Coastal Engineering 1994

Proceedings of the
twenty-fourth international conference

October 23-28; 1994
Kobe, Japan

Conference held under the auspices of the
Coastal Engineering Research Council of the
American Society of Civil Engineers

Organized by the Japan Society of Civil Engineers

Edited by Billy L. Edge

Property of Alumni Library
Wentworth Institute of Technology

Published by the
American Society of Civil Engineers
345 East 47th Street
New York, New York 10017-2398

ABSTRACT

This proceedings contains over 200 papers presented at the 24th International Conference on Coastal Engineering which was held in Kobe, Japan, October 23-28, 1994. The book is divided into six parts: 1) Characteristics of coastal waves and currents; 2) long waves and storm surges; 3) coastal structures; 4) coastal processes and sediment transport; 5) coastal, estuarine and environmental problems; and 6) case studies. The individual papers include such topics as the effects of wind, waves, storms and currents as well as the study of sedimentation and beach nourishment. Special emphasis is given to case studies of completed engineering projects. With the inclusion of both theoretical and practical information, these papers provide the civil engineer and related fields with a broad range of information on coastal engineering.

Coastal engineering 1994: proceedings of the twenty-fourth international conference, October 23-28, 1994, Kobe, Japan
edited by Billy L. edge.
  p.   cm.
  "Conference held under the auspices of the Coastal Engineering Research Council of the American Society of Civil Engineers; organized by the Japan Society of Civil Engineers (JSCE)."Papers presented at the 24th International Conference on Coastal Engineering.
  Includes index.
  ISBN 0-7844-0089-X
  1. Coastal engineering—Congresses. 2. Ocean waves—Congresses. 3. Shore protection-Congresses. I. Edge, Billy L. II. Coastal Engineering Research Council (U.S.) III. Doboku Gakkai. IV. International Conference on Coastal Engineering (24th: 1994: Kobe, Japan)
TC203.5.C6184 1995   95-18653
627'.58—dc20   CIP

The Society is not responsible for any statements made or opinions expressed in its publications.

Photocopies. Authorization to photocopy material for internal or personal use under circumstances not falling within the fair use provisions of the Copyright Act is granted by ASCE to libraries and other users registered with the Copyright Clearance Center (CCC) Transactional Reporting Service, provided that the base fee of $2.00 per article plus $.25 per page copied is paid directly to CCC, 222 Rosewood, Drive, Danvers, MA 01923. The identification for ASCE Books is 0-7844-0089-X/95 $2.00 + $.25. Requests for special permission or bulk copying should be addressed to Permissions & Copyright Dept., ASCE.

Copyright © 1995 by the American Society of Civil Engineers,
All Rights Reserved.
Library of Congress Catalog Card No: 95-18653
ISBN 0-7844-0089-X
Manufactured in the United States of America.

FOREWORD

The 24th International Conference on Coastal Engineering was held in Kobe, Japan. The 24th ICCE, like the ones before it, was organized and managed by volunteers from within Japan representing private, industrial and governmental contributors. This Conference represented an opportunity to share scientific and engineering information and provided a forum for interaction with other engineers and scientists working on similar coastal problems. The time and effort contributed to the development and organization of each conference becomes more complex and difficult as they continue to grow in size and content. The Local Organizing Committee worked on the planning for this Conference for four years. The 24th ICCE was a tremendous success in every way. The **Proceedings** of this Conference will represent a major step forward in the field of coastal engineering.

The chapters in this **Proceedings** have been prepared by the authors who were selected to make presentations at the 24th International Conference on Coastal Engineering. The authors were asked to make presentations at the Conference based upon review of the abstracts which were submitted well in advance of the Conference. The Technical Review Committee included six professionals who are active in the field of coastal engineering. One of the members is a representative of the Local Organizing Committee; the other five members were selected for their broad understanding and recognition in the field. The papers included in this volume are eligible for discussion in the **Journal of the Waterways, Port, Coastal and Ocean Division** of the ASCE. All papers are eligible for ASCE awards.

Venues for the upcoming conferences are listed below:

25th ICCE - Orlando, FL USA	1996
26th ICCE - Copenhagen, Denmark	1998
27th ICCE - Sydney, Australia	2000

Coastal engineers who would desire to host a future conference in their country should contact the Secretary of the Coastal Engineering Research Council to receive information about submitting a proposal.

The continuing coordination of the International Conferences on Coastal Engineering is through the Coastal Engineering Research Council of the ASCE. The Research Council began at the instigation of Professor Boris Bakhmeteff who as chairman of the Research Committee of the Engineering Foundation suggested the formation of the Council on Wave Research. The Council was established in June 1950 under the Engineering Foundation. In 1963 the Council was transferred from the Foundation to the ASCE and was renamed the Coastal Engineering Research Council which better described its expanded function.

Members of the coastal engineering community recognized that the problems they faced required broad based research to better define the coastal and ocean phenomena with which they dealt. The Foundation felt that it was important that all disciplines working in the coastal area should have an opportunity and be encouraged to communicate with one another through the mechanism of interdisciplinary conferences.

The first conference was held in Long Beach, CA in 1950. The papers which were delivered at the conference were published and became the first coastal engineering conference **Proceedings.** Although the conferences began with a national focus they quickly became international in scope. After planning and conducting 24 conferences on coastal engineering the Series has been established as the principal conference on coastal engineering in the world. Contributors to the conference represent nearly all coastal nations and the numbers of abstracts which are submitted for consideration are generally twice as large as the available opportunity for presentations. The **Proceedings** of the conferences are all available from the ASCE.

Billy L. Edge, Secretary
Coastal Engineering Research Council
American Society of Civil Engineers

ACKNOWLEDGMENTS

LOCAL ORGANIZING COMMITTEE

    Chairman:
        K. Sasayama - Mayor of Kobe
    Vice Chairman:
        Y. Iwagaki - Professor, Meijo University
        K. Horikawa - President, Saitama University
        N. Shiraishi - Nippon Tetrapod Co., Ltd. (deceased)
        T. Sawaragi - Professor, Osaka University
        Y. Tsuchiya - Professor, Meijo University

EXECUTIVE COMMITTEE

    Co-Chairman:
        T. Sawaragi - Professor, Osaka University
    Co-Chairman:
        Y. Tsuchiya - Professor, Meijo University
        M. Hattori - Professor, Chuo University
        M. Inoue - Professor, Kansai University
        K. Oda - Professor, Osaka City University
        A. Watanabe - Professor, University of Tokyo
        T. Sakai - Professor, Kyoto University
        Y. Kawata - Professor, Kyoto University
        I. Deguchi - Professor, Osaka University

SUPPORTING ORGANIZATIONS

    International Association for Hydraulic Research
    Japanese Association for Coastal Zone Studies
    Japan Society for Natural Disaster Science
    The Society of Naval Architects of Japan
    Architectural Institute of Japan
    Meteorological Society of Japan
    The Oceanographic Society of Japan
    The Ministry of Agriculture, Forestry and Fisheries
    The Ministry of Transport
    The Ministry of Construction

FINANCIAL SUPPORT

    City of Kobe
    Hyogo Prefectural Government
    Osaka Prefectural Government
    Wakayama Prefectural Government
    City of Osaka
    City of Himeji
    City of Sakai
    Japan Ocean Development Construction Association
    The Federation of Electric Power Companies
    Japan Association for Wave-Dissipating and Foot Protection Blocks
    Japan Iron and Steel Foundation
    The Japan Port and Harbor Association
    Osaka Bay Regional Offshore Environmental Improvement Center
    All Japan Fishing Ports Association
    Fishery Infrastructure Development Center
    The Japanese Institute of Technology on Fishing Ports and Communities
    Japan Consulting Engineers Cooperation
    Osaka Gas Co., Ltd.
    The Portopia 81 Foundation
    The Kajima Foundation
    Commemorative Association for the Japan World Exposition (1970)
    International Exchange Fund, JSCE

COASTAL ENGINEERING RESEARCH COUNCIL (ASCE)

    Chairman:
        R. G. Dean
    Vice Chairman:
        O. T. Magoon
    Secretary:
        B. L. Edge
    Members:
        K. Horikawa
        T. Saville, Jr.
        R. M. Noble
        J. W. Kamphuis
        R. A. Dalrymple
        K. d'Angremond
        S. A. Hughes

CONTENTS

PART I

Plenary Session
THE PRESENT AND FUTURE OF COASTAL ENGINEERING IN JAPAN....... 1
Yuichi Iwagaki

Chapter 1
ON THE CHARACTERISTICS OF ONE-DIMENSIONAL SPECTRA
AND NON-DIMENSIONAL PARAMETERS OF WIND WAVES.................. 12
Toshio Aono, Chiaki Goto

Chapter 2
SWASH MOTION DUE TO OBLIQUELY INCIDENT WAVES...................... 27
Toshiyuki Asano

Chapter 3
LABORATORY COMPARISON OF DIRECTIONAL WAVE
MEASUREMENT SYSTEMS AND ANALYSIS TECHNIQUES 42
Michel Benoit, Charles Teisson

Chapter 4
ACCURACY OF WIND AND WAVE EVALUATION IN COASTAL
REGIONS.. 57
Luciana Bertotti, Luigi Cavaleri

Chapter 5
A SPECTRAL MODEL FOR WAVES IN THE NEAR SHORE ZONE 68
R.C. Ris, L.H. Holthuijsen, N. Booij

Chapter 6
WIND VARIABILITY AND EXTREMES STATISTICS.............................. 79
Luigi Cavaleri, Luciana Bertotti

Chapter 7
MEAN FLUX IN THE FREE SURFACE ZONE OF WATER WAVES
IN A CLOSED WAVE FLUME... 86
Witold Cieslikiewicz, Ove T. Gudmestad

Chapter 8
VERTICAL VARIATIONS OF FLUID VELOCITIES AND SHEAR
STRESS IN SURF ZONES 98
Daniel T. Cox, Nobuhisa Kobayashi, Akio Okayasu

Chapter 9
VORTICITY EFFECTS IN COMBINED WAVES AND CURRENTS 113
I. Cummins, C. Swan

Chapter 10
WAVES IN AN ANNULAR ENTRANCE CHANNEL 128
Robert A. Dalrymple, J.T. Kirby

Chapter 11
WAVE DAMPING BY KELP VEGETATION 142
Alfonse Dubi, Alf Tørum

Chapter 12
NONLINEAR COUPLING IN WAVES PROPAGATING OVER A BAR 157
Y. Eldeberky, J.A. Battjes

Chapter 13
AN ABSORBING WAVE-MAKER BASED ON DIGITAL FILTERS 168
Peter Frigaard, Morten Christensen

Chapter 14
WAVE CLIMATE STUDY IN WADDEN SEA AREAS 181
Ralf Kaiser, Günther Brandt, Joachim Gärtner, Detlef Glaser,
Frerk Jensen, H.D. Niemeyer, Joachim Grüne

Chapter 15
FALSE WAVES IN WAVE RECORDS AND NUMERICAL SIMULATIONS 192
Marcos H. Giménez, C.R. Sánchez-Carratalá, J.R. Medina

Chapter 16
MEASURING WAVES WITH MANOMETER TUBES 207
David J. Hanslow, P. Nielsen, K. Hibbert

Chapter 17
QUANTITY OF SPRAY TRANSPORTED BY THE STRONG WIND
OVER BREAKING WAVES 219
Nobuhiro Matsunaga, Misao Hashida, Hiroyuki Mizui, Yuji Sugihara

Chapter 18
EXTENSION OF THE MAXIMUM ENTROPY PRINCIPLE METHOD
FOR DIRECTIONAL WAVE SPECTRUM ESTIMATION 232
Noriaki Hashimoto, Toshihiko Nagai, Tadashi Asai

Chapter 19
A REGRESSION MODEL FOR ESTIMATING SEA STATE PERSISTENCE 247
Yoshio Hatada, Masataka Yamaguchi

Chapter 20
THE MAXIMUM SIGNIFICANT WAVE HEIGHT IN THE SOUTHERN
NORTH SEA 261
*L.H. Holthuijsen, J.G. de Ronde, Y. Eldeberky, H.L. Tolman,
N. Booij, E. Bouws, P.G.P. Ferier, J. Andorka Gal*

Chapter 21
IMPROVED BOUNDARY CONDITIONS TO A TIME-DEPENDENT
MILD-SLOPE EQUATION FOR RANDOM WAVES 272
Toshimasa Ishii, Masahiko Isobe, Akira Watanabe

Chapter 22
TIME-DEPENDENT MILD SLOPE EQUATIONS FOR RANDOM WAVES 285
Masahiko Isobe

Chapter 23
MODELLING MOVEABLE BED ROUGHNESS AND FRICTION FOR
SPECTRAL WAVES 300
L.M. Kaczmarek, J.M. Harris, B.A. O'Connor

Chapter 24
DIFFERENCE BETWEEN WAVES ACTING ON STEEP AND
GENTLE BEACHES 315
Kazumasa Katoh

Chapter 25
WAVE BREAKING UNDER STORM CONDITION 330
Yoshiaki Kawata

Chapter 26
APPLICATION OF MAXIMUM ENTROPY METHOD TO THE REAL
SEA DATA 340
Taerim Kim, Li-Hwa Lin, Hsiang Wang

Chapter 27
PROBABILITY OF THE FREAK WAVE APPEARANCE IN A
3-DIMENSIONAL SEA CONDITION 356
Akira Kimura, Takao Ohta

Chapter 28
ON THE JOINT DISTRIBUTION OF WAVE HEIGHT, PERIOD &
DIRECTION OF INDIVIDUAL WAVES IN A 3-DIMENSIONAL
RANDOM SEAS.. 370
   J.G. Kwon, Ichiro Deguchi

Chapter 29
SPECTRAL WAVE-CURRENT BOTTOM BOUNDARY LAYER FLOWS........ 384
   Ole Secher Madsen

Chapter 30
TIME DOMAIN MODELLING OF WAVE BREAKING, RUNUP AND
SURF BEATS 399
   Per A. Madsen, O.R. Sørensen, H.A. Schäffer

Chapter 31
ORTHONORMAL WAVELET ANALYSIS FOR DEEP-WATER
BREAKING WAVES 412
   Nobuhito Mori, Takashi Yasuda

Chapter 32
A FULLY DISPERSIVE-NONLINEAR WAVE MODEL AND ITS
NUMERICAL SOLUTIONS 427
   Kazuo Nadaoka, Serdar Beji, Yasuyuki Nakagawa

Chapter 33
A GENERALIZED GREEN-FUNCTION METHOD FOR WAVE
FIELD ANALYSIS 442
   Hitoshi Nishimura, Michio Matsuoka, Akira Matsumoto

Chapter 34
COUPLED VIBRATION EQUATIONS FOR IRREGULAR WATER
WAVES.. 455
   Masao Nochino

Chapter 35
NONLINEAR EVOLUTION OF DIRECTIONAL WAVE SPECTRA
IN SHALLOW WATER 467
   Okey Nwoġu

Chapter 36
NON-GAUSSIAN PROBABILITY DISTRIBUTION OF COASTAL WAVES..... 482
   M.K. Ochi, K. Ahn

Chapter 37
PROBABILITY CHARACTERISTICS OF ZERO-CROSSING WAVE
HEIGHT 497
  Takao Ohta, Akira Kimura

Chapter 38
NUMERICAL SIMULATION AND VALIDATION OF PLUNGING
BREAKERS USING A 2D NAVIER-STOKES MODEL............................. 511
  H.A.H. Petit, P.Tönjes, M.R.A. Van Gent, P. Van Den Bosch

Chapter 39
WAVE VELOCITY FIELD MEASUREMENTS OVER A SUBMERGED
BREAKWATER 525
  Marco Petti, Paul A. Quinn, Gianfranco Liberatore, William J. Easson

Chapter 40
VELOCITY FIELD MEASUREMENTS AND THEORETICAL
COMPARISONS FOR NON-LINEAR WAVES ON MILD SLOPES 540
  Paul A. Quinn, Marco Petti, Michele Drago, Clive A. Greated

Chapter 41
THE CONCEPT OF RESIDENCE TIME FOR THE DESCRIPTION OF
WAVE RUN-UP, WAVE SET-UP AND WAVE RUN-DOWN 553
  Holger Schüttrumpf, Hendrik Bergmann, Hans-Henning Dette

Chapter 42
BOTTOM SHEAR STRESSES UNDER RANDOM WAVES WITH
A CURRENT SUPERIMPOSED.. 565
  Richard R. Simons, Tony J. Grass, Wameidh M. Saleh,
  Mehrdad M. Tehrani

Chapter 43
EFFECTS FROM DIRECTIONALITY & SPECTRAL BANDWIDTH
ON NON-LINEAR SPATIAL MODULATIONS OF DEEP-WATER
SURFACE GRAVITY WAVE TRAINS 579
  Carl Trygve Stansberg

Chapter 44
SHEAR STRESSES AND MEAN FLOW IN SHOALING AND
BREAKING WAVES 594
  Marcel J.F. Stive, Huib J. De Vriend

Chapter 45
PREDICTION OF THE MAXIMUM WAVE ON THE CORAL FLAT 609
  Dede M. Sulaiman, Shigeaki Tsutsui, Hiroshi Toshioka,
  T. Yamashita, S. Oshiro, Y. Tsuchiya

Chapter 46
DEVELOPMENT OF A SUBMERGED DOPPLER-TYPE DIRECTIONAL
WAVE METER 624
  Tomotsuka Takayama, Noriaki Hashimoto, Toshihiko Nagai,
  Tomoharu Takahashi, Hiroshi Sasaki, Yoshiki Ito

Chapter 47
BRAGG SCATTERING OF WAVES OVER POROUS RIPPLED BED 635
  Hijime Mase, Ken Takeba

Chapter 48
NON-REFLECTIVE MULTI-DIRECTIONAL WAVE GENERATION
BY SOURCE METHOD 650
  Masahiro Tanaka, Takumi Ohyama, Tetsushi Kiyokawa, Kazuo Nadaoka

Chapter 49
THE GROWTH OF WIND WAVES IN SHALLOW WATER 665
  L.A. Verhagen, I.R. Young

Chapter 50
ESTIMATION OF TYPHOON-GENERATED MAXIMUM WAVE
HEIGHT ALONG THE PACIFIC COAST OF JAPAN BASED ON
WAVE HINDCASTING 674
  Masataka Yamaguchi, Yoshio Hatada

Chapter 51
RUN-UP OF IRREGULAR WAVES ON A GENTLY SLOPING BEACH 689
  Yoshimichi Yamamoto, Katsutoshi Tanimoto, Karunarathna Harshinie

Chapter 52
SOLITON-MODE WAVEMAKER THEORY AND SYSTEM FOR
COASTAL WAVES 704
  Takashi Yasuda, Takeshi Hattori, Seirou Shinoda

Chapter 53
ON A METHOD FOR ESTIMATING REFLECTION COEFFICIENT
IN SHORT-CRESTED RANDOM SEAS 719
  Hiromune Yokoki, Masahiko Isobe, Akira Watanabe

Chapter 54
THE DIRECTIONAL WAVE SPECTRUM IN THE BOHAI SEA.................... 731
    Yuxiu Yu, Shuxue Liu

Chapter 55
IRREGULAR WAVES OVER AN ELLIPTIC SHOAL 746
    Xiping Yu, Hiroyoshi Togashi

Chapter 56
PERFORMANCE OF A SPECTRAL WIND-WAVE MODEL IN
SHALLOW WATER 761
    Gerbrant Ph. van Vledder, John G. de Ronde, Marcel J.F. Stive

PART II

Chapter 57
THE GENERATION OF LOW-FREQUENCY WAVES BY A SINGLE
WAVE GROUP INCIDENT ON A BEACH... 776
    Gary Watson, Timothy C.D. Barnes, D. Howell Peregrine

Chapter 58
INFLUENCE OF LONG WAVES ON SHIP MOTIONS IN A
LAGOON HARBOUR 791
    Volker Barthel, Etienne Mansard

Chapter 59
RESONANT FORCING OF HARBORS BY INFRAGRAVITY WAVES 806
    Gordon S. Harkins, Michael J. Briggs

Chapter 60
NUMERICAL SIMULATION OF THE 1992 FLORES TSUNAMI IN
INDONESIA: DISCUSSION ON LARGE RUNUP HEIGHTS IN THE
NORTHEASTERN FLORES ISLAND... 821
    Fumihiko Imamura, Tomoyuki Takahashi, Nobuo Shuto

Chapter 61
A COMPARATIVE EVALUATION OF WAVE GROUPING MEASURES 832
    E.P.D. Mansard, S.E. Sand

Chapter 62
RELATIONSHIP OF A MOORED VESSEL IN A HARBOUR AND
LONG WAVE CAUSED BY WAVE GROUPS...................................... 847
    Toshihiko Nagai, N. Hashimoto, T. Asai, I. Tobiki, K. Ito,
    T. Toue, A. Kobayashi, T. Shibata

Chapter 63
COHERENT STRUCTURE OF TIDAL TURBULENCE IN A ROTATING
SYSTEM OF OSAKA-BAY 861
  Tsukasa Nishimura, Tomonao Kobayashi, Goichi Furuta

Chapter 64
DEVELOPMENT OF A PARTIALLY THREE-DIMENSIONAL MODEL
FOR SHIP MOTION IN A HARBOR WITH ARBITRARY BATHYMETRY 871
  Takumi Ohyama, M. Tsuchida

Chapter 65
THE MEASURED AND COMPUTED HOKKAIDO NANSEI-OKI
EARTHQUAKE TSUNAMI OF 1993 886
  Tomoyuki Takahashi, Nobuo Shuto, Fumihiko Imamura,
  Hideo Matsutomi

Chapter 66
QUASI-THREE-DIMENSIONAL MODEL FOR STORM SURGES AND
ITS VERIFICATION 901
  Takao Yamashita, Yoshito Tsuchiya, Hiroshi Yoshioka

PART III

Chapter 67
ANALYSIS OF PRACTICAL RUBBLE MOUNDS 918
  N.W.H. Allsop, R.J. Jones, P. Besley, C. Franco

Chapter 68
FRICTION AND CLAMPING FORCES IN WAVE LOADED PLACED
BLOCK REVETMENTS 932
  Adam Bezuijen

Chapter 69
A FIELD EXPERIMENT ON THE INTERACTION
WAVES-REFLECTING WALL 945
  Paolo Boccotti

Chapter 70
THE APPLICATION OF LOAD-CELL TECHNIQUE IN THE STUDY
OF ARMOUR UNIT RESPONSES TO IMPACT LOADS 958
  Hans F. Burcharth, Zhou Liu

Chapter 71
STEEP WAVE DIFFRACTION BY A SUBMERGED CYLINDER.................. 973
J.T. Aquije Chacaltana, A.F.T. da Silva

Chapter 72
WAVE STRESSES ON RUBBLE-MOUND ARMOUR 986
Andrew M. Cornett, Etienne Mansard

Chapter 73
DAMAGE ANALYSIS FOR RUBBLE-MOUND BREAKWATERS 1001
Michael H. Davies, Etienne P.D. Mansard, Andrew M. Cornett

Chapter 74
NUMERICALLY MODELING PERSONNEL DANGER ON A
PROMENADE BREAKWATER DUE TO OVERTOPPING WAVES 1016
Kimihiko Endoh, Shigeo Takahashi

Chapter 75
WAVE OVERTOPPING ON VERTICAL AND COMPOSITE
BREAKWATERS..1030
L. Franco, M. de Gerloni, J.W. van der Meer

Chapter 76
AN INVESTIGATION OF THE WAVE FORCES ACTING ON
BREAKWATER HANDRAILS 1046
Atsushi Hujii, Shigeo Takahashi, Kimihiko Endoh

Chapter 77
RUBBLE MOUND BREAKWATER STABILITY UNDER OBLIQUE
WAVES: AN EXPERIMENTAL STUDY 1061
J.-C. Galland

Chapter 78
WAVE LOADS ON SEADYKES WITH COMPOSITE SLOPES & BERMS 1075
Joachim Grune, Hendrik Bergmann

Chapter 79
COMPUTERISED METHODOLOGY TO MEASURE RUBBLE MOUND
BREAKWATER DAMAGE... 1090
Bruno Chilo, Franco Guiducci

Chapter 80
WAVE BREAKING OVER PERMEABLE SUBMERGED BREAKWATERS 1101
Masataro Hattori, Hiroyoki Sakai

Chapter 81
WAVE FORCES ACTING ON A VORTEX EXCITED VIBRATING
VERTICAL CYLINDER IN WAVES 1115
Kenjirou Hayashi, Futoshi Higaki, Koji Fujima,
Toshiyuki Sigemura, John R. Chaplin

Chapter 82
OVERTOPPING OF SEA WALLS UNDER RANDOM WAVES 1130
D.M. Herbert, N.W.H. Allsop, M.W. Owen

Chapter 83
STABILITY OF HIGH-SPECIFIC GRAVITY ARMOR BLOCKS 1143
Masahiro Ito, Y. Iwagaki, H. Murakami, K. Nemoto, M. Yamamoto,
M. Hanzawa

Chapter 84
ROCK ARMOURED BEACH CONTROL STRUCTURES ON
STEEP BEACHES 1157
R.J. Jones, N.W.H. Allsop

Chapter 85
EFFECT ON ROUGHNESS TO IRREGULAR WAVE RUN-UP..................... 1169
Jea-Tzyy Juang

Chapter 86
WAVE OVERTOPPING OF BREAKWATERS UNDER OBLIQUE WAVES 1182
Jørgen Juhl, Peter Sloth

Chapter 87
CONSTRUCTION OF OFFSHORE FISHING PORT FOR PREVENTION
OF COASTAL EROSION 1197
Takeshi Kawaguchi, O. Hashimoto, T. Mizumoto, A. Kamata

Chapter 88
TOPOGRAPHICAL CHANGE AROUND MULTIPLE LARGE
CYLINDRICAL STRUCTURES UNDER WAVE ACTIONS......................... 1212
Chang-Je Kim, Koichiro Iwata, Yoshihito Miyaike, Hong-Sun Yu

Chapter 89
STABILITY OF RUBBLE MOUND FOUNDATIONS OF COMPOSITE
BREAKWATERS UNDER OBLIQUE WAVE ATTACK 1227
Katsutoshi Kimura, Shigeo Takahashi, Katsutoshi Tanimoto

Chapter 90
ANALYSIS OF NONLINEAR COEFFICIENTS OF REFLECTION
AND TRANSMISSION OF WAVES PROPAGATING OVER A
RECTANGULAR STEP 1241
Wudhipong Kittitanasuan, Yoshimi Goda

Chapter 91
OSCILLATORY MOTIONS AND PERMANENT DISPLACEMENTS
OF CAISSON BREAKWATERS SUBJECT TO IMPULSIVE
BREAKING WAVE LOADS.. 1255
P. Klammer, H. Oumeraci, H.-W. Partenscky

Chapter 92
HYDRAULIC CHARACTERISTICS AND FIELD EXPERIENCE OF
NEW WAVE DISSIPATING CONCRETE BLOCKS (ACCROPODE) 1269
Masanori Kobayashi, Sumio Kaihatsu

Chapter 93
EXPERIMENTAL STUDY ON DEVELOPING PROCESS OF LOCAL
SCOUR AROUND A VERTICAL CYLINDER.. 1284
Tomonao Kobayashi, Kenji Oda

Chapter 94
WAVE-INDUCED UPLIFT LOADING OF CAISSON BREAKWATERS 1298
Andreas Kortenhaus, Hocine Oumeraci, Søren Kohlhase, Peter Klammer

Chapter 95
LINEAR AND NONLINEAR WAVE FORCES EXERTED ON A
SUBMERGED HORIZONTAL PLATE... 1312
Haruyuki Kojima, A. Yoshida, T. Nakamura

Chapter 96
MODERN FUNCTIONAL DESIGN OF GROIN SYSTEMS......................... 1327
Nicholas C. Kraus, Hans Hanson, Sten H. Blomgren

Chapter 97
RESHAPING BREAKWATERS IN DEEP AND SHALLOW WATER
CONDITIONS... 1343
A. Lamberti, G.R. Tomasicchio, F. Guiducci

Chapter 98
DESIGN OF BREAKWATERS AND BEACH NOURISHMENT 1359
Christian Laustrup, Holger Toxvig Madsen

Chapter 99
CIRCULAR CHANNEL BREAKWATER TO REDUCE WAVE
OVERTOPPING AND ALLOW WATER EXCHANGE 1373
Dal Soo Lee, Woo Sun Park, Nobuhisa Kobayashi

Chapter 100
INCREASED DOLOS STRENGTH BY SHAPE MODIFICATION 1388
S.A. Luger, D.T. Phelp, A. Van Tonder, A.H. Holtzhausen

Chapter 101
INFLUENCE OF WAVE DIRECTIONALITY ON STABILITY
OF BREAKWATER HEADS... 1397
Y. Matsumi, E.P.D. Mansard, J. Rutledge

Chapter 102
COST-EFFECTIVENESS OF D-ARMOR BREAKWATER.......................... 1412
Josep R. Medina

Chapter 103
THE CORE-LOC: OPTIMIZED CONCRETE ARMOR.............................. 1426
Jeffrey A. Melby, George F. Turk

Chapter 104
STABILITY OF ARMOR STONES OF A SUBMERGED WIDE-CROWN
BREAKWATER 1439
Norimi Mizutani, Teofilo Monge Rufin, Jr., Koichiro Iwata

Chapter 105
WAVE INDUCED FLOW AROUND SUBMERGED SLOPING PLATES 1454
Hitoshi Murakami, Sadahiko Itoh, Yoshihiko Hosoi, Yoshiyuki Sawamura

Chapter 106
SECOND-ORDER WAVE INTERACTION WITH ARRAYS OF
VERTICAL CYLINDERS OF ARBITRARY CROSS SECTION 1469
Keisuke Murakami, Akinori Yoshida

Chapter 107
HYDRODYNAMIC FORCES ON BOTTOM-SEATED HEMISPHERE
IN WAVES AND CURRENTS.. 1484
Hidenori Nishida, A. Tada, F. Nishihira

Chapter 108
THE PRESSURE FIELD DUE TO STEEP WATER WAVES INCIDENT
ON A VERTICAL WALL 1496
D.H. Peregrine, M.E. Topliss

Chapter 109
RESULTS OF EXTENSIVE FIELD MONITORING OF DOLOS
BREAKWATERS... 1511
D. Phelp, S. Luger, A. Van Tonder, A. Holtzhausen

Chapter 110
FAILURE OF RUBBLE MOUND STRUCTURES DUE TO THE STORM
DURATION AND THE IRREGULARITY OF OCEAN WAVES 1526
Cheong-Ro Ryu, Hyeon-Ju Kim

Chapter 111
BLOCK SUBSIDENCE UNDER PRESSURE AND FLOW 1541
T. Sakai, H. Gotoh, T. Yamamoto

Chapter 112
STABILITY OF ROCK ON BEACHES... 1553
*Gerrit J. Schiereck, Henri L. Fontijn, Wout V. Grote,
Paul G.J. Sistermans*

Chapter 113
SHORT TERM WAVE OVERTOPPING RATE OF BLOCK
ARMORED SEAWALL 1568
Tsunehiro Sekimoto, Hiroshi Kunisu, Tsuyoshi Yamazaki

Chapter 114
ESTIMATING THE SLIDING DISTANCE OF COMPOSITE BREAKWATERS
DUE TO WAVE FORCES INCLUSIVE OF IMPULSIVE FORCES 1580
Kenichiro Shimosako, S. Takahashi, K. Tanimoto

Chapter 115
BED SHEAR STRESS AND SCOUR AROUND COASTAL STRUCTURES 1595
B.M. Sumer, J. Fredsøe, N. Christiansen, S.B. Hansen

Chapter 116
LABORATORY MEASUREMENT OF OBLIQUE IRREGULAR
WAVE REFLECTION ON RUBBLE-MOUND BREAKWATERS.................... 1610
Charles Teisson, Michel Benoit

Chapter 117
STONE MOVEMENT ON A RESHAPED PROFILE 1625
G. Roberto Tomasicchio, A. Lamberti, F. Guiducci

Chapter 118
THE LARGE SCALE DOLOS FLUME STUDY 1641
George F. Turk, Jeffrey A. Melby

Chapter 119
H_O PARAMETER FOR PRELIMINARY DESIGN OF CONVENTIONAL
BREAKWATER STRUCTURAL HEAD: DATA ANALYSIS OF
SPANISH NORTH COAST HARBOURS 1657
Vincente Negro Valdecantos, Ovidio Varela Carnero

Chapter 120
NUMERICAL MODELLING OF BREAKING WAVE IMPACTS
ON A VERTICAL WALL 1672
N.T. Wu, H. Oumeraci, H.-W. Partenscky

Chapter 121
WIND EFFECTS ON RUNUP AND OVERTOPPING 1687
Donald L. Ward, Christopher G. Wibner, Jun Zhang,
Billy L. Edge

Chapter 122
MODELLING OF WAVE OVERTOPPING OVER BREAKWATER 1700
F. Zhuang, C. Chang, J.J. Lee

Chapter 123
STRESSES IN TETRAPOD ARMOR UNITS INDUCED BY WAVE ACTION 1713
Kees d'Angremond, J.W. van der Meer, C.P. van Nes

Chapter 124
PORE PRESSURES IN RUBBLE MOUND BREAKWATERS 1727
M.B. de Groot, H. Yamazaki, M.R.A. van Gent, Z. Kheyruri

Chapter 125
WAVE ACTION ON AND IN PERMEABLE STRUCTURES 1739
M.R.A. van Gent, P. Tönjes, H.A.H. Petit, P. Van Den Bosch

Chapter 126
PROBABILISTIC CALCULATIONS OF WAVE FORCES ON
VERTICAL STRUCTURES 1754
J.W. van der Meer, K. d'Angremond, J. Juhl

PART IV

Chapter 127
VELOCITY AND PRESSURE BOUNDARY CONDITIONS FOR FLOW
OVER THE PERMEABLE BOUNDARY OF A POROUS MEDIUM 1770
  Ismail Aydin

Chapter 128
PHYSICAL EXPERIMENTS ON THE EFFECTS OF GROINS ON
SHORE MORPHOLOGY 1782
  Peyman Badiei, J. William Kamphuis, David G. Hamilton

Chapter 129
LINE-MODELING OF SHOREFACE NOURISHMENT 1797
  Willem T. Bakker, Nico F. Kersting, Hanz D. Niemeyer

Chapter 130
STATISTICAL VARIATIONS IN BEACH PARAMETER CHANGE RATES
FOR WALLED & NON-WALLED PROFILES AT SANDBRIDGE, VA 1812
  John M. Hazelton, David R. Basco, D. Bellomo, G. Williams

Chapter 131
A BOTTOM BOUNDARY LAYER SEDIMENT RESPONSE TO
WAVE GROUPS 1827
  J. Lee, S. O'Neil, K. Bedford, R. Van Evra

Chapter 132
GEOMORPHOLOGICAL ANALYSIS OF A BEACH AND SANDBAR
SYSTEM..1837
  Chen-Shan Kung, Marcer Stive, Geffery Toms

Chapter 133
SUSPENDED SEDIMENT TRANSPORT IN INNER SHELF WATERS
DURING EXTREME STORMS..1849
  O.S. Madsen, T.A. Chisholm, L.D. Wright

Chapter 134
SEA BED STABILITY ON A LONG STRAIGHT COAST............................1865
  E. Christensen, R. Deigaard, J. Fredsoe

Chapter 135
THE RESPONSE OF GRAVEL BEACHES IN THE PRESENCE
OF CONTROL STRUCTURES 1880
  T.T. Coates, N. Dodd

Chapter 136
THE ROLE OF ROLLERS IN SURF ZONE CURRENTS 1895
William R. Dally, Daniel A. Osiecki

Chapter 137
THREE DIMENSIONAL MORPHOLOGY IN A NARROW WAVE TANK:
MEASUREMENTS AND THEORY 1906
Robert G. Dean, Tae-Myoung Oh

Chapter 138
NUMERICAL SIMULATION OF FINITE AMPLITUDE SHEAR
WAVES & SEDIMENT TRANSPORT 1919
Rolf Deigaard, Erik Damgaard Christensen,
Jesper Svarrer Damgaard, Jørgen Fredsøe

Chapter 139
BEACH NOURISHMENT AND DUNE PROTECTION 1934
Hans H. Dette, Arved J. Raudkivi

Chapter 140
PROFILE CHANGE OF A SHEET FLOW DOMINATED BEACH 1946
Mohammad Dibajnia, Takuzo Shimizu, Akira Watanabe

Chapter 141
A NONLINEAR SURF BEAT MODEL 1961
Zhili Zou, Nicholas Dodd

Chapter 142
PIV MEASUREMENTS OF OSCILLATORY FLOW OVER A RIPPLED BED 1975
H.C. Earnshaw, T. Bruce, C.A. Greated, W.J. Easson

Chapter 143
SHEAR INSTABILITY OF LONGSHORE CURRENTS: EFFECTS OF
DISSIPATION AND NON-LINEARITY 1983
Albert Falqués, Vincente Iranzo, Miquel Caballería

Chapter 144
INTERPRETATION OF SHORELINE POSITION FROM AERIAL
PHOTOGRAPHS 1998
John S. Fisher, Margery F. Overton

Chapter 145
SETTLING COLUMNS PARAMETRIC TESTS APPLIED TO
COASTAL SEDIMENT CONSOLIDATION 2004
Stephane Gallois, Alain Alexis, Pierre Thomas

Chapter 146
SEDIMENT-CLOUD BASED MODEL OF SUSPENSION OVER RIPPLE
BED DUE TO WAVE ACTION 2013
Hitoshi Gotoh, Tetsuro Tsujimoto, Hiroji Nakagawa

Chapter 147
WAVE OVERTOPPING AND SEDIMENT TRANSPORT OVER DUNES 2028
Mark W. Hancock, Nobuhisa Kobayashi

Chapter 148
SEDIMENT TRANSPORT OVER RIPPLES IN WAVES AND CURRENT 2043
*Gilles Perrier, Erik Asp Hansen, C. Villaret, Rolf Deigaard,
Jorgen Fredsøe*

Chapter 149
IN-SITU DETERMINATION OF THE CRITICAL BED-SHEAR
STRESS FOR EROSION OF COHESIVE SEDIMENTS 2058
Erik-Jan Houwing, Leo C. van Rijn

Chapter 150
CONTROL OF CROSS-SHORE SEDIMENT TRANSPORT BY A
DISTORTED RIPPLE MAT 2070
Isao Irie, N. Ono, S. Hashimoto, S. Nakamura, K. Murakami

Chapter 151
HOW MUCH VELOCITY INFORMATION IS NECESSARY TO
PREDICT SEDIMENT SUSPENSION IN THE SURF ZONE? 2085
Bruce E. Jaffe, David M. Rubin, Asbury Sallenger, Jr.

Chatper 152
BEACH PROFILE SPACING: PRACTICAL GUIDANCE FOR
MONITORING NOURISHMENT PROJECTS 2100
Timothy W. Kana, Christopher J. Andrassy

Chapter 153
WATERTABLE OVERHEIGHT DUE TO WAVE RUNUP ON A
SANDY BEACH 2115
Hong-Yoon Kang, P. Nielsen, D.J. Hanslow

Chapter 154
A MODEL FOR CROSS SHORE SEDIMENT TRANSPORT 2125
Irene Katopodi, Nikos Kitou

Chapter 155
NUMERICAL MODELLING OF FLOW OVER RIPPLES USING
SOLA METHOD..2140
Hyoseob Kim, Brian A. O'Connor, Youngbo Shim

Chapter 156
SWASH DYNAMICS UNDER OBLIQUELY INCIDENT WAVES 2155
Nobuhisa Kobayashi, Entin A. Karjadi

Chapter 157
WAVE RUN-UP AND SEA CLIFF EROSION 2170
S.-M. Shih, P. Komar, K.J. Tillotson, W.G. McDougal, P. Ruggiero

Chapter 158
MEASUREMENT OF PARAMETERS, DIRECTION AND RATE OF
BEDFORM MIGRATION 2185
Ruben D. Kos'yan, I.S. Podymov

Chapter 159
BREACH-GROWTH RESEARCH PROGRAMME AND ITS PLACE
IN DAMAGE ASSESSMENT FOR A POLDER 2197
Arie W. Kraak, W.T. Bakker, J. van de Graaff, H.J. Steetzel,
P.J. Visser

Chapter 160
SWASH ZONE WAVE CHARACTERISTICS FROM SUPERTANK 2207
David L. Kriebel

Chapter 161
MORPHOLOGICAL MONITORING OF A SHOREFACE NOURISHMENT
NOURTEC EXPERIMENT AT TERSCHELLING, THE NETHERLANDS.........2222
Aart Kroon, Piet Hoekstra, Klass Houwman, Gerben Ruessink

Chapter 162
NUMERICAL MODEL FOR LONGSHORE CURRENT DISTRIBUTION
ON A BAR-TROUGH BEACH 2237
Yoshiaki Kuriyama

Chapter 163
PREDICTION OF BEACH PROFILE CHANGE AT MESOSCALE
UNDER RANDOM WAVES 2252
Magnus Larson

Chapter 164
A QUASI-3D SURF ZONE MODEL 2267
Jung Lyul Lee, Hsiang Wang

Chapter 165
QUANTIFICATION OF LONGSHORE TRANSPORT IN THE SURF
ZONE ON MACROTIDAL BEACHES 2282
Franck Levoy, Olivier Monfort, Helene Rousset, Claude Larsonneur

Chapter 166
A THREE DIMENSIONAL MODEL FOR WAVE INDUCED CURRENTS 2297
Bin Li, Roger J. Maddrell

Chapter 167
INFLUENCE OF OFFSHORE BANKS ON THE ADJACENT COAST 2311
N.J. MacDonald, B.A. O'Connor

Chapter 168
WAVE RUNUP ON COMPOSITE-SLOPE AND CONCAVE BEACHES 2325
Robert H. Mayer, D.L. Kriebel

Chapter 169
SIMULATION OF NEARSHORE WAVE CURRENT INTERACTION
BY COUPLING A BOUSSINESQ WAVE MODEL WITH A 3D
HYDRODYNAMIC MODEL 2340
Roberto Mayerle, Andreas Schröter, Werner Zielke

Chapter 170
EXPERIMENTAL RESULTS OF WAVE TRANSFORMATION ACROSS
A SLOPING BEACH 2350
Constantine D. Memos

Chapter 171
A RELATIVE INTERCOMPARISON BETWEEN VARIABLE WAVE
SHOALING, BREAKING AND TRANSITION ZONE FORMULATIONS 2365
G.P. Mocke, F. Smit

Chapter 172
ANALYSIS OF COASTAL PROCESSES AT TORONTO ISLANDS................2380
R.B. Nairn, R.D. Scott, C.D. Anglin, P.J. Zuzek

Chapter 173
STABILITY AND MANAGEMENT OF AN ARTIFICIAL BEACH.................2395
Naofumi Shiraishi, Hoiku Ohhama, Taiji Endo,
Patricia G. Pena-Santana

Chapter 174
SUSPENDED SEDIMENT PARTICLE MOTION IN COASTAL FLOWS..........2406
Peter Nielsen

Chapter 175
LONG-TERM MORPHODYNAMICAL DEVELOPMENT OF THE
EAST FRISIAN ISLANDS AND COAST 2417
Hanz D. Niemeyer

Chapter 176
FIELD OBSERVATION AND NUMERICAL SIMULATION OF BEACH
AND DUNE SCARPS...2434
Ryuichiro Nishi, Michio Sato, Hsiang Wang

Chapter 177
EFFECTS OF CONTROLLED WATER TABLE ON BEACH
PROFILE DYNAMICS 2449
Tae-Myoung Oh, R.G. Dean

Chapter 178
LABORATORY EXPERIMENTS ON 3-D NEARSHORE CURRENTS
AND A MODEL WITH MOMENTUM FLUX BY BREAKING WAVES...........2461
Akio Okayasu, Koji Hara, Tomoya Shibayama

Chapter 179
SUSPENDED SEDIMENT CAUSED BY WAVES AND CURRENTS 2476
Masanobu Ono, Kyu Han Kim, Toru Sawaragi, Ichiro Deguchi

Chapter 180
DEVELOPMENT OF A DUNE EROSION MODEL USING
SUPERTANK DATA...2488
Margery F. Overton, John S. Fisher, Kyu-Nam Hwang

Chapter 181
NUMERICAL MODELLING OF THREE-DIMENSIONAL
WAVE-DRIVEN CURRENTS IN THE SURF-ZONE..................................2503
Philippe Péchon, Charles Teisson

Chapter 182
SEDIMENT TRANSPORT IN VARIOUS TIME SCALES 2513
Zbigniew Pruszak, Ryszard B. Zeidler

Chapter 183
SEDIMENT TRANSPORT UNDER (NON)-LINEAR WAVES
AND CURRETNS 2527
*Jan S. Ribberink, Irene Katopodi, Khaled A.H. Ramadan,
Ria Koelewijn, Sandro Longo*

Chapter 184
WAVES AND CURRENTS AT THE EBRO DELTA SURF ZONE:
MEASUREMENTS AND MODELLING 2542
*A. Rodriguez, A. Sánchez-Arcilla, F.R. Collado, V. Gracia,
M.G. Coussirat, J. Prieto*

Chapter 185
A NUMERICAL SIMULATION OF BEACH EVOLUTION BASED
ON A NONLINEAR DISPERSIVE WAVE-CURRENT MODEL 2557
Shinji Sato, Michael B. Kabiling

Chapter 186
AN EXPERIMENTAL STUDY ON BEACH TRANSFORMATION DUE
TO WAVES UNDER THE OPERATION OF COASTAL DRAIN SYSTEM........2571
Michio Sato, Sadakatsu Hata, Masahiro Fukushima

Chapter 187
WAVE BREAKING AND INDUCED NEARSHORE CIRCULATIONS 2583
Ole R. Sørensen, Hemming A. Schäffer, Per A Madsen, Rolf Deigaard

Chapter 188
ACCURACY AND APPLICABILITY OF THE SPM LONGSHORE
TRANSPORT FORMULA...2595
J.S. Schoonees, A.K. Theron

Chapter 189
FIELD VERIFICATION OF A NUMERICAL MODEL OF BEACH
TOPOGRAPHY CHANGE DUE TO NEARSHORE CURRENTS,
UNDERTOW AND WAVES 2610
Takuzo Shimizu, Masahito Tsuru, Akira Watanabe

Chapter 190
AN ATTEMPT TO MODEL LONGSHORE SEDIMENT TRANSPORT
ON THE CATALONIAN COAST 2625
J.P. Sierra, A.I. Presti, A. Sánchez-Arcilla

Chapter 191
EROSION AND OVERTOPPING OF A GRASS DIKE LARGE
SCALE MODEL TESTS 2639
G.M. Smith, J.W.W. Seijffert, J.W. van der Meer

Chapter 192
CALCULATION OF TOMBOLO IN SHORELINE NUMERICAL MODEL 2653
Kyung Duck Suh, C. Scott Hardaway, Jr.

Chapter 193
BEACH IMPROVEMENT SCHEMES IN FALSE BAY 2668
D.H. Swart, J.S. Schoonees

Chapter 194
BEACH EROSION IN KUTA BEACH, BALI & ITS STABILIZATION 2683
Abdul R. Syamsudin, Yoshito Tsuchiya, Takao Yamashita

Chapter 195
FUNDAMENTAL CHARACTERISTICS OF A NEW WAVE
ABSORBING SYSTEM UTILIZING SAND LIQUEFACTION 2698
Shigeo Takahashi, S. Yamamoto, H. Miura

Chapter 196
LABORATORY STUDY OF SURF-ZONE TURBULENCE ON
A BARRED BEACH 2712
Francis C.K. Ting

Chapter 197
BEACH EROSION AROUND A SAND SPIT—AN EXAMPLE OF
MIHONO-MATSUBARA SAND SPIT 2726
Takaaki Uda, Koji Yamamoto

Chapter 198
SHORECIRC: A QUASI 3-D NEARSHORE MODEL 2741
    A.R. Van Dongeren, F.E. Sancho, I.A. Svendsen, U. Putrevu

Chapter 199
A MODEL FOR BREACH GROWTH IN SAND-DIKES..............................2755
    Paul J. Visser

Chapter 200
SCALING EFFECTS ON BEACH RESPONSE PHYSICAL MODEL 2770
    Xu Wang, Li-Hwa Lin, Hsiang Wang

Chapter 201
A NUMERICAL MODEL OF BEACH CHANGE DUE TO SHEET-FLOW 2785
    Akira Watanabe, Kazuhiko Shiba, Masahiko Isobe

Chapter 202
FIELD TESTS OF RADIATION-STRESS ESTIMATORS OF LONGSHORE
SEDIMENT-TRANSPORT 2799
    Thomas E. White

Chapter 203
SUSPENDED SEDIMENT CONCENTRATION PROFILES UNDER
NON-BREAKING AND BREAKING WAVES...2813
    Rattanapitikon Winyu, Tomoya Shibayama

Chapter 204
MASS TRANSPORT AND ORBITAL VELOCITIES WITH
LAGRANGEIAN FRAME OF REFERENCE 2828
    Stefan Woltering, Karl-Friedrich Daemrich

Chapter 205
CROSS-SHORE PROFILE MODELLING UNDER RANDOM WAVES............2843
    Yongjun Wu, H-H Dette, H. Wang

Chapter 206
MULTIPLE BAR FORMATION BY BREAKER-INDUCED VORTICES:
A LABORATORY APPROACH...2856
    Da Ping Zhang, Tsuguo Sunamura

Chapter 207
IDENTIFICATION OF SOME RELEVANT PROCESSES IN
COASTAL MORPHOLOGICAL MODELLING..2871
    Hakeem Johnson, Ida Brøker, Julio A. Zyserman

Chapter 208
EXPERIMENTAL SHOREFACE NOURISHMENT, TERSCHELLING, (NL) 2886
J.P.M. Mulder, J. van de Kreeke, P. van Vessem

PART V

Chapter 209
IMPORTANCE OF PERMEABILITY IN THE SEDIMENTATION
CONSOLIDATION PROCESS 2902
Alain Alexis, P. Thomas

Chapter 210
WAVE-CURRENT INTERACTION WITH MUD BED 2913
Nguyen Ngoc An, T. Shibayama

Chapter 211
ON RESIDUAL TRANSPORT IN SHALLOW TIDAL BASINS 2928
Andrea Balzano

Chapter 212
THE EXTENT OF INLET IMPACTS UPON ADJACENT SHORELINES 2943
Kevin R. Bodge

Chapter 213
THE SPREADING OF DREDGING SPOILS DURING CONSTRUCTION
OF THE DENMARK-SWEDEN LINK 2958
*Ida Brøker, John Johnsen, Morten Lintrup, Anders Jensen,
Jacob Steen Møller*

Chapter 214
EXPERIMENTAL STUDIES ON THE EFFECT OF THE DREDGING
ON CHANG-HWA RECLAMATION AREA, TAIWAN 2972
Tai-Wen Hsu, Hsien-Kuo Chang

Chapter 215
OSCILLATIONS INDUCED BY IRREGULAR WAVES IN HARBOURS 2987
C.R. Chou, W.Y. Han

Chapter 216
MECHANISM AND ESTIMATION OF SEDIMENTATION IN
BANGKOK BAR CHANNEL 3002
*Ichiro Deguchi, Toru Sawaragi, Masanobu Ono,
Sucharit Koontanakulvong*

Chapter 217
TURBIDITY & SUSPENDED SEDIMENT ASSOCIATED WITH
BEACH NOURISHMENT DREDING 3016
Daniel M. Hanes

Chapter 218
SOIL MECHANICS OF SHIP BEACHING 3030
N.-E. Ottesen Hansen, B.C. Simonsen, M.J. Sterndorff

Chapter 219
THE DILUTION PROCESSES OF ALTERNATIVE HORIZONTAL
BUOYANT JETS IN WAVE MOTION 3045
*Hwung-Hweng Hwung, Jih-Ming Chyan, Chen-Yue Chang,
Yih-Far Chen*

Chapter 220
STUDY ON THE BEHAVIORS OF THE COHESIVE SEDIMENT
IN THE YANGTZE ESTUARY 3060
Xiaochuan Zeng, Yixin Yan, Kai Yen

Chapter 221
DISPERSION OF POLLUTION IN A WAVE ENVIRONMENT 3071
R. Koole, C. Swan

Chapter 222
MODEL OF BIVALVE ON/OFFSHORE MOVEMENT BY WAVES 3086
Hisami Kuwahara, Junya Higano

Chapter 223
INITIAL GAP IN BREAKOUT OF HALF-BURIED SUBMARINE
PIPE DUE TO WAVE ACTION 3099
Adrian W.K. Law, M.A. Foda

Chapter 224
ANALYTICAL SOLUTION FOR THE WAVE-INDUCED EXCESS
PORE-PRESSURE IN A FINITE-THICKNESS SEABED LAYER 3111
Waldemar Magda

Chapter 225
RESPONSE CHARACTERISTICS OF RIVER MOUTH TOPOGRAPHY
IN WIDE TIME SCALE RANGE 3126
A. Mano, M. Sawamoto, M. Nagao

Chapter 226
HALF-LIFE PERIOD OF SEDIMENTATION—MODEL TEST
ON A SCALE 1:1 3139
  Helmut Manzenrieder

Chapter 227
RISK ASSESSMENT FOR COASTAL AND TIDAL DEFENCE SCHEMES 3154
  I.C. Meadowcroft, P.H. von Lany, N.W.H. Allsop, D.E. Reeve

Chapter 228
WATER OXYGENATION IN THE VICINITY OF COASTAL
STRUCTURES DUE TO WAVE BREAKING 3167
  C.I. Moutzouris, E.I. Daniil

Chapter 229
ENVIRONMENTAL ASSESSMENT OF HYPOTHETICAL
LARGE-SCALE RECLAMATION IN OSAKA BAY, JAPAN 3178
  Keiji Nakatsuji, Toshiaki Sueyoshi, Kohji Muraoka

Chapter 230
FLOOD AND EROSION CONTROL IN THE CONTEXT OF
SEA-LEVEL RISE 3193
  E. Bart Peerbolte, Herman G. Wind

Chapter 231
OFFSHORE BREAKWATERS VERSUS BEACH NOURISHMENTS—
A COMPARISON 3208
  M. Pluijm, J.C. van der Lem, A.W. Kraak, J.H.W. de Ruig

Chapter 232
MORPHOLOGICAL MODELLING OF KETA LAGOON CASE 3223
  J.A. Roelvink, D.J.R. Walstra, Z. Chen

Chapter 233
COST-BENEFIT ANALYSIS OF SHORE PROTECTION INVESTMENTS 3237
  L. Felipe Vila Ruiz, Fernando Bernaldo de Quirós

Chapter 234
PARAMETRIZATION FOR CONCEPTUAL MORPHODYNAMIC
MODELS OF WADDEN SEA AREAS 3251
  Ernst Schroeder, R. Goldenbogen, H. Kunz

Chapter 235
MUD TRANSPORT AND MUDDY BOTTOM DEFORMATION
BY WAVES 3266
Daoxian Shen, Masahiko Isobe, Akira Watanabe

Chapter 236
WAVE-INDUCED SEDIMENT RESUSPENSION AND MIXING IN
SHALLOW WATERS...3281
Y. Peter Sheng, X. Chen, E.A. Yassuda

Chapter 237
A NUMERICAL MODEL FOR BEACH DEFORMATION AROUND
RIVER MOUTH DUE TO WAVES AND CURRENTS..................3295
Tomoya Shibayama, Akiko Yamada

Chapter 238
DEVELOPMENT OF A NUMERICAL SIMULATION METHOD FOR
PREDICTING THE SETTLING BEHAVIOR & DEPOSITION
CONFIGURATION OF SOIL DUMPED INTO WATERS 3305
Kazuki Oda, Takaaki Shigematsu

Chapter 239
PLANE DESIGN OF "SPAC"; COUNTERMEASURE AGAINST SEABED
SCOUR DUE TO SUBMERGED DISCHARGE AND LARGE WAVES............3320
Takao Shimizu, Masaaki Ikeno, Hisayoshi Ujiie, Kazuaki Yamauchi

Chapter 240
MODELLING AND ANALYSIS TECHNIQUES TO AID MINING
OPERATIONS ON THE NAMIBIAN COASTLINE 3335
G.G. Smith, G.P. Mocke, D.H. Swart

Chapter 241
ENGINEERING APPROACH TO COASTAL FLOW SLIDES 3350
Theo P. Stoutjesdijk, Maarten B. de Groot, Jaap Lindenberg

Chapter 242
EROSION OF LAYERED SAND-MUD BEDS IN UNIFORM FLOW...............3360
Hilde Torfs

Chapter 243
STANDING WAVE INDUCED SOIL RESPONSE IN A POROUS SEABED 3369
Ching-Piao Tsai, Tsong-Lin Lee

Chapter 244
EROSION CONTROL BY CONSIDERING LARGE SCALE COASTAL
BEHAVIOR 3378
    Yoshito Tsuchiya, Takao Yamashita, Tatsuhisa Izumi

Chapter 245
SHORELINE EROSION DUE TO OFFSHORE TIN MINING 3393
    Suphat Vongvisessomjai

Chapter 246
BEHAVIORS OF FLUID MUD UNDER OSCILLATORY FLOW 3408
    Hiroyuki Yamanishi, Tetsuya Kusuda

Chapter 247
FORMATION OF HABITATS FOR BIVALVES BY PORT AND
HARBOR STRUCTURES 3420
    Kenji Yano, S. Akeda, Y. Miyamoto, S. Kuwabara

Chapter 248
USE OF THREE-DIMENSIONAL HYDRODYNAMICS MODEL FOR
TIDAL INLETS STUDIES 3432
    E.A. Yassuda, Y.P. Sheng

Chapter 249
STUDY OF UPWELLING PHENOMENA OF ANOXIC WATER "A-OSHIO" 3447
    Jong Seong Yoon, Keiji Nakatsuji, Kouji Muraoka

Chapter 250
SEA LEVEL RISE AND COAST EVOLUTION IN POLAND 3462
    Ryszard B. Zeidler

PART VI

Chapter 251
FORMATION OF DYNAMICALLY STABLE SANDY BEACHES
ON AMANOHASHIDATE COAST BY SAND BYPASSING 3478
    Ikuo Chin, Minoru Yamada, Yoshito Tsuchiya

Chapter 252
THE RECONSTRUCTION OF FOLLY BEACH 3491
    Billy L. Edge, Millard Dowd, Robert G. Dean, Patrick Johnson

Chapter 253
THE COMPLEMENTARY INTERACTION BETWEEN BEACH
NOURISHMENT AND HARBOUR MANAGEMENT: FOUR CASES
IN SPAIN 3507
  Gregorio Gómez-Pina, Jose L. Ramírez

Chapter 254
WAVE IMPACTS ON THE EASTERN SCHELDT BARRIER
EVALUATION OF 5 YEARS FIELD MEASUREMENTS 3522
  Leo Klatter, Hans Janssen, Michiel Dijkman

Chapter 255
SALINITY & WATER LEVELS IN THE WESER ESTUARY DURING
THE LAST HUNDRED YEARS—ANTHROPOGENIC INFLUENCES
ON THE COASTAL ENVIRONMENT 3533
  Hans Kunz

Chapter 256
REVIEW OF SOME 30 YEARS BEACH REPLENISHMENT
EXPERIENCE AT DUNGENESS NUCLEAR POWER STATION, UK 3548
  Roger Maddrell, Bill Osmond, Bin Li

Chapter 257
PROJECT, WORKS AND MONITORING AT BARCELONA
OLYMPIC BEACHES 3564
  Carlos Peña, Manuel F. Covarsi

Chapter 258
SANTA CRISTINA BEACH NOURISHMENT WORKS AND
MONITORING PROGRAM 3579
  Eduardo Toba, G. Gomez-Pina, J. Alvarez

Chapter 259
DESIGN & CONSTRUCTION OF AN EXTENDED BERM
BREAKWATER AT PORT OF HAINA, DOMINICAN REPUBLIC 3594
  David W. Yang, Mark H. Lindo, Edward J. Schmeltz,
  Joaquin Fernandez, Daniel Gomez

Chapter 260
REHABILITATION OF THE WEST BREAKWATER—
PORT OF SINES 3608
  Orville T. Magoon, J.R. Weggel, B.L. Edge, E. Mansard,
  R.W. Whalin, W.F. Baird

Subject Index .... .3615
Author Index .... .3621

The Present and Future of Coastal Engineering in Japan

Yuichi Iwagaki[1]

The subject I am going to talk on is the present and future of Coastal Eng. in Japan, which is rather large-scale topics both in time and space. Because of the limits of my abilities and time appointed for this speech, I should like to scale down the subject in space, and focus on the present and future of Coastal Eng. in Kansai and Chubu regions, mainly in Osaka Bay Area including here, City of Kobe.

Change in Coastal Engineering in Japan

I should like to start with describing very briefly a change in Coastal Eng. in Japan. As you all may know, the 1st International Conference on Coastal Eng. took place at Long Beach in 1950. Four years later, in 1954, the 1st Japanese Conference on Coastal Eng. was held here at Kobe. At this stage, we experienced frequent attacks of Typhoon that caused serious disasters. That is, if we count Typhoons that caused more than 100 death and missing for every ten years, 13 Typhoons occurred during the period from 1945 to 1954. But the number reduced very drastically to 5 from 1955 to 1964, and 4 from 1965 to 1974 (see, Table 1). This change reflected our research topics as shown in Table 2. Until 1964, Japanese Conference on Coastal Eng., which takes place every year, covered mainly the topics and problems of waves, storm surge, tsunami and coastal disasters. Around 1965, new study fields were introduced in the Conference; those include tidal currents, density currents, and sea water exchange. These new fields were motivated by the increased demand for reclamation in the coastal zone. At this stage, Japan was just entering a high-growth period of the economy. Eventually, the high level of

Table 1 Typhoon which generated over one hundred of killed and missing (after 1945)

date of occurrence	name of typhoon	killed and missing (persons)
1945. 9. 17	Makurazaki	3,756
1945. 10. 10	Akune	877
1947. 9. 13	Catharine	1,930
1947. 9. 15	Ion	838
1949. 6. 15	Delia	468
1949. 8. 15	Judith	179
1949. 8. 31	Kitty	160
1950. 9. 3	Jane	509
1951. 10. 13	Ruth	943
1952. 6. 22	Dinah	135
1953. 9. 24	No. 13	478
1954. 9. 10	No. 12	107
1954. 9. 25	Toyamaru	1,761
1958. 9. 26	Kanogawa	1,175
1959. 8. 12	No. 7	235
1959. 9. 26	Isewan	5,041
1961. 9. 15	Daini Muroto	202
1961. 10. 25	No. 26	114
1966. 9. 17	No. 24, 26	314
1968. 8. 10	Hidagawa	135
1971. 8. 27	No. 23, 25, 26	147
1974. 5. 29	No. 8	143
1975. 8. 5	No. 5, 6	143
1976. 9. 7	No. 17	169
1979. 10. 14	No. 20	111

---

[1] Dean, Faculty of Science and Technology, Meijo University, Shiogamaguchi 1-501, Tempaku-ku, Nagoya 468, Japan

Table 2   Change in topics of Coastal Engineering in Japan

Period	Special topics in Proc. of Coastal Eng., JSCE
1954 - 1964	Coastal disaster, Waves, Storm surge, Tunami
1965 - 1979	Tidal current, Density current, Sea water exchange, Heated water discharge
1980 - at present	Water pollution, Environmental problem, Sea level rise

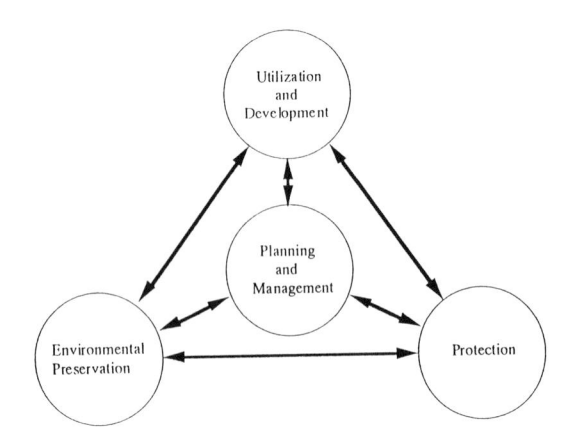

Fig.1   Mutual relation between coastal activities

economic growth brought various pollution problems. Water pollutants are one of the coastal problems. In 1976, we started to carry out study on the water quality, for example, the problems connected with dissolved oxygen. Since then, about 20 percent of the papers presented at the Japanese conference aims at the improvement of the water quality. I looked over the 93's proceedings, and found that 50 out of 236 papers, that comes up to 21 percent, still treats the environmental problems including the water pollution. The ratio seems to be unchanged for the last 25 years.

For the last 15 years, the preservation of coastal environments is highlighted. Global environment issues, especially sea level rise and its impact on the coastal zone become our problems. In addition, we become more and more aware of conservation of ecosystem and natural environment. Future coastal works have to be undertaken in a coordinated manner with the nature conservation. Fig.1 shows a conception of the coastal management. Further utilization of coastal resources has to be carefully planned satisfying the sometimes conflicting requirements of protection, development and conservation.

Coastal Protection

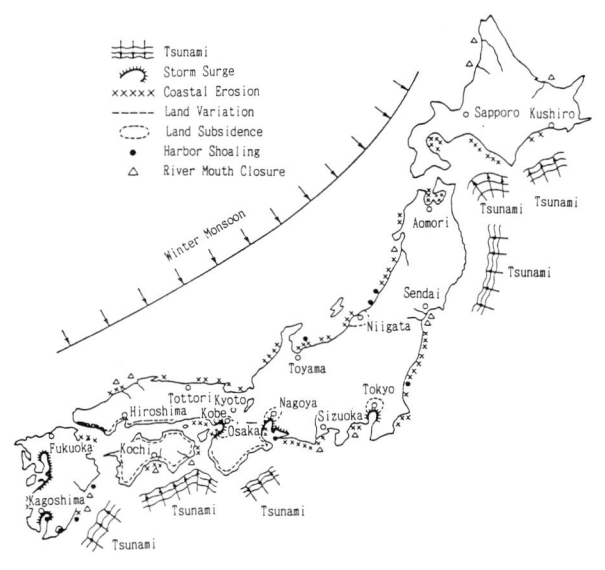

Fig.2 Illustrative map of usual coastal disaster places in Japan

Table 3 Storm surges of over 2m anomaly by typhoons during 1900-1987

date	place	max. anomaly (m)	cause
1917. 10. 1	Tokyo Bay	2.1	Typhoon
1930. 7. 18	Ariake Bay	2.5	Typhoon
1934. 9. 21	Osaka Bay	3.1	Ty. Muroto
1938. 9. 1	Tokyo Bay	2.2	Typhoon
1950. 9. 3	Osaka Bay	2.4	Ty. Jane
1956. 8. 17	Ariake Bay	2.4	Ty. No. 9
1959. 9. 26	Ise Bay	3.5	Ty. Isewan
1961. 9. 16	Osaka Bay	2~2.5	Ty. Daini Muroto
1964. 9. 25	Osaka Bay	2	Ty. No. 20
1965. 9. 10	East of Seto Island Sea	2.2	Ty. No. 23
1970. 8. 21	Tosa Bay	2.4	Ty. No. 10
1972. 9. 16	Ise Bay	2.0	Ty. No. 20

Next we will look briefly at the respective fields; coastal protection, utilization and environmental preservation in Japan, first at the coastal protection. Natural conditions surrounding the nation are very severe as one can tell from the geographical location of Japan. There is constant danger of the tsunami attack due to earthquake, the storm surge due to Typhoon, and the erosion due to waves (see, Fig.2). Since Japan belongs to Circum-Pacific earthquake belt, earthquakes and tsunamis associated with large earthquakes often occur. Many earthquakes caused severe damage, and tsunami sometimes caused still more loss of life and property damage along the coastline. Japan also lies on the course of Typhoon born in Southern Ocean, and so suffers various disasters. The average number of Typhoons is 25 to 30 per year, 31 for this year up to

Table 4 Some examples of design waves

Name of Coast	Prefecture	Design Waves Wave Height(m)	Design Waves Wave Period(sec)	note
Tsugaru	Aomori	5.70	–	monsoon
Sakata	Yamagata	8.30	12.0	monsoon
Kaetsu	Fukui	8.30	14.0	monsoon
Kochi	Kochi	10.10	15.2	model Typhoon
Fuji-Yoshihara	Sizuoka	20.00	20.0	model Typhoon

now, and about 4 out of them strike Japan. Major coastal disasters caused by Typhoon are due to storm surges. Storm surge disasters often occur especially when Typhoon passes on the western side of large bay opening south, for example, Tokyo Bay, Ise Bay, Osaka Bay, Ariake Bay and Kagoshima Bay. Table 3 shows the area of storm surge and the name of Typhoon for period of 1900 through 1987 for the cases where the maximum meteorological tide anomaly exceeds 2 m. The largest anomaly was observed at the Nagoya Port due to the Ise Bay Typhoon in September 1959. It was recorded 3.5 m. Until now, for the past 90 years, this record is the largest. I myself carried out statistical analysis on the return period using the observed data for the past 42 years. My analysis gives that the return period of the value of 3.5m is 200 to 400 years.

By contrast, the coast of Japan Sea is most threatened with attack of the monsoon during winter. According to the record at Hajikizaki in Sado Island, the maximum velocity of the wind was 31.7 m/s, and the north-northwest wind continued to blow at more than 10m/s for 21 hours. High wind waves are generally developed, and thus the design wave height adopted for coastal dikes and revetments is typically 8 to 9 m and the period of 12 to 14 s. The design waves for the Pacific coast are generally based on the waves generated by Typhoons. For example, the design wave height and period for the Kochi Coast is 10.1m and 15.2s, and 20m and 20s for the Yoshihara Coast in Suruga Bay which is the maximum design wave in Japan (see, Table 4). Problems caused by action of such high waves are the beach erosion. Coastal erosion is influenced by various factors, but chiefly caused by a shortage of beach material supplied from the river, and by blocking the movement of the coastal sediment with the construction of port and harbor structures, and the reclamation. In other words, the beach erosion is a fatal problem associated with the utilization and the development of river and coastal zone.

We now look at the social environment of the coastal zone in Japan. There are 47 Prefectures in Japan. Among them, only 8 Prefectures, are not adjacent to the sea. These account for 14 percent in the total area and 15 percent in the total population. That is to say, 85 percent of the Japanese people are adjacent to the sea and receive benefit from the sea. The total population of Tokyo, Chiba and Kanagawa Prefectures along Tokyo Bay accounts for 20 percent, Aichi and Mie Prefectures occupy 7 percent in the total population, and Osaka, Hyogo and Wakayama Prefectures occupy 12 percent as summarized in Table 5. The total population of these three major bay areas comes up to 39 percent of the national population. A large city which has a population of more than one million is also distributed along these major bay areas; Tokyo, Yokohama and

Table 5 Concentration of population and cities in three bays

Name of bay	Population (million)	Name of city (over 1 million)
Tokyo	25.11 (20%)	Tokyo, Yokohama, Kawasaki
Ise	8.41 (7%)	Nagoya
Osaka	15.01 (12%)	Osaka, Kobe
Total	48.53 (39%)	

Kawasaki along Tokyo Bay, Nagoya along Ise Bay, and Osaka and Kobe along Osaka Bay. The population highly concentrates along these three bay areas. Major ports are also located along these bays; Port of Kobe, Chiba, Nagoya, Yokohama and Kawasaki. The volume of cargo handled at these five ports ranks top five in Japan. The ports along three major bays handle 75 percent of the total exports and 60 percent of the total imports.

Measures against storm surges were well planned and taken in three major bays, Tokyo Bay, Ise Bay and Osaka Bay where population and industry highly concentrate. Coastal dikes, seawalls and watergates were constructed to prevent sea water penetration. In designing these structures, the most disastrous Ise Bay Typhoon is generally used to determine the sea level and wave height assuming the typhoon route which generates the worst effect at the location of interest.

Tsunami countermeasures are very difficult to establish because the tsunami frequently attacks small towns located at the back of a deeply indented rias bay. The past highest tsunami record at the location of interest is used to determine the design sea level. Tsunami seawall 10m in height was constructed at Taro-cho in Iwate Prefecture, and tsunami breakwaters were constructed at Ofunato Bay in Iwate Prefecture and elsewhere. Recent tsunami disaster occurred in Okusiri Island due to the South West off Hokkaido in July 1993, however, indicates that even after the experience of the tsunami disaster due to the 1983 Japan Sea earthquake we were not able to prevent the disastrous damage and lost 200 lives. I believe that the best measures are to establish a speedy refuge system and a speedy and accurate warning system.

Needless to say, the forecasting and observation of incoming waves are essential to the coastal protection. As we all realize, ocean waves are not monochromatic but random, so in designing coastal structures the theory for monochromatic waves cannot be always applied to random wave transformation, random wave forces, random wave runup and overtopping. Dr. Yoshimi Goda, professor of Yokohama National University, has been investigating this problem for a long time, and successfully established a method for introducing the random wave property; for example, he introduced the directional spectrum into the design forces for port and harbor structures. His works on this problem should be recognized as a great success in coastal, port and harbor engineering research in Japan. We are indeed proud of Dr. Goda. Recently, the coastal engineering committee of JSCE organized the subcommittee for the purpose of reviewing existing papers on the coastal waves, including the computing techniques of wave transformation, tsunami transmission, storm surge generation, wave-structure interaction and wave-soil interaction. The first work of the subcommittee was done and the result was published, in this July but unfortunately in Japanese.

I think that one of the most difficult problem we are faced with is beach erosion. As I said before, the beach erosion is mostly fatalistic in retaliation for revolution of nature by human. As countermeasures for the beach erosion, we used to adopt jetties, offshore breakwaters, sand nourishment works, bypassing works, artificial reefs, headland works and so on. These measures are, however, not permanent but temporal in most cases, which means we just gain time. We should try to enlarge time and space in our view, and reconsider the human activity once more. I believe that a fundamental countermeasure is tightly connected with our philosophy of nature in the human life.

Utilization and Development of the Coastal Zone

There is not enough time to go into details, but I like to call your attention now to the utilization and development of the coastal zone. Japan extends 2,600 km from northern Hokkaido to southern Okinawa. Four islands, Hokkaido, Honsyu, Shikoku and Kyushu are connected each other by the long-span bridges and or submarine tunnels. At present, we have the vision of the national land axis, which is defined as the

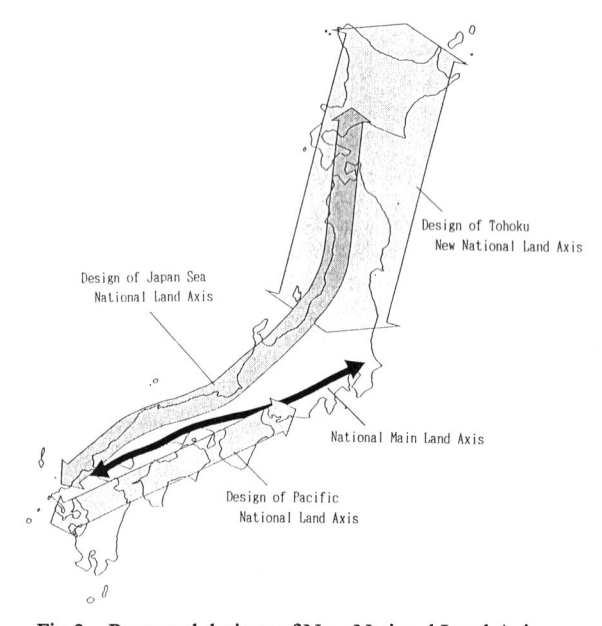

Fig.3   Proposed designs of New National Land Axis

common space providing information and communication, and the extensive economical area as sketched in Fig.3 . Along this axis, highways, railways, cities and industries are planned to develop. More recently, in addition to the main axis, various designs of new national land axis are proposed as shown in Fig.3 . Among the proposed axises, the most expected one is the Pacific new national land axis, which runs 800km from the western Kyushu to Chukyo Area through the central Kyushu, Shikoku, Awaji Island and Kii Peninsula. If this new design axis will be put into practice in the next century, long-span bridges across the channels or submarine tunnels will be constructed at Irako channel of the Ise Bay mouth, Tomogashima channel of the Osaka Bay mouth, and Hayasui channel between Kyushu and Shikoku Islands. These big projects will make coastal engineering researches more active , and contribute the further development of the coastal zones along the new axis.

One of the large-scale utilization at the coastal zone is by a man-made island. A typical example is an offshore airport. The first one in Japan is the Nagasaki Airport in Ohmura Bay, which was constructed on the small island, Minoshima, but needed some reclamation works. Nagasaki Airport was opened with the runway of 3,000m length in 1980. The second offshore airport is the Kansai International Airport, which many of you arrived at for this conference. The airport was open on September 4th of this year as the nation's first airport available for 24 hours. The present airport was constructed on the basis of the first stage plan. The construction of man-made island of 511ha needed seven years since starting the work. The expense was as much as 1 trillion 5 hundred billion yen and the soil for reclamation was as much as 180 million cubic meters. The airport is located 5km off the coast to keep the craft noise on the land less than a critical value 70 according to the WECPNL index (see, Fig.4) . The average water depth of the airport site is 18m . The geological survey of the seabed showed the alluvial clay layer of 20m thick, and below that, the Pleistocene clay layer. So, the foundation improvement was necessary to accelerate the consolidation settlement. The improvement by using over one million sand piles was actually performed for the alluvial clay layer. Expected average settlement of the whole airport is estimated 10m at the time of opening and 11.5m in 50 years as shown in Fig.5 . For the time being, additional reclamation of the offshore side is being investigated for the overall plan for 3 runways and 1,200 ha. area. As the currently planned offshore airports, we have Chubu International Airport in Ise Bay (see, Fig.6), which is expected open in 2005, Kobe Airport expected open in 2003, and New Kitakyushu Airport in 2005. At the site of the Chubu International Airport, the Observation

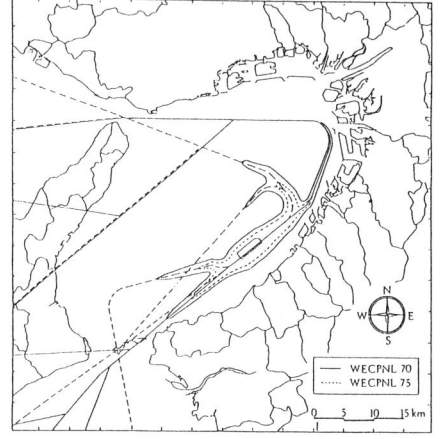

Fig.4 Aircraft noise estimation
[WECPNL:Weighted Equivalent Continuous Perceived Noise Level]

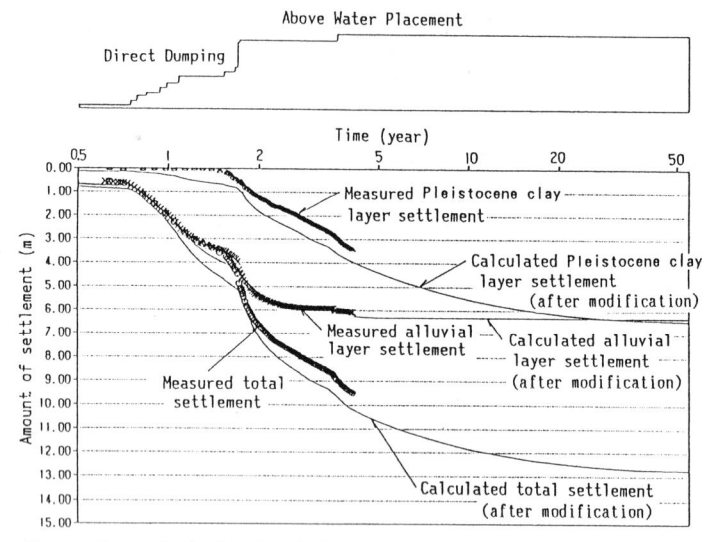

Fig.5 Comparison of calculated and observed settlement in the pilot area of the airport island   (After Kansai International Airport Project, JSCE, Sept. 1992)

Tower for measuring various environmental factors has been already constructed and in operation.

I should like to make special mention of enactment of the special law for Osaka Bay Area Development in 1992. By this law, Osaka Bay Area is expected to develop greatly in the 21st century, and at present a grand design of the development by 2025 is under review.

Environmental Preservation

There are some other facilities for the utilization of coastal zone, such as reclamation for the waterfront industrial area and for the agricultural farm, electric power stations, oil storage station, seaports, fishery ports, waste disposal facilities, marinas and so on. The

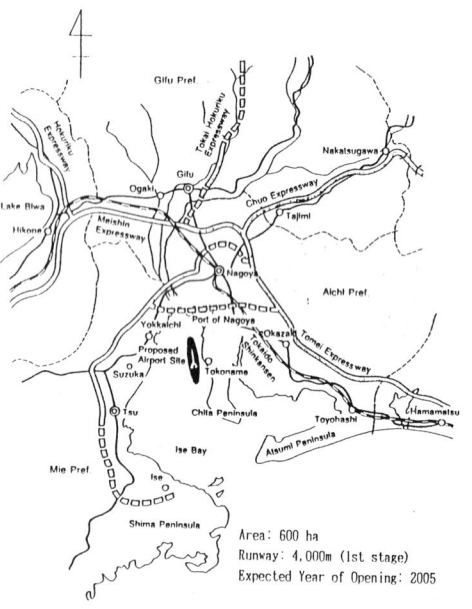

Fig.6 Proposed site of Chubu International Airport

Area: 600 ha
Runway: 4,000m (1st stage)
Expected Year of Opening: 2005

construction of such facilities has much possibility of destroying natural environment and worsening living environment. The plan for the projects for the coastal zone is sometimes stopped and withdrawn by the opposition of residents, fishermen and nature-conservation groups. Whenever a big civil engineering project is planned, the environmental impact assessment has to be generally made by law. The further advancement of coastal engineering is definitely required to make a more precise assessment. For Seto Inland Sea including Osaka Bay, the special law named as Act on Special Measures for the Seto Inland Sea Environmental Conservation was enacted in 1978, which covers the various regulations concerning the environmental conservation. The development of Osaka Bay Area is subject to the regulations by this special law.

As for the improvement of water quality and landscape of the coastal zone, the methods of artificial beach nourishment, sand banking on the sea bottom, artificial formations of tideland and seaweed bed have been proposed. The effects of these improvement methods for the sea water quality have not been cleared yet.

More global environment issues arise from the fact that the atmospheric concentration of carbon dioxide has increased rapidly since the industrial revolution in the second half of the 18th century, and is still increasing at the rate of 1.5ppm per year. The rapid concentrations induce the greenhouse effect and warm the surface temperature of the earth. The mass added to oceans due to deglaciation and the thermal expansion of oceans' volume due to heat diffusion will result in the global sea level rise. According to IPPC, the sea level rise reaches 0.3m to 1.1m, in average 0.65m by the end of the 21st century. This problem is very important to coastal engineers. Recently, the subcommittee under the coastal engineering committee of JSCE has investigated the possible effects of sea level rise on the coasts in Japan and the possible countermeasures. The final report was published in July of this year. I think this report will stimulate the further progress of coastal engineering research.

Concluding Remarks

In closing my speech, I should like to summarize my opinion to promote a sustainable development of the coastal zone. First of all, in planning coastal projects we should consider the harmony of mutual relations between the utilization and development, the protection and preservation, and the environmental conservation. This means Harmony with Nature, the key word of this conference. Secondly, from a viewpoint of the paradigm of coastal engineering, we should include not only the field based on the mechanics but also the field related to environmental problems in the research topics. Third, in planning coastal zone projects, we should consider the multifarious values, that is, not only the economical value but also the environmental and the social-cultural values. Finally, we should establish the mitigation engineering in the near future.

PART I

Characteristics of Coastal Waves and Currents

CHAPTER 1

On The Characteristics of One-Dimensional Spectra
and Non-Dimensional Parameters of Wind Waves

Toshio Aono*   Chiaki Goto**

Abstract
Similarity relationship among non-dimensional significant wave parameters are discussed which is based upon the 3/2 power law. The characteristics of the wind wave spectra in deep water are investigated by using the parameters of JONSWAP spectrum and the 3/2 power law. From theoretical and empirical arguments, it is confirmed that a f^{-4} power law exists at high frequency range, that JONSWAP spectrum parameter γ and σ are varied with fetch, and that parameter α and γ satisfied a $-1/3$ power law. In shallow water region, spectral form of wind waves is varied with the shoaling coefficient. Through the analysis of the wind wave spectra, a new spectral formula is obtained.

1.Introduction
Relationship between the marine surface wind and wind waves gives us the basic knowledge in field as diverse as air-sea interaction, wave hindcasting and engineering design of maritime structures.

Many researchers pointed out that the frequency spectra of wind-generated gravity waves shows similarity. It is well known that a similarity law is applicable to the spectral form of wind waves. Phillips (1959) derived the f^{-5} power law using the similarity and dimensional analysis argument. Many functional forms of the wind wave spectrum have been proposed in the past, such as Pierson-Moskowitz (1964), Bretshneider-Mitsuyasu (1970) and JONSWAP (1973). However, such spectral forms represent the fully developed condition of wind waves, except the

---
* Senior Research Engineer, Technical Research Institute, Toa Corporation, 1-3 Anzen-cho, Tsurumi-ku, Yokohama 230, Japan
** Professor, Department of Civil Engineering, Tokai Univ., 1117 Kita-kaname, Hiratsuka city, Kanagawa 259-12, Japan

JONSWAP, so that in the case of rapid development of wind waves by typhoon, observed and calculated spectral forms do not show good agreement, especially near the peak frequency. Recently, Toba(1973) derived the f^{-4} power law based on 3/2 power law. This argument of spectral gradient was related to the resistance law between the wind and the wind waves.

These characteristics of wind waves are investigated in deep water, but field observation data is limited at shallow water region, especially under high wave condition. The standard form of shallow water spectrum is proposed by Thornton (1977) and Bouse et. al (1985). These spectrum is based upon the f^{-5} power law.

The present work reveals the quantitative relationship linking marine surface wind and wind waves. The paper discusses the significant wave parameters and the frequency spectrum of wind waves by using the 3/2 power law relation.

2.Observation of wave data

In this study, field observation wave data are used. To get long fetch wave data, 13 wave observation points around the coasts of Japan are used, and to get short fetch wave data, Meteorological station or MT station at Osaka bay is used.

(1) Ocean wind waves

Figure. 1 shows the wave observation system called as NOWPHAS operated by Port and Harbour Research Institute, Ministry of Transport. There are 41 wave observation points. In this study, ocean wave data obtained at 13 major points around the Japan coast are used. These points are located at relatively deep water and there are no topography effects. Table 1 shows the observation systems and

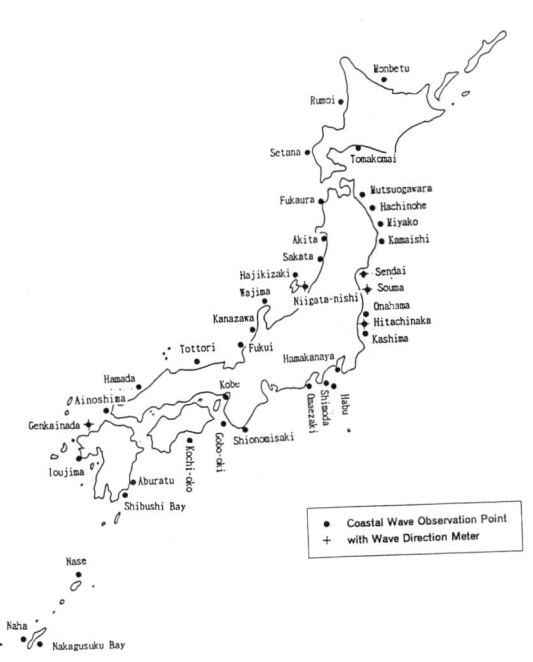

Fig. 1 The NOWPHAS system

Table 1 Observation points

Location	Wavegage Type	Depth(m)	Latitude	Longitude
Hajikizaki	U S W	- 5 4	38° 20'39" N	138° 30'25" W
Wajima Port	U S W	- 5 0	37° 25'40"	136° 54'19"
Fukaura Port	U S W	- 4 9 . 6	40° 39'25"	139° 54'57"
Mutuogawara Port	U S W	- 4 9	40° 55'20"	141° 25'40"
Hitachinaka Port	U S W	- 3 0	36° 23'24"	140° 39'36"
Habu Port	U S W	- 4 9	34° 40'23"	139° 27'18"
Kouchi-oki	A W	- 1 2 0	33° 15'24"	133° 30'06"
Hamada Port	U S W	- 5 1	34° 54'07"	132° 02'21"
Aburatu Port	U S W	- 4 8 . 5	31° 33'27"	131° 26'32"
Setana Port	U S W	- 5 2 . 9	42° 26'30"	139° 49'16"
Monbetu Port	U S W	- 5 2	44° 24'58"	143° 26'00"
Naha Port	U S W	- 5 1	26° 15'19"	127° 38'56"
Nakagusuku Bay	U S W	- 5 0	26° 14'15"	127° 58'10"

condition of each points. The sampling frequency of wave data is 2 Hz. Waves are defined by using the zero up crossing method. Frequency spectra are calculated by using the FFT method. The number of total observation cases are 2546.

These observation points have no wind profile. Thus, a data of the wind waves are determined by using the following criteria.
a) From the time series of significant parameters, significant wave height should be in its developed stage.
b) The swell component is not included in wave data clearly. Thus, a spectral form should be a single peak spectrum.
c) The JONSWAP spectrum parameter γ should be greater than 1.

(2) Wind waves of short fetch

Short fetch waves are obtained at MT station in Osaka Bay. Figure 2 shows the location of MT station. The water depth is 17m. The wave data and wind velocity at 10m high from sea surface are measured at MT station. Sampling frequency is 10 Hz. The analysis method is the same as for ocean waves. The duration of observation used is from 1984 to 1991. The FFT analysis is applied

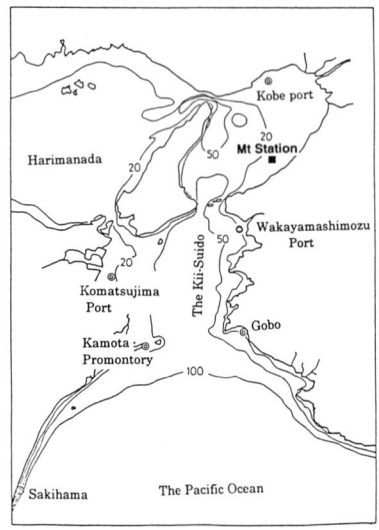

Fig. 2 Location of MT station

to the duration of storm or typhoon. The number of total observation data is 70080 and the data analyzed by FFT is 246.

3. Non-dimensional parameters of wind waves

The non-dimensional parameters that describe the characteristics of wind waves are as follows.

$$\frac{gH_{1/3}}{u_*^2}, \frac{gT_{1/3}}{u_*}, \frac{g^2E}{u_*^4}, \frac{gF}{u_*^2}, \frac{f_m u_*}{g}, \frac{C}{u_*}, \frac{H_{1/3}}{L_{1/3}} \quad (1)$$

where $H_{1/3}$ is the significant wave height, $T_{1/3}$ the significant wave period, E the wave energy, F the fetch, f_m the peak frequency of spectrum, C the wave celerity, $L_{1/3}$ the wave length corresponding to the significant wave period, u_* the friction velocity of wind, and g the gravity acceleration. Equation (2) is the 3/2 power law relation of Toba (1972):

$$\left(\frac{gH_{1/3}}{u_*^2}\right) = B\left(\frac{gT_{1/3}}{u_*}\right)^{3/2}, B = 0.062 \quad (2)$$

Goto (1990) modified the coefficient B to 0.067. Toba (1992) also showed that if wind waves include components of swell, the exponent of the power law is changed from 3/2 to 2. The friction velocity of wind is calculated by using Eq. (3):

$$u_* = \sqrt{C_D} U_{10} \quad (3)$$

where U_{10} is wind speed at a 10m high from the sea surface. To determine the drag coefficient C_D of sea surface, the following Mitsuyasu's formula (1980) is applied:

$$C_D = \begin{cases} (1.290 - 0.024 U_{10}) \times 10^{-3} & (U_{10} < 8m/s) \\ (0.581 + 0.063 U_{10}) \times 10^{-3} & (U_{10} \geq 8m/s) \end{cases} \quad (4)$$

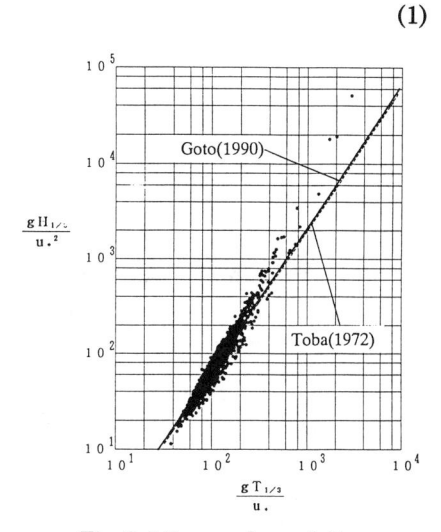

Fig. 3 3/2 power law relation

Figure 3 shows the relation of $gH_{1/3}/u_*^2$ and $gT_{1/3}/u_*$ by using the data at MT station. The solid line shows Toba's 3/2 power law while the dotted line shows the Goto's. The deviation of the data from the corresponded line represents the effects of swell components. It is confirmed that the measured wind waves data satisfied the 3/2 power law.

If the relationship linking non-dimensional fetch and energy, significant wave height and wave energy, peak frequency and significant

Table 2 Relationship among the non-dimensional parameters
$(Y = aX^n)$

X Y	gH/U^2 a	n	gT/U a	n	g^2E/U^4 a	n	gF/U^2 a	n	C/U a	n	f_mU/g a	n	H/L a	n
gH/U^2	1	1	B	$\frac{3}{2}$	a	$\frac{1}{2}$	$aA^{1/2}$	$\frac{1}{2}$	$(2\pi)^{3/2}B$	$\frac{3}{2}$	$\dfrac{B}{b^{3/2}}$	$-\frac{3}{2}$	$(2\pi)^3 B^4$	-3
gT/U	$\dfrac{1}{B^{2/3}}$	$\frac{2}{3}$	1	1	$\dfrac{a^{2/3}}{B^{2/3}}$	$\frac{1}{3}$	$\dfrac{a^{2/3}A^{1/3}}{B^{2/3}}$	$\frac{1}{3}$	2π	1	$\dfrac{1}{b}$	-1	$(2\pi)^2 B^2$	-2
g^2E/U^4	$\dfrac{1}{a^2}$	2	$\dfrac{B^2}{a^2}$	3	1	1	A	1	$\dfrac{(2\pi)^3 B^2}{a^2}$	3	$\dfrac{B^2}{a^2 b^3}$	-3	$(2\pi)^6 \dfrac{B^8}{a^2}$	-6
gF/U^2	$\dfrac{1}{a^2 A}$	2	$\dfrac{B^2}{a^2 A}$	3	$\dfrac{1}{A}$	1	1	1	$\dfrac{(2\pi)^3 B^2}{a^2 A}$	3	$\dfrac{B^2}{b^3 a^2 A}$	-3	$(2\pi)^6 \dfrac{B^8}{a^2 A}$	-6
C/U	$\dfrac{1}{2\pi B^{2/3}}$	$\frac{2}{3}$	$\dfrac{1}{2\pi}$	1	$\dfrac{a^{2/3}}{2\pi B^{2/3}}$	$\frac{1}{3}$	$\dfrac{a^{2/3}A^{1/3}}{2\pi B^{2/3}}$	$\frac{1}{3}$	1	1	$\dfrac{1}{2\pi b}$	-1	$2\pi B^2$	-2
f_mU/g	$\dfrac{B^{2/3}}{b}$	$-\frac{2}{3}$	$\dfrac{1}{b}$	-1	$\dfrac{B^{2/3}}{ba^{2/3}}$	$-\frac{1}{3}$	$\dfrac{B^{2/3}}{ba^{2/3}A^{1/3}}$	$-\frac{1}{3}$	$\dfrac{1}{2\pi b}$	-1	1	1	$\dfrac{1}{(2\pi)^2 bB^2}$	2
H/L	$2\pi B^{4/3}$	$-\frac{1}{3}$	$2\pi B$	$-\frac{1}{2}$	$2\pi \dfrac{B^{4/3}}{a^{1/3}}$	$-\frac{1}{6}$	$2\pi \dfrac{B^{4/3}}{a^{1/3}A^{1/6}}$	$-\frac{1}{6}$	$(2\pi)^{1/2}B$	$-\frac{1}{2}$	$2\pi b^{1/2}B$	$\frac{1}{2}$	1	1

wave period are revealed, all non-dimensional parameter are related to each other by using 3/2 power law.

Table 2 shows the relationship among the non-dimensional parameters. In this table, a is the coefficients of the relationship between wave height and wave energy (**Fig. 4**), A the coefficients of the relationship between fetch and energy (**Fig. 5**), b the coefficients of the relationship between peak frequency and wave periods (**Fig. 6**), and B the coefficient of 3/2 power law. These relation indicate that if the friction velocity and other only one parameter of wind waves are determined, all significant wave parameters can be calculated. The coefficients determine from observed data, are as follows:

$A = 0.00016, \quad a = 3.86, \quad B = 0.067, \quad b = 1.13$ \hfill (5)

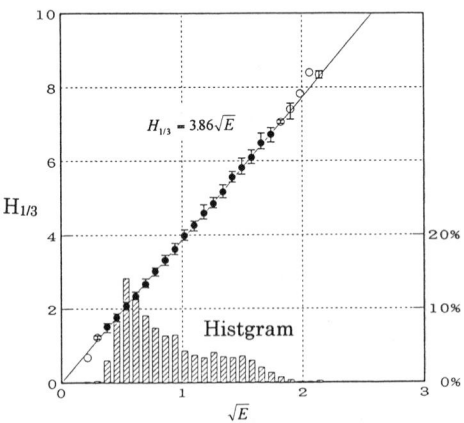

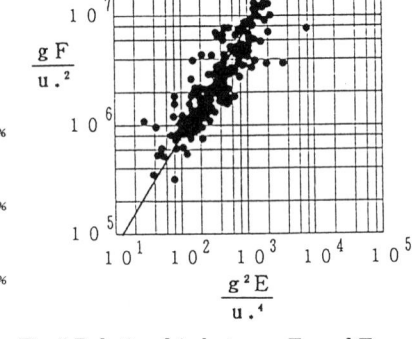

Fig. 4 Relationship between $H_{1/3}$ and E

Fig. 5 Relationship between E* and F*

Table 3 Comparison between observed and calculated results

Y \ X		gH/u_*^2	gT/u_*	g^2E/u_*^4	gF/u_*^2	C/u_*	$f_m u_*/g$	H/L
gH/u_*^2	observed calculated ratio	1 1 1	0.067 0.067 1	3.86 3.86 1.	0.0475 0.0518 0.9170	1.0675 1.0552 1.0117	0.0628 0.0556 1.1295	0.006 0.005 1.2
gT/u_*	observed calculated ratio	6.091 6.062 1.005	1 1 1	14.97 14.91 1.004	0.7941 0.8098 0.9806	6.281 6.283 0.999	0.949 0.883 1.0747	0.185 0.177 1.0452
g^2E/u_*^4	observed calculated ratio	0.0676 0.0671 1.0075	3.10×10⁻⁴ 3.01×10⁻⁴ 1.0299	1 1 1	1.6×10⁻⁴ 1.6×10⁻⁴ 1	0.0786 0.0747 1.0522	2.74×10⁻⁴ 2.07×10⁻⁴ 1.3237	1.46×10⁻⁶ 1.68×10⁻⁶ 0.8690
gF/u_*^2	observed calculated ratio	523 419 1.25	2.2267 1.8830 1.1825	6250 6250 1	1 1 1	551 467 1.18	1.9214 1.2964 1.4821	0.0136 0.0105 1.2952
C/u_*	observed calculated ratio	0.9696 0.9648 1.0050	0.1592 0.1590 1.0013	2.3840 2.3741 1.0042	0.1264 0.1289 0.9806	1 1 1	0.1510 0.1405 1.0747	0.0295 0.0282 1.0461
$f_m u_*/g$	observed calculated ratio	0.1575 0.1457 1.0810	0.9489 0.8830 1.0746	0.0640 0.0592 1.0811	1.2191 1.0904 1.1180	0.1511 0.1405 1.0754	1 1 1	5.6241 4.9829 1.1287
H/L	observed calculated ratio	0.1745 0.1710 1.0205	0.4259 0.4210 1.0116	0.1113 0.1090 1.0211	0.4755 0.4678 1.0165	0.1700 0.1679 1.0125	0.4377 0.4480 0.9770	1 1 1

Table 3 shows the comparison between observed and calculated results. The results show good agreement. Thus, table 2 shows similarity of wind waves parameter.

4. Wave spectrum in deep and shallow water

On the basis of the non-dimensional significant wave parameter, frequency spectra of wind waves in the deep and shallow water region are discussed and a new functional form of wind wave spectrum is proposed.

The JONSWAP spectrum is an extension form of Pierson-Moskowitz spectrum and is applicable to cases ranging from developing wave to fully developed wave. Equation (6) is the generalized JONSWAP spectrum.

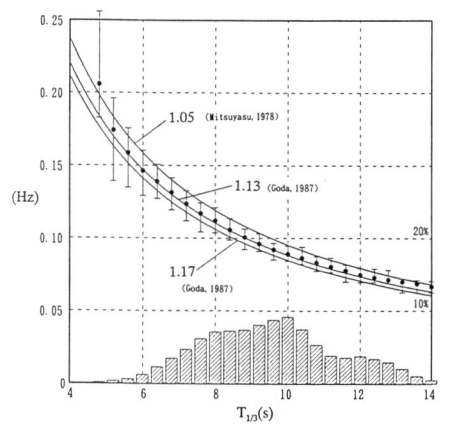

Fig. 6 Relationship between f_{m*} and $T_{1/3}$

$$S(f) = \alpha(2\pi)^{-m+1} g u_*^{5-m} f^{-m} \exp\left\{-\frac{m}{4}\left(\frac{f}{f_m}\right)^{-4}\right\} \gamma^\beta \quad (6)$$

$$\beta = \exp\{-(1 - f/f_m)^2 / 2\sigma^2\}$$

where α is scale factor, γ the peak enhancement factor, σ the band width near peak frequency. By using this equation, the similarity and the characteristics of slope of high frequencies and parameter of spectrum are discussed.

Figure 7 shows the slope of wind wave spectra at high frequencies by using least-squares estimation. The high frequency range is the same as Donelan's (1985). The x-axis is significant wave height. In this figure, the vertical lines are the standard deviations of the data to indicate the amount of scatter. From Fig. 7, the mean slope of wind wave spectra at high-frequency range is approximated by f^{-4} power law. The amount of scatter decreases with increasing wave height. This fact implies that the 3/2 power law is applicable in the frequency domain and the development of the wind waves is essentially a strongly nonlinear phenomenon. Hence, the JONSWAP spectrum follows the f^{-4} law, expressed as follows:

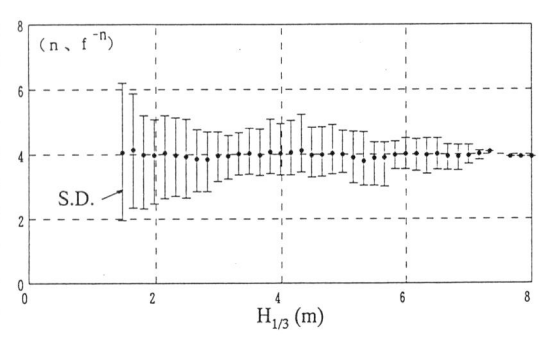

$$S(f) = \alpha (2\pi)^{-3} g u_* f^{-4} \exp\left\{-(f/f_m)^{-4}\right\} \gamma^\beta \qquad (7)$$

Fig. 7 Slope of the spectrum in high frequency range

The functional forms of the peak enhancement factor γ are proposed by Donelan (1985) and Mitsuyasu (1980). They use the non-dimensional peak frequency as variable. Figure 8 shows the relationship linking γ and non-dimensional peak frequency $f_{m*}=f_m u_*/g$. The vertical lines are standard deviations. There is a large scatter in γ, but the mean value is increasing with non-dimensional peak frequency and has log-linear relation. It means the spectral form is change to Pierson-Moskowitz spectrum with the increase of the fetch. Hence, the following empirical formula for γ is obtained:

$$\gamma = 6 f_{m*}^{0.15} \qquad (8)$$

The scale factor α is varied with spectral form. In this study, the relationship between α and γ is discussed. The total energy of JONSWAP spectrum is

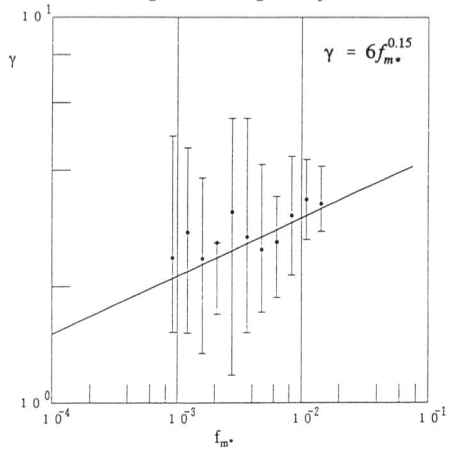

Fig. 8 Relationship between γ and f_{m*}

$$E = \alpha(2\pi)^{-3} g u_* f_m^{-4} M_0 \tag{9}$$

where, M_0 is expressed as Eq. (10)

$$M_0 = \int_0^\infty \left(\frac{f}{f_m}\right)^{-4} \exp\left\{-\left(\frac{f}{f_m}\right)^{-4}\right\} \gamma^\beta df \tag{10}$$

therefore, M_0 is approximated by (Goto and Aono, 1993)

$$M_0 = 0.30 \gamma^{1/3} \tag{11}$$

By using this approximate formula for M_0 and the relation of non-dimensional energy and peak frequency, and the 3/2 power law, this equation is derived:

$$\alpha \approx \gamma^{-1/3} \tag{12}$$

From Eq. (12), α is related to γ. Table 4 shows the relationship linking α and γ derived using the 3/2 power law for various spectral forms.

Table 4 Relationship between α and γ

Power	Pierson·Moskowitz	JONSWAP
-4	C o n s t .	$\alpha \sim \gamma^{-1/3}$
-5	$\alpha \sim \left[\dfrac{g^2 E}{u_*^4}\right]^{-1/3}$	$\alpha \sim \gamma^{-1/3} \left[\dfrac{g^2 E}{u_*^4}\right]^{-1/3}$

Figure 9 shows the relation of α and γ, while solid line correspond to the least squares estimate. The coefficient is determined from the line and yields

$$\alpha = 0.17 \gamma^{-1/3}. \tag{13}$$

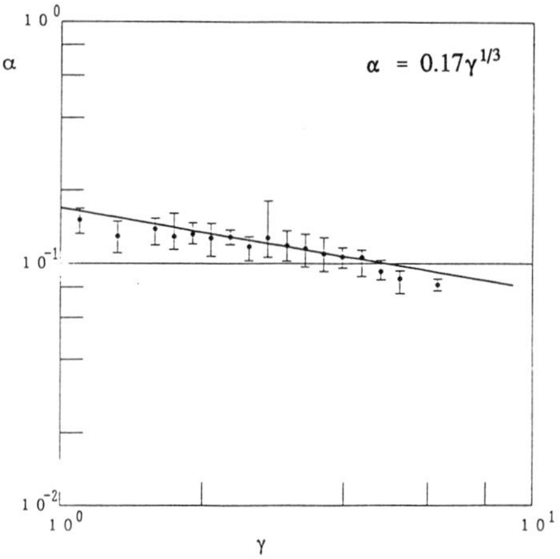

Figure 10 and Fig. 11 show the relation of σ_1, σ_2, and the non-dimensional peak frequency. The symbol σ_1 is low frequency side while σ_2 is high frequency side. Because the low frequency side include the weak swell components, there are no trends in σ_1. In the high frequency side, σ_2 varies inversely as the non-dimensional frequency on a log-log axis. This means the spectral form near peak frequency is varied from sharp to mild in the case of developing wind waves. Hence the band width near peak frequency is determined:

Fig. 9 Relation between α and γ

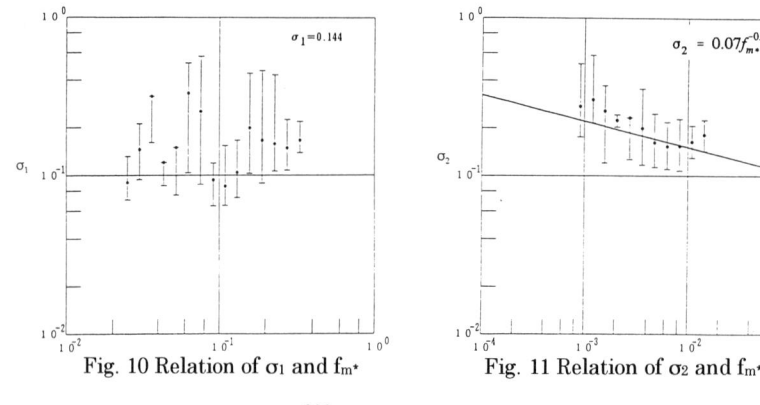

Fig. 10 Relation of σ_1 and f_{m*} Fig. 11 Relation of σ_2 and f_{m*}

$$\sigma_1 = 0.144, \quad \sigma_2 = 0.07 f_{m*}^{-0.16} \tag{14}$$

These characteristics of spectral parameters are in deep water region. Next, the spectrum in shallow water region is discussed.

Figure 12 show the change of the coefficient of 3/2 power law in shallow water. The x-axis is relative depth, the shaded bar is the histogram of observed data and the solid line is the linear shoaling coefficient K_S. The mean value of B/B_0 shows good agreement with K_S. The 3/2 power law is also extended in a shallow water region by using the shoaling coefficient.

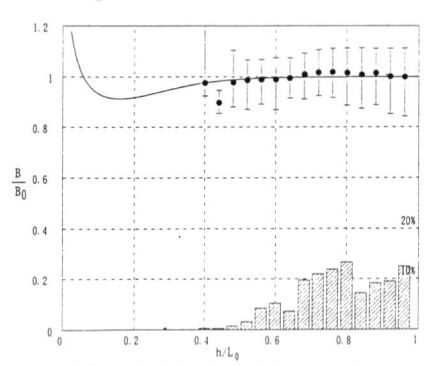

Fig. 12 Relation of B and h/L_0

The spectral form in shallow water is derived by using 3/2 power law. In this derivation, subscript 0 indicates values in deep water and s indicates there in shallow water. The integrated form of the JONSWAP spectrum is

$$E_0 = \alpha_0 g u_* f_m^{-3} M_1 \tag{15}$$

where M_1 is

$$M_1 = \int_0^\infty \left(\frac{f}{f_m}\right)^{-4} \exp\left\{-\left(\frac{f}{f_m}\right)^{-4}\right\} \gamma_0^\beta d\left(\frac{f}{f_m}\right). \tag{16}$$

From the analogy of M_0, M_1 is approximated by Eq. (17)

$$M_1 = c \gamma_0^{1/3} \tag{17}$$

Thus, Eq. (15) is expressed as
$$E_0 = c\alpha_0 u_* \gamma_0^{1/3} f_m^{-3} \qquad (18)$$
Using the relationship of $u_* f_m/g$ and gT_0/u_* and the relationship of gH_0/u_*^2 and $g^2 E_0/u_*^4$, Eq. (19) is derived.
$$\left[\frac{gH_0}{u_*^2}\right] = a\left[\alpha_0 b^{-3} c\gamma_0^{1/3}\right]^{1/2}\left[\frac{gT_0}{u_*}\right]^{3/2} \qquad (19)$$
Equation (19) is the 3/2 power law which derived from the spectral form. The coefficient B_0 of 3/2 power law in deep water region is expressed as Eq. (20).
$$B_0 = \left[ab^{-3/2}\right]\left[\alpha_0 c\gamma_0^{1/3}\right]^{1/2} \qquad (20)$$
If the coefficients a, b and B_0 is constant, the relationship between α_0 and γ_0 is
$$\alpha_0 = \left[\frac{B^2}{ab^{-3}}\right]\frac{1}{c\gamma_0^{1/3}} \qquad (21)$$
The same argument applies in shallow water. The total energy and the coefficient of 3/2 power law in shallow water is expressed by the following equations:
$$E_s = c\alpha_{sy}\gamma_s^{1/3} u_* f_m^{-3} \qquad (22)$$
$$\left[\frac{gH_s}{u_*^2}\right] = a\left[\alpha_s b^{-3} c\gamma_s^{1/3}\right]^{1/2}\left[\frac{gT_s}{u_*}\right]^{3/2} \qquad (23)$$
The coefficient of 3/2 power law is
$$B_s = B_0 K_s = a\left[\alpha_s b^{-3} c\gamma_s^{1/3}\right]^{1/2} \qquad (24)$$
The combination of Eq. (20) and Eq. (24) lead to Eq. (25).
$$\left[\frac{\gamma_s}{\gamma_0}\right]^{1/6} = K_s \qquad \alpha_0 = \alpha_s \qquad (25)$$

Equation (25) determine the spectral form in shallow water. **Figure 13** shows the change of γ with relative depth h/L_0. The mean value of γ and K_s show good agreement.

The characteristics of the wind waves spectrum in deep water are invest-

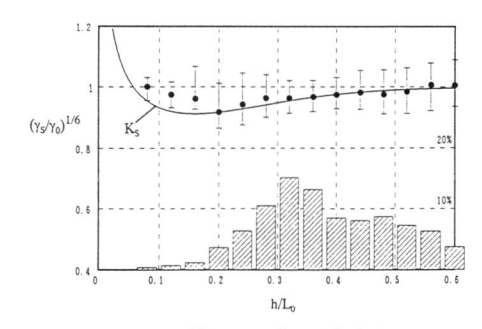

Fig. 13 Change of γ with h/L_0.

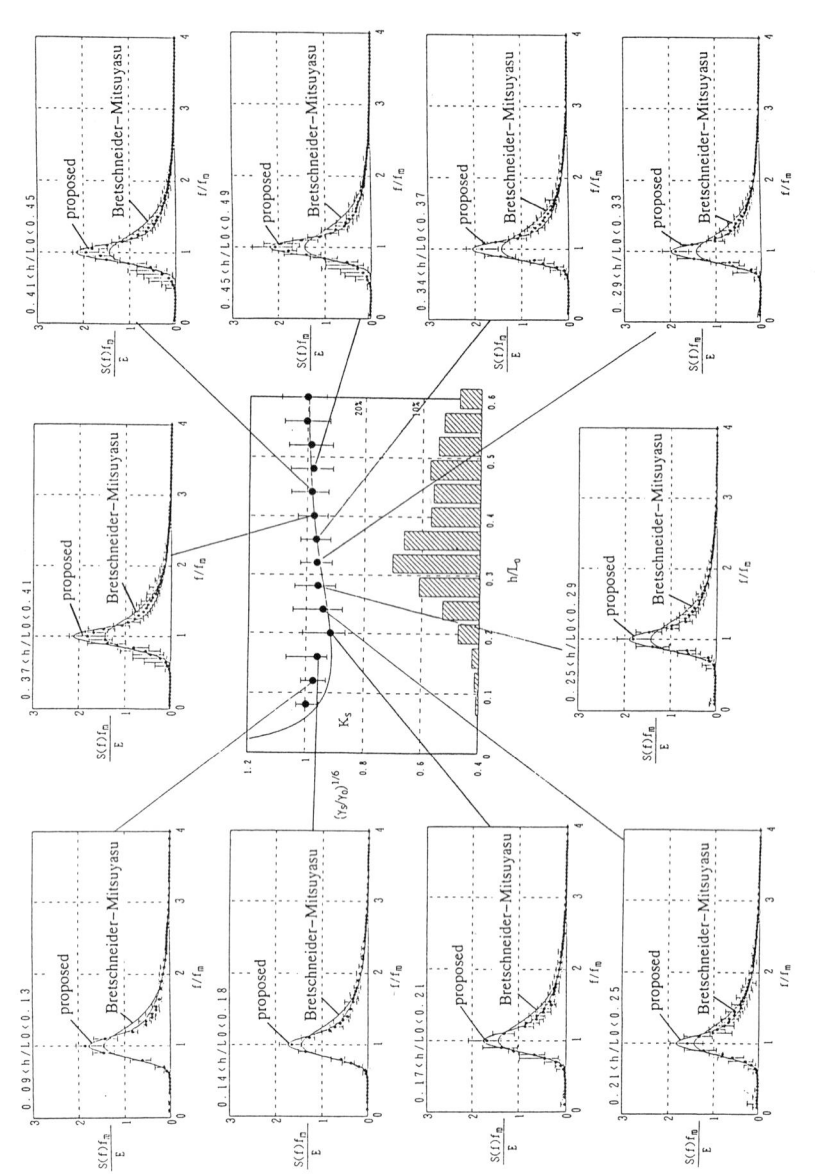

Fig. 14 Change of the spectral form with h/L₀.

gated by using the parameters of JONSWAP spectrum. From theoretical and empirical arguments, it is confirmed that a f⁻⁴ power law exists at high frequency range, that γ and σ are varied with fetch, and that a -1/3 power law, based on 3/2 power law, also exists in relation between α and γ. Wave energy is concentrated near the peak frequency during the development stage, and approaches the Pierson-Moskowitz spectrum gradually with the increase of fetch. In shallow water region, the spectral form of wind waves varies with the shoaling coefficient. From these characteristics of the wind wave spectrum, the following new spectral formula is obtained:

$$S(f) = \alpha(2\pi)^{-3} g u_* f^{-4} \exp\left[-(f/f_m)^{-4}\right]\gamma^{\beta}$$
$$\beta = \exp\left[1-(f/f_m)^2/2\sigma^2\right] \quad (26)$$

$$\gamma = 6 f_{m*}^{0.15}, \quad \alpha = 0.17\gamma^{-1/3}, \quad f_m = 1/1.136 T_{1/3}$$
$$\sigma_1 = 0.144, \quad \sigma_2 = 0.07 f_{m*}^{-0.16}, \quad f_{m*} = f_m u_*/g \quad (27)$$
$$u_* = H_{1/3}^2/gB^2 T_{1/3}, \quad B = 0.067$$

In shallow water region, the spectral form of wind waves is expressed as
$$S_S(f) = K_S^{6\beta} S(f) \quad (28)$$
In this spectrum, the input variables are significant wave height and period only. **Figure 14** shows the comparison of the newly-proposed spectral model with the observed spectral data and Bretshneider-Mitsuyasu spectrum. The spectral form varies with h/L_0 and the proposed model and observed data show good agreement.

5. Discussion

Table 4 shows the relationship between non-dimensional wave height and period derived from selected spectral forms. Where B is coefficients and β_1 is power. It is confirmed that 3/2 power law relation applies to f⁻⁴ power law and 2 power law relation applies to f⁻⁵ power law. 2 power law relation is revealed under the condition that the swell components are included in the wind waves. This is the reason of such change of power.

Table 4 Relationship between H· and T· derived from spectral forms.

	Pierson·Moskowitz		JONSWAP	
Power	B	β1	B	β1
-5	a b⁻²[0.2α]¹ᐟ²	2	a b⁻²[0.2α γ¹ᐟ³]¹ᐟ²	2
-4	0.5a[α b⁻³]¹ᐟ²	3/2	a[α b⁻³ c γᵐ]¹ᐟ²	3/2

The 3/2 power law is not dependent on the fetch and shows local equilibrium between wind and wind waves. In the frequency domain, such local equilibrium is revealed by the f^{-4} power law. It implies that the spectral form depends on the characteristics of the high frequency range. However, spectral parameters depend on the fetch, so that the spectral form cannot be determined from the 3/2 power law directly. The implication in the case of developing wind waves is that α varies with γ, which depends on the fetch, in order to satisfy the 3/2 power law.

To apply Eq. (28) for actual work, it is necessary to investigate the error between the observed data and Eq. (28). The error of the calculated spectrum is defined as

$$E_r = \frac{\int_0^\infty |S_c(f) - S_o(f)| df}{\int_0^\infty S_c(f) df} \quad (29)$$

Figure 15 shows the characteristics of the error of proposed spectrum and Bretschneider-Mitsuyasu spectrum. The horizontal axis is the U_{10} calculated by using the 3/2 power law and Mitsuyasu's C_D law. The error of the proposed spectrum is relatively small in the entire range of wind velocity. The error is being large in the range of low wind velocity and the wind velocity is greater than the 22m/s. These characteristics of error indicates the mixture of wind waves and swell at low wind velocity range. The increase of error at high wind velocity range suggest the change of the wind resistance law, but there is not enough the data such high wind velocity range.

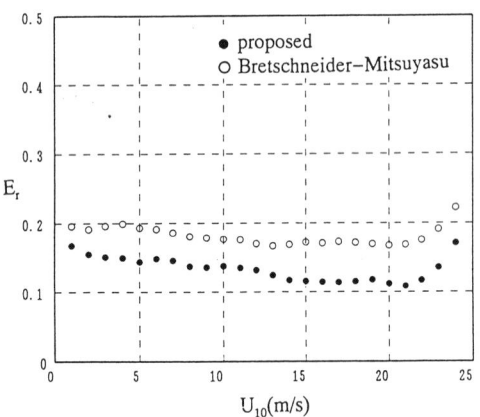

Fig. 15 Error of the spectrum

Figure 16 shows the results of comparison of the observed and proposed spectra. The spectrum of Bretschneider-Mitsuyasu is given in this figure for reference. Observation data was taken from Kochi-Oki on July, 1981. Time series of $H_{1/3}$, $T_{1/3}$, h/L_0, and error which defined by Eq. (29) is also shown in Fig. 16. It can be seen that the proposed spectrum shape agrees well with the observed one.

6. Conclusions

The major conclusions of this study are as follows:
(1) It was confirmed that the measured wind waves data satisfied the 3/2

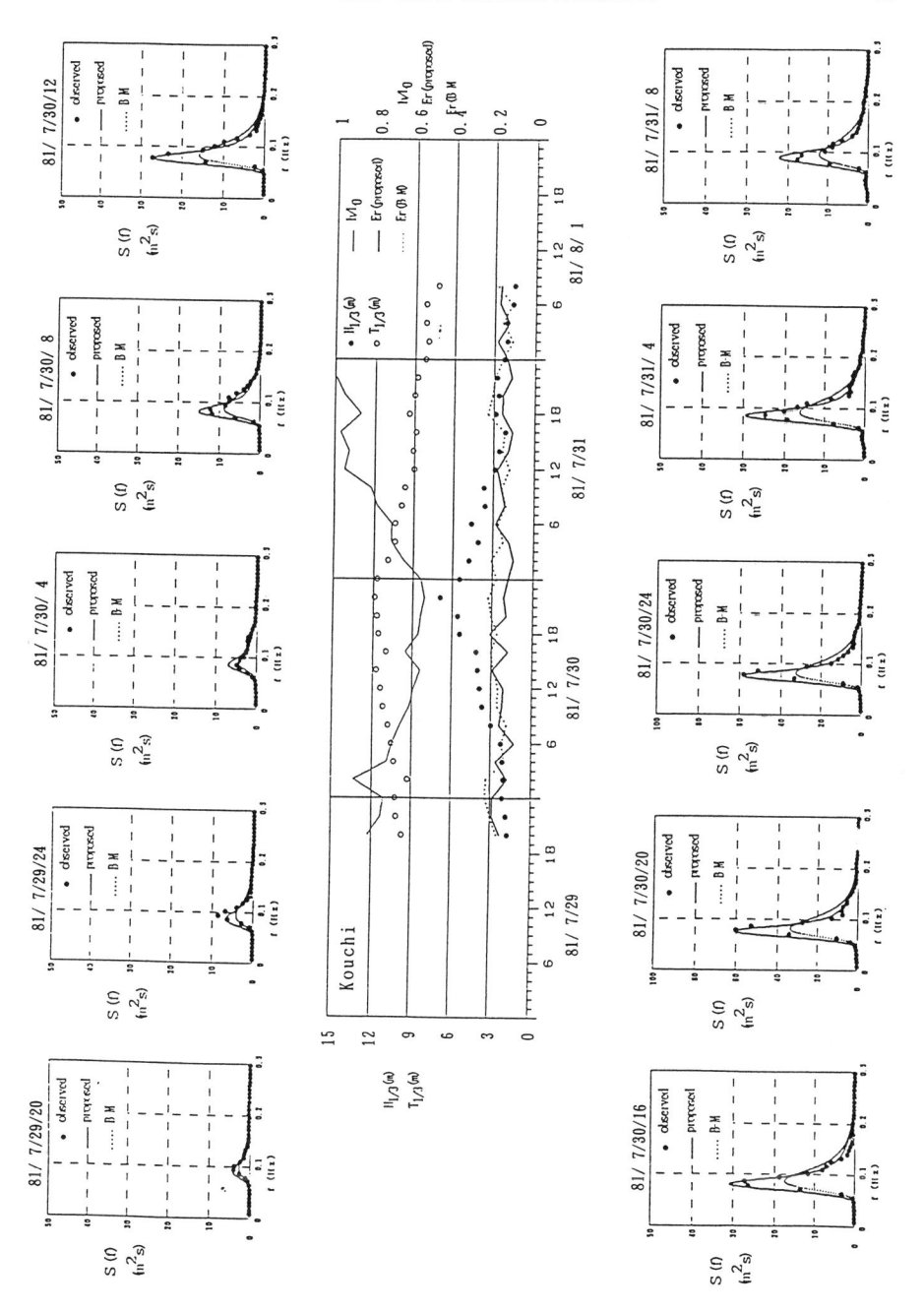

Fig. 16 Results of the comparison of the observed and proposed spectra.

power law which is an empirical formula between non-dimensional wave height and period.
(2) By using the 3/2 power law relation, non-dimensional parameters of wind waves are all related.
(3) The new spectral model, which is based on the 3/2 power law, is proposed. This spectral model is evaluated from $H_{1/3}$ and $T_{1/3}$, and calculated result is very accurate. The applicability of this new spectrum is very wide, ranging from the case of developing wind to that of fully developed wind, and from deep to shallow water region.

Reference
(1) Bouws, E., H. Gunther, W. Rosenthal and C. L. Vincent (1985): Similarity of the wind wave spectrum in finite depth water 1.Spectral form, J. Geophys. Res., Vol.90 No.C1, pp.975-986.
(2) Donelan, M.A., J. Hamilton and W. H. Hui (1985): Directional spectra of wind-generated waves, Phil. Trans. R. Soc. Lond. A315,pp.509-562.
(3) Ebuchi, N., Y. Toba and H. Kawamura (1992): Statistical study on the local equilibrium between wind and wind waves by using data from ocean data buoy stations, J. Oceanogr., Vol.48, pp.77-92.
(4) Goto, C., K. Suetsugu and T. Nagai (1990): Wave Hindcast Model for Short Fetch Sea, Rep. of P.H.R.I. Vol.29, No.3, pp. 3-26.
(5) Goto, C. and T. Aono (1993) : On the Characteristics of One-Dimensional Spectra and Non-Dimensional Parameters of Wind Waves - Wave Hindcast Model Using the Hybrid-parameter Methods (2nd report) -, Rep. of P.H.R.I, Vol. 32, No.1, pp. 53-99 (in Japanese)
(6) Hasselman, K. et. al(1973):Measurements of wind wave growth and swell decay during the Joint North Sea Wave Project (JONSWAP), Deut. Hydrgr. Z.,Suppl. 8, pp.1-95.
(7) Mitsuyasu, H (1968): On the growth of the spectrum of wind generated waves (1), Rep. Res. Inst. Mech. Kyushu Univ., Vol.16, pp.459-482.
(8) Mitsuyasu, H., R. Nakamura and T. Komori (1971): Observation of the wind and waves in Hakata Bay, Rep. Res. Inst. Mech. Kyushu Univ., Vol.19, pp.37-74.
(9) Phillips, O. M. (1958): The equilibrium range in the spectrum of wind-generated waves, J. Fluid Mech., 4, pp. 426-434.
(10) Pierson, W. J. and L. Moskowitz (1964): A proposed spectral form for fully developed wind seas based on the similarity theory of S. A. Kitaigorodskii, J. Geophys. Res., 69, pp.5181-5190.
(11) Thornton, E. B. (1977): Rederivation of the saturation range in the frequency spectrum of wind-generated gravity waves, J. Phys. Oceanogr., Vol.7, pp.137-140.
(12) Toba, Y. (1972): Local balance in the air-sea boundary processes, I. On the growth process of wind waves, J. Oceanogr Soc. Japan, 28, pp. 109-120.
(13) Toba, Y. (1973): Local balance in the air-sea boundary processes, III. On the spectrum of wind waves, J. Oceanogr. Soc. Japan, 29, pp. 209-220.

CHAPTER 2

SWASH MOTION DUE TO OBLIQUELY INCIDENT WAVES

Toshiyuki Asano[1]

abstract

A numerical model is developed to predict the flow characteristics in a swash zone for obliquely incident wave trains. The two-dimensional shallow water equations are de-coupled into independent equations each for on-off shore motion and for longshore motion. A front of swash wave train is treated as a moving boundary which makes the solution predictable for landward zone of the still water level. The results show non-vanishing longshore velocities and volume flux at the still water shoreline. These quantities are found to increase with the beach slope. The two-dimensional uprush and downrush motion near the front of swash waves shows skew figures, which may cause zig-zag longshore sediment transport inherent in swash zone.

1.INTRODUCTION

Swash zone is the most familiar area that we can easily observe its motion while walking along a sandy beach, but it is one of the unsolved area from the hydrodynamic point of view. The difficulty lies on that waves in swash zone are highly nonlinear and show such a unique behaviour that the seabed is immersed during run-up and dried during run-down alternatively. Lately, the swash motion has received much attention because the sediment process in this zone provides the important boundary condition for the beach evolution. And also, recent studies have reported an important new finding that longshore sediment transport takes two major peaks located at the breaking point and on the foreshore in the swash zone(Bodge - Dean, 1987; Kamphuis, 1991;b).

Under obliquely incident waves, sediment near shoreline moves in a zig-zag way which results in the inherent longshore transport in the swash zone.

---
[1]Dept. of Ocean Civil Engrg., Kagoshima Univ., Korimoto, Kagoshima, 890, JAPAN

Intensive turbulence generated in the uprush and backwash waves causes a large volume of sediment to be suspended. Moreover, infragravity wave motion might be more influential on the sediment dynamics in this very shallow water region(Thornton - Abdelrahman, 1991).

Although such complicated dynamics needs to be investigated for full descriptions, this study, as the first step, focuses on the velocity field as a basic forcing function of the sediment motion under obliquely incident monochromatic waves. The horizontally two-dimensional water particle velocities are computed based on Ryrie(1983)'s analysis. Moreover, the velocity measurements in the swash zone under obliquely incident waves are conducted using tracer method. Through comparisons between the numerical and experimental results, the hydrodynamic properties related to longshore sediment transport are discussed.

2. NUMERICAL ANALYSIS

2.1 Formulation of Shallow Water Wave Equations

Based on Ryrie's formulation, the following two dimensional wave and topographic system is assumed. The incident monochromatic waves with straight parallel crests are assumed to arrive at the seaward boundary with an angle θ_B (Fig. 1). The x' - and y' - axis, in which the prime indicates the dimensional variables, is chosen to be in the normal and the parallel direction to the shoreline, respectively. The z' - axis is taken positive upward with $z' = 0$ at the still water level(SWL). The beach slope S' is herein restricted to be uniform and its contours are assumed to be straight and parallel to the shoreline. The water depth at the toe of the slope is given as d'_B where the offshore boundary conditions for the incident waves are provided. The free surface is located at $z' = \eta'$, so that the instantaneous water depth h' is given by $h' = \eta' + (d'_B - S'x')$.

The governing equations for the mass and momentum conservations may be expressed as

$$\frac{\partial h'}{\partial t'} + \frac{\partial}{\partial x'}(h'u') + \frac{\partial}{\partial y'}(h'v') = 0 \tag{1}$$

$$\frac{\partial}{\partial t'}(h'u') + \frac{\partial}{\partial x'}(h'u'^2) + \frac{\partial}{\partial y'}(h'u'v') =$$
$$-gh'\frac{\partial \eta'}{\partial x'} - \frac{1}{2}f' \mid u' \mid u' \tag{2}$$

$$\frac{\partial}{\partial t'}(h'v') + \frac{\partial}{\partial x'}(h'u'v') + \frac{\partial}{\partial y'}(h'v'^2) =$$
$$-gh'\frac{\partial \eta'}{\partial y'} - \frac{1}{2}f' \mid u' \mid v' \tag{3}$$

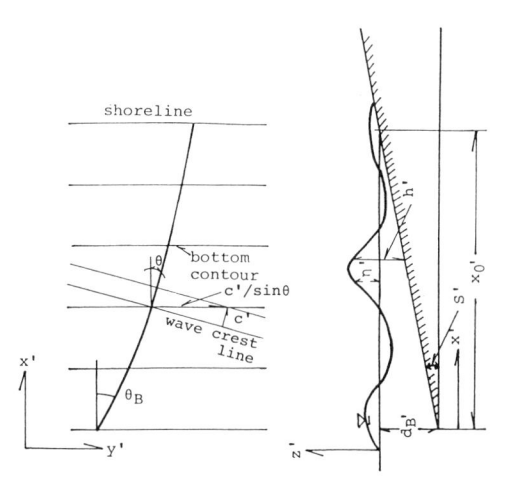

Fig. 1: Two dimensional plane wave on a uniform beach

in which, t' is the time, u', v' is the on-offshore and longshore velocity respectively, g is the gravitational acceleration and f' is the bottom friction factor. Eqs.(2) and (3) are the two dimensional nonlinear shallow water equations, where the vertical pressure distribution is assumed to be hydrostatic.

According to Kobayashi et al.(1987), the following dimensionless variables using the characteristic period T' and height H' associated with the incident wave train are introduced.

$$u = \frac{u'}{\sqrt{gH'}}; \quad \eta = \frac{\eta'}{H'}; \quad h = \frac{h'}{H'}; \quad t = \frac{t'}{T'};$$
$$x = \frac{x'}{T'\sqrt{gH'}}; \quad c = \frac{c'}{\sqrt{gH'}}; \quad f = \frac{1}{2}\sigma f';$$
$$S = T'\sqrt{\frac{g}{H'}}S'; \quad d_B = d'_B/H'; \quad \sigma = T'\sqrt{g/H'}; \qquad (4)$$

in which, S is expressed by the surf similarity parameter ξ as $S = \sqrt{2\pi}\xi$, d_B corresponds to the inverse number of relative water depth at the offshore boundary and σ/h means the ratio of wave length to water depth L'/h'.

Since we have assumed that the wave crest is straight parallel and bottom topography does not vary in the y - direction, the observed wave motion moving along the alongshore direction at the speed $C'/\sin\theta$ (which remains constant throughout the surf zone by Snell's law) becomes independent in the y- direction. Consequently, a new independent variable referred to "pseudotime" $\hat{t}'$ is introduced to unify two independent variables t' and y'.

$$\hat{t}' = t' - \frac{\sin\theta_B}{C'_B}y' \qquad (5)$$

Providing the incident wave angle θ_B is small, the following small parameter ϵ can be used for the scaling of the governing equations

$$\epsilon = \frac{\sin\theta_B}{C'_B}\sqrt{gH'} \tag{6}$$

Since the length scale of variations in the y' direction is much greater than that in the x' direction and also the velocity v' is much smaller than u', we introduce the following transformations in order to make these variables of order unity

$$v = \frac{v'}{\epsilon\sqrt{gH'}}; \quad y = \frac{\epsilon y'}{T'\sqrt{gH'}} \tag{7}$$

Using the above scaling parameters and transformations, the normalized pseudotime is given by,

$$\hat{t} = t - y \tag{8}$$

and the differentiations with the dimensional variables are replaced by as follows.

$$\frac{\partial}{\partial t} = \frac{\partial}{\partial \hat{t}} \qquad \frac{\partial}{\partial y} = -\frac{\partial}{\partial \hat{t}} \tag{9}$$

After these deductions, the original equations (1) $\sim$ (3) are considerably simplified into the following non-dimensional form.

$$\frac{\partial h}{\partial \hat{t}} + \frac{\partial (uh)}{\partial x} = \epsilon^2 \frac{\partial (vh)}{\partial \hat{t}} \tag{10}$$

$$\frac{\partial u}{\partial \hat{t}} + u\frac{\partial u}{\partial x} + \frac{\partial h}{\partial x} + S + \frac{fu|u|}{h} = \epsilon^2 v\frac{\partial u}{\partial \hat{t}} \tag{11}$$

$$\epsilon\left\{\frac{\partial v}{\partial \hat{t}} + u\frac{\partial v}{\partial x} - \frac{\partial h}{\partial \hat{t}} + \frac{f|u|v}{h}\right\} = \epsilon^3 v\frac{\partial v}{\partial \hat{t}} \tag{12}$$

If we neglect higher terms than $O(\epsilon^2)$, that is, the right hand sides of Eqs. (10), (11) are set to be zero, the resultant equations are the same as the usual one dimensional shallow water wave equations normally incident to the shoreline. If we consider Eq. (12) in the order $O(\epsilon)$, the equation becomes as

$$\frac{\partial v}{\partial \hat{t}} + u\frac{\partial v}{\partial x} - \frac{\partial h}{\partial \hat{t}} + \frac{f|u|v}{h} = 0 \tag{13}$$

After all, the leading order in ϵ of perturbation equations yields decoupled equations each for on-offshore motion and for longshore motion. Thus, the longshore velocity $v(x,\hat{t})$ can be solved like one-dimensional analysis once $h(x,\hat{t})$ and $u(x,\hat{t})$ is known.

2.2 Numerical Method

The basic equations were solved using an explicit Lax-Wendroff finite difference method. To attenuate numerical oscillations in the vicinity of wave front, a so-called artificial-viscosity term was included (Hibberd-Peregrine, 1979). The computational domain in the cross shore direction was discretized into 100 grid points so that one wave length can be represented at least 20 grid points. The time discretization was set as $\Delta t = T/4000$.

The initial condition for free water surface η is still water condition; $\eta = 0$. Accordingly, the flow velocities u and v are set to be zero for all the computational domain. The offshore boundary was set at around one wave length offshore from the breaking point so that the boundary value for time averaged longshore velocity V may be given by zero. The offshore boundary should be devised to make the reflected waves from the on-shore side transmit through the boundary freely. The total water depth at the offshore boundary is expressed as

$$h_B = d_B + \eta_i(t) + \eta_r(t), \quad at \quad x = 0 \tag{14}$$

in which, η_i, η_r denotes the water surface fluctuation due to the incident and reflected waves, respectively. The time variation η_r was evaluated using the retreat characteristic variable β as follows (Kobayashi et al., 1987)

$$\eta_r(t) = \sqrt{d_B}\beta(t)/2 - d_B, \quad at \quad x = 0 \tag{15}$$

The offshore boundary condition for u_B is easily determined by the boundary values of β and h_B, and that for v_B was assumed to be $u_B \tan\theta$. The onshore boundary was treated as a moving boundary in which the front of wet waterline node was determined as such a location that the total water depth is less than a given threshold small depth (Hibbert-Peregrin, 1979).

2.3 Numerical Results

Calculations were carried out under similar conditions of Kamphuis's experiments(1991; a, b); that is, the water depth at the offshore boundary $d'_B = 0.50$m, the incident wave height $H' = 12.4$cm, wave period $T' = 1.15$s, incident wave angle $\theta_B = 10°$, beach slope $S' = 1/10$.

Fig. 2 shows the computed spatial variations of free surface elevation η' (bottom), on-offshore velocity u' (middle) and longshore velocity v' (top). Fig. 2(a) and (b) shows the results under the bottom friction factor $f' = 0.01$ and $f' = 0.10$, respectively. Comparison of these figures indicates that the results of free surface elevation η' and on-offshore velocity u' are little affected by the change of the friction factor, whereas the results of the longshore velocity v' shows the considerable difference by f' especially around the still water shoreline(S.W.S.L.).

Fig. 2(c) and (d) are the results when the beach slope S' is changed from the original input condition. Although, Eq. (13) for the longshore velocity v' does not include S' term, the effects of beach slope are indirectly involved in v'

through the alterations of h' and u' by S'. Under the mild slope case $S' = 1/20$, the longshore velocity around S.W.S.L. takes small value as shown in Fig. 2(c). On the other hand, under steep slope case $S' = 1/5$, the longshore current velocity becomes large even in the landward region of S.W.S.L. The longshore velocity v' near S.W.S.L. is found to be positive not only up-rush phase but also down rush phase.

In order to discuss the properties of calculated longshore velocity, the following characteristics are investigated; the peak value of temporal velocity fluctuation at a specific point; v'_{peak}, the time average value; v'_{mean} and the volume flux of longshore velocity; Q', defined as

$$Q' = \frac{1}{T} \int_0^T h'v' dt \tag{16}$$

Fig. 3 illustrates the on-offshore variations of the above characteristics. As the beach slope S' becomes steeper, all the characteristics increase and the peaks of the distributions move toward S.W.S.L. (Fig. 3(a)). Non-vanishing velocities and volume flux are found at the still water shoreline and in further landward region. Fig. 3(b) indicates the effect of incident wave period T', where the increase of T' has the same effects as the increase of S'. Fig. 3 (c) illustrates the results when the friction factor f' is varied. The characteristics decrease with the friction factor, but the differences are not so significant.

Fig. 4 shows the temporal variations of free surface elevation η' (bottom), velocity vector $\mathbf{v}'$ (middle) and volume flux vector $\mathbf{Q}'$ (top) at the locations from offshore to onshore. In order to stress the difference between run-up and downwash motion, this calculation was carried out under relatively steep slope $S' = 1/5$ and large wave incident angle $\theta_B = 20°$. The stick-diagrams of $\mathbf{v}'$ show that the water mass runs up on the beach with a certain angle to the shoreline, then runs down approaching the right angle. The diagram of volume flux $\mathbf{Q}'$ shows more predominant difference in magnitude between run-up and run-down phases, because the water depth h' becomes larger in run-up phase than in run-down phase. This property is more evident as the location $(x - x_0)/x_0$ ($x_0 = d'_B/S'$: the position of still water shoreline) is in the direction of landward. These properties seem to be useful to explain the zig-zag sediment transport in a swash zone. The velocity vector $\mathbf{v}'$ or the volume flux vector $\mathbf{Q}'$ may govern the sediment movement corresponding whether the transport mode is bed load or suspended load, respectively.

OBLIQUELY INCIDENT WAVES 33

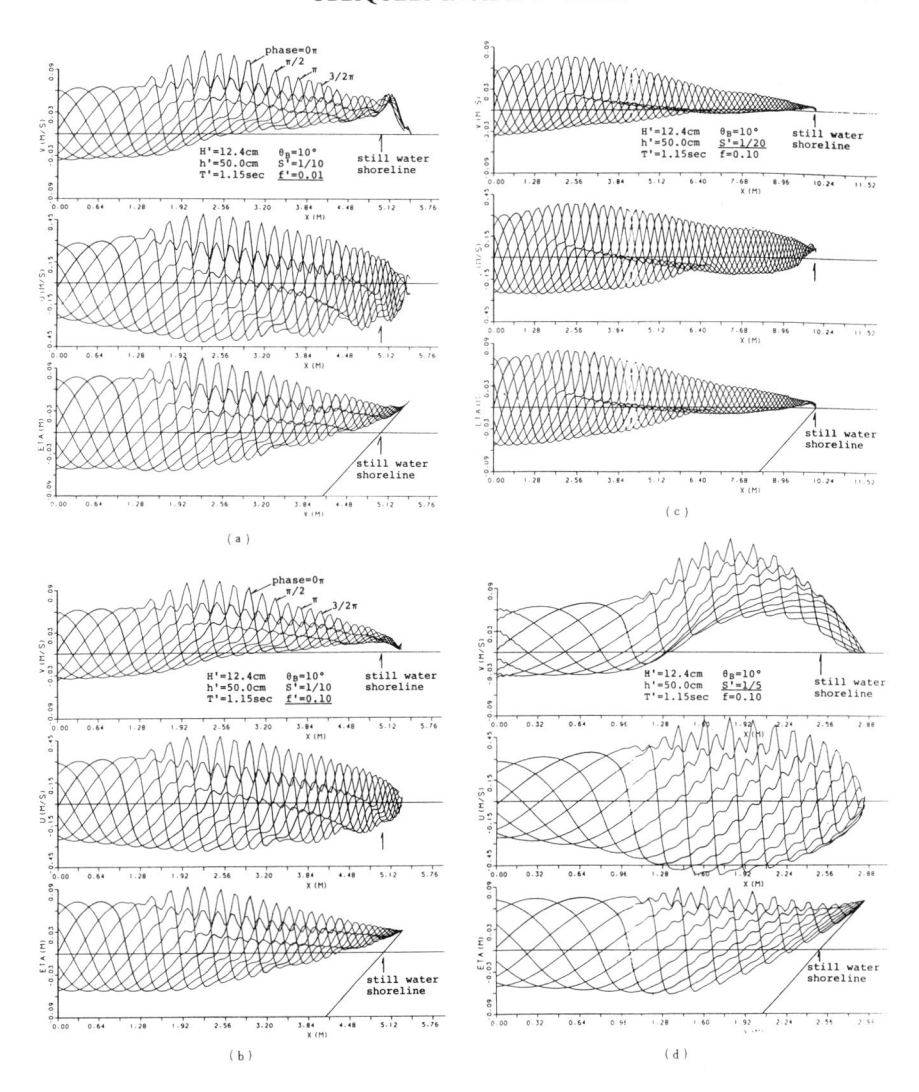

Fig. 2: Spatial variations of longshore velocity v' (top), on-offshore velocity u' (middle) and free surface elevation η' (bottom)

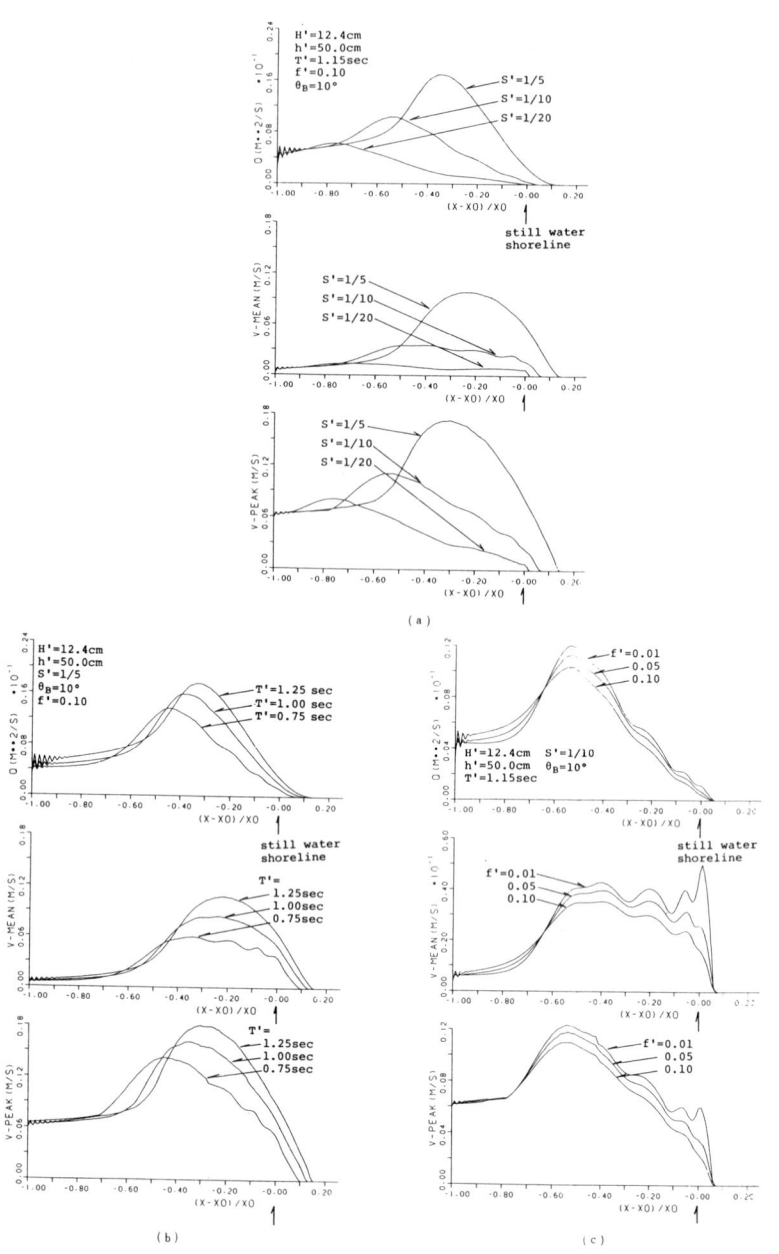

Fig. 3: On-offshore variations of the peak longshore velocity v'_{peak} (bottom), time averaged longshore velocity v'_{mean} (middle) and time averaged longshore volume flux Q' (top)

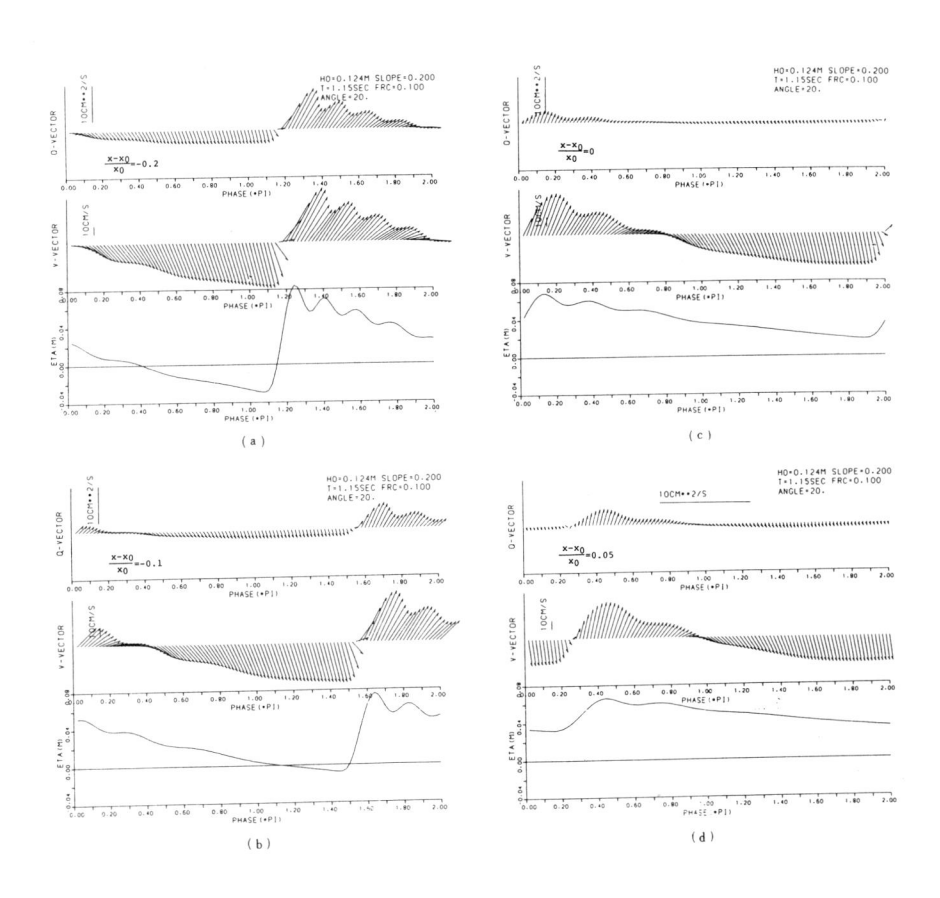

Fig. 4: Temporal variations of free surface elevation η' (bottom), velocity vector $\mathbf{v}'$ (middle) and longshore volume flux vector $\mathbf{Q}'$ (top)

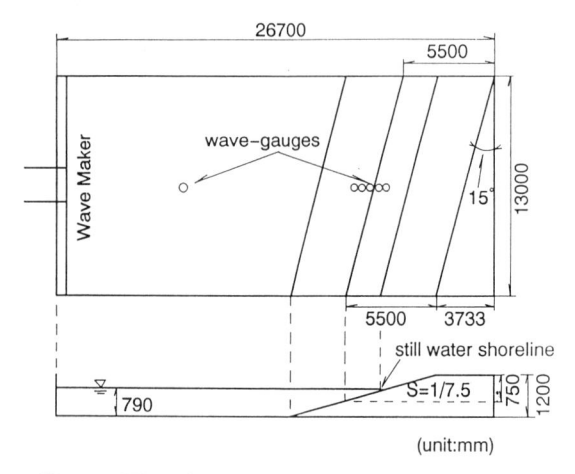

Fig. 5: Wave basin and experimental set-up

3. EXPERIMENTAL STUDY

3.1 Experimental Arrangements and Procedure

A wave basin of 26.7m long, 13m wide and 1.2m deep was used. A uniform plane slope of the gradient S=1/7.5 was set up at an angle $\theta = 15°$ with a wave generator which was equipped at the other end of the basin. The slope was carefully built up by covering 30mm thick concrete mortar with reinforced mesh over a sandy mound. The water depth in the offshore uniform depth region was kept at 79cm constant throughout the experiments. Water surface fluctuations were measured with an array of five capacitance type wave gauges at 12.5cm interval on the slope. The experimental set-up is illustrated as Fig. 5.

In order to measure very shallow flows in a swash zone including landward region of the still water shoreline, tracer method was adopted. An important demand for the tracer is to represent the fluid velocity accurately in the swash zone. Another demand is visibility because the tracer movements were recorded with a bird's-eye video camera locating $5 \sim 6$ m above the water surface. After several trials, the following three type tracers were chosen: a fluorescent color sphere float of styrene form with 3cm in diameter(type-A), a 2.5mm thick circular plate made of plywood with 110mm diameter(type-B) and a 6.6mm thick octagonal plate made of plywood with 93mm in diagonal length(type-C). A small electro-magnetic current meter with a cylindrical shape sensor 16mm in length and 5mm in diameter was also used in the preliminary measurements.

For each test run, the velocity measurement was started two minutes after the start of wave generation to attain the steady state condition. The velocity measuring area for the tracer method should be chosen where the uniformity of the longshore current is confirmed. In order to detect the position of a moving tracer throughout the surf zone and swash zone, a 20cm-mesh reference

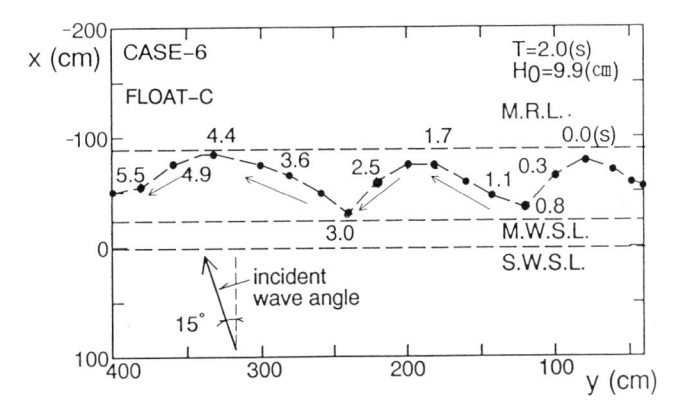

Fig. 6: Trajectories of tracers

frame of 3.6m × 3.6m was installed. The trajectories of the tracers over the reference frame were recorded at an interval of 1/30 second. Frame by frame analysis of the video film yields Lagrangian velocities both in on-offshore and alongshore directions. For the detailed descriptions of the experiment, see Asano et al.(1994).

3.2 Experimental Results on Longshore Velocity in Surf and Swash Zone

Fig. 6 shows examples of plane trajectories of tracers. Each tracer runs up obliquely on the slope, then tends to run-down vertically. This property corresponds to the numerical results illustrated in Fig. 4.

The longshore velocity of a fluid particle was herein evaluated from the alongshore displacement of the tracer during one wave cycle. Fig. 7 shows an example of the on-offshore distributions of longshore current velocity V. In plotting the data V, the x-position was determined by the central point of a tracer over one wave cycle. Some data which the position after one wave cycle were clearly shifted in the on-offshore direction were discarded. Fewer data are obtained in the negative x-region because the plate type tracers (type-B and C) are frequently thrown up on the slope. Also fewer data are available slightly onshore of the breaking point because white-caps generated by breaking waves often make tracer invisible. Since no obvious differences by the tracer types are noticed in Fig. 7, results will be shown without distinguishing the tracer types in the following. One important point in this result is that the longshore current velocity at the still water shoreline x=0 has the same order of magnitude as that in the surf zone.

Fig. 8 shows the results where the incident wave heights are almost the same. Here, the measured longshore current velocities including the case shown in Fig. 7 are averaged over every 10cm segment in the on-offshore direction, then plotted in the non-dimensional form V/V_{max} against normalized alongshore

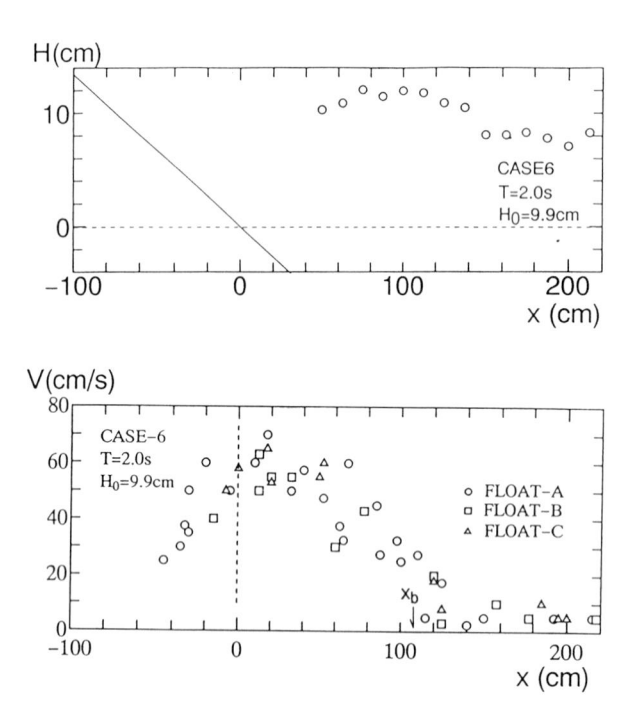

Fig. 7: On-offshore distributions of wave height and longshore current velocity

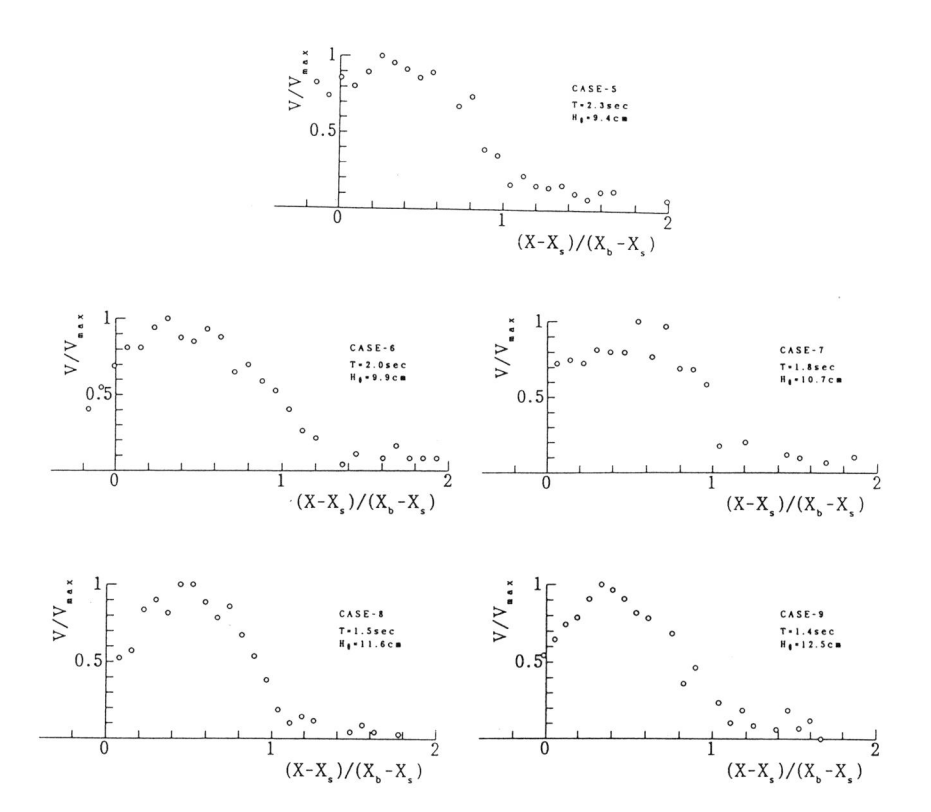

Fig. 8: On-offshore distributions of longshore current velocity

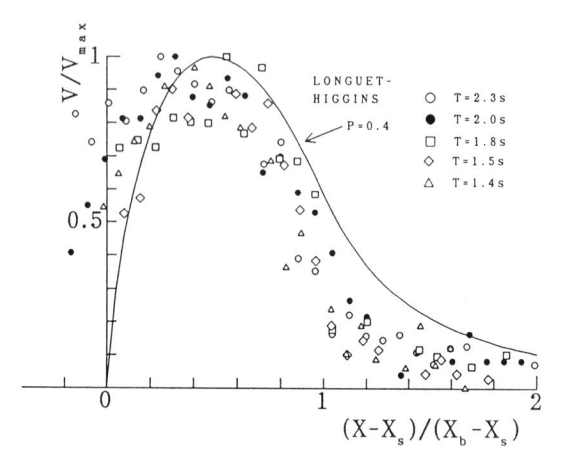

Fig. 9: On-offshore distributions of longshore current velocity and comparison with Longuet-Higgins'(1970) analytical solution

position.

Fig. 9 shows the comparison of all the data shown in Fig. 8 with Longuet-Higgins'(1970) analytical solution, in which the ratio between the horizontal mixing term and the friction term P is given by 0.4. It should be noted that little attention has been paid on the longshore current velocity at and beyond the shoreline landward, and conventional time averaged models inevitably predict the longshore current at the shoreline V_s as 0. Whereas, Fig. 8 and Fig. 9 show that V_s has a substantial value even at the mean water shoreline $x = x_s$ after wave set-up. It is also noticed that the longshore velocity at the mean water shoreline V_s increases with the incident wave period. This property agrees with the numerical results shown in Fig. 3 (b).

Further investigation reveals that the present results as well as a part of Visser(1991)'s results on V_s/V_{max} are well arranged by the surf similarity parameter (See Asano et al.(1994)).

4. CONCLUSIONS

A numerical model is developed to predict the flow characteristics in a swash zone for obliquely incident wave trains. The two dimensional shallow water equations are de-coupled into independent equations each for on-off shore motion and for longshore motion. The numerical results on the longshore velocity in a swash zone are compared with measured results by the tracer method. Both the numerical and experimental studies reveal the following properties.

(1) Substantial longshore velocities have been obtained even in the landward region of the still water shoreline.
(2) The magnitude of the longshore current velocities at the mean water shoreline is found to increase with the beach slope and incident wave period.

(3) The swash wave which rushed up obliquely on a slope tend to run down approaching to the right angle with the shoreline. The property may cause zig-zag longshore sediment transport inherent in swash zone.

Quantitative comparison between numerical and experimental results should be conducted with considering the effects of the vertical velocity profiles and turbulence in swash waves.

REFERENCES

- Asano T., H. Suetomi and J. Hoshikura (1994): Velocity measurements in swash zone generated by obliquely incident waves, Coastal Engineering in Japan (in Printing).

- Asano T. and T. Nakano (1992): Numerical Analysis on obliquely incident run-up waves, Proc. of Coastal Engrg., JSCE, Vol. 39, pp.26-30 (in Japanese).

- Bodge K. R. and R. G. Dean (1987): Short-term impoundment of longshore transport, Proc. of Coastal Sediment '87, pp.468-483.

- Hibberd, S. and Peregrine, D.H. (1979): Surf and run-up on a beach; A uniform bore, J. of Fluid Mech., Vol.95, pp.323-345.

- Kamphuis J. W. (1991, a): Wave transformation, Coastal Engineering, Vol.15, pp.173-184.

- Kamphuis J. W. (1991, b): Alongshore sediment transport rate distribution, Proc. of Coastal Sediment '91, pp.170-183.

- Kobayashi N., A. K. Otta and I. Roy (1987): Wave reflection and run-up on rough slopes, J. of Waterway, Port, Coastal and Ocean Div., ASCE, Vol.113, No.3, pp.282-298.

- Longuet-Higgins, M. S. (1970) : Longshore current generated by obliquely incident sea waves, J. Geophys. Res. , Vol.75, pp.6778-6801.

- Ryrie S. C. (1983): Longshore motion generated on beaches by obliquely incident bores, J. Fluid Mech., Vol.129, pp.193-212.

- Thornton E. B. and S. Abdelrahman (1991): Sediment transport in the swash due to obliquely incident wind-waves modulated by infragravity waves ; Proc. Coastal Sediment '91, pp. 100-113.

- Visser, P. J.(1991) : Laboratory measurements of uniform longshore currents, Coastal Engineering, Vol. 15, pp.563-593.

CHAPTER 3

LABORATORY COMPARISON OF DIRECTIONAL WAVE MEASUREMENT SYSTEMS AND ANALYSIS TECHNIQUES

Michel BENOIT [1] and Charles TEISSON [2]

Abstract

In order to define a directional wave sensor for laboratory experiments, three measuring systems as well as seven directional analysis methods are combined, applied and compared on three different tests performed in a directional wave basin. "Single-point" gauges are found to accurately analyse unimodal spectra when associated to advanced methods (Fit to bimodal model, Iterative Maximum Likelihood Method, Maximum Entropy Method, Bayesian Method). The heave-pitch-roll gauge used in this study is in particular very simple and shows promising capabilities. For bimodal spectra (two directional peaks at same frequency) however, only the wave probe array combined with the Maximum Entropy Method or the Bayesian Method appears to be able to produce reliable estimates.

1. INTRODUCTION — SCOPE OF THE STUDY

The measurement of directional wave spectrum may be performed through various systems, including co-located gauges (directional buoys, pressure sensor combined with a 2D currentmeter,...), arrays of gauges (wave probe arrays or mixed instruments arrays) or remote-sensing systems (satellite synthetic aperture radar, aerial stereo-photography techniques,...). Each of these measuring devices delivers a rather limited amount of information and the estimation of directional wave spectrum is then an awkward inverse problem, mathematically speaking. In order to get an estimate from the data anyway, various directional analysis methods have been proposed : Fourier Series Decomposition, Fit to parametric models, Maximum Likelihood Methods, Maximum Entropy Methods, Bayesian Methods,...

From practical point of view these methods exhibit different behaviours and characteristics for instance in mathematical complexity, directional accuracy, directional spectrum shape dependency, computing time, numerical convergence,... A good number of these methods have recently been implemented at Laboratoire National d'Hydraulique (LNH) and tested quite extensively on numerical simulations using heave-pitch-roll data (Benoit, 1992) as well as gauge array data (Benoit, 1993). The present study aims to proceed a step further in this comparative analysis by evaluating the capabilities of the methods on laboratory data.

---
1 Research Engineer — Maritime Group
2 Head of Maritime Group
   EDF - Laboratoire National d'Hydraulique, 6, quai Watier 78400 CHATOU, FRANCE

We briefly recall that the main unknown of the problem is the directional wave spectrum S(f,θ), a function of wave frequency f and direction of propagation θ. The following conventional decomposition is used :   S(f,θ) = E(f).D(f,θ)

E(f) is the classical variance or 1D-spectrum that may be estimated by a single record of free-surface elevation. and D(f,θ) is the Directional Spreading Function (DSF) satisfying two important properties :

$$D(f,\theta) \geq 0 \text{ over } [0, 2\pi] \quad \text{and} \quad \int_0^{2\pi} D(f,\theta) \, d\theta = 1$$

The directional analysis procedure may be roughly decomposed into three steps :
a. record simultaneously one or several wave properties (elevation, velocities, pressure, slopes,...) at one or more locations : $X_i(t), \ldots, X_N(t)$ (N ≥ 3)
b. compute the cross-spectra between each pair of recorded signals :

$$G_{ij}(f) = \int_{-\infty}^{+\infty} R_{ij}(\tau) \, e^{-i2\pi f\tau} \, d\tau \quad \text{with} \quad R_{ij}(\tau) = \lim_{T \to \infty} \frac{1}{T} \int_0^T X_i(t).X_j(t+\tau) \, dt$$

c. estimate the directional spectrum by inverting the following set of equations :

$$G_{ij}(f) = \int_0^{2\pi} H_i(f,\theta) . \overline{H_j}(f,\theta) . S(f,\theta) . \exp(-\vec{k}.\vec{x_{ij}}) \, d\theta$$

In this study, three measuring systems are set up in LNH directional wave basin (see section 2) and seven directional analysis methods are selected (see section 3). The experimental lay-out is described in section 4 and the three test-cases are presented in section 5. The comparative analysis of results is reported in section 6.

2. DIRECTIONAL MEASURING DEVICES

Three measuring systems are considered for laboratory measurements (figure 1) :

— <u>a wave probe array</u> : the array is composed of five probes (numbered from 1 to 5) laid out on the same configuration as the one used by Nwogu (1989). The wave probes are resistive-type wires mounted on a frame that allows a precise positionning. The radius R of the array is 0.40 m. As it will be presented in section 5, the wavelength corresponding to peak frequency is Lp = 2.42 m, and thus the ratio R/Lp is about 16 %.

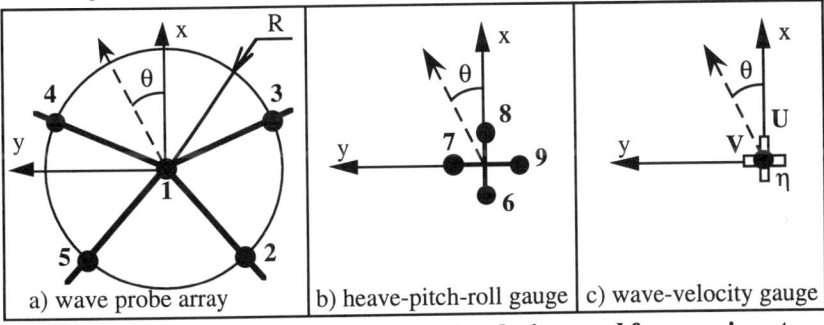

Figure 1 : the three directional measuring devices used for experiments.

— a "heave-pitch-roll" gauge : this gauge aims to deliver the same type of signals as the heave-pitch-roll buoy used in the field. To that extent, four wave probes are set up close to each other in a very simple way (see figure 1-b). From the four recorded free-surface elevation time series, the elevation and two orthogonal slopes of free-surface at the center of the gauge are computed :

$\eta(t) = (\eta_6 + \eta_7 + \eta_8 + \eta_9)/4.$

$\frac{\partial \eta}{\partial x}(t) = \frac{\eta_8(t) - \eta_6(t)}{d_{6-8}}$

$\frac{\partial \eta}{\partial y}(t) = \frac{\eta_7(t) - \eta_9(t)}{d_{7-9}}$ with : $d_{6-8} = 11.4$ cm and $d_{7-9} = 13.2$ cm

— a wave-velocity gauge : as the previous one, this gauge is also a "single-point" gauge, recording at the same location the free-surface elevation (through a wave probe) and the two horizontal components of velocity (through a 3D acoustic velocimeter, from which only the two velocity signals U and V are kept).

3. DIRECTIONAL ANALYSIS METHODS

Among the methods available at LNH, the seven following ones are considered, because each of them may be used both for "single-point" systems and arrays :

— **Weighted Fourier Series (WFS)** : the directional spreading function is expressed as a truncated Fourier series whose first coefficients are computed from the cross-spectra (Borgman, 1969). A weighting function is used to avoid possible negative values taken by this estimate (Longuet-Higgins et al., 1963).

— **Fit to bimodal Gaussian model (2MF2)** : A bimodal parametric model obtained from linear combination of two unimodal Gaussian-type models is used. Its five unknown parameters are determined from the cross-spectra. In the case of "single-point" systems, the problem becomes awkward because there are only four information avalaible and additional constraints are thus needed (Benoit, 1992).

— **Maximum Likelihood Method (MLM)** : By this method the directional spectrum is regarded as a linear combination of the cross-spectra. The weighting coefficients are calculated with the condition of unity gain of the estimator in the absence of noise (Oltman-Shay and Guza, 1984 ; Krogstad, 1988). Recently, Haug and Krogstad (1993) proposed a modified version of MLM (the constrained MLM) for gauge arrays which is not taken into account here.

— **Iterative Maximum Likelihood Method (IMLM2)** : The estimate obtained from the former method is not consistent with the measured cross-spectra. It may be iteratively modified to let its cross-spectra become closer to the ones obtained from the data (Oltman-Shay and Guza, 1984).

— **Maximum Entropy Method (MEM2)** : The method is based on the definition of Shannon for entropy : $\chi = -\int_0^{2\pi} D(f,\theta).\ln(D(f,\theta))\, d\theta$.

This entropy is maximized under the constraints given by the cross-spectra. The application to single-point systems or gauge arrays is described by Kobune and Hashimoto (1986), Nwogu et al. (1987) and Nwogu (1989).

This entropy definition is different from the one used by other authors (e.g. Lygre and Krogstadt, 1986) which generally appears to be less powerfull.

— **Bayesian Directional Method (BDM)** : No *a priori* assumption is made about the spreading function which is considered as a piecewise-constant function over [0 , 2π]. The unknown values of D(f,θ) on each of the K segments dividing [0 , 2π] are obtained by considering the constraints of the cross-spectra and an additional condition on the smoothness of D(f,θ) (Hashimoto *et al.*, 1987).

— **Variational Fitting Technique - Long-Hasselmann Method (LHM)** : Long and Hasselmann (1979) developed this method by which an initial simple estimate is iteratively modified to minimize a "nastiness" function that takes into account the various conditions on the spreading function. The application to buoy data is described in detail by Long (1980).

When referring to previous LNH numerical comparative surveys (Benoit, 1992), the Fit to unimodal model method (e.g. Borgman, 1969) and the Eigenvector Method (Mardsen and Juszko, 1987) have been dropped. The former is definitely unable to analyse bimodal cases and the latter has not been applied to gauge arrays.

4. EXPERIMENTAL LAY-OUT

The LNH multirectional wave facility is a rectangular wave basin of 50 m by 30 m used for coastal studies. The segmented wavemaker is composed of 56 piston-type paddles. The width of the paddles is 0.40 m. The total wavemaker length is thus 22.4 m. It is movable along the main side of the basin. The maximum water depth in the basin is 0.80 m. The facility is equipped with numerous mobile upright progressive wave absorbers. Each absorber unit measures 2.8 m by 2 m, allowing variable and adpatable absorber configurations in the basin. Tidal currents may also be simuleted in addition to waves.

For the present experiments only the eastern part of the basin was used as there was a breakwater model set up in the remaining part of the basin. Only the first 30 paddles were activated, giving an effective wavemaker length of 12 m. The test area was then a rectangle of 12 m by 25 m (figure 2) limited by wave absorbers. Fully reflective sheets over a length of 6 m were set-up at each side of the wavemaker in order to increase the work area through the corner reflection method. The bottom was flat over the whole test area. The water depth was kept constant at 0.60 m.

The three measuring devices were located 6 m apart from the wavemaker. They were set up every 0.40 m on a line parallel to the wavemaker (see figure 2).

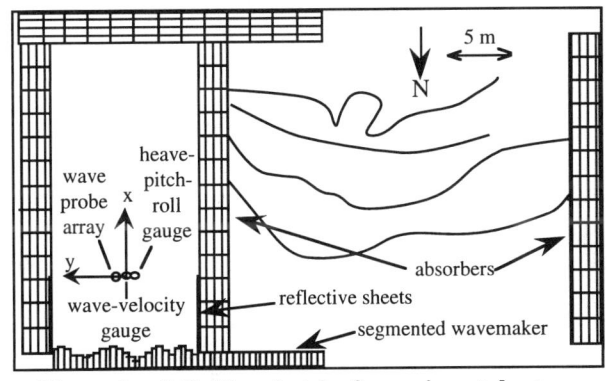

Figure 2 : definition sketch of experimental setup.

5. CHARACTERISTICS OF LABORATORY TEST-CASES

5.1 Wave simulation characteristics

The directional wave simulation is achieved by using a "single summation" method (also called "single direction per frequency method") (Miles, 1989) :

$$\eta(x,y,t) = \sum A_n \cos(2\pi.f_n.t - k_n(x.\cos\theta_n + y.\sin\theta_n) + \varphi_n)$$

The sea surface elevation is obtained through a linear superposition of numerous elementary components. The amplitude of each component is related to the target spectrum through : $A_n = \sqrt{2 S(f_n, \theta_n) \Delta f_n \Delta\theta_n}$

The phases are uniformly and randomly distributed over $[0, 2\pi]$. The directions θ_n are of the form $k.\Delta\theta$, but randomly distributed over $[0, 2\pi]$.

The frequency spectrum $E(f)$ is a classical JONSWAP spectrum with a significant wave height of 0.10 m, a peak period of 1.3 s and a peak-factor $\gamma = 5$. The simulated directional spreading function (DSF) is frequency independent :

$$\Pi_{s1,\alpha1,s2,\alpha2,\lambda}(\theta) = \lambda.\Pi_{s1,\alpha1}(\theta) + (1-\lambda).\Pi_{s2,\alpha2}(\theta) \qquad \text{with } 0 \le \lambda \le 1$$

with : $\Pi_{s,\alpha}(\theta) = \Delta(s) \cos^{2.s}\left[(\theta-\alpha)/2\right]$ if $\theta-\alpha \in [-\theta_m; \theta_m]$ $(\theta_m = 60°)$

Three DSF are simulated with the following characteristics (see figure 3) :

Test	Description	s1	α1	s2	α2	λ
A1	Unimodal Broad DSF	1	0.			1.
A3	Unimodal sharp DSF	15	0.			1.
A5	Bimodal DSF	25	-40.°	5	30.°	0.5

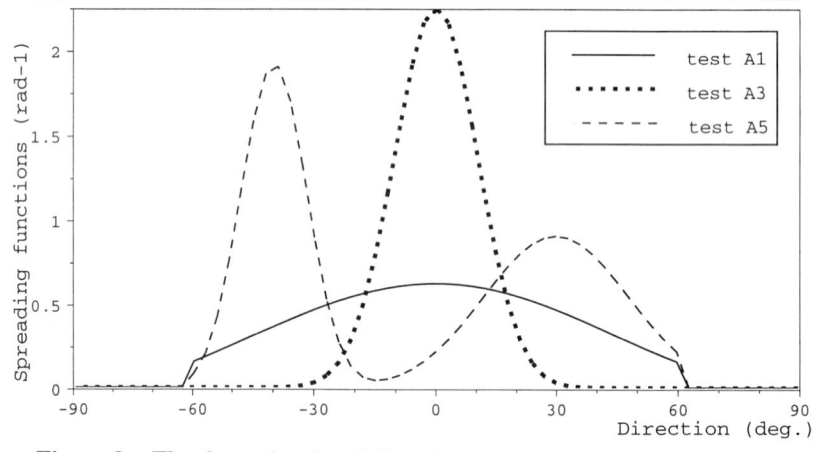

Figure 3 : The three simulated directional spreading functions.

5.2 Signal recording characteristics

The signals are recorded with a time step of 0.05 s over a duration of 819.2 s.

5.3 Cross-spectral analysis characteristics

The spectral analysis procedure is based on the technique of the averaged periodogram on the whole recorded signals partitioned in segments of 512 points. An overlapping of 25% between adjacent segments is used. The resulting frequency resolution is 0.039 Hz. Directional analysis is carried out between 0.5 and 1.25 Hz.

6. PRESENTATION AND DISCUSSION OF LABORATORY TESTS RESULTS

The directional spectra analysed on the three laboratory test-cases are presented using 2D-plots on figures 4-a and 4-b (test A1), 5-a and 5-b (test A3), 6-a and 6-b (test A5). Figure 7 gives a 3D-view of directional spectra analysed by MEM2 and BDM for the three measuring devices on test A5.

6.1 Analysis of test A1 — Unimodal broad spectrum — Figures 4-a and 4-b :

— Wave probe array : Reliable estimates are obtained by the BDM and MEM2 methods only. The WFS, MLM and LHM estimates are broader than the target spectrum. The 2MF2 and IMLM2 estimates are either too sharp or quite bimodal and reveal some instable behaviour of the methods.

— Heave-pitch-roll gauge : Very accurate and similar estimates are obtained from the 2MF2, IMLM2, MEM2 and BDM methods. This similarity allows a certain confidence in the results of this gauge. The spectra analysed by the WFS, MLM and LHM methods are too broad.

— Wave-velocity gauge : the behaviour of analysis methods is very similar to the heave-pitch-roll gauge, but the estimated spectra are a little bit sharper than the former ones. Correct estimates are again obtained from the 2MF2, IMLM2, MEM2 and BDM methods.

6.2 Analysis of test A3 — Unimodal sharp spectrum — Figures 5-a and 5-b :

— Wave probe array : Best estimates are obtained from MEM2 method. The BDM and IMLM2 methods produce acceptable results, but the latter one shows some numerical instabilities out of peak region. The 2MF2 method also exhibits some numerical instabilties, resulting in spurious peaks of the spectrum. The WFS, MLM and LHM methods are unable to model the sharpness of the spectrum and appear to be unefficient for sea-states with narrow angular spreading of energy.

— Heave-pitch-roll gauge : the best estimates are given by IMLM2, BDM and MEM2 methods. The spectrum analysed by the 2MF2 method is clearly too sharp, while the spectra analysed by the WFS method especially, but also by the MLM and LHM methods, are far too broad.

— Wave-velocity gauge : As for test A1, the observations for this measuring device are very close to the ones of heave-pitch-roll gauge. Again the most accurate estimates are achieved by the IMLM2, BDM and MEM2 methods. On this second case however the directional widths of the estimates are very close to those obtained from heave-pitch-roll gauge.

6.3 Analysis of test A5 — Bimodal spectrum — Figures 6-a, 6-b and 7 :

— Wave probe array : Reliable estimates are obtained from the 2MF2, MEM2 and BDM methods. For these three methods the bimodal nature of the sea-state (with a difference in the shapes of the two peaks) is clearly reproduced. On figure 7 it may be seen that the spectra given by MEM2 and BDM agree quite well with the theoretical spectrum. One must emphasis that this bimodal case with two peaks at the same frequency only separeted by 70 degrees is very severe. Bimodality of spectrum is harly detected by the LHM and the IMLM2. The WFS and MLM methods only produce an unimodal and very broad spectrum.

— Heave-pitch-roll gauge : The results given by the various methods are definitely worse than for the wave probe array. The quite low number of information recorded by the single point-system undoubtedly limits here the resolution capability of the analysis methods. The bimodal nature of the spectrum is

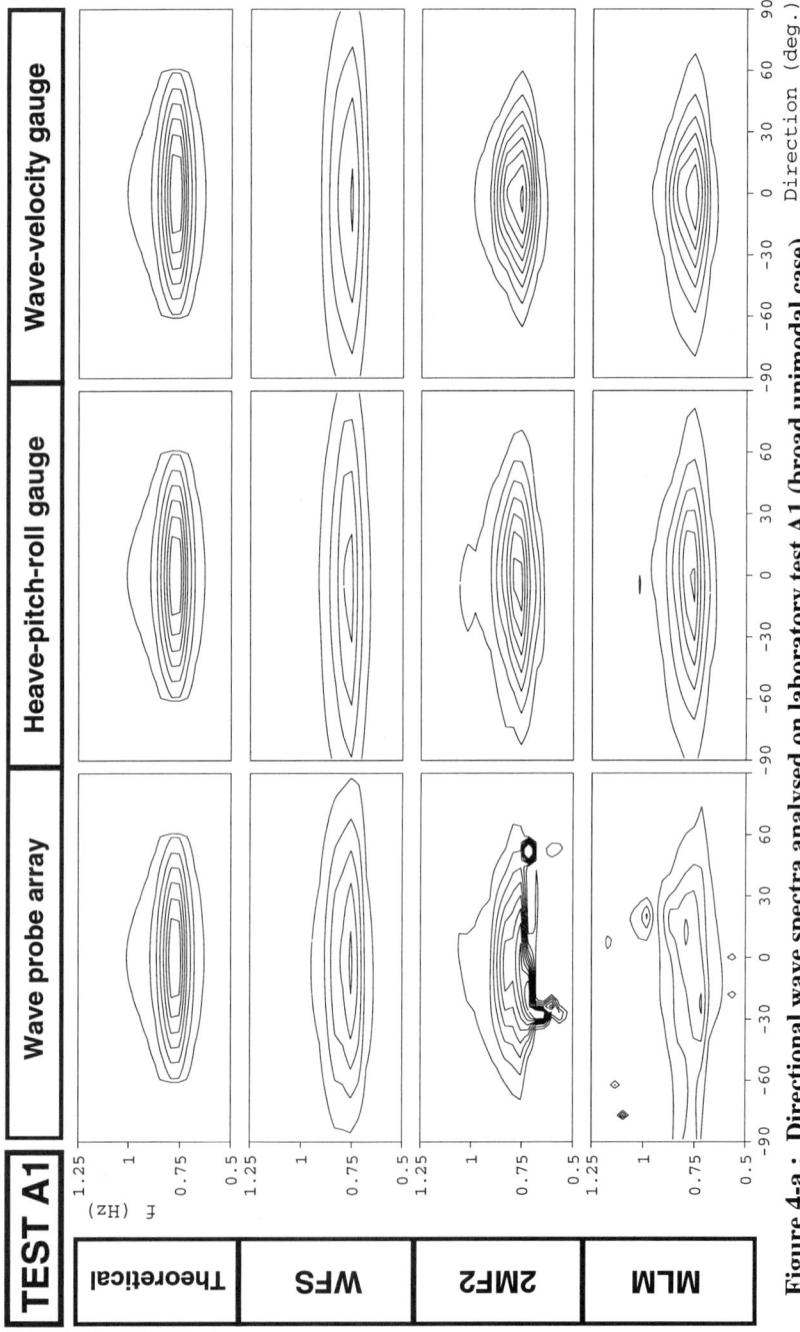

Figure 4-a : Directional wave spectra analysed on laboratory test A1 (broad unimodal case) Comparison of measuring systems and analysis methods (WFS, 2MF2, MLM)

DIRECTIONAL WAVE MEASUREMENT SYSTEMS 49

Figure 4-b: Directional wave spectra analysed on laboratory test A1 (broad unimodal case) Comparison of measuring systems and analysis methods (IMLM2, MEM2, BDM, LHM)

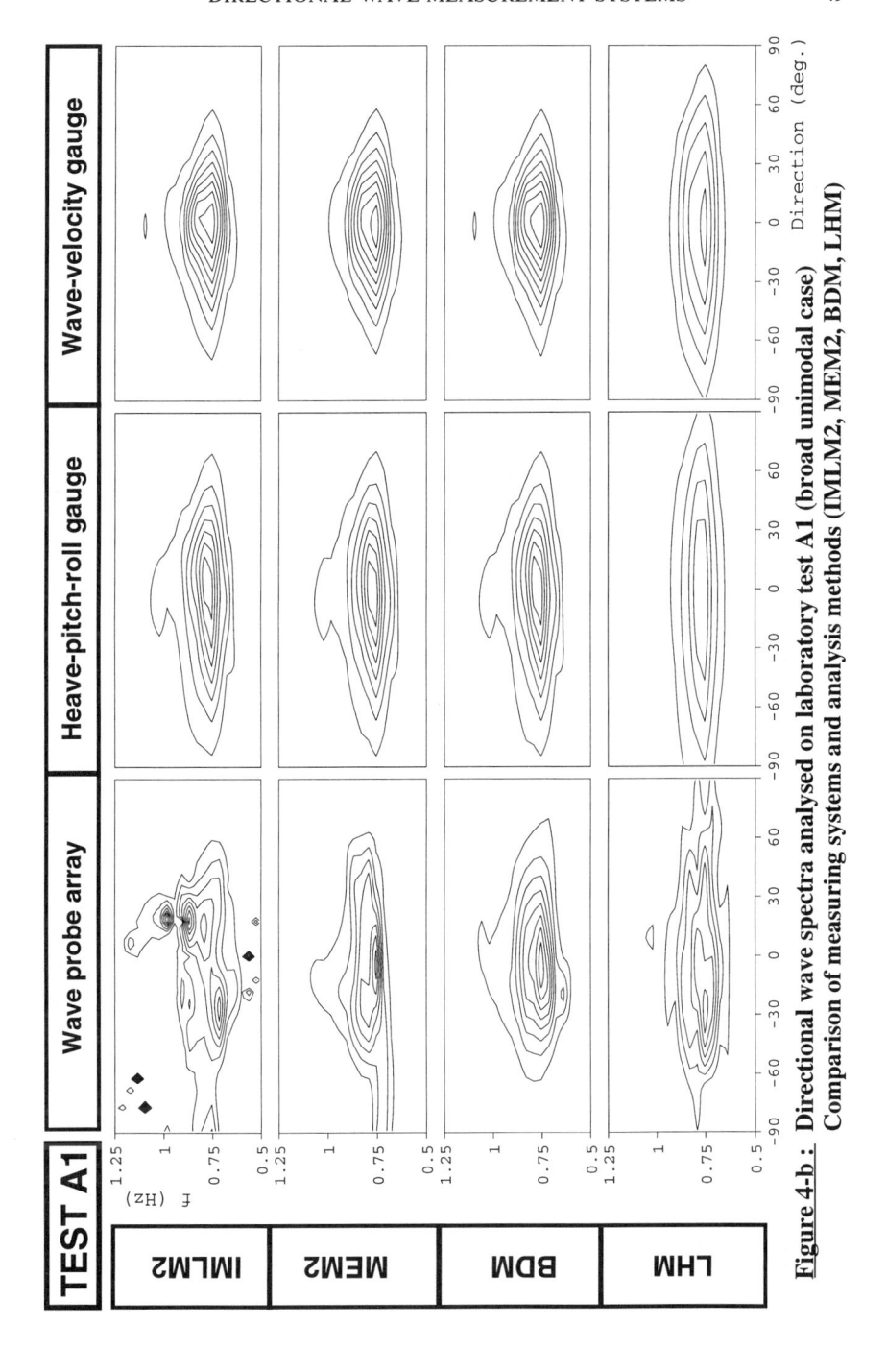

Figure 5-a : Directional wave spectra analysed on laboratory test A3 (Sharp unimodal case) Comparison of measuring systems and analysis methods (WFS, 2MF2, MLM)

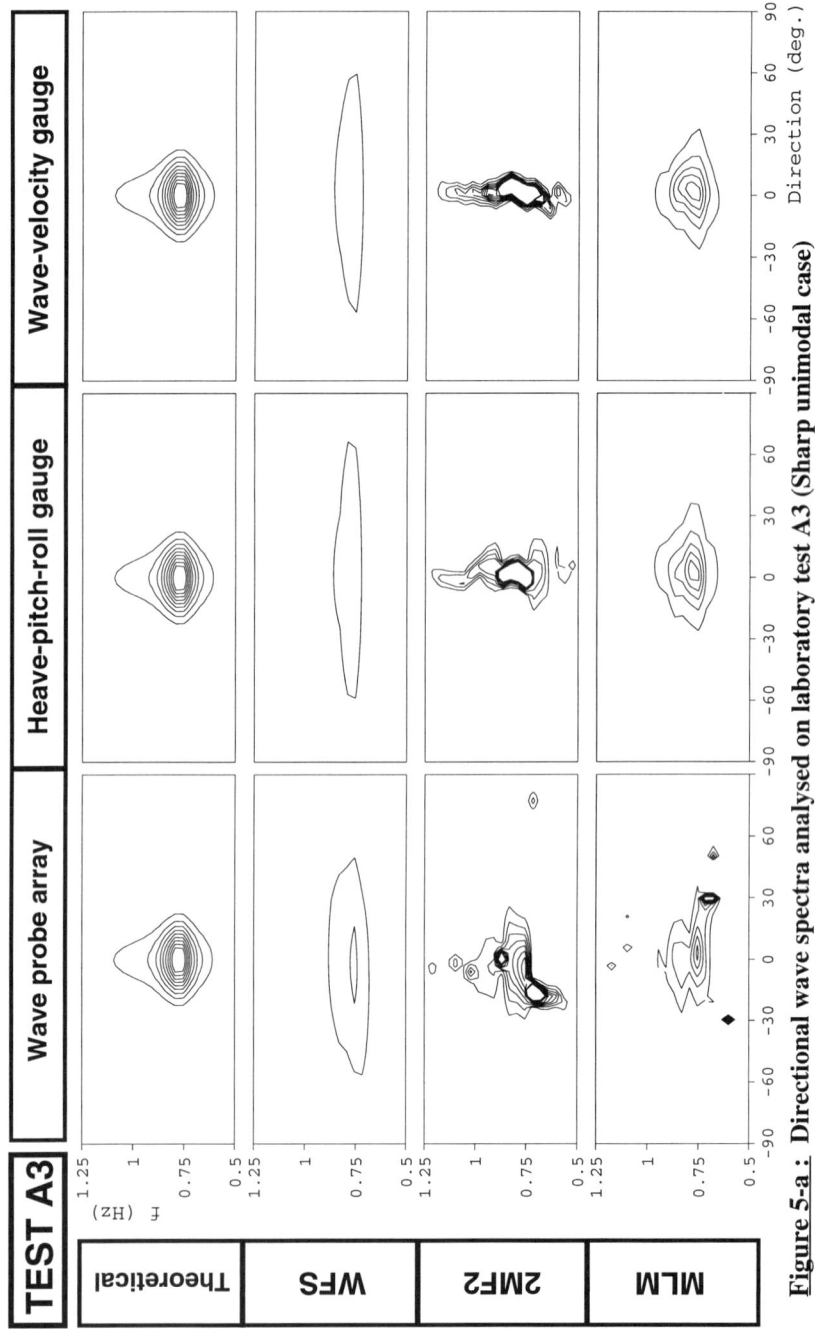

DIRECTIONAL WAVE MEASUREMENT SYSTEMS

Figure 5-b : Directional wave spectra analysed on laboratory test A3 (Sharp unimodal case) Comparison of measuring systems and analysis methods (IMLM2, MEM2, BDM, LHM)

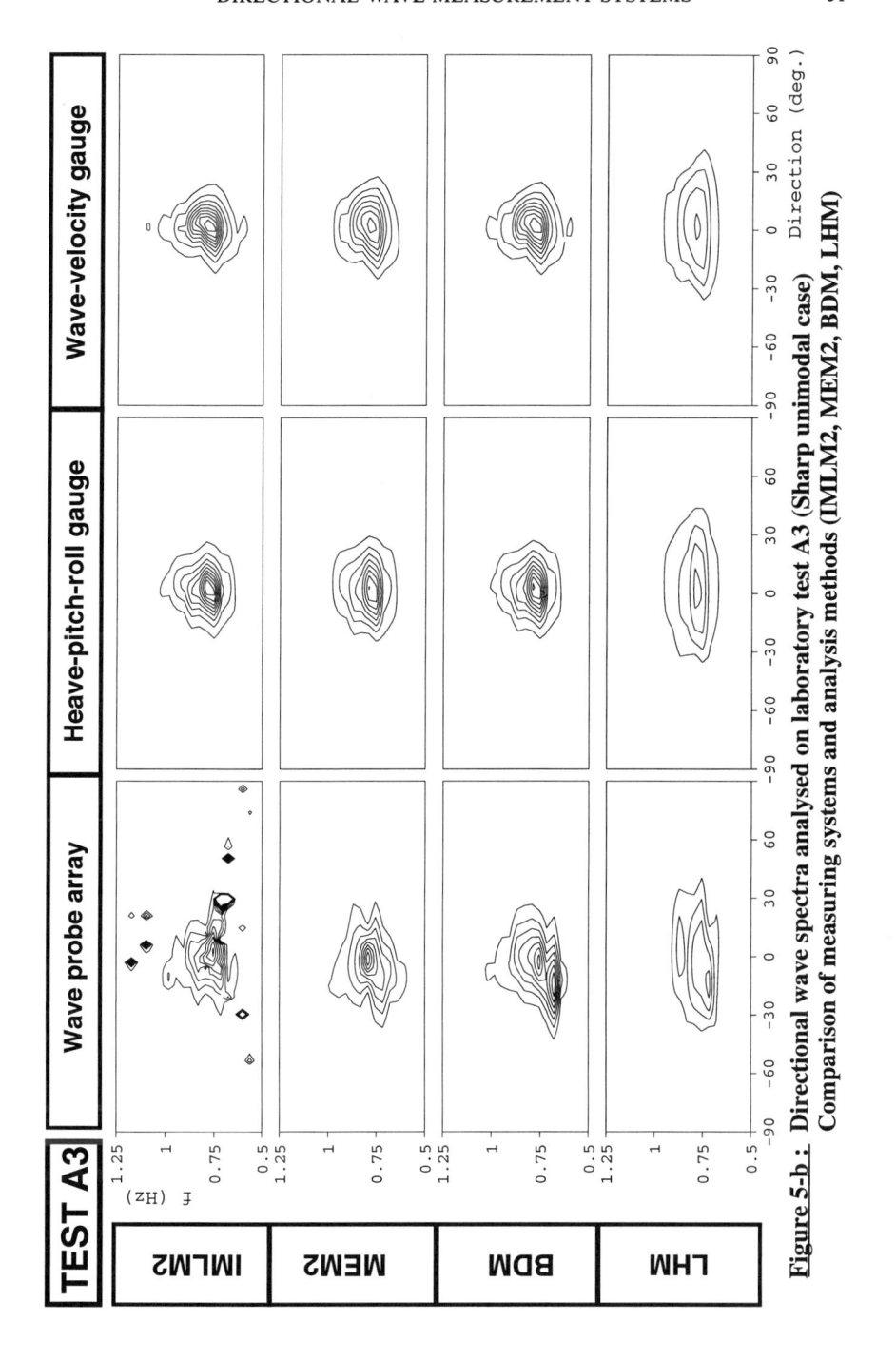

Figure 6-a : Directional wave spectra analysed on laboratory test A5 (Bimodal case)
Comparison of measuring systems and analysis methods (WFS, 2MF2, MLM)

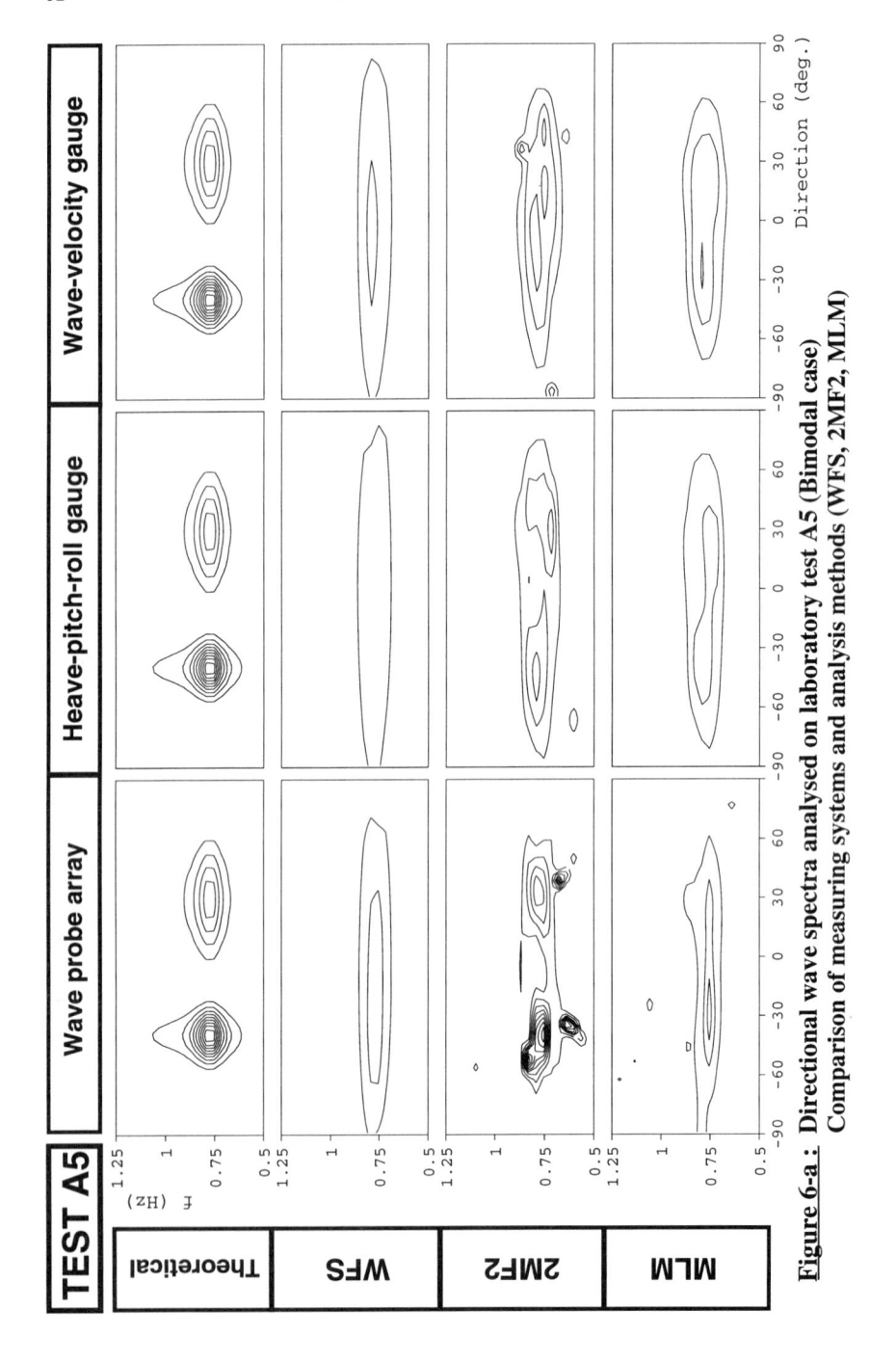

Figure 6-b.: Directional wave spectra analysed on laboratory test A5 (Bimodal case) Comparison of measuring systems and analysis methods (IMLM2, MEM2, BDM, LHM)

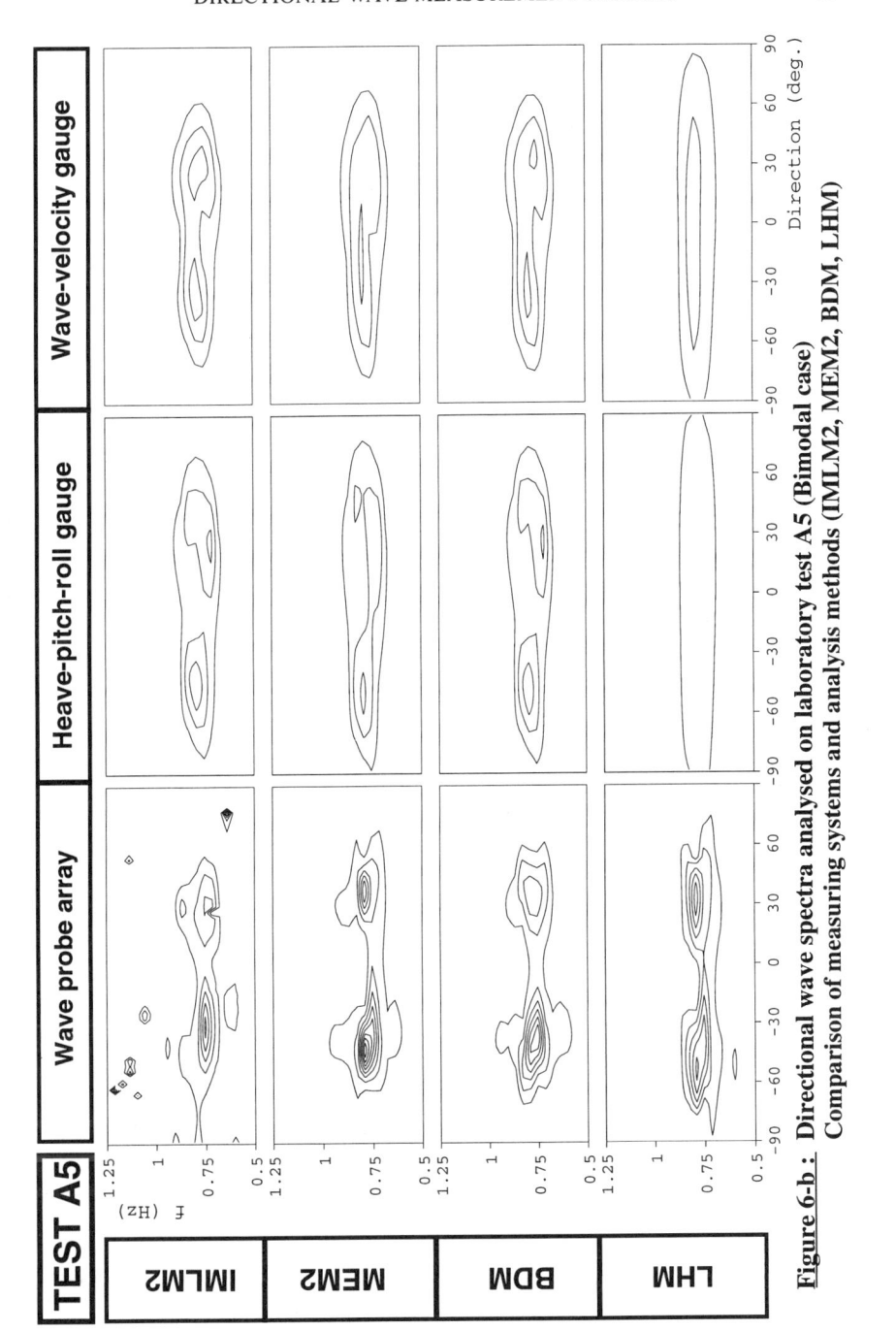

slightly noticeable for the 2MF2, IMLM2, MEM2 and BDM methods, but it is almost impossible to get detailed information about the relative shapes of the two peaks. Again these four methods give very similar and concordant results.

— Wave-velocity gauge : Most of the estimates are of same quality —and generally even a little bit worse— than the ones obtained from the heave-pitch-roll gauge. An exception is maybe the IMLM2 method which shows some better results than for the heave-pitch-roll gauge. Figure 7 however indicates that on this test-case the heave-pitch-roll gauge is superior to the the wave-velocity gauge when associated with sophisticated methods (MEM2 and BDM).

7. CONCLUSIONS — FUTURE WORK

Based on the comparative analysis of the various laboratory experiments carried out during this study, the following conclusions may be expressed :
- Great care should be given to the preliminary steps of signal recording and spectral analysis. There is a strong need to record rather long time series in order to get minimum variance spectral estimates. It seems worthwhile to increase the number of degrees of freedom of the cross-spectra as much as possible.
- The analysis of <u>unimodal directional sea-states</u> may be quite efficiently achieved by "single-point" (or "co-located") measuring systems, recording only three wave signals : heave-pitch-roll gauge or wave-velocity gauge.
- Each of these gauges usually produces concordant estimates from the 2MF2, IMLM2, MEM2 and BDM methods. When associated to one of these analysis methods, the "single-point" gauges exhibit good resolution capabilities for unimodal spectra, whatever the directional spreading of energy is.
- In addition, the comparison of tests results seems to indicate that the heave-pitch-roll gauge is somewhat superior to the wave-velocity gauge. This point needs however to be confirmed on more extensive experiments with velocity measurement at various depths. Furthermore, the heave-pitch-roll gauge used for these laboratory experiments is very simple and may be set up at moderate cost. Although these performances have to be confirmed on additional test-cases, the present experimental results indicate promising capabilities for operational laboratory measurements.
- For <u>complex bimodal sea-states,</u> the "single-point" systems usually fail to reproduce the shape of the target spectra. The bimodal nature of sea-state is hardly detected by these systems, even by using sophisticated analysis methods (MEM2, IMLM2, BDM). The output resolution seems too low to allow physical interpretation of results. For such bimodal cases, the amount of input information must be extended (up to five or more). The gauge array composed of five wave probes used in these experiments has proven to be able to produce reliable estimates on the three tests.
- The gauge array however usually requires more refined and complex numerical treatment because of the additional assumption it needs about spatial homogeneity of sea-state. For this type of measuring system, one must emphasis the need of advanced methods (MEM2 or BDM) in order to get reliable results .

Additional laboratory experiments with differents conditions will be performed in order to confirm or modify the present conclusions. The extension of these methods for the measurement of directional wave spectra close to a reflective structure is also a major research field for the next future.

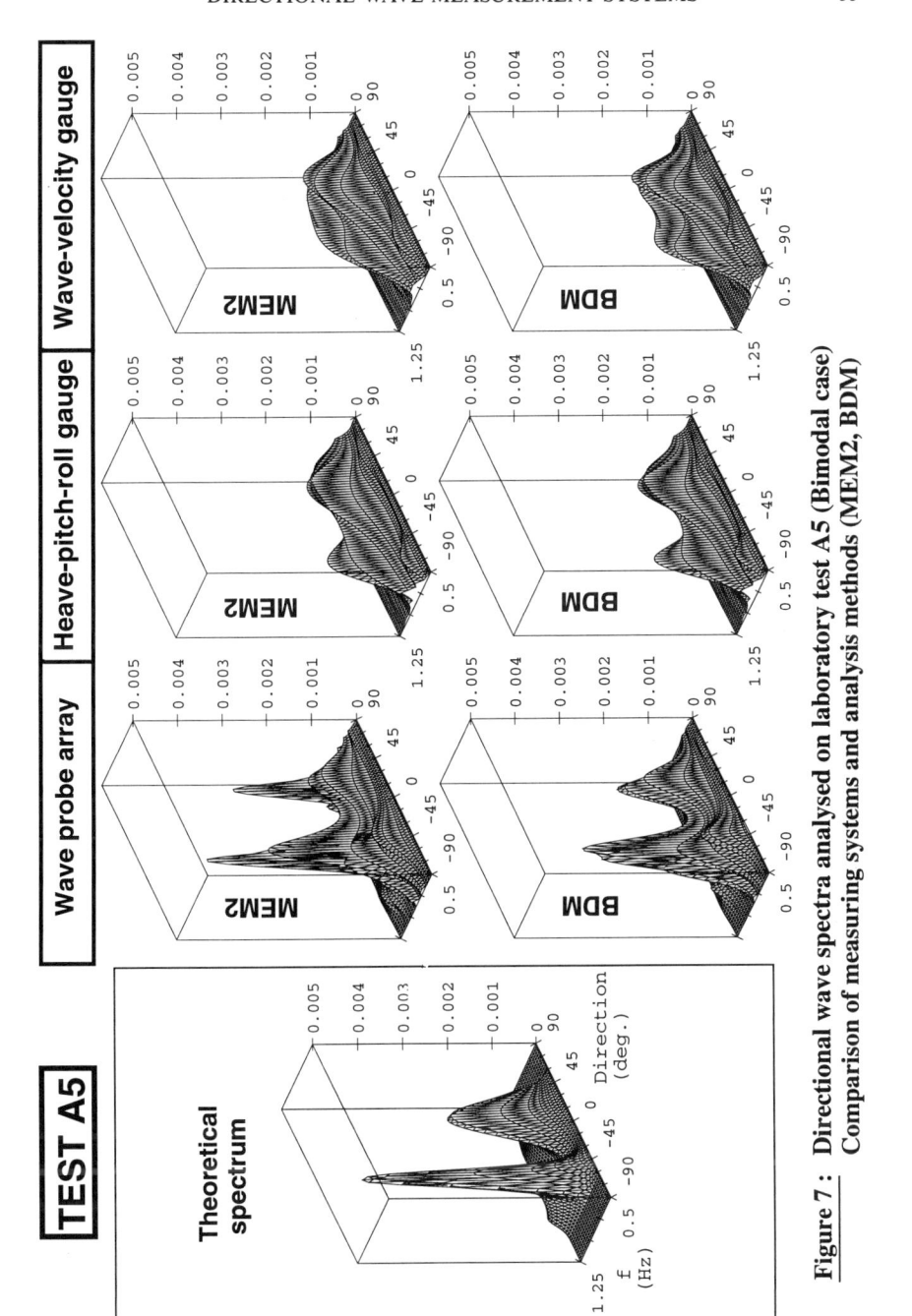

DIRECTIONAL WAVE MEASUREMENT SYSTEMS 55

Figure 7: Directional wave spectra analysed on laboratory test A5 (Bimodal case) Comparison of measuring systems and analysis methods (MEM2, BDM)

8. ACKNOWLEDGEMENTS

This study is a joint research program between EDF-Laboratoire National d'Hydraulique and the French Ministry of the Sea (Service Technique de la Navigation Maritime et des Transmissions de l'Equipement — STNMTE).

9. REFERENCES

BENOIT M. (1992) : Practical comparative performance survey of methods used for estimating directional wave spectra from heave-pitch-roll data. *Proc. 23rd Int. Conf. on Coastal Eng. (ASCE), pp 62-75, Venice (Italy).*

BENOIT M. (1993) : Extensive comparison of directional wave analysis methods from gauge array data. *Proc. 2nd Int. Symp. on Ocean Wave Measurement and Analysis (ASCE), pp 740-754, New-Orleans (USA).*

BORGMAN L.E. (1969) : Directional spectra models for design use. *Offshore Technology Conference, Houston, Texas*

HASHIMOTO N., KOBUNE K., KAMEYAMA Y. (1987) : Estimation of directional spectrum using the Bayesian approach and its application to field data analysis. *Report of the Port and Harbour Research Institute., vol 26, pp 57-100.*

HAUG O., KROGSTAD H. E. (1993) : Estimation od directional spectra by ML/ME methods. *Proc. 2nd Int. Symp. on Ocean Wave Measurement and Analysis (ASCE), pp 394-405, New-Orleans (USA).*

KOBUNE K., HASHIMOTO N. (1986) : Estimation of directional spectra from maximum entropy principle. *Proc. 5th OMAE Symp., pp. 80-85, Tokyo (Japan).*

KROGSTAD H.E. (1988): Maximum likelihood estimation of ocean wave spectra from general arrays of wave gauges. *Modelling, Identification and Control, vol 9, pp 81-97*

LONG R.B., HASSELMANN K. (1979) : A variational technique for extracting directional spectra from multicomponent wave data. *J. Phys. Ocean., vol 9, pp 373-381.*

LONG R.B. (1980) : The statistical evaluation of directional estimates derived from pitch/roll buoy data. *J. Phys. Oceanogr., vol 10, pp 944-952.*

LONGUET-HIGGINS M.S., CARTWRIGHT D.E., SMITH N.D. (1963) : Observations of the directional spectrum of sea waves using the motions of a floating buoy. *Ocean Wave Spectra, Prentice-Hall, pp 111-136.*

LYGRE A., KROGSTAD H.E. (1986) : Maximum entropy estimation of the directional distribution in ocean wave spectra. *J. Phys. Oceanogr., vol 16, pp 2052-2060.*

MARDSEN R.F., JUSZKO B.A. (1987) : An eigenvector method for the calculation of directional spectra from heave, pitch and roll buoy data. *J. Phys. Oceanogr., vol 17, pp 2157-2167.*

MILES M.D. (1989) : A note on directional random wave synthesis by the Single Summation Method. *Proc. 23rd IAHR Congress,Vol C, pp243-250, Ottawa (Canada).*

NWOGU O.U., MANSARD E.P.D., MILES M.D., ISAACSON M. (1987) : Estimation of directional wave spectra by the maximum entropy method. *Proc. 17th IAHR Seminar - Lausanne (Switzerland).*

NWOGU O.U. (1989) : Maximum entropy estimation of directional wave spectra from an array of wave probes. *Applied Ocean Research, vol 11, N°4, pp 176-182*

OLTMAN-SHAY J., GUZA R.T. (1984) : A data-adaptative ocean wave directional-spectrum estimator for pitch-roll type measurements. *J. Phys. Oceanogr., vol 14, pp 1800-1810.*

CHAPTER 4

Accuracy of Wind and Wave Evaluation in Coastal Regions

Luciana Bertotti[1] and Luigi Cavaleri[1]

Abstract

We have made a critical analysis of the processes and the parameters that affect the accuracy with which wind and waves can be evaluated close to coast. For each process we quantify the possible error, whenever possible complementing this with numerical tests and practical cases.

1. Introduction

The standard evaluation of the performance of a wave model is usually done off the coast, in the open sea. Typically the analysis fields (wave fields obtained using as input the wind provided by the analysis of 3-D meteorological models) are compared with the measured data available at certain locations. This provides a fair estimate of the overall performance. More detailed analyses, possibly referred to some test cases, can provide information on specific aspects of the model. The forecast fields are compared with the analysis fields to assess the reliability of the results in the forecast mode (in doing so we effectively check the forecast wind fields).

The related statistics are commonly available (see e.g., Günther et al., 1992 and Komen et al., 1994). Particularly during the last three years, with most meteorological centers moved to high resolution meteorological models, the results are quite satisfactory. The average bias for the significant wave height Hs is about 0.10-0.15 m or less, the rms error is limited to a few tens of centimeters.

However, the condition is not similarly satisfactory in coastal areas. Here a number of problems arise. First, the orography of the coast strongly affects the wind field, hence the local evolution of the wave fields. Then, the shallow waters bring to relevance a number of processes, some of them being of difficult evaluation, but

---

[1]Istituto Studio Dinamica Grandi Masse-CNR, San Polo 1364, 30125 Venice, Italy

The practical application of wave modeling in coastal areas requires therefore a careful analysis of the local conditions, to assess, even if only on a qualitative basis, the relevance of each process. This will tell us where to focus our attention, following the principle of "larger corrections first".

We have done an evaluation of the potential relevance of the processes active in coastal areas and of the model parameters that affect the accuracy of the results. Whenever possible, this has been done with numerical tests, supporting the results with wave measurements at suitable locations. In section 2 we discuss the evaluation of the wind field and the related consequences on the evaluation of the wave fields. Then (section 3) we analyze the relevance of the conservative processes arising from the interaction of the waves with the bottom. The dissipative processes are analyzed in section 4. In 5 we turn our attention to the resolution of the grid and of the wave model. Interactions with currents are briefly mentioned in 6. The overall results are summarized and commented in the final section 7.

2. Wind in Coastal Areas

The wind is the source of the whole energy present in the sea in the wind wave frequency range. The sensitivity of waves to even limited variations imply a careful attention to the modifications of the wind fields in coastal areas.

Wind can be modified both at large and local scales. Analyzing a very severe storm in the Mediterranean Sea, Cavaleri et al. (1991) report an increase of the wind speed of about 30% by increasing the resolution from 150 to 70 km. In another case in the same area (Cavaleri et al., 1993), a further increase of resolution to 40 km succeeded in revealing an otherwise unnoticed local turn of the wind, strictly associated to the local orography, that produced a 5 m significant wave height in the Gulf of Genoa, duly found in the measured data. It is not possible to specify the characteristics of a meteorological model that are required for a sufficient accuracy. As a practical rule, we can say that, given a characteristic length D of the local orography (the dimension of a bay, or of an island or a promontory), a good wind requires a resolution of D/5 or better. The same applies if D is the minimal distance from a coast with a complicated orography at which we want to evaluate the fields.

3. Wave Conservative Bottom Processes

We discuss the following processes: refraction, shoaling, bottom scattering. While the first two are a standard part of any shallow water wave model, the third one is rarely considered, but it can become dominant in certain conditions.

Refraction. Well established, both with grid and ray techniques. If no particular complication arises (e.g., caustics), the accuracy for the single spectral component is of the order of a few percents and a few degrees in direction.

More care is required when dealing with a full 2-D spectrum, from which we extract the mean direction (we anticipate here a result connected to the subject of section 5). Hubbert and Wolf (1991) have considered a narrow swell approaching at

60° with respect to the isobaths a one-dimensional 1:10⁴ sloping coast. Figure 1 shows the resulting turning of the swell mean direction while approaching the beach, as a function of the directional resolution used in the model. While all the results are good, and excellent for a resolution of 15° or better, we need to go down at 5° before being able to reproduce the result obtained with Snel's law (note: Komen et al., 1994, p. 345, point out that, contrarily to the common use, the correct spelling of this Dutch mathematician requires a single "l").

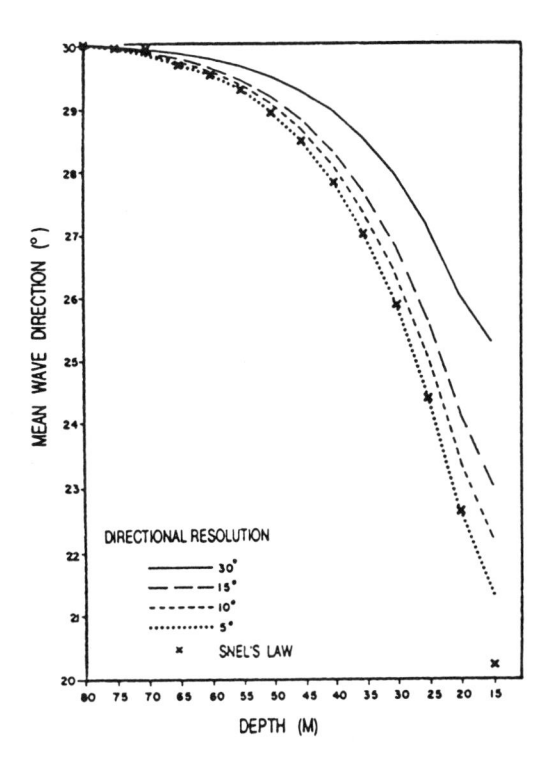

Figure 1. Refraction on a sloping bottom as a function of directional resolution (after Hubbert and Wolf, 1991).

Shoaling. When shoaling a wave spectrum towards the coast, two basic approaches are possible: to use linear theory for each component separately, or to summarize the spectrum into a representative wave of given height and period, and to use one of the several nonlinear theories available. To our knowledge no general method to deal with nonlinear shoaling of the whole 2-D spectrum has been published. We expect some substantial improvement not far in the future. For the time being we call the attention to one result of strong interest for the coastal engineer. Starting from recorded data and by numerical integration of the KdV equation, Osborne (1993) has analyzed the shoaling of a heavy swell case in the Northern Adriatic Sea,

shows the same system of waves at 6 meters of depth. The relevant point is the development by nonlinear interactions of long period components, formally appearing as a train of solitons. These long components are important for beach shaping and harbor management. They appear for large Ursell numbers in the field. In this case the results of standard theories should be taken with care.

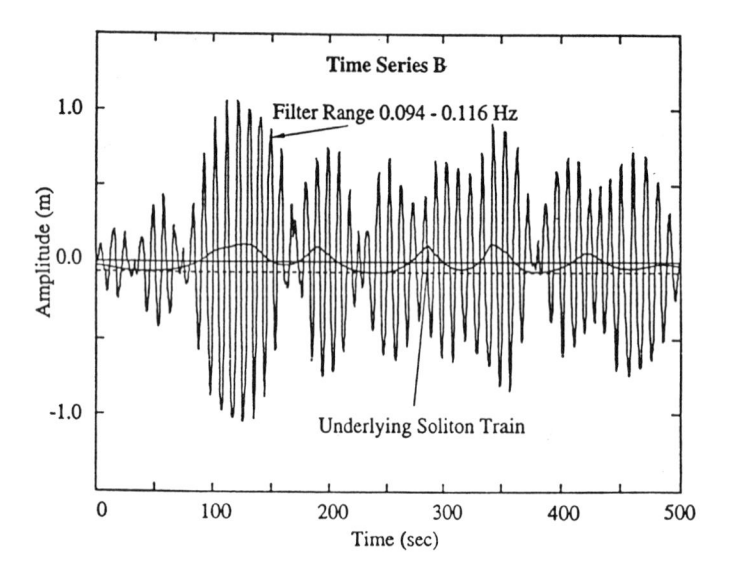

Figure 2. Non linear shoaling of a heavy swell case in the Northern Adriatic Sea. Surface profile evaluated at 6 m of depth. Note the development of long period wave components.

Bottom-scattering. We refer here to the interaction of a surface wave spectrum with the oscillations of the bottom. The theory is well established (see Long, 1973), but it has rarely been applied for the practical difficulty to have the necessary data available (the 2-D spectrum of the depth variations is required) and because of the very large computer power requirements. However, some laboratory experiments have clearly confirmed the theory and provided spectacular results. Davies and Heathershaw (1983) have shown that four oscillations of the bottom (wavelength half of that of the surface wave) are sufficient to reflect in the opposite direction 80% of the incoming wave height. Ten oscillations reflect 90%.

The wavelength is critical, which makes the application problematic. However, because of its potential dominant role, this process should be kept in mind whenever a series of transversal parallel bars or reefs is present in front of a coast.

4. Dissipative Bottom Processes

We discuss bottom friction, percolation, breaking and bottom elasticity.

Bottom friction. It can be evaluated with both linear and nonlinear (fully spectral) approaches. The non linearity leads to more correct evaluations, but this is paid with two orders of magnitude in computer power. A rather comprehensive treatment of the subject is given by Weber (1991). The linear approach suffices till water particles velocity at the bottom of about 0.15 m/s, above which it underestimates the energy loss at an increasing rate. A first hand estimate of the expected wave conditions at a given location and of the associated orbital velocity will tell the user which approach is to be followed.

A good example of the possible difference between the two approaches is given in figure 3, showing a 1-D spectrum at an oceanographic tower located on 16 m of depth at the far north of the Adriatic Sea, in front of Venice (Cavaleri et al., 1989). The tower is at the upper end of a long, slowly sloping continental platform, and the swell represented in the figure has been propagating in shallow water for many tens of kilometers. Clearly, the linear approach (WAM in the figure) fails to dissipate the low frequency energy at a sufficiently high rate.

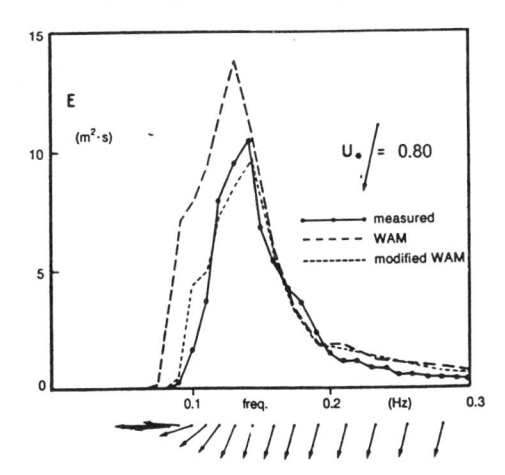

Figure 3. Comparison between measured and evaluated 1-D spectra in the Northern Adriatic Sea. WAM, evaluated with linear theory for bottom friction; modified WAM, with non linear theory (after Cavaleri et al., 1989).

Percolation. Of little importance offshore, it becomes important when the bottom is composed of shingles or very coarse sand, which are usually found close to the beach. Its role is never dominant. If to be considered, its proper evaluation requires laboratory tests to measure the transmission coefficient necessary for the estimate of the related energy budget (see, e.g., Shemdin et al., 1978).

Breaking. It is the most dominant factor for wave height in shallow water. Its consideration is essential whenever Hs>0.4 depth.

The breaking is not fully understood. Notwithstanding this, well-devised approaches provide very good results (see Battjes and Beji, 1993, for a clear example). In general, also simple methods provide acceptable results, simply because all of them essentially follow the basic principle of limiting the wave height to a certain percentage of depth. Larger differences, in percent terms, are found very close to shore.

Bottom elasticity. This phenomenon is rarely considered because in the very large majority of cases the energy involved is negligible. Besides, the bottom material (e.g., sand) is practically elastic, with a negligible absorption of energy. In a few special cases (the Mississippi Delta and the Bay of Bengal are the best known examples), the bottom is locally composed of viscoelastic mud. In this case tremendous absorptions of energy can be experienced in heavy storm. Figure 4 shows the evolution of a shoaling wave spectrum during hurricane Federic (Forristall et al., 1990). In 30 km the wave height passed from 8.6 m (in deep water) to 2.4 m (in 19 m of depth), a loss of energy of more than 90%. It is obvious then, whenever present, this process must be considered and it is going to be the dominant one.

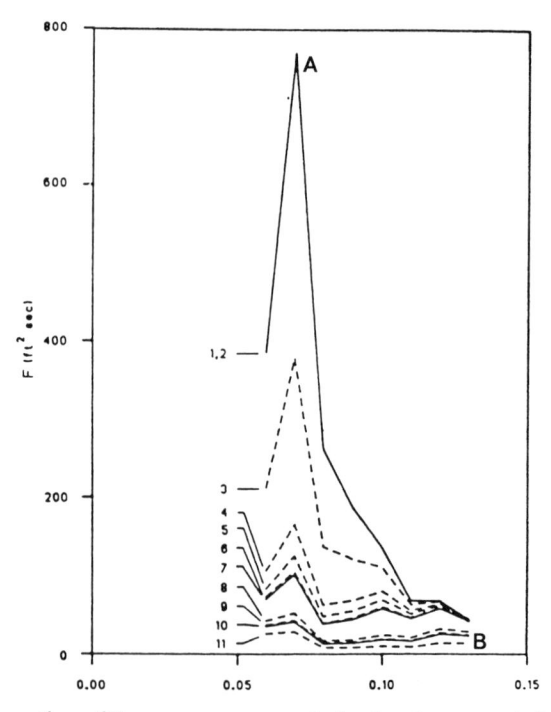

Figure 4. Attenuation of the wave spectrum during hurricane Frederic in the Gulf of Mexico. A and B locations are only 30 km apart (after Forristall et al., 1990).

5. Grid and Model Resolution

Grid resolution. It is essential in establishing the scale at which we want to analyze the phenomenon. The resolution affects the results of a wave model, particularly when strong gradients are present in the field. In this case a doubling of the resolution (say from 40 to 20 km) can increase the estimate of the peak wave height by 10-20%.

The grid resolution establishes also the accuracy with which we describe the coast. An uncertainty of half the grid step size on the actual position of the coast must be considered. This becomes critical in slanting fetch conditions or, e.g., with waves coming towards the coast after going around a promontory enclosing a gulf. To avoid errors larger than 10%, the point of interest should be at a distance from the coast at least five times the uncertainly in its exact location.

The overall effect of the grid resolution is exemplified in Figure 5. A very severe storm in the Mediterranean Sea (Hs>11 m between Tunisia, Sardinia and Sicily, 8 m in the Sicily Channel) has been hindcast using the same input wind, but two different resolutions, namely 0.5° and 0.25° (see Cavaleri et al., 1991). The figure shows the differences (with 0.5 m isolines) between the two fields, at the peak of the storm. They are partly due to a better description of the wave generation, and partly to coastal effects.

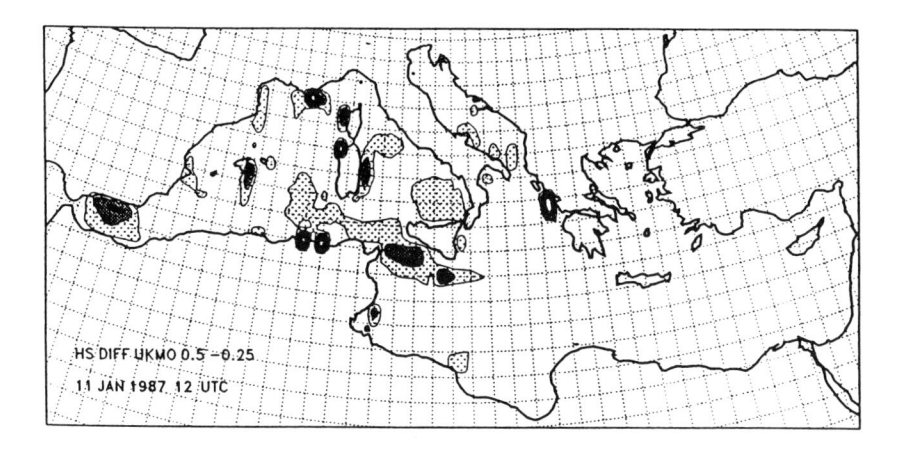

Figure 5. Wave height comparison between the hindcasts of a severe storm in the Mediterranean Sea done with different grid resolution. The differences are indicated as isolines at 0.5 m interval (after Cavaleri et al., 1991).

Model resolution. The integration time step ΔT is connected to the grid step size and to the time scale of the phenomenon we want to describe. Therefore ΔT must be equal or smaller than the time required by the minimal change we want to detect in the evolution of the storm. If larger, the phenomenon will be smoothed, and we must expect a likely underestimate of the peak conditions.

The resolution in frequency is usually not a problem. We have formerly discussed in section 3 the implications for refraction. Some particular cases, e.g., the proper evaluation of swell on the Pacific Ocean, can require an extension towards the low frequency range. De La Heras (1990) gives a nice example of this.

A more critical aspect for coastal engineers is the resolution in direction. The usual 24 or 30 degree resolution suffices for most of the cases. However, when approaching a complicated shallow water topography or a winding coastal shape, an increased resolution will provide a substantially better description of the wave distribution. Errors of 15-20% on Hs can easily be found at some location, if a coarse resolution in direction is used.

6. Interactions with Current

The wave-current interactions are usually neglected by the wave modeller for two reasons. First, in the large majority of cases the currents are not strong enough to affect a developed wave field in an appreciable way. Second, very rarely a detailed distribution of the current field is available. In any case many wave models (see Tolman, 1991 and Komen et al., 1994) are built to face the problem.

In practical terms, till when the current speed is below a few tens of centimeters per second, there is no strict need of taking it into consideration. Rather the problem for the coastal engineer is the eventual, if necessary, availability of a detailed description of the current field. Particularly in coastal areas, with a strong spatial variability, this can be a serious problem that deserves a particular attention. Besides, to properly evaluate the interactions, the grid resolution of the wave model must be better than that required for a proper description of the current field.

7. Summary

In the previous sections we have highlighted the possible relevance of each single process and model parameter in the modeling of wind waves in coastal areas. The difficulty in so doing is that the influence of the single factor can span a wide range of values, depending on the conditions in the area of interest.

Some processes, like bottom scattering and bottom elasticity, require special conditions for their appearance. They are usually not considered. However, when the conditions are present, their role becomes dominant.

Even if only on a qualitative basis, we have summarized in figure 6, the relevance of the single physical processes. The figure provides an "expected" level of influence at 20 m of depth, at 5 m of depth, and the maximum possible relevance of each process in the local energy budget.

SHALLOW WATER EFFECT	INFLUENCE ON WAVES
REFRACTION	
SHOALING	
BOTTOM FRICTION	
BOTTOM SCATTERING	
PERCOLATION	
BOTTOM ELASTICITY	
INTERACTION WITH CURRENT	

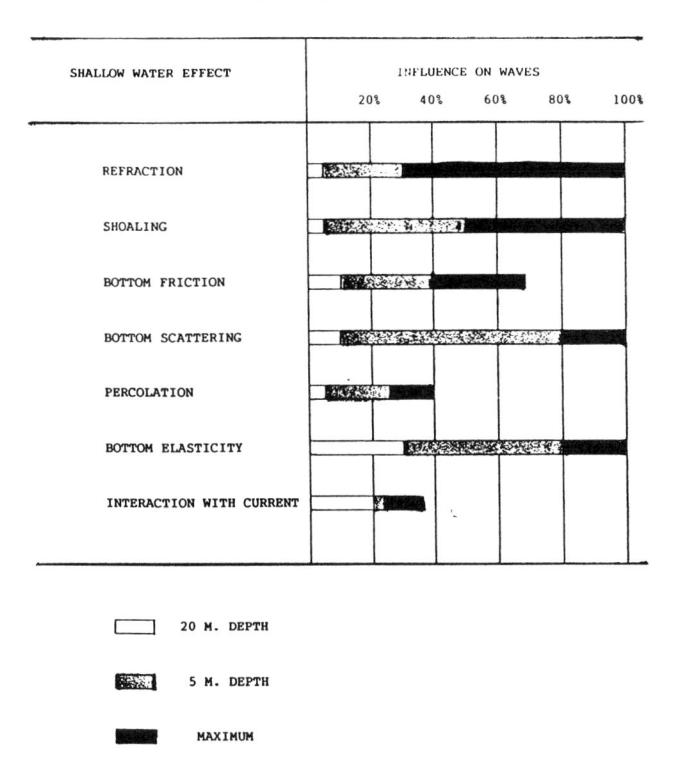

☐ 20 M. DEPTH

▨ 5 M. DEPTH

■ MAXIMUM

Figure 6. Possible influence of the single physical processes affecting waves in shallow water.

Acknowledgments

This research has been partially funded by the *Progetto Salvaguardia Laguna Venezia*.

References

Battjes J.A. and S. Beji, 1993. Breaking waves propagation over a shoal, ICCE 1992, **3**, 42-51.

Cavaleri, L., L. Bertotti and P. Lionello, 1989. Shallow water application of the third generation WAM wave model, *J. Geophys. Res.*, **C94**, 8111-8124.

Cavaleri, L., L. Bertotti and P. Lionello, 1991. Wind wave-cast in the Mediterranean Sea, *J. Geophys. Res.*, **C96**,10739-10764.

Cavaleri, L., L. Bertotti, C. Koutitas, S. Christopoulos, G. Komen, G. Burgers, K. Mastenbroek, J.M. Lefevre, A. Guillaume, J.C. Carretero, A. Guerra, L. Iovenitti and P. Cherubini, 1993. MAST Contract 0042, Final Report, 211 pp.

Davies, A.G. and A.D. Heathershaw, 1983. Surface wave propagation over sinusoidally varying topography: Theory and observation, I.O.S. Report no. 159, Part 1, Wormley (UK), 88 pp.

Forristall, G.Z., E.H. Doyle, W. Silva and M. Yoshi, 1990. Verification of a soil interaction model (SWIM), p. 41-68, In: *Modeling Marine Systems*, **II**, A.M. Davies (ed), CRC Press, Boca Raton, Florida (USA).

Günther, H., P. Lionello, P.A.E.M. Janssen, L. Bertotti, C. Brüning, J.C. Carretero, L. Cavaleri, A. Guillaume, B. Hanssen, S. Hasselmann, K. Hasselmann, M. de las Heras, A. Hollingworth, M. Holt, J.M. Lefevre and R. Portz, 1992. Implementation of a third generation ocean wave model at the European Centre for Medium-Range Weather Forecasts, Final Report for EC Contract SC1-0013-C(GDF), ECMWF, Reading (UK).

Heras, M. de las, 1990. WAM hindcast of long period swell, KNMI Afeling Oceanografisch Onderzoek Memo, OO-90-09, De Bilt (NL), 10 pp.

Hubbert, K.P. and J. Wolf, 1991. Numerical investigation of depth and current refraction of waves, *J. Geophys. Res.*, **C96**(C2), 2737-2748.

Komen, G.J., L. Cavaleri, M. Donelan, K. Hasselmann, S. Hasselmann and P.A.E.M. Janssen, 1994. *Dynamics and Modelling of Ocean Waves*, Cambridge University Press, 532 pp.

Long, R.B., 1973. Scattering of surface waves by an irregular bottom. *J. Geophys. Res.*, **78**, 7861-7870.

Osborne, A.R., 1993. Behavior of solitons in random-function solutions of the periodic Korteweg-de Vries equation, *Physical Review Letters*, **71** (19), 3115-3118.

Shemdin, P., K. Hasselmann, S.V. Hsiao and K. Herterich, 1978. Non-linear and linear bottom interaction effects in shallow water, p. 347-372, In: *Turbulent*

Fluxes Through the Sea Surface, Wave Dynamics and Prediction, A. Favre and K. Hasselmann (eds.), Plenum Press, New York, 677 pp.

Tolman, H.L., 1991. A third-generation model for wind on slowly varying, unsteady and inhomogeneous depths and currents, *J. Phys. Oceanogr.*, **21**, 782-797.

Weber, S.L., 1991. Eddy-viscosity and drag-law models for random ocean wave dissipation, *J. Fluid Mech.*, **232**, 73-98.

CHAPTER 5

A SPECTRAL MODEL FOR WAVES IN THE NEAR SHORE ZONE

R.C. Ris[1], L.H. Holthuijsen[1] and N. Booij[1]

Abstract

The present paper describes the second phase in the development of a fully spectral wave model for the near shore zone (SWAN). Third-generation formulations of wave generation by wind, dissipation due to whitecapping and quadruplet wave-wave interactions are added to the processes of (refractive) propagation, bottom friction and depth-induced wave breaking that were implemented in the first phase (Holthuijsen et al., 1993). The performance and the behaviour of the SWAN model are shown in two observed cases in which waves are regenerated by the wind after a considerable decrease due to shallow water effects. In the case of the Haringvliet (a closed branch of the Rhine estuary, the Netherlands) reasonable results in terms of significant wave heights were obtained. In the case of Saginaw Bay (Lake Huron, USA), the SWAN model underestimates the significant wave height deep inside the bay (as did two other models). Due to the absence of triad wave-wave interactions in the model the mean period is not properly shifted to the higher frequencies in shallow water. Adding these triads is planned for the next phase of developing the SWAN model.

Introduction

In conventional wave models (at least in coastal engineering) wave components are traced from deep water into shallow water along wave rays to obtain realistic estimates of wave parameters in coastal areas, lakes and estuaries. However, this technique often results in chaotic wave ray patterns which are difficult to interpret. Moreover, nonlinear processes cannot be calculated efficiently. Using a spectral wave model that represents the evolution of the waves on a grid is superior in several respects. The inherent spatial smoothing of such a model ensures a realistic smooth representation of the wave pattern, and it allows an efficient representation of the random, short-crested waves with their generation and dissipation. Models of

---

[1] Delft University of Technology, Department of Civil Engineering, Stevinweg 1, 2628 CN, Delft, The Netherlands.

this type are fairly common for oceans and shelf seas. However in coastal applications several orders of magnitude more computer effort is required due to the very high spatial resolution that is needed and the numerical techniques that are used. We reduced the required computer effort greatly by considering stationary situations only and by developing an unconditionally stable propagation scheme. The model is described and results are shown of tests for observed conditions in the Haringvliet and in Saginaw Bay with effects of wind, whitecapping, quadruplet wave-wave interactions, bottom friction and depth-induced wave breaking included.

Model formulation

The SWAN model (Simulation of WAves in Near shore areas) is conceived to be a third-generation, stationary wave model that is discrete spectral in both frequencies and directions. The qualification "third-generation" implies that the spectrum in the model evolves free from any a priori restraints (Komen et al., 1994). The model is formulated in terms of action density N (energy density divided by relative frequency: $N = E/\sigma$). From the possible representations of the spectrum in frequency, direction and wave number space, we choose a formulation in terms of relative frequency σ and direction of propagation θ (normal to the wave crest of a wave component). This is convenient for two reasons. Because of efficiency and accuracy, an implicit numerical scheme should be chosen for energy transport across the directions (refraction) and an explicit scheme for energy transport across the frequencies or wave numbers (frequency or wave number shift). The latter is subject to a Courant stability criterion which is more restrictive for the wave number formulation than the formulation in terms of frequency. Choosing the relative frequency σ rather than absolute frequency ω has the advantage that the relationship between the relative frequency and wave number remains unique when currents are added.

In general the evolution of the spectrum can be described by the spectral action balance equation (e.g., Phillips, 1977):

$$\frac{\partial}{\partial t}N(\sigma,\theta) + \nabla_{x,y} \cdot [(\underline{c}_g + \underline{U})N(\sigma,\theta)] + \frac{\partial}{\partial \sigma}[c_\sigma N(\sigma,\theta)] + \frac{\partial}{\partial \theta}[c_\theta N(\sigma,\theta)] = \frac{S(\sigma,\theta)}{\sigma} \qquad (1)$$

The first term in the left-hand side is the rate of change of action density in time, the second term is the rectilinear propagation of action in geographical x,y-space. The third term describes the shifting of the relative frequency due to currents and unsteadiness of depths with propagation velocity c_σ in σ-space. The fourth term represents the propagation in θ-space (depth and current induced refraction) with propagation velocity c_θ. This action balance equation implicitly takes into account the interaction between waves and currents through radiation stresses. The term $S(\sigma,\theta)$ at the right hand side of the action balance equation is the source term representing the growth by wind, the wave-wave interactions and the decay by bottom friction, whitecapping and depth-induced wave breaking.

To reduce computer time, we removed time from the action balance equation (i.e., $\partial/\partial t = 0$). This is acceptable for most coastal conditions since the residence time of the waves is usually far less than the time scale of variations of the wave boundary condition, the ambient current or the tide. For cases in which the time scale of these variations becomes important, i.e., variable incoming waves at the boundary, or variable winds or currents, a quasi-stationary approach can be taken by repeating the computations for predefined time intervals. For the wind effects the formulations of Cavaleri and Malanotte-Rizzoli (1981) and Snyder et al. (1981) are used. For the bottom friction effects the formulation from JONSWAP (Hasselmann et al., 1973; WAMDI group, 1988) is taken with the friction coefficient Γ equal 0.067 m^2s^{-3} (wind sea conditions; Bouws and Komen, 1983). Whitecapping is represented by the formulation of Hasselmann (1974) and Komen et al. (1984) in which the dissipation is controlled primarily by the steepness of the waves. Depth-induced wave breaking is modelled by a spectral version of the Battjes/Janssen wave breaking model (Battjes and Janssen, 1978) resulting in a dissipation which does not affect the shape of the spectrum itself (Beji and Battjes, 1993; Battjes et al., 1993). For the breaking coefficient in this model we use $\gamma=0.73$ which is the average value in the field experiments summarized in Table 1 of Battjes and Stive (1985). The quadruplet wave-wave interactions of Hasselmann (1962) are calculated with the Discrete Interaction Approximation (DIA) of Hasselmann et al. (1985) as in the WAM model (WAMDI group, 1989). A complete version of the SWAN model is planned to include triad wave-wave interactions. In very shallow water these triad wave-wave interactions transfer energy to higher frequencies. However, the triad interactions are still poorly understood and no suitable expression in terms of energy density has yet been derived (only for non-dispersive waves, Abreu et al., 1992). Efforts are presently being made to model the triad interactions based on the work of other authors, e.g., Beji and Battjes (1993), Battjes et al. (1993) and Madsen et al. (1991).

The numerical algorithm

In coastal wave models it is customary to propagate waves from deeper water towards the shore. In the HISWA model (Holthuijsen et al., 1989) this is exploited by propagating the waves line by line roughly parallel with the crests from deeper water to shallower water over a regular grid. This is computationally very efficient and unconditionally stable from a numerical point of view but waves can only propagate within a directional sector of about 120°. Complicated wave conditions with extreme refraction, reflections or initial cross seas cannot be properly accommodated. In the SWAN model we retain the unconditionally stable character of this technique but we expand it to accommodate these complicated conditions (wave from all directions). This modified version of the technique of the HISWA model (which in turn was borrowed from parabolic refraction/diffraction models) is a forward marching scheme in geographic x,y-space in a sequence of four 90° sectors of wave propagation (quadrants). In the first quadrant the state in a gridpoint (x_i, y_j) is determined by its up-wave gridpoints (x_{i-1}, y_j) and (x_i, y_{j-1}).

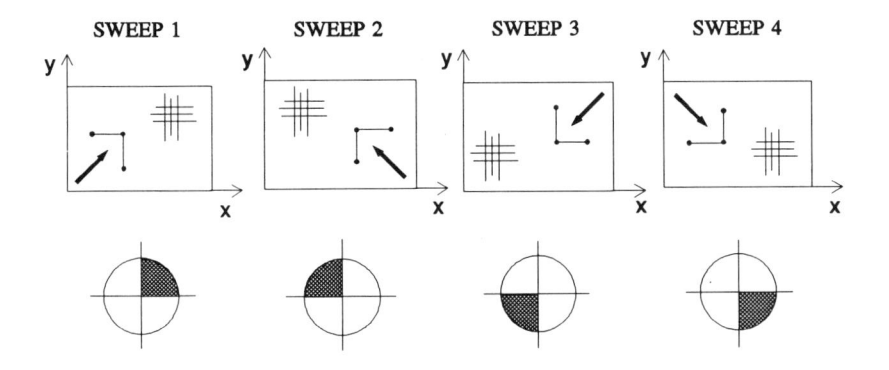

Fig. 1 Numerical scheme for wave propagation in geographic space in SWAN with the appropriate directional quadrant indicated per sweep for which the waves are propagated.

The computation is therefore unconditionally stable for all wave propagation directions in the 90° quadrant between the up-wave x- and y-direction because the wave characteristics lie within this quadrant. The waves in this quadrant are propagated with this scheme over the entire geographical region on a rectangular grid (sweep 1, Fig. 1). By rotating the stencil over 90°, the next quadrant (90°-180°) is propagated (sweep 2). Rotating the stencil twice more ensures propagation within all four quadrants. This allows waves to propagate from all directions with an unconditionally stable scheme. In cases with current- or depth induced refraction, action density can shift from one quadrant to another. This is taken into account through the boundary conditions of the directional quadrants and by repeating the computations with converging results. Hence the method is characterized as an iterative four-sweep technique (Holthuijsen et al., 1993). Typically we choose a change of less then 1% in significant wave height and mean wave period in 99% of the wetted geographic gridpoints to terminate the iteration. The propagation in θ-direction (refraction) is computed with an implicit scheme to achieve numerical stability for large bottom gradients (central second-order scheme). The corresponding tri-diagonal matrix is solved with a Thomas algorithm (Abbott and Basco, 1989). Preliminary results of propagation tests (also with currents induced wave blocking) for analytical and real cases show an accurate and stable behaviour of the wave model. In all of the next cases, however, no currents are present.

The integration of the source terms is straightforward for all frequencies of the discretized spectrum (the prognostic part of the spectrum, $\sigma < \sigma_{max}$). For frequencies higher than σ_{max}, a diagnostic spectral tail is added to the spectrum. To ensure a stable integration of the source terms we have used explicit schemes for the input source terms and semi-implicit and fully implicit schemes for the sink terms. To suppress the development of numerical instabilities, the maximum growth of energy density per sweep in a spectral bin is limited to a fraction of 10% (Tolman, 1992) of the fully developed equilibrium level (Phillips, 1958). Decay is not similarly limited to allow realistic rapid decrease near the shore. In SWAN the frequency σ is exponential distributed ($\sigma_{i+1} = \alpha\sigma_i$, with α constant) so that the calculation of the nonlinear transfer scales with frequency and can be integrated economically.

Field measurements: the Haringvliet and Saginaw bay

We will concentrate on a comparison with observations in two observed cases in which regeneration after decay is dominant. Both cases are in shallow water with an initial decrease of wave energy due to dissipation and refraction and a subsequent increase of wave energy due to wind. One is taken from the Haringvliet in the Netherlands (a closed branch of the Rhine estuary; Holthuijsen et al. 1989; Holthuijsen et al., 1993). Here a shoal protects the branch (5 km length scale) from the open sea (see Fig. 2). The computations have been carried out for a situation which occurred on October 14, 1982 at 23.00 h (1.95 m depth over the top of the shoal). The waves are locally generated in the southern North Sea and approach the estuary from NW direction and travel across and around the shoal. The observations show that a considerable fraction of wave energy is dissipated over the shoal. Behind the shoal the waves are regenerated by the wind. The wind speed was 16.5 m/s from NW. During this period currents were practically absent. The other case is taken from Saginaw Bay (USA; Bondzie and Panchang, 1993). Here a shallow region with an island protects the bay (25 km length scale) from Lake Huron (see Fig. 3). During a storm event in May, 1981 wave conditions were recorded with wave gauges at three locations (A, B and C). The wind velocity was 11.2 m/s, blowing along the main axis of the bay from Lake Huron. The observed wave height initially decreases between location A and B and then increases between location B and C.

Model results for the Haringvliet

The resolution of the bottom grid for the Haringvliet is $\Delta x = 500$ m and $\Delta y = 500$ m. The directional resolution in the spectrum is $\Delta\theta = 10°$ and the frequency resolution is $\Delta f = 0.1045 \cdot f$ between 0.055 Hz and 0.66 Hz. The observed wave boundary conditions at location 1 (see Fig. 2) are a significant wave height $H_s = 3.54$ m and a peak period $T_P = 8.3$ s. A JONSWAP spectrum is assumed at this upwave boundary with a $\cos^2(\theta)$ directional distribution since the observed width of the directional energy distribution is about 31°. We carried out two calculations with the SWAN model to show the effects of the regeneration of wave energy by wind: one with and one without wind. Fig. 2 shows the pattern of the significant wave height. The results indicated in Table 1 as "wind" are obtained with all mechanisms of generation and dissipation activated (the same station identification as in Holthuijsen et al. (1989) is used). Note that no measurements are available for location 2 at

23.00 h. The significant wave height thus computed agrees fairly well with the observations. The results indicated in Table 1 as "no wind" are obtained with only depth-induced breaking and bottom friction activated.

Table 1   Measurements and SWAN results at various locations in the Haringvliet and Saginaw Bay of significant wave height H_S and mean wave period T_M.

Location	Measurements H_S (m)	T_M (s)	SWAN results "wind" H_S (m)	T_M (s)	"no wind" H_S (m)	T_M (s)
Haringvliet						
1	3.54	6.6	3.54	6.6	3.54	6.6
2	—	—	3.22	6.8	3.32	6.6
3	2.63	6.4	2.85	6.9	2.85	6.8
4	2.71	6.3	2.82	6.9	2.81	6.8
5	0.79	3.2	0.95	6.0	0.82	6.8
6	1.41	4.9	1.54	6.5	1.42	6.8
7	1.84	6.0	1.77	6.7	1.66	6.9
8	1.08	3.7	1.00	5.2	0.65	6.4
Saginaw Bay						
A	1.90	7.3	1.91*	7.3*	1.92*	7.3*
B	0.94	4.2	0.80	3.7	0.23	6.9
C	1.30	—	0.78	3.5	0.07	5.3

* fitted.

A comparison of the significant wave height between "wind" and "no wind" shows the relative importance of wind effects on the significant wave height deep inside in the branch (1.00 m versus 0.65 m at location 8). It is obvious that the added wind input, whitecapping and the quadruplet wave-wave interactions were essential to obtain the better results ("wind") deep inside the Haringvliet. The regeneration of waves is clearly visible as a second, high frequency peak in the spectrum (see Fig. 4). This secondary peak shifts the mean frequency to higher values but not sufficiently (compare the computed and observed mean wave period T_M in Table 1), most probably because the triad interactions are absent in the present version of SWAN.

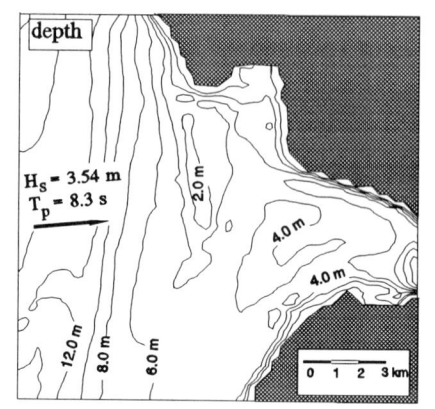

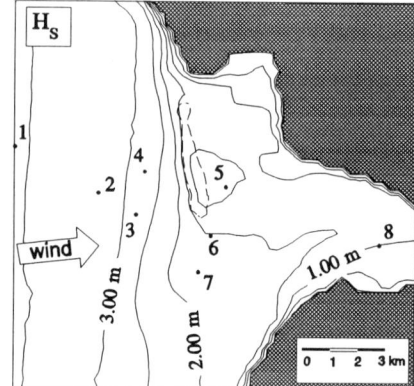

Fig. 2 Left-hand panel: the bathymetry of the Haringvliet area (contour line interval 2 m). Right-hand panel: results of the SWAN computation in terms of significant wave heights (all mechanisms activated) and location of the eight buoys. The dashed line indicates the location of the shoal.

Model results for Saginaw Bay

We carried out the calculations for Saginaw Bay using the same bottom grid resolution of 1200 m by 1200 m as used by Bondzie and Panchang (1993). To assume an up-wind boundary that is as homogeneous as possible, we choose this boundary at a line just outside the bay (the right-hand boundary of Fig. 3). Since the observations were taken during a storm, we assume a $\cos^2(\theta)$ distribution and a JONSWAP spectrum at this up-wave boundary. For such conditions, a directional resolution of $\Delta\theta = 10°$ is sufficient. We used a frequency resolution $\Delta f = 0.1225 \cdot f$ between 0.555 Hz and 1. Hz. The significant wave height and peak period at the up-wave boundary was chosen such that at location A the model reproduces the observed wave height and period (Table 1). The calculated pattern of the significant wave height is shown in Fig. 3. The wave height gradually decreases between the up-wave boundary and the entrance of the bay. Due to depth effects, more wave energy penetrates into the deeper entrance than into the shallower entrance of the bay (roughly 20 % lower wave height near location A than in the other entrance). As the waves penetrate into the bay they refract laterally to the shallower parts of the bay. The calculated significant wave heights and mean periods at the three wave gauges are shown in Table 1. In contrast with the observations, the calculated wave height shows no growth between gauge B and C at all. Calculations with a 25% higher wind speed, which is fairly realistic since the wind speed was recorded on land, also did not show the observed growth (similarly for a 50% higher wind speed). Additional computations with another third-generation wave model

(WAVEWATCH; Tolman, 1991, 1992) also failed to reproduce the observed net growth. Also the HISWA model failed in this (Bondzie and Panchang, 1993; who used an uniform wave boundary condition along a straight line across the entrance through location A). Still the effect of wind on the significant wave height deep inside the bay is clearly visible in the model as shown in Table 1 (0.78 m versus 0.07 m at location C) and Fig. 4. Apparently the decay of the low-frequency part of the spectrum is compensated by the growth of the high-frequency part.

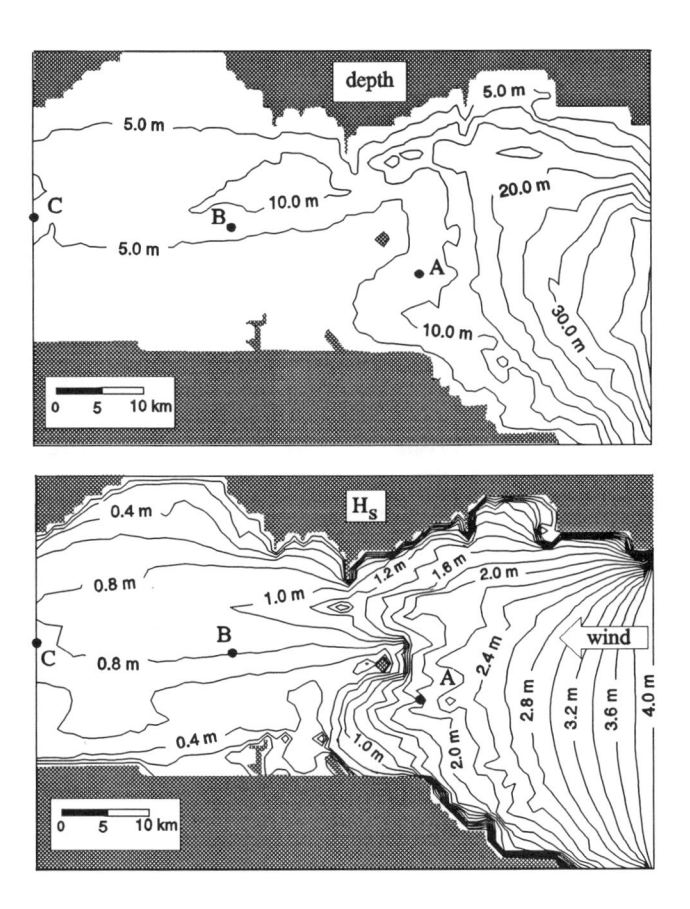

Fig. 3 Top panel: the bathymetry of Saginaw Bay (contour line interval 5 m). Bottom panel: results of the SWAN computation in terms of significant wave heights (all mechanisms activated) and the locations of the three wave gauges.

Haringvliet

Location 4
H_S = 2.82 m

Location 5
H_S = 0.95 m

Location 8
H_S = 1.00 m

Saginaw Bay

Location A
H_S = 1.91 m

Location B
H_S = 0.80 m

Location C
H_S = 0.78 m

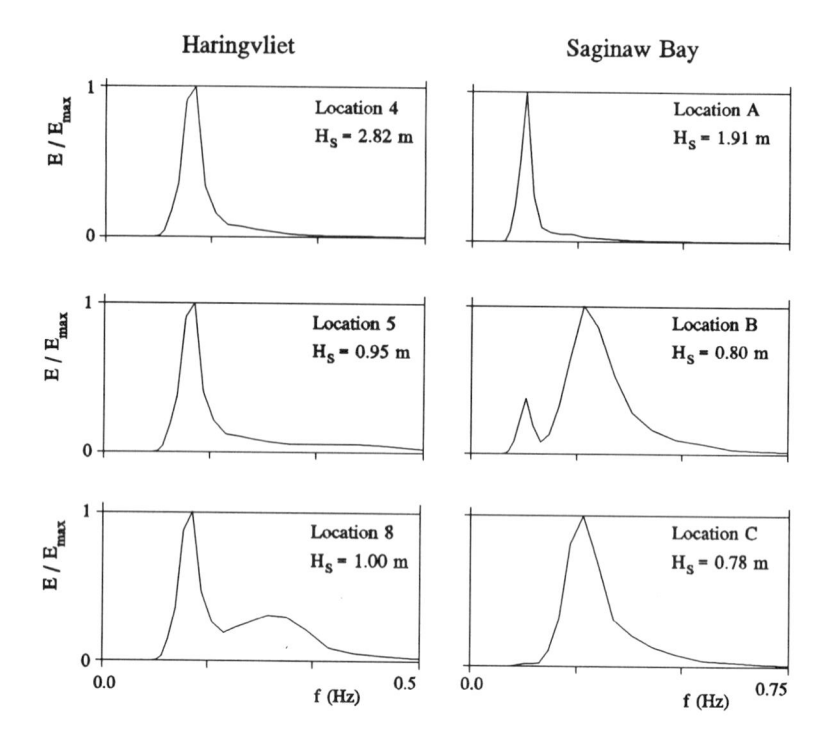

Fig. 4 Left-hand panels: Computed, normalized frequency spectra at location 4, 5 and 8 of the Haringvliet. Right-hand panels: Computed, normalized frequency spectra at locations A, B, and C of Saginaw Bay.

Conclusions and future work

We have described the second phase in the development of a third-generation fully spectral wave model for near-shore applications (SWAN) that is stationary and unconditionally stable. This permits economically feasible, high-resolution computations with all relevant effects of propagation, generation and dissipation included without a priori restraints on the development of the wave spectrum. The present version includes refractive propagation (currents included), wind generation, quadruplet wave-wave interactions, whitecapping, bottom friction, depth-induced breaking and wave blocking.

We have performed a number of computations with the SWAN model in two cases: the Haringvliet and Saginaw Bay. In the Haringvliet case the agreement of the computed significant wave heights with the observations is reasonable in spite of the absence of the triad interactions. We have found in Saginaw Bay a significant difference between the calculated and observed wave heights deep inside the bay. Although the effect of regeneration of waves by wind is clearly visible in the model, we could not reproduce the observed wave growth between location B and C, even when we increased the wind speed. The wave periods are not properly computed, in particular in the Haringvliet case. This underscores the importance of the next phase in the development of the SWAN model in which a parameterized formulation for the triad wave-wave interactions will be implemented.

Acknowledgements

We thank our colleagues C. Bondzie and V.G. Panchang of the University of Maine for providing us with the data of Saginaw Bay. We also thank H.L. Tolman for his permission to use the WAVEWATCH II model for the Saginaw Bay computation.

References

Abbott, M.B. and D.R. Basco (1989). *Computational Fluid Dynamics*, Jonh Wiley & Sons, Inc., New York, 425 p.

Abreu, M., A. Larraza and E. Thornton (1993). Nonlinear transformation of directional spectra in shallow water. *J. Geophysical Res.*, 97 (C10), 15,579-15,589.

Battjes, J.A. and J.P.F.M. Janssen (1978). Energy loss and set-up due to breaking of random waves. *Proc. 16th Int. Conf. Coastal Engineering*, Hamburg, 569-587.

Battjes, J.A., Y. Eldeberky and Y. Won (1993). Spectral Boussinesq modelling of random, breaking waves. *Proc. of 2nd Int. Symposium on Ocean Wave Measurement and Analysis*, New Orleans, 813-820.

Beji, S and J.A. Battjes (1993). Experimental investigation of wave propagation over a bar. *Coastal Engineering*, 19, 151 - 162.

Bondzie, C. and V.G. Panchang (1993). Effect of bathymetric complexities and wind generation in a coastal wave propagation model. *Coastal Engineering*, 21, 333-366.

Bouws, E. and G.J. Komen (1983). On the balance between growth and dissipation in an extreme depth-limited wind-sea in the southern North sea. *J. Phys. Oceanography*, 13, 9, 1653-1658.

Cavaleri, L. and P. Malanotte-Rizzoli (1981). Wind wave prediction in shallow water: Theory and applications. *J. Geophys. Res.*, 86, No. C11, 10, 961-973.

Hasselmann, K. (1962). On the non-linear energy transfer in a gravity wave spectrum. Part 1. General theory. *J. Fluid Mech.*, 12, 481-500.

Hasselmann, K. (1963). On the non-linear energy transfer in a gravity wave spectrum. Part 2. Conservation theorems; wave-particle analogy; irreversibility. *J. Fluid Mech.*, 15, 273-281.

Hasselmann, K., T.P. Barnett, E. Bouws, H. Carlson et al. (1973). Measurements of wind-wave growth and swell decay during the Joint North Sea Wave Project (JONSWAP). *Ergänzungsheft zur Deutschen Hydrographischen Zeitschrift*, 12.

Hasselmann, K. (1974). On the spectral dissipation of ocean waves due to whitecapping. *Boundary-Layer Meteorology*, Vol. 6, No. 2, 200-228.

Hasselmann, K. and S. Hasselmann (1985). Computations and parameterizations of the nonlinear energy transfer in a gravity-wave spectrum. Part I: A new method for efficient computations of the exact nonlinear transfer integral. *J. Phys. Oceanography.*, *15*, 1369-1377.

Holthuijsen, L.H., N. Booij and T.H.C. Herbers (1989). A prediction model for stationary, short crested waves in shallow water with ambient currents. *Coastal Engineering*, *13*, 23 - 54.

Holthuijsen, L.H., N. Booij and R.C. Ris (1993). A spectral wave model for the coastal zone. *Proc. of 2nd Int. Symposium on Ocean Wave Measurement and Analysis*, New Orleans, 630-641.

Komen, G.J., S. Hasselmann and K. Hasselmann (1984). On the existence of a fully developed wind-sea spectrum. *J. Phys. Oceanography.*, *14*, 1271-1285.

Komen, G.J., L. Cavaleri, M. Donelan, K. Hasselmann, S. Hasselmann and P.A.E.M. Janssen (1994). *Dynamics and modelling of ocean waves.* Cambridge University Press, UK, 560 p.

Madsen, P.A., R. Murray and O.R. Sørensen (1991). A new form of the Boussinesq equations with improved linear dispersion characteristics. *Coastal Engineering*, *15*, 4, 371-388.

Phillips, O.M. (1958). The equilibrium range in the spectrum of wind-generated waves. *J. Fluid Mech.*, 4, 426-434.

Phillips, O.M. (1977). *The dynamics of the upper ocean, 2nd edition*, Cambridge University Press, 261 p.

Snyder, R.L., F.W. Dobson, J.A. Elliott and R.B. Long (1981). Array measurements of atmospheric pressure fluctuations above surface gravity waves. *J. Fluid Mech.*, 102, 1-59.

Tolman, H.L. (1991). A third-generation model for wind waves on slowly varying unsteady, and inhomogeneous depths and currents. *J. Phys. Oceanography.*, 21, no. 6, 782-797.

Tolman, H.L. (1992). Effects of numerics on the physics in a third generation wind-wave model. *J. Phys. Oceanography.*, 22, no. 10, 1095-1111.

WAMDI group (Hasselmann et al.), 1988, The WAM model - a third generation ocean wave prediction model, *J. Phys. Oceanography*, 18, 1775-1810.

CHAPTER 6

Wind Variability and Extremes Statistics

Luigi Cavaleri[1] and Luciana Bertotti[1]

Abstract

Continuous records of wind speed and direction show a high variability both in the high and low frequency ranges. This variability is usually not considered in the numerical hindcast of a storm. We discuss the related implications for the maximum wave heights and for the values from the statistics of extremes.

1. Introduction

Extremes statistics at a given location are based on the availability of extended time series of the parameter of interest. Notwithstanding the recent increase in their number, measured wave data are still scarce and not sufficiently representative closer to coast where the wave conditions exhibit a strong variability. If proper data are not available at the location of interest, the usual solution is the hindcast, with suitable mathematical models, of all the relevant storms of the last 10 or 20 years, using their output as a basis for the extremes statistics. In this paper we analyze one aspect of this reconstruction relevant for the final results.

Meteorological models provide a smooth description of the atmosphere, their filtering characteristics depending on the grid step size and on the time integration step. In a model, the representation of the passage of a storm at a given location is characterized by a smooth growth of the wind speed and a similarly smooth decay. This is not what is experienced in the field. Cavaleri and Burgers (1992, henceforth referred to as CB) point out that levels of turbulence with rms percentage variability $\sigma = 0.10$, up to values $\sigma = 0.30$, are common in nature. This turbulence leads to a substantial increase of the maximum significant wave height Hs in a storm.

In this paper, first we briefly describe (in section 2) the physics of the process and the effects on the evolution of a storm. Then (section 3) we focus on the statistics of the extremes, showing how the related results are affected by data derived from "turbulent" storms. The overall findings are summarized in section 4.

---

[1]Istituto Studio Dinamica Grandi Masse-CNR, San Polo 1364, 30125 Venice, Italy

2. Turbulent Wave Growth

Figure 1 shows four records of wind speed U with different degrees of turbulence taken from an oceanographic platform located in the Northern Adriatic Sea (Cavaleri, 1979). The turbulence ranges from periods of seconds, where it interacts with the basic wave generation process, till one hour and beyond, shifting gradually into the synoptic variability. Here we focus our attention on the part from one minute upwards. In practice we do not deal with the frequency range connected to the pure generation.

Within its range of variability, the wind speed happens to be for part of the time lower than the phase speed of part of the spectral frequencies. CB show that, through a rectification of the Miles generation process (1957), the relatively fast turbulence (i.e., with period approximately between one and twenty minutes) leads to an enhancement of the actual significant wave height Hs, the enhancement increasing with the level σ of the turbulence. A second order effect, but acting also on the low speed waves, i.e., on the high frequency range of the wave spectrum, is associated with the non linear relationship between friction velocity and wind speed.

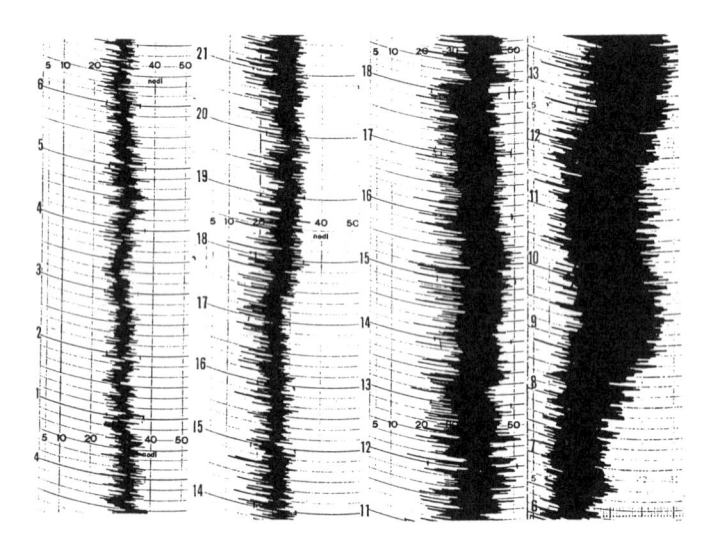

Figure 1. Records of wind speed with different degrees of turbulence.
Wind speed in knots. Time in hours (after Komen et al., 1994).

Figure 2 shows the classical case of time limited wave growth, with different turbulence σ ranging from 0.0 (uniform wind) till 0.30 (very strong turbulence). We see that the latter value leads to an increase of the final Hs of more than 30%.

A further increase of the maximum wave height in a storm derives from the relatively long period of turbulence of the wind field (periods from twenty minutes till several hours). These oscillations are clearly recognizable by direct inspection of a record lasting one day or more. Obviously the wave field reacts to this variability with a related Hs variability throughout the field. To simulate such a variability we need first to simulate the turbulent wind. CB show that this can be done with a Markov chain, where, for a given σ, the time scale of the turbulence is dictated by the correlation α between the sequential U values. Actual turbulent records seem to be well reproduced by this approach, provided the correct σ and α are used. The α =0.90 seems to be a good value for data taken at one minute intervals. CB have introduced a Markov chain turbulence with these characteristics in a uniform wind field, then repeating the test of figure 2. The results are shown in figure 3, where the test has been extended also to the cases of α=0.95 and 0.99. The σ was equal to 0.25. The effect of air turbulence is clear. The "turbulent" growth curve follows the smooth one (already enhanced by fast turbulence, compare with figure 2), waving around it. Note that in figure 3 each couple of lines has been shifted up by 2 meters for the sake of clarity.

By direct inspection of the diagram in figure 3 (but similar results are often found in recorded Hs time series, even if obscured by the usual 3-hour intervals), we recognized immediately the further increase of the maximum Hs value, the increase being directly dependent on σ and α.

We can summarize the present situation as follows. Standard numerical wave hindcasts are based on wind fields obtained from meteorological models. Turbulence is usually not considered, and the field evolves smoothly in time. The introduction of wind turbulence affects the wave field in two ways. On one hand, it increases the actual Hs values. On the other, it forces the wave field to oscillate around the otherwise smooth growth curve, reaching in the process still higher wave heights. Note that, while the first effect is fully determined by σ, and it can therefore be correctly evaluated, for the latter we come across statistics. The highest Hs in a turbulent storm is crudely a matter of chance. The consequence of this on the statistics of the extreme wave heights is the subject of the next section.

3. The Uncertainty in the Extremes Statistics - The Probability of a Probability

The classical procedure of extremes statistics starts from a long term time series of the parameter of interest, typically available as a regular sequence of single values at 3-hour intervals. Then, a subset of values is selected, according to one of two principles: pick up (a) all the values above a certain threshold, (b) the highest value for each pre-established time interval (for wave height a month, a semester, or, one year are a regular choice). If an extended time series is not available, the wave hindcast of a large number of storms is performed, retaining as input information for the extremes statistics the highest Hs in each storm. The selected data are then best-fitted by some extremal distribution, Weibull, Gumbel or FT-1, exponential being among the common ones. Given the distribution and the number of data in the subset, we can then estimate the probability to overcome a given value at the next event or

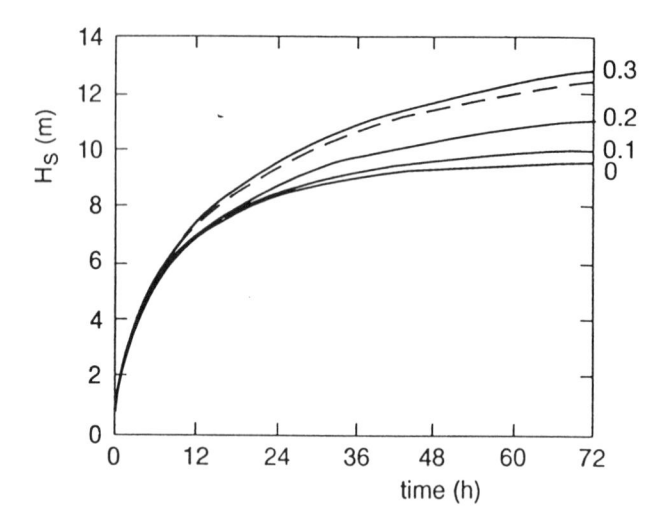

Figure 2. Time growth of the significant wave height under a 20 m/s wind with different degree of turbulence (after Komen et al., 1994).

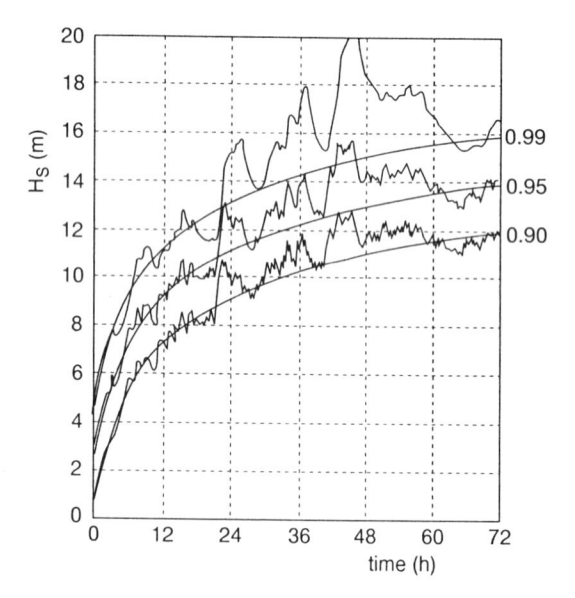

Figure 3. Oscillations of the significant wave height in the time growth curve as a function of the degree of correlation in the sequential wind values (after Komen et al., 1994).

during the next time interval, depending on how the data have been selected. Given the period covered by the input time series, the statistics can be usefully referred to time. In practice, we can reply to the following question. What is the probability P^+ to overcome a certain value H within T years? More generally, given two of the quantities P^+, H, T, we can immediately deduce the third one.

The analytical expression relating the three quantities is

$$P^+ = 1 - [\, p^-(H)\,]^{nT} \qquad (1)$$

where p^- is the no-exceeding probability at the next event deduced from the extremal distribution, and n is the average number of storms per year. A full discussion of the subject is found in Gumbel (1958) and practical applications illustrated in Cavaleri et al. (1986).

The graphical representation of the results is particularly enlightening. Figure 4 shows the exceedance probability for a certain area of the Tyrrhenian Sea (see Cavaleri et al. 1986). The enhancement of the wave height H due to the somehow more efficient generation described in the previous section means that we are dealing with higher wave heights. This crudely shifts all the lines in figure 4 to the right, the shift depending on the σ of turbulence typical of the area and on the kind of storms we are considering.

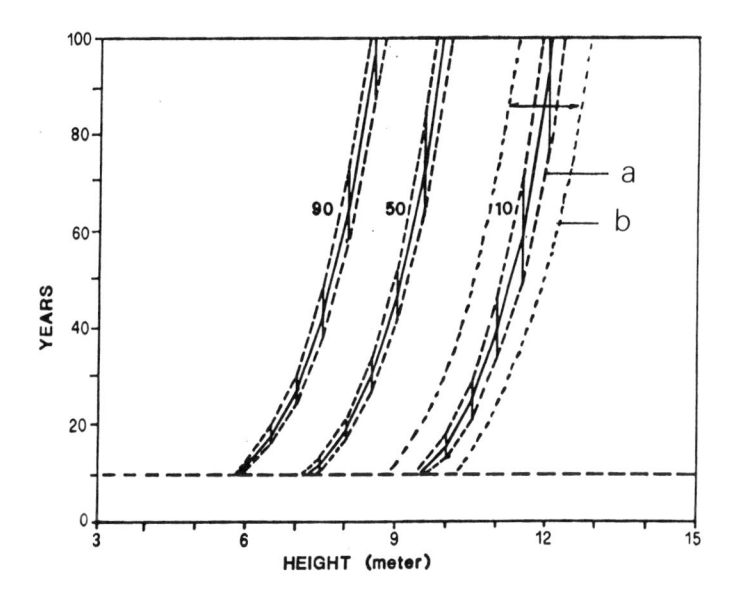

Figure 4. Exceedance probability (given by the number close to each continuous curve) with respect to wave height and elapsed period. The broken lines represent the corresponding confidence limits due to a) choice of the storms and b) wind turbulence.

Three points must be stressed. First, Resio (1978) warns that an extremal statistic produces meaningful results only if applied to a consistent data set, i.e., including only data of the same kind. We cannot mix data associated with substantially different kinds of storms, e.g., southern swell and extra-tropical storms in the North Atlantic Ocean. Clearly the presence of turbulence stresses further this point.

Second, the confidence we have in our input data must depend on the actual source. If derived from a hindcast, we must be aware that they are likely to be underestimated (if turbulence is likely to be present and it has not been considered). But also recorded data have problems. Our usual 20 minute sample chosen for the record is just a random choice in a waving time series as the ones in figure 3. As such, the confidence limits on the actual average representative value (the smooth growth curves in figure 3) are much larger than those associated only to the sampling variability connected with the randomness of the surface, typically 15% instead of 6-8%.

Third, after estimating, by means of diagrams as the one in figure 4 or the related expression (1), the extremal conditions, i.e., a certain H_e, we must warn the user of our results that the maximum H he will be likely to come across in such conditions is going to be higher than H_e. Our numerical results suggest an increase between 5 and 10%. Starting from the extremes in figure 4, we have evaluated the associated confidence limits by a combined used of the Jack-knife and Montecarlo techniques.

A full description of the Jack-knife technique can be found in Cavaleri et al. (1986) who also describe practical applications. Basically, given a subset of N data and the extremal distribution fitted to them, the technique estimates the reliability of the results by checking how much they depend on the single datum. This is done by excluding in turn each datum and getting a new extremal fit on the remaining N-1 data. This produces N new estimates of the extremes that are statistically analyzed to provide an estimate of the confidence limits of the overall extremal evaluation. Acting on the extremes shown in figure 4, we have so obtained the limits given by the two (a) curves lying close to the original ones.

This procedure does not account for the uncertainty on the single data due to sampling variability and/or turbulence effects. This can be obtained by Montecarlo technique. Each datum has been left oscillating randomly around its original value, following a Gaussian distribution with $\sigma=10\%$, and the fitting procedure repeated for each realization. Similarly to the Jack-knife technique, the results have been statistically analyzed providing a new estimate of the confidence limits. These are given in figure 4 by the wider (b) limits associated with each extremal curve. It is obvious that the turbulence introduces into the estimate of the extremes an uncertainty much larger than derived from the sampling of the input data.

4. Conclusions

We summarize here the relevant points for extremes statistics when dealing with measured or hindcast data.

Measured data.
- enhancement of wave height: naturally present in the data, hence automatically considered.
- confidence limits for choice of input data: to be evaluated by Jack-knife technique (or similar one).
- enhancement of estimated extreme values, because of oscillations in the growth curve, estimated to be 5-10%.

Hindcast data.
- enhancement of wave height: its consideration requires introduction of turbulence into the input wind data.
- confidence limits for choice of input data to be evaluated by the Jack-knife technique (or similar one).
- confidence limits for turbulent records: not required.
- enhancement of estimated extreme values, because of oscillations in the growth curve, estimated to be 5-10%.

Acknowledgments

This research has been partially supported by the *Progetto Salvaguardia Laguna Venezia*.

References

Cavaleri, L., 1979. An instrumental system for detailed wind wave study, *Il Nuovo Cimento*, Serie 1, **2C**, 288-304.

Cavaleri, L. and G.J.H. Burgers, 1992. Wind gustiness and wave growth, KNMI Afdeling Oceanografish Onderzoek Memo, 00-92-18, De Bilt, 38 pp.

Cavaleri, L., P. De Filippi, G. Grancini, L. Iovenitti and R. Tosi, 1986. Extreme wave conditions in the Tyrrhenian Sea, *Ocean Engng.*, **13**, 2, 157-180.

Gumbel, E.J., 1958. *Statistics of Extremes*, Columbia University Press, NY, 375 pp.

Komen, G., L. Cavaleri, M. Donelan, K. Hasselmann, S. Hasselmann and P.A.E.M. Janssen, 1994. *Dynamics and Modelling of Ocean Waves*, Cambridge University Press, 532 pp.

Miles, J.W., 1957. On the generation of surface waves by shear flows, *J. Fluid Mech.*, **3**, 185-204.

Resio, D.T., 1978. Some aspects of extreme waves prediction related to climatic variations, In: Proc. 10th Offshore Technology Conference, Houston, Texas, **3**, 1967-1980.

CHAPTER 7

MEAN FLUX IN THE FREE SURFACE ZONE OF WATER WAVES IN A CLOSED WAVE FLUME

Witold Cieślikiewicz [1] and Ove T. Gudmestad [2]

Abstract

This paper presents some results of research relating to the theoretical predictions of mass-transport velocity within the free surface zone of water waves in intermediate water depth. The theoretical results are compared with measurements made in a wave flume.

The theoretical estimate of a mean drift has allowed for a better estimation of the return flow in the wave flume. Examples of such estimation are given and graphically presented in the paper. Finally, the stability of the obtained mean velocity profiles throught the experiments is examined.

Introduction

The occurrence of a second-order mean drift is one of the more interesting, and by the same time, important non-linear features of a progressive water gravity wave. This drift, an apparent mass transport, influences such a phenomena like migration of sediments and pollutant particles in the water, and it can also result in the pilling up of water at a beach, with an associated increase in the local mean water level.

---

[1] Institute of Hydroengineering, Polish Academy of Sciences, Kościerska 7, 80-953 Gdańsk, Poland. Presently at British Maritime Technology, 7 Ocean Way, Ocean Village, Southampton SO14 3TJ, England

[2] Statoil, P.O. Box 300 Forus, 4001 Stavanger, Norway

The formulae for the mass-transport velocity and the total mean flux are usually derived in the Lagrangian frame. The total mean flux in Eulerian frame was developed by Starr (1947) for a small-amplitude wave train and by Phillips (1960) for a random wave field. These traditional approaches, in the Eulerian frame, allow us only to treat the total flux like a physical quantity "existing on a subset of zero measure," namely, *exactly* at the free surface of the wave. Within such an approach we are not able to discuss the distribution of the mean horizontal velocity in the free surface zone. We would then have the artificial situation that the mean velocity along the vertical is everywhere equal to zero except at the free surface.

Tung (1975) has shown a positive mean value of the horizontal orbital velocity of random waves in the near surface zone, but to the authors' knowledge, this has not been discussed and interpreted as a current induced by waves until the works of Cieślikiewicz and Gudmestad (1993, 1994). In those works the modified particle velocity $\bar{u}$ of a wave field is introduced by following the approach of Tung (1975):

$$\bar{\mathbf{u}}(\mathbf{x}, z, t) = \begin{cases} \mathbf{u}(\mathbf{x}, z, t) & \text{for } z \leq \zeta(\mathbf{x}, t) \\ 0 & \text{for } z > \zeta(\mathbf{x}, t) \end{cases} \quad (1)$$

in which $\mathbf{u}$ is unmodified water wave orbital velocity, ζ is the free surface elevation, $\mathbf{x}$ is the location vector on the horizontal plane, z-axis is directed vertically upwards and t is the time.

Random waves

Tung (1975) had derived the probability density function and the first three statistical moments of this modified velocity. The mean value of the horizontal velocity component u is given as

$$\langle \bar{u}(z) \rangle = r(z)\sigma_u(z)Z(z') \quad (2)$$

where $Z(\gamma) = (2\pi)^{-1/2} \exp(-\gamma^2/2)$, σ_ζ and σ_u are the standard deviations of ζ and u, respectively, and $z' = z/\sigma_\zeta$. Assuming that the wave is unidirectional and denoting the frequency spectrum by $S(\omega)$, the cross-correlation coefficient of u and ζ is given as

$$r(z) = \frac{1}{\sigma_u(z)\sigma_\zeta} \int_0^\infty \frac{gk}{\omega} \frac{\cosh k(z+h)}{\cosh kh} S(\omega)\, d\omega \quad (3)$$

in which g denotes the gravitational acceleration, h is the water depth, and the wavenumber k is related to the angular frequency ω by the dispersion relation.

Cieślikiewicz and Gudmestad (1993) developed the formula for total mean flux of random waves using the modified velocity (1) in the following form:

$$q = \int_0^\infty \frac{gk}{\omega \cosh kh} W(-h; \sigma_\zeta, k) S(\omega) \, d\omega \qquad (4)$$

where the function W is defined as

$$W(z_a; \sigma, k) = \int_{z_a/\sigma}^\infty \cosh k(z+h) Z(z) \, dz$$

$$= \frac{1}{2} \exp\left[\frac{\sigma^2 k^2}{2}\right] \left\{ e^{kh} Q(z/\sigma + k\sigma) + e^{-kh} Q(z/\sigma - k\sigma) \right\} \qquad (5)$$

and where $Q(z) = \int_z^\infty Z(\gamma) \, d\gamma$

The approximation of that formula leads to the result obtained by Phillips (1960). Phillips' formula for total mean flux, $q = \int_{-h}^\infty \langle u(z) \rangle \, dz$, may be easily obtained by using the modified velocity (1) and formulae (2) and (3) (see Cieślikiewicz 1994):

$$q = \frac{1}{\sigma_\zeta} \int_{-h}^\infty Z(z') \left[\int_0^\infty \frac{gk}{\omega} \frac{\cosh k(z+h)}{\cosh kh} S(\omega) \, d\omega \right] dz =$$

$$\int_0^\infty \frac{gk}{\omega} S(\omega) \left[\int_{-h}^\infty \frac{\cosh k(z+h)}{\cosh kh} Z(z') \, dz' \right] d\omega \approx$$

$$\int_0^\infty \frac{gk}{\omega} S(\omega) \left[\int_{-h}^\infty Z(z') \, dz' \right] d\omega \qquad (6)$$

since we assume in practice that $h \gg 3\sigma_\zeta$, then for $|z| < 3\sigma_\zeta$ we obtain in the above integral an approximation $\cosh k(z+h)/\cosh kh \approx 1$. A large error in this approximation outside the region $|z| < 3\sigma_\zeta$ is nonessential since the value of $Z(z')$ is close to zero here. As we have assumed $h \gg 3\sigma_\zeta$, we have $\int_{-h}^\infty Z(z') \, dz' \approx \int_{-\infty}^\infty Z(z') \, dz' = 1$. Thus

$$q \approx \int_0^\infty \frac{gk}{\omega} S(\omega) \, d\omega \qquad (7)$$

Note that for deep water waves above formula can be rewritten as $q = \int_0^\infty \omega S(\omega) \, d\omega$ which is the value of the spectral moment of the first order m_1.

Deterministic wave

In the paper of Cieślikiewicz and Gudmestad (1994), the same approach as described above for random waves, is adapted to deterministic small-amplitude waves. The result for the total mean flux $q^{(d)}$ in the unidirectional wave case is equal to

$$q^{(d)} = \frac{ga}{\omega} I_1(ak) \tag{8}$$

where $I_1(\cdot)$ is the modified Bessel function of the first order. The above formula in approximation gives a result first obtained by Starr (1947): $M^{(d)} = \rho q^{(d)} = E/C$ where E is the average energy per unit surface area and C is the phase velocity. It should be emphasised that this approximation may be easily obtained by assuming that

$$\frac{\cosh k(z+h)}{\cosh kh} \approx 1 \quad \text{for} \quad |z| \leq a \tag{9}$$

where a is the wave amplitude. Consider a unidirectional progresive small-amplitude wave of the form

$$\zeta(x,t) = a \cos(kx - \omega t) \tag{10}$$

The associated horizontal velocity under the wave is given by

$$u(x,z,t) = \frac{gak}{\omega} \frac{\cosh k(z+h)}{\cosh kh} \cos(kx - \omega t) \quad \text{for} \quad z \in [-h, \zeta(x,t)] \tag{11}$$

Introduce the extension of u on the z-domain $[-h, \infty)$ by the definition

$$\overline{u}(x,z,t) = \begin{cases} u(x,z,t) & \text{for} \quad z \leq \zeta(x,t) \\ 0 & \text{for} \quad z > \zeta(x,t) \end{cases} \tag{12}$$

The mean value of u over a wave period T of a deterministic wave is

$$m_{\overline{u}}^{(d)}(z) = <\overline{u}(x,z,t)>^{(d)} = \frac{1}{T} \int_{-T/2}^{T/2} \overline{u}(x,z,t)\,dt \tag{13}$$

In view of (11) and (12)

$$m_{\overline{u}}^{(d)}(z) = \begin{cases} \dfrac{1}{T} \displaystyle\int_{t_1(z)}^{t_2(z)} u(x,z,t)\,dt & \text{for} \quad |z| \leq a \\ 0 & \text{for} \quad |z| > a \end{cases} \tag{14}$$

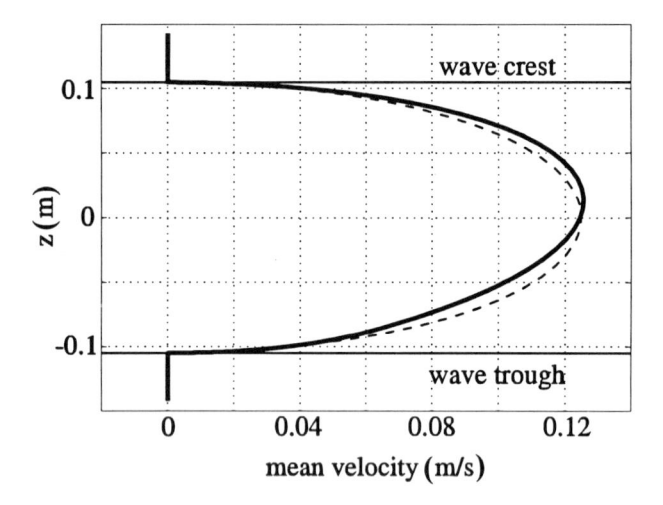

Fig. 1. Theoretical Eulerian mean velocity profile for deterministic small-amplitude wave (solid line) and its approximation (dashed line).

where t_1 and t_2 are such that

$$\begin{cases} z = \zeta(x,t_1) = \zeta(x,t_2) \\ z \leq \zeta(x,t) \quad \text{for} \quad t_1 \leq t \leq t_2 \end{cases} \quad (15)$$

Carrying out the integration in (14) yields

$$m_{\overline{u}}^{(d)}(z) = \begin{cases} \dfrac{gak}{\pi\omega} \dfrac{\cosh k(z+h)}{\cosh kh} \sin\left(\arccos \dfrac{z}{a}\right) & \text{for} \quad |z| \leq a \\ 0 & \text{for} \quad |z| > a \end{cases} \quad (16)$$

Taking into account the approximation (9) we obtain

$$m_{\overline{u}}^{(d)}(z) \approx \begin{cases} \dfrac{gak}{\pi\omega} \sin\left(\arccos \dfrac{z}{a}\right) & \text{for} \quad |z| \leq a \\ 0 & \text{for} \quad |z| > a \end{cases} \quad (17)$$

The mean horizontal velocity profiles according to above expressions are presented in Fig. 1.

To obtain the total mean flux $q^{(d)}$ at a fixed position x (Eulerian frame) we perform the following integration

$$q^{(d)} = \int_{-h}^{\zeta(x,t)} m_{\overline{u}}^{(d)}(z)\,dz \quad (18)$$

In view of (17)

$$q^{(d)} \approx \int_{-a}^{a} \frac{gak}{\pi\omega} \sin\left(\arccos\frac{z}{a}\right) dz \qquad (19)$$

By substituting $\theta = \arccos(z/a)$ we obtain

$$q^{(d)} \approx \frac{ga^2 k}{\pi\omega} \int_0^{\pi} \sin^2\theta\, d\theta = \frac{ga^2}{2}\frac{k}{\omega} \qquad (20)$$

Therefore, the flow of mass $M^{(d)}$ in approximation is equal to

$$M^{(d)} = \rho q^{(d)} \approx \frac{\rho g a^2}{2}\frac{k}{\omega} = \frac{E}{C} \qquad (21)$$

which is the well-known result usually derived in the Lagrangian frame. The quantitative understanding of mass transport and return flow in the closed wave flume plays an important role in experimental studies of water wave kinematics. A review of recent research relating to the problem of return flow may be found in Gudmestad (1993).

Return flow

A theoretical prediction of the mean horizontal velocity (in the Eulerian frame) allows for a better estimation of the return flow in the wave flume. We suggest in the present study that the difference between the predicted and the measured mean values of mean horizontal velocity gives an estimate of the return current in the wave flume. Figs. 2, 3, and 4 show the results of such calculations applied to laboratory data. These data were collected in the Norwegian Hydrotechnical Laboratories' 33 m long, 1.02 m wide and 1.8 m deep wave channel by a two-component Laser Doppler Velocimeter (LDV). The LDV allowed wave velocity measurements from wave crest down to tank bottom but at one point in space only during one run. In order to obtain the distributions for the statistical properties of the velocity along the vertical axis it was necessary to repeat the experiment with exactly the same free surface elevation spectrum but locating the LDV station at different vertical positions. The first series I18, consisting of 12 runs is given by the significant wave height $H_S = 0.21$ m and peak period $T_p = 1.8$ s, while the second series I24 consisting of 13 runs is given by the significant wave height $H_S = 0.25$ m and peak period $T_p = 2.4$ s. The third series R15B consisting of 10 runs represens a deterministic wave given by the value of wave height $H = 0.26$ m and period $T = 1.5$ m. Digitisation of the free surface elevation and velocity time series was carried at a rate of 40 Hz and

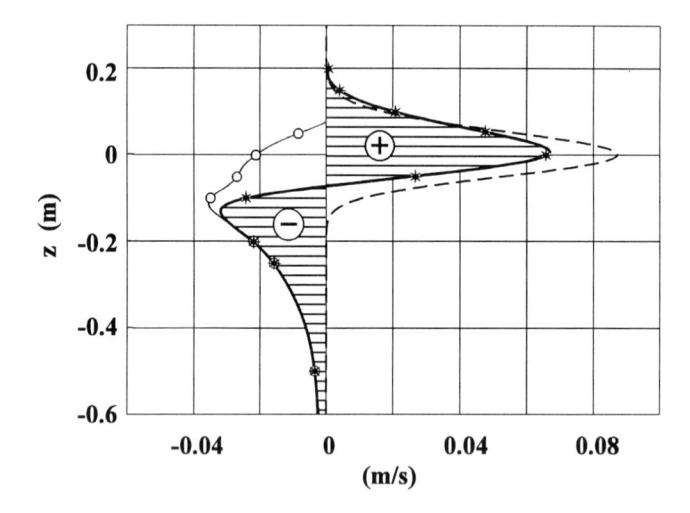

Fig. 2. Mean value of horizontal velocity and estimate of return flow: ∗ observed mean values, – – – – theoretical mean drift, ○ estimated return flow. Wave case I18.

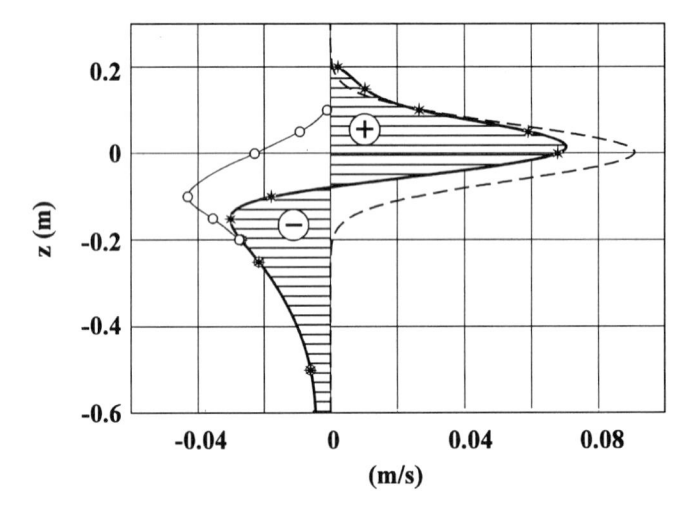

Fig. 3. Mean value of horizontal velocity and estimate of return flow: ∗ observed mean values, – – – – theoretical mean drift, ○ estimated return flow. Wave case I24.

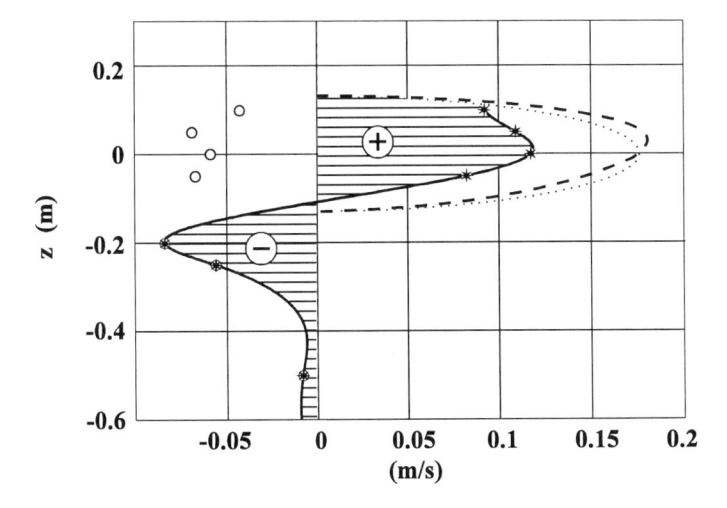

Fig. 4. Mean value of horizontal velocity and estimate of return flow for regular wave: ∗ observed mean values, – – – – theoretical mean drift and its approximation (dotted line), ○ estimated return flow.

samples of 32,768 measuring points were collected. The water depth was 1.3 m. The experimental arrangement is described in detail in papers by Skjelbreia *et al.* (1989, 1991).

In Figs. 2 and 3 the measured mean horizontal velocity is marked with stars for the irregular wave cases I18 and I24, respectively, from Skjelbreia's measurements. The dashed line presents the theoretical mean value of the modified (according to equation (1)) horizontal velocities. Open circles show the estimated values of the return flow.

Data for a deterministic case have been examined through analysis of data series R15B from Skjelbreia's experiments. The full line in Fig. 4 presents the measured (in the Eulerian frame) mean flow for this deterministic wave case.

The return flow profiles presented in Figs. 2, 3, and 4 were averaged over the whole velocity data collected during each run. In order to examine the stability of the mean velocity profiles obtained, each time series has been divided into four equal parts. We believe that all resulting sub-series were long enough for calculation of a statistical estimate of their mean values and standard deviations. We believe also that a comparison of those four mean values provides us with at least an indication of the stability of the mean velocity profiles. The results of these estimations are presented on Figs. 5 and 6 for the wave cases I18 and I24, respectively. It can be noted that the mean values as well as standard deviations of horizontal velocity calculated for each of four parts of time series are very much the same. Closer examination of the plots shows some similarities between I18 and I24 wave cases indicating that some trends in the mean horizontal velocity may exist. For example in both cases, for elevations below $z = -0.4$ m the mean values in the first quarter of the experiment series have the smallest values, while, on the other hand, above that level it has the largest values.

Conclusions

In the approach of Cieślikiewicz and Gudmestad (1993) we are able, in the Eulerian frame, to discuss the distribution of the mean velocity along the vertical axis—the mean horizontal velocity is "stretched out" from the exact location at the surface onto the free surface zone. Moreover, we are able to calculate not only the total mean water flux but also the flux between two given z-elevations. The theoretical results relating to the current in the direction of the wave advance can be used for better estimation of a return current in the wave flume.

REFERENCES

Cieślikiewicz, W. and Gudmestad, O. T., (1993). Stochastic characteristics of orbital velocities of random water waves. *J. Fluid Mech.*, 255, 275-299.

Cieślikiewicz, W. and Gudmestad, O. T., (1994). Mass transport within the free surface zone of water waves. *Wave Motion*, 19, pp. 275-299.

Cieślikiewicz, W., (1994). Influence of free surface effects on wave kinematics. *Schriftenreihe des Vereins der Freunde und Förderer des GKSS-Forschungszentrums Geesthacht e.V. Heft Nr. 5*, Geesthacht.

Gudmestad, O. T., (1993). Measured and predicted deep water wave kinematics in regular and irregular seas. *Marine Structures*, 6, 1-73.

Phillips, O. M., (1960). The mean horizontal momentum and surface velocity of finite-amplitude random gravity waves. *J. Geophys. Res.*, 65, No. 10, 3473-3476.

Skjelbreia, J., Tørum, A., Berek, E., Gudmestad, O. T., Heideman, J. and Spidsøe, N., (1989). Laboratory measurements of regular and irregular wave kinematics. E&P Forum Workshop on Wave and Current Kinematics and Loading IFP, Rueil Malmaison, France, 45-66.

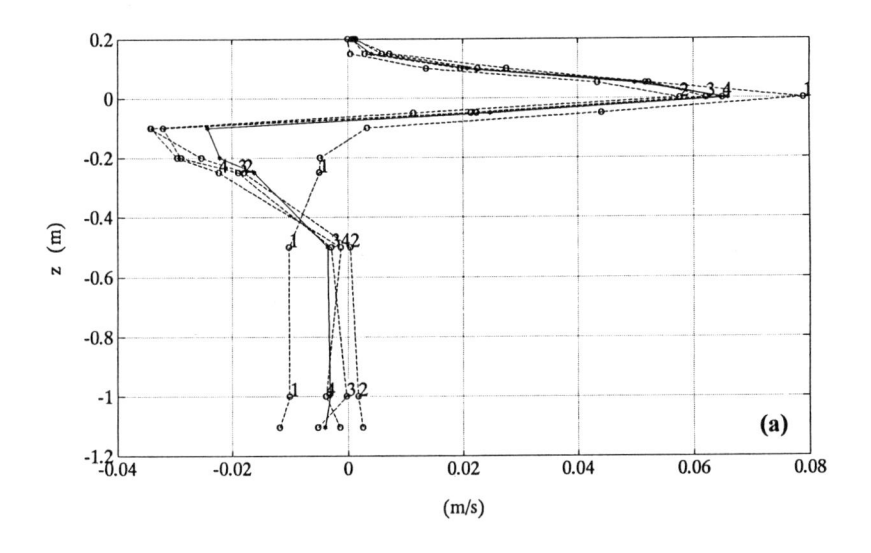

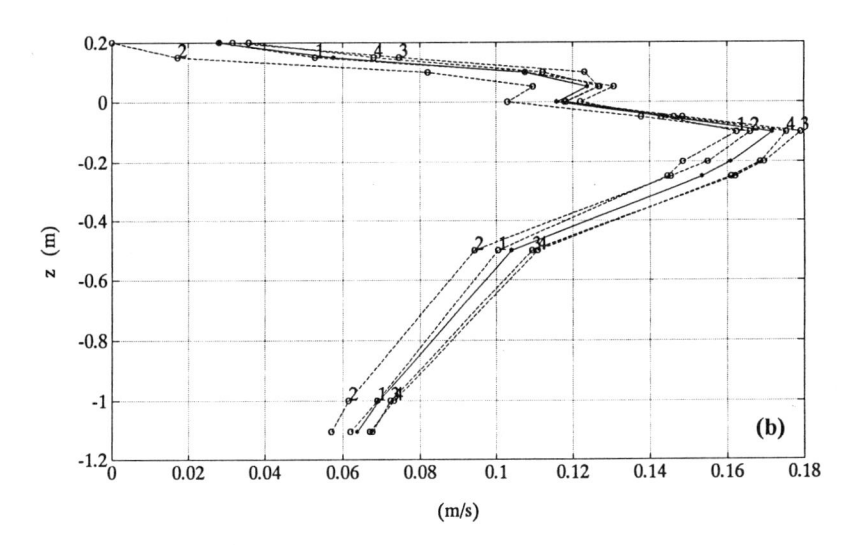

Fig. 5. Mean values (a) and standard deviations (b) of horizontal velocity calculated for each sub-series (dashed lines with circles) against the one calculated for the whole data series (solid line with stars). Wave case I18. The numbers of successive quarters of the experiment series are indicated.

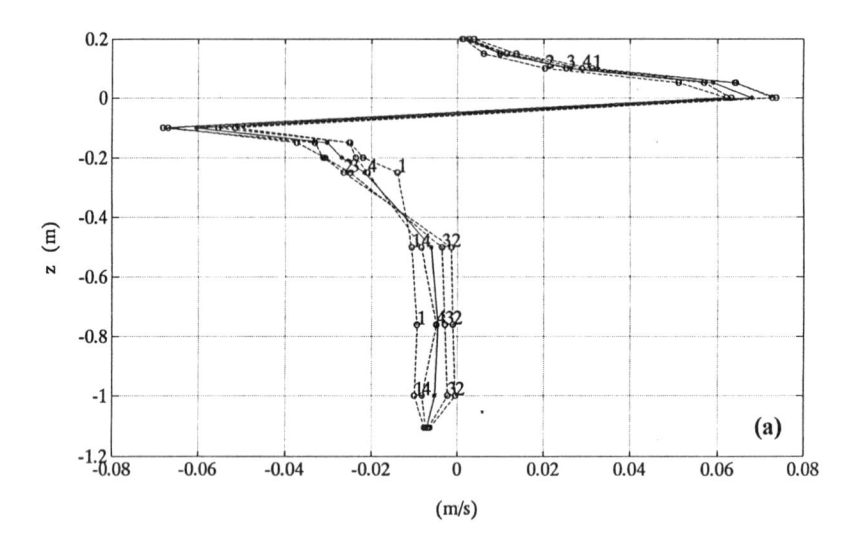

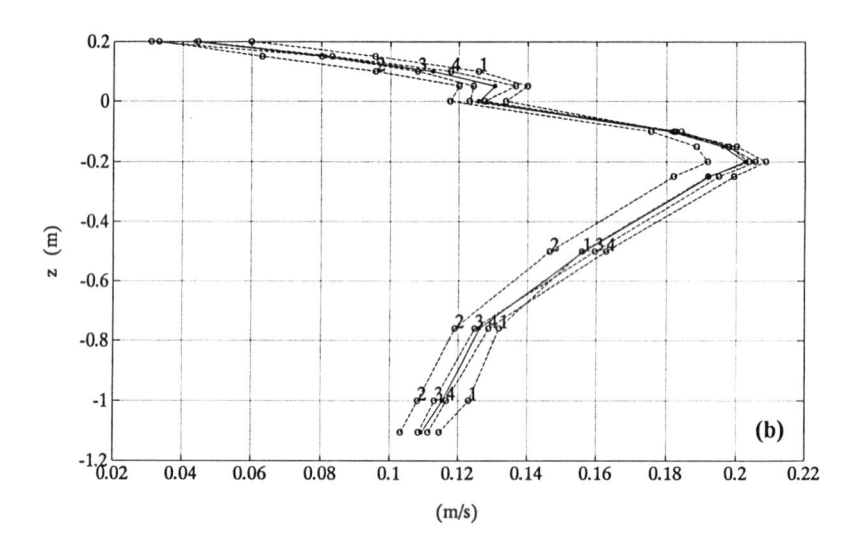

Fig. 6. Mean values (a) and standard deviations (b) of horizontal velocity calculated for each sub-series (dashed lines with circles) against the one calculated for the whole data series (solid line with stars). Wave case I24. The numbers of successive quarters of the experiment series are indicated.

Sjelbreia, J. E., Berek, E., Bolen, Z. K., Gudmestad, O. T., Heideman, J. C., Ohmart, R. D., Spidsøe, N. and Tørum, A., (1991). Wave kinematics in irregular waves. Proc. 10th Intern. Conf. on Off-shore Mech. and Arctic Engng., Stavanger, Norway, 223-228.

Starr, V. P., (1947). A momentum integral for surface waves in deep water. *J. Marine Res.*, 6, No. 2, 126-135.

Tung, C. C., (1975). Statistical properties of the kinematics and dynamics of a random gravity wave field. *J. Fluid Mech.*, 70, 251-255.

CHAPTER 8

Vertical Variations of Fluid Velocities and Shear Stress in Surf Zones

Daniel T. Cox[1], Nobuhisa Kobayashi[2], and Akio Okayasu[3]

ABSTRACT: Detailed laboratory measurements are made of the velocity fluctuations to investigate the processes of the turbulence generation, advection, diffusion and dissipation in the surf zone. An order of magnitude analysis of the transport equation of the turbulent kinetic energy using the normalization adopted by Kobayashi and Wurjanto (1992) indicates an approximate local equilibrium of turbulence for shallow water waves in the surf zone. Estimates are found for common surf zone turbulence parameters. The calibrated values are used to show that the eddy viscosity varies gradually over depth and is nearly time-invariant and that the local equilibrium of turbulence is a reasonable approximation for spilling waves in the inner surf zone.

INTRODUCTION

The spatial and temporal variations of fluid velocities, shear stress, and turbulence intensity are required for a detailed analysis of sediment transport in the surf zone (Deigaard et al., 1986). Field measurements of turbulent velocity fluctuations in the surf zone are difficult due to the harsh conditions for hot film anemometers and problems of calibration and voltage drift (George et al., 1994). Laser-Doppler anemometry has been used in the laboratory (e.g., Stive, 1980; Nadaoka and Kondoh, 1982) to measure turbulent velocity fluctuations. However, no detailed analysis has been made of the turbulent kinetic energy transport equation in the nearshore region with laboratory data. Turbulence measurements in the surf zone are presented herein and are used to show that the local equilibrium of turbulence is a reasonable approximation for spilling waves in the inner surf zone. In addition,

---

[1]Graduate Student, Center for Applied Coastal Research, University of Delaware, Newark, DE 19716 USA. Email: dtc@coastal.udel.edu Tel: +1 302 831 8477

[2]Professor and Associate Director, Center for Applied Coastal Research, University of Delaware, Newark, DE 19716 USA. Email: nk@coastal.udel.edu

[3]Assistant Professor, Department of Civil Engineering, Yokohama National University, 156 Tokiwadai, Yokohama 240, Japan. Email: okayasu@coast.cvg.ynu.ac.jp

the present analysis will be used to calibrate coefficients for the simple turbulence model for the surf zone.

TURBULENCE MODEL

The transport equation of the turbulent kinetic energy, k, is normally written as

$$\frac{\partial k}{\partial t} + u_j \frac{\partial k}{\partial x_j} = \nu_t \left(\frac{\partial u_i}{\partial x_j} + \frac{\partial u_j}{\partial x_i}\right) \frac{\partial u_i}{\partial x_j} + \frac{\partial}{\partial x_j}\left(\frac{\nu_t}{\sigma_k}\frac{\partial k}{\partial x_j}\right) - C_d^{3/4}\frac{(k)^{3/2}}{\ell} \quad (1)$$

where use is made of the repeated indices, $i = 1, 2$; t is time, $x_1 = x$ is the onshore directed horizontal coordinate; $x_2 = z$ is the vertical coordinate, positive upward with $z = 0$ at the still water level (SWL); $u_1 = u$ and $u_2 = w$ are the horizontal and vertical velocities; and σ_k is an empirical constant associated with the diffusion of k. The turbulent eddy viscosity, ν_t, may be expressed as (e.g., ASCE, 1988)

$$\nu_t = C_d^{1/4} \ell \sqrt{k} \quad (2)$$

in which ℓ is the turbulent mixing length, and C_d is an empirical coefficient.

The typical values of C_d and σ_k for *steady turbulent flow* are $C_d \simeq 0.08$ and $\sigma_k \simeq 1.0$ (Launder and Spalding, 1972). The value of C_d is determined herein for normally incident waves on a rough, impermeable slope under the assumption of the approximate local equilibrium of turbulence. The equation for the dissipation rate of k might be used to estimate the turbulence length scale (k-ϵ model) but this equation is more empirical than (1) and gives only slightly better results for the case of bed shear stress calculations (Fredsøe and Deigaard, 1992). Alternatively, the mixing length ℓ in (2) may be specified simply as

$$\ell = \begin{cases} \kappa(z - z_b) & \text{for } z < \left(\overline{C_\ell} h/\kappa + z_b\right) \\ \overline{C_\ell} h & \text{for } z \geq \left(\overline{C_\ell} h/\kappa + z_b\right) \end{cases} \quad (3)$$

where κ is the von Karman constant ($\kappa \simeq 0.4$); z_b is the bottom elevation; h is the instantaneous water depth; and $\overline{C_\ell}$ is an empirical coefficient related to the eddy size. $\overline{C_\ell}$ is written with an overbar to show that it is time-invariant and to differentiate it from C_ℓ used later. Eq. (3) is similar to that used by Deigaard et al. (1986) for their analysis of suspended sediment in the surf zone in which use was made of $\overline{C_\ell} = 0.07$ and the mean water depth, $\bar{h}$, instead of the instantaneous depth, h. The use of h should be more appropriate in the swash zone in light of the limited field data of Flick and George (1990). Svendsen (1987) suggested $\overline{C_\ell} = 0.2$–0.3 for the steady undertow. The value of C_ℓ for the *unsteady* flow and the time-averaged value, $\overline{C_\ell}$, will also be determined for the present data.

The dimensionless variables are introduced following Kobayashi and Wurjanto (1992):

$$t' = \frac{t}{T}; \quad x' = \frac{x}{T\sqrt{gH}}; \quad z' = \frac{z}{H}; \quad u' = \frac{u}{\sqrt{gH}}; \quad w' = \frac{w}{H/T} \quad (4)$$

$$p' = \frac{p}{\rho g H}; \quad \nu'_t = \frac{\nu_t}{H^2/T}; \quad k' = \frac{k}{gH/\sigma}; \quad \ell' = \frac{\ell}{H/\sqrt{\sigma}}; \quad \sigma = \frac{T\sqrt{gH}}{H} \qquad (5)$$

where the primes indicate dimensionless quantities, T and H are the characteristic wave period and height of the shallow water waves, and σ is the ratio between the horizontal and vertical length scales. The order of magnitude of k, ℓ, and ν_t is estimated such that the resulting normalized equations become consistent with the measured data as explained later.

Substitution of (4) and (5) into (1) under the assumption of $\sigma^2 \gg 1$ yields

$$\sigma^{-1}\left(\frac{\partial k'}{\partial t'} + u'\frac{\partial k'}{\partial x'} + w'\frac{\partial k'}{\partial z'}\right) = \tau'\frac{\partial u'}{\partial z'} + \sigma^{-1}\frac{\partial}{\partial z'}\left(\frac{\nu'_t}{\sigma_k}\frac{\partial k'}{\partial z'}\right) - C_d^{3/4}\frac{k'^{3/2}}{\ell'} \qquad (6)$$

where the first and third terms on the right-hand-side are the production and dissipation terms, respectively. For their analysis of suspended sediment in the surf zone, Deigaard et al. (1986) used (6) in which the advection terms were neglected and the production of k' was estimated empirically. In short, they attempted to predict the variation of k' without analyzing u', w' and τ'. Eq. 6 indicates that the production and dissipation of k' are dominant under the assumption of $\sigma^2 \gg 1$. This is qualitatively consistent with the findings of Svendsen (1987) who concluded that only a very small portion of the energy loss in the breaker (2–6% for the cases considered) was dissipated below trough level.

Considering the empirical nature of (6) with the coefficients σ_k and C_d as well as the uncertainty of the free surface boundary condition of k' even for steady turbulent flow (Rodi, 1980), (6) may be simplified further by neglecting the terms of the order σ^{-1} and the resulting equation is expressed in dimensional form as

$$\frac{\tau}{\rho}\frac{\partial u}{\partial z} \simeq C_d^{3/4}\frac{k^{3/2}}{\ell} \qquad (7)$$

which implies the local equilibrium of turbulence. Substitution of $\tau/\rho = \nu_t \partial u/\partial z$ and (2) into (7) yields

$$k = |\tau|/(\rho\sqrt{C_d}) \qquad (8)$$

$$\nu_t = \ell^2\left|\frac{\partial u}{\partial z}\right| \qquad (9)$$

With these assumptions, (8) is used to determine the appropriate value of C_d. Eq. (9) corresponds to the standard mixing length model (ASCE, 1988) and is used with (2) to determine C_ℓ and $\overline{C_\ell}$ in (3). The degree of the local equilibrium of turbulence is assessed using (7) with the calibrated coefficients C_d and $\overline{C_\ell}$.

EXPERIMENT and DATA REDUCTION

The experiment was conducted in the 33 m long, 0.6 m wide and 1.5 m deep wave flume at the University of Delaware. A hydraulically actuated piston wavemaker with a 1 m stroke was at the far end; and a rough, uniform 1:35 slope was emplaced at the near end of the flume. The water depth was 0.4 m in the constant depth section. Regular cnoidal waves were specified at the wavemaker, and the waves broke by spilling on the impermeable slope. The rough slope consisted of a layer of natural sand grains with median diameter $d_{50}=1$ mm glued to Plexiglas sheets and mounted on the entire slope. This was used to increase the bottom boundary layer thickness for estimating the bottom shear stress. A detailed analysis of the bottom shear stress outside and inside the surf zone is given in Cox, et al. (1995).

The free surface elevations were measured using capacitance-type wave gages with a sampling rate of 100 Hz. The velocities were measured using a two-component laser-Doppler anemometer with a pair of burst spectrum analyzers. The effective sampling rate was in excess of 1×10^3 data points per second, and the sampling rate was later reduced by band averaging to 100 Hz before the phase averaging procedure described below. The free surface and velocity fluctuations were measured at six vertical lines and are denoted L1, L2, ..., L6 for brevity. The horizontal spacing of the measuring lines was on the order of 1 m, and the vertical spacing of the measuring points was on the order of 1 cm except near the bottom where measurements were made on the order of a fraction of the grain height, i.e. less than 1 mm. Details of the experiment are provided in Okayasu and Cox (1995).

The free surface and velocity measurements were reduced by a standard phase averaging procedure over 50 waves. The sampling interval was $\Delta t = 0.01$ s and the wave period was $T = 2.2$ s which gave $J = T/\Delta t = 220$ as the number of data points or phases per wave. The phase-averaged free surface elevations, η_a, were computed from the measured free surface, η_m, where the subscripts a and m refer to the phase-averaged and measured quantities. The variance of the free surface elevation, σ_η^2, and the standard deviation, σ_η, were also computed. For the figures presented here, the phases are aligned with zero-upcrossing of the free surface elevation at $t = (T/4) = 0.55$ for the six measuring lines (Cox, 1995).

The normalization parameters for the turbulent quantities in (5) are given in Table 1 with the range of values for the measured data as explained later. The range is found by taking the minimum and maximum values for the phase-averaged quantites between the trough level and the bottom boundary layer which is defined simply as 1 cm above the impermeable bottom and consistent with the analysis of Cox (1995). The ranges given in parentheses are for the bottom boundary layer. The cross-shore locations of the measuring lines are given in Table 2 and are characterized as follows: L1 is seaward of the break point; L2 is at the break point which is defined as the start of aeration in the tip of the wave; L3 is in the transition region where the wave form goes from organized motion to a turbulent bore; and L4, L5, and L6 are in the inner surf zone where the saw-toothed wave shape is a well-developed turbulent bore (Cox, 1995). Table 1 indicates that the scaling of k, ℓ, and ν_t in (5) is appropriate inside the surf zone for L3 to L6.

Table 1: Range of k, ℓ, and ν_t for L1 to L6 and Normalization Quantities.

Line No.	k (cm^2/s^2)	gH/σ (cm^2/s^2)	ℓ (cm)	$H/\sqrt{\sigma}$ (cm)	ν_t (cm^2/s)	H^2/T (cm^2/s)
L1	0.2 – 2.8 (0.3 – 18.8)	684	0.60 – 1.20 (0.01 – 0.40)	3.04	0.11 – 0.61 (0.00 – 0.27)	79.4
L2	0.4 – 3.8 (0.4 – 35.9)	1007	0.60 – 1.91 (0.01 – 0.40)	4.19	0.17 – 1.03 (0.01 – 0.28)	132.9
L3	14.4 – 297 (4.9 – 45.6)	645	0.60 – 3.23 (0.01 – 0.40)	2.89	0.65 – 18.4 (0.01 – 0.76)	73.4
L4	20.7 – 559 (6.8 – 65.7)	337	0.61 – 4.26 (0.02 – 0.41)	1.68	0.81 – 34.8 (0.01 – 0.68)	30.9
L5	17.4 – 206 (4.1 – 35.6)	268	0.60 – 2.64 (0.01 – 0.40)	1.39	0.97 – 12.9 (0.01 – 0.60)	22.8
L6	13.9 – 179 (4.4 – 76.7)	162	0.61 – 2.00 (0.02 – 0.41)	0.91	0.81 – 11.1 (0.02 – 0.83)	11.6

Table 2 lists the free surface statistics for L1 to L6 where x is the onshore directed horizontal coordinate with $x = 0$ cm at L1; d is the distance below the still water level to the top of the Plexiglas sheet, i.e. the bottom of the 1 mm sand layer; H is the local wave height given by $H = [\eta_a]_{max} - [\eta_a]_{min}$ where the subscripts min and max indicate the minimum and maximum values of a phase-averaged quantity; $\overline{\eta_a}$ is the setup or setdown; $\overline{\sigma_\eta}$ is the time-average of the standard deviation of η_a; and $[\sigma_\eta]_{min}$ and $[\sigma_\eta]_{max}$ are the minimum and maximum of the standard deviation values over the wave period. The cross-shore variations of $\overline{\eta_a}$, $[\eta_a]_{min}$, and $[\eta_a]_{max}$ have been well studied; however, less mention has been made of $\overline{\sigma_\eta}$, $[\sigma_\eta]_{min}$, and $[\sigma_\eta]_{max}$. From Table 2, $[\sigma_\eta]_{max}$ is very small for L1 indicating repeatability of the wave form. For L2, $[\sigma_\eta]_{max}$ increases slightly due to irregularities of wave breaking. For L3 in the transition region, $[\sigma_\eta]_{max}$ is at a maximum. For L4 to L6, $[\sigma_\eta]_{max}$ decreases with increasing distance to the shore. It is interesting to note the cross-shore variation of $[\sigma_\eta]_{max}$ because it could be used to better quantity the transition region of the surf zone (e.g., Nairn et al., 1990 and references therein).

The signal dropouts are excluded in the phase averaging of the measured horizontal and vertical velocities. The phase-averaged horizontal and vertical velocities are denoted u_a and w_a, and the horizontal and vertical velocity variances are denoted σ_u^2 and σ_w^2. The turbulent normal stresses may be assumed to be equal to $-\rho\sigma_u^2$ and $-\rho\sigma_w^2$ in the horizontal and vertical directions, where ρ is the fluid density. The phase-averaged covariance of the measured horizontal and vertical velocities is denoted σ_{uw}, and the turbulent shear stress, τ, may be assumed to

Table 2: Phase-Averaged Free Surface Statistics for L1 to L6.

Line No.	x (cm)	d (cm)	H (cm)	$\overline{\eta_a}$ (cm)	$[\eta_a]_{min}$ (cm)	$[\eta_a]_{max}$ (cm)	$\overline{\sigma_\eta}$ (cm)	$[\sigma_\eta]_{min}$ (cm)	$[\sigma_\eta]_{max}$ (cm)
L1	0	28.00	13.22	-0.30	-3.88	9.34	0.10	0.05	0.22
L2	240	21.14	17.10	-0.44	-3.60	13.50	0.14	0.06	0.98
L3	360	17.71	12.71	-0.05	-2.82	9.89	0.41	0.19	2.06
L4	480	14.29	8.24	0.20	-2.33	5.91	0.38	0.17	1.37
L5	600	10.86	7.08	0.75	-1.60	5.48	0.28	0.15	1.03
L6	720	7.43	5.05	1.13	-0.82	4.23	0.22	0.11	0.92

be equal to $-\rho\sigma_{uw}$.

Figure 1 compares the vertical variation of the Froude-scaled time-averaged horizontal turbulent intensity for the present measurements L3 to L6 with the data of George et al. (1994), Stive (1980), and Nadaoka and Kondoh (1982). The data of George et al. (1994) from their Figure 8a are for the natural surf zone and include random waves of both plunging and spilling type. The frozen turbulence assumption was used to extract the turbulent signal. The middle curve of George et al. (1994) indicates the mean value in several vertical bins and the envelope is this mean ±1 standard deviation plus the uncertainty in the data reduction. The data of Nadaoka and Kondoh (1982) from their Figure 7 are for Case 1, spilling waves on a 1:20 slope, and include only the measuring lines inside the surf zone, i.e. P1 to P5. A frequency filter was used to extract the turbulent signal. It is noted that these data are plotted in Figure 1 using d rather than $\overline{h}$. The data of Stive (1980) are also taken from Figure 8a of George et al. (1994) and are presumably for Test 1, spilling waves on a 1:40 slope, and include the measuring lines in the transition region as well as the inner surf zone. Phase-averaging was used to extract the turbulent signal.

Only the data of George et al. (1994) are for multidirectional random waves measured in the field. The other three data sets are for normally incident, regular waves measured in the laboratory. The comparison of the present data set with that of Stive (1980) shows that the phase averaging method gives consistent results for laboratory waves of similar type. The comparison with Nadaoka and Kondoh (1982) indicates that the frequency filter may underestimate the turbulent signal as noted by other researchers (e.g., George et al., 1994). Nevertheless, it would be useful to have a simple relation between the turbulent signals from the two methods since phase averaging cannot be used for random waves in a natural surf zone. Interpretation of the data of George et al. (1994) is difficult

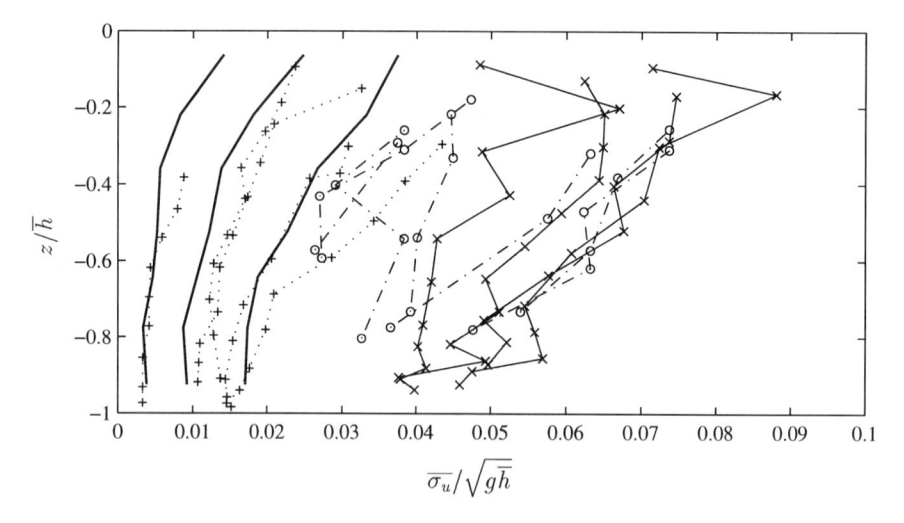

Figure 1: Comparison of Vertical Variation of Froude-Scaled Horizontal Turbulence Intensity with George et al. (1994) (———); Stive (1980) (o – · –); Nadaoka and Kondoh (1982) (+ · · · ·); Present Data L3 to L6 (× ———).

because of the method used to extract the turbulent signal and because the waves were random and multidirectional. Clearly, more work is necessary in this area.

Figure 2 shows the temporal variations of the phase-averaged horizontal and vertical velocity variances, σ_u^2 and σ_w^2, and velocity covariance, σ_{uw}, for five vertical elevations for L2. The five vertical locations for L2a to L2e are -5.04, -13.04, -17.04, -20.04, and -20.94 cm, respectively, where the still water level is $z = 0.0$ cm and the still water depth is $d = 21.14$ cm. Figure 2 shows almost no turbulence in the interior, and the turbulence seems to be confined to the bottom boundary layer. Also, as indicated in the caption, the proposed scaling may not be appropriate in the boundary layer outside the surf zone. The same quantities of Figure 2 for L2 are shown in Figure 3 for L4. The five vertical locations for L4a to L4e are -2.19, -6.19, -10.19, -13.19, and -14.09 cm, respectively, where the still water depth is $d = 14.29$ cm. This figure shows the spread and decay of turbulence generated by wave breaking. Also, the peak of the turbulence shifts downward. For L4a, the horizontal velocity variance is greater than the vertical variance over most of the wave period except at $t = 0.6$ s when the phase-averaged horizontal and vertical velocities are approximately the same (Cox, 1995). The proposed scaling indicated in the caption seems appropriate here. For L4c, the horizontal and vertical variances are approximately the same since the turbulence becomes more isotropic even though the vertical velocity is much smaller than the horizontal velocity at this elevation. For L4e, the horizontal variance is again greater than the vertical variance since the vertical turbulent fluctuations may be limited by the solid boundary. Also, the covariance is negative for L4a to L4c.

Figure 4 shows the detail of the cross-shore variations of time-average horizontal and vertical variances, $\overline{\sigma_u^2}$ and $\overline{\sigma_w^2}$, and the time-averaged covariance, $\overline{\sigma_{uw}}$. Comparison of $\overline{\sigma_u^2}$ and $\overline{\sigma_w^2}$ for L3 to L6 shows that they are about the same magnitude below trough level and decay linearly downward except in the lower portion of the water column where $\overline{\sigma_u^2}$ remains approximately constant over depth and $\overline{\sigma_w^2}$ tends to zero near the bottom.

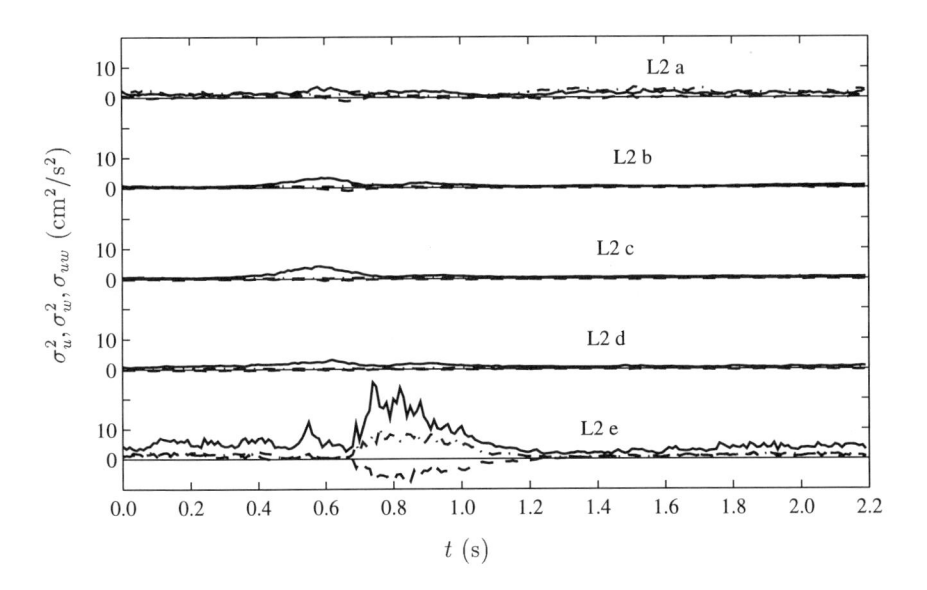

Figure 2: Temporal Variations of Phase-Averaged Horizontal Velocity Variance, σ_u^2 (——); Vertical Velocity Variance, σ_w^2 (-··-); and Covariance, σ_{uw} (- -) for Five Vertical Elevations for L2 with $gH/\sigma = 1007$ cm^2/s^2.

ANALYSES OF WAVE GENERATED TURBULENCE

The dimensional shear stresses, τ_{ij}, are written in tensor notation as (e.g., Rodi, 1980)

$$\tau_{ij} = \rho \left[\nu_t \left(\frac{\partial u_i}{\partial x_j} + \frac{\partial u_j}{\partial x_i} \right) - \frac{2}{3} k \delta_{ij} \right] \qquad (10)$$

where δ_{ij} is the Kronecker delta and k is the turbulent kinetic energy per unit mass which can be expressed in terms of the normal stresses as

$$k = -\frac{1}{2\rho} (\tau_{11} + \tau_{22} + \tau_{33}) \qquad (11)$$

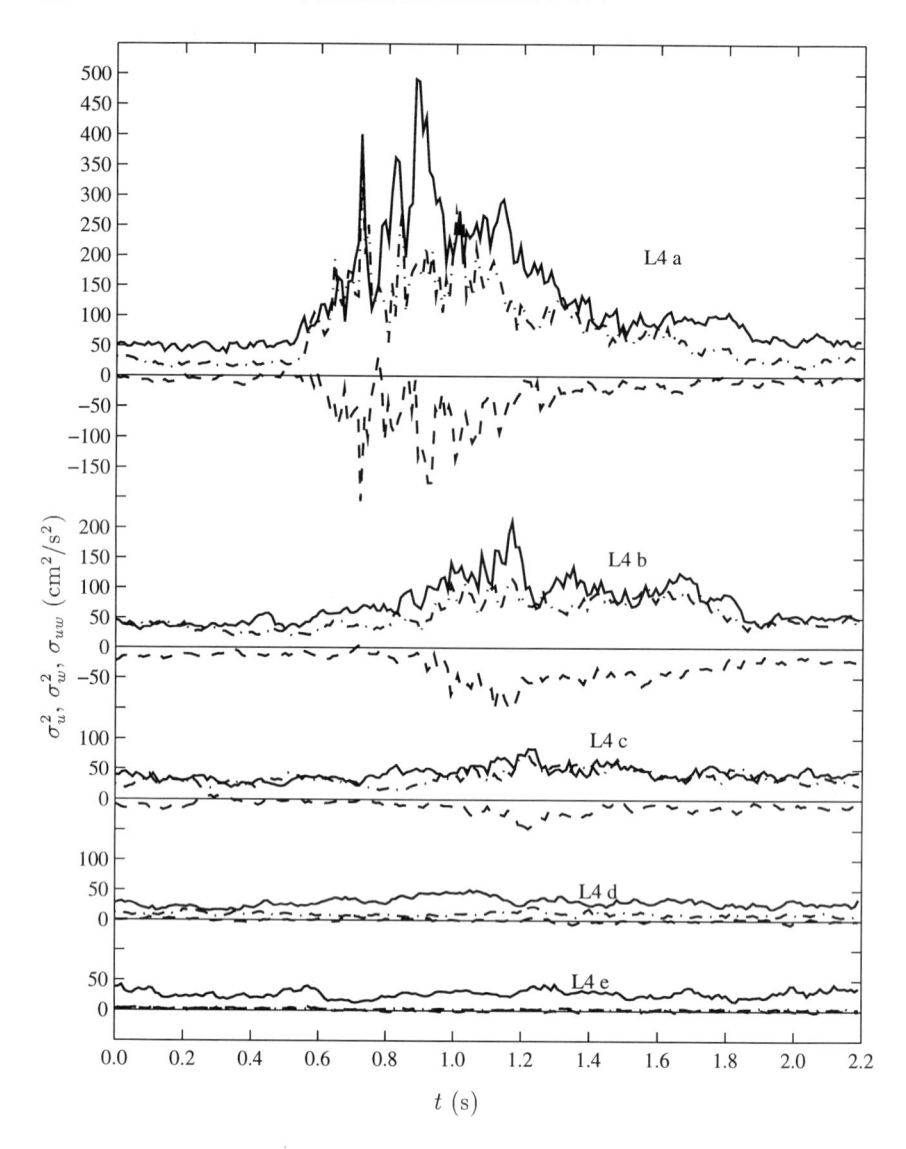

Figure 3: Temporal Variations of Phase-Averaged Horizontal Velocity Variance, σ_u^2 (——); Vertical Velocity Variance, σ_w^2 (-·-); and Covariance, σ_{uw} (- -) for Five Vertical Elevations for L4 with $gH/\sigma = 337$ cm^2/s^2.

VERTICAL VARIATIONS OF FLUID VELOCITIES   107

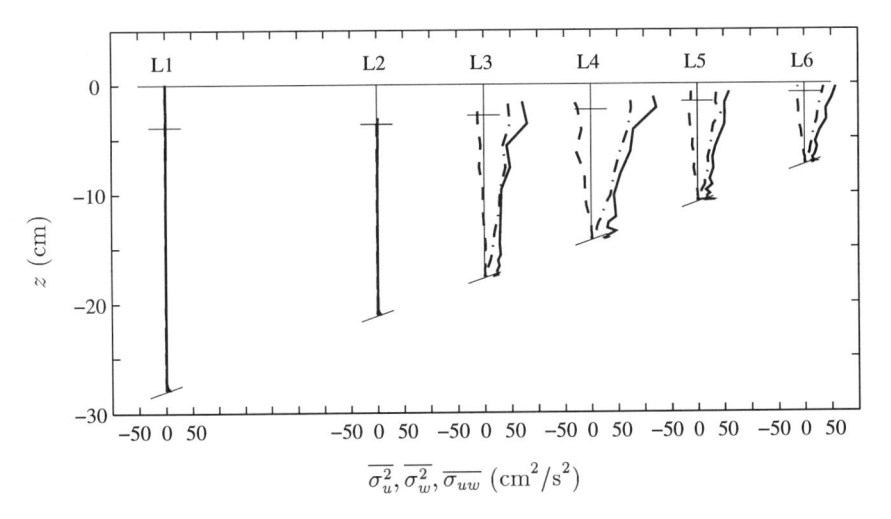

Figure 4: Cross-Shore Comparison of Mean Horizontal and Vertical Velocity Variances and Mean Covariance with $\overline{\sigma_u^2}$ (——); $\overline{\sigma_w^2}$ (-·-); and $\overline{\sigma_{uw}}$ (- -) for L1 to L6.

Assuming that Reynolds averaging is the same as the phase averaging used here, the standard definition of k in terms of the variances is given as

$$k = \frac{1}{2}\left(\sigma_u^2 + \sigma_w^2 + \sigma_v^2\right) \tag{12}$$

The transverse velocity variance, σ_v^2, was not measured for this experiment.

For idealized two-dimensional turbulent flow, $\partial u_3/\partial x_3 = 0$ so that $\tau_{33} = -\frac{2}{3}\rho k$ and then $\sigma_v^2 = \frac{2}{3}k$. This reduces (12) to

$$k = -\frac{3}{4\rho}(\tau_{11} + \tau_{22}) = \frac{3}{4}\left(\sigma_u^2 + \sigma_w^2\right) \tag{13}$$

The use of (10) results in $\frac{1}{2}(\sigma_v^2/k) = \frac{1}{3}$, corresponding to homogeneous isotropic turbulence. For steady turbulent flow, the ratios of the normal stresses to the turbulent kinetic energy have been tabulated by Svendsen (1987). This table indicates that the range is $0.21 \leq \frac{1}{2}(\sigma_v^2/k) < \frac{1}{3}$ so that σ_v^2 may be overestimated slightly here.

Having measured σ_u^2 and σ_w^2 directly, Cox (1995) determined whether the ratio of the vertical to horizontal velocity variance, $C_w = \sigma_w^2/\sigma_u^2$, is constant over a wave period. The results show that the values lie in the range $0.06 \leq C_w \leq 0.86$ for the variances below trough level whereas the range for the types of flows listed in Svendsen (1987) is $0.16 \leq C_w \leq 1.00$. For L3 to L6, C_w is fairly constant

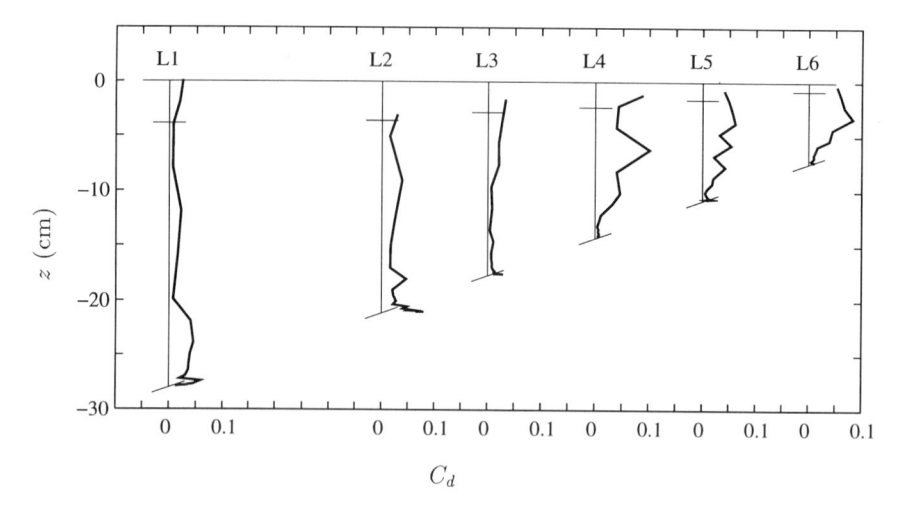

Figure 5: Cross-Shore Variation of C_d for L1 to L6.

over depth with $C_w \simeq 0.7$ until the lower portion of the water column where it decreases linearly toward the rough bottom. Further comparisons of the temporal variations of $C_w \sigma_u^2$ and σ_w^2 similar to Figures 2 and 3 show that it is appropriate to assume that C_w is constant over a wave period in the boundary layer outside the surf zone and below trough level inside the surf zone (Cox, 1995).

A least-squares error method is used to calibrate C_d following (8). Assuming that C_d is independent of time, the least-squares equation is

$$\sqrt{C_d} = \frac{\sum_{j=1}^{J} |\sigma_{uw}|_j k_j}{\sum_{j=1}^{J} k_j^2} \qquad (14)$$

where j indicates the phase out of $J = 220$ phases. Figure 5 shows the cross-shore variation of C_d for L1 to L6 using (14). The vertical variation of C_d is distinctly different for the three regions: L1 and L2 seaward of breaking, L3 in the transition region, and L4 to L6 in the inner surf zone. For L1 and L2, $C_d \simeq 0.06$ in the bottom boundary layer whereas a similar value $C_d = 0.08$ has been used for steady flows and for oscillatory flows in nonbreaking waves. Above the bottom boundary layer for L1 and L2, the values on the right-hand-side of (14) are near zero so that the estimated values in this region are not useful. For L3 in the transition region, the magnitude of C_d is less than 0.03 over most the water column even though the values for k and σ_{uw} are non-zero. For L4 to L6, a typical value is $C_d \simeq 0.05$ below trough level except in the lower portion where it decreases to a small value.

Eqs. (9) and (2) with the calibrated values of C_d are used to determine C_ℓ in

VERTICAL VARIATIONS OF FLUID VELOCITIES 109

(3). For this procedure, an error term is computed for a range of C_ℓ by summing the absolute value of the difference of (9) and (2) over the water column at each of the 220 phases. The error term is given as

$$Err(j) = \frac{1}{I}\sum_{i=1}^{I}\left|(\ell^2 \mid \frac{\partial u}{\partial z}\mid) - (C_d^{1/4}\ell\sqrt{k})\right|_i, \quad j = 1, 2, \ldots, 220 \qquad (15)$$

where the index i refers to points in the vertical measuring line. The ranges of C_ℓ were $0.01 \leq C_\ell \leq 0.20$ for L1 and L2 and $0.05 \leq C_\ell \leq 0.45$ for L3 to L6. The value of C_ℓ that gave the least error in (15) was adopted at that phase. Figure 6 shows the temporal variation of the adopted value of C_ℓ at each of the 220 phases for L4. This figure shows the amount of scatter expected for the calibrated C_ℓ and shows that there is a slight variation over the wave period. The bore arrives at $t \simeq 0.6$ s (see also Figure 8 for the relative phases of the free surface elevation in the inner surf zone). The time-average values, $\overline{C_\ell}$, computed for all the measuring lines L1 to L6 are $\overline{C_\ell} = 0.032$ (.021); 0.055 (.041); 0.117 (.065); 0.211 (.105); 0.162 (.081); and 0.172 (.089), respectively, where the standard deviation is given in parentheses (Cox, 1995). This gives an overall value of $\overline{C_\ell} \simeq 0.04$ (0.03) outside the surf zone, and $\overline{C_\ell} \simeq 0.12$ (0.07) in the transition region, and $\overline{C_\ell} \simeq 0.18$ (0.09) for the inner surf zone.

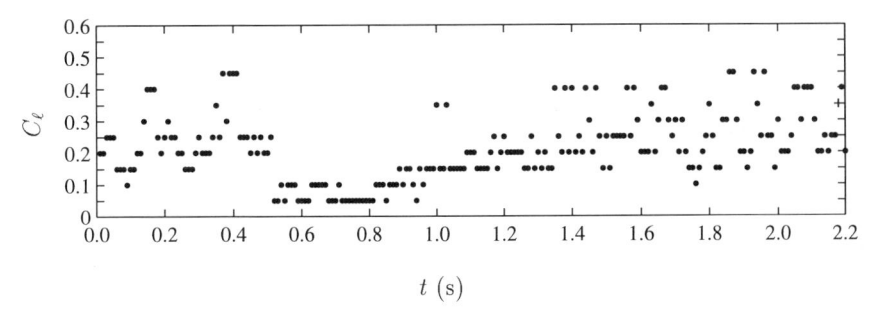

Figure 6: Temporal Variation of Adopted C_ℓ Value at Each of 220 phases for L4 with $\overline{C_\ell} = .211$.

Figure 7 shows the vertical and temporal variations of the eddy viscosity, ν_t, given in (2) computed using the calibrated values of C_d and $\overline{C_\ell}$ for L4. The light vertical lines in the upper figure indicate the extent of the water column at the given phase. The two light horizontal lines in the lower left corner of the top figure indicate the vertical range plotted in detail in the bottom figure. In the bottom figure, z_m is the vertical coordinate from the bottom where $z_m = 0$ on the top of the Plexiglass sheet, i.e. the bottom of the 1 mm sand layer. From both figures, it is clear that ν_t at a given phase increases gradually from the bottom until about the middle of the water column where it is more or less constant over depth. Also, it is reasonable to assume that ν_t is time-invariant except near trough level with the passing of the bore.

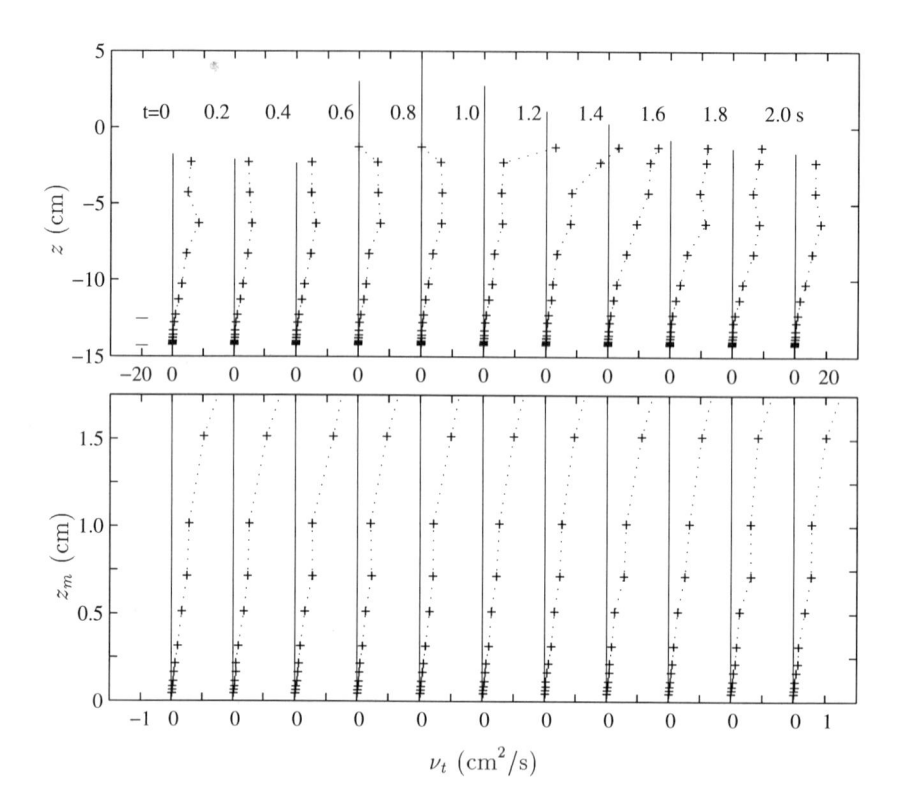

Figure 7: Vertical Variations of Eddy Viscosity, ν_t, at Eleven Phases for L4 with $H^2/T = 30.9$ cm²/s.

Figure 8 shows the temporal variation of the dissipation term $C_d^{3/4}(k^{3/2}/\ell)$ and the production term $\tau(\partial u/\partial z)$ using the calibrated C_d and $\overline{C_\ell}$ values for L5. Smoothing was used for the final plot since the contour lines of the unsmoothed values are difficult to discern in black and white (Cox, 1995). Also, only the measuring points above the bottom boundary layer are plotted. This figure shows that the approximate local equilibrium of turbulence is a reasonable assumption for spilling waves in the inner surf zone. It is noted that the numerical derivatives for additional terms in the dimensional equivalent of (6) were computed and that the noise level was on the same order as the quantities of interest.

CONCLUSIONS

Turbulence measurements of spilling waves were presented and used to show that

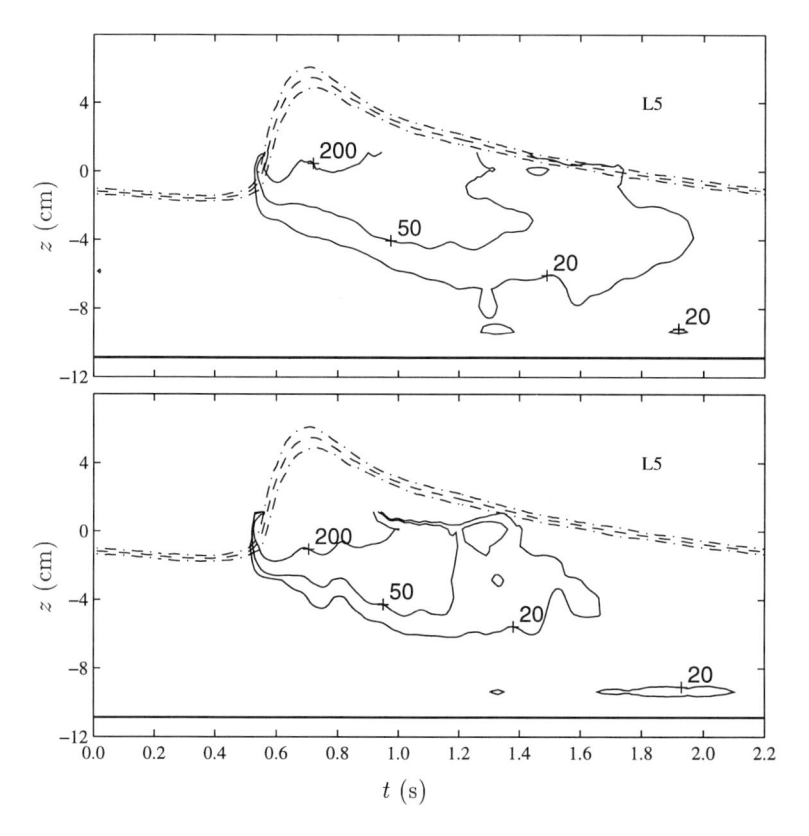

Figure 8: Contour Plot of Temporal Variation of Dissipation Term, $C_d^{3/4}(k^{3/2}/\ell)$, (Top) and Production Term, $\tau(\partial u/\partial z)$, (Bottom) using Calibrated C_d and $\overline{C_\ell}$ with η_a (– –) and $\eta_a \pm \sigma_\eta$ (–·–) for L5.

the local equilibrium of turbulence is a reasonable approximation for spilling waves in the inner surf zone. Further, the empirical coefficient for the mixing length was shown to be roughly constant over the wave period but varied in the cross-shore direction. The typical values were of $\overline{C_\ell} \simeq 0.04$ (0.03) outside the surf zone, and $\overline{C_\ell} \simeq 0.12$ (0.07) in the transition region, and $\overline{C_\ell} \simeq 0.18$ (0.09) for the inner surf zone. The coefficient related to the dissipation of k was found to be $C_d \simeq 0.06$ in the bottom boundary layer outside the surf zone. In the transition region, the magnitude of C_d was less than 0.03 over most of the water column. In the inner surf zone a typical value was $C_d \simeq 0.05$ over most of the water column except in the lower portion where it decreased to a small value. The eddy viscosity was also shown to increase approximately linearly from the bottom to the middle of the water column where the value became more or less constant over depth. The eddy viscosity was fairly constant over the wave period except near trough level with passing of the bore.

ACKNOWLEDGMENTS

This work was sponsored by the U.S. Army Research Office, University Research Initiative under contract No. DAAL03-92-G-0116 and by the National Science Foundation under grand No. CTS-9407827. Dr. Okayasu was supported by the Foundation for International Exchanges, Yokohama National University, during his stay at the University of Delaware.

REFERENCES

ASCE Task Committee on Turbulence Models in Hydraulic Computations (1988). "Turbulence modeling of surface water flow and transport: Part I to V." *J. Hydraulic Engrg.*, ASCE, 114(9), 970–1073.

Cox, D. T. (1995). "Experimental and numerical modeling of surf zone hydrodynamics." Ph.D. Dissertation, University of Delaware, Newark.

Cox, D. T., Kobayashi, N. and Okayasu, A. (1995). "Bottom shear stress in the surf zone." in preparation for *J. Geophys. Res.*

Deigaard, R., Fredsøe, J. and Hedegaard, I. B. (1986). "Suspended sediment in the surf zone." *J. Wtrway. Port Coast. and Oc. Engrg.*, ASCE, 112(1), 115–127.

Flick, R. E. and George, R. A. (1990). "Turbulence scales in the surf and swash." *Proc. 22nd Coast. Engrg. Conf.*, ASCE, 557–569.

Fredsøe, J. and Deigaard, R. (1992). *Mechanics of Coastal Sediment Transport*, Advanced Series on Ocean Engineering, Volume 3, World Scientific, New Jersey.

George, R. A., Flick, R. E. and Guza, R. T. (1994). "Observations of turbulence in the surf zone." *J. Geophys. Res.*, 99(C1), 801–810.

Launder, B. E. and Spalding, D. B. (1972). *Mathematical Models of Turbulence*. Academic Press, New York, NY.

Nadaoka, K. and Kondoh, T. (1982). "Laboratory measurements of velocity field structure in the surf zone by LDV." *Coast. Engrg. in Japan*, 25, 125–145.

Nairn, R.B, Roelvink, J.A. and Southgate, H.N. (1990). "Transition zone width and implications for modelling surfzone hydrodynamics." *Proc. 22nd Coast. Engrg. Conf.*, ASCE, 68–81.

Okayasu, A. and Cox, D. T. (1995). "Laboratory study on turbulent velocity field in the surf zone over a rough bed." in preparation for *Coast. Engrg. in Japan*.

Kobayashi, N. and Wurjanto, A. (1992). "Irregular wave setup and run-up on beaches." *J. Wtrway. Port Coast. and Oc. Engrg.*, ASCE, 118(4), 368–386.

Rodi, W. (1980). *Turbulence Models and Their Application in Hydraulics*. Int'l. Assoc. for Hydraul. Res., Delft, The Netherlands.

Stive, M. J. F. (1980). "Velocity and pressure field of spilling breakers." *Proc. 17th Coast. Engrg. Conf.*, ASCE, 547–566.

Svendsen, I. A. (1987). "Analysis of surf zone turbulence." *J. Geophys. Res.*, 92(C5), 5115–5124.

CHAPTER 9

Vorticity effects in combined waves and currents.

I. Cummins[1] & C. Swan[2].

Abstract.

This paper concerns the interaction of waves and currents, and in particular the effect of the time-averaged vorticity distribution associated with a sheared current. Laboratory data describing both the "initial interaction" of waves and currents, and the "equilibrium" conditions arising within an established wave-current combination are presented. These results are compared to both the existing irrotational solutions and a multi-layered numerical model capable of describing an arbitrary current profile. The interaction of regular waves and sheared currents is shown to be in good agreement with this latter solution. However, a similar description of random waves on sheared currents is limited by the wave-induced changes in both the mean current profile and the associated turbulent structure.

1. Introduction.

In general, the interaction of waves and currents may be sub-divided into two distinct stages. The first corresponds to the "initial interaction" which arises when a given wave train (specified in the absence of a current) propagates onto a pre-determined current profile. This stage is usually solved in terms of a "gradually varying flow", and describes the initial changes in the wave height, the wave length, and (under some circumstances) the current profile. In contrast, the second stage concerns the description of the so-called "equilibrium conditions" arising from an established wave-current combination. It is this stage which seeks to define the fluid motion appropriate to the wave height, the wave period, the water depth, and the current profile determined in stage 1.

---

[1] Research student & [2] Lecturer. Department of Civil Engineering, Imperial College of Science Technology and Medicine, London, United Kingdom. SW7 2BU.

In its simplest form the combination of waves and currents involves a series of regular waves propagating on a depth-uniform current. In this case the current has no associated vorticity, and laboratory measurements (Thomas, 1990 and Swan, 1990) suggest that the "initial interaction" is well defined by the conservation of wave action and the "equilibrium conditions" are in good agreement with a Doppler shifted solution (Fenton, 1985). Although this case appears very simplistic it is, in fact, valid for a wide range of flow conditions. For example, in the case of a logarithmic current profile (typical of many tidal flows) the vorticity distribution is largely confined to a lower layer adjacent to the sea bed. In this case the vorticity has little effect beyond the near-bed region, and an approximate uniform current provides a reasonable description of both the "initial interaction" and the resulting oscillatory flow. Furthermore, if the current profile is weakly sheared, an "equivalent uniform current" based upon the flow conditions at the water surface (Hedges and Lee, 1992) may be sufficient to define the resulting flow field.

Unfortunately, the success of these simplistic solutions, and in particular the ease with which they can be incorporated within large-scale coastal models, has tended to detract from those situations in which a uniform approximation is inappropriate. For example, in the case of a wind generated current (or, indeed, a wind modified current) offshore measurements suggest that the current profile is strongly sheared in the vicinity of the water surface. In this case a non-uniform vorticity distribution arises (ie. the current is not linearly sheared), and consequently the wave motion may be very different to that which is predicted by the existing irrotational wave-current solutions. Indeed, Tsao (1959) suggested that the wave motion would itself become rotational; while Swan (1992) provided explicit analytical calculations (within a truncated series expansion), and confirmed that the near-surface vorticity distribution altered the wave kinematics over the entire water depth.

The present paper will consider the interaction of waves with depth varying currents, and will examine the importance of the vorticity distribution. Section 2 commences with a brief outline of the experimental apparatus. Laboratory measurements describing the interaction of regular waves and random waves with a variety of current profiles are presented in sections 3 and 4. Section 3 concerns the "equilibrium conditions" and compares the measured data with a variety of solutions including a five-layered numerical model similar to that outlined by Dalrymple and Heideman (1989) and previously discussed by Cummins and Swan (1993). In section 4 this model is, in turn, used within an iterative procedure to solve the energy transfer equation first outlined by Longuet-Higgins and Stewart (1960). Although this latter approach is computationally intensive, it is applicable to both regular and random waves, and (unlike the conservation of wave action) provides an explicit description of the spectral changes arising when random waves propagate onto strongly sheared currents. This is of particular importance from a practical point of view. Finally, some concluding remarks concerning the importance of the vorticity distribution and the applicability of the various solutions are made in section 5.

2. Laboratory apparatus.

The experimental measurements were undertaken in a purpose built wave flume which allows the simultaneous generation of waves and co-linear currents. This facility is 25m long, 0.3m wide, and has a working depth of 0.7m. It is equipped with a numerically controlled random wave paddle located at one end of the wave flume, and a large passive absorber (consisting of poly-ether foam) at the other. The current is introduced via three loops of re-circulating pipework which are pumped individually to give a total volume flow of 0.45m^3/s. Each loop is fully reversible, and the inlets and outlets are adjustable (in height) to give a variety of both "favourable" (in the same direction as the phase velocity) and "adverse" (in the opposite direction to the phase velocity) current profiles. With this arrangement it is possible to generate a uniform current of approximately 0.2m/s, or a highly sheared current in which the near-surface velocities may be as large as 0.6m/s. A sketch showing the layout of this apparatus is given on figure 1.

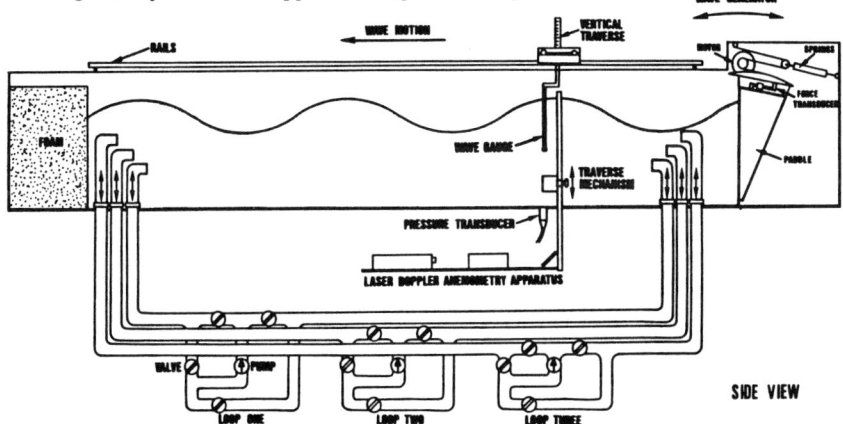

Figure 1. Laboratory apparatus.

Within this study measurements of the water surface elevation were obtained from surface piercing wave gauges which were mounted above the wave flume. Each gauge consists of two vertical wires and provides a time history of the water surface elevation at one point fixed in space. These probes cause a minimal disturbance of the water surface, and have a measuring accuracy of ± 1mm. The velocity field was measured using laser Doppler anemometry. This was based upon a 35mW helium-neon laser, used to create a three beam arrangement with cross polarisation. The intersection of the beams was located along the centre-line of the wave flume, and produced a measuring volume which was estimated to be 0.5mm^3. This intersection was observed in forward scatter using two photomultipliers positioned on the opposite side of the wave flume. This arrangement provides the optimal signal to noise ratio, with no disturbance of the flow field. After seeding the flow with milk, added in the ratio of 100ppm., a data rate of 2.5 khz was achieved with a measuring accuracy of $\pm 2\%$.

3. Equilibrium conditions.

The interaction of regular waves (T=0.75s and H=0.083m) with a "favourable" uniform current is considered on figures 2a and 2b. The first of these figures describes the current profile measured both before and after the interaction with the wave train; while the second figure describes the depth variation in the horizontal component of the wave velocity measured beneath a wave crest. In this case, and indeed in all other cases with zero vorticity, there is virtually no change in the current profile, and the measured wave kinematics are in good agreement with the fifth order Doppler shifted solution proposed by Fenton (1985).

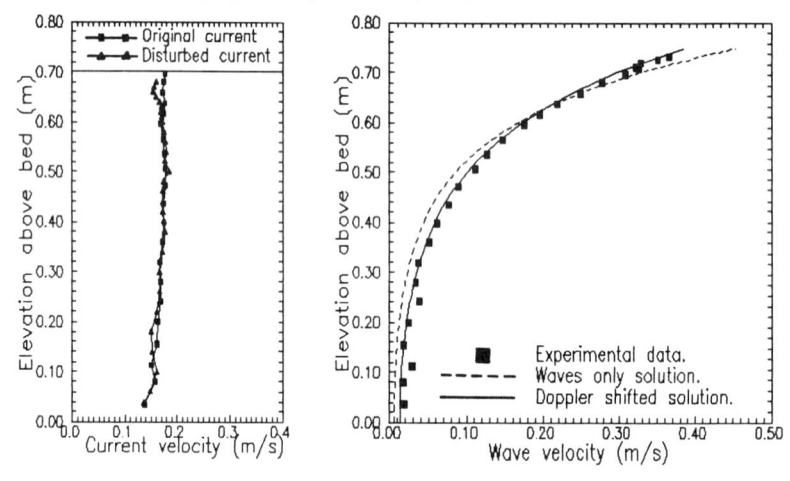

Figures 2a-2b. Regular waves on a "favourable" uniform current.

Figures 3a-3b present a similar sequence of results describing the interaction of regular waves on an "adverse" sheared current; while figures 4a-4b correspond to waves on a "favourable" sheared current. In both of these cases there is some evidence of a change in the current profile (ΔU). This is particularly apparent in the "favourable" case (figure 4a) where the interaction with the wave motion appears to reduce the time-averaged near-surface vorticity (ie. the current becomes less sheared). In contrast, figure 3a suggests that in an "adverse" current the magnitude of the near-surface vorticity increases.

If the disturbed current profile (or that measured in the presence of waves) is used to calculate the wave-induced kinematics, a Doppler shifted solution based on the magnitude of the near-surface current is in poor agreement with the laboratory data (figures 3b and 4b). However, a non-linear numerical model, which incorporates the effects of the vorticity distribution, provides a good description of the measured data. The numerical solution referred to in figures 3b and 4b is based upon a five layered approximation in which the measured current profile is described by five linear segments of variable depth. This approach represents an extension of the bi-linear model originally proposed by Dalrymple (1974), and provides a satisfactory compromise between the description of the current profile (particularly the vorticity distribution) and the computational effort required for convergence. This model has been rigorously tested against other wave-current models (Chaplin, 1990), and is described in detail by Cummins and Swan (1993). Figures 3b and 4b confirm the importance of the vorticity distribution, and suggest that this must be taken into account if the wave kinematics are to be correctly predicted.

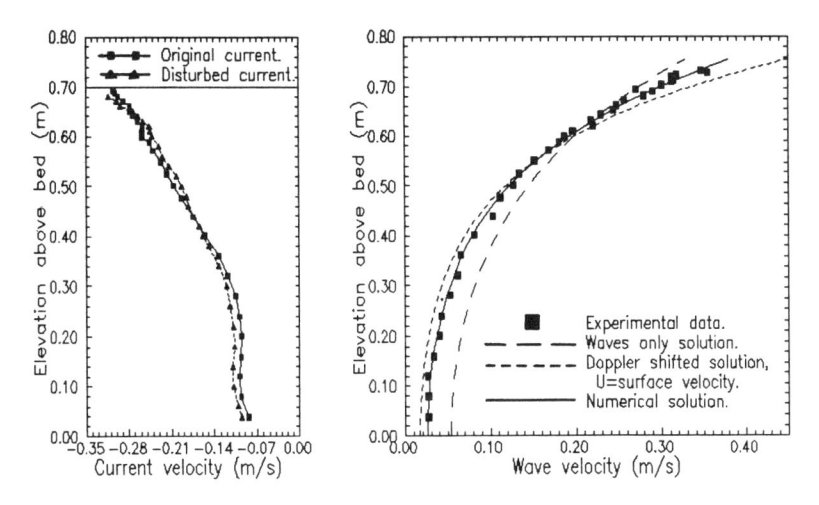

Figures 3a-3b. Regular waves on an "adverse" sheared current.

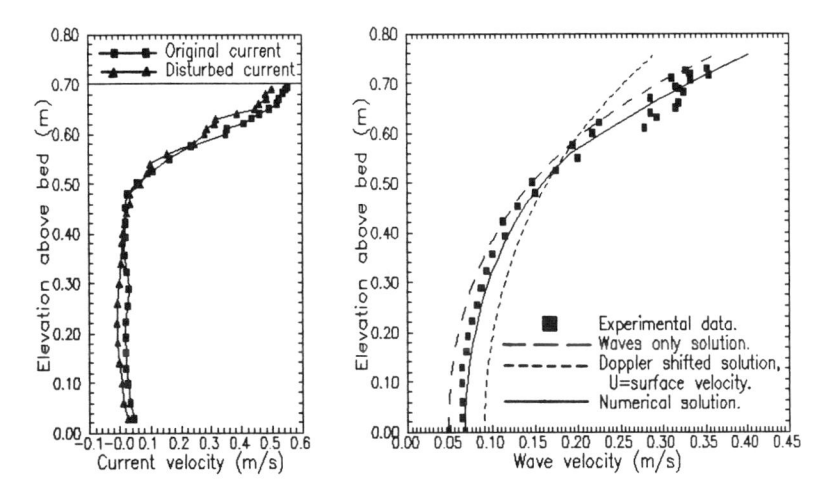

Figures 4a-4b. Regular waves on a "favourable" sheared current.

Previous work by Hedges and Lee (1992) suggests that the interaction of waves with sheared currents may be described by an "equivalent uniform current". This is defined as the uniform current which produces the same wave number as the measured current for a given wave period, wave height, and water depth. In other words, it ensures that the dispersive characteristics of the waves are correctly modelled. However, this does not imply that the underlying kinematics will be correctly predicted. Indeed, Hedges and Lee comment that the solution may be inappropriate if there are regions of very strong shear; while Swan (1992) suggests that if this is indeed the case, an additional rotational term arises within the description of the wave kinematics. The measured data appears to confirm this effect. Figure 5 compares the measured kinematics on a "favourable" sheared current with the numerical model (discussed above), and a Doppler shifted solution based upon an "equivalent uniform current". In this case (and indeed, in several other cases involving layers of strong current shear) the "equivalent uniform current" does not provide a good description of the wave kinematics.

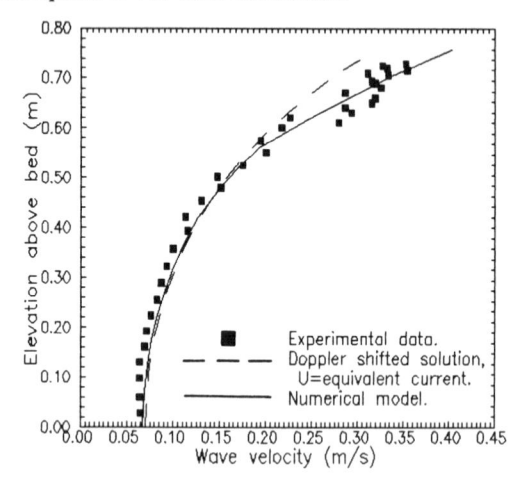

Figure 5. Comparison with an "equivalent uniform current".

4. Initial interaction.

When a wave train first propagates onto a current there are changes in the wave number (k), the wave height (H), and under some circumstances the current profile, U(z). The numerical model (discussed above) is able to predict the wave number for a given wave period, wave height, water depth, and current profile. If we assume that the water depth is known, and (at present) that the current profile remains unchanged, the energy transfer equation first identified by Longuet-Higgins and Stewart (1960) may be applied in conjunction with the numerical model to define an iterative solution for the overall change in the wave train (ie. Δk and ΔH).

If R defines the mean rate of energy transfer across a fixed surface (S), Longuet-Higgins and Stewart (1960) give:

$$R = \int_S \left(P + \frac{1}{2}\rho \underline{u}^2 + \rho gz\right)\underline{u}.\underline{n}\ dS \qquad (1)$$

where $\underline{u}$ is the velocity vector, P is the pressure, ρ is the density, g is the gravitational constant and $\underline{n}$ a unit vector normal to the surface S.

If we consider a control volume, and assume that there is no reflection of wave energy, then the sum of the mean energy transfer rates due to the current (acting alone) and the wave train (also acting alone), must exactly balance the total mean rate of energy transfer associated with the combined wave-current motion. This approach is entirely consistent with the original analysis outlined by Longuet-Higgins and Stewart (1960) in which they introduced the concept of radiation stress. Indeed, Longuet-Higgins and Stewart suggested in a footnote (page 574) that the vorticity may be taken into account by supposing U to be dependent upon z (ie. U(z)). This is exactly what we have done in the present analysis.

Regular waves.

Having coupled the energy transfer equation with the numerical model the interaction of regular waves with a uniform current was considered. In this case the present calculations were shown to be in good agreement with the fifth order solution (based upon the conservation of wave action) proposed by Thomas (1990). In a second test-case the interaction of regular waves with a linearly sheared current was considered, and the results compared with the second order solution outlined by Jonsson et al. (1978). In this case there was again good agreement between the two solutions provided (as one might expect) the comparisons were restricted to small amplitude waves. However, in those cases involving larger wave amplitudes, the difference between these solutions emphasises the importance of the higher order non-linear interactions. These results appear to be consistent with the laboratory observations presented by Swan (1990) and, in particular, the comparison with the third order kinematics predicted by Kishida and Sobey (1988). Further comments concerning the importance of the non-linear interactions are provided by Cummins and Swan (1993).

The experimental measurements presented on tables 1a and 1b concern a total of seven cases involving the interaction of regular waves and depth varying currents. Cases 1-4 correspond to a "favourable" sheared current (table 1a); while cases 5-7 correspond to an "adverse" sheared current (table 1b). In each case the initial wave properties (measured in the absence of a current) are denoted by H_o and T_o; while the wave height measured in the presence of a current is indicated by H. Three comparative solutions are al o presented. The first corresponds to a uniform current based upon the near surface velocity ($U=U_s$); the second corresponds to an "equivalent uniform current" ($U=U_e$); and the third represents the present solution. In each case the numerical model provides the best description of the measured data.

Initial state.		Lab. data	Predicted wave height		
T_0	H_0	H	$U=U_s$	$U=U_e$	Model.
.75	.083	.079	.051	.061	.077
.90	.088	.082	.055	.066	.081
1.1	.102	.096	.065	.083	.091
1.2	.107	.104	.071	.093	.096

Table 1a: Wave height change on a "favourable" sheared current.

Initial state.		lab. data	Predicted wave height		
T_0	H_0	H	$U=U_s$	$U=U_e$	Model.
0.9	.075	.086	.107	.111	.094
1.0	.075	.096	.111	.110	.098
1.1	.075	.097	.113	.108	.098

Table 1b: Wave height change on an "adverse" sheared current.

In figures 6a-6b the wave lengths associated with cases 1-7 are again compared to three potential solutions. The first corresponds to a waves only solution and thus describes the waves in the absence of the current. The second corresponds to a Doppler shifted solution based upon the near surface current ($U=U_s$), while the third represents the present numerical results. In each case the latter solution provides the best description of the measured data. In general, the measured changes in both the wave height and the wave length suggest that the vorticity distribution acts to reduce the effect of the Doppler shift associated with the surface current. This is consistent with the second order approximation presented by Swan (1992), but because of the non-linearity of the wave-current interaction a higher order numerical solution is required to provide a good fit to the measured data.

Random waves.

Having demonstrated that the present approach is able to describe the changes in a regular wave train propagating onto a strongly sheared current, a similar approach will be applied to investigate the spectral changes in a random wave train. The laboratory apparatus described in section 2 was used to generate a random wave train with a Pierson-Moskowitz input spectrum. Both the water surface elevation ($\eta(t)$) and the horizontal velocity component (u(t)) at z=-0.1m were sampled at 25Hz for 200 minutes, and the resulting data was analysed using a ten point moving average. Figures 7a-7b concern the interaction with a "favourable" uniform current. In figure 7a the uppermost curve (indicated by a dashed line) describes the spectrum of the

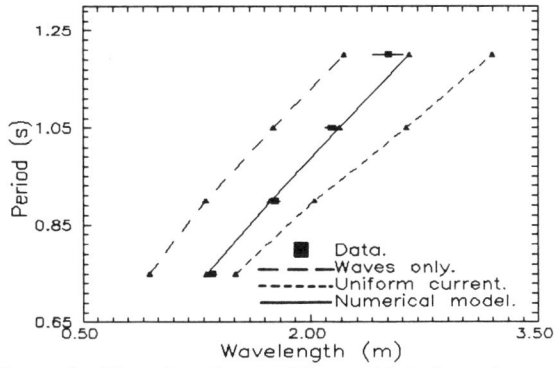

Figure 6a. Wave length on a "favourable" sheared current.

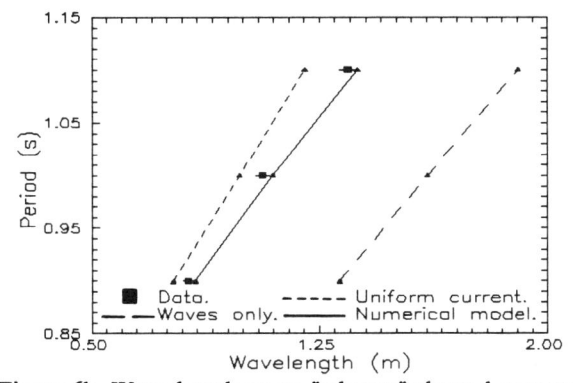

Figure 6b. Wave length on an "adverse" sheared current.

water surface elevation (S$\eta\eta$) measured in the absence of a current; the solid squares represent the data measured in the combined wave-current motion; and the solid line corresponds to the present numerical model. A similar sequence of results is presented on figure 7b, but in this case they correspond to the spectrum of the measured horizontal velocity (Suu) at z=-0.1m. Together, these figures suggest that the proposed numerical model provides a good description of the changes in both the water surface elevation and the underlying kinematics.

Figures 8a-8b present a similar sequence of plots concerning the interaction with an "adverse" uniform current. In this case the steepening of the high frequency components induces wave breaking, and consequently the experimental measurements within the range ($\omega > 7$ rad/s) diverge from the predicted behaviour. This result was also noted by Hedges et al. (1985), and an experimental correction (referred to as the equilibrium range constraint) was proposed. Having incorporated this correction, the present model once again provides a good description of the laboratory data.

(a) Water surface elevation.

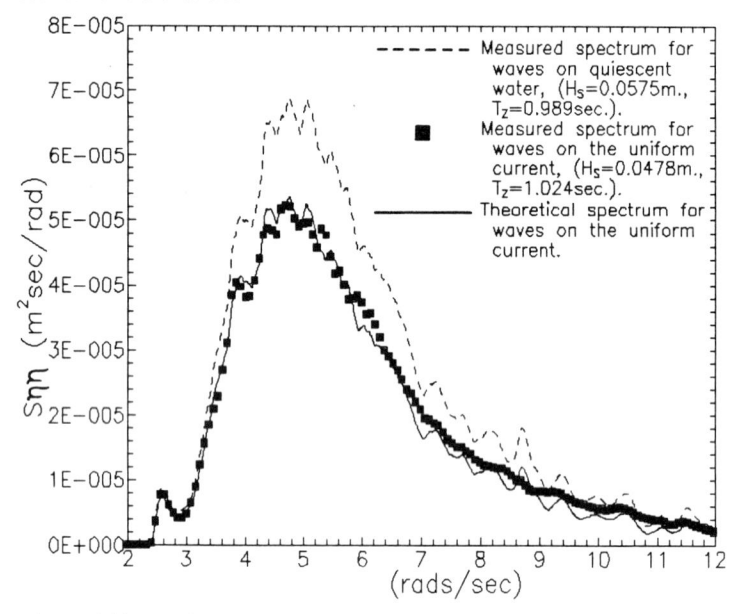

- - - - - Measured spectrum for waves on quiescent water, (H_S=0.0575m., T_Z=0.989sec.).
■ Measured spectrum for waves on the uniform current, (H_S=0.0478m., T_Z=1.024sec.).
—— Theoretical spectrum for waves on the uniform current.

$S\eta\eta$ (m²sec/rad) vs (rads/sec)

(b) Horizontal kinematics (z=-0.1m.)

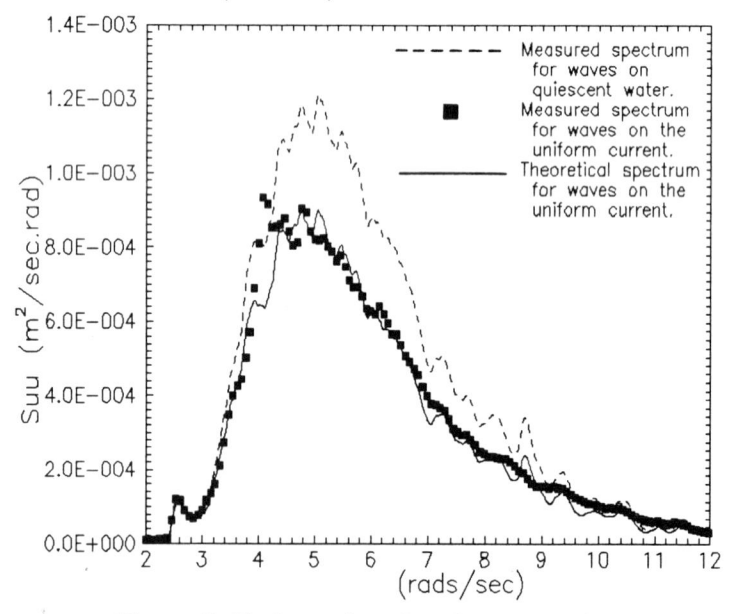

- - - - - Measured spectrum for waves on quiescent water.
■ Measured spectrum for waves on the uniform current.
—— Theoretical spectrum for waves on the uniform current.

S_{uu} (m²/sec.rad) vs (rads/sec)

Figures 7a-7b. Interaction of random waves with a "favourable" uniform current.

(a) Water surface elevation.

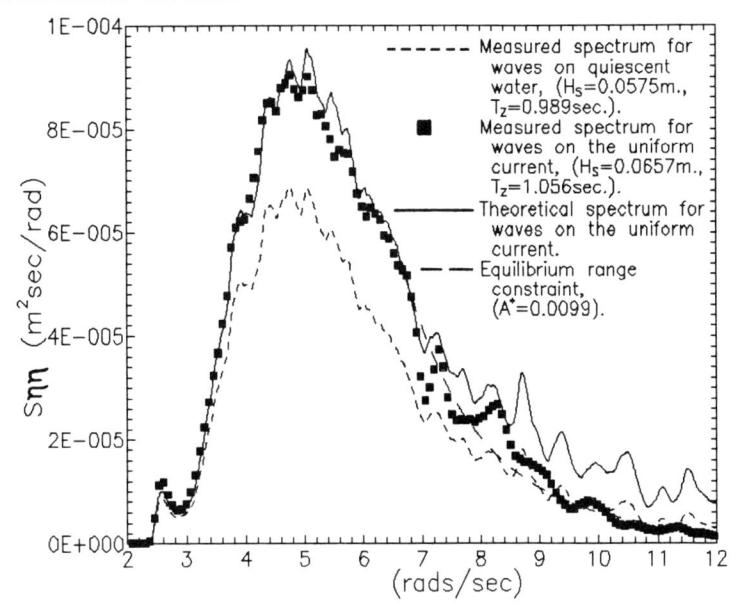

Legend:
- Measured spectrum for waves on quiescent water, (H_S=0.0575m., T_z=0.989sec.).
- ■ Measured spectrum for waves on the uniform current, (H_S=0.0657m., T_z=1.056sec.).
- Theoretical spectrum for waves on the uniform current.
- Equilibrium range constraint, (A^*=0.0099).

$S_{\eta\eta}$ (m²sec/rad) vs (rads/sec)

(b) Horizontal kinematics (z=-0.1m).

Legend:
- ■ Measured spectrum for waves on the uniform current.
- Theoretical spectrum for waves on quiescent water.
- Theoretical spectrum for waves on the uniform current.
- Equilibrium range constraint, (A^*=0.0099).

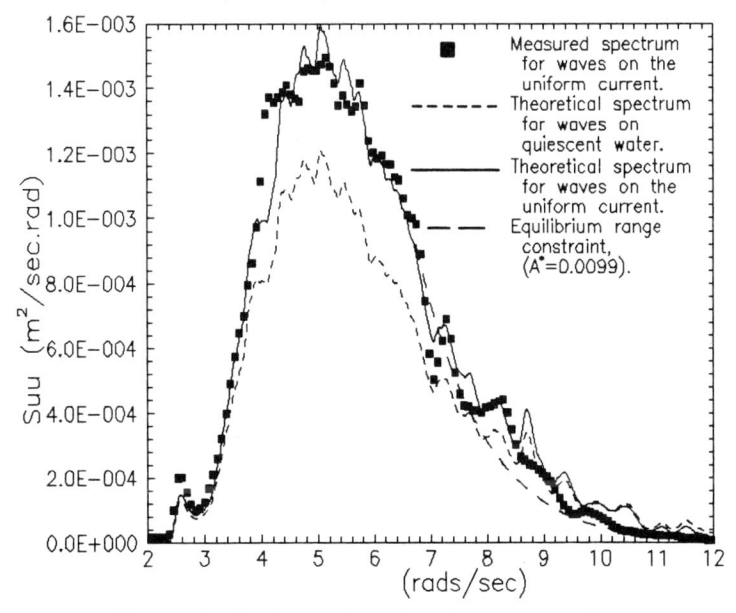

S_{uu} (m²/sec·rad) vs (rads/sec)

Figures 8a-8b. Interaction of random waves with an "adverse" uniform current.

These examples (involving the interaction with a uniform current) provide further validation of the proposed numerical scheme. However, in practice, the existing analytical solutions are also appropriate to these cases, and provide a simpler calculation procedure. Indeed, if the individual waves are linear, the first order approximation originally proposed by Longuet-Higgins and Stewart (1960) may be applied. Alternatively, if the waves are steeper (and, in particular, if the interaction involves an "adverse" current) the fifth order solution proposed by Thomas (1990) will be appropriate.

In contrast, if the current profile is strongly sheared and involves a non-uniform vorticity distribution, the analytical solutions are invalid. In this case the present numerical scheme provides the only method of determining the change in the wave spectra. Figures 9a-9b concern exactly this case and compare the measured data with both a uniform current approximation and the numerical solution. The power spectrum of the water surface elevation ($S\eta\eta$) is considered in figure 9a. Although, in this case, the numerical model provides the best description of the measured data, there remain significant differences between the observed and predicted behaviour. These discrepancies are probably associated with a change in the current profile similar to that noted on figures 3a and 4a. The present formulation assumes that the current velocity is unchanged by the wave-current interaction. Indeed, if the fluid is inviscid and the flow laminar, the vorticity must remain constant along a streamline. However, in the present investigation the turbulent intensity (or the root-mean-square velocity fluctuations expressed as a ratio of the mean current velocity) was of the order of 8-10%. This provides an effective transport mechanism capable of re-distributing the vorticity profile. Without a detailed description of the turbulent structure, the change in the current profile remains indeterminate. This represents an important limitation to the present model. Nevertheless, figure 9a suggests that the numerical calculations provide a significant improvement over the uniform current approximation based upon the near-surface velocity.

Figure 9b describes the spectrum of the horizontal velocity at $z=-0.1m$. In this case the measured data, including an apparent bi-modal peak, does not correspond to either of the existing solutions. Indeed, the data lies mid-way between the uniform current approximation ($U=U_s$) and the numerical predictions. These results may, once again, reflect the importance of the current change (ΔU). However, the wave spectrum indicated on figure 9b, has been derived by subtracting the power spectrum of the turbulent fluctuations (measured in the absence of waves) from the total velocity spectrum measured in combined waves and currents. If, as discussed above, the interaction of waves and currents produces a modification of both the vorticity distribution and the turbulent structure, the present results may also reflect the uncertainty in the turbulence spectrum arising in combined waves and currents.

VORTICITY EFFECTS 125

(a) Water surface elevation.

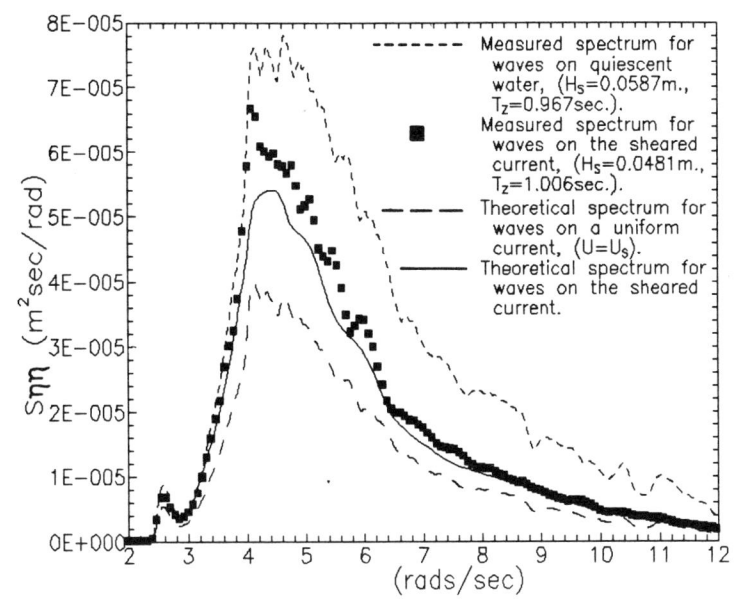

(b) Horizontal kinematics (z=-0.1m).

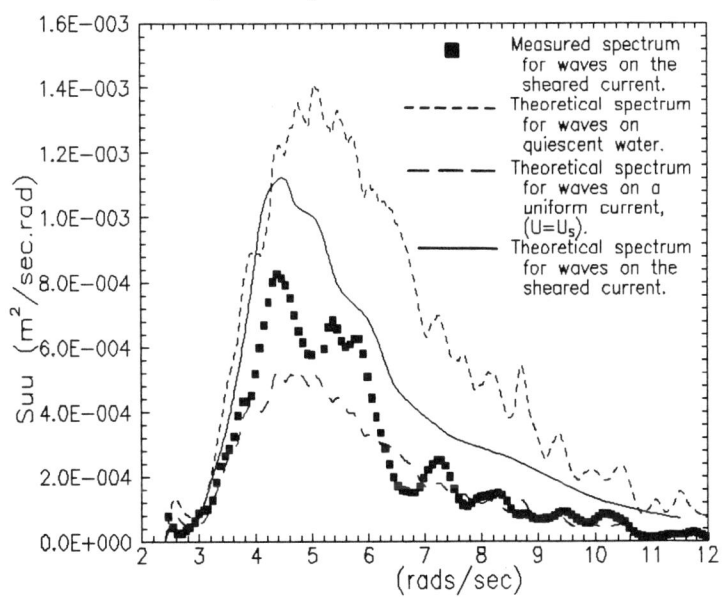

Figures 9a-9b. Interaction of random waves with a "favourable" sheared current.

5. Concluding remarks.

The present paper has considered the interaction of waves and currents, and has presented the results of a new experimental study. This has considered both regular and random waves, and has sought to identify the importance of the time-averaged vorticity distribution. Preliminary measurements concerning the interaction with a uniform current (zero vorticity) confirm that the initial changes in both the wave height and the wave length are consistent with the fifth order solution proposed by Thomas (1990). Furthermore, in the absence of vorticity, the underlying kinematics are in good agreement with the Doppler shifted solution proposed by Fenton (1985).

In contrast, if the current profile is strongly sheared, and in particular if there is significant vorticity at the water surface, the present data suggest that the resulting flow field cannot be predicted by an "equivalent uniform current". However, a multi-layered numerical model, capable of describing an arbitrary current profile, provides a good description of the kinematics beneath a regular wave train. Furthermore, if the solution is coupled with an energy transfer equation the initial change in both the wave height and the wave number can be satisfactorily predicted. Finally, a similar approach was applied to the propagation of random waves on a strongly sheared current. In this case the predicted power spectrum of the water surface elevation was in reasonable agreement with the laboratory data. However, there were important differences between the observed and predicted spectrum of the underlying wave motion. At present, these discrepancies are believed to reflect the wave-induced changes in both the mean current profile and its associated turbulent structure. Until these changes are clarified, the exact nature of these important wave-current interactions will remain indeterminate.

Acknowledgements.

The authors gratefully acknowledge the financial support provided by the Engineering and Physical Sciences Research Council (EPSRC) and The Royal Society.

References.

Chaplin, J.R. (1990) Computation of non-linear waves on a current of arbitrary non-uniform profile. OTH 90 327, HMSO.

Cummins, I. & Swan, C. (1993) Non-linear wave-current interactions. Soc. for Underwater Tech., 29, 35-51.

Dalrymple, R.A. (1974) Water waves on a bi-linear shear current, Proc. 14th. Int. Conf. Coastal Eng., 1, 626-641.

Dalrymple, R.A. & Heideman, J.C. (1989) Non-linear water waves on a vertically sheared current. Proc. I.M.P. Forum. Paris, 1989.

Fenton, J.D. (1985) A fifth-order Stokes theory for steady waves. J. Waterway, Port, Coastal and Ocean Eng., 111, 216-234.

Hedges, T.S., Anastasiou, K. & Gabriel, D. (1985) Interaction of random waves and currents. J. Waterway, Port, Coastal and Ocean Eng. ASCE 111(2), 275-288.

Hedges, T.S. & Lee, B.W. (1992) The equivalent uniform current in wave-current computations. Coastal Eng., 16, 301-311.

Jonsson, I.G., Brink-Kjaer, O. & Thomas, G.P. (1978) Wave action and set-down for waves on a shear current. J. Fluid Mech. 87, 227-250.

Kishida, N. & Sobey, R.J. (1988) Stokes theory for waves on linear shear current, J. Eng. Mech., ASCE, Vol. 114, 1317-1334.

Longuet-Higgins, M.S. & Stewart, R.W. (1960) Changes in the form of short gravity waves on long waves and tidal currents, J. Fluid Mech., 8, 565-583.

Swan, C. (1990) An experimental study of waves on a strongly sheared current profile. Proc. 22nd. Int. Conf. Coastal Eng. Delft (1990). 1, 489-502.

Swan, C. (1992). A stream function solution for waves on a strongly sheared current. Proc. 23rd. Int. Conf. Coastal Eng. Venice (1992). 1, 684-697.

Thomas, G.P. (1990) Wave-current interactions: An experimental and numerical study. Part 2. Non-linear waves. J. Fluid Mech., 216, 505-536.

Tsao, S. (1959) Behaviour of surface waves on a linearly varying flow. Moskow. Fiz.-Tech. Inst. Issl. Mekh. Prikl. Mat., 3, 66-84.

CHAPTER 10

Waves in an Annular Entrance Channel

Robert A. Dalrymple, F. ASCE and James T. Kirby, M. ASCE[1]

Abstract

Waves propagating in a curved channel are examined analytically and with a variety of parabolic and spectral models. The results show that the wide angle parabolic method is reasonably robust, but not exact, while a spectral method based on trigonometric functions is superior to a Chebyshev polynomial method. Angular spectrum models are discussed and shown to be equivalent to an eigenfunction expansion in the cross-channel direction.

Introduction

Waves propagating in curved narrow (with respect to the wave length) channels behave as if the channel were straight. However, when the curved channel is wide, then wave reflection from the outer wall and the diffraction of waves around the channel bend at the inner wall lead to a complicated sea state within the channel.

The prediction of the transmission of waves into harbors with curved entrance channels and the use of annular wave tanks (in small laboratories) depend on our ability to model waves in curved domains. Here waves propagating in a curved channel, taken to be a section of a circular annulus, are studied. An analytical and three different types of numerical solutions are obtained and compared. The analytical solution is used to verify the validity of the numerical techniques developed in a conformally mapped domain, resulting from mapping the curved channel into a straight channel.

An annular channel can be classified as narrow or wide with respect to the incident wave depending on the dimensionless width, defined by Dalrymple, Kirby, and Martin (1995) as $k(r_2 - r_1) = kw$, where r_2 is the outer radius of the channel, r_1 is the inner radius, w is the width of the channel, and k is the wavenumber of the progressive wave train found from the dispersion equation,

$$\sigma^2 = gk \tanh kh$$

---

[1]Center for Applied Coastal Research, Ocean Engineering Laboratory, Univ. Delaware, Newark, DE 19716

where the angular frequency of the wave is $\sigma = 2\pi/T$, the wavenumber $k = 2\pi/L$, h is the water depth, and g the acceleration of gravity. The parameter kw is 2π times the number of waves that fit <u>across</u> the channel width; wide channels can fit numerous waves across their widths.

In acoustics, the problem of sound propagation in curved ducts has been studied by Rostafinski in a variety of papers (see References). The mathematical problem is analogous, although now the computing tools make the problem much simpler.

Analytical Solution

For linear wave theory, with irrotational motion and an incompressible fluid, the governing equation for the velocity potential of the wave motion is the Laplace equation. The Laplace equation is written conveniently for this problem in polar form; therefore, we will take the velocities to be given by

$$u = -\frac{\partial \Phi}{\partial r} \tag{1}$$

$$v = -\frac{1}{r}\frac{\partial \Phi}{\partial \theta} \tag{2}$$

$$w = -\frac{\partial \Phi}{\partial z} \tag{3}$$

and the Laplace equation is:

$$\frac{\partial^2 \Phi}{\partial r^2} + \frac{1}{r}\frac{\partial \Phi}{\partial r} + \frac{1}{r^2}\frac{\partial^2 \Phi}{\partial \theta^2} + \frac{\partial^2 \Phi}{\partial z^2} = 0 \tag{4}$$

Figure 1 shows a layout of an annular channel with a maximum θ value of π, corresponding to the positive y axis.

Taking the depth h as constant, we assume $\Phi(r, \theta, z) = \phi(r, \theta) \cosh k(h + z)$, where the wavenumber k is fixed from the dispersion relationship. Substituting into (4), gives the polar form of the Helmholtz equation:

$$\frac{\partial^2 \phi}{\partial r^2} + \frac{1}{r}\frac{\partial \phi}{\partial r} + \frac{1}{r^2}\frac{\partial^2 \phi}{\partial \theta^2} + k^2 \phi = 0 \tag{5}$$

The boundary conditions for vertical no-flow channel sidewalls are

$$\frac{\partial \phi}{\partial r} = 0, \text{ at } r = r_1, \text{ the inner wall} \tag{6}$$

$$\frac{\partial \phi}{\partial r} = 0, \text{ at } r = r_2, \text{ the outer wall.} \tag{7}$$

The general solution for ϕ that satisfies these boundary conditions is

$$\phi = \sum_{n=0}^{N} a_n \left[Y'_{\gamma_n}(kr_1) J_{\gamma_n}(kr) - J'_{\gamma_n}(kr_1) Y_{\gamma_n}(kr) \right] e^{i\gamma_n \theta} \tag{8}$$

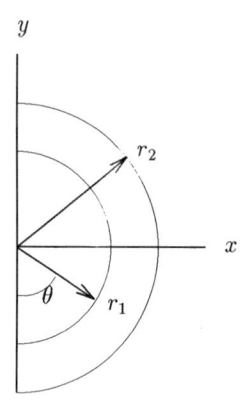

[Figure 1: Schematic Diagram of Circular Channel]

$$\phi = \sum_{n=0}^{N} a_n F_n(r)\, e^{i\gamma_n \theta} \tag{9}$$

where k is the wavenumber of the incident wave, r_1 is the inner radius of the channel, and γ_n must satisfy

$$Y'_{\gamma_n}(kr_1) J'_{\gamma_n}(kr_2) - J'_{\gamma_n}(kr_1) Y'_{\gamma_n}(kr_2) = 0, \ n=1, 2, \ldots, N$$

to enforce a no-flow boundary condition on the outer wall $(r = r_2)$ (see, e.g. Kirby et al., 1994). There are only N real values of γ_n that satisfy this equation, which are ranked in descending order of magnitude. The largest γ_n is less than kr_2.

Each of the terms in the summation (Eq. 9) is a wave mode, propagating in the θ direction with $1/\gamma_n$ waves per 2π radians, and a cross-channel variation given by the radial term $F_n(r)$. These terms are orthogonal to each other with weight $(1/r)$ from the Sturm-Liouville theorem and provide a method for determining the a_n values.

The values of the unknown coefficients a_n in Eq. 9 are found from the initial condition at $\theta = 0$, which is prescribed as a function of r. Here we assume that the wave height is uniform across the mouth of the channel. In actuality, there is an interaction between the channel and the ocean that leads to variations across the channel. These are assumed to be small. For $\phi(r,0) = 1$, we use the expression for the velocity potential (9) and the orthogonality of the r terms to find the a_n.

$$a_n = \frac{\int_{r_1}^{r_2} r^{-1} \phi(r,0) F_n(r)\, dr}{\int_{r_1}^{r_2} r^{-1} F_n^2(r)\, dr} \tag{10}$$

WAVES IN ANNULAR ENTRANCE 131

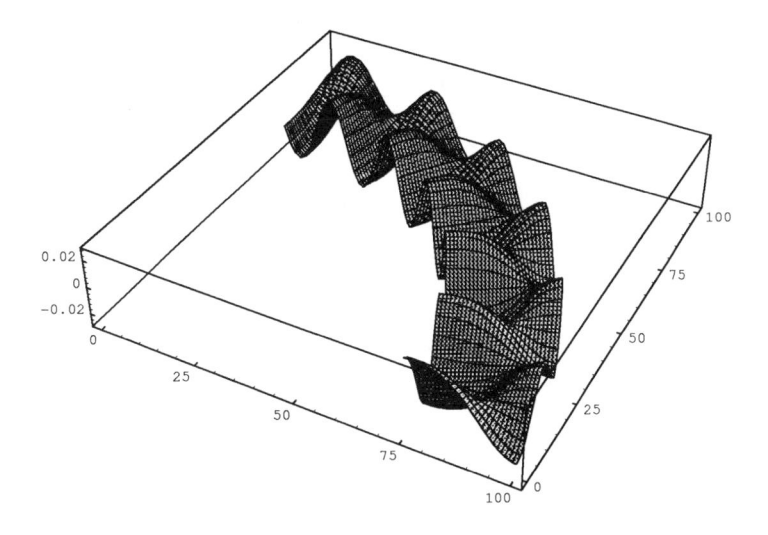

Figure 3: First Mode of Analytical Solution

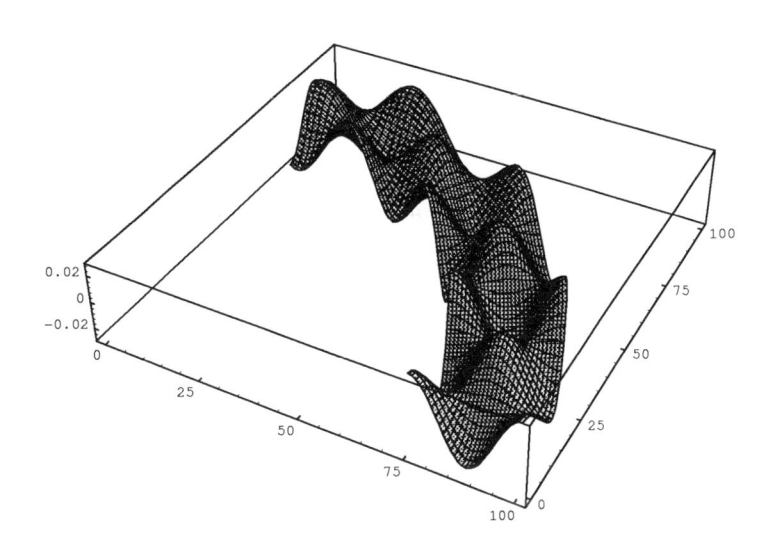

Figure 4: Second and Last Mode for Analytical Solution

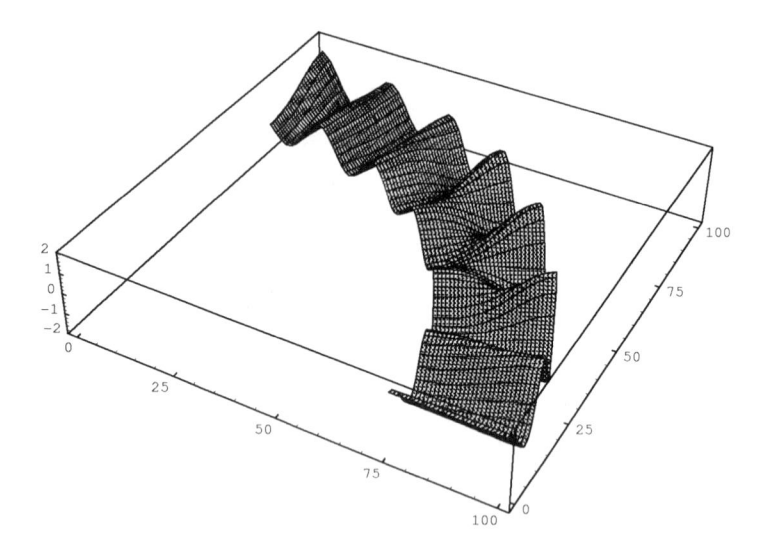

Figure 5: Analytic Solution (Sum of All Three Modes)

seiching problem was studied by Campbell (1953), who had to deal with waves in a circular (ship) testing channel.

Parabolic Modelling

A simple parabolic model can be easily derived from the Helmholtz equation (5). Since the waves will travel in the azimuthal direction (particularly for narrow channels), most of the oscillation in the wave form can be described by a periodic function in θ and we write

$$\phi(r,\theta) = A(r,\theta)\, e^{i\Gamma\theta} \qquad (11)$$

where Γ is a dimensionless constant and A is likely to vary slowly in the θ direction. Substituting into the Helmholtz equation, we find, after neglecting a second derivative of A with respect to θ (assumed to be small).

$$2ikr_0\frac{\partial A}{\partial \theta} + (k^2r^2 - k^2r_0^2)A + r\frac{\partial}{\partial r}\left(r\frac{\partial A}{\partial r}\right) = 0 \qquad (12)$$

The value of Γ was taken as kr_0, where r_0 is a reference radius arbitrarily taken as the mid-point of the channel, $\bar{r}$, or $r_0 = \bar{r} = (r_1 + r_2)/2$. This model will work for narrow channels.

We can obtain more correct values of r_0 by comparing to the γ_1 angular wavenumbers from the analytical solution. By comparing to a number of cases, we find an approximate linear trend (with unit slope) between $(r_0 - \bar{r})/\bar{r}$ with the relative width of the channel $k(r_2 - r_1)$. Therefore for narrow channels, $r_0 = \bar{r}$, but for wide channels, $r_0 = \bar{r}(1 + k(r_2 - r_1))$.

Kirby, Dalrymple, and Kaku (1994) provide a general form for small and large angle parabolic models in arbitrary mapped and conformal domains. Here we need only the mapping for conformal domains.

The constant depth Helmholtz equation with constant coefficients is conformally mapped into a variable coefficient Helmholtz equation:

$$\frac{\partial^2 \phi}{\partial u^2} + \frac{\partial^2 \phi}{\partial v^2} + k^2 J(u,v)\phi = 0 \tag{13}$$

Here, $J(u,v)$ is the Jacobian of the transformation. Assuming that $\phi(u,v)$ varies rapidly in the propagation direction u, we write

$$\phi(u,v) = Re\left\{ A(u,v) e^{i \int k_0 J_0^{1/2} du} \right\} \tag{14}$$

where $k_0 = k(u, v_0)$ and $J_0 = J(u, v_0)$ and v_0 is a fixed reference distance. Substituting into Eq. 13 and neglecting a second derivative of A with respect to u, Kirby et al. find a small angle parabolic model:

$$2ik_0 J_0^{1/2} \frac{\partial A}{\partial u} + i\frac{\partial(K_0 J_0^{1/2})}{\partial u} A + (K^2 J - K_0^2 J_0)A + \frac{\partial^2 A}{\partial v^2} = 0 \tag{15}$$

In wide channels, the turning of the channel leads to large angle discrepancies between the assumed azimuthal propagation direction and the actual wave direction. To allow for these wave directions differing from the azimuthal direction to a greater extent than permitted by the small angle parabolic model, a wide angle model was developed:

$$2ikJ^{1/2}\frac{\partial A}{\partial u} + 2kJ^{1/2}\left(kJ^{1/2} - k_0 J_0^{1/2}\right) A + i\frac{\partial(kJ^{1/2})}{\partial u} A + \left\{\frac{3}{2} - \frac{1}{2}\left(\frac{k_0^2 J_0}{k^2 J}\right)^{1/2}\right\}\frac{\partial^2 A}{\partial v^2}$$
$$- \frac{3i}{4k^2 J}\frac{\partial(kJ^{1/2})}{\partial u}\frac{\partial^2 A}{\partial v^2} + \frac{i}{2kJ^{1/2}}\frac{\partial^3 A}{\partial u v v} = 0 \tag{16}$$

Kirby et al. use a conformal mapping to convert a circular channel into a straight channel in the (u,v) domain. One such map is $w = u + iv = \pi/2 - i\ln(z/r_m)$, where $r_m = \sqrt{r_1 r_2}$. For this case, the $kJ_0^{1/2}$ is not a function of u and the equations simplify. For the small angle parabolic models, it is easily shown that Eq. 12 and 15 are equivalent. They compared their models to the exact solution for a wide angle case; the simple parabolic began to fail 60° from the channel mouth, while the wide angle model compared reasonably well for the full 180°test channel. They also went further, adding nonlinear effects to the models.

Spectral Modelling

Fourier-Galerkin Modelling

Angular spectrum modelling has been used to model waves propagating over sloping and irregular bathymetry in regions which are bounded by a rectangular box (cf. the review by Dalrymple and Kirby, 1992). The angular spectrum model was defined by Booker and Clemmow (1950) in a Cartesian coordinate system as the decomposition of the initial condition into progressive plane waves. The subsequent wave field is found by summing the plane waves within the computational domain. In an (x,y) coordinate system, the angular spectrum corresponds to a Fourier-Galerkin spectral method of solution, as the plane waves in this system are described by trigonometric functions. In the polar coordinate system, the angular spectrum model corresponds to a modal decomposition, based on the analytical modes shown above. Therefore the angular spectrum model in any coordinate system corresponds to an eigenfunction expansion for that coordinate system.

The Fourier-Galerkin model (using trigonometric functions in the cross-channel (radial) direction) is a natural extension of the angular spectrum model in Cartesian coordinates (e.g., Dalrymple et al., 1989): it is based on the Fourier transform of the wave field in the lateral direction. In the Fourier domain, the wave equations are split into forward and backward propagating waves. Only the forward waves are kept as back-reflection is assumed to be negligible.

Dalrymple, Kirby and Martin (1995) have used Fourier-Galerkin modelling to examine waves in conformally mapped channels. In the (u,v) conformal domain, the Fourier transform pair is:

$$f_n(u) = \mathcal{T}_F[\phi(u,v)] = \frac{1}{2v_b}\int_{-v_b}^{v_b} \phi(u,v)\cos[\lambda_n(v+v_b)]\,dv, \quad n=0,1,2,\ldots \quad (17)$$

$$\phi(u,v) = \mathcal{T}_F^{-1}[f(u)] = \sum_{n=0}^{\infty} \epsilon_n f_n(u)\cos[\lambda_n(v+v_b)] \quad \text{for } -v_b < v < v_b. \quad (18)$$

where $v_b = \frac{1}{2}\ln(r_2/r_1)$, which is half the channel width in the conformal domain.

Defining

$$\overline{k^2 J}(u) = \frac{1}{2v_b}\int_{-v_b}^{v_b} k^2 J(u,v)\,dv. \quad (19)$$

and substituting this into the governing Helmholtz equation gives

$$\frac{\partial^2 \phi}{\partial u^2} + \frac{\partial^2 \phi}{\partial v^2} + \overline{k^2 J}\,(1-\nu)\phi = 0, \quad (20)$$

where

$$\nu(u,v) = 1 - k^2 J/\overline{k^2 J} \quad (21)$$

Fourier transforming the above Helmholtz equation yields

$$\frac{d^2 f_n}{du^2} + \gamma_n^2 f_n - \overline{k^2 J}\, T_F[\nu\, T_F^{-1}[f]] = 0, \qquad n = 0, 1, 2, \ldots, \qquad (22)$$

where

$$\gamma_n^2(u) = \overline{k^2 J} - \lambda_n^2 \quad \text{and} \quad \lambda_n = \tfrac{1}{2} n\pi/v_b.$$

Splitting the problem into forward and backward propagating modes, they find the following equation for the forward propagating modal amplitudes, f_n, in the Fourier-transformed conformal domain:

$$\frac{df_n(u)}{du} = i\gamma_n f_n - \frac{i\,\overline{k^2 J}}{2\gamma_n} T_F[\nu\, T_F^{-1}[f]], \qquad n = 0, 1, 2, \ldots \qquad (23)$$

where

$$\overline{k^2 J} = \frac{k^2 (r_2^2 - r_1^2)}{2 \ln(r_2/r_1)} \quad \text{and} \quad \nu(v) = 1 - \frac{2 \ln(r_2/r_1)\, r_1 r_2 e^{-2v}}{r_2^2 - r_1^2},$$

This set of first order ordinary differential equations is solved numerically.

After the conformal mapping, the cosines used in the Fourier transform are no longer an optimal basis, as the forms of the lateral eigenfunctions $F_n(r)$ are far more complicated, as shown in Eq. 9. The errors increase with the width of the channel. This has implications also for application of the angular spectrum model for open coast cases where the bathymetry is varying significantly in the longshore direction.

Chebyshev-tau Modelling

Chebyshev polynomials provide another set of orthogonal bases with which to expand the wave potential across the channel. Panchang and Kopriva (1989) have used these polynomials for the solution of the mild-slope equation. Here a cross-channel Chebyshev transform is used to develop the governing equation in the transform domain, which must be scaled to be in the range of -1 to 1, so we define $\zeta = v/v_b$, so that the lateral boundaries are located at $\zeta = \pm 1$. The appropriate Chebyshev-transform pair is:

$$c_n(u) = T_c[\phi(u,v)] = \frac{\epsilon_n}{\pi} \int_{-1}^{1} \frac{\phi(u, v_b \zeta)\, T_n(\zeta)}{\sqrt{1-\zeta^2}}\, d\zeta, \qquad n = 0, 1, 2, \ldots, \qquad (24)$$

$$\phi(u,v) = T_c^{-1}[c(u)] = \sum_{n=0}^{\infty} c_n(u)\, T_n(\zeta) \qquad \text{for } -1 < \zeta < 1. \qquad (25)$$

As the Chebyshev polynomials do not satisfy the lateral boundary conditions, the tau method, which forces the Chebyshev sum to satisfy these conditions, Is used.

Introducing $\overline{k^2 J}$ and ν, defined by (19) and (21), respectively, we find that the Chebyshev transform of (20) is

$$\frac{d^2 c_n(u)}{du^2} + \overline{k^2 J}\, c_n(u) + \frac{1}{v_b^2} c_n^{(2)} - \overline{k^2 J} \mathcal{T}_c[\nu\, \mathcal{T}_c^{-1}[c]] = 0.$$

The splitting in the transform domain yields:

$$\frac{dc_n^+}{du} = i\gamma_0 c_n^+(u) + \frac{i}{2\gamma_0}\left(\frac{1}{v_b^2} c_n^{(2)} - \overline{k^2 J}\, \mathcal{T}_c[\nu\, \mathcal{T}_c^{-1}[c^+]]\right) = 0. \qquad (26)$$

There are significant disadvantages of the Chebyshev approach. First all modes are progressive, while for the Fourier-Galerkin method, only those modes for which γ_n are positive propagate (thus reducing the number of simultaneous differential equations to solve). Also, the splitting of the Chebyshev transformed equation introduces an error, which does not occur with the Fourier-Galerkin model; it in fact reduces the Chebyshev method to an equivalent of the small angle parabolic model (Dalrymple et al., 1995).

Results

For narrow channels, all of the methods, parabolic and spectral, work well. As the channel width kw increases, then the errors begin increase for all the methods. As an example of these errors, we increase the width in the previous example to $w = 125$ m ($r_1 = 75$ m, $r_2 = 200$ m); now $kw = 37.625$, or almost 6 wavelengths can fit across the channel.

The exact solution, given by the analytical model, is shown in Figure 6. The waves entering the channel from the right, begin to diffract around the channel bend and to reflect from the outer radius. At about 120°, the wave field is dominated by reflected waves from the outer side wall. Finally at 180°, the wave field is reasonable complex, indicating that a 180°bend in a laboratory wave channel will not result in planar waves.

The surface elevations along the outer wall ($r = r_2$) predicted by the exact solution and the small and wide angle parabolic models are shown in Figure 7. The wide angle parabolic model does a reasonable job for the 180°length of channel shown, while the small angle begins to fail at 50°. Note that the phase of the waves is well predicted by the parabolic model.

For the Fourier-Galerkin model, a comparison to the exact solution for water surface elevations at the outer radius is shown in Figure 8 for 90°, while the Chebyshev result is shown in Figure 9. Both have problems with the phase speed of the wave as evidenced by the mis-matching of the wave crests. The Fourier-Galerkin results have the biggest discrepancy about 52°, but the Chebyshev model fairs worse, with mismatches over the entire sector from 50-90 °.

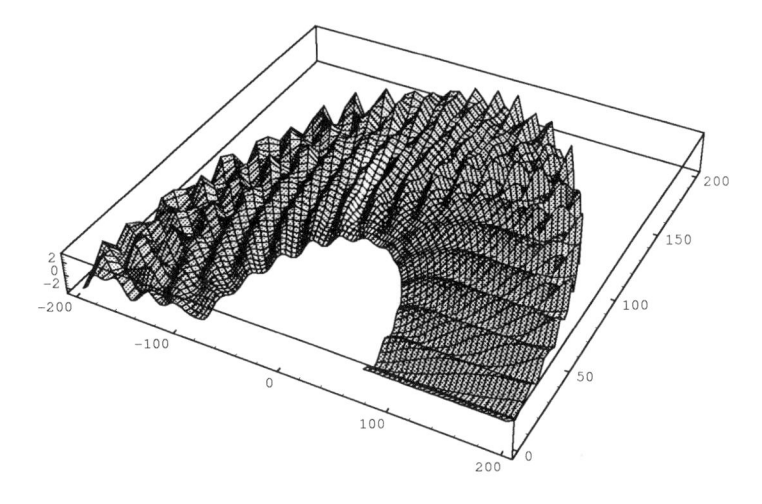

Figure 6: Exact Solution for Waves in a Wide Circular Channel

Conclusions

Waves propagating in curved channels are shown for constant depth channels that are annular in planform. For a given wave height at the mouth of the channel, the wave field within the channel is predicted by an analytic solution and parabolic and spectral (Fourier-Galerkin and Chebyshev-tau) methods.

The analytic solution, by separation of variables, shows the modal structure of the wave field. The lowest mode, with no zero crossings across the channel, combines with higher order modes to form the total wave field. The wider the channel the more modes are present with the largest mode having an angular wavenumber γ_1 less than kr_2. Higher modes have smaller values of angular wavenumber.

All of the numerical models work well for narrow channels (say $kw < 8$) and become more inaccurate as the dimensionless width of the channel increases, with the wide-angle parabolic model doing better than the Fourier-Galerkin model, which out-performed the Chebyshev-tau model for the wide channel case ($kw = 37.6$).

Acknowledgments

Both authors were supported in part by the NOAA Office of Sea Grant,

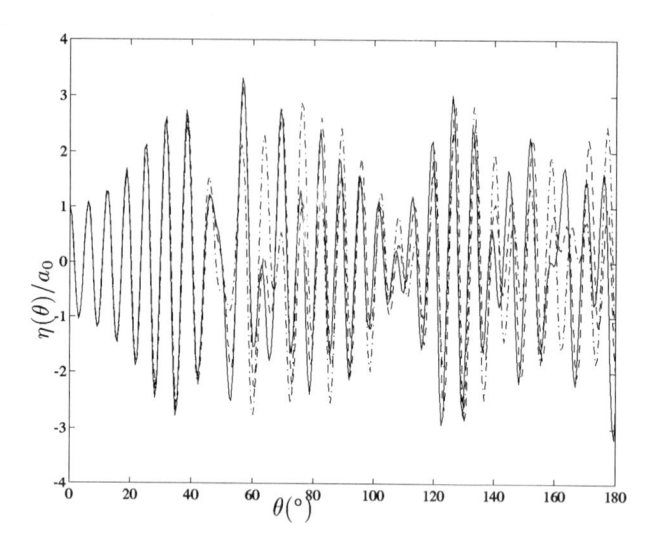

Figure 7: Parabolic model Results. Analytical solution is the solid line; dashed line, wide angle parabolic model and the dashed dot line is the small angle parabolic model.

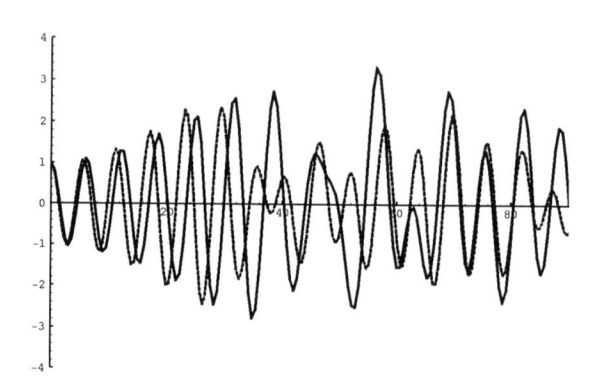

Figure 8: Comparison of the Water Surface Variation Along Outer Wall Between the Exact Solution (solid line) and the Fourier-Galerkin Model (dashed line)

As an example, Figures 4, 4, 2 show the only three wave modes that comprise the total solution shown in Figure 5. The first mode ($n = 0$) has no zero crossing across the channel and the wave action is confined to the outer channel wall. This is the equivalent "whispering gallery" mode. The next mode ($n = 1$) has one zero crossing and a longer angular wave length (defined as $2\pi/\gamma_2$ radians). The last mode has two zero crossings across the channel and an even longer angular wave length. For this case, the channel radii are $r_1 = 75$ m, $r_2 = 100$ m. The channel depth is 4 m and the wave period is 4 s. The corresponding wavenumber k is 0.301 m^{-1}, and therefore the dimensionless channel width is $kw = 7.53$, or 1.2 wave lengths fit across the channel. It is neither a narrow nor a very wide channel. The three values of γ_n are 27.6676, 22.8667, 14.2745 from the gravest to the highest mode.

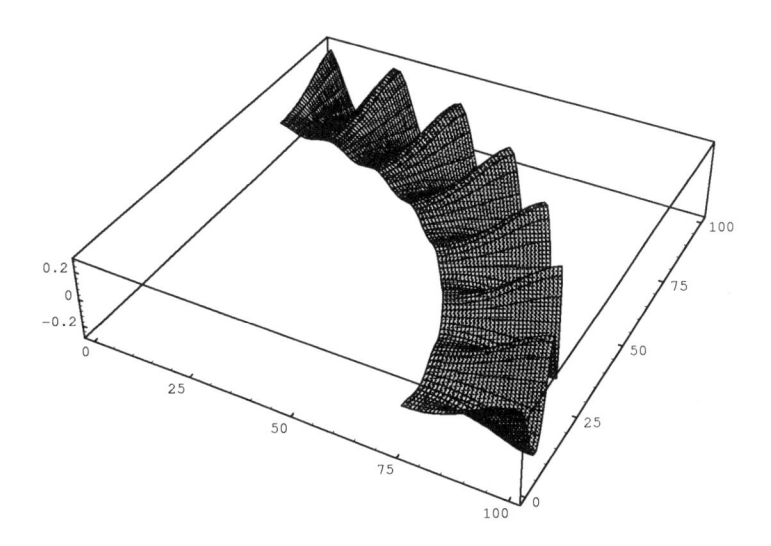

Figure 2: Zeroth (Whispering Gallery) Mode for Analytical Solution

One interesting phenomena of wide channels is the 'loss' of waves within the channel. Since the length of the outer circumference is greater than the inner, there are more waves around the outer circumference than the inner, which means that waves become short-crested across the channel and singular points in the wave phase occur, similar to that discussed by Nye and Berry (1974), Radder (1992), and Dalrymple and Martin (1994).

This problem of progressive waves entering the channel with a given wavenumber k is different than the seiching problem in an annular channel where an integer number of waves fit along the centerline of the circular channel. The

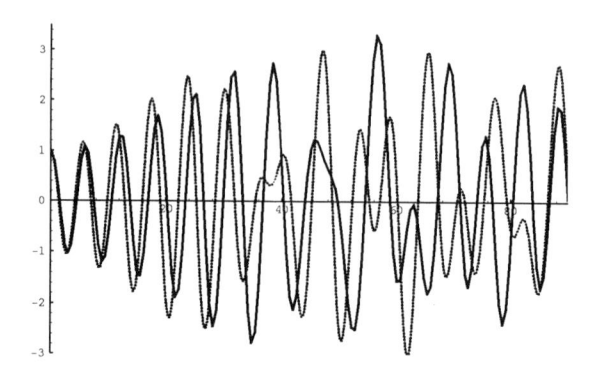

Figure 9: Comparison of the Water Surface Variation Along Outer Wall Between the Exact Solution (solid line) and the Chebyshev-tau Model (dashed line)

Department of Commerce, under Award No. NA 16 RG 0162 (University of Delaware). The U.S. Government is authorized to produce and distribute reprints for government purposes notwithstanding any copyright notation that may appear herein.

References

Booker, H.G. and Clemmow, P.C., "The Concept of an Angular Spectrum of Plane Waves, and its Relation to that of Polar Diagram and Aperture Distribution," *Proc. Inst. Electr. Engineering*, 3, 97, 11, 1950.

Campbell, I.J., "Wave Motion in an Annular Tank," *Phil. Magazine*, Series 7, 44, 355, 845-853.

Dalrymple, R.A. and J.T. Kirby, "Angular Spectrum Modelling of Water Waves," *Reviews in Aquatic Sciences*, CRC Press, 6, Iss. 5 and 6, 383-404, 1992.

Dalrymple, R.A., K.D. Suh, J.T. Kirby, and J.W. Chae, "Models for Very Wide Angle Water Waves and Wave Diffraction. Part 2. Irregular Bathymetry," *J. Fluid Mechanics*, 201, 299-322, 1989.

Dalrymple, R.A., J.T. Kirby and P.A. Martin, "Spectral Methods for Forward Propagating Water Waves in Conformally-mapped Channels," *Applied Ocean Research*, in press.

Kirby, J.T., R.A. Dalrymple, and H. Kaku, "Parabolic Approximations for Water Waves in Conformal Coordinate Systems,"

Coastal Engineering, 23, 185-213, 1994.

Nye, J. F. and M. V. Berry, "Dislocations in wave trains," *Proc. Roy. Soc. London,* A, 336, 165–190, 1974.

Martin, P.A. and R.A. Dalrymple"On Amphidromic Points," *Proc. Roy. Soc. London, A*, 444, 91–104, 1994.

Panchang, V.G. and D. A. Kopriva, "Solution of two-dimensional water-wave propagation problems by Chebyshev collocation," *Mathl. Comput. Modelling*, 12, 625–640, 1989.

Radder, A. C., "Efficient elliptic solvers for the mild-slope equation using the multigrid technique, by B. Li and K. Anastasiou: comments" *Coastal Engineering*, **18**, 347–350, 1992.

Rostafinski, W., "On Propagation of Long Waves in Curved Ducts," *J. Acoustical Soc. of America,"* 52, 5, Pt. 2, 1411-1420, 1972.

Rostafinski, W., "Analysis of Propagation of Waves of Acoustic Frequencies in Curved Ducts," *J. Acoustical Soc. of America,* 56, 1, 11-15, 1974.

Rostafinski, W., "Transmission of Energy in Curved Ducts," *J. Acoustical Soc. of America,"* 56, 3, 1005-1007, 1974.

Rostafinski, W., "Acoustic Systems Containing Curved Duct Sections," *J. Acoustical Soc. of America,"* 60, 1, 23-28, 1976.

CHAPTER 11

WAVE DAMPING BY KELP VEGETATION

[1]Alfonse Dubi and [2]Alf Tørum

1. INTRODUCTION

Aquatic vegetation, like seagrasses, macroalgae and trees whether submerged or subaerial are an important feature of a coastal ecosystem. In addition to the structural and functional aspects to the environment, they are known to reduce wave and current energies propagating through them. The reduction of energy would then influence sediment motion and thus render an impact on coastal sediment transport. The dissipative character of large stands of kelp has been studied for instance by Jackson and Winant, (1983), Dalrymple et al. (1984) and for artificial seaweed as material for shore protection Price et al. (1968).

Kelp is a macroalga which grows on hard rock and stone and extracts all of its nutrients from the water column. The plant consists of a root-like holdfast organ, a stipe and a frond (Fig.1). The general properties of a fully grown (4-8 years) kelp are summarized as follows:
Length of stipe: 1-2 meters. Fronds have the same length as the stipe.
Specific gravity: 1.18 kg/cu. m; Biomass: 10-30 kg/sq. m
Growth density: 10-15 per sq. meter of horizontal area

---

[1]Dept. of Structural Engineering, The Norwegian Inst. of Technology, (NTH), N-7034 Trondheim, Norway

[2]Dept. of Structural Engineering, NTH/SINTEF-NHL, N-7034 Trondheim, Norway

Specimen found at water depths 2-20 metes.

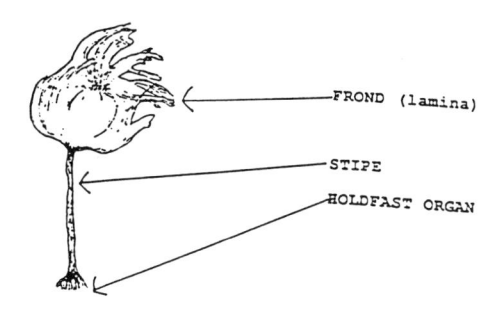

Figure 1: Kelp *Laminaria hyperborea*

Kelp is harvested at several places along the Norwegian coast and is used as a raw material for the manufacturing of various chemicals. In some areas the harvesting of the seaweed has become a controversial issue in which there is suspicion among coastal zone managers that the harvesting results in beach erosion. This study is actually a consequence of the controversy.

Basing on the most recent work by Asano et al. (1992), a new analysis is developed for the flow model and the vegetation motion using field and experimental results carried out on kelp fronds and kelp plant models. The theoretical model is compared with experimental results. The influence of kelp vegetation on beach erosion is not included in this paper because of space limitation.

2. BASIC FORMULATION FOR THE FLOW MODEL

Let us consider small amplitude waves propagating in the x- direction in water of depth h above submerged vegetation of mean height d. We employ cartesian coordinates (x,z) fixed on the mean free surface, z=0, where z is positive upwards (see Fig.2). The surface displacement at the free surface is given by $\eta_1 = a_0 e^{i(kx-\omega t)}$ and displacement at the interface is $\eta_2 = b_0 e^{i(kx-\omega t)}$. Let us assume flat bottom, potential flow in the water layer, frictional flow in the vegetation zone. At the interface the viscous shear stresses and the corresponding layer, δ, are initially neglected. The bottom shear stress is considered to be negligible in comparison with the frictional resistance of the vegetation. Further, let us assume known a priori the wave amplitude a_0, the angular frequency $\omega = 2\pi/T$ both of which are real and positive. k is a wave number and T is the wave period.

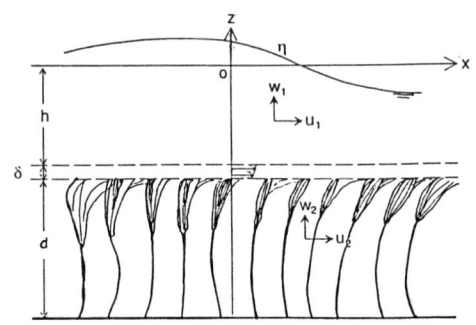

Figure 2: Definition sketch of the flow model

The equations of motion employed are the linearized momentum equations for the water and vegetation zones. For a unit volume

$$\frac{\partial U_1}{\partial t} = -\frac{1}{\rho}\nabla P_1 \tag{1}$$

$$\frac{\partial U_2}{\partial t} = -\frac{1}{\rho}\nabla P_2 - F \tag{2}$$

and the equation of continuity

$$\nabla U = 0 \tag{3}$$

where the subscripts 1 and 2 denote the water and the vegetation zones respectively; t = time; $U = (u,w)$ = water particle velocity vector; ρ is the fluid density and P is the dynamic pressure. $F = (Fx, Fz)$ = force vector acting on vegetation given as

$$Fx = \frac{\rho}{2} N [C_{Dx} A]_e |u_2 - \dot{\xi}| (u_2 - \dot{\xi}) + \rho N |C_{ax} V|_e (\dot{u}_2 - \ddot{\xi}) \tag{4}$$

and

$$Fz = \frac{\rho}{2} N [C_{Dz} A]_e |w_2| w_2 + \rho N [C_{az} V]_e \dot{w}_2 \tag{5}$$

where N is the number of vegetation per unit horizontal area, C_{Dx} and C_{Dz} are drag force coefficients in the x- and z- directions respectively; C_a is the added mass coefficient, u_2 and w_2 are the horizontal and vertical velocities of the fluid particles. ξ is the horizontal displacement of the vegetation stipe with the dot denoting the derivative with respect to time. The subscript e indicates the equivalent value taking into account of both the stipe and the frond. A and V are the total projected area and volume of the plants in this unit volume. More details on the equivalent values are given in the section on the solution for the vegetation motion.

Let us assume that the particle velocities and the dynamic pressure are sinusoidal such that

$$U = U(z)\, e^{i(kx-\omega t)} \tag{6}$$

and

$$P = P(z)\, e^{i(kx-\omega t)} \tag{7}$$

where $i^2 = -1$.

Substituting equations (4), (5), (6) and (7) into equations (1) and (2) we get a new set of equations. For the upper layer we have

$$\frac{\partial u_1}{\partial t} = -\frac{1}{\rho}\frac{\partial P_1}{\partial x} \tag{8}$$

$$\frac{\partial w_1}{\partial t} = -\frac{1}{\rho}\frac{\partial P_1}{\partial z} \tag{9}$$

and for the lower layer we have

$$\frac{\partial P_2}{\partial x} = -\rho\, \omega\, f_x\, u_2 \tag{10}$$

$$\frac{\partial P_2}{\partial z} = -\rho\, \omega\, f_z\, w_2 \tag{11}$$

where fx and fz are the horizontal and vertical force components due to the presence of vegetation and expressed by

$$f_x = f_{Dx} + i\, f_{Ix} \tag{12}$$

and

$$f_z = f_{Dz} + if_{Iz} \tag{13}$$

in which the horizontal and vertical drag force terms are defined by

$$f_{Dx} = \frac{1}{2} [C_{Dx}A]_e \left|1 - \frac{\dot{\xi}}{u_2}\right| \left(1 - \frac{\dot{\xi}}{u_2}\right) |u_2|/\omega \tag{14}$$

and

$$f_{Dz} = \frac{1}{2} [C_{Dz}A]_e |w_2|/\omega \tag{15}$$

The inertial force terms are expressed by

$$f_{Ix} = [C_{mx}V]_e \left|1 - \frac{\dot{\xi}}{u_2}\right| \tag{16}$$

Here $C_{mx} = 1 + C_{ax}$

$$f_{Iz} = -[C_{mz}V]_e \tag{17}$$

We impose the following linearized boundary conditions at the free surface, interface and bottom boundaries on the momentum equations (8), (9) (10) and (11):

$$\eta(x,t) = \frac{1}{g}\frac{\partial \phi_1}{\partial t} \quad at \quad z = 0 \tag{18}$$

$$\frac{\partial^2 \phi_1}{\partial t^2} + g\frac{\partial \phi}{\partial z} = 0 \quad at \quad z = 0 \tag{19}$$

$$\rho \frac{\partial \phi_1}{\partial t} = P_1 = P_2 \quad at \quad z = -h \tag{20}$$

$$-\frac{\partial \phi_1}{\partial z} = w_1 = -\frac{1}{\rho \omega f_z}\frac{\partial P_2}{\partial z} = w_2 \quad at \quad z = -h \tag{21}$$

$$-\frac{1}{\rho \omega f_z}\frac{\partial P_2}{\partial z} = 0 \quad at \quad z = -(h+d) \tag{22}$$

$$\frac{\partial P_2}{\partial z} = 0 \quad at \quad z = -(h+d) \tag{23}$$

where $\eta = a_0 e^{i(kx-\omega t)}$ is the free surface elevation at SWL, a_0 is the wave amplitude at the origin, g is the acceleration due to gravity. k is complex = $k_r + k_i$ in the subscripts r and i denote real and imaginary values. Substituting k into the surface elevation, the local wave amplitude is found to decay exponentially as $a = a_0 \exp(-k_i x)$.

The solutions for the flow model can be shown to be

$$\Phi_1 = i\frac{C}{\omega}[\cosh(\alpha kd)\cosh(k(h+z)) - \frac{i}{\alpha fx}\sinh(\alpha kd)\sinh(k(h+z))]\exp[i(kx-\omega t)] \tag{24}$$

$$P_2 = \rho C \cosh(\alpha k(h+d+z))\exp[i(kx-\omega t)] \tag{25}$$

where

$$C = \frac{ga_0}{\cosh(\alpha kd)\cosh(kh)[1 - \frac{i}{\alpha fx}\tanh(\alpha kd)\tanh(kh)]} \tag{26}$$

and

$$\alpha = \sqrt{\frac{|fz|}{|fx|}} \leq 1 \tag{27}$$

is the force ratio. In the upper layer a velocity potential Φ_1 exists which satisfy Laplace equation

$$\nabla^2 \Phi = 0 \tag{28}$$

where the particle velocities are expressed as

$$u_1 = -\frac{\partial \phi_1}{\partial x}; \quad w_1 = -\frac{\partial \phi_1}{\partial z} \tag{29}$$

In the vegetation zone, however, the particle velocities can be obtained by substituting equation (25) into equations (10) and (11) to give

$$u_2 = -i \frac{Ck}{\omega fx} \cosh(\alpha k(h+d+z)) \exp[i(kx-\omega t)] \tag{30}$$

$$w_2 = -\frac{Ck}{\omega fx} \sinh\alpha k(h+d+z) \exp[i(kx-\omega t)] \tag{31}$$

We remark that the horizontal and vertical wave numbers in the upper layer are the same when the fluid is inviscid and homogeneous. In the vegetation zone they are different due to the different horizontal and vertical resistance forces. At the interface, the horizontal particle velocities are discontinuous, i.e $u_1 \neq u_2$, thus a shear stress is present which is accounted for by a boundary-layer type of solution.

Finally, from the combined kinematic and dynamic free surface boundary conditions we derive the dispersion relationship given by

$$\omega^2 = gk \frac{\tanh kh - \frac{i}{\alpha fx} \tanh \alpha kd}{1 - \frac{i}{\alpha fx} \tanh \alpha kd \tanh kh} \tag{32}$$

For given α, ω, h, d and fx the unknown complex wave number k can be found by solving equation (32) by iteration. This is done in the coming section on calculated results.

3. SOLUTION FOR THE VEGETATION MOTION

In order to solve the flow field described above we need the knowledge of the kelp motion. The basic approach is the Morison equation in which the forces resisting the fluid flow are the sum of the drag and inertial forces. Following Asano et al. (1992), the motion of the vegetation is regarded as a forced vibration wih one degree of freedom. Let the horizontal displacement of a single kelp plant be denoted by ξ while the differentiation with respect to time t be denoted by the over dot. For a unit length of kelp, the equation of motion is given by

$$m_0\ddot{\xi} + c_1\dot{\xi} + k_0\xi = \frac{1}{2}\rho C_{Dx}A|u_2-\dot{\xi}|(u_2-\dot{\xi}) + \rho V\dot{u}_2 + \rho VC_{ax}(\ddot{u}_2-\ddot{\xi})$$ (33)

where m_0 = mass of kelp per unit length, c_1 = structural damping, k_0 = spring constant, C_{Dx} and C_{ax} are drag and added mass coefficients respectively, A is the projected area and V is the volume per unit length of kelp. Neglecting the structural damping on the assumption that it is small compared to the frictional forces and rearranging we get

$$m\ddot{\xi} + \frac{1}{2}\rho C_{Dx}A|u_2-\dot{\xi}|\dot{\xi} + k_0\xi = \frac{1}{2}\rho C_{Dx}|u_2-\dot{\xi}|u_2 + \rho(1+C_{ax})\ddot{u}_2$$ (34)

The general solution of equation (34) requires iterations involving volumetric integration of unknown variables. For our particular case of kelp we shall simplify the equation of motion before attempting to solve the coupled system. The kelp plant consists of a stipe and a frond which together make a total height d. The stipe can be represented as a slender vertical cylinder of height $d-l_k$ with uniformly distributed mass and the frond is taken as a concentrated mass at the top of the stipe. Here l_k is the half length of the frond. Integrating over depth gives the equation that represents the integrated effect over the water column by the motion at the top of the stipe. Now the equation of motion (34) becomes

$$m_e\ddot{\lambda} + \frac{1}{2}\rho[C_{Dx}A]_e|u_\lambda-\dot{\lambda}|\dot{\lambda} + k_0 = \frac{1}{2}[C_{Dx}A]_e|u_\lambda-\dot{\lambda}|u_\lambda + \rho[C_{mx}V]_e\dot{u}_\lambda$$ (35)

where we have assumed that the velocities of both the stipe and fluid can be treated as varying linearly from the bottom to the top of the stipe, that is

$$\dot{\xi} = \frac{h+d+z}{d-l_k}\dot{\lambda} \; ; \quad u_2 = \frac{h+d+z}{d-l_k}u_\lambda$$ (36)

where λ refers to the level at the top of the stipe and the subscript e stands for equivalent values for the stipe and frond such that

$$m_e = \frac{1}{2}m_0 + m_f + \rho(\frac{1}{2}C_{as}V_s + C_{af}V_f)$$ (37)

Here subscripts s and f denote the stipe and frond properties respectively. The mass per unit length, m_0, of the stipe and its equivalent added mass is weighted by ½ to imply the conversion of the entire kelp motion to the top. This is valid also for the drag and inertial force coefficients which are given by

$$[C_{Dx}A]_e = \frac{1}{2}C_{Ds}A_s + C_{Df}A_f$$ (38)

$$[C_m V]_e = \frac{1}{2}(1+C_{as})V_s + (1+C_{af})V_f \tag{39}$$

Prior to linearization of equation (35) we need to establish the relationship between the drag forces and the flow velocity. One complication with kelp is that the projected area of the frond varies in a flow field. When the velocity (or the relative velocity as the case may be) is zero, the kelp will assume an upright position and the projected area is largest. As the velocity increases, the plant tends to bend over and the fronds tend to streamline in the direction of the flow thereby reducing the projected area. A field experiment has been carried out by the authors in collaboration with prof. Martin Mork and dr. scient. student Kjersti Sjøtun of the University of Bergen (UiB) using a research vessel " Hans Brattstrøm" belonging to UiB. Drag forces were measured on 7 fronds of different sizes of the Norwegian kelp by ship towing. In the laboratory, the forces have been measured using a shear plate on which 95 model kelp plants were fixed. Details of the procedure follow in the section on experimental set-up. Results from the two experiments as shown in Fig.3 show that the drag force does not follow the normal quadratic relationship with velocity, instead, the force is linearly proportional to the velocity.

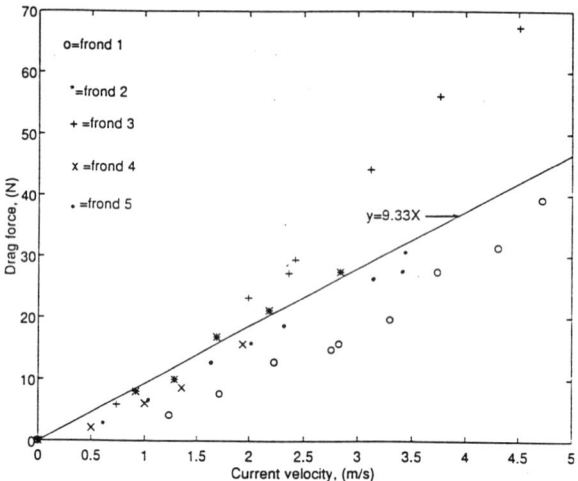

Figure 3: Variation of the drag force on kelp fronds with current velocity

Basing on the two experiments, the following general relationship is proposed between the projected area and the flow velocity:

$$\frac{\rho}{2}[C_{Dx}A]_e = F_\lambda |u_\lambda - \dot\lambda|^{-m} = constant \tag{40}$$

where F_λ = equivalent drag force coefficient evaluated at elevation λ, or at

$z = -(h + l_k)$. Substituting equation (40) into equation (35) gives

$$m_e \ddot{\lambda} + N_D \dot{\lambda} + k_0 \lambda = N_D u_\lambda + N_I \dot{u}_\lambda \qquad (41)$$

where

$$N_D = F_\lambda \left| u_\lambda \left(1 - \frac{\dot{\lambda}}{u_\lambda}\right) \right|^{1-m} \qquad (42)$$

Both F_λ and m are empirically determined constants. The experimental results shown in figure 3, suggest that m=1 and F_λ=9.3 and 30.9 for currents and waves (fig. 7) respectively. With this information we can now proceed with the linearization of equation (35) to get an analytical solution.

Assuming small amplitudes for the vegetation and fluid particle motion and that the particle velocity amplitude is much greater than the maximum velocity amplitude of vegetation, equation (35) now becomes

$$m_e \ddot{\lambda} + \frac{1}{2}\rho[C_{Dx}A]_e |u_\lambda| \dot{\lambda} + k_0 \lambda = \frac{1}{2}\rho[C_{Dx}A]_e |u_\lambda| u_\lambda + \rho[C_{mx}V]_e \dot{u}_\lambda \qquad (43)$$

which represents a linear system provided k_0 is also constant. The solution to this equation gives a ratio known as the velocity amplification factor

$$A_m = \frac{\dot{\lambda}}{u_2|_{z=-(h+l_k)}} = \frac{1 - i\omega \frac{N_I}{N_D}}{1 + i\omega\left(\frac{\omega_n^2}{\omega^2} - 1\right)\frac{m_e}{N_D}} \qquad (44)$$

where $\omega_n = \sqrt{(k_0/m_e)}$ in which k_0 is the spring constant which was determined experimentally to be 20 N/m for deflections up to 55 cm at the top of the plant. From equation (44) we can derive the quantity

$$|1 - A_m| = \frac{\omega\left[\left(\frac{\omega_n^2}{\omega^2} - 1\right) + N_I\right]}{\sqrt{N_D^2 + \omega^2\left(\frac{\omega_n^2}{\omega^2} - 1\right)^2}} \qquad (45)$$

The linearized damping force coefficient used in the solution for the flow model can be established by applying the principle of equivalent work which the energy dissipation of the actual system to that of a linear system (Wang and Tørum,1994). The time averaged work done by the actual system per unit surface area is given by

$$W_a = \alpha_k \, N \, F_\lambda (1 - \frac{\lambda}{u_\lambda}) \overline{u_\lambda^2} \tag{46}$$

where N is the number of plants per unit surface area, α_k is a force reduction factor due to group effect and the over bar denotes the time averaging. For the linearized flow system

$$W_L = \rho f_{Lx} \overline{\int_{-(h+d)}^{-(h+l_k)} u_2^2 \, dz} \tag{47}$$

Equating equation (46) and (47) gives

$$f_{Lx} = \frac{\alpha_k \, N \, F_\lambda (1 - A_m) \overline{u_\lambda^2}}{\rho \overline{\int_{-(h+d)}^{-(h+l_k)} u_2^2 \, dz}} \tag{48}$$

Substituting equation (30) into equation (48) gives

$$f_{Lx} = \frac{\alpha_k \, N \, F_\lambda (1 - A_m) \cosh^2 k_s (d - l_k)}{\rho (\frac{1}{2} + \frac{\sinh 2k_s d}{4 k_s d})} \tag{49}$$

where $k_s = \alpha k$. Then the linearized damping force coefficient becomes

$$fx = \frac{f_{Lx}}{\omega} + [C_m V]_e (1 - A_m) \tag{50}$$

4. EXPERIMENTAL RESULTS AND DISCUSSION

4.1 Experimental setup

The experiment was carried out in a 33 m long, 1 m wide and 1.6 m high wave tank as shown in Fig.4. Five thousand models (scale 1:10) of typical Norwegian kelp plants were fixed in the wave flume bottom over a span of 9.3 meters. This represented a density of about 12 plants per horizontal square meter in the field. Eight capacitance wave gauges were used to measure surface elevations, one shear plate to measure the horizontal force and a mini current meter was inserted in the plants 4 centimeters above the shear plate. One of the wave gauges was fixed above the shear plate. The shear plate was fixed flush with the bottom. The location of the first wave gauge, taken to be x=0 and the last taken to be x = 7m were fixed about 1.2 meters inward from the outer boundaries. In total, 50 tests were carried out for different wave periods (6-14 s, full scale) and wave heights in water depth of 60 cm. Analysis of the results is done within $0 \le x \le 7$ m.

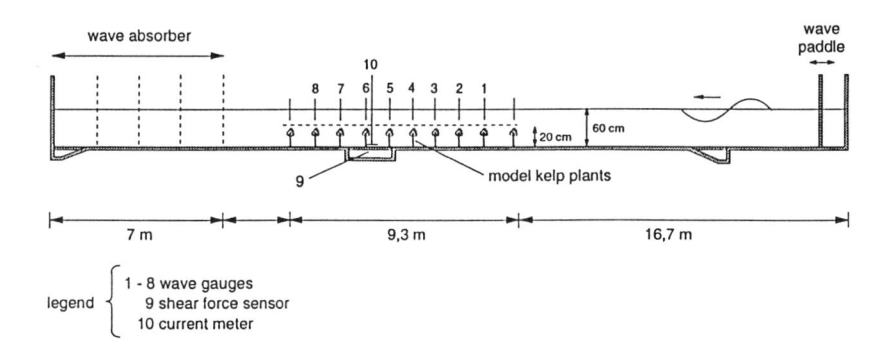

Figure 4: Test setup

4.2 Comparison of theoretical and experimental results

The wave heights measured at the eight locations along the wave channel are fitted to an exponential decay curve H/Ho = exp(-0.00327X) as shown in Fig. (5) whereby H is the local wave height and Ho is the incident wave height measured at x=0, ki =0.00327/m was found by regression based on the least squares in the MATLAB environment.

Basing on the relationship proposed in equations (40) and (42), the force can then be generalized as

$$F = F_\lambda \, |1-A_m| \, u_\lambda \tag{51}$$

where $|1-A_m|$ is given by equation (45). Fig. 6 shows this function fitted to the measured force for given wave heights. Fig. 7 and Fig. 8 show the linear variation of the measured force with the horizontal particle velocity and wave height respectively.

Finally, inserting equation (50) into the dispersion relationship given by equation (32) for a given water depth ,(h+d) wave period, T, and number of vegetation per unit surface area, the damping coefficient, ki, is found by iteration in the MATLAB environment . This solution, however, does not include the contribution of the shear stress to damping.

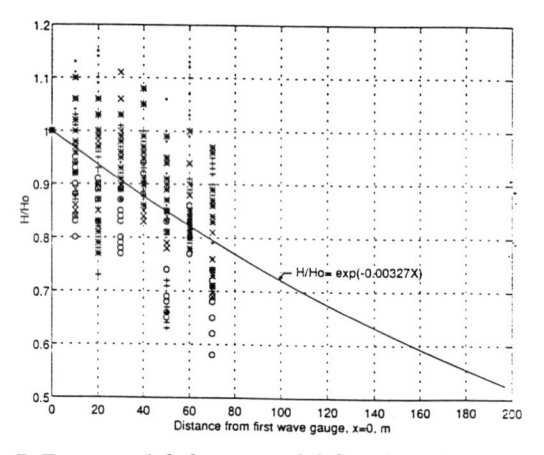

Figure 5. Exponential decay model fitted to data

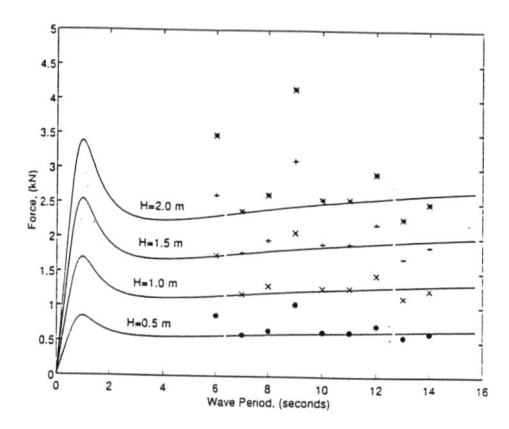

Figure 6. Theoretical force model (Eq. (51)) fitted to measured force

WAVE DAMPING

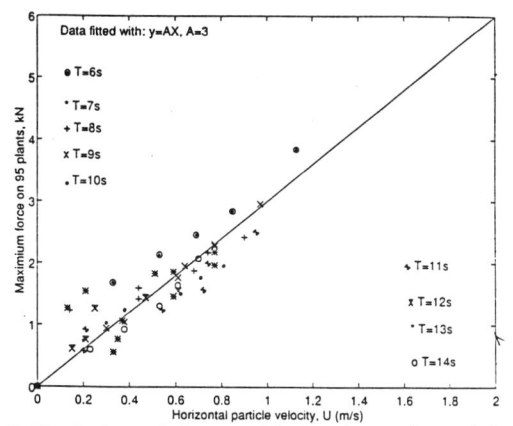

Figure 7: Variation of force with horizontal particle velocity amplitude

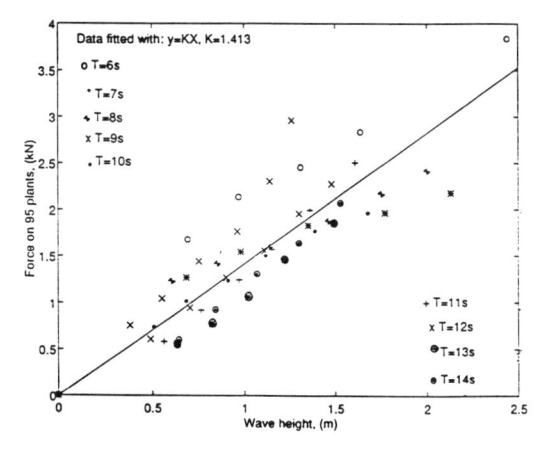

Figure 8: Variation of force with wave height

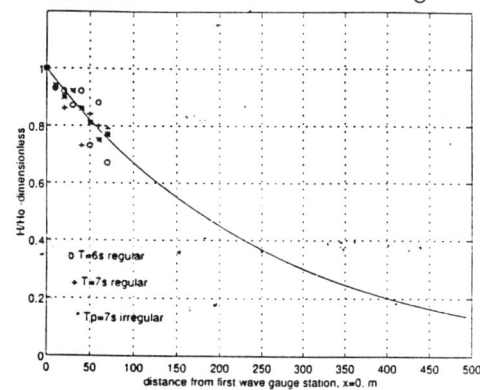

Figure 9: Comparison of scatter for regular and irregular waves

5. CONCLUSION

The present study has given an analytical solution for water waves propagating over submerged vegetation taking account of the vegetation motion using the field experimental results. The linear relationship of the damping force with velocity has been applied effectively to obtain an analytical solution for the otherwise iterative equation of the vegetation motion. The average damping coefficient has been found to be ki=0.00327/m for the type of kelp we considered. However there has been a large scatter of about 30% which may have been due to reflection of the waves from the wave absorber (Fig. 4). A trial run with irregular waves has revealed a smaller scatter (Fig.9). In this study only regular waves were used. It is our intention to use irregular waves to investigate further on the damping force and the damping coefficient.

6. ACKNOWLEDGEMENT

We acknowledge the financial support granted to us by The Norwegian Directorate of Nature Management and The Norwegian Universities Committee for Development, Research and Education (NUFU).

7. REFERENCES

1. Asano, T., Deguchi, H., and Kobayashi, N. (1992): Interaction between water waves and vegetation. *Proc. 23rd International Conference on Coastal Engineering,* Venice, Italy

2. Dalrymple R.A., Kirby, J.T., and Hwang, P.A. (1984): Wave diffraction due to areas of energy dissipation. *Journal of Waterway, Port, Coastal and Ocean Engineering.* ASCE, Vol.110(1)

3. Jackson, G.A., and Winant, C.D. (1983): NOTE Effect of a kelp forest on coastal currents. *Continental Shelf Research,* **2(1)**, pp75-80

4. Price, W.A., Tomlinson, K.W., and Hunt, J.N., (1968): The effect of artificial seaweed in promoting the build up of beaches. Proc. 11th International Conference on Coastal Engineering, London, pp 570-578

5. Wang, H and Tørum, A. (1994): A numerical model on beach response behind coastal kelp. SINTEF-NHL Report No. STF60 A 94092

CHAPTER 12

NONLINEAR COUPLING IN WAVES PROPAGATING OVER A BAR

Y. Eldeberky[1] and J.A. Battjes[2]

Abstract

The degree of nonlinear coupling in a random wavefield propagating over and beyond a bar is examined using a physical wave flume as well as numerical simulations based on time-domain extended Boussinesq equations and their frequency-domain counter-part. The nonlinear phase speed is computed from the evolution of the nonlinear part of the phase function inherent in the frequency-domain model. Over the bar, the phase speeds of the higher harmonics are larger than the linear estimates due to the nonlinear couplings, resulting in virtually dispersionless propagation, while beyond the bar crest, nonlinear effects on the phase speed vanish rapidly, implying full release of bound harmonics. Quantitative measures of nonlinearity such as the skewness and asymmetry have also been determined. They have near-zero values in the deep-water region on either side of the bar and a pronounced peak over the bar. On the downwave side, the random wave field is found to be spatially homogeneous. This implies that it can be fully described by the energy density spectrum without additional phase information related to the bar location.

1. Introduction

The research described in this contribution deals with the propagation of nonbreaking waves over a shallow bar. It is aimed at investigating the degree of nonlinear coupling as waves evolve over and beyond a bar region. It is a continuation of previous related work by the authors on this subject (Eldeberky and Battjes, 1994).

---

[1]Research Assistant, [2]Professor at Delft University of Technology, Department of Civil Engineering, Section Hydraulics, Stevinweg 1, P.O. Box 5048, 2628 CN Delft, Netherlands

The conventional viewpoint is that on the seaside, the harmonics, bound to the primary, are amplified because of the increasing nonlinearity in the shoaling region, and that they are released on the shoreside, at least partially, because of the decreasing nonlinearity in the deepening region. Strictly speaking, however, even in the shoaling region free components are generated as a result of the nonhomogeneity, whereas conversely some degree of phase coupling may remain in the deepening region.

The phenomena mentioned above can with reasonable accuracy be modelled with Boussinesq-type equations, but these are unwieldy in applications involving wave propagation over large distances, for which phase-averaged, energy-based models with linear propagation are better suited. It is then convenient to switch to such a model at some distance downwave from the bar. Two questions then arise:
(1) How far downwave do the nonlinearities extend with non-negligible intensities? The answer to this question determines whether and from where the switch to a linear propagation model is justified.
(2) Are the fixed phase relations between harmonics, which are induced in the bar region, noticeable on the downwave side of the bar? If so, the wavefield on the downwave side of the bar would be spatially nonhomogeneous, with a "memory" of what happened over the bar (although the local nonlinear exchanges may already have vanished). This phenomenon is to be expected for wavefields with a discrete spectrum of only a few harmonics, but we expect that these effects cancel out in case of a continuous spectrum. If this is indeed the case, the wavefield shoreward of the bar would again be statistically homogenous. Knowledge of the energy spectrum on the downwave side (including the harmonics generated over the bar) would then be sufficient to characterize the wave field in a statistical sense, without additional phase information from "upwave" regions expressing the distance downwave from the bar.

The purpose of this paper is to investigate the questions raised above in a quantitative manner. This is done by determining the values and spatial distributions of the nonlinear contributions to the phase speed and of the skewness and asymmetry parameters, which are quantitative measures of nonlinearity.

The paper is arranged as follows. In section 2, the experimental arrangement is described. In section 3, the model equations used in the numerical simulations are presented. The analyses using the measures of nonlinearity followed by some results are given in section 4. Finally, section 5 presents the summary and conclusions.

2. Experimental Approach

Experiments with random waves reported by Beji and Battjes (1993) have yielded time series of surface elevation at a number of stations over a bar (still-water depth 0.10 m) and on either side of it (still-water depth 0.40 m) (Fig. 1). A mechanical wave maker was used to generate wave signals according to a prescribed spectrum. The target spectrum was narrow-banded with peak frequency of 0.40 Hz and variance of 0.35 cm^2. At the downwave side, a beach with 1:20 slope was used as an absorbing boundary. Surface elevations were measured at stations 1 to 8 (the other stations refer to the numerical simulations). Station 1 is at the beginning of the upslope side of the bar, station 2 is 5.0 meters from station 1, and stations 3 to 8 are positioned every 1.0 meter.

These measurements have been analyzed to address the questions raised above. The analyses are mainly in the frequency-domain. Power spectra for these observations indicated significant transfer of energy from the spectral peak to higher frequencies (Beji and Battjes, 1993). Bispectral analysis showed the intensity and the spectral distribution of nonlinear couplings (Eldeberky and Battjes, 1994). Here we focus on the total measures of nonlinearity such as the skewness and asymmetry in order to investigate the spatial variation in the intensity of nonlinear couplings.

Due to the fact that the spatial coverage of the experimental data (up to station 8) is not enough to examine nonlinearity beyond the bar, a numerical "wave flume" has been employed to obtain results farther downwave. This is done using the time- and frequency-domain extended Boussinesq equations. The model equations are presented in the next section.

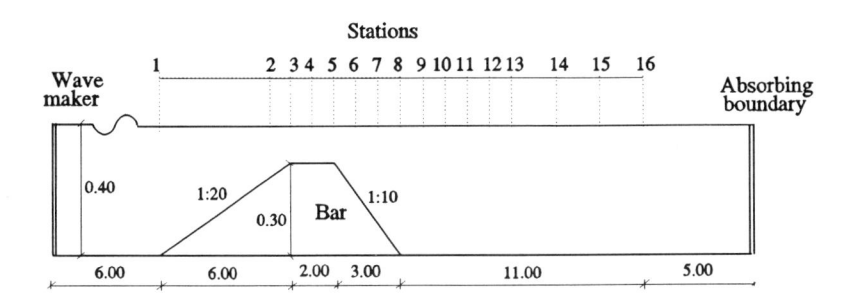

Fig. 1. Definition sketch of wave flume and location of wave gauges. All distances are expressed in meters

3. Numerical Approach

3.1 Time-domain model

Numerical simulations were performed using a 1-D extended Boussinesq model (Beji and Battjes, 1994) with improved dispersion characteristics, as in Madsen and Sørensen (1992), describing relatively long, small amplitude waves propagating in water of slowly varying depth:

$$\frac{\partial u}{\partial t} + u\frac{\partial u}{\partial x} + g\frac{\partial \xi}{\partial x} = \frac{2}{5}h^2\frac{\partial^3 u}{\partial x^2 \partial t} + h\frac{\partial h}{\partial x}\frac{\partial^2 u}{\partial x \partial t} + \frac{1}{15}gh^2\frac{\partial^3 \xi}{\partial x^3} , \quad (1)$$

$$\frac{\partial \xi}{\partial t} + \frac{\partial}{\partial x}[(h+\xi)u] = 0 \quad (2)$$

Here, ξ is the surface displacement, u the depth-averaged horizontal velocity, h the still water depth, and g the gravitational acceleration. This set of equations has been integrated numerically using a difference scheme as described by Beji and Battjes (1994).

In the computation, the initial condition used is the unperturbed state. At the upwave boundary (station 1), the surface elevation is set equal to the experimental values; velocity values are derived from these using the long-wave approximation. At the outgoing boundary, an absorbing boundary condition has been used to ensure that the disturbances leave the computational domain without reflection.

3.2 Frequency-domain model

Numerical simulations were also performed using the frequency-domain counter part of the extended Boussinesq equations mentioned above. For one-dimensional propagation considered so far, the time (t) variation of the surface elevation (ξ) at each location (x) is expanded in a Fourier series as in

$$\xi(t;x) = \sum_{p=-\infty}^{\infty} A_p(x) \exp\{i(\omega_p t - \psi_p(x))\} \quad (3)$$

with A_p denoting a complex amplitude, p indicating the rank of the harmonic, $\omega_p = p\omega_1$, and $d\psi_p/dx = k_p$, the wave number corresponding to ω_p according to the dispersion equation for the linearised Boussinesq equations. By substituting (3) into the time-domain Boussinesq equations, and by neglecting certain higher-order terms on the assumption of a sufficiently gradual evolution of the wave field, Madsen and Sørensen (1993) develop a set of coupled evolution equations for the set of complex amplitudes A_p which in abbreviated form can be written as

$$\frac{dA_p}{dx} = L_p\frac{dh}{dx}A_p + \sum_{m=1}^{p-1} Q^+_{m,p}A_m A_{p-m} + \sum_{m=1}^{\infty} Q^-_{m,p}A^*_m A_{p+m} \quad (4)$$

The first term on the right represents linear shoaling, proportional to the bottom slope dh/dx, the second term the triad sum interactions and the third the triad difference interactions. Complete expressions for the coefficients L, Q^+ and Q^- can be found in Madsen and Sørensen (1993). For the numerical integration of the evolution equation (4) and the applications to random waves with a given energy spectrum refer to Battjes et al. (1993).

4. Analysis

4.1 Phase speed

The evolution equation of the frequency-domain model is formulated in terms of complex Fourier amplitudes that contain the nonlinear part of the phase function. This equation (4) is used to obtain an evolution equation for the nonlinear part of the phase function in order to derive and evaluate a nonlinear correction to the linear phase speed. To this end, we express the complex amplitude $A_p(x)$ in its magnitude $a_p(x)$ and its phase $\alpha_p(x)$:

$$A_p(x) = a_p(x) e^{i\alpha_p(x)} \tag{5}$$

After straightforward algebra, we obtain the following phase evolution equation for each harmonic after omitting the x-dependency for abbreviation,

$$a_p \frac{d\alpha_p}{dx} = \Im\left[\frac{dA_p}{dx}\right] \cos\alpha_p - \Re\left[\frac{dA_p}{dx}\right] \sin\alpha_p \tag{6}$$

The values of a_p, α_p, and dA_p/dx at each location can be obtained from the numerical integration of equation (4). The phase speed of each harmonic, linear $(c_p)_l$ and nonlinear $(c_p)_{nl}$, can be obtain as follows

$$(c_p)_l = \frac{\omega_p}{(k_p)_l} = \frac{\omega_p}{d\psi_p/dx} \tag{7}$$

$$(c_p)_{nl} = \frac{\omega_p}{(k_p)_{nl}} = \frac{\omega_p}{d\psi_p/dx + d\alpha_p/dx}, \tag{8}$$

where $(k_p)_l$ and $(k_p)_{nl}$ are the linear and nonlinear phase change per unit length.

To examine the intensity of nonlinear coupling, the nonlinear phase speeds are compared to the linear predictions obtained from the extended Boussinesq equations.

For random incident waves, assumed to have independent, random phases at station 1, Fig. 2 shows the spectral densities, the linear and the nonlinear phase speeds at different locations over the bar. Over the upslope of the bar (station 2), the nonlinear phase speeds of the higher harmonics are larger than the linear estimates due to the nonlinear couplings to the primary. Over the bar crest (station 4), the nonlinear phase speeds are nearly constant and equal to $\sqrt{gh}$ (nondispersive shallow-waves). Beyond the bar (station 8), the nonlinear predictions of phase speed agree with the linear estimates, implying full release of bound harmonics. The results farther beyond the bar (not shown here) did not show any deviation in the nonlinear phase speeds from the linear predictions, implying the absence of nonlinear couplings. Likewise, the energy spectrum of the waves did not evolve in the region beyond the bar. This in turn means that nonlinear models are not needed to characterize the wavefield beyond the bar.

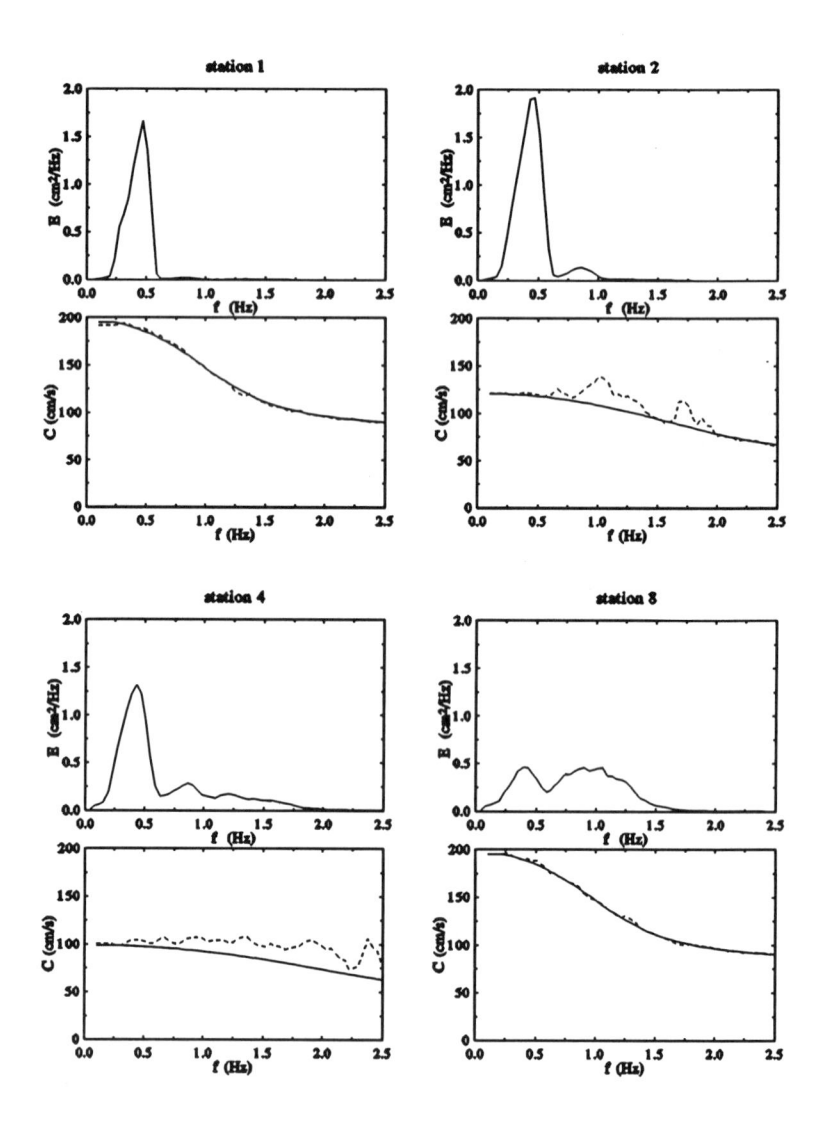

Fig. 2. Computed energy density spectra (upper panels), linear (solid line) and total (dashed line) phase speeds (lower panels) at stations 1, 2, 4, and 8

4.2 Skewness and Asymmetry

To measure the nonlinearities associated with the nonlinear couplings, higher-order moments are needed such as the skewness and asymmetry. These are measures of asymmetry of the wave profile around the horizontal (crest to trough asymmetry) and the vertical (front to back asymmetry) plane respectively. Skewness and asymmetry values have been computed according to the definitions given by Hasselmann *et al.* (1963) and Elgar and Guza (1985) respectively.

The original wave flume and its numerical simulation

The skewness and asymmetry have been computed for both the measured and time-domain computed surface elevations at various locations over the bar. Their variations are given in Fig. 3. The comparison shows the ability of the numerical model to reproduce the nonlinear evolution of waves propagating in varying water depth with sufficient accuracy for the present purpose.

The variations indicate a significant increase on the upslope to a maximum over the bar. To the lee of the bar crest, the skewness and asymmetry decrease rapidly to near-zero values, comparable to those on the exposed side of the bar. (For skewness values less than 0.2, Ochi and Wang (1984) found virtually no deviation from the Gaussian probability density of the sea surface elevation.) This in turn means absence of significant nonlinear interactions at the downwave side of the bar. We expect that for random waves with a continuous spectrum, this will imply spatial homogeneity. This is checked below by using computed signals in the region beyond that of the measurements.

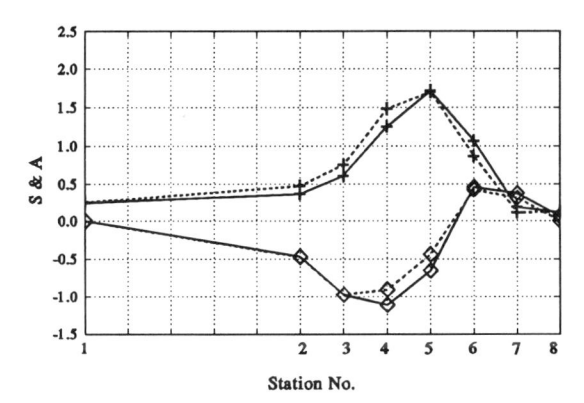

Fig. 3. Skewness (+) and asymmetry ($\diamond$): comparison between results from the physical wave flume (solid lines) and the numerical wave flume (dashed lines)

Numerical simulation for extended region
Time-domain numerical computations have been performed extending to distances farther downwave in order to examine the homogeneity of the wavefield in that region. The computational domain now extends to station 16; the distances between stations 9 to 13 are 1.0 m and those between stations 14 to 16 are 2.0 m.

Computations are done for two different upwave boundary conditions, corresponding to sinusoidal and irregular waves. The former is to demonstrate the spatial nonhomogeneity associated with the interference between a primary wave and its harmonics. The latter is to investigate the matter for the case of a continuous spectrum, where the innumerable interferences are expected to cancel, resulting in a homogeneous wavefield.

To demonstrate the contrast between the two cases, the nonlinearity parameter (a/h) is kept constant by imposing the same surface elevation variance at the upwave boundary. The computations are performed for the same record duration as in the physical experiment, and the computed signals at stations 1 to 16 have been analyzed in the frequency-domain in the same manner as the experimental records.

Sinusoidal waves
Fig. 4 shows the spatial variations of the amplitudes of the primary wave and its harmonics. Over the bar, a significant energy transfer takes place into the second, third and fourth harmonics. Beyond the bar, the amplitudes do not vary because of absence of nonlinear interactions.

The corresponding variations in the skewness and asymmetry are shown in Fig. 5. They indicate a significant increase on the upslope to a maximum over the bar as a result of harmonic generation. On the downslope side of the bar, the skewness and asymmetry decrease rapidly to values between ± 0.5. Beyond the bar, although the amplitudes of the harmonics are nearly constant, the skewness and asymmetry vary significantly as a result of the varying phase lags between the freely propagating component-waves, resulting in a spatially nonhomogenous wavefield.

Irregular waves
Fig. 6 shows the variations of skewness and asymmetry over the upslope, the bar crest, the downslope and farther downwave; the latter is of particular interest here. It shows that over the horizontal region, and in contrast to the case of sinusoidal waves, the skewness and asymmetry remain at near-zero values (less than 0.2), comparable to those on the exposed side of the bar, without any significant spatial variations. This implies that there is no memory of the bar location, in contrast to the discrete case. The practical implication of this is that the waves downwave from the bar can again be assumed to have independent, random phases.

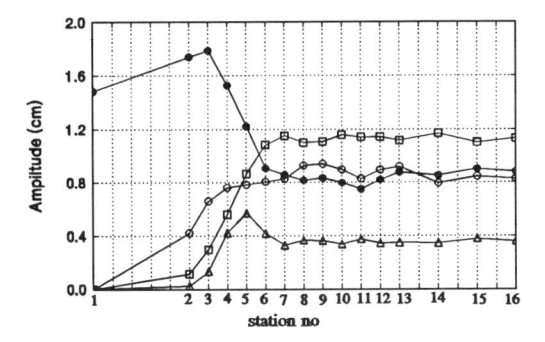

Fig. 4. Spatial variations of the amplitudes of the primary wave and its harmonics primary wave (●), second harmonic (○), third harmonic (□), and fourth harmonic (△)

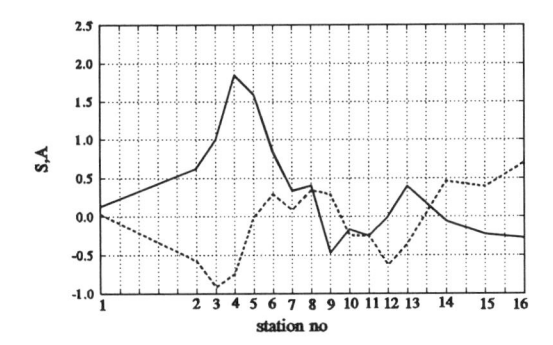

Fig. 5. The spatial variations of skewness (solid line) and asymmetry (dashed line) for sinusoidal incident waves (at station 1)

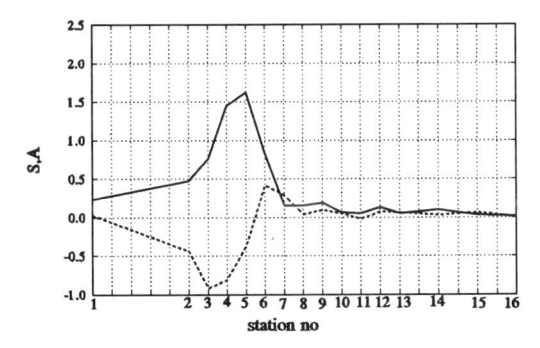

Fig. 6. The spatial variations of skewness (solid line) and asymmetry (dashed line) for irregular waves

5. Summary and Conclusions

The degree of nonlinear coupling as waves evolve over and beyond a bar is examined. The extended Boussinesq equations that had shown success in this kind of application are used in this investigation together with experimental data. The situation considered was such that significant harmonic generation took place on the upslope leading to the bar crest. The following indicators of nonlinearity were used: the nonlinear phase speed, the skewness and asymmetry. Inspection of their spatial variations has led to the following conclusions.

- The comparison between linear and nonlinear estimates of the phase speeds indicates that over the bar, the bound harmonics travel faster than their corresponding free waves that have linear phase speeds. Beyond the bar, they propagate with the linear phase speed implying that they are fully released.

- The skewness and asymmetry have near-zero values in the deep-water region on either side of the bar and a pronounced peak over the bar. On the downwave side, the wavefield is found to be spatially homogeneous for irregular waves without memory of the phase couplings which existed over the bar crest. This is in contrast to the case of a discrete, finite set of wave components.

Summarizing, the wavefield on the downwave side is virtually linear and statistically homogeneous. It can be fully described by the energy density spectrum with linear propagation, and without the need for additional phase information reflecting the site-dependent distance downwave from the bar.

Acknowledgements

The work presented here is part of a project sponsored by the National Institute for Coastal and Marine Management and the Road and Hydraulics Engineering Division of Rijkswaterstaat, the Dutch Department of Public Works.

References

Beji, S., and Battjes, J.A. (1993). Experimental investigation of wave propagation over a bar. *Coastal Eng.*, 19, 151-162.
Beji, S., and Battjes, J.A. (1994). Numerical simulation of nonlinear waves propagating over a bar. *Coastal Eng.*, 23, 1-16.
Battjes, J.A., Eldeberky, Y., and Won, Y. (1993). Spectral Boussinesq modelling of breaking waves. *Proc. Int. Conf. WAVES '93*, New Orleans, ASCE, pp. 813-820, New York.
Eldeberky, Y., and Battjes, J.A. (1994). Phase lock in waves passing over a bar. *Proc. Int. Sym. WAVES-PHYSICAL AND NUMERICAL MODELLING*

'94, Vancouver, pp. 1086-1095.

Elgar, S. and Guza, R.T. (1985). Observations of bispectra of shoaling surface gravity waves. *J. Fluid Mech.*, Vol. 161, 425-448.

Hasselmann, K., Munk, W., and McDonald, G. (1963). Bispectra of ocean waves. In: *Time Series Analysis* (edited by M. Rosenblatt), 125-139, Wiley, New York.

Madsen, P.A., and Sørensen, O.R. (1992). A new form of the Boussinesq equations with improved linear dispersion characteristics. Part 2: a slowly-varying bathymetry. *Coastal Eng.*, 18, 183-205.

Madsen, P.A., and Sørensen, O.R. (1993). Bound waves and triad interactions in shallow water. *J. Ocean Eng.*, 20 (4), 359-388.

Ochi, M.K. and Wang, W.C. (1984). Non-Gaussian characteristics of coastal waves. *Proc. 19th Int. Conf. on Coastal Eng.*, Houston, Vol.1, pp.516-531

CHAPTER 13

AN ABSORBING WAVE-MAKER
BASED ON DIGITAL FILTERS

Peter Frigaard and Morten Christensen
Hydraulics & Coastal Engineering Laboratory
Aalborg University, Sohngaardsholmsvej 57, DK-9000 Aalborg, Denmark

ABSTRACT

An absorbing wave maker operated by means of on-line signals from digital FIR filters is presented.

Surface elevations are measured in two positions in front of the wave maker. The reflected wave train is separated from the sum of the incident and rereflected wave trains by means of digital filtering and subsequent superposition of the measured surface elevations. The motion of the wave paddle required to absorb reflected waves is determined and added to the original wave paddle control signal.

Irregular wave tests involving test structures with different degrees of reflection show that excellent absorption characteristics have been achieved with the system.

INTRODUCTION

Coastal engineering problems are often solved by means of physical models. Physical modelling of coastal engineering phenomena requires the capability of reproducing natural conditions in the laboratory environment.
One of the problems associated with the physical modelling of water waves in laboratory wave channels is the presence of rereflected waves.
In nature the sea constitutes an open boundary which absorbs waves reflected by the coastal system.
A wave channel is a closed system: waves reflected from a model structure will be rereflected at the wave paddle, thus altering the characteristics of the wave train incident to the model structure. Consequently, the reproduction of a specified

incident wave train will often be impossible when a reflective structure is being tested.

The problem of rereflection can be reduced by applying a so-called absorbing wave maker: a combined wave generator and active wave absorber, which, in addition to generating incident waves, absorbs waves reflected from the test structure. The construction of an absorbing wave maker requires (Gilbert (1978)):

1. A means of detecting reflected waves as they approach the wave maker

2. A means of making the paddle generate waves that are, in effect, equal and opposite to the reflected waves so that the reflected waves are cancelled out as they reach the paddle. This requirement is over and above the need to generate the primary incident waves.

Milgram (1970) presented a system in which waves in a channel were absorbed by means of a moving termination at the end of the channel. The motion of the termination needed for absorption was determined by analog filtering of a surface elevation signal measured in front of the termination. This active wave absorber was not used in a combined generation and absorption mode.

Bullock and Murton (1989) described the conversion of a conventional wedge-type wave maker to an absorbing wave maker. The system developed by Bullock and Murton was based on analog filtering of a surface elevation signal measured on the face of the wave paddle. Good absorption characteristics were achieved with a less-than-perfect circuit design.

Recently, an absorbing wave maker has been installed in a wave channel at Aalborg Hydraulic Laboratory, Aalborg University.

In the following the design of this absorbing wave maker is presented and its performance is evaluated based on the results of physical model tests.

PRINCIPLE

The absorbing wave maker is operated by digital FIR filters working in real time. The relation between the input η and the output x of a digital FIR filter of length N is given by the discrete convolution integral:

$$x^k = \sum_{i=-M}^{i=M} h^i \eta^{k-i} \quad , \quad M = \frac{N-1}{2} \tag{1}$$

where h denotes the filter impulse response (filter operator). Given a desired frequency response, the corresponding FIR filter operator is obtained by computing the inverse discrete Fourier transform of the complex frequency response function, see e.g. Karl (1989). Notice, that the filter output is delayed $\frac{N-1}{2}$ time steps relative to the input. For FIR filters operating in real time, this time delay must be removed.

The paddle displacement correction signal needed for absorption of reflected waves is determined by means of digital filtering and subsequent superposition of surface elevation signals measured in two positions in front of the wave maker (fig. 1).

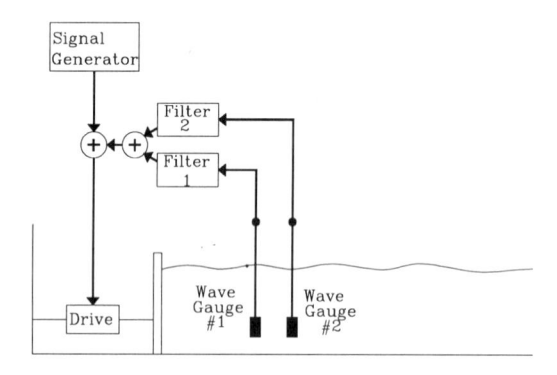

Figure 1: Principle of absorbing wave maker.

When active absorption is applied, the paddle displacement correction signal is added to the input paddle displacement signal read from the signal generator, causing the wave maker to operate in a combined generation/absorption mode. Having outlined the principle of the system, the only remaining problem in the design process is to specify the frequency response of the FIR filters applied.

FREQUENCY RESPONSE OF FIR FILTERS

In fig. 2, a wave channel equipped with two wave gauges is shown.

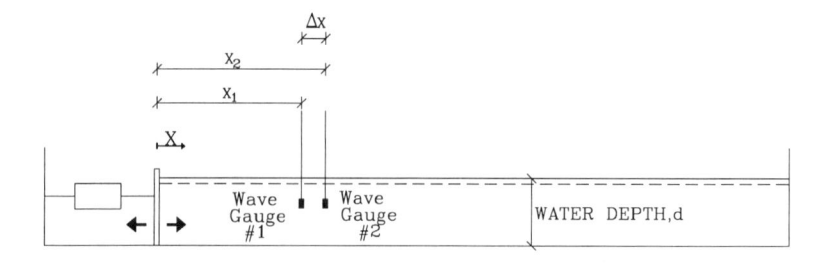

Figure 2: Wave channel with piston-type wave maker.

The surface elevation signal at a position x may be regarded as a sum of harmonic components. Considering an isolated component of frequency f, the surface elevation arising from this component may be written as the sum of the corresponding incident and reflected wave components:

$$\begin{aligned}\eta(x,t) &= \eta_I(x,t) + \eta_R(x,t) \\ &= a_I cos(2\pi f t - kx + \phi_I) + a_R cos(2\pi f t + kx + \phi_R)\end{aligned} \quad (2)$$

where
- f : frequency
- $a = a(f)$: wave amplitude
- $k = k(f)$: wave number
- $\phi = \phi(f)$: phase

and indices I and R denote incident and reflected, respectively.

Provided a linear relation exists between a given paddle displacement signal and its corresponding surface elevation signal, the paddle displacement correction signal, $X_{corr}(t)$, which cancels out the reflected component without disturbing the incident component, is given by

$$X_{corr}(t) = B \cdot a_R cos(2\pi f t + \phi_R + \phi_B + \pi) \quad (3)$$

where
- B : piston stroke/wave height relation
- ϕ_B : phaseshift between paddle displacement and surface elevation on the face of the paddle

In the following it is shown that it is possible to amplify and phase shift the surface elevation signals from the two wave gauges in such a way that their sum is identical to the paddle correction signal corresponding to absorption of the reflected component as given by eq. (3).

At the two wave gauges (fig. 2) we have:

$$\begin{aligned}\eta(x_1,t) &= a_I cos(2\pi f t - kx_1 + \phi_I) + a_R cos(2\pi f t + kx_1 + \phi_R) \quad (4)\\ \eta(x_2,t) &= a_I cos(2\pi f t - kx_2 + \phi_I) + a_R cos(2\pi f t + kx_2 + \phi_R) \\ &= a_I cos(2\pi f t - kx_1 - k\Delta x + \phi_I) + \\ &\quad a_R cos(2\pi f t + kx_1 + k\Delta x + \phi_R)\end{aligned} \quad (5)$$

where $x_2 = x_1 + \Delta x$ has been substituted into eq. (5).

An amplification of C and a theoretical phase shift ϕ^{theo} are introduced into the expressions for $\eta(x,t)$. The modified signal is denoted η^*. For the i'th wave gauge signal the modified signal is defined as:

$$\begin{aligned}\eta^*(x_i,t) &= Ca_I cos(2\pi f t - kx_i + \phi_I + \phi_i^{theo}) + \\ &\quad Ca_R cos(2\pi f t + kx_i + \phi_R + \phi_i^{theo})\end{aligned} \quad (6)$$

This gives at wave gauges 1 and 2:

$$\eta^*(x_1,t) = C a_I \cos(2\pi f t - kx_1 + \phi_I + \phi_1^{theo}) + \\ C a_R \cos(2\pi f t + kx_1 + \phi_R + \phi_1^{theo}) \quad (7)$$

$$\eta^*(x_2,t) = C a_I \cos(2\pi f t - kx_1 - k\Delta x + \phi_I + \phi_2^{theo}) + \\ C a_R \cos(2\pi f t + kx_1 + k\Delta x + \phi_R + \phi_2^{theo}) \quad (8)$$

The sum of $\eta^*(x_1,t)$ and $\eta^*(x_2,t)$, which is denoted $\eta^{calc}(t)$, is:

$$\eta^{calc}(t) = \eta^*(x_1,t) + \eta^*(x_2,t)$$

$$= 2C a_I \cos\left(\frac{k\Delta x + \phi_1^{theo} - \phi_2^{theo}}{2}\right)$$

$$\cos\left(2\pi f t - kx_1 + \phi_I + \frac{-k\Delta x + \phi_1^{theo} + \phi_2^{theo}}{2}\right) +$$

$$2C a_R \cos\left(\frac{-k\Delta x + \phi_1^{theo} - \phi_2^{theo}}{2}\right)$$

$$\cos\left(2\pi f t + kx_1 + \phi_R + \frac{k\Delta x + \phi_1^{theo} + \phi_2^{theo}}{2}\right) \quad (9)$$

It is seen that $\eta^{calc}(t)$ and $X_{corr}(t) = B a_R \cos(2\pi f t + \phi_R + \phi_B + \pi)$ are identical signals in case:

$$2C \cos\left(\frac{k\Delta x - \phi_1^{theo} + \phi_2^{theo}}{2}\right) = B \quad (10)$$

$$kx_1 + \frac{k\Delta x + \phi_1^{theo} + \phi_2^{theo}}{2} = \phi_B + \pi + n \cdot 2\pi, \; n \in (0,\pm 1,\pm 2,..) \quad (11)$$

$$\frac{k\Delta x + \phi_1^{theo} - \phi_2^{theo}}{2} = \frac{\pi}{2} + m \cdot \pi, \; m \in (0,\pm 1,\pm 2,..) \quad (12)$$

Solving eqs. (10)-(12) with respect to $\phi_1^{theo}, \phi_2^{theo}$ and C with $n = m = 0$ gives

$$\phi_1^{theo} = \phi_B - k\Delta x - kx_1 + 3\pi/2 \quad (13)$$

$$\phi_2^{theo} = \phi_B - kx_1 + \pi/2 \quad (14)$$

$$C = \frac{B}{2\cos(-k\Delta x + \pi/2)} \quad (15)$$

Eqs. (13)-(15) specify the frequency responses, i.e. the amplification factors and phase shifts, of FIR filters 1 and 2 in fig. 1.
Even though the theoretical frequency response of the filters easily can be calculated from the eqs. (13)-(15) one should notice that an actual realization of such

a theoretical frequence response in FIR filters might be rather difficult to obtain. The aim of this paper is not to describe design of Fir filters, but notice that the actual frequency response of the filters (read: performance of the absorbing system) are strongly dependent upon: Type of wave-maker, water depth, location of wave gauges, number of filter coefficients, sample frequency in filter etc.

PHYSICAL MODEL TEST

In order to determine the performance of the active absorption method described above, the method was implemented in the control system of a piston-type wave maker placed in a small laboratory wave channel at Aalborg Hydraulic Laboratory, Aalborg University and the method was implemented in the control system of the wedge type wave maker placed in the CIEM at LIM/UPC, Catalonia University of Technology.
The geometry of the Aalborg Hydraulics Laboratory wave channel and the wave gauge positions are given in fig. 3.

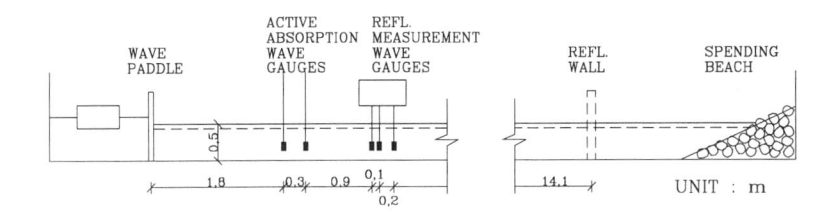

Figure 3: Wave channel and wave gauge positions.

The active absorption system was based on surface elevation measurements obtained in wave gauges positioned at distances of $x_1 = 1.80$ m and $x_2 = 2.10$ m from the wave paddle. The phase ϕ_B and gain B (see eq. (3)) were determined using the linear transfer functions derived by Biesel (1951).
When active absorption was applied, the surface elevation time series were recorded and digitized by means of a PC equipped with an A/D-D/A-card, digital filtering and superposition were performed, and the resulting paddle displacement correction signal was added to the input signal read from the signal generator.
At the far end of the channel, a spending beach was situated. In order to be able to perform tests with different degrees of reflection from the channel termination, provision was made for mounting a vertical reflecting wall in front of the spending beach.

The channel is equipped with three pairs of wave gauges mounted on a beam at distances of 3.0 m, 3.1 m and 3.3 m from the wave paddle. These gauges are used for reflection measurements.
A water depth of $d = 0.5$ m was maintained throughout the test series.

In order to evaluate the efficiency of the absorbing wave maker when applied to irregular wave tests involving test structures with different degrees of reflection, tests covering all four permutations of the alternatives

- Either with or without active absorption applied

- Either with the spending beach or the reflecting wall at the far end of the channel

were performed.
All tests were performed with exactly the same input from the signal generator: a wave paddle displacement signal corresponding to a JONSWAP-spectrum with significant wave height $H_s = 0.04$ m, peak frequency $f_p = 0.6$ Hz and peak enhancement factor $\gamma = 3.3$ sampled at a frequency of $f_s = 40$ Hz and generated by means of digital filtering of Gaussian white noise in the time domain. In each test the incident and reflected spectra were resolved as described by Mansard and Funke (1980). The incident spectra are given in fig. 4.

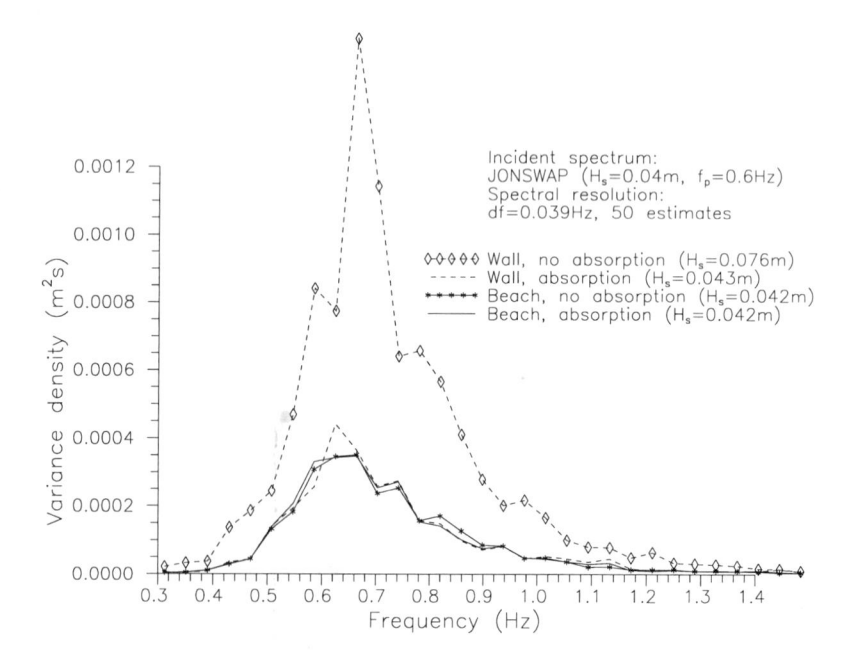

Figure 4: Incident wave spectra.

The tests showed that the spending beach reflected only 5-10% of the incoming wave energy in the frequency range of the input spectrum. Consequently, the "beach, no absorption" incident spectrum in fig. 4 may be regarded as the target spectrum: the disturbances introduced by rereflection are negligible.
The "beach, absorption" incident spectrum is almost identical to the target spectrum. This implies, that applying the active absorption system to tests involving test structures with little reflection will not introduce disturbances in the incident spectrum.
The efficiency of the active absorption system is demonstrated by the test results obtained with the reflecting wall installed at the far end of the channel. When active absorption is applied, the incident spectrum is in excellent agreement with the target spectrum, whereas the incident spectrum obtained without active absorption is significantly distorted by rereflection (see fig. 4.).

In order to visualize the effect of active absorption in the time domain, the following test was performed: the reflecting wall was installed in the far end of the channel, and irregular waves were generated. After 60 seconds, wave generation was terminated, and active absorption was applied. A surface elevation time series recorded at $x = 3.0$ m is given in fig. 5 a.
For comparison, a time series recorded in a similar test in which active absorption was not applied is given in fig. 5 b.

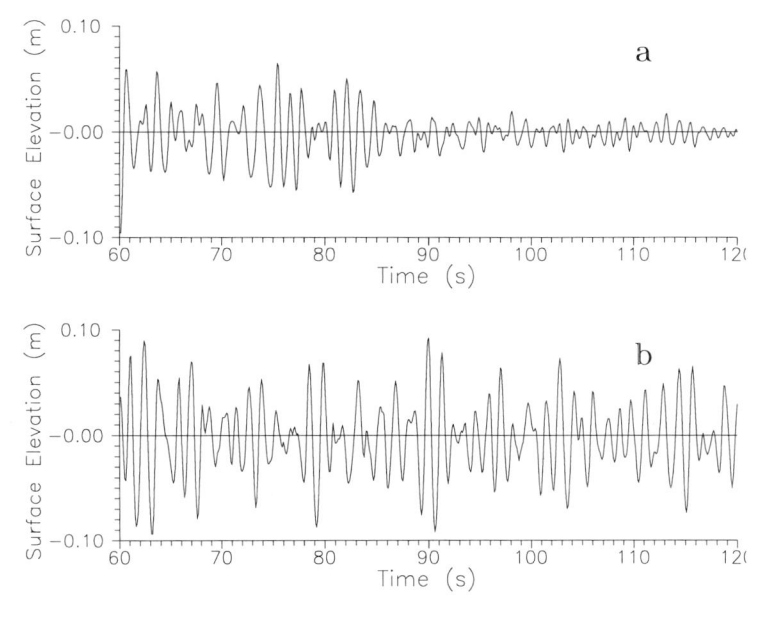

Figure 5: Time series obtained with (a) and without absorption (b).

Furthermore, the stability of the system was tested.
Again, the reflecting wall was mounted at the far end of the channel. The active absorption system was applied, and a paddle displacement time series of length $T = 51.2\,\text{s}$ was generated, and sent repeatedly to the wave maker.
Two surface elevation time series of length T were recorded starting from $t = T$ and $t = 25\,T$, respectively, and the incident and reflected spectra were resolved. The resulting incident spectra are given in fig. 6.

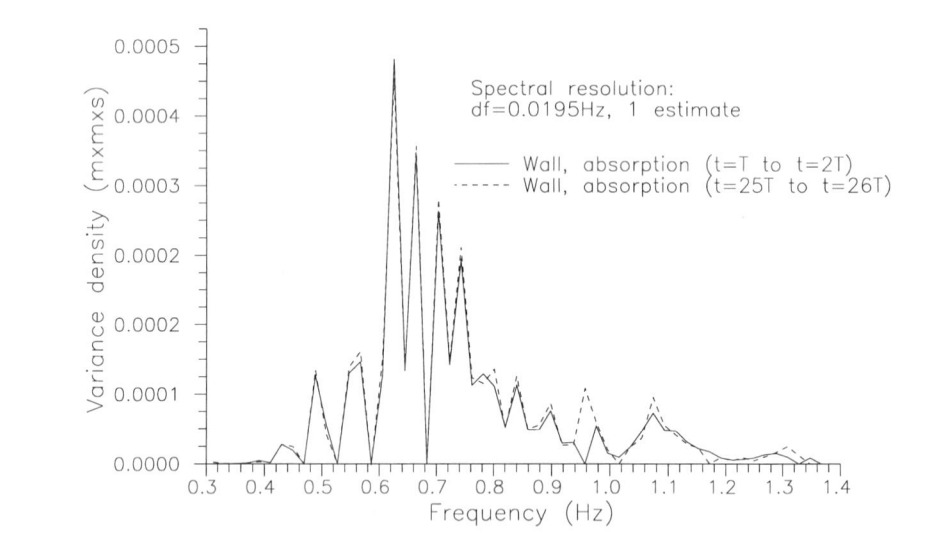

Figure 6: Incident wave spectra.

Fig. 6 indicates that the system is stable. Apparently, 25 repetitions of the input signal (approx. 20 minutes of wave generation) have not caused significant disturbance in the incident spectrum despite reflection from the wall at the channel termination.

The geometry of the wave channel in the CIEM at LIM/UPC and the wave gauge positions are given in fig. 7.

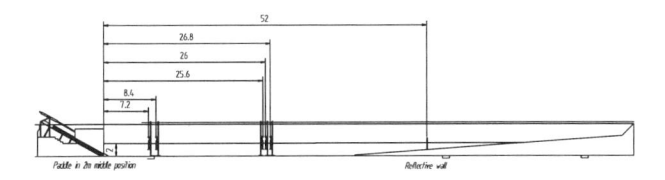

Figure 7: Wave channel and wave gauge positions.

In this test, the reflection compensation system was based on surface elevation measurements obtained in wave gauges positioned at distances of $x_1 = 7.20m$ and $x_2 = 8.40m$ from the wave paddle. The phase ϕ_B and gain B were determined using the linear transfer functions derived by Biesel (1951). An additional phase shift was introduced in order to compensate for a measured time delay of $0.1s$ between demand and feedback signals in the wave maker control system.

When reflection compensation was applied, the surface elevation time series were recorded and digitised by means of a PC equipped with an A/D-D/A-card, digital filtering and superposition was performed, and the resulting paddle displacement correction signal was added to the input signal read from the signal generator.

At the far end of the channel, a spending beach was situated. In order to be able to perform tests with different degrees of reflection from the channel termination, a vertical reflecting wall could be placed in front of the spending beach.

The channel was equipped three wave gauges mounted at distances of 25.6m, 26.0m and 26.8m from the wave paddle. These gauges were used for reflection measurements.

A water depth of $d = 2.0m$ was maintained throughout the test series.

All tests were performed with exactly the same input from the signal generator: a wave paddle displacement signal corresponding to a JONSWAP-spectrum with significant wave height $H_s = 0.25m$, peak period of $T_p = 3s$ and peak enhancement factor $\gamma = 3.3$ sampled at a frequency of $f_s = 20Hz$ and generated by means of digital filtering of Gaussian white noise in the time domain.

In each test, the incident and reflected spectra were resolved as described by Mansard and Funke (1980).

The spending beach had a reflection coefficient of only 6-8 % in the frequency range of the input spectrum. Consequently, the incident spectrum measured with the spending beach at the far end of the channel and no reflection compensation applied may be regarded as the target spectrum: the disturbances introduced by rereflection are negligible.

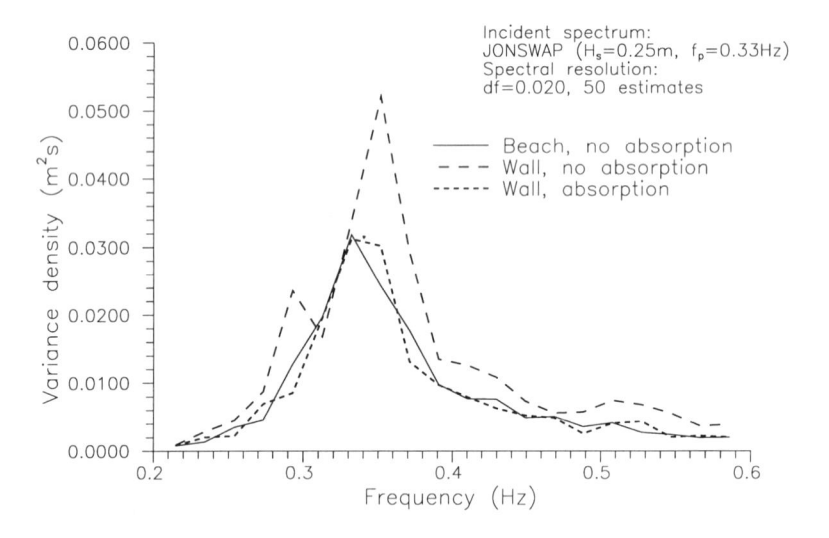

Figure 8: Incident wave spectra.

In fig. 8 the efficiency of the active absorption system is demonstrated. The figure shows that the tests in the large flume are very comparable to the tests in the small flume.

The performance of the reflection compensation system developed at Aalborg University (AU) appears to be excellent. When the system is applied to the test with a spending beach at the channel termination, the measured incident spectrum is almost identical to the target spectrum (fig. 4 and fig. 8). This implies, that applying this reflection compensation system to tests involving test structures with little reflection will not introduce disturbances in the incident spectrum.

LONG WAVES

Laboratory tests with irregular waves often give problems with absorption of long waves. Absorption of long waves requires an enormous stroke of the paddle. This means, that the designer of the filters always have to limit the low frequent performance of the system in order to have enough stroke in the wave maker system. Fig. 9 shows the performance of the system used in the tests at Aalborg Hydraulics Laboratory. Even though the figure indicates a rather poor performance of the system for long waves, it should be noted that as long as the system absorbs a part of the re-reflected waves it will prevent the growth of long waves.

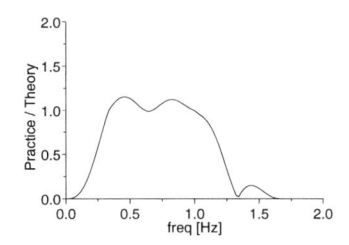

Figure 9: Performance of system.

CONCLUSION

A method for active absorption of reflected waves in wave channels by means of an absorbing wave maker has been presented. (Notice, that the system also can be very usefull in numerical models).
The motion of the wave maker, which is needed for absorption, is determined by means of digital filtering and subsequent superposition of surface elevations measured in two fixed positions in front of the wave maker.
The method has been implemented in the control system of a piston-type wave maker installed in a wave channel, and irregular wave tests involving test structures with different degrees of reflection have been carried out in order to determine the performance of the absorbing wave maker.
The tests performed imply that excellent absorption characteristics have been achieved. The absorbing wave maker is capable of reducing the problem of rereflection considerably even at very high levels of reflection. Furthermore, the active absorption system appears to be stable.
Converting a conventional wave maker to an absorbing wave maker based on the method presented above is relatively inexpensive considering the improvements achieved: the only requirements are two conventional wave gauges and a PC

equipped with an A/D-D/A-card. These facilities will normally already be available in most laboratory environments (if a PC equipped with an A/D-D/A-card is used as signal generator for the wave maker, the wave gauges can be connected to this computer, allowing the computer to perform signal generation and correction signal calculation simultaneously).

ACKNOWLEDGEMENTS

The authors would like to thank the Danish Technical Research Council for financial support and the LIM/UPC for the use of the large flume.

REFERENCES

Biesel, F., 1951. Les Apparails Generateurs de Houle en Laboratorie. *La Houille Blanche,* Vol. 6, nos. 2, 4 and 5.

Bullock, G.N. and Murton, G.J., 1989. Performance of a Wedge Type Absorbing Wave Maker. *Journal of Waterway, Port, Coastal and Ocean Engineering,* Vol. 115, No. 1.

Christensen, M. and Høgedal, M., 1994. Reflection Compensation in the CIEM, LIM/UPC – Test Report. Aalborg University

Frigaard, P. and Brorsen, B., 1993. A Time Domain Method for Separating Incident and Reflected Irregular Waves. To be published in *Coastal Engineering.*

Gilbert, G., 1978. Absorbing Wave Generators. Hydr. Res. Station notes, Hydr. Res. Station, Wallingford, Oxon, United Kingdom, 20, 3-4.

John H. Karl, 1989. *An Introduction to Digital Signal Processing.* Academic Press, San Diego.

Mansard, E. and Funke, E., 1980. The Measurement of Incident and Reflected Spectra Using a Least Squares Method. *Proceedings, 17th International Conference on Coastal Engineering,* Vol. 1, pp 154-172, Sydney, Australia.

Milgram, J.S., 1970. Active Water-Wave Absorbers. *J. Fluid Mech.,* 43(4), 845-859.

CHAPTER 14

Wave Climate Study in Wadden Sea Areas

Ralf Kaiser[1], Günther Brandt[1], Joachim Gärtner[2], Detlef Glaser[1], Joachim Grüne[3], Frerk Jensen[4], Hanz D. Niemeyer[1]

Abstract

Significant features on Wadden Sea wave climate are evaluated in respect of the state of the art. Main emphasis was laid on an analysis of the governing boundary conditions of local wave climate in island sheltered Wadden Sea areas with extensions being sufficient for local wind wave growth. Explanatory for significant wave heights a reliable parametrization of local wave climate has been evaluated by using generally available data of water level and wind measurements.

Introduction

In the German Bight comprehensive wave climate investigations in six distinct Wadden Sea areas have been carried out in recent years (fig. 1). Each of them represents a significant type of Wadden Sea coastal areas: Mesotidal barrier island coasts where the protective islands are seaborne or remnants of former mainland and macrotidal estuarine coasts [NIE-

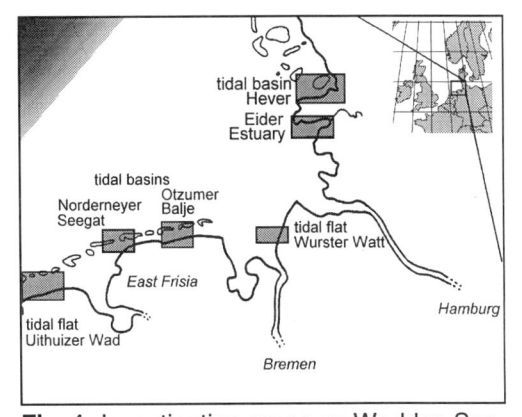

Fig. 1: Investigation areas on Wadden Sea Wave climate in the German Bight

---

[1]Coastal Research Station of the Lower Saxonian Central State Board for Ecology, Norderney/East Frisia, Germany

[2]Regional Board for Water Management (ALW), Heide, Germany

[3]Joint Research Facility Large Wave Channel, Hannover, Germany

[4]Regional Board for Water Management (ALW), Husum, Germany

MEYER 1990]. The measuring locations within each of these typical areas have been chosen in such a way, that each location itself represents the conditions of a significant part of the total area in respect of the typical morphological boundary conditions. Main aim of these investigations was to establish a parametrization for local wave climate in dependence of the distinct types of morphological boundary conditions of each area as a basis for a more general approach.

Wave climate in Wadden Seas is characterized by strong hydro-dynamical-morphological interactions due to restriction of water depths and an often very complex three-dimensional underwater topography. Waves propagating from the offshore shelf via ebb deltas towards the flats break partly or completely experience significant energy dissipation. The generation of strong waves by local windfields on the intertidal flats depends remarkably on tidal elevation and on the wind-induced surge set-up. Therefore this effect is restricted to those limited time intervals for which these necessary boundary conditions occur. On the one hand a reliable forecasting of wave parameters is mostly not achievable by using generally known forecasting procedures, which have been evaluated in areas for less differentiated boundary conditions. On the other hand as useful field data are still very poor the need for such data is tremendous.

Wave climate in island sheltered Wadden Sea areas is characterized by three different types of origin [NIEMEYER 1984, 1991]:

1. Local wind waves,
2. offshore swell entering via the tidal inlets,
3. offshore generated wave systems being generated by onshore directed strong winds and storms and propagating via the tidal inlets.

• = measurements on tidal flats
+ = measurements on flooded salt marshes

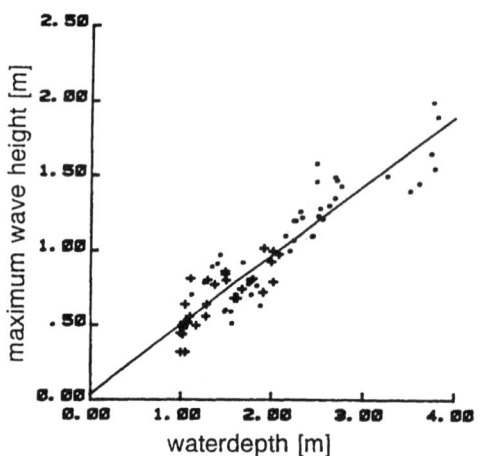

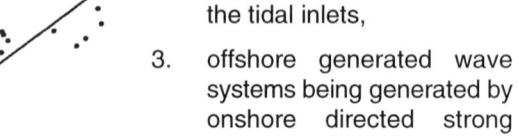

Fig. 2: Wave height/water depth relation for island sheltered tidal flat areas and salt marshes at the East Frisian coast [NIEMEYER 1991]

Due to previous investigations it was established knowledge that the latter type is the most important one as well for design of coastal structures as for impacts on tidal flat and salt marsh morphology [NIEMEYER 1983]: The dynamical equilibrium

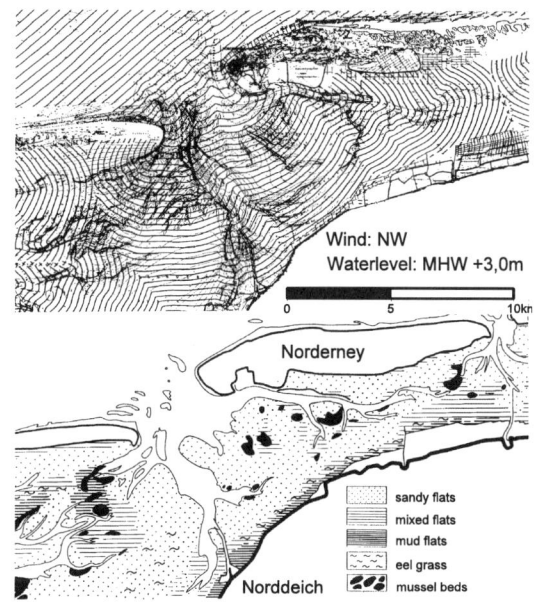

Fig. 3: Combination of refraction diagram and map of surface sediments for the Norderneyer Seegat [NIEMEYER 1987b]; map of surface sediments by RAGUTZKI [1982]

between morphologically stable tidal flats above mean sea-level and waves of type 3 is characterized by a strong linear relationship of local wave heights and water depths which is also valid for adjacent supratidal salt marshes (Fig. 3). The relationship is strictly fitting for onshore directed strong winds and storm but is only valid for areas above MSL.

A combination of refraction diagram and a map of surface sediments has been used by NIEMEYER [1987b] in order to demonstrate the interaction of waves and morphology and the resulting surface sediment distribution for the tidal basin of the tidal inlet Norderneyer Seegat (Fig. 3). The landward succession of sandy, mixed and muddy flats follows the wave propagation from the inlet to the main land. The sector of wave propagation with only modest refraction coincides with the smallest band of mixed and muddy flats in front of the mainland. Due to the dominant changes the wave climate experiences on the tidal inlet bar [NIEMEYER 1987a], wave propagation landward of the inlet is mainly independent from offshore wind direction and setup above mean high water level [NIEMEYER 1986]. There are also small muddy and sandy flats landwards in wave direction of the mussel beds.

Although the data of Wadden Sea waves for the areas above MSL fits generally to the linear relationship H_s/h one has to be aware that different boundary conditions generate distinct spectral energy distributions which is evident by examples of wave spectra (fig. 4) taken from NIEMEYER et al. [1992]. For spectrum No. 107 the still water level is about 1m above MHW and there are wind velocities of 11 m/s from NNW. Energy is distributed over the range from 15-3 seconds, with even higher energy for lower frequencies. For the same water level occurring in coincidence with higher wind velocities from more westerly directions spectrum no. 43 shows a more pronounced peak energy concentration, but its total energy is only about 20% higher than that

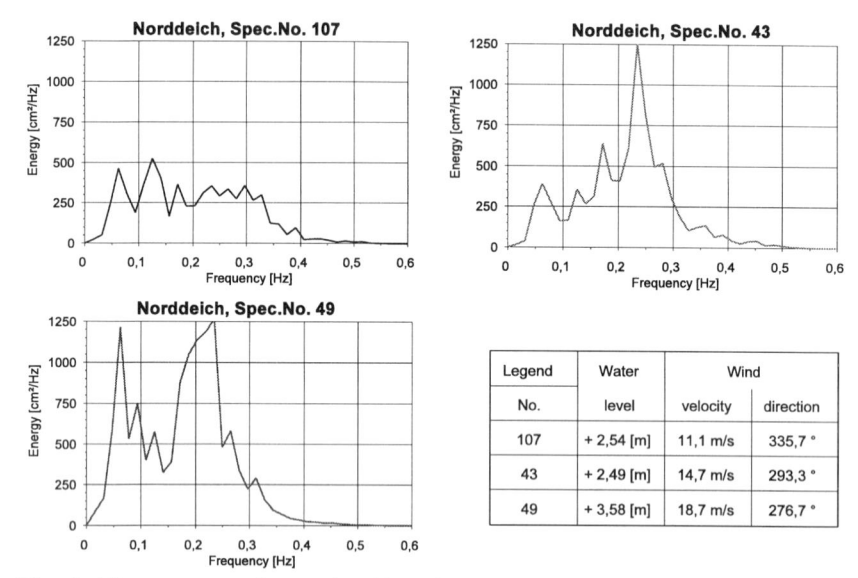

Fig. 4: Wave spectra for station Norddeich (see fig. 3) for different boundary conditions [NIEMEYER et al. 1992]

Legend	Water	Wind	
No.	level	velocity	direction
107	+ 2,54 [m]	11,1 m/s	335,7 °
43	+ 2,49 [m]	14,7 m/s	293,3 °
49	+ 3,58 [m]	18,7 m/s	276,7 °

one of no. 107. Waterdepth and fetch for the spectrum No. 107 were more convenient in respect of generating higher waves. The high peak in spectrum No. 43 is due to waves entering from the North-Sea experiencing an energy shift to higher frequencies. An increase in water level of about 1m and higher wind velocities from west produced perform the boundary conditions for spectrum No. 49 with a significant low frequent high peak. For water levels with this height even longer waves with periods of about 15 s can penetrate into the Wadden Sea.

Regional characteristics of Wadden Sea wave climate

An detailed study on wave climate in distinct German Wadden Sea areas [NIEMEYER et al. 1992] reflects their remarkable morphodynamical features in the different regions causing specific characteristic interactions everywhere. Regional wave climate in a particular Wadden Sea area is therefore often very remarkably distinct from that one occurring in another one. Explanatory a comparison of the wave climate in the Hever tidal basin (fig. 1 + 5) and the East Frisian Wadden Sea coast is carried out in order to make differences evident as explained before. The East Frisian Wadden Sea is protected by sea built barrier islands and with relative small tidal basins. The Hever tidal basin has comparatively larger extensions (e. g. width up to 25 km), a wider spreaded tidal inlet and a lower ebb delta with shoals laying 3m below MSL.

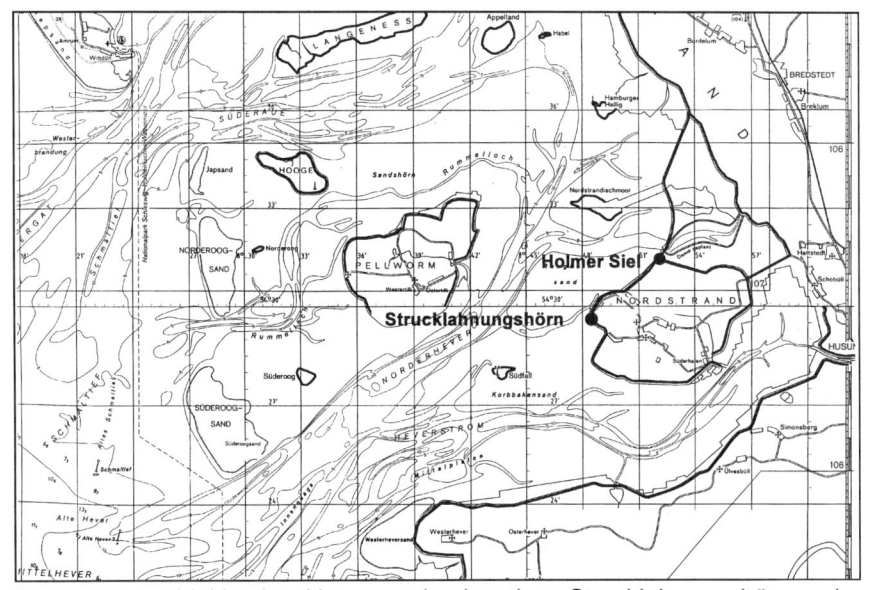

Fig. 5: Hever tidal basin with measuring locations Strucklahnungshörn and Holmer Siel

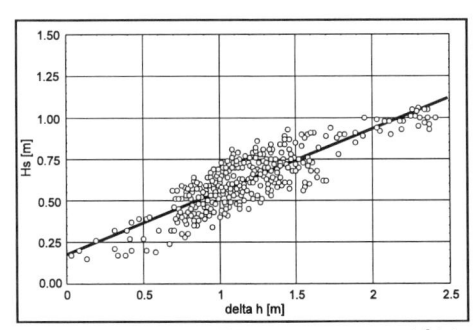

Fig. 6: Norddeich [Hs=0.177+0.378*Δh]

The islands in this system are remnants of the former mainland. According to the hydrodynamical classification of tidal inlets [HAYES 1979] both areas are mesotidal and characterized by mixed energy, tide dominated [NIEMEYER 1990, NIEMEYER et al. 1992].

In the figures 6 to 8 significant wave heights versus wind-induced set-up is plotted for the stations Norddeich (fig. 3) at the mainland coast of the tidal basin of the Norderneyer Seegat and the stations Strucklahnungshörn in the tidal basin of the Hever (fig. 5). At the station Norddeich data represent strictly fitting linear relationship for the whole range of positive Δh-values with relative small scattering. The data of the station Strucklahnungshörn (fig. 7) reflect a more differentiated correlation of both parameters: There is no pronounced relationship between significant wave height and smaller values of Δh. With increasing Δh two distinct values of both parameters can be distinguished which identify

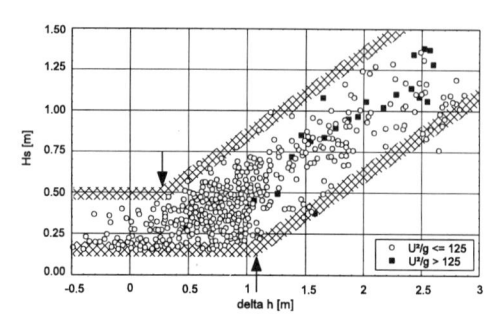

Fig. 7: Significant wave height as function of wind-set-up; Strucklahnungshörn

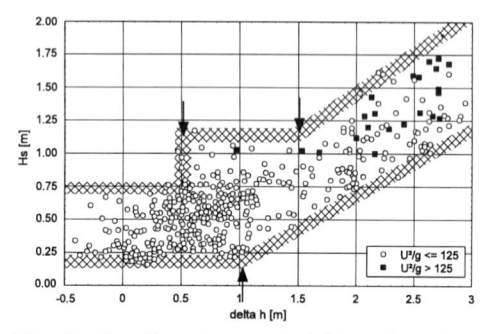

Fig. 8: Significant wave height as function of wind-set-up; Holmer Siel

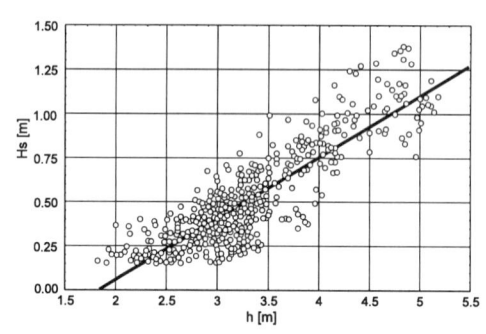

Fig. 9: Strucklahnungshörn [Model: Hs = -0.633 + 0.347 h]

the transition to a relationship for an upper and a lower value. Higher waves increase already for a lower Δh. Smaller waves need a larger critical value of Δh in order to fulfill the same linear relationship. The scattering of the data is significantly higher than for that one of the station Norddeich, which is also valid if only higher wind velocities are taken into consideration.

The station Holmer Siel (fig. 8) is situated landward of Strucklahnungshörn. Regarding the correlation of the same parameters three distinct values of Δh mark significant changes. Similar to the data from Strucklahnungshörn the significant wave heights cover a certain range of values for set-ups in the same order of magnitude. For a wind-set-up of about 0.5 m a rapid increase of the upper value of significant wave height variation is obvious but it remains constant for increasing Δh until the value of 1.5 m. Beyond that figure as well the upper limit as the lower limit of significant wave height variation increase with the set-up whereas the lower one starts already to increase from a set-up of about 1.0 m.

It is of great importance to find a reliable parametrization of local wave climate in order to make the results of field data analysis for a broader range of applications suitable and easily adaptable for coastal managers. Therefore relations are

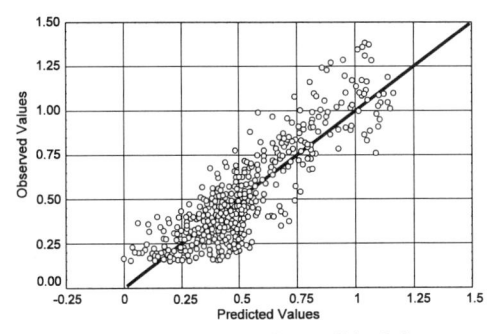

Fig. 10: Strucklahnungshörn [Model: Hs = -0.633 + 0.347 h]; r = 0.859

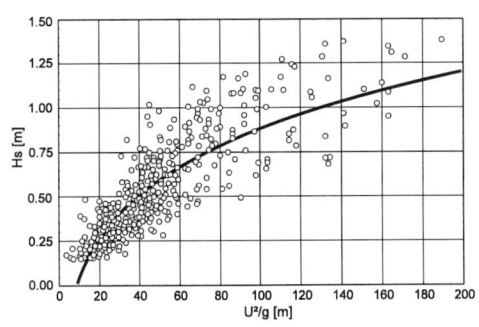

Fig. 11: Strucklahnungshörn [Model: Hs=1.757 + 1.231 $(U^2/g)^{0.166}$]

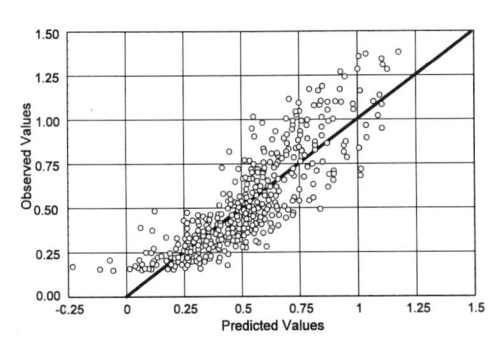

Fig. 12: Strucklahnungshörn [Model: Hs = -1.76 + 1.23 $(U^2/g)^{0.166}$]; r = 0.84994

required which on the one hand vary only to a small extent and can on the other derived from a small number of parametrizated boundary conditions. Comparing the correlation of the significant wave-height and local water depth for the station Strucklahnungshörn (Fig. 9) with the same one island sheltered tidal flat areas and salt marshes at the East Frisian coast, it becomes evident that the first one in respect of its statistical quality must be regarded as insufficient for a reliable parametrization of local wave climate in the basin of the Hever inlet. Correlating significant wave heights with local water-depths and plotting observed against predicted values (Fig. 10), there are deviations of more than ± 35%. The strength of the relationship given by the coefficient of determination is r^2 = 0.74. Doing the same for significant wave heights and local wind velocities (Fig. 11) for the station Strucklahnungshörn the relationship is in agreement with the physical process nonlinear. The results of the nonlinear regression model scatter as well as that one for the linear wave height / water-depth regression (fig. 9) in relation to measured data. Evaluating the fit of the model by the correlation of predicted versus observed values there are also large deviations and the coefficient of determination has a similar value: r^2 = 0.72 (Fig. 12).

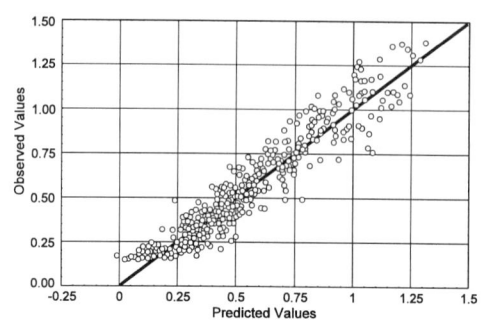

Fig. 13: Strucklahnungshörn [Model: Hs = $-0.81 + 0.21 h + 0.16 (U^2/g)^{0.372}$]; r = 0.950

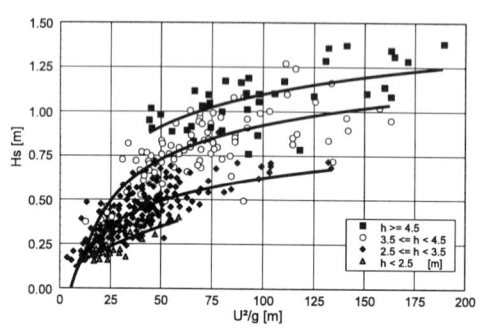

Fig. 14: Strucklahnungshörn, Models: Hs = $f(U^2/g)$ differentiated for four ranges of water depth

The boundary conditions waterdepth and local windfields effect the local wave climate interactively by superimposing each other. These combined effects are described by the following equation: $H_s = f(h, U)$. According to the physical background the model for the estimation of the relationship has a linear term for waterdepth and a nonlinear one for the wind velocity (fig. 13). The scatterplot of values predicted by this model versus observed wave heights show much smaller deviations of about 25% in comparison with those ones which estimate significant wave heights by considering only one boundary condition. The coefficient of determination is $r^2 = 0.903$ which can directly be compared with those ones gained previously, because they are calculated for the same data set. But for a reliable parametrization of local wave climate this result is still insufficient. In order to improve the empirical relationship by a more differentiated consideration of boundary conditions we have to break down the data of water levels and wind into different groups. The creation of four data subsets for different ranges of water depths allow a deeper insight into the processes governing local Wadden Sea wave climate (fig. 14): The significant wave heights are plotted versus the local wind velocities for four distinct water level subsets. Incorporated are additionally the nonlinear regression models for the different groups, but the scattering is significantly large. Due to the limiting condition of the water depth even higher wind velocities result in lower wave heights for lower water depth. The limitation of local wave heights due to water depth are stronger than could be expected in respect of the shallow water breaking criteria. This effect has already been detected for East and West Frisian Wadden Sea areas [NIEMEYER 1983] and is obviously a typical feature of wave climate in island sheltered Wadden Sea areas.

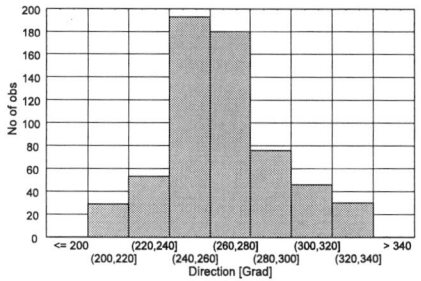

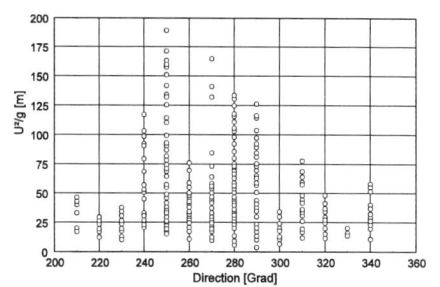

Fig. 15: Wind Strucklahnungshörn; numbers of observations for wind directions from 200° to 340°

Fig. 16: Wind Strucklahnungshörn; distribution of velocities in respect of directions

For coastal areas and particularly regions like the Wadden Sea with their complex morphology the orientation to the open sea and the wind direction are of great importance because they determine both wave growth and energy dissipation. Therefore a differentiation in respect of wind directions seems to be an appropriated approach. The distribution of both the numbers of observations and the wind velocities in respect of the relevant sectors for the Hever inlet are documented in the figures 15 and 16. Due to measurement procedures the distribution of wind observations in these graphs is dependent of wave measurements, because these measurements were triggered by distinct wave height or water depth exceedence levels and reflect therefore situations with higher water levels and higher waves. Highest waves could be expected in the Hever tidal basin for wind from the sector South-West to North-West. Grouping the data sets for the wind directions from 240° to 290° for sectors of 10° is used to produced differentiated data sets for the nonlinear model $H_s = f(h, U)$. The fit of the nonlinear model $H_s = f(h, U)$ after grouping for different wind directions has generally improved (fig. 17-22). Particularly for the sectors 240°, 250° and 270° the relationship has a very high significance.

Fig. 17: Strucklahnungshörn [Model: $H_s = -1.32 + 0.27\,h + 0.27\,(U^2/g)^{0.34}$]
$r = 0.98453$

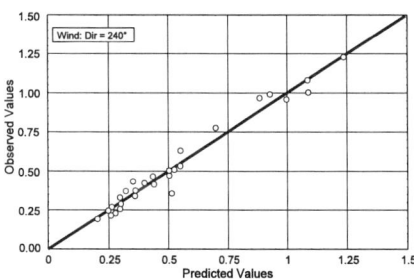

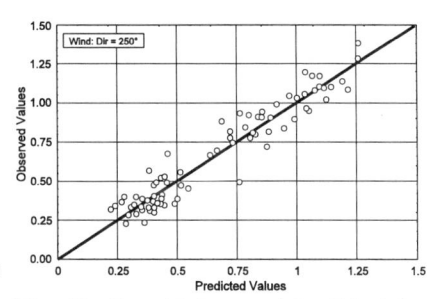

Fig. 18: Strucklahnungshörn [Model: $H_s = -1.84 + 0.27\,h + 0.93\,(U^2/g)^{0.13}$]
$r = 0.96238$

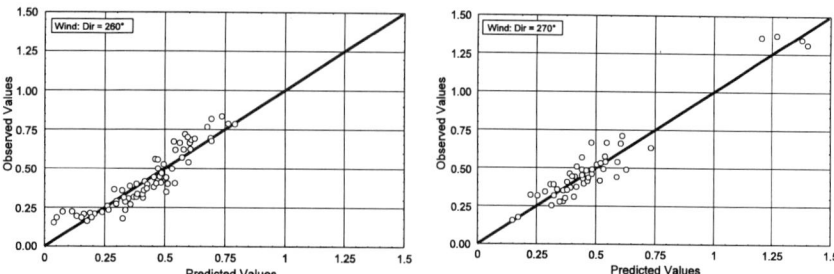

Fig. 19: Strucklahnungshörn [Model: $H_s=1.25+0.22\ h+(-2.58)\ (U^2/g)^{-0.16}$]
r = 0.92363

Fig. 20: Strucklahnungshörn [Model: $H_s= -0.24+0.16\ h+0.003\ (U^2/g)^{1.09}$]
r = 0.95980

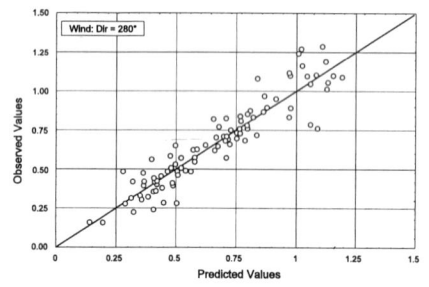

Fig. 21: Strucklahnungshörn [Model: $H_s= -0.63+0.23\ h+0.06\ (U^2/g)^{0.50}$]
r = 0.93208

Fig. 22: Strucklahnungshörn [Model: $H_s= -0.76+0.24\ h+0.11\ (U^2/g)^{0.40}$]
r = 0.93705

Conclusions:

- In island sheltered Wadden Sea areas with large onshore-offshore extensions local wind effects on wave climate cannot be neglected. The simple wave height / waterdepth relation is therefore no longer sufficient for a parametrization of Wadden Sea wave climate.

- The maximum wave heights in Wadden Sea areas are limited by water depths by a critical value below the shallow water breaking limit. They are independent from wind velocities beyond a certain level but depend on wind direction in respect of both wave growth and energy dissipation.

References

HAYES, M.O. [1979]: Barrier island morphology as a function of tidal and wave regime. in: S. P. Leatherman: Barrier islands, Academic Press, New York, pp. 1-27

NIEMEYER, H.D. [1983]: On the Wave Climate at Island Sheltered Wadden Sea Coasts (in German). BMFT-Forschungsbericht MF 0203

NIEMEYER, H.D. [1984]: Hydrographische Untersuchungen in der Leybucht zum Bauvorhaben Leyhörn. Jber. 1983 Forsch.-Stelle f. Insel- u. Küstenschutz, Bd. 35

NIEMEYER, H.D. [1986]: Ausbreitung und Dämpfung des Seegangs im See- und Wattengebiet von Norderney. Jber. 1985 Forsch.-Stelle Küste, Bd. 37

NIEMEYER, H.D. [1987a]: Changing of wave climate due to breaking on a tidal inlet bar. Proc. 20th Intern. Conf. o. Coastal Eng. Taipei, ASCE, New York

NIEMEYER, H.D. [1987b]: Seegang und Biotopzonierung in Wattgebieten. in: Niedersächsischer Umweltminister: Umweltvorsorge Nordsee - Belastungen - Gütesituation - Maßnahmen -. Hildesheim

NIEMEYER, H.D. [1990]: Morphodynamics of tidal inlets. Civ. Eng. Europ. Course Prog. o. Cont. Educ. Coast. Morph., Syll. Delft Univ. o. Tech. Int.-Int. Civ. Eng.

NIEMEYER, H.D. [1991]: Case study Ley Bay: an alternative to traditional enclosure. Proc. 3rd Conf. Coast & Port Eng. i. Devel. Countr., Mombasa/Kenya

NIEMEYER, H.D. ; GÄRTNER, J. & GRÜNE, J. [1992]: Naturuntersuchungen von Wattseegang an der deutschen Nordseeküste. Schlußbericht zum BMFT-Forschungsvorhaben MTK 464 B

RAGUTZKI, G. [1982]: Verteilung der Oberflächensedimente auf den niedersächsischen Watten. Jber. 1980 Forsch.-Stelle f. Insel- u. Küstenschutz, Bd. 32

CHAPTER 15

FALSE WAVES IN WAVE RECORDS AND NUMERICAL SIMULATIONS

Marcos H. Giménez[1], Carlos R. Sánchez-Carratalá[2] and Josep R. Medina[3]

ABSTRACT

It is common practice to consider the random waves as a succession of discrete waves characterized by individual amplitudes and periods. The zero-up-crossing criterion isolates some discrete waves that are not physical waves. The orbital criterion avoids these "false waves". As a result, the orbital criterion proves to be more consistent and robust, and to have a less variability. The selection of the discretization criterion results in some significant differences in the wave statistics, which are analyzed. As an example, while the mean period for the zero-up-crossing criterion is T_{02}, the mean period for the orbital criterion is T_{01}.

INTRODUCTION

Regular waves can be characterized by amplitude and period, and random waves may be described by the energy spectrum. However, it is common practice to consider the random waves as a succession of "discrete waves" characterized by individual amplitude and period.

---

[1] Ass. Prof., Dept. of Applied Physics, EUITI, Universidad Politécnica de Valencia, Camino de Vera s/n, 46022, Valencia, SPAIN.

[2] Ass. Prof., Dept. of Applied Physics, ETSICCP, Universidad Politécnica de Valencia, Camino de Vera s/n, 46022, Valencia, SPAIN.

[3] Professor, Director of the Lab. of Ports and Coasts, Dept. of Transportation, ETSICCP, Universidad Politécnica de Valencia, Camino de Vera s/n, 46022, Valencia, SPAIN.

Unfortunately, a variety of reasonable criteria for discretizing waves have been proposed by different authors. In fact, any method used to define a discrete wave in regular waves could be extended to the case of random waves.

A number of papers are related to wave statistics and may be affected by the wave discretization procedure. Moreover, a variety of subjective criteria are used for neglecting small waves in the analysis.

Giménez et al. (1994) have proposed an orbital criterion for discretizing waves. Using numerical simulations, the authors have proved that this method is more consistent and robust than the zero-up-crossing criterion. These results are in good agreement with the observations given by Pires-Silva and Medina (1994) analyzing wave records off the coast of Portugal.

This paper describes first the most common wave discretization methods, and summarizes the concepts and properties of "orbital wave" and "false wave". The advantages of the orbital criterion are presented, including consistency, robustness and a less variability. Finally, the influence of the wave discretization criteria on the wave statistics is analyzed using numerical simulations.

WAVE DISCRETIZATION CRITERIA

The more commonly used wave discretization criteria are the following:

* The ZUC criterion

In the zero-up-crossing criterion, a discrete wave is limited by two consecutive up-crossings of the mean level. Following Rice (1954), Longuet-Higgins (1958) showed that for linear random waves the mean period using the ZUC criterion is T_{02}, where T_{ij} is given by:

$$T_{ij} = \sqrt[j-i]{\frac{m_i}{m_j}} \qquad (1)$$

where m_n is the nth moment of the energy spectrum $S(f)$,

$$m_n = \int_0^\infty f^n S(f) df \qquad (2)$$

* The ZDC criterion

In the zero-down-crossing criterion, a discrete wave is limited by two consecutive down-crossings of the mean level. For linear random waves, the ZUC and the ZDC criteria are statistically equivalent. Therefore, the mean period using the ZDC criterion is also T_{02}.

* The crest-to-crest criterion

In the crest-to-crest criterion, a discrete wave is limited by two consecutive maxima of the surface displacement function. From Rice (1954), it can be proved that the mean period using the crest-to-crest criterion is T_{24}.

THE ORBITAL CRITERION

For linear waves, the free surface elevation in a fixed point, $\eta(t)$, can be modeled by:

$$\eta(t) = \sum_{i=1}^{M} c_i \cos(2\pi f_i t + \varphi_i) \tag{3}$$

where the frequencies f_i are $i\Delta f$, the phases φ_i are random variables distributed uniformly over the interval $[0, 2\pi[$, and the amplitudes c_i are such that over any frequency interval $[f_i, f_i + \Delta f[$ is:

$$\frac{1}{2} c_i^2 = \int_{f_i}^{f_i + \Delta f} S(f) \, df \tag{4}$$

The Hilbert transform of $\eta(t)$ is:

$$\hat{\eta}(t) = \sum_{i=1}^{M} c_i \sin(2\pi f_i t + \varphi_i) \tag{5}$$

The functions $\eta(t)$ and $\hat{\eta}(t)$ can be taken as the real and the imaginary part, respectively, of the analytical function $AF(t)$:

$$AF(t) = \sum_{i=1}^{M} c_i \exp[j(2\pi f_i t + \varphi_i)] = \eta(t) + j\hat{\eta}(t) \tag{6}$$

where $j = \sqrt{-1}$ is the imaginary unit. $AF(t)$ can be expressed in the form:

$$\left. \begin{array}{l} AF(t) = A(t) \exp[j\theta(t)] \\ A(t) = \sqrt{\eta^2(t) + \hat{\eta}^2(t)} \\ \theta(t) = \arctan \dfrac{\hat{\eta}(t)}{\eta(t)} \end{array} \right\} \tag{7}$$

where $A(t)$ is the wave envelope and $\theta(t)$ is the phase angle. As noted by Medina and Hudspeth (1987), $AF(t)$ represents the orbital movement of a point floating on the sea surface.

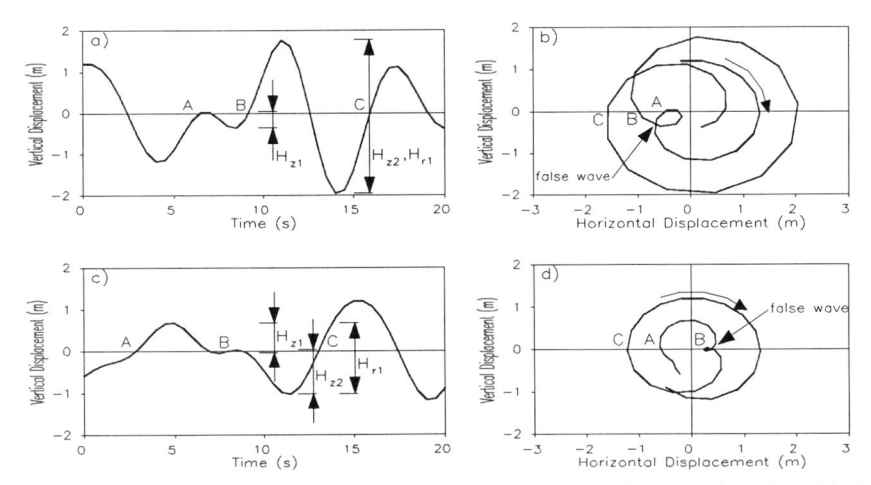

Figure 1.- ZUC criterion vs. orbital criterion: (a) example of time series; (b) orbital analysis of a); (c) example of time series; (d) orbital analysis of c).

According to Giménez et al. (1994), the orbital criterion defines a discrete wave as corresponding to a 2π advance of the phase angle in the complex plane. Figures 1(a) and 1(c) show two pieces of numerical simulation from a JONSWAP spectrum. Figures 1(b) and 1(d) represent the corresponding analytical functions AF(t). The ZUC and the orbital waves are denoted by H_z and H_r respectively. In both pieces two ZUC waves (AB and BC), but only one orbital wave (AC), are present. The first small wave in Figure 1(a) is not considered by the orbital criterion. Furthermore, the two ZUC waves in Figure 1(c) are only one, and higher, wave in the orbital criterion.

Giménez et al. (1994) define false wave as any discrete wave that does not correspond to a 2π advance in the complex plane. Examples of these false waves are shown in Figures 1(b) and 1(d). In the same reference, the authors prove (both, mathematically and numerically) that the mean period using the orbital criterion is T_{01}. Furthermore, the discrepancy between T_{01} and T_{02}, is completely due to the presence of false waves.

ADVANTAGES OF THE ORBITAL CRITERION

Because of its dependence of m_4, the crest-to-crest criterion is very sensitive to the cut-off frequency. On the other hand, the ZUC and the ZDC are statistically equivalent for linear random waves. Therefore, the ZUC criterion has been used as reference for analyzing the advantages of the orbital criterion. Giménez et al. (1994) have carried out that comparison, and have come to the next conclusions:

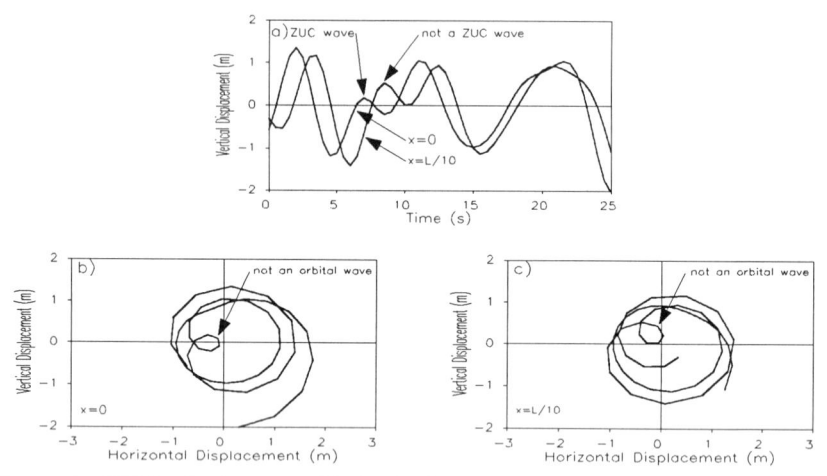

Figure 2.- ZUC criterion vs. orbital criterion: (a) time series of two close points; (b) orbital analysis of the first point; (c) orbital analysis of the second point.

* Consistency

Figure 2(a) shows a numerical simulation of two time series corresponding to two points in the sea surface separated by 10% of the mean wavelength L. There is a perturbation that is a ZUC wave in x=0, but not in x=L/10. This physical inconsistency can be solved using the orbital criterion. Figures 2(b) and 2(c) show that the perturbation is a false wave.

Another problem of the ZUC criterion is the wide variety of subjective thresholds for neglecting small invalid waves (see Rye, 1974; van Vledder, 1983; Thompson and Seeling, 1984; Mansard and Funke, 1984; Mase and Iwagaki, 1986). For the orbital criterion, any wave that does not correspond to a 2π advance of the phase angle is not an actual wave. No additional thresholds are required.

* Robustness

Figure 3(a) shows a piece of simulation, and the same record when a 5% of white noise is added. Additional ZUC waves appear due to the presence of noise. However, Figure 3(b) shows that these waves are not orbital waves. In fact, most of the additional ZUC waves due to noise are false waves. Therefore, the ZUC criterion is more sensitive to noise than the orbital criterion.

Giménez et al. (1994) have analyzed the sensitivity to noise using numerical simulations. The mean period of orbital waves is underestimated about 2% when a 2% of white noise is added. On the contrary, the resulting error in the mean period of ZUC waves is about 10%. The underestimations with a 5% of white noise are about 5% for orbital waves, and 20-25% for ZUC waves.

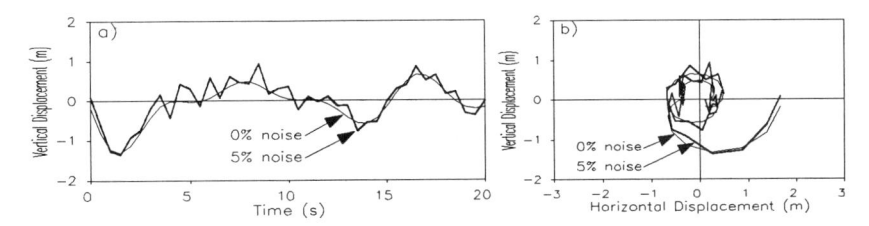

Figure 3.- Influence of white noise: (a) time series; (b) orbital analysis.

* Less variability

As proved in Appendix A, the variabilities of the mean periods T_{01} and T_{02} are:

$$CV^2[T_{01}] = \frac{\int_0^\infty S^2(f)[T_{01}f - 1]^2 df}{m_0^2 T_R} \qquad (8)$$

$$CV^2[T_{02}] = \frac{\int_0^\infty S^2(f)[(T_{02}f)^2 - 1]^2 df}{4 m_0^2 T_R} \qquad (9)$$

where CV[.] is the coefficient of variation and T_R is the length of the record. The result depends on the spectral shape. Table 1 shows the variability for JONSWAP type spectra (see Goda, 1985) with a peak frequency of 0.1 Hz and different values of the peak enhancement parameter γ. The mean period T_{01} proves to have a less variability than T_{02}.

γ	$CV[T_{01}]$	$CV[T_{02}]$
1	$0.677/\sqrt{T_R}$	$0.695/\sqrt{T_R}$
2	$0.686/\sqrt{T_R}$	$0.744/\sqrt{T_R}$
3.3	$0.678/\sqrt{T_R}$	$0.771/\sqrt{T_R}$
5	$0.655/\sqrt{T_R}$	$0.776/\sqrt{T_R}$
7	$0.623/\sqrt{T_R}$	$0.762/\sqrt{T_R}$
10	$0.577/\sqrt{T_R}$	$0.728/\sqrt{T_R}$

Table 1.- Variability of T_{01} and T_{02}.

DISTRIBUTIONS OF WAVE HEIGHTS AND PERIODS

The mean period is not the only statistical parameter that is affected by the selected wave discretization criterion. The parameters related with discrete waves are altered in two ways. First of all, the presence of false waves varies significantly the number and characteristics of small waves. As a second effect, the total number of orbital waves in a record is less than the number of ZUC waves. Therefore, the relative number of higher waves is slightly greater in the orbital criterion than in the ZUC criterion.

A comparison between the statistics for orbital and ZUC waves has been developed using numerical simulations. These simulations were obtained using a DSA-FFT algorithm (see Tuah and Hudspeth, 1982) and a JONSWAP type spectrum (see Goda, 1985) with $N=8192$ points and a time interval $\Delta t=0.2$ sec. 1000 simulations were carried out for different values of the peak enhancement parameter ($\gamma=1,2,3.3,5,7,10$). The number of simulated ZUC waves varies from about 180000 (for $\gamma=10$) to 210000 (for $\gamma=1$). The results for all the indicated values of γ have been analyzed, and figures corresponding to the extremal values $\gamma=1$ and $\gamma=10$ are included in this paper.

The methodology developed by Sobey (1992) was used for the statistical analysis of the simulations. Therefore, the wave periods were obtained by linear interpolation in the neighborhood of the zero-up-crossings. The wave heights were determined from quadratic interpolation at the crest and the trough. The wave heights were normalized by H_{rms}, and the wave periods by T_{01}. Finally, the waves were aggregated, according to their normalized height and period, into a joint histogram in 0.025x0.025 dimensionless bins.

Figures 4(a) and 4(b) show the joint distributions of wave heights and periods, for $\gamma=1$, corresponding to the orbital and the ZUC criteria respectively. Figures 5(a) and 5(b) show the same distributions for $\gamma=10$. For both criteria, the distribution is bimodal. The external contour corresponds to points with a probability of 0.01. The other contours correspond to probabilities of 0.20, 0.40, 0.60 and so on. The values in the thicker lines are 0.01, 1.00 and 2.00.

A maximum is located near the point corresponding to H_{rms} and T_{01}. This maximum becomes higher for narrower spectra, and is slightly greater in the orbital criterion. On the other hand, the maximum corresponding to small waves has a less value for orbital waves.

Figures 6(a) and 6(b) show the distributions of wave heights for $\gamma=1$ and $\gamma=10$ respectively. The probability of both the orbital and the ZUC waves is underestimated by the Rayleigh distribution near its mode, and overestimated in the ranges of small and high waves. As expected according to the narrow band assumption, the fitting is slightly better for narrower spectra.

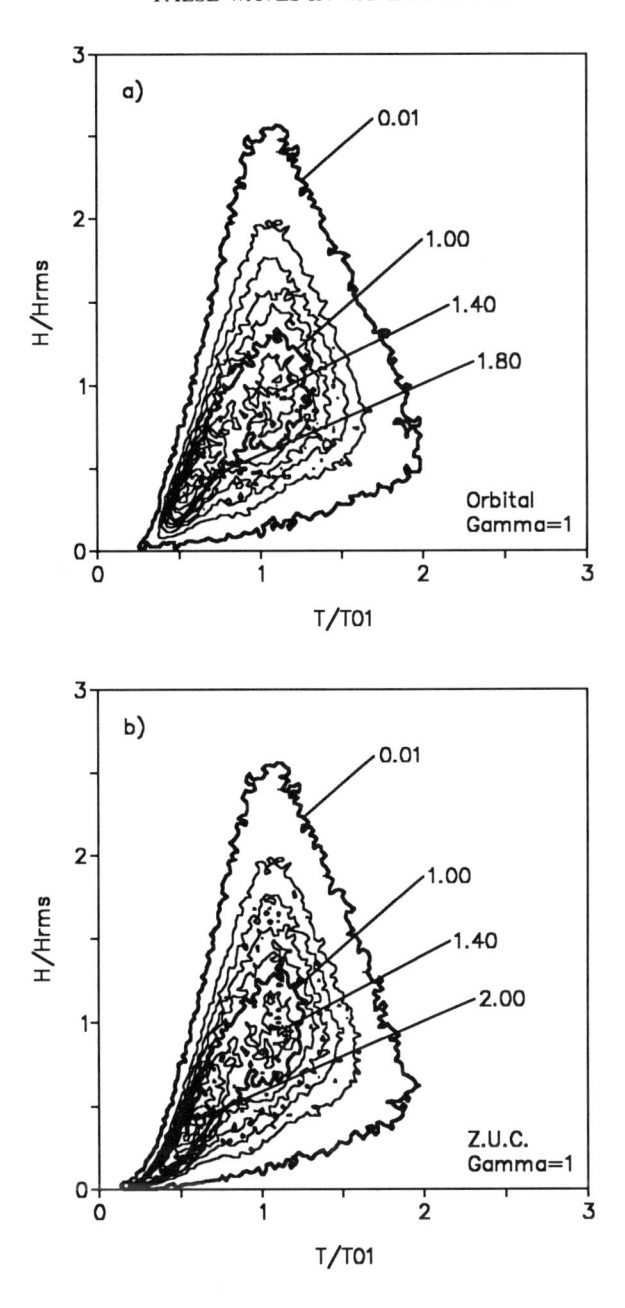

Figure 4.- Joint distributions of wave heights and periods for $\gamma = 1$: (a) orbital waves; (b) ZUC waves.

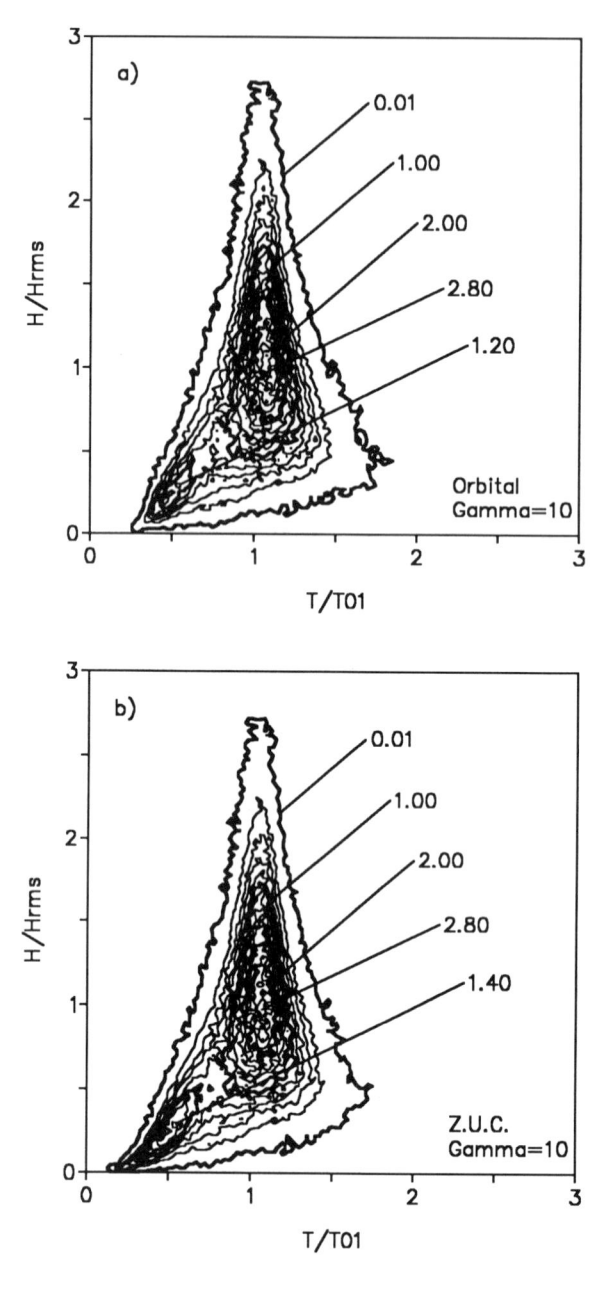

Figure 5.- Joint distributions of wave heights and periods for $\gamma=10$: (a) orbital waves; (b) ZUC waves.

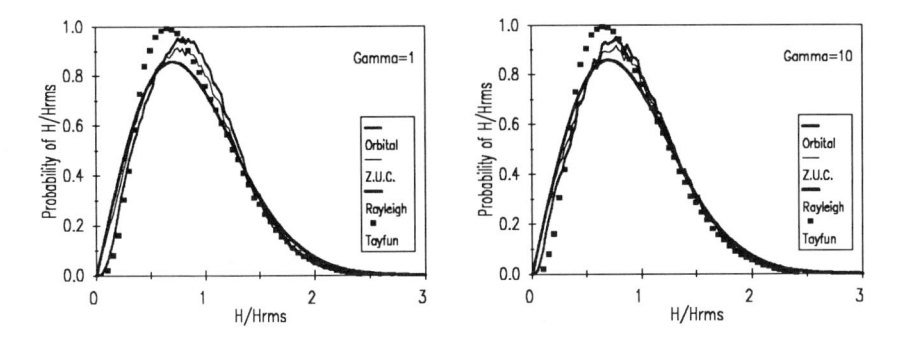

Figure 6.- Distributions of wave heights: (a) $\gamma=1$; (b) $\gamma=10$.

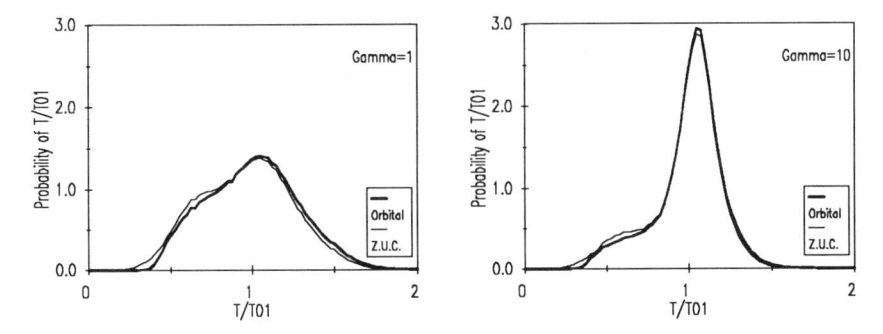

Figure 7.- Distributions of wave periods: (a) $\gamma=1$; (b) $\gamma=10$.

A comparison between the distributions of orbital and ZUC wave heights leads to the next conclusions: in the orbital criterion, the probability is less in the range of small waves, greater around the mode, and nearly the same in the range of hight waves. This behaviour could be expected due to the presence of false waves in the ZUC criterion.

Figures 6(a) and 6(b) also show the distribution of Tayfun (1981) corresponding to $\nu=0.2$. This distribution is based on the values of the envelope with a lag of $T_{o1}/2$, and therefore is supposed to be more appropriate for orbital waves than for ZUC waves.

Figures 7(a) and 7(b) show the distributions of wave periods for $\gamma=1$ and $\gamma=10$ respectively. As noted above, the mean period is T_{01} for orbital waves and T_{02} for ZUC waves. It can be concluded from these distributions that the orbital waves have higher periods than the ZUC waves, as expected.

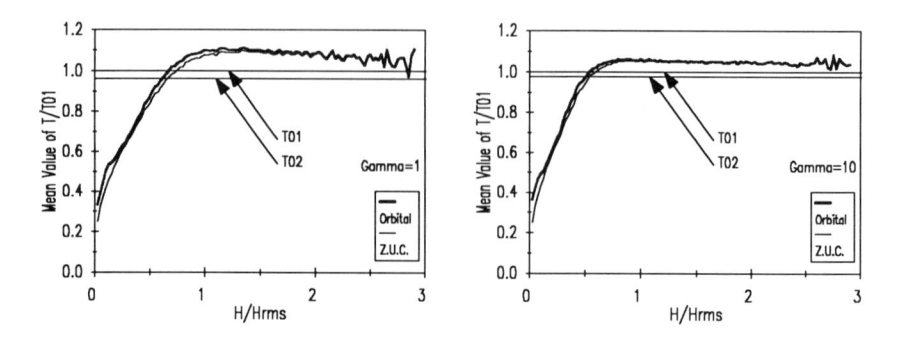

Figure 8.- Variation of the mean wave period as a function of the wave height: (a) $\gamma=1$; (b) $\gamma=10$.

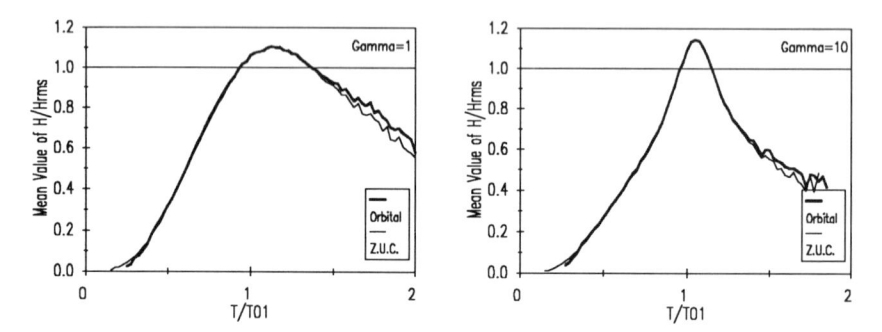

Figure 9.- Variation of the mean wave height as a function of the wave period: (a) $\gamma=1$; (b) $\gamma=10$.

Figures 8(a) and 8(b) show the variation of the mean wave period for $\gamma=1$ and $\gamma=10$, respectively, as a function of the wave height. The mean value of the orbital waves is greater for every wave height, especially in the range of small waves. Waves with $H/H_{rms}>0.5$-0.7 have a mean period over T_{01}, with a certain tendency to this value for higher waves.

Figures 9(a) and 9(b) show the variation of the mean wave height for $\gamma=1$ and $\gamma=10$, respectively, as a function of the wave period. The greatest mean wave heights correspond to periods around T_{01}. The orbital waves have a larger mean wave height than the ZUC waves in the range of high periods.

CONCLUSIONS

The ZUC criterion commonly used to discretize wave records generates discrete waves that do not correspond to physical waves. The orbital criterion avoids these "false waves", and proves to be more consistent and robust, and to have a less variability. Finally, the wave statistics are altered by the selection of the wave discretization criteria. The differences are basically located in the range of small waves, but some slight variations are also found for larger waves.

ACKNOWLEDGEMENTS

The authors gratefully acknowledge the financial support provided by the Dirección General de Investigación Científica y Técnica (PB92-0411).

APPENDIX A: VARIABILITY OF THE MEAN PERIODS

Random waves, taken as a Gaussian ergodic stochastic process, are described by the energy spectrum $S(f)$. When this spectrum is estimated from records of finite length T_R, the resulting M components are independent random variables. If T_R is large enough, then every component is distributed as a chi-squared law with two degrees of freedom. The mean and variance of this distribution are:

$$\left. \begin{array}{l} E[S_{m,R}] = S_m \\ \sigma^2[S_{m,R}] = S_m^2 \end{array} \right\} \quad (A.1)$$

where $S_m = S(m\Delta f)$ with $\Delta f = 1/T_R$, and $S_{m,R}$ is the corresponding estimation. The ith moment of the estimated spectrum is:

$$m_{i,R} = \sum_{m=1}^{M} f_m^i S_{m,R} \Delta f \quad (A.2)$$

From (A.1) and (A.2), it can be proved that the mean of the random variable $m_{i,R}$ is:

$$E[m_{i,R}] = m_i \quad (A.3)$$

and the covariance between the moments $m_{i,R}$ and $m_{j,R}$ is:

$$C[m_{i,R}, m_{j,R}] = \frac{1}{T_R} \int_0^\infty f^{i+j} S^2(f) df \quad (A.4)$$

The square of the coefficient of variation of $m_{i,R}$ is:

$$CV^2[m_{i,R}] = \frac{C[m_{i,R}, m_{i,R}]}{E^2[m_{i,R}]} = \frac{1}{m_i^2 T_R} \int_0^\infty f^{2i} S^2(f) df \qquad (A.5)$$

The period $T_{ij} = (m_i/m_j)^k$, where $k = 1/(j-i)$, cannot be determined when $S(f)$ is unknown. On the other hand, $T_{ij,R} = (m_{i,R}/m_{j,R})^k$ can be obtained from the estimated spectrum. However, $T_{ij,R}$ is a random variable, and presents a certain variability. The objective of this appendix is to obtain an expression for that variability.

A Taylor expansion of $T_{ij,R}$ and an integration lead to:

$$E[T_{ij,R}] = \int_0^\infty \int_0^\infty T_{ij,R} p(m_{i,R}, m_{j,R}) dm_{i,R} dm_{j,R} \approx$$
$$\approx T_{ij}\left[1 + \frac{k(k-1)}{2} CV^2[m_{i,R}] + \frac{k(k+1)}{2} CV^2[m_{j,R}] - k^2 \frac{C[m_{i,R}, m_{j,R}]}{m_{i,R} m_{j,R}}\right] \qquad (A.6)$$

The same method leads to:

$$E[T_{ij,R}^2] = \int_0^\infty \int_0^\infty T_{ij,R}^2 p(m_{i,R}, m_{j,R}) dm_{i,R} dm_{j,R} \approx$$
$$\approx T_{ij}^2\left[1 + k(2k-1)CV^2[m_{i,R}] + k(2k+1)CV^2[m_{j,R}] - 4k^2 \frac{C[m_{i,R}, m_{j,R}]}{m_{i,R} m_{j,R}}\right] \qquad (A.7)$$

Therefore:

$$CV^2[T_{ij,R}] = \frac{E[T_{ij,R}^2] - E^2[T_{ij,R}]}{E^2[T_{ij,R}]} \approx k^2 \left[CV^2[m_{i,R}] + CV^2[m_{j,R}] - 2\frac{C[m_{i,R}, m_{j,R}]}{m_{i,R} m_{j,R}}\right] \qquad (A.8)$$

From (A.4) and (A.5), if T_R is large enough, then $CV[m_{i,R}]$, $CV[m_{j,R}]$ and $C[m_{i,R}, m_{j,R}]$ are much smaller than one, and, according to (A.6):

$$E[T_{ij,R}] = T_{ij} \qquad (A.9)$$

Therefore, the random variable $T_{ij,R}$ can be used for estimating T_{ij}, and the expression (A.8) gives the variability of the estimation. For T_{02} the result is:

$$CV^2[T_{02,R}] = \frac{1}{4}\left[CV^2[m_{0,R}] + CV^2[m_{2,R}] - 2\frac{C[m_{0,R}, m_{2,R}]}{m_{0,R}m_{2,R}}\right] =$$

$$= \frac{1}{4}\left[\frac{1}{m_0^2 T_R}\int_0^\infty S^2(f)df + \frac{1}{m_2^2 T_R}\int_0^\infty f^4 S^2(f)df - \frac{2}{m_0 m_2 T_R}\int_0^\infty f^2 S^2(f)df\right] = \quad (A.10)$$

$$= \frac{1}{4m_0^2 T_R}\int_0^\infty S^2(f)\left[1 + \frac{m_0^2}{m_2^2}f^4 - 2\frac{m_0}{m_2}f^2\right]df = \frac{1}{4m_0^2 T_R}\int_0^\infty S^2(f)\left[(T_{02}f)^2 - 1\right]^2 df$$

This expression was obtained by Cavanié (1979). On the other hand, the variability of the estimation of T_{01} is:

$$CV^2[T_{01,R}] = CV^2[m_{0,R}] + CV^2[m_{1,R}] - 2\frac{C[m_{0,R}, m_{1,R}]}{m_{0,R}m_{1,R}} =$$

$$= \frac{1}{m_0^2 T_R}\int_0^\infty S^2(f)df + \frac{1}{m_1^2 T_R}\int_0^\infty f^2 S^2(f)df - \frac{2}{m_0 m_1 T_R}\int_0^\infty f S^2(f)df = \quad (A.11)$$

$$= \frac{1}{m_0^2 T_R}\int_0^\infty S^2(f)\left[1 + \frac{m_0^2}{m_1^2}f^2 - 2\frac{m_0}{m_1}f\right]df = \frac{1}{m_0^2 T_R}\int_0^\infty S^2(f)\left[T_{01}f - 1\right]^2 df$$

REFERENCES

Cavanié, A.G. (1979). Evaluation of the Standard Error in the Estimation of Mean and Significant Wave Heights as well as Mean Period from Records of Finite Length. *Proc. Int. Conf. Sea Climatology*, 73-88.

Giménez, M.H., Sánchez-Carratalá, C.R. and Medina, J.R. (1994). Analysis of False Waves in Numerical Sea Simulations. *Ocean Engineering*, 21 (8), 751-764.

Goda, Y. (1985). *Random Seas and Design of Maritime Structures.* University of Tokyo Press.

Longuet-Higgins, M.S. (1958). On the Intervals between Successive Zeros of a Random Function, *Proc. Roy. Soc.*, Ser.A, 246, 99-118.

Mansard, E.P.D. and Funke, E.R. (1984). Variabilité Statistique des Paramètres des Vagues, *Proc. Int. Symp. Marit. Struct. in the Mediterranean Sea*, Athens, Greece, 1.45-1.59.

Mase, H. and Iwagaki, Y. (1986). Wave Group Analysis from Statistical Viewpoint, *Proc. Ocean Struct. Dynamics Symp.*, Corvallis, Oregon, 145-157.

Medina, J.R. and Hudspeth, R.T. (1987). Sea States Defined by Wave Height and Period Functions, *Seminar on Wave Analysis and Generation in Laboratory Basins*, 22nd IAHR Congress, Laussanne, Switzerland, 249-259.

Pires-Silva, A.A. and Medina, J.R. (1994). False Waves in Wave Records. *Ocean Engineering*, 21 (8), 765-770.

Rice, S.O. (1954). Mathematical Analysis of Random Noise, *Selected Papers on Noise and Stochastic Processes*, Nelson Wax, ed., Dover Publications, Inc., New York.

Rye, H. (1974). Wave Group Formation among Storm Waves, *Proc. 14th ICCE*, Copenhagen, Denmark, 164-183.

Sobey, R.J. (1992). The Distribution of Zero-Crossing Wave Heights and Periods in a Stationary Sea State. *Ocean Engineering*, 19 (2), 101-118.

Tayfun, M.A. (1981). Distribution of Crest-to-Trough Wave Heights. *Journal of the Waterway, Port, Coastal and Ocean Division*, 107 (WW3), 149-158.

Thompson, E.F. and Seeling, W.N. (1984). High Wave Grouping in Shallow Water, *J. Waterway, Port, Coastal and Ocean Engng.*, 110 (2), 139-157.

Tuah, H. and Hudspeth, R.T. (1982). Comparisons of Numerical Random Sea Simulations, *J. Waterway, Port, Coastal and Ocean Division*, 108 (WW4), 569-589.

van Vledder, G.Ph. (1983). Verification of the Kimura Model for the Description of Wave Groups, Part 2, M.Sc. Thesis, Tech. Univ. Delft.

CHAPTER 16

MEASURING WAVES WITH MANOMETER TUBES

David J Hanslow[1], Peter Nielsen[2] & Kevin Hibbert[1]

Abstract
A new type of wave gauge has been developed for the measurement of waves near a beach or near an existing coastal structure. It consists of a nylon tube with diameter between 0.5 and 1cm and length up to 500m. The seaward end of the tube is open so that the wave induced pressure fluctuations can be transmitted through the water in the tube. The landward end, which is conveniently above the water, is fitted with a pressure transducer. This is simple and reliable technology well suited for use in developing countries. Maintenance and running coasts are also very low. The frequency response function of the system is somewhat complicated but a workable formula is presented and "once and for all" calibration of the system can be done very easily.

Background
The authors were prompted to look for a new type of nearshore wave gauge by the difficulty encountered with getting representative wave data for the Brunswick Heads field site during storm conditions. The problem with that particular site is that the nearest offshore waverider, which is off Cape Byron, tends to go a drift during "interesting" weather conditions. This leads to increased difficulty with interpreting the most interesting data.

The above mentioned example is not isolated. There is a considerable general need for a simple, inexpensive method of measuring nearshore wave heights, i e within 50 to 500 metres of a beach or an existing structure.

The existing devices for nearshore wave measurements include surface piercing gauges, "Schwartz poles", and bottom mounted current meters and pressure transducers. All of these are well proven but not without problems.

---

[1] Coast and Floods Branch, New South Wales Public Works Department, Po Box 5280 Sydney 2001, Australia.

[2] Department of Civil Engineering, University of Queensland, Brisbane, Australia 4072. Fax +61 7 365 4599, e-mail: nielsen@civil.uq.oz.au.

Not all locations allow the installation of a "Schwartz pole" because the supporting structure becomes very expensive with increasing depth and it may present a navigation hazard.

One problem with bottom mounted pressure transducers/current meters is that they tend to get lost to trawlers etc. Secondly, the fact that it is impossible to check their performance during the deployment has lead to many diappointments when the instruments, upon recovery, have been found to contain no useful data. The tubes of the present system are usually buried in the sand and thus protected from "trawler attack", see Figure 1.

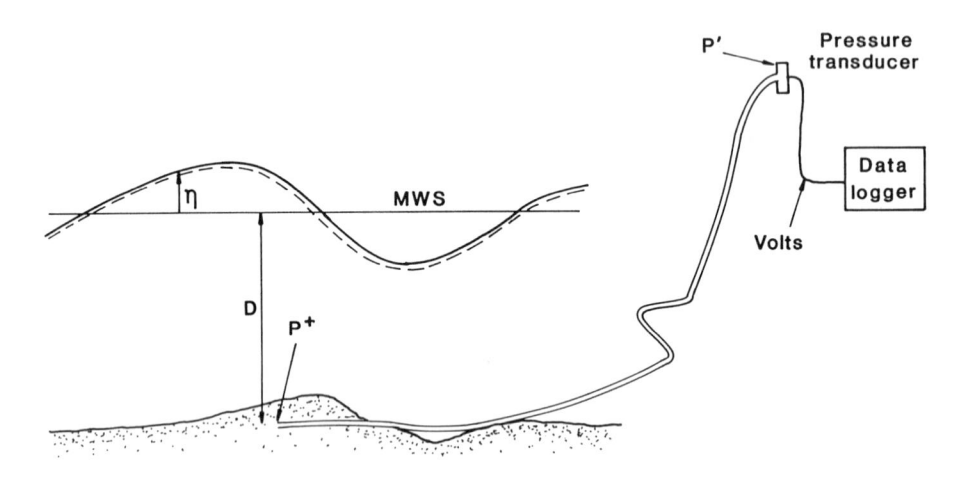

Figure 1: The idea is to estimate the water surface elevation time series $\eta(t)$ on the basis of measured pressure fluctuations $p'(t)$ at the landward end of a water filled nylon tube.

A new type of cheap and reliable nearshore wavegauge

The present study has investigated the possibility of measuring waves by monitoring the pressure fluctuations $p'(t)$ at the landward end of nylon tubes of *10mm* OD and lengths between 50 and 500 metres. The seaward ends of the tubes are exposed to the wave induced pressure fluctuations $p^+(t)$, See Figure 1.

The main problem addressed here is that of estimating of p^+ on the basis of p'. However, the second step in the process of getting wave data from p', namely determining the surface elevation η from p^+ is also discussed briefly.

Initial field testing

Initial field testing has been performed with the new tube-transducer system in order to assure that the transmitted pressure signal $p'(t)$ is of adequate strength and that the system is easy to operate in the field. The system performed very satisfactorily in these tests. The transducer connected to the tubes which are permanently installed at Brunswick Heads. Connection and evacuation of the sytem takes only a few minutes.

The pressure signal was initially recorded by a chart recorder in the field. The $p'(t)$-signal has a low noise level and is of adequate strength to be recorded with the chart recorders 0-$50mV$ range. The transducers applied in the present study are Model AB Pressure Transducers from Data Instruments Inc, Ma, USA. In later field tests the data was recorded in digital form on a portable PC. Various combinations of pressure transducers and recording equipment can be used but it should be noted that the working range for the pressure transducer should be about 0.5 - 1 atmosphere absolute, i e, the working pressure is below atmospheric pressure.

Pressure transducer on the bed

Relationships between the dynamic bottom pressure p^+ ($p(t) = \bar{p} + p^+(t)$) and the surface elevation η may be taken from linear wave theory for monocromatic small amplitude waves

$$\eta = \frac{p^+}{\rho g} \cosh kh \qquad (1)$$

where k is the wave number $2\pi/\lambda$, λ is the wave length and h is the water depth.

For irregular, non-linear waves, local approximations may be used. Nielsen (1989) recommended the formula

$$\eta_n = \frac{p_n^+}{\rho g} \exp\left\{\frac{2}{3} \frac{-p_{n-1}^+ + 2p_n^+ - p_{n+1}^+}{p_n^+ g \delta_t^2}(h + \frac{p_n^+}{\rho g})\right\} \qquad (2)$$

based on a measured time series $p_1^+, p_2^+, p_3^+, \cdots$.

As an alternative to the local approximations approach, the classical spectrum transformation approach may be applied. With this method a Fourier transform is applied to the pressure record. Then each spectral estimate is transformed in accordance with Equation (1). Finally, the inverse Fourier transform is applied to the transformed spectral estimates to obtain an estimate of the surface elevation time series. The relative merits of the two methods has been discussed by Nielsen (1989).

Transducer burried in the bed

It is possible that a layer of sand on top of the "open" end of the tube can cause extra damping. Most likely however, the effect is negligible for typical beach sand and typical wave frequencies.

Sleath (1970) and Maeno & Hasagawa (1987) measured pore pressures inside the bed simultaneously with pressures at the bed surface. The pressure amplitude ratios agreed reasonably with the formula

$$|p^+| = \rho g \frac{H}{2} \frac{\cosh k(z+h_1)}{\cosh kh \; \cosh kh_1} \qquad (3)$$

where z is the transducer elevation measured from the sand surface h is the water depth and h_1 is the thickness of the sand bed, see e g Sleath (1984).

According to this formula it makes very little difference whether a transducer, at a fixed depth below the water surface, is covered with sand or not. Consider for example, the situation in Figure 2.

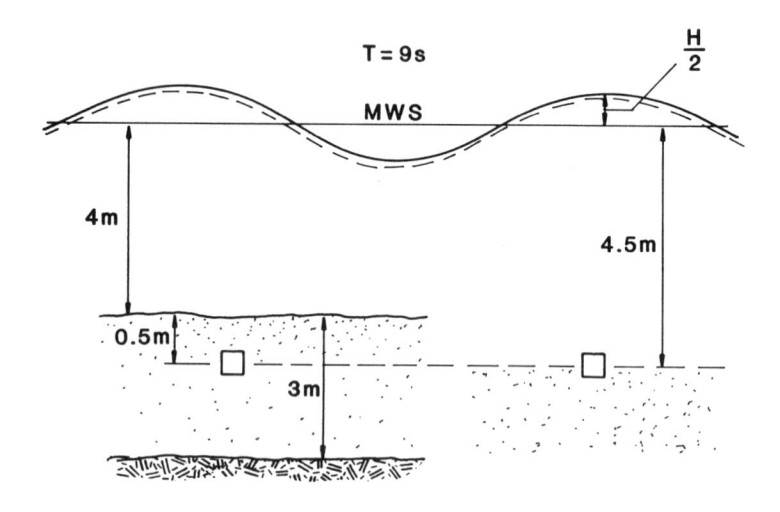

Figure 2: The pressure felt by a transducer *4.5m* below the MWS under a *9s* wave changes very little due to a sand cover of *0.5m*.

A pressure transducer is placed *4.5m* below the MWS. The wave period is 9s, and we consider the situation where the sand level is at the transducer level as well as the situation where the sand level is 0.5m above the transducer. The bed is assumed impermeable below -7m.

Linear wave theory plus Equation (3) gives $|p^+| = 0.889\rho gH/2$ for the "uncovered case" and $|p^+| = 0.885\rho gH/2$ for the "covered" case. This difference is negligible compared to the accuracy of linear wave theory. It should be noted however that finer material like silt or mud will provide a stronger damping of the pressure signal than 0.2mm sand.

The speed of pressure waves in a flexible tube

The frequency response of the tube-transducer system in Figure 1 depends crically on the speed of pressure waves in the tube. This speed, in turn, is a function of the compressibility of the water, the rigidity of the tube walls and, if air bubbles are present, of the bubble concentration.

In an infinite fluid, the speed c of a plane sound wave is determined by the density ρ and the compressibility K

$$c = \frac{1}{\sqrt{\rho K}} \qquad (4)$$

The speed of sound in sea water is approximately 1500m/s, corresponding to a compressibility K of $4.4 \cdot 10^{-10} Pa^{-1}$.

If the fluid is contained in a flexible tube, the speed of sound will be reduced in accordance with the formula

$$c = \frac{1}{\sqrt{\rho (K+D)}} \qquad (5)$$

where the distensibility D of the tube is defined in terms of the normal cross sectional area A_o and the excess pressure p_e by

$$D = \frac{1}{A_o} \frac{dA}{dp_e}\bigg|_{p_e=0} \qquad (6)$$

If the tube cross section is circular, the distensibility can be due to stretching of the wall only. However, if the normal tube cross section is not circular a greater distensibility may be partly due to this non-circularity. In this case, an area increase may be obtained by bending the wall towards the circular shape. Thus, the speed of sound in an oval shaped tube will be lower than in a perfectly circular tube for the same wall thickness.

Experimental determination of the distensibility

Experiments were performed to determine the distensibility of *10mmOD* Nylex tubing by monitoring the volume increase in *60m* of tube as function of excess pressure.

The results are shown in Figure 3 and the best-fit distensibility $(\frac{1}{V_o}\frac{dV}{dp})$ was found to be $2.0 \cdot 10^{-8} Pa^{-1}$. We note that the behaviour of the tube material is linear up to excess pressures of at least *120 kPa*, corresponding to *12m* excess head of water.

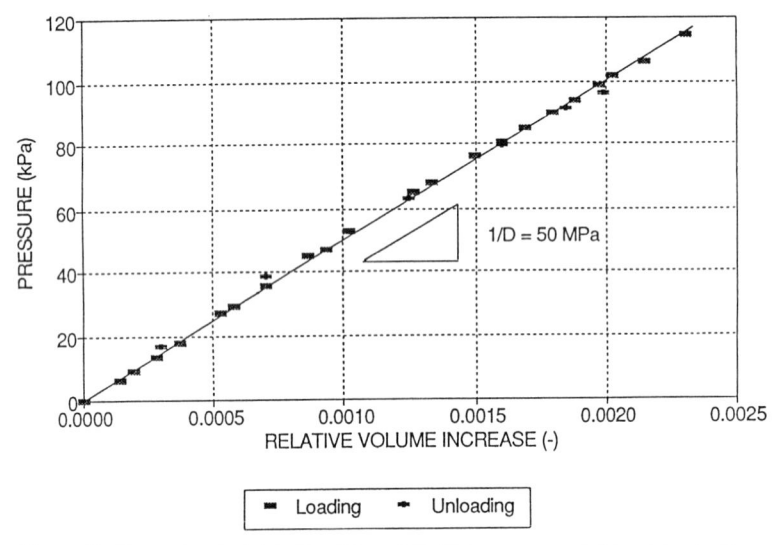

Figure 3: Expansion test data for *10mm OD* Nylex pressure tubing (standard).

According to Equation (5) this corresponds to the speed of sound c = 224m/s for a *10mm OD* Nylex tube with no air bubbles.

A complementary streching test was performed on a short (*150mm*) length of tube to determine Young's modulus E for the tube material. Based on a measured ID of *6.7mm* and wall thickness δ of *1.67mm* the result was $E = 2.03 \cdot 10^8 Pa$.

Through the simple relationship

$$D \approx \frac{d}{\delta E} \qquad (7)$$

this gives a distensibility of $1.95 \cdot 10^{-8} Pa^{-1}$ in close agreement with the directly measured value above. The manufacturer's value for E is $35 kg/mm^2$ corresponding to $3.4 \cdot 10^8 Pa$ for standard tubing (all colours) and $100 kg/mm^2$ (They must be thinking in terms of "kg force") corresponding to $9.8 \cdot 10^8 Pa$ for "semi rigid tubing (only black). The discrepancy (3.4 versus 2.0) being due to uncertainty of tube dimensions in test and to variable humidity. The laboratory tests were perfomed with fully wet tubes.

The effect of air bubbles on the speed of sound

Small, isolated air bubbles can also slow down the pressure waves in the tube. They do this by effectively increasing the distensibility. To quantify this effect, consider for simplicity an air bubble of volume V_o at the ambient pressure p_o, which is compressed isothermally. Its volume is then given by $V(p) = V_o p_o/p$ and hence,

$$\frac{dV}{dp} = -V_o \frac{p_o}{p^2} \approx -\frac{V_o}{p_o} \qquad (8)$$

The presence of air bubbles with concentration C_{air} (vol/vol) will therefore increase the distensibility by the amount

$$D_{air} \approx \frac{C_{air}}{p_o} \qquad (9)$$

leading to the reduced speed of sound

$$c = \frac{1}{\sqrt{\rho\,(K + D + C_{air}/p_o)}} \qquad (10)$$

Test for linearity with regular waves

A series of measurements were conducted at the University of Queensland in the period August to October 1993 to establish the possible existence and importance of nonlinearity of the systems response to regular waves.

Regular but not quite simple harmonic pressure waves with "heights" in the range $0.5m < H < 4.5m$ and periods in the range $2s < T < 7s$ were generated by moving a small reservoir with mercury up and down in a quasi simple-harmonic fashion. The test tube was approximately $100m$ of $10mm$ OD Nylex standard tubing.

The data indicate that the gain is only weakly dependent upon the amplitude and hence, the use of a linear frequency response model (Equation (15)) developed below for the system is reasonably well justified.

Frequency response for the tube-transducer system

As a working hypothesis, it was assumed that the tube-transducer system can be modelled in analogy with a dampened "quarter length resonator". That is, it has resonnant pressure wave modes of the form indicated in Figure 4. Such a system has the approximate frequency response function

$$F(f) = \frac{1}{\cos\left(\frac{\pi}{2}\frac{f}{f_o}\right) + i D_E\left(\frac{f}{f_o}\right)} \qquad (11)$$

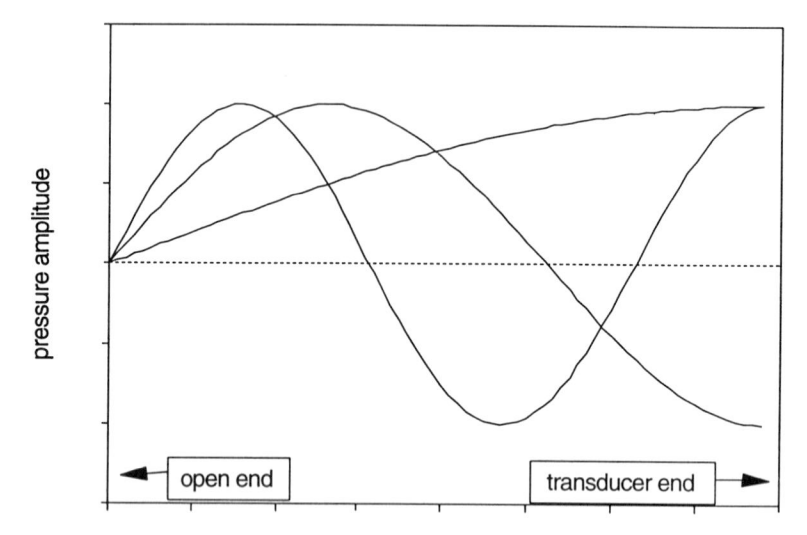

Figure 4: Assuming that the pressure transducer forms a hard, reflecting boundary, the tube/transducer system will behave as a dampened quarter length resonater and have infinitely many resonant wave modes of which the first three are shown here.

where $D_E\ (f/f_o)$ is an energy dissipation function. The corresponding gain function is

$$G(f) = |F(f)| = \frac{1}{\sqrt{\cos^2\left(\frac{2}{\pi}\frac{f}{f_o}\right) + D_E^2\left(\frac{f}{f_o}\right)}} \qquad (12)$$

In these expressions, f_o is the lowest resonance frequency corresponding to the resonance period T_o. It is seen from Equation (12) that the gain function has peaks for all odd multiples of the resonance frequency f_o. This model is in reasonable agreement with experiments see Figure 5

The experiments show the first two peaks of the gain function rather clearly.

The mode of resonnance has a pressure antinode at the transducer end and a node at the open end of the tube, see Figure 4. Hence, the resonant pressure wave in the tube resembles a seiche in a bay. The length of the tube must in that case be 1/4 of the wavelength of the first mode: $L = \lambda_o/4 = c\,T_o/4$, corresponding to the resonance frequency

$$f_o = \frac{c}{4L} \qquad (13)$$

the frequencies of the higher resonant modes are all the odd multiples of f_o.

MEASURING WAVES 215

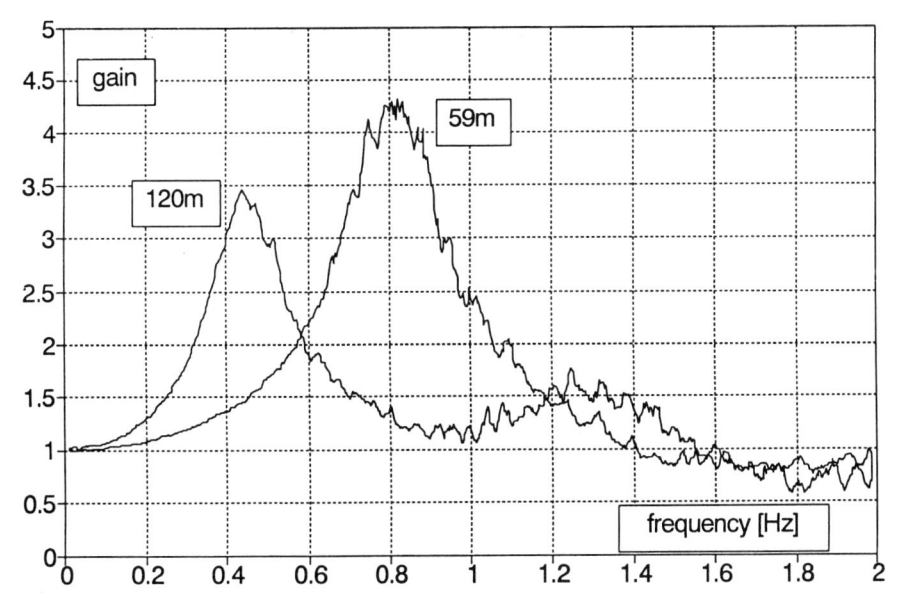

Figure 5: Gain functions measured in the laboratory for a 120m and a 59m OD10mm tube (Nylex standard pressure tubing). The frequency response data was obtained by comparing the output of two transducers. One at the closed end of the tube and one at the "open" end where irregular pressure focing was provided by moving a water filled open ended tube up and down,

Energy dissipation and damping

The maximum gain values observed with *10mm* Nylex tubes of lengths *60m* to *120m* are of the order *4.5* and *3.5* respectively, see Figure 4.

These finite gain values indicate some damping in the system. The nature of this damping i e, the loss of energy is not completely understood. - The energy may be turned into heat in the fluid and in the tube walls. Alternatively, it may be radiated away. Radiation may occur along the full length of the tube or mainly from the open end.

Energy loss in the form of heating of the fluid is caused by the viscosity ν and may be estimated as follows. The rate of heat generation in a boundary layer is $\overline{u_\infty \tau}$ where u_∞ is the velocity at the edge of the boundary layer and τ is the wall shear stress. The wall shear stress may, under the assumption $\sqrt{\nu T} \ll d$, be estimated by the formula $|\tau| = \rho \sqrt{2\pi f \nu} \; |u_\infty|$ which holds for a plane, oscillatory boundary layer, see e g Nielsen 1992, p 21. The velocity amplitude $|u_\infty|$ is related to the pressure amplitude by $u = p/\rho c$, see Lighthill 1978, p 4. Hence the energy dissipation due to fluid viscosity in a tube of length L and diameter d over one period can be estimated by

$$DE_{fluid} \approx \rho u^2 \sqrt{2\pi f \nu}\ L\,d\,T.$$

The loss of energy as heat in the tube walls may be estimated as follows. For a given pressure amplitude $|p|$, the deformation of the tube wall is of the magnitude $\dfrac{|p|\,d^2}{\delta E}$ where δ is the wall thickness and E is Young's modulus for the tube material. Hence, the work done on the tube wall per unit length through one cycle is of the magnitude $\dfrac{|p|^2 d^3}{\delta E}$ and the work done on a tube of length L in one period is $\dfrac{|p|^2 d^3}{\delta E}\,L$. Since the general magnitude of $|p|$ along the tube is $|p'|$, this may be written

$$DE_{wall} \sim \frac{|p'|^2 d^3}{\delta E} L \approx \frac{|p'|^2 L\,d^2}{\rho c^2},\quad \text{since } \rho c^2 \approx E\frac{\delta}{d}\quad \text{for relatively flexible tubes,}$$

cf Equations (5) and (7).

The energy flux through the tube cross section is of the order $|p|cd^2$ corresponding to an energy input of $|p^+|\,c\,d^2 T$ at the open end during one wave period.

Based on these considerations, and with $p' = G\,p^+$, we find that the relative energy loss D_E can be quantified approximately by

$$D_E = \frac{\text{energy loss}}{\text{energy input}} \sim \frac{\dfrac{|p'|^2 d^2 L}{\rho c^2} + \dfrac{|p'|^2}{\rho c^2}\sqrt{2\pi f \nu}\,T\,d\,L}{|p^+|\,c\,d^2\,T}$$

$$D_E = \frac{\text{energy loss}}{\text{energy input}} = G\frac{|p'|}{\rho c^2}\frac{L}{c}[C_1 f + C_2 \frac{\sqrt{\nu}}{d}\sqrt{f}\,] \qquad (14)$$

where C_1 and C_2 are dimensionless coefficients.

A semi enmpirical gain function

Based on the analysis above, which indicates the existence of some damping terms proportional to the square root of the frequency and some which are proporttional to the frequency it seems reasonable to suggest a semi empirical gain function of the form

$$G = \frac{1}{\sqrt{\cos^2\left(\dfrac{\pi}{2}\dfrac{f}{f_o}\right) + [B_1 \dfrac{fL}{c}(1 + B_2 \sqrt{\dfrac{\nu}{fd^2}})\,]^2}} \qquad (15)$$

Based on the data shown in Figure 5 the values of the dimensionless constants were found to be $(B_1, B_2) = (0.58, 5.0)$. The fact that $B_2 \gg B_1$ indicates that the loss due to fluid viscosity is dominant compared to the loss due to deformation of the tube walls for these tubes. The formula (15) with these values of B_1 and B_2 is compared with the data in Figure 6.

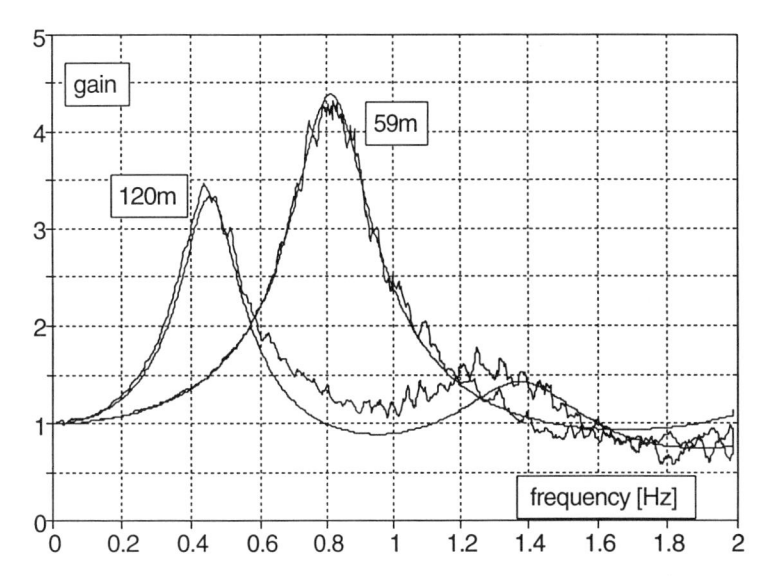

Figure 6: The semi empirical gain function (15) compared with laboratory measurements.

The matching of the shape of the gain function is not perfect, but the discrepancies seem to be due to the cosine function not having the right shape.

Calibration
The values of the constants B_1 and B_2 which were determined on the basis of the data in Figure 5 may not be universal. Hence, when working with systems with different lengths, diameters and tube materials it would be wise to calibrate the system before deployment. This calibration is best done in the way described in connection with Figure 5.

Evaluation of the system
The indication of the tests carried out so far is that the new wave gauge offers a cheap and reliable alternative to existing gauges for measuring the wave conditions in shallow (< 5-6m) depths near (0-500m) beaches or existing structures. The main advantages of the new system are that it is cheap and easy to service and interrogate

because all electronics are kept "high and dry".

Based on the field, it can be concluded that using the sytem in the field with tubes which are already in place is easy (installation time about 5 minutes in fair weather). Earlier tests also show that deployment of the tubes for a one or two day experiment on a beach is manageable, provided a 4 wheel drive vehicle is at hand to help pull the tubes back ashore.

The pressure signal is of adequate strength, and the noise level is very low provided the tubes are prevented from moving with the waves. This is normally achieved by tying the tubes to an 8mm steel chain.

Variable degrees of sand cover over the seaward end of the tube seem not to cause problems with the translation of dynamic bottom pressures into water surface elevations. However, thich layers of silt or mud may have a very strong dampening effect.

The frequency response function for the system is fairly complicated. It has several peaks and the dampening is non-linear. However, with the use of Nylex standard pressure tubing the second peak of the gain function will generally be well outside the frequency range of ocean waves. For example, a *600m 10mmOD* tube system will have its second peak at $f = 0.45$Hz *(fo = 0.15Hz)* and most of the energy in the pressure spectra from wind waves on a beach lies at freqencies below 0.25Hz.

References

Lighthill, M J (1978): *Waves in fluids*. Cambridge University Press.

Maeno, Y & T Hasegawa (1987): In situ measurements of wave induced pore pressure for predicting properties of seabed deposits. *Coastal Engineering in Japan, Vol 30, No 1*, pp 99-115.

Nielsen, P (1989): Analysis of natural waves by local approximations. *J Waterway, Port,Coastal and Ocean Engineering, Vol 115, No 3*, ASCE, pp 384-396.

Nielsen, P (1992): *Coastal bottom boundary layers and sediment transport*. World Scientific, 324pp.

Sleath, J F A (1970): Wave induced pressures in beds of sand. *J Hydraulics Division A S C E, Vol 96, No Hy2*, pp 367-378.

Sleath, J F A (1984): *Sea Bed Mechanics*, John Wiley and Sons, UK, 335pp.

CHAPTER 17

Quantity of Spray Transported by Strong Wind over Breaking Waves

Nobuhiro Matsunaga[1], Misao Hashida[2], Hiroyuki Mizui[3] and Yuji Sugihara[4]

Abstract

The quantity of spray transported by strong wind has been measured experimentally when the wind blows over waves propagating on a sloping bed. The leeward variation of spray concentration is much smaller than that in the vertical direction because of the quasi uniform supply of spray from the water surface. In the theoretical development, therefore, the balance has been analyzed between the upward flux of the concentration by turbulent diffusion and the downward flux due to the spray precipitation. Three characteristic quantities introduced in the analysis have been obtained by superposing the experimental data on the theoretical solutions. They are related to the wave and wind parameters to predict quantitatively the spray concentration supplied from sea to shore.

1. Introduction

Japan is well known as a country which many typhoons pass through every year. Two big typhoons hit successively the western

---

1. Associate Professor, Department of Earth System Science and Technology, Kyushu University, Kasuga 816, Japan
2. Professor, Department of Civil Engineering, Nippon Bunri University, Oita 870-03, Japan
3. Civil Engineering Department, Nagoya Regional Office, Fujita Corporation, Nagoya 460, Japan
4. Research Associate, Department of Earth System Science and Technology, Kyushu University, Kasuga 816, Japan

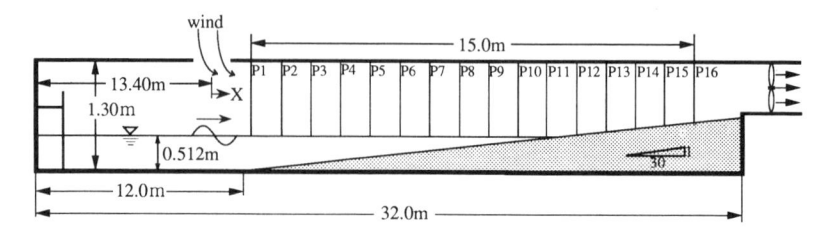

Fig. 1. Schematic diagram of experimental set-up.

part of Japan on September 1991. They gave extensive severe damages along the coastal region, such as the stoppage of electric current over more than four days and poor harvest of crops. A report of the investigation showed that a large quantity of spray of sea water distributed over a wide area and it caused these disasters. To predict the magnitude of the salt damage, it is necessary to evaluate how much the quantity of spray is transported from sea to shore. Some studies (e. g., Toba, 1959; Toba & Tanaka, 1967; Hama & Takagi, 1970) have been already made on the generation and transport of sea-salt particles. However, many fundamental problems have been left unsolved.

In this study, the quantity of spray which is transported when strong wind blows over breaking waves has been measured experimentally, and the discussion has been made about vertical and leeward profiles of the spray concentration.

2. Experimental arrangements

Experiments were carried out by using a wave tank. It was equipped with an inhalation-typed wind tunnel. Its schematic diagram is shown in figure 1. The tank was 32 m long, 0.60 m wide and 1.30 m high. It was covered with a semicircular cylindrical ceiling whose radius was 0.37 m. As a model beach, a sloping flat bed of 1/30-grade was fixed at one side of the tank. The mean depth of water was 0.52 m at the horizontal bed section. Spilling breakers were formed by making two-dimensional regular waves propagate on the sloping bed. The wave parameters are given in table 1, where T is the wave period and H_0, L_0 and C_0 are the wave height, wavelength and wave velocity in deep water, respectively.

The wind was taken into the tank by opening a part of the ceiling. Water spray entrained into the air was captured by arranging vertically ten or twelve cylindrical containers filled with cotton. Their diameter and length were 3.0 m and 5.0 cm, respectively. The vertical interval

Table 1. Experimental parameters.

Run	T(s)	H_0(cm)	L_0(cm)	C_0(m/s)	U_m(m/s)	H_0/L_0	U_m/C_0	γ
1					12.0		6.15	
2	1.25	11.6	243.7	1.95	16.8	0.048	8.63	0.2
3					19.4		9.95	
4					12.0		7.69	
5	1.00	10.9	156.0	1.56	16.9	0.070	10.8	0.4
6					17.5		11.2	
7					12.1		7.76	
8	1.00	14.0	156.0	1.56	16.9	0.090	10.8	0.6
9					19.1		12.2	

between the containers was 5.0 cm. As shown in figure 1, the test section was 15.0 m long. The positions from p1 to p16 were data-taking stations. Their leeward interval was 1.0 m except for Run 6, in which it was 1.2 m. At each station, wave height, wave set-up and spray quantity were measured. The concentration of spray in the air C (g/cm^3) was obtained by dividing the mass of spray transported per square centimeter and second by the cross-sectionally averaged wind velocity U. The wind velocities U_m given in table 1 are ones obtained by averaging U in the leeward direction. The values of non-dimensional parameters H_0/L_0, U_m/C_0 and γ are also given in table 1. The parameter γ will be mentioned in §3.2 and 3.3.

In the following discussion, the horizontal coordinate axis x is taken in the leeward direction from the center of the wind intake. The vertical axis z is taken upwards from wave crest at each data-taking station.

3. Results and discussion

3.1 Transport of spray by wind

Photo 1 shows the generation of spray from a spilling breaker under strong wind. It was taken at x = 10 m in Run 6. A large amount of spray is occurring from the breaking crest.

Photos 2(a) and (b) show vertical profiles of spray size obtained by using droplet sampling paper, t being the time for which the paper had been exposed to spray. The paper is made of filter paper, and a solution of aniline blue dye in benzine is applied to it. From these profiles, it is seen that the spray size decreases remarkably in the

Photo 1. Generation of spray from spilling breaker under strong wind.

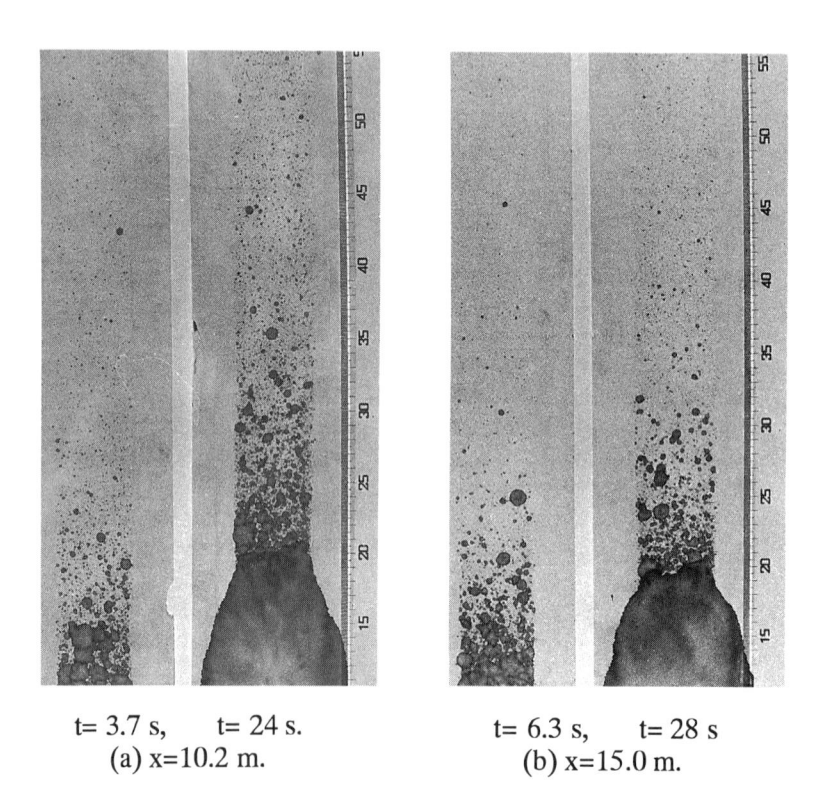

t= 3.7 s,   t= 24 s.
(a) x=10.2 m.

t= 6.3 s,   t= 28 s
(b) x=15.0 m.

Photo 2. Vertical profiles of spray size.

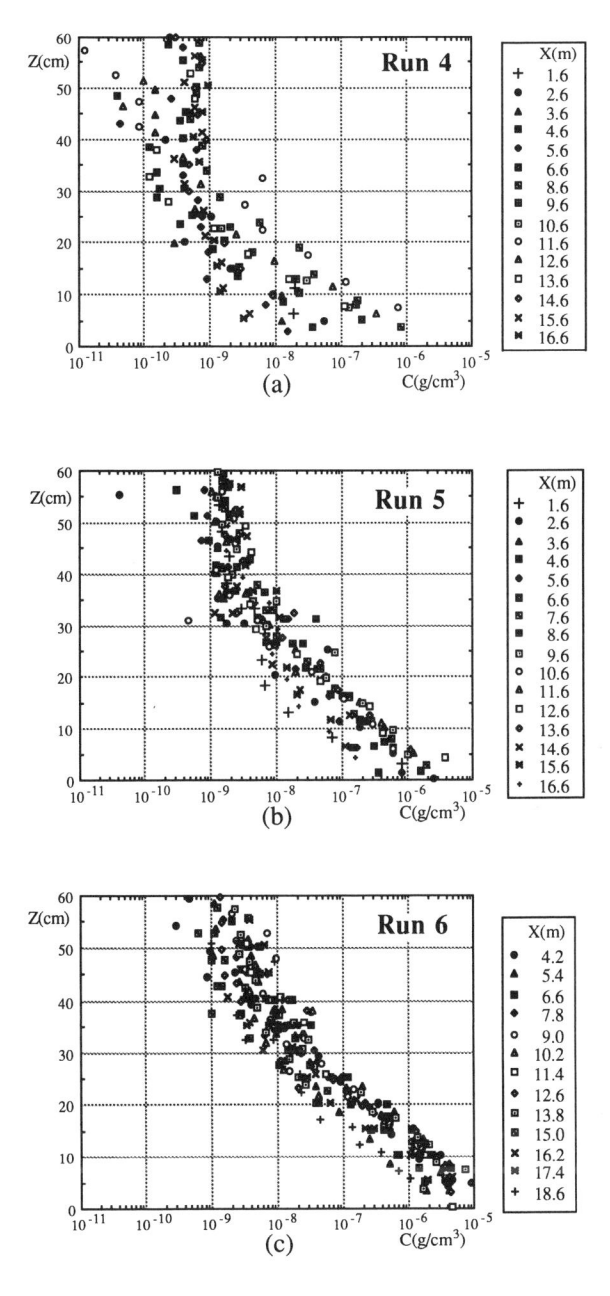

Fig. 2. Vertical profiles of spray concentration.

vertical direction but the vertical profiles do not change so rapidly in the leeward direction.

Figures 2 (a) to (c) show some examples of vertical profiles of spray concentration C. The values of C increase with the increase of x. The rate of increase in the leeward direction is one or two orders of magnitude over ten and several meters. On the other hand, the change in the vertical direction is much larger than that in the leeward direction. The values of C decrease vertically upwards to three or four orders of magnitude over about 40 cm. It means that the supply of spray from the water surface is almost uniform in the leeward direction.

3.2 Theoretical development

The following assumptions have been made in the theoretical development.
- The field is steady and two-dimensional.
- The leeward variation is much smaller than the vertical one.
- Vertical mean velocity is nearly equal to zero.

Under these assumptions, the governing equation for the spray concentration becomes

$$\frac{\partial}{\partial z}\left(D \frac{\partial C}{\partial z}\right) + \frac{\partial}{\partial z}(w_0 C) = 0, \tag{1}$$

where D is the coefficient of turbulent diffusion and w_0 the settling velocity of spray. Boundary conditions are given by

$$C = C_*(x, 0) \quad \text{at } z = 0$$
$$\text{and} \quad C \to 0 \quad \text{as } z \to \infty \tag{2}$$

Let us introduce the dimensionless quantities defined by

$$f(\tilde{z}) = C/C_*(x, 0), \quad g(\tilde{z}) = D/w_0 \, l_*(x, 0) \quad \text{and} \quad \tilde{z} = z/l_*(x, 0), \tag{3}$$

where $l_*(x, 0)$ is a characteristic length scale defined as a value of D/w_0 at $z = 0$. Substituting equation (3) into equation (1), we obtain

$$df/d\tilde{z} + g^{-1} f = 0 \tag{4}$$

Considering that the diffusion coefficient increases generally with z, w_0 decreases oppositely because of the decrease of the spray size with z and $g(0)=1$, we can assume the form of g as follows.

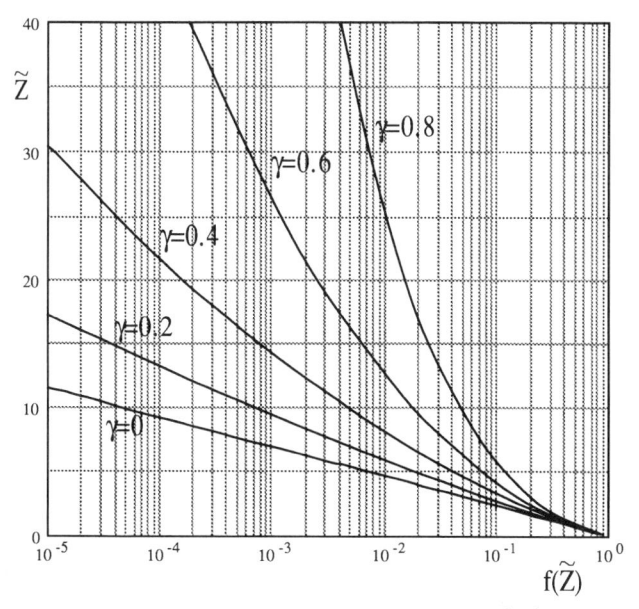

Fig. 3. Vertical profiles of theoretical solutions for various values of γ.

$$g = (1 + \tilde{z})^{\gamma} . \qquad (\gamma > 0) \qquad (5)$$

Substituting equation (5) into equation (4) and using $f(0) = 1$, we obtain the solution

$$f(\tilde{z}) = \exp\left[\frac{1}{1-\gamma}\left\{1 - (1+\tilde{z})^{1-\gamma}\right\}\right]. \qquad (6)$$

Since $f(\infty) \to 0$, γ must take a value from 0 to 1.

Figure 3 shows the vertical profiles of the theoretical solutions for various values of γ. As γ increases, spray diffuses to higher elevation and the vertical profile C becomes uniform. Therefore, it is the result which can be accepted easily.

3.3 Quantification of spray concentration

To predict quantitatively the spray concentration, the quantities introduced in the theoretical development, C_*, l_* and γ should be related with the wave and wind parameters. For their evaluations, the method

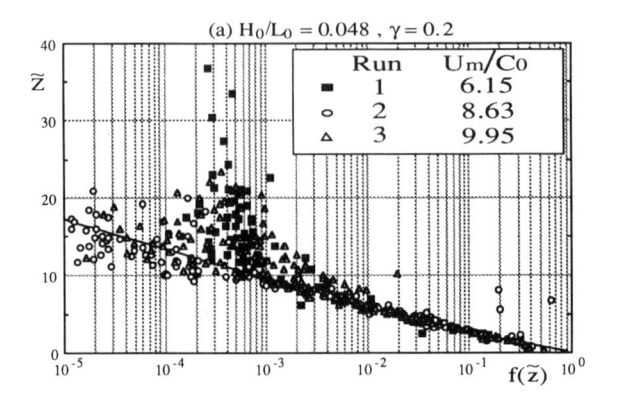

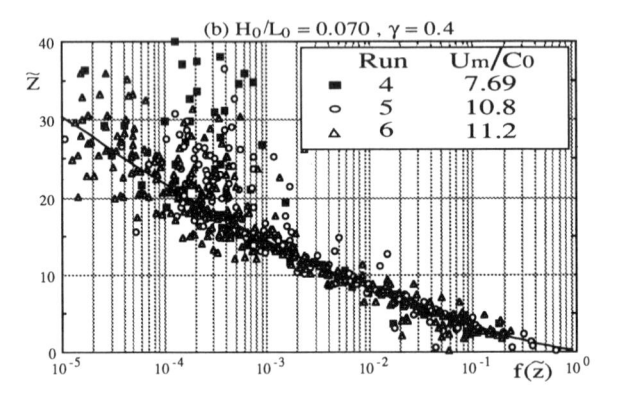

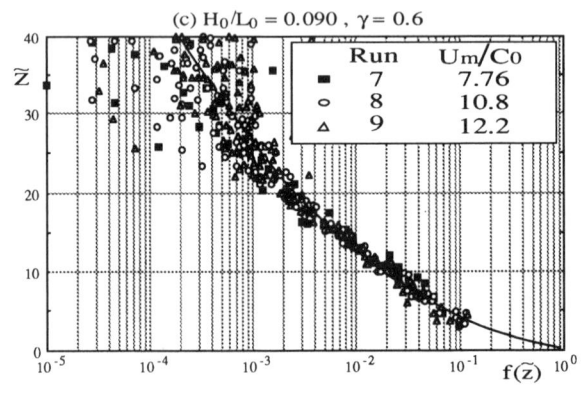

Fig. 4. Degree of agreement between the non-dimensionalized experimental data and theoretical solutions.

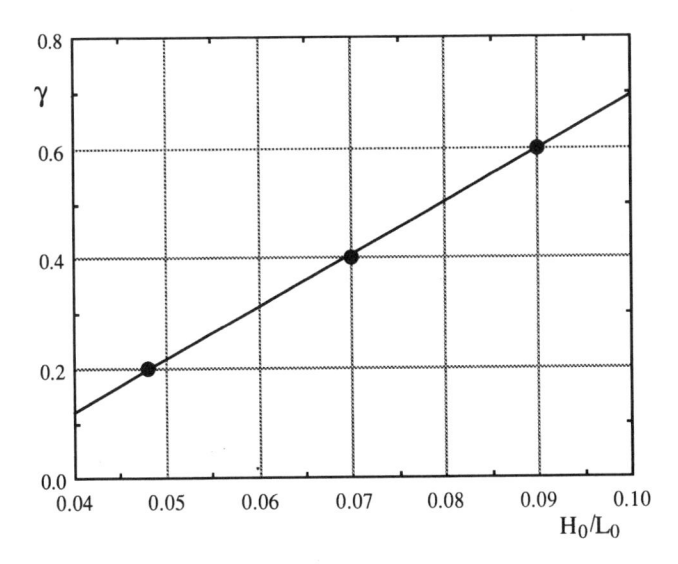

Fig. 5. Relation between γ and H_0/L_0.

was used of superposing the experimental data on the theoretical solutions in log-log graph paper, and the theoretical curve fitting best to the experimental data was determined by the method of trial and error. The values of γ determined by this method are given in table 1. Figures 4 (a) to (c) show the degree of agreement between the experimental data and theoretical curves. Though some scattering can be seen in the upper region, the agreement between the both is very good in the lower region. From table 1, it is seen that γ does not depend on U_m/C_0 but H_0/L_0. In figure 5, the values of γ are plotted against H_0/L_0. The values of γ increase linearly with the increase of H_0/L_0.

The leeward variations of $C_*(x, 0)$ are given in figures 6 (a) to (c). The values of C_* increase gradually with x. After reaching the maximum value $C_{*\,max}$, it decrease rapidly. It results from the decrease of the supply of spray due to the wave propagation into a swash zone. $C_{*\,max}$ is an important quantity to evaluate the maximum quantity of spray transported from sea to shore. In figure 7, the values of $C_{*\,max}/\rho_0$ are plotted against U_m/C_0, where ρ_0 is the density of water. They increase exponentially with the wind velocity.

Figures 8 (a) to (c) give the leeward variations of l_*. The values

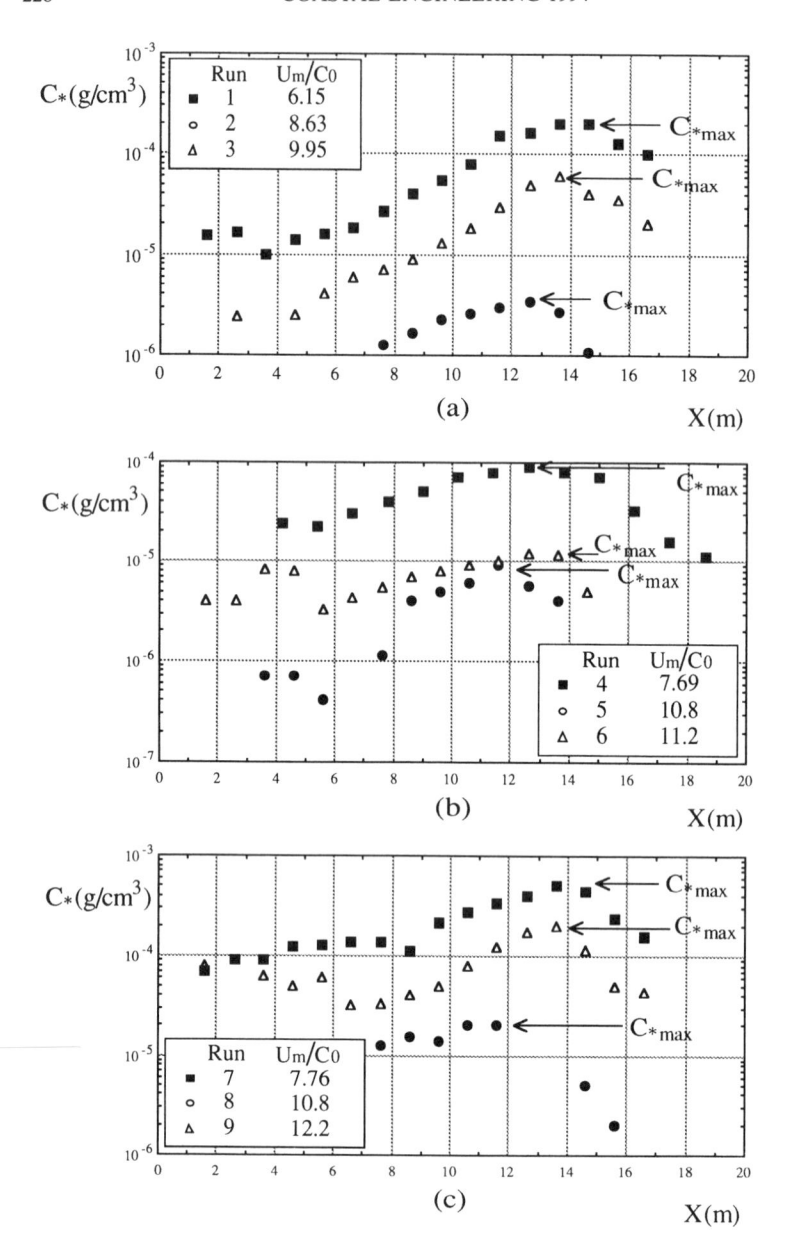

Fig. 6. Variations of C_* in the leeward direction.

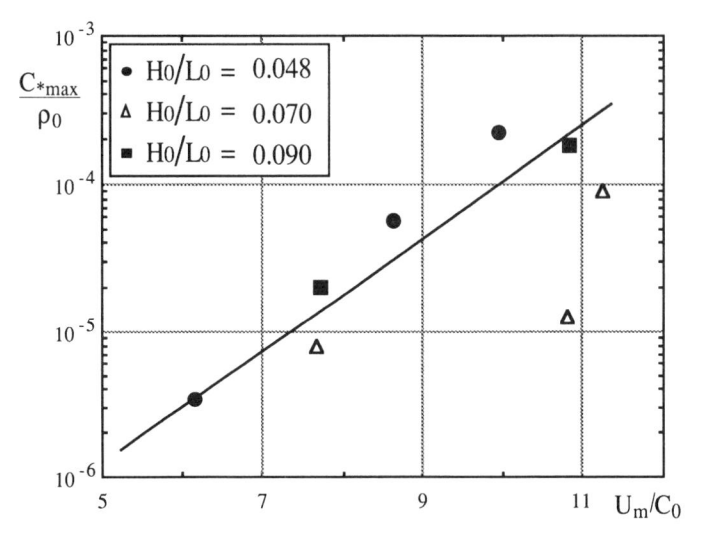

Fig. 7. Relation between $C_{*\,max}/\rho_0$ and U_m/C_0.

of l_* are approximately constant in the leeward direction and tend to decrease with the increase of the wave steepness. Here, let us define $\overline{l_*}$ as a value of l_* averaged in the leeward direction. Figure 9 shows the relation between $\overline{l_*}/H_0$ and H_0/L_0. The values of $\overline{l_*}/H_0$ decrease linearly with the increase of H_0/L_0 without depending on U_m/C_0.

It should be noted that the relatively small leeward-variations of C_* and l_* guarantee the validity of the second assumption made in §3.2.

4. Conclusions

In the case when strong wind blows over breaking waves, the vertical and leeward profiles of spray concentration are investigated experimentally and theoretically. The leeward variation is much smaller than that in the vertical direction. The non-dimensional vertical profiled is expressed by a exponential function. The three characteristic quantities to determine the spray concentration supplied from sea to shore, i. e., the maximum concentration at the elevation of wave crest $C_{*\,max}$, a characteristic length scale of the vertical profiles $\overline{l_*}$ and the non-dimensional parameter determining the vertical profile of the ratio of the diffusion coefficient to the settling velocity γ are related to the wave and wind parameters. The values of $C_{*\,max}$ increase exponentially with the increase of wind velocity. The values of γ

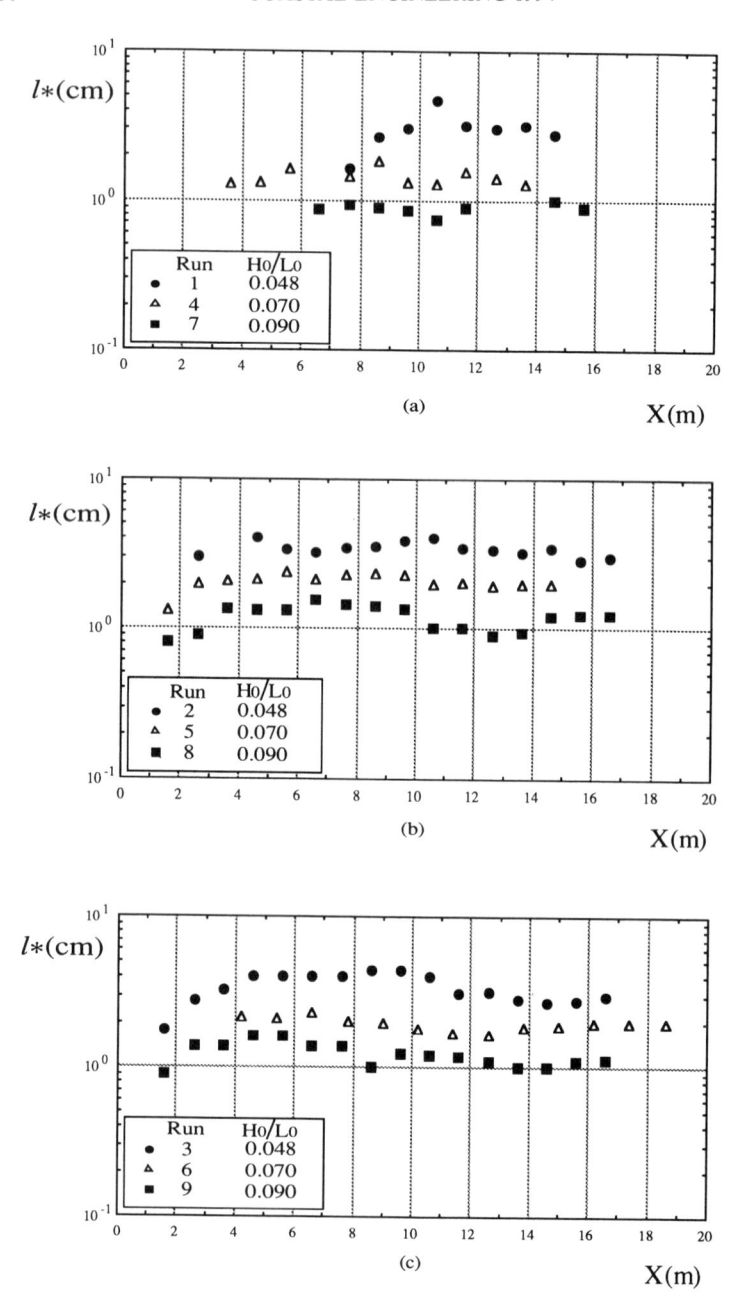

Fig. 8. Variations of l_* in the leeward direction.

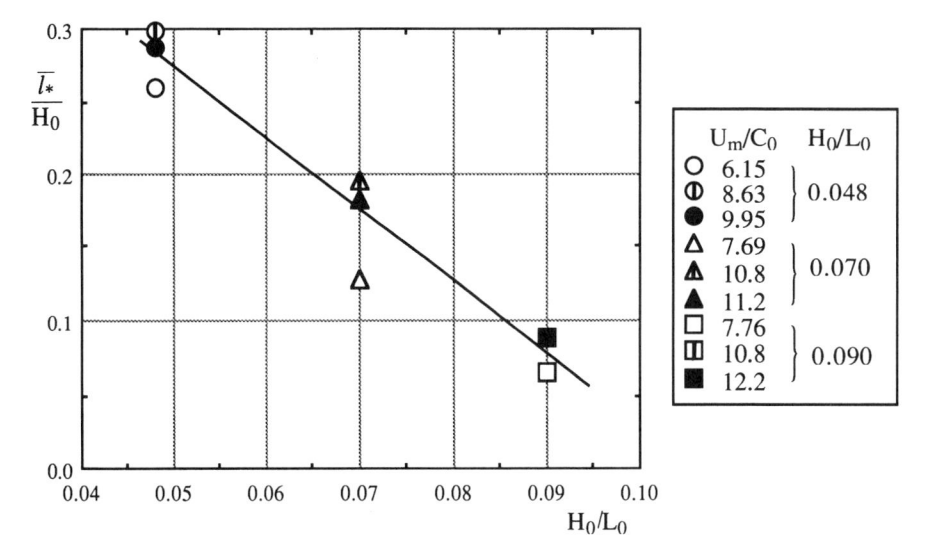

Fig. 9. Relation between $\bar{l}_*/H_0$ and H_0/L_0.

depend only on the wave steepness and increase with its increase. On the other hand, the ratios of $\bar{l}_*$ to the wave height decrease linearly as the wave steepness increases. The non-dimensional vertical profile of spray concentration and the evaluation of $C_{*\,max}$, $\bar{l}_*$ and γ enable us to know how much the spray concentration is supplied from sea to shore. It offers us the boundary condition on the sea side in calculating the spray concentration on land.

The authors are grateful to Ms Y. Arizumi and K. Uzaki for their help.

REFERENCES

HAMA, K. and TAKAGI, N. (1970) Measurement of sea-salt particles on the coast under moderate winds. Papers in Meteorology and Geophysics, **21**, 449-458.

TOBA, Y. (1959) Drop production by bursting of air bubbles on the sea surface (II): Theoretical study on the shape of floating bubbles, J. Oceanogr. Soc. Japan, **15**, 121-130.

TOBA, Y. and TANAKA, M. (1967) Basic study on salt damage (I): Production sea-salt particles and a model of their transport inland, Disaster Prevention Research Institute Annuals, **10**, 331-342 (in Japanese).

CHAPTER 18

Extension of the Maximum Entropy Principle Method for Directional Wave Spectrum Estimation

Noriaki Hashimoto[1], Toshihiko Nagai[2] and Tadashi Asai[3]

Abstract

This paper presents an extension of the maximum entropy principle method (MEP), named the extended MEP (EMEP), as a general and practical method for estimating directional wave spectra. Since the EMEP is formulated to consider errors in the cross-power spectra, it proves to be an accurate, reliable, and robust method against such errors. In addition, we also examine the EMEP using numerical simulation and field wave data, with its validity being subsequently discussed.

1. Introduction

The directional spectrum of ocean waves expresses their fundamental properties by describing the energy distribution as a function of wave frequency and wave propagation direction. Many methods have been proposed for estimating the directional spectra of various types of ocean wave measurements, e.g., the direct Fourier transformation method (DFTM), parametric method, maximum likelihood method (MLM), extended maximum likelihood method (EMLM), maximum entropy principle method (MEP), and Bayesian directional spectrum estimation method (BDM).

The MEP (Hashimoto and Kobune, 1985) can estimate the directional spectra using three simultaneously measured quantities related to random wave motion, for instance, pitch, roll, and heave data, and when applied as such, estimates of directional spectra have better directional resolution than other existing methods. On the other

---
[1] Chief, Ocean Energy Utilization Laboratory, Hydraulic Engineering Division.
[2] Chief, Marine Observation Laboratory, Marine Hydrodynamics Division.
[3] Member, Environmental Hydraulics Laboratory, Marine Hydrodynamics Division.
Port and Harbour Research Institute, Ministry of Transport,
1-1 Nagase 3-chome, Yokosuka 239, Japan

hand, the MEP is not a general method because it is restricted to applications involving three-quantity measurements.

In comparison, the BDM (Hashimoto, 1987) can handle more than three arbitrary-mixed instrument array measurements and has the highest resolution among existing methods for estimating the directional spectrum under such conditions. This method is a robust method for estimating the directional spectrum using cross-power spectra contaminated by estimation errors, though it is also a general method and requires the use of time-consuming iterative calculations.

Consequently, a method needed to be developed that can be applied to arbitrary general measurements, while at the same time easily yielding an accurate and reliable estimate of the directional spectrum.

The present paper discusses inherent drawbacks of several existing methods for estimating the directional spectrum, and then describes a new method, the Extended MEP (EMEP), which correspondingly functions as a general yet practical method. Since the EMEP is formulated in the same manner as the BDM, i.e., considering errors in the cross-power spectra, it demonstrates both robustness and reliability. Here we also examine the EMEP using numerical simulation and field wave data, with its validity being subsequently discussed.

2. Fundamental Equation Related to Directional Spectrum

The general relationship between the cross-power spectrum for a pair of wave properties and the wave-number frequency spectrum was introduced by Isobe et al. (1984) as follows:

$$\Phi_{mn}(\omega) = \int_k H_m(k,\omega)H_n^*(k,\omega)\exp\{-ik(x_n - x_m)\}S(k,\omega)dk, \qquad (1)$$

where ω is the angular frequency, k the wave number vector, $\Phi_{mn}(\omega)$ the cross-power spectrum between the m-th and n-th wave properties, $H_m(k,\omega)$ the transfer function from the water surface elevation to the m-th wave property, i the imaginary unit, x_m the location vector of the wave probe for the m-th wave property, $S(k,\omega)$ the wave-number frequency spectrum, and " * " denotes the conjugate complex. The wave number k is related to the frequency f by the following dispersion relationship:

$$\omega^2 = (2\pi f)^2 = gk \tanh kd, \qquad (2)$$

hence, the wave-number frequency spectrum can be expressed as a function of the frequency f and wave propagation direction θ. Equation (1) can therefore be rewritten as

$$\Phi_{mn}(f) = \int_0^{2\pi} H_m(f,\theta)H_n^*(f,\theta)[\cos\{k(x_{mn}\cos\theta + y_{mn}\sin\theta)\} \\ - \sin\{k(x_{mn}\cos\theta + y_{mn}\sin\theta)\}]S(f,\theta)d\theta, \qquad (3)$$

where $x_{mn} = x_n - x_m$, $y_{mn} = y_n - y_m$, and $S(f,\theta)$ is the directional spectrum.

The directional spectrum is commonly expressed as a product of the frequency spectrum $S(f)$ and the directional spreading function $G(\theta|f)$, i.e.,

$$S(f,\theta) = S(f)G(\theta|f), \tag{4}$$

with $S(f,\theta)$ taking a non-negative value and satisfying the following relationship:

$$\int_0^{2\pi} S(f,\theta)d\theta = S(f). \tag{5}$$

Substitution of Eq. (4) into (5) yields

$$\int_0^{2\pi} G(\theta|f)d\theta = 1. \tag{6}$$

The transfer function $H_m(f,\theta)$ in Eq. (3) is generally expressed as

$$H_m(f,\theta) = h_m(f)\cos^{\alpha_m}\theta \sin^{\beta_m}\theta, \tag{7}$$

where the function $h_m(f)$ and parameters α_m and β_m obtained from small amplitude wave theory are specified for each measured quantity in Isobe et al. (1984).

Equations (1) and (3) are the fundamental equations for estimating the directional spectrum on the basis of simultaneous measurements of various wave properties. If the function $S(k,\omega)$ or $S(f,\theta)$ is determined which respectively satisfies Eq. (1) or (3), and it has a non-negative value, then this function is termed as the directional spectrum.

3. Existing Methods for Estimating the Directional Spectrum

If an infinite number of wave properties are measured, i.e., the cross-power spectra are known for infinite pairs of m and n in Eq. (1) or (3), the directional spectrum can be uniquely determined. However, only a limited number of wave properties can actually be measured at a limited number of locations; thus, the directional spectrum cannot be uniquely determined since the number of component waves is infinite. Existing methods for directional spectral estimation therefore attempt to determine the unique function by introducing some type of idea and/or trick, and as a result, their advantages and disadvantages depend on the employed concept. The most prominent methods and their advantages/disadvantages are briefly discussed as follows.

(1) Direct Fourier Transformation Method (DFTM)

This method was first proposed by Barber (1961) and uses the following estimation equation:

$$\hat{S}(f,\theta) = \alpha \sum_m \sum_n \Phi_{mn}(f)\exp\{ik(x_n - x_m)\}, \qquad (8)$$

where α is a proportionality constant such that the estimate of the directional spectrum satisfies Eq. (5). Although the method's computation is easy, the directional resolution is low and a negative energy distribution sometimes occurs.

(2) Parametric Method

Parametric methods employed by Longuet-Higgins et al. (1963), Panicker and Borgman (1974), and Hasselman et al. (1980) assume a specific formulation for the directional spectrum, such the following truncated Fourier series or a cosine-powered function:

$$S(f,\theta) = a_0(f) + \sum_{n=1}^{N}\{a_n(f)\cos\theta + b_n(f)\sin\theta\} \qquad (9)$$

$$S(f,\theta) = S(f)\sum_n \alpha_n(f)\cos^{2s_n(f)}\{\frac{\theta - \theta_n(f)}{2}\}. \qquad (10)$$

It should be realized that these methods are only able to approximate the real directional spectrum when it suitably fits the model.

(3) Extended Maximum Likelihood Method (EMLM)

This method was proposed by Isobe et al. (1984) who extended the maximum likelihood method (MLM) developed by Capon (1969) so as to handle an arbitrary combination of wave properties. The EMLM formula is derived using a window theory, namely,

$$S(k,\omega) = \alpha \Big/ [\sum_m \sum_n \Phi_{mn}^{-1}(\omega)H_m^*(k,\omega)H_n(k,\omega)\exp\{ik(x_n - x_m)\}], \qquad (11)$$

where $\Phi_{mn}^{-1}(\omega)$ is the mn element of inverse matrix $\Phi^{-1}(\omega)$, while α is a proportionality constant such that the directional spectrum $\hat{S}(k,\omega)$ satisfies Eq. (5).

The EMLM has high-directional resolution and versatility, and consequently, has been widely employed in directional wave analysis. On the other hand, when the layout of the probe array is not proper or the cross-power spectra are contaminated by errors, this method sometimes estimates erroneous peaks or negative values, while also in some cases failing to yield a smooth and continuous estimation of the directional spectrum.

(4) Maximum Entropy Principle Method (MEP)

We previously developed the MEP (Hashimoto and Kobune, 1985), which provides a powerful means for estimating the directional spectrum when three-quantity point measurements are available. Such data can be obtained, for example, using a discus buoy or a two-axis current meter and a wave gauge. The estimation equation of the MEP's directional spreading function is expressed by maximizing the entropy with an assumption that the directional spreading function is a type of probability density function, i.e.,

$$G(\theta|f) = \exp[a_0(f) + \sum_{n=1}^{2}\{a_n(f)\cos n\theta + b_n(f)\sin n\theta\}], \quad (12)$$

where the coefficients a_n and b_n are the Lagrange multipliers. The advantage of this expression is that even though the Fourier series in the power of the exponential function is truncated by $n = 2$, it has non-negative values and yields a wide range of shapes for $G(\theta|f)$.

It should be noted that although the MEP was originally developed to be used with three-quantity point measurements, Nwogu (1989) expanded it for application to array measurements. Since this expansion was carried out using the same procedure as the original one, the directional spreading function results in a complex form including Bessel functions; thereby making the computation more difficult than the original MEP.

(5) Bayesian directional spectrum estimation method (BDM)

The BDM (Hashimoto, 1987) provides an accurate and reliable estimate of the directional spectrum for array measurements consisting of more than three arbitrary quantities. The assumed estimation equation of the BDM's directional spreading function is not a formal mathematical function, instead being a piece-wise constant function expressed as

$$\hat{G}(\theta|f) = \sum_{k=1}^{K} \exp\{x_k(f)\} I_k(\theta), \quad (13)$$

where
$$x_k(f) = \ln\{G(\theta_k|f)\} \quad (14)$$

$$I_k(\theta) = \begin{cases} 1 & : (k-1)\Delta\theta \le \theta < k\Delta\theta \\ 0 & : \text{otherwise} \end{cases} \quad (15)$$

$(k = 1, \cdots, K)$.

Equation (13) can be determined by minimizing Akaike's Bayesian information criterion (ABIC) (Akaike, 1980) and applying an additional condition that the directional spreading function be smooth and continuous, namely,

$$\{x_k - 2x_{k-1} + x_{k-2}\}^2 \to \text{small}$$
$$(k = 1, \cdots, K \text{ and } x_0 = x_K, \ x_{-1} = x_{K-1}). \quad (16)$$

The advantage of Eq. (13) is that even though the additional condition of Eq. (16) is imposed, it generates non-negative values and yields a wide range of arbitrary shapes for $G(\theta|f)$. However, since it consists of many unknown parameters $x_k(f)$; $(k = 1,\cdots,K)$, the BDM involves the use of time-consuming iterative calculations.

After reviewing the various methods, it is clear that the each formulates a tailored model for approximating the directional spectrum, being characterized by some unknown parameters. Though the principle and method of deriving each model are different, inherent advantages and disadvantages arise due to the model's different characteristics. When considering the MEP and BDM, since they are characterized by an exponential function incorporating a power function, an advantage exists in that these models result in non-negative values and flexibly yield a wide range of arbitrary shapes of the directional spectrum.

4. Formulation of Extended Maximum Entropy Principle Method (EMEP)

To simplify the nomenclature in the equations, Eq. (3) is rewritten using the upper triangular components of the hermitian matrix $\Phi(f)$, such that

$$\phi_i(f) = \int_0^{2\pi} H_i(f,\theta) G(\theta|f) d\theta \qquad (i = 1,\cdots,K), \tag{17}$$

where K is the number of equations, and

$$H_i(f,\theta) = H_m(f,\theta) H_n^*(f,\theta) [\cos\{k(x_{mn}\cos\theta + y_{mn}\sin\theta)\} \\ - \sin\{k(x_{mn}\cos\theta + y_{mn}\sin\theta)\}]/W_{mn}(f) \tag{18}$$

$$\phi_i(f) = \Phi_{mn}(f)/\{S(f) W_{mn}(f)\} \tag{19}$$

$$G(\theta|f) = S(f,\theta)/S(f), \tag{20}$$

with $W_{mn}(f)$ being a weighting function introduced for normalizing and non-dimensionalizing the errors of the cross-power spectra. This function is represented by the following equation (standard deviations of error of co-spectrum and quadrature-spectrum, Bendat and Pirsol, 1986) for the real and imaginary part of Eq. (17), respectively:

$$\sigma[\hat{C}_{mn}(f)] \approx [\{\Phi_{mm}(f)\Phi_{nn}(f) + C_{mn}(f)^2 - Q_{mn}(f)^2\}/2N_a]^{1/2} \tag{21}$$

$$\sigma[\hat{Q}_{mn}(f)] \approx [\{\Phi_{mm}(f)\Phi_{nn}(f) - C_{mn}(f)^2 + Q_{mn}(f)^2\}/2N_a]^{1/2}, \tag{22}$$

where $\Phi_{mn}(f) = C_{mn}(f) - iQ_{mn}(f)$ and N_a is the number of the ensembled average.

The directional spreading function normally takes values greater than or equal to zero. However, in the EMEP, the function is treated as a function which always takes positive values. Then, as a general expression for $G(\theta|f)$, we can extend Eq. (12) (or simplify Eq. (13)) to obtain

$$G(\theta|f) = \frac{\exp[\sum_{n=1}^{N}\{a_n(f)\cos n\theta + b_n(f)\sin n\theta\}]}{\int_0^{2\pi} \exp[\sum_{n=1}^{N}\{a_n(f)\cos n\theta + b_n(f)\sin n\theta\}]d\theta} \quad (23)$$

where $a_n(f)$, $b_n(f)$; $(n = 1, \cdots, N)$ are unknown parameters.

For the sake of convenience, the complex values $\phi_i(f)$ and $H_i(f,\theta)$ are rewritten as

$$\left.\begin{array}{ll} \phi_i = \text{Re}\{\phi_i(f)\} & \phi_{K+i} = \text{Im}\{\phi_i(f)\} \\ H_i(\theta) = \text{Re}\{H_i(f,\theta)\} & H_{K+i}(\theta) = \text{Im}\{H_i(f,\theta)\} \end{array}\right\} \quad (24)$$

so that $\phi_i(f)$ and $H_i(f,\theta)$ are real. For simplicity, the frequency f is omitted on the LHS of Eq. (24), and is also omitted hereafter.

When Eq. (17) is applied to the observed data, the errors contained in the cross-power spectra must be taken into account. Thus, after making the substitution of Eq. (23) into (17), it can be modified by considering the existence of errors ε_i, i.e.,

$$\varepsilon_i = \frac{\int_0^{2\pi}\{\phi_i - H_i(\theta)\}\exp\{\sum_{n=1}^{N}(a_n\cos n\theta + b_n\sin n\theta)\}d\theta}{\int_0^{2\pi}\exp\{\sum_{n=1}^{N}(a_n\cos n\theta + b_n\sin n\theta)\}d\theta} \quad (25)$$

$$(i = 1, \cdots, M)$$

where M is the number of independent equations left after eliminating the meaningless equations such as the zero co-spectrum and zero quadrature-spectrum.

Now, if ε_i; $(i = 1, \cdots, M)$ are assumed to be independent of each other, and the probability of their occurrence is expressed by the normal distribution having a zero mean and variance σ^2, the optimal $G(\theta|f)$ is one which minimizes $\sum \varepsilon_i^2$. However, Eq. (25) is nonlinear with respect to a_n, b_n; $(n = 1, \cdots, N)$, and is therefore difficult to solve. Consequently, let us apply Newton's technique of local linearization and iteration to solve the problem.

If approximate solutions $\tilde{a}_n$, $\tilde{b}_n$; $(n = 1, \cdots, N)$ are known, the solution a_n, b_n; $(n = 1, \cdots, N)$ can be written as

$$a_n = \tilde{a}_n + a'_n \\ b_n = \tilde{b}_n + b'_n \Biggr\} ,\qquad (26)$$

where a'_n, b'_n are the residuals between the solution a_n, b_n and the approximate solution $\tilde{a}_n$, $\tilde{b}_n$.

Substitution of Eq. (26) into (25) and rearranging yields the following linearized equation with respect to a'_n, b'_n:

$$\varepsilon_i = Z_{N,i} - \sum_{n=1}^{N}(a'_n X_{n,i} + b'_n Y_{n,i}) \qquad (i = 1,\cdots,M), \qquad (27)$$

where

$$Z_{N,i} = \frac{\int_0^{2\pi}\{\phi_i - H_i(\theta)\}F_N(\theta)d\theta}{\int_0^{2\pi}F(\theta)_N d\theta} \qquad (28)$$

$$X_{n,i} = Z_{N,i}\left\{\frac{\int_0^{2\pi}F_N(\theta)\cos n\theta d\theta}{\int_0^{2\pi}F_N(\theta)d\theta} - \frac{\int_0^{2\pi}\{\phi_i - H_i(\theta)\}F_N(\theta)\cos n\theta d\theta}{\int_0^{2\pi}\{\phi_i - H_i(\theta)\}F_N(\theta)d\theta}\right\} \qquad (29)$$

$$Y_{n,i} = Z_{N,i}\left\{\frac{\int_0^{2\pi}F_N(\theta)\sin n\theta d\theta}{\int_0^{2\pi}F_N(\theta)d\theta} - \frac{\int_0^{2\pi}\{\phi_i - H_i(\theta)\}F_N(\theta)\sin n\theta d\theta}{\int_0^{2\pi}\{\phi_i - H_i(\theta)\}F_N(\theta)d\theta}\right\} \qquad (30)$$

$$F_N(\theta) = \exp\{\sum_{n=1}^{N}(\tilde{a}_n \cos n\theta + \tilde{b}_n \sin n\theta)\}. \qquad (31)$$

Equation (27) can be solved iteratively by assuming a proper approximate solution $\tilde{a}_n$, $\tilde{b}_n$; $(n = 1,\cdots,N)$.

Minimizing $\sum \varepsilon_i^2$ for a particular data set also introduces an additional problem of choosing the optimal finite order N for the model (Eq. (23)); hence to overcome this, the minimum Akaike's Information Criterion (AIC) procedure (Akaike, 1973) is incorporated into the above iterative calculations to yield a reasonable and smooth estimate of $G(\theta|f)$. The AIC is given by

$$\text{AIC} = M(\ln 2\pi + 1) + M\ln\hat{\sigma}^2 + 2(2N+1), \qquad (32)$$

where M is the number of independent equations (Eq. (27)) and $\hat{\sigma}^2$ is the estimate of the variance of ε_i : $(i = 1,\cdots,M)$.

5. Numerical computation of the EMEP

To estimate the directional spectrum using the EMEP, the computation must be performed from lower ($N = 1$) to higher orders. During the iterative computation, however, the computation occasionally becomes unstable in special cases. If so, a control parameter δ is introduced into Eq. (26) to under-relax the iterative computation:

$$\left. \begin{array}{l} a_n = \tilde{a}_n + \delta \; a_n' \\ b_n = \tilde{b}_n + \delta \; b_n' \end{array} \right\} \tag{33}$$

that is, when the iterative computation is unstable, the control parameter δ is changed to a smaller value, followed by $\delta = (0.5)^k : (k = 0, \cdots, 4)$.

The numerical computation procedure including the minimization of the AIC is summarized as follows:

1) Select the lowest model order $N = 1$ and compute $X_{n,i}$, $Y_{n,i}$, and $Z_{N,i}$ of Eq. (28) $\sim$ (30) assuming the initial approximate solutions of $\tilde{a}_1$ and $\tilde{b}_1$ being equal to zero. Then, to obtain solutions a_1 and b_1, carry out the iterative computation of Eq. (27) and perform the least square method for $\sum \varepsilon_i^2$ until the absolute values of residuals $|a'_1|$ and $|b'_1|$ become small enough ($|a'_1|$, $|b'_1| < 0.01$).

2) Compute the AIC by Eq. (32).

3) Substitute a_1 and b_1 obtained in step 1) into Eq. (31) vice $\tilde{a}_1$ and $\tilde{b}_1$, leaving $\tilde{a}_2$ and $\tilde{b}_2$ equal to zero, and carry out the same procedure as step 1) to obtain the solutions a_i, $b_i : (i = 1,2)$ of the 2nd order model ($N = 2$). If during the iterative computation one of the absolute values of residuals $|a'_i|$, $|b'_i| : (i = 1,2)$ has a value greater than 30, terminate the computation and adopt $N = 1$ as the optimal model order, with the values a_1 and b_1 obtained in step 1) being chosen as the solutions. If the iterative computation does not converge after 100 iterations, terminate the computation and similarly choose the solutions obtained in step 1). If $|a'_i|$, $|b'_i| : (i = 1,2)$ become less than 0.01, compute the AIC using Eq. (32) and proceed to the next step.

4) If the AIC obtained in step 3) is greater than that of step 2), or if the absolute value of the difference between the two AICs is much less than 1, adopt $N = 1$ as the optimal model order and choose the values a_1 and b_1 of step 1) as the solutions. If the above cases do not hold, proceed to the next step.

5) Change the model order into a higher one (3rd, 4th, ...) and repeat in the same manner as steps 3) and 4).

6) During the computation of step 5), if the AIC of the model order $N+1$ is greater than that of the previous order N, or when the absolute value of the difference between the AIC of the model order $N+1$ and that of the previous order N is much less than 1, or when the iterative computation does not converge at the model order $N+1$, then terminate the computation and choose the model order

MAXIMUM ENTROPY PRINCIPLE

N and $a_i, b_i : (i = 1, \cdots, N)$ as the optimal solutions.

7) Substitute the optimal solutions $a_i, b_i : (i = 1, \cdots, N)$ into Eq. (23) to get the optimal directional spreading function $\hat{G}(\theta|f)$.

In addition, since the number of unknown parameters should be less than or equal to the number of independent equations, i.e., $2N \leq M$, steps 1) ~ 6) should be performed at most within this order.

6. Examination of EMEP by Numerical Simulation

Here we will employ numerical simulation to examine the validity of using the EMEP to estimate the directional spectrum. The procedure is the same as that used for examining the EMLM (Isobe et al., 1984), where the employed directional spreading function is a cosine-powered type function expressed as

$$G(\theta) = \sum_i \alpha_i \cos^{2S_i}\left(\frac{\theta - \theta_i}{2}\right). \qquad (34)$$

When $i = 1$ only, Eq. (34) yields a unimodal directional spreading function, while a bimodal function is formulated by the superposition of two unimodal directional spreading functions, i.e., $i = 1$ and 2, having a different coefficient α_i, mean direction θ_i, and spreading parameter S_i. The cross-power spectra utilized for the numerical simulation are obtained by numerically integrating Eq. (3).

Figure 1 compares the given directional spreading function (TRUE) and the estimated ones by the EMEP, MEP, BDM, and EMLM, where the three measured quantities (sea surface elevation and two orthogonal slopes on the surface at the same location) are assumed to be the simulated observation condition. The ordinate is normalized by dividing the value of the directional spreading function by the maximum value of the TRUE directional spreading function. As indicated, the EMEP yields the same estimate as the MEP and can detect the small peak in proper direction in **Fig. 1(e)** and **(f)**, though it overestimates the main peak and underestimates the

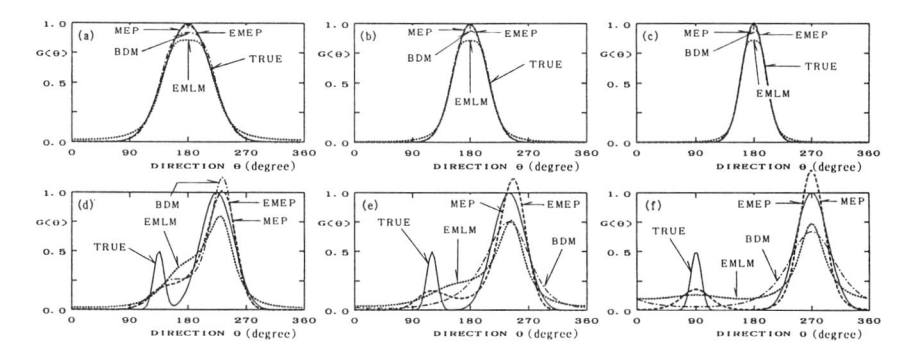

Fig. 1 Comparison of EMEP, MEP, BDM and EMLM (Three-quantity measurement)

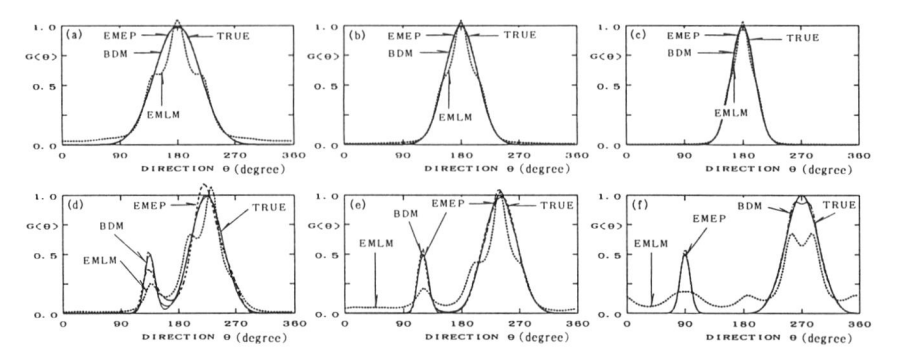

Fig. 2 Comparison of EMEP, MEP, BDM and EMLM (Star array measurement)

small peak. Also note that the EMEP (MEP) shows the minimum leakage of the wave energy into neighboring directional bands. On the other hand, the EMLM appears to recognize the existence of the small peak in **Fig. 1(e)**, but its estimated direction is not proper, while the BDM does not even recognize the small peak.

Figure 2 shows results when a star array consisting of four wave gauges is assumed as the simulated observation condition. The minimum distance D between the wave gauges is assumed as $D/L = 0.2$, where L is the wave length. In comparison with **Fig. 1**, the directional resolution of the EMEP, BDM, and EMLM is improved in **Fig. 2(d), (e)**, and **(f)**. In particular, the EMEP is very close to the BDM and shows good agreement with TRUE. In contrast, the EMLM underestimates the small peak and shows some energy leakage around the peaks, especially in **Fig. 2(e)** and **(f)**.

7. Field Data Analysis

Here we apply the EMEP to analyze wave records acquired at an offshore oil rig (Iwaki-oki Station) located 42 km off the Iwaki coast, the northeastern coast of the main island of Japan (**Fig. 3**), where the water depth is 155 m. **Figure 4** shows the oil rig, which has installed on its legs, four step-resistance wave gauges, a two-axis ultrasonic current meter, and a pressure sensor (**Fig. 5**). The simultaneous measurement of seven elements is performed over 20 min every 2 h. The Onahama Port Construction Office, Second Port Construction Bureau, Ministry of Transport, has been conducting this multi-element measurement of directional waves since 1986.

Figure 6 shows typical directional spectra of a swell having a significant wave height $H_{1/3} = 3.81$ m, period $T_{1/3} = 12.3$ s, and directional spreading parameter $S_{max} = 75$, which are estimated by the EMEP, BDM, and EMLM. Note that the shape of the EMEP estimate is very similar to that of the BDM, but different from the EMLM's. The major difference between them is that the EMLM is more diffuse and does not show a concentration around the peak. In addition, the peak of the EMLM is obviously much lower than that of the EMEP and BDM.

MAXIMUM ENTROPY PRINCIPLE 243

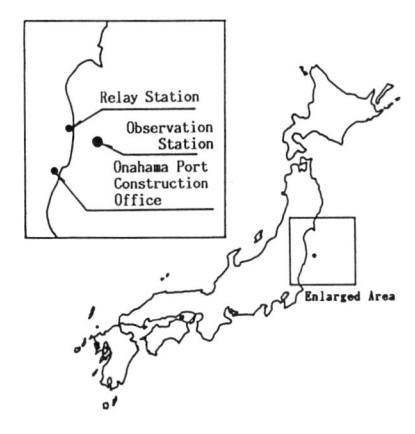

Fig. 3 Location of the wave observation station.

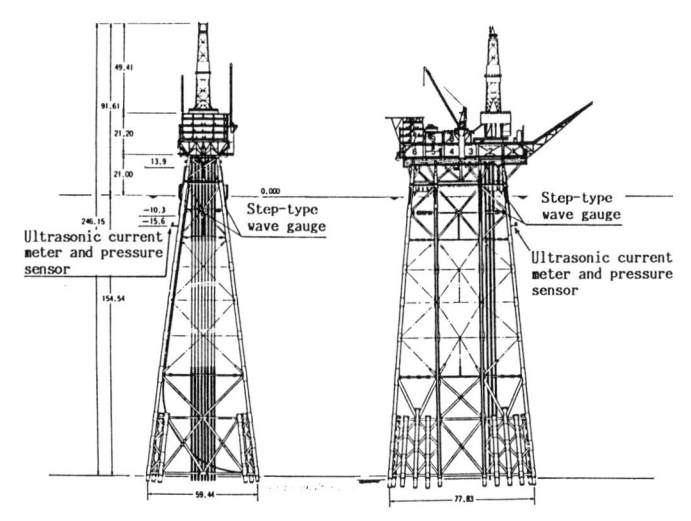

Fig. 4 Offshore oil rig and the location of wave instruments.

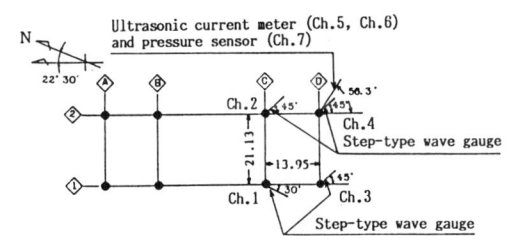

Fig. 5 Wave instrument array.

244 COASTAL ENGINEERING 1994

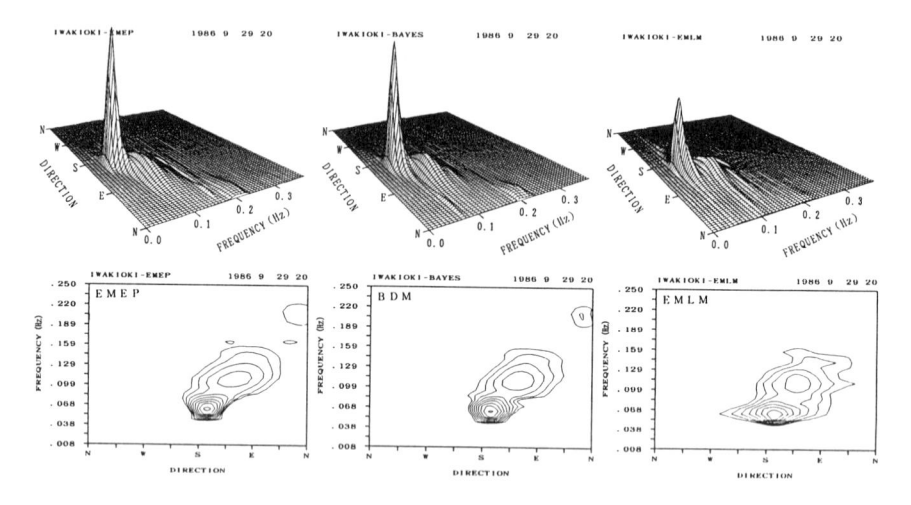

Fig. 6 Typical directional spectra of a swell estimated by EMEP, BDM and EMLM

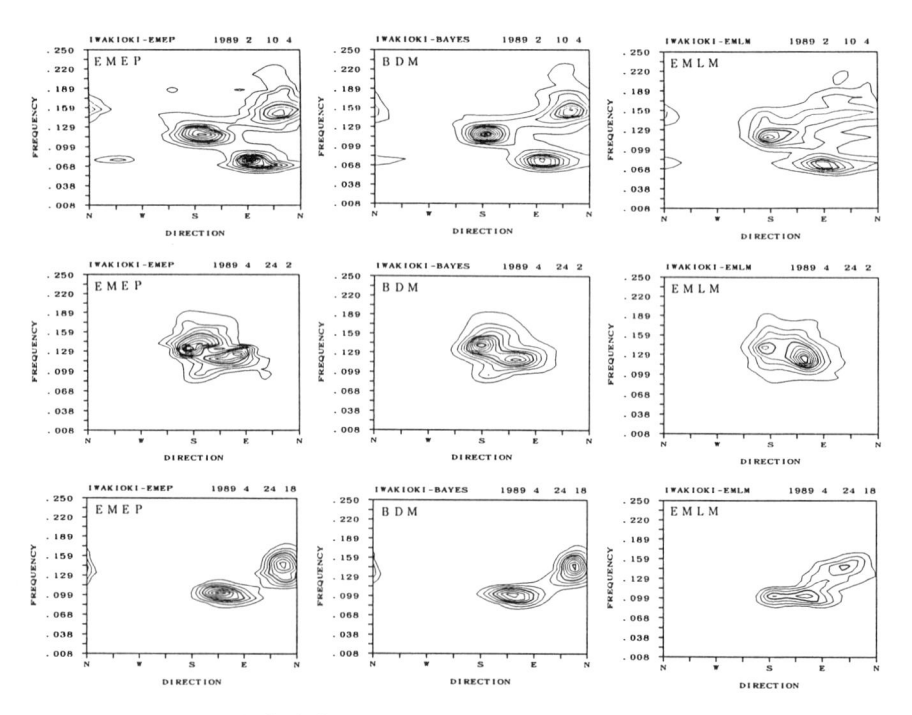

Fig. 7 Typical directional spectra of multiple wave systems estimated by EMEP, BDM, and EMLM.

Sometimes several wave systems exist from different sources, with **Fig. 7** showing such an example measured at the Iwaki-oki Station for the EMEP, BDM, and EMLM. The contour lines of the relative spectral density are plotted by direction versus frequency domain. The upper figures show directional spectra having $H_{1/3}$ = 2.18 m and $T_{1/3}$ = 9.0 s, while the middle ones are $H_{1/3}$ = 2.02 m and $T_{1/3}$ = 8.2 s and the lower ones $H_{1/3}$ = 3.94 m and $T_{1/3}$ = 8.8 s. In the EMEP and BDM, several wave groups can be clearly seen coming from different directions with different peak frequencies, whereas some of the EMLM energy peaks are diffused and not as clear.

8. Conclusions

Precise determination of the directional spectral characteristics of ocean waves is an essential step in planning and designing of coastal and offshore structures. Since field measurements of directional wave spectra require deployment of multiple sensors, all of which must be maintained in good condition for successful data recording, the obtained data must be analyzed using an accurate, reliable, and robust method to estimate the directional wave spectrum. We believe the proposed EMEP meets these criteria and can be successfully employed for carrying out research on directional seas.

Our major conclusions are as follows:
1) The EMEP can be applied to handle arbitrary-mixed instrument array measurements.
2) When the EMEP is applied to three-quantity measurements, it yields the same estimate as the MEP and has higher resolution than the EMLM.
3) When the EMEP is applied to more than three arbitrary-mixed instrument array measurements, it yields almost the same estimate as the BDM and has the highest resolution among other existing methods.
4) A personal computer can be used to perform the EMEP; thereby obtaining real-time estimation of directional spectra.

References

Akaike, H. (1973): Information theory and an extension of the maximum likelihood principle, 2nd Inter. Symp. on Information Theory (Petrov, B. N. and Csaki, F. eds.), Akademiai Kiado, Budapest, pp. 267−281.

Akaike, H. (1980): Likelihood and Bayes procedure, Bayesian statistics (Bernardo, J. M., De Groot, M. H., Lindley, D. U., and Smith, A. F. M. eds.), University Press, Valencia, pp. 143−166.

Barber, N. F. (1961): The directional resolving power of an array of wave detectors, Ocean Wave Spectra, Prentice Hall, Inc., pp. 137−150.

Bendat, J. S. and A. G. Piersol (1986) : Random Data Analysis and Measurement Procedures, 2nd ed., John Wiley & Sons, 566 p.

Capon, J. (1969): High-resolution frequency-wavenumber spectrum analysis, Proc. IEEE, Vol. 57, pp. 1480−1418.

Hashimoto, N. and K. Kobune (1985): Estimation of directional spectra from the

Maximum Entropy Principle, Rept. of P.H.R.I., Vol. 23, No. 3, pp. 123 – 145 (in Japanese); or Kobune, K. and N. Hashimoto (1986): Estimation of directional spectra from the Maximum Entropy Principle, Proc. 5th Inter. OMAE Symp., Tokyo, pp. 80 – 85.

Hashimoto, N. (1987): Estimation of directional spectra from a Bayesian approach, Rept. of P.H.R.I., Vol. 26, No. 2, pp. 97 – 125 (in Japanese); or Hashimoto, N., K. Kobune and Y. Kameyama (1987): Estimation of directional spectrum using the Bayesian Approach, and its application to field data analysis, Rept. of P.H.R.I., Vol. 26, No. 5, pp. 57 – 100; or Hashimoto, N. and K. Kobune (1988): Directional spectrum estimation from a Bayesian approach, Proc. 21st ICCE, Spain, Vol. 1, pp. 62 – 76.

Hashimoto, N. and K. Kobune (1987): Estimation of directional spectra from a Bayesian approach in incident and reflected wave field, Rept. of P.H.R.I., Vol. 26, No. 4, pp. 3 – 33 (in Japanese).

Hasselman, D. E., M. Dunckel, and J. A. Ewing (1980): Directional wave spectra observed during JONSWAP 1973, Journal of Physical Oceanography, Vol. 10, pp. 1264 – 1280.

Isobe, M., K. Kondo and K. Horikawa (1984): Extension of MLM for estimating directional wave spectrum, Proc. Symp. on Description and Modeling of Directional Seas, Paper No. A-6, 15 p.

Longuet-Higgins, M. S., D. E. Cartwright, and N. D. Smith (1963): Observation of the directional spectrum of sea waves using the motions of a floating buoy, Ocean Wave Spectra, Prentice Hall Inc., New Jersey, pp. 111 – 136.

Nwogu, O. (1989): Maximum Entropy estimation of directional wave spectra from an array of wave probes, Applied Ocean Res., Vol. 11, No. 4, pp. 176 – 193.

Panicker, N. N. and L. E. Borgman (1974): Enhancement of directional wave spectrum estimate, Proc. 13th ICCE, Copenhagen, pp. 258 – 279.

CHAPTER 19

A Regression Model for Estimating Sea State Persistence

Yoshio Hatada[1] and Masataka Yamaguchi[2]

Abstract

This paper deals with regression models for estimating probability distributions of long term wave height, sea state persistence above a wave height threshold and peak wave height during a storm, and yearly-averaged occurrence rate of peak wave height event. The models, in which input condition is the variance of long term wave height at a concerned sea area, were constructed on the basis of statistical analysis of the long term wave height data obtained around the Japanese coast. First, the effect of observation interval on the above-mentioned wave climate statistics was investigated and a method to remove the effect by use of FFT was proposed. Second, Applicability of each regression model was confirmed from close agreement between estimation and observation for the wave climate statistics. Finally, return periods of extreme wave height with prescribed sea state persistence and wave height threshold at wave observation points around the Japanese coast were estimated with a combination of the present regression models.

1. Introduction

Sea state persistence is an important factor to be considered in maritime activity and design of coastal structures. Although many studies on its statistical properties (Lawson and Abernethy, 1975; Graham, 1982; Takahashi et al., 1982; Kuwashima and Hogben, 1984; Smith, 1988; Yamaguchi et al., 1989, 1993; Teisson, 1990; Mathiesen, 1994) have been conducted using the observed
---
[1] Research Assistant, Dept. of Civil and Ocean Eng., Ehime Univ., Bunkyocho 3, Matsuyama 790, Ehime Pref., Japan
[2] Prof., Dept. of Civil and Ocean Eng., Ehime Univ., Bunkyocho 3, Matsuyama 790, Ehime Pref., Japan

wave data, probability distribution of sea state persistence of high waves which is crucial to the design of coastal structures and a statistical model for estimating sea state persistence of high waves still remain open to question, because of the lack of long term observed wave data with good quality.

The aim of this study is to present a regression model for estimating return period of extreme wave height with prescribed sea state persistence and wave height threshold on the basis of statistical analysis for distributions of significant wave height and sea state persistence and peak wave height during a storm using long term data of coastal waves around Japan acquired over several years by the Japan Meteorological Agency and the Bureaus of Harbor Construction, Ministry of Transport.

2. Wave Data and Analysis Method

Wave data used in the analysis are those measured for 20 minutes every 1 to 3 hours with a sonic-type wave gauge installed in water of 30 to 50 m depth. Fig. 1 shows location of 14 wave observation points around the Japanese coast. The longest and shortest observation periods are 14 years at Irouzaki in the Pacific coast and Kyougamisaki in the Japan Sea coast and 5 years at Shirihamisaki in the Pacific coast of North Japan. The

Fig. 1 Location of wave observation points.

most prominent feature of the wave data is that breaks in the data record are rare, in which even the highest ratio of breaks is less than 5 % of the total run at Sakihama in the Pacific coast. One or two missing data points in the time series of wave height were filled using linear interpolation but larger gaps due to breaks were not filled.

Fig. 2 illustrates the definitions of sea state persistence τ and peak wave height H_p. Individual sea state persistence τ is defined as linearly-interpolated time span above a wave height threshold $H_{1/3c}$. Wave height threshold is prescribed every 0.25 m for the range of 0.5 to 5 m. Peak wave height H_p is obtained by fitting of parabolic curve to the three largest wave height data to recover the reduction of wave height associated with observation of discrete interval and application of FFT.

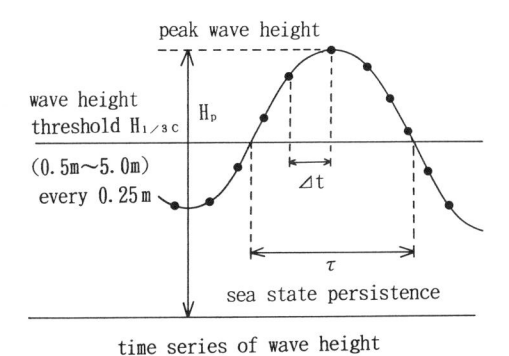

Fig. 2 Definitions of sea state persistence and peak wave height.

3. Fitting of Probability Distribution to Wave Data

Four kinds of probability distributions such as 3-parameter and 2-parameter Weibull distributions, 3-parameter lognormal distribution and hypergamma distribution (Suzuki, 1964) or generalized gamma distribution (Ochi, 1992) were used for data fitting. As a result of goodness of fit test, the 3-parameter Weibull distribution was chosen for fitting to the long term wave height data and peak wave height data, and the 2-parameter Weibull distribution was used for fitting to persistence data above a wave height threshold given every 0.25 m. The method applied for the estimation of the parameters is the maximum likelihood method. Each parameter is distinguished with suffix "H", or "τ" or "e". Non-exceedance probability of the 3-parameter Weibull distribution

F(x) is written as

$$F(x) = 1 - \exp[\{(x-b)/(x_0-b)\}^k],\qquad(1)$$

in which k, b and x_0 are the shape parameter, location parameter and scale parameter respectively. If the location parameter b is set to zero, Eq. (1) reduces to 2-parameter Weibull distribution.

Fig. 3 shows an example of the probability distributions fitted to the wave height data. Goodness of fit of the 3-parameter Weibull distribution to the wave height data is satisfactory in this case. The other distributions produce poor fit in the higher or lower wave height region and the parameters estimated for hypergamma distribution are not always statistically stable.

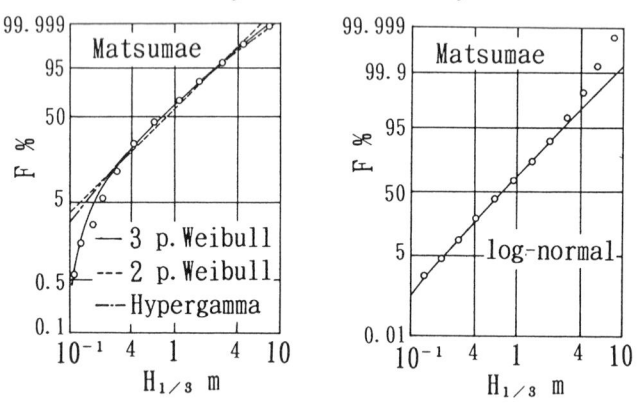

Fig. 3 Fitting of probability distributions to wave height data.

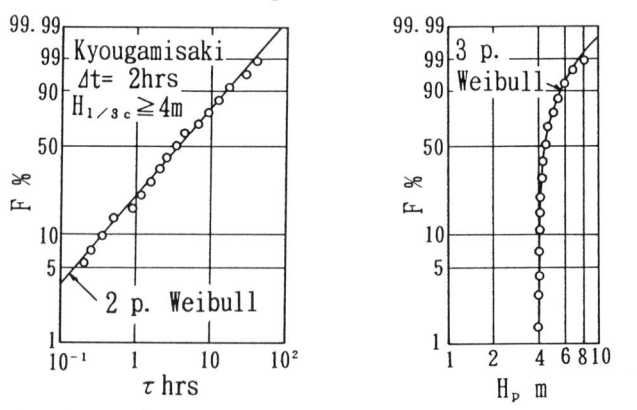

Fig. 4 Fitting of 2-parameter Weibull distribution to sea state persistence data and fitting of 3-parameter Weibull distribution to peak wave height data.

Examples of the fitting of the 2-parameter Weibull distribution to the persistence data and the fitting of the 3-parameter Weibull distribution to the peak wave height data are indicated in Fig. 4. Goodness of fit of these distributions is satisfactory. When sea state persistence data are well approximated with the 2-parameter Weibull distribution, use of the 2-parameter distribution rather than the 3-parameter distribution is preferable, because the estimates of parameters in this case are statistically more stable.

4. Effect of Sampling Interval on Wave Climate Statistics

In order to investigate the effect of sampling interval on wave climate statistics, time series of wave height data were resampled with sampling intervals of 2, 3, 4, 5, 6, 8, 10 and 12 hours from those observed every 1 hour and with sampling intervals of 4, 6, 8, 10 and 12 hours from those observed every 2 hours. Then, wave climate characteristics were analyzed using the above-mentioned methods. The left side of Fig. 5 shows an example of the relation between shape parameter k_τ of the 2-parameter Weibull distribution fitted to sea state persistence data and wave height threshold $H_{1/3c}$ in the case of sampling time of $\Delta t=2$ hours, and the right side of Fig. 5 is the relation between shape parameter k_e of the 3-parameter Weibull distribution fitted to peak wave height data and wave height threshold $H_{1/3c}$. Relations between the Weibull parameters and wave height threshold are approximated well with power functions such as $k_\tau = a_{k\tau}(H_{1/3c})^{b_{k\tau}}$ and $k_e = a_{ke}(H_{1/3})^{b_{ke}}$. These results hold for the other parameters $x_{0\tau}$, b_e and x_{0e} of the Weibull distribution.

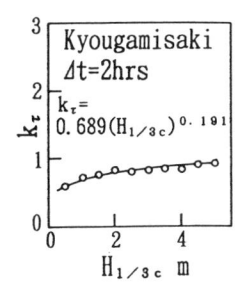

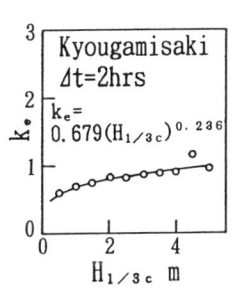

Fig. 5 Relation between Weibull parameter and wave height threshold.

The relations between the coefficients $a_{k\tau}$, a_{ke} in the power function and sampling interval Δt are indicated in Fig. 6. These coefficients, and consequently, the probability distributions of sea state persistence and peak wave height are strongly dependent on sampling interval Δt. But the parameters of the 3-parameter Weibull distribution fitted to long term wave height data are almost independent of sampling interval.

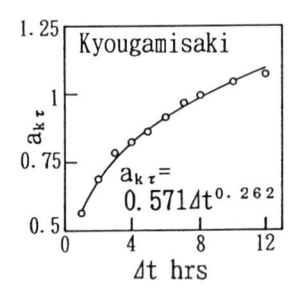

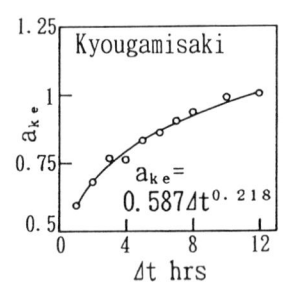

Fig. 6 Dependence of coefficient in power function on observation interval.

In order to remove the effect of the sampling interval on wave climate statistics, higher frequency components in time series of wave height data were filtered out by making use of FFT, and then the same methods as before were applied for the estimation of the wave climate statistics. Fig. 7 describes the effect of cut-off period on the coefficient in the power function, and an almost constant value of the coefficient can be found for the range of sampling interval shorter than half of a prescribed cut-off period. But it should be noted that filtering of time series of wave height data produces longer sea state persistence data and lower peak wave height data. In the following analysis, the cut-off period of $\Delta t=6$ hours is chosen in order to avoid the smoothing effect as much as possible and to make quality of wave height data similar with respect to the sampling interval, when the sampling interval used in actual observation is taken into account.

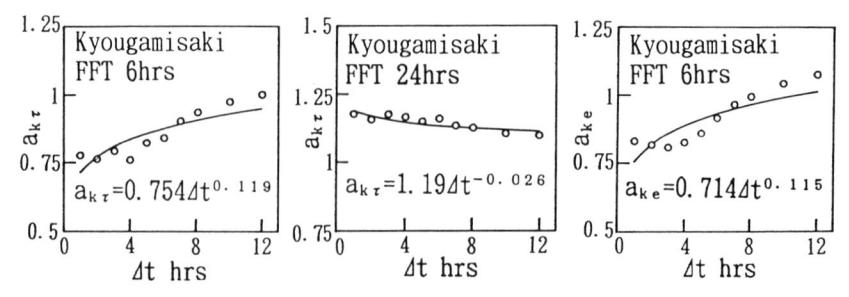

Fig. 7 Effect of cut-off period on coefficient in power function.

5. Construction of Regression Models

(1) Regression model for wave height distribution

Fig. 8 shows the relations between parameters of the 3-parameter Weibull distribution fitted to long term wave height data and variance H_σ^2 of wave height data in each wave observation point. The relations are approximated well by power functions such as

$$k_H = 1.10(H_\sigma^2)^{-0.246}, \quad b_H = 0.088(H_\sigma^2)^{-0.894},$$
$$x_{0H} = 1.19(H_\sigma^2)^{0.088} \tag{2}$$

These relations are approximate ones, because the parameters of the Weibull distribution theoretically depend on mean, variance and skewness of the population. But long term wave height distribution can be approximately estimated from these regression equations by giving wave height variance in the sea area as an input condition.

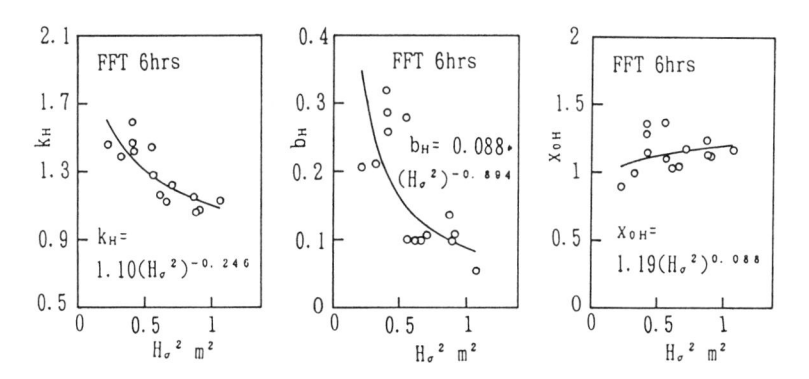

Fig. 8 Relation between Weibull parameter for wave height distribution and variance of wave height.

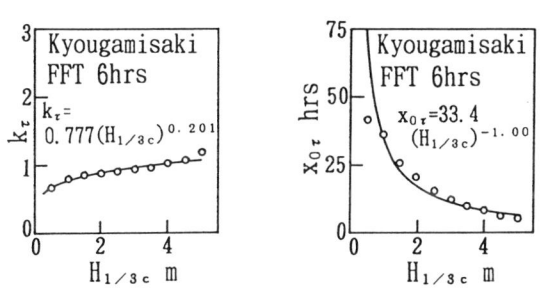

Fig. 9 Relation between sea state persistence parameter and wave height threshold.

(2) Regression model for sea state persistence

Relations between sea state persistence parameters and wave height threshold are indicated in Fig. 9. The relations are approximated well by regression equations such as

$$k_\tau = a(H_{1/3c})^b, \quad x_{0\tau} = c(H_{1/3c})^d \tag{3}$$

Fig. 10 shows the relations between coefficients in the above regression equations and variance of wave height at 14 wave observation points around the Japanese coast. Relatively high correlation is found between these variables and the relations are approximated with power functions such as

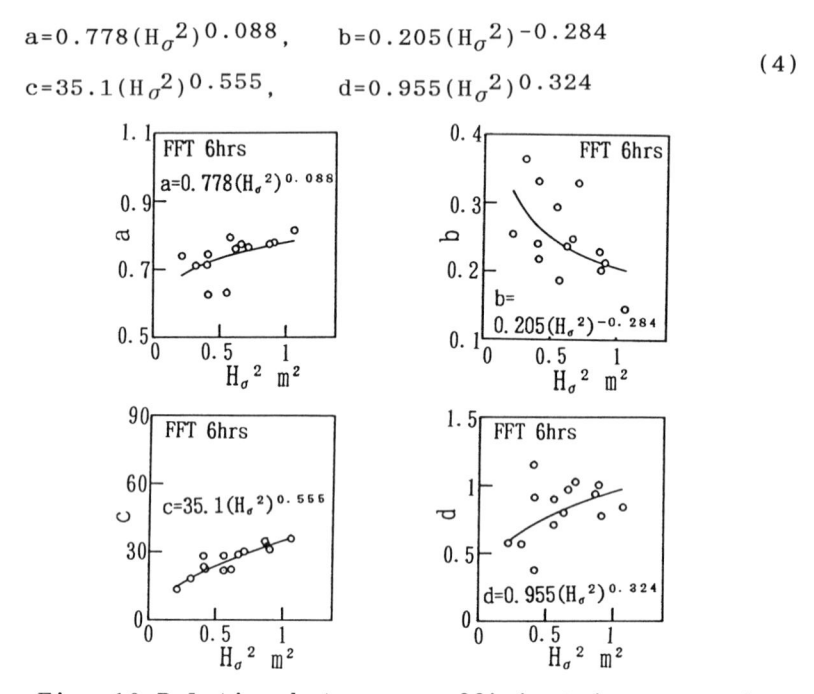

$$a = 0.778(H_\sigma^2)^{0.088}, \quad b = 0.205(H_\sigma^2)^{-0.284}$$
$$c = 35.1(H_\sigma^2)^{0.555}, \quad d = 0.955(H_\sigma^2)^{0.324} \tag{4}$$

Fig. 10 Relation between coefficient in regression equation and variance of wave height.

Thus, probability distribution of sea state persistence above a prescribed wave height threshold can be estimated by using these regression models, when wave height variance in the sea area is given as an input condition. Fig. 11 illustrates the comparison between estimation and observation for mean $\bar{\tau}$ and standard deviation τ_σ of sea state persistence and probability distribution. Good correspondence is observed between estimation and observation.

ESTIMATING SEA STATE PERSISTENCE

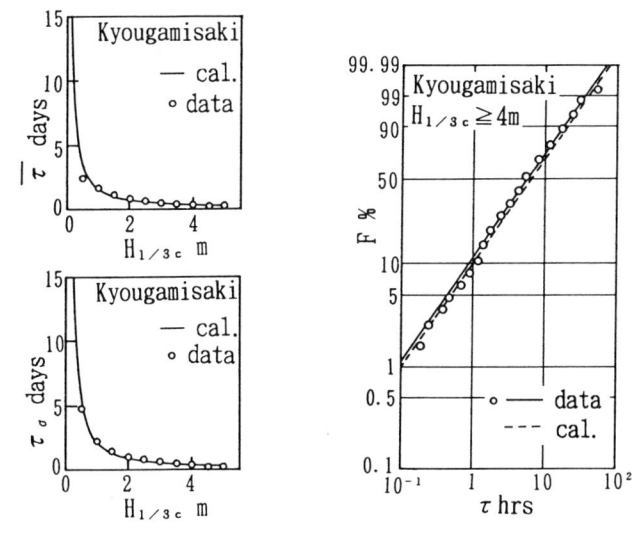

Fig. 11 Comparison between estimation and observation for sea state persistence.

(3) Regression models for peak wave height and its occurrence rate

Fig. 12 shows the relation between peak wave height parameters in the Weibull distribution k_e, b_e, x_{0e} and yearly-averaged occurrence rate of peak wave height $\overline{N}$, and wave height threshold $H_{1/3c}$. The former three relations are approximated with power functions such as

$$k_e = a(H_{1/3c})^b, \quad b_e = c(H_{1/3c})^d, \quad x_{0e} = e(H_{1/3c})^f \tag{5}$$

and the last relation is approximately expressed with exponential function as

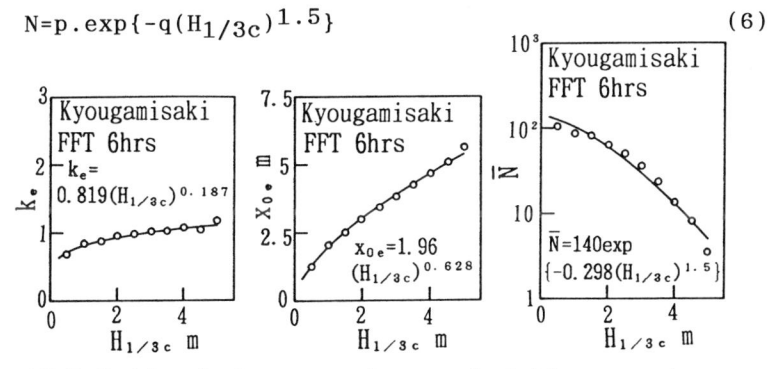

$$N = p \cdot \exp\{-q(H_{1/3c})^{1.5}\} \tag{6}$$

Fig. 12 Relation between peak wave height parameters and occurrence rate, and wave height threshold.

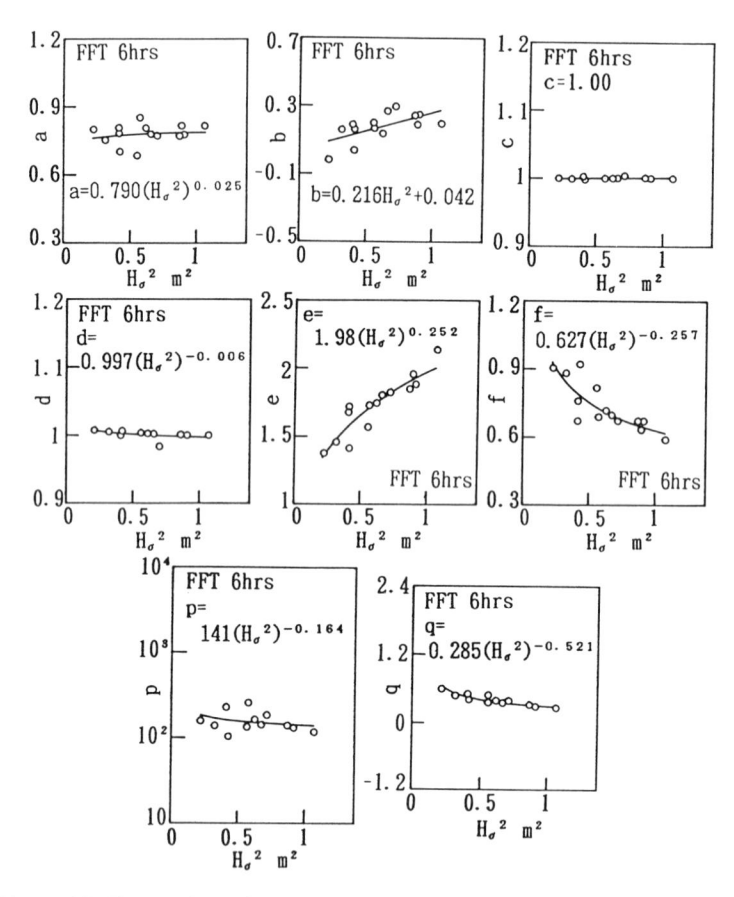

Fig. 13 Relation between coefficients in regression equations and wave height variance.

Then relations between coefficients in regression equations and variance of wave height estimated individually at 14 wave observation points are shown in Fig. 13. These relations are also well approximated with power functions as

$$a=0.790(H_\sigma^2)^{0.025}, \quad b=0.216H_\sigma^2+0.042$$
$$c=1.000, \quad d=0.997(H_\sigma^2)^{-0.006}$$
$$e=1.98(H_\sigma^2)^{0.252}, \quad f=0.627(H_\sigma^2)^{-0.257} \quad (7)$$
$$p=141(H_\sigma^2)^{-0.164}, \quad q=0.285(H_\sigma^2)^{-0.521}$$

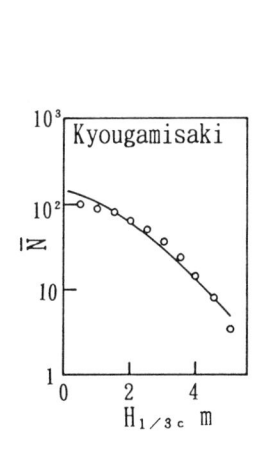

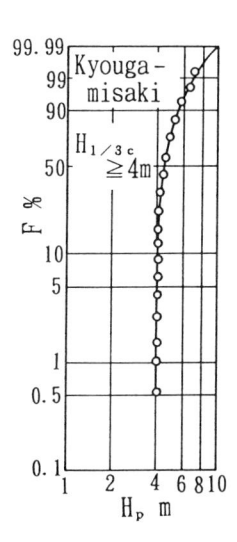

Fig. 14 Comparison between estimation and observation for yearly-averaged occurrence rate of peak wave height and probability distribution of peak wave height.

Probability distributions of peak wave height and yearly-averaged occurrence rate of peak event at a prescribed wave height threshold can be estimated by using these regression models for given wave height variance. Fig. 14 is an example of the estimation for yearly-averaged occurrence rate of peak wave height and probability distribution of peak wave height. High correlation between estimation and observation can be seen.

6. Estimation of Return Period of Extreme Wave Height

Assuming independence between sea state persistence and peak wave height, return period R of peak wave height H_p with prescribed sea state persistence τ above a wave height threshold $H_{1/3c}$ can be estimated from the following equation

$$1/R = N(H_{1/3c})\{1-F(H_p;H_{1/3c})\}\{1-F(\tau;H_{1/3c})\} \quad (8)$$

in which $F(H_p;H_{1/3c})$ and $F(\tau;H_{1/3c})$ are non-exceedance probabilities of peak wave height H_p and sea state persistence τ above a wave height threshold $H_{1/3c}$. When variance of long term wave height at a location around the coastal area of Japan is given as an input condition, the Weibull parameters and yearly-averaged occurrence rate of peak event above a prescribed wave height threshold required in Eq. (8) are determined by using all regression equations mentioned before.

Table 1 illustrates examples of return periods of peak wave height H_p=8 m with sea state persistence τ=12 hours above a wave height threshold $H_{1/3c}$=4 m which are estimated around the coastal area of Japan. The return periods of extreme wave height under these conditions range from about 160 to 270 years. The longest return period is obtained at Atsumi facing Northern Japan Sea. This is due to low exceedance probability of a prescribed extreme wave height and the resulting longer return period. The second longest return period is evaluated at Satamisaki facing Western Pacific Ocean. This is due to geographical situation around Satamisaki. That is to say, two islands, Tanegashima Island and Yakushima Island are located near Satamisaki, which tends to be sheltered from incoming waves by these islands.

Table 1 Return period R for peak wave height H_p with sea state persistence τ above wave height threshold $H_{1/3c}$ for H_p=8 m, τ=12 hours and $H_{1/3c}$=4 m.

location	H_σ^2 (m^2)	$\bar{N}(H_{1/3c})$	$1-F(\tau)$	$1-F(H_P)$	R (years)
Matsumae	.620	8.2	.246	.0029	169
Atsumi	1.067	15.4	.276	.0009	269
Kyougamisaki	.885	12.7	.268	.0014	211
Kashima	.663	9.0	.251	.0026	173
Fukuejima	.565	7.2	.240	.0035	165
Shirihamisaki	.406	4.3	.214	.0066	165
Enoshima	.555	7.0	.239	.0036	164
Irouzaki	.407	4.3	.215	.0066	164
Sakihama	.320	2.7	.194	.0105	179
Satamisaki	.216	1.1	.157	.0223	249
Kiyanmisaki	.415	4.4	.216	.0064	164
Wajima	.869	12.4	.267	.0015	207
Kanazawa	.906	13.0	.269	.0013	216
Tottori	.708	9.8	.255	.0023	178

7. Conclusions

Main results obtained in this study are summarized as follows;

(1) Effects of wave observation interval on the distributions of sea state persistence and peak wave height during a storm are relatively strong, whereas the effect on the long term wave height distribution is negligibly weak.

(2) Effect of observation interval on wave climate statistics can be removed by use of FFT.

(3) Probability distribution of sea state persistence above a prescribed wave height threshold can be estimated with a proposed regression model, when variance of long term wave height is given as an input condition.

(4) At the coastal sea area of Japan, return period of extreme wave height with prescribed sea state persistence and wave height threshold can be estimated with a combination of regression models for probability distributions of sea state persistence and peak wave height, and yearly-averaged occurrence rate of peak wave height, when wave height variance in the sea area is given as an input condition.

8. Acknowledgment

The authors thanks the Japan Ocean Data Center of Maritime Safety Agency, the Maritime Section of Japan Meteorological Agency and the Bureaus of Harbor Construction, Ministry of Transport for kindly offering the valuable wave data. Thanks are also due to Mr. M. Ohfuku, Technical Officer of Civil and Ocean Engineering, Ehime University and Mr. Arai, former student of Department of Ocean Engineering, Ehime University for their sincere assistance during the study.

References

Graham, C.(1982): The parameterization and prediction of wave height and wind speed persistence statistics for oil industry operational planning purposes, Coastal Eng., Vol. 6, pp.303-329.

Kuwashima, S. and N. Hogben(1984): The estimation of persistence statistics from cumulative probability of wave height, Rept. No. R183, NMI Ltd., 72p..

Lawson, N. V. and C. L. Abernethy(1975): Long term wave statistics off Botany Bay, Proc. 2nd Austr. Conf. on Coastal and Ocean Eng., pp. 167-176.

Mathiesen, M.(1994): Estimation of wave height duration statistics, Coastal Eng., Vol. 23, pp.167-181.

Ochi, M. K.(1992): New approach for estimating the severest sea state from statistical data, Proc. 23rd ICCE, Vol. 1, pp.512-525.

Smith, O. P.(1988): Duration of extreme wave condition, J. Waterway, Port, Coastal, and Ocean Eng., ASCE, Vol. 114, No. 1, pp. 1-17.

Suzuki, E.(1964): Hypergamma distribution and its fitting to rainfall data, Papers in Meteorol. and Geophys., Vol. 15, pp.31-51.

Takahashi, T. et al.(1982): Statistical properties of coastal waves, Proc. 29th Japanese Conf. on Coastal Eng., pp.11-15 (in Japanese).

Teisson, C.(1990): Statistical approach of duration of extreme storms, Consequences on break water damages, Proc. 22nd ICCE, Vol. III, pp.1851-1860.

Yamaguchi, M. et al.(1989): Analysis of wave climate around the coastal area of Japan, Natural Disas. Sci., 8-2, pp.23-45 (in Japanese).

Yamaguchi, M. et al.(1993): Statistics of duration of severe sea state and its evaluation model, Proc. Coastal Eng., Vol. 40, pp.116-120 (in Japanese).

CHAPTER 20

THE MAXIMUM SIGNIFICANT WAVE HEIGHT IN THE SOUTHERN NORTH SEA

L.H. Holthuijsen[1], J.G. de Ronde[2], Y. Eldeberky[1], H.L. Tolman[3], N. Booij[1], E. Bouws[4], P.G.P. Ferier[1], J. Andorka Gal[2]

Abstract

The maximum possible wave conditions in the southern North Sea are estimated with synthetic storms and a third-generation wave model. The storms and the physics in the wave model have been chosen within the uncertainty of the state-of-the-art to have maximum effect. The wave conditions appear to be limited by the presence of the bottom and to some extent by the assumed maximum wind speed of 50 m/s. The maximum significant wave height thus determined for the southern North Sea is about 0.4 times the local water depth.

Introduction

Extrapolations of observations usually provide fair estimates of extreme conditions as long as the physical regime of the waves does not change dramatically. Such a change may occur because the wind speed is limited and in shallow seas the water depth is limited. An extrapolation of the significant wave height should take this into account, possibly as an upper limit of the significant wave height. In the present study, the existence of such an upper limit in the southern North Sea is investigated.

Although the maximum wind speed in the North Sea is not well known, consultation with meteorologists suggested the 50 m/s wind speed as an upper limit. We investigate the sensitivity of the our results for this assumption. The physical phenomena of waves in these conditions are not well known either but we use a state-of-the-art wave model. For deep water this type of model has shown to be reliable in hurricane conditions.

---

[1] Delft University of Technology, the Netherlands, [2] Ministry of Public Works and Transport, the Netherlands, [3] NOAA, National Meteorological Center, USA, [4] Royal Netherlands Meteorological Institute, the Netherlands

The wave model

We use the numerical wave model WAVEWATCH II (Tolman, 1991, 1992). It is based on the discrete spectral action balance equation with a number of optional formulations for wave generation and dissipation. First we use it with initial choices for these formulations and then we investigate the alternatives.

The wave model

The action balance equation is a generalization of the energy balance equation in the presence of currents (e.g., Phillips, 1977). In the model this balance is formulated for propagation over a sphere with coordinates longitude λ and latitude ϕ:

$$\frac{\partial N(\omega,\theta)}{\partial t} + (\cos\phi)^{-1}\frac{\partial C_\phi \cos\phi N(\omega,\theta)}{\partial \phi} + \frac{\partial C_\lambda N(\omega,\theta)}{\partial \lambda} + \frac{\partial C_\omega N(\omega,\theta)}{\partial \omega} + \frac{\partial C_\theta N(\omega,\theta)}{\partial \theta} = S(\omega,\theta)$$

in which $N(\omega,\theta)$ is the action density of the waves, defined as the energy density $E(\omega,\theta)$ divided by the relative frequency σ, as function of absolute frequency ω and direction θ. The left-hand-side represents the local rate of change of the action density, propagation along great circles, shifting of the absolute frequency due to time variations in depth and currents, and refraction. The expressions are taken from linear wave theory (e.g. Mei, 1983). We will not consider currents in the present study. The right-hand-side of the above balance (the net production of wave action) represents all effects of generation and dissipation of the waves. The processes which are included in the model are: wave generation by wind, nonlinear quadruplet wave-wave interactions and dissipation (white-capping and bottom friction). The WAVEWATCH II model incorporates more than one formulation for each of these processes except for depth-induced wave breaking, which we added. For this we used the formulation of Battjes and Janssen (1978) and Battjes et al. (1993). As an alternative for Battjes and Janssen (1978) we added the formulation of Roelvink (1993).

In discretizing the spectrum we used a logarithmic frequency distribution from .022 Hz (a rather low frequency, chosen to cover extreme conditions) to .75 Hz with a 10% resolution. The directional resolution is 15°. From the numerical options in WAVEWATCH II (see Tolman, 1992) we choose to use the up-wind propagation scheme with either the static integration of the source terms (constant-wind cases) or dynamic integration (all other cases). The spatial resolution of the bottom grid is about 8 km except near the Dutch coast where it is 3 km. The bathymetry is indicated in Fig. 1. In view of the type of storms that we will consider, we imposed a uniform increase of 5 m in water depth to simulate the corresponding storm surge. We verified with sensitivity computations that the wave conditions at the northern boundary of the model (at 62° N) are not relevant for the southern North Sea in the extreme conditions considered here.

Maximum wave physics

We initially use formulations that are identical to those of the published WAM model (WAMDI group, 1988; to investigate the sensitivity of the results for the assumed maximum wind speed). However, we replaced the bottom friction

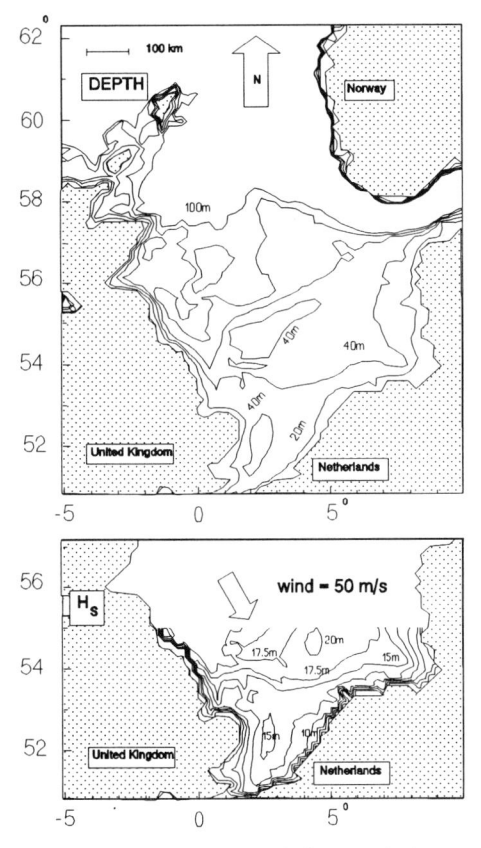

Fig. 1 The bathymetry of the North Sea (8 km resolution; upper panel) and the maximum significant wave height in the southern North Sea computed with 16 km resolution in a uniform wind field (50 m/s from 330°; lower panel). Physics selected for maximum effect.

formulation of Hasselmann et al. (1973) with the formulation of Madsen et al. (1988) and we added the depth-induced breaking of Battjes and Janssen (1978) and Battjes et al. (1993). This setting of the physics is summarized in Table 1 under the heading "initial".

The maximum significant wave height in given wind conditions is achieved by minimizing the effect of dissipation and maximizing the effects of generation (within the uncertainty of the state-of-the-art). We therefore varied the formulations of the physics in the model to each of the available options in WAVEWATCH II (Table 1).

WAVEWATCH II		
physics	initial	options
wind input	Snyder et al. (1981) + Komen et al. (1984)	Janssen (1991)
wave-wave interactions	Hasselmann and Hasselmann (1985)	-
white-capping	Komen et al. (1984)	Komen et al. (1984) + Janssen (1991)
bottom friction	Madsen et al. (1988)	Hasselmann et al. (1973)
depth-induced breaking	Battjes and Janssen (1978) + Battjes et al. (1993)	Roelvink (1993) + Battjes et al. (1993)

Table 1.   *The formulations of the physics of wave generation and dissipation in WAVEWATCH II (Tolman, 1991, 1992). The formulations that generate the maximum significant wave height are indicated with shading.*

We hindcasted for each variation independently the significant wave height at all stations indicated in Fig. 4 for wind speed 50 m/s from direction 330°. After all variations were considered and selected (as indicated in Table 1), we hindcasted the significant wave height in the southern North Sea with the selected formulations of the physics combined. The results are shown in Fig. 1. The significant wave height thus obtained is typically 1 m higher than obtained with the initial model setting.

Verification
We verified results of the model with the selected formulations for generation and dissipation (Table 1) with WAVEC buoy observations in a severe storm (December 12, 1990). The time series of the significant wave height for the station with best agreement between observed and computed maximum significant wave height and the station with the worst agreement are given in Fig. 2. The maximum significant wave height at these and other stations is given in Table 2. The agreement between observation and computation is generally reasonable except at station SON which is planned to be investigated further (the buoy seems to be located between two ship wrecks).

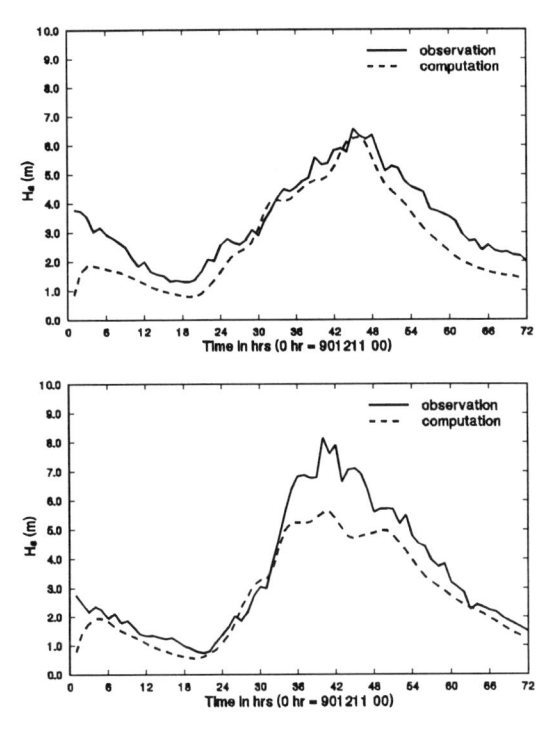

Fig. 2. The observed and measured time series of the significant wave height in the storm of December 12, 1990 at stations EUR (top panel) and SON (bottom panel).

station	observations	computations	difference
AUK	12.20 m	12.47 m	-0.27 m (2%)
SON	7.70 m	5.60 m	+2.10 m (26%)
ELD	7.70 m	7.25 m	+0.45 m (6%)
K13	7.70 m	7.60 m	+0.10 m (1%)
YM6	6.70 m	6.40 m	+0.30 m (4%)
EUR	6.25 m	6.30 m	-0.05 m (1%)
LEG	6.00 m	5.42 m	+0.58 m (10%)

Table 2. Observed and computed maximum significant wave height at various locations in the storm of December 12, 1990.

The wave model for the parametric storms
To find the storm that would generate the maximum significant wave height in the southern North Sea, we used a search procedure with a large number of synthetic storms (see below). For economic reasons these hindcasts were carried out with the second-generation wave model DOLPHIN-B described by Holthuijsen and de Boer (1988). This wave model has been adapted for shallow water and tuned to resemble the behaviour of the WAVEWATCH II model (initial setting) at station K13 in the storm of Feb. 1953. Dedicated computations showed that the storms that generated the largest significant wave heights at station K13 also generated the largest values at the other locations along the Dutch coast. The computations for the selected synthetic storm were repeated with the WAVEWATCH II model.

The wind field

In search for the existence of a physical upper-limit of the significant wave height we assume, in consultation with meteorologists of the Royal Netherlands Meteorological Institute, a wind speed of 50 m/s to be the maximum realizable sustained wind speed over the North Sea (at 10 m elevation). This is only a crude estimate and we will therefore determine the sensitivity of the maximum significant wave height for this assumption.

Uniform wind
The waves are hindcasted in a uniform wind field over the entire North Sea until a stationary situation is achieved for various wind speeds and directions. For these hindcasts a 16 km spatial resolution was used (sensitivity tests showed practically no effect of refining the resolution to 8 km). Fig. 3 shows that between northerly and westerly wind directions the significant wave height is weakly dependent of the direction with most (but not very pronounced) maxima at wind direction 330°. To find the sensitivity for the wind speed, we repeated these hindcasts for wind speeds in this direction increasing from 20 m/s to 60 m/s. As shown in Fig. 3 the sensitivity is rather weak (practically absent at the shallower stations) around the assumed maximum wind speed of 50 m/s.

Results
To obtain high resolutions results along the Dutch coast, we refined the computations in the southern North Sea with nested computations (from a 16 km via an 8 km to a 3 km grid resolution). The results are shown in Fig. 4.

Synthetic storm
To investigate whether the maximum wave conditions found in the above constant wind field are physically realizable, we hindcasted the waves in an extreme storm that we synthesized from historic storms. We selected this storm with search procedures involving wave hindcasts in a large number of synthetic storms.

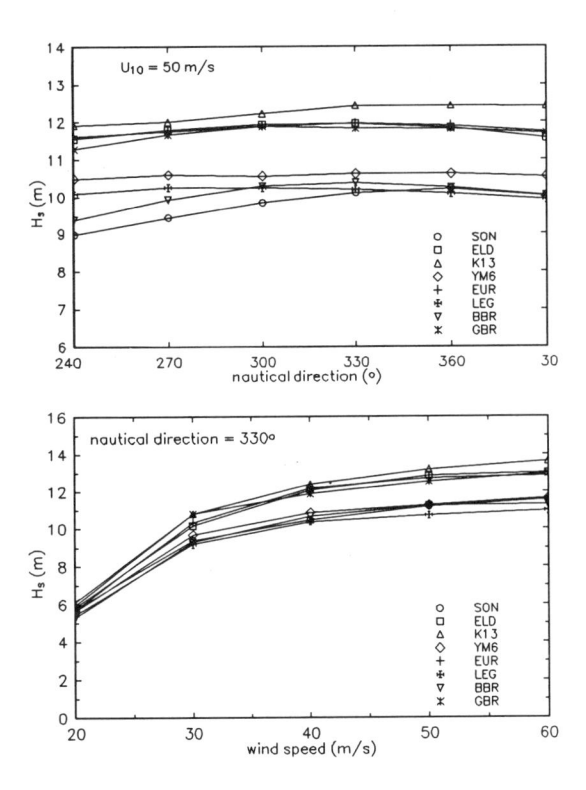

Fig. 3 The significant wave height as a function of wind direction (50 m/s) at various stations (top panel). The significant wave height as a function of wind speed from 330° at various stations (bottom panel).

Storm parameterization

To represent the atmospheric pressure in the synthetic storms we used a spatial Gaussian distribution with the radius to maximum wind different along the four major (orthogonal) axes of the storm. This created an elliptical asymmetric pressure field. From this pressure field we computed the geostrophic wind which we reduced to 65% and turned counter-clockwise by 15° to estimate the surface wind at 10 m elevation. Storms with surface wind speeds exceeding 50 m/s were removed from the search (due to the incremental nature of the search, small overshoots of about 2 m/s were permitted). The parameters of these synthetic storms were all assumed to vary linearly in time, characterized by one value at the moment when the centre of the storm is located at 10° W and one value 72 hours later.

We varied these time histories in the following search procedure within limits obtained from historic storms. To that end we analyzed five storms that are considered by meteorologists to be the severest storms in the southern North Sea over the last decennia.

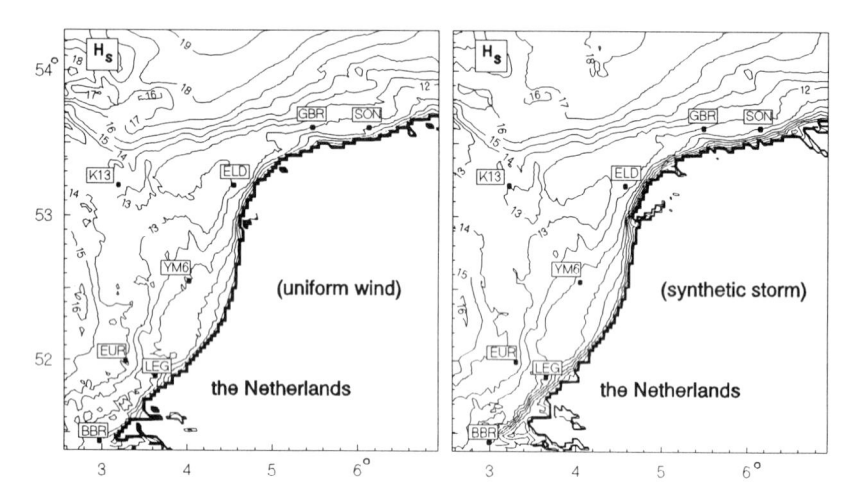

Fig. 4 The maximum significant wave height along the Dutch coast computed with 3 km resolution in a uniform wind field (wind speed = 50 m/s from 330°) and in a synthetic storm (maximum wind speed = 51.8 m/s).

These are all storms from westerly or north-westerly directions: 1st Feb '53, 21st Dec '54, 3rd Jan '76, 19th-25th Nov '81 and 26th Feb - 2nd March '90. (We verified with extra hindcasts that storms with tracks from more northerly directions were irrelevant.) We followed these storms on standard weather maps after they passed 10° W longitude and we visually estimated as a function of time: the forward speed, the central pressure, the orientation of the major axes and the radii along these axes. We thus obtained for each of these parameters five time histories. By roughly approximating the upper and lower envelope of these time histories with straight lines. From this we estimated the limits of the parameter values at the start of the storm and 72 hours later. For the start and end positions of the storms we used the results of an earlier and more extensive analysis of historic storms by Zwart (1993).

The search procedure

To determine which synthetic storm would lead to the largest significant wave height in the southern North Sea, we have used a sequential binary search with the storm parameters varying within the limits obtained from the above analysis of historic storms. First a reference hindcast is carried out with the value of all storm parameters (start and end values considered separately) set at their mid-range value. In sequence each storm parameter is then investigated: it is set at two values, one at the centre of the upper half-range and one at the centre of the lower half-range. All other parameters retain their reference value. Two hindcasts then decide which of these two values produces the largest significant wave height. The reference value of this parameter is then replaced by this selected value (it retains this value during the continuation of the search). Then the next storm parameter is modified similarly. After all storm parameters are thus investigated and selected the procedure is repeated twice, each time cutting the range of the storm parameter in half and centering it at the last selected reference value (three iterations in all). To increase the probability that the proper storm has been selected, we also carried out a synoptic binary search (replacing the reference values only after each of the three iterations has been completed) and a random search (shifting the mean to the selected value and reducing the widths of the assumed distributions by 50% after each of three iterations). A total of about 800 hindcasts was thus carried out. As the searches are carried out with three iterations, the resolution of the result is 1/8 of the original parameter range.

Results

The storm with the maximum significant wave height was selected by the sequential binary search. It is a fairly small but intense storm (300 to 400 km radius) tracking across the southern North Sea from a westerly direction. An inspection of the results suggests that to achieve the extreme wave heights, the wave field in the southern North Sea requires a locally high wind speed to compensate for local, bottom- induced dissipation (particularly breaking). Within the permitted range of atmospheric pressure, this locally high wind speed can be achieved only with a fairly small radius of the storm. The results of the other searches (synoptic binary and random) were storms that were similar in pattern to the one found with the sequential binary search but with somewhat lower significant wave heights.

To obtain the results with third-generation formulations, we carried out the hindcast for the selected synthetic storm with the WAVEWATCH II model with the formulations selected for maximum effect of the physics. This hindcast was nested to 3 km along the Dutch coast. The resulting maximum significant wave height (at each location) is given in Fig. 4.

Discussion

Comparing the two wave fields in Fig. 4, it is obvious that the maximum significant wave height obtained with a uniform wind field is practically equal to that obtained with the extreme synthetic storm. Apparently the wave field in these conditions (maximum wind speed of about 50 m/s) is dominated by the local water depth. This is supported by the similarity between the pattern of the maximum significant wave height and the water depth (Figs. 1 and 4). In fact, in the region considered with depth between 10 and 55 m, the ratio between maximum significant wave height and depth varies only between 0.35 and 0.45. This suggests a local equilibrium between generation on the one hand (for 50 m/s wind speed) and dissipation on the other. An uncertainty analysis based on sensitivity computations and on the uncertainty range of the coefficients in the selected formulations of the physics of wave generation and dissipation (not presented here) indicates that (a) the mechanism of bottom-induced breaking dominates the estimate of the maximum significant wave height and (b) the uncertainty of the estimated maximum significant wave height is about 5% upward and 25% downward.

Conclusions

The significant wave height in the southern North Sea seems to be limited by the local water depth and to some extent by an upper limit of the wind speed. The physics of wave generation and dissipation in these extreme conditions are not well known but with the available formulations selected for maximum effect, an estimate of the physical maximum of the significant wave height has been made. This has been done for selected wind fields using the third-generation model WAVEWATCH II (Tolman, 1991, 1992). The wind speed was limited to 50 m/s.

The results in a uniform wind field over the entire North Sea and in a selected extreme synthetic storm (based on historic information) are almost equal, indicating a local balance between wind generation and bottom induced dissipation. The local ratio between maximum significant wave height and water depth is nearly constant in the southern North Sea and equal to about 0.4.

References

Battjes, J.A. and J.P.F.M. Janssen, 1978, Energy loss and set-up due to breaking of random waves, *Proc. 16th Int. Conf. Coastal Engineering*, Hamburg, 569 - 587

Battjes, J.A., Eldeberky, Y., and Won, Y., 1993, Spectral Boussinesq modelling of random, breaking waves, *Proc. Int. Symposium WAVES '93*, New Orleans, ASCE, pp.813-820, New York

Hasselmann, K., T.P. Barnett, E. Bouws, H. Carlson, D.E. Cartwright K. Enke, J.A.Ewing, H. Gienapp, D.E. Hasselmann, P. Kruseman, A. Meerburg, P. Müller, D.J. Olbers, K. Richter, W. Sell and H. Walden, 1973, Measurements of wind-wave growth and swell decay during the Joint North Sea Wave Project (JONSWAP). *Ergänzungsheft zur Deutschen Hydrographischen Zeitschrift*, Reihe A (8) Nr. 12, 95 pp.

Hasselmann, S. and K. Hasselmann, 1985, Computations and parameterizations of the nonlinear energy transfer in a gravity-wave spectrum, Part I: A new method for efficient computations of the exact nonlinear transfer integral, *J. Phys. Oceanogr.*,15, 1369-1377

Holthuijsen, L.H., and de Boer, S., 1988, Wave forecasting for moving and stationary targets, In: *Computer Modelling in Ocean Engineering,* Eds. B.Y. Schrefler and O.C. Zienkiewicz, Balkema, Rotterdam, 231-234

Janssen, P.A.E.M., 1991, Quasi-linear theory of wind-wave generation applied to wave forecasting, *J. Phys. Oceanogr.*, 21, 1631 - 1642

Komen, G.J., S. Hasselmann and K. Hasselmann, 1984, On the existence of a fully developed wind-sea spectrum. *J. Phys. Oceanogr.*, **14**, 1271-1285.

Madsen, O.S., Y.-K. Poon and H.C. Graber, 1988b, Spectral wave attenuation by bottom friction: theory. *Proc. 21st Int. Conf. Coastal Eng.*, ASCE, Malaga, 492-504.

Mei, C.C., 1983, *The applied dynamics of ocean surface waves*, Wiley, New York

Phillips, O.M., 1977, *The dynamics of the upper ocean*, Cambridge University Press

Roelvink, J.A., 1993, Dissipation in random wave groups incident on a beach, *Coastal Engineering*, 19, 127 - 153

Snyder, R.L., F.W. Dobson, J.A. Elliott and R.B. Long, 1981, Array measurements of atmospheric pressure fluctuations above surface gravity waves. *J. Fluid Mech.*, **102**, 1-59.

Tolman, H.L., 1991, A third-generation model for wind waves on slowly varying, unsteady, and inhomogeneous depths and currents, *J. Phys. Oceanogr.*, 21 (6), 782 -797

Tolman, H.L., 1992, User manual for WAVEWATCH-II, NASA/GSFC, Laboratory for Oceans

WAMDI group, 1988, The WAM model - a third generation ocean wave prediction model. *J. Phys. Oceanogr.*, **18**, 1775-1810.

Zwart, B., 1993, De stormvloed van 1 februari 1953, Memorandum KNMI, VEO 93 -01, pp. 20, in Dutch

CHAPTER 21

Improved Boundary Conditions to a Time–Dependent Mild–Slope Equation for Random Waves

Toshimasa Ishii[1], Masahiko Isobe[2], and Akira Watanabe[2]

Abstract

The offshore and side boundary conditions to a time–dependent mild slope equation for random waves are improved to introduce given incident waves into and extract reflected waves from the computational domain with reduced computational time and storage size. The resulting numerical model is applied to calculate the wave field, nearshore current, and bottom topography change around a detached breakwater.

1 Introduction

A time–dependent mild–slope equation for random waves was derived from the mild slope equation by approximating frequency–independent expressions to the frequency–dependent coefficients (Kubo *et al.*, 1991; Kotake *et al.*, 1992). In the numerical solution, however, the peripheral region for absorbing the outgoing wave energy becomes large in comparison with the calculation domain of interest. In addition, because the velocity potential due to the incident waves must be calculated at each point in the peripheral region, extensive computational time is required.

The present study deals with the incident wave boundary condition and the open boundary condition in more detail and improves them, so that the deformation of multi–directional irregular waves may be calculated with much less computational time and storage size. The numerical model developed is applied to calculate the wave field, nearshore current, and bottom topography change around a detached breakwater on a uniform slope.

---
[1]Tokyo Electric Power Co., Ltd.
[2]Dept. of Civil Eng., Univ. of Tokyo, Bunkyo–ku, Tokyo 113, Japan

2 A time–dependent mild–slope equation for random waves

A time–dependent mild–slope equation for random waves (Kubo et al., 1992) is given by

$$\nabla(\bar{\alpha}\nabla\tilde{\eta}) + i\nabla\left[\bar{\beta}\nabla\left(\frac{\partial\tilde{\eta}}{\partial t}\right)\right] + \bar{k}^2\bar{\alpha}(1+if_D)\tilde{\eta} + i\bar{\gamma}(1+if_D)\frac{\partial\tilde{\eta}}{\partial t} = 0 \qquad (1)$$

$$\bar{\alpha} = \bar{C}\bar{C}_g \qquad (2)$$

$$\bar{\beta} = \frac{\bar{C}}{\bar{k}}[-2(1-\bar{n}) + \frac{1}{2\bar{n}}(2\bar{n}-1)\{1-(2\bar{n}-1)\cosh 2\bar{k}d\}] \qquad (3)$$

$$\bar{\gamma} = \bar{k}\bar{C}[2\bar{n} + \frac{1}{2\bar{n}}(2\bar{n}-1)\{1-(2\bar{n}-1)\cosh 2\bar{k}d\}] \qquad (4)$$

$$\bar{n} = (1/2)(1 + 2\bar{k}d/\sinh 2\bar{k}d) \qquad (5)$$

where $\bar{C}$ is the wave celerity, $\bar{C}_g$ the group velocity, $\bar{k}$ the wave number, d the water depth, t the time, ∇ the horizontal gradient operator, i the unit of imaginary number, and f_D an energy absorbing coefficient. The symbol ¯ denotes quantities at the representative frequency $\bar{\omega}$. The relation between $\tilde{\eta}$ and the temporal water surface variation η is :

$$\eta = Re[\tilde{\eta}e^{-i\bar{\omega}t}] \qquad (6)$$

3 Improvement of boundary condition for incident waves

3.1 Layer boundary method

In the numerical solution of horizontal two–dimensional problems, we introduce the incident waves at the offshore and side boundaries. In addition, the outgoing waves should propagate out at the boundaries without reflection.

Kubo et al. (1992) presented a method that satisfies the above conditions, whereby the incident waves are introduced through a layer boundary. This technique is illustrated in Figure 1. The energy of the outgoing waves is absorbed in the energy absorbing layer of a sufficient width, and the incident irregular waves must be prescribed as an excitation force at all grid points in the layer by a superposition of component waves. Consequently, the method requires considerable computational time and large storage, and therefore the number of component waves is restricted to about 200. Also, as seen from results of computations performed by Kubo *et al.* (1992) and the authors for multi-directional irregular waves around a detached breakwater, the distribution of the wave height was asymmetrical due to the statistical variation.

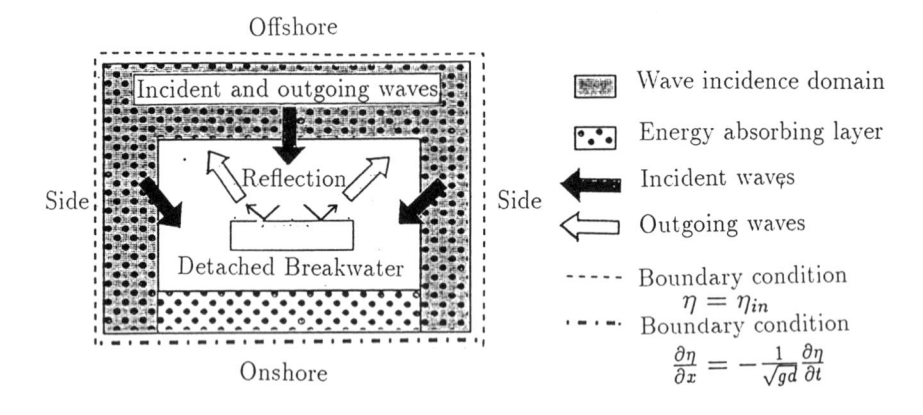

Fig.1: Method for introducing incident waves through a layer boundary (Kubo et al., 1992)

3.2 Line boundary method

The incident wave boundary condition is improved in the present study by introducing the incident waves through a line boundary as shown in Figure 2. The incident wave potential is only specified along the line boundary, which significantly reduces computational time and storage. With the new method wave transformation can be calculated for irregular incident waves even with as many as 1,000 components, whereas the previous method has a practical limit of about 200 components. As a result, the deformation of multi-directional irregular waves can be calculated with a much higher accuracy because the statistical variation is reduced.

The method is summarized in the following. First we set up a line boundary where the incident waves are introduced. Inside the line boundary, incident and outgoing waves are to be dealt with, whereas only outgoing waves exist outside the line boundary as shown in Figure 2. For this purpose, we only have to adjust the incident wave component $\tilde{\eta}_{in}$ in the finite difference equation for which the central grid point is located adjacent to the line boundary. The incident waves are introduced into the calculation domain through this operation.

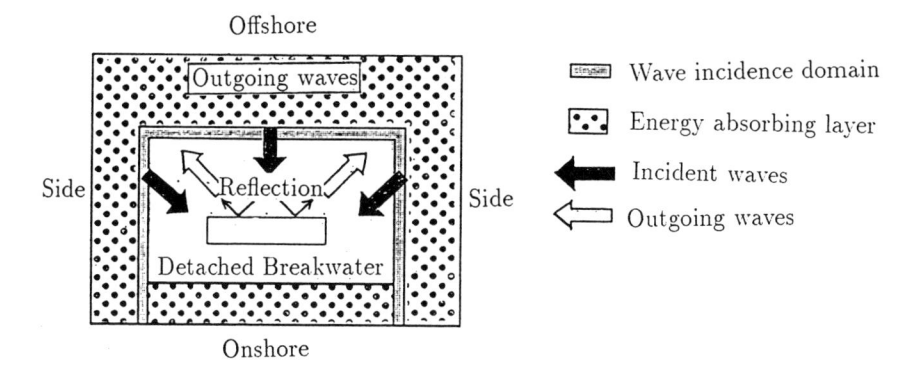

Fig.2: Method for introducing incident waves through a line boundary

To describe the method in more detail, we apply it to a horizontal two–dimensional problem. The ADI method is employed in the numerical calculations to solve the present equation. The calculations are carried out alternately in the x and y directions, implying that the application of the present method must be considered separately for each direction. Figure 3 shows the application in the x direction. For a horizontal two–dimensional model on a uniform slope, the finite difference equation of the time–dependent mild–slope equation for random waves in the x direction is written

$$\frac{\bar{\alpha}_{i+1} - \bar{\alpha}_{i-1}}{2\Delta x} \left\{ \frac{\tilde{\eta}_{i+1,j}^{t+1} - \tilde{\eta}_{i-1,j}^{t+1}}{2\Delta x} \right\}$$
$$+ \bar{\alpha}_i \left\{ \frac{\tilde{\eta}_{i-1,j}^{t+1} - 2\tilde{\eta}_{i,j}^{t+1} + \tilde{\eta}_{i+1,j}^{t+1}}{\Delta x^2} + \frac{\tilde{\eta}_{i,j-1}^{t} - 2\tilde{\eta}_{i,j}^{t} + \tilde{\eta}_{i,j+1}^{t}}{\Delta y^2} \right\}$$
$$+ i\frac{\bar{\beta}_{i+1} - \bar{\beta}_{i-1}}{2\Delta x} \left\{ \frac{(\tilde{\eta}_{i+1,j}^{t+1} - \tilde{\eta}_{i-1,j}^{t+1}) - (\tilde{\eta}_{i+1,j}^{t} - \tilde{\eta}_{i-1,j}^{t})}{2\Delta x \Delta t} \right\}$$
$$+ i\bar{\beta}_i \left\{ \frac{(\tilde{\eta}_{i-1,j}^{t+1} - 2\tilde{\eta}_{i,j}^{t+1} + \tilde{\eta}_{i+1,j}^{t+1}) - (\tilde{\eta}_{i-1,j}^{t} - 2\tilde{\eta}_{i,j}^{t} + \tilde{\eta}_{i+1,j}^{t})}{\Delta x^2 \Delta t} \right.$$
$$+ \left. \frac{(\tilde{\eta}_{i,j-1}^{t} - 2\tilde{\eta}_{i,j}^{t} + \tilde{\eta}_{i,j+1}^{t}) - (\tilde{\eta}_{i,j-1}^{t-1} - 2\tilde{\eta}_{i,j}^{t-1} + \tilde{\eta}_{i,j+1}^{t-1})}{\Delta y^2 \Delta t} \right\}$$
$$+ \bar{k}^2 \bar{\alpha}_i (1 + if_D) \frac{\tilde{\eta}_{i,j}^{t+1} + \tilde{\eta}_{i,j}^{t}}{2} + i\bar{\gamma}_i (1 + if_D) \frac{\tilde{\eta}_{i,j}^{t+1} - \tilde{\eta}_{i,j}^{t}}{\Delta t} = 0 \qquad (7)$$

By rearranging the equation under the condition that $\Delta x = \Delta y = \Delta l$, the left-hand side contains only unknown values (time step $t+1$) and the right-hand side

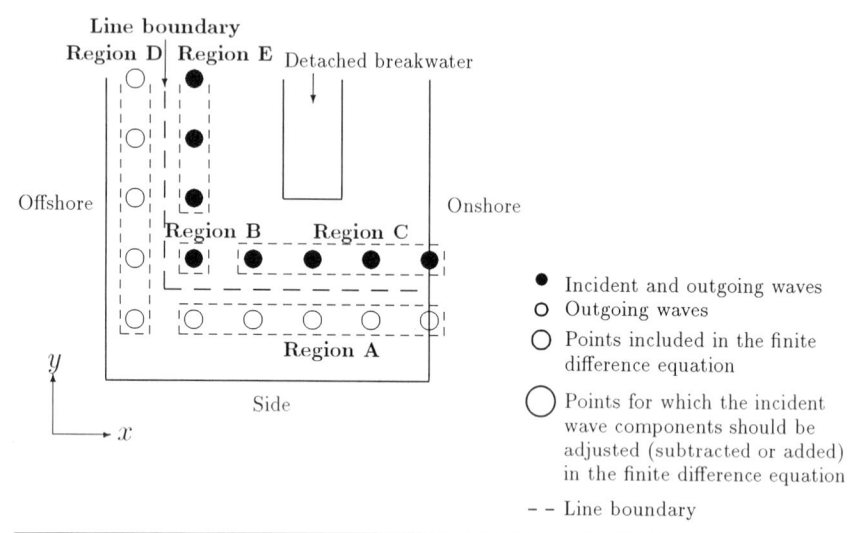

Fig.3: Illustration of the method of introducing incident waves through a line boundary for the case of calculation in the x direction

only known values (time steps t and $t-1$), and the above equation can be written

$$\begin{aligned}
& A1_i \tilde{\eta}_{i-1,j}^{t+1} + A2_i \tilde{\eta}_{i,j}^{t+1} + A3_i \tilde{\eta}_{i+1,j}^{t+1} \\
= {} & B1_i \tilde{\eta}_{i-1,j}^{t} + B2_i \tilde{\eta}_{i,j}^{t} + B3_i \tilde{\eta}_{i+1,j}^{t} + B4_i \tilde{\eta}_{i,j-1}^{t} + B5_i \tilde{\eta}_{i,j+1}^{t} \\
& + C1_i \tilde{\eta}_{i,j-1}^{t-1} + C2_i \tilde{\eta}_{i,j}^{t-1} + C3_i \tilde{\eta}_{i,j+1}^{t-1}
\end{aligned} \qquad (8)$$

where (i, j) is the grid number in the (x, y) coordinate system, t the time,

$A1_i \sim A3_i$, $B1_i \sim B5_i$ and $C1_i \sim C3_i$ the coefficients determined by $\bar{\alpha}_i, \bar{\beta}_i, \bar{\gamma}_i$ etc. Equation (8) includes grid points indicated with a double circle in Figure 3. There are three such points at time $t+1$, five points at time t and three points at time $t-1$ around the central grid point (i, j). These grid points are divided into two types according to their position relative to the line boundary. In the first type incident and outgoing waves coexist, and these grid points are shown with solid circles. In the second type only outgoing waves exist, which is indicated by empty circles. As an illustration, let us consider region A. The point$(i, j+1)$ at times t and $t-1$ are points including both incident and outgoing waves whereas the double circle points are points with outgoing waves only. Hence, we apply an operation which subtracts the incident wave term $\tilde{\eta}^t_{i,j+1,\text{in}}$, $\tilde{\eta}^{t-1}_{i,j+1,\text{in}}$ from point$(i, j+1)$ at times t and $t-1$. For this particular case, the finite difference equation (8) applied to outgoing waves may be written

$$\begin{aligned}
& A1_i \tilde{\eta}^{t+1}_{i-1,j} + A2_i \tilde{\eta}^{t+1}_{i,j} + A3_i \tilde{\eta}^{t+1}_{i+1,j} \\
=\ & B1_i \tilde{\eta}^t_{i-1,j} + B2_i \tilde{\eta}^t_{i,j} + B3_i \tilde{\eta}^t_{i+1,j} + B4_i \tilde{\eta}^t_{i,j-1} + B5_i \tilde{\eta}^t_{i,j+1} \\
& + C1_i \tilde{\eta}^{t-1}_{i,j-1} + C2_i \tilde{\eta}^{t-1}_{i,j} + C3_i \tilde{\eta}^{t-1}_{i,j+1} \\
& - B5_i \tilde{\eta}^t_{i,j+1,\text{in}} - C3_i \tilde{\eta}^{t-1}_{i,j+1,\text{in}}
\end{aligned} \quad (9)$$

In regions B to E, the incident waves are introduced along the line boundary using the same method. In the case of the y direction, the position of the double circles at $t+1$ and $t-1$ time step should be reversed.

The computational time required by the present method was compared with the previous method for the case of a calculation region with 80×100 grid points, 100 component waves, and 3000 time steps. Multi-directional irregular waves were expressed using the single summation method. In this case, the computational time required by the present method was one-third that of the previous method.

4 Improvement of open boundary condition

4.1 One-dimensional open boundary condition

The boundary condition at the outer edge of the energy absorbing layer used in the numerical calculations presented by Kubo et al. (1992) is shown in Figure 1. The one-dimensional Sommerfeld radiation condition is employed at the onshore boundary with the celerity C approximated by the long wave celerity in the same way as Ohyama et al. (1990). This condition is motivated by the fact that the wave direction becomes perpendicular to the onshore boundary due to wave refraction. The condition at the offshore and side boundary is that the water surface elevation is equal to that of the incident waves. This condition is based on the assumption that the outgoing waves are absorbed perfectly in the energy absorbing layer. In the numerical calculation presented by Kubo et al. (1992), a wide energy absorbing layer was needed to achieve perfect absorption.

In the present study, a new open boundary condition is derived from the Sommerfeld–type boundary condition by applying a Taylor expansion in terms of the angular frequency and wave direction. The aim is to reduce the energy absorbing layer. We present the result for the one–dimensional case first and then for the two-dimensional case.

The new open boundary condition for the one-dimensional case may be written

$$\eta(x_b, t + \Delta t) = \eta(x_b - \bar{C}\Delta t, t)$$
$$+\bar{C}(1 - \bar{n})\Delta t[\frac{\partial \eta(x_b - \bar{C}\Delta t, t)}{\partial x} - i\bar{k}\eta(x_b - \bar{C}\Delta t, t)] \quad (10)$$

where $\eta(x_b, t + \Delta t)$ is the water surface elevation at the outer edge of the energy absorbing layer, and $\bar{C}$, $\bar{n}$ and $\bar{k}$ the wave celerity, shallowness factor and wave number at the representative frequency, respectively.

Equation(10) consists of coefficients which are independent of the frequency of component waves as a result of applying the Taylor expansion to the wave celerity C, which is a coefficient in the Sommerfeld–type boundary condition. Equation (10) is, therefore, applicable to irregular waves. The physical meaning of equation (10) is illustrated in Figure 4.

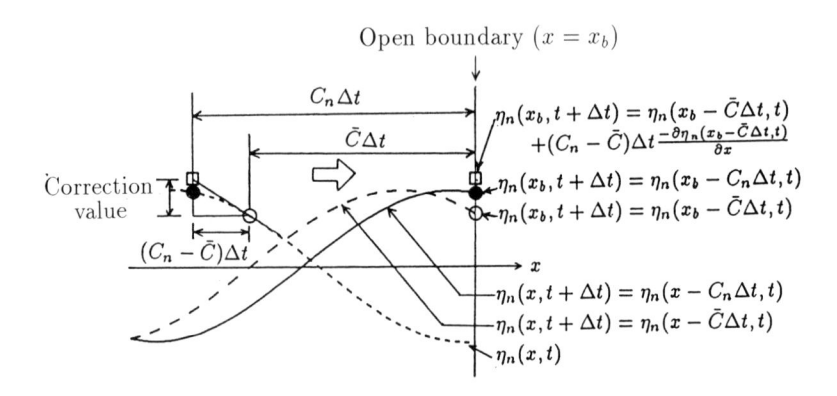

Fig.4: Physical interpretation of the improved Sommerfeld–type boundary condition for the one-dimensional case

The first term on the right–hand side of equation (10) expresses that the water surface elevation of the wave at time t and at a point which is $\bar{C}\Delta t$ away from the outer edge of the energy absorbing layer is equal to that at the outer edge after Δt, because the wave moves to the outer edge. $\bar{C}$ is the wave celerity at the representative angular frequency, however, the wave celerity of irregular waves is different from $\bar{C}$. This is because irregular waves consist of many component waves with various angular frequencies. The second term on the right-hand side is a correction for the phase of the water surface elevation at the outer edge of the absorbing layer due to the difference in wave celerity.

Equation(10) is derived as follows. First we have

$$\eta(x_b, t + \Delta t) = \eta_n(x_b - \bar{C}\Delta t, t) - (C_n - \bar{C})\Delta t \frac{\partial \eta(x_b - \bar{C}\Delta t, t)}{\partial x} \quad (11)$$

Expanding C_n into a Taylor series of $\Delta\omega_n$ and truncating the series at the first order gives

$$C_n = \bar{C} + \left(\overline{\frac{dC}{d\omega}}\right)\Delta\omega_n \quad (12)$$

where $\left(\overline{\frac{dC}{d\omega}}\right)$ and $\Delta\omega_n$ are written as

$$\left(\overline{\frac{dC}{d\omega}}\right) = \frac{1}{\bar{k}}\left(1 - \frac{1}{\bar{n}}\right) \quad (13)$$

$$\Delta\omega_n = \left(\overline{\frac{d\omega}{dk}}\right)\Delta k_n = \bar{n}\bar{C}\Delta k_n \quad (14)$$

Next, we assume that $\eta_n(x,t)$ in equation (11) is expressed as $\eta_n(x,t) = a_n e^{i(k_n x - \omega_n t)}$ and differentiate $\eta_n(x,t)$ with respect to x. Expanding k_n in $\eta_n(x,t)$ into a Taylor series and truncating it at the first order gives

$$\frac{\partial \eta_n(x - \bar{C}\Delta t, t)}{\partial x} = i(\bar{k} + \Delta k_n)\eta_n(x - \bar{C}\Delta t, t) \quad (15)$$

Substituting equations (12) to (15) into the second term of equation (11) and neglecting second-order terms of the resulting equation yield

$$-(C_n - \bar{C})\Delta t \frac{\partial \eta_n(x - \bar{C}\Delta t, t)}{\partial x} = i\bar{C}(1 - \bar{n})\Delta t \Delta k_n \eta_n(x - \bar{C}\Delta t, t) \quad (16)$$

Equation (15) is rewritten as

$$i\Delta k_n \eta_n(x - \bar{C}\Delta t, t) = \frac{\partial \eta_n(x - \bar{C}\Delta t, t)}{\partial x} - i\bar{k}\eta_n(x - \bar{C}\Delta t, t) \quad (17)$$

Substituting equation (17) into equation (16) gives

$$-(C_n - \bar{C})\Delta t \frac{\partial \eta_n(x - \bar{C}\Delta t, t)}{\partial x}$$
$$= \bar{C}(1 - \bar{n})\Delta t \left(\frac{\partial \eta_n(x - \bar{C}\Delta t, t)}{\partial x} - i\bar{k}\eta_n(x - \bar{C}\Delta t, t)\right) \quad (18)$$

Equation (18) consists only of coefficients that are independent of the frequency and hence determined only by the representative frequency. Superposition of equation (18) for an infinite number of component waves gives the second term on the right–hand side of equation (10). The present open boundary condition combines equation (10) and the energy absorbing layer where the energy absorbing coefficient f_D is increased linearly towards the outer edge of the layer in the same way as Ohyama et al. (1990).

In the present study, we determine the width of the energy absorbing layer and the coefficients $\bar{C}$, $\bar{n}$, and $\bar{k}$ in equation (10) as follows. The width is given as $0.6L_1$, where L_1 is the longest wavelength in the irregular waves. This width is chosen because reflection due to friction of the energy absorbing layer begins to increase rapidly at a width of $0.6L$ for a regular wave with wavelength L as obtained from the numerical calculations. Moreover, the coefficients $\bar{C}$, $\bar{n}$, and $\bar{k}$ in equation (10) are determined from the component wave having the longest wavelength. The energy absorption in the layer is the lowest for this component wave, implying that the energy at the outer edge of the layer is the highest. The behavior of the open boundary condition is shown in Figure 5. f_{Dmax} is the

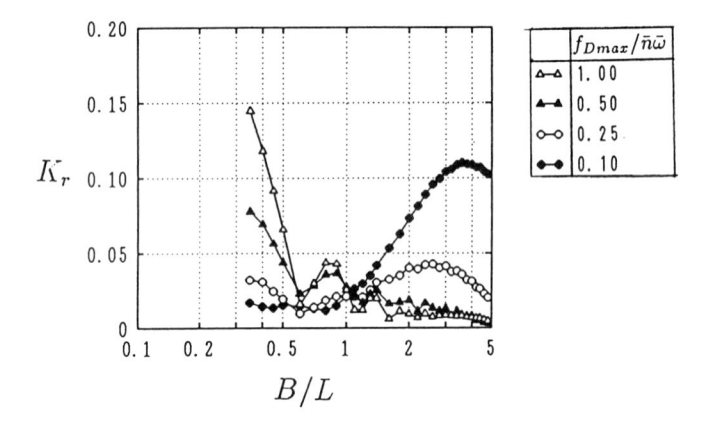

Fig.5: Reflection coefficient for the improved open boundary condition in the one–dimensional case

maximum value of the energy absorbing coefficient, given at the outer edge of the layer. Figure 5 displays the reflection coefficient for component waves in an irregular wave field at a water depth of 10m that has a representative period of 6s and a longest period of 12s. It is seen that if B/L is greater than 0.6 and a proper value of f_{Dmax} is selected, the reflection coefficient becomes sufficiently small. The previous method needed a width of $2L$ for a regular wave of wavelength L. Thus, a significant improvement is achieved with the present method.

4.2 Two–dimensional open boundary condition

In the horizontal two–dimensional case, it is necessary to approximate both the wave angle and the angular frequency by expanding them in Taylor series. For the two–dimensional case, equation (10) is written

$$\begin{aligned}
&\eta_n(x,y,t+\Delta t) \\
=\ &\eta_n(x-\bar{C}\Delta t\cos\theta_n, y-\bar{C}\Delta t\sin\theta_n, t) \\
&+\bar{C}(1-\bar{n})\Delta t\cos\theta_n \frac{\partial\eta_n(x-\bar{C}\Delta t\cos\theta_n, y-\bar{C}\Delta t\sin\theta_n, t)}{\partial x} \\
&+\bar{C}(1-\bar{n})\Delta t\sin\theta_n \frac{\partial\eta_n(x-\bar{C}\Delta t\cos\theta_n, y-\bar{C}\Delta t\sin\theta_n, t)}{\partial y} \\
&-i\bar{C}(1-\bar{n})\Delta t\bar{k}\eta_n(x-\bar{C}\Delta t\cos\theta_n, y-\bar{C}\Delta t\sin\theta_n, t)
\end{aligned} \quad (19)$$

The wave angle θ_n is expressed as the sum of a representative angle $\bar{\theta}$ and a deviation $\Delta\theta_n$:

$$\theta_n = \bar{\theta} + \Delta\theta_n \quad (20)$$

We then substitute equation (20) into equation (19) and apply the additional theorem. Neglecting second order terms gives

$$\Delta t\cos\theta_n = \Delta t\cos\bar{\theta} \quad (21)$$

$$\Delta t\sin\theta_n = \Delta t\sin\bar{\theta} \quad (22)$$

Substituting equations (21) and (22) into equation (19) gives an equation in which the coefficients are independent of component angular frequencies and wave angles. Superimposing the equation gives an improved Sommerfeld–type boundary condition for the two–dimensional case.

$$\begin{aligned}
&\eta(x,y,t+\Delta t) \\
=\ &\eta(x-\bar{C}\Delta t\cos\bar{\theta}, y-\bar{C}\Delta t\sin\bar{\theta}, t) \\
&+\bar{C}(1-\bar{n})\Delta t\cos\bar{\theta} \frac{\partial\eta(x-\bar{C}\Delta t\cos\bar{\theta}, y-\bar{C}\Delta t\sin\bar{\theta}, t)}{\partial x} \\
&+\bar{C}(1-\bar{n})\Delta t\sin\bar{\theta} \frac{\partial\eta(x-\bar{C}\Delta t\cos\bar{\theta}, y-\bar{C}\Delta t\sin\bar{\theta}, t)}{\partial y} \\
&-i\bar{C}(1-\bar{n})\Delta t\bar{k}\eta(x-\bar{C}\Delta t\cos\bar{\theta}, y-\bar{C}\Delta t\sin\bar{\theta}, t)
\end{aligned} \quad (23)$$

where the representative wave angle $\bar{\theta}$ is determined from the gradient of the water surface elevation in the x and y directions at the outer edge of the energy-absorbing layer :

$$\bar{\theta} = \tan^{-1}\left(\frac{\partial \eta(x,y,t)/\partial y}{\partial \eta(x,y,t)/\partial x}\right) \tag{24}$$

5 Calculation of wave field, nearshore current and bottom topography change

With the boundary conditions developed above, we carried out the calculation of deformation of multi–directional irregular waves around a detached breakwater on a uniform slope. The waves consisted of 958 components and were given by the single summation method. Also, the nearshore current and the bottom topography change were calculated by using the results of the calculation of the wave field. These results were compared with calculations for regular waves.

The method of calculating the nearshore current and the bottom topography change was similar to that of Kubo et al. (1992). Figure 6 shows the distribution of wave height for multi–directional irregular waves (($H_{1/3})_0 = 1$m, $T_{1/3} = 6$s, $(\theta_p)_0 = 0°$) around the breakwater, and Figure 7 shows the distribution for regular waves ($H_0 = 0.706$m, T=6s) which has the same wave energy as the irregular waves. In the results for regular waves, a node and antinode occur in front of the detached breakwater, but for the multi–directional irregular waves, the variation in the wave height is small except just in front of the breakwater. This is because the multi–directional irregular waves consist of component waves with various wave direction and frequency, which affects the position of the node and antinode. Furthermore, the unreasonable value on the wave height due to the statistical variation that can be seen in the front of the breakwater in the results of Kubo et al. (1992) does not appear in the results of the present study. Figures 8 and 9 show the calculation results for the nearshore current under the above wave field. Overall, the current shows almost the same tendency, but the current velocity due to the multi–directional irregular waves is smaller than that due to the regular waves. This is because the radiation stress gradients become smaller around the average break point due to the variability in the break–point location and the large width of the surf zone for multi–directional irregular waves. In addition, in front of the breakwater the variation in the wave height, and hence the radiation stress, is large, however, a current does not occur since the gradients of the radiation stress and the mean water level are balanced here. Figure 10 shows the result of the calculation of the bottom topography change after 24 hours, computed by using the wave field in Figure 6 and the nearshore current

IMPROVED BOUNDARY CONDITIONS 283

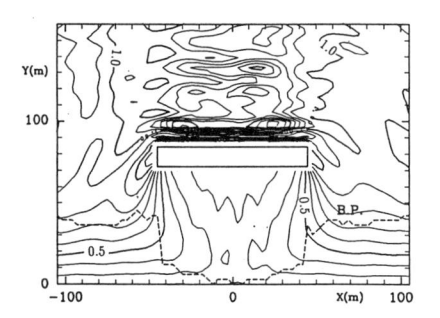

Fig.6: Wave height distribution
(Multi-directional irregular waves)

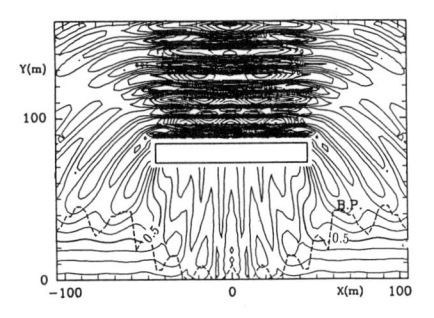

Fig.7: Wave height distribution
(Regular waves)

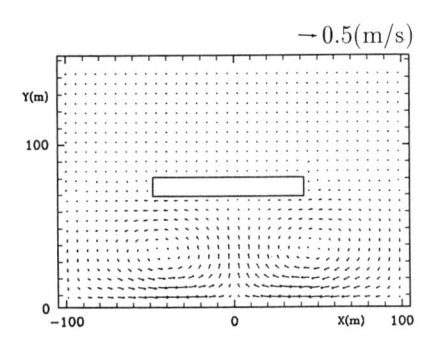

Fig.8: Velocity distribution of nearshore current
(Multi-directional irregular waves)

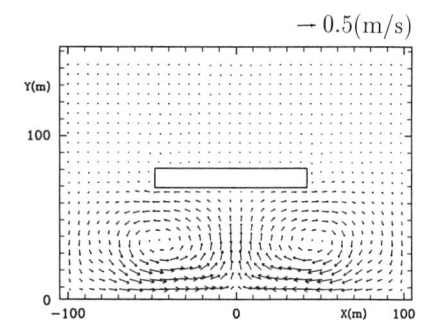

Fig.9: Velocity distribution of nearshore current
(Regular waves)

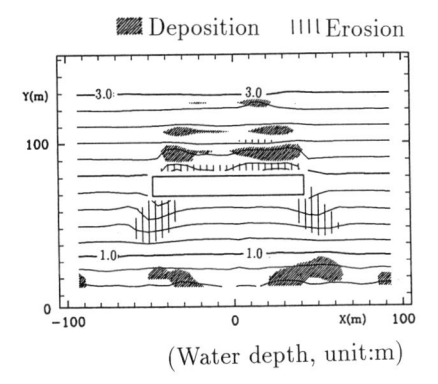

Fig.10: Bottom topography change
(Multi-directional irregular waves)

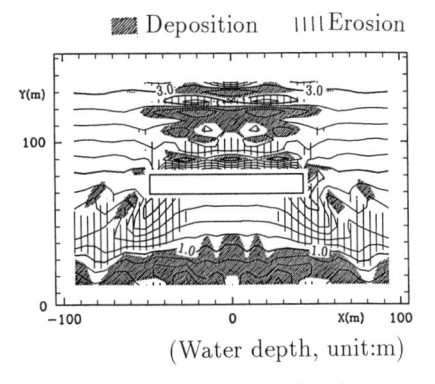

Fig.11: Bottom topography change
(Regular waves)

field in Figure 8 as a steady external force. Figure 11 shows the result of the calculation of the bottom topography change after 24 hours for regular waves. By comparing these two figures, it is seen that the bottom topography change due to multi–directional irregular waves is smaller than that caused by regular waves with equal energy. This is because the bottom wave velocity and nearshore current are smaller for the multi–directional irregular waves.

6 Conclusion

A new incident wave boundary condition which is more efficient than that proposed previously is presented. In addition, the width of the energy absorbing layer is reduced by improving the non-reflective boundary condition for irregular waves. As a result, the accuracy in predicting the wave height distribution was considerably improved by reducing the statistical variation of the wave characteristics. The numerical model developed was applied to calculate the wave field, nearshore current and bottom topography change around a detached breakwater on a uniform slope. The results from calculations with regular and irregular waves were compared.

Acknowledgement

The authors wish to thank Dr. Larson, Associate Professor, University of Lund, Sweden for his assistance in improving the English expressions of this paper.

References

Berkhoff, J. C. W.(1972) : Computation of combined refraction–diffraction, Proc. 13th Int. Conf. on Coastal Eng., pp. 191-203.

Kotake, Y., M. Isobe and A. Watanabe (1992): On the high–order time–dependent mild–slope equation for irregular waves, Proc. 39th Japanese Conf. on Coastal Eng., pp. 91-95 (in Japanese).

Kubo, Y., Y. Kotake, M. Isobe and A. Watanabe (1992) : Time–dependent mild–slope equation for random waves, Proc. 23rd Int. Conf. on Coastal Eng., ASCE, pp. 419-431.

Ohyama, T. and N. Nadaoka (1990) : Modeling the transformation of nonlinear waves passing over a submerged step, Proc. Japanese Conf. on Coastal Eng., JSCE, Vol. 37, pp. 16-20 (in Japanese).

CHAPTER 22

Time-dependent mild-slope equations for random waves

Masahiko Isobe[1]

Abstract

Linear and nonlinear governing equations are derived to calculate the time evolution of random waves subject to refraction and diffraction.

In the linear theory, the frequency-dependent coefficients in the mild-slope equation (Berkhoff, 1972) are approximated by a rational function of the frequency, and then a time-dependent and frequency-independent expression of the mild-slope equation is derived. The resulting equation is applicable to simulate the transformation of random waves in the nearshore zone. Results of numerical calculation agree well with experimental results for random wave shoaling in the offshore zone.

A set of nonlinear governing equations is also derived to simulate the nonlinear wave transformation. The velocity potential for the wave motion is expressed as a series in terms of a given set of vertical distribution functions. Then, the Lagrangian is integrated vertically and the variational principle is applied to yield a set of nonlinear, time-dependent, two-dimensional governing equations for the nonlinear random wave tranformation. Comparison between the results of numerical calculation and flume experiment shows good agreement for the random wave shoaling near the breaking point and for wave disintegration due to a submerged breakwater.

1 Introduction

The mild-slope equation derived by Berkhoff (1972) has widely been used in the numerical calculation of refraction and diffraction of regular waves. However, the randomness of sea waves has a significant effect on the wave transformation especially due to refraction and diffraction. In this paper, linear and nonlinear governing equations are derived to calculate the time evolution of random waves subject to refraction and diffraction. In the linear equation, a term for the energy dissipation due to breaking is added to simulate the random wave field in the near shore zone. Results of numerical calculations are compared with those of laboratory experiments in wave flumes.

2 Linear Theory

2.1 Derivation

2.1.1 rational approximation

The mild-slope equation derived by Berkhoff (1972) is written as

$$\nabla(cc_g \nabla \hat{\eta}) + k^2 cc_g \hat{\eta} = 0 \qquad (1)$$

---

[1]Dept. of Civil Eng., Univ. of Tokyo, Tokyo 113, Japan.

where $\hat{\eta}$ is the complex amplitude of the water surface elevation, c the wave celerity, c_g the group velocity and k the wave number, and ∇ denotes the differential operator in the horizontal two directions. To simplify the equation, the transformation by Radder (1979) is employed:

$$\hat{\phi} = \hat{\eta}/\sqrt{cc_g} \tag{2}$$

Then, within the accuracy up to the first order in the bottom slope, the resultant equation becomes a Helmholtz equation:

$$\nabla^2 \hat{\phi} + k^2 \hat{\phi} = 0 \tag{3}$$

The time-dependent quantity, ϕ, corresponding to $\hat{\phi}$ is expressed as

$$\phi = \hat{\phi} e^{-i\omega t} \tag{4}$$

where ω is the angylar frequency and t the time. For the random wave analysis, ϕ is composed of an infinite number of component waves and the angular frequency differs from component to component; however, a unique value must be chosen to express ϕ of random waves. Thus a slowly varying amplitude, $\tilde{\phi}$, is defined from ϕ as

$$\phi = \tilde{\phi} e^{-i\bar{\omega} t} \tag{5}$$

where $\bar{\omega}$ is a certain representative angular frequency such as the average frequency. Comparison between Eqs. (4) and (5) gives

$$\tilde{\phi} = \hat{\phi} e^{-i\omega' t} \tag{6}$$

where ω' is the deviation from the representative frequency and defined as

$$\omega' = \omega - \bar{\omega} \tag{7}$$

Equation (6) implies that $\tilde{\phi}$ is a slowly varying function of time.

Since the Helmholtz equation (3) is independent of time, the governing equation for $\tilde{\phi}$ has the same form. When an energy dissipation term which is expressed in terms of the energy dissipation coefficient, f_D, is added, the equation is expressed as

$$\nabla^2 \tilde{\phi} + k^2(1 + if_D)\tilde{\phi} = 0 \tag{8}$$

Equation (8) cannot be used to calculate $\tilde{\phi}$ of random waves directly since the coefficients included vary with the frequency. Linear approximation (Kubo et al., 1992) and parabolic approximation (Kotake et al., 1992) to the coefficients were employed in the previous studies. To improve the accuracy of approximation, a rational function is used in the present study.

Consider the following approximation to Eq. (8):

$$\nabla^2 \tilde{\phi} - ia_1 \nabla^2 \left(\frac{\partial \tilde{\phi}}{\partial t}\right) + (b_0 + ic_0)\tilde{\phi} + i(b_1 + ic_1)\frac{\partial \tilde{\phi}}{\partial t} - b_2 \frac{\partial^2 \tilde{\phi}}{\partial t^2} = 0 \tag{9}$$

where the coefficients a_1, b_0, b_1, b_2, c_0 and c_1 are constants and independent of the frequency. Theoretical consideration on stability condition requires that the highest orders of approximation for b and c should be second and first, respectively. The order for a is lower than that for b so that the ADI method may be available in the numerical calculation.

For monochromatic progressive waves expressed as

$$\tilde{\phi} = ae^{i(\hat{k}x\cos\theta + \hat{k}y\sin\theta - \omega't)} \tag{10}$$

Eq. (9) becomes

$$-\hat{k}^2 + a_1\hat{k}^2\omega' + (b_0 + ic_0) + (b_1 + ic_1)\omega' + b_2\omega'^2 = 0 \tag{11}$$

from which the approximated dispersion relation is obtained as

$$\hat{k}^2 = (b_0 + b_1\omega' + b_2\omega'^2)/(1 - a_1\omega') \tag{12}$$

The values of the coefficients should be determined so that the error in the approximation (12) may become minimum without causing numerical instability.

2.1.2 determination of coefficients

Equation (11) can be solved for ω' as

$$\omega' = \{-(a_1\hat{k}^2 + b_1 + ic_1) \pm \sqrt{(a_1\hat{k}^2 + b_1 + ic_1)^2 - 4b_2(-\hat{k}^2 + b_0 + ic_0)}\}/(2b_2) \tag{13}$$

To avoid numerical divergence, $Im\{\omega'\} \leq 0$. This requires that the magnitude of the imaginary part for $\sqrt{}$ should not exceed c_1. Let the real and imaginary parts in the $\sqrt{}$ be denoted by X and Y, respectively, then the condition is written as

$$X \geq 0 \qquad (c_1 = 0) \tag{14}$$
$$X \geq (Y/2c_1)^2 - c_1^2 \qquad (c_1 > 0) \tag{15}$$

The above condition should be satisfied for an arbitrary $\hat{k}$, which yields

$$b_1^2 - 4b_0b_2 \geq 0, \quad c_0 = 0 \qquad (c_1 = 0) \tag{16}$$

$$\left(\frac{c_0}{c_1}\right)^2 - \left(\frac{b_1}{b_2}\right)\left(\frac{c_0}{c_1}\right) + \left(\frac{b_0}{b_2}\right) \leq 0 \qquad (c_1 > 0) \tag{17}$$

Within the above restrictions, we take the equal sign for the sake of convenience. Then,

$$b_1 = 2\sqrt{b_0b_2} \tag{18}$$

$$c_1 = (2b_2/b_1)c_0 \tag{19}$$

By considering the above two equations, independent parameters are a_1, b_0, b_2 and c_0.

When we determine the values of these parameters, we compensate for the error included in the finite difference form of the equation. For waves progressive in the x-direction, the central finite difference expressions for each term in Eq. (9) are related with the corresponding derivatives as

$$\left.\frac{\partial^2 \tilde{\phi}}{\partial x^2}\right|_{F.D.} = \alpha_2\beta_0 \frac{\partial^2 \tilde{\phi}}{\partial x^2}, \quad \left.\frac{\partial^2}{\partial x^2}\left(\frac{\partial \tilde{\phi}}{\partial t}\right)\right|_{F.D.} = \alpha_2\beta_1 \frac{\partial^2}{\partial x^2}\left(\frac{\partial \tilde{\phi}}{\partial t}\right), \quad \left.\tilde{\phi}\right|_{F.D.} = \beta_0\tilde{\phi}$$
$$\left.\frac{\partial \tilde{\phi}}{\partial t}\right|_{F.D.} = \beta_1 \frac{\partial \tilde{\phi}}{\partial t}, \quad \left.\frac{\partial^2 \tilde{\phi}}{\partial t^2}\right|_{F.D.} = \beta_2 \frac{\partial^2 \tilde{\phi}}{\partial t^2} \Bigg\} \tag{20}$$

where $|_{F.D.}$ denotes the finite difference expressions and

$$\beta_0 = (2/3)\cos\omega'\Delta t + (1/3), \qquad \beta_1 = \{(\sin\omega'\Delta t)/(\omega'\Delta t)\}^2$$
$$\beta_2 = \left\{\left(\sin\frac{\omega'\Delta t}{2}\right)/\left(\frac{\omega'\Delta t}{2}\right)\right\}^2, \quad \alpha_2 = \left\{\left(\sin\frac{\hat{k}\Delta x}{2}\right)/\left(\frac{\hat{k}\Delta x}{2}\right)\right\}^2 \tag{21}$$

are correction factors. Then, instead of Eq. (11), the finite difference equation for Eq. (9) implies

$$-\alpha_2\hat{k}^2\beta_0 + a_1\alpha_2\hat{k}^2\beta_1\omega' + b_0\beta_0 + b_1\beta_1\omega' + b_2\beta_2\omega'^2 = 0 \tag{22}$$

for $c_0 = c_1 = 0$. Three independent parameters can be determined from three sets of ω' and k which satisfy the dispersion realation exactly:

$$-b_2^*\alpha_2 k_l^2\beta_0 + a_1^*\alpha_2 k_l^2\beta_1\omega_l' + \xi^2\beta_0 + 2\xi\beta_1\omega_l' + \beta_2\omega_l'^2 = 0 \quad (l = 1, 2, 3) \tag{23}$$

where
$$b_2^* = 1/b_2, \quad \xi = \sqrt{b_0/b_2}, \quad a_1^* = a_1/b_2 \tag{24}$$

Since Equations expressed by (23) are linear in b_2^* and a_1^*, these parameters can be eliminated to yield a parabolic equation in terms of ξ. After solving for ξ, we can determine b_2, b_0, a_1 and b_1 by Eqs. (18) and (24).

2.1.3 breaking wave model

Breaking wave model used is the same as Isobe (1987). First, the relative wave amplitude is defined by

$$\gamma = |\eta|/h \tag{25}$$

The critical relative amplitude, γ_b, for breaking of an individual wave in the random wave train is given as

$$\gamma_b = 0.8 \times \gamma_b' \tag{26}$$

$$\gamma_b' = 0.53 - 0.3\exp(-3\sqrt{h/\bar{L}_o}) + 5\tan^{3/2}\beta \exp\{-45(\sqrt{h/\bar{L}_o} - 0.1)^2\} \tag{27}$$

After breaking, the energy dissipation coefficient is introduced as follows:

$$f_D = \frac{5}{2}\tan\beta\sqrt{\frac{1}{k_o h}}\sqrt{\frac{\gamma - \gamma_r}{\gamma_s - \gamma_r}} \tag{28}$$

$$\gamma_s = 0.4 \times (0.57 + 5.3\tan\beta) \tag{29}$$

$$\gamma_r = 0.135 \tag{30}$$

From f_D determined at the representative frequency, $c_0 = \bar{k}^2 f_D$ and c_1 is calculated by Eq. (19). Thus all the coefficients in Eq. (9) are determined and $\tilde{\phi}$ can be calculated.

2.1.4 water surface elevation

In Eq. (2) which determines the water surface elevation from the calculated $\tilde{\phi}$, the coefficient $\sqrt{cc_g}$ is also a function of frequency but can accurately be approximated by a second-order polynomial function. Therefore, Eq. (2) is approximated by

$$\tilde{\eta} = d_0\tilde{\phi} + d_1\frac{\partial\tilde{\phi}}{\partial t} + d_2\frac{\partial^2\tilde{\phi}}{\partial t^2} \tag{31}$$

where the constants, d_0, d_1 and d_2, are determined from the values of $\sqrt{cc_g}$ at three different frequencies:

$$d_0 + d_1\beta_1\omega_l' + d_2\beta_2\omega_l'^2 = (1/\sqrt{cc_g})_l \quad (l = 1, 2, 3) \tag{32}$$

2.2 Error evaluation

Since $1/\sqrt{cc_g}$ is constant for low frequency and proportional to the frequency for high frequency, the second-order approximation has a high accuracy. On the other hand, since k^2 is proportional to ω^2 and ω^4 for low and high frequency, respectively, even the rational approximation may not have a sufficient accuracy.

Figure 1 shows the interval from $\omega_{\min}$ to $\omega_{\max}$ within which the relative error of k^2 is less than 1%. The three frequencies for determining the coefficients are denoted

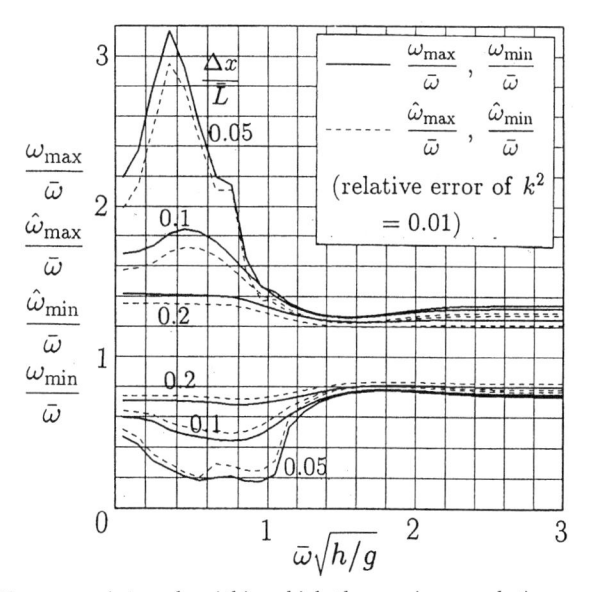

Figure 1: Frequency intervals within which the maximum relative error of k^2 is 1%

by $\hat{\omega}_{\min}$, $\bar{\omega}$ and $\hat{\omega}_{\max}$, and therefore ω'_1, ω'_2 and ω'_3 in Eq. (23) are $\hat{\omega}_{\min} - \bar{\omega}$, 0 and $\hat{\omega}_{\max} - \bar{\omega}$, respectively. The horizontal axis is the nondimensionalized representative angular frequency $\bar{\omega}\sqrt{h/g}$ (h: the water depth; g: the gravitational acceleration). Lines are drawn for various relative grid size $\Delta x/\bar{L}$ ($\bar{L}$: the wavelength at $\bar{\omega}$). Figure 2 shows the same interval for various relative errors. From these figures, the interval becomes narrowest at about $\bar{\omega}\sqrt{h/g} = 1.4$. Finally, Fig. 3 shows the narrowest interval as a function of the relative error. The relative grid size, $\Delta x/\bar{L}$, is assumed to be 0.1 but does not have much influence.

From Fig. 3, if the relative error of k^2 is permitted up to 10%, most of the energy in random waves will be included in the interval and therefore the transformation of random waves can be analyzed by Eq. (9). This may usually be the case because the wave energy at the frequency far different from the representative frequency is usually small. However, for random waves with a very wide banded spectrum and a small relative error of k^2, the frequency interval have to be divided into several sections and the results of calculation for each section are superimposed.

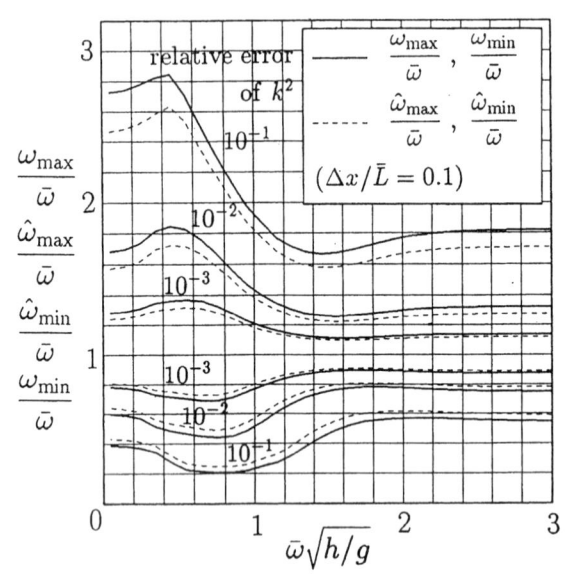

Figure 2: Frequency intervals for various maximum relative errors of k^2

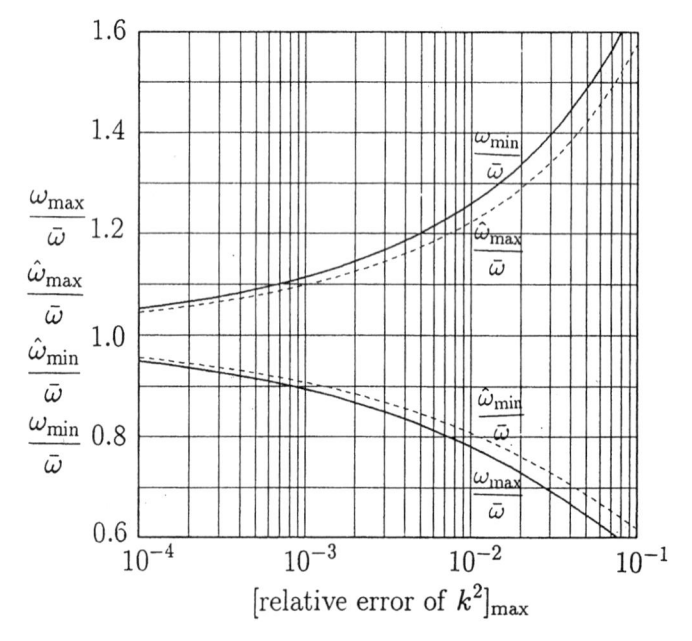

Figure 3: Frequency intervals in terms of the allowable relative error of k^2

2.3 Sample calculations

Once the representative frequency and the grid size are fixed, the gird size to wavelength ratio may become large for high frequency. Figure 4 examines the effect of grid size for analyzing wave shoaling on a uniformly sloping bottom. The angular frequency, ω, of the waves analyzed is $0.8\bar{\omega}$ for which the error of k^2 becomes almost maximum. In the upper figure, the agreement with the analytical solution is very good. Even in the lower figure for which the relative grid size is as large as 0.321, the agreement is not bad, which may be acceptable in analyzing far side frequency band component.

(a) $\Delta x/\bar{L}_o = 0.04, \Delta t/\bar{T} = 0.1$ $\quad (\omega/\bar{\omega} = 0.8, \hat{\omega}_{\min}/\bar{\omega} = 0.6, \hat{\omega}_{\max}/\bar{\omega} = 1.4)$

0.05	←	$h/\bar{L}_o$	→	0.01
0.074	←	h/L	→	0.032
0.059	←	$\Delta x/L$	→	0.129

(b) $\Delta x/\bar{L}_o = 0.1, \Delta t/\bar{T} = 0.2$ $\quad (\omega/\bar{\omega} = 0.8, \hat{\omega}_{\min}/\bar{\omega} = 0.6, \hat{\omega}_{\max}/\bar{\omega} = 1.4)$

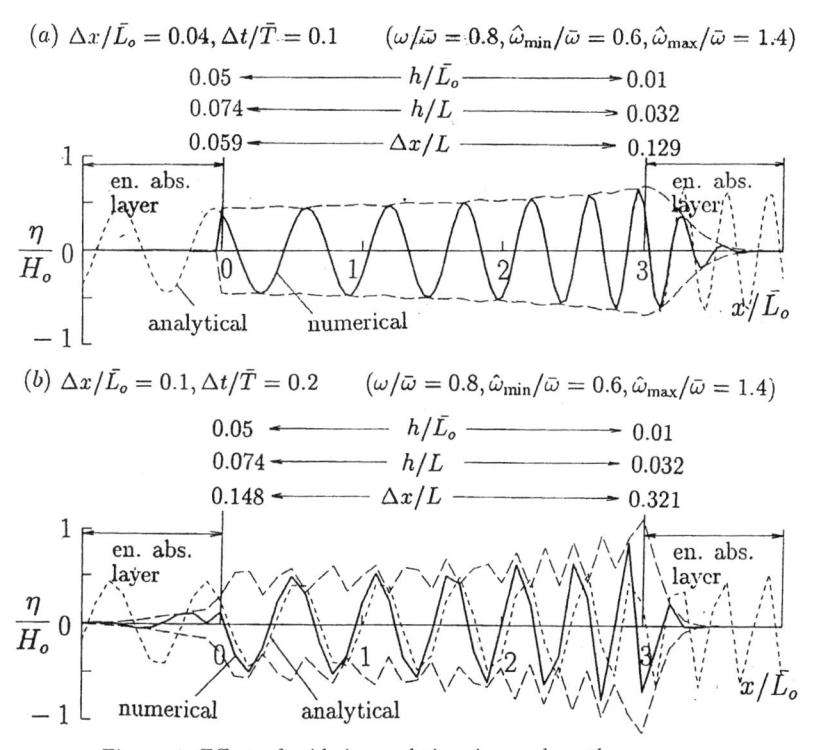

0.05	←	$h/\bar{L}_o$	→	0.01
0.074	←	h/L	→	0.032
0.148	←	$\Delta x/L$	→	0.321

Figure 4: Effect of grid size and time interval on the accuracy

Figure 5 compares calculated and measured water surface elevation due to shoaling random waves. The incident wave profile which is shown in the upper figure was measured on a horizontal bottom with water depth of 40cm. From the point, a horizontal bottom with 0.4m in length, a 1/10 slope with 1m, and a 1/30 slope are installed. The onshoreward measuring point is located 2.6m from the beginning of the 1/30 slope and the water depth there is 21cm. The frequency interval was divided into four sections in the numerical calculation. The agreement is seen to be very good. However, near the breaker zone where nonlinearity of waves is strong, steepening of wave crests can not be reproduced by the present linear theory, even though energies of wave groups are fairly well reproduced. This implies that the present linear theory can be used to

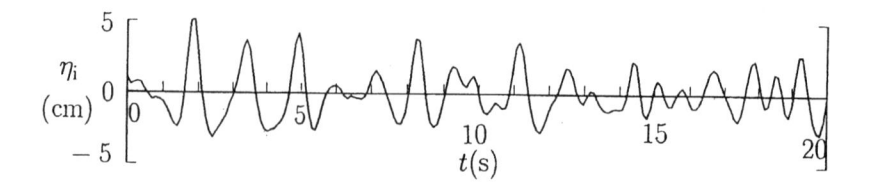

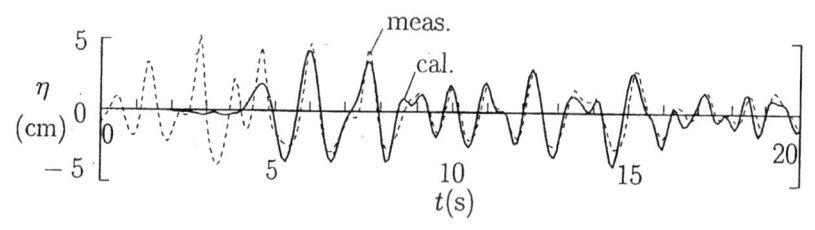

Figure 5: Comparison between calculated and measured water surface elevation in shoaling water

predict the distribution of integral properties such as the wave energy and radiation stress. To predict the wave profile in the nearshore zone, nonlinear theory must be employed.

3 Nonlinear Theory

3.1 Derivation

3.1.1 definition of Lagrangian

A Lagrangian L which is euivalent to the basic equation and boundary conditions for water surface waves is given as follows (Luke, 1967):

$$L[\phi,\eta] = \int_{t_1}^{t_2} \iint_A \int_{-h}^{\eta} \left\{ \frac{\partial \phi}{\partial t} + \frac{1}{2}(\nabla\phi)^2 + \frac{1}{2}\left(\frac{\partial \phi}{\partial z}\right)^2 + gz \right\} dz \, dA \, dt \qquad (33)$$

where unknown functions are the velocity potential ϕ and the water surface elevation η, and t_1 and t_2 denote the beginning and end of time, A the area of concern in (x, y) plane, h the water depth, g the gravitational acceleration, $\nabla = (\frac{\partial}{\partial x}, \frac{\partial}{\partial y})$ the differential operator in the horizontal directions, $(x, y) = \mathbf{x}$ the horizontal coordinates, z the vertical coordinates, and t the time.

The variation of L due to small variations of ϕ and η is obtained from Eq. (33):

$$\begin{aligned}\delta L = &-\int_{t_1}^{t_2} \iint_A \int_{-h}^{\eta} \left(\nabla^2\phi + \frac{\partial^2 \phi}{\partial z^2} \right) \delta\phi \, dz \, dA \, dt \\ &- \int_{t_1}^{t_2} \iint_A \left[\left\{ \frac{\partial \phi}{\partial t} + \frac{1}{2}(\nabla\phi)^2 + \frac{1}{2}\left(\frac{\partial \phi}{\partial z}\right)^2 + g\eta \right\} \delta\eta \bigg|_{z=\eta} \right. \\ &\left. + \left\{ \frac{\partial \eta}{\partial t} + \nabla\eta\nabla\phi - \frac{\partial \phi}{\partial z} \right\} \delta\phi \bigg|_{z=\eta} + \left\{ \nabla h \nabla\phi + \frac{\partial \phi}{\partial z} \right\} \delta\phi \bigg|_{z=-h} \right] dA \, dt\end{aligned}$$

$$+ \int_{t_1}^{t_2} \oint_C \int_{-h}^{\eta} \frac{\partial \phi}{\partial n} \delta\phi \, dz \, ds \, dt + \iint_A \int_{-h}^{\eta} \left[\delta\phi\right]_{t_1}^{t_2} dz \, dA \tag{34}$$

where C denotes the boundary of A. To terminate L for small variations of ϕ and η in an arbitrary point, all the integrands in the above equation must vanish. The Laplace equation for ϕ can be obtained from the first integral, and the dynamic and kinematic free surface boundary conditions and the bottom boundary condition, respectively, from the first, second and third terms in the second integral. Therefore the application of the variational principle to L results in the basic equation and boundary conditions for water surface waves. The third and fourth integrals are, respectively, related with the lateral and initial conditions which are given in each specific problem.

3.1.2 vertical distribution functions

Wave equations such as the mild-slope equation and Boussinesq equation are two-dimensional equations which are obtained by integrating vertically the govering three-dimensional equations. For the integration, vertical distribution functions are introduced theoretically or a priori. Massel (1993) derived an extended mild-slope equation by introducing a vertical distribution function of hyperbolic cosine type and integrating the governing equation. A clear and generalized consept of this procedure was proposed by Nadaoka and Nakagawa (1993) and Nadaoka et al. (1994) in deriving a strongly-nonlinear, strongly-dispersive wave equation by applying the Galerkin method to the Euler equations of motion. Nochino (1994) used another set of vertical distribution functions to derive a nonlinear dispersive equation. The present theory also introduces vertical distribution functions and integrate the Lagrangian to yield a nonlinear mild-slope equation.

First, the three-dimensional dependent variable, ϕ, is expanded into a series in terms of a certain set of vertical distribution functions, $Z_\alpha(z)$, given a priori:

$$\phi(\mathbf{x}, z, t) = \sum_{\alpha=1}^{N} Z_\alpha(z; h(\mathbf{x})) f_\alpha(\mathbf{x}, t) \equiv Z_\alpha f_\alpha \tag{35}$$

where the function, Z_α, may change according to the water depth h, and f_α is the coefficient for Z_α and a function of x and t but not of z. The summation convention will be applied hereafter.

Then, after substituting Eq. (35) into Eq. (33), the integration is carried out in the vertical direction:

$$L[f_\alpha, \eta] = \int_{t_1}^{t_2} \iint_A \xi(f_\alpha, \frac{\partial f_\alpha}{\partial t}, \eta, \frac{\partial \eta}{\partial t}) \, dA \, dt \tag{36}$$

$$\xi(f_\alpha, \frac{\partial f_\alpha}{\partial t}, \eta, \frac{\partial \eta}{\partial t}) = \frac{g}{2}(\eta^2 - h^2) + \tilde{Z}_\beta \frac{\partial f_\beta}{\partial t} + \frac{1}{2} A_{\gamma\beta} \nabla f_\gamma \nabla f_\beta + \frac{1}{2} B_{\gamma\beta} f_\gamma f_\beta$$
$$+ C_{\gamma\beta} f_\gamma \nabla f_\beta \nabla h + \frac{1}{2} D_{\gamma\beta} f_\gamma f_\beta (\nabla h)^2 \tag{37}$$

where

$$\tilde{Z}_\alpha = \int_{-h}^{\eta} Z_\alpha \, dz \tag{38}$$

$$A_{\alpha\beta} = \int_{-h}^{\eta} Z_\alpha Z_\beta \, dz \qquad (39)$$

$$B_{\alpha\beta} = \int_{-h}^{\eta} \frac{\partial Z_\alpha}{\partial z} \frac{\partial Z_\beta}{\partial z} \, dz \qquad (40)$$

$$C_{\alpha\beta} = \int_{-h}^{\eta} \frac{\partial Z_\alpha}{\partial h} Z_\beta \, dz \qquad (41)$$

$$D_{\alpha\beta} = \int_{-h}^{\eta} \frac{\partial Z_\alpha}{\partial h} \frac{\partial Z_\beta}{\partial h} \, dz \qquad (42)$$

The above coefficients, $\tilde{Z}_\alpha$, $A_{\alpha\beta}$, $B_{\alpha\beta}$, $C_{\alpha\beta}$ and $D_{\alpha\beta}$, are obtained from given vertical distribution functions and then the Lagrangian is expressed by Eq. (36) as an integral in the horizontal two-dimensional plane.

3.1.3 variational principle

Application of the variational principle to Eq. (36) in terms of η and f_α yields Euler equations which are expressed in general forms:

$$\frac{\partial \xi}{\partial f_\alpha} = \frac{\partial}{\partial t}\left[\frac{\partial \xi}{\partial(\partial f_\alpha/\partial t)}\right] + \nabla\left[\frac{\partial \xi}{\partial(\nabla f_\alpha)}\right] \qquad (43)$$

$$\frac{\partial \xi}{\partial \eta} = \frac{\partial}{\partial t}\left[\frac{\partial \xi}{\partial(\partial \eta/\partial t)}\right] + \nabla\left[\frac{\partial \xi}{\partial(\nabla \eta)}\right] \qquad (44)$$

Substituting Eq. (37) into Eqs. (43) and (44), a set of nonlinear partial differential equations is obtained for analyzing nonlinear water wave transformation:

$$Z_\alpha^\eta \frac{\partial \eta}{\partial t} + \nabla(A_{\alpha\beta}\nabla f_\beta) - B_{\alpha\beta}f_\beta + \nabla(C_{\beta\alpha}f_\beta\nabla h) - C_{\alpha\beta}\nabla f_\beta \nabla h - D_{\alpha\beta}f_\beta(\nabla h)^2 = 0 \qquad (45)$$

$$g\eta + Z_\beta^\eta \frac{\partial f_\beta}{\partial t} + \frac{1}{2}Z_\gamma^\eta Z_\beta^\eta \nabla f_\gamma \nabla f_\beta + \frac{1}{2}\frac{\partial Z_\gamma^\eta}{\partial z}\frac{\partial Z_\beta^\eta}{\partial z}f_\gamma f_\beta + \frac{\partial Z_\gamma^\eta}{\partial h}Z_\beta^\eta f_\gamma \nabla f_\beta \nabla h$$

$$+\frac{1}{2}\frac{\partial Z_\gamma^\eta}{\partial h}\frac{\partial Z_\beta^\eta}{\partial h}f_\gamma f_\beta (\nabla h)^2 = 0 \qquad (46)$$

where

$$Z_\alpha^\eta = Z_\alpha|_{z=\eta} \qquad (47)$$

The above equations includes terms up to the second order in the bottom slope; however, vertical distribution functions given will usually be consistent only with a horizontal or mild-slope bottom. Therefore, on assuming that the bottom slope is mild, the terms of the second order are neglected to yield a set of nonlinear mild-slope equations:

$$Z_\alpha^\eta \frac{\partial \eta}{\partial t} + \nabla(A_{\alpha\beta}\nabla f_\beta) - B_{\alpha\beta}f_\beta + (C_{\beta\alpha} - C_{\alpha\beta})\nabla f_\beta \nabla h + \frac{\partial Z_\beta^\eta}{\partial h}Z_\alpha^\eta f_\beta \nabla \eta \nabla h = 0 \qquad (48)$$

$$g\eta + Z_\beta^\eta \frac{\partial f_\beta}{\partial t} + \frac{1}{2}Z_\gamma^\eta Z_\beta^\eta \nabla f_\gamma \nabla f_\beta + \frac{1}{2}\frac{\partial Z_\gamma^\eta}{\partial z}\frac{\partial Z_\beta^\eta}{\partial z}f_\gamma f_\beta + \frac{\partial Z_\gamma^\eta}{\partial h}Z_\beta^\eta f_\gamma \nabla f_\beta \nabla h = 0 \qquad (49)$$

The total number of equations is $(N+1)$ since Eqs. (48) and (49) contain 1 and N components, respectively. On the other hand, the total number of unknowns, η and f_α ($\alpha = 1$ to N), is $(N+1)$. Therefore, with appropriate boundary conditions, the equations can be solved numerically. Then, the velocity is obtained through the velocity potential expressed by Eq. (35).

3.2 Sample vertical distribution functions

A set of vertical distribution functions should be selected so that the velocity potential may accurately be expressed by Eq. (35) with a small number of terms. As understood from the small amplitude wave theory, hyperbolic cosine functions may be effective for deep to intermediate water, whereas polynomial functions for very shallow water. Here, for the sake of simplicity, polynomial functions are chosen and analytical expressions for the coefficients are shown.

As inferred from shallow water wave theory, we select a set of even-order polynomial functions:
$$Z_\alpha = \left(\frac{h+z}{h}\right)^{2(\alpha-1)} \tag{50}$$

Then, Eqs. (47) and (39) to (41) give
$$Z_\alpha^\eta = \zeta^{2\alpha_1} \tag{51}$$

$$A_{\alpha\beta} = h \frac{\zeta^{2(\alpha_1+\beta_1)+1}}{2(\alpha_1+\beta_1)+1} \tag{52}$$

$$B_{\alpha\beta} = \frac{4\alpha_1\beta_1}{h} \frac{\zeta^{2(\alpha_1+\beta_1)-1}}{2(\alpha_1+\beta_1)-1} \tag{53}$$

$$C_{\alpha\beta} = -2\alpha \left[\frac{\zeta^{2(\alpha_1+\beta_1)+1}}{2(\alpha_1+\beta_1)-1} - \frac{\zeta^{2(\alpha_1+\beta_1)}}{2(\alpha_1+\beta_1)}\right] \tag{54}$$

where
$$\zeta = (h+z)/h \tag{55}$$
$$\alpha_1 = \alpha - 1 \tag{56}$$
$$\beta_1 = \beta - 1 \tag{57}$$

To check the effectiveness of the polynomial functions, the dispersion relation of the linearized equation is examined for a horizontal bottom. In the linear theory, the coefficients expressed by Eqs. (51) to (53) are evaluated at $z=0$ instead of $z=\zeta$. By denoting the quantities at $z=0$ by superscript $^\circ$, the following equation can be obtained by eliminating η from the linearized forms of Eqs. (48) and (49) on a horizontal bottom:
$$-\frac{1}{g} Z_\alpha^\circ Z_\beta^\circ \frac{\partial^2 f_\beta}{\partial t^2} + \nabla(A_{\alpha\beta}^\circ \nabla f_\beta) - B_{\alpha\beta}^\circ f_\beta = 0 \tag{58}$$

where from Eqs. (51) to (53)
$$Z_\alpha^\circ = 1 \tag{59}$$
$$A_{\alpha\beta}^\circ = h/\{2(\alpha_1+\beta_1)+1\} \tag{60}$$

$$B^o_{\alpha\beta} = 4\alpha_1\beta_1/[h\{2(\alpha_1+\beta_1)-1\}] \tag{61}$$

For progressive waves,
$$f_\alpha = a_\alpha e^{i(\hat{k}-\omega t)} \tag{62}$$

By substituting the above expression into Eq. (58), the following homogenious equations are obtained:

$$\sum_{\beta=1}^{N}\left(\frac{\omega^2}{g}-\frac{1}{h}\frac{4\alpha_1\beta_1}{2(\alpha_1+\beta_1)-1}\right)a_\beta = \hat{k}^2\sum_{\beta=1}^{N}\frac{h}{2(\alpha_1+\beta_1)+1}a_\beta \tag{63}$$

For a given ω, $\hat{k}^2$ is obtained as an eigenvalue which gives a non-trivial solution to the above equations. At least up to $N=4$, it was confirmed numerically that only one eigenvalue is positive and the others are negative. A positive value of $\hat{k}^2$, i.e., a real value of $\hat{k}$, corresponds to progressive waves, and a negative value to evernescent waves. The relationship between the frequency and wave celerity of the progressive waves is shown in Fig. 6 for various N. As can be seen, agreement with the linear wave theory

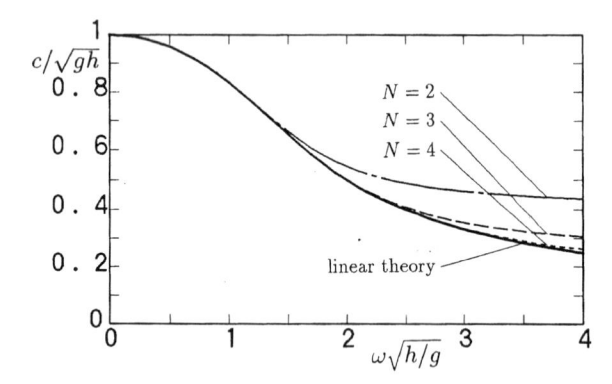

Figure 6: Dispersion relation for vertical distribution functions of even-order polynomials

is good for shallow to deep water even with small N.

3.3 Sample calculations

Figure 7 compares calculated and measured water surface elevation η and bottom velocity u_b in shoaling water. Even for Case 2-2 in which nonlinearity is strong at the measuring point, agreement is good for the bottom velocity as well as the water surface elevation.

Figure 8 compares calculated and measured water surface elevation around a submerged breakwater. Nonlinearity and dispersion are significant on the breakwater and on the horizontal bottom, respectively. Agreement is good even with a small number of N ($N=3$).

TIME-DEPENDENT MIND-SLOPE EQUATIONS

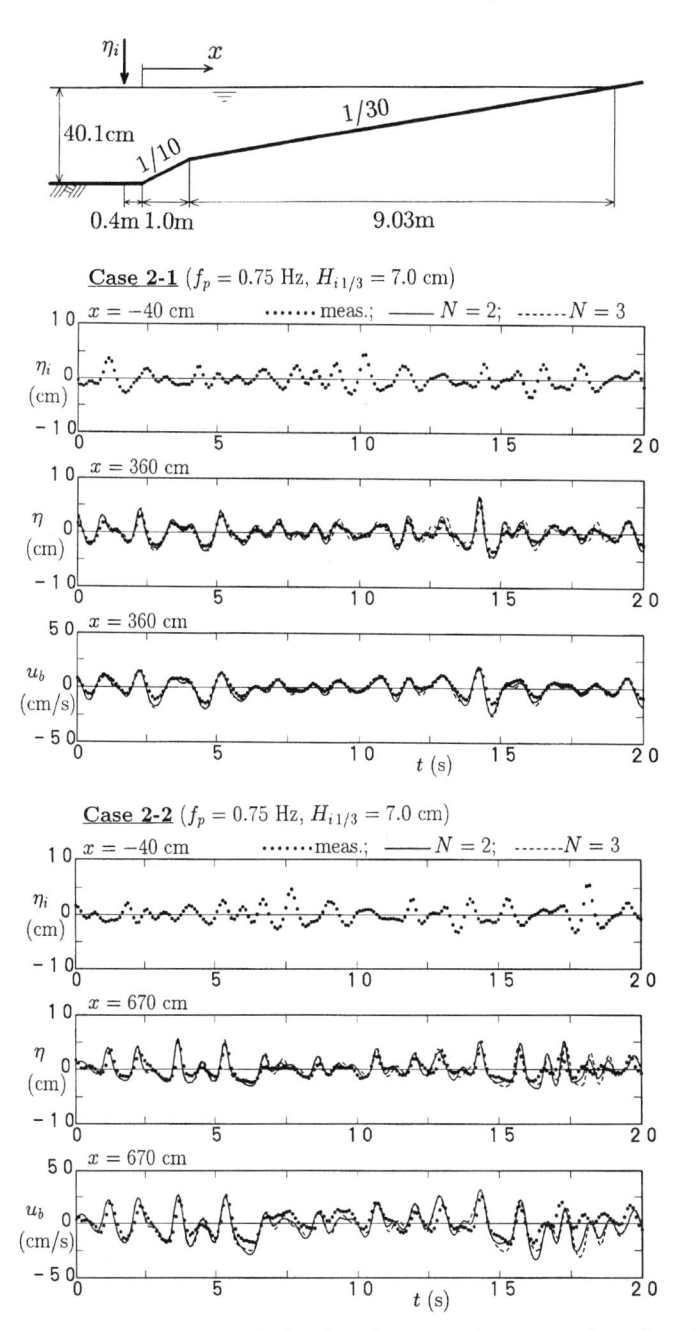

Figure 7: Comparison between calculated and measured water surface elevation and bottom velocity in shoaling water

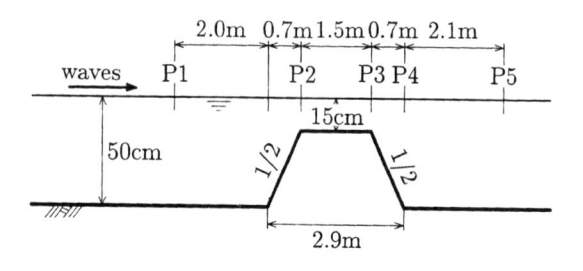

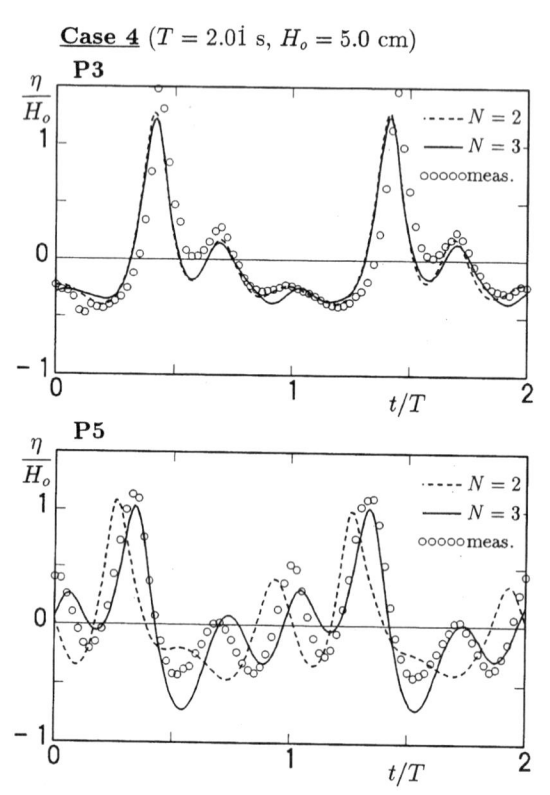

Figure 8: Comparison between calculated and measured water surface elevation on and behind the submerged breakwater

4 Conclusions

A time-dependent mild-slope equation for random waves is derived from the mild-slope equation by approximating rational function to the frequency-dependent coefficients. This equation allows to simulate the time evolution of short-crested random waves in the nearshore area. Agreement between calculated and measured water surface elevation in the offshore zone is good because wave nonlinearity is not essential.

A nonlinear mild-slope equation is derived by expanding the velocity potential into a series in terms of vertical distribution functions and then applying the variational principle to a Lagrangian. Comparison between calculated and measured quantities confirms the validity of the theory even for a strongly nonlinear and dispersive wave field.

References

[1] Berkhoff, J. C. W. (1972): Computation of combined refraction-diffraction, Proc. 13th Int. Conf. on Coastal Eng., ASCE, pp. 471–490.

[2] Isobe, M. (1987): A parabolic equation model for transformation of irregular waves due to refraction, diffraction and breaking, Coastal Eng. in Japan, Vol. 30, pp. 33–47.

[3] Kotake, Y., M. Isobe and A. Watanabe (1992): On a high-order approximation of the time-dependent mild-slope equation for irregular waves, Proc. 39th Japanese Conf. on Coastal Eng., pp. 91–95 (in Japanese).

[4] Kubo Y., M. Isobe and A. Watanabe (1992): Time-dependent mild slope equation for random waves, Proc. 23rd Int. Conf. on Coastal Eng., pp. 419–431.

[5] Luke, J. C. (1967): A variational principle for a fluid with a free surface, J. Fluid Mech., Vol. 27, pp. 395–397.

[6] Massel, S. R. (1993): Extended refraction-diffraction equation for surface waves, Coastal Eng., Vol. 19, pp. 97–126.

[7] Nadaoka, K. and Y. Nakagawa (1993): Simulation of nonlinear wave fields with the newly derived nonlinear dispersive wave equations, Proc. 40th Japanese Conf. on Coastal Eng., pp. 6–10 (in Japanese).

[8] Nakaoka, K., S. Beji, Y. Nakagawa and O. Ohno (1994): A fully-dispersive nonlinear wave model and its numerical solution, Proc. 24th Int. Conf. on Coastal Eng., ASCE (in press).

[9] Nochino, M. (1994): Coupled oscillation equation for strongly nonlinear random wave field and its basic characteristics, Proc. 41st Japanese Conf. on Coastal Eng., pp. 11–15 (in Japanese).

[10] Radder, A. C. (1979): On the parabolic equation method for water-wave propagation, J. Fluid Mech., Vol. 95, pp. 159–176.

CHAPTER 23

Modelling Moveable Bed Roughness and Friction for Spectral Waves

Kaczmarek, L.M.[1], Harris, J.M.[2], O'Connor, B.A.[2]

Abstract
The present paper is concerned with the simulation of turbulent boundary layer dynamics over a moveable seabed in random waves. A new theoretical approach for the evaluation of moveable bed roughness in spectral waves based on the grain-grain interaction idea is presented and tested against data from the laboratory and field. The new approach is combined with the methodology which assumes that the spectral wave condition can be represented by a monochromatic representative wave. Good results have been obtained, although further testing against data gathered in the North Sea is required.

1. Introduction

For a moveable sandy bed, one may distinguish three general seabed conditions due to the action of surface gravity waves: a flat bed, rippled bed and sheet flow. If we consider the latter condition, the need to study sediment transport under wave-induced sheet flow conditions is necessary in the understanding of beach profile changes in the surf zone. The understanding of nearbed sediment dynamics is also of great importance for the mathematical description of cross-shore sediment transport.

To understand the effect of changing bed roughness by the hydrodynamic forces requires knowledge of the dynamic behaviour of sand grains in the collision-dominated, high concentration nearbed region. At high shear stresses and sediment transport intensities, the nearbed sediment transport appears to take place in a layer with a thickness that is large compared to the grain size. It is therefore not possible to properly describe flow in this layer by conventional engineering models which assume

---

[1] Polish Academy of Sciences, Institute of Hydro-Engineering, IBW PAN, 7 Koscierska, 80-953, Gdansk, Poland.

[2] Department of Civil Engineering, University of Liverpool, Brownlow St., P.O. Box 147, Liverpool L69 3BX. UK.

that bed load transport occurs in a layer that has a thickness of the order of one or two grain diameters.

The present paper is concerned with developing an iterative procedure for the estimation of the effective bed roughness for a monochromatic wave, as characterised by the roughness parameter k_s, and extending this to the case of a spectral sea. The nearbed sediment dynamics are modelled in two regions with continuous profiles of stress and velocity. Namely (i) a granular fluid region and (ii) a wall bounded turbulent fluid shear region.

In sheet flow conditions it is assumed that the external drag produced by the boundary layer flow is related to the particle interactions within the sub-bed layer and hence to the effective roughness at the boundary.

2. The sheet flow model
2.1 Formulation of the problem

A typical velocity distribution with depth of a rough bed is supposed to be characterised (Kaczmarek & O'Connor 1993a,b) by a sub-bottom flow and a main or outer flow, as shown in Figure 1.

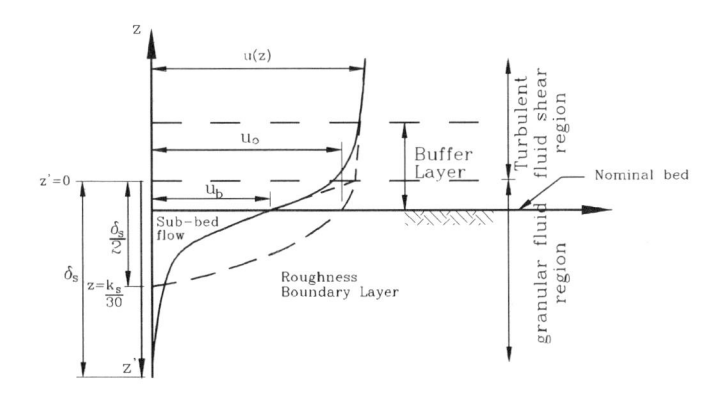

Figure 1: Definition sketch of turbulent flow over a moveable bed.

The velocity distribution is supposed to be continuous. Its intersection with the nominal bottom is the apparent slip velocity u_b. The downward extension of the velocity distribution in the outer zone of the main flow yields a fictitious slip velocity, u_o at the nominal bed, which is necessarily larger than u_b because of the supposed asymptotic transition in the buffer layer between the sub-bed flow and the fully turbulent flow in the turbulent-fluid shear region.

The velocity distribution in the roughness layer depends on the type of geometric roughness pattern and the bed permeability. There must be some transition between both parts of the velocity distribution bridged by the buffer zone. However, for present purposes it is assumed that the velocity distribution in the turbulent-fluid shear region can be determined by parameters dependent on the geometric roughness properties of the bed and the outer flow parameters, such as the free-stream orbital velocity. It is proposed to extend the sub-bed granular-fluid flow region to the matching point with the velocity distribution in the turbulent-fluid shear region. Thus, shear stress velocities in the two layers are set equal at the theoretical bed level, as it is shown in Figure 1, point A.

The sub-bed flow region has a high sediment concentration. For sheet flow conditions in this layer, chaotic collisions of grains are the predominant mechanism. In this case water does not really transfer shear stresses at all. The dynamic state of such a mixture is characterised by stresses σ_{ij} which are the sum of dynamic σ^*_{ij} and plastic σ°_{ij} stresses.

The first problem, therefore, is to determine the velocity profile distribution in the upper turbulent layer, which means determining the effective roughness height of the bed k_s as well as the lower grain-fluid flow. The intersection of these two profiles will determine point A (See Figure 1).

2.2 Mathematical description of flow in the turbulent upper region

It is assumed that flow in the upper layer is governed by the simplified equation of motion:

$$\frac{\partial u}{\partial t} = \frac{\partial U}{\partial t} + \frac{1}{\rho}\frac{\partial \tau}{\partial z} \qquad (2.1)$$

in which u(z,t) is, in general, a combined wave-period-averaged "steady" current and wave velocity and U(t) is the free-stream wave velocity at the top of the wave boundary layer.

The present work uses an eddy viscosity model, which is an extension of Kajiura's (1968) and Brevik's (1981) model. Thus the eddy viscosity over the flow depth is assumed to be given by the equations.

$$v_t(z) = \kappa u_{fmax} z \qquad \text{for} \quad \frac{k_s}{30} \le z \le \frac{\delta_m}{4} + \frac{k_s}{30} \qquad (2.2)$$

$$v_t(z) = \kappa u_{fmax}\left(\frac{\delta_m}{4} + \frac{k_s}{30}\right) \qquad \text{for} \quad \frac{\delta_m}{4} + \frac{k_s}{30} < z \le 2\delta_m + \frac{k_s}{30} \qquad (2.3)$$

in which κ is von Karmen's constant; u_{fmax} is the maximum value of bed shear velocity ($u_f(\omega t)$) during the wave period that is max [$u_f(\omega t)$]; δ_m is the maximum value of δ_1 and δ_2, that is, max (δ_1, δ_2) where δ_1 and δ_2 are the boundary layer thickness at the moments corresponding to maximum and minimum velocity (of the combined wave and current flow) at the top of the turbulent boundary layer.

The quantities u_{fmax}, δ_m are determined from the solution of the integral equation derived from equation (2.1) as used by Fredsøe (1984):

$$\frac{\tau(\delta)}{\rho} - \frac{\tau_o}{\rho} = -\int_{\frac{k_s}{30}}^{\delta+\frac{k_s}{30}} \frac{\partial}{\partial t}(U-u)dz \qquad (2.4)$$

Fredsøe (1984) assumed a logarithmic velocity profile in the boundary layer

$$\frac{u}{u_f} = \frac{1}{\kappa} \ln \frac{30z}{k_s} \qquad (2.5)$$

The solution of equation (2.4) using Fredsøe's (1984) approach enables the value of u_{fmax} to be determined, if k_s is specified. Equation (2.1) can then be solved to provide the velocity distribution in the wave boundary layer.

2.3 Mathematical description of the flow in the granular-fluid region

Particle interactions in the shear-grain-fluid flow are assumed to produce two distinct types of behaviour. The Coulomb friction between particles gives rise to rate-independent stresses (of the plastic type) and the particle collisions give rise to stresses that are rate-dependent (of the viscous type). We assume the co-existence of both types of behaviour and the stress tensor is divided into two parts.

$$\sigma_{ij} = \sigma_{ij}^0 + \sigma_{ij}^* \qquad (2.6)$$

Where σ_{ij}^0 is the plastic stress and σ_{ij}^* is the viscous stress.

For two-dimensional deformation in the rectangular Cartesian co-ordinates x' and z' the Coulomb yield criterion is satisfied by employing the following stress relations:

$$\sigma_{x'x'}^0 = -\sigma'(1+\sin\varphi\cos 2\psi) \qquad (2.7)$$
$$\sigma_{z'z'}^0 = -\sigma'(1-\sin\varphi\cos 2\psi) \qquad (2.8)$$
$$\sigma_{x'z'}^0 = -\sigma'\sin\varphi\cos 2\psi \qquad (2.9)$$

Where φ is the quasi-static angle of internal friction, while ψ, denoting the angle between the major principal stress and the x'-axis is equal to:

$$\psi = \frac{\pi}{4} - \frac{\varphi}{2} \qquad (2.10)$$

For the average normal stress:

$$\sigma' = -\left(\frac{\sigma_{x'x'}^0 + \sigma_{z'z'}^0}{2}\right) \qquad (2.11)$$

we employ the following approximate expression (Sayed and Savage 1983).

$$\sigma' = \alpha^0 \left(\frac{c - c_0}{c_m - c} \right) \qquad (2.12)$$

where α^0 is a constant and c_0 and c_m are the solid concentrations corresponding to fluidity and closest packing respectively.

The viscous part of the stress tensor according to Sayed and Savage (1983) is assumed to have the following form:

$$\sigma^*_{x'x'} = \sigma^*_{z'z'} = -(\mu_0 + \mu_2)\left(\frac{\partial u}{\partial z'}\right)^2 \qquad (2.13)$$

$$\sigma^*_{x'z'} = \sigma^*_{z'x'} = \mu_1 \left|\frac{\partial u}{\partial z'}\right| \frac{\partial u}{\partial z'} \qquad (2.14)$$

in which the viscous stress coefficients μ_0, μ_1 and μ_2 are functions of the solids concentration c:

$$\frac{\mu_1}{\rho_s d^2} = \frac{0.03}{(c_m - c)^{1.5}} \qquad (2.15)$$

$$\frac{\mu_0 + \mu_2}{\rho_s d^2} = \frac{0.02}{(c_m - c)^{1.75}} \qquad (2.16)$$

Considering steady fully developed two-dimensional shear-grain-flow, the balance of linear momentum according to Kaczmarek & O'Connor (1993a,b) yields:

$$\alpha^0 \left[\frac{c - c_0}{c_m - c}\right] \sin\varphi \sin 2\psi + \mu_1 \left[\frac{\partial u}{\partial z'}\right]^2 = \rho u_f^2 \qquad (2.17)$$

$$\alpha^0 \left[\frac{c - c_0}{c_m - c}\right](1 - \sin\varphi \cos 2\psi) + (\mu_0 + \mu_2)\left[\frac{\partial u}{\partial z'}\right]^2 =$$
$$\left[\frac{\mu_0 + \mu_2}{\mu_1}\right]_{c=c_0} \rho u_f^2 + (\rho_s - \rho)g \int_0^{z'} c \, dz \qquad (2.18)$$

where ρ is the density of the fluid.

Eliminating $(\partial u / \partial z')$ from equations (2.16) and (2.17) allows the calculation of the profiles of the sub-bed sediment concentration c and velocity u in relation to known maximum shear stress ($\rho u_{f\,max}^2$) at the theoretical bed level ($z'=0$).

In Kaczmarek & O'Connor (1993a,b) equation (2.18) was solved for c as a function of depth (z') by using an iteration method in conjunction with numerical integration. Integration started at the theoretical bed level ($z'=0$) with $c=c_0$. Proceeding downwards at each step the iteration method was used to evaluate c. The integration was stopped when c was equal to c_{ms}. For the calculations the following numerical values were recommended for the various sand beds.

$$\frac{\alpha^0}{\rho_s g d} = 1 \; ; \; c_0 = 0.32 \; ; \; c_m = 0.53 \; ; \; c_{ms} = 0.50 \; ; \; \varphi = 24.4°$$

3. Results for monochromatic waves
3.1 Plane bed

The above procedure was used to compare computations for the model with the experimental results of Horikawa *et al.* (1982). The conditions for Horikawa *et al.*'s test 1 were used for the model calculations: d = 0.2mm, s = ρ_s/ρ = 2.66, φ = 24.4°, T = 3.64s and U = 127 cm/s. A value of k_s = 7.3mm was found for the roughness parameter.

Having obtained the roughness parameter it is then easy to obtain the instantaneous profiles both in the turbulent layer and the sub-bottom flow zone without reference to empirical formulas of any kind. Knowing u_f and solving equations 2.17 and 2.18 the velocity and concentration distributions at any time inside the entire sub-bottom layer can be found.

The results are shown in Figure 2. A reasonable agreement is obtained between the model and the laboratory data.

The model was then run for a range of conditions including those outside its range of application. The results of these tests are shown in Figure 3. The calculations were obtained using the simplified iteration procedure to determine k_s by introducing a simple logarithmic distribution (2.5) instead of the numerical solution of equation (2.1). Such a simplification makes the calculations much more efficient. It is seen that the roughness parameter, k_s, decreases with increasing dimensionless bed shear stress θ_{max} and k_s is seen to attain its greatest value for small dimensionless shear stresses where $\theta \approx 1$ (the transition from plane bed to ripples).

The trend shown in the present results, that is, that the roughness parameter increases drastically with decreasing dimensionless maximum shear stress, is similar to that shown by Nielsen (1992). Nielsen (1992) showed that the hydraulic roughness for equilibrium ripple formations is of the order $100 d_{50}$ to $1000 d_{50}$. However, for artificial flat beds where measurements were taken before ripples had time to form Nielsen (1992) found that the hydraulic roughness decreased with decreasing grain roughness Shields parameter.

Next, calculations were carried out for a moveable sandy bed (d = 0.2mm, s = 2.66, φ = 24.4°) with a variety of wave heights with a mean water depth of 5.0m. The wave period was kept constant at T = 3.6s. The maximum shear stresses were calculated on the basis of equation (2.4) and using the simplified iteration procedure to determine k_s.

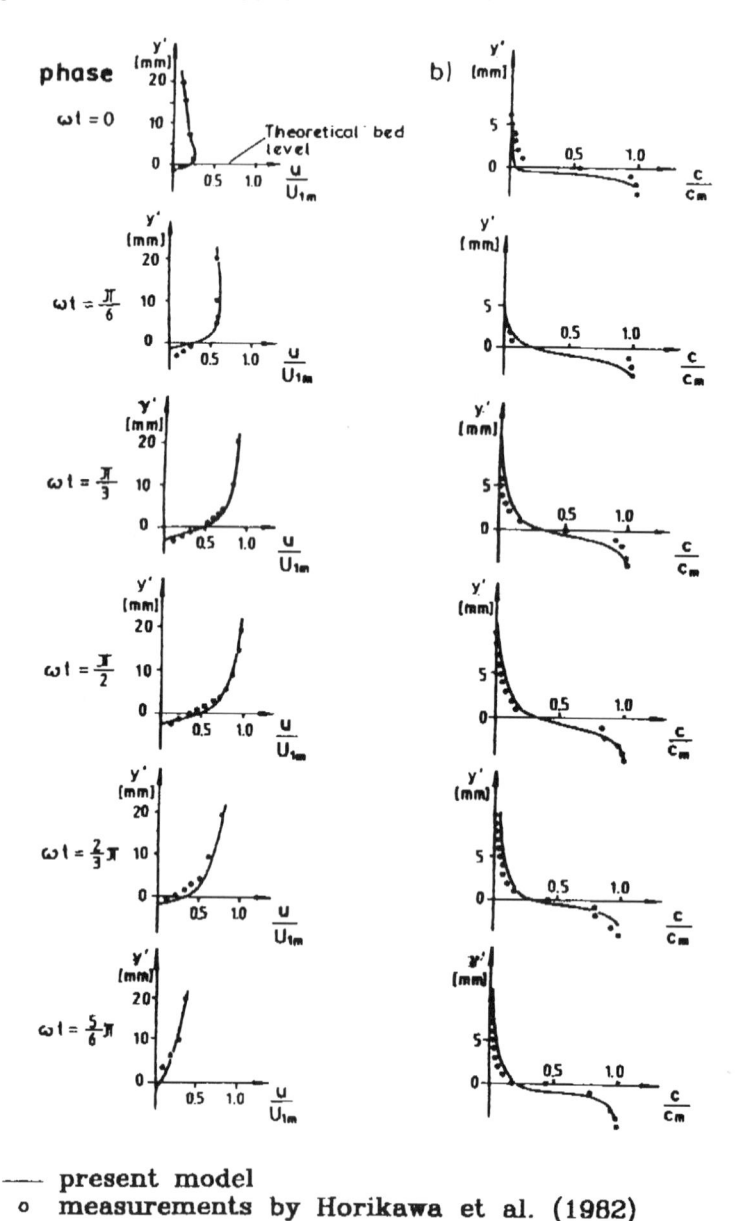

——— present model
o measurements by Horikawa et al. (1982)

Figure 2: Theoretical and experimental distributions of velocity (a) and concentration (b) below and above the bed.

The results of the analysis of friction for wave-induced sheet flow, shown in Figure 4, suggest that the present approach restricted to the sheet flow regime may be extended to lower flow regimes and on the basis of analogy used to investigate lower flow conditions involving bed ripples.

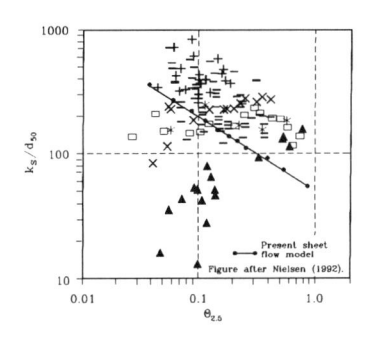

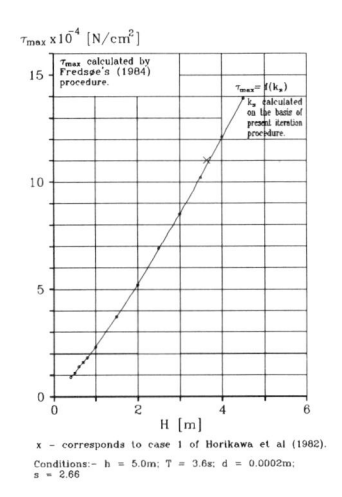

Figure 3: Nikuradse sand roughness by present theory along with results of Nielsen.

Figure 4: Calculation of maximum shear stresses

3.2 Rippled bed

Calculations for a rippled bed were performed for two different sediment sizes (0.2mm and 0.12mm diameter quartz sands). The calculations were carried out in two steps. Firstly, the values of the bed roughness k_s were obtained using the proposed iterative scheme. Then, the friction factors were calculated on the basis of an adjusted version of the semi-empirical formula of Jonsson and Carlsen (1976) in order to include the effects of the vortices formed in the lee of the roughness element crest due to turbulent mixing.

The theoretical results are shown in Figure 5. Presented alongside these results are the experimental results over a moveable bed reported by Madsen et al. (1990). The values of wave friction factor f_w are plotted against the representative value of a fluid-sediment interaction parameter, defined as:

$$S_r = \frac{\theta'}{\theta_c} \tag{3.1}$$

in which the skin Shields parameter is defined for a monochromatic wave as:

$$\theta' = \frac{u'^2_{f\,max}}{(s-1)gd} \qquad (3.2)$$

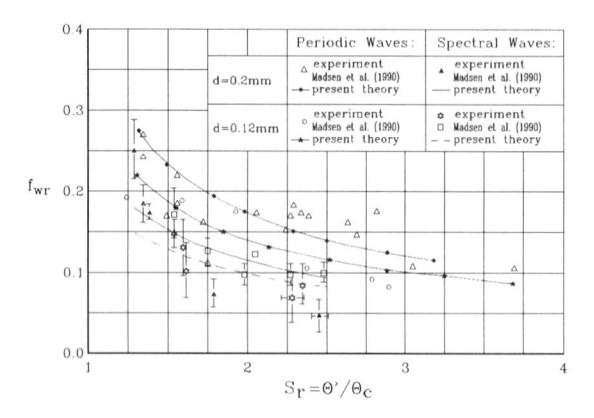

Figure 5: Moveable bed friction factors.

The agreement between theoretical and experimental results appears quite reasonable. It therefore appears that the sheet flow model can be used to investigate rippled bed conditions.

If the model can be used to investigate rippled bed conditions then it might also be possible to extend the analogy to include spectral wave conditions.

4 Spectral sheet flow model
4.1 Introduction

In the real world the Sea's motion is a random process. To describe a real sea it is usual to use spectral methods. However, it is possible to simplify the process by using appropriate representative values for the spectral components (See O'Connor *et al.* 1992).

The effect of random waves on bed roughness needs to be studied, since it is known that the bed friction changes between mono-frequency and random wave conditions. It is hypothesized by Madsen *et al* (1990) that the larger waves in a spectral simulation shave off the sharp ripple crests thereby causing the observed reduction in dissipation and friction factors for spectral waves. In an attempt to explain this reduction of spectral wave friction factors a new theoretical approach for predictive evaluation of moveable bed roughness for spectral waves is proposed. The new approach is based on the methodology which assumes that the spectral wave condition can be represented by a monochromatic wave and is combined with the theoretical grain-grain interaction ideas.

4.2 Modified iterative method

Following on from the iterative method used for monochromatic waves, a modified iterative procedure to evaluate the moveable bed roughness under spectral waves is proposed as shown in Figure 6.

Representative values are used in the calculation routine for the free stream velocity and the angular frequency. Previously for monochromatic waves, the maximum value of shear stress was the maximum value of shear stress during a wave period. For spectral waves the maximum value of the random shear stress time series is used:

$$\tau_{max} = \frac{3\tau_{rms}}{\sqrt{2}} = 3\sigma_\tau \qquad (4.1)$$

The choice of this maximum value of the random shear stress time series was checked using the simple Rayleigh Method as well as a through running a more sophisticated one dimensional through depth (1DV) k-ε boundary layer model.

4.3 Spectral shear stress

Using the Rayleigh method, it is possible to quickly determine a value for the shear stress for a random time series. Assuming a Rayleigh distribution then:

$$\frac{\tau_{max}}{\tau_{rms}} = [\ln(N)]^{\frac{1}{2}} = R \qquad (4.2)$$

T_z /sec	10	10	10
Time /min	10	20	40
N	60	120	240
R	2.02	2.18	2.34

The assumed value of R is:
$$3/\sqrt{2} = 2.12$$

The 1DV k-ε boundary layer model provides a method to directly simulate a random from shear stress from a known random velocity field. The method is based on the previous work of O'Connor *et al.* (1992) where a zero equation mixing length model was used to simulate a random sea.

The two equation k-ε model uses the standard equations to represent the momentum, the turbulent energy, k and the dissipation rate, ε.

Momentum:

$$\frac{\partial u}{\partial t} = \frac{\partial u_f}{\partial t} + \frac{\partial}{\partial z}\left(\nu_t \frac{\partial u}{\partial z}\right) \qquad (4.3)$$

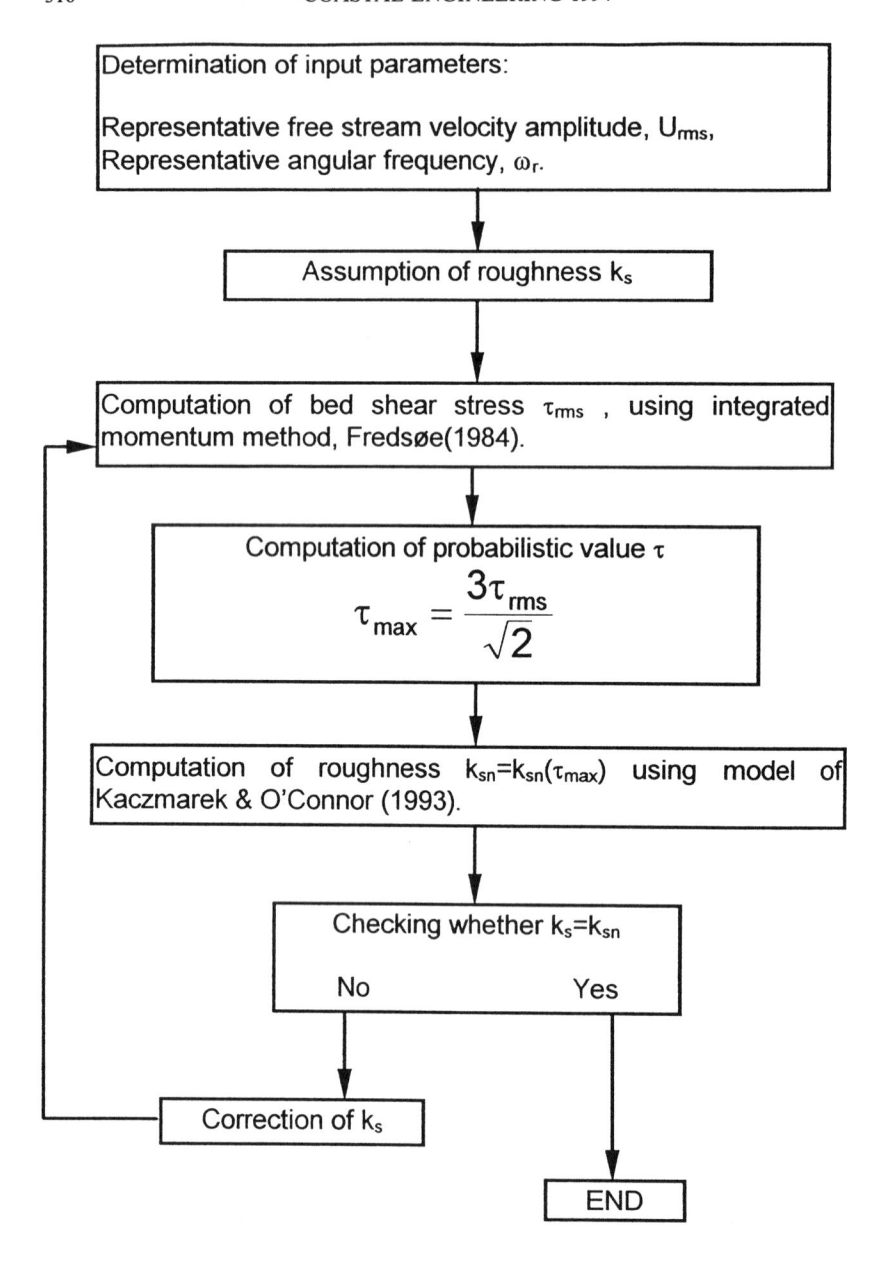

```
┌─────────────────────────────────────────────────┐
│ Determination of input parameters:              │
│                                                 │
│ Representative free stream velocity amplitude, U_rms, │
│ Representative angular frequency, ω_r.          │
└─────────────────────────────────────────────────┘
                        │
                        ▼
        ┌───────────────────────────────┐
        │ Assumption of roughness k_s   │
        └───────────────────────────────┘
                        │
                        ▼
┌─────────────────────────────────────────────────┐
│ Computation of bed shear stress τ_rms, using integrated │
│ momentum method, Fredsøe(1984).                 │
└─────────────────────────────────────────────────┘
                        │
                        ▼
        ┌───────────────────────────────┐
        │ Computation of probabilistic value τ │
        │                               │
        │         τ_max = 3τ_rms / √2   │
        └───────────────────────────────┘
                        │
                        ▼
┌─────────────────────────────────────────────────┐
│ Computation of roughness k_sn=k_sn(τ_max) using model of │
│ Kaczmarek & O'Connor (1993).                    │
└─────────────────────────────────────────────────┘
                        │
                        ▼
        ┌───────────────────────────────┐
        │ Checking whether k_s=k_sn     │
        │      No           Yes         │
        └───────────────────────────────┘
              │              │
              ▼              ▼
    ┌──────────────────┐   ┌─────┐
    │ Correction of k_s│   │ END │
    └──────────────────┘   └─────┘
```

Figure 6: Modified iteration scheme for spectral waves.

Turbulent Energy, k:

$$\frac{\partial k}{\partial t} = \frac{\partial}{\partial z}\left(\frac{v_t}{\sigma_k}\frac{\partial k}{\partial z}\right) + v_t\left(\frac{\partial u}{\partial z}\right)^2 - \varepsilon \qquad (4.4)$$

Dissipation Rate, ε:

$$\frac{\partial \varepsilon}{\partial t} = \frac{\partial}{\partial z}\left(\frac{v_t}{\sigma_\varepsilon}\frac{\partial \varepsilon}{\partial z}\right) + c_{1\varepsilon}\frac{\varepsilon}{k}v_t\left(\frac{\partial u}{\partial z}\right)^2 - c_{2\varepsilon}\frac{\varepsilon^2}{k} \qquad (4.5)$$

Turbulent Eddy Viscosity, v_t:

$$v_t = c_1 \frac{k^2}{\varepsilon} \qquad (4.6)$$

The upper boundary condition for the k-ε model is given by :-

$$u_o(t) = \sqrt{\frac{2M_{oc}}{N}} \sum_{n=1}^{N}\left(\frac{\omega_n}{\sinh(k_n d)}\right)\cos(-\omega_n t + \delta_n) \qquad (4.7)$$

Results from the model appear to indicate that the shear stress time series is not necessarily Rayleigh in its distribution. A typical model value for R was 2.6.

5. Results
5.1 Spectral bed roughness

The ability of the present iteration procedure, shown in Figure 6 to evaluate moveable bed roughness, k_s, under spectral waves was checked for a sandy bed: s = ρ_s/ρ = 2.66 φ = 24.4° with different grain size and various wave conditions. The results of the computations plotted in Figure 7 are for both irregular and regular waves.

In an attempt to explain the reduction of spectral wave friction factors the present theoretical approach was compared with Madsen et al. (1990) laboratory data. The results are shown in Figure 5 with the previous results for a monochromatic wave. The parameters are defined as before except that for spectral waves as the skin Shields parameter is given by:

$$\theta' = \frac{u_{fr}'^2}{(s-1)gd} = \frac{\tau_{rms}'}{\rho(s-1)gd} \qquad (5.1)$$

The calculation of the friction factors were carried out in two steps. First, the values of the bed roughness k_s were obtained using the modified iterative scheme (Figure 6) with Fredsøe's (1984) model used to determine the bed shear stress, τ_{rms}. Then the friction factors were calculated on the basis of adjusted the semi-empirical formula of Jonsson

& Carlsen (1976), as for monochromatic waves, in order to include the contribution of vortex formation in the lee of the roughness crests on the shear stress. Here, Jonsson & Carlsen's (1976) formulae were proposed for the calculations of both the friction factors and the dimensionless skin shear stresses.

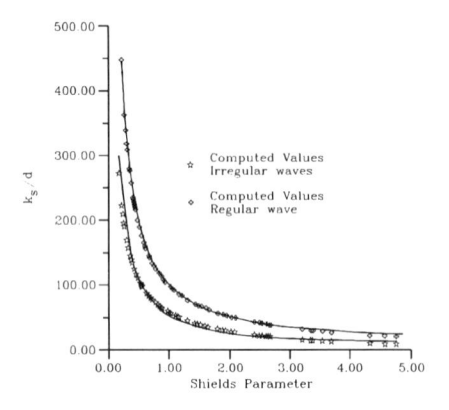

Figure 7: **Results of sheet flow model for regular and random waves.**

Similarly as for monochromatic waves, the calculations were performed for two different sediments (0.2mm and 0.12mm diameter quartz sands). Again the agreement between theoretical and experimental results appears quite satisfactory.

6. Conclusions

The sheet flow model appears to produce reasonable results for the conditions tested. However further testing is required.

The use of the model for a range of flow conditions and grain sizes produces a trend of large bed roughnesses at low flow regimes. According to this trend it is suggested that the sheet flow model provides a simple method, or rather an analogy, for the investigation of rippled bed conditions.

Using τ_{rms} to represent mono-frequency waves and τ_{max} to represent spectral waves produces a reasonable agreement with laboratory data.

The simple model results for k_s / d may be of use in preliminary engineering estimates although further testing is required. The present findings can be summarized for both the plane and rippled bed by the equation:

$$\tau_{rms1,2} = F_{1,2}\left[U_{rms}, T_p, k_s = f\left(\frac{3\tau_{rms1}}{\sqrt{2}}, s, d\right)\right] \qquad (6.1)$$

The subscripts 1 and 2 refer to the plane and rippled bed respectively.
The function f is described by the proposed iterative procedure and may be represented by the approximating formula:

$$\log\left[\frac{k_s}{d}\right] = -1.05 \log[\theta_{rms1}] + 4.00 \tag{6.2}$$

where the Shields parameter is calculated using Fredsøe's (1984) model.
The above approximation differs from that given for monochromatic waves due to the largest waves causing a reduction in the roughness parameter.
To calculate the function F in the case of a plane bed, Fredsøe's (1984) model is recommended ($F_1 \rightarrow \tau_{rms1}$). For the rippled bed case, the empirical formula of Kamphuis (1975) or semi-empirical formula of Jonsson & Carlsen (1976) have been used ($F_2 \rightarrow \tau_{rms2}$) in order to include the effects of the vortices formed in the lee of the roughness crest on the turbulent mixing.
Based on experimental data, it was found that the representative period equals the peak period. It appears as though the proposed method of predicting bed roughness in spectral waves by using ideas derived for sheet-flow modelling and a representative design wave is capable of providing realistic values for effective bed roughness height. Further work is in progress on the application of the model to additional North Sea data.

7. References

Brevik, I. (1981). 'Oscillatory rough turbulent boundary layers.' J. Waterways, Port, Coastal and Ocean Eng. Div., ASCE, Vol. 107, No.WW3, pp175-188.

Fredsøe, J. (1984). 'The turbulent boundary layer in combined wave and current motion.' J. Hydraulic Eng., ASCE, Vol. 110, No. HY8, pp1103-1120.

Horikawa, K., Watanabe, A. and Katori, S. (1982). 'Sediment transport under sheet flow condition.' Proc. 18th Int. Conf. on Coastal Eng., ASCE, Cape Town, South Africa, pp1335-1352.

Jonsson, I.G. and Carlsen, N.A. (1976). 'Experimental and theoretical investigations in an oscillating turbulent boundary layer.' J. Hydr. Res., Vol. 14, No. 1, pp45-59.

Kaczmarek, L.M. and O'Connor, B.A. (1993a). 'A new theoretical approach for predictive evaluation of wavy roughness on a moveable-flat bed.' Part I, Report No. CE/14/93, Department of Civil Engineering, University of Liverpool, 31pp.

Kaczmarek, L.M. and O'Connor, B.A. (1993b). 'A new theoretical approach for predictive evaluation of wavy roughness on a moveable-rippled bed.' Part II, Report No. CE/15/93, Department of Civil Engineering, University of Liverpool, 29pp.

Kamphuis, J.W. (1975). 'Friction factor under oscillatory waves.' J. Waterways, Port, Coastal and Ocean Eng. Div., ASCE, Vol. 101, No. WW2, pp135-144.

Kajiura, K. (1968). 'A model of the bottom boundary layer in water waves.' Bull. Earthq. Res. Inst., Univ. Tokyo, Vol. 46, pp75-123.

Madsen, O.S., Mathison, P.P. and Rosengaus, M.M. (1990). 'Moveable bed friction factors for spectral waves.' Proc. 22nd Int. Conf. on Coastal Eng., ASCE, pp420-429.

Nielsen, P. (1992). 'Coastal bottom boundary layers and sediment transport'. Advanced Series on Ocean Engineering, Vol. 4, World Scientific, Singapore, 324pp.

O'Connor, B.A., Harris, J.M., Kim, H., Wong, Y.K., Oebius, H.U. and Williams, J.J. (1992). 'Bed boundary layers.' Proc. 23rd Int. Conf. on Coastal Eng., ASCE, pp2307-2320.

Savage, S.B. (1984). 'The mechanics of rapid granular flows.' Advances in Applied Mechanics, Vol. 24, pp289-367.

Sayed, M. and Savage, S.B. (1983). 'Rapid gravity flow of cohesionless granular materials down inclined chutes.' J. Applied Mathematics and Physics (ZAMP), Vol.34, pp84-100.

CHAPTER 24

Difference between Waves Acting on Steep and Gentle Beaches

Kazumasa KATOH [1]

Abstract

The physical difference of beach erosions between on the steep and on the gentle beaches has been discussed, based on the field data and the theory of generation of the infragravity waves. As a result, it is shown that the incident waves are predominant in the wave run-up phenomena on the steeper beach, while on the gentler beach the infragravity waves are predominant.

Dear Prof. Dalrymple,

Based on the data obtained at the Hazaki Oceanographical Research Facility (**HORF**), I reported at the 23nd ICCE held in Venice, Italy, that the foreshore erodes under the action of infragravity waves of one to several minutes in a period (Katoh and Yanagishima, 1992). This conclusion, however, was distressed with the question made by one of participants, you Prof. Dalrymple, in the conference. Your question was; "There are many examples of beach erosion in experimental flumes with the regular waves, where the infragravity waves cannot exist. How do you explain away the experimental facts of erosion with your conclusion?". Your question was very excellent, because I could not answer on that time.

Since then, I have been studying on the physical difference between the beach erosions in the experimental flume and in the field. Here, I am going to explain the advanced conclusion on the role of the infragravity waves in the process of beach erosion, in the following.

---

1) Chief of Littoral Drift Laboratory, Port and Harbour Research Institute, Ministry of Transport, 3-1-1, Nagase, Yokosuka, 239 JAPAN

Critical Level of Foreshore Profile Change

At the HORF which is located on the sandy beach facing to the Pacific Ocean, Katoh and Yanagishima(1990) carried out the field observations on the foreshore berm erosion and formation. By analyzing the small scale sand deposition on the higher elevation when the berm eroded (see Figure 1), Katoh and Yanagishima (1992) presented that the critical levels of foreshore profile change in the processes of both erosion and accumulation are expressed by the significant wave run-up level, R_{max};

$$R_{max} = (\overline{\eta})_0 + 0.96(H_L)_0 + 0.31 \quad (m), \tag{1}$$

where $(\overline{\eta})_0$ is the mean sea level and $(H_L)_0$ is the height of infragravity waves at the shoreline, respectively, and the third constant term is considered to represent the run-up effect of incident wind waves.

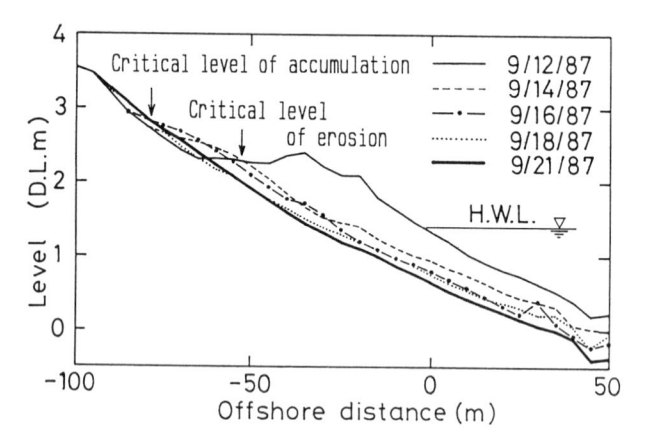

Figure 1 Typical example of berm erosion.

Equation (1) is the empirical relation which has been derived from the field data. The observation spans wide range of incident waves (offshore significant wave heights 0.39 -5.11 m and significant wave periods 4.5-15.6 s), $(H_L)_0 = 0.15$ -1.23 m, and $(\overline{\eta})_0 = 0.79$ -1.88 m. There is, however, a serious problem due to a fixation of observation site, that is to say, the constancy of profile slope. Namely, the effect of beach slope is excluded from equation (1), which restricts its general application.

For making up for this insufficiency of excluding the effect of slope, the empirical equation (1) has been compared with the previous results of experimental and theoretical studies. As the comparison of the results of field observation at the HORF with the Goda's theory (1975) has been already done in regard to the wave set-up at the shoreline (Yanagishima and Katoh,1990), the run-up heights of incident waves and the heights of infragravity waves at the shoreline have been examined in this study.

Run-up Heights of Incident Waves

Mase and Iwagaki (1984) carried out experiments as to the run-up of irregular waves on the uniform slope (1/5 to 1/30). As a result, they showed the relationship between the run-up height and the Iribarren number ξ, or a surf similarity parameter, as follow;

$$\frac{R_{1/3}}{H_{1/3}} = 1.378 \, \xi^{0.702}, \qquad (2)$$
$$\xi = \tan\beta / \sqrt{H_{1/3} / L_o} \, ,$$

where $R_{1/3}$ is the significant run-up height which is defined by the crest method, $H_{1/3}$ and L_o are a significant wave height and a wavelength in deep water respectively, and $\tan\beta$ is a bottom slope. As Mase et al. used a still water level as a reference level for the wave run-up height, the effect of wave set-up is included in the run-up height in their analysis. On the other hand, as the reference level in equation (1) is a mean water level at the shoreline under the action of waves on beach, the wave set-up is included in $(\overline{\eta})_o$, not in the wave run-up height. In short, a direct comparison of equation (1) with equation (2) is impossible. Then, a magnitude of wave set-up at the shoreline, $\overline{\eta}_{max}$, has been estimated from the data of offshore significant waves by using the Goda's theory (1975). In the calculation, the bottom slope is 1/60, which is the mean slope in the surf zone at the HORF. By adding $\overline{\eta}_{max}$ to the third constant term in equation (1), the run-up height above the still water level, Rs, is evaluated as follow;

$$Rs = 0.31 + \overline{\eta}_{max}, \quad (m) \qquad (3)$$

Figure 2 shows the relation between the non-dimensional run-up height normalized by the offshore significant wave height and ξ, where the bottom slope is fixed to be 1/60. The data plotted by circles and by triangles are obtained in the processes of berm erosion and berm formation respectively (Katoh and Yanagishima, 1993,1993). From this figure, we have a linear relationship between two parameters as

$$\frac{Rs}{H_{1/3}} = 2.50 \, \xi \, , \qquad (4)$$

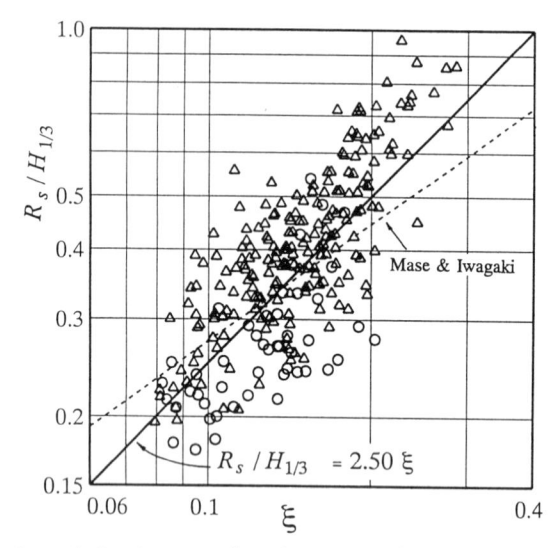

Figure 2 Relation between ξ and run-up heights of incident waves.

which is shown by a solid line closed to the equation (2).

Equation (2) proposed by Mase et al. is the significant run-up height by the crest method, while equation (3) is the critical run-up height where the significant profile change occurs. Although the physical meaning are different each other, it can be said that both run-up heights depend on ξ. Here, I have to give a supplementary explanation concerning equation (4). Since the data plotted in Figure 2 do not include the effect of bottom slope which is fixed to be 1/60 in analysis, it must be properly said that the non-dimensional run-up height is inversely proportional to the square root of wave steepness. However, equation (4) is the similar form to equation (2) which is based on the experimental data obtained on the bottom slope from 1/10 to 1/30. By taking this similarity into account, it can be concluded that the non-dimensional run-up height is proportional to ξ.

It is recognized in Figure 2 that the data plotted by the triangles, in the process of berm formation, are scattered mainly above the solid line, while the circles in the process of berm erosion are below the line. The leading cause of these inclined properties is considered to be due to the disregard of the effect of bottom slope in the area from the shallow water depth to the foreshore. If we consider the profile changes in the processes of berm formation, being steeper, and berm erosion, being gentle, the data by the triangles are shifted to the right and the data by circles to the left. As a result, all data will be plotted closer to the solid line.

Heights of Infragravity Waves at the Shoreline

Statistics of infragravity waves

In order to observe the waves near the shoreline, an ultra-sonic wave gauge was installed to the pier deck at the location where the mean water depth was about 0.4 meter in M.W.L. The wave measurement was carried out during 20 minutes of every hour with the sampling interval of 0.3 seconds (Katoh and Yanagishima, 1990). By utilizing the wave profile data, the wave heights and periods of infragravity waves were calculated by the following equations based on the result of spectra analysis;

$$H_L = 4.0 \sqrt{m_0} \quad , \tag{5}$$

$$T_L = \sqrt{m_0 / m_2} \quad , \tag{6}$$

$$m_n = \int_0^{f_c} f^n S(f) df, \tag{7}$$

where H_L and T_L are the wave height and the period of infragravity waves respectively, f is the frequency, $S(f)$ is the spectral energy density, f_c is the threshold frequency of 0.33 Hz. The wave height of infragravity waves at the shoreline where the water depth was zero, $(H_L)_0$, has been estimated by the following transformation equation (Katoh and Yanagishima, 1990);

$$(H_L)_0 = H_L \sqrt{(1 + h / H_{1/3})} \quad , \tag{8}$$

where h is the water depth at the observation point which can be evaluated from the bottom level and the mean water surface level.

A statistical analysis has been done for the data obtained during 4 years from 1989 to 1990, provided that some data have been excluded according to the following criteria;
(a) The data obtained when the water depth was shallower than 0.5 meter should be excluded, because the sea bottom sometimes emerged when the waves ran down offshoreward.
(b) The data obtained when the water depth was deeper than 1.1 meters is not preferable, because the relative distance from the observation point to the shoreline was large.

Figures 3 and 4 show the frequency distributions of the heights and the periods of infragravity waves at the shoreline, respectively. The highest frequency of wave heights is in the rank of 0.2 to 0.3 meter, while the mean height is 0.38 meter. The number of cases is 45 (4.1%) for the waves higher than 0.8 meter, and 15 (

1.4%) for the waves higher than 1.0 meter. The maximum wave height is 1.34 meters. The wave periods distribute in the range from 40 to 100 seconds, being 62.4 seconds in average.

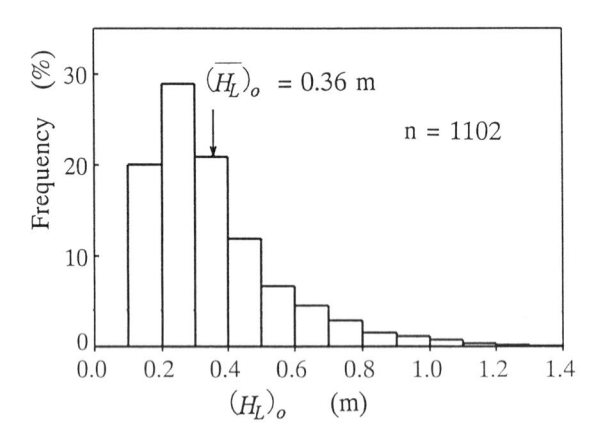

Figure 3 Frequency distribution of heights of infragravity waves.

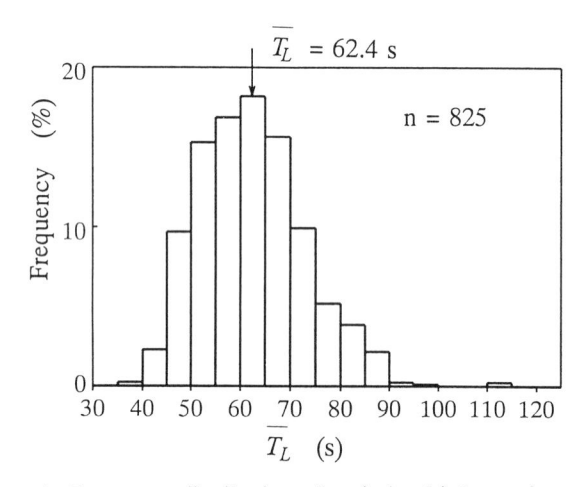

Figure 4 Frequency distribution of periods of infragravity waves.

Figure 5 shows the relation between the heights of infragravity waves at the shoreline and the offshore wave energy flux. On the upper, the significant wave height is shown as an indicator for the case that the wave period is 8.2 seconds. Although there is a little scattering of data, by means of the least square method we have a following relation;

$$(H_L)_0 = 0.23 \ E_f^{0.51} \tag{9}$$

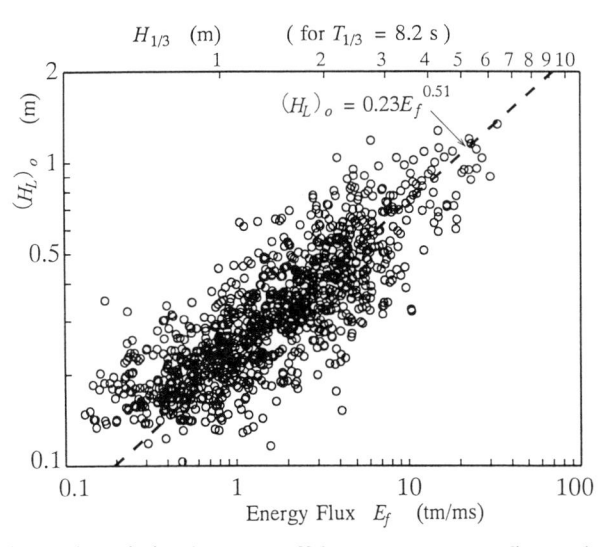

Figure 5 Relation between offshore wave energy flux and the height of infragravity waves at the shoreline.

where E_f is the offshore wave energy flux. Yanagishima and Katoh(1990) reported that the wave set-up at the shoreline, $\bar{\eta}_{max}$, at the HORF can be well explained only by the offshore wave energy flux,

$$\bar{\eta}_{max} = 0.14 \, E_f^{0.4} \tag{10}$$

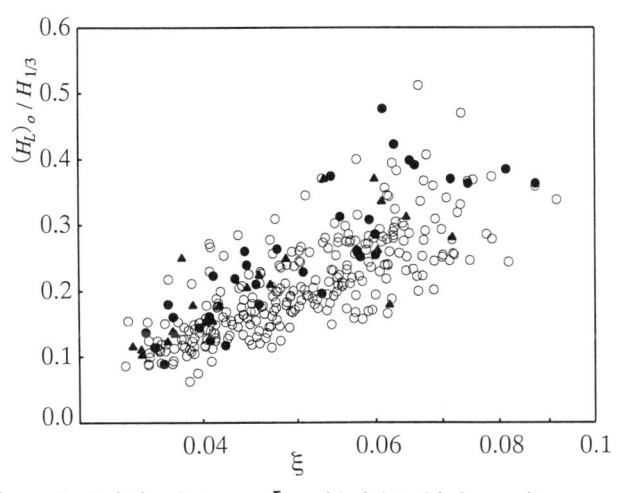

Figure 6 Relation between ξ and height of infragravity waves.

By substituting equations (9) and (10) into the first and second terms of equation (1) respectively, the critical level of foreshore change is expressed only by the wave energy flux. This result corresponds to another result that the daily changes of shoreline position can be well predicted by the offshore wave energy flux (Katoh and Yanagishima, 1988).

Figure 6 shows the relation between the non-dimensional height of infragravity waves at the shoreline, which are normalized by the offshore significant wave heights, and the Iribarren number ξ, provided that the mean slope in the area of 5 to 8 meters in depth where the incident waves break in a storm is utilized, which is 1/140. In this figure, the data obtained when the offshore significant wave height was larger than 2 meters is plotted. The closed triangles and the closed circles are the data which have been analyzed with respect to the berm formation and erosion (Katoh and Yanagishima,1992). Figure 6 shows that the non-dimensional height of infragravity waves increases with ξ. However, since the bottom slope is fixed to be 1/140 in the analysis, it must be properly said that the non-dimensional height increases with a decrease of wave steepness.

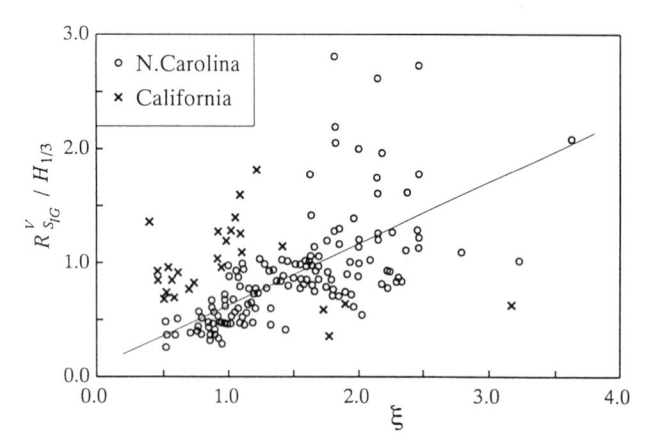

Figure 7 Relation between ξ and significant vertical swash excursion of infragravity waves (ξ is defined with the foreshore slope, from Guza et al., 1984).

In the same way, Guza et al.(1984) analyzed the two sets of wave run-up data, which were obtained on the beaches facing to the Pacific Ocean and the Atlantic Ocean. Figure 7 shows their result, where R^V_{SIG} is the significant vertical swash excursion of infragravity waves. We should remark that the foreshore slope was considered in their calculation of ξ. Figure 7 shows that two data sets are systematically different in this parameter space. While the data obtained in the

Carolina beach show a clear trend as shown by the best fit solid line, the data obtained in the California beach show no significant slope when ξ is large. Guza et al.(1984) said that the apparent discrepancy between the data sets lay in the rather arbitrary choice of a cutoff frequency for the infragravity band. However, I think as explained later that it depends on the difference of bottom slope in the wave breaking zone.

Effect of bottom slope on the infragravity waves

As the dependence of infragravity waves at the shoreline on ξ is different from beach to beach, the theory on the generation of infragravity waves by Symonds et al.(1982) has been examined. They used the non-dimensional, depth-integrated, and linearized shallow water equations, that is,

$$\chi \frac{\partial U}{\partial t} + \frac{\partial \zeta}{\partial x} = -\frac{\partial (a^2)}{2x \partial x} \qquad (11)$$

$$\frac{\partial \zeta}{\partial t} + \frac{\partial (xU)}{\partial x} = 0 \qquad (12)$$

$$\chi = \frac{\sigma^2 X}{g \tan\beta}, \quad U = \frac{2U'}{3\gamma\sigma X}, \quad t = \sigma t',$$

$$\zeta = \frac{2\zeta'}{3\gamma^2 X \tan\beta}, \quad x = \frac{x'}{X}, \quad a = \frac{a'}{\gamma X \tan\beta}, \qquad (13)$$

where $\sigma = 2\pi / \overline{T_R}$; $\overline{T_R}$ is a repetition period of wave group, x' is a distance offshore with the origin at the shoreline, X is the mean position of break point, g is the gravitational acceleration, U' is the depth-integrated velocity, γ is the ratio of the incident wave height to the water depth in the surf zone, t' is the time, ζ' is the level of sea surface, and a' is the amplitude of incident waves.

Nakamura and Katoh(1992) pointed out that the Symonds' theory overestimates the height of infragravity waves in comparison with the field data. They modified the Symonds' theory by taking a time delay of small wave breaking due to propagation into consideration, which well predicts the wave height of infragravity waves. According to the modified theory, the non-dimensional wave profile of infragravity waves at the shoreline, ζ_0, is as follow (see, Nakamura and Katoh, 1992);

$$\zeta_0 = \sum_{n=1} C_n \sin(nt + \varepsilon_n), \qquad (14)$$

where C_n are the coefficients which have been expressed in complicated forms, and ε_n are the phase lags.

In order to examine the characteristic of wave profile expressed by equation (14), relatively simple, but acceptable, assumptions will be introduced, although they make a little sacrifice of quantity. By assuming that the ratio of the wave amplitude to the water depth, γ, is constant, we have,

$$X = \frac{\overline{H}}{2\gamma\tan\beta}, \qquad (15)$$

where $\overline{H}$ is the mean wave height which is correlated with $H_{1/3}$ as

$$H_{1/3} = 1.6\,\overline{H}. \qquad (16)$$

Next, let us assume that the height of incident waves in groups varies sinusoidally (Nakamura and Katoh,1992), that is,

$$H = \overline{H} + \sqrt{2}\,H_{1/3}\cos(\sigma t)\,/\,3. \qquad (17)$$

By utilizing equations (15),(16), and (17), the ratio of the amplitude of wave break point varying to X is 0.75 ($= \Delta a$).

Figure 8 shows the theoretical relation between C_n (; n=1 to 4) and χ, which have been obtained based on the above assumption. In Figure 8, the amplitude of infragravity waves at the shoreline is also shown, which is calculated by means of $\sqrt{2}\,(\zeta_o)_{ms}$ from the wave profile composed by equation (14) for n=1 to 4.

In regard to χ, an interesting rewriting will be possible. By referring the empirical relation between T_R and $T_{1/3}$ ($T_R = 9.24 T_{1/3}$, Nakamura and Katoh, 1992), we have

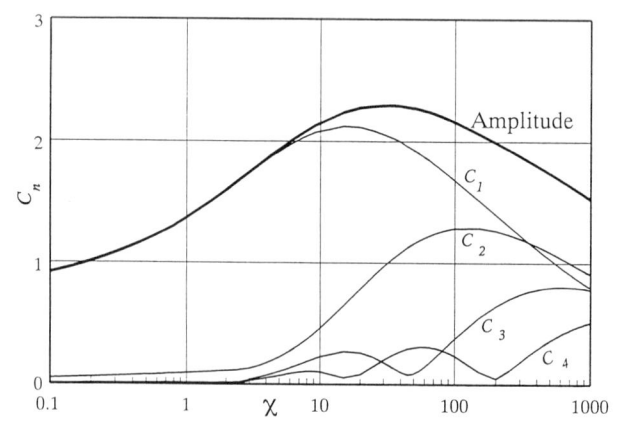

Figure 8 Relation between χ and C_n or amplitude of infragravity waves at the shoreline.

$$\sigma = \frac{2\pi}{9.24\, T_{1/3}} = \frac{1}{9.24}\sqrt{\frac{2\pi g}{L_o}}\ . \tag{18}$$

Then, by substituting equations (15),(16), and (18) into equation (13), we have

$$\chi = \frac{0.023}{\gamma}\, \frac{H_{1/3}/L_o}{\tan^2\beta} = \frac{0.023}{\gamma \xi^2}\ , \tag{19}$$

that is to say, χ is the function of ξ. Furthermore, another relation in equation (13) can be rewritten for the condition at the shoreline as follow;

$$\sqrt{2}\,(\zeta_o)_{rms} = \frac{(H_L)_o}{3\gamma^2 X \tan\beta}\ . \tag{20}$$

According to the parameter definitions, we have

$$\overline{H} = 2\gamma X \tan\beta\ . \tag{21}$$

By substituting equation (21) into equation (20) and taking equation (16) into account, we finally have

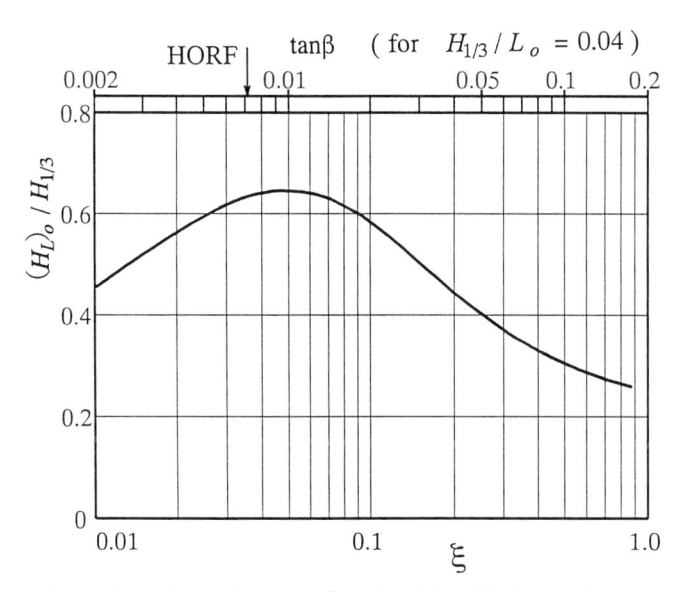

Figure 9 Relation between ξ and height of infragravity waves at the shoreline (theory).

$$\frac{(H_L)_0}{H_{1/3}} = \frac{3\gamma}{3.2} \sqrt{2} \, (\zeta_0)_{rms} \, . \tag{22}$$

By utilizing equations (19) and (22), Figure 8 can be represented by using new parameters.

Figure 9 is the transformed relation between new parameters, by assuming $\gamma = 0.3$. On the upper side, a scale is marked for the bottom slope when the mean value of offshore wave steepness at the HORF, 0.04, is adopted. The mean bottom slope in the area of wave breaking in a storm is 1/140 at the HORF, which is indicated by an arrow on the upper side. In a range of bottom slope up to 0.01 (=1/100), the non-dimensional wave height increases with ξ, of which tendency is the similar as that shown in Figure 6. On the other hand, it decreases with ξ in a range of bottom slope steeper than 0.01.

In the analyses by Guza et al.(1984), they defined ξ by the foreshore slope, not by the bottom slope in the wave breaking area. Then, I have inspected the bottom slopes of two beaches in literature. The mean bottom slope in the area from −2 to −6 meters is 1/127 in the North Carolina beach (Holman and Shallenger,1985), which is too much gentler than the foreshore slope. The bottom slope in the Torrey Pines Beach, California, is almost constant up to the water depth of 7 meters, being 1/45 (Guza and Thornton,1985). By utilizing these slopes, the values of ξ have been calculated, provided that the foreshore slopes are 6 degrees

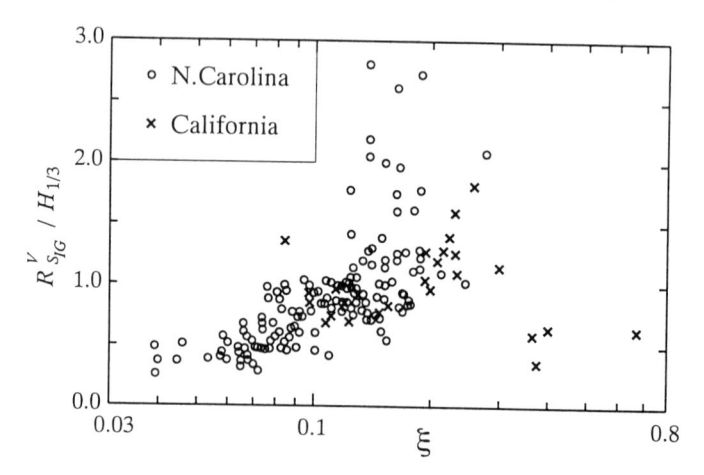

Figure 10 Relation between ξ and significant vertical swash excursion of infragravity waves (ξ is defined with the mean bottom slope in the wave breaking area).

in both the beaches. Figure 10 is the result of calculation, in which the data are rearranged to give the qualitative agreement with the theoretical relation in Figure 9.

Discussion and Conclusions

Figure 11 shows the result of present study. A thick line is the theoretical relation between the non-dimensional height of infragravity waves at the shoreline and ξ, which has been qualitatively verified with three sets of field data. The linear relationship between the non-dimensional run-up height of incident waves and ξ, equation (4), is superimposed on this figure by a thin line, which is curved in a semi-log space. There is an intersection of two lines, which is roughly corresponding to the bottom slope of 1/25 when the wave steepness is 0.04. In the right-hand side from the intersection, on the steeper beach, the incident waves are predominant in the wave run-up phenomena, while on the gentler beach the infragravity waves are predominant.

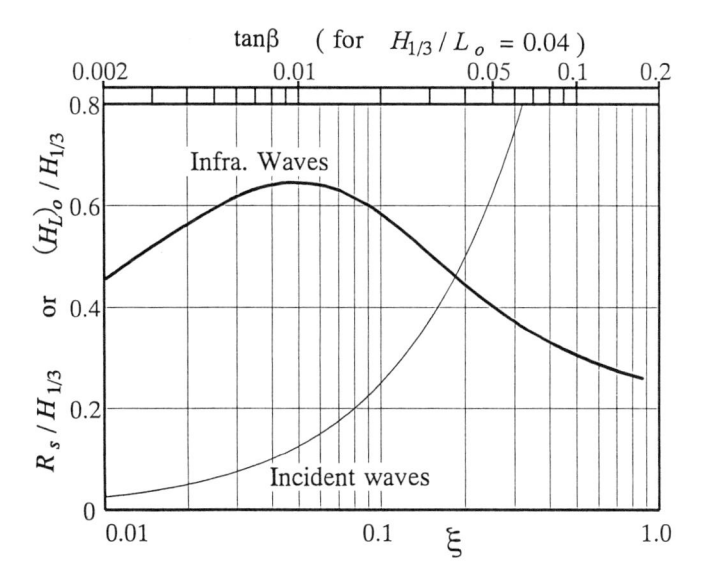

Figure 11 Dependences of infragravity waves and incident waves on ξ.

Almost all of the model experiments were conducted on the model beach steeper than 1/30, probably due to the limited length of flume. Under this condition, the dominating external force for profile changes is the incident waves. Then, the foreshore erodes even though only the incident waves are reproduced in the experiment. On the other hand, the infragravity waves exist on the gentle beach in the field. However, the evidences which show the cross relation between the beach erosion and the infragravity waves are scarce. One of reasons is that many field observations in connection with the wave run-up were carried out on the steep beach (according to literature survey by Kubota ;1991). Another reason is that many field observations were done in the relatively calm wave conditions. The smaller waves break closer to the beach where the bottom slope is steeper because the profile of nearshore topography is usually concave upward. In short, the development of infragravity waves is weak when the incident waves break on the steep bottom. Therefore, in order to have quantitative information on the relation between the infragravity waves and the beach erosion, the further field observations are required concerning the phenomena on the gentle beaches in storms.

Finally, the author is grateful to Mr. Satoshi Nakamura, a member of the littoral drift laboratory, for his courtesy in using the computer program which he developed to predict the height of infragravity waves in the surf zone.

References

Goda,Y.(1975):Deformation of irregular waves due to depth controlled wave breaking, Rep. of PHRI, Vol.14, No.3, pp.56-106(in Japanese, see Goda, 1985).

Goda,Y.(1985):Random seas and design of maritime structures, University of Tokyo Press, 323p.

Guza,R.T., E.B.Thornton and R.A.Holman(1984):Swash on steep and shallow beaches, Proc. of 19th ICCE, pp.708-723.

Guza,R.T. and E.B.Thornton(1985):Observation of surf beat, J.G.R., Vol.90, No.C2, pp.3161-3172.

Holman,R.A. and A.H.Sallenger(1985):Setup and swash on a natural beach, J.G.R., Vol.90, No.C1, pp.945-953.

Katoh,K. and S.Yanagishima(1988):Predictive model for daily changes of shoreline, Proc. of 21st ICCE, pp.1253-1264.

Katoh,K. and S.Yanagishima(1990):Berm erosion due to long period waves, Proc. of 22nd ICCE, pp.2073-2086.

Katoh,K. and S.Yanagishima(1992):Berm formation and berm erosion, Proc. of 23rd ICCE, pp.2136-2149.

Katoh,K. and S.Yanagishima(1993):Beach erosion in a storm due to infragravity waves, Rep. of PHRI, Vol.31, No.5, pp.73-102.

Kubota,S.(1991):Dynamics of field wave swash and its prediction, a doctor's thesis, Chuo University, 232p.(in Japanese)

Mase,H. and Y.Iwagaki(1984):Run-up of random waves on gentle slopes, Proc. of 19th ICCE, pp.593-609.

Nakamura,S. and K.Katoh(1992):Generation of infragravity waves in breaking process of wave groups, Proc. of 23rd ICCE, pp.990–1003.

Symonds,G., D.A.Huntley and A.J.Bowen(1982):Two-dimensional surf beat:Long wave generation by a time-varing breakpoint, J.G.R., Vol.87, No.C1, pp.492–498.

Yanagishima,S. and K.Katoh(1990):Field observation on wave set-up near the shoreline, Proc. of 22nd ICCE, pp.95–108.

CHAPTER 25

Wave Breaking under Storm Condition

Yoshiaki Kawata, M.ASCE[1]

Abstract

The effects of strong wind on wave breaking were discussed with the field data. At around U =23m/s the maximum wave steepness of 0.034 was obtained. At the wind velocity of more than 23m/s, the wave steepness becomes small once because reformed waves which were already broken at the offshore were measured. This corresponds with the changes of the ratio of wave breaking. The relationship between the mean period of SIWEH and significant wave period was made clear. It was found that pulsating behavior of undertow influences wave breaking in a shallow water.

Introduction

In spite of many trials to explain the mechanism of wave transformation and wave breaking, in a shallow water, we have still obscure phenomena in the nearshore zone. For example, under high wind, it is very difficult to predict the location of breaking points of incoming waves. Most of coastal structures have been designed to avoid impulsive shock pressure due to wave breaking, but usually breaker index has been developed on the basis of experimental data conducted under no wind condition. Some coastal disasters have been generated by underestimation or miscalculation of wave forces due to wave breaking.

On a mild sloping beach, high waves under strong wind break two or more times in the crosshore direction. In the shoaling process, we observed frequently that a lower wave just after a higher wave breaks easily. This mechanism was not well understood yet. Moreover, at the wind velocity of more than 20m/s, the applicability of breaker index and plausible maximum wave steepness have not made clear due to lack of field data. Moreover, high wind waves usually come to the shore in accompany with high wind.

The effects of strong wind on wave breaking were discussed with some traditional methods. Firstly, changes of wave steepness and ratio of wave breaking with wind velocity were studied. Secondly, applicability of breaker index was investigated. In order to make clear the effect of undertow on wave breaking, the relationship between the dominant period of SIWEH and significant wave period was also justified. The measurement was conducted on 18 November, 1981 on which developing low pressure moved eastward in the Japan Sea and high wind blew around the coast.

---

1 Professor, Disaster Prevention Research Institute, Kyoto University Gokasho, Uji, Kyoto 611, Japan

Field Observation Site and Method

The field experiments were conducted on the Ogata coast which is a straight sandy beach of 20km long facing the Japan Sea. At the measuring site (water depth:7m, bottom slope:1/130, offshore distance:200m), capacitance wave gauges were installed as shown in Fig. 1. Two wave gauges were set respectively in the crosshore direction at the distances of 12.9m landward and 19.6m seaward from the measuring site. A unit of data is each 100 sec in length and 40 units were analyzed as every averaged data. VTR systems focused on sea surface at Ch. 1 Wind characteristics were measured with the anemometer installed at the 35m offshore from Ch. 1 and its height was 15m above mean sea level (tidal range is about 30cm). At the observatory, the anemometer was installed at the height of 10m above ground level. Dominant wave direction was determined through the analysis of directional spectra with the data obtained by eleven wave gauges. The undertow was measured with the ultrasonic type current meters at the height of 1m above the bottom.

Observation Results

Firstly, we analyzed the data with surf similarity parameter ξ to divide into breaking and non-breaking waves. The wave length was calculated with small amplitude wave theory and second-order Stokes wave theory. Figure 2 shows one of the results in which η_d means the distance between mean sea level and wave trough. From this figure, it was found that it is impossible to clarify breaking and non-breaking with the traditional expression. In the similar expression such as η_u / η_d (η_u: distance between mean sea level and wave crest), $H / \eta_d T$ (T: wave period), the combination of and the variables defined here can not express the criterion of breaking and non-breaking. In these relationship, we introduced the wind velocity as the parameter, but we could not get good results to explain wave breaking phenomena.

Effects of High Wind on Wave Breaking

In this chapter, we firstly tried to justify the applicability of breaker index and changes of wave steepness and rate of wave breaking with wind velocity. On the Ogata coast, wind waves develop in accompany with the movement of low pressure and the wave direction changes from west, northwest to east sequentially. It is generally

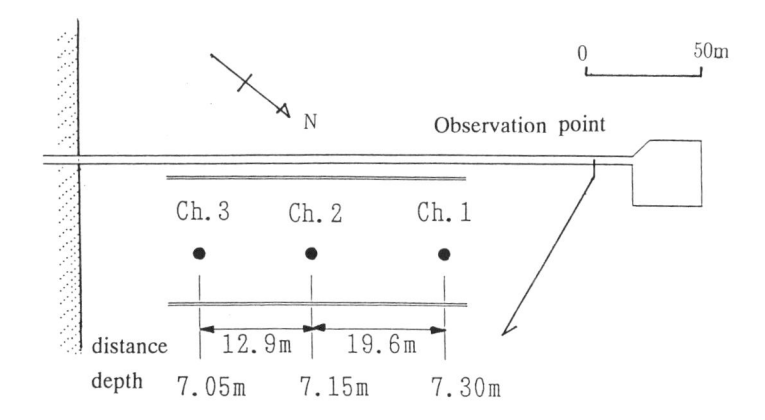

Fig. 1 Observation pier and point on the Ogata coast

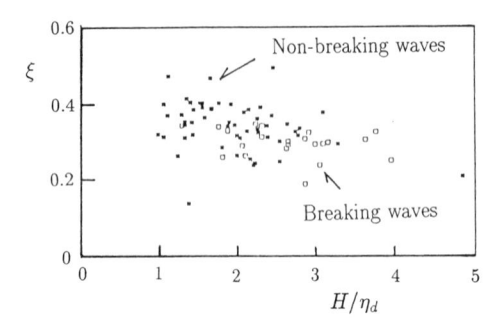

Fig. 2 Relationship between H / η_d and surf similarity parameter

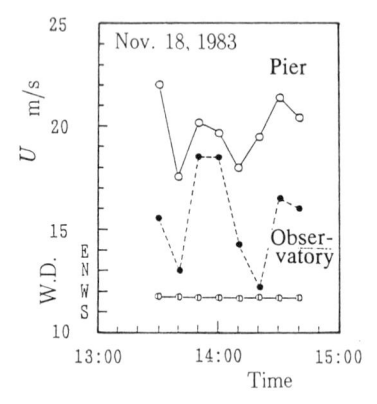

Fig. 3 Changes of wind characteristics

recognized that wave height increases until wave direction becomes northwest. After this stage, waves change to swell.

Figure 3 shows an example of sequential changes of wind characteristics. Every point is 10 min. averaged data. At the pier, we had very high wind and wave breaking with spilling type frequently occurred. Figure 4 shows the changes of the wave steepness H/L_o and ratio of wave breaking R with wind velocity U. Wave steepness was measured at the point of Ch. 3. Individual wave steepness is calculated and averaged in the period of 10 min. The ratio R is defined as the number of wave breaking divided by the total wave number. In the case of (a) (wind direction :WSW) the increase of wave steepness with wind velocity is clear and also the increase of R show same tendency. On the contrary, in the case of (b)(wind direction:WNW), the following results were observed: At around U =23m/s the maximum wave steepness of 0.034 was obtained and at the wind velocity of more than 23m/s, the wave steepness becomes small once because reformed waves which were already broken at the offshore were measured. Furthermore, more than 26m/s, the reformed waves also begin to break. The difference of wind velocity of

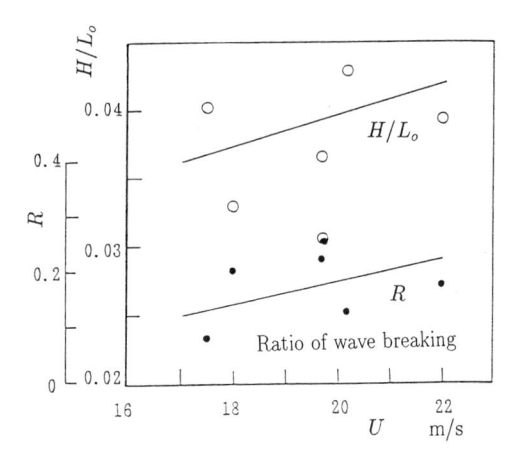

(a) wind direction: WSW

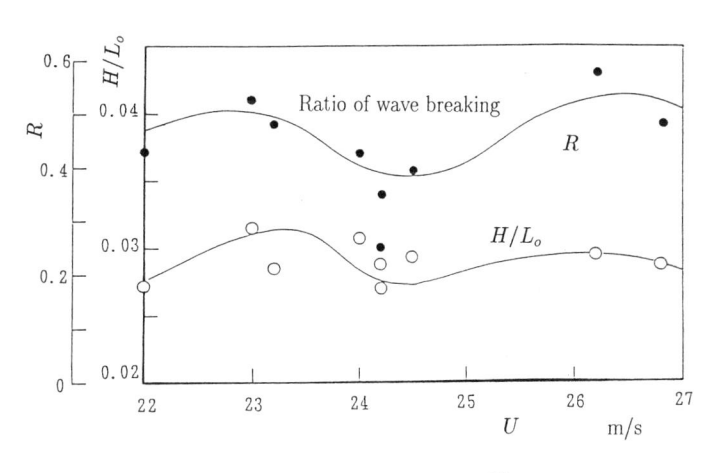

(b) wind direction: WNW

Fig. 4 Changes of wave steepness and ratio of wave breaking with wind velocity

only 3m/s is corresponds to the enlargement of width of breaker zone. The effect of wind direction on the changes of wave steepness is reflected the difference not only fetch and duration of wind waves but also of wave refraction characteristics. For example, the westerly waves changes from long crested to short crested waves due to wave refraction.

The changes of the rate of wave breaking is similar to that of wave steepness as follows:
a) wind velocity is less than 23m/s, the ratio R increases up to 0.5
b) 23 to 25m/s, R decreases to 0.4.
c) more than 25m/s, $R \approx 0.5$
They correspond well with the changes of wave steepness under high wind conditions.

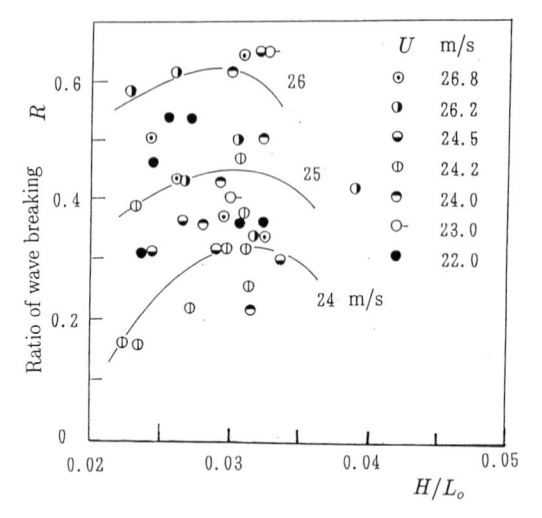

(a) wind velocity is more than 22m/s

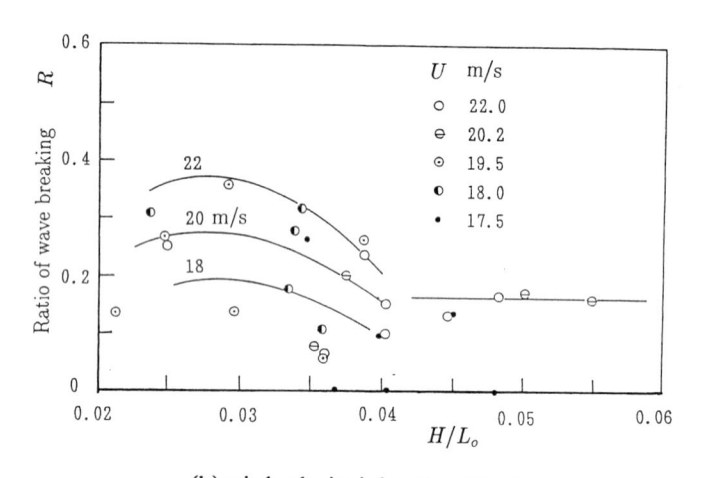

(b) wind velocity is less than 22m/s

Fig. 5 Changes of ratio of wave breaking and wave steepness

Figure 5 shows the relationship between wave steepness and ratio of wave breaking. The following remarks were observed:
a) wind velocity is less than 22m/s, ratio of wave breaking increases with wind velocity. When wave steepness becomes more than 0.04, ratio of wave breaking keeps 0.16.
b) wind velocity ranges from 22 to 24m/s, ratio of wave breaking slightly decreases.
c) more than 24m/s, waves whose steepness is more than 0.04 can not be found.

Figure 6 shows the applicability of breaker index proposed by Goda(1970). The data were recorded at $U = 26.7$ m/s and $R = 0.457$. It was found that the breaking height was classified into some groups in each wave age and the breaker index was the upper limit of the field data.

Effect of Undertow on Wave Breaking

In the field, a following lower wave behind a higher wave is sometimes prone to breaking. The following mechanism may be related with the wave breaking phenomena on a mild slope in a shallow water:
a) high waves have low trough level, therefore, following low waves may break.
b) undertow due to high wave breaking promotes breaking of following waves.

Under storm wave condition on the Ogata coast, wave grouping is clearly formed and trough level of carrier waves is almost constant in every storm conditions as shown in Fig. 7. This may be only applicable to the shallow water at the depth of 7m, but the If

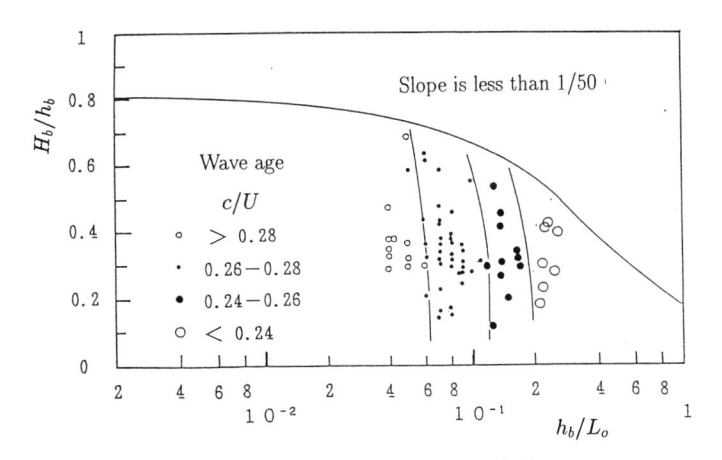

Fig. 6 Applicability of breaker index

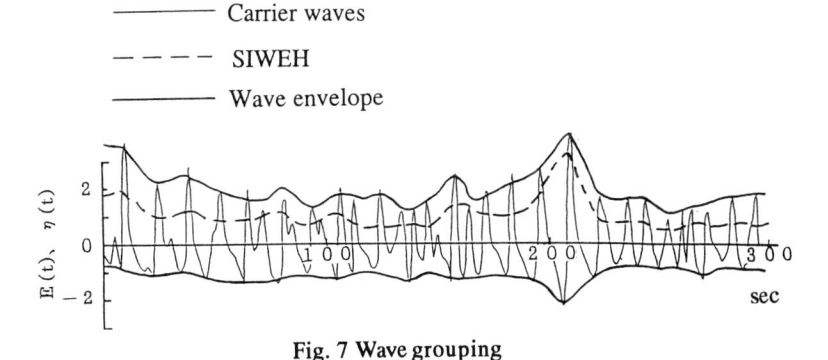

Fig. 7 Wave grouping

hypothesis a) is inadequate. Secondly, we discuss about the plausibility of hypothesis b). If we assume the control volume whose rectangular frame includes one wave length in the crosshore direction and one wave height above the bottom with moving coordinate, the bore model gives undertow due to energy dissipation at the moment of wave breaking. In this case, local water level changes due to changes of radiation stress and its gradient. However, a shoreline is the fixed boundary and flow structure can not be described without consideration about conservation of momentum or energy. Moreover, time scale in the analysis has to be comparable to spatial scale of width of nearshore zone. Therefore, formation of wave grouping and its related phenomena are focused on explanation of wave breaking.

As already pointed out, traditional methods can not predict the wave breaking. The reason of the pulsating occurrence of wave breaking is firstly discussed. If the pulsation exists, the oscillation of wave energy level with nearly same period should be found. Figure 8 shows the changes of wave height along Chs. 1, 2 and 3. The increase of wave height depends on wave transformation or existence of reverse flow. The averaged data show that the increase of wave height is larger than that predicted by wave transformation with small amplitude theory (in the calculation, wave height of 2m and wave period of 8s). In order to make wave height twice, it is necessary to add the reverse flow at the velocity of 2.75m/s. This is very large, but in the field the breaking phenomena is unsteady and the increase of wave height due to breaking also exist, this result suggests that the reverse flow may influence the wave breaking process.

Long Period Fluctuation of Current Velocity

Figure 9 shows the power spectrum of current velocity in the Northwest direction (it is almost in the crosshore direction) at the height of 1m above the bottom (water depth is about 6m)(Tsuchiya et al., 1989). In this figure, the first peak with the frequency of about 0.1Hz responds to that of incoming dominant waves. In the low frequent range, it is found to be some primary peaks.

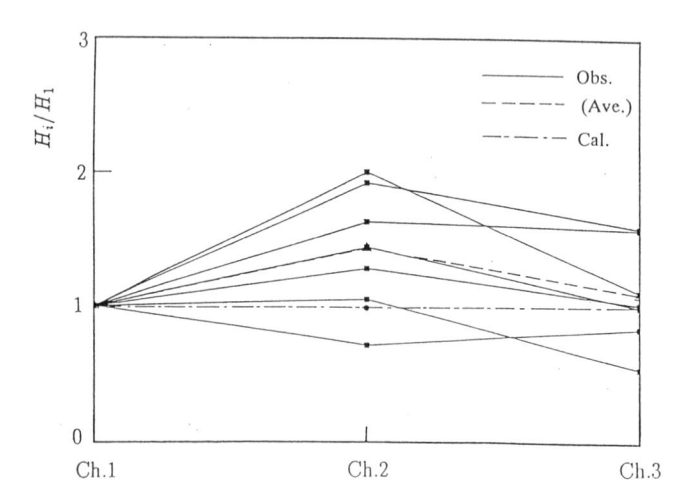

Fig. 8 Changes of wave height in the crosshore direction

Figure 10 shows the long period fluctuation of the current velocity in which the higher frequent component were cut off with the low pass filter of 0.03Hz. The current was recognized with the amplitude of 50cm/s. By this pulsating fluctuation, bursting of the bottom sediment were observed as shown in Fig. 11. The optical suspended sediment sensors are mounted in a vertical steel pipe of 3cm in diameter by 4m long. The lowest sensor was set at the height of 10cm above the bottom.(kawata et al., 1989). This bursting occurred at the moment of the concurrence of undertow and maximum water particle velocity in the offshore direction. The results reveals the importance of undertow to control bottom sediment concentration.

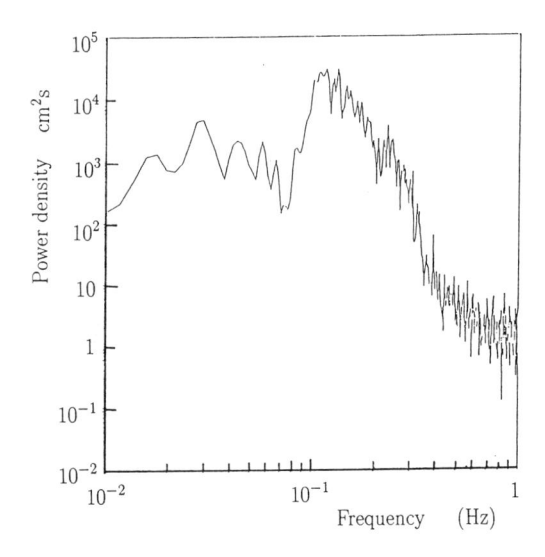

Fig. 9 Power spectrum of current velocity in the crosshore direction

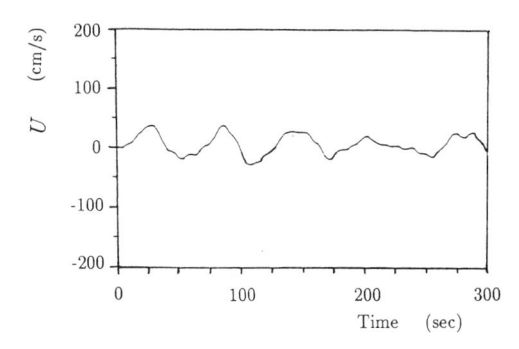

Fig. 10 Long period oscillation of undertow

Correspondence of Undertow and Wave Breaking

It is necessary to make clear the generation mechanism of undertow whose period is several to around ten times of those of incoming waves. Firstly, we checked the wave energy levels of wave grouping on the Ogata coast. Figure 12 shows that the mean period of SIWEH is proportional to that of significant wave period. This figure shows that the increase of wave period well correlate to the long period variation in wave grouping. Secondly, we checked the relationship between the time variation of undertow and the occurrence of wave breaking. The procedure is as follows:
  a) In every 200 waves recorded at Chs. 1, 2 and 3, significant wave height and its period are calculated. We get a series of them during 600 waves.
  b) In every breaking wave, the period T_{sh} (the period of SIWEH in Fig. 12, given by 5.81 $T_{1/3}$ +1) is shifted behind it and check the certain wave breaking or not.
  c) This checking is applied to every individual 30 waves.

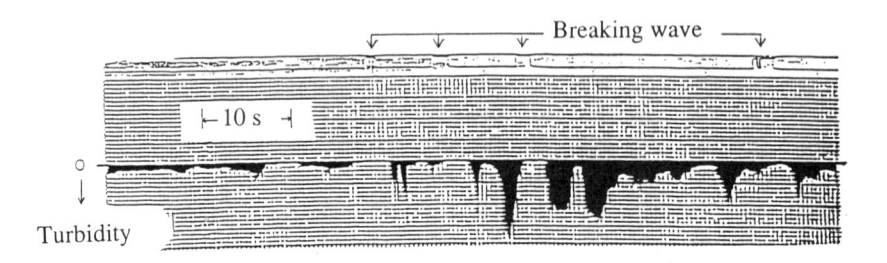

Fig. 11 Suspended sediment bursting due to wave breaking

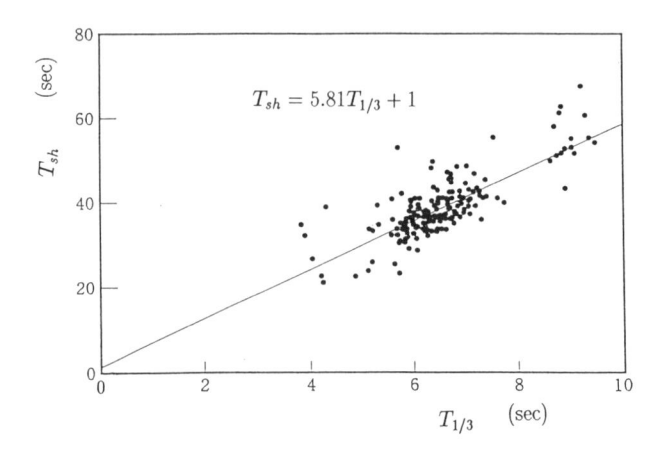

Fig. 12 Relationship between mean period of SIWEH and significant wave period

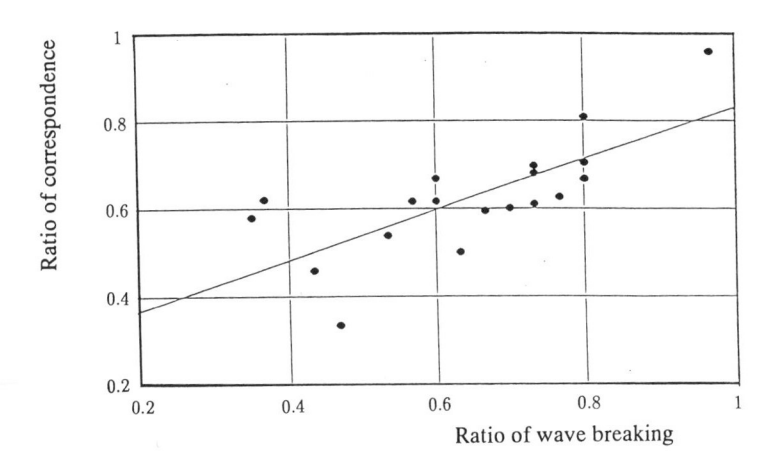

Fig. 13 Changes of ratio of correspondence with ratio of wave breaking

Figure 13 shows the relationship the ratio of correspondence and the ratio of wave breaking. It was found that the increase of the former is roughly predicted by the increase of the latter. Therefore, it was concluded that wave breaking in a shallow water is well depend on the pulsation of the energy level of wave grouping.

Conclusions

The effects of high wind on wave breaking in a shallow water were discussed with the field data. In rough sea state under strong wind, the wave breaking conditions in the nearshore zone can not be predicted with some traditional expressions. At around U =23m/s the maximum wave steepness reaches to 0.034. At the wind velocity of more than 23m/s, the wave steepness becomes small once because reformed waves which were already broken at the offshore site were measured. This corresponds with the changes of the ratio of wave breaking. The relationship between the mean period of variation of SIWEH and significant wave period was made clear. It was found that pulsating behavior of undertow influences wave breaking in a shallow water.

References

Goda, Y. : A synthesis of breaking indices, Trans. JSCE, Vol. 2, Part 2, 1970, pp. 227-230.
Kawata, Y. et al.: Field observation of sediment transport under storm wave conditions, Proc. Coastal Engineering, JSCE, Vol. 36, 1989, pp. 269-273 (in Japanaese).
Tsuchiya, Y. et al.: Longterm nearshore current observation in surf zones with ultra-sonic current meters, Annuals, DPRI, Kyoto University, No. 32B-2, 1989, pp. 847-879 (in Japanese).

CHAPTER 26

APPLICATION OF MAXIMUM ENTROPY METHOD TO THE REAL SEA DATA

Taerim Kim[1], Li-Hwa Lin[2], and Hsiang Wang[3]

Abstract

Two different versions of maximum entropy methods(MEM) were compared with two conventional methods for analyzing directional ocean wave spectra. The two MEMs were originally proposed by Lygre and Krogstad (1986) and Kobune and Hashimoto (1986), respectively, and the two conventional methods are the truncated Fourier series method(TFS) and Longuet-Higgins parametric model(LHM). The comparisons included hypothetical idealized cases and actual measured data. For the hypothetical cases, the MEM by Kobune and Hashimoto clearly performed better, particularly for dual-peaked spectra. As for the comparisons from measured data, the MEM generally yielded narrower directional spreading than the two conventional methods but all methods gave nearly identical main direction information. However, this MEM does have occasional convergence problem in real sea data analysis. The problem is removed with the aid of an approximation scheme. This modified scheme is employed in the automated directional spectral analysis of measured sea data.

Introduction

In coastal engineering applications, directional ocean wave information becomes increasingly important owing to the advancement of technology and the demand of better design information. In order to acquire more accurate information of the directional sea waves, much efforts have been devoted to the development of measuring system and the method of analyzing the data. There are several different measuring systems utilized today. For instance, a heave/pitch/roll buoy has been used in the open ocean while wave gage array or pressure transducer and bi-axial current meter are often deployed in coastal water to collect directional wave data. Since ocean waves can be treated as random signals in both time and space, the information derived from all these measuring systems are truncated partial statistical properties, such as in the

---

[1]Graduate Assistant,[2]Research Scientist,[3]Professor, Coastal and Oceanographic Engineering Department, University of Florida, Gainesville, FL 32611, USA.

form of moments or in the expression of a finite number of Fourier coefficients. Based on the limited information, different methods have been developed to estimate the true ocean wave properties which were often expressed as directional ocean wave spectra. The most direct method clearly is to express the directional spectrum by a finite Fourier series presentation known as truncated Fourier series (TFS) method. However, this method is found to often produce the unreasonable results of negative energy components in the directional domain. When this situation occurs, the estimate is evidently badly biased. For other methods which were developed to analyze directional spectrum, the Longuet-Higgins' parametric model(LHM) is presently the most popular one owing to its concise form and the guaranteed non-negative spectral values. However, the model always gives symmetrical single-peaked directional distribution for each frequency band. Hence, LHM is not suitable for waves with frequency bands containing multi-directional peaks. The maximum entropy directional spectrum estimator developed recently has a major improvement. The method is capable of showing both multiple peaks and asymmetric distribution in direction (Kim, et al., 1993). Therefore, the MEM is particularly useful for shallow water application where waves can be very asymmetric in nature.

Two different entropy definitions have been utilized in finding the corresponding maximum entropy directional spectra. One is from Lygre and Krogstad (1986) who applied MEM under the assumption of a complex Gaussian, stationary process of directional waves and the other is from Kobune and Hashimoto (1986) regarding the directional distribution of wave spectrum as a probability density function. The maximum entropy estimator derived by Lygre and Krogstad was designated here as MEM I and the one by Kobune and Hashimoto as MEM II. Benoit (1992) compared twelve different methods for estimating the directional wave spectrum based on numerical simulations and showed that MEM II gives more reliable estimates but the computational time is rather long. Brissette et al. (1992) also compared several methods and pointed out that MEM I often overpredicts the energy at the distribution peaks. Kim et al. (1993) tested TFS, LHM, MEM I, and MEM II using three different types of target spectrum and concluded that MEM II is more reliable than the other methods compared. They also suggested an approximation scheme of MEM II to avoid the occasional problem in MEM II and, hence, to significantly reduce the computational time for the practical use. Up to the present, most of the tests for the reliability of the directional spectrum estimator were done based on either artificially simulated target spectra or just one sample set of data. However, simulation test of different directional methods may also be influenced by the target spectra chosen and, therefore, does not provide sufficient evidence for the better method.

In this paper, MEM II was compared and evaluated with two classical methods, TFS and LHM, using two sets of real time series data. The comparisons include the spectral pattern and statistical parameters of dominant frequency peak direction and mean direction, which are important for many different applications in the coastal and ocean engineering.

Methods Estimating Directional Spectrum

Mathematically, a true directional spectrum $E(\sigma,\phi)$, with σ and ϕ denoting the frequency and direction, respectively, can be expressed in terms of an infinite Fourier series as

$$E(\sigma,\phi) = \frac{A_0(\sigma)}{2} + \sum_{n=1}^{\infty}[A_n(\sigma)\cos(n\phi) + B_n(\sigma)\sin(n\phi)], \quad |\phi| \le \pi,$$

where A_0, A_n and B_n are the frequency dependent Fourier coefficients, which can be determined based on the measured or simulated sea data. Although several different techniques analyzing the directional spectrum have been developed in the past, only two classical and two versions of MEM methods are discussed here. The two classical methods are the truncated Fourier series (TFS) and the Longuet-Higgins' parametric model (LHM). The two MEM methods (MEM I and II) have different entropy definitions. To be consistent, the four methods are summarized below based upon the first five Fourier coefficients, A_0, A_1, B_1, A_2, and B_2.

(1) TFS (truncated Fourier series)

The directional estimator expressed by the truncated five-term Fourier series is

$$E(\sigma,\phi) = \frac{A_0(\sigma)}{2} + \sum_{n=1}^{2}[A_n(\sigma)\cos(n\phi) + B_n(\sigma)\sin(n\phi)] \quad |\phi| \le \pi.$$

(2) LHM (Longuet-Higgins parametric model)

The parametric model proposed by Longuet-Higgins (1963) is

$$E(\sigma,\phi) = E(\sigma)\frac{2^{2s-1}}{\pi}\frac{\Gamma^2(s+1)}{\Gamma(2s+1)}\cos^{2s}\frac{\phi-\phi_o}{2},, \quad |\phi_o| \le \pi,\ s > 0,$$

where Γ denotes the Gamma function, $s = s(\sigma)$ and $\phi = \phi_o(\sigma)$ are the directional spreading parameter and the symmetric center direction, respectively. As a first order approximation, the parameters s and ϕ_o can be determined from

$$s = \frac{C_1}{1-C_1}, \quad \phi_o = \tan^{-1}\frac{B_1}{A_1}, \quad C_1 = \frac{\sqrt{A_1^2 + B_1^2}}{A_0}.$$

(3) MEM I (Maximum Entropy Approach, Method I)

By defining the entropy of directional sea as

$$M = -\int_0^{2\pi} \ln H(\sigma,\phi)d\phi.$$

and maximizing M, Lygre and Krogstad(1986) showed that

$$E(\sigma,\phi) = E(\sigma)H(\sigma,\phi), \quad H(\sigma,\phi) = \frac{1 - d_1 c_1^* - d_2 c_2^*}{2\pi|1 - d_1 e^{-i\phi} - d_2 e^{-2i\phi}|^2},$$

with

$$c_1 = \frac{A_1}{A_0} + i\frac{B_1}{A_0}, \quad c_2 = \frac{A_2}{A_0} + i\frac{B_2}{A_0}, \quad d_1 = \frac{(c_1 - c_2 c_1^*)}{(1 - |c_1|^2)}, \quad d_2 = c_2 - c_1 d_1,$$

where $H(\sigma, \phi)$ is the directional distribution function and the asterisk indicates a complex conjugate.

(4) MEM II (Maximum Entropy Approach, Method II)

By maximizing the entropy defined as

$$M = -\int_0^{2\pi} \overline{H}(\sigma, \phi) \ln H(\sigma, \phi) d\phi,$$

Kobune and Hashimoto (1986) showed that

$$H(\sigma, \phi) = \exp[-\sum_{j=0}^{4} \lambda_j(\sigma) \alpha_j(\phi)],$$

where $\alpha_0(\phi) = 1$, $\alpha_1(\phi) = \cos\phi$, $\alpha_2(\phi) = \sin\phi$, $\alpha_3(\phi) = \cos 2\phi$, $\alpha_4(\phi) = \sin 2\phi$, and λ_j's are the Lagrange's multipliers. The λ_j's are determined by iteration method solving a set of nonlinear equations:

$$\int_0^{2\pi} [\beta_i(\sigma) - \alpha_i(\phi)] \cdot \exp[-\sum_{j=1}^{4} \lambda_j(\sigma) \alpha_j(\phi)] d\phi = 0, \quad i = 1, 2, 3, 4$$

with

$$\lambda_0 = \ln\{\int_{-\pi}^{\pi} \exp[\sum_{j=1}^{4} \lambda_j(\sigma) \alpha_j(\phi)] d\phi\},$$

where $\beta_1(\sigma) = A_1/A_0$, $\beta_2(\sigma) = B_1/A_0$, $\beta_3(\sigma) = A_2/A_0$, and $\beta_4(\sigma) = B_2/A_0$. It is noted here that, based upon the first five Fourier coefficients measured, the directional spreading function $H(\sigma, \phi)$ determined from both MEM I and II can have at most two directional peaks, which can be understood by taking $\partial H/\partial \phi = 0$. Using the maximum entropy technique to estimate the directional spectrum is also attractive in that the wave spectrum computed does not have to be symmetrical in direction. This certainly indicates a better approach than the conventional LHM method of which the directional distribution is modelled by a symmetrical function. On the other hand, the TFS is known to often yield a biased estimate to the directional spectrum and the computed directional spectrum may have negative energy components.

Fig. 1 displays several comparisons of four methods shown above for the simulated directional spectra which include unimodal, bimodal, and asymmetric distributions. It is seen that TFS can result in non-positive spectral densities, LHM always gives a symmetric, single-peaked distribution, and MEM I generally produces two peaks and overestimates the peak. The overall comparisons show that MEM II generates the closest estimates to the target spectra for all cases tested.

WAVE DIRECTIONAL SPECTRUM

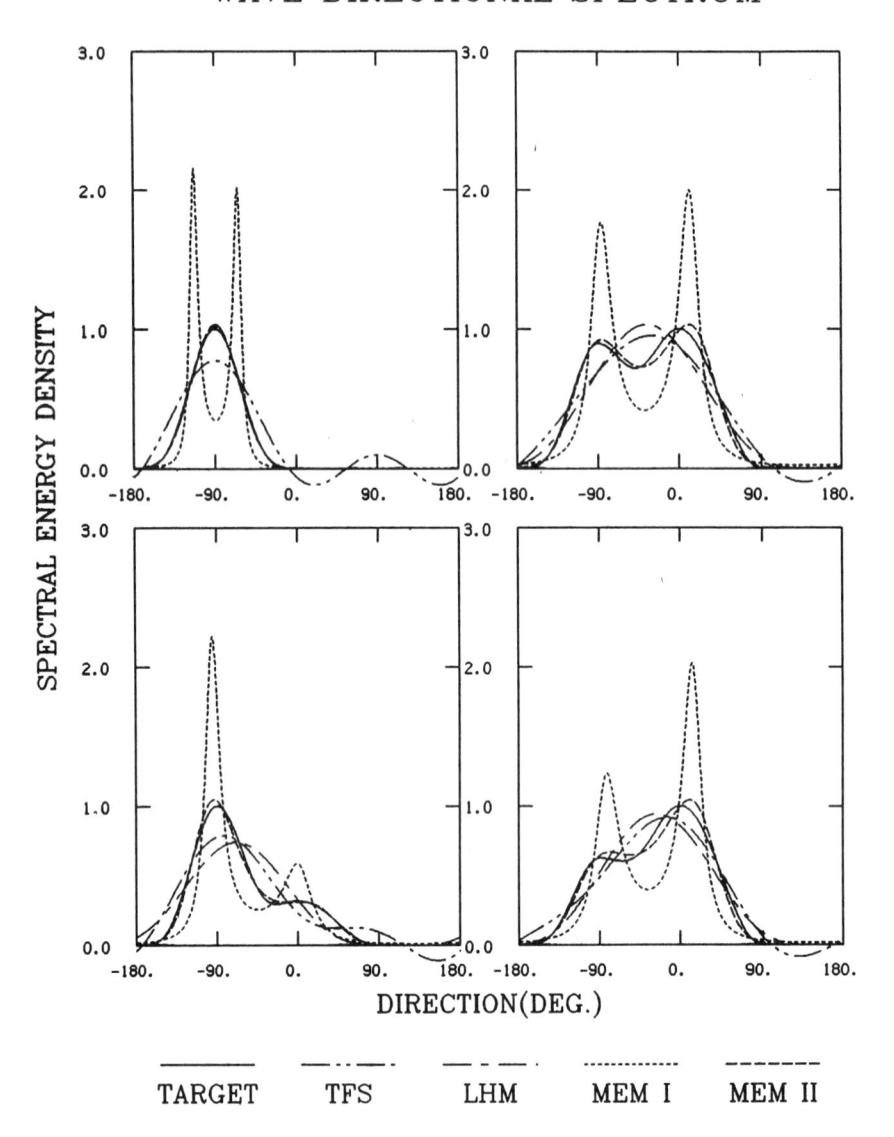

Figure 1: Comparison of simulation results with target spectra.

Application of MEM II

In general, there are no difficulties in computation of directional spectra based on TFS, LHM, and MEM I. However, when MEM II is applied, a nonconvergence problem may occur due to numerical iterations. This problem can be overcome by using an approximation scheme for solving the Lagrange's multipliers. It can be shown that by expanding the exponential term appearing in the nonlinear equations solving for λ_j's to the second order as

$$\int_0^{2\pi} [\beta_i(\sigma) - \alpha_i(\phi)] \cdot \{1 - \sum_{j=1}^{4} \lambda_j(\sigma)\alpha_j(\phi) + \frac{[\sum_{j=1}^{4} \lambda_j(\sigma)\alpha_j(\phi)]^2}{2}\} = 0,$$

an approximation of solutions of λ_i, $i = 1, 2, 3, 4$, can be obtained as

$$\lambda_1 = 2\beta_1\beta_3 + 2\beta_2\beta_4 - 2\beta_1(1 + \sum_{i=1}^{4} \beta_i^2), \quad \lambda_2 = 2\beta_1\beta_4 - 2\beta_2\beta_3 - 2\beta_2(1 + \sum_{i=1}^{4} \beta_i^2),$$

$$\lambda_3 = \beta_1^2 - \beta_2^2 - 2\beta_3(1 + \sum_{i=1}^{4} \beta_i^2), \quad \lambda_4 = 2\beta_1\beta_2 - 2\beta_4(1 + \sum_{i=1}^{4} \beta_i^2).$$

This approximation scheme is designated as MEM AP2 in the present paper. Fig. 2 shows some numerical simulations comparing the original MEM II and MEM AP2 spectra along with the target spectra. Although the MEM AP2 is not identical to MEM II, it generally yields reasonably good result to the unimodal, bimodal, and asymmetric target spectra.

Extended MEM II

As an extension of the MEM II based on the five Fourier coefficients measured, the directional distribution function may be also estimated by combining the first and any J-th directional modes as

$$H(\sigma, \phi) = \exp[-\lambda_0 - a_1 \cos(\phi - \phi_1) - a_J \cos J(\phi - \phi_J)].$$

For example, when $J = 3$,

$$H(\sigma, \phi) = \exp[-\lambda_0 - \lambda_1 \cos(\phi) - \lambda_2 \sin(\phi) - \lambda_5 \cos(3\phi) - \lambda_6 \sin(3\phi)].$$

The solution of λ_j's from the above equation can be obtained either by iteration method or from an approximation scheme as followings:

$$\lambda_1 = -\beta_1 \frac{2\sum_{i=1}^{4}\beta_i^2 + 2.5[\sum_{i=1}^{4}\beta_i^2 - (\sum_{i=1}^{4}\beta_i^2)^2]^2}{\beta_1^2 + \beta_2^2 + 2\beta_1\beta_2\beta_4 + \beta_1^2\beta_3 - \beta_2^2\beta_3}, \quad \lambda_2 = \lambda_1\beta_2/\beta_1,$$

$$\lambda_5 = \frac{\lambda_1(3\beta_2^2 - \beta_1^2) - 4(\beta_1\beta_3 - \beta_2\beta_4)}{2(\beta_1^2 + \beta_2^2)}, \quad \lambda_6 = \frac{\lambda_2(\beta_2^2 - 3\beta_1^2) - 4(\beta_3\beta_2 + \beta_1\beta_4)}{2(\beta_1^2 + \beta_2^2)}.$$

The above approximation scheme is designated here as MEM AP3. Fig. 3 shows some comparisons between the MEM II and MEM AP3 with the target spectra. It is seen that MEM AP3 can still generate good estimate to the target spectrum and

WAVE DIRECTIONAL SPECTRUM

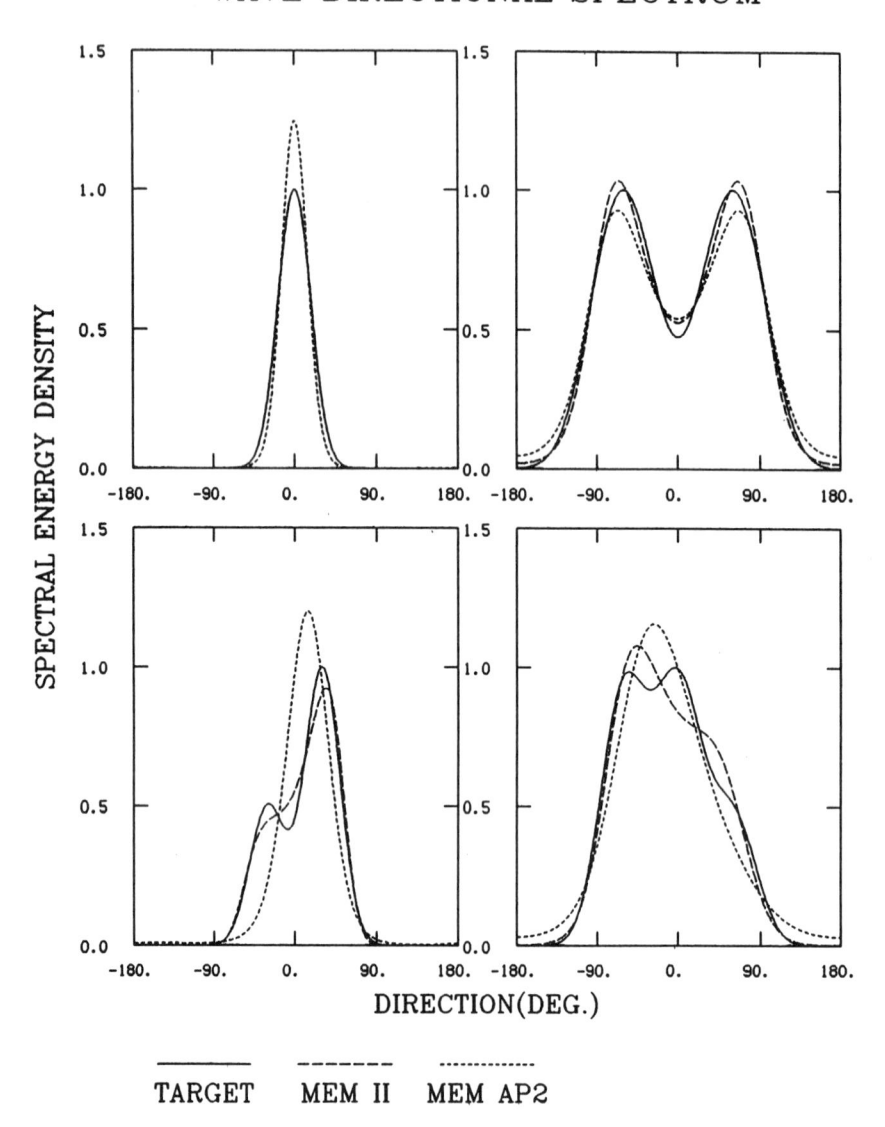

Figure 2: Comparison of the MEM II and MEM AP2 with target spectra.

WAVE DIRECTIONAL SPECTRUM

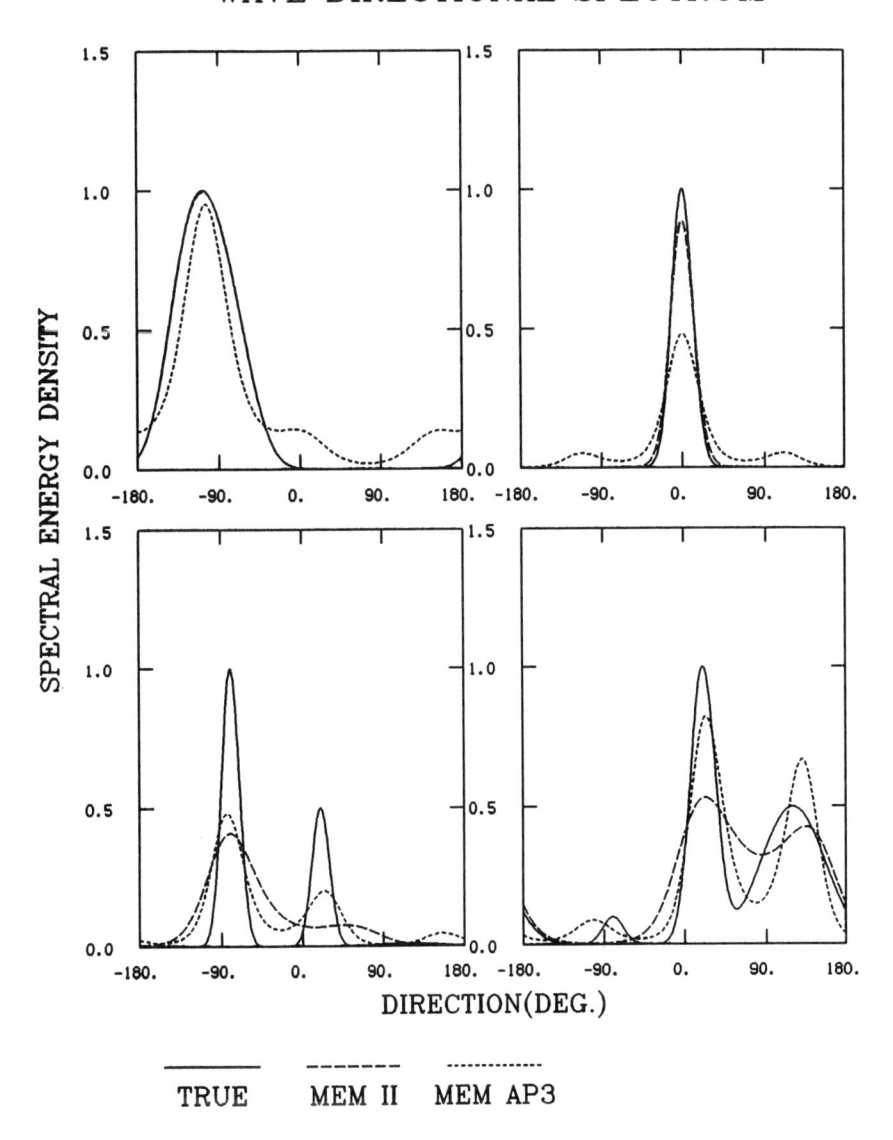

Figure 3: Comparison of the MEM II and MEM AP3 with target spectra.

sometimes show better result than MEM II depending on the target spectrum tested. However, similar to MEM II, the extended model combining the first and limited higher directional modes can result false, although small, side lobe(s) in directions.

Real Sea Data Analysis

Although MEM II shows attractive advantage in simulation test over TFS, LHM, and MEM I, it is more important to see how the method corresponds when applied to the real sea data. The effort here was to test MEM II for the measured time series sea data and compare the results with those from TFS and LHM. Two sets of real data representing two different sea states near coast were chosen for the test. One is a typical storm event of high wind and large waves. The other is for an event of combined swell and local waves due to moderate wind. Both data sets composed of two-day time series. In addition to the wave data, the wind information was also collected from the nearby coastal weather station. In most cases, the computations in MEM II converged rapidly after about five iterations. When they did not converge, the approximation scheme of MEM AP2 automatically took over the calculation.

Fig. 4 shows the computed results of directional spectra for the large wind and wave event which occurred at the Perdido Key, Florida, in the Gulf of Mexico from January 16th to 17th, 1994. The results displayed that new short waves were developed in the beginning when small wind started over the calm sea. As the wind strengthened, the waves were seen to grow steadily and, meanwhile, extend the spectral pattern toward the low frequency region. During the high wind stage, the waves appeared to have reached a state of equilibrium as the spectral pattern remained nearly stationary. As the wind gradually died out, spectra exhibited energy dissipation near the high frequency end. The spectral estimates for this case mostly have single directional peak. The estimates from TFS and MEM II occasionally yielded asymmetric distribution with two peaks, but mostly in frequency bands with little energy content. In terms of the directional dispersion, the directional spectra computed by MEM II displayed much narrower distribution than the results from TFS and LHM methods. Since the MEM II is deemed to predict better directional properties than TFS and LHM, the narrow distribution of directional spectra as estimated by MEM II shall be more representative to the real sea waves.

Fig. 5 shows the results for the event of combined swell and moderate wind waves. The measurement was taken at Cape Canaveral on the Atlantic coast from December 20th to the 21st, 1993. In this event, westbound sea swell was observed throughout the two-day time data used in analysis. At first, only small short waves in scattered directions were observed as the wind was nearly absent. Later on as a moderate southerly wind started, local waves were rapidly developed heading to the north. It was noticed that although local wind waves and existing swell have different directions, as the local wind waves grew, the directions of wind waves and swell began to merge in the middle frequency region. This result seems to suggest that interaction took place between wind waves and swell in this frequency range. Again, the general pattern of directional distribution is narrower as obtained from MEM II than the other two methods.

Ideally, if both wind waves and swell are present at sea, there may be good chance

MAXIMUM ENTROPY METHOD 349

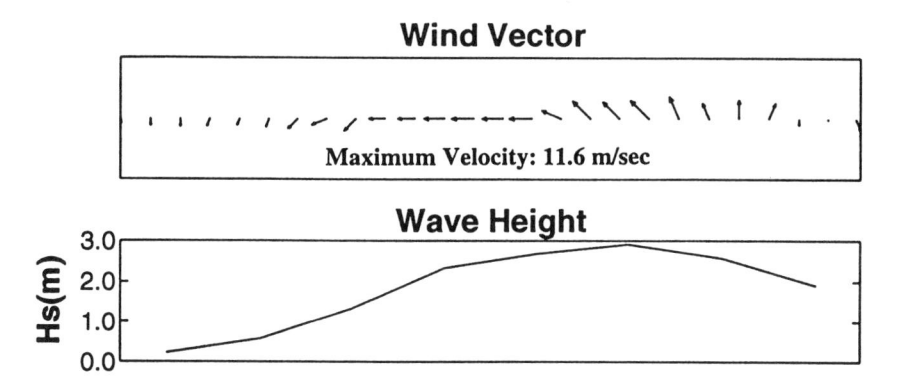

Wind Vector

Maximum Velocity: 11.6 m/sec

Wave Height

Directional Spectrum

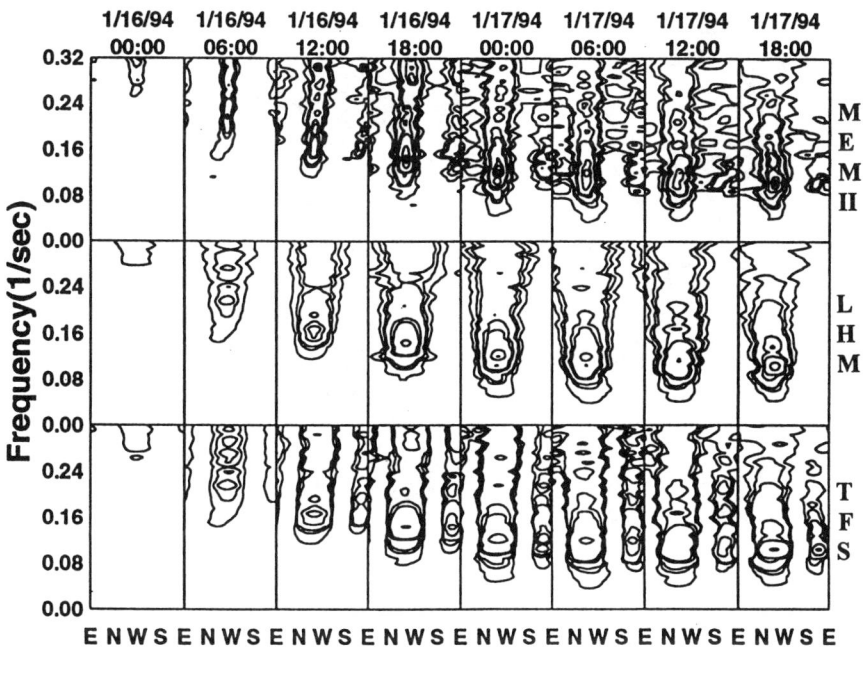

Figure 4: Comparison of MEM II, TFS, and LHM with real time series sea data measured at Perdido Key, Florida (contours in log scale).

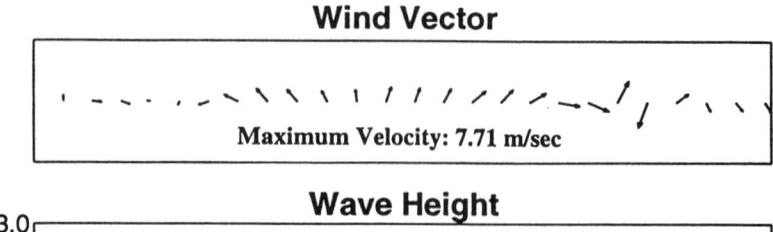

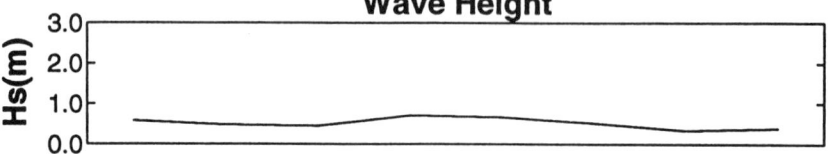

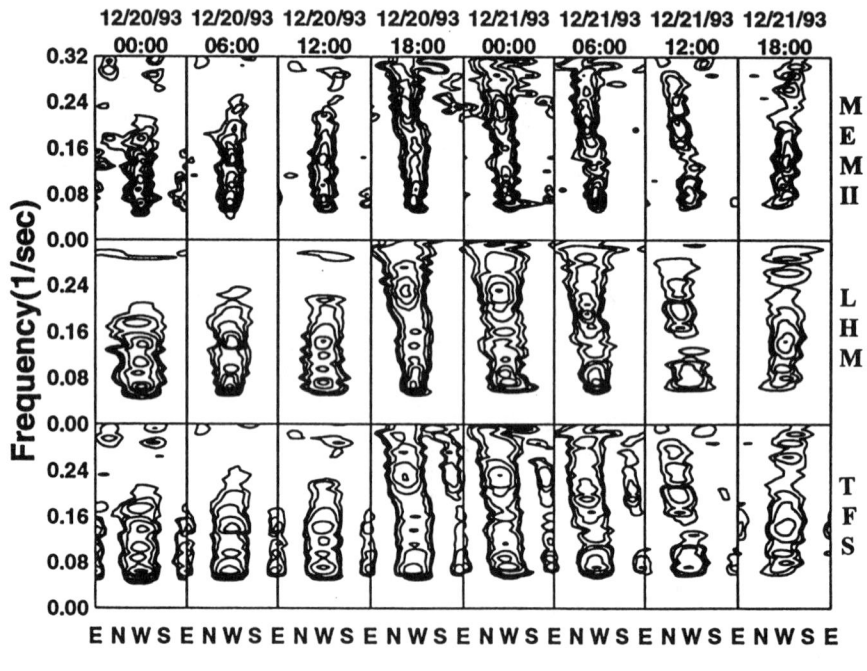

Figure 5: Comparison of MEM II, TFS, and LHM with real time series sea data measured at Cape Canaveral, Florida (contours in log scale).

to have two peaks in directional distribution at the same frequency. However, based on the results of real sea data from MEM II, two directional peaks at the same frequency were rare events, excluding the cases with false secondary peak which usually occurred in the opposite direction of the main peak direction. One possible explanation is the interaction between wind wave and swell components in the same frequency band. However, outside the main energy content range where spectral energy is relatively low, a few cases with two directional peaks were found. A few examples with the presence of directional spectra with dual peaks or asymmetric distribution from the combined wind sea and swell event are shown in Fig. 6.

In order to compare the different methods more specifically for real sea data analysis, three statistical parameters, namely, the peak direction, the mean direction, and the standard deviation at the dominant frequency, which corresponds to the largest spectral energy in frequency domain, were computed. Figs. 7 and 8 show the three parameters computed for the two sets of real sea data tested earlier. For the peak direction at the dominant frequency, all the LHM, TFS, and MEM II results are almost identical. For the mean direction at the dominant frequency, the LHM gives the same direction as the peak direction but both the TFS and MEM II show different directions from the peak directions because of the asymmetry of the directional distribution. In terms of standard deviation, the MEM II exhibits much narrower distribution than both LHM and TFS. In other words, wave energy is more concentrated around the main peak direction as resulted by MEM II.

Conclusions

Four different methods analyzing directional wave spectrum were compared using numerical simulation and actual measured sea data. The four methods include the Truncated Fourier series(TFS), the Longuet-Higgins parametric model(LHM), two different maximum entropy methods (MEM I and II) that utilize different definitions of entropy. The numerical simulation consisted of a variety of target spectra with different properties. And the test results showed that the maximum entropy method with the entropy defined as a statistical probability density function, named here as MEM II, is clearly performed better than the other methods in estimating the target spectra. From the real sea data analysis, it is also concluded that MEM II is most suitable as the method could differentiate dual peaks in the same frequency component and detect the evolution of directional interactions of each frequency component.

The specific findings from the study were summarized below:

(1) For the four different methods compared as candidates for analyzing the measured directional waves, the LHM is restricted to a symmetrical single peak distribution, the TFS has the disadvantage producing negative energy component, the MEM I often overestimates the peak, and the MEM II may have a convergence problem.

(2) For applications to both simulated and real sea data, the MEM II is considered superior to the other methods compared in the paper. The convergence problem of the MEM II in numerical iterations can be overcome by using an approximation scheme.

Wave Directional Spectrum

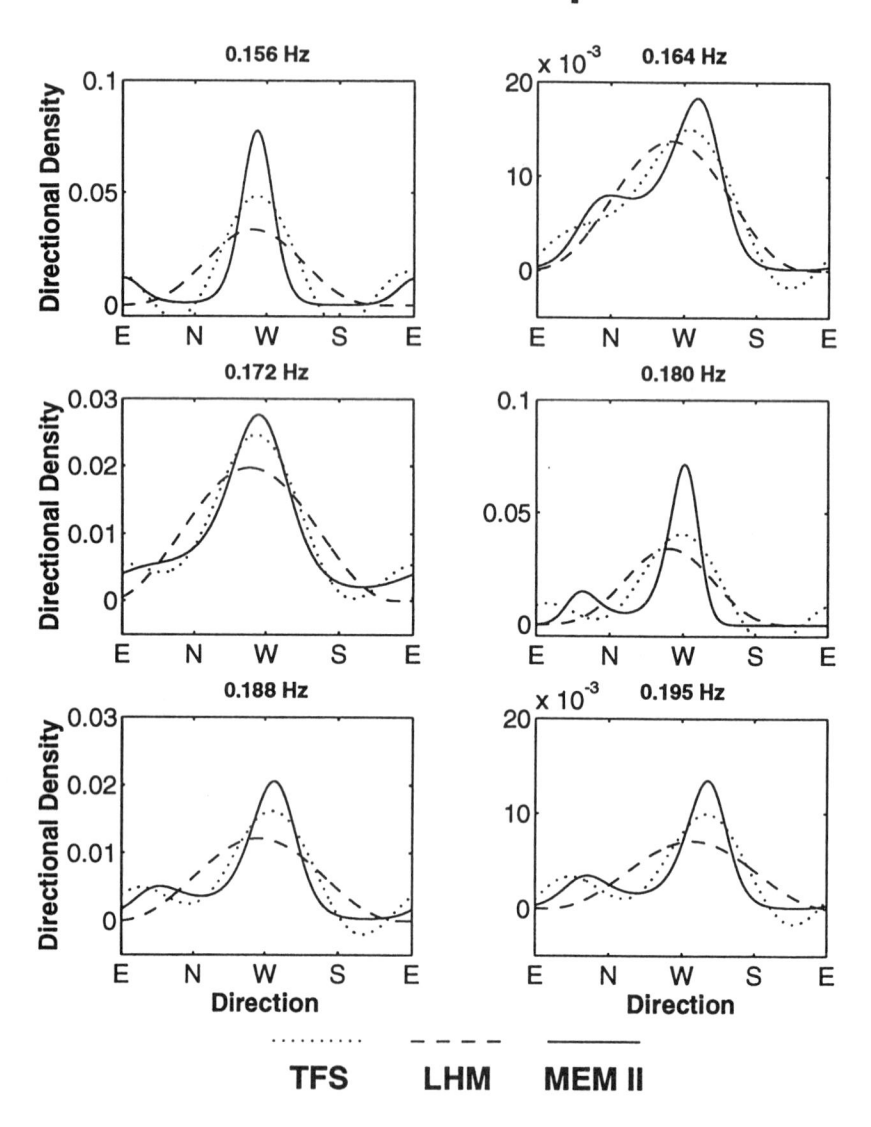

Figure 6: Examples of bimodal and asymmetric directional spectra computed by MEM II along with those from LHM and TFS, based on measured real sea data.

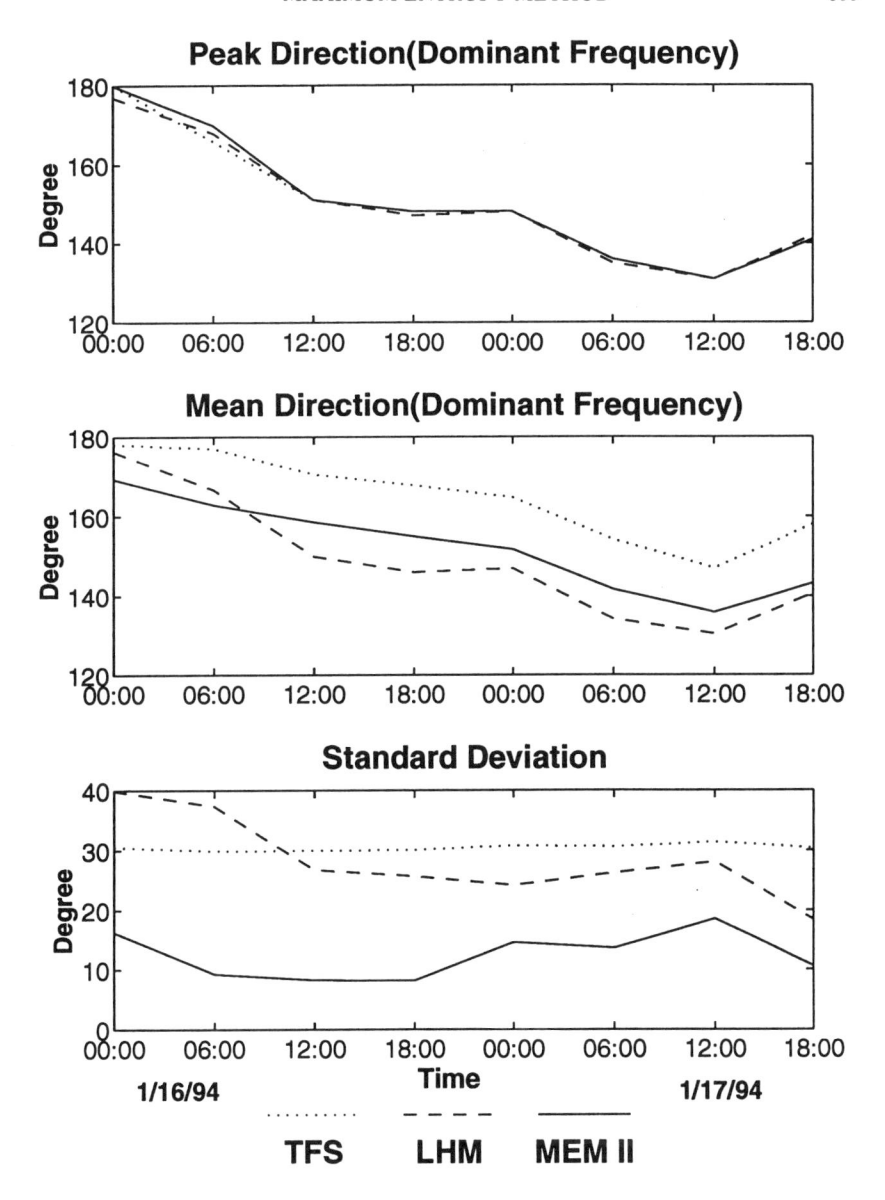

Figure 7: Comparison of dominant frequency peak and mean directions with standard deviations computed based on Perdido Key data.

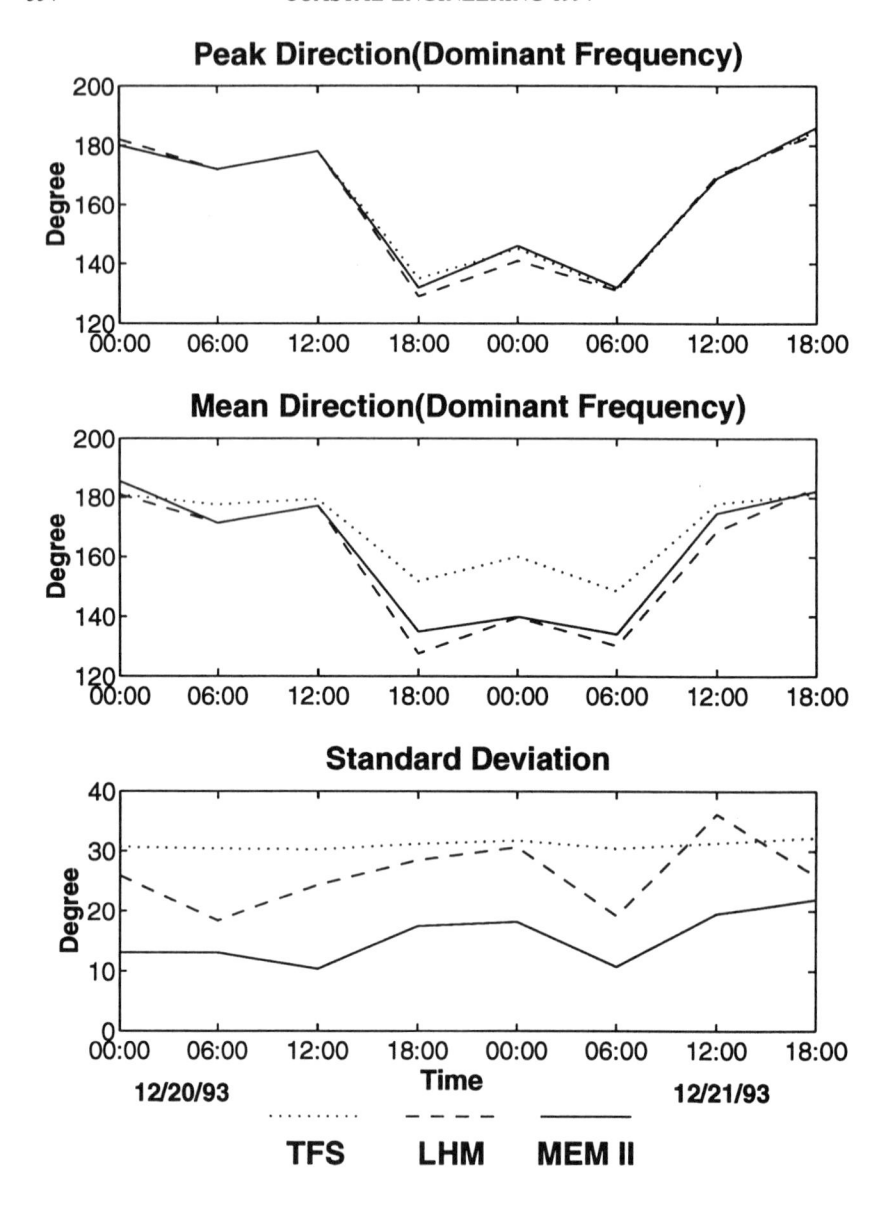

Figure 8: Comparison of dominant frequency peak and mean directions with standard deviations computed based on Cape Canaveral data.

(3) Applying TFS, LHM, and MEM II to the real sea time series data shows similar patterns of directional spectrum. The TFS, LHM, and MEM II are all seen to result almost identical peak direction for the dominant frequency component. However, they all yield different mean directions for dominant frequency component. This is because LHM produces symmetrical directional distribution whereas TFS and MEM II give asymmetrical distribution of directional spectrum. The MEM II, in general, produces narrower directional distribution than the other two methods.

(4) Even with the presence of both local wind waves and swell, it is generally rare to have two distinguished directional peaks at the same frequency. It appears that in the mid frequency range that contains most of the wave energy, the directional components from wind waves and swell tend to merge.

(5) With the aid of an approximation scheme, the MEM II can be programmed for practical applications such as automated directional spectrum analysis from real time data with significant reduction of computational time.

References

Benoit, M., 1992, "Practical Comparative Performance Survey of Methods Used for Estimating Directional Wave Spectra from Heave-Pitch-Role Data.",Proc. 23rd ICCE, ASCE, pp.62-75.

Brissette, F.P. and Tsanis, I.K.,1992, "Estimation of Wave Directional Spectra from Pitch-Roll Buoy Data.", J. WPCOE, ASCE, Vol. 120, No. 1., pp 93-115.

Krogstad, H.E., 1989, "Reliability and Resolution of Directional Wave Spectra from Heave, Pitch, and Roll Data Buoys", Directional Ocean Wave Spectra, The Johns Hopkins University Press, Baltimore and London, pp.66-71.

Longuet-Higgins, M. S., Cartwright, D. E., and Smith, N. D., 1963, "Observations of the Directional Spectrum of Sea Waves Using The Motion of a Floating Buoy", in Ocean Wave Spectra, Prentice Hall, Englewood Cliffs, N. J., pp.111-136.

Kim, T., Lin, L. W., and Wang, H., 1993, "Comparisons of Directional Wave Analysis Methods.", Ocean Wave Measurement and Analysis, Proc. of the Second International Symposium, New Orleans, pp.554-568.

Kobune,K., and Hashimoto, N., 1986, "Estimation of Directional Spectra from the Maximum Entropy principle", in Proc. 5th International Offshore Mechanics and Arctic Engineering (OMAE) Symposium, Tokyo, pp.80-85.

Lygre, A., and Krogstad, H. E., 1986, "Maximum Entropy Estimation of the Directional Distribution in Ocean Wave Spectra", J. Phys. Oceanogr. 16, pp.2052-2060.

CHAPTER 27

Probability of the freak wave appearance in a 3-dimensional sea condition

Akira Kimura[1] and Takao Ohta[2]

Abstract

In this study, the appearance probability of the freak waves is theoretically introduced applying the definitions by Klinting and Sand (1987). Their three conditions are formulated theoretically applying the probability distributions for the run of wave heights (Kimura,1980) and the distance of a mean point of the zero-crossing wave crest and trough from mean water level (Kimura and Ohta, 1992b). Its appearance probability in a uni-directional irregular wave condition is studied first, the theory is extended then to the 3-dimensional wave condition, and the definition is discussed in terms of the probability in the last.

1. Introduction

The term "freak waves" may be used to express a huge wave in height. Freak waves have been seen and reported in many places in the world. Many fishing boats, even a man of wars have been destroyed by exceptionally huge waves. And the possibilities have been pointed out that the recent disasters on break waters at port of Sines (Portugal), Bilbao (Spain), (per Bruun, 1985) are also due to the freak waves. Although a common recognitions "what is the freak wave" may not have been established yet, it may have following properties as described by per Bruun (1985).
It is a single "mammoth" short crested wave with, apparently, little relation to its neighboring waves. It has a high crest but not necessarily a corresponding pronounced trough. It does not stay long but break down in small waves.
In 1987, clear definition was made by Klinting and Sand as,

----
[1] Professor and [2] Research associate of the Faculty of Eng., Tottori Univ., Koyama Minami 4-101, Tottori, 680, Japan

(1) it has a wave height higher than twice the significant wave height,
(2) its wave height is larger than 2 times of the fore-going and the following wave heights,
(3) its wave crest height is larger than 65% of its wave height.

Recently, only the first condition may be used for the definition (Sand, 1990). However very large appearance probability is given if only the first condition is used. Furthermore the idea that there is a big jump in wave heights, is not realized in the recent definition. The present study applies three all conditions by Klinting and Sand theoretically in the cases of uni-directional and directional random sea conditions. The importance of the conditions is compared and examined through the individual probability.

2. Definition for freak wave

Three conditions given by Klinting and Sand (1987) are as follows.

If we have a following time series of wave height,

$$\ldots\ldots, H_{j-1}, H_j, H_{j+1}, \ldots\ldots$$

and if H_j is the freak wave (Fig.1), the definition by them is expressed as,

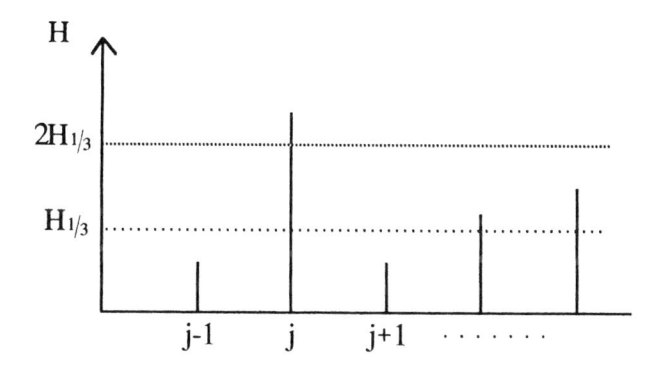

(1) $H_j > H_{1/3}$, (Condition 1)
(2) $H_j > 2H_{j-1}$, (Condition 2A)
 and $H_j > 2H_{j+1}$, (Condition 2B)
(3) $\eta_j > 0.65 H_j$, (Condition 3)

Fig.1 Time series of wave heights. (H_j is the freak wave)

in which $H_{1/3}$ is a significant wave height and η_j is a crest height of H_j.

3. Probability distribution of wave heights

The wave height distribution for the zero-crossing deep water irregular waves agrees well with the Rayleigh distribution. However the agreement of data and distribution in a very large part of wave height has not been sufficiently investigated yet. Goda (1985) reported the measured data show slightly larger probability of appearance than the Rayleigh distribution in a large part of the distribution. Kimura (1981), Mase (1986) reported that increasing non-linearity on wave profiles brings narrower wave height distribution. However Yasuda (1992) showed the non-linearity brings about no significant difference on wave height distribution when the waves are those of fully saturated in a deep sea condition.

If a physical mechanism in which the freak waves are brought about differs from other wind waves, there is a possibility that the freak waves do not follow the statistical law for irregular wave heights. Per Bruun (1985) pointed several phenomena such as orthogonal crossing of waves, overtaking of waves. However numerical simulations for the waves from two separate wind wave sources showed no significant difference in the wave height distribution from the Rayleigh distributions. Therefore we apply the Rayleigh distribution for the wave height distribution in this study. Further assumption used in this study was that waves are those of fully saturated in deep water condition.

4. Formulation of the condition 2A and 2B by Klinting and Sand

If the time series of wave height,

$$\ldots\ldots, H_{j-1}, H_j, H_{j+1}, \ldots\ldots$$

forms a Markov chain and its transition probability is given by the normalized 2-dim. Rayleigh distribution (Kimura, 1980), the probability for the condition 2A is given as follows.

The probability of the first "jump" from H_{j-1} to H_j ($H_j > 2H_{j-1}$) is given as

$$p_{12}(H_j)\,dH = \frac{\int_0^{H_j/2} dH_1 \int_{H_j}^{H_j+dH} p(H_1,H_2) dH_2}{\int_{H_j}^{H_j+dH} p(H_1)\,dH_1}, \qquad (1)$$

in which $p(H_1,H_2)$ is the 2-dim. Rayleigh distribution and

$p(H_1)$ is the Rayleigh distribution which are given by

$$p(H_1,H_2) = \frac{\pi^2}{4(1-\kappa^2)} H_1 H_2 \exp\left(-\frac{\pi}{4(1-\kappa^2)}(H_1^2 + H_2^2)\right) I_0\left(\frac{\pi \kappa H_1 H_2}{2(1-\kappa^2)}\right), \quad (2)$$

and

$$p(H_1) = \frac{\pi}{2} H_1 \exp\left(-\frac{\pi}{4} H_1^2\right). \quad (3)$$

H is a normalized wave height by the mean wave height. Correlation parameter (κ) between H_1 and H_2 is correlated to the wave spectrum (Battjes and van Vledder, 1984) as,

$$\kappa = (\rho^2 + \lambda^2)^{1/2} / m_0, \quad (4)$$

where

$$\rho = \int_{f_d}^{f_u} S(f) \cos\left(2\pi (f - f_m) T_m\right) df,$$

$$\lambda = \int_{f_d}^{f_u} S(f) \sin\left(2\pi (f - f_m) T_m\right) df,$$

$$f_m = m_1 / m_0,$$

$$T_m = 1 / f_m,$$

$$m_n = \int_{f_d}^{f_u} f^n S(f) df,$$

$$f_d = (-0.186 / r + 0.735) f_p \quad : (4 \leq r \leq 20)$$
$$f_u = (1.61 / r + 1.62) f_p \quad : (4 \leq r \leq 20) \quad (4)'$$

in which S(f) is a power spectrum, f_p is its peak frequency, r is a shape factor of the spectrum.

$$S(f) = (f/f_p)^{-r} \exp\left[\left(\frac{r}{4}\right)\left\{1 - (f/f_p)^{-4}\right\}\right]. \quad (5)$$

Narrow integration range from f_d to f_u instead of 0 and ∞ respectively, in the calculations for ρ, λ and m_n is used to improve the value of correlation parameter (Kimura and Ohta, 1992a).

The probability of the second jump from H_j to H_{j+1} ($H_j > 2H_{j+1}$) is also given by

$$p_{21}(H_j)\,dH = \frac{\int_0^{H_j/2} dH_2 \int_{H_j}^{H_j+dH} p(H_1,H_2)\,dH_1}{\int_{H_j}^{H_j+dH} p(H_1)\,dH_1} \tag{6}$$

in which $P(H_1,H_2)$ and $p(H_1)$ are given by eqs.(2), (3) respectively. Combining eqs.(1) and (6), the condition 2A and 2B is given as

$$p_{f1}(H)\,dH = p_{12}(H)\,dH\,p_{21}(H). \tag{7}$$

5. Formulation of the condition 3

Two waves in Fig.2 have the same zero-down-cross wave height and period but different crest heights. To clarify the difference between these two waves, Kimura and Ohta (1992b) introduced the new parameter : the mean point between wave crest and trough (Fig.2). Applying this parameter, the condition 3 is formulated as,

$$d/H_* > 0.15, \tag{8}$$

in which d is a distance from the mean water level to the mean point between wave crest and trough, H_* is a wave height.

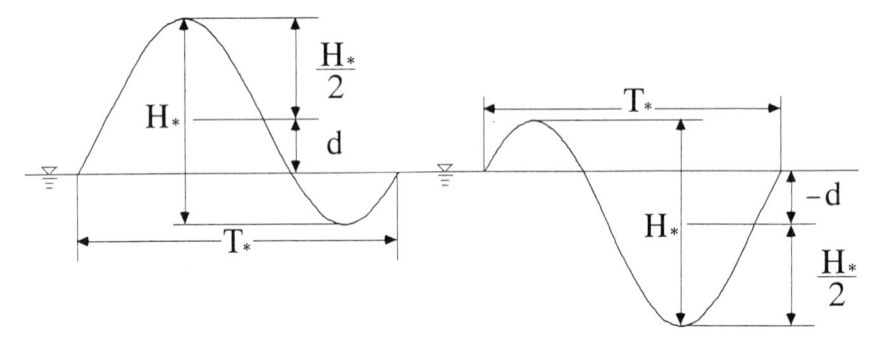

Fig.2 Zero-crossing waves with the same wave height and period but different crest heights.

Probability of eq.(8) is theoretically given as follows. Putting $\varepsilon = d/H_*$, combined distribution of ε and H is given as (Kimura and Ohta, 1992b),

$$p(\varepsilon,H) = \frac{\pi^2 H^3(1-4\varepsilon^2)}{1-\kappa_2^2}\exp\left(-\frac{\pi H^2(1-4\varepsilon^2)}{2(1-\kappa_2^2)}\right)I_0\left(-\frac{\pi\kappa_2 H^2(1-4\varepsilon^2)}{2(1-\kappa_2^2)}\right) \tag{9}$$

in which H is the normalized wave height with its mean ($H=H_*/H_m$). The condition 3 (eq.8) is also expressed as

$$p_{f2}(H) = \int_{0.15}^{0.5} p(\varepsilon|H)\, d\varepsilon , \qquad (10)$$

in which $p(\varepsilon|H)$ is determined as

$$p(\varepsilon|H) = p(\varepsilon,H)/p^*(H) , \qquad (11)$$

where

$$p^*(H) = \int_0^{2H} p(A_1,H)\, dA_1 , \qquad (12)$$

and

$$p(A_1,H) = \frac{\pi^2 A_1(2H-A_1)}{2(1-\kappa_2^2)} \exp\left(-\frac{\pi\{A_1^2+(2H-A_1)^2\}}{4(1-\kappa_2^2)}\right) I_0\left(\frac{\pi\kappa_2 A_1(2H-A_1)}{2(1-\kappa_2^2)}\right) . \qquad (13)$$

In eqs.(9) and (13), κ_2 is the correlation parameter. This value is calculated using the values 0, ∞ and $T_m/2$ instead of f_d, f_u and T_m respectively in eqs.(4), (4)'.

Solid line in Fig.3 shows the calculated p_{f2} (for a fully saturated sea condition, for example, r=5).

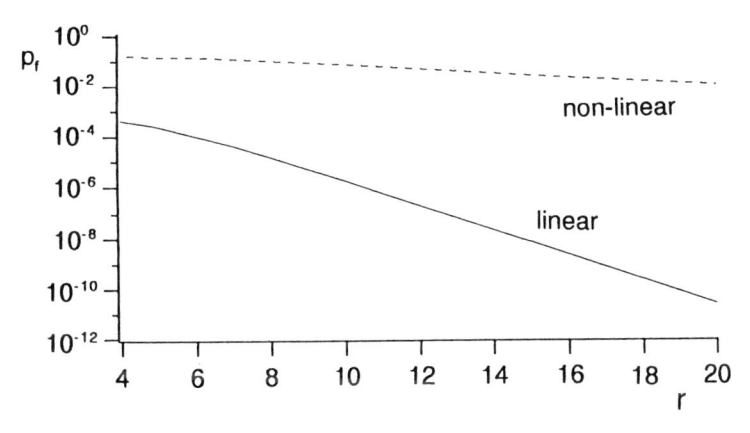

Fig.3 p_{f2} with non-linearity (dotted line) without non-linearity (solid line)

If we take the non-linearity of wave profile into account, p_{f2} increases considerably as follows.
In a deep sea condition, the 3rd order wave profile is

given as,

$$\eta = a \cos(2\pi\theta) + \frac{\pi a^2}{L}\cos(4\pi\theta) + \frac{3\pi^2 a^3}{2L^2}\cos(6\pi\theta),\quad (14)$$

in which θ is a phase, L is a wave length and the relation between wave height H_* and a is given by

$$H_* = 2a + 3\frac{\pi^2}{L^2}a^3.\quad (15)$$

Wave steepness of a significant wave for fully saturated wind waves is about 0.04 ~ 0.05 (Goda, 1975). Since $H_f > 2H_{1/3}$ (H_f : freak wave height), wave steepness of the freak wave may be larger than 0.1, and d/H_* may be approximately,

$$\pi a^2 / L H_* = \pi/4 \cdot H/L.\quad (16)$$

Putting H/L in eq.(16) equals to 0.1, we obtain ε to be about 0.08. Therefor taking the non-linearity into account, the condition 3 may be able to change as,

$$p_{f2}(H) = \int_{0.15-0.08}^{0.5} p(\varepsilon|H)\,d\varepsilon.\quad (17)$$

6. Formulation of the condition 1

Assuming the conditions 2A and 2B and the condition 3 to be independent, the condition 1 together with 2A, 2B and 3 is formulated as,

$$p_f = \int_{2H_{1/3}}^{\infty} p_{f1}(H)\,p_{f2}(H)\,dH\quad (18)$$

7. Result of the calculation

Result of the calculations are listed in Table-1. If we apply eq.(10) for the condition 3, p_f is about 0.155×10^{-4} when r=5 in eq.(5). Narrower spectrum brings far smaller value for p_f. The effect of non-linearity on the condition 3 is compared in Fig.3. Broken line show p_f from eqs.(17) instead of eq.(10). p_f with non-linearity gives a far larger value. ε does not distribute widely when wave height is very large (Kimura and Ohta, 1992b) and large waves have non-linearity on their profiles, the condition 3 may not be important.

Furthermore, the second jump ($H_j > H_{j+1}$) in the condition 2 may not be an important also. To realize only a hazardous property of the freak wave, consecutive wave height after the freak wave may not be important.

Table-1 Appearance probability of the freak waves

Conditions considered	uni-directional	directional
1	0.321×10^{-3} (1/3,100)	0.477×10^{-3} (1/2,100)
1, 2A, 2B, 3	0.155×10^{-4} (1/65,000)	-
1, 2A, 2B	0.106×10^{-3} (1/9,400)	-
1, 2A	0.198×10^{-3} (1/5,000)	0.278×10^{-3} (1/3,600)

The calculated result is listed in Table-1 when the conditions 2B and 3 are neglected.

8. 3-dimensional sea condition

If we use a single wave gauge in the measurements, the wave gauge can not always record the local maximum in a directional sea condition as shown in Fig.4. If wave gauges can selectively record wave profiles at the local maxima of short crested waves, we may have larger appearance probability of large waves. In this section, the change in the appearance probability of the freak wave is introduced when the maximum

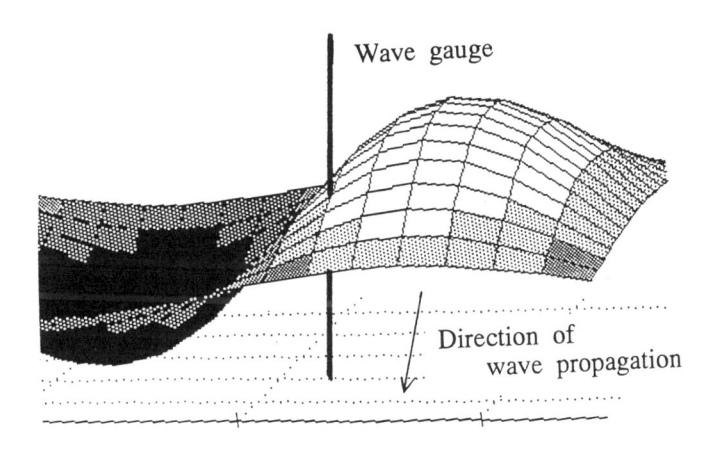

Fig.4 Short crested wave and wave gauge

wave height within a certain distance from a fixed point is applied.

We place a plane Q which is vertical to the horizontal still water plane (x'- y') and is perpendicular to the dominant wave direction. x' is taken in the dominant direction of waves and y' is taken on Q. Figure 5 shows schematically a wave envelope for the cross section of short crested wave profile on Q at a certain instance. A wave gauge is placed at point A and this point is taken as an origin (x'=0,y'=0). Using the Taylor series expansion, the envelope R(y') is expanded around A as,

$$R(y') = R_A + R'_A(y') + R''_A(y')^2/2 + \cdots . \tag{19}$$

R_A' and $R_A"$ are the first and second derivatives of R at point A. Since we only discuss the wave height in the vicinity around A, we apply three terms in eq.(19) : R is approximated with a quadratic function around A. The value of R at the local maximum (R_B) is given as

$$R_B = \left\{ 2R_A R''_A - R'^2_A \right\} / 2R''_A . \tag{20}$$

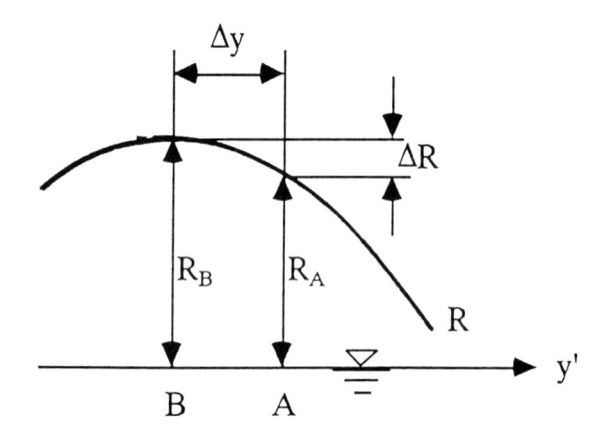

Fig.5 Wave envelope on Q

A distance from A to B is given as

$$\Delta y = \left| R'_A / R''_A \right| . \tag{21}$$

The probability that the value $R_B - R_A$ exist within $\Delta R \sim \Delta R + dR$ is given by

$$P_F = \int_S P(R', R''; R_A) \, dR' \, dR'' \tag{22}$$

in which $P(R_A', R_A''; R_A)$ is the conditional probability distribution of R_A' and R_A'' for the given value of R_A. S is the region of integration. Figure 6 shows the region S schematically. Solid lines shows the relations,

$$\Delta R = -R_A'^2/2R_A'' , \qquad (23)$$

and

$$\Delta R = -R_A'^2/2R_A'' + dR , \qquad (24)$$

respectively where $\Delta R = R_B - R_A$. Dotted line shows the relation,

$$\Delta y = |RA'/RA''| = const. \qquad (25)$$

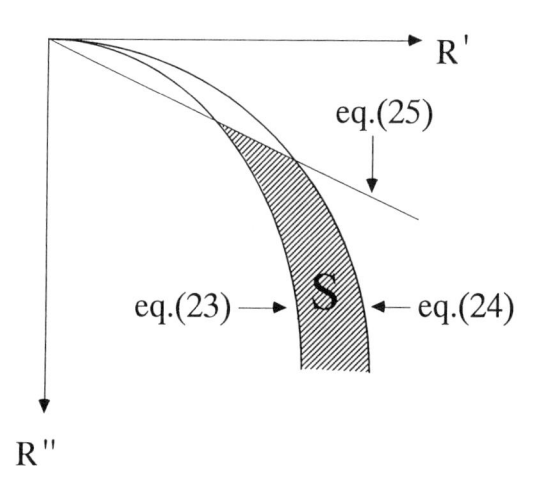

Fig.6 Region of integration S

If the shadowed part is taken as a region S, eq.(22) gives a probability of $R_B - R_A = \Delta R \sim \Delta R + dR$ within a distance Δy on Q from A.

$P(R_A', R_A''; R_A)$ is introduced as follows. The combined distribution of R_A, R_A', R_A'' : $P(R_A', R_A'', R_A)$, is derived theoretically by Rice(1945) as

$$p(R, R', R'') = 2\alpha \int_0^\infty \exp\left(-\beta\phi'^4 - \gamma\phi'^2\right) d\phi' , \qquad (26)$$

in which

$$\alpha = \frac{R^2}{(2\pi)^{3/2}\sqrt{B_4}} \exp\left\{-\frac{1}{2B^2}(B_0 R^2 - 2B_2 RR'' + B_{22} R'^2 + B_4 R''^2)\right\}, \quad (27)$$

$$\beta = B_4 R^2 / 2B^2,$$

$$\gamma = (B_{22} R^2 - 2B_4 RR'' + 2B_2 R^2) / (2B^2),$$

where

$$B = b_0 b_2 b_4 + 2b_1 b_2 b_3 - b_2^3 - b_0 b_3^2 - b_4 b_1^2 \quad (28)$$

$$B_0 = (b_2 b_4 - b_3^2) B, \qquad B_{22} = (b_0 b_4 - b_2^2) B,$$

$$B_1 = -(b_1 b_4 - b_2 b_3) B, \qquad B_2 = (b_1 b_3 - b_2^2) B,$$

$$B_3 = -(b_0 b_3 - b_1 b_2) B, \qquad B_4 = (b_0 b_2 - b_1^2) B.$$

b_i (i=0,1,2,3,4) is given as follows,

$$b_0 = \overline{(l_{c1}^2)} = \overline{(l_{s1}^2)}, \qquad b_1 = \overline{(l_{c1} l_{s2})} = \overline{(l_{c2} l_{s1})}, \quad (29)$$

$$b_2 = \overline{(l_{c2}^2)} = \overline{(l_{s2}^2)}, \qquad b_3 = \overline{(l_{s2} l_{c3})} = \overline{(l_{c2} l_{s3})},$$

$$b_4 = \overline{(l_{c3}^2)} = \overline{(l_{s3}^2)},$$

where

$$l_{c1} = \sum_{n=1}^{\infty} C_n \cos\left(u_n' x' - u_m' x' + \varepsilon_n\right)$$

$$l_{s1} = \sum_{n=1}^{\infty} C_n \sin\left(u_n' x' - u_m' x' + \varepsilon_n\right) \quad (30)$$

and

$$l_{c2} = (l_{c1})', \quad l_{s2} = (l_{s1})', \quad l_{c3} = (l_{c1})'', \quad l_{s3} = (l_{s1})''. \quad (31)$$

()' and ()" are the first and the second derivatives, C_n is calculated by the relation (Longuet-Higgins,1957),

$$\sum_{u,v}^{u+du, v+dv} \frac{C_n^2}{2} = E(u,v)\, du\, dv, \quad (32)$$

where E(u, v) is the directional wave spectrum in which u and v are the wave number of component wave in x' and y' directions respectively. Σ means to take the total of $C_n^2/2$ in the region u ~ u+du and v ~ v+dv. The conditional distribution $P(R_A', R_A''; R_A)$ is given as,

$$P(R', R''; R_A) = P(R', R'', R) / p(R)\Big|_{R=R_A}, \quad (33)$$

in which p(R) is the Rayleigh distribution.

Rice (1945) gave a theoretical expression for eq.(26), the integration was made numerically, for the simplicity.

In the calculation of E(u, v), Bretschneider-Mitsuyasu spectrum with the significant wave of $H_{1/3}$ =5.5m, $T_{1/3}$ =10s is used. For the directional function, Mitsuyasu type directional function (Goda,1985) with S_{max} =10 is used. When the sea is in a fully saturated condition, above value for S_{max} is recommended (Goda,1985). The power spectrum and the directional function is multiplied and transformed into E(u, v), applying the dispersion relation of component waves. Above power spectrum brings same statistical properties of freak waves when r=5 in eq.5 and directionality is ignored.

Figure 7 shows the probability that R_B exceed 2 times of $R_{1/3}$ (1/3 highest amplitude) in terms of $R_A/(2R_{1/3})$. Considerable probability exists in the region $R_A/(2R_{1/3})$ > 0.85.

If we take the waves of $R_A < 2R_{1/3}$ but $R_B > 2R_{1/3}$ into account as freak waves, appearance probability of the freak wave can be calculated as follows.

When only the condition 1 is used for the freak wave definition,

$$p = \int_0^\infty p_F p(H) \, dH , \qquad (34)$$

in which

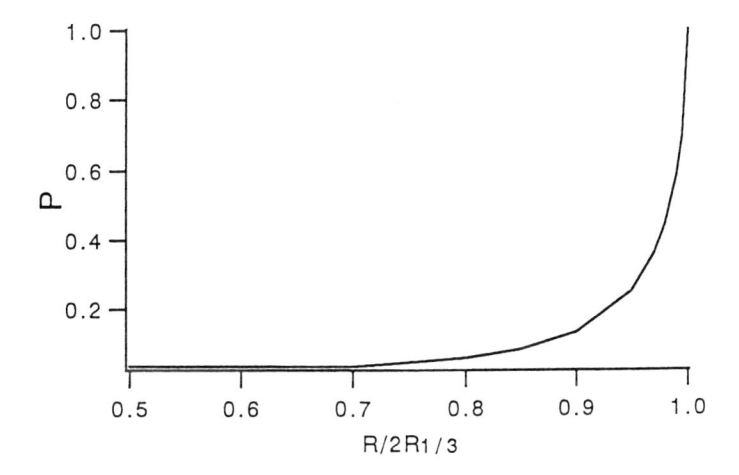

Fig.7 Appearance probability of the freak wave within the vicinity 0.1 $L_{1/3}$.

$$p_F = \begin{cases} p_F & : 0 < R < 2R_{1/3} \\ 1 & : R > 2R_{1/3} \end{cases} \qquad (35)$$

p_F is given by eq.(22).
If the conditions 2B and the 3 are neglected, the appearance probability is given by,

$$p = \int_0^\infty p_F p_{12}(H)\, dH \qquad (36)$$

Equation (35) is also applied in eq.(36) for p_F.
The results from eqs.(34) and (36) are also listed in Table-1. If we take the directional property of waves into account, the appearance probability increases about 45% when the sea condition is fully saturated and the condition 1 and 2A are applied in the freak wave definition.

9. Conclusion

If only the conditions 1 and 2A given by Klinting and Sand are applied, the appearance probability of the freak wave is about 2.78×10^{-4}. This means a freak wave appears once every 3,600 waves on the average. Considering the extremely disastrous properties of this wave, this is slightly too frequent. We may have to use a higher wave height for the freak wave. We also have to continue looking for the possibility that freak wave may not follow the ordinary statistical law but comes from other physical mechanism.

References

Battjes, J. A. and G. Ph. van Vledder (1984) : Verification of Kimura's theory for wave group statistics, Proc. 19th ICCE, pp.642-648.
Goda, Y. (1985): Random seas and design of maritime structures, Univ. of Tokyo Press, 323p.
Kimura, A. (1980) : Statistical properties of random wave groups, Proc. 17th ICCE, pp.2974-2991.
Kimura, A. (1981) : Joint distribution of the wave heights and periods of random sea waves, Coastal Eng. in Japan, Vol.24 pp.77-92.
Kimura, A. and T. Ohta (1992a) : On probability distributions of the zero-crossing irregular wave height, Proc. 6th Intl. Symp. on Stochastic Hydraulics, pp.291-298.
Kimura, A. and T. Ohta (1992b) : An additional parameter for the zero crossing wave definition and its probability distribution, Proc. 23rd ICCE, pp.378-390.
Klinting, P. and S. Sand (1987) : Analysis of prototype freak wave, Coastal Hydrodynamics (Ed. Darlymple), ASCE, pp.618-632.

Longuet-Higgins, M.S. (1957) : The statistical analysis of a random, moving surface, Phil. Trans. Roy. Soc. London, Ser.A (966), Vol.249, pp.321-387.

Mase, H., A. Matsumoto and Y. Iwagaki (1986) : Calculation model of random wave transformation in shallow water, Proc. JSCE, No.375/II-6, pp.221-230. (in Japanese)

per Bruun (1985) : Design and construction of mounds for breakwater and coastal protection, Elsevier, 938p.

Rice S.O. (1945) : Mathematical analysis of random noise, reprinted in selected paper on noise and stochastic processes, Dover Pub. Inc. pp.133-294.

Sand, S. (1990) : Report from the working group on breaking and freak waves, Water wave kinematics (Ed. Torum and Gudmestad), NATO ASI series, pp.17-21.

Yasuda, T., K. Ito and N. Mori (1992) : Non-linearity effect on the statistical property of uni-directional irregular waves, Proc. JSCE, No.443, pp.83-92. (in Japanese)

CHAPTER 28

On the Joint Distribution of Wave Height,
Period and Direction of Individual Waves
in a Three-Dimensional Random Seas

J.G.Kwon[1] and Ichiro Deguchi[2]

Abstract

A theoretical expression for the joint disribution of wave height, period and direction is derived based on the hypothesis that sea surface is a Gaussian stochastic process and that a band-width of energy spectra is sufficiently narrow. The derived joint distribution is found to be an effective measure to investigate characteristics of three-dimensional random wave fields in shallow water through field measurements.

I. Introduction

A variety of studies has been conducted on the sediment transport in shallow water regions and a lot of formula has been proposed on the rate of sediment transport. However, these formulas do not always predict the same estimation of the rate of sediment transport even under the same conditions. This discrepancy can be explained by the following reasons: i) the dynamics and kinematics of sediment movement are not fully understood yet and each formula contains empirical coefficients which have to be fixed through experiments or field measurements, and ii) the accurate measurement of sediment transport rate is extremely difficult. In the case of applying these formula to the sediment transport in the fields, further difficulties such as how to take into account the effect of irregularities in wave heights, periods and directional spreading of incident waves arise.

On the other hand, a wave transformation including wave breaking in the shallow water reigion is a

---

[1] Assistant Professor, Dept. of Environmental Eng., Pusan Univ., Changjun-dong, Kumjeoung-ku, Pusan, 609-735, Korea

[2] Assistant Professor, Dept. of Civil Eng., Osaka Univ., 2-1 Yamadaoka, Suita City, Osaka 565, Japan

non-linear and discontinuous phenomenon. Therefore, a so-called individual wave analysis (or a wave-by-wave analysis) rather than spectral approach seems to be adequate to investigate the wave transforation in such regions.

In this study, a joint distribution of wave height, period and direction of zero-down crossing waves which is required in the individual wave analysis in the shallow water region is derived theoretically. The applicability of the derived joint distribution, which is hereafter referred to as H-T-θ joint distribution, to the shallow water waves in the fields is examined through fields observations.

II. Derivation of H-T-θ Joint Distribution

2.1 Expression of Waves in Three-dimensional Random Seas

In a deep water region where a dispersive property of wave is strong, three-dimensional random seas waves are expressed by a directional spectrum. A transformation of irregular waves is also analyzed as the transformation of directional spectrum. While wave transformations in the shallow water region, where a significant sediment transport takes place, are usually investigated by applying the individual wave analysis (or the wave-by-wave analysis) of zero-down(or up)-crossing waves due to the non-linearity and discontinuity caused by wave breaking. The applicability of this approach has already been verified through experiments in two-dimensional wave tanks (for example, Mase et al., 1982).

The authors aim at applying the individual wave analysis to the three-dimensional random sea waves in shallow water with directional spreading. To do so, the joint distribution of wave height, period and direction of individual zero-down(or up)crossing waves has to be given.

Theoretical investigations on the joint probability density functions of H-T and H-θ have been conducting assuming that the band width of the frequency spectra of surface displacement $\eta(t)$ is sufficiently narrow so that $\eta(t)$ can be expressed by the envelope function. In this paper, referring to these results, the joint distribution of H-T-θ is derived from envelpe functions of surface displacements $\eta(t)$, bi-directional water particle velocities u(t), v(t) and time derivatives of surface displacements $\dot{\eta}(t)$.

The surface displacement in three-dimensional

random sea is usually expressed by the sum of an infinite number of sine-waves of amplitudes a_{ij} and periods T_i, each of which has different wave direction θ_j, in the following form:

$$\eta(t) = \sum_{i=1}^{\infty}\sum_{j=1}^{\infty} a_{ij}\cos\phi_{ij} \tag{1}$$

$$\phi_{ij} = k_i(x\cos\theta_j + y\sin\theta_j) - 2\pi f_i t - \varepsilon_{ij} \tag{2}$$

where k_i are the wave numbers and f_i are the frequencies, both of which correspond to the periods T_i, θ_j are the wave directions and ε_{ij} represent the phase differences of the waves whose amplitudes are a_{ij}. A coordinate system used in this study is shown in Fig.1.

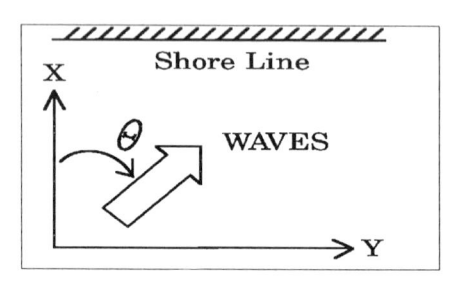

Fig. 1   Coordinate system

In the same way, water particle velocities in x- and y-directions u(t), v(t) are expressed as follows:

$$u(t) = \sum_{i=1}^{\infty}\sum_{j=1}^{\infty} b_i\cos\theta_j\, a_{ij}\cos\phi_{ij} \tag{3}$$

$$v(t) = \sum_{i=1}^{\infty}\sum_{j=1}^{\infty} b_i\sin\theta_j\, a_{ij}\cos\phi_{ij} \tag{4}$$

$$b_i = 2\pi f_i \frac{\cosh k_i z}{\sinh k_i h} \tag{5}$$

where, z is the height from the bottom where water particle velocities were measured..
    Besides these three time series, a time derivative of the surface elevation $\dot{\eta}(t)$ which is expressed by Eq.(6) is required to obtain the H-T-θ joint distribution.

$$\dot{\eta}(t) = \sum_{i=1}^{\infty}\sum_{j=1}^{\infty} 2\pi f_i \, a_{ij} \sin \phi_{ij} \qquad (6)$$

2.2 Envelope Functions

To express envelopes of time series of the quantities given by Eqs.(1), (3), (4) and (6), the phase function given by Eq.(2) is rewritten by using a representative frequency $\bar{f}$ as,

$$\dot{\phi}_{ij} = k_i(x\cos\theta_j + y\sin\theta_j) - 2\pi(f_i - \bar{f})t - \varepsilon_{ij} \qquad (7)$$

Substituting Eq.(7) into Eqs.(1), (3), (4) and (6), the following envelope functions of time series of $\eta(t)$, u(t), v(t) and $\dot{\eta}(t)$ are obtained:

$$\begin{aligned}
\eta(t) &= \eta_c(t)\cos 2\pi \bar{f} t + \eta_s(t)\sin 2\pi \bar{f} t \\
\dot{\eta}(t) &= \dot{\eta}_c(t)\cos 2\pi \bar{f} t + \dot{\eta}_s(t)\sin 2\pi \bar{f} t \\
u(t) &= u_c(t)\cos 2\pi \bar{f} t + u_s(t)\sin 2\pi \bar{f} t \\
v(t) &= v_c(t)\cos 2\pi \bar{f} t + v_s(t)\sin 2\pi \bar{f} t
\end{aligned} \qquad (8)$$

where,

$$\begin{aligned}
\eta_c(t) &= \sum_{i=1}^{\infty}\sum_{j=1}^{\infty} a_{ij} \cos \dot{\phi}_{ij} \\
\eta_s(t) &= \sum_{i=1}^{\infty}\sum_{j=1}^{\infty} a_{ij} \sin \dot{\phi}_{ij} \\
\dot{\eta}_c(t) &= \sum_{i=1}^{\infty}\sum_{j=1}^{\infty} 2\pi(f_i - \bar{f}) a_{ij} \sin \dot{\phi}_{ij} \\
\dot{\eta}_s(t) &= \sum_{i=1}^{\infty}\sum_{j=1}^{\infty} 2\pi(f_i - \bar{f}) a_{ij} \cos \dot{\phi}_{ij} \\
u_c(t) &= \sum_{i=1}^{\infty}\sum_{j=1}^{\infty} b_i \cos\theta_j \, a_{ij} \cos \dot{\phi}_{ij} \\
u_c(t) &= \sum_{i=1}^{\infty}\sum_{j=1}^{\infty} b_i \cos\theta_j \, a_{ij} \cos \dot{\phi}_{ij} \\
u_s(t) &= \sum_{i=1}^{\infty}\sum_{j=1}^{\infty} b_i \cos\theta_j \, a_{ij} \sin \dot{\phi}_{ij} \\
v_c(t) &= \sum_{i=1}^{\infty}\sum_{j=1}^{\infty} b_i \sin\theta_j \, a_{ij} \cos \dot{\phi}_{ij} \\
v_s(t) &= \sum_{i=1}^{\infty}\sum_{j=1}^{\infty} b_i \sin\theta_j \, a_{ij} \sin \dot{\phi}_{ij}
\end{aligned} \qquad (9)$$

The amplitude of each carrier wave and the phase relation between carrier waves and water particle velocities are determined by these envelope functions.

2.3 Joint Probability Density Function of Envelope Amplitudes

As can be understood from Eq.(8), these eight envelope amplitudes are stationary Gaussian stochastic processes with zero mean by virtue of the central limit theorem. Therefore, the probability density function for eight envelope amplitudes is expressed as :

$$P(\eta_c, \eta_s, u_c, u_s, v_c, v_s, \dot{\eta}_c, \dot{\eta}_s)$$

$$= \frac{1}{(2\pi)^4 [M]^{1/2}} \text{EXP}\left[-\frac{1}{2}\sum_{i=1}^{8}\sum_{j=1}^{8}\frac{M_{ij}}{[M]}\zeta_i\zeta_j\right]$$

$$= \frac{1}{(2\pi)^4 (m_{00} m_{02} m_{20} m_{22})} * \text{EXP}\left[-\frac{1}{2*\Delta} * \left\{A_{11}\left(\frac{\dot{\eta}_c^2 + \dot{\eta}_s^2}{m_{00}}\right)\right.\right.$$

$$+ A_{22}\left(\frac{u_c^2 + u_s^2}{m_{20}}\right) + A_{44}\left(\frac{\dot{\eta}_c^2 + \dot{\eta}_s^2}{m_{22}}\right) + A_{33}\left(\frac{v_c^2 + v_s^2}{m_{02}}\right) +$$

$$2A_{12}\left(\frac{\eta_c u_c + \eta_s u_s}{\sqrt{m_{00} m_{20}}}\right) + 2A_{13}\left(\frac{\eta_c v_c + \eta_s v_s}{\sqrt{m_{00} m_{02}}}\right) + 2A_{23}\left(\frac{u_c v_c + u_s v_s}{\sqrt{m_{02} m_{20}}}\right) +$$

$$2A_{14}\left(\frac{\eta_c \dot{\eta}_s + \dot{\eta}_c \eta_s}{\sqrt{m_{00} m_{22}}}\right) + 2A_{24}\left(\frac{u_c \dot{\eta}_s - u_s \dot{\eta}_c}{\sqrt{m_{22} m_{20}}}\right) + 2A_{34}\left(\frac{v_c \dot{\eta}_s - v_s \dot{\eta}_c}{\sqrt{m_{22} m_{02}}}\right)\right]$$

$$\text{............(10)}$$

where,

$$[M] = \begin{pmatrix} M_0 & 0 \\ 0 & M_0' \end{pmatrix}, \quad M_0 = \begin{pmatrix} m_{00} & m_{10} & m_{01} & 0 \\ m_{10} & m_{20} & m_{11} & m_{12} \\ m_{01} & m_{11} & m_{02} & m_{21} \\ 0 & m_{12} & m_{21} & m_{22} \end{pmatrix},$$

$$M_0' = \begin{pmatrix} m_{00} & m_{10} & m_{01} & 0 \\ m_{10} & m_{20} & m_{11} & -m_{12} \\ m_{01} & m_{11} & m_{02} & -m_{21} \\ 0 & -m_{12} & -m_{21} & m_{22} \end{pmatrix}$$

is the determinant of covariance matrix whose elements $\langle \zeta_i, \zeta_j \rangle$ are defined as :

$$\langle \eta_c \eta_c \rangle = \langle \eta_s \eta_s \rangle = \langle \eta^2 \rangle = m_{00}, \quad \langle \eta_c v_c \rangle = \langle \eta_s v_s \rangle = \langle \eta v \rangle = m_{01}$$

$$\langle u_c u_c \rangle = \langle u_s u_s \rangle = \langle u^2 \rangle = m_{20}, \quad \langle u_c v_c \rangle = \langle u_s v_s \rangle = \langle uv \rangle = m_{11}$$

$$\text{............(11)}$$

$$\langle v_c v_c \rangle = \langle v_s v_s \rangle = \langle v^2 \rangle = m_{02}, \quad \langle \dot{\eta}_c \dot{\eta}_c \rangle = \langle \dot{\eta}_s \dot{\eta}_s \rangle = \langle \dot{\eta}^2 \rangle = m_{22}$$
$$\langle \eta_c u_c \rangle = \langle \eta_s u_s \rangle = \langle \eta u \rangle = m_{10}, \quad \langle u_c \dot{\eta}_s \rangle = -\langle u_s \dot{\eta}_c \rangle = m_{12}$$
$$\langle v_c \dot{\eta}_s \rangle = -\langle v_s \dot{\eta}_c \rangle = m_{21} \quad \text{\dotfill (11)}$$

M_{ij} is the co-factor of $\langle \zeta_i, \zeta_j \rangle$ and

$$\Delta = (1 + 2\gamma_{11}\gamma_{12}\gamma_{21} - \gamma_{12}^2 - \gamma_{21}^2 - \gamma_{11}^2 - \gamma_{10}^2 - 2\gamma_{01}\gamma_{10}\gamma_{12}\gamma_{21}$$
$$+ 2\gamma_{10}\gamma_{01}\gamma_{11} + \gamma_{10}^2\gamma_{21}^2 + \gamma_{01}^2\gamma_{12}^2 - \gamma_{01}^2)$$

$$A_{11} = (1 + 2\gamma_{11}\gamma_{12}\gamma_{21} - \gamma_{12}^2 - \gamma_{21}^2 - \gamma_{11}^2)$$

$$A_{12} = (\gamma_{21}^2\gamma_{10} + \gamma_{01}\gamma_{11} - \gamma_{10} - \gamma_{01}\gamma_{12}\gamma_{21})$$

$$A_{13} = (\gamma_{10}\gamma_{11} + \gamma_{12}^2\gamma_{01} - \gamma_{10}\gamma_{12}\gamma_{21} - \gamma_{01})$$

$$A_{14} = (\gamma_{10}\gamma_{12} + \gamma_{01}\gamma_{21} - \gamma_{10}\gamma_{11}\gamma_{21} - \gamma_{01}\gamma_{11}\gamma_{12})$$

$$A_{22} = (1 - \gamma_{21}^2 - \gamma_{01}^2)$$

$$A_{23} = (\gamma_{01}\gamma_{10} + \gamma_{12}\gamma_{21} - \gamma_{11})$$

$$A_{24} = (\gamma_{11}\gamma_{21} + \gamma_{01}^2\gamma_{12} - \gamma_{12} - \gamma_{10}\gamma_{01}\gamma_{21})$$

$$A_{33} = (1 - \gamma_{12}^2 - \gamma_{10}^2)$$

$$A_{44} = (1 + 2\gamma_{10}\gamma_{01}\gamma_{11} - \gamma_{01}^2 - \gamma_{11}^2 - \gamma_{10}^2)$$

where,
$$\gamma_{10} = m_{10}/\sqrt{m_{00}m_{20}}, \quad \gamma_{01} = m_{01}/\sqrt{m_{00}m_{02}}, \quad \gamma_{11} = m_{11}/\sqrt{m_{20}m_{02}}$$
$$\gamma_{12} = m_{12}/\sqrt{m_{20}m_{22}}, \quad \gamma_{21} = m_{21}/\sqrt{m_{02}m_{22}}, \quad \text{\dotfill (12)}$$

Among these nine covariances from m_{00} to m_{21} of preceding description, seven covariance except for m_{12} and m_{21} can be directly calculated from the time series of surface elevation $\eta(t)$ and horizontal water particle velocities u(t) and v(t). While, m_{12} and m_{21} can be calculated from the directional spectra $s(f, \theta)$ as:

$$m_{12} = -\int_0^\infty \int_{-\pi}^{\pi} 2\pi b(f)(f-\bar{f}) \cos\theta \; s(f,\theta) d\theta df$$
$$m_{21} = -\int_0^\infty \int_{-\pi}^{\pi} 2\pi b(f)(f-\bar{f}) \sin\theta \; s(f,\theta) d\theta df$$
(13)

2.4 Joint Probability Density Function of Wave Height, Period and Direction

To derive the joint probability density function for wave height, period and direction, the following series of variable transformation are carried out:

1) Normalization of the envelope functions.

$$N_c = \eta_c/\sqrt{m_{00}}, \quad N_s = \eta_s/\sqrt{m_{00}}, \quad V_c = v_c/\sqrt{m_{02}}, \quad V_s = v_s/\sqrt{m_{02}},$$
$$U_c = u_c/\sqrt{m_{20}}, \quad U_s = u_s/\sqrt{m_{20}}, \quad \dot{N}_c = \dot{\eta}_c/\sqrt{m_{22}}, \quad \dot{N}_s = \dot{\eta}_s/\sqrt{m_{22}},$$
(14)

2) Introduction of the amplitude R and phase angle δ of carrier waves.

$$N_c = R\cos\delta \quad , \quad N_s = R\sin\delta$$
$$\dot{N}_c = \dot{R}\cos\delta - R\dot{\delta}\sin\delta, \quad \dot{N}_s = \dot{R}\sin\delta + R\dot{\delta}\cos\delta$$
(15)

where, $R^2 = N_c^2 + N_s^2$. $\delta = \tan^{-1}(N_s/N_c)$ (16)

3) Transformation of the phase angle of water particle velocities so that the phase of the surface displacement becomes a standard and introduction of the polar coordinate system after defining the wave direction of each wave θ(after Isobe, 1987).

$$U_c = u_p\cos\delta - u_q\sin\delta, \quad U_s = u_p\sin\delta + u_q\cos\delta$$
$$V_c = v_p\cos\delta - v_q\sin\delta, \quad V_s = v_p\sin\delta + v_q\cos\delta$$
(17)

$$\theta = \tan^{-1}(v_p/u_p)$$
$$u_p = W\cos\theta \quad , \quad v_p = W/\Gamma\sin\theta$$
(18)

After conducting these transformation of variables of Eq.(10), the following joint probability density function for R, δ and θ is obtained by integrating with respect to u_q, v_q, $\dot{R}$, δ and W which have nothing to do with H-T-θ joint distribution:

$$P(R,\theta,\delta) = \frac{1}{2\pi^{3/2}*\Gamma} *R^2 * \mathrm{EXP}\ [\frac{R^2}{2\Delta}*(A_{11}+A_{44}\dot{\delta}^2+2A_{14}\dot{\delta})]\ *$$
$$[\frac{\sqrt{\Delta}}{A} + \frac{RB}{A^{3/2}}*\frac{\sqrt{\pi}}{\sqrt{2}}*\mathrm{EXP}\{\frac{R^2B^2}{2A\Delta}\}*\{1-\mathrm{Pr}(-\frac{RB}{\sqrt{A\Delta}})\}]$$
(19)

where, $\Gamma = \sqrt{m_{02}/m_{20}}$ is a longcrestedness parameter and

$$\Pr(\zeta) = \frac{1}{\sqrt{2\pi}} * \int_{-\infty}^{\zeta} \exp\left(-\frac{t^2}{2}\right) dt$$
$$A = (A_{22}\cos\theta^2 + A_{33}\sin\theta^2/\Gamma^2 + 2A_{23}\cos\theta\sin\theta/\Gamma)$$
$$B = -(A_{12}\cos\theta + A_{13}\sin\theta/\Gamma + A_{24}\dot{\delta}\cos\theta + A_{34}\dot{\delta}\sin\theta/\Gamma)$$

When the spectral bandwidth is sufficiently narrow, R and $\dot{\delta}$ in Eq.(19) can be related to the wave height $\overline{H}$ and the period $\overline{T}$ (Longuet - Higgins, 1975) as :

$$R = H/2, \quad \dot{\delta} = 2\pi(\overline{f}-f) = 2\pi(1/\overline{T} - 1/T) \tag{20}$$

By using these relations together with the zero-th and first special moments m_0 and m_1, wave heights and periods are normalized as follows :

$$\tau = T/\overline{T} = 2\pi/(2\pi\overline{f}-\dot{\delta})*m_1/m_0$$
$$x = H/\overline{H} = 2R/(2\pi m_0)^{1/2} \tag{21}$$

Substituting, Eq.(21) into Eq.(19), the following joint probability density function for wave heights(x), period (τ) and direction (θ) is obtained :

$$P(x,\tau,\theta) = \frac{x^2 * \overline{\sigma}}{2^3 * \Gamma * \tau^2} * \text{EXP}\left[-\frac{\pi}{4\Delta} * x^2(A_{11} + A_{44}\overline{\sigma}^2(1-1/\tau)^2\right.$$
$$\left. + 2A_{14}\overline{\sigma}(1-1/\tau))\right] * \left[\frac{\sqrt{\Delta}}{A} + \frac{B'}{A^{3/2}} * \frac{\sqrt{\pi}}{\sqrt{2}} * x * \frac{\sqrt{\pi}}{\sqrt{2}}\right. \tag{22}$$
$$\left. \text{EXP}\left\{\frac{B'^2\pi}{4A\Delta}x^2\right\} * \left\{1 - \Pr\left(-B'\frac{\sqrt{\pi}}{\sqrt{2A\Delta}}x\right)\right\}\right]$$

where, $\overline{\sigma} = 2\pi m_1 m_0$ and
$$B' = -\{A_{12}\cos\theta + A_{13}\sin\theta/\Gamma + A_{24}\overline{\sigma}(1-1/\tau)\cos\theta$$
$$+ A_{34}\overline{\sigma}(1-1/\tau)\sin\theta/\Gamma\}$$

Akai et al.(1988) also derived the joint distribution of wave height, frequency and direction. They omitted cross correlation term(γ_{21}) for influence of the asymmetrical property of directional spreading, but in our derivation, all correlation are taken into account.

III. H-T-θ Joint Distribution of Measured Waves in Shallow Water

3.1 Field Observation

Field observation were carried out at two coasts to verify the proposed joint probability density function for wave height, period and direction under the condition of wind waves in a winter. One observation site was the Keinomatsubara Beach located in the west coast of the Awaji Island and the other was the Nishikinohama Beach on the southern part of OSAKA Bay. These locations are shown in Fig. 2.

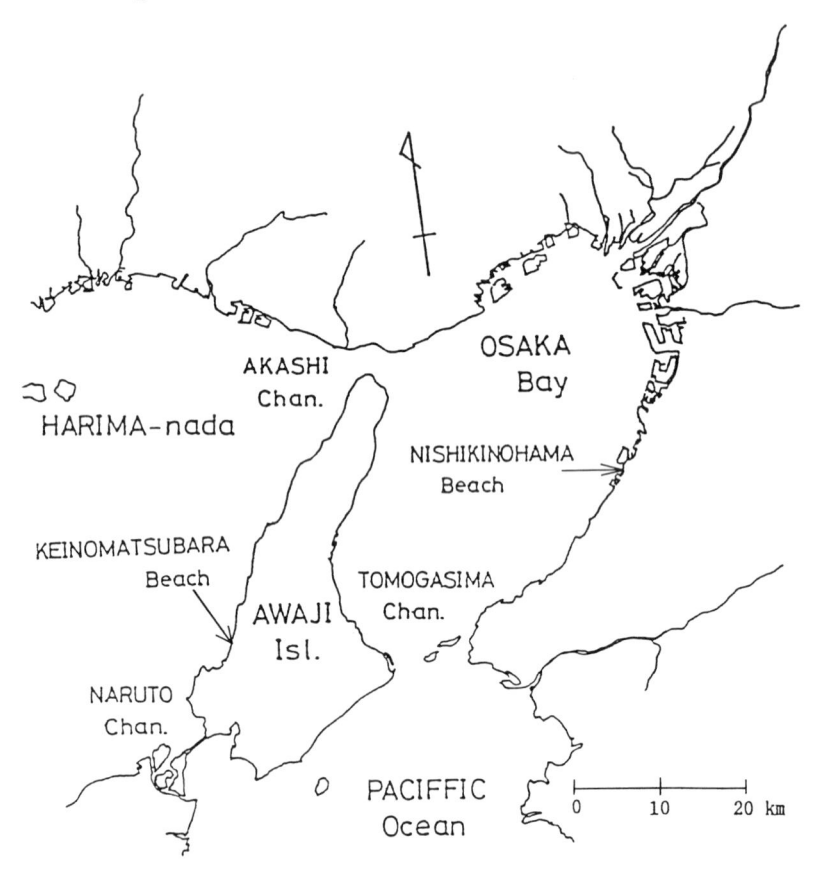

Fig. 2 Location of observation sites

Wave heights were measured by capacipitance type wave gages. Longshore and cross-shore water particle velocities were measured by bi-directional electromagnetic current meters at the same place of wave gauges and about 25cm above the bottom to consist a so-caled 3-element array. Analogue data from these installments were first recorded by an analogue data recorder and then digitized by an A-D converter at a sampling time of 0.1sec for the data processing. Very small waves whose frequencies were greater than $4* f_p$ (f_p : peak frequency) were disregarded in the data processing.

3.2 Characteristics of Measured Waves

Before discuss the joint distribution, characteristics of measured waves are examined briefly. Table 1 shows the statistical characteristics obtained from the time series in which more than 500 waves were recorded at the two coasts. In the following analysis, the x-axis is rotated to be the principle direction of incident waves.

Table 1 Statistical characteristics of measured waves

Case No.	Dep. (cm)	r_{10}	r_{01}	r_{11}	r_{12}	r_{21}	γ	U_r	Kurt.	Skew.	Q_p
1-1	82	0.806	0.103	-0.004	0.042	0.204	0.305	21.4	3.19	0.162	2.37
1-2	48	0.827	0.101	-0.021	0.148	0.717	0.257	30.7	3.14	0.193	1.84
1-3	97	0.821	0.037	-0.015	0.003	0.412	0.301	6.0	3.27	0.048	2.04
1-4	115	0.819	0.103	-0.020	0.051	0.632	0.392	2.6	2.94	0.023	2.32
1-5	121	0.693	0.161	-0.020	0.019	0.163	0.317	4.6	3.36	0.052	1.90
1-6	134	0.783	0.125	-0.012	0.066	0.680	0.355	4.3	3.07	0.033	1.89
1-7	67	0.596	0.142	-0.022	-0.020	0.656	0.296	24.7	2.71	-0.251	2.72
1-8	113	0.808	0.115	-0.009	-0.291	-0.009	0.299	9.4	2.95	0.016	2.22
2-1	80	0.933	0.026	-0.003	-0.000	0.649	0.307	66.6	3.32	0.144	3.46
2-2	1320	0.906	-0.006	0.000	0.276	0.433	0.431	10.0	2.99	0.107	2.39

Results of Case-number from 1-1 to 1-6 and Case-number from 2-1 to 2-2 correspond to the data obtained at Nishikinohama Beach and Keinomatsubara Beach, respectively. r_{ij} in the table is the covariance defined by Eqs. (11) and (12), γ is the longcrestedness parameter, ν

is the band width parameter and Q_P is the peakedness parameter. Ursell number U_r is calculated by using the significant wave height and the period. The integration of the directional spetrum, which was estimated by EMLM(Isobe et al., 1984), was carried out between $0.5 f_p$ and 3 f_p to obtain m_{12} and m_{21}.

It is found from Table 1 that a large part of measured waves has a significant non-linear property (U_r =2.6-66.6 , Kurtosis=2.939-3.266 , skewness=-0.251-0.193) and had a wide directional spreading(γ=0.2565-0.4306). The value of r_{21} which represents asymmetrical property of directional spreading of incident waves is relatively large when compared with other covariances in the table. However, any correlation between parameters of directional spreading of incident waves (r_{10} and γ) and those of asymmetry of directional spreading(r_{12} and r_{01}) can not be seen.

3.3 Joint Distribution of Wave Height, Period and Direction

In this paper, Case 2-1 whose significant wave height and r_{21} are relatively large and more than 1000 waves were recorded is analyzed as an example to examine the applicability of theoretical joint distribution for H, T and θ derived in this study.

Figure 3 (a)~(c) shows nondimensional scatter diagram of wave period($T/\overline{T}$) and direction(θ) of measured 1000 waves(Case 2-1) under the condition of wave height shown in the figure. The class bands of nondimensional wave heights and directions are 0.25 and 22.5°, respectively. Numerals in the figure show the frequency of zero-down crossing waves. In the figure, predicted isolines of frequency obtained from multiplying the integrated probability density (Eq.(21)) between the range of wave height shown in the figure by the total number of measured waves(1000) are also illustrated by solid lines.

Figure 4 is the joint distribution of non-dimensional wave height and directions under the conditions of 0.25< $T/\overline{T}$<0.75 (Fig(a)), 0.75< $T/\overline{T}$<1.25 (Fig(b)) and 1.25< $T/\overline{T}$<1.75 (Fig(c)). Solid lines in the figures are the predicted isolines of frequency calculated in the same way as those in Fig.3.

In Fig. 3, the predicted isolines exhibit almost symmetric profile with respect to the direction when

JOINT DISTRIBUTION OF WAVE HEIGHT 381

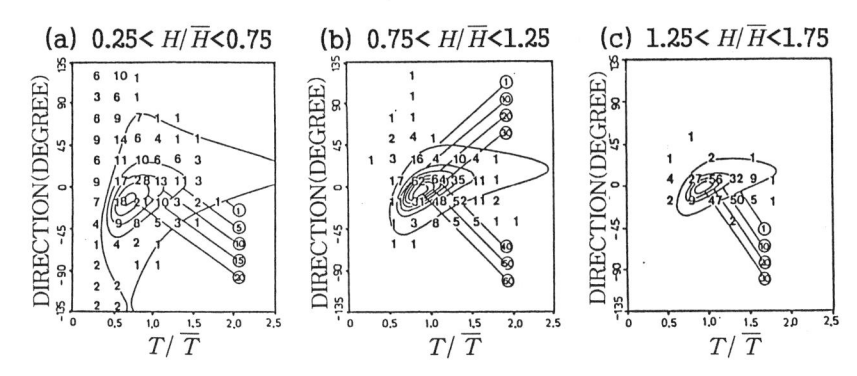

Figure 3 Joint distribution of wave directions and periods

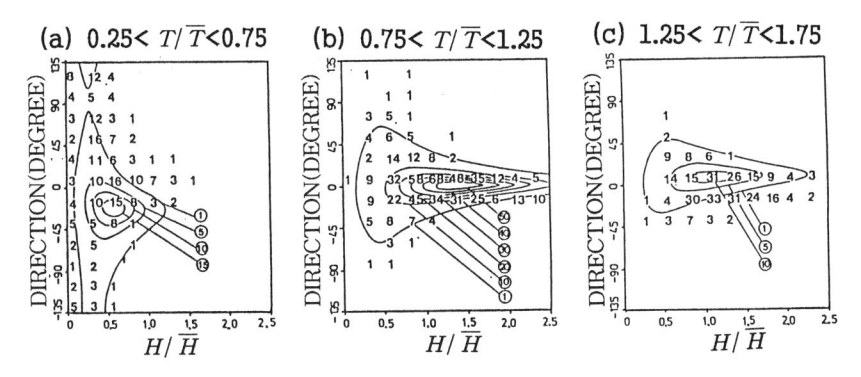

Figure 4 Joint distribution of wave directions and heights

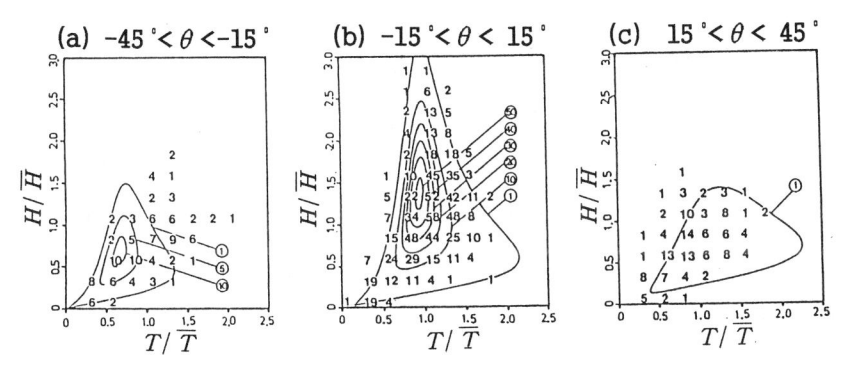

Figure 5 Joint distribution of wave heights and periods

$T/\overline{T}$<0.75 . On the other hand, the predicted isolines in Fig.4 show asymmetry around the principle direction (θ =0.0°), ie, wave periods distribute asymmetrically around the principle direction. This difference is partly explained by the fact that wave refraction does not depend deeply on wave heights but mainly on wave periods. It is also found from Figs. 3 and 4 that the predicted isolines do not coincide well with the measured frequency in the region where $T/\overline{T}$<0.5 or $H/\overline{H}$<0.5 or $|\theta|$>45°. However, in the region of $H/\overline{H}$>0.75, $T/\overline{T}$>0.75 and $|\theta|$<15°, where a large part of measured waves is included, a relatively good agreement is seen.

Figure 5 is a scatter diagram of non-dimensional wave heights and periods under the conditions of -45°<θ <-15° (Fig(a)), -15°<θ<15° (Fig (b)) and 15°<θ<45° (Fig(c)). Solid lines in the figures are the isolines of frequency calculated in the same way as the former two figures. In the region of -15°<θ<15°, which is shown in Fig.(b), the predicted isolines coincide well with the measured frequency indicating the tail toward the origin in the regions of $T/\overline{T}$<0.75 and $H/\overline{H}$<0.75 due to the correlation between the wave heights and periods. The correlation coefficient between wave heights and periods of Case 2-1 is 0.57.

Although the tail can be seen in Figs.(a) and (c), the agreement between the predicted and measured frequency is not good in these regions. It is also found that the joint distribution of wave heights and periods is not symmetrical with respect to the wave direction by comparing Figs.(a) and (c).

IV. CONCLUSIONS

Summing up the results of the study described above, the followings are the major conclusions:
(1) The joint probability density function of wave height, period and direction is derived by using time series of surface displacement $\eta(t)$, its time derive $\dot{\eta}(t)$, and horizontal two component water particle velocities u(t), v(t) assuming that $\eta(t)$ is a Gaussian process with a narrow band power spectra. It is found that the marginal joint distribution of wave heights and direction has less influence of the asymmetrical property of directional spreading (directional spectra) than that of wave period and direction.
(2) The derived joint distribution are compared with the measured joint distribution in the field of shallow

water. In the region where the appearance frequency of waves is high (ie. $T/\overline{T}>0.75$, $H/\overline{H}>0.75$, $|\theta|<15°$) , a relatively high agreement is obtained between measured and predicted frequencies. Especially, the marginal distribution of wave heights and periods in the region of $|\theta|<15°$, the predicted frequency coincides fairley well with the measured one. However, in the region of low appearance frequency, both of them do not agree well with each other.

The measured waves had a strong asymmetric property of directional spreading with respect to the principle direction. They also exhibited wide directional spreading and a little nonlinearity.

References

Akai, S., et al., (1988), Joint distribution of wave height, frequency and direction in multi-directional wave fields, Proc. Japanese Conf. on Coastal Engineering, pp. 142~147, in Japanese.

Isobe, M., K. Kondo and K. Horikawa, (1984), Extension of MLM for estimating directional wave spectrum, Proc. Sympo. on Dispersion and Modeling of Directional Seas, Paper No. A-6, 15p.

Isobe, M., (1988), On the joint distribution of wave heights and directions, Proc. 21st Conf. on Coastal Eng., pp. 524~538.

Longuet-Higgins, M. S., (1975), On the joint distribution of the periods and amplitudes of sea waves, Jour. Geophysical Res. Vol. 80, No. 18, pp. 2688~2694.

Mase, H. and Y. Iwagaki, (1982), Wave heights distribution and wave grouping in surf zone, Proc. 18th Conf. on Coastal Eng., pp. 58~76.

CHAPTER 29

Spectral Wave-Current Bottom Boundary Layer Flows

Ole Secher Madsen[1]

Abstract

　　Based on the linearized governing equations, a bottom roughness specified by the equivalent Nikuradse sand grain roughness, k_N, and a time-invariant eddy viscosity analogous to that of Grant and Madsen (1979 and 1986) the solution is obtained for combined wave-current turbulent bottom boundary layer flows with the wave motion specified by its near-bottom orbital velocity directional spectrum. The solution depends on an *a priori* unknown shear velocity, u_{*r}, used to scale the eddy viscosity inside the wave boundary layer. Closure is achieved by requiring the spectral wave-current model to reduce to the Grant-Madsen model in the limit of simple periodic plane waves. To facilitate application of the spectral wave-current model it is used to define the characteristics (near-bottom orbital velocity amplitude, u_{br}, radian frequency, ω_r, and direction of propagation, ϕ_{wr}) of a representative periodic wave which, in the context of combined wave-current bottom boundary layer flows, is equivalent to the wave specified by its directional spectrum. Pertinent formulas needed for application of the model are derived and their use is illustrated by outlining efficient computational procedures for the solution of wave-current interaction for typical specifications of the current.

Introduction

　　Over the past couple of decades several theoretical models for turbulent bottom boundary layers associated with combined wave-current flows have been proposed. In view of the vastly different levels of sophistication with which these different models represent turbulence, their end result, most

---

[1] R.M. Parsons Laboratory, Massachusetts Institute of Technology, Cambridge, MA 02139 USA

notably their predicted effect of the presence of waves on the current, is surprisingly similar. Thus, in the absence of solid experimental evidence as to which of the many models is truly *the* best, choosing a particular model for applications becomes a matter of personal preference and convenience. However, since all existing wave-current interaction models have been derived for a wave motion corresponding to a simple periodic plane progressive wave, one additional choice – the choice of simple periodic wave characteristics to represent a wave motion which more realistically is described by its directional spectrum – must be made prior to model applications.

The objective of this study is therefore to derive a simple theoretical model for turbulent wave-current bottom boundary layer flows for a wave motion described by its directional spectrum and to use this model to determine the characteristics of the periodic wave which, in the context of wave-current interaction, is equivalent to the directional sea.

Theoretical Model

The theoretical model for spectral wave-current boundary layer flows is based on the linearized boundary layer equation which, in standard notation, reads

$$\frac{\partial \mathbf{u}}{\partial t} = -\frac{1}{\rho}\frac{\partial p}{\partial x_i} + \frac{\partial}{\partial z}\left[\nu_t \frac{\partial \mathbf{u}}{\partial z}\right] \qquad (1)$$

in which the turbulent eddy viscosity, ν_t, is assumed to be scaled by an *a priori* unknown shear velocity, u_{*r}, which inside the wave boundary layer, $z < \delta_{wc}$, reflects the combined wave-current flow and outside the wave boundary layer is a function of only the average, *i.e.* the current, shear velocity, u_{*c}. To keep the analysis relatively simple and because the universally made *assumption* of a *single* roughness scale, the equivalent Nikuradse sand grain roughness, k_N, for currents and waves alone as well as for combined wave-current flows has been experimentally validated (Mathisen and Madsen, 1993) only for this model, we adopt the Grant-Madsen (1979 and 1986) eddy viscosity

$$\nu_t = \begin{cases} \kappa u_{*r} z & \text{for } z < \delta_{wc} \\ \kappa u_{*c} z & \text{for } z > \delta_{wc} \end{cases} \qquad (2)$$

Assuming u_{*r} and therefore ν_t to be time-invariant, resolving velocity and pressure into their mean and time-varying components, *i.e.*

$$\mathbf{u} = \bar{\mathbf{u}} + \tilde{\mathbf{u}} = \mathbf{u_c} + \mathbf{u_w} \quad ; \quad p = \bar{p} + \tilde{p} = \bar{p} + \tilde{p}_b \qquad (3)$$

and introducing these expressions in the governing equation, two separate equations are obtained. One for the wave motion

$$\frac{\partial(\mathbf{u_w} - \mathbf{u_b})}{\partial t} = \frac{\partial}{\partial z}\left[\nu_t \frac{\partial(\mathbf{u_w} - \mathbf{u_b})}{\partial z}\right] \qquad (4)$$

in which the relationship between $\tilde{p}_b$ and the near-bottom velocity $\mathbf{u_b}$, predicted by potential theory, has been invoked; and another equation for the current

$$\nu_t \frac{\partial \mathbf{u_c}}{\partial z} = \tau_\mathbf{c}/\rho = u_{*c}^2\{\cos\phi_c, \sin\phi_c\} \qquad (5)$$

in which the law-of-the-wall arguments have been used, and ϕ_c denotes the current angle with the x-axis.

Wave Solution

For the wave solution only the eddy viscosity formulation assumed for $z < \delta_{wc}$ is pertinent. Hence, the wave portion of the wave-current problem is completely analogous to the pure wave boundary layer problem discussed and solved by Madsen et al. (1988) for a wave motion described by its directional spectrum, i.e. the solution to (4) may be written as a sum of wave components, each being the real part of

$$\mathbf{u_{w}}_{nm} = \left[1 - \frac{\text{ker}2\sqrt{\zeta_n} + i\text{kei}2\sqrt{\zeta_n}}{\text{ker}2\sqrt{\zeta_{no}} + i\text{kei}2\sqrt{\zeta_{no}}}\right]\mathbf{u_b}_{nm}e^{i\omega_n t} \qquad (6)$$

in which the velocity is related to the directional near-bottom orbital velocity spectrum, $S_{u_b}(\omega, \theta)$, through,

$$\mathbf{u_b}_{nm} = \sqrt{2S_{u_b}(\omega_n, \theta_m)d\theta d\omega}\{\cos\theta_m, \sin\theta_m\} \qquad (7)$$

ker and kei denote the zeroth order Kelvin functions and

$$\zeta_n = z\omega_n/(ku_{*r}) \qquad (8)$$

with ζ_{no} denoting the value of ζ_n at $z = z_o = k_N/30$ where k_N is the equivalent Nikuradse sand grain roughness of the bottom.

Current Solution

The solution for the current, obtained from (5) and (2), is for $z < \delta_{wc}$

$$\mathbf{u_c} = \frac{u_{*c} u_{*c}}{u_{*r} \kappa} \ell n \frac{z}{z_o} \{\cos\phi_c, \sin\phi_c\} \quad (9)$$

and for $z > \delta_{wc}$

$$\mathbf{u_c} = \frac{u_{*c}}{\kappa} \ell n \frac{z}{z_{oa}} \{\cos\phi_c, \sin\phi_c\} \quad (10)$$

in which z_{oa} – the apparent bottom roughness experienced by the current in presence of waves – is obtained by matching the currrent velocities at $z = \delta_{wc}$, i.e.

$$\ell n \frac{\delta_{wc}}{z_{oa}} = \frac{u_{*c}}{u_{*r}} \ell n \frac{\delta_{wc}}{z_o} \quad (11)$$

Closure

In the preceeding analysis we have formally obtained the solution for the near-bottom turbulent flow associated with a wave motion described by its directional frequency spectrum and a superimposed steady current. However, the solution can be evaluated only when the value of the representative wave-current shear velocity, u_{*r}, is known. To determine this unknow, i.e. to close the problem, we obtain the bottom shear stress for each wave component from

$$\tau_{\mathbf{w}nm} = \rho\kappa u_{*r} z \frac{\partial \mathbf{u_w}_{nm}}{\partial z}\bigg|_{z=z_o} =$$

$$\rho\kappa u_{*r}\sqrt{\zeta_n}\frac{\partial \mathbf{u_w}_{nm}}{\partial(2\sqrt{\zeta_n})}\bigg|_{\zeta_n=\zeta_{no}} =$$

$$\rho\kappa u_{*r}\sqrt{\zeta_{no}}\frac{-\mathrm{ker}'2\sqrt{\zeta_{no}} - i\,\mathrm{kei}'2\sqrt{\zeta_{no}}}{\mathrm{ker}2\sqrt{\zeta_{no}} + i\mathrm{kei}2\sqrt{\zeta_{no}}} \mathbf{u_b}_{nm} e^{i\omega_n t} \quad (12)$$

in which "prime" denotes the derivative of the zeroth order Kelvin function with respect to its argument, and $\mathbf{u_b}_{nm}$ is given by (7).

This equation may alternatively be interpreted as an expression for the directional spectrum of the wave-associated bottom shear stress

$$S_{\tau_w}(\omega,\theta) = K^2(\omega) S_{u_b}(\omega,\theta) \quad (13)$$

$$K(\omega) = \rho\kappa u_{*r}\sqrt{\zeta_o} \left| \frac{-\text{ker}'2\sqrt{\zeta_o} - i\text{kei}2\sqrt{\zeta_o}}{\text{ker}2\sqrt{\zeta_o} + i\text{kei}2\sqrt{\zeta_o}} \right| \quad (14)$$

is a function of ω since

$$\zeta_o = \frac{z_o \omega}{\kappa u_{*r}} = \frac{k_N \omega}{30\kappa u_{*r}} \quad (15)$$

From (13) we may obtain the representative amplitude of the bottom shear stress of a simple harmonic wave with the same variance as the spectral representation, *i.e.* the root-mean-square amplitude,

$$\tau_{wr}^2 = 2 \iint S_{\tau_w}(\omega,\theta) d\theta d\omega = \int_0^\infty \int_0^{2\pi} K^2(\omega) [2S_{u_b}(\omega,\theta)] d\theta d\omega \quad (16)$$

with a direction given by

$$\tan\theta_{wr} = \frac{\iint S_{\tau_w}(\omega,\theta) \sin\theta d\omega d\theta}{\iint S_{\tau_w}(\omega,\theta) \cos\theta d\omega d\theta} \quad (17)$$

In principle (16) and (17) may be evaluated if the bottom roughness, k_N, and the near-bottom orbital velocity spectrum, $S_{u_b}(\omega,\theta)$, are known. This would lead to a representative wave-associated bottom shear stress amplitude vector

$$\boldsymbol{\tau_{wr}} = \tau_{wr}\{\cos\theta_{wr}, \sin\theta_{wr}\} \quad (18)$$

which depends on the unknown representative wave-current shear velocity, u_{*r}.

Final closure is obtained by requiring the spectral wave-current solution to reduce to that of Grant and Madsen (1986) in the limit of simple periodic waves, *i.e.*

$$u_{*r}^2 = \frac{1}{\rho}|\boldsymbol{\tau_{wr}} + \boldsymbol{\tau_c}| \quad (19)$$

The bottom shear stress in (12) is evaluated at $z = z_o = k_N/30$ rather than by taking the limit $z \to 0$, which was used in Madsen et al. (1988). For small values of ζ_o which, by (15), is seen to correspond to small values of the bottom roughness, this difference is of negligible importance. However, for larger roughness values and hence larger values of ζ_o this difference has important implications that will be discussed in Appendix A.

The Representative Periodic Wave

Actual application of the theoretical spectral wave-current model as presented in the preceeding section would be extremely cumbersome, particularly since the evaluation of the integrals in (16) and (17) presumes u_{*r} to be known. For this reason a further simplification is achieved by introducing the concept of a representative periodic wave which, in the context of wave-current interaction, is equivalent to the spectral wave representation.

This representative periodic wave is characterized by its near-bottom orbital velocity amplitude, u_{br}, radian frequency, ω_r, and direction of propagation, ϕ_{wr}. To obtain the representative wave characteristics we start by writing (16) in the form

$$\tau_{wr}^2 = 2\int_0^\infty K^2(\omega)\int_0^{2\pi} S_{u_b}(\omega,\theta)d\theta d\omega =$$
$$\int_0^\infty \left[K^2(\omega_r) + \left.\frac{\partial K^2}{\partial \omega}\right|_{\omega=\omega_r}(\omega-\omega_r) + \cdots\right]\int_0^{2\pi} 2S_{u_b}(\omega,\theta)d\theta d\omega \quad (20)$$

Neglecting higher order terms in the expansion of the transfer function $K^2(\omega)$ around $\omega = \omega_r$, the expression given by (20) reduces to that for a simple periodic wave

$$\tau_{wr} = \tau_{wm} = K(\omega_r)u_{br} \quad (21)$$

in which τ_{wm} is the maximum bottom shear stress of the periodic wave with the representative wave orbital velocity amplitude given by

$$u_{br} = \sqrt{2\iint S_{u_b}(\omega,\theta)d\omega d\theta} \quad (22)$$

and the representative wave radian frequency taken as

$$\omega_r = \frac{\iint \omega S_{u_b}(\omega,\theta)d\omega d\theta}{\iint S_{u_b}(\omega,\theta)d\omega d\theta} \quad (23)$$

Finally, the direction of propagation of the representative periodic wave may be obtained from (17) as

$$\tan \phi_{wr} = \frac{\iint S_{u_b}(\omega,\theta) \sin\theta \, d\omega d\theta}{\iint S_{u_b}(\omega,\theta) \cos\theta \, d\omega d\theta} \qquad (24)$$

when it is assumed either that $K(\omega)$ is sufficiently accurately represented by $K(\omega_r)$ or that the directional spreading function for $S_{u_b}(\omega,\theta)$ is independent of radian frequency.

For completeness it is noted that the expression for ω_r, (23), differs from that obtained by Madsen et al. (1988). Since Madsen et al. (1988) based their representative radian frequency on rather intuitive arguments whereas (23) is based on more rigorous considerations the expression given here should be considered the correct definition of ω_r. The effects of small variations of ω_r on the predicted wave-current interaction are, however, insignificant compared to other uncertainties so this point is made primarily to avoid confusion.

Application of Spectral Wave-Current Model

Pertinent Formulas

To illustrate the application of the spectral wave-current model we assume that the current is specified and that the characteristics of the representative periodic wave are known.

We first define the angle between current direction, ϕ_c, and direction of wave propagation, ϕ_{wr}, as

$$\phi_{cw} = \phi_c - \phi_{wr} \qquad (25)$$

With this definition, the representative shear velocity is obtained from (19) as

$$u_{*r}^2 = \frac{1}{\rho}\left|\tau_{wr}\{1,0\} + \tau_c\{\cos\phi_{cw}, \sin\phi_{cw}\}\right| = C_\mu u_{*wm}^2 \qquad (26)$$

in which $u_{*wm} = \sqrt{\tau_{wr}/\rho}$ is the shear velocity based on the maximum representative wave-associated shear stress,

$$C_\mu = \left(1 + 2\mu|\cos\phi_{cw}| + \mu^2\right)^{1/2} \qquad (27)$$

and

$$\mu = \tau_c/\tau_{wr} = \left(\frac{u_{*c}}{u_{*wm}}\right)^2 \qquad (28)$$

expresses the ratio of current and wave bottom shear stresses; a ratio which generally is much smaller than unity and therefore results in values of C_μ, given by (27), close to unity.

To obtain the maximum wave shear stress, we introduce the wave friction factor concept in the presence of a current through the definition

$$\frac{1}{\rho}\tau_{wr} = u_{*wm}^2 = \frac{1}{2}f_{wc}u_{br}^2 \qquad (29)$$

Introducing (14) and (15), with $\omega = \omega_r$, in (21) and using (29) with (26) to replace shear stresses and shear velocities lead to an implicit equation for the wave friction factor in the presence of a current

$$\sqrt{f_{wc}/C_\mu} = \kappa\sqrt{2\zeta_{ro}}\left|\frac{-\text{ker}'2\sqrt{\zeta_{ro}} - i\text{kei}'2\sqrt{\zeta_{ro}}}{\text{ker}2\sqrt{\zeta_{ro}} + i\text{kei}2\sqrt{\zeta_{ro}}}\right| \qquad (30)$$

in which

$$\zeta_{ro} = \frac{k_N\omega_r}{30\kappa u_{*r}} = \frac{(k_N\omega_r)/(u_{br}C_\mu)}{(30\kappa/\sqrt{2})\sqrt{f_{wc}/C_\mu}} \qquad (31)$$

Written in this form clearly brings out the feature that the wave friction factor, in the presence of a current, is a function of the relative magnitude of the current shear stress, expressed through the factor C_μ defined by (27) and (28), and the bottom roughness, $(C_\mu u_{br}/\omega_r)/k_N = C_\mu A_{br}/k_N$, relative to the representative wave near-bottom orbital excursion amplitude, $A_{br} = u_{br}/\omega_r$, modified by the factor C_μ to account for the presence of a current. It is particularly interesting to note that (30) and (31) reduce to the pure wave case in the absence of any current since then $C_\mu = 1$. Thus, (30) and (31) may be regarded as the generalized wave friction factor relationship valid both in the presence and absence of a current.

To evaluate the wave friction factor as a function of relative roughness, series expansions for zeroth order Kelvin functions and their derivatives, given in Abramowitz and Stegun (1972, Chapter 9), are used to obtain $\sqrt{f_{wc}/C_\mu}$ from (30) with von Karman's constant $\kappa = 0.4$ for a chosen value of ζ_{ro}. Once $\sqrt{f_{wc}/C_\mu}$ is obtained from (30) the corresponding relative roughness, $C_\mu u_{br}/(\omega_r k_N)$, is obtained from (31). The resulting relationships between wave friction factor and relative roughness is shown in Figure 1, and may be approximated by the following explicit formulas

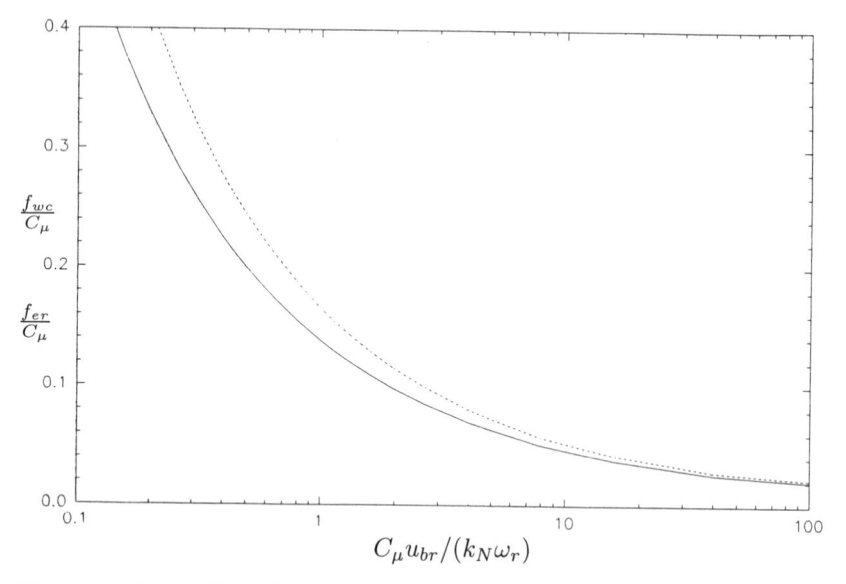

Figure 1: Generalized Friction Factor Diagram for Waves in the Presence of a Current. Wave Friction Factor, f_{wc}, (dashed line), Wave Energy Dissipation Factor, f_{er} (full line), for the Representative Periodic Wave.

$$f_{wc} = C_\mu \exp\left\{7.02\left(\frac{C_\mu u_{br}}{k_N \omega_r}\right)^{-0.078} - 8.82\right\} \qquad (32)$$

for $\qquad 0.2 < C_\mu u_{br}/(k_N \omega_r) < 10^2;$

and

$$f_{wc} = C_\mu \exp\left\{5.61\left(\frac{C_\mu u_{br}}{k_N \omega_r}\right)^{-0.109} - 7.30\right\} \qquad (33)$$

for $\qquad 10^2 < C_\mu u_{br}/(k_N \omega_r) < 10^4.$

These expressions are accurate to within about 1% for the indicated ranges of relative roughness and are far superior to the approximate implicit friction factor equations that may be drived from (30) and (31) under the assumption of small values ζ_{ro}, i.e. large values of $C_\mu u_{br}/(k_N \omega_r)$. Thus the wave-current friction factor equation given e.g. by Madsen et al. (1988) is not only far more cumbersome to use but it is also less accurate than (33)! The author is indebted to Prof. Stephen R. McLean (U.C. Santa Barbara, personal communication) for pointing out the advantages of friction factor formulas of this type (originally suggested by Swart, 1974).

The Wave Boundary Layer Thickness, δ_{wc}

The matching level for the current profile, here referred to as the wave boundary layer thickness and denoted by δ_{wc}, has up to this point not been defined in quantitative terms.

Considerations of δ_{wc} as the level required for the wave orbital velocity to approach its free stream value within a certain percentage were used by Grant and Madsen (1979 and 1986) to arrive at

$$\delta_{wc} = \alpha \frac{\kappa u_{*r}}{\omega_r} \qquad (34)$$

with α-values in the range of 1 to 2. Madsen and Wikramanayake (1991) concluded from a comparison of current velocity profiles predicted by the Grant-Madsen model with limited experimental result as well as predictions afforded by more sophisticated turbulence closure models that reasonable agreement was obtained for α-values in the range of 1 to 1.5.

If therefore one accepts (34) as the appropriate expression for δ_{wc} one is faced with the problem of predicting a wave boundary layer thickness smaller than the equivalent Nikuradse sand grain roughness, k_N, when the roughness is large. This apparent inconsistency of the wave-current theory may be removed by requiring that δ_{wc} should be at least some fraction of the Nikuradse roughness, k_N. Choosing, somewhat arbitrarily, $\delta_{wc} \geq k_N$ for the current profile matching level leads to a limiting value of (34) given by

$$\frac{k_N \omega_r}{\kappa u_{*r}} = 30 \zeta_{ro} \leq \alpha \qquad (35)$$

in which (31) was used.

Choosing, in honor of Bill Grant, $\alpha = 2$ use of (35) in (30) and subsequent use of (31) leads to a limiting value of the relative roughness for which δ_{wc} is determined from (34), and results in the following definition:

$$\delta_{wc} = \begin{cases} 2\kappa u_{*r}/\omega_r & \text{for } C_\mu u_{br}/(k_N \omega_r) > 8 \\ k_N & \text{for } C_\mu u_{br}/(k_N \omega_r) < 8 \end{cases} \qquad (36)$$

Whereas there is evidence in support of (34) for small relative roughness, the "prediction" of $\delta_{wc} = k_N$ should be regarded as tentative (at best). Since large roughness values, for naturally occuring wave-current flows over a movable bed, generally are associated with ripples and since $k_N \simeq 4\times$ (ripple height) for two-dimensional equilibrium-range ripples, e.g. Grant and Madsen (1982), the prediction from (36) of $\delta_{wc} = 4\times$ (physical scale of two-dimensional roughness features) does, however, appear reasonable.

Application

Assuming the bottom roughness ($k_N = 30z_o$) and the representative periodic wave characteristics (u_{br}, ω_r, and ϕ_{wr}) known the method of solution depends on the specification of the current.

Current Specified by u_{*c}^2 and ϕ_c. This specification may physically arise in near-shore waters with a strong wind-driven shore-parallel velocity. In this situation the bottom shear stress may be approximated by the shore-parallel wind stress.

Obtaining ϕ_{cw} from (25), solution is started by initially taking $\mu = \mu^{(o)} = 0$ in (28), to obtain $C_\mu = C_\mu^{(o)} = 1$ from (27). With $C_\mu = C_\mu^{(o)}$ and known wave and bottom roughness conditions $f_{wc} = f_{wc}^{(o)}$ is obtained from (32) or (33), whichever is appropriate. Now, $f_{wc}^{(o)}$ and (29) provide a first estimate of $u_{*wm}^{(o)}$, so that (28) may be revisited to obtain an improved estimate of $\mu = \mu^{(1)}$. With $\mu = \mu^{(1)}$ the procedure is repeated and iteration is terminated when $f_{wc}^{(n+1)} = f_{wc}^{(n)}$ within 1%.

Convergence is generally achieved within a few iterations and from knowledge of u_{*c} and u_{*wm} the represenatative shear velocity, u_{*r}, is obtained from (26). The current velocity profile may now be evaluated from (9) and (10) with δ_{wc} specified by (36) and the apparent bottom roughness, z_{oa}, may be determined from (11).

Current Specified by ϕ_c and $u_c(z_r) = u_{cr}$. This specification of the current corresponds to a current velocity and direction measured at a given elevation, z_r, above the bottom. (It is assumed that $z_r > \delta_{wc}$.)

For this specification of the current the solution procedure is somewhat more cumbersome. Again, after use of (25) the iterations are started by $\mu = \mu^{(o)} = 0$ in (28) and $C_\mu = C_\mu^{(o)} = 1$ from (27) resulting in $f_{wc} = f_{wc}^{(0)}$ from (32) or (33) and then $u_{*wm} = u_{*wm}^{(o)}$ from (29) followed by $u_{*r} = u_{*r}^{(o)}$ from (26). Since it is assumed that $z_r > \delta_{wc}$, the specified current velocity is given by (10), which with the aid of (11) may be written as a quadratic equation in the current shear velocity, u_{*c},

$$u_c(z_r) = u_{cr} = \frac{u_{*c}}{\kappa} \ell n \frac{z_r}{\delta_{wc}} + \frac{u_{*c}^2}{\kappa u_{*r}} \ell n \frac{\delta_{wc}}{z_o} \qquad (37)$$

Since δ_{wc} is given by (36) this equation, with $\kappa = 0.4$, $z_o = k_N/30$ and the latest value of u_{*r} may be solved to give

$$u_{*c} = \frac{u_{*r}}{2} \frac{\ell n(z_r/\delta_{wc})}{\ell n(\delta_{wc}/z_o)} \left(-1 + \sqrt{1 + \frac{4\kappa \ell n(\delta_{wc}/z_o)}{(\ell n(z_r/\delta_{wc}))^2} \frac{u_{cr}}{u_{*r}}} \right) \qquad (38)$$

For the initial iteration $u_{*r} = u_{*r}^{(o)}$ in (36) and (38) yields the first approximation for the current shear velocity $u_{*c} = u_{*c}^{(o)}$. With $u_{*c} = u_{*c}^{(o)}$ and $u_{*wm} = u_{*wm}^{(o)}$ the value of μ may be updated by use of (28) and the procedure may be repeated until convergence is achieved ($f_{wc}^{(n+1)} = f_{wc}^{(n)}$ within 1%). Again, convergence is generally achieved after a couple of iterations and the current velocity profile is determined from (9) and (10) with the apparent bottom roughness obtained from (11).

Discussion and Conclusions

Based on the linearized governing equations and adopting a simple time-invariant eddy viscosity a solution for combined wave-current turbulent bottom boundary layer flows was obtained for a wave motion specified by its near-bottom orbital velocity directional spectrum, $S_{u_b}(\omega, \theta)$. Closure was achieved by requiring the model to reduce the Grant-Madsen model in the limit of simple periodic plane waves. The spectral wave-current interaction model was used to determine the characteristics of a representative periodic wave which, in the context of combined wave-current bottom boundary layer flows, was equivalent to the directional sea. This representative periodic wave is specified by u_{br}, its near-bottom orbital velocity amplitude,

$$u_{br} = \iint 2 S_{u_b}(\omega, \theta) d\omega d\theta,$$

i.e. the root-mean-square bottom velocity amplitude of the directional sea; ω_r, its radian frequency,

$$\omega_r = \frac{\iint \omega S_{u_b}(\omega, \theta) d\omega d\theta}{\iint S_{u_b}(\omega, \theta) d\omega d\theta},$$

i.e. the mean frequency of the directional sea; and ϕ_{wr}, its direction of propagation,

$$\phi_{wr} = \arctan \frac{\iint \omega S_{u_b}(\omega, \theta) \sin \theta d\omega d\theta}{\iint S_{u_b}(\omega, \theta) \cos \theta d\omega d\theta},$$

i.e. the mean direction of the directional sea.

Although based on the simple Grant-Madsen eddy viscosity formulation it is believed that the representative periodic wave characteristics determined here can be adopted with any wave-current interaction model to determine the effect of a directional sea on the near-bottom flow associated with a superimposed current. This transferability of the present results to any model that is based on a time-invariant eddy viscosity is assured on theoretical grounds and is not likely to seriously affect the practical significance of results obtained from elaborate numerical turbulent-closure models.

The most severe limitation of the spectral wave-current model's ability to predict current velocity profiles in the presence of waves is associated with the uncertainty in the determination of matching level for the current profile segments, *i.e.* the wave boundary layer thickness, δ_{wc}, for large values of the equivalent bottom roughness. This limitation is not unique to the class of models presented here. It is inherent in all theoretical formulations of turbulent boundary layer flows that apply the turbulent no-slip condition at $z = z_o$, and may, for large roughness, lead to the nonsensical result of a boundary layer thickness less than the physical scale of bottom roughness elements. Clearly, when δ_{wc} is not large relative to the physical scale of the bottom roughness, assuming a horizontally uniform flow is a poor assumption. This issue is particularly important in the context of wave-current interaction in the coastal environment where wave-generated bottom bedforms (ripples) create a large bottom roughness. Theoretical and especially experimental studies are required to shed some light on this problem whose existence is acknowledged here by tentatively suggesting $\delta_{wc} \geq k_N$.

In retrospect these characteristics of the representative periodic wave can hardly be considered surprising. The most important wave characteristic in terms of wave-current interaction is the magnitude of the bottom orbital velocity since this has the greatest effect on the bottom shear stress which, in turn, dominates the turbulence intensity within the wave boundary layer. Since the wave-associated bottom shear stress is proportional to the square of the near-bottom orbital velocity amplitude, u_{bm}^2, with the "constant" of proportionality, the wave friction factor, being "essentially constant" it follows directly that the near-bottom velocity of a representative periodic wave should be $(\overline{u^2}_{bm})^{1/2}$, *i.e.* exactly the form of u_{br} that we obtained through laborious, albeit theoretically rigorous, considerations. Despite the "intuitively obvious" nature of our rigorously derived expression for the representative periodic wave's bottom orbital velocity amplitude it is interesting to note that several investigators have chosen to represent a random sea by a "significant" bottom velocity amplitude, *i.e.* greater than our u_{br} by a factor of $\sqrt{2}$. The theoretically rigorous derivation of the representative periodic wave characteristics presented here therefore firmly establishes the *significance* of the root-mean-square near-bottom wave orbital velocity (and the *insignificance* of the "significant" velocity) in the context of combined wave-current bottom boundary layer flows.

Acknowledgements

The research presented in this paper was sponsored by the MIT Sea Grant College Program under Grant NA90-AA-D-SG424 from the Office of Sea Grant of NOAA during the early stages, and continued under sponsorship from NSF's Marine Geology and Geophysics Program Grant Nos. OCE-9017878 and 9314366 and NSF's Ocean Sciences CoOP Grant No. OCE-9123513.

References

Abramowitz, M., I.A. Stegun (1972) *Handbook of Mathematical Functions.* National Bureau of Standards Applied Math Series, No. 55, pp. 379-509.

Grant W.D., O.S. Madsen (1979) Combined wave and current interaction with a rough bottom. *Journal of Geophysical Research* 84(C4):1979-1808.

Grant, W.D., O.S. Madsen (1982) Moveable bed roughness in oscillatory flow. *Journal of Geophysical Research* 87(C1):469-481.

Grant, W.D., O.S. Madsen (1986) The continental shelf bottom boundary layer, In *Annual Review of Fluid Mechanics* (M. Van Dyke, ed.), 18:265-305.

Kajiura, K. (1968) A model of the bottom boundary layer in water waves. *Bulletin of the Earthquake Research Institute* 46:75-123.

Madsen, O.S., Y.-K. Poon, H.C. Graber (1988) Spectral wave attenuation by bottom friction: Theory. *Proceedings 21st International Conference on Coastal Engineering*, ASCE, Torremolinos. 1:492:504.

Madsen, O.S., P.N. Wikramanayake (1991) Simple models for turbulent wave-current bottom boundary layer flow. U.S. Army Corps of Engineers, WES, Report No. DRP-91-1.

Mathisen, P.P., O.S. Madsen (1993) Bottom roughness for wave and current boundary layer flows over a rippled bed. MIT Sea Grant College Program. Tech. Rept. MITSG93-27.

Swart, D.H. (1974) Offshore sediment transport and equilibrium beach profiles. Delft Hydraulic Laboratory. Publication No. 131.

Appendix A: Spectral Wave Energy Dissipation

As mentioned earlier the effect of evaluating the bottom shear stress at $z = z_o(\zeta = \zeta_{ro})$ rather than by taking the limit $z \to 0 (\zeta \to 0)$ has pronounced effects for larger roughness values. Thus the wave friction factors predicted by (32) are significantly lower than those obtained if the limit $z \to 0$ were used to define the bottom shear stress. Furthermore, the phase difference, φ_τ, between bottom shear stress and free stream wave orbital velocity, which is important for the computation of wave energy dissipation in the wave bottom boundary layer, is extremely sensitive to the choice of definition, $z = z_o$ or $z \to 0$, used to evaluate the bottom shear stress for large values of the bottom roughness.

The phase difference, φ_τ, is given by the argument of the complex fraction of Kelvin functions and their derivatives in (12), (14) and (30). The phase difference, $\varphi_{\tau r}$, for the representative periodic wave, obtained from (30), may be approximated by the explicit relationship

$$\phi_{\tau r}^\circ = 33 - 6.0 \log_{10} \frac{C_\mu u_{br}}{k_N \omega_r} \quad \text{for} \quad 0.2 < C_\mu u_{br}/(k_N \omega_r) < 10^3 \quad (39)$$

which is accurate within $1°$ for the range indicated.

For each wave component the rate of energy dissipation in the bottom boundary layer, D_{nm}, is obtained from Kajiura (1968) with bottom shear stress given by (12) and free stream velocity by (6), evaluated for $\zeta_n \to \infty$. The resulting formula is written in compact form by using (26) and (29) to express u_{*r} and generalizing (30) and (39) for arbitrary choice of ω. The end result is

$$D_{nm} = \overline{\tau_{\mathbf{w}nm} \cdot (\mathbf{u}_{\mathbf{w}nm})_{\zeta_n \to \infty}} = \frac{1}{4} \rho \sqrt{f_{wc}} \sqrt{f_{wc,n}} \cos \varphi_{\tau n} u_{br} \mathbf{u}_{\mathbf{b}nm}^2 \quad (40)$$

in which $f_{wc,n}$ and $\varphi_{\tau n}$ are obtained from (32) or (33) and (39), respectively, with ω_r replaced by ω_n and $\mathbf{u}_{\mathbf{b}nm}$ is given by (7).

Madsen et al. (1988) used the $z \to 0$ definition to obtain the wave friction factor and neglected the influence of the phase difference, φ_τ, in the evaluation of the energy dissipation rate. Strictly speaking this limits the applicability of their results to the range of relative roughness associated with (33). For this range $f_{wc,n}$ and $\cos \varphi_{\tau n}$ depend weakly on ω_n and may be replaced by their values for $\omega_n = \omega_r$. In this way (40) becomes equivalent to Eq. (26) in Madsen et al. (1988) if their "f_{wr}" is replaced by f_{er} = "the representative wave energy dissipation factor" = $f_{wc} \cos \varphi_{\tau r}$. For small bottom roughness $\cos \varphi_{\tau r} \simeq 1$ and $f_{er} = f_{wc}$ makes (40) identical to Eq. (26) of Madsen et al.. However, as seen in Figure 1, f_{er} may be significantly different from f_{wc} for large bottom roughness.

CHAPTER 30

TIME DOMAIN MODELLING OF
WAVE BREAKING, RUNUP, AND SURF BEATS

P.A. Madsen[*], O.R. Sørensen[*] and H.A. Schäffer[*]

Abstract

In this paper we study wave breaking and runup of regular and irregular waves, and the generation of surf beats. These phenomena are investigated numerically by using a time-domain primary-wave resolving model based on Boussinesq type equations. As compared with the classical Boussinesq equations the ones adopted here allow for improved linear dispersion characteristics, and wave breaking is incorporated by using a roller concept for spilling breakers. The swash zone is represented by incorporating a moving shoreline boundary condition and radiation of short and long period waves from the offshore boundary is allowed by the use of absorbing sponge layers. The model results presented include wave height decay, mean water setup, depth-averaged undertow, shoreline oscillations and the generation and release of low frequency waves.

Introduction

Previous work on the modelling of surf beats and low frequency waves in the surf zone has mainly been based on the wave-averaged approach for the primary waves combined with linear or nonlinear equations for the long waves (Symonds et al., 1982; Schäffer, 1993; Roelvink, 1993).

In the analytical work of Symonds et al. (1982) a sinusoidal variation of the break point was assumed corresponding to weakly modulated short waves. Inside the surf zone the modulation or groupiness of these waves was assumed to vanish and the wave height was taken to be a fixed proportion of the local water depth. The low frequency waves were described by the linear shallow water equations driven by radiation stresses according to linear wave theory. Incoming bound long waves and frictional effects were neglected.

---

[*] International Research Centre for Computational Hydrodynamics (ICCH). Located at the Danish Hydraulic Institute, Agern Alle 5, DK-2970 Hørsholm, Denmark

Schäffer (1990, 1993) improved this analytical work in a number of ways of which the inclusion of the incoming bound long waves and their transformation on a sloping beach turned out to be most important for the results. He also considered different types of surf zone models for the incident modulated short waves: The first one was the saturation approach by Symonds et al. (1982), neglecting groupiness inside the surf zone. The second one assumed a fixed break point position but allowed for full transmission of groupiness into the surf zone. The model finally adopted was a hybrid between the two allowing the transmission of groupiness to be zero, partial, full or even to be reversed.

Roelvink (1993) developed a numerical model using a nonlinear description of the low frequency waves based on the nonlinear shallow water equations including bottom friction and radiation stress terms. The primary waves were described by a wave energy balance equation including dissipation terms similar to the formulation by Battjes and Janssen (1978).

As an alternative to the wave-averaged approaches Watson and Peregrine (1992) used the nonlinear shallow water (NSW) equations to resolve the short wave motion as well as the long wave motion. A shock-capturing method allowed for an automatic treatment of bores and shocks without the need for any special tracking algorithm. The drawback of this method is, however, the lack of frequency dispersion, which forces the high frequency waves to move with shallow water celerities and to steepen into bores at a certain distance from the seaward boundary. Another problem is the prediction of the fluctuating surf zone width. It is well known that the NSW equations are not able to predict the break point position in regular waves. Hence it is unlikely that they should be able to give a good estimate of the time varying break point in wave groups and irregular waves.

In the present work we have studied and modelled the nonlinear interaction processes described above on the basis of a special type of Boussinesq equations. The equations were derived by Madsen et al. (1991) and Madsen and Sørensen (1992) and have proved to incorporate improved linear dispersion characteristics and shoaling properties, which are important for a correct representation of the nonlinear energy transfer (Madsen and Sørensen, 1993). Incorporation of wave breaking is an essential part of the model complex and it is based on the concept of surface rollers following the formulation by Schäffer et al. (1992,1993). The model has proved to be able to represent a variety of processes such as the initiation and cessation of spilling wave breaking, and the evolution of wave profiles before, during and after wave breaking.

The present paper presents a further extension of the Boussinesq model by including the swash zone and the moving shoreline. The new extension of the model makes it possible to simulate wave breaking and runup of irregular waves and to study the generation of surf beats and low frequency waves induced by short-wave groups.

Only 1D results will be presented in this paper, while the extention to 2D horizontal problems is described by Sørensen et al. (1994).

A Simple Model for Spilling Breakers

The incorporation of wave breaking for spilling breakers in the Boussinesq model is based on the concept of surface rollers as described by Schäffer et al. (1992,1993). A brief summary of the concept will be given in the following. First of all it can be split up into two parts:
- Determination of the spatial and temporal variation of surface rollers.
- Determination of the effect of the rollers on the wave motion.

The determination of the rollers is based on the heuristic geometrical approach by Deigaard (1989). First of all wave breaking is initiated when the local slope of the surface elevation exceeds a critical value, $\tan\phi$. Due to the transition from initial breaking to a bore-like stage in the inner surf zone this angle, ϕ, is assumed to vary in time. Breaking is assumed to be initiated for an angle of 20 Degrees, which then gradually changes (with an exponential decay) to a smaller terminal angle of 10 Degrees. Locally, the roller is defined as the water above the tangent of $\tan\phi$ and wave breaking is assumed to cease when the maximum of the local slope becomes less than $\tan\phi$.

The determination of the effect of the surface rollers on the wave motion is inspired by the simple model suggested by Svendsen (1984). The basic principle is that the surface roller is considered as a volume of water being carried by the wave with the wave celerity and this is assumed to result in the vertical distribution of the horizontal particle velocity shown in Fig 1. Although this is a very simple approximation it allows for the description of some important physical phenomena:
- It leads to additional convective terms in the depth-integrated momentum equations. This results in a conversion of potential energy into forward momentum flux immediately after breaking and while the wave height starts to decay the radiation stress may keep constant for a while. This leads to the well known horizontal shift between the break point and the point where the setup in the mean water level is initiated.
- The inclusion in the mass balance of the net mass transport due to the roller has a significant effect on the time-average of the depth-averaged particle velocities and improves the prediction of the depth-averaged undertow in the surf zone.

These two features are illustrated in the following by Fig 2, dealing with wave height decay and setup, and by Fig 3 dealing with wave height decay and undertow.

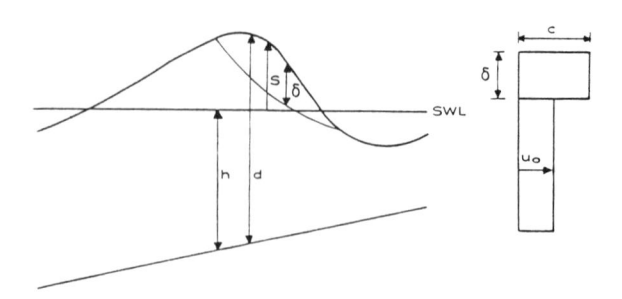

Figure 1 Cross-section and assumed velocity profile of a breaking wave with a surface roller.
Definitions: S=elevation, d=total water depth, δ=roller thickness, c=wave celerity, Uo=particle velocity.

Fig 2 shows a comparison with the measurements of test 1 presented by Stive (1980). In this test incident monochromatic waves shoal and break as spilling breakers on a plane slope of 1:40. The wave period is 1.79 s and at the seaward boundary the depth is 0.70 m and the wave height is 0.145 m. From the variation in wave heights wave breaking is seen to occur at a distance of 24.5 m from the wavemaker, while the setup in mean water level starts at 25.5 m. The numerical results obtained by the present Boussinesq model are slightly underestimating the last part of the shoaling, breaking occurs at 24.0 m and setup starts at 24.7 m. Except for the discrepancy near the break point the overall agreement with the measurements of wave heigth and setup is acceptable.

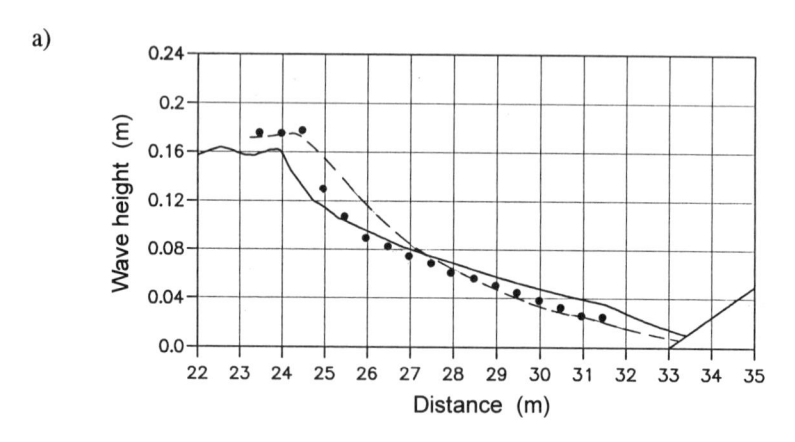

a)

b)

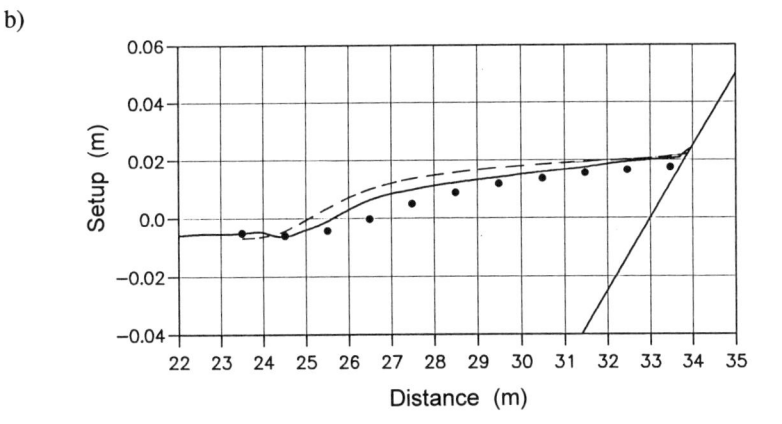

[Figure: Setup (m) vs Distance (m), showing computed and measured MWL variation from 22 to 35 m]

Figure 2   Monochromatic waves on a mild slope
              a) Computed and measured variation of the wave height
              b) Computed and measured variation of MWL
              —— Present model ---- Kobayashi et al. (1989)
              • Measurements by Stive (1980)

Kobayashi et al. (1989) solved the nonlinear shallow water NSW equations by using a dissipative shock-capturing method in which bores or shock fronts are frozen to cover only a few grid points. Due to the lack of dispersion this type of model will predict breaking to occur very near the seaward boundary. For this reason the NSW calculations were started very near the observed breaking point at a depth of 0.2375 m with an incoming wave heigth of 0.172 m. From Fig 2 we notice that for the computed results the positions of the break point and the start of setup coincide. Furthermore, the wave height decay is clearly underestimated initially. However, the overall performance of the NSW model is very good.

Fig 3 shows a comparison with the measurements by Hansen and Svendsen (1984). In this test incident monochromatic waves shoal and break as spilling breakers on a plane slope of 1:34.25. The wave period is 2.0 s and at the seaward boundary the depth is 0.36 m and the wave height is 0.12 m. The Boussinesq model was started at the toe of the slope with a cnoidal input. The variation of the wave height is shown in Fig 3a. We notice that the position of the break point is quite well predicted, and also the rate of energy decay after breaking is satisfactory. However, again the maximum wave heights occurring just before breaking are underestimated in the simulation. The depth-averaged undertow is obtained by determining the time-average of the particle velocity U_o, defined in Fig. 1. In Fig. 3b this is compared to the measurements of Hansen and Svendsen (1984). The agreement is seen to be much improved as compared with the NSW results by Kobayashi et al. (1989).

a)

b)

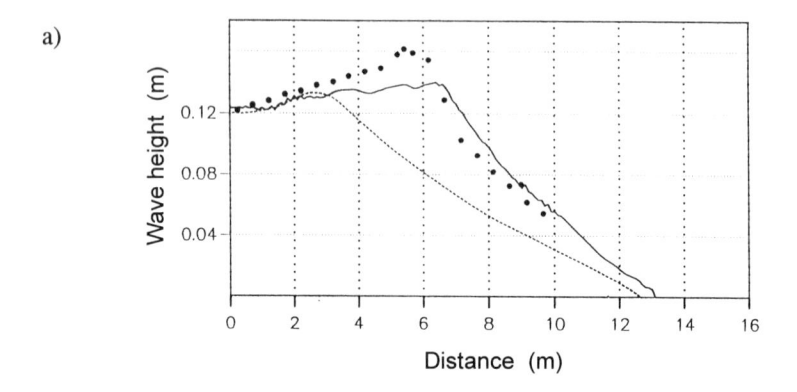

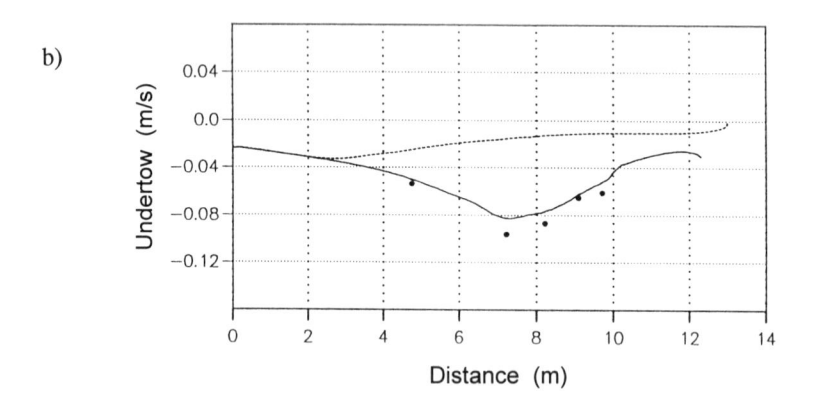

Figure 3 Monochromatic waves on a mild slope
a) Computed and measured variation of the wave height
b) Computed and measured depth-averaged undertow
——— Present model ---- Kobayashi et al. (1989)
• Measurements by Hansen and Svendsen (1984)

For this case measured wave profiles immediately seaward of the observed break point were not available, and consequently the NSW model was started at the toe of the slope. This position being quite far from the break point makes it a very difficult test for the NSW model. Clearly, the NSW model fails to predict the break point and the wave height starts decaying much too early. This generally leads to a significant underestimation of the wave height. Also, the undertow is clearly underestimated, partly due to the discrepancy in the wave height and partly due to the fact that the NSW equations do not include the important net mass transport in the rollers.

Swash Oscillations and Runup

One of the difficult points in simulating runup of regular and irregular waves is the treatment of the moving shoreline. In the present work we have adopted a modified version of the slot-technique described by Tao (1983). A brief summary of this concept will be given in the following.

First of all the computational domain is extended artificially to cover the solid beach by introducing narrow channels or slots in which the waves can propagate. The width of the slot will enter the depth-integrated mass and momentum equations in two ways:
- Generally the water depth will be replaced by a cross sectional area which will include the slot area.
- The time derivative of the surface elevation will be multiplied by the slot width.

The artificial slots have to be very narrow to avoid a distortion of the mass balance and a disturbance of the flow in the physical domain. On the other hand, the numerical solution will break down when the width becomes too small. In practice the width is chosen in the interval between 0.01 and 0.001 times the grid size. An alternative way to interpret this technique is to consider the solid beach to be replaced by a porous beach with a very low porosity factor (the relative slot width).

In order to make this technique operational in connection with Boussinesq type models a couple of problems call for special attention:
- The Boussinesq terms are switched off at the still water shoreline where their relative importance is extremely small anyway.
- The convective terms are treated by central differences in the physical flow domain and by upwind differences inside the slots.
- An explicit filter is introduced near the still water shoreline to remove short wave instabilities during uprush and downrush.

Some of the advantages of the method is that it works very well in combination with implicit numerical schemes and it is quite easy to generalize and implement in two horizontal dimensions. The drawback is that the method will always introduce minor errors in the mass balance and these will generally lead to an underestimation of the maximum uprush and downrush on impermeable slopes.

To check the accuracy of the method we have tested the numerical model against the analytical solution for non-breaking shallow water waves on a sloping beach (Carrier and Greenspan, 1958). The analytical solution is based on the nonlinear shallow water equations, hence for this test case we have switched off the Boussinesq terms everywhere.

The test case considered is an initial water depth of 0.5 m, a wave period of 10 s, a slope of 1/25, and a wave height of 50% of the limiting value (giving breaking at rundown). Fig 4 shows the horizontal motion of the shoreline computed with three different values of the relative slot width (0.01, 0.005 & 0.001). The numerical solution is clearly converging towards the analytical solution, but even for a value of 0.001 the maximum runup is still underestimated by 8%. The reason is that the upper part of the swash zone is containing only a thin film of water and this is quite sensitive to even small portions of water entering the porous beach.

Further testing and a more complete description of the technique will appear in Madsen et al. (1995).

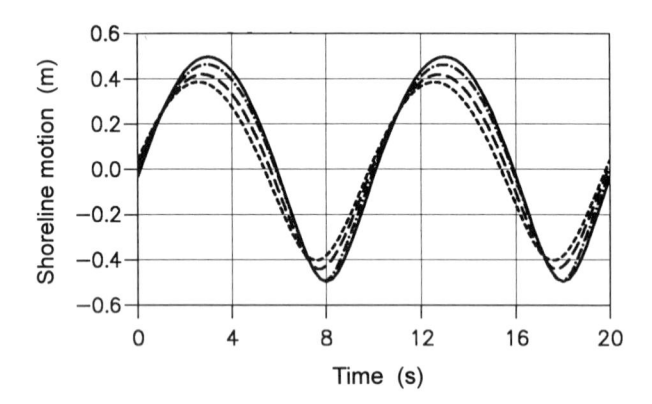

Figure 4 Horizontal motion of the shoreline
——— Analytical solution by Carrier & Greenspan (1958)
--------- Present model with a slot width of 0.01
- - - -   -"-   slot width of 0.005
-·-·-·-   -"-   slot width of 0.001

Low Frequency Waves in the Surf Zone

The extension of the Boussinesq model to include wave breaking of spilling breakers and a moving shoreline formulation makes it possible to simulate the generation of surf beats due to shoaling, breaking and runup of irregular waves. The simplest possible study of this phenomenon can be made with a bichromatic wave group composed of a superposition of two sine waves with slightly different frequencies.

For simplicity we consider a test case where linear boundary conditions can be applied at the seaward boundary. The bathymetry consists of a horizontal section of 40 m with a water depth of 2.0 m and a section of 75 m with a constant slope of 1/34.25 (Fig 5). Waves are generated internally and re-reflection of waves from the seaward boundary is avoided by using a 5 m wide sponge layer. As input we apply linear bichromatic waves with frequencies of 0.4 hz and 0.5 hz and with amplitudes of 0.03 m on both frequencies. This leads to a *linear* wave group in the horizontal section of the flume, and it has several advantages: One is that higher order boundary conditions are not neccessary and the other is that the determination of the amplitude of the outgoing free low frequency wave due to the surf beat can be determined accurately by simple means. This is not the case if the seaward boundary is placed in shallow water. In order to resolve the superharmonics in shallow water we use a grid size of 0.1 m and a time step of 0.02 s.

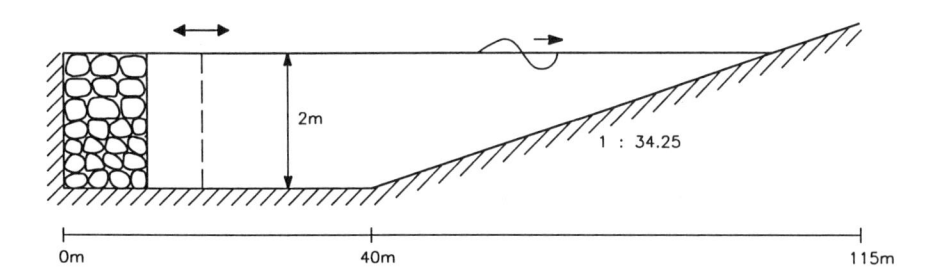

Figure 5   Bichromatic waves on a mild slope. Sketch of model setup.

Fig 6 shows the instantaneous surface elevation as a function of the distance from the seaward boundary. The individual waves can be seen to steepen and break in shallow water. The figure also contains two curves, showing the maximum and minimum elevation obtained during the last two wave groups of the simulation. The maximum curve has three spikes indicating three different break points. The outermost point is seen to be at 101 m and the envelope drops steeply landwards of this point due to the roller dissipation. The second break point occurs at 102.5 m and a new steep descent follows. This pattern indicates that the highest waves decay to become smaller than the second highest waves before the latter start decaying. In other words, the modulation of the primary waves or the groupiness is reversed inside the surf zone. This phenomenon was also discussed by Schäffer (1993).

Fig 7 shows the resulting surf beat obtained by low-pass filtering the surface elevation time series in each grid point with a cutoff at 0.1 hz. The envelope plot shows a combination of bound and free long waves running in the onshore direction and reflected free waves running in the offshore direction. Furthermore, the stationary setup is included in the plot. An almost perfect nodal point is occurring at 94 m, but further offshore the nodes become more and more open indicating that the outward free long waves become dominant compared to the inward bound long waves. The overall picture for the long waves shows the same behaviour as the analytical solution by Schäffer (1993) for weakly modulated primary waves.

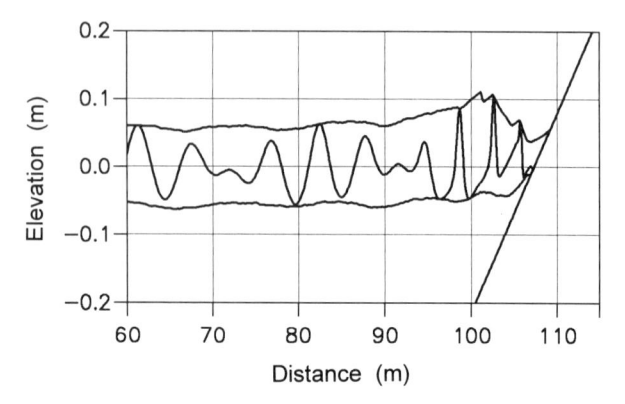

Figure 6 Bichromatic waves on a mild slope. Spatial variation of the surface elevation.

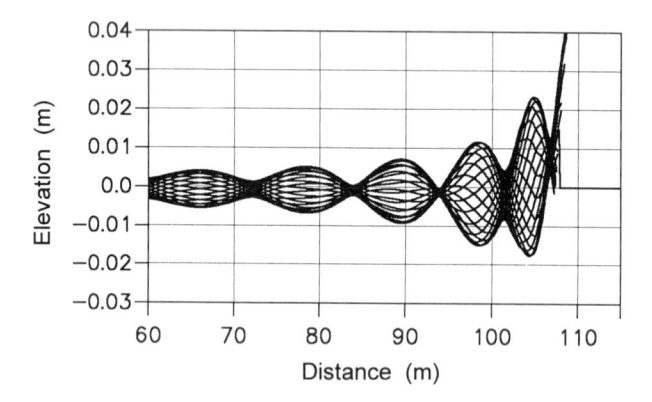

Figure 7 Bichromatic waves on a mild slope. Envelope of lowpass filtered surface elevations.

Fig 8 shows the temporal variation of the shoreline and the spatial and temporal evolution of the individual rollers identified by the model. The outermost break point is again spotted at 101 m and we clearly see how the break point changes in time. The slope of the trajectories of the rollers indicate the local speed of the breaking waves and especially close to the shoreline this is clearly influenced by the swash oscillations. The temporal variation of the shoreline shows a low frequency variation superimposed by individual swash oscillations. Unfortunately, the use of the slot-technique for handling the moving boundary has the undesired effect, that the individual wave runups disappear too rapidly due to the thin film of water penetrating the porous beach. In reality we expect the shoreline motion to be less spiky because the thin film of water will stay on the slope until the next wave arrives. This aspect requires further investigation.

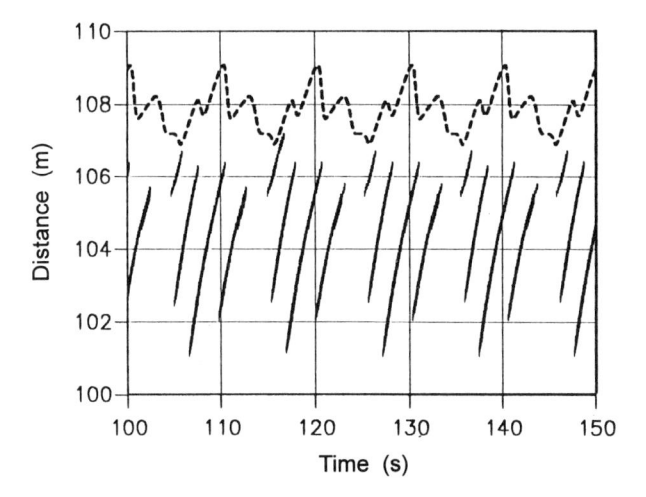

Figure 8 A time-space plot of the horizontal motion of the shoreline (---) and the tracks of detected surface rollers (—).

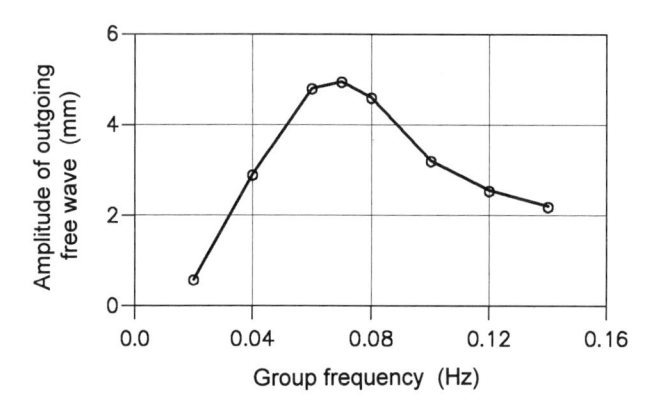

Figure 9 The amplitude of the outgoing free long wave as a function of the group frequency, Δf.

To investigate the sensitivity of the surf beat to the variation of the group frequency a number of simulations were made with different fully modulated bichromatic wave inputs: The first frequency was fixed to be 0.5 hz while the second was varied in the range of 0.36 hz - 0.48 hz, leading to group frequencies in the interval of 0.14 hz to 0.02 hz. Input amplitudes of 0.03 m on both frequencies were used in all cases. Fig 9 shows the amplitude of the resulting outgoing free long wave as a function of the group frequency, Δf. A clear local maximum is obtained for a value of 0.07 hz indicating that the surf beat mechanism is indeed sensitive to the ratio between the surf zone width and the wave length of the low frequency waves. The local maximum in the resulting amplitude occurs when the long waves reflected from the shoreline are in phase with the long waves generated to move in the offshore direction by the oscillating break point. This confirms the mechanism first described by Symonds et al. (1982).

Further investigations and comparisons with e.g. the measurements by Kostense (1984) will appear in Madsen et al. (1995).

Conclusion

A Boussinesq type model with improved linear dispersion characteristics and with a roller concept for spilling breakers is extended to include the swash zone and the moving shoreline. The model is capable of representing a variety of processes such as the initiation and cessation of breaking, and the evolution of wave profiles before, during and after wave breaking. In this work we have concentrated on resulting wave-averaged quantities such as wave height decay, mean water setup and depth-averaged undertow.

The new extension of the model to include the swash zone makes it possible to simulate wave breaking and runup of irregular waves and to study the generation of surf beats and low frequency waves induced by short-wave groups. The test cases presented is this paper are only first examples of what can be achieved by the use of the Boussinesq model. The results obtained so far are qualitatively correct and most promising. However, further work is necessary to quantify the accuracy of the results.

Acknowledgement

This work has been financed mainly by the Danish National Research Foundation and partly by the Commission of the European Communities, Directorate General for Science, Research, and Development under MAST contract No. MAS2-CT92-0027. Their financial support is greatly appreciated.

References

Battjes, J.A. and J.P.F.M. Janssen (1978), "Energy loss and setup due to breaking in randow waves". Proc. 16th Int. Conf. on Coastal Eng., ASCE, pp 569-587.

Carrier, G.F. and H.P. Greenspan (1958),"Water waves of finite amplitude on a sloping beach". Journal of Fluid Mechanics, Vol 4, pp 97-109.

Deigaard, R. (1989), "Mathematical Modelling of waves in the surf zone". Prog. Report 69, pp 47-59, ISVA, Technical University, Lyngby, Denmark.

Hansen, J.B. and I.A. Svendsen (1984), "A theoretical and experimental study of undertow", In Proc. of the 19th Coastal Eng. Conf., pp 2246-2262.

Kobayashi, N., G.S. De Silva and K.D. Watson (1989), "Wave transformation and swash oscillation of gentle and steep slopes". Journal of Geophysical Research, Vol. 94, No. C1, pp 951-966.

Kostense, J.K. (1984),"Measurements of surf beat and set-down beneath wave groups". In Proc. of the 19th Coastal Eng. Conf., pp 724-740.

Madsen, P.A., R. Murray and O.R. Sørensen (1991), "A new form of the Boussinesq equations with improved linear dispersion characteristics. Part 1", Coastal Engineering, Vol 15, pp 371-388.

Madsen, P.A. and O.R. Sørensen (1992), "A new form of the Boussinesq equations with improved linear dispersion characteristics. Part 2: A slowly-varying bathymetry", Coastal Engineering, Vol 18, pp 183-204.

Madsen, P.A. and O.R. Sørensen (1993), "Bound waves and Triad Interactions in shallow water", Ocean Engineering, Vol 20, pp 359-388.

Madsen, P.A., O.R. Sørensen and H.A. Schäffer (1995), "Wave breaking, swash oscillations and surf beats simulated by a Boussinesq type model". Submitted for publication.

Roelvink, J.A. (1993)," Surf beat and its effect on cross-shore profiles". Technical University of Delft, 117 pp.

Schäffer, H.A. (1990), "Infragravity water waves induced by short-wave groups". Technical University of Denmark, Series paper 50, 168 pp.

Schäffer, H.A., R. Deigaard and P.A. Madsen (1992), "A two-dimensional surf zone model based on the Boussinesq equations". In Proc. of the 23th Coastal Eng. Conf., Chapter 43, pp 576-590.

Schäffer, H.A. (1993), "Infragravity waves induced by short-wave groups". J. Fluid Mech. (1993), vol. 247, pp. 551-588.

Schäffer, H.A., P.A. Madsen and R. Deigaard (1993), "A Boussinesq model for waves breaking in shallow water". Coastal Engineering, Vol 20, pp 185-202.

Stive, M.J.F. (1980),"Velocity and pressure field of spilling breakers".In Proc. of the 17th Coastal Eng. Conf., pp 547-566.

Svendsen, I.A. (1984),"Wave Heights and setup in a surf zone". Coastal Engineering, Vol 8 (4), pp 303-329.

Symonds, G., G.A. Huntley and A.J. Bowen (1982), "Two dimensional surf beat: Long wave generation by a time-varying break point". J. Geophys. Res, 87, C1, pp 492-498.

Sørensen, O.R., H.A. Schäffer, P.A. Madsen and R. Deigaard (1994), "Wave breaking and induced nearshore circulations modelled in the time domain". In Proc. of the 24th Coastal Eng. Conf., Kobe.

Tao, J. (1983). "Computation of wave run-up and wave breaking". Internal Report, 40 pp, Danish Hydraulic Institute.

Watson,G. and D.H. Peregrine (1992), "Low frequency waves in the surf zone". In Proc. of the 23th Coastal Eng. Conf., Chapter 61, pp 818-831.

CHAPTER 31

Orthonormal Wavelet Analysis for Deep-Water Breaking Waves

Nobuhito MORI [*] and Takashi YASUDA [†]

ABSTRACT

This study aims to develop a method using wavelet transform to detect wave breaking event from temporal water surface elevation data. Sudden surface jump associated with breaking wave is regarded as shock wave and shock wavelet spectrum is defined to detect the occurrence of the surface jump. The visual observation of breaking wave crest shows that this method can almost completely detect the occurrence of breaking wave in random wave trains.

1. INTRODUCTION

Wave breaking plays important roles in numerous aspects of horizontal and vertical momentum transfer from surface waves to current and mixing of surface layer. Since breaking waves are associated with steep and giant waves, wave breaking is very important phenomenon to estimate their upper limit and the occurrence probability of their wave height. Furthermore, breaking waves exert the strong wave induced force, which occasionally bring about impact pressure to structures.

A great deal of effort has been made on direct observations of wave breaking in the ocean. One of them is direct visual observation of white caps to detect the breaking waves[Holthuijsen & Herbers(1986)], because the wave breaking is related to some sort of instability near the crests. Since they required much efforts, the visual observations are not adequate to ordinary routine observation. Another approach is to detect the breaking events in the time series of surface elevation, directly. Longuet-Higgins & Smith(1983) observed breaking waves by using a surface jump meter. Recently, Su & Cartmill(1993) developed a

---

[*]Graduate student, Graduate School of Gifu University, Yanagido, Gifu 501-11, Japan.
[†]Professor, Dept. of Civil Eng., Gifu University, Yanagido, Gifu 501-11, Japan.

new detection method of breaking wave by a void fraction technique. These direct measurement methods are quite well but require special and sophisticated instrument, so that they are not also in general use. Weissman et al.(1984) measured energy backscattering of the high frequency components due to wave breaking with the running Fourier spectrum method. Their detection method is empirical and has not objectivity, nevertheless their method need not any special instruments. They use the Fourier spectrum analysis that is generally effective to analyze an energy change, periodicity and a power law of data. However, it is not suitable to use unsteady process such as breaking. The reason why the Fourier analysis has not temporally or spatially local information on data is that its integral basis consists of periodic function.

Recently, a new method of aperiodic and unsteady data analysis which has a locally confined integral basis, so-called *'wavelet analysis'*, is getting well known[for example, Meyer(1991) and Farge(1992)]. Shen et al.(1994) studied local energy characteristics of wind generated waves using a continuous wavelet transform. Since the continuous wavelets, however, have overcomplete basis which causes formal relations among expansion coefficients, they are not suitable to analyze local energy properties. Meyer(1989) studied and formulated the orthogonal analyzing wavelet system. This orthogonal wavelet transform is known as adequate analysis of the local energy characteristics of the data[Mori et al.(1993)].

In this study we make a rational breaking wave detection scheme to indicate and to measure small jumps and discontinuities in surface elevation associated with breaking waves. Further, we check the validity of the method by experimental data and analyze the local energy properties of breaking wave using the orthonormal wavelet analysis.

2. PRINCIPLE OF MEASUREMENT

2.1 Orthonormal wavelet expansion

Since the kernel functions of continuous wavelet transforms are not mutually orthogonal, we have the redundancy of wavelet coefficients independent of data. The excellent mathematical formulation of orthogonal analyzing wavelet, *multi-resolution analysis*, was developed by Mallat(1989). Accordingly, in this paper, we employ the orthonormal wavelet expansion to analyze a water surface elevation.

The orthogonal wavelet expansion of an arbitrary function $\eta(t)$ is written as

$$\eta(t) = \sum_{j=1}^{\infty} \sum_{k=1}^{\infty} \alpha_{j,k} \psi_{j,k}(t), \quad (j, k \in Z) \tag{1}$$

in which $\alpha_{j,k}(j,k \in Z$, where Z is the set of all integers) is an expansion coefficient and $\psi_{j,k}(t) \in L(\mathbf{R}^2)$ is a complete orthonormal set of wavelets generated from an analyzing wavelet, which is sometimes called mother wavelet, by discrete translations k and is corresponding to temporal position of time $k/2^j$, and discrete dilations j corresponds to frequency. It is conventional to take the discrete dilation as

$$\psi_{j,k}(t) = 2^{\frac{j}{2}}\psi(2^j t - k), \quad (j,k \in Z) \tag{2}$$

with the orthonormality condition. From the orthogonality, the wavelets allow us to obtain the expansion coefficient $\alpha_{j,k}$ in Eq.(1) by taking the innerproduct of $\eta(t)$ and $\psi_{j,k}^*(t)$ as

$$\alpha_{j,k} = \int_{-\infty}^{\infty} \eta(t)\psi_{j,k}^*(t)dt. \tag{3}$$

There are typically three types of the orthonormal analyzing wavelets as Meyer's, Daubechies' and Battle-Lemariés's. To investigate the complete relation to the Fourier analysis for the purpose of this study, we follow Meyer's method(1989) which has the properties that; i) $\psi(t)$ is a real analytic function, ii) $\psi(t)$ and its derivatives of any order are rapidly decreasing functions, iii) the moments of any order are zero, and iv) the Fourier transform of $\psi(t)$ has a compact supported in the Fourier space.

We define the mother function $\tilde{\phi}(\omega)$ of analyzing wavelet that is an infinitely differentiable real function satisfying the following conditions,

$$a) \quad \left. \begin{array}{l} \tilde{\phi}(\omega) \geq 0, \\ \tilde{\phi}(\omega) = \tilde{\phi}(-\omega), \\ \tilde{\phi}(\omega) \text{ is monotonically decreasing for } \omega \geq 0, \end{array} \right\} \tag{4}$$

$$b) \quad \left. \begin{array}{ll} \tilde{\phi}(\omega) = 1 & (|\omega| \leq 2\pi/3), \\ = 0 & (|\omega| \geq 4\pi/3), \end{array} \right\} \tag{5}$$

$$c) \quad \{\tilde{\phi}(\omega)\}^2 + \{\tilde{\phi}(\omega - 2\pi)\}^2 = 1 \quad (2\pi/3 \leq |\omega| \leq 4\pi/3). \tag{6}$$

The conditions of a) to c) do not uniquely determine $\tilde{\phi}(\omega)$, so that we can make arbitrary functions $\tilde{\phi}(\omega)$, if $\tilde{\phi}(\omega)$ satisfies the above conditions of a) to c). We employ here the mother function $\tilde{\phi}(\omega)$ defined as

$$\tilde{\phi}(\omega) = \sqrt{g(\omega)g(-\omega)}, \tag{7}$$

where

$$g(\omega) = \frac{h(4\pi/3 - \omega)}{h(\omega - 2\pi/3) + h(4\pi/3 - \omega)}, \tag{8}$$

$$h(\omega) = \begin{cases} \exp\left(-1/\omega^2\right), & (\omega > 0) \\ 0, & (\omega \leq 0) \end{cases} \tag{9}$$

(a) Meyer's analyzing wavelet $\psi(t)$

(b) The Fourier spectrum $\tilde{\psi}(\omega)$ of Meyer's analyzing wavelet $\psi(t)$

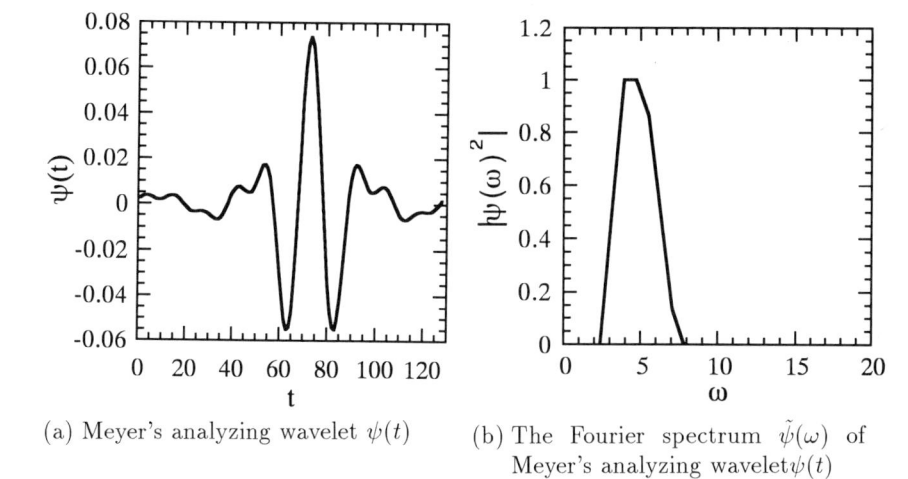

Figure 1: Illustrations of Meyer's analyzing wavelet.

which is the same definition of Yamada & Ohkitani(1990).

From the conditions of a) to c) and Eqs.(7) to (9), the Fourier coefficients of analyzing wavelet $\tilde{\psi}(\omega)$ are defined as

$$\tilde{\psi}(\omega) = e^{(-i\omega/2)}\sqrt{\{\tilde{\phi}(\omega/2)\}^2 - \{\tilde{\phi}(\omega)\}^2}. \tag{10}$$

Note that $\tilde{\phi}(\omega)$ has a compact support in $\{\omega \,|\, 2\pi/3 \leq |\omega| \leq \pi/3\}$. Therefore, the wavelet coefficient is directly connected to the Fourier coefficient.

The inverse Fourier transform of $\tilde{\psi}(\omega)$ gives the following desired analyzing wavelet,

$$\psi(t) = \frac{1}{2\pi}\int_{-\infty}^{\infty} \tilde{\psi}(\omega)e^{i\omega t}d\omega. \tag{11}$$

Fig.1 shows Meyer's analyzing wavelet and its Fourier spectrum. The analyzing wavelet is very regular. However, it is not very well localized in physical space but is supported compact in the Fourier space. The fast algorithm for wavelet transform with Meyer's analyzing wavelet using FFT algorithm is formulated by Yamada & Ohkitani(1991).

2.2 Relation between the wavelet and the Fourier spectra

It could be worth pointing out that the relations between the wavelets and the Fourier spectra. Meyer's analyzing wavelet has a useful property that is the compactness of the support in the Fourier space. Eq.(6) shows that $\tilde{\psi}(\omega)$ is only included in $[-2^{j+3}\pi/3, -2^{j+1}\pi/3] \cup [2^{j+1}\pi/3, 2^{j+3}\pi/3]$. Since the square of wavelet coefficient is corresponding to the energy, the direct relation between

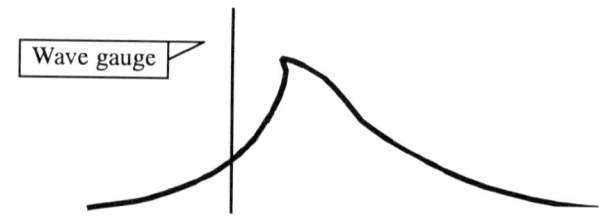

Figure 2: Illustration of breaking wave passing a wave gauge.

the wavelet and the Fourier spectrum is expected as

$$E_j = \sum_{k=1}^{\infty} |\alpha_{j,k}|^2 \sim \omega S(\omega) \quad (\omega \sim 2^{j+2}\pi/3) \qquad (12)$$

where ω is angular frequency, $S(\omega)$ the Fourier spectrum and E_j the wavelet spectrum. Particularly, the relationship of a power law of the energy spectrum between the wavelet and the Fourier spectrum can be expressed as

$$E_j \sim 2^{-j(p-1)} \iff S(\omega) \sim \omega^{-p}. \qquad (13)$$

Eq.(13) gives the relation that ω^{-p} in the Fourier spectrum is equivalent to $2^{-j(p-1)}$ in the wavelet spectrum.

2.3 Local and shock wavelet spectra

The purpose of this study is to detect jumps in water surface elevation associated with breaking. We suppose that the sea surface elevation $\eta(t)$ is measured as a function of time t with a discrete sampling time Δt at a fixed horizontal position as illustrated in Fig.2. The wave passing the sensor will generally show a smooth rise $\partial \eta/\partial t$. But, if a just breaking wave passes the sensor, we can expect a sudden jump of the surface elevation. The magnitude of the sudden jump is associated with the scale of breaker type: *i.e.* for a plunging breaker this is relatively large and it is smaller for spilling breaker.

The Fourier series of shock wave represented as

$$y(t) = \begin{cases} At, & t \leq 0.5, \\ A(t-1), & t \geq 0.5, \end{cases} \quad (0 \leq t \leq 1) \qquad (14)$$

where A is the height of shock is easily given by $\sum(-4/n)\sin(n\omega t)$. Therefore, a power law of the energy spectrum of the breaking wave with the sudden jump is expected to ω^{-2} or 2^{-j}.

To investigate the local energy information among the scale, we define the local wavelet spectrum $L_{j,k'}$ for the scale $j \geq j_s$ as

$$L_{j,k'} = \sum_{k'} |\alpha_{j,k}|^2, \quad (0 \leq k' \leq 2^{j_s}) \qquad (15)$$

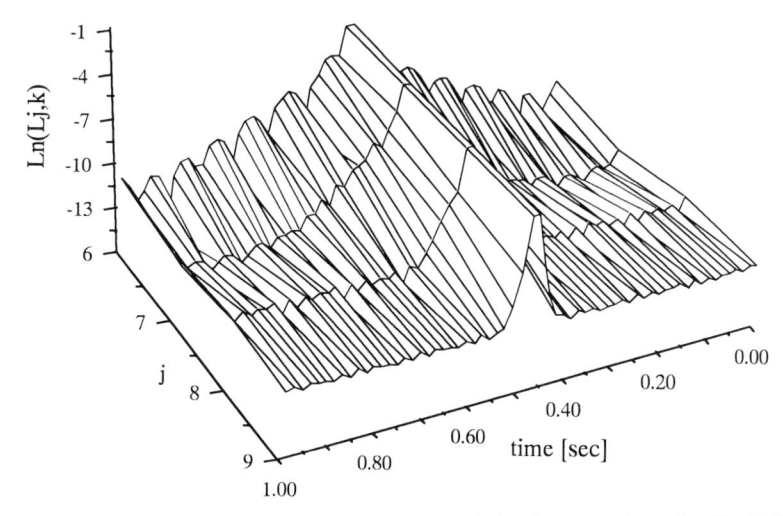

Figure 3: The local wavelet spectrum $L_{j,k'}$ of shock wave given by Eq.(14).

where, j_s is minimum dilation mode of the local wavelet spectrum which determine the resolution and $\sum$ denotes the special summation over k satisfying: $(k/2^{j_s} \leq k'/2^j \leq (k+1)/2^{j_s})$. The local wavelet spectrum $L_{j,k'}$ can analyze characteristics of microscopic or local energy properties for data. Besides, we introduce the shock wavelet spectrum to detect the surface jump of the shock wave described by Eq.(14). The shock wavelet spectrum $M_{j,k'}$ is defined by following as

$$M_{j,k'} = 2^j \times \sum_{k'} |\alpha_{j,k}|^2. \quad (0 \leq k' \leq 2^{j_s}) \tag{16}$$

Since power law of shock wave is 2^{-j} for the wavelet spectrum, that is corresponding to power law of ω^{-2} for the Fourier spectrum, the shock wavelet spectrum detects the shock as a constant power whenever jumps observed in the surface elevation.

Fig.3 shows the local wavelet spectrum(j_s=6) for the shock wave given by Eq.(14). The number of points to discretize the shock wave is 512. The local wavelet spectrum shows the energy distribution of *time-frequency* space and indicates existence of the high crest corresponding the time of shock at t=0.5. This result implies the effectiveness of the local wavelet spectrum for local energy analysis. To make clear the local energy properties of the shock wave, the shock wavelet spectrum(j_s=6) for shock wave of Eq.(14) is shown in Fig.4. We can easily detect the shock from the characteristic structures of shock wavelet spectrum $M_{j,k'}$ both of the occurrence time and their magnitudes.

To say nothing of the accuracy of detection of shock depend on discretization of data. The relation sampling frequency Δt and the local power law is shown

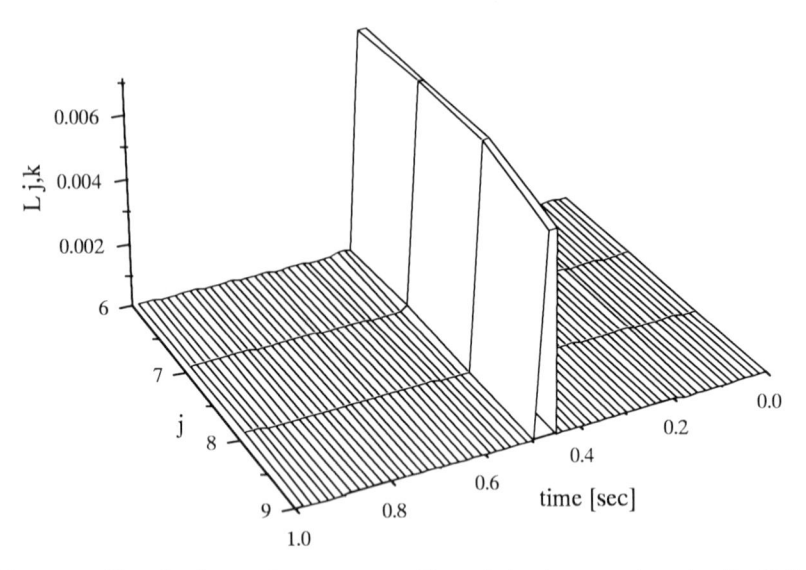

Figure 4: The shock wavelet spectrum $M_{j,k'}$ of shock wave given by Eq.(14).

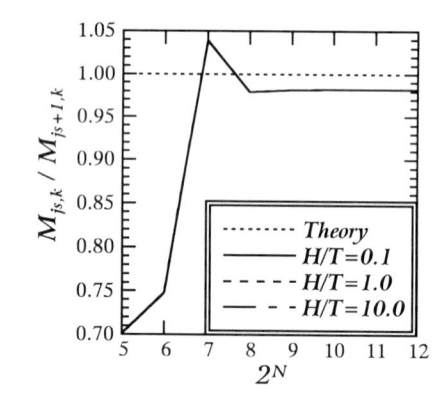

Figure 5: Comparison of the local power law of shock wave between the numerical and analytical one.

Figure 6: Location of wave gages.

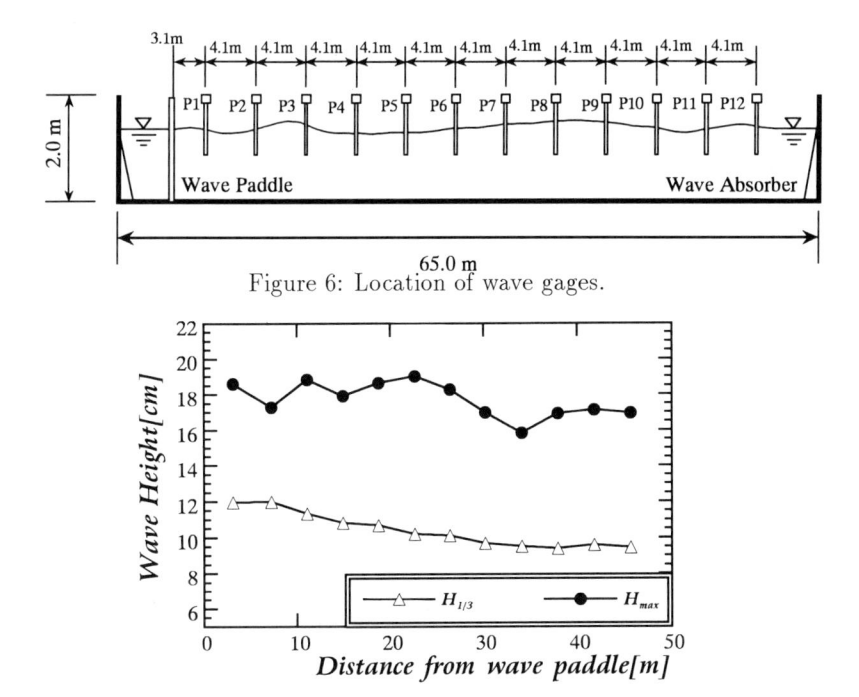

Figure 7: Spatial variation of the wave height statistics.

in Fig.5, where H is the height of the shock wave and T is the wave period. The value of j_s is set equal to $j_{max}-2$ and the local power law of the wavelet spectra are calculating the scales between the j_s and $j_s + 1$. The error of estimating of power law is 2.7% for $N=256$ to 4096, 4% for $N=128$, 25% for $N=64$ and 30% for $N=32$. We conclude that the number of point $N \geq 64$ or 128 per one wave is required to accurate estimation the jump from the data. Note that the accuracy of estimation is independent of the amplitude of the shock wave.

3. EXPERIMENTS

3.1 Experimental condition

The experiments were conducted in the glass channel installed at Technical Research & Development Institute of Nishimatsu construction Co.,Ltd. The channel is 65m long, 1m wide, 2m high and was filled to a depth of 0.98m. Waves were generated by computer-controlled piston type wave paddle. The initial spectra of the surface elevations are composed of the Wallop type spectra with the band width $m=10$ and the peak frequency $f_p=1$Hz, giving a wavenumber $k_p=4.072$m^{-1} and characteristic water depth $k_p h=3.99$, so that the waves

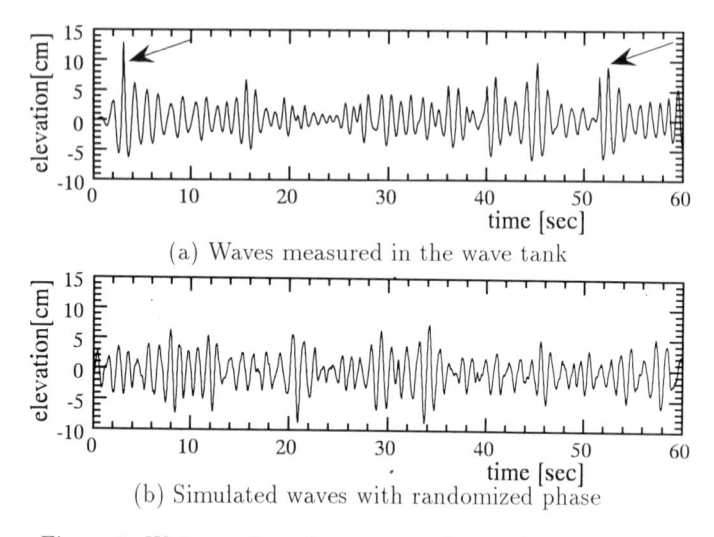

[Figure 8: Water surface elevations of the random wave trains.
(a) Waves measured in the wave tank
(b) Simulated waves with randomized phase]

were deep water waves. Water surface displacements were measured with twelve capacitance type wave gages as shown in Fig.6. The measurements were performed at a sampling frequency of 100Hz for over 100sec. At the same, spatial surface profiles were recorded by video camera to examine the breaking event.

Figure 7 shows the spatial variation of H_{max} and $H_{1/3}$. Although, the value of H_{max} is fluctuated, the spatial variation of $H_{1/3}$ shows that there is some energy dissipation due to the wave breaking. In the following, we only focus the surface elevations at P5.

3.2 Local energy characteristics of nonlinear wave

The values of skewness and kurtosis at P5 are 0.245 and 3.473, so that the waves are found to have weekly nonlinear characteristics. For comparison with the experimental data, we calculate artificial random phase wave data(linear waves) which is obtained by the inverse Fourier transform of the original surface elevations of P5 after randomizing their phases uniformly over$[0,2\pi]$ with their amplitudes unchanged. The surface elevations of experimental data and simulated wave with randomized wave are shown in Fig.8, respectively. The two arrows in the Figure indicate the time when breaking event just occurs.

Weissman et al.(1984) developed the detection method of breaking waves based on the singularities of the high frequencies, which is the intrinsic frequency of gravity-capillary waves, by a trial and error method with the running Fourier transform. The wave profile band-passed of high frequencies(10-12Hz) shown in Fig.8(a) is illustrated in Fig.9. The bursts of energy in the high frequency com-

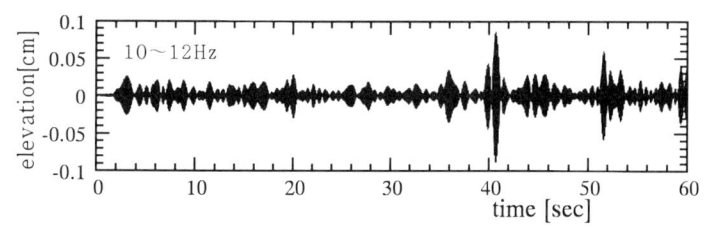

Figure 9: Water surface profile of band-pass filtered wave with the high frequency components(10-12Hz) of Fig.8(a).

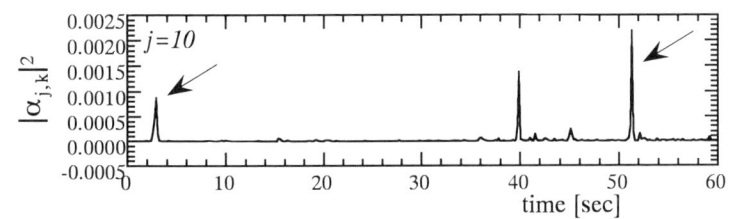

Figure 10: Temporal distribution of the square value of wavelet coefficient, $|\alpha_{j,k}|^2$, of the experimental wave

ponents are indicative of breaking events. He detected wave breaking from burst of the temporal energy distributions in this frequency band. Fig.10 shows the square value of wavelet coefficients $\alpha_{j,k}$ at the scale $j=10$ which is corresponding to about 8Hz. The distributions of square values of $\alpha_{j,k}$ clearly indicate the singularities of the high frequencies, which are corresponding to the time of breaking waves, in comparison with the result of the Fourier band filtered method(Fig.9). The same empirical method to detect the breaking waves can be applied by the wavelet analysis and will be given better result rather than the running Fourier transform, but let us then considered here the energy structure and energy cascade process among lower and higher frequency components.

It was already shown that the local wavelet spectrum is effective to analyze the temporal energy structures in previous section. Fig.11 shows the local wavelet spectra with $j_s=7$ for the experimental wave and the simulated wave, respectively. The experimental data indicate that characteristic structure of energy distribution is shown running in parallel with the j-axis. Particularly, some big crests can be observed at the large scale into small scale. Although the Fourier spectrum is the same as experimental wave, there is no pattern or structure in the local wavelet spectrum as the experimental wave in the simulated wave(b) and they seem to distributing uniformly. This implies that the high frequency components of experimental wave, nonlinear wave, are not constant in amplitude and that there are sharp peaks sporadically distributed along the time series. In other words, the high frequency components are intermittent in

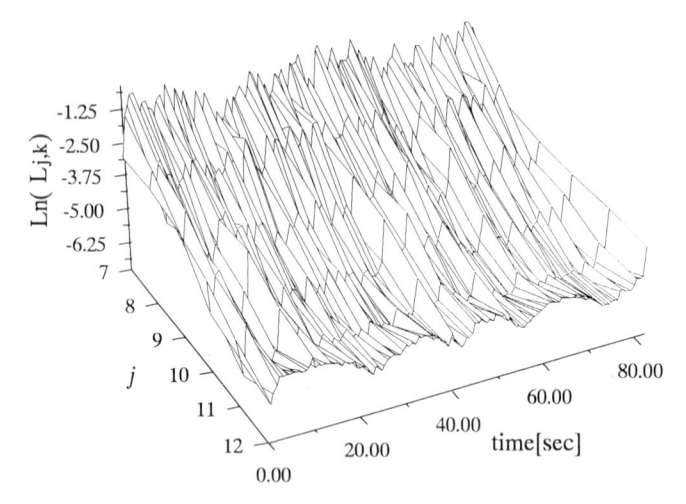

(a) Waves measured in the wave tank

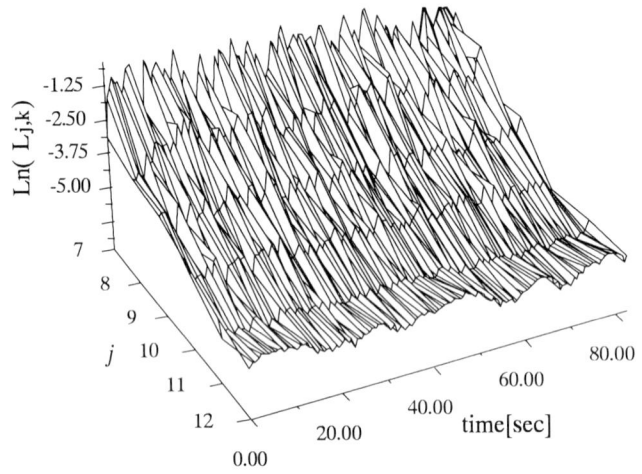

(b) Simulated waves with randomized phase

Figure 11: Local wavelet spectra for the measured and simulated waves.

the nonlinear wave.

The sharp peaks and structures in wavelet space(j,k) are related with nonlinear wave-wave interaction but we may leave the details to further studies.

3.3 Detection of breaking wave

The nonlinear waves have the local energy characteristics as already shown in Fig.11. This will lead us further into a consideration of detection of breaking

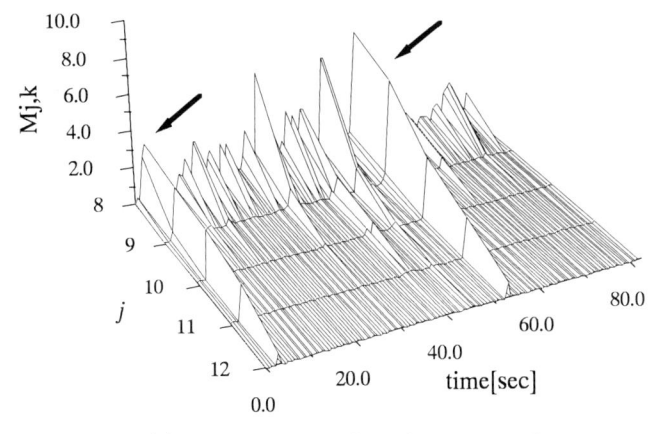

(a) Waves measured in the wave tank

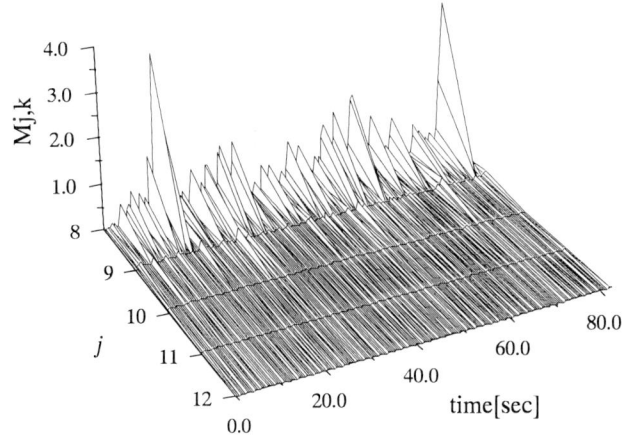

(b) Simulated waves with randomized phase

Figure 12: Shock wavelet spectra for the measured and simulated wave.

wave by shock wavelet spectra. The shock wavelet spectra of experimental wave and simulated wave are shown in Fig.12. The sharp peaks are shown intermittently in the experimental wave, corresponding to the time when breaking wave passes. Therefore, we need some detector function to judge whether they break or not. Thus, we define the detector function a_i is calculated as

$$a_i = \frac{M_{j_s+m,k_i}}{M_{j_s,k_i}}, \qquad (17)$$

where subscript i denotes temporal position ($0 \leq k' \leq 2^{j_s}$), j_s is resolution of scale parameters and m is distance between the j_s. The illustration of the

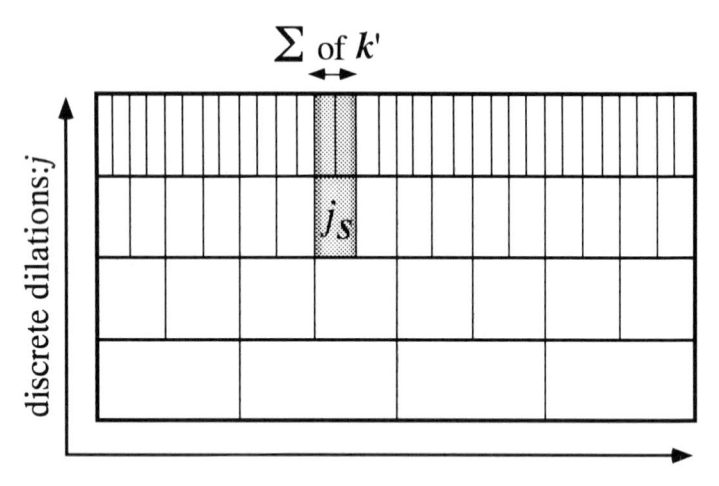

Figure 13: Illustration of wavelet space and relationship with parameters

Table 1: Accuracy of the present detection method

$a \backslash n$		$j_s=7$ $m=1$	$j_s=7$ $m=2$	$j_s=8$ $m=1$	$j_s=8$ $m=2$
0.9	(-1)	2	1	3	2
0.8		3	1	4	2
0.6		4	1	5	2
0.5	(-2)	5	1	5	3

wavelet space and the relationship with parameter is shown in Fig.13. The experimental result of detection of the method for the experimental wave is shown Table 1($j_s=7$ is corresponding peak frequency of the spectrum). The actual number of breaking waves is two, therefore, $j_s=8$, $m=2$, $a_i=0.9$ and 0.8, 0.6 and $j_s=7$, $m=1$, $a_i=0.9$ gives quite nice value when we selected. The fine resolution scale of $j_s=8$ accurately detect the breaking waves rather than $j_s=7$ and the $m=2$ gives more accurate results than $m=1$. The reason for this result is sampling frequency of shock and fluctuation of the local power law due to the external noise of data.

It could be concluded that the present method can detect the breaking wave from the temporal water surface elevation data, if the adequate parameters a and m are selected as $a_i \geq 0.8$ and $m=2$.

4. CONCLUSION

We applied the orthonormal wavelet expansion to the water surface elevations of random waves and studied their local characteristics. It is found that the sudden surface jumps associated with the breaking waves were well reflected in the shock wavelet spectra. Thus, we developed the rational detection method of breaking wave as following procedure:

1. The wavelet coefficients $\alpha_{j,k}$ of the surface elevations are calculated Eq.(3).

2. The appropriate resolution with the scale j_s is selected.

3. The shock wavelet spectra are calculated with Eq.(16).

4. The local power law of the surface elevations are calculated Eq.(17).

5. Breaking waves are expected to have the local power a law 2^{-j} of the wavelet spectrum

Furthermore, the sudden surface jumps can be well detected, if they are described by using sufficiently many discretized points. We demonstrated the validity of the method by comparing with the experimental data. Note that this method can be applied to not only deep water waves but also shallow water.

ACKNOWLEDGEMENT

The authors are grateful to Dr.Yamada at University of Tokyo for valuable advice and wish to thank Technical Research & Development Institute of Nishimatsu construction Co.,Ltd. for their cooperation. One of the authors is supported by fellowships of the Japan Society for the Promotion of Science for Japanese Junior Scientists. Further, this research is supported in part by Grant-in-Aid for Scientific Research(JSPS-0440), The Ministry of Education, Science and Culture.

REFERENCES

Farge, M. (1992) Wavelet transforms and their appllications to turbulence, *Annual.Review.Fluid Mech.*, 24, pp.395-457.

Holthuijsen, L.H. & T.H.C.Herbers(1986) Statistics of breaking waves observed as whitecaps in the open sea, *J.Phys.Oceanogr*, 16, pp.290-297.

Kennedy, R.M. & Snyder, R.L.(1983) On the formation of whitecaps by threshold mechanism Part II: Monte Carlo experiments, *J.Phys.Oceanogr*, 13, pp.1493-1504.

Longuet-Higgins, M.S. & N.D.Smith(1983) Measurement of breaking waves by a surface jump meter, *J.Geo.Res.*, 88, C14, pp.9823-9831.

Mallat, S.(1989) Multiresolution approximations and wavelet orthonormal bases of $L^2(R)$, Tans.Am.Math.Soc., 315, pp.69-89.

Meyer, Y.(1989) Wavelets, eds. J.M. Combes et al., (Springer).

Meyer, Y.(1991) Wavelets and Applications, (Masson-Springer).

Mori, N., T.Yasuda & M.Yamada(1993) Orthonormal wavelet expansion and its application to water waves, Proc.40th Coastal Eng.Conf., JSCE, 40-1, pp.141-145(in Japanese).

Shen, Z., W.Wang & K.Mei(1994) Finestructure of wind waves analyzed with wavelet transform, J.Phys.Oceanogr., 24, pp.1085-1094.

Su, M.Y. & J.Cartmill(1993) Breaking wave measurement by a void fraction technique, WAVE'93, pp.951-962.

Weissman, M.A., S.S.Ataktürk and K.B.Katsaros(1984) Detection of breaking events in a wind-generated wave field, J.Phys.Oceanogr., 14, pp.1608-1619.

Yamada, M. & K.Ohkitani(1990) Orthonormal wavelet expansion and its application to turbulence, Prog.Theor.Phys., 83, No.5, pp.819-823.

Yamada, M. & K. Ohkitani(1991) An identification of energy in turbulence by orthonormal wavelet analysis, Prog.of Theoretical Phys., 86, No.4, pp.799-815.

CHAPTER 32

A Fully-Dispersive Nonlinear Wave Model and its Numerical Solutions

Kazuo Nadaoka[1], Serdar Beji[2] and Yasuyuki Nakagawa[3]

Abstract

A set of fully-dispersive nonlinear wave equations is derived by introducing a velocity expression with a few vertical-dependence functions and then applying the Galerkin method, which provides an optimum combination of the vertical-dependence functions to express an arbitrary velocity field under wave motion. The obtained equations can describe nonlinear non-breaking waves under general conditions, such as nonlinear random waves with a wide-banded spectrum at an arbitrary depth including very shallow and far deep water depths. The single component forms of the new wave equations, one of which is referred to here as "*time-dependent nonlinear mild-slope equation*", are shown to produce various existing wave equations such as Boussinesq and mild-slope equations as their degenerate forms. Numerical examples with comparison to experimental data are given to demonstrate the validity of the present wave equations and their high performance in expressing not only wave profiles but also velocity fields.

INTRODUCTION

Although evolution of non-breaking waves is principally governed by their nonlinearity and dispersivity, there exist no wave equations which can express these two effects under general conditions. For example, the Boussinesq-type equations are *weakly* nonlinear-dispersive equations and can describe only shallow water waves. Although several successful attempts for extending their applicable range in relative water depth have been reported (Madsen, et al., 1991; Nwogu, 1993, etc.), even such an improved model cannot be relied on if the depth becomes comparable with the wave length or more.

---

[1] Department of Civil Engineering, Tokyo Institute of Technology, 2-12-1 O-okayama, Meguro-ku, Tokyo 152, Japan
[2] Faculty of Naval Architecture and Marine Technology, Istanbul Technical University, Maslak 80626, Istanbul, Turkey
[3] Hydrodynamics Laboratory, Port and Harbour Research Institute, Nagase 3-1-1, Yokosuka 239, Japan

The mild-slope equation of Berkhoff (1972) has no restriction on depth; but it can be used only for linear monochromatic (and hence *non-dispersive*) waves. The time-dependent forms of the mild-slope equation (e.g., Smith and Sprinks, 1975) can describe the dispersive evolution of linear random waves; but their band-width of spectrum is restricted to be narrow. (In this sense, they may be called "*narrow-banded* mild-slope equations".)

To break through all these restrictions, in the present study, new fully-dispersive nonlinear wave equations have been developed. Unlike the Boussinesq equations, which are derived with an asymptotic expansion procedure, the new equations are obtained by introducing a velocity expression with a few vertical-dependence functions and then applying the Galerkin method, which provides an optimum combination of the vertical-dependence functions to express an arbitrary velocity field of waves. The derived equations can express nonlinear non-breaking waves under general conditions, such as nonlinear random waves with a *wide-banded* spectrum at an *arbitrary depth* including very shallow and far deep water depths.

In the following sections, the principal idea and derivation procedure of the new wave equations[*] as well as their simplified forms are presented with some numerical examples and their comparison to experimental data to demonstrate the validity of the equations and the performance especially in expressing velocity fields. Besides theoretical relationships of the present theory to various existing wave equations such as Boussinesq and mild-slope equations are also shown.

THEORY

Principal Idea :

Generally speaking, any mathematical procedure to obtain a water-wave equation is a conversion process from original basic equations defined in a 3-D (x,y,z) space to wave equations to be defined in a horizontal 2-D (x,y) space. For this conversion, we must introduce an assumption on the vertical dependence of the velocity field.

For example, the Boussinesq equations are obtained by introducing the following expression with polynomials of z on the velocity potential Φ (e.g., Mei, 1983*)*:

$$\Phi(x,y,z,t) = \sum_{m=0}^{\infty} \Phi_m(x,y,t)(z+h)^m, \qquad (1)$$

where h is the water depth and the vertical coordinate z is taken upward from the still water level. With the Laplace equation of Φ and the boundary condition at the horizontal bottom, the above equation may be expressed as

$$\Phi(x,y,z,t) = \Phi_0 - \frac{(z+h)^2}{2!}\nabla^2\Phi_0 + \frac{(z+h)^4}{4!}\nabla^2\nabla^2\Phi_0 - \cdots, \qquad (2)$$

where $\nabla = (\partial/\partial x, \partial/\partial y)$. Usually only the first two terms in the above equation are retained to derive the Boussinesq equations.

This procedure is a kind of asymptotic expansion of Φ around the long wave limit,

---

[*] The fundamental idea of the present theory and numerical examples only for linear random waves with wide-band spectrum have been given in Nadaoka and Nakagawa (1991, 1993a,b). The extension to nonlinear waves but in more complicated form of equations has been reported in Nadaoka and Nakagawa (1993c).

and hence the Boussinesq equations can be applied only to shallow water waves. This restriction is related to the fact that the asymptotic approximate form of eq.(2) is not enough to express a velocity field under deeper waves. This in turn suggests that derivation of new wave equations with much wider applicability may be achieved by providing a more reasonable way to express the vertical dependence of a velocity field for more general cases including random waves in deep water.

In the present study, the following assumption is introduced to express the horizontal velocity vector, $q = (u,v)$:

$$q(x,y,z,t) = \sum_{m=1}^{N} U_m(x,y,t) F_m(z), \qquad (3)$$

where

$$F_m(z) = \frac{\cosh k_m(h+z)}{\cosh k_m h}. \qquad (4)$$

The choice of cosh functions in the above as the vertical-dependence functions is based on the general 2-D solution of Laplace equation of Φ on the horizontal bottom (e.g., Nadaoka and Hino, 1983),

$$\Phi(x,z,t) = \int_{-\infty}^{\infty} A(k,t) \frac{\cosh k(h+z)}{\cosh kh} \exp(ikx) dk, \qquad (5)$$

where k is the wavenumber and $A(k,t)$ is a time-varying wavenumber spectrum. It should be noted that eq.(5) is valid also for nonlinear waves and hence the use of eq.(4) as the vertical-dependence function $F_m(z)$ is not restricted to linear waves.

In the discrete form of eq.(5),

$$\Phi(x,z,t) \cong \sum_{i=1}^{i\max} A(k_i,t) \exp(ik_i x) \Delta k \frac{\cosh k_i(h+z)}{\cosh k_i h}, \qquad (6)$$

we need a large number of the spectral component $A(k_i,t)$ in case of broad-banded random waves. However this fact does not necessarily mean that N in eq.(3) should be a large number, in spite of the resemblance between eqs.(3) and (6). This is true if each function, $\cosh k_i(h+z)/\cosh k_i h$, in eq.(6) can be expressed by eq.(3) with a *few* prescribed $F_m(z)$.

Galerkin Expression of a Velocity Field :

To examine this, the following approximation has been attempted:

$$\frac{\cosh k(h+z)}{\cosh kh} \cong \sum_{m=1}^{N} Q_m F_m(z), \qquad (7)$$

where k is an arbitrary wavenumber and $F_m(z)$ is as defined in (4). For this approximation we need a mathematical procedure to determine the unknown coefficients Q_m ($m=1,\cdots,N$). For this purpose, in the present study, the Galerkin method has been employed.

Figure 1 shows the results of the approximation for five values of kh, covering very shallow to deep water depths. The number of components in eq.(7), N, is only 4 in this case and the prescribed values of $k_m h$ for $F_m(z)$ are 1.6, 3.5, 6.0, 10.5, respectively. The fact that the remarkably good agreements between the exact and approximated values are obtained for any arbitrary kh means that the

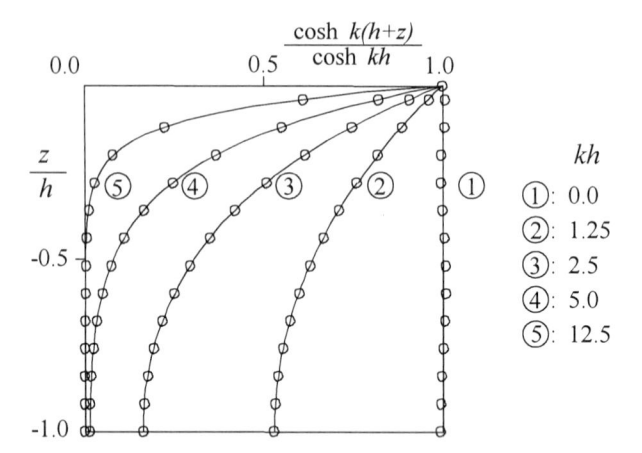

Fig. 1 Comparisons of the exact and approximated vertical distribution functions.

velocity expression by eqs.(3) and (4) with a small N may be applied to wave field under general conditions, such as random waves at an arbitrary depth including very shallow and far deep water depths. This is the most important finding to provide a basis of the new formulation of wave equations described in what follows.

Derivation of Fully-Dispersive Nonlinear Wave Equations :

With this basis of formulation, we are now ready to proceed to the derivation of new wave equations (for details, see Nadaoka et al., 1994).

The basic equations defined in 3-D (x,y,z) space are the continuity equation,

$$\nabla \cdot q + \frac{\partial w}{\partial z} = 0, \tag{8}$$

and an alternative exact form of the Euler equation for irrotational flow (Beji, 1994),

$$\frac{\partial q}{\partial t} + \nabla\left[g\eta + \int_z^\eta \frac{\partial w}{\partial t}dz + \frac{1}{2}\left(q_s \cdot q_s + w_s^2\right)\right] = 0, \tag{9}$$

where q_s and w_s are the velocity components at the free surface $z=\eta$.

The vertical velocity w is obtained from the continuity equation (8) by substituting eqs.(3) and (4) and integrating from the bottom to an arbitrary depth z:

$$w(x,y,z,t) = -\sum_{m=1}^{N} \nabla \cdot \left[\frac{\sinh k_m(h+z)}{k_m \cosh k_m h} U_m(x,y,t)\right]. \tag{10}$$

The vertical integration of the continuity equation (8) over the entire depth gives

$$\frac{\partial \eta}{\partial t} + \nabla \cdot \left(\int_{-h}^{\eta} q dz\right) = 0, \tag{11}$$

which then with the substitution of eqs.(3) and (4) yields

$$\frac{\partial \eta}{\partial t} + \sum_{m=1}^{N} \nabla \cdot \left[\frac{\sinh k_m(h+\eta)}{k_m \cosh k_m h} U_m \right] = 0. \tag{12}$$

To obtain the evolution equations of U_m ($m = 1,\cdots,N$), on the other hand, we may apply the Galerkin method to the momentum equation (9). Namely, after substituting eqs.(3) and (4) into eq.(9), the resulting equation is multiplied by the depth dependent function $F_m(z)$ and vertically integrated from $z=-h$ to η. Since the depth-dependence function has N different modes, we obtain a total of N vector equations corresponding to each mode:

$$\sum_{m=1}^{N} a_{nm} \frac{\partial U_m}{\partial t} + b_n \nabla \left[g\eta + \frac{1}{2}(q_s \cdot q_s + w_s^2) \right] =$$
$$\sum_{m=1}^{N} \left[c_{nm} \nabla (\nabla \cdot U_m)_t + d_{nm} \cdot (\nabla \cdot U_m)_t \right], \quad (n = 1, 2, \cdots, N) \tag{13}$$

where

$$a_{nm} = a_{mn} = \frac{1}{2\cosh k_m h \cosh k_n h} \left\{ \frac{\sinh(k_m + k_n)(h+\eta)}{k_m + k_n} + \frac{\sinh(k_m - k_n)(h+\eta)}{k_m - k_n} \right\},$$

$$b_n = -g \frac{\sinh k_n(h+\eta)}{k_n \cosh k_n h},$$

$$c_{nm} = c_{mn} = \frac{1}{k_m^2 \cosh k_m h \cosh k_n h} \left[\frac{\cosh k_m(h+\eta)\sinh k_n(h+\eta)}{k_n} \right. \tag{14}$$
$$\left. -\frac{1}{2}\left\{ \frac{\sinh(k_m + k_n)(h+\eta)}{k_m + k_n} + \frac{\sinh(k_m - k_n)(h+\eta)}{k_m - k_n} \right\} \right].$$

The coefficients d_{nm} in eq.(13) have rather complicated mathematical forms, but may be evaluated as being nearly equal to D_{nm} shown in eq.(19) later. In this evaluation the neglected terms are $O(\varepsilon \cdot \nabla h)$.

Equations (12) and (13) constitute a solvable set of equations for $2N+1$ unknowns, η, U_m ($m = 1,\cdots,N$), and describe their evolution as wave equations. It should be noted that no approximation has been introduced on the nonlinearity and that the full-dispersivity can be attained by taking only a few components, as demonstrated later; hence eqs.(12) and (13) may be referred to as "fully-dispersive nonlinear wave equations".

It should be further noted that k_m in eqs.(12) and (14) are *not* the wavenumbers in a usual sense like the spectral wavenumbers k_i in eq.(6), but they are the parameters to prescribe $F_m(z)$ so as to approximate a velocity field well enough. The wavenumber parameters k_m ($m = 1,\cdots,N$) are to be specified with the linear dispersion relation, $\omega_m^2 = gk_m \tanh k_m h$, by prescribing the angular frequencies ω_m ($m = 1,\cdots N$) as a set of input data for the computation to properly cover the wave spectrum concerned. Therefore k_m must be treated as spatially varying quantities, according to the variation in $h(x,y)$.

Weakly Nonlinear Version of Fully-Dispersive Wave Equations :

Although equations (12) and (13) may express both full nonlinearity and dispersivity, they have disadvantages in computational aspects; i.e., the coefficients (14) includes many hyperbolic functions, besides the arguments of them have the unknown variable η. These points are undesirable in terms of computational time and robustness of the numerical algorithm.

Therefore, in the present study, a simplified version of eqs.(12) and (13) has been also developed by introducing a weakly-nonlinear formulation. By invoking a Taylor series expansion of q around $z=0$, and keeping only the first-order nonlinear contributions both in eqs.(9) and (11), we obtain

$$\frac{\partial \eta}{\partial t} + \nabla \cdot \left(\int_{-h}^{0} q\, dz + \eta q_0 \right) = 0, \tag{15}$$

$$\frac{\partial q}{\partial t} + \nabla \left[g\eta + \int_z^0 \frac{\partial w}{\partial t} dz + \eta \frac{\partial w_0}{\partial t} + \frac{1}{2}\left(q_0 \cdot q_0 + w_0^2 \right) \right] = 0, \tag{16}$$

in which q_0 and w_0 are the velocities at the still water level $z=0$.

With the corresponding change of the upper limit of the vertical integration from η to 0 in the Galerkin procedure, we get the following simultaneous equations as the weakly-nonlinear version of eqs. (12) and (13).

$$\frac{\partial \eta}{\partial t} + \sum_{m=1}^{N} \nabla \cdot \left[\left(\frac{C_m^2}{g} + \eta \right) U_m \right] = 0, \tag{17}$$

$$\sum_{m=1}^{N} A_{nm} \frac{\partial U_m}{\partial t} + B_n \nabla \left[g\eta + \eta \frac{\partial w_0}{\partial t} + \frac{1}{2}\left(q_0 \cdot q_0 + w_0^2 \right) \right] =$$

$$\frac{\partial}{\partial t} \sum_{m=1}^{N} \left[C_{nm} \nabla (\nabla \cdot U_m) + D_{nm} (\nabla \cdot U_m) \right], \quad (n=1,2,\cdots,N) \tag{18}$$

where,

$$A_{nm} = \frac{\omega_n^2 - \omega_m^2}{k_n^2 - k_m^2}, \qquad A_{nn} = \frac{g\omega_n^2 + h\left(g^2 k_n^2 - \omega_n^4\right)}{2gk_n^2}, \qquad \omega_m^2 = gk_m \tanh k_m h,$$

$$B_n = \frac{\omega_n^2}{k_n^2}, \qquad C_{nm} = \frac{B_n - A_{nm}}{k_m^2}, \qquad D_{nn} = \nabla C_{nn}, \tag{19}$$

$$D_{nm} = \frac{2}{k_m^2 - k_n^2} \left[\frac{2\nabla k_m}{k_m} \left\{ A_{nm} - \left(k_m^2 - k_n^2 \right) C_{nm} \right\} + \frac{g\nabla h}{\cosh k_n h \cdot \cosh k_m h} \right],$$

q_0 and w_0 in eq.(18) may be evaluated as

$$q_0 = \sum_{m=1}^{N} U_m, \qquad w_0 = -\sum_{m=1}^{N} \nabla \cdot \left(\frac{B_m}{g} U_m \right). \tag{20}$$

As shown in (19), the coefficients of the weakly nonlinear version of the equations are considerably simplified as compared with those defined in (14).

Linear Dispersion Characteristics of New Wave Equations :

The linear dispersion characteristics of the present wave equations can be examined by solving the eigenvalue problem defined with the linearized equation of the fully-dispersive equations and with the prescribed values of $k_m h$ ($m=1,...,N$). An example of the computed dispersion curve is shown in Fig. 2, where $N=4$ and the same values as those for Fig.1 are assigned to $k_m h$ ($m=1,...,N$). The computed values show perfect agreements with the theoretical linear dispersion curve over wide wavenumber domain extending from very shallow to far deep water. This remarkable feature of the present wave equations becomes more prominent by comparing with the dispersion curves of the classic Boussinesq equations and of the improved Boussinesq equations (Madsen et al. 1991), as shown in Fig.2.

The reason why the present equations can possess fully dispersive characteristics may be found by examining the dispersive characteristics of the linearized single-component ($N=1$) equation. In this case, the following analytical expression of the

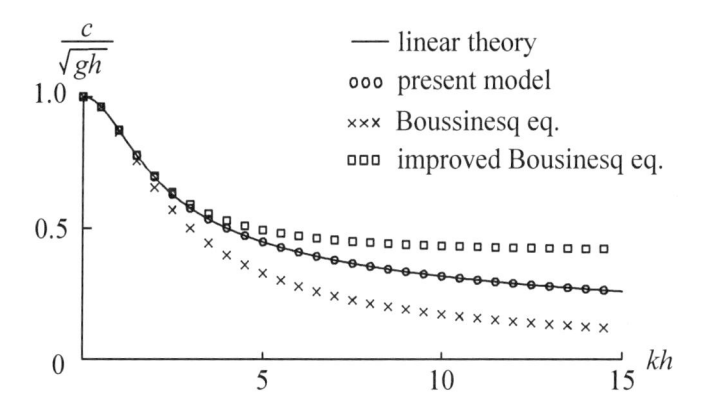

Fig.2 Linear dispersion characteristics of the fully-dispersive equations.

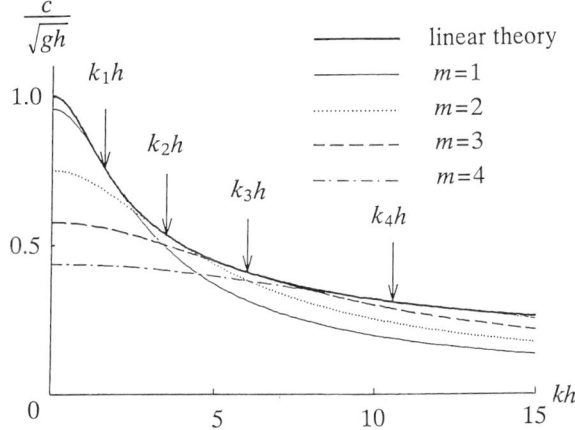

Fig.3 Dispersion curves of the single-component equation with each $k_i h$ for Fig.2.

dispersion relation can be obtained by solving the corresponding eigenvalue problem :

$$C^2 = \frac{C_p^3}{C_g + \frac{k^2}{k_p^2}(C_p - C_g)}, \tag{21}$$

where C_p and C_g are the theoretical linear phase and group velocities corresponding to the prescribed wavenumber k_p, while k and C denote an arbitrary incident wavenumber and the corresponding phase celerity dictated by the dispersion relation (21). Figure 3 depicts the dispersion curves described by eq.(21), in which one of the values of $k_m h$ ($m=1,...,N$) for Fig.2 is given respectively for each curve as $k_p h$. As it is seen, each selected component describes a dispersion curve which is tangent to the exact curve at the selected wavenumber k_p. Hence combining all these contributions by the Galerkin procedure, we obtain the perfect agreement in the expression of the dispersive characteristics as shown in Fig.2.

Single-Component (*N*=1) Forms :

"*Narrow-banded* nonlinear wave equations"

The fact that as shown in Fig.3 each selected component describes a dispersion curve which is tangent to the exact curve at the selected wavenumber k_p means that if the waves in concern have a narrow-band spectrum centered at k_p, the single-component (*N*=1) versions of the wave equations (12) and (13) or (17) and (18) may be employed as "*narrow-banded* nonlinear wave equations".

For example, the single-component forms of eqs.(17) and (18) may be written as:

$$\frac{\partial \eta}{\partial t} + \nabla \cdot \left[\left(\frac{C_p^2}{g} + \eta \right) q_0 \right] = 0, \tag{22}$$

$$C_p C_g \frac{\partial q_0}{\partial t} + C_p^2 \nabla \left[g\eta + \eta \frac{\partial w_0}{\partial t} + \frac{1}{2}(q_0 \cdot q_0 + w_0^2) \right] =$$
$$\frac{\partial}{\partial t} \left\{ \frac{C_p(C_p - C_g)}{k_p^2} \nabla(\nabla \cdot q_0) \right\} + \nabla \left[\frac{C_p(C_p - C_g)}{k_p^2} \right] (\nabla \cdot q_0) \right\}, \tag{23}$$

where C_p and C_g denote the phase and group velocities corresponding to k_p as defined by the linear theory.

By specifying C_p and C_g in these equations, we can show that various existing wave equations may be reproduced as the degenerate forms. For example, Airy's shallow water equations and Boussinesq equations can be obtained as follows.

(1) Airy's shallow water equations: $C_p = C_g = \sqrt{gh}$

$$\frac{\partial \eta}{\partial t} + \nabla \cdot [(h + \eta) q_0] = 0, \tag{24}$$

$$\frac{\partial q_0}{\partial t} + \nabla \left(g\eta + \frac{1}{2} q_0 \cdot q_0 \right) = 0. \tag{25}$$

(2) Boussinesq equations:

$$C_p = \sqrt{gh}\left(1 - \frac{k_p^2 h^2}{6}\right), \quad C_g = \sqrt{gh}\left(1 - \frac{k_p^2 h^2}{2}\right)$$

$$\frac{\partial \eta}{\partial t} + \nabla \cdot [(h+\eta)q_0] + \frac{h^3}{3}\nabla^2(\nabla \cdot q_0) = 0, \tag{26}$$

$$\frac{\partial q_0}{\partial t} + \nabla\left(g\eta + \frac{1}{2}q_0 \cdot q_0\right) = 0, \tag{27}$$

where all the higher-order terms have been neglected.

Combined Form of the Single-Component Equations :
"Time-dependent nonlinear mild-slope equation"

The single-component equations (22) and (23) may be combined, with the introduction of the mild-slope assumption, to give the following equation of η (Beji and Nadaoka, 1994):

$$C_g \eta_{tt} - C_p^3 \nabla^2 \eta - \frac{(C_p - C_g)}{k_p^2}\nabla^2 \eta_{tt} - C_p \nabla(C_p C_g) \cdot (\nabla \eta)$$

$$-\frac{3}{2}gC_p\left(3 - 2\frac{C_g}{C_p} - \frac{k_p^2 C_p^4}{g^2}\right)\nabla^2(\eta^2) = 0. \tag{28}$$

By further manipulations, the linearized equation of (28),

$$C_g \eta_{tt} - C_p^3 \nabla^2 \eta - \frac{(C_p - C_g)}{k_p^2}\nabla^2 \eta_{tt} - C_p \nabla(C_p C_g) \cdot (\nabla \eta) = 0, \tag{29}$$

can be found to lead to the time-dependent (or "*narrow-banded*") mild-slope equation proposed by Smith and Sprinks (1975),

$$\eta_{tt} + \omega_p^2\left(\frac{C_p - C_g}{C_p}\right)\eta - \nabla(C_p C_g \nabla \eta) = 0, \quad (\omega_p = C_p k_p) \tag{30}$$

and also to Berkhoff's (1972) elliptic equation as an original steady form of the mild-slope equation,

$$k_p^2 C_p C_g Z + \nabla \cdot (C_p C_g \nabla Z) = 0, \tag{31}$$

in which Z denotes a spatially varying wave amplitude. Therefore, eq.(28) can be regarded as an extension of the mild-slope equations to nonlinear waves. In this sense, eq.(28) may be called "*time-dependent nonlinear mild-slope equation*". However its linear dispersion characteristics are not the same as those of the time-dependent mild-slope equation (30), since the latter equation approximates more limited region around ω_p in the dispersion curve (Beji and Nadaoka, 1994). This means that even in the linear version of eq.(28), eq.(29), the new mild-slope equation has an advantage as compared with the conventional one.

Unidirectional Simplified Form of Nonlinear Mild-Slope Equation :

Equation (28) may be further elaborated for case of unidirectional propagation of waves in the positive x-direction only. The reason of taking up the analysis of such a simplified case lies in the attractive form of the KdV equation, which will be recovered as a special case. Skipping the derivation procedure (see Beji and Nadaoka, 1994) the equation we obtain is

$$C_g \eta_t + \frac{1}{2} C_p (C_p + C_g) \eta_x - \frac{(C_p - C_g)}{k_p^2} \eta_{xxt} - \frac{C_p(C_p - C_g)}{2k_p^2} \eta_{xxx}$$
$$+ \frac{1}{2}\left[C_p(C_g)_x + (C_p - C_g)(C_p)_x\right]\eta + \frac{3}{4} g \left(3 - 2\frac{C_g}{C_p} - \frac{k_p^2 C_p^4}{g^2}\right)(\eta^2)_x = 0, \quad (32)$$

which describes the weakly-nonlinear wave evolution of a narrow-banded unidirectional wave field centered at the primary wave frequency $\omega_p = k_p C_p$.

The specification of C_p and C_g in eq.(32) yields again some degenerate forms. For weakly-dispersive shallow water waves, the specification,

$$C_p = \sqrt{gh}\left(1 - \frac{k_p^2 h^2}{6}\right), \quad C_g = \sqrt{gh}\left(1 - \frac{k_p^2 h^2}{2}\right)$$

leads to the KdV equation for a gently varying depth :

$$\eta_t + C_0\left[\eta_x + \frac{h_x}{4h}\eta + \frac{h^2}{6}\eta_{xxx} + \frac{3}{4h}(\eta^2)_x\right] = 0, \quad (33)$$

in which $C_0 = \sqrt{gh}$.

For deep water, on the other hand, $C_p = \sqrt{g/k_p}$, $C_g = C_p/2$, then we have

$$\eta_t + \frac{3}{2} C_p \eta_x - \frac{1}{k_p^2} \eta_{xxt} - \frac{C_p}{2k_p^2}\eta_{xxx} + \frac{3}{2}\frac{g}{C_p}(\eta^2)_x = 0, \quad (34)$$

which can be shown to admit the second-order Stokes' waves in deep water as an analytical solution.

NUMERICAL EXAMPLES

The forms of the single-component equations are in perfect correspondence with those of the Boussinesq equations. This is an important advantage because it allows the adoption of an efficient implicit scheme which has been developed for solving the Boussinesq equations. The numerical schemes for the combined forms of the single-component equations, (28) and (32), are of course much simpler and need shorter computational time.

The fully-dispersive equations, on the other hand, require a more complicated scheme to solve the N momentum equations. In the present study, a generalized Thomas algorithm, or the so-called block elimination method, is used for solving the linear algebraic equations resulting from an implicit three-time-level, centered discretization of the momentum equations (Nadaoka, et al., 1994). The values of

U_m ($m = 1, \cdots, N$) at the boundary may be prescribed by applying a Galerkin procedure similar to that for eq.(7). The angular frequencies ω_m to specify the corresponding k_m ($m = 1, \cdots, N$) are chosen so as to properly cover the frequency spectrum concerned. It has been found through various numerical computations, some of which will be shown later, that N required is usually no more than 3 and the use of slightly different set of ω_m ($m = 1, \cdots, N$) yields no appreciable difference in the computational results and hence the apparent ambiguity in selecting ω_m ($m = 1, \cdots, N$) does not affect the validity of the present model.

The following parts are devoted to show some typical numerical examples.

(1) Linear random waves

To examine the fundamental performance of the fully-dispersive equations, their linearized equations were applied to a case of linear random waves in deep water with a Bretschneider-type spectrum, which has a broad band-width in comparison with, e.g., JONSWAP. The relative water depth to the wave length corresponding to the mean period T_m is one ($h/L_m=1$). Figure 4 shows the comparisons with the predictions of linear theory for surface displacement and horizontal velocity at two different depths after 20 wave periods elapsed over a distance of five wavelengths. Good agreement with the theory is observed for both the surface displacement and velocity profiles. In the computation three components were used: $k_1h=2\pi$, $k_2h=3\pi$, $k_3h=5\pi$ with $\Delta x=L_m/90$ and $\Delta t= T_m/90$. The relatively fine resolutions were deemed necessary for the accurate representation of higher frequency components with shorter wavelengths and periods. No sponge layer was needed to improve the absorption at the outgoing boundary; the computational domain was not longer than shown. Good absorption of the radiating waves is attributable to the fact that the outgoing waves are radiated at three different wavenumbers instead of one. This is an important advantage especially in long time simulation of random waves.

(2) Nonlinear regular and irregular waves propagating over a bar

Further examinations on the fully-dispersive equations have been made through the comparisons with the laboratory data obtained by the experiment on the nonlinear wave deformation over a submerged trapezoidal bar as shown in Fig.5, which is similar to that of Beji and Battjes (1994). The two-component form of eqs. (17) and (18) was used for the computation by selecting the corresponding angular frequencies as $\omega_1=2\pi/T$ and $\omega_2=4\pi/T$. The experimental data compared is that for which the incident wave height H and period T are 2.0cm and 1.5s, respectively. Note that in this case the largest relative depth h/L observed was 0.35 at most. Figure 6(a) shows the comparisons for the water surface profiles at station 3, 5 and 7, while Fig.6(b) represents the velocity comparison at three depths at station 7, where an appreciable wave-decomposition phenomenon was observed. On the other hand, Figs.7(a) and (b) show the comparisons in which the improved Boussinesq equations of Madsen et al. (1991) were used for the computation. From these results, it is found that in the water surface profiles the present wave equations show good but nearly the same degree of agreements as compared with the improved Boussinesq equations. In the velocity profiles, on the contrary, the agreements for the improved Boussinesq equations deteriorate, although for the present equations the agreements are comparable to those in the surface profiles.

As a test for nonlinear random waves traveling over a submerged trapezoidal bar, the experimental data of Beji and Battjes (1994) was compared with the computational results by the two-component wave equations (17) and (18). The incident wave field has a JONSWAP type random wave spectrum with a peak period $T_p=2s$. The first four stations are in the upslope region where the nonlinear shoaling takes place. The remaining three stations are in the downslope region, where harmonic wave decomposition becomes appreciable. In the computations k_1 and k_2 for each component are selected to be k_p and πk_p, respectively, where k_p denotes the wavenumber corresponding to the peak period T_p. Figure 8 shows the results of the comparison at six different stations, indicating good agreements at all the stations. (It has been also found that even in case of using the simplest single-component equation (32) for the computations the agreements are slightly worse but still comparable to those indicated in Fig.8.)

(3) Stokes and cnoidal waves

The comparisons with the theories of steady nonlinear wave train have been also conducted by using the various versions of the present wave equations. The wave theories compared are of Stokes, cnoidal and solitary waves. As an example, Fig.9 show the comparisons with the second-order Stokes and cnoidal wave theories. The computations were carried out with the unidirectional simplified form of the nonlinear mild-slope equation (32). It is found that even this simplest version of equations can describe steady nonlinear wave trains with remarkably good accuracy under wide conditions including very shallow and far deep water waves. Solitary waves are also found to be well predicted by the present models, although the results are not presented here (see Nadaoka, et al.,1994; Beji & Nadaoka, 1994).

CONCLUSIONS

The major conclusions of the present study are summarized as follows :

1. Fully-dispersive nonlinear wave equations are presented which can express nonlinear non-breaking waves under general conditions, such as nonlinear random waves with *wide-spectrum* at an *arbitrary depth* including very shallow and far deep water depths.

2. The single-component forms of the new wave equations, one of which is referred to as "*time-dependent mild-slope equation*", are shown to produce various existing wave equations like Boussinesq and mild-slope equations as their degenerate forms.

3. Even under relatively shallow wave condition, present wave model can evaluate more precisely the velocity field than the improved Boussinesq equations.

Acknowledgement

The authors would like to thank Prof. J.A. Battjes for his permission to use the experimental data and a graduate student O. Ohno for his helpful cooperation in laboratory works.

References

Beji, S.(1994): Note on conservation equations for nonlinear surface waves. *Submitted to Coastal Eng.*

Beji, S. and Battjes, JA.(1994): Numerical simulation of nonlinear wave propagation over a bar, Coastal Eng., Vol.23, pp.1-16.

Beji, S. and Nadaoka, K.(1994): A time-dependent nonlinear mild-slope equation. *Submitted to Proc. Roy. Soc. London A*

Berkhoff, J.C.W.(1972): Computation of combined refraction-diffraction. Proc. 13th Int. Conf. Coastal Eng., ASCE, pp.471-490.

Madsen, P.A., Murray, R. and Sørensen, O.R.(1991): A new form of the Boussinesq equations with improved linear dispersion characteristics, Coastal Eng., Vol.15, pp. 371-388.

Mei, C.C.(1983): The applied dynamics of ocean surface waves, John Wiley & Sons, 740pp.

Nadaoka, K. and Hino, H.(1984): Conformal mapping solution of a wave field on the

arbitrary shaped sea bottom, Proc. 19th Int. Conf. Coastal Eng., ASCE, pp.1192-1208.
Nadaoka, K and Nakagawa, Y.(1991): A Galerkin derivation of a fully-dispersive wave equation and its background, Tech. Rep. No.44, Dept. Civil Eng., Tokyo Inst. of Tech., pp.63-75. (*in Japanese*)
Nadaoka, K. and Nakagawa Y.(1993a): Fully-dispersive wave equations derived by a Galerkin formulation, *Meet'n '93*, ASCE/ASME/SES Abstracts, p.724.
Nadaoka, K. and Nakagawa, Y.(1993b): Derivation of fully-dispersive wave equations for irregular wave simulation and their fundamental characteristics, J. Hydraulics, Coastal and Environmental Eng., JSCE, No.467, pp.83-92. (*in Japanese*)
Nadaoka, K. and Nakagawa, Y.(1993c): Simulation of nonlinear wave fields with the newly derived nonlinear dispersive wave equations, Proc. Coastal Eng., JSCE, Vol.40, pp.6-10. (*in Japanese*)
Nadaoka, K., Beji, S. and Nakagawa, S.(1994): A fully-dispersive weakly-nonlinear wave model. *Submitted to Proc. Roy. Soc. London A*
Nwogu, O.(1993): Alternative form of Boussinesq equations for nearshore wave propagation, J. Waterway, Port, Coastal, and Ocean Eng., ASCE, Vol.119, No.6, pp.618-638.
Smith, R. and Sprinks, T.(1975): Scattering of surface waves by a conical island, J. Fluid Mech., Vol.72, pp.373-384.

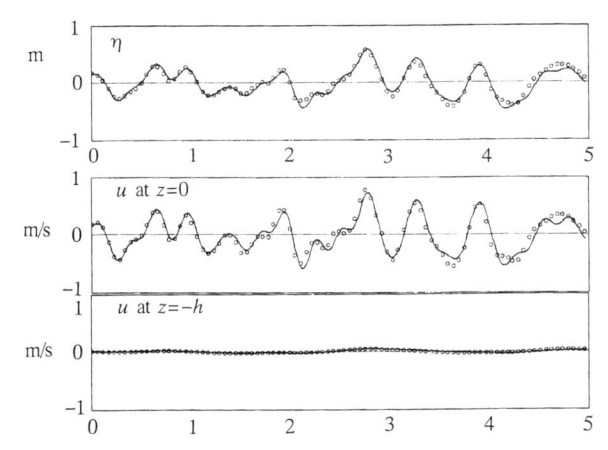

Fig.4 Linear random wave simulation ; linear theory (—) vs. computational results (O).

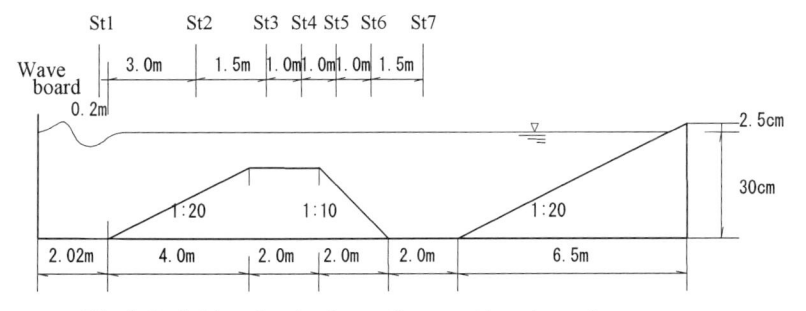

Fig.5 Definition sketch of wave flume and locations of wave gauges.

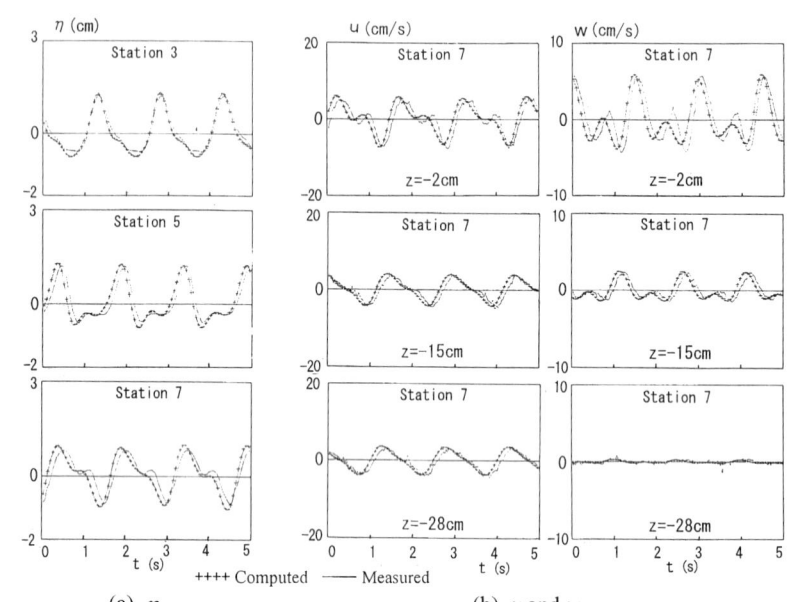

(a) η (b) u and w
Fig.6 Comparisons in η, u and w for the present fully-dispersive equations.

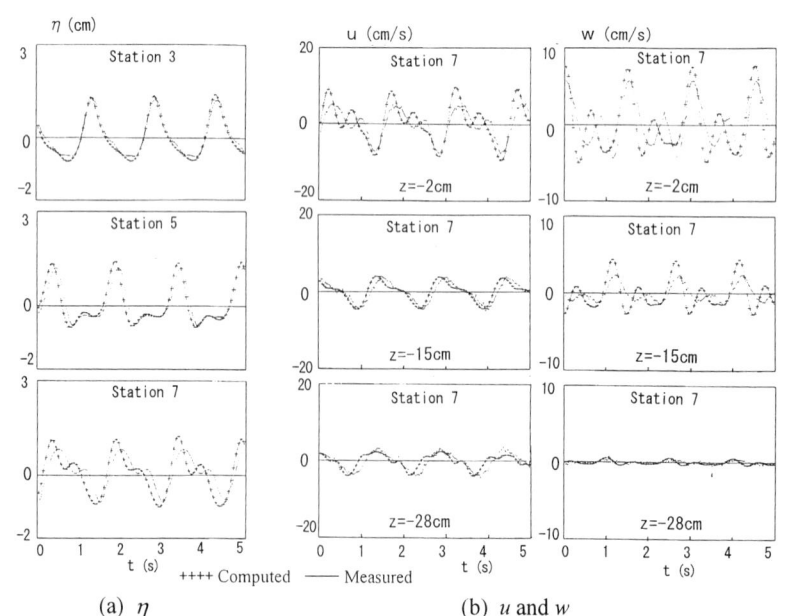

(a) η (b) u and w
Fig.7 Comparisons in η, u and w for the improved Boussinesq equations.

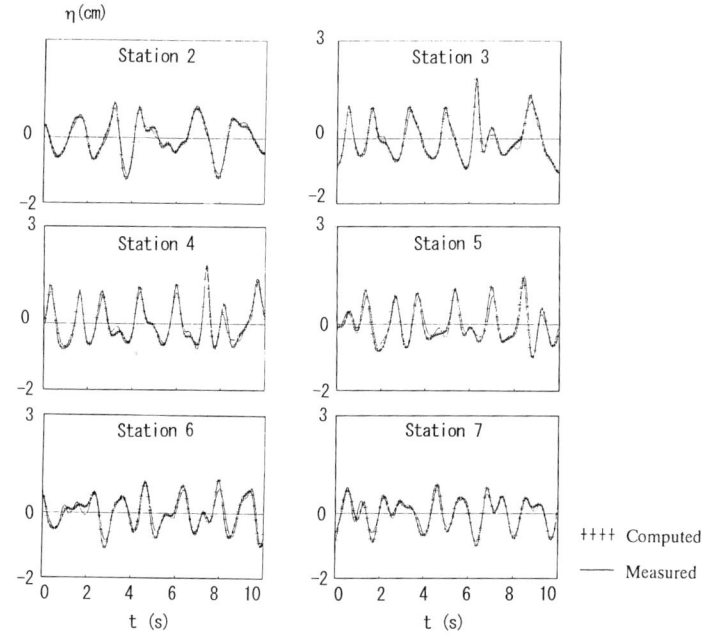

Fig.8 Comparison with laboratory data for nonlinear random waves propagating over a bar.

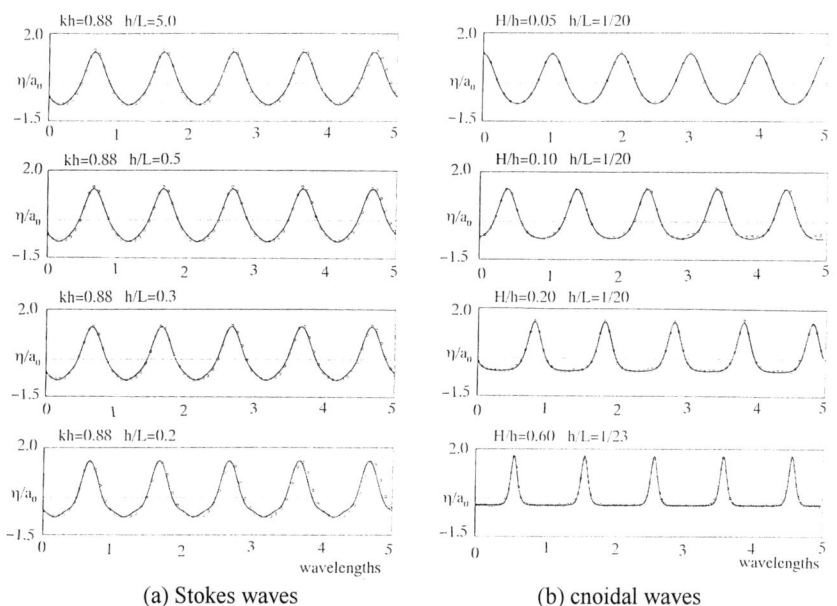

(a) Stokes waves (b) cnoidal waves
Fig.9 Comparison with Stokes and cnoidal wave theories; theory (—) vs. computed (O).

CHAPTER 33

A Generalized Green-Function Method for Wave Field Analysis

Hitoshi Nishimura [1], Michio Matsuoka [2] and Akira Matsumoto [3]

Abstract

The Green-function method for analyzing wave refraction, diffraction and reflection is improved by deriving rational explicit formulations of boundary conditions. Possibilities and limitations of the method are discussed. Trial computations and their comparisons with experiments demonstrate the validity and usefulness of the numerical model.

Introduction

To analyze refraction, diffraction and reflection of simple sinusoidal waves is one of the very fundamental problems in the field of coastal engineering. If an accurate and simple method is provided for this purpose, deformation of irregular waves may also be analyzed through superposition of solutions for constituent waves. In this context, numerical modeling based on the Green-function method is quite promising since it describes the diffraction and multiple reflection of waves more accurately than any other methods. This kind of numerical model was first proposed by Barailler and Gaillard (1967), and has been widely used in particular for simulation of waves in semi-closed sea basins. In numerical models presently used, however, continuation of solutions are often insufficient at artificial boundaries.

In the following sections, rational explicit formulations of boundary conditions is derived for more rigorous computation of two-dimensional wave fields. Validity of the numerical model thus improved is examined through trial computations and their comparisons with experiments.

---

[1] Professor, Inst. of Eng. Mech., Univ. of Tsukuba., Tsukuba, Ibaraki, 305, Japan.
[2] Senior Researcher, Applied Hydraulic Laboratory, Nippon Tetrapod Co., Ltd., 2-7 Nakanuki, Tsuchiura, Ibaraki, 300, Japan.
[3] Ditto.

Principle of the Green-Function Method

If we express the wave profile $\zeta(x, y, t)$ in a water area with a uniform depth as

$$\zeta = f(x, y) \cdot e^{i\omega t} \qquad (1)$$

the complex amplitude $f(x, y)$ satisfies the following Helmholtz equation:

$$\frac{\partial^2 f}{\partial x^2} + \frac{\partial^2 f}{\partial y^2} + k^2 f = 0 \qquad (2)$$

where (x, y) is the Cartesian coordinate in a horizontal plane, t is the time, ω is the angular frequency, and k is the wave number.

Suppose the existence of a perfectly reflective wall along the x-axis and a wave source at a point (ξ, η), as shown in Fig.1. It is well known that the resulting wave field in the semi-finite region of $y > 0$ is then

$$f(x, y) = -\frac{i\Gamma}{4}[H_0^{(1)}(kr^+) + H_0^{(1)}(kr^-)] \qquad (3)$$

$$r^{\pm} = \sqrt{(x - \xi)^2 + (y \mp \eta)^2} \qquad (4)$$

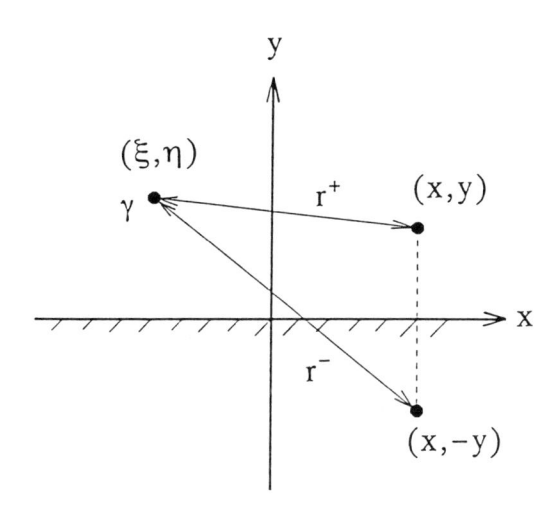

Figure 1: Definition sketch.

in which $H_0^{(1)}$ denotes the zeroth-order Hankel function of the first kind. Note that the complex source intensity Γ represents the wave phase as well as the amplitude. The solution (3) satisfies the basic equation (2) and, at the same time, the boundary condition $\partial(f_1+f_2)/\partial y = 0$ along the x-axis. It also satisfies Sommerfeld's radiation condition toward the infinity.

For a particular case that the wave source is located on the x-axis ($\eta = 0$), the expressions (3) and (4) are simplified as follows:

$$f(x,y) = -\frac{i\Gamma}{2} H_0^{(1)}(kr) \tag{5}$$

$$r^\pm = \sqrt{(x-\xi)^2 + y^2} \tag{6}$$

Distribution of such wave sources along the boundary of a water area will cause a composite wave field, which is expressed by the integration of the unit solution:

$$f(x,y) = \int_C \gamma(s) \cdot H_0^{(1)}(kr)\,ds \tag{7}$$

where s is the coordinate taken along the boundary C, r is the distance from the boundary point (s) to the point (x,y), and $\gamma(s)ds$ corresponds to the wave source intensity Γ in Eq.(5). The wave source distribution has to be determined so that the resulting wave field satisfies all the boundary conditions imposed, as correctly suggested by Lee (1969, 1971) for harbor oscillation analysis.

Relationship between the Source Intensity and Wave Amplitude on a Boundary

Consider the situation that wave sources in the semi-infinite region of $y > 0$ produce a wave field $f_1(x,y)$. As to the waves propagating across the x-axis, the following relationship proves from Green's theorem between the complex amplitude and its gradient in the y-direction:

$$f_1(x,0) = -\frac{i}{2}\int_{-\infty}^{\infty} \left.\frac{\partial f_1(\xi,\eta)}{\partial \eta}\right|_{\eta=0} \cdot H_0^{(1)}(k\,|x-\xi|)\,d\xi \tag{8}$$

On the other hand, a similar relationship is obtained for a wave field $f_2(x,y)$ which is produced in the same region by the source distribution $\gamma_2(x)$ along the x-axis:

$$f_2(x,y) = -\frac{i}{2}\int_{-\infty}^{\infty} \left.\frac{\partial f_2(\xi,\eta)}{\partial \eta}\right|_{\eta=0} \cdot H_0^{(1)}(kr)\,d\xi \tag{9}$$

It can thus be concluded that the source intensity is directly proportional to the local gradient of resulting wave amplitude in the normal direction to the boundary:

$$\gamma_2(\xi) = -\frac{i}{2}\frac{\partial f_2(\xi,\eta)}{\partial \eta}\bigg|_{\eta=0} \tag{10}$$

Condition of Reflective Boundary

If x-axis is a perfectly reflective boundary in the above discussion, then the incident waves $f_1(x,y)$ and reflected waves $f_2(x,y)$ appear in the semi-infinite region. The boundary condition in this case is

$$\frac{\partial(f_1+f_2)}{\partial y}\bigg|_{y=0} = 0 \tag{11}$$

which leads to the following simple expression of the source intensity:

$$\gamma_2(\xi) = -\frac{i}{2}\frac{\partial f_2(\xi,\eta)}{\partial \eta}\bigg|_{\eta=0} \tag{12}$$

In rigorously formulating a condition of partial reflection, hydraulic mechanism of wave reflection has to be known. A conventional method for simulating the partial reflection is to simply reduce the source intensity by multiplying the reflection coefficient β as follows:

$$\gamma_2(\xi) = -\frac{i\beta}{2}\frac{\partial f_1(\xi,\eta)}{\partial \eta}\bigg|_{\eta=0} \tag{13}$$

Continuation of Solutions along an Open Boundary

For the convenience of numerical computation, the whole region in question is often divided into subregions by supposing imaginary boundaries between them. A breakwater gap illustrated in Fig.2 is a typical example of such a boundary. In the figure, waves $f_I(x,y)$ are incident to the imaginary boundary C on the x-axis. Note that $f_I(x,y)$ represents all the incident waves to C including those from the boundaries immediately beside C.

If the boundary is either finite or semi-infinite, waves are partly reflected as they are transmitted across the boundary. In other words, two kinds of wave sources are in general to be distributed along a transmissive boundary. On one side of the boundary, the reflected waves $f_R(x,y)$ is superposed on the incident waves $f_I(x,y)$, whereas only transmitted waves $f_T(x,y)$ propagate on the other side. It is thus feasible for these two wave fields to coincide in terms of both the amplitude and its gradient along the boundary C:

$$(f_I+f_R)\big|_C = f_T\big|_C \tag{14}$$

$$\left.\frac{\partial(f_I + f_R)}{\partial y}\right|_C = \left.\frac{\partial f_T}{\partial y}\right|_C \qquad (15)$$

These conditions lead to the following integral equations for the source intensities γ_R and γ_T for reflected and transmitted waves respectively:

$$2\int_C \gamma_R(\xi) \cdot H_0^{(1)}(k\,|x-\xi|)\,\mathrm{d}\xi = f_I^*(x,0) - f_I(x,0) \qquad (16)$$

$$2\int_C \gamma_T(\xi) \cdot H_0^{(1)}(k\,|x-\xi|)\,\mathrm{d}\xi = f_I^*(x,0) + f_I(x,0) \qquad (17)$$

where

$$f_I^*(x,0) = -\frac{i}{2}\int_C \left.\frac{\partial f_I(\xi,\eta)}{\partial \eta}\right|_{\eta=0} \cdot H_0^{(1)}(k\,|x-\xi|)\,\mathrm{d}\xi \qquad (18)$$

and

$$f_R(x,y) = \int_C \gamma_R(\xi) \cdot H_0^{(1)}(kr)\,\mathrm{d}\xi \qquad (19)$$

$$f_T(x,y) = \int_C \gamma_T(\xi) \cdot H_0^{(1)}(kr)\,\mathrm{d}\xi \qquad (20)$$

For solution of the above equations, they are discretized by dividing the boundary into a number of segments. The problem is then ascribed to linear systems of simultaneous equations, and the source intensities are obtained through matrix operations. Since the coefficient matrices of the systems are fixed for each boundary of this type regardless of wave conditions, inverse matrices once calculated can be repeatedly used throughout the computation.

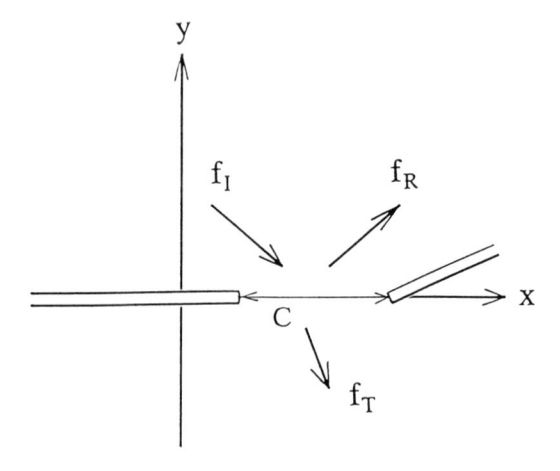

Figure 2: Imaginary boundary.

In the actual computation, it is not necessary to solve both the above equations. After solving one of them for γ_R or γ_T, we can easily evaluate the other from the following relationship:

$$\gamma_I(x) + \gamma_R(x) + \gamma_T(x) = 0 \tag{21}$$

Even if incident waves with uniform amplitude are incident to the boundary, wave sources thus obtained are not uniform with high peaks at the boundary ends. These peaks represent the singularity of the boundary ends, reproducing fields of diffracted waves with good approximation. It is important to note here that the source intensity cannot necessarily be related to the local energy density of incident waves.

Outline of the Numerical Model

When the configuration of a sea area in question is complicated, the whole region for computation is divided into several convex polygonal subregions. A semi-infinite open sea area is regarded as one of the subregions. The other subregions are totally enclosed by boundaries, at least one of them being transmissive. All the boundaries are subdivided into a number of segments and the effect of each segment is represented by a wave source. A transmissive boundary serves as two boundaries at once for two subregions on its both sides, where two sources defined are for either reflected or transmitted waves depending on the subregion currently treated.

The distribution of source intensities is calculated for each boundary. The flow of computation starts from the the subregion of wave incidence. It moves from one subregion to the next, and from one boundary to another in a subregion. Since the source intensities are interdependent, these boundarywise calculations over the whole region are repeated until all the intensities reach an equilibrium.

It is somewhat difficult to formulate the condition of a partially transmissive boundary. Such a boundary may be reasonably regarded as a fully reflective boundary in evaluating reflected waves, and as a fully open boundary for transmitted waves. Prior to these calculations, the incident wave amplitudes are to be reduced by multiplying reflection or transmission coefficient. In this case, the source intensities for reflected and transmitted waves have to be memorized separately on both the sides of the boundary.

The present method for wave field analysis can be applied even to a water area with arbitrary bathymetry by numerically obtaining unit solutions to replace the Hankel function. For the calculation of unit solutions, a relatively simple method for analyzing wave refraction may be employed, since the main part of wave diffraction is included in principle in the process of superposition of point source waves. Nonlinear wave deformation, however, can never be analyzed by means of the Green-function approach as it is essentially based on the superposition of unit solutions.

Physical Model Experiments

A series of experiments on waves in a harbor were conducted (Photo 1) to provide wave distribution data to examine performance of the numerical model described above. The model harbor configuration shown in Fig.3 and experimental conditions listed in Table 1 were determined along one of the model cases specified for trial simulation works by a subcommittee of the Coastal Engineering Conference, JSCE. The model was installed on a horizontal bed in a narrow wave basin. Gaps between the harbor and basin walls were filled by wave absorbing material to avoid the elevation of the mean water level.

In some cases of the experiments, tetrapod mounds were arranged on the outer sea sides of the breakwaters, but all the other walls were vertical walls. As a matter of fact, arrangement of absorbing facility inside the harbor significantly narrows the harbor area in such a small-scale model. The reflection coefficient for each part of the model was separately estimated by applying Healy, Goda (1976) and Isaacson's (1991) methods, as summarized in Table 2.

Wave heights were measured using servo-type gauge array at every 10cm grid point over the whole area near and inside the harbor. The measurements were repeated more than twice for each case, but no significant scattering was observed in the data obtained. The incident wave heights listed in Table 1 were obtained at the location of the harbor entrance prior to the model installation.

Photo 1: Experimental setup.

Figure 3: Experimental setup.

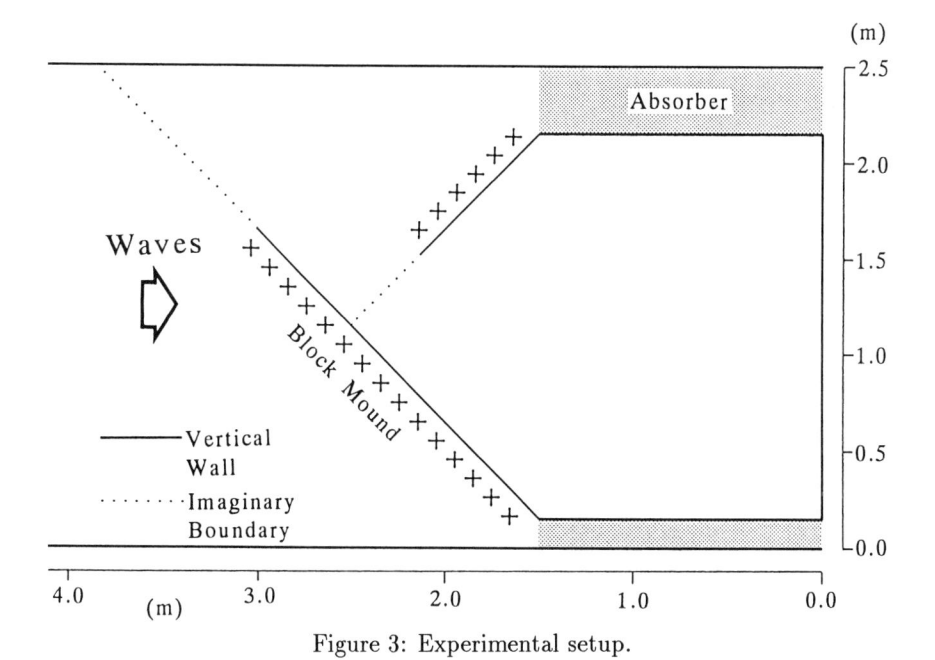

Table 1: Experimental conditions.

Case	Water depth (cm)	Period (s)	Incident wave height (cm)	Breakwaters
1	12.0	0.70	2.62	with block mounds
2	12.0	0.72	2.06	with block mounds
3	12.0	0.70	1.96	without block mounds

Table 2: Reflection coefficient.

	Measured value	For computation
Vertical wall	0.95-1.00	0.95
Block mound	0.35-0.40	0.40
Wave absorber	0.30-0.35	0.30

Comparison of Numerical Computations and Experiments

In the numerical computations, the whole region was divided into four subregions. Two imaginary boundaries are shown in by dotted lines in Fig.3. Another boundary for wave incidence was assumed at 4m from the outer breakwater tip, and further offshore area was treated as a semi-infinite region. All the boundaries were divided into segments with length of roughly 1/20 wavelength. The reflection coefficient values used are also shown in Table 2, and the boundarywise computations were repeated until the relative accuracy of 1/1000 at maximum was attained.

Figure 4 compares measured and calculated distributions of relative wave heights normalized with the incident wave height for Case-1 with a period of 0.70s. The numerical model well simulates the field of standing waves formed inside the harbor, although the wave heights calculated are somewhat smaller than those measured. The incident wave height may have been substantially increased involving reflected waves from the wave generator. In numerical analyses, waves in the innermost area of the harbor are apt to be underestimated as they are subject to multiple diffractions. This sort of tendency, however, is not apparent at all here.

Figure 5 presents a similar comparison for Case-2 with a slightly longer period of 0.72s. It is seen that the wave period sensitively affect the wave field in such a system of multiple reflection, as is well simulated by the numerical model. These periods may be close to one of the resonant oscillation periods of the harbor water.

The effects of the absorbing mounds on the breakwater fronts first appears on the wave height distribution outside the harbor. Then the change in wave heights along the harbor entrance indirectly influences the wave field inside the harbor. Figure 6 shows wave fields for Case-3, where the block mounds were eliminated. The computation again reasonably reproduces significant differences in the wave height distribution which is noticed through comparison with Fig.4 for the Case-1 experiment.

Concluding Remarks

The Green-function method provides a powerful tool for analyzing wave diffraction and multiple reflection of coastal and harbor waves. The rational treatment of imaginary boundaries allows the arbitrary division of water area with a complicated configuration without deteriorating the accuracy of total computation. The present model is rather simple and minimizes empirical factors for its actual application.

Since computer memory and computational labor required are not so large, even irregular waves can be treated so far as the linear superposition of constituent waves are acceptable. The model may be extended for arbitrary bathymetry, but cannot contribute to analyses of nonlinear wave deformation.

GENERALIZED GREEN-FUNCTION METHOD 451

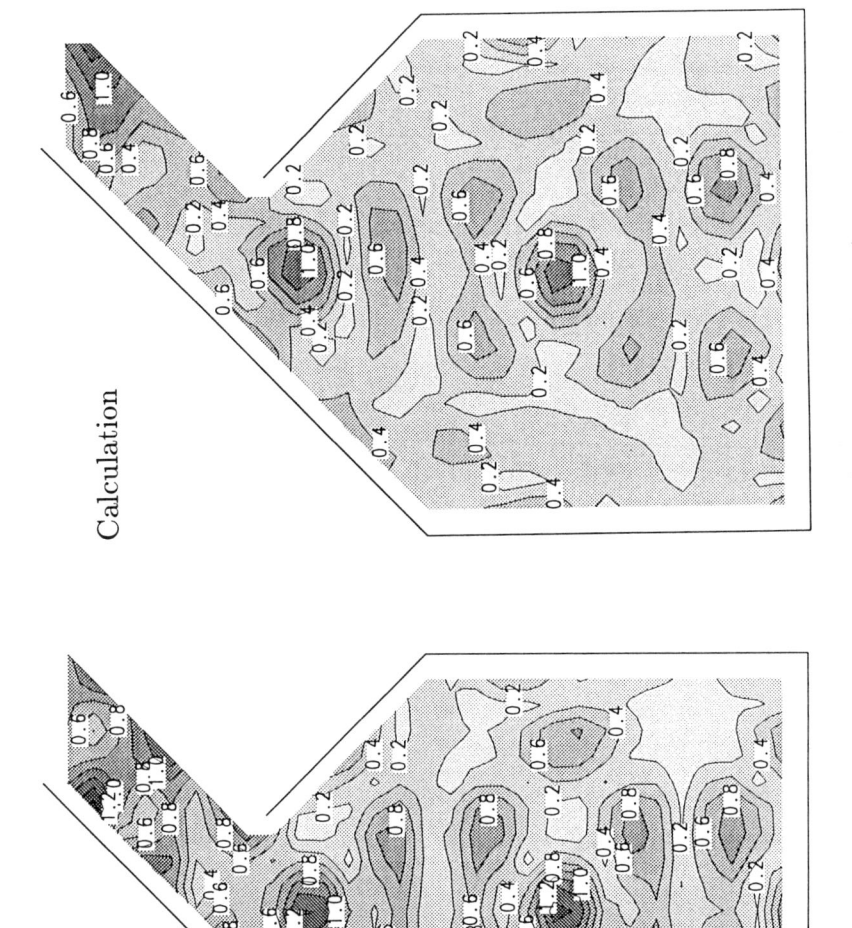

Fig. 4: Relative wave height distribution (Case 1, period: 0.70s).

452 COASTAL ENGINEERING 1994

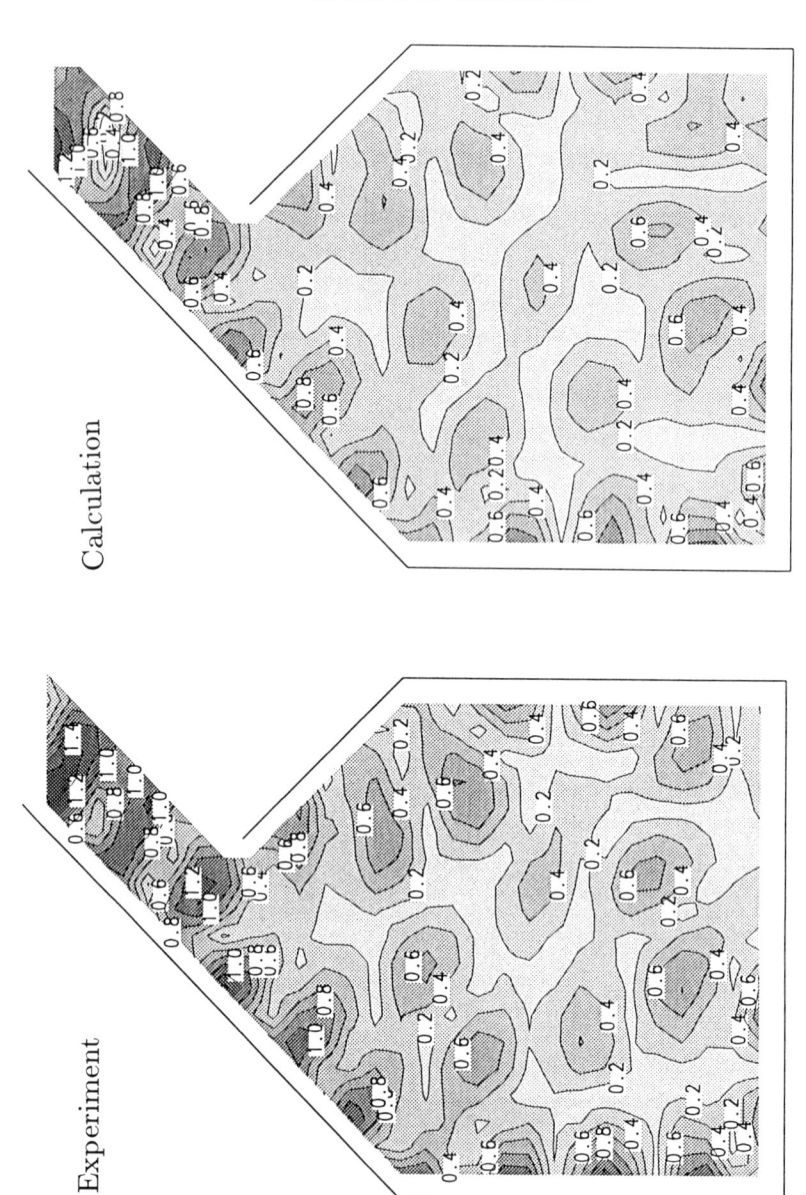

Fig. 5: Relative wave height distribution (Case 2, period: 0.72s).

GENERALIZED GREEN-FUNCTION METHOD 453

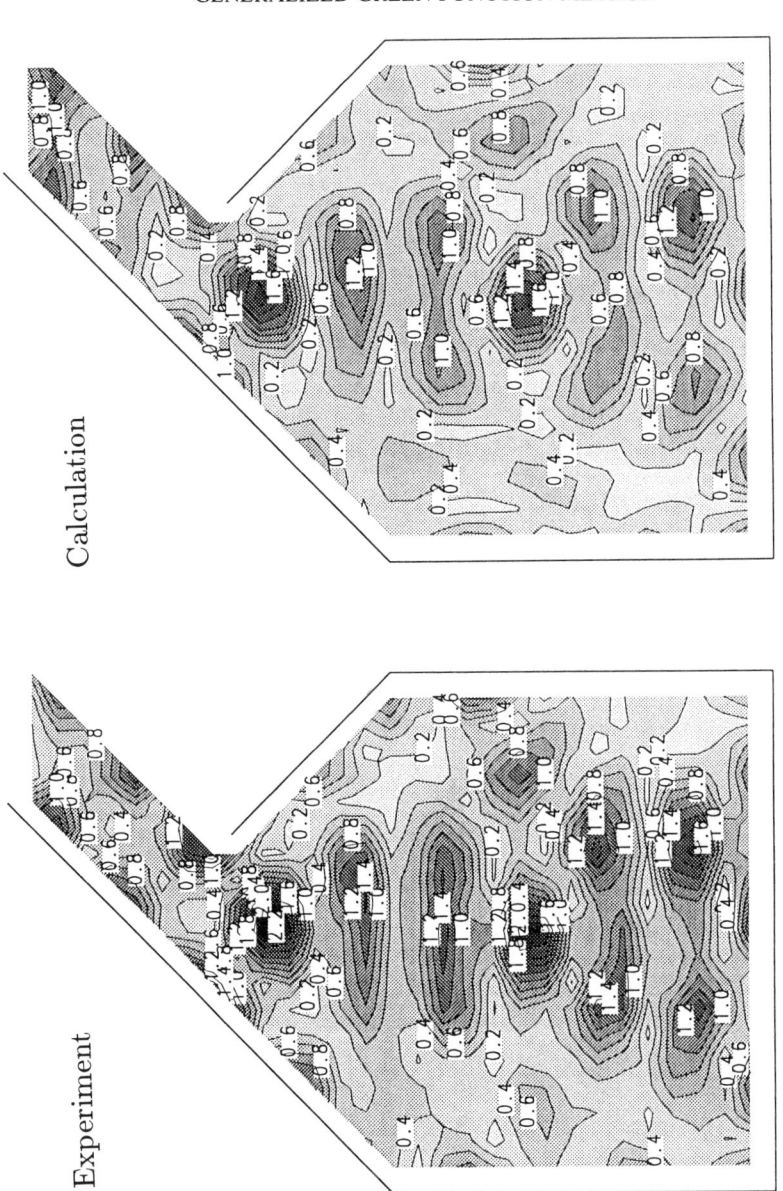

Fig. 6: Relative wave height distribution (Case 3, period: 0.70s, without block mounds).

Acknowledgment

The authors' gratitude is due to Mr. Nobumasa Shinoda, Toyota System Research Co., for his good help in the computer work for this study.

References

Barailler, L., and P. Gaillard (1967). Évolution récente des modèles mathématiques dagitation due à la houle. La Houile Blanche, Vol.22, No.8, pp.861-869.

Goda, Y. (1976). Estimation of incident and reflected waves in random wave experiments. Tech. Note of Port and Harbour Res. Inst., Min. of Transport, Japan, No.248, 24p. (in Japanese)

Isaacson, M (1991). Measurement of regular wave reflection, J. Waterway, Port, Coastal and Ocean Eng., ASCE, Vol.117, pp.553-509.

Lee, J.-J., and F. Raichlen (1969). Wave induced oscillations in harbors of arbitrary shape. W. M. Keck Lab., Calif. Inst. of Tech., Rep. No.KH-R-20, 266p.

Lee, J.-J., and F. Raichlen (1971). Wave induced oscillations in harbors with connected basins. W. M. Keck Lab., Calif. Inst. of Tech., Rep. No.KH-R-26, 135p.

CHAPTER 34

COUPLED VIBRATION EQUATIONS FOR IRREGULAR WATER WAVES

Masao Nochino*

Abstract

A new system of equations is proposed for calculating wave deformation of irregular waves under the assumption of small amplitude wave and of mild slope of bottom configuration. The system is composed of a vibration equation for water surface elevation and three elliptic equations defined in horizontal plane. These equations are coupled each other. The coupled vibration equations are capable of calculating water surface elevation of irregular waves in a time domain.

1 Introduction

Some equations have been proposed for the deformation of water waves; the mild slope equation by Berkhoff(1972), the unsteady mild slope equation considering wave-current interaction by Liu(1983), Boussinesq equation by Pregrine(1978). The mild slope equations are able to apply only for a simple harmonic wave. The Boussinesq equation is derived for nonlinear long waves. The equation to calculate the deformation of irregular waves including long waves is expected for the coastal structure design and the analysis of coastal process.

Kubo et. al.(1992) modified the time-dependent mild slope equation by applying the Fourier expansion technique to the coeffecnt in the mild slope equation. The modified mild equation is capable of simulating random waves without long waves. Nadaoka et. al.(1993) proposed the fully-dispersive wave equation capable of simulating random waves with long waves.

In this paper, a new system of equations is proposed for calculating the deformation of the linear and irregular water waves including long waves. The waves are expressed as motions of coupled vibrations. The system of equations are applicable for the waves in the range of deep water depth to very shallow water depth.

---
*Associate Professor, Osaka Institute of Technology, Department of Civil Engineering 5-16-1 Omiya Asahi-ku, Osaka, 565, Japan

2 Theory

2.1 Basic equation

The motion equations and the equation of continuity are expressed as follows under the assumption that the motion due to waves be small and the nonlinear terms in the motion equations be negligible.

$$\frac{\partial u}{\partial t} + \frac{\partial \tilde{p}}{\partial x} = 0 \qquad (1)$$

$$\frac{\partial v}{\partial t} + \frac{\partial \tilde{p}}{\partial y} = 0 \qquad (2)$$

$$\frac{\partial w}{\partial t} + \frac{\partial \tilde{p}}{\partial z} = 0 \qquad (3)$$

$$\frac{\partial u}{\partial x} + \frac{\partial v}{\partial y} + \frac{\partial w}{\partial z} = 0 \qquad (4)$$

where, $\tilde{p}$; the fluctuating pressure defined as

$$\tilde{p} = p/\rho + gz, \qquad (5)$$

x,y ;the coordinates in horizontal axes , z ; the vertical coordinate in upward direction with the origin at still water surface, p ; the pressure, ρ; the water density, g; the gravitational acceleration, u,v,w; the water particle velocity in x, y, z direction, respectively. Taking divergence of the motion equation, the Laplace equation of the fluctuating pressure $\tilde{p}$ are given as follows.

$$\triangle \tilde{p} = 0 \qquad (6)$$

In this paper, Equation(6) is treated as the basic equation and $\tilde{p}$, the main variable as same as the water surface elevation η.

2.2 Boundary conditions

There are three boundary conditions; two boundary conditions at free surface and one at bottom. These boundary conditions should be expressed by the main variable and linearized as the motion equations are linearized.

The dynamic and kinematic free surface boundary conditions are expressed as follows.

$$\tilde{p} = g\eta \qquad \text{at } z = 0 \qquad (7)$$

$$\frac{\partial \eta}{\partial t} = w \qquad \text{at } z = 0 \qquad (8)$$

The motion equation should be satisfied at free surface. Eliminating the vertical velocity w in above equation by using the motion equation in vertical direction, Eq. (3), the kinematic boundary condition, Eq.(8), can be expressed as follow.

$$\frac{\partial^2 \eta}{\partial t^2} = -\frac{\partial \tilde{p}}{\partial z} \qquad \text{at } z = 0 \qquad (9)$$

The boundary condition at bottom ($z = -h$) is commonly expressed as follow.

$$u\frac{\partial h}{\partial x} + v\frac{\partial h}{\partial y} + w = 0 \quad \text{at } z = -h \qquad (10)$$

where, $h = h(x, y)$ indicates the water depth. Derivating above equation with respect to time and applying the motion equations (1)-(3), the following equation is given as the boundary condition at bottom expressed by the fluctuation pressure.

$$\left(\nabla \tilde{p} \cdot \nabla h + \frac{\partial \tilde{p}}{\partial z}\right) = 0 \quad \text{at } z = -h \qquad (11)$$

where $\nabla = (\partial/\partial x, \partial/\partial y)$ indicate the differential operator in horizontal plane.

2.3 Expansion series of fluctuating pressure

The fluctuating pressure $\tilde{p}$ is expanded by using the series of Legendre's Polynomials $P_n(z)$ as follows.

$$\tilde{p} = \sum_{m=1}^{\infty} q_m P_{2(m-1)}(\tilde{z}) \qquad (12)$$

where $q_m = q_m(x, y, t)$ is the coefficient of mth term, and

$$\tilde{z} = 1 + z/h. \qquad (13)$$

The variable $\tilde{z}$ is defined in the interval of $[0, 1]$ as the coordinate z is defined in $[-h, 0]$ for the linearized theory. The Legendre's Polynomials $P_m(\tilde{z})$ are defined as follows.

$$P_0(\tilde{z}) = 1, \quad P_m(\tilde{z}) = \frac{1}{2^m m!}\frac{d^m}{d\tilde{z}^m}(\tilde{z}^2 - 1)^m, \quad m = 1, 2, \cdots \qquad (14)$$

For example,

$$P_2(\tilde{z}) = \frac{1}{2}(3\tilde{z}^2 - 1), \quad P_4(\tilde{z}) = \frac{1}{8}(35\tilde{z}^4 - 30\tilde{z}^2 + 3), \quad \text{etc..} \qquad (15)$$

The series of Legendre's Polynomials $P_{2(m-1)}(z)$ have the property of the series of the orthogonal functions.

$$\int_0^1 P_{2(m-1)}(z) P_{2(n-1)}(z) dz = \begin{cases} 0, & \text{for } m \neq n \\ \frac{1}{4m-3}, & \text{for } m = n \end{cases}$$
$$(m, n = 1, 2, \cdots) \qquad (16)$$

The above relation implies that all coefficients q_m in Eq.(12) be uniquely determined.

The expression by the infinite series in Eq.(12) is not suitable for a numerical calculation. Let's suppose, in this paper, that the fluctuating pressure $\tilde{p}$ be expressed by the series of the first 4 terms in the right side of Eq.(12), such as

$$\tilde{p} = q_1 P_0(\tilde{z}) + q_2 P_2(\tilde{z}) + q_3 P_4(\tilde{z}) + q_4 P_6(\tilde{z}). \tag{17}$$

The dynamic and kinematic boundary condition at free surface are rewritten as follows by substituting Eq.(17) into Eqs.(7) and (9).

$$g\eta = q_1 + q_2 + q_3 + q_4 \tag{18}$$

$$\frac{\partial^2 \eta}{\partial t^2} = -\frac{1}{h}(3q_2 + 11q_3 + 21q_4) \tag{19}$$

2.4 Depth integrated equation

There are five unknown variables; $\eta(x, y, t), q_m(x, y, t), (m = 1, 2, 3, 4)$. All variables are the functions defined in horizontal plane. There are two equations, Eqs.(18) and (19) which relate the five unknown variables. Therefore, three more new equations are required in order to solve the five unknown variables, instead of Eq.(6) defined in three dimensional space.

The Galerkin method is applied to Eq.(6) to make up the three new equations.

$$\int_{-h}^{0} P_{2(m-1)}(\tilde{z}) \Delta \tilde{p} dz + P_{2(m-1)}(0) \left(\nabla h \cdot \nabla \tilde{p} + \frac{\partial \tilde{p}}{\partial z}\right)\bigg|_{z=-h} = 0 \tag{20}$$

$$(\text{for} \quad m = 1, 2, 3)$$

The second term in left hand side of the above equation is added so that the boundary condition at bottom, Eq.(11) is satisfied. Integrating the above equation and eliminating the coefficient q_4 by using Eq(18), the following equations are given.

$$\nabla^2 q_1 + \frac{\nabla(88q_1 + 24q_2 - 88q_3)}{128h}\nabla h - \frac{21q_1 + 18q_2 + 11q_3}{h^2}$$
$$= \frac{40g\nabla\eta\nabla h}{128h} - \frac{21g\eta}{h^2} \tag{21}$$

$$\nabla^2 q_2 + \frac{\nabla(-190q_1 + 258q_2 + 330q_3)}{128h}\nabla h - \frac{90q_1 + 90q_2 + 55q_3}{h^2}$$
$$= -\frac{130g\nabla\eta\nabla h}{128h} - \frac{90g\eta}{h^2} \tag{22}$$

$$\nabla^2 q_3 + \frac{\nabla(99q_1 - 693q_2 - 205q_3)}{128h}\nabla h - \frac{99q_1 + 99q_2 + 99q_3}{h^2}$$
$$= \frac{333g\nabla\eta\nabla h}{128h} - \frac{99g\eta}{h^2} \tag{23}$$

When deriving the above equations, the assumption for the mild slope of the bottom configuration is applied; the terms related $\nabla^2 h$ and $|\nabla h|^2$ are neglected.

The Eq(19) can be rewritten as follows by eliminating q_4 in a same manner.

$$\frac{\partial^2 \eta}{\partial t^2} = -\frac{1}{h}(21g\eta - 21q_1 - 18q_2 - 10q_3) \tag{24}$$

Eq.(24) shows the form of a vibration equation for the water surface elevation η with the exciting force as a function of q_1, q_2 and q_3. The variables q_1, q_2 and q_3 are determined by the Eqs. (21)-(23) which has the form of the elliptic equation defined in the horizontal plane with the exciting term as a function of η. Eqs.(21)-(24) are coupled each other. The variables η and q_m are in phase. Therefore, the system of Eqs.(21)-(24) expresses the water waves as the system of motions of the coupled vibrations, not as a system of wave equations.

The system of coupled vibration equations are expressed by the water depth h and the constant coefficients. The system is independent to both the wave frequency and the wave length L.

3 Dispersion Relation of Coupled Vibration Equation

The system of coupled vibration equations satisfy the dynamic and kinematic boundary condition which are linearized. Applying the coupled vibration equations for water waves, the dispersion relation of the coupled vibration equation should be coincide with the one of the small amplitude theory (Airy's wave theory).

Suppose the monochromatic wave progressing in the x direction with angular frequency of σ, wave number of k on the water of uniform depth h. The water surface elevation η and the coefficients q_m can be expressed as follows,

$$\eta = \Re\left\{\hat{\eta} e^{i(kx-\sigma t)}\right\} \tag{25}$$

$$q_m = \Re\left\{\hat{q_m} e^{i(kx-\sigma t)}\right\} \qquad m = 1, 2, 3, 4 \tag{26}$$

as η and q_m are in phase, where, $\Re$ denotes the real part and $\hat{}$ the complex amplitude. Substituting above equations to Eqs.(21)-(23), the coefficients q_m are obtained as function of η.

$$q_1 = \frac{21g\eta\left(495 + 60(kh)^2 + (kh)^4\right)}{10395 + 4725(kh)^2 + 210(kh)^4 + (kh)^6} \tag{27}$$

$$q_2 = \frac{45g\eta(kh)^2\left(77 + 2(kh)^2\right)}{10395 + 4725(kh)^2 + 210(kh)^4 + (kh)^6} \tag{28}$$

$$q_3 = \frac{99g\eta(kh)^4}{10395 + 4725(kh)^2 + 210(kh)^4 + (kh)^6} \tag{29}$$

$$q_4 = \frac{g\eta(kh)^6}{10395 + 4725(kh)^2 + 210(kh)^4 + (kh)^6} \tag{30}$$

Substituting the above equations into Eq.(24), the relation between the angular frequency σ and the wave number k is finally obtained as follows.

$$\frac{\sigma^2 h}{g} = \frac{21(kh)^2(495 + 60(kh)^2 + (kh)^4)}{10395 + 4725(kh)^2 + 210(kh)^4 + (kh)^6} \quad (31)$$

This relation is the dispersion relation for the system of coupled vibration equations and corresponds to the one in the Airy's wave theory.

$$\frac{\sigma^2 h}{g} = kh \tanh kh \quad (32)$$

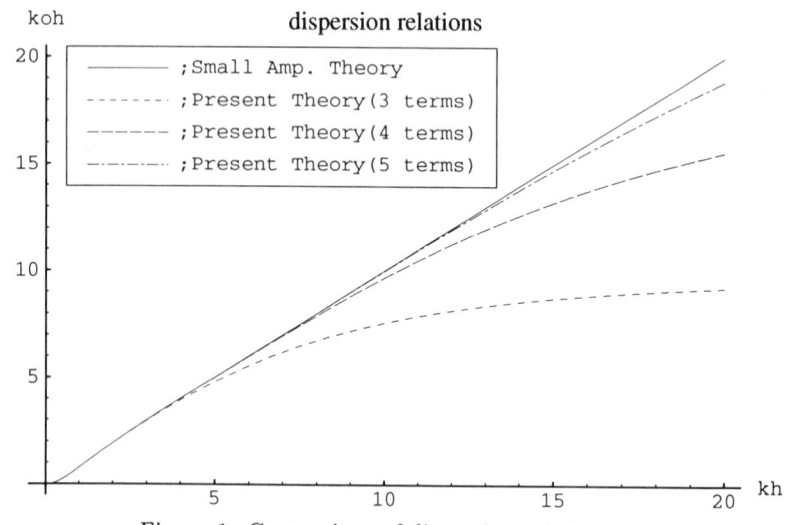

Figure 1: Comparison of dispersion relations

Figure 1 shows the relation of Eq.(32) by the solid line and of Eq.(31) by the broken line. The left hand side of these equations $\sigma^2 h/g$ are expressed by $k_o h$ in the figure. The left two lines in the figure indicate the dispersion relation in the case that the first 3 and 5 terms are adopted in the Eq.(12)(see Appendix).

The dispersion relation of Eq.(31) coincide very well with the one of Airy's wave theory in the rage of kh less than $7(h/L \simeq 1)$. This results implies that the system of coupled vibration equations be capable of expressing the waves on the very shallow water depth to the deep water depth.

Group velocity is important factor for waves deforming in a shallow water region and wave groups progressing in a deep water region. The group velocity C_g is defined by the gradient of the angular frequency σ with respect to the wave number k, that is,

$$C_g = \frac{\partial \sigma}{\partial k}. \quad (33)$$

The Eq.(31) gives the C_g/C of the coupled vibration equation by differentiating the equation with respect to k, where C denotes the phase velocity.

$$\frac{C_g}{C} = \frac{30}{495 + 60kh^2 + kh^4} \frac{343035 + 83160kh^2 + 14049kh^4 + 564kh^6 + 10kh^8}{10395 + 4725kh^2 + 210kh^4 + kh^6} \quad (34)$$

The Figure-2 shows the C_g/C due to the Airy's wave theory and to the present

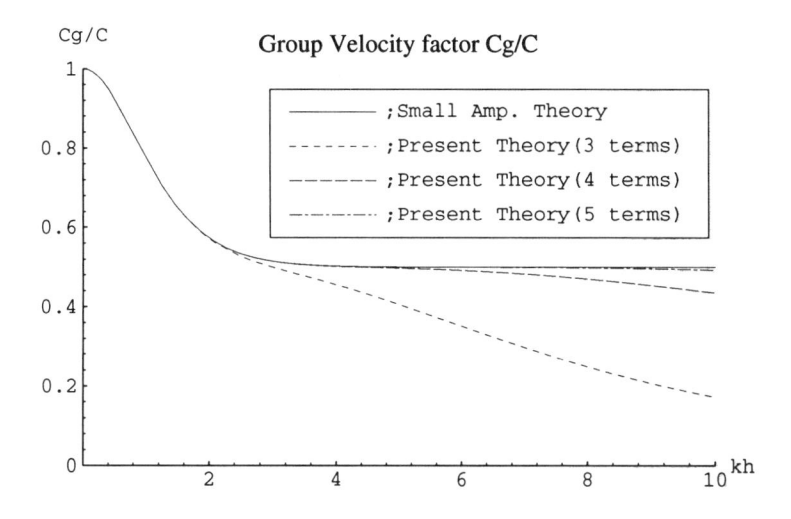

Figure 2: Comparison of group velocity

theory with the first 3, 4 and 5 terms as same as Figure-1. The group velocities due to the present theories in Figure-2 show less accuracy as comparing to the Airy's theory than the dispersion relation in Figure-1. This decrease of the accuracy is caused by differential calculus of the dispersion relation, Eq.(31) in order to obtain the group velocity.

Eq.(31) indicates the form of a *Padé* approximation for the right side of Eq.(32). The accuracy of differential coefficient of the approximated function is generally less than of the approximated function.

4 Calculation Method and Results

4.1 Incident boundary

The time series of water surface elevation is commonly measured in both field measurement and laboratory test. The position at measuring point of off-shore waves is generally treated as the incident boundary in most of calculations. In the calculation of the coupled vibration equations for irregular water waves, the water surface elevation η and the coefficients q_m should be given at the incident boundary. The measured data give the water surface elevation η_{in} at incident boundary. The coefficient q_m is calculated by Eqs.(27)-(30) when the wave number k is given.

The wave number k is determined by the following method. The wave number k and the angular frequency σ for irregular waves are assumed to be a function of time and satisfy the following equations.

$$\{\sigma(t)\}^2 = gk(t)\tanh k(t)h \tag{35}$$

$$\frac{d^2\eta_{in}}{dt^2} = -\{\sigma(t)\}^2 \eta_{in} \tag{36}$$

The second equation gives the angular frequency changing with time and the wave number is determined by the first equation. The angular frequency, however, becomes infinity when the η_{in} is nearly equals to zero in the computer calculation. In order to avoid this problem, the complex water surface elevation ζ is used instead of η. the complex water surface elevation ζ is defined as

$$\zeta(t) = \eta_{in} + i\tilde{\eta}_{in}, \tag{37}$$

where, $\tilde{\eta}_{in}$ is the Hilbert transform of η_{in} and i is the imaginary unit. $\tilde{\eta}_{in}$ is calculated by the following relation.

$$i\tilde{H}(\omega) = \begin{cases} H(\omega), & \text{for } \omega > 0 \\ 0, & \text{for } \omega = 0 \\ -H(\omega), & \text{for } \omega < 0 \end{cases}, \tag{38}$$

where $H(\omega)$ and $\tilde{H}(\omega)$ are the complex Fourier transforms of η_{in} and $\tilde{\eta}_{in}$, respectively. Finally, the angular frequency $\sigma(t)$ is redefined by the complex water surface elevation $\zeta(t)$.

$$\{\sigma(t)\}^2 = -\Re\left\{\frac{d^2\zeta/dt^2}{\zeta}\right\} \tag{39}$$

4.2 Calculation results

Eqs.(21)-(24) are applied to the waves progressing in one direction on the constant water depth, $7m$. The waves is composed of the two wave groups; WAVE-A of which the dominant wave period is $9s$ and WAVE-B, $3s$. Each wave group has a very narrow spectra, but is not monochromatic waves. That is, the waves calculated in this paper are parts of the components of the uni-directional irregular waves.

The finite difference method is applied for the equations. The time interval is $0.183s$, the distance of calculation nodes; $0.73m$, the number of nodes; 2800. The water channel is $2km$ long and one end of the channel is a reflection boundary.

The calculated results are shown in Figure 3 as the snapshots of the water surface elevation per $37.5s$. In the Figure, each wave group progresses with the each group velocity. Each wave group stretches with progress because of the frequency dispersion. The WAVE-A started later catches up with, passes the WAVE-B started earlier, then, reflects at the end, again meets with, and passes through the WAVE-B.

These results are a matter of course as the liner wave theory. A point is that the results are calculated in the time domain by the system of the coupled vibration equations.

5 Discussion

In the present theory, the fluctuation pressure $\tilde{p}$ is expanded to 4 terms in Eq.(17) with the coefficients from q_1 to q_4. There is, however, no q_4 appeared in the final form of the coupled vibration equation in Eqs.(21)-(24). The evolution problem is calculated without q_4.

The vertical distribution of fluctuation pressure $\tilde{p}$ in Airy's theory is given as follows.
$$\tilde{p} = g\eta \frac{\cosh k(h+z)}{\cosh kh} = g\eta \frac{\cosh kh\tilde{z}}{\cosh kh} \tag{40}$$
The right side of above equation can be exressed by the Taylor expansion.
$$\tilde{p} = \frac{g\eta}{\cosh kh}\left(1 + \frac{(kh)^2}{2!}\tilde{z}^2 + \frac{(kh)^4}{4!}\tilde{z}^4\right) + O\left(g\eta\frac{(kh)^6}{6!}\right) \tag{41}$$

Comparing Eq.(17) and above equation, the coefficient q_4 corresponds to the resident term in the right side of above equation. The coefficient q_4 indicates the truncation error of the coupled vibration equations.

The combined kinematic-dynamic free surface boundary condition is given as
$$\frac{\partial^2 \Phi}{\partial t^2} = -g\frac{\partial \Phi}{\partial z} \tag{42}$$
in the linear water wave problem based on the velocity potential theory. Assuming that the velocity potential $\Phi(x, y, z, t)$ is expressed as the following form,
$$\Phi = \phi(x, y, t)\frac{\cosh k(h+z)}{\cosh kh}, \tag{43}$$
Eq.(42) is rewritten as follow.
$$\frac{\partial^2 \phi}{\partial t^2} = -gk \tanh kh \phi \tag{44}$$

This equation has the form of a vibration eqauiton. Expressing water waves as the form of vibration equation is a traditional and natural method. The coupled vibration equations, however, express the any shape of vertical distribution of the fluctuating pressure instead of an unique shape of vertical distribution such like the velocity potential in Eq.(43).

6 Conclusion

The system coupled vibration equations is derived for water waves progressing on the water with mild slope of the bottom configuration. The motion of water surface elevation is expressed by the vibration equation with the coupled exciting force. The new system of equations is able to be applied for the random waves composed by the monochoromatic waves whose relative depth h/L is less than 1.

REFERENCES

Berkhoff, J. C. W.(1972): Computation of combined refraction diffraction. *Proc. 13th ICCE*, ACSE, New York, pp.471-490.

Kubo, Y. , Y. Kotake, M. Isobe and A. Watanabe(1992): Time dependent mild slope Equation for random waves., Proc. 23rd ICCE, pp.419-431.

Nadaoka, K. and Y. Nakagawa(1993): Fully nonlinear-dispersive wave equations derived by a Galerkin formulation., *Meet'n '93, ASCE/ASME/SES Abstracts*, p.724

Liu, P. L.-F.(1983): Wave-current interactions on a slowly varying topography. *J. Geophys. Res.*, Vol.*88(C7)*, pp4421-4426.

Pregrine, D. H.(1967). Long waves on a beach. *J. Fluid Mech.*, Vol.*27*, pp.815-827

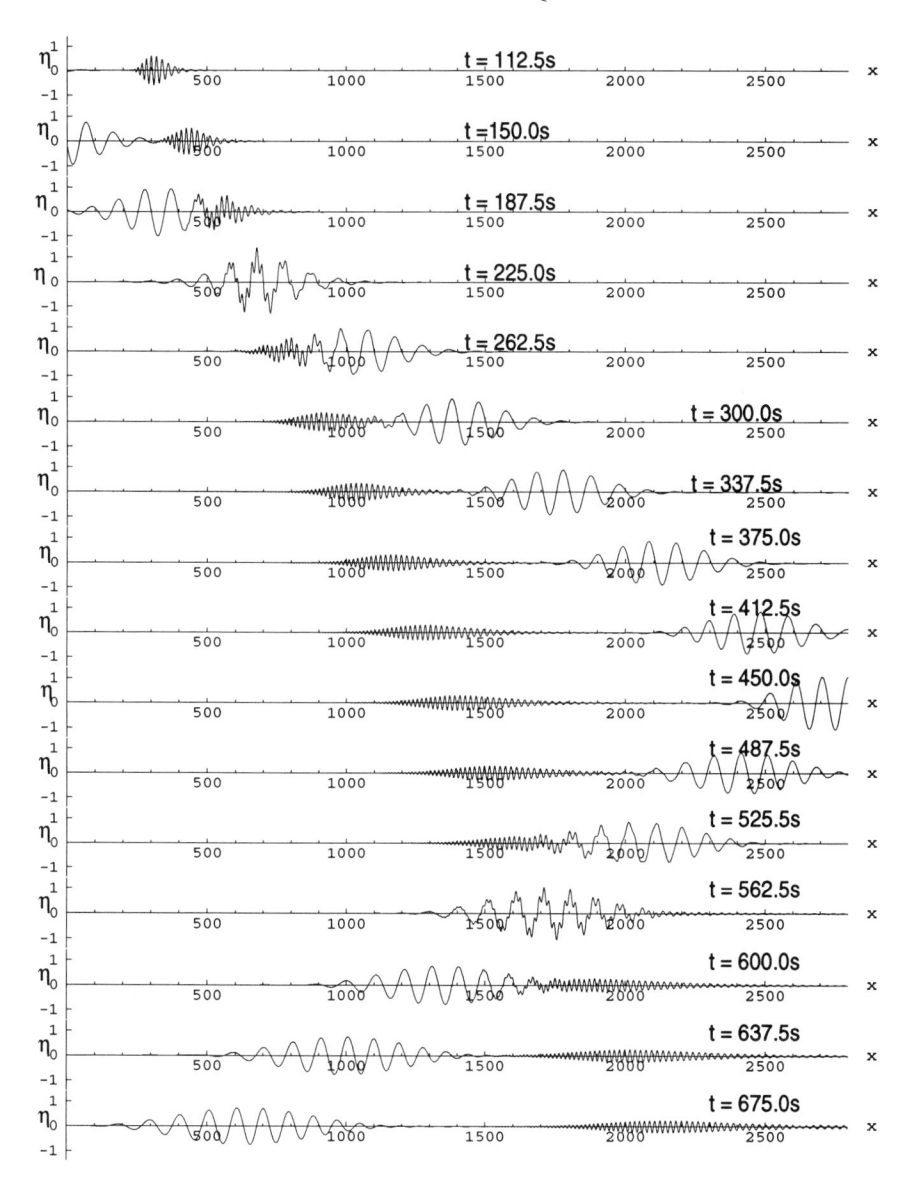

Figure 3: Propagation of two wave groups

APPENDIX

Equations for 3 Terms

$$g\eta = q_1 + q_2 + q_3 \tag{45}$$

$$\nabla^2 q_1 + \frac{\nabla(3q_1 + 7q_2)}{4h}\nabla h - \frac{10q_1 + 7q_2}{h^2} = \frac{3g\nabla\eta\nabla h}{4h} - \frac{10g\eta}{h^2} \tag{46}$$

$$\nabla^2 q_2 + \frac{\nabla(-10q_1 - 11q_2)}{4h}\nabla h - \frac{35q_1 + 35q_2}{h^2} = -\frac{5g\nabla\eta\nabla h}{2h} - \frac{35g\eta}{h^2} \tag{47}$$

$$\frac{\partial^2 \eta}{\partial t^2} = -\frac{1}{h}(10g\eta - 10q_1 - 7q_2) \tag{48}$$

Equations for 5 Terms

$$g\eta = q_1 + q_2 + q_3 + q_4 + q_5 \tag{49}$$

$$\nabla^2 q_1 + \frac{\nabla(35q_1 + 99q_2 - 13q_3 + 75q_4)}{64h}\nabla h \\ - \frac{36q_1 + 33q_2 + 26q_3 + 15q_4}{h^2} = \frac{35g\nabla\eta\nabla h}{64h} - \frac{36g\eta}{h^2} \tag{50}$$

$$\nabla^2 q_2 + \frac{\nabla(-95q_1 - 111q_2 + 65q_3 - 210)}{64h}\nabla h \\ - \frac{165q_1 + 165q_2 + 130q_3 + 75q_4}{h^2} = -\frac{95g\nabla\eta\nabla h}{64h} - \frac{165g\eta}{h^2} \tag{51}$$

$$\nabla^2 q_3 + \frac{\nabla(639q_1 + 351q_2 + 571q_3 + 1575q_4)}{256h}\nabla h \\ - \frac{234q_1 + 234q_2 + 234q_3 + 135q_4}{h^2} = \frac{639g\nabla\eta\nabla h}{256h} - \frac{234g\eta}{h^2} \tag{52}$$

$$\nabla^2 q_4 - \frac{\nabla(1222q_1 + 1066q_2 + 1794q_3 + 1291q_4)}{256h}\nabla h \\ - \frac{195q_1 + 195q_2 + 195q_3 + 195q_4}{h^2} = -\frac{1222g\nabla\eta\nabla h}{256h} - \frac{195g\eta}{h^2} \tag{53}$$

$$\frac{\partial^2 \eta}{\partial t^2} = -\frac{1}{h}(36g\eta - 36q_1 - 33q_2 - 26q_3 - 15q_4) \tag{54}$$

CHAPTER 35

NONLINEAR EVOLUTION OF DIRECTIONAL WAVE SPECTRA IN SHALLOW WATER

Okey Nwogu[1]

ABSTRACT

Nonlinear aspects of the transformation of a multidirectional wave field in shallow water are investigated using Boussinesq-type equations. Second-order interactions between different frequency components in an irregular sea state produce lower and higher harmonic components at the sum and difference frequencies of the primary waves. For water of constant depth, expressions are derived from the Boussinesq equations for the magnitude of the second-order waves induced by bidirectional, bichromatic waves. These are used to investigate the effect of the direction of wave propagation on the near-resonant interactions that occur in shallow water. For waves propagating in water of variable depth, a numerical model based on a time-domain solution of the governing equations is used to the predict the spatial evolution of the directional wave spectrum. The results of the numerical model are compared to experimental results for the propagation of bidirectional, bichromatic waves and irregular, multidirectional waves on a constant slope beach.

1. INTRODUCTION

Surface waves in the ocean are short-crested or multidirectional with components of different amplitudes and frequencies, propagating in different directions. As surface waves propagate from deep to shallow water, the directional wave spectrum is transformed due to both linear and nonlinear processes. Linear refraction leads to a narrower directional distribution in shallow water, with the principal direction more closely aligned to the beach contours. Changes to the directional spectrum due to linear effects can be accurately predicted by linear refraction models (e.g. Longuet-Higgins, 1957). However, Freilich et al. (1990) found that linear theory could not predict the amplitudes and directions observed in certain frequency bands in field studies. This is due to the near-resonant amplification of wave components induced at the sum and difference frequencies of the primary waves. These wave harmonics could also propagate in directions quite different from those of the primary waves, resulting in substantial changes to the frequency and directional distribution of wave energy.

---
[1] Institute for Marine Dynamics, National Research Council, Ottawa, Canada K1A 0R6

In shallow water depths, Boussinesq-type equations (e.g. Peregrine, 1967) are able to describe the near-resonant quadratic interactions that occur in shoaling multidirectional waves. Various time and frequency domain methods have been used to solve the two-dimensional form of the Boussinesq equations. Liu *et al.* (1985) developed a parabolic model for the shoreward propagation of individual wave components while Kirby (1990) proposed an angular spectrum model. Abreu *et al.* (1992) developed a shallow water spectral model in which near-resonant interactions were only considered between colinear waves. In intermediate water depths, third-order interactions become the dominant nonlinear mechanism for the cross-spectral transfer of energy. Suh *et al.* (1990) have developed an angular spectrum model of the mild-slope equation for Stokes waves which includes cubic interactions.

In this paper, the Boussinesq model of Nwogu (1993) is used to investigate the effect of near-resonant nonlinear wave-wave interactions on the transformation of directional wave spectra in shallow water. Compared to the Boussinesq model of Peregrine (1967), the model proposed by Nwogu (1993) can be applied over a wider range of water depths due to improved frequency dispersion characteristics obtained by changing the velocity variable from the depth-averaged velocity to the velocity at an arbitrary distance from the still water level. Analytical expressions are derived from the Boussinesq equations for the bidirectional quadratic transfer function of the second-order waves induced in water of constant depth. These are used to evaluate the effect of the direction of propagation of the wave components on the near-resonant quadratic interactions that occur in shallow water. For irregular multidirectional waves propagating in water of variable depth, a numerical model based on a time domain solution of the equations is used to predict the spatial evolution of the directional wave spectrum. The results of the numerical model are compared to experimental results for the shoaling of multidirectional waves on a constant slope beach.

2. THEORETICAL MODEL

2.1 Governing Equations

Boussinesq equations represent the depth-integrated equations for the conservation of mass and momentum for weakly nonlinear and mildly dispersive waves, propagating in water of variable depth. By assuming a quadratic variation of the velocity potential over depth, the governing equations of fluid motion can be integrated over the depth, reducing the three-dimensional problem to a two-dimensional one. The continuity and momentum equations can be expressed in terms of the water surface elevation, $\eta(\mathbf{x}, t)$, and the horizontal velocity, $\mathbf{u}_\alpha(\mathbf{x}, t)$, at an arbitrary distance z_α from the still water level as (see Nwogu, 1993):

$$\eta_t + \nabla \cdot [(h+\eta)\mathbf{u}_\alpha] + \nabla \cdot \left[\left(\frac{z_\alpha^2}{2} - \frac{h^2}{6}\right) h\nabla(\nabla \cdot \mathbf{u}_\alpha) + \left(z_\alpha + \frac{h}{2}\right) h\nabla[\nabla \cdot (h\mathbf{u}_\alpha)]\right] = 0 \quad (1)$$

$$\mathbf{u}_{\alpha t} + g\nabla\eta + (\mathbf{u}_\alpha \cdot \nabla)\mathbf{u}_\alpha + \left[\frac{z_\alpha^2}{2}\nabla(\nabla \cdot \mathbf{u}_{\alpha t}) + z_\alpha\nabla[\nabla \cdot (h\mathbf{u}_{\alpha t})]\right] = 0 \quad (2)$$

where $\nabla = (\partial/\partial x, \partial/\partial y)$, $h(\mathbf{x})$ is the water depth and g is the gravitational acceleration. The elevation of the velocity variable z_α is a free parameter and is chosen to minimize the differences between the linear dispersion characteristics of the Boussinesq model and linear theory. An optimum depth for the velocity variable, $z_\alpha = -0.53h$, gives errors of less than 2% in the phase speed from shallow water depths up to the deep water depth limit.

2.2 Second-Order Forced Waves

As surface waves propagate, the different frequency components interact to produce wave components at the harmonics of the primary wave frequencies. The simplest example of this nonlinear phenomenon is the combination of two wave trains with different frequencies. The nonlinear boundary conditions at the free surface result in the generation of forced waves at the sum and difference frequencies of the primary waves, $\omega_1 \pm \omega_2$. The forced waves are bound or phase-locked to the primary waves or wave groups, and propagate in directions given by the sum and difference of the wavenumber vectors, $\mathbf{k}_1 \pm \mathbf{k}_2$. The magnitude of the forced harmonics can be determined by solving the governing equation of motion, with the free surface boundary condition satisfied at second-order in wave steepness. Starting from the Laplace equation, Dean and Sharma (1981) derived expressions for the magnitude of the forced harmonics for bidirectional, bichromatic waves, i.e. two waves with different frequencies, propagating in different directions. We shall now derive corresponding expressions from the Boussinesq equations. Consider a wave train consisting of two periodic waves with frequencies, ω_1 and ω_2, amplitudes, a_1 and a_2, propagating in directions, θ_1 and θ_2 respectively. The water surface elevation is given by:

$$\eta^{(1)}(\mathbf{x},t) = a_1 \cos(\mathbf{k}_1 \cdot \mathbf{x} - \omega_1 t) + a_2 \cos(\mathbf{k}_2 \cdot \mathbf{x} - \omega_2 t), \quad (3)$$

where $\mathbf{k} = (k\cos\theta, k\sin\theta)$. The angle, θ, is defined relative to the positive x axis. The individual waves are assumed to satisfy the first-order or linearized form of the Boussinesq equations, i.e.

$$\eta_t^{(1)} + h\nabla \cdot \mathbf{u}_\alpha^{(1)} + \left(\alpha + \frac{1}{3}\right)h^3 \nabla \cdot \left[\nabla(\nabla \cdot \mathbf{u}_\alpha^{(1)})\right] = 0, \quad (4)$$

$$\mathbf{u}_{\alpha t}^{(1)} + g\nabla\eta^{(1)} + \alpha h^2 \nabla\left(\nabla \cdot \mathbf{u}_{\alpha t}^{(1)}\right) = 0, \quad (5)$$

where $\alpha = (z_\alpha/h)^2/2 + (z_\alpha/h)$. The wavenumber, k, is thus related to the wave frequency by the following dispersion relation:

$$\frac{\omega^2}{k^2} = gh\left[\frac{1 - \left(\alpha + \frac{1}{3}\right)(kh)^2}{1 - \alpha(kh)^2}\right] \quad (6)$$

The first-order horizontal velocities corresponding to the water surface elevation given by equation (3) can be determined from equations (4) and (5) as:

$$u_\alpha^{(1)}(\mathbf{x},t) = \frac{\omega_1}{k_1'h}\cos\theta_1 a_1 \cos(\mathbf{k}_1\cdot\mathbf{x}-\omega_1 t) + \frac{\omega_2}{k_2'h}\cos\theta_2 a_2 \cos(\mathbf{k}_2\cdot\mathbf{x}-\omega_2 t), \quad (7)$$

$$v_\alpha^{(1)}(\mathbf{x},t) = \frac{\omega_1}{k_1'h}\sin\theta_1 a_1 \cos(\mathbf{k}_1\cdot\mathbf{x}-\omega_1 t) + \frac{\omega_2}{k_2'h}\sin\theta_2 a_2 \cos(\mathbf{k}_2\cdot\mathbf{x}-\omega_2 t), \quad (8)$$

where $k' = k[1-(\alpha+1/3)(kh)^2]$. At second-order in wave steepness, the surface elevation and velocities have to satisfy the following set of equations:

$$\eta_t^{(2)} + h\nabla\cdot\mathbf{u}_\alpha^{(2)} + \left(\alpha+\frac{1}{3}\right)h^3\nabla\cdot\left[\nabla(\nabla\cdot\mathbf{u}_\alpha^{(2)})\right] = -\nabla\cdot\left(\eta^{(1)}\mathbf{u}_\alpha^{(1)}\right) \quad (9)$$

$$\mathbf{u}_{\alpha t}^{(2)} + g\nabla\eta^{(2)} + \alpha h^2\nabla\left(\nabla\cdot\mathbf{u}_{\alpha t}^{(2)}\right) = -\left(\mathbf{u}_\alpha^{(1)}\cdot\nabla\right)\mathbf{u}_\alpha^{(1)} \quad (10)$$

The second-order wave will consist of a sub-harmonic at the difference frequency, $\omega_1-\omega_2$, and super-harmonics at the sum frequencies, $2\omega_1$, $2\omega_2$ and $\omega_1+\omega_2$. It can be written as:

$$\eta^{(2)}(\mathbf{x},t) = \frac{a_1^2}{2}G_+(\omega_1,\omega_1,\theta_1,\theta_1)\cos(2\mathbf{k}_1\cdot\mathbf{x}-2\omega_1 t)$$

$$+ \frac{a_2^2}{2}G_+(\omega_2,\omega_2,\theta_2,\theta_2)\cos(2\mathbf{k}_2\cdot\mathbf{x}-2\omega_2 t)$$

$$+ a_1 a_2 G_\pm(\omega_1,\omega_2,\theta_1,\theta_2)\cos(\mathbf{k}_\pm\cdot\mathbf{x}-\omega_\pm t), \quad (11)$$

where $\mathbf{k}_\pm = \mathbf{k}_1\pm\mathbf{k}_2$, and $G_\pm(\omega_1,\omega_2,\theta_1,\theta_2)$ is a bidirectional quadratic transfer function that relates the amplitude of the second-order forced wave to the first-order amplitudes. The quadratic transfer function can be determined by substituting equations (3), (7) and (8) for the first-order surface elevation and velocities into the second-order equations (9) and (10), and solving for the amplitudes of the second-order surface elevation and velocities. This leads to:

$$G_\pm(\omega_1,\omega_2,\theta_1,\theta_2) = \frac{\omega_1\omega_2(k_\pm h)^2\cos\Delta\theta[1-(\alpha+\frac{1}{3})(k_\pm h)^2]}{2\lambda k_1' k_2' h^3} +$$

$$\frac{\omega_\pm[1-\alpha(k_\pm h)^2][\omega_1 k_2'h[k_1 h\pm k_2 h\cos\Delta\theta]+\omega_2 k_1'h[k_1 h\cos\Delta\theta\pm k_2 h]]}{2\lambda k_1' k_2' h^3}, \quad (12)$$

where $\Delta\theta = \theta_1-\theta_2$,

$$k_\pm = |\mathbf{k}_1\pm\mathbf{k}_2| = \sqrt{k_1^2+k_2^2\pm 2k_1 k_2\cos\Delta\theta}, \quad (13)$$

and

$$\lambda = \omega_\pm^2[1-\alpha(k_\pm h)^2] - gk_\pm^2 h[1-(\alpha+1/3)(k_\pm h)^2]. \quad (14)$$

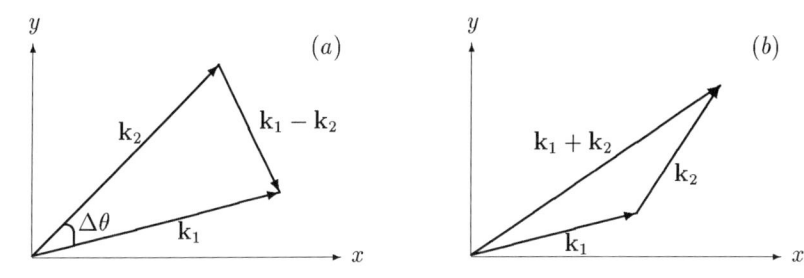

Figure 1. Directions of the sum and difference of the wavenumber vectors.

$\lambda = 0$ is actually the dispersion relation for first-order free waves in the Boussinesq model (Eqn. 6). The second-order forced waves do not satisfy the linear dispersion relation because of the forcing terms on the right hand side of equations (9) and (10), unlike second-order free waves that satisfy the homogeneous part of the equations. The second-order interactions could be resonant ($\lambda \to 0$) if the frequency and wavenumber of the forced wave corresponds to that of a free wave. This occurs to unidirectional wave trains in shallow water depths where the phase velocities of the forced waves are nearly equal to the phase velocities of free waves, leading to a near-resonant amplification of the second-order waves. Wave breaking and cross-spectral energy transfers, however, limit the amplitude of the second-order waves in shallow water.

We shall now examine in detail, the differences between the magnitudes and directions of propagation of the sub-harmonics and super-harmonics in unidirectional and bidirectional seas. The sub-harmonic component, which is commonly referred to as the set-down component, travels at the velocity of the wave group, ω_-/k_-, along a direction defined by the difference of the wavenumber vectors as shown in Figure 1(a). This direction could be quite different from the direction of propagation of the primary waves. If k_1 is nearly equal to k_2, the second-order long wave would travel in a direction almost perpendicular to the average direction of the first-order short waves. On a sloping beach, this might excite the edge (transverse) wave modes as discussed by Gallagher (1971). In contrast, the super-harmonics travel in a direction nearly coincident with the average direction of the primary waves as shown in Figure 1(b).

The quadratic transfer function was evaluated for bidirectional waves with different angles of separation between the individual wave components. The transfer function of the sub-harmonic wave component is plotted in Figure 2 for an example where the frequency difference, $\Delta\omega/\omega = 0.1$, with $\omega_1 = \omega - \Delta\omega/2$ and $\omega_2 = \omega + \Delta\omega/2$. Three crossing angles, $\Delta\theta = 0°$, $10°$ and $20°$ were considered. Also shown in Figure 2 are results obtained from expressions based on a second-order solution of the Laplace equation, derived by Dean and Sharma (1981).

A number of interesting observations can be made from Figure 2. The first one is that for unidirectional waves, the Boussinesq model underestimates the magnitude of the set-down wave, particularly in intermediate water depths. However, for bidirectional

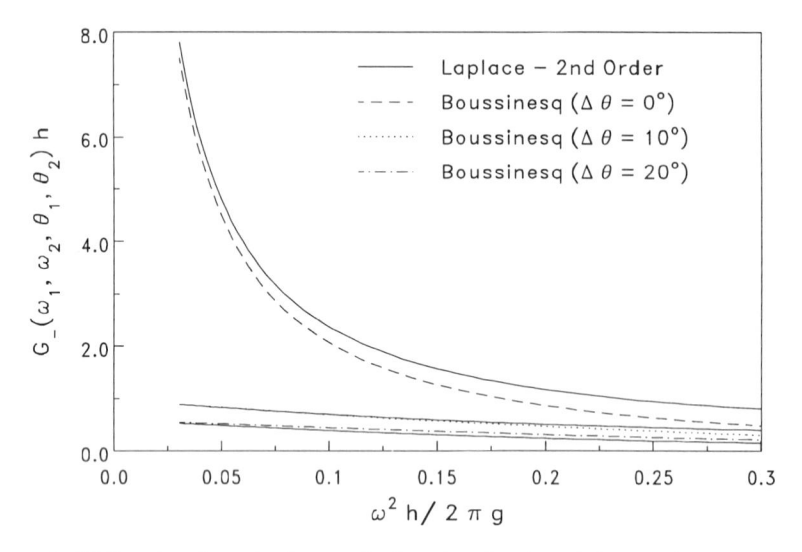

Figure 2. Bidirectional quadratic transfer function for the sub-harmonic component ($\Delta\omega/\omega = 0.1$).

waves, the differences between the Boussinesq and Laplace models becomes negligible. The second one is that the directionality of the waves leads to a significant reduction of the set-down wave, as was noted by Sand (1982). In shallow water with $\omega^2 h/2\pi g < 0.05$, the amplitude of the forced long wave for $\Delta\theta = 10°$ is reduced by at least a factor of 5. This is because forced long waves in directional wave fields are not resonantly amplified in shallow water. Near-resonant amplification occurs for unidirectional waves because the wavenumber of the forced long wave $|k_1 - k_2|$ is nearly equal to that of a free long wave $k(\omega_-)$, or equivalently, ω_- and $|k_1 - k_2|$ nearly satisfy the linear dispersion relation with $\lambda \to 0$ in equation (12). In bidirectional seas, the magnitude of the wavenumber vector of the forced long wave, $|k_1 - k_2|$, is significantly larger than that of the corresponding free wave $k(\omega_-)$ for small angles of separation. The forced waves are, thus, no longer close to satisfying the dispersion relation for free waves.

The quadratic transfer function for the super-harmonics is plotted in Figure 3 for an example with $\Delta\omega/\omega = 0.1$, and $\Delta\theta = 0°$ and $40°$. The forced higher harmonics are slightly reduced in bidirectional waves, but not as significantly as the sub-harmonic component. For $\omega^2 h/2\pi g = 0.05$ and $\Delta\theta = 20°$, the super-harmonic component is reduced by 12%, compared to 90% for the sub-harmonic component. In contrast to the sub-harmonics, the super-harmonics in directional wave fields are resonantly amplified in shallow water. This is because with the addition of the wavenumber vectors, the magnitude of of the wavenumber vector of the forced super-harmonic is closer to that of the corresponding free wave. The directional wave transformation model proposed

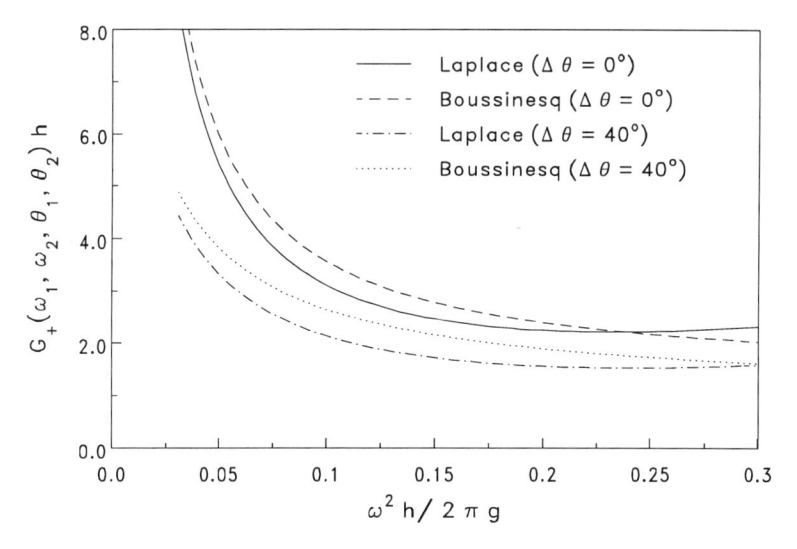

NONLINEAR EVOLUTION OF DIRECTIONAL WAVE SPECTRA 473

Figure 3. Bidirectional quadratic transfer function for the super-harmonic component ($\Delta\omega/\omega = 0.1$).

by Abreu *et al.* (1992) assumes that near-resonant, second-order interactions only occur between colinear waves in shallow water. The present analysis, does, however, show that second-order interactions between non-colinear waves in shallow water are near-resonant for the super-harmonic component, although the strength of the interaction decreases with increasing angular separation.

3. Time Domain Solution

For waves propagating in water of variable depth, free second and higher-order waves are generated in addition to the forced waves. The bidirectional quadratic transfer function (Eqn. 12) can no longer be used to determine the amplitude of the wave harmonics. In this paper, we employ the time-domain model of Nwogu (1995) for irregular multidirectional wave propagation in water of variable depth. The model solves the governing set of Boussinesq-type equations using an iterative Crank-Nicolson finite difference method, with a predictor-corrector scheme used to provide the initial estimate. The computational domain is discretized using a rectangular grid, with the dependent variables η, u_α and v_α defined at the grid points in a staggered manner. The numerical solution procedure consists of solving an algebraic expression for η at all grid points, tridiagonal matrices for u_α along lines in the x direction and tridiagonal matrices for v_α along lines in the y direction at every time step. Details of the third-order accurate scheme can be found in Nwogu (1995).

The boundaries of the computational domain may be specified as wave input boundaries or solid walls. Along incident wave boundaries, time series of u_α, $u_{\alpha,xx}$ and $v_{\alpha,xy}$ or v_α, $v_{\alpha,yy}$ and $u_{\alpha,xy}$, corresponding to regular or irregular, unidirectional or multidirectional sea states are input at the grid points. The time histories may be derived from target directional wave spectra using the random phase, single direction per frequency model of wave synthesis (see Miles and Funke, 1987). Waves propagating out of the domain are artificially absorbed in damping regions placed next to solid wall boundaries. Artificial damping of wave energy is accomplished by introducing terms out of phase with the water surface velocity and fluid acceleration into the continuity and momentum equations respectively (see Nwogu, 1995). The output of the numerical model are time histories of η, u_α and v_α at desired grid points in the computational domain. Directional wave spectra estimates are obtained from the time records by using the high resolution, maximum entropy method (Nwogu et al., 1987)

4. NUMERICAL AND EXPERIMENTAL RESULTS

Laboratory experiments were also conducted to investigate the propagation of multidirectional waves on a constant slope beach. The experiments were carried out in the three-dimensional wave basin of the Hydraulics Laboratory, National Research Council of Canada. The basin is 30 m wide, 20 m long and 3 m deep and is equipped with a 60-segment directional wave generator. The individual wave boards are 0.5 m wide and 2 m high. Wave energy absorbers made of perforated metal sheets are installed along the other sides of the basin not occupied by the wave generator. A 1:25 constant slope beach with an impermeable concrete cover was constructed in the basin, parallel to the wave generator. The toe of the slope was located 4.6 m away from the wave boards. Bidirectional bichromatic waves and irregular multidirectional sea states were generated in the basin. The water depth in the constant depth portion of the basin was 0.56 m. The water surface elevation along the centerline of the basin was measured with a linear array of 23 water level gauges. The experimental set-up is discussed in greater detail in Nwogu (1993).

4.1 Shoaling of Bidirectional, Bichromatic Waves

Consider the shoaling of a bichromatic wave train with component wave periods, $T_1 = 1.65$ s, $T_2 = 1.5$ s, and heights, $H_1 = 0.041$ m, $H_2 = 0.037$ m. The tests were carried for both a unidirectional version with $\theta_1 = \theta_2 = 0°$, and a bidirectional version with $\theta_1 = 30°$ and $\theta_2 = -15°$. The spectral density of the measured surface elevation time history at location ($h = 0.134$ m) is shown in Figure 4 for the bidirectional wave train. In addition to second-order wave harmonics at $2f_1$, $2f_2$ and $f_1 \pm f_2$, third-order wave components are also observed at $2f_1 \pm f_2$, $2f_2 \pm f_1$, $3f_1$, and $3f_2$, as well as some fourth-order components.

The numerical model was used to simulate the propagation of the unidirectional and

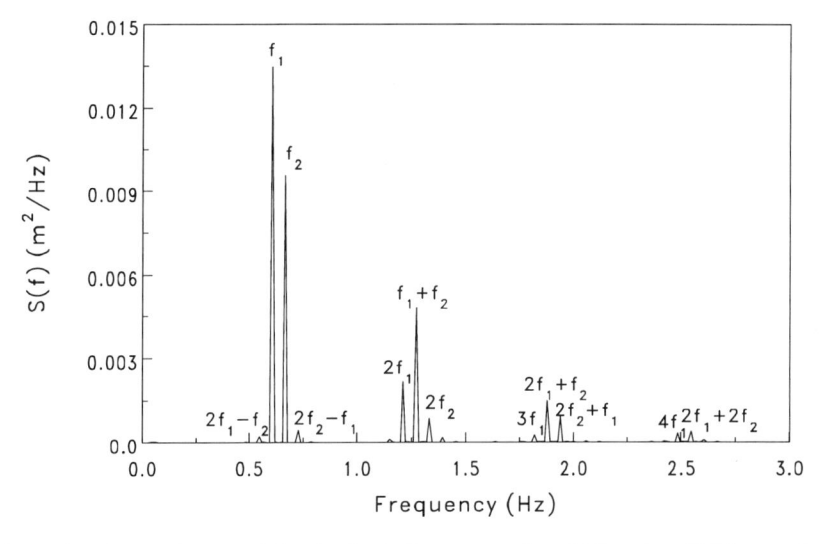

Figure 4. Measured spectral density in shallow water for a bidirectional bichromatic wave train ($\Delta\theta = 45°$, $h = 0.134$ m).

bidirectional wave trains on the 1:25 beach. The simulations were carried out using a time step size $\Delta t = 0.05$ s, and spatial grid sizes $\Delta x = 0.1$ m and $\Delta y = 0.2$ m. A contour plot of the instantaneous water surface elevation in the basin for the bidirectional wave train is shown in Figure 5. The variation in amplitude of a select number of wave harmonics along the centerline of the basin are plotted in Figure 6 for the unidirectional wave train, and Figure 7 for the bidirectional wave train. The experimental results are also shown in the figures. The Boussinesq model is observed to reasonably predict the measured variation in amplitude of the first-order waves and higher harmonics. The model, however, underestimates the magnitude of the set-down component for the unidirectional wave in shallow water. The simulated set-down wave has an amplitude of 0.003 m at $h = 0.134$ m, while the measured wave has an amplitude of 0.006 m. The numerical model underestimates the magnitude of the long period wave partly because it does not simulate the reflected free waves that are generated after wave breaking and runup.

The ability of the numerical model to predict the change in wave direction between the deep and shallow water depths was also examined. The maximum entropy method, described by Nwogu et al. (1987), was used to estimate the directional distributions at different frequency bands from the simulated η, u_α and v_α time series. The directions in the deep ($h = 0.56$ m) portion of the basin are $\theta(f_1) = 30°$ and $\theta(f_2) = -15°$. At $h = 0.134$ m, the Boussinesq model predicts $\theta(f_1) = 18°$ and $\theta(f_2) = -13°$ while Snell's law predicts $\theta(f_1) = 27°$ and $\theta(f_2) = -13°$. Good agreement is observed for the $-15°$ wave but not the $30°$ wave. The differences are primarily due to diffraction

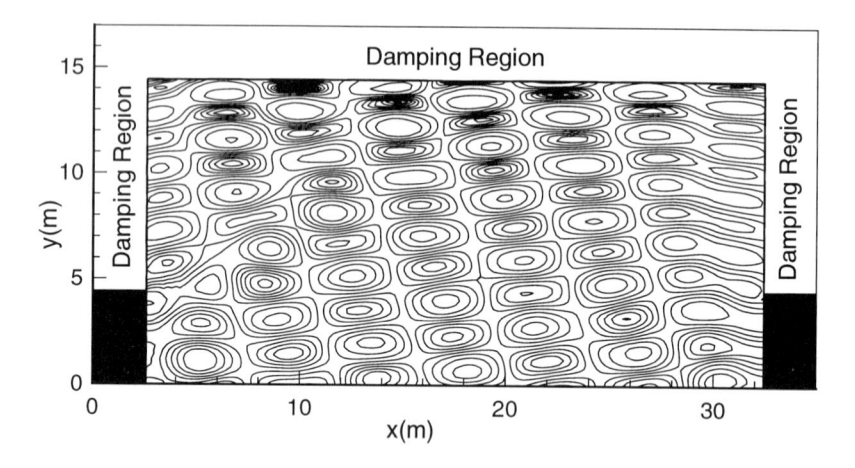

Figure 5. Contour plot of instantaneous water surface elevation for a bidirectional bichromatic wave train shoaling on a 1:25 beach ($\theta_1 = 30°$, $\theta_2 = -15°$).

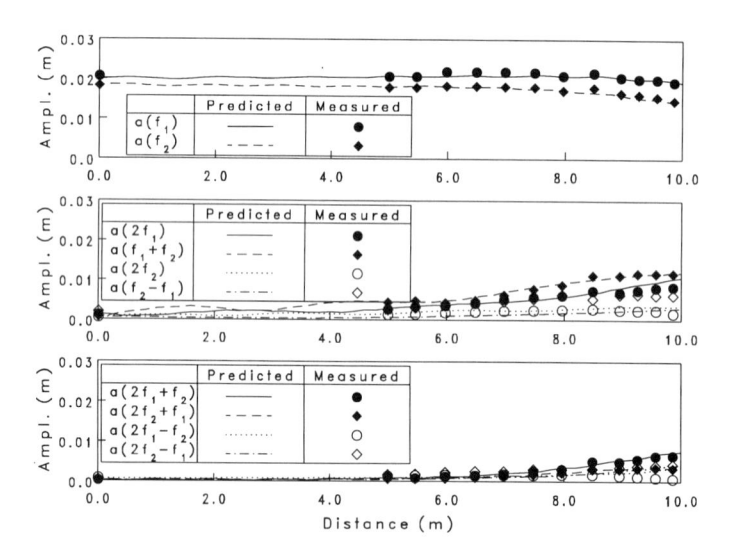

Figure 6. Variation in amplitude of wave harmonics for a unidirectional bichromatic wave train shoaling on a 1:25 beach ($\Delta\theta = 0°$).

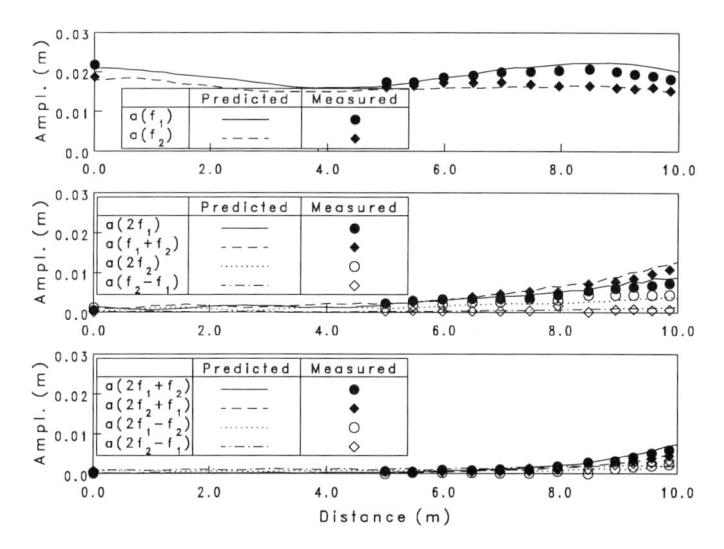

NONLINEAR EVOLUTION OF DIRECTIONAL WAVE SPECTRA 477

Figure 7. Variation in amplitude of wave harmonics for a bidirectional bichromatic wave train shoaling on a 1:25 beach ($\Delta\theta = 45°$).

effects caused by the finite width of the wave generator. The effect of diffraction is more important in the shallow portion of the basin for oblique waves with large angles of incidence. For the second-order higher harmonics, the numerical model predicts $\theta(2f_1) = 19°$, $\theta(f_1 + f_2) = 2°$, and $\theta(2f_2) = -12°$. The predicted direction for the sum frequency component is within 2° of that determined from the sum of the wavenumber vectors.

4.2 Shoaling of Irregular Multidirectional Waves

Consider the shoaling of a bimodal sea state with local sea and swell components on the 1:25 beach from deep ($h = 0.56$ m) to shallow water ($h = 0.18$ m). The incident sea states were synthesized using the random phase, single direction per frequency model of wave synthesis (see Miles and Funke, 1987). The JONSWAP spectrum was used to describe the frequency distribution of wave energy while the parametric cosine power function was used for the directional distribution. The local sea component has a significant wave height, $H_{mo} = 0.062$ m, peak period, $T_p = 1.5$ s, peak enhancement factor, $\gamma = 3.3$, and a directional distribution defined by $D(\theta) = \cos^{12}(\theta - 22.5°)$. The swell component is characterized by $H_{mo} = 0.068$ m, $T_p = 2$ s, $\gamma = 10$, and $D(\theta) = \cos^{44}(\theta + 22.5°)$. The incident directional wave spectrum is shown in Figure 8.

The Boussinesq model was used to simulate the shoaling of the bimodal sea state with

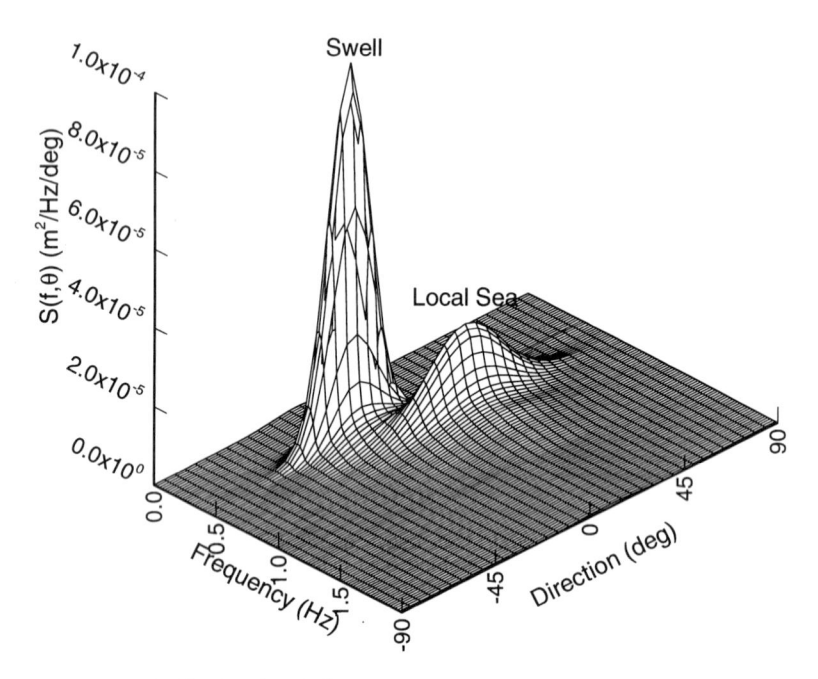

Figure 8. Numerically simulated directional spectrum of a bimodal sea state in deep water ($h = 0.56$ m).

time step size, $\Delta t = 0.05$ s and spatial grid sizes, $\Delta x = 0.1$ m, $\Delta y = 0.2$ m. Figure 9 shows a comparison of the wave spectrum at $h = 0.56$ m with the measured and predicted spectral densities at $h = 0.18$ m. Reasonably good agreement is observed between the measured and predicted wave spectra. Directional spectral estimates were also obtained from the simulated η, u_α and v_α time histories using the maximum entropy method (Nwogu et al., 1987). The predicted directional spectrum at $h = 0.18$ m is shown in Figure 10. Nonlinear wave-wave interaction effects in the shoaling process substantially change the directional wave spectrum, with the transfer of energy across frequency and direction bands. There is a growth of the second, third and fourth harmonics of the swell component, and the generation of components at the vector sum of the local sea and swell components, and the vector sum of the local sea and second harmonic of the swell component. Such modifications of the directional wave spectrum can only be obtained with the use of a model that simultaneously treats nonlinearity and directionality in shoaling waves.

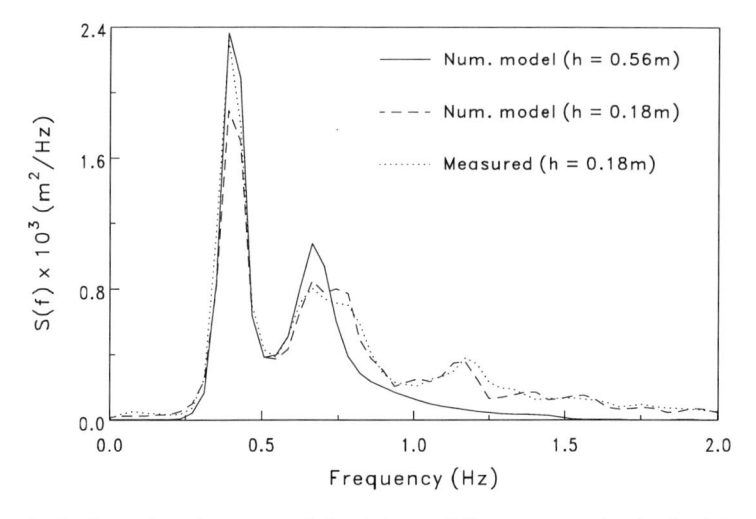

Figure 9. Surface elevation spectral densities at different water depths for bimodal sea state.

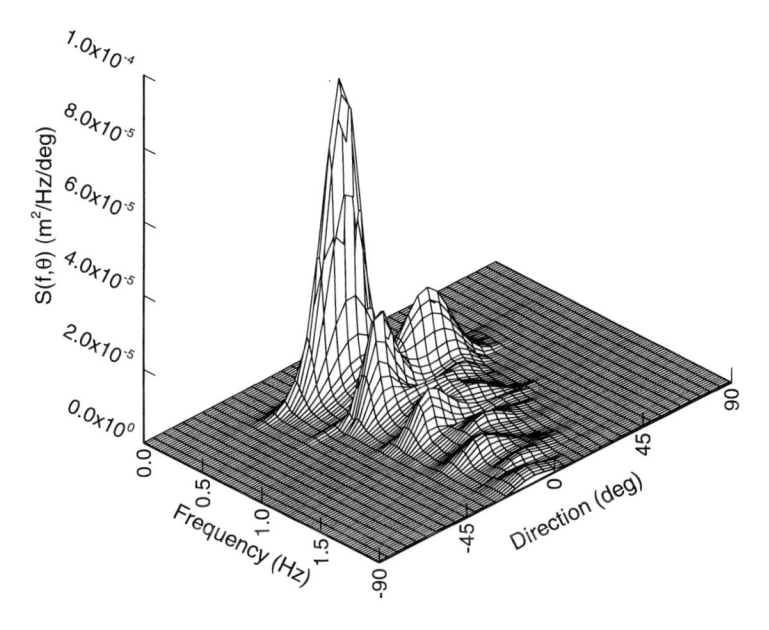

Figure 10. Numerically simulated directional spectrum of a bimodal sea state in shallow water ($h = 0.18$ m).

5. CONCLUSIONS

A Boussinesq model has been used to investigate the effect of near-resonant nonlinear interactions on the transformation of directional wave spectra in shallow water. The amplification of second-order components induced at the sum and difference frequencies of the primary waves is near-resonant for unidirectional waves in shallow water. In multidirectional sea states, however, the second-order interactions are near-resonant for the higher harmonics but non-resonant for the lower harmonics. This leads to a significant reduction in magnitude of the long period waves induced by shoaling multidirectional waves. The spatial evolution of the directional wave spectrum in water of variable depth was predicted using a time domain Boussinesq model, and the maximum entropy method for directional wave analysis. The numerical model, which includes the effects of shoaling, refraction, diffraction and reflection, was able to predict a substantial modification of the directional wave spectrum in shallow water due to near-resonant nonlinear interactions.

REFERENCES

Abreu, M., Larraza, A., and Thornton, E. 1992. Nonlinear transformation of directional wave spectra in shallow water. *Journal of Geophysical Research*, **97**, 15579-15589.

Dean, R.G., and Sharma, J.N. 1981. Simulation of wave systems due to nonlinear directional spectra. *Proc. Int. Symposium on Hydrodynamics in Ocean Engineering*, Trondheim, Norway, 1211-1222.

Freilich, M.H, Guza, R.T., and Elgar, S.L. 1990. Observations of nonlinear effects in directional spectra of shoaling gravity waves. *Journal of Geophysical Research*, **95**, 9645-9656.

Gallagher, B. 1971. Generation of surf beat by non-linear wave interactions. *Journal of Fluid Mechanics*, **49**, 1-20.

Kirby, J.T. 1990. Modelling shoaling directional wave spectra in shallow water. *Proc. 22nd Int. Conference on Coastal Engineering*, Delft, The Netherlands, 109-122.

Liu, P.L.-F., Yoon, S.B., and Kirby, J.T. 1985. Nonlinear refraction-diffraction of waves in shallow water. *Journal of Fluid Mechanics*, **153**, 185-201.

Longuet-Higgins, M.S. 1957. On the transformation of a continuous spectrum by refraction. *Proc. Cambridge Phil. Soc.*, **53**, 226-229.

Miles, M.D., and Funke, E.R. 1987. A comparison of methods for the synthesis of directional seas. *Proc. 6th Int. Offshore Mechanics and Arctic Engineering Symposium*, Houston, II, 247-255.

Nwogu, O. 1993. Alternative form of Boussinesq equations for nearshore wave propagation. *Journal of Waterway, Port, Coastal and Ocean Engineering*, ASCE, **119**(6), 618-638.

Nwogu, O. 1995. Time domain simulation of shoaling multidirectional waves. *submitted to Coastal Engineering*.

Nwogu, O., Mansard, E.P.D., Miles, M.D, and Isaacson, M. 1987. Estimation of directional wave spectra by the maximum entropy method. *Proc. IAHR Seminar on Wave Analysis and Generation in Laboratory Basins*, XXII IAHR Congress, Lausanne, 363-376.

Peregrine, D.H. 1967. Long waves on a beach. *Journal of Fluid Mechanics*, **27**, 815-827.

Sand, S.E. 1982. Long waves in directional seas. *Coastal Engineering*, **6**, 195-208.

Suh, K.D., Dalrymple, R.A., and Kirby, J.T. 1990. An angular spectrum model for propagation of Stokes waves. *Journal of Fluid Mechanics*, **221**, 205-232.

CHAPTER 36

NON–GAUSSIAN PROBABILITY DISTRIBUTION OF COASTAL WAVES

M.K. Ochi[*] and K. Ahn[**]

Abstract

This paper presents the development of a probability density function applicable to waves in finite water depth (which can be considered to be a nonlinear, non–Gaussian random process) in closed form. The derivation of the density function is based on the Kac-Siegert solution developed for a nonlinear mechanical system, but the parameters involved in the solution are evaluated from the wave record only. Further, the probability density function is asymptotically expressed in closed form. Comparisons between the presently developed probability density function and histograms constructed from wave records show good agreement.

Introduction

The wind–generated wave profile observed in a sea of finite water depth is significantly different from that observed in deep water in that there is a definite excess of high crests and shallow troughs in contrast to those of waves in deep water. This is attributed to nonlinear wave–wave interaction (energy transfer) between component waves, and such waves are considered to be a typical non–Gaussian random process.

Probability distributions applicable for presenting non–Gaussian random waves have been derived through three different approaches; (a) application of the orthogonal polynomials to the probability density function [Longuet–Higgins 1963], (b) application of Stokes wave theory [Tayfun 1980, Huang, et al., 1983], and (c) application of the Kac–Siegert solution [Langley 1987].

Derivation of the non–Gaussian probability density function by applying the concept of orthogonal polynomials is well–known as the

---
[*] Professor, Coastal and Oceanographic Engineering Department, 336 Weil Hall, University of Florida, Gainesville, Florida, 32611, U.S.A.
[**] Assistant professor, Civil Engineering Department, Handong, University, Pohang City, Korea.

Gram–Charlier series probability density function. The probability density function, however, is given in series form. Therefore, the density function usually has a negative value at some part of the distribution caused by the finite number of waves used in computing the density function.

The probability density function derived based on the application of Stokes waves imposes a preliminary form on the wave profile such that the individual waves are expressed as a Stokes expansion to the 2nd or 3rd order components. It is highly desirable to justify the validity of the assumption involved for random waves, particularly waves in shallow water. On the other hand, the probability density function derived by application of the Kac–Siegert solution appears to be pertinent for waves in finite water depth, since the solution represents a nonlinear, non–Gaussian random process. The probability density function, however, cannot be presented in closed form. Hence, it is not possible to develop a probability density function applicable to wave height therefrom.

This paper presents a probability density function applicable to waves in finite water depth which can be considered to be a nonlinear, non–Gaussian random process. The density function is derived based on the Kac–Siegert solution through spectral analysis. During the course of analysis, the wave spectrum is decomposed into linear and nonlinear components in order to clarify the degree of nonlinearity involved in shallow water waves. Furthermore, the probability density function is derived in closed form so that it can be used for derivation of the distribution functions applicable to peaks and troughs of waves in finite water depth.

Basic Concept

The basic concept of the analysis applied in the present study is that the stochastic characteristics of waves in finite water depth may be considered to be the same as those of the output of a nonlinear mechanical system. In order to elaborate on this concept, let us examine the variation of wave profiles observed from deep to shallow water.

Figure 1 shows portions of wave records obtained by the Coastal Engineering Research Center, US Army, during the Atlantic Ocean Remote Sensing Land–Ocean Experiment (ARSLOE) project in 1980 at Duck, North Carolina. Included in each figure is the distance from the shoreline at the location where data were taken. As can be seen, the wave profile recorded by Gage 710 (obtained at the deepest of the three locations) is almost the same as that observed in deep water waves. As the water depth decreases, the wave profile shows a definite excess of high crests and shallow troughs which is a typical feature of a non–Gaussian random process. In other words, the wave profile transforms from a Gaussian random process to a non–Gaussian random process as they move from deep to shallow wave areas.

Let us compare the transformation of wave profiles from deep to shallow water with the output of a nonlinear mechanical system having a Gaussian random input with various degrees of nonlinearity. The

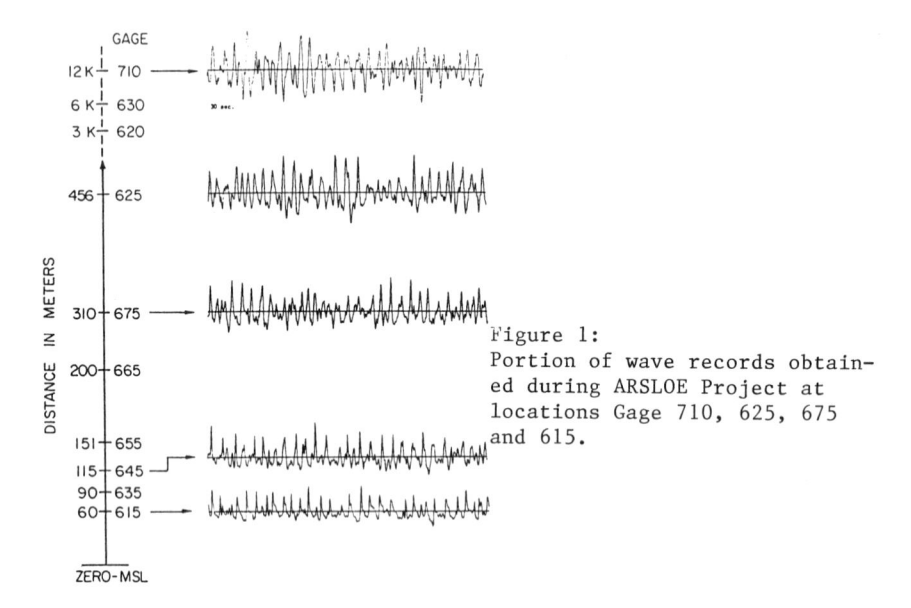

Figure 1:
Portion of wave records obtained during ARSLOE Project at locations Gage 710, 625, 675 and 615.

output of a system with very weak nonlinearity may be considered as a Gaussian random process. However, the statistical properties of the system's output show increasing non-Gaussian characteristics with increase in the intensity of system's nonlinearity. Thus, we may consider the augmentation of the non-Gaussian characteristics of waves with decreasing water depth to be analogous to that observed in the output of a nonlinear system as its nonlinearity increases.

With this basic concept in mind, we may apply the unique solution of the output of a nonlinear system developed by Kac-Siegert [1947] to nonlinear waves. That is, the output (response) of a nonlinear system which can be presented by Volterra's stochastic series expansion can be presented in terms of the standardized normal random variable as follows:

$$y(t) = \sum_{j=1}^{N} \left(\beta_j z_j + \lambda_j z_j^2 \right) \tag{1}$$

where $y(t)$ = output of a nonlinear system

   z_i = standardized normal variate.

The parameters β_j and λ_j are evaluated by finding the eigenfunction and eigenvalues of the integral equation given by

$$\int K(\omega_1,\omega_2)\psi_j(\omega_2)d\omega_2 = \lambda_j \psi_j(\omega_1) \tag{2}$$

where $K(\omega_1,\omega_2)$ = $H(\omega_1,\omega_2)\sqrt{S(\omega_1)\,S(\omega_2)}$

$\psi_j(\omega)$ = orthogonal eigenfunction

$H(\omega_1,\omega_2)$ = 2nd order frequency response function

$S(\omega)$ = output spectral density function.

As can be seen, knowledge of the 2nd order frequency response function, $H(\omega_1, \omega_2)$ is necessary in order to solve the integral equation given in Eq.(2), but there is no way to evaluate it for random waves. Recently, however, the authors developed a probability density function applicable to the response of a nonlinear mechanical system in closed form based on Kac-Siegert's solution without knowledge of the second order frequency response function [Ochi and Ahn 1994]. Based on the analogy between the nonlinear mechanical system and transformation of wave profiles as they move from deep to shallow water, this paper presents the application of the authors' method in the above reference to the analysis of waves in finite water depth.

An approach to evaluate the nonlinear properties of random waves by applying the Kac-Siegert solution was considered earlier by Langley [1987]. Our approach, however, differs from Langley's approach in that (i) the wave spectrum is decomposed into linear and nonlinear components in order to examine how notably the nonlinearity increases with decreasing water depth, (ii) spectral analysis methodology instead of wave potential theory is applied in evaluating the parameters β_j and λ_j, and (iii) the probability density function applicable to wave profile y(t) is presented in closed form.

Presentation of Nonlinear Waves

Let us write the surface profile of nonlinear waves as follows:

$$y(t) = \mathrm{Re} \sum_{k=1}^{N} c_k \, e^{i(\omega_k t + \varepsilon_k)} + \mathrm{Re} \sum_{k=1}^{N} \sum_{\ell=1}^{N} c_k\, c_\ell \left[q_{k\ell}\, e^{\{(\omega_k+\omega_\ell)t+\varepsilon_+\}} \right.$$
$$\left. + r_{k\ell}\, e^{i\{(\omega_k-\omega_\ell)t+\varepsilon_-\}} \right] \quad (3)$$

where ω = frequency, ε = phase lag.

$q_{k\ell}$ is a coefficient associated with $(\omega_k + \omega_\ell)$ which is the sum of the interaction between two frequency components ω_k and ω_ℓ, while $r_{k\ell}$ is associated with the interaction differential between ω_k and ω_ℓ. Since there is a phase shift between the frequencies, ω_k and ω_ℓ, in general, these interaction coefficients are complex numbers.

Equation (3) has often been considered for simulation studies as well as for analysis of nonlinear ocean waves in which the interaction coefficients are evaluated from second order nonlinear wave theory [Hasselmann 1962, Hudspeth and Chen 1979, Sharma and Dean 1979, Anastasiou et al. 1982, etc.].

Let us write the first term of Eq.(3) as follows:

$$y_1(t) = \sum_{k=1}^{N} c_k e^{i(\omega_k t + \varepsilon_k)} = \sum_{k=1}^{N} c_k \{\cos(\omega_k t + \varepsilon_k) + i\sin(\omega_k t + \varepsilon_k)\} \quad (4)$$

We assume that the linear component $y_1(t)$ is a narrow-band Gaussian random process. By ignoring the factor ρg, $(1/2)c_k^2$ represents the spectral density, $S_L(\omega_k)\Delta\omega$, where the subscript L stands for the spectral density of the linear wave components. That is, the spectral density at $\omega = \omega_k$ is written by

$$s_k = \{S_L(\omega_k)\Delta\omega\}^{1/2} = \left\{ \int_{\omega_k - (\Delta\omega/2)}^{\omega_k + (\Delta\omega/2)} S_L(\omega) d\omega \right\}^{1/2}. \quad (5)$$

Then by defining

$$c_k \cos(\omega_k t + \varepsilon_k) = s_k \cdot u_k, \quad \text{and} \quad c_k \sin(\omega_k t + \varepsilon_k) = s_k \cdot v_k, \quad (6)$$

the linear component can be written as

$$y_1(t) = \operatorname{Re} \sum_{k=1}^{N} s_k (u_k + i v_k). \quad (7)$$

The second term of Eq.(3) can be similarly written as

$$y_2(t) = \operatorname{Re} \sum_{k=1}^{N} \sum_{\ell=1}^{N} s_k s_\ell \{ q_{k\ell}(u_k + iv_k)(u_\ell + iv_\ell)$$
$$+ r_{k\ell}(u_k + iv_k)(u_\ell - iv_\ell) \}, \quad \text{where} \quad s_\ell = \{S_L(\omega_\ell)\Delta\omega\}^{1/2} \quad (8)$$

Then, by taking the real-part of Eqs.(7) and (8), the nonlinear wave profile can be presented as follows:

$$y(t) = y_1(t) + y_2(t) = \sum_{k=1}^{N} s_k u_k$$
$$+ \sum_{k=1}^{N} \sum_{\ell=1}^{N} \{s_k s_\ell (q_{k\ell} + r_{k\ell}) u_k u_\ell + s_k s_\ell (q_{k\ell} - r_{k\ell}) v_k v_\ell\} \quad (9)$$

Separation of Linear and Nonlinear Components of Wave Spectrum

Several methods for decomposing a wave spectrum into linear and nonlinear components have been developed to date. These include Tick [1959, 1961], Hamada [1965], Hudspeth and Chen [1979], Masuda, et al. [1979] and Anastasiou et al. [1982], among others. Almost all of these methods for evaluating the nonlinear part are based on the second order interaction kernel of a weakly nonlinear solution. Kim and Power [1979], on the other hand, have developed a method using the bicoherent spectrum to separate the nonlinear wave-wave interaction of coherent waves in plasma fluctuation data. Their method may be applied for separation of the spectral energy density of a random process with strong nonlinear characteristics.

Since the Kim-Power method was developed to evaluate the wave-wave interaction associated with two arbitrarily chosen constant frequencies, the method is extended in the present study so that any two frequencies associated with wave-wave interaction are not constant; instead they are variables. This implies that the non-linear component of the spectral density at a frequency ω_m is equal to the accumulation of nonlinear interaction associated with various pairs of frequency components ω_k and ω_ℓ under the condition that $\omega_k + \omega_\ell = \omega_m$. Furthermore, we consider the interaction not only at the frequency $(\omega_k + \omega_\ell)$ but also at the frequency $(\omega_k - \omega_\ell)$, where $\omega_k > \omega_\ell$. The latter is equivalent to the sum interaction between ω_ℓ and $(\omega_k - \omega_\ell)$. In evaluating the interaction at the frequency $(\omega_k - \omega_\ell)$, it is assumed that the spectral energy density at frequencies smaller than the minimum frequency of the main energy density in the spectrum (ω_s in Figure 2) is due to nonlinear interaction associated with the difference between various combinations of the two frequency components at ω_k and ω_ℓ. The spectral density for frequencies greater than ω_s is due to the nonlinear interaction associated with the sum of various combinations of two frequency components.

Let us evaluate the interaction due to the sum of two frequency components. We may write the Fourier transform of $y(t)$ at the frequency ω_m as follows:

$$Y(\omega_m) = Y_L(\omega_m) + \sum_{\omega_k + \omega_\ell = \omega_m} A_L(\omega_k, \omega_\ell) \, Y_L(\omega_k) \, Y_L(\omega_\ell), \qquad (10)$$

where $Y_L(\omega_j)$ = Fourier transform of the linear component $y_1(t)$ at the frequency j,

$A_L(\omega_k, \omega_\ell)$ = coupling coefficient.

The summation in Eq.(10) is for various combinations of components at ω_k and ω_ℓ where $\omega_k + \omega_\ell = \omega_m$, and that the second term is a convolution in discrete form. The coupling coefficient $A_L(\omega_k, \omega_\ell)$ can be obtained by multiplying each side of Eq.(10) by the product of the conjugates of $Y_L(\omega_k)$ and $Y_L(\omega_\ell)$, $Y_L^*(\omega_k) \, Y_L^*(\omega_\ell)$, and by taking the expected value. That is,

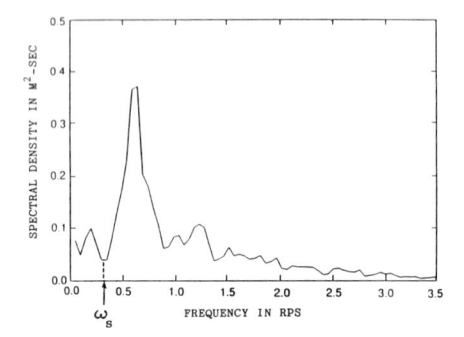

Figure 2: Definition of frequency ω_s.

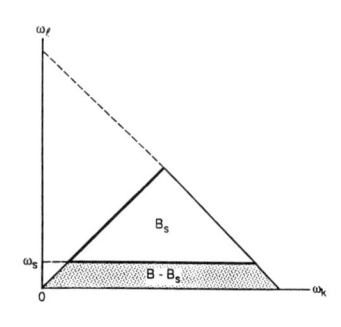

Figure 3: Domains for computing bispectra $B(\omega_k, \omega_\ell)$ and $B(\omega_\ell, \omega_k-\omega_\ell)$.

$$A_L(\omega_k,\omega_\ell) = \frac{B^*(\omega_k,\omega_\ell)}{E\left[|Y_L(\omega_k) Y_L(\omega_\ell)|^2\right]}, \quad (11)$$

where $B^*(\omega_k,\omega_\ell)$ = conjugate of the bispectrum $B(\omega_k, \omega_\ell)$. It should be noted that the domain of the bispectrum in this case is limited to that in B_s as indicated in Figure 3.

The spectral density function at the frequency ω_m can be obtained from Eqs.(10) and (11) as

$$S(\omega_m) = E\left[|Y_L(\omega_m)|^2\right] + \sum_{\omega_k+\omega_\ell=\omega_m} |A_L(\omega_k,\omega_\ell)|^2 \cdot E\left[|Y_L(\omega_k) Y_L(\omega_\ell)|^2\right]$$

$$= E\left[|Y_L(\omega_m)|^2\right] + \sum_{\omega_k+\omega_\ell=\omega_m} \{b(\omega_k,\omega_\ell)\}^2 \cdot S(\omega_m), \quad (12)$$

where
$$b(\omega_k,\omega_\ell) = \left\{\frac{|B(\omega_k,\omega_\ell)|^2}{E\left[|Y_L(\omega_k) Y_L(\omega_\ell)|^2\right] E\left[|Y(\omega_m)|^2\right]}\right\}^{1/2}. \quad (13)$$

The second term of Eq.(12) represents the accumulation of energy densities associated with interactions which occur at the frequency $\omega_k + \omega_\ell$. It is noted that $E[|Y_L(\omega_k) Y_L(\omega_\ell)|^2]$ in Eq.(12) is unknown in advance; hence we may evaluate $E[|Y(\omega_k) Y(\omega_\ell)|^2]$ for a given spectrum, and use it as an initial value in finding $b(\omega_k, \omega_\ell)$ by iteration.

For the nonlinear components which occur at the frequency $(\omega_k - \omega_\ell)$, $B^*(\omega_k, \omega_\ell)$ as well as $B(\omega_k, \omega_\ell)$ in Eq.(12) are replaced by

$B^*(\omega_\ell, \omega_k - \omega_\ell)$ and $B(\omega_\ell, \omega_k - \omega_\ell)$, respectively. Note that the domain of the bispectrum $B(\omega_\ell, \omega_k - \omega_\ell)$ is limited to that in $(B - B_s)$ as illustrated in Figure 3.

As an example of application of the method for separating the spectrum into linear and nonlinear components, Figures 4 through 6 show the results of computations carried out on three wave records, Gages 710, 625 and 615; a portion of each record is shown in Figure 1. It can be clearly seen in these figures that the ratio of nonlinear energy to total energy increases as water depth decreases. That is, for the Gage 710 wave record (depth 21.4 m), there exists no appreciable nonlinear energy component in its spectrum. On the other hand, for the Gage 625 record (depth 9.94 m), an appreciable amount of nonlinear energy exists at low and high frequencies, and the same trend can be observed on the Gage 615 wave record (depth 2.28 m) with substantial increase of nonlinear energy. It is noted that no nonlinear component exists in the neighborhood of the frequency where the spectrum peaks irrespective of sea severity.

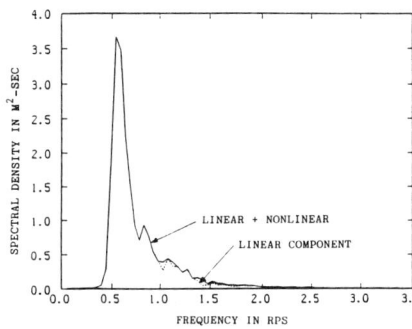

Figure 4: Separation of linear & nonlinear components of spectrum for wave record Gage 710.

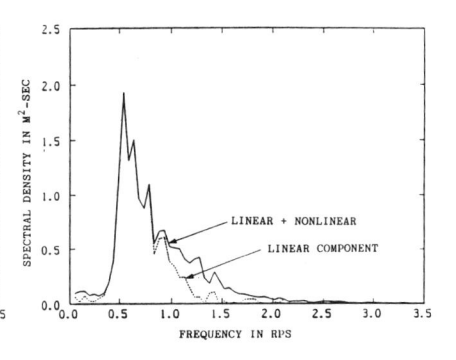

Figure 5: Separation of linear & nonlinear components of spectrum for wave record Gage 625.

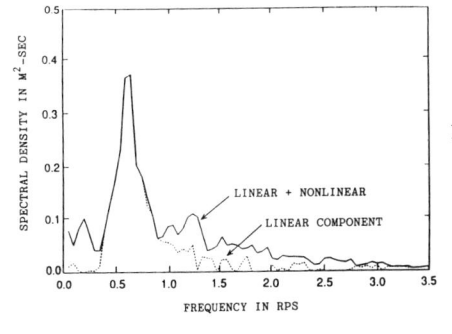

Figure 6: Separation of linear and nonlinear components of spectrum for wave record Gage 615.

Evaluation of Interaction Coefficients $q_{k\ell}$ and $r_{k\ell}$

The interaction coefficients $q_{k\ell}$ and $r_{k\ell}$ in Eq.(9) can be evaluated through bispectral analysis of the wave $y(t)$.

The bispectrum $B(\omega_k, \omega_\ell)$ is most commonly evaluated in the fundamental region, which is the octant, the domain defined by $0 \leq \omega_\ell \leq \omega_k$ and $0 \leq \omega_k \leq \infty$ as shown in Figure 3. The volume under the bispectrum is equal to the 3rd moment of $y(t)$ if the mean value of $y(t)$ is zero. That is,

$$E\left[\{y(t)\}^3\right] = 6 \sum_{k=1}^{N} \sum_{\ell=1}^{N} \text{Re}\{B(\omega_k, \omega_\ell)\}. \tag{14}$$

On the other hand, the 3rd moment of $y(t) = y_1(t) + y_2(t)$ can be obtained as

$$E\left[\{y(t)\}^3\right] = 3 E\left[\{y_1(t)\}^2 y_2(t)\right] + E\left[\{y_2(t)\}^3\right]. \tag{15}$$

Note that $y_1(t)$ is a normal distribution with zero mean, therefore $E[\{y_1(t)\}^3]$ and $E[y_1(t)\{y_2(t)\}^2]$ are zero.

Since the linear component $y_1(t)$ is usually much greater than the nonlinear component $y_2(t)$, the second term of Eq.(15) may be neglected in comparison with the first term. Hence, we have from Eq.(9)

$$E\left[\{y(t)\}^3\right] = 3 E\left[\{y_1(t)\}^2 y_2(t)\right] = 6 \sum_{k=1}^{N} \sum_{\ell=1}^{N} s_k^2 s_\ell^2 (q_{k\ell} + r_{k\ell}). \tag{16}$$

Then, Eqs.(14) and (16) yield

$$\sum_{k=1}^{N} \sum_{\ell=1}^{N} (q_{k\ell} + r_{k\ell}) = \sum_{k=1}^{N} \sum_{k=1}^{N} \frac{\text{Re}\{B(\omega_k, \omega_\ell)\}}{s_k^2 s_\ell^2}. \tag{17}$$

As stated in earlier, the domain of the bispectrum applicable for the interactions associated with the sum of two frequency components is B_s as shown in Figure 3. Therefore, from Eq.(17), the interaction coefficient $q_{k\ell}$ can be obtained as

$$q_{k\ell} = \frac{1}{s_k^2 s_\ell^2} B_s(\omega_k, \omega_\ell). \tag{18}$$

Similarly, the interaction coefficient $r_{k\ell}$ can be evaluated as

$$r_{k\ell} = \frac{1}{s_k^2 s_\ell^2} \left\{ B(\omega_\ell, \omega_k - \omega_\ell) - B_s(\omega_k, \omega_\ell) \right\}. \tag{19}$$

Application of the Kac-Siegert Solution

The functional relationship between parameters β_j and λ_j in the Kac-Siegert solution and the spectral densities of the linear wave components, s_k and s_ℓ, as well as the interaction coefficients $q_{k\ell}$ and $r_{k\ell}$ are presented by Langley [1987] in a concise matrix form. Although our methods of deriving the linear wave component as well as the interaction coefficients are quite different from Langley's methods, the functional relationship between (β_j, λ_j) and $(s_k, s_\ell, q_{k\ell}, r_{k\ell})$ can still be applied to the present problem. That is, Langley presents Eq.(9) as follows:

$$y(t) = \underline{s}' \underline{u} + \underline{u}' (\underline{Q} + \underline{R}) \underline{u} + \underline{v}' (\underline{Q} - \underline{R}) \underline{v}. \tag{20}$$

where $\underline{Q}$ and $\underline{R}$ are real symmetric matrices whose $k\ell$-th components are $s_k s_\ell q_{k\ell}$ and $s_k s_\ell r_{k\ell}$, and $\underline{s}$, $\underline{u}$ and $\underline{v}$ are vectors whose k-th components are s_k, u_k and v_k, respectively. We may write the two matrices in Eq.(20) as

$$\begin{aligned} \underline{Q} + \underline{R} &= \underline{w}_1' \underline{\Lambda}_1 \underline{w}_1 \\ \underline{Q} - \underline{R} &= \underline{w}_2' \underline{\Lambda}_2 \underline{w}_2 \end{aligned} \tag{21}$$

where the column vectors of $\underline{w}_j$ (j = 1, 2) are normalized eigenvectors. They are orthonormal vectors satisfying the condition $\underline{w}_j' \underline{w}_j = \underline{I}$, where $\underline{I}$ is the identity matrix. The elements of the diagonal matrices $\underline{\Lambda}_j$ (j = 1, 2) are eigenvalues of matrix $\underline{Q} + \underline{R}$ and $\underline{Q} - \underline{R}$, respectively. Since $\underline{Q} + \underline{R}$ and $\underline{Q} - \underline{R}$ are N by N symmetric matrices, there is a total of 2N eigenvalues.

Let us write

$$z_j = \begin{cases} (\underline{w}_1' \underline{u})_j, & j = 1, 2, \ldots \ldots N \\ (\underline{w}_2' \underline{v})_{j-N} & j = (N+1), \ldots \ldots 2N \end{cases} \tag{22}$$

where z_j are independent standardized Gaussian random variables. Then, Langley has shown that Eq.(20) can be reduced to the Kac-Siegert formulation given in Eq.(1) with the following relationships:

$$\beta_j = \begin{cases} (\underline{s}' \underline{w}_1)_j, & j = 1, 2, \ldots \ldots N \\ 0 & j = (N+1) \ldots \ldots 2N \end{cases}$$

$$\lambda_j = \begin{cases} (\Delta_1)_j, & j = 1, 2, \ldots\ldots N \\ (\Delta_2)_j, & j = (N+1), \ldots\ldots 2N \end{cases} \quad (23)$$

Since the parameters of the Kac-Siegert solution are obtained through spectral analysis, the probability density function applicable to nonlinear, non-Gaussian waves can be derived numerically with the aid of the characteristic function as suggested by Kac-Siegert.

As an example, the method is applied to the Gage 615 wave record which indicates strong nonlinear characteristics. A comparison between the probability density function and the histogram constructed from the data is shown in Figure 7. Excellent agreement between them can be seen in the figure. This result implies that the Kac-Siegert solution evaluated from the time history of waves yields a probability density function representing the statistical properties of waves in finite water depth which have strong nonlinear characteristics. It is unfortunate, however, that the probability density function cannot be obtained in closed form.

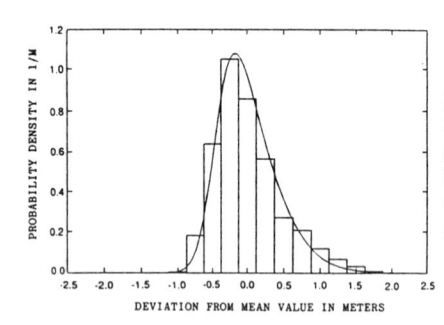

Figure 7:
Comparison of probability density function obtained from the Kac-Siegert solution and histogram constructed from data for wave record Gage 615.

Asymptotic Probability Distribution for Non-Gaussian Waves

One method for deriving the probability density function in closed form is to present the Kac-Siegert solution (Eq.1) as a function of a single random variable instead of the summation of the standardized normal distribution and its squared quantity. For this, let us present Eq.(1) as

$$Y = U + aU^2 \quad (24)$$

where "a" is a constant (unknown) and U is a normal variate with mean μ_* and variance σ_*^2, both of which are also unknown. The value of these unknowns will be determined from the following three equations which are derived by equating the cumulant generating function of Eq.(24) with that of Kac-Siegert's solution given in Eq.(1) (see Ochi and Ahn 1994).

$$\begin{cases} a\sigma_*^2 + a\mu_*^2 + \mu_* = 0 \\ \left(\sum_{j=1}^{2N} \beta_j^2\right) + 2\left(\sum_{j=1}^{2N} \lambda_j^2\right) = \sigma_*^2 - 2a^2\sigma_*^4 \\ 3\left(\sum_{j=1}^{2N} \beta_j^2\lambda_j\right) + 4\left(\sum_{j=1}^{2N} \lambda_j^3\right) = 3a\sigma_*^4 - 8a^3\sigma_*^6. \end{cases} \quad (25)$$

It should be noted that the left side of Eq.(25) can be presented in terms of cumulants, k_1, k_2 and k_3. If we have a wave record obtained for a sufficiently long time (on the order of 20 minutes) and if we let the mean value be the zero line, we have $k_1 = 0$, and thereby k_2 and k_3 are equal to the sample moments $E[y^2]$ and $E[y^3]$, respectively. Thus, we can determine the unknown parameters a, μ_* and σ_*^2 by simply evaluating the sample moments from the wave record y(t) and by applying the following relationship:

$$a\sigma_*^2 + a\mu_*^2 + \mu_* = 0$$
$$\sigma_*^2 - 2a^2\sigma_*^4 = E[x^2] \quad (26)$$
$$2a\sigma_*^4(3 - 8a^2\sigma_*^2) = E[x^3].$$

Since the random variable U in Eq.(24) is now a normal variate with known mean μ_* and variance σ_*^2, the probability density function of Y can be derived by applying the technique of change of random variables from U to Y. Unfortunately, however, the density function thusly derived vanishes at a point $y = -(1/4a)$ due to a singularity involved in the density function.

In order to circumvent this drawback, let us present the functional relationship between Y and U given in Eq.(24) inversely such that the random variable U is expressed as a function of Y as follows:

$$U = \frac{1}{\gamma a}\left(1 - e^{-\gamma aY}\right), \quad (27)$$

where γ is a constant; 1.28 for $y \geq 0$ and 3.00 for $y \leq 0$. Justification for selecting these constant values of γ is given in Ochi and Ahn 1994. It may suffice here to say that these constants are valid even for a random process with very strong nonlinear characteristics.

It is noted that the values of γ are different for positive and negative y-values. This results in a slight difference in the slope of the probability density function at $y = 0$, although the density function is continuous at this point.

By using the functional relationship given in Eq.(27), we may derive the probability density function of Y from the random variable U which obeys the normal distribution with mean μ_* and variance σ_*^2. The change of random variables technique yields the probability density function of Y as follows:

$$f(y) = \frac{1}{\sqrt{2\pi}\,\sigma_*} e^{-\frac{1}{2(\gamma a \sigma_*)^2}(1 - \gamma a \mu_* - e^{-\gamma a y})^2 - \gamma a y}, \qquad (28)$$

where $\gamma = \begin{cases} 1.28 & y \geq 0 \\ 3.00 & y < 0. \end{cases}$

By applying the method for determining the limit of an indefinite function, it can be easily proved that Eq.(28) reduces to a normal probability density function with mean μ_* and variance σ_*^2 if a = 0; namely, for a linear system.

Figure 8 shows a comparison of the newly developed asymptotic probability density function with the histogram of the Gage 615 wave record. The parameters a, μ_* and σ_*^2 are determined from Eq.(26). Included also in the figure is the normal probability density function with zero mean and the variance evaluated from the record. As can be seen in the figure, the histogram deviates from the normal density function to a great extent; but, the agreement between the histogram and the newly developed probability density function is excellent.

Figures 9 and 10 show comparisons between the probability density function given in Eq.(28) and histograms for the Gage 625 (water depth 9.94 m) and the Gauge 710 (water depth 21.4 m), respectively. In the latter case, the probability density function is normally distributed and agrees very well with the histogram.

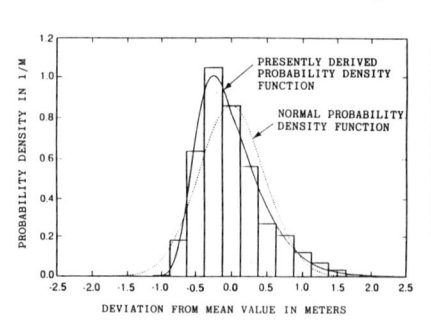

Figure 8: Comparison of presently developed probability density function and histogram constructed from data for wave Gage 615.

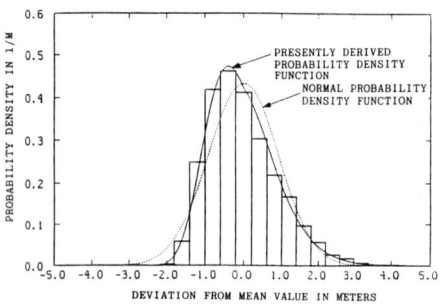

Figure 9: Comparison of presently developed probability density function and histogram constructed from sata for wave Gage 625.

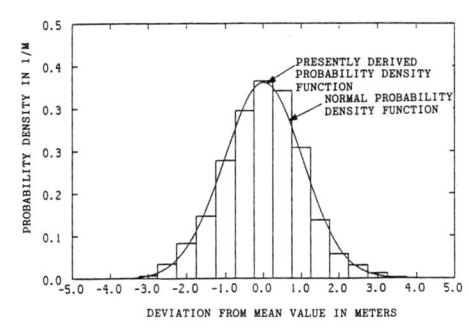

Figure 10:
Comparison of presently developed probability density function and histogram constructed from data for wave record Gage 710.

From the results of these comparisons, it can be concluded that the probability density function covers the non–Gaussian distribution observed for random waves in shallow water as well as the Gaussian distribution observed for random waves in deep water, and that the probability density function agrees very well with the histograms constructed from wave data.

Conclusions

This paper presents the results of a study on a probability density function applicable to waves in finite water depth developed based on the concept of Kac–Siegert's solution for the output of a nonlinear mechanical system. Although the Kac–Siegert solution requires knowledge of the second order frequency response function of the system, we developed a method to evaluate the parameters involved in the solution only from measured waves through spectral analysis. Further, the probability density function associated with the Kac–Siegert solution is asymptotically expressed in closed form (Eq.28). The presently developed probability density function covers the non–Gaussian distribution observed for random waves in shallow water as well as the Gaussian distribution observed for random waves in deep water. Comparison between the probability density function and histograms constructed from wave records show good agreement.

Acknowledgements

This study was carried out in connection with the project on probability functions of nonlinear systems sponsored by the Naval Civil Engineering Laboratory through contract N 47408–91–C–1204 to University of Florida. The authors would like to express their appreciation to Mr. Paul Palo for his valuable discussions received during the course of this project. The authors are also grateful to Ms. Laura Dickinson for typing the manuscript.

References

Anastasiou, K., et al. (1982), The non-linear properties of random wave kinematics, Proc. 3rd Int. Confer. on Behaviour Offshore Struct., pp.493-515.

Hamada, T. (1965), The secondary interactions of surface waves, Rep. Port and Harbor Res. Inst., No.10, pp.1-28.

Hasselmann, K. (1962), On the nonlinear energy transfer in a gravity wave spectrum, part I, general, J. Fluid Mechanics, Vol.12, pp.481-500.

Huang, N.E., Long, S.R., Tung, C.C. and Yuan, Y. (1983), A non-Gaussian statistical model for surface elevation of nonlinear random wave fields, J. Geophy. Res., Vol.88, No.C12, pp.7597-7606.

Hudspeth, R.T. and Chen, M-C. (1979), Digital simulation of nonlinear random waves, J. Waterway, Port, Coastal & Ocean, Vol.105, No.WW1, pp.67-85.

Kac, M. and Siegert, A.J.F. (1947), On the theory of noise in radio receivers with square law detectors, J. Applied Physics, Vol.8, pp.383-397.

Kim, Y.C. and Powers, E.J. (1979), Digital bispectral analysis and its applications to nonlinear wave interactions, IEEE Trans. Plasma Science, Vol.PS-7, No.2, pp.120-131.

Langley, R.S. (1987), A statistical analysis of non-linear random waves, Ocean Eng., Vol.14, No.5, pp.389-407.

Longuet-Higgins, M.S. (1963), The effect of non-linearities on statistical distribution in the theory of sea waves, J. Fluid Mech. Vol.17, Part 3, pp.459-480.

Masuda, A., Kuo, Y. and Mitsuyasu, H. (1979), On the dispersion relationship of random gravity waves. Part 1, Theoretical framework, J. Fluid Mech., Vol.92, Part 4, pp.717-730.

Ochi, M.K. and Ahn, K. (1994), Probability distribution applicable to non-Gaussian random processes, J. Prob. Eng. Mechanics, Vol.9, No.4, pp.255-264.

Sharma, J.N. and Dean, R.G. (1979), Second order directional seas and associated forces, Proc. 11th Offshore Tech. Confer., Paper OTC 3645.

Tayfun, M.A. (1980), Narrow-banded nonlinear sea waves, J. Geoph. Res., Vol.85, No.C3, pp.1548-1552.

Tick, L.J. (1959), Nonlinear probability models of ocean waves I, J. Math and Mech., Vol.8, pp.184-196.

Tick, L.J. (1961), Nonlinear probability models of ocean waves, Proc. Conf. Ocean Wave Spectra, pp.163-169.

CHAPTER 37

PROBABILITY CHARACTERISTICS OF
ZERO-CROSSING WAVE HEIGHT

by T. Ohta[1] and A. Kimura[2]

Abstract

This study deals with the probability distribution of zero-crossing wave height applying the definition of the zero-up(down)-cross method faithfully. In this study, gap between wave crest or trough and the envelope at the same location, which has been neglected in the ordinary studies is taken into account in the wave height definition. Its probability distribution is approximated with the Weibull distribution. The probability distribution of the zero-crossing wave height is, then, introduced theoretically together with the theory by modified Tayfun Method and the statistical properties of the gaps. The numerically simulated irregular wave height distributions agree well with the theoretical distribution.

Introduction

The Rayleigh distribution has been used as a probability distribution of zero-crossing wave height. Although this distribution agrees very well for the almost irregular wave height distributions, it was derived theoretically as a probability distribution of "wave amplitude" in the case of narrow band spectrum by Longuet—Higgins(1952). Therefore, the Rayleigh distribution is not the theoretical probability distribution for "wave height", even when the wave spectrum is narrow. Tayfun(1981,1983) tried to derive the probability distribution of the zero-crossing wave height on the basis of its definition faithfully. However, his distribution is considerably different from the Rayleigh distribution when the wave spectrum is wide. We know the Rayleigh distribution
----
1 Research Associate, Dept. of Social Systems Eng., Faculty
  of Eng., Tottori Univ., 4-101 Koyama Minami Tottori, Japan
2 Prof., Dept. of Social Systems Eng., Faculty of Eng.,
  Tottori Univ., 4-101 Koyama Minami Tottori, Japan

can be applied to the probability distribution of zero-crossing "wave height" sufficiently. The agreements of the theory and data are mainly around its mean. Very few studies examined the wave height distribution in the very large part, larger than twice the mean wave height, for example. In the design of structural durability and reliability, however, reliable probability in a range over 2.5 times of the mean wave height may become necessary. This study aims at deriving the probability distribution of zero-crossing "wave height", applying the basic definition for the zero-crossing wave height and considering small errors which is inevitably introduced in the ordinary theory.

Definition of Zero-Crossing Wave Height

In a spectrum theory, the envelope for irregular wave profile has been used instead of the amplitude at crest and trough of zero-crossing wave. Fig.1 shows irregular wave profile η, its envelope R and their enlargements around $t=t_2$. $\eta(t_1)$, $\eta(t_2)$ and $\eta(t_3)$ show maxima or minima of η. R_{m_1}, R_{m_2} and R_{m_3} are the simultaneous envelope amplitudes respectively. The zero-crossing wave height is defined, in principle, as a sum of consecutive maximum and minimum of η between two zero-up or down-crossing points. For example, the wave height of the first zero-down-crossing wave in fig.1 is described as $H_1 = \eta(t_1) + \eta(t_2)$. In the ordinary theory, maxima and minima of zero-crossing waves are approximated by the simultaneous envelope amplitudes, then H_1 is given by eq.1.

$$H_1 = R_{m_1} + R_{m_2} \tag{1}$$

When the wave spectrum is very narrow, the envelope changes gradually. Longuet-Higgins(1952) assumed that wave amplitudes are equal to the envelope amplitudes, and we have been applying the Rayleigh distribution as the probability distribution of zero-crossing "wave height", putting $H_1=2R_{m1}$. However, if the wave spectrum is wide, this assumption brings considerable errors. On the basis of eq.1, Tayfun(1981) derived probability distributions of zero-crossing wave height. Since the gaps between $\eta(t_j)$ and R_{m_j}, which are shown by δ_j in fig.1, are order of ν^2 (Tayfun, 1989, $\nu^2 = m_0 m_2/m_1^2 - 1$, m_n : n-th order spectral moment), he neglected them. However the zero-crossing wave height should be defined, in principle as

$$H_1 = \left(R_{m_1} + R_{m_2} \right) - \left(\delta_1 + \delta_2 \right) \tag{2}$$

To derive the probability distribution of zero-crossing wave height which basis on eq.2, it is necessary to make clear the additional probability distribution for R_{m_j} and δ_j. The envelope

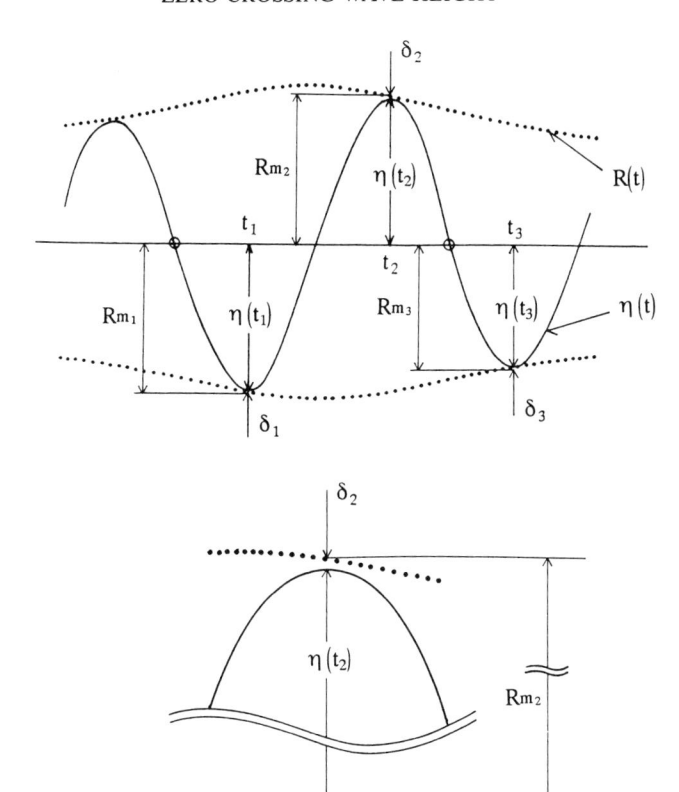

Fig. 1 Components of wave height

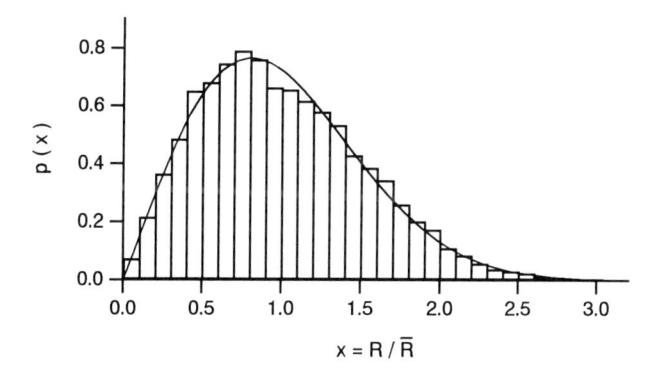

Fig. 2 Probability distribution of R (r = 5)

amplitudes R(t) follow the Rayleigh distribution, however, R_{m_j} is not statistically uniform samples from the population of envelope amplitude. In other words, because of the uneven interval between consecutive R_{m_j}, its distribution may be depart from the Rayleigh distribution. And there is no theoretical probability distribution for δ_j. In this study, the probability distribution for R_{m_j} and δ_j are investigated experimentally through numerical simulations.

Numerical Simulations

The irregular wave profiles η were simulated by FFT method (8192 points, Δt=0.05s). Next wave spectrum S(f) with variable shape factor r (r=4,5,6,7,8,9,10,15,20) is used in the simulations.

$$S(f) = \left(\frac{f}{f_p}\right)^{-r} \exp\left[\frac{r}{4}\left\{1 - \left(\frac{f}{f_p}\right)^{-4}\right\}\right]$$
(3)

where, f_p is a peak frequency of S(f). The envelope R(t) was calculated by the following equation.

$$R(t) = \sqrt{\eta^2(t) + \hat{\eta}^2(t)}$$
(4)

where, $\hat{\eta}$ is the Hilbelt transformation of η. The histgram in fig.2 shows the frequency distribution of R(t). The theoretical probability distribution for R(t) is the Rayleigh distribution, however, we attempted to apply the Weibull distribution as a more general distribution (eq.5).

$$p(x) = \frac{\alpha}{2\gamma} x^{\alpha-1} \exp\left(-\frac{x^\alpha}{2\gamma}\right)$$
(5)

where, $x = R/\bar{R}$

$$\gamma = \frac{1}{2}\left[\Gamma\{(1+\alpha)/\alpha\}\right]^{-\alpha}$$

α is the shape parameter, γ is the scale parameter and Γ is the Gamma function. In fig.2, the solid line shows the Weibull distribution whose shape parameter α=2.027 and scale parameter γ=0.639. In the case of α=2.0 and γ=0.637, it is the Rayleigh distribution. Therefore the probability distribution for R(t) agrees well with the Rayleigh distribution. The histgram in fig.3 illustrates the frequency distribution of R_{m_j}. In this case, the spectral shape parameter r is 5. Applying the Weibull distribution to this frequency distribution, we obtained a result that the shape parameter α_2=1.864 and the scale parameter γ_2=0.697. The solid line in fig.3 shows that Weibull distribution. In fig.4, the histgram

ZERO-CROSSING WAVE HEIGHT 501

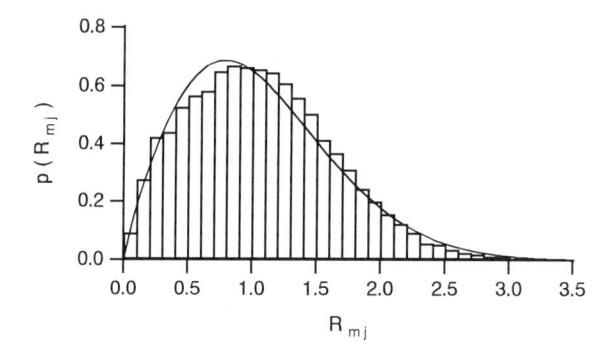

Fig. 3 Probability distribution of R_{mj} ($r=5$)

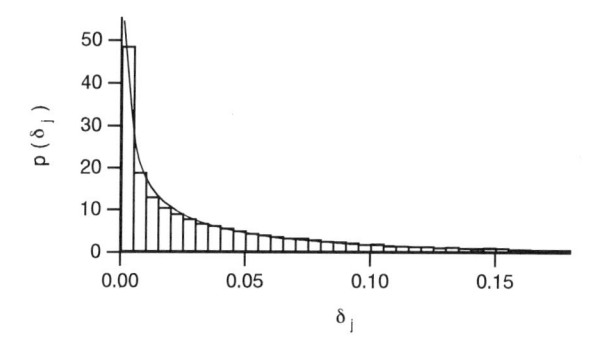

Fig. 4 Probability distribution of δ_j ($r=5$)

Tab. 1 Parameters for Weibull distribution

r	ν	α_1	γ_1	α_2	γ_2	$\overline{\delta_j}$	$\overline{R_{mj}}$
4	0.5571	0.6325	0.08649	1.829	0.8065	0.08469	1.154
5	0.4041	0.6394	0.06427	1.864	0.6975	0.05435	1.062
6	0.3247	0.6418	0.05081	1.898	0.6223	0.03802	0.9959
7	0.2768	0.6430	0.04274	1.939	0.5666	0.02891	0.9459
8	0.2444	0.6394	0.03767	1.975	0.5193	0.02356	0.9037
9	0.2205	0.6467	0.03262	1.921	0.4792	0.01955	0.8677
10	0.2018	0.6546	0.02900	1.941	0.4495	0.01701	0.8395
15	0.1514	0.6690	0.01988	1.941	0.3548	0.01043	0.7432
20	0.1256	0.6942	0.01462	1.988	0.2999	0.00775	0.6854

shows the frequency distribution of δ_j when the spectral shape parameter r is 5. Solid line shows the Weibull distribution with shape parameter $\alpha_1=0.639$ and scale parameter $\gamma_1=0.0643$. The results for other spectral shape parameters, whose range

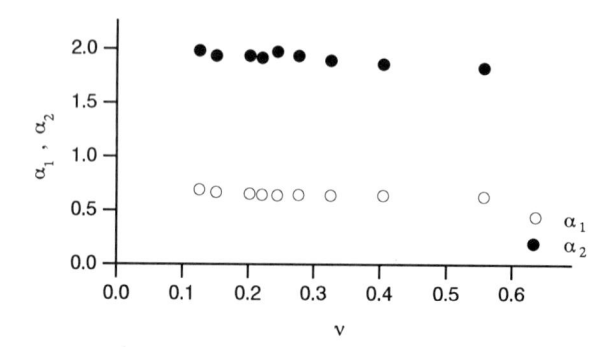

Fig. 5 Shape parameters of Weibull distribution

is from 4 to 20, are given in tab.1. Fig.5 shows shape parameters of the Weibull distribution. The open circles show the shape parameters of the Weibull distribution for δ_j and the closed circles show those for R_{m_j}. Horizontal axis shows the spectral band width parameter ν. As ν increases, the shape parameter of the Weibull distribution has tendency to decrease. The distributions obtained become a little flatter shape compared with the Rayleigh distribution. We concluded that uneven sampling from the population of R(t) causes this deviation from the Rayleigh distribution.

Probability Distribution of Wave Height

It is confirmed so far that the probability distributions for R_{m_j} and δ_j are well approximated by the Weibull distribution. To derive the probability distribution of wave height with the definition as eq.2, it is necessary to investigate the following points.

1. The joint probability distribution between consecutive envelope amplitudes R_{m_j} and $R_{m_{j+1}}$; $p(R_{m_j},R_{m_{j+1}})$
2. The joint probability distribution between δ_j and δ_{j+1} ; $p(\delta_j,\delta_{j+1})$
3. The joint probability distribution between $R_m=(R_{m_j}+R_{m_{j+1}})$ and $\delta=(\delta_j+\delta_{j+1})$; $p(R_m,\delta)$

First of all, the probability distribution for R_{m_j} and $R_{m_{j+1}}$ are also approximated by the Weibull distribution. The shape parameter is denoted by α_2. Since the interval between R_{m_j} and $R_{m_{j+1}}$ is about half of the mean period, R_{m_j} and $R_{m_{j+1}}$ must have correlation. With only above limited conditions, however, it is not possible to determine the joint probability distribution theoretically. In this study, we use the 2-dimensional Weibull distribution (Kimura,1981) as the joint probability distribution for the consecutive envelope amplitudes, $p(R_{m_j}, R_{m_{j+1}})$.

$$p\left(R_{m_j}, R_{m_{j+1}}\right) = \frac{\alpha_2^2}{4\left(\gamma_2^2-\rho^2\right)} R_{m_j}^{\alpha_2-1} R_{m_{j+1}}^{\alpha_2-1}$$

$$\cdot \exp\left\{-\frac{\gamma_2}{2\left(\gamma_2^2-\rho^2\right)}\left(R_{m_j}^{\alpha_2}+R_{m_{j+1}}^{\alpha_2}\right)\right\} \cdot I_0\left(\frac{R_{m_j}^{\alpha_2/2} R_{m_{j+1}}^{\alpha_2/2}}{\gamma_2^2-\rho^2}\rho\right) \quad (6)$$

where, I_0 is the modified Bessel function of the first kind (0-th order), ρ is the correlation parameter between R_{m_j} and $R_{m_{j+1}}$.

$$\rho = \kappa \gamma_2 \quad (7)$$

and κ is given as following (Kimura and Ohta, 1992).

$$\kappa = \sqrt{\mu_{13}^2 + \mu_{14}^2} / m_0 \quad (8)$$

$$\mu_{13} = \int_{f_d}^{f_u} S(f) \cos\left\{2\pi(f-\bar{f})t_m\right\} df$$

$$\mu_{14} = \int_{f_d}^{f_u} S(f) \sin\left\{2\pi(f-\bar{f})t_m\right\} df$$

$$f_d = \left(-0.186/r + 0.735\right) f_p \quad (4 \leq r \leq 20)$$

$$f_u = \left(1.61/r + 1.62\right) f_p \quad (4 \leq r \leq 20)$$

$$\bar{f} = m_1/m_0 \quad , \quad t_m = \sqrt{m_0/m_2}$$

Because R_{m_j} is not normalized by its mean, the scale parameter γ_2 is given as follows.

$$\gamma_2 = \frac{1}{2}\left[\Gamma\left\{(1+\alpha_2)/\alpha_2\right\}\right]^{-\alpha_2} \overline{R_{m_j}}^{\alpha_2} \quad (9)$$

Fig.6 shows that the 2-dimensional Weibull distribution agrees well with the simulated joint frequency distribution between R_{m_j} and $R_{m_{j+1}}$. Three cases of r=5,10 and 20 were shown in fig.6. Similar agreements are obtained in other cases (r=4,6,7,8,9,15).

Second, the probability distribution for δ_j and δ_{j+1} are also approximated by the Weibull distribution. The shape parameter is described by α_1. Open circles in fig.7 show the correlation coefficients between δ_j and δ_{j+1}, and fig.8 illustrates the distribution of δ_j and δ_{j+1} in the case r=5. Although calculated

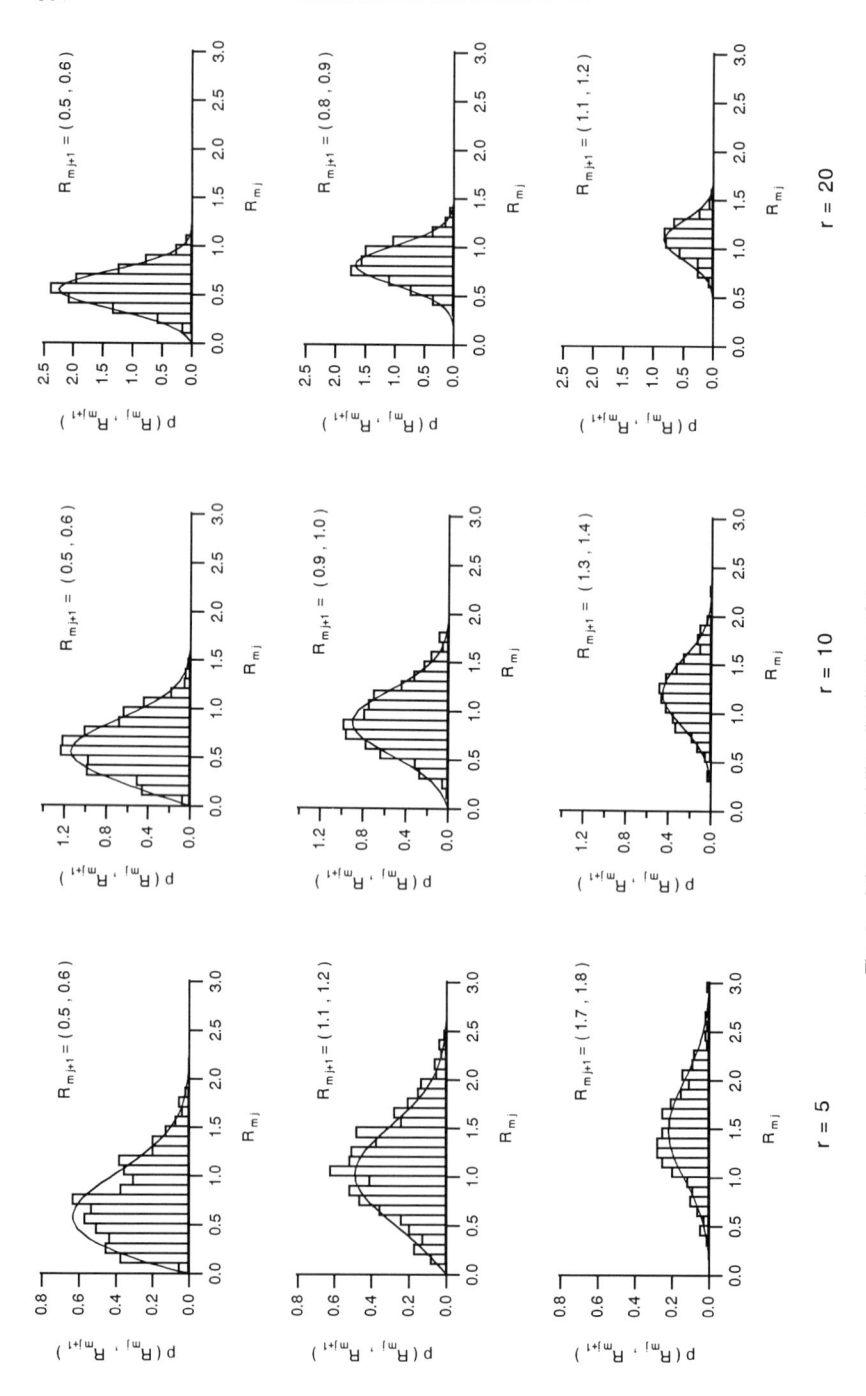

Fig. 6 Joint probability distribution of R_{mj} and R_{mj+1}

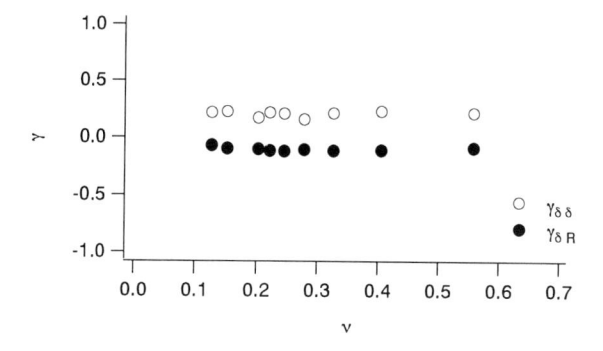

Fig. 7 Correlation coefficients

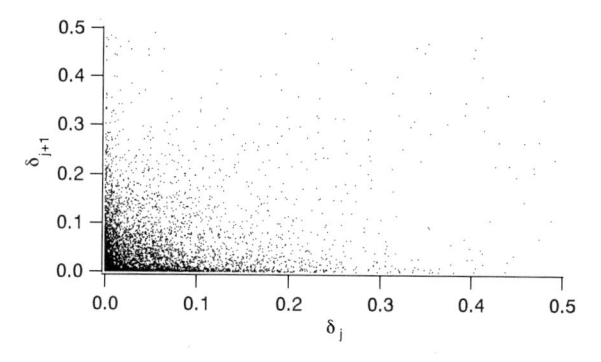

Fig. 8 Distribution of δ_j and δ_{j+1} (r = 5)

correlation coefficient is about 0.2 in this case, we can not see apparent correlation as shown in fig.8. Considering δ_j and δ_{j+1} are independent, then, we tried to apply the product of Weibull distribution as a joint probability distribution between δ_j and δ_{j+1}, $p(\delta_j, \delta_{j+1})$.

$$p(\delta_j, \delta_{j+1}) = \frac{\alpha_1^2}{4\gamma_1^2} \delta_j^{\alpha_1 - 1} \delta_{j+1}^{\alpha_1 - 1} \exp\left\{-\frac{1}{2\gamma_1}\left(\delta_j^{\alpha_1} + \delta_{j+1}^{\alpha_1}\right)\right\} \tag{10}$$

Since δ_j is not normalized by its mean, the scale parameter γ_1 is given as

$$\gamma_1 = \frac{1}{2}\left[\Gamma\left\{(1+\alpha_1)/\alpha_1\right\}\right]^{-\alpha_1} \overline{\delta_j}^{\alpha_1} \tag{11}$$

Fig.9 illustrates a comparison between the product of the Weibull distribution and the simulated joint frequency distribution of δ_j and δ_{j+1} in the cases of r=5,10 and 20.

506   COASTAL ENGINEERING 1994

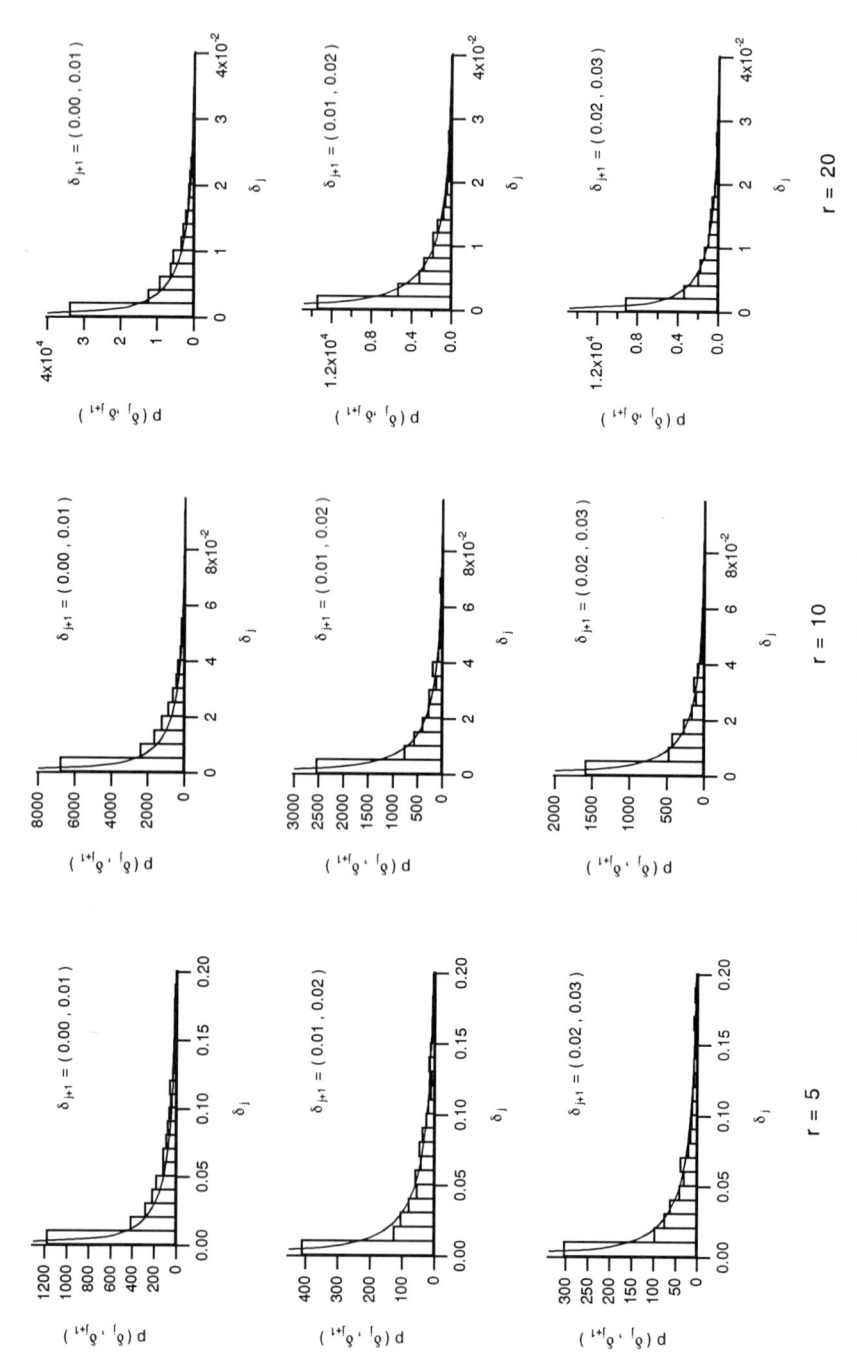

Fig. 9 Joint probability distribution of δ_j and δ_{j+1}

Third, the joint probability distribution between $R_m = (R_{m_j} + R_{m_{j+1}})$ and $\delta = (\delta_j + \delta_{j+1})$ is determined by using above results. Closed circles in fig.7 show the correlation coefficients between R_m and δ. Although the calculated values are about -0.1, we consider R_m and δ to be independent. Therefore, the joint probability distribution between R_m and δ is given by the product of the probability distribution for R_m and it for δ.

Using above results, we derive the probability distribution of zero-crossing wave height. First, the probability distribution for R_m, which is denoted by $p(R_m)$, is given as eq.(12).

$$p(R_m) = \int_0^{R_m} \frac{\alpha_2^2}{4(\gamma_2^2 - \rho^2)} R_{m_j}^{\alpha_2 - 1} (R_m - R_{m_j})^{\alpha_2 - 1}$$

$$\cdot \exp\left[-\frac{\gamma_2}{2(\gamma_2^2 - \rho^2)} \left\{ R_{m_j}^{\alpha_2} + (R_m - R_{m_j})^{\alpha_2} \right\} \right]$$

$$\cdot I_0\left[\frac{R_{m_j}^{\alpha_2/2} (R_m - R_{m_j})^{\alpha_2/2}}{\gamma_2^2 - \rho^2} \rho \right] dR_{m_j} \quad (12)$$

Second, the probability distribution for δ, which is described by $p(\delta)$, is given as following.

$$p(\delta) = \int_0^{\delta} \frac{\alpha_1^2}{4\gamma_1^2} \delta_j^{\alpha_1 - 1} (\delta - \delta_j)^{\alpha_1 - 1}$$

$$\cdot \exp\left[-\frac{1}{2\gamma_1} \left\{ \delta_j^{\alpha_1} + (\delta - \delta_j)^{\alpha_1} \right\} \right] d\delta_j \quad (13)$$

The zero-crossing wave height is defined as $H = R_m - \delta$, and the probability distribution of H can be obtained by

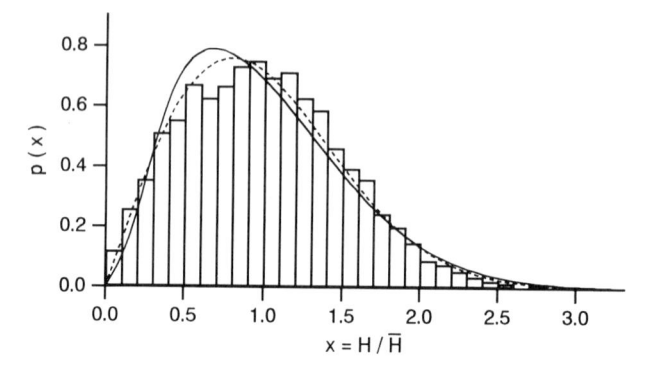

Fig.10 (a) Wave height distribution (r = 5)

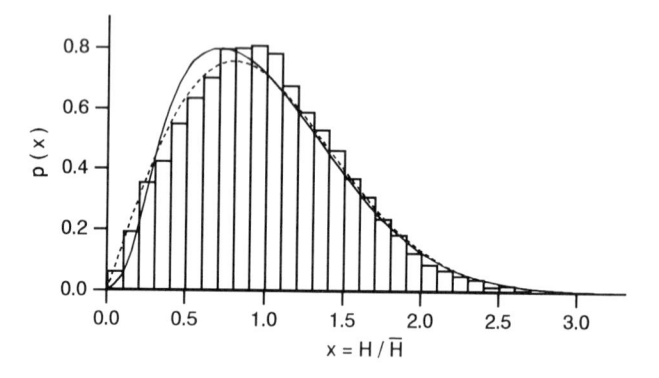

Fig.10 (b) Wave height distribution (r = 10)

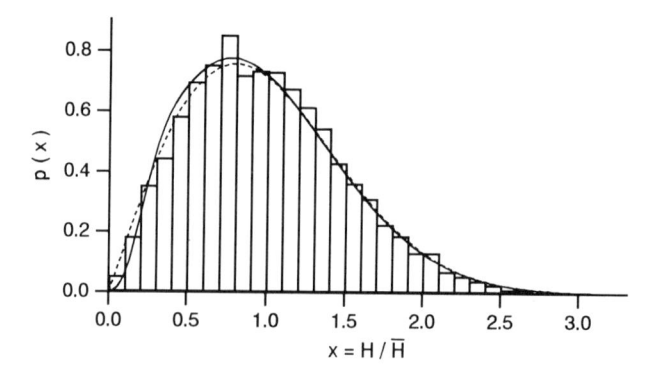

Fig.10 (c) Wave height distribution (r = 20)

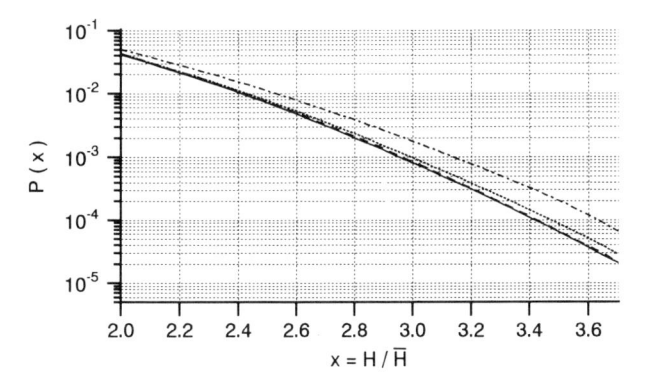

Fig.11 Exceedance probability
(chain line : r=5, dotted line : r=10, broken line : r=20, solid line : Rayleigh)

$$p(H) = \int_H^\infty \int_0^{R_m} \frac{\alpha_2^2}{4\left(\gamma_2^2 - \rho^2\right)} R_{m_j}^{\alpha_2 - 1} \left(R_m - R_{m_j}\right)^{\alpha_2 - 1}$$

$$\cdot \exp\left[-\frac{\gamma_2}{2\left(\gamma_2^2 - \rho^2\right)} \left\{R_{m_j}^{\alpha_2} + \left(R_m - R_{m_j}\right)^{\alpha_2}\right\}\right]$$

$$\cdot I_0 \left[\frac{R_{m_j}^{\alpha_2/2} \left(R_m - R_{m_j}\right)^{\alpha_2/2}}{\gamma_2^2 - \rho^2} \rho\right] dR_{m_j}$$

$$\cdot \int_0^{R_m - H} \frac{\alpha_1^2}{4\gamma_1^2} \delta_j^{\alpha_1 - 1} \left(R_m - H - \delta_j\right)^{\alpha_1 - 1}$$

$$\cdot \exp\left[-\frac{1}{2\gamma_1} \left\{\delta_j^{\alpha_1} + \left(R_m - H - \delta_j\right)^{\alpha_1}\right\}\right] d\delta_j \, dR_m$$

(14)

As results of the numerical calculation of eq.(14), the probability distributions of zero-crossing wave height are obtained as shown in fig.10(a)-(c). (a) is the case when the spectral shape parameter r=5, (b) r=10 and (c) r=20. The solid line shows the present theory and the broken line shows the Rayleigh distribution. Fig.11 shows the exceedance probability of wave height. The chain line, dotted line and broken line show the distribution when r=5,10 and 20, and solid line shows the Rayleigh distribution respectively. We can see considerably larger probability than the Rayleigh distribution when r=5.

Conclusion

The probability distributions of zero-crossing wave height, when the gap between maximum(minimum) of wave profile and the simultaneous envelope amplitude is taken into account, were derived. As the result, larger probability of exceedance than the Rayleigh distribution was obtained in the range of larger wave height, when the spectrum is wide.

References

Kimura, A. (1981): Joint distribution of the wave heights and periods of random sea waves, Coastal Eng. in Japan, Vol.24, pp.77-92.

Kimura, A. and T. Ohta (1992): On probability distributions of the zero-crossing irregular wave height, Proc. 6th IAHR Int'l Symp. on Stochastic Hydr., pp.291-298.

Longuet-Higgins, M.S. (1952): On the statistical distributions of the heights of sea waves, J. Marine Res., Vol.11, No.3, pp.245-265.

Tayfun, M.A. (1981): Distribution of crest-to-trough wave height, J. Wtrway., Port, Coast. and Oc. Eng., ASCE, 107(WW3), pp.149-156.

Tayfun, M.A. (1983): Effects of spectrum band width on the distribution of wave heights and periods, Ocean Eng., Vol.10, No.2, pp.107-118.

Tayfun, M.A. (1989): Envelope, phase, and narrow-band models of sea waves, J. Wtrway, Port, Coast., and Oc. Eng., ASCE, Vol.115, No.5, pp.594-613.

CHAPTER 38

NUMERICAL SIMULATION AND VALIDATION OF
PLUNGING BREAKERS USING A 2D NAVIER-STOKES MODEL

H.A.H. Petit[1], P. Tönjes[2], M.R.A. van Gent[3], P. van den Bosch[1]

ABSTRACT

The numerical model SKYLLA, developed for simulation of breaking waves on coastal structures is described. The model is based on the Volume Of Fluid method and solves the two-dimensional (2DV) Navier-Stokes equations. Weakly reflecting boundary conditions allow waves to enter and leave the computational domain. Impermeable boundaries can be introduced to simulate a structure. A two-model approach can be used to simulate overtopping over a low crested structure. Results obtained with the model are compared with those obtained with physical model tests for waves on a 1:20 slope of a submerged structure.

INTRODUCTION

Traditionally, wave motion on coastal structures was studied by means of physical small-scale model tests. Some phenomena can be studied quite well on a small scale whereas others, like those which involve effects of viscosity, cannot.
Numerical models do not have the disadvantage of scaling however, they have the disadvantage that the equations they solve represent a simplification of reality.
Most models used to simulate wave motion on structures either solve the shallow water equations or potential flow formulations. For examples of the first we refer to Kobayashi et al.(1987) and Van Gent (1994). For examples of methods based on potential flow we refer to Klopman (1987) for the two dimensional case and to Broeze (1993) for a solver for three-dimensional flow. The shallow water equation

---

[1] Delft Hydraulics, P.O.Box 152, 8300 AD, Emmeloord, The Netherlands.
[2] Ministry of Transport, Public Works and Water Management. Directorate-General of Public Works and Water Management (Rijkswaterstaat), Road and Hydraulic Engineering Division, P.O. Box 5044, 2600 GA, Delft, The Netherlands.
[3] Delft University of Technology, Dept. Civil Engrg., P.O. Box 5048 GA Delft, The Netherlands.

solvers cannot directly simulate wave breaking but need to add extra dissipation to simulate the wave height reduction caused by breaking. The potential flow solvers can solve the flow very accurately up to the moment where the flow domain becomes multiply connected as a result of the breaking process. After that moment these methods become unstable and the calculation breaks down. Solvers based on the MAC (Marker And Cell) or the VOF (Volume Of Fluid) method can solve the Navier-Stokes equation for breaking waves.

THE 2D-NAVIER-STOKES MODEL

The Volume Of Fluid method (Hirt and Nichols, 1981) has been made applicable for simulation of wave and flow phenomena on coastal structures (Petit & Van den Bosch, 1992 and Van der Meer et al., 1992). The model solves the two-dimensional incompressible Navier-Stokes equations with a free surface.
For the treatment of the free surface a redistribution of water contained in the cells of the computational grid has to take place once the velocity is known. The method called FLAIR (Ashgriz and Poo, 1991) has been adopted for this purpose. Arbitrary free-slip boundaries can be introduced in the model to simulate breaking waves on impermeable coastal structures. The numerical simulation of the breaking process is not limited to the moment where the fluid domain becomes multiply connected.

IMPROVEMENTS

Recent improvements of the model involve the use of weakly reflecting boundary conditions that allow nonlinear waves based on a Rienecker and Fenton (1981) (R&F) formulation to enter the domain. Further improvements allow the simulation of overtopping at a dike, not only with respect to the volume of water, but also a detailed simulation of water running down the rear of the dike (Petit et al. 1994). Furthermore, the simulation of flow through permeable structures has been made possible for the model (Van Gent et al. 1993) which, however, is beyond the framework of this paper.

WEAKLY REFLECTING BOUNDARY CONDITIONS

In Figure 1 we show a situation where weakly reflective boundary conditions are needed at both sides of the model.
The waves are assumed to enter the domain at the left. They are given by the free surface elevation $\eta_{in}(x_0,t)$, and the velocity components $u_{in}(x_0,y,t)$ and $w_{in}(x_0,y,t)$ in x- and y direction respectively. At the right boundary the incoming waves are set to zero, although the model allows waves to be sent in from both sides. The equations which prescribe the weakly reflecting boundary conditions at the left boundary are:

$$\frac{\partial}{\partial t}(\eta - \eta_{in}) - C\frac{\partial}{\partial x}(\eta - \eta_{in}) = -r(\eta - \overline{\eta}) \tag{1}$$

$$\frac{\partial}{\partial t}(u-u_{in}) - C\frac{\partial}{\partial x}(u-u_{in}) = -r(u-\bar{u}) \qquad (2)$$

$$\frac{\partial}{\partial t}(w-w_{in}) - C\frac{\partial}{\partial x}(w-w_{in}) = 0 \qquad (3)$$

Here, it is assumed that at the boundary the free surface elevations and the velocity components can be decomposed as the sum of a wave travelling to the right and a wave travelling to the left. For the surface elevation at the left boundary this becomes:

$$\eta(x,t) = \eta_{out}(x+Ct) + \eta_{in}(x-Ct)$$

In the case where $r=0$ this signal satisfies the weakly reflecting boundary condition (1) perfectly.

If, again for the case of $r=0$, η is replaced by $\eta + d$ where d is a constant, equation (1) will still be satisfied. This means that a change in the time averaged water level caused by inaccuracies in the code will not be corrected by the boundary conditions. The same problem occurs for the weakly reflective boundary condition for the x- velocity component (2). We have experienced that in using free slip boundary conditions at the bottom unrealistic average velocities can develop during lengthy computations. By choosing r equal to a small positive constant a time averaged value for the free surface elevation $\bar{\eta}$ and for the velocity in x direction $\bar{u}$ can be prescribed. Although the amplitude of the incoming signals will be reduced for positive values of r, small values like $r = \omega/5$ which theoretically reduce the amplitude by a factor of 0.995 prove to work quite well.

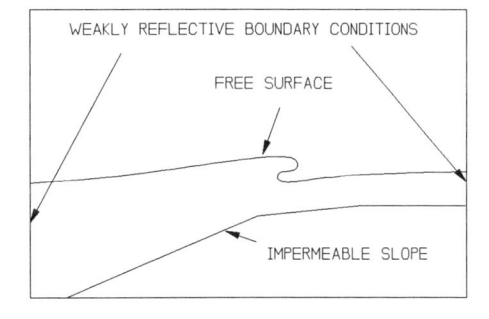

Figure 1 Model application with two weakly reflective boundary conditions

In order to test the quality of the weakly reflecting boundary conditions with incoming nonlinear waves, the velocities and the free surface elevation from the R&F solutions were used. At the left boundary of a numerical wave flume with a constant water depth, these waves were generated using a weakly reflecting boundary condition. At the right boundary again a weakly reflecting boundary condition was used to allow the waves to leave the domain undisturbed. At both boundaries of the Navier-Stokes model the velocities and the surface elevation were calculated and compared with the incoming signal. For a flume with the length of one wave length (wave height 0.2 m, period 3.0 s) the time series of the free surface elevation at the right and the left boundary are shown in Figure 2. Here we can see that, once the initial disturbances have left the domain, incoming and outgoing signals match nicely.

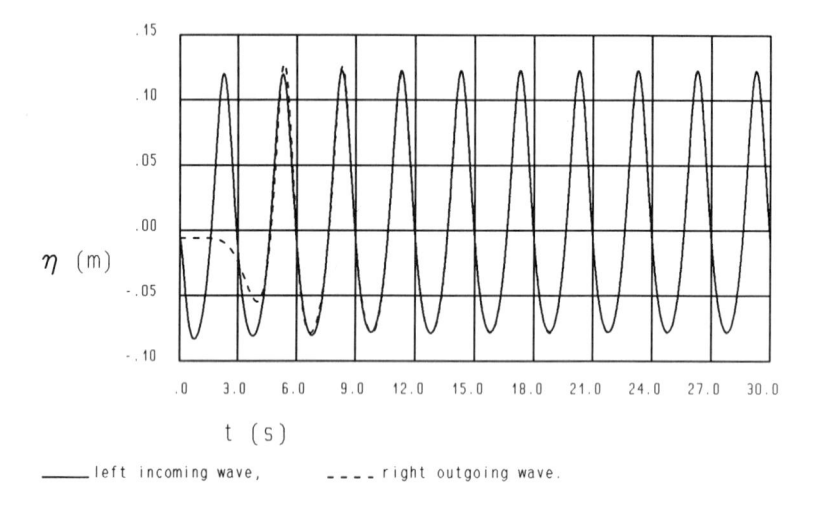

Figure 2 Time series of free surface elevation in a numerical wave flume in order to test the weakly reflecting boundary conditions at both sides

IMPERMEABLE BOUNDARIES

The velocity components used in the VOF method are defined at the centres of the cell faces. In order to discretize the spacial derivatives in the Navier-Stokes equations, velocity components at several locations are used. They are indicated by the arrow in Figure 3 for the case of the momentum equations in the horizontal direction.

In order to model an impermeable boundary as indicated by the line, one could choose to change the stencil of velocity components such that none of the velocity components needed in the discretization is beneath the impermeable boundary. The disadvantage of this approach is that on a vector computer the vectorization of the computational process would be frustrated by the different treatment of the equations inside the fluid and at the boundaries. We wanted to avoid this problem and decided to define virtual velocities at those positions beneath the impermeable boundary. They are indicated by the dotted vectors in Figure 3. In Figure 4 an example of a submerged structure is shown where only the virtual velocities are given.
The virtual velocities which are to be defined beneath the surface of the structure are determined by the boundary conditions at the surface.

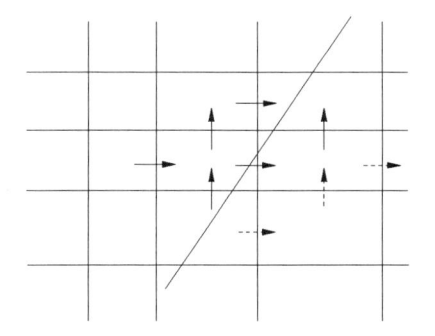

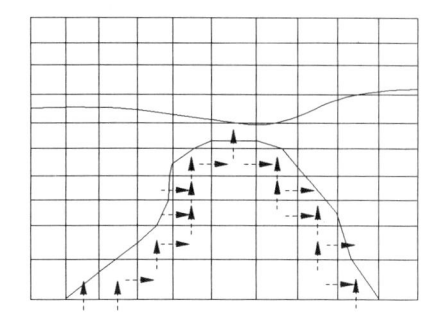

Figure 3 Velocity components needed to discretize the horizontal momentum equations

Figure 4 Virtual velocity components beneath the surface of the structure

In the program only the free slip boundary condition was implemented. Both conditions at the impermeable surface now become:

$$\frac{\partial u_\tau}{\partial n} = 0 \qquad (4)$$

$$u_n = 0 \qquad (5)$$

where u_n is the velocity component in normal direction to the impermeable surface, n the coordinate in this direction and u_τ the velocity component along the surface.

The 14 cell categories which are identified in the program are shown in Figure 5. As can be seen here the impermeable boundary is to be modelled as a straight line inside each cell. For the case of category 4 we will examine how the virtual velocities can be determined. The velocity components shown in this figure are those which are used to discretize the impermeability and free slip condition.

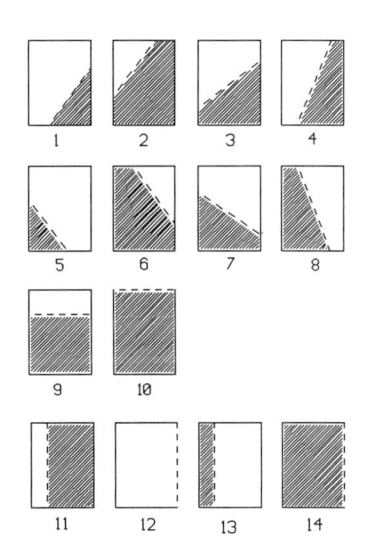

Figure 5 Examples of cell categories used in SKYLLA

By using the components of the normal unit vector at the part of the slope in this cell, n_x and n_y, the equations (4) and (5) can be rewritten as:

$$n_x^2 \frac{\partial w}{\partial x} - n_y^2 \frac{\partial u}{\partial y} - n_x n_y \left(\frac{\partial u}{\partial x} - \frac{\partial w}{\partial y} \right) = 0 \tag{6}$$

$$n_x u + n_y v = 0 \tag{7}$$

The velocity component shown in Figure 6 can be used to find a first order accurate approximation of the derivatives in the free slip condition at the collocation point indicated by the small circle in Figure 6. At his same position the impermeability of the slope can be approximated second order accurately by using linear interpolation. In this way two linear equations are found from which the virtual velocities can be determined. For each cell category the two virtual velocities involved are chosen such that all velocities needed for the discretization of the Navier-Stokes equation are available. Furthermore, the virtual velocities determined for one cell do not coincide with the virtual velocities of another cell.

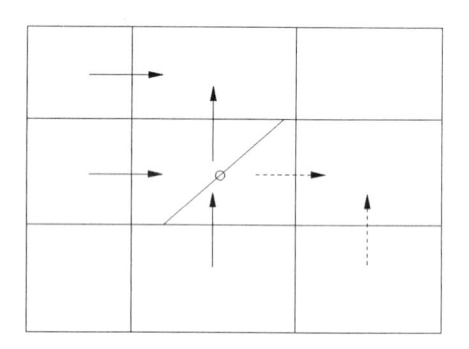

Figure 6 Virtual velocity components for a cell of category 4

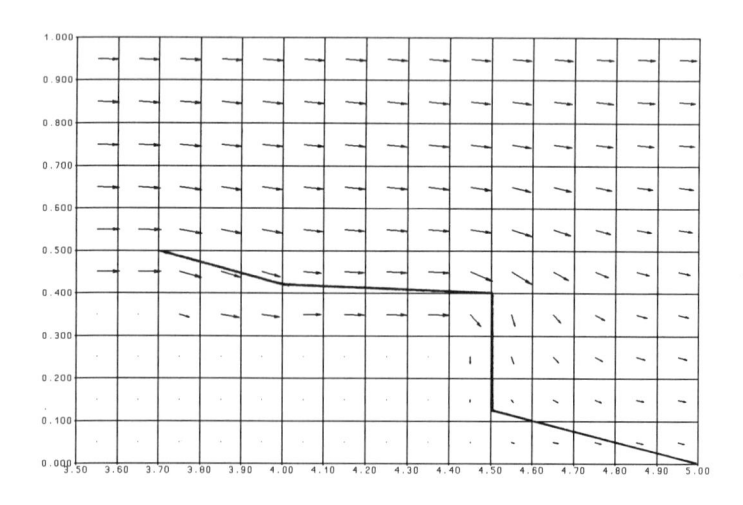

Figure 7 Test for impermeable free-slip boundaries

In Figure 7 we show the result of a computation with a falling slope. The components of the velocity vectors shown here were determined as the averaged values of the velocity components at the boundaries of each cell. The velocities beneath the impermeable boundary are partially determined by the virtual velocities. As can be seen in this figure, the resulting flow near the structure is well aligned with the surface of the structure.

OVERTOPPING BOUNDARY CONDITIONS

Computations with the VOF method are very costly. Especially if the cell sizes are small the explicit time solver will need very small time steps to keep the computations stable. In each time step a Poisson pressure equation needs to be solved to ensure the incompressibility of the fluid. This leads to a set of equations to be solved for the pressure in each cell. The computational effort to solve the pressure equations is roughly proportional to $(N*M)^{2.5}$ where N is the number of cells in horizontal direction and M the number in vertical direction. In cases where a low-crested structure is to be modelled the computation can be carried out applying two separate computational domains, provided that the flow at the top of the crest has supercritical velocity.

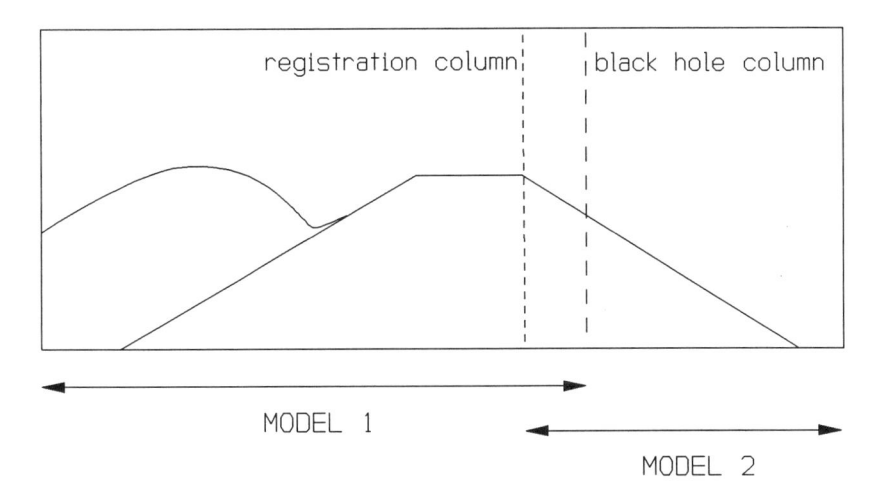

Figure 8 Registration and black hole column at different locations used in the two model approach

The first part of the computation takes place in model 1 as indicated in Figure 8. At the right boundary indicated by 'black hole column' we use the boundary conditions $u_x = 0$ and $w_x = 0$. Furthermore, we set the F value equal to zero in this column each time step. The F values and velocity components are registered during the computation in the 'registration column' each time step. The registered quantities are used during the second computation which involves the flow in model 2 as indicated in Figure 9.

In a strict sense the incompressibility condition can only be satisfied in a simply connected domain by solving the pressure Poisson equation. We expect, however, that in the case where free slip boundary conditions are used and the layer of water at the crest of the structure flows with a supercritical velocity, the errors introduced by using this method will be small.

VALIDATION OF WAVES ON A 1:20 BAR

In order to gain insight in the performance of the numerical model, physical model tests were performed with waves on a submerged bar with a front slope of 1:20. Here we did not use the two-model approach as the velocities at the top of the structure would not be supercritical. Incident regular waves broke on this bar as weakly plunging breakers. Figure 9 shows the experimental set-up used. The numerical set-up used in the verification runs was simpler because at the time the verification took place the falling slope option had not been implemented. Figure 10 shows the left part of the slope used in the experiment.

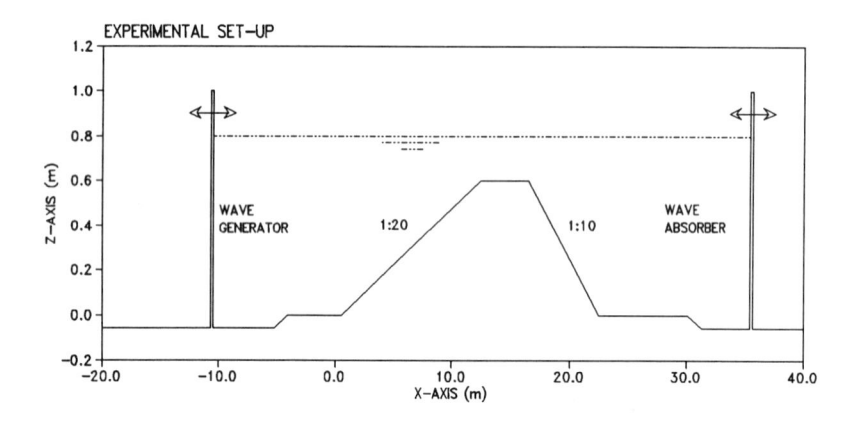

Figure 9 Bottom topography used in wave flume

Velocity profiles at a large number of locations were measured with Laser-Doppler Velocity meters. Those positions are indicated by the blocks in Figure 10. The wave profile was recorded using a video camera. The position of the free surface was determined electronically from the video registration, which resulted in two or more lines in regions with much air entrainment. As the position of the free surface is not a variable in the VOF method the free surface had to be defined using the

F-function. This quantity is the volume fraction of the cell which is filled with fluid (which explains the name Volume Of Fluid). We chose the value $F = 1/2$ to define the location of the free surface. To send in waves into the numerical model we used solutions obtained by the R&F (1981) method. The parameters used to get this solution were obtained by comparing the free surface as prescribed by R&F with the measured free surface of the incoming waves assuming the reflected waves to be negligibly small.

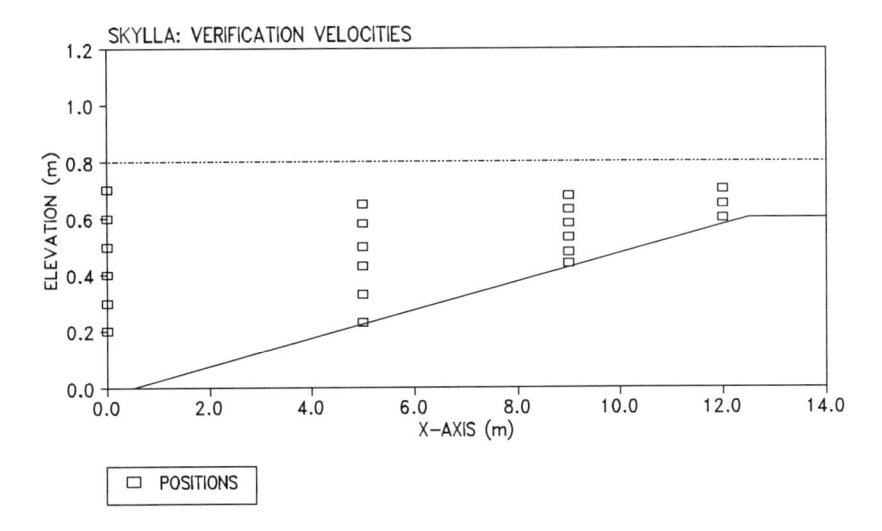

Figure 10 Schematized bottom used in numerical simulation

The following wave parameters were found:

Wave height	:	0.29 m
Wave period	:	1.80 s
Still water level	:	0.80 m

For the R&F solution we used 16 Fourier components. The mean Eulerian velocity was set to zero m/s. The resulting wave length of the incoming waves was 4.41 m. Comparison of the velocity profiles of the R&F solutions and the measured velocities showed that the crest velocities were somewhat too large wheras the trough velocities were underestimated in an absolute sense. We expect this to be caused by the fact that the undertow is assumed to be uniformly distributed over the vertical in the potential model solved by R&F. In practice, however, smaller velocities occur near the bottom and higher velocities more upward in the vertical.

For the computation 480 cells were used in horizontal direction and 50 in the vertical. The kinematic viscosity was set to 0.001 m^2/s.

After the numerical solutions had become periodic we started the comparison. Figure 11 shows the first of the comparisons. Here we see that once the waves start climbing the slope the wave length of the numerical waves become smaller than the measured value.

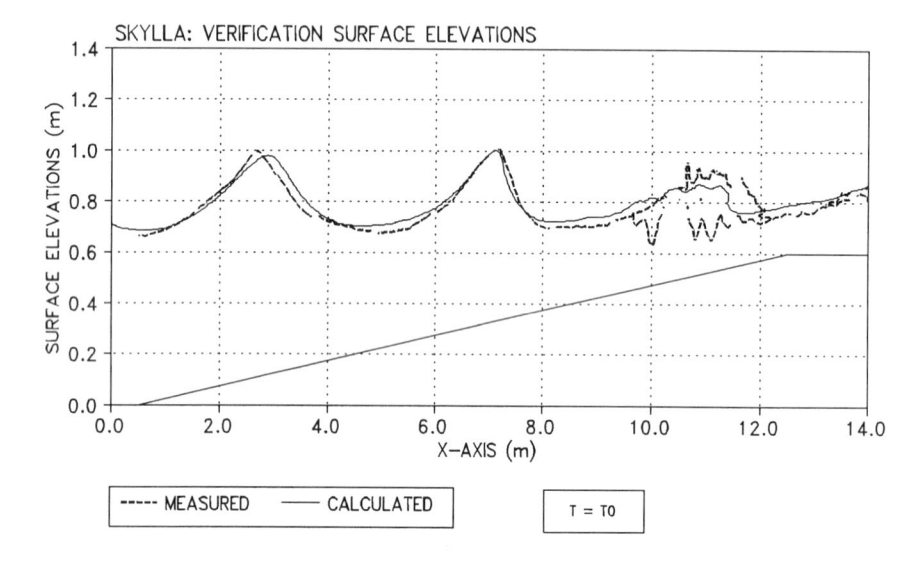

Figure 11

Here we can also see that, the breaking process in the numerical model takes place at the right position. Figures 12 and 13 show the comparison at time intervals of 0.48 s.

The effect of shoaling which is clearly visible in Figure 12 for the measured spilling wave at 8.5 m is not represented well in the numerical simulation. Furthermore, it can be seen in Figure 13 that the breaking process itself develops faster in the numerical model as the decrease in wave height is faster. The transmitted waves at the right boundary, however, were found to be rather accurate.

In Figure 14 we show the measured and computed horizontal and vertical velocities at the left boundary of the model. The problems which arise when in using R&F solutions regarding undertow which were mentioned earlier, reduce the absolute velocity at the trough of the waves. In Figures 15 and 16 we show the comparison of these velocities at 5 and 9 m from the left boundary of the computational domain. All velocities shown here were measured at about 0.5 m from the zero level of the wave flume.

PLUNGING BREAKERS USING 2D NAVIER-STORES

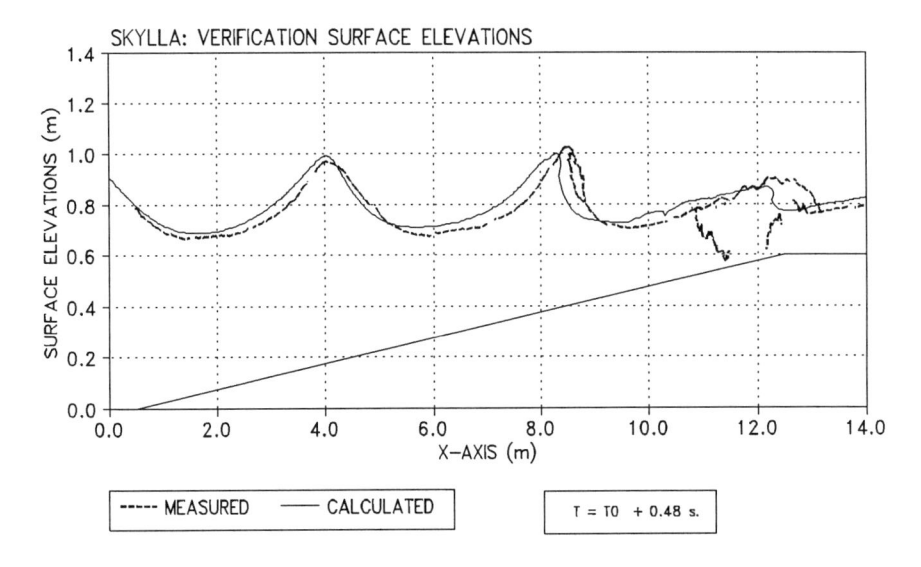

Figure 12

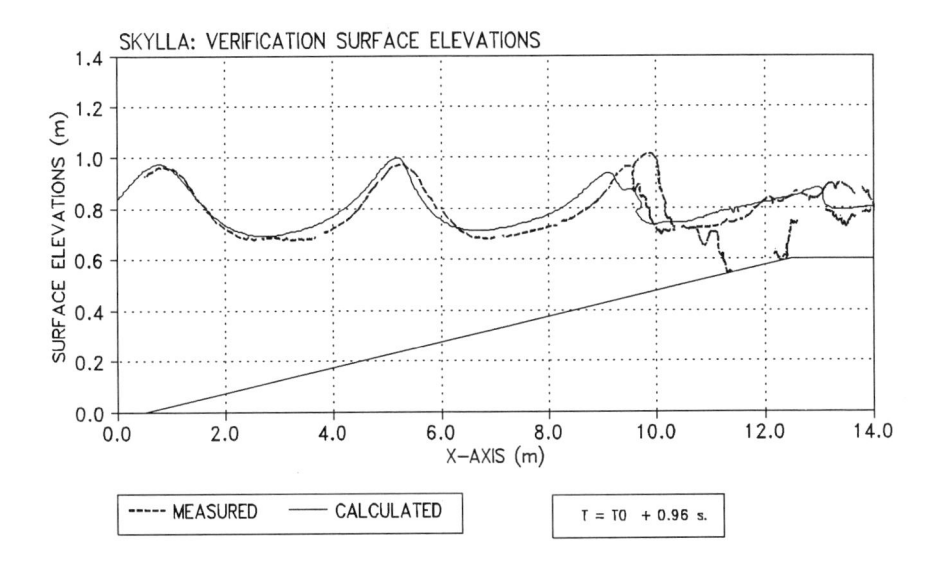

Figure 13

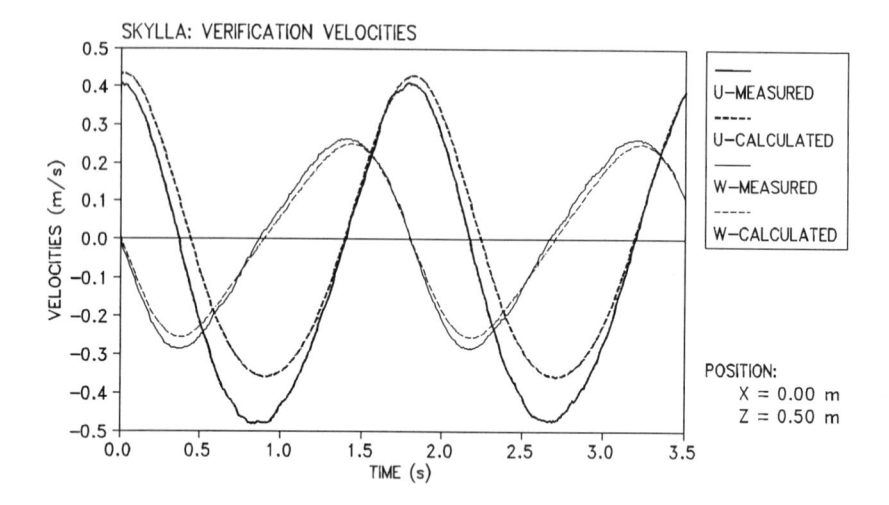

Figure 14 Comparison of measured and calculated velocities; u denotes horizontal velocities, w denotes vertical velocities

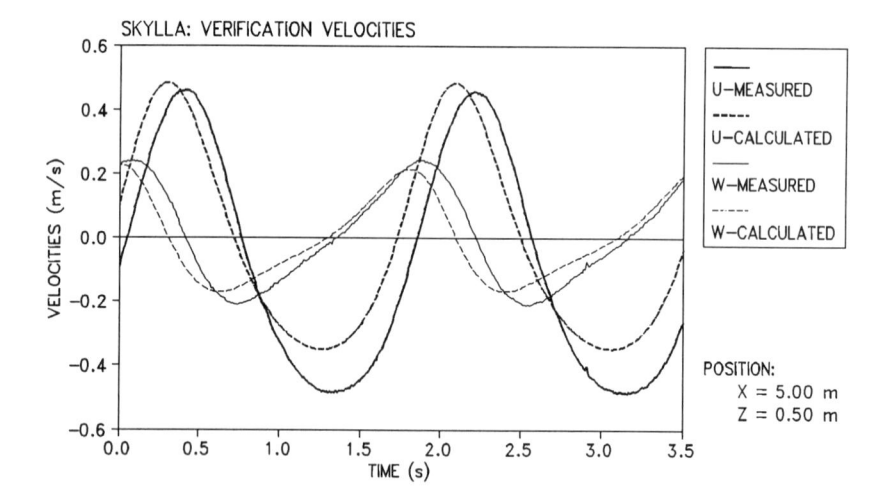

Figure 15 Comparison of measured and calculated velocities; u denotes horizontal velocities, w denotes vertical velocities

SKYLLA: VERIFICATION VELOCITIES

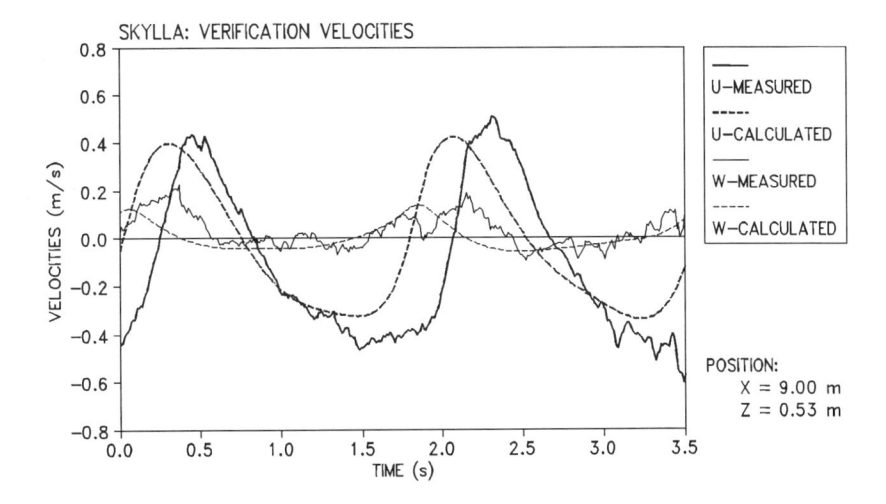

Figure 16 Comparison of measured and calculated velocities; u denotes horizontal velocities, w denotes vertical velocities.

CONCLUSIONS

The VOF method has been made applicable for the computation of breaking waves on coastal structures. Verification with measurements has shown that the program can fairly well simulate waves on a structure. The differences with measurements found in the comparison can partly be explained by the way the boundary conditions were used to define the incoming waves.

ACKNOWLEDGEMENT

This work was funded by the Ministry of Public Works of the Netherlands, Rijkswaterstaat, Road and Hydraulic Engineering Division.

REFERENCES

Ashgriz, N. and J.Y. Poo (1991). "FLAIR: Flux Line-Segment Model for Advection and Interface Reconstruction", J. of Comp. Physics 93.

Broeze, J. (1993). "Numerical Modelling of Nonlinear Free Surface Waves with a 3D Panel Method", Ph.D. Thesis, Twente University, Enschede.

Hirt, C.W. and Nichols, B.D. (1981). "Volume Of Fluid method for the dynamics of free boundaries". J. of Comp. Physics 39.

Klopman, G. (1987). "Numerical simulation of breaking waves on steep slopes", Coastal Hydrodynamics, ed. R.A. Dalrymple.

Kobayashi, N., Otta A.K. and Roy I. (1987). Wave reflection and run-up on rough slopes, J. of Water ways, Ports, Coastal and Ocean Engrg., ASCE, Vol. 113, no. 3.

Petit, H.A.H. and Van den Bosch, P. (1992). SKYLLA: Wave motion in and on coastal structures. Numerical analysis of program modifications. Delft Hydraulics Report H1351.

Petit, H.A.H, Van den Bosch, P and Van Gent, M.R.A. (1994). SKYLLA: Wave motion in and on coastal structures, Implementation of impermeable slopes and overtopping-boundary conditions. Delft Hydraulics, report no H1780.11.

Rienecker, M.M. and Fenton, J.D. (1981). "A Fourier Method for Steady Water Waves", J. of Fluid Dynamics, Vol. 104.

Van Gent, M.R.A., De Waal, J.P., Petit, H.A.H. and Van den Bosch, P. (1994). SKYLLA: Wave motion in and on coastal structures, Verification of kinematics of breaking waves on an offshore bar. Delft Hydraulics, report no H1780.03.

Van Gent, M.R.A. (1994). "The modelling of wave action on and in coastal structures", Coastal Engineering, Elsevier, Amsterdam. Vol. 22.

Van Gent, M.R.A., Petit, H.A.H., Van den Bosch, P. (1993). SKYLLA: Wave motion in and on coastal structures, Implementation and verification of flow on and in permeable structures. Delft Hydraulics, report no H1780.

Van der Meer, J.W., Petit, H.A.H., Van den Bosch, P., Klopman, G. and Broekens, R.D. (1992). "Numerical Simulation of Wave Motion on and in Coastal Structures". ASCE, Proc. 22nd ICCE, Venice, Italy.

CHAPTER 39

Wave Velocity Field Measurements over a Submerged Breakwater.

Marco Petti,[1] Paul A. Quinn,[2] Gianfranco Liberatore[3] & William J. Easson[4]

Abstract

The main focus of the experiment was to observe and measure large-scale vortices generated by breaking waves over a submerged breakwater. These flow structures are important in sediment transport due to their ability to trap sediment particles negating their normal settling velocity. The technique of Particle Image Velocimetry (PIV) was used to measure the spatial distribution of velocities at an instant; an approach which is essential for measuring coherent structures in the flow. Experiments were conducted in the 50m flume at The University of Florence which was fitted with a 1:100 beach, wave elevations, as well as velocity fields, were measured.

The paper deals with the experimental details and a display of the velocity and vorticity maps obtained using PIV.

Introduction

There is considerable interest in the use of submerged breakwaters for coastal protection purposes, but there is little information on the efficiency of such structures due to the poor knowledge of the local hydrodynamics. Consequently there are no analytical or numerical models (Kobayashi & Wurjanto, 89) which can fully describe the processes involved, and so experimental data is still valuable (Arhens, 89, Van der Meer, 90, Petti & Ruol, 91, 92). This study attempts to examine the velocity fields of waves breaking over submerged breakwaters, paying particular attention to the formation of large scale vortices formed by the breaking waves. These play a major part in sediment transport because sediment

---

[1] Lecturer, Department of Civil Engineering, The University of Florence, Via di S. Marta, 3, I-50139 Florence, Italy.

[2] Research Associate, Fluid Dynamics Unit, Dept. of Physics & Astronomy, The University of Edinburgh, Edinburgh, EH9 3JZ, UK.

[3] Professor, University of Udine, Via Cotonificio, 114, 33100 Udine, Italy.

[4] Senior Lecturer, Department of Mechanical Engineering, The University of Edinburgh, Edinburgh, EH9 3JL, U.K.

particles can become trapped in vortices thereby losing their normal settling rates (Nielsen, 92, 94). Consequently how far these vortices travel and the time they take to decay is of significant interest. A further aspect to consider is the effect localised scour, due to large scale vortices, has on the stability of the structure. This paper concentrates on the experiments carried out and the initial results found, which are presented in the form of velocity and vorticity maps.

It is the technique of Particle Image Velocimetry (PIV) which is possibly the most novel aspect of this work; as it provides a spatial distribution of velocities at an instant it is ideal for measuring coherent structures. Previous experiments have used point measuring systems such as directional micro-propellers (Petti & Ruol, 1992) or electromagnetic velocimeters (Mizutani et al, 1992). This approach has the drawback that as only a time history of the wave velocity is obtained at a single point, coherent structures in the flow, such as large scale vortices, tend to get averaged out.

Particle Image Velocimetry (PIV)

A review of PIV is given by Adrian (1991) and an introduction to the process in a coastal engineering context is given by Greated et al (1992). Further references to work using PIV in this publication include Quinn et al (1994) and Earnshaw et al (1994). In order to avoid duplication I refer you to those papers for a brief overview of the processes involved.

The only additional point relating to these experiments was that the technique of image shifting was used (Bruce et al, 1992, Earnshaw et al, 1994). Without going into detail this allows us to measure zero and near-zero velocities and resolves the 180° directional ambiguity inherent with the autocorrelation method of PIV analysis; it is analogous to frequency shifting in Laser Doppler Anemometry (LDA).

Experiments

Experiments were carried out in the 50m flume at The University of Florence, which was fitted with a 1:100 sloping beach. For this study a submerged breakwater was installed about 26m from the wave maker and 22m from the foot of the beach. The breakwater was constructed of perspex with an Aluminium support structure and was 13.6cm high and 26 cm wide at the top, with an offshore facing slope of 1:3.5 and a shoreward facing slope of 1:1.5. The wave flume is shown in Figure 1 and the positions of the breakwater, PIV measurement section and wave gauges are indicated.

PIV was used to measure the velocity fields and a set of 16 parallel-wire resistance-type wave gauges were used to measure the surface elevations of the waves. The 15W Argon ion laser, PIV illumination system and photographic equipment were brought over from The University of Edinburgh for this collaborative study. The laser light sheet required for PIV has to be introduced into the

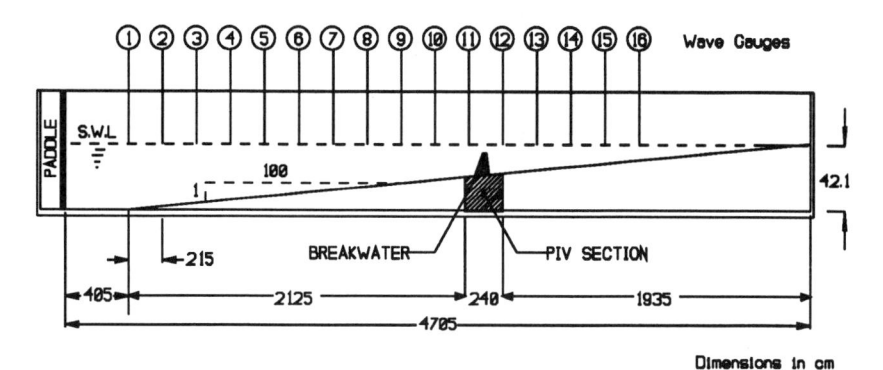

Figure 1: The Wave Flume at The University of Florence.

Test No.	Period, T_0 (s)	Wave Height, H_0 (m)	Ursell No.
1	2	0.10	19.2
2	3	0.10	47.0
3	4	0.14	89.3

Table 1: Wave Parameters

water from below the beach surface. As the bottom of the flume is not made of glass, a special section of the beach had to be built. This allowed the laser sheet in through the side of the flume, below the level of the beach and reflected it off an underwater mirror vertically up through a transparent strip in the beach into the flow. This is shown in Figure 2.

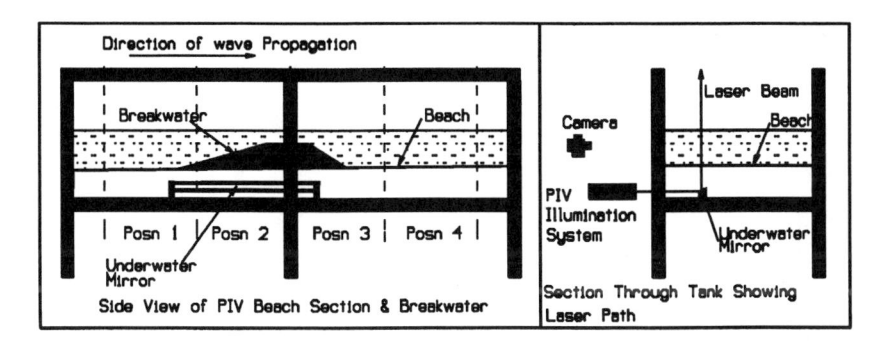

Figure 2: The PIV Beach Section

Three regular waves were used whose parameters are shown in Table 1. These were chosen to break before, over and behind the structure. The waves were generated by a absorbing piston-type wave maker, in a water depth of 0.42m. PIV measurements were made at four positions around the breakwater and four phases of each wave, were recorded at each position, the phases being separated by $90°$.

Wave Gauge Results

Figure 3 shows typical surface elevation records from gauges 1, 3, 5 & 7 for the 3s wave; Figure 4 shows the same for gauges 9, 11, 13 & 15. There are a couple of points worth noting here, firstly that gauge No. 1 is in constant depth water ($h_0 = 0.42$m) and secondly that gauge No. 11 is just in front of the breakwater.

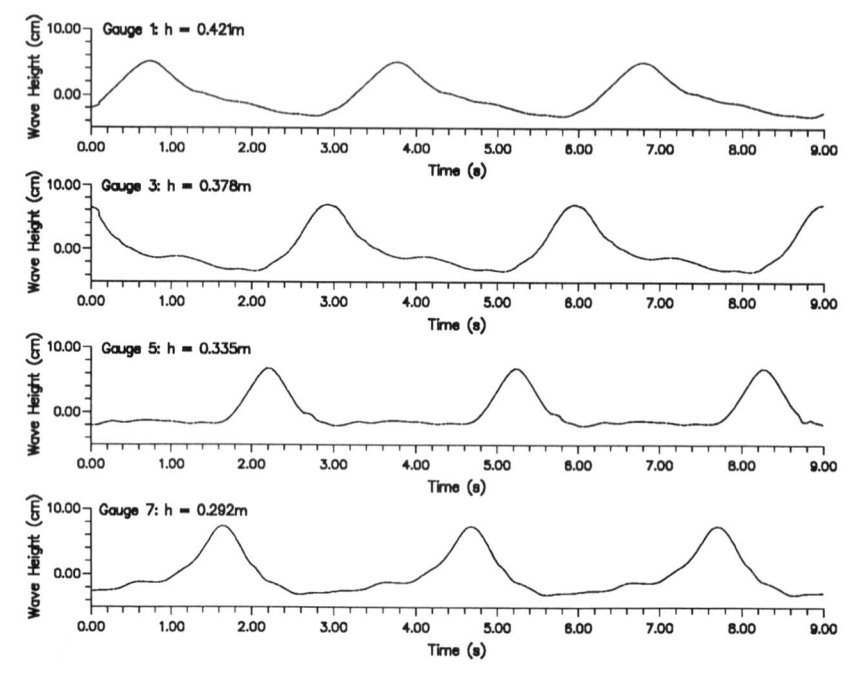

Figure 3: Wave Heights (T = 3s)

The first thing to notice from gauge 1 is that solely monochromatic waves are not being produced. There is definite evidence of a second wave being generated, and judging from the subsequent gauges this is a free wave, although it is at least of a much smaller amplitude than that of the main one. One can ascribe this generation of a free wave to the relatively high non-linearity of these waves. The Ursell numbers, calculated from Equation 1 are shown in Table 1; the values for the 3s & 4s waves being particularly high. It is perhaps not too surprising to see

Figure 4: Wave Heights (T = 3s)

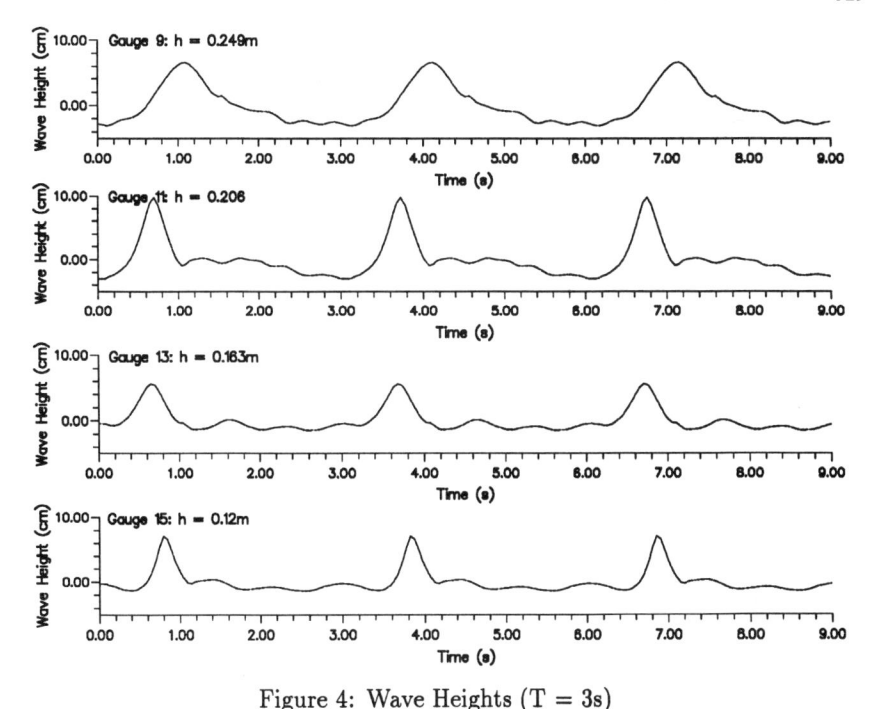

this kind of instability from a wave maker trying to generate sinusoidal waves, Osborne & Petti (1994) give a more detailed description of this phenomenon.

$$U_0 = (H_0 L_0^2)/h_0^3 \tag{1}$$

Gauges 9 and 11 show significant reflected high-frequency waves, particularly in the troughs, with gauge 11 also showing the maximum wave elevation, of about 0.1m, due to its position just in front of the breakwater. In addition to this, one can also see (from Figures 3 & 4) how the phase celerity decreases as the waves approach the breakwater. Gauges 13 & 15 show diminished wave height and increased high frequency components following breaking, with evidence from gauge 15 of some frequency recombination as the wave crests have increased in height and sharpened in profile.

With the generation of a free wave and a measured reflection coefficient value of about 10%, there was a noticeable variation of wave heights at all the wave gauge positions. A zero-up-crossing significant wave height analysis from all 16 wave gauges was carried out and this gives a good indication of the scatter in wave heights. This zero-crossing analysis is given for all three waves in Figure 5

Although there is some significant scatter of the wave heights one can see the general trend of increasing wave height towards the breakwater, located just

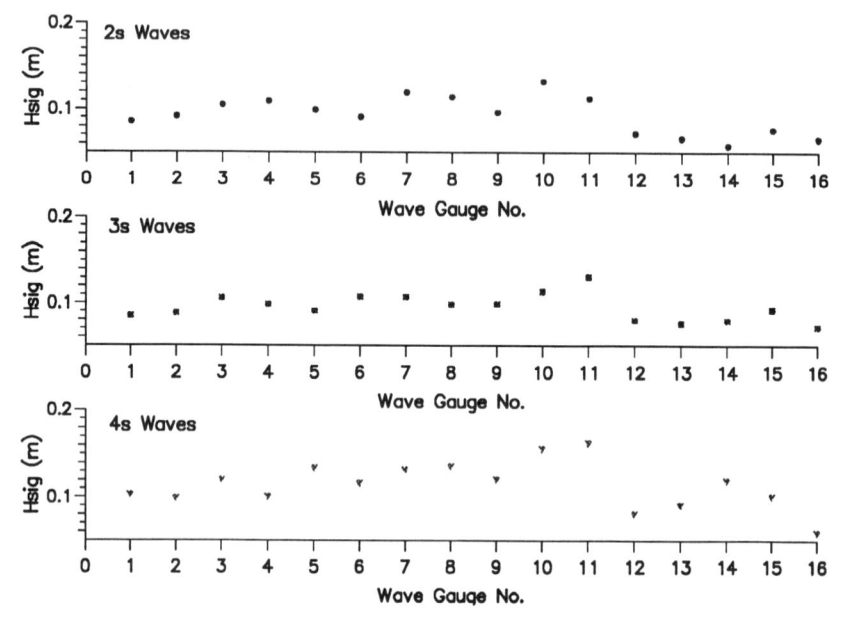

Figure 5: Zero up-crossing significant wave height distribution.

after gauge 11, and a decrease following breaking. The maximum waveheight was recorded at gauge number 10 for the 2s wave, this is due to the fact that the wave has already started to break by the time it reaches gauge 11. There is further evidence of post-breaking recombination of frequency components with the increase in significant wave height after the initial decrease due to breaking. This is particularly so for the 4s waves.

PIV Results

Due to the large number of measurements taken only a small set of the 48 velocity vector maps can be shown here. As the focus of the paper is the measurement of vortices generated by the breaking waves, only one example of a wave breaking on the front face of the breakwater is shown. The other results are presented, by means of velocity and vorticity maps, for the positions behind the breakwater. The convention of positive vorticity indicating an anti-clockwise rotation has been adopted throughout.

Figure 6 shows a 2s wave breaking on the front face of the breakwater. It is interesting to notice that the wave is breaking where the wave meets the fast flowing backwash, returning over the breakwater, and not at the top of the wave as one might expect. This formation of a localised bore on the structure gives rise to small waves which travel in the offshore direction.

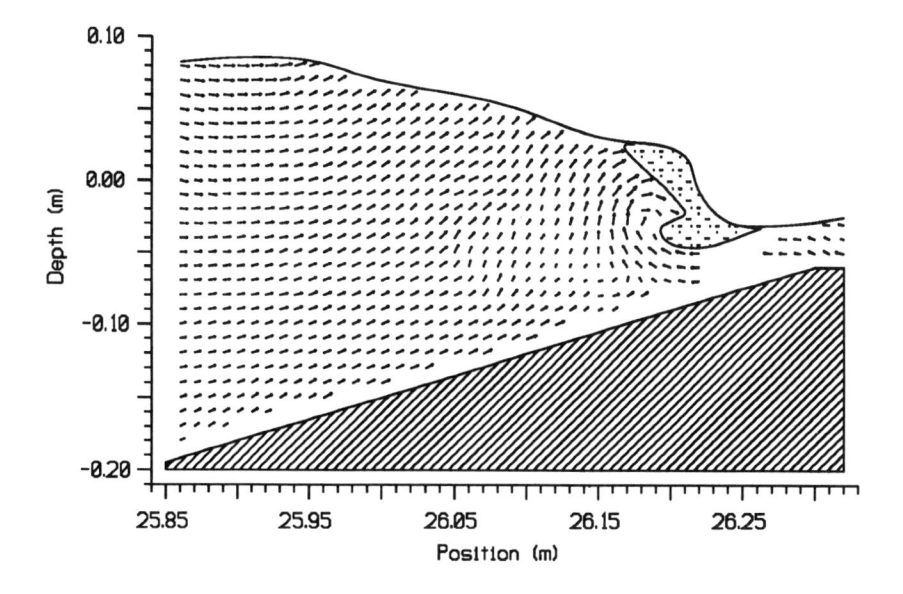

Figure 6: Vector map: 2s wave: Phase 4: Position 2.

Considering now the 3s wave, Figures 7, 9 & 11 show velocity maps for the first three phases at position three, with Figures 8, 10 & 12 showing the corresponding vorticity maps.

Phase 1 shows a positive vortex down near the bottom of the breakwater in an area particularly sensitive to scour, (Liberatore, 92) . Phase 2 shows a large region of negative vorticity near the surface to the right caused by the overturning of the wave that has just broken, sweeping upwards the positive vortex that was located near the foot of the structure in the previous phase. A small negative vortex is also being shed by the top corner of the breakwater. The third phase shows that the main negative vortex of the previous phase has reduced to a ring of smaller vortices and the positive vortex has now either dissipated or been carried back over the structure.

If we now look at position 4 (shorewards of position 3) for the 3s wave we can see from Figures 13 & 14 that at phase 4 there is still a significant negative vortex. This is most likely to have been generated by the previous wave and has, therefore, persisted for about 5s and moved about 0.55m from the rear of the breakwater.

Turning our attention now to the 4s waves, Figures 15 & 17 show velocity maps for phases 2 & 3 at position 3, with Figures 16 & 18 showing their corresponding vorticity maps.

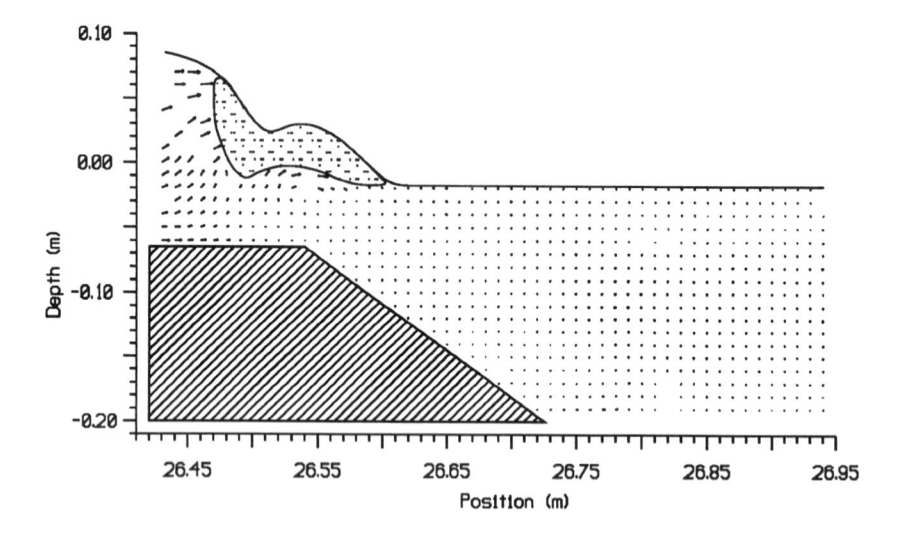

Figure 7: Vector map: 3s wave: Phase 1: Position 3.

Phase 2 shows negative vorticites near the top right and on the top corner of the breakwater, the former being generated by the overturning breaker. There is a positive vortex above the rear toe of the structure in a similar position to the positive vortex shown at the same phase and position for the 3s wave. In phase 3, however, we can see that the large negative vortex has been swept into the next position and all that remains are two positive vortices near the bed and surface and a positive vortex being generated at the top corner of the structure as the flow returns back over the breakwater.

Moving on to the final position, Figure 19 shows the velocity maps for phase 2 again and Figure 21 shows the same for phase 4. Figures 20 & 22 show their vorticity maps, respectively. Phase 2 occurs just after the wave has broken and there is still some aeration of the flow near the surface, particularly near the centre of the figure. It has the greatest magnitude of vorticity values of any of the waves, with large negative vortices near the surface on the right of Figure 20 and near the bed on the left. Two phases later we can see that both negative vortices have persisted with the main one, centred about 27.09m in Figure 22 and the second one located on the left hand side of the flow.

WAVE VELOCITY FIELD MEASUREMENTS 533

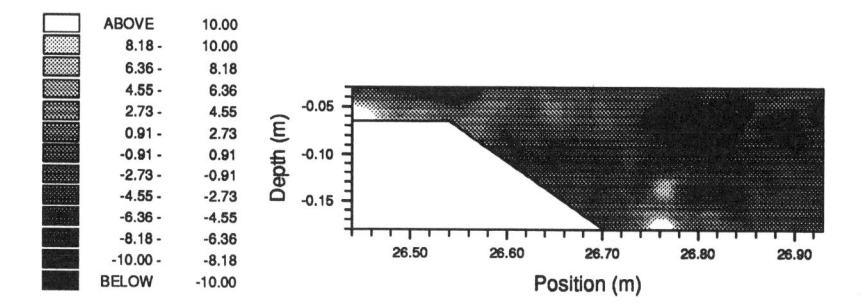

Figure 8: Vorticity map: 3s wave: Phase 1: Position 3.

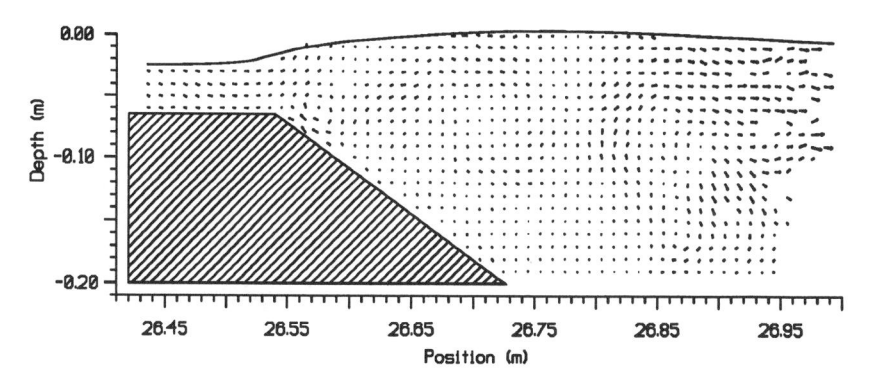

Figure 9: Vector map: 3s wave: Phase 2: Position 3.

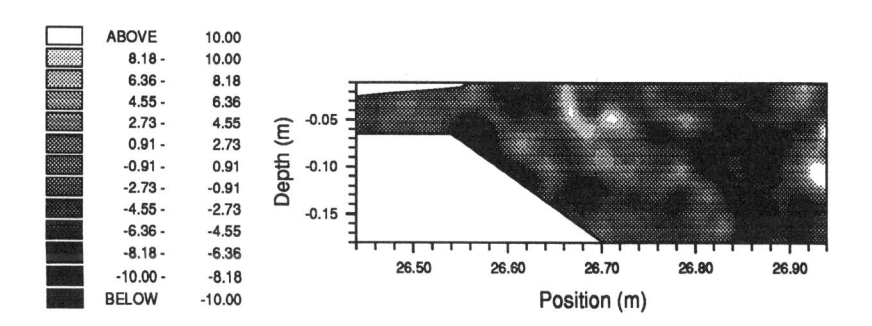

Figure 10: Vorticity map: 3s wave: Phase 2: Position 3.

534 COASTAL ENGINEERING 1994

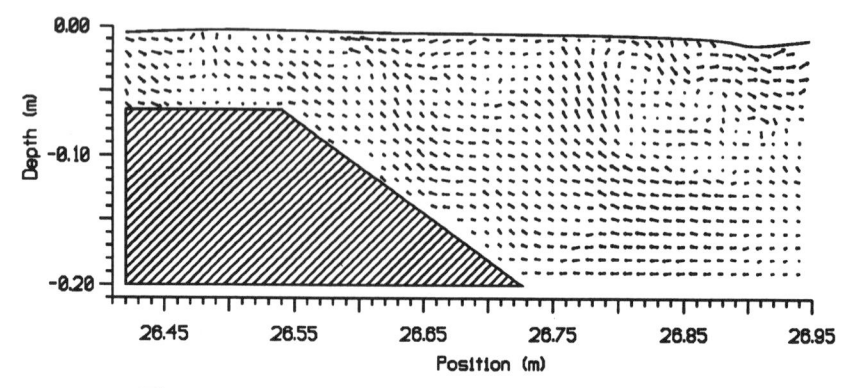

Figure 11: Vector map: 3s wave: Phase 3: Position 3.

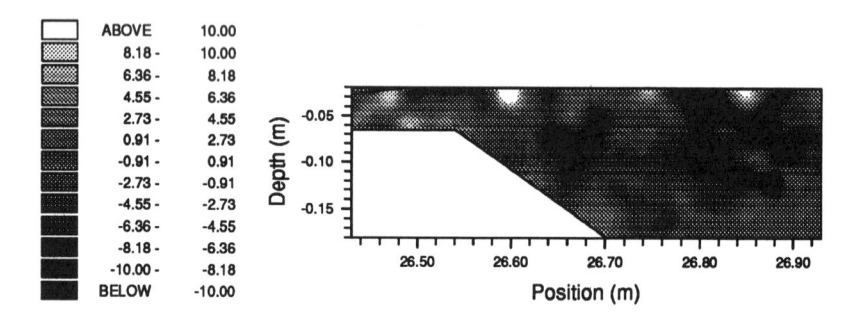

Figure 12: Vorticity map: 3s wave: Phase 3: Position 3.

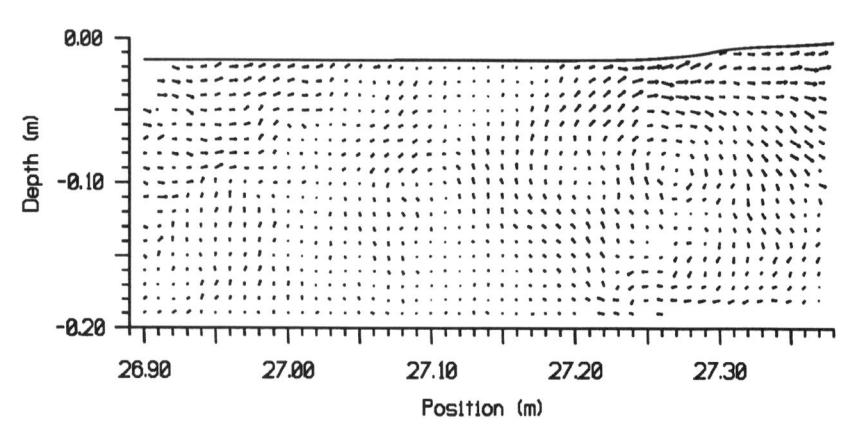

Figure 13: Vector map: 3s wave: Phase 4: Position 4.

WAVE VELOCITY FIELD MEASUREMENTS 535

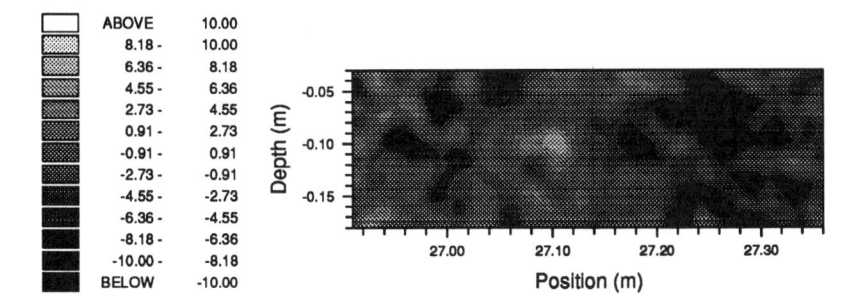

Figure 14: Vorticity map: 3s wave: Phase 4: Position 4.

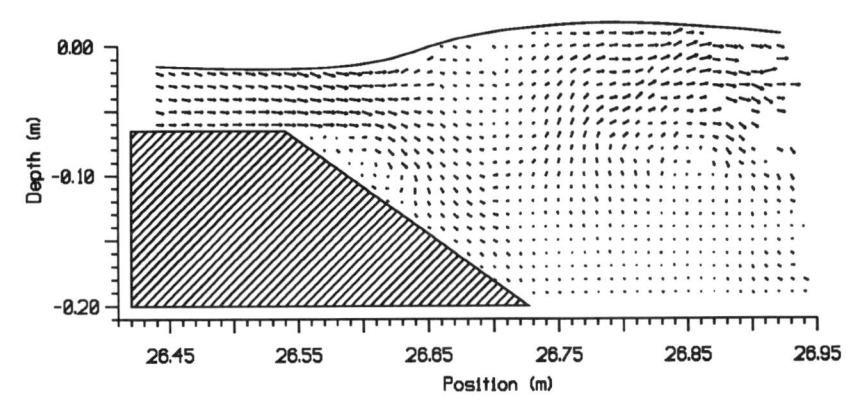

Figure 15: Vector map: 4s wave: Phase 2: Position 3.

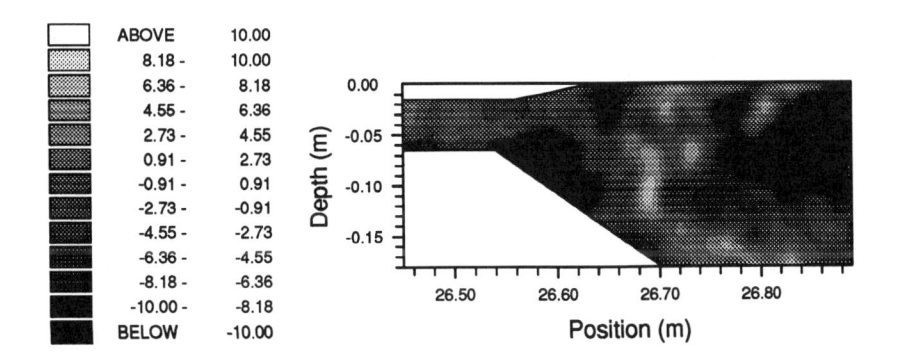

Figure 16: Vorticity map: 4s wave: Phase 2: Position 3.

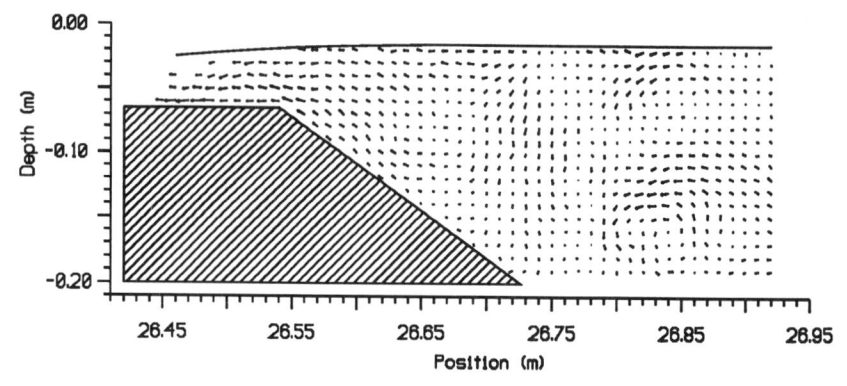

Figure 17: Vector map: 4s wave: Phase 3: Position 3.

Figure 18: Vorticity map: 4s wave: Phase 3: Position 3.

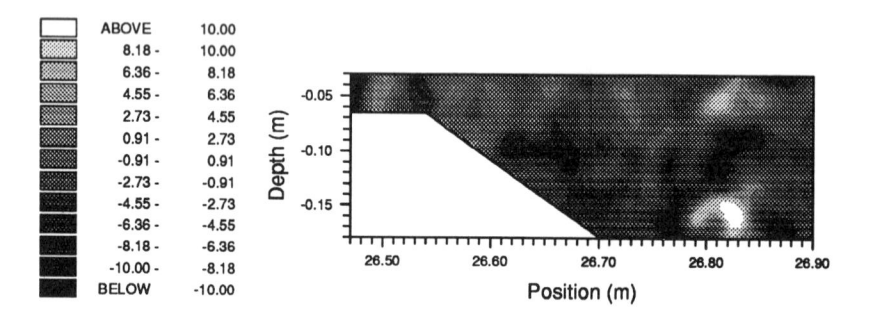

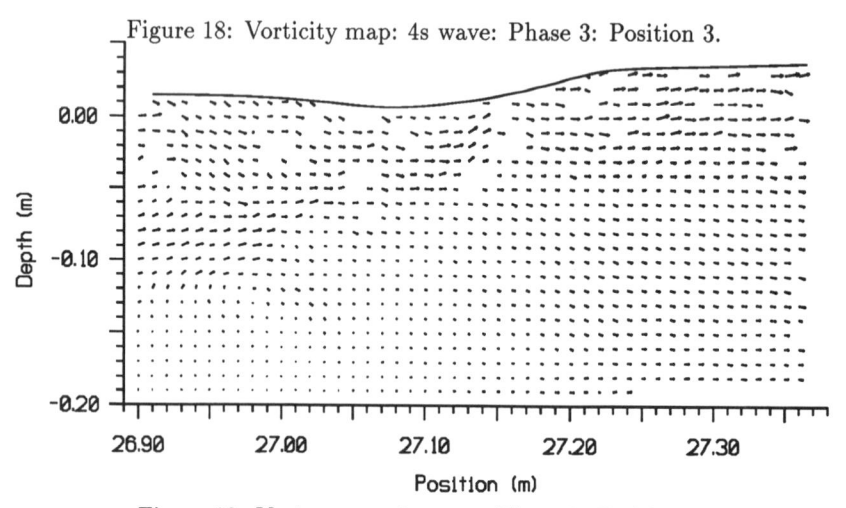

Figure 19: Vector map: 4s wave: Phase 2: Position 4.

WAVE VELOCITY FIELD MEASUREMENTS 537

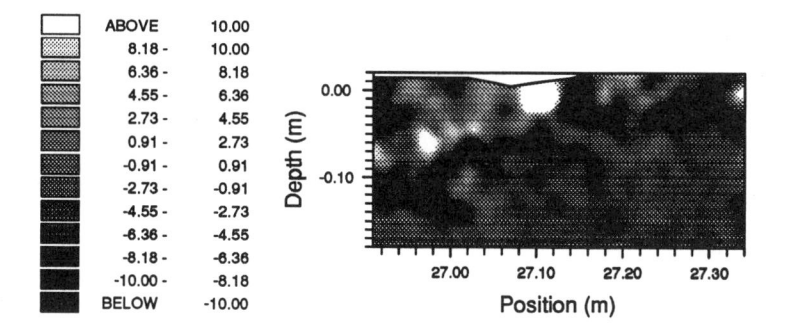

Figure 20: Vorticity map: 4s wave: Phase 2: Position 4.

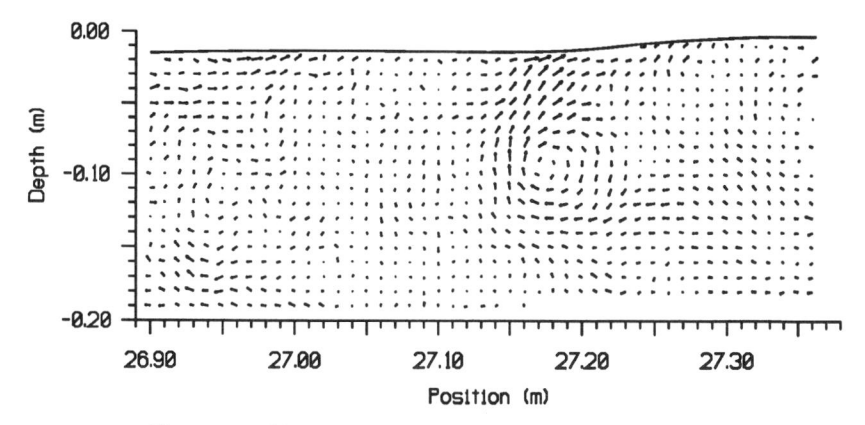

Figure 21: Vector map: 4s wave: Phase 4: Position 4.

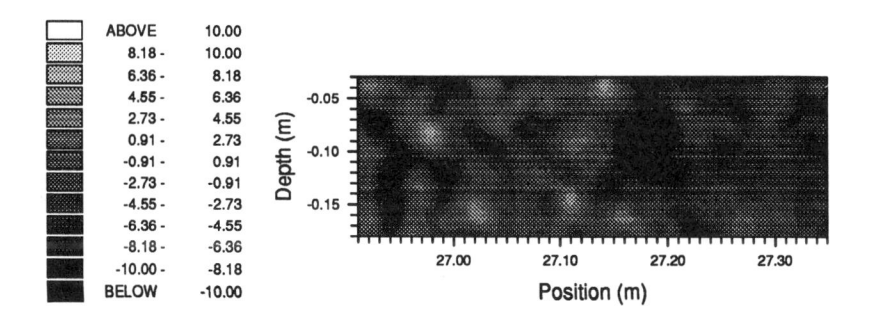

Figure 22: Vorticity map: 4s wave: Phase 4: Position 4.

Conclusions

The main focus of this study is to demonstrate the scale and significance of large scale vortices generated by waves breaking over a submerged breakwater. The ability to do this lies in the use of PIV as the measurement system, due to the fact that these flow structures are only measurable with a technique which records the spatial distribution of velocities at an instant. The detail of these measurements is apparent and a thorough analysis of the data is now called for. However, one can get an initial impression, from the results presented (eg. Figures 7-10 showing phases 1 & 2 of the 3s waves at position 3), of the importance of including vortex generation and interaction in the formulation of sediment transport numerical models. One such model is the Discrete Vortex Model described in Pedersen et al, 1992.

Acknowledgements

This work was undertaken as part of the MAST G8-M Coastal Morphodynamics research programme. It was funded by the Commission of the European Communities Directorate General for Science, Research and Development under contract N^o. EC MAST-II 0CT 92 0027; their support is greatly appreciated.

The authors also wish to extend their gratitude to the technicians, Mauro Gioli, Muzio Mascherini and Frank Morris, who played an essential role in conducting the experments in Italy with such a tight time limit.

References

Adrian, R.J. (1991) *Particle Imaging Techniques for Experimental Fluid Dynamics*. Ann. Rev. Fluid Mechanics, 23:261-304.

Arhens, R.J. (1991) *Stability of Reef Breakwaters*. J. Waterway, Port, Coastal and Ocean Engineering, ASCE, 1729-1738.

Bruce, T. & Easson, W.J., (1992) *The Kinematics of Wave Induced Flows Around Near-bed Pipelines*. Proc. 23rd Int. Conf. Coastal Eng. 229:2990-2998.

Earnshaw, H.C., Bruce, T, Greated, C.A. & Easson, W.J. (1994) *PIV Measurements of Oscillatory Flow over a Rippled Bed*. Proc. 24th Int. Conf. Coastal Eng., Kobe, Japan.

Greated, C.A., Skyner, D.J. and Bruce, T., (1992) *Particle Image Velocimetry (PIV) in the Coastal Engineering Laboratory*. Proc. 23rd Int. Conf. Coastal Eng. 15:212-225.

Kobayashi, N. & Wurjanto, K. (1989) *Wave Transmission over Submerged Break-*

waters. J. Waterway, Port, Coastal and Ocean Eng., ASCE, (115) 5, 662-680.

Liberatore, G. (1992) *Detatched Breakwaters and their use in Italy.* Proc. Short Course on Design and Reliability of Coastal Structures, 23rd Int. Conf. Coastal Eng.

Mizutani, N., Iwata, K., Rufin, T.M. and Kurata, K. (1992) *Laboratory Investigation on the Stability of a Spherical Armour Unit as a Submerged Breakwater.* Proc. 23rd Int. Conf. Coastal Eng. pp 1400-1413.

Nielsen, P. (1992) *Coastal Bottom Boundary Layers and Sediment Transport.* World Scientific, Singapore.

Nielsen, P. (1994) *Suspended Particle Motion in Coastal Flows.* Proc. 24th Int. Conf. Coastal Eng., Kobe, Japan.

Osborne, A.R. & Petti, M. (1994) *Laboratory Generated Shallow-Water Surface Waves: Analyses using the Periodic, Inverse Scattering Transform.* Physics of Fluids, Vol. 6, No. 5, 1727-1744.

Pedersen, C., Deigaard, R., Fredsøe, J. & Hansen, E.A. (1992) *Numerical Simulation of Sand in Plunging Breakers.* Proc. 23rd Int. Conf. Coastal Eng. pp 2344-2357.

Petti, M. and Ruol, P. (1991). *Experimental Studies on the Behaviour of Submerged Breakwaters.* Proc. 3rd Int. Conf. Coastal and Port Eng. in Developing Countries, 167-178.

Petti, M. and Ruol, P. (1992). *Laboratory Tests on the Interaction between Nonlinear Long-waves and Submerged Breakwaters.* Proc. 23rd Int. Conf. Coastal Eng. pp 792-803.

Quinn, P.A., Petti, M., Drago, M. & Greated, C.A. (1994) *Velocity Field Measurements and Theoretical Comparisons for Non-Linear Waves on Mild Slopes.* Proc. 24th Int. Conf. Coastal Eng., Kobe, Japan.

Van der Meer, J.W. & Pilaczyk, K.W. (1990) *Stability of Low-Crested and Reef Breakwaters.* Proc. 22nd Int. Conf. Coastal Eng., ASCE, 1375-1388.

CHAPTER 40

Velocity Field Measurements and Theoretical Comparisons For Non-Linear Waves on Mild Slopes.

Paul A. Quinn,[1] Marco Petti,[2] Michele Drago[3] & Clive A. Greated[4]

Abstract

The way in which the Boussinesq and Serre models deal with the internal kinematics of waves on mildly sloping beaches is examined. The equation normally used on the depth-averaged horizontal velocity to impose a parabolic profile vertically up through the wave is studied using experimentally and theoretically obtained values. In general, it is found that for near-bed regions the equation provides satisfactory comparisons, however, as one approaches the surface the theoretical values can substantially exceed the experimental ones, particularly in the crest of the wave.

Introduction

A great deal of emphasis has been placed recently on the use of the Boussinesq and Serre models in determining near-shore wave motion (Dingemans, 94a, 94b), mainly due to its recently improved frequency dispersion capabilities. It has some shortcomings, however, in its treatment of the internal wave kinematics based mainly on the fact that it only predicts explicitly the depth-averaged horizontal velocity and relies on an equation to provide a vertical profile of velocity up through the wave, and the divergence of that equation to provide a distribution of the vertical component of velocity.

A description of the Boussinesq and Serre models used in this comparison can be found in Brocchini et al, (1992). They also give the equation for the parabolic profile of velocity as:

---

[1] Research Associate, Fluid Dynamics Unit, Department of Physics & Astronomy, The University of Edinburgh, Edinburgh, EH9 3JZ, U.K.

[2] Lecturer, Department of Civil Engineering, The University of Florence, Via di S. Marta, 3, I-50139 Florence, Italy.

[3] ENVI, Snamprogetti SpA., Fano, Italy.

[4] Professor of Fluid Dynamics, Department of Physics & Astronomy, The University of Edinburgh, Edinburgh, EH9 3JZ, U.K.

$$u(z) = \bar{u} - (h/2)(h\bar{u})_{xx} + (h^2/6)\bar{u}_{xx} - z(h\bar{u})_{xx} - (z^2/2)\bar{u}_{xx} \qquad (1)$$

Where u is the horizontal velocity component, $\bar{u}$ is its depth-average, $\bar{u}_{xx}$ is its second derivative with respect to x and h is the local depth from the still water level (SWL). If the bottom topography is known then all that is required to calculate $u(z)$ is a spatial distribution of $\bar{u}$. As Particle Image Velocimetry provides a 2-D spatial distribution of the velocity field it is ideally suited for use in Equation 1.

The first part of this paper uses the values for $\bar{u}$ and $\bar{u}_{xx}$ from PIV results in Equation 1 to calculate the vertical profile of velocity, $u(z)$, and then compares this to the actual profile measured in the experiments. This is based upon the supposition that the best fit to the actual velocity profile will be obtained using the experimental values of $\bar{u}$ and $\bar{u}_{xx}$ in Equation 1. Alternatively, if the model predicted the depth-averaged velocity distribution perfectly will it then provide the correct vertical profile of velocity?

The second part of the paper is then to compare the distribution of the depth-averaged horizontal velocity predicted by the Boussinesq and Serre models with the experiments and then calculate their vertical profiles of velocity.

Particle Image Velocimetry (PIV)

Although there are a number of institutions around the world who now use PIV as a matter of course, it is still relatively new in the field of coastal engineering when compared to techniques like Laser Doppler Anemometry (LDA), and as such warrants a brief description. A comprehensive review of PIV is given by Adrian (1991) and an introduction to the process in a coastal engineering context is given by Greated et al (1992). Quinn et al (1993) gives an account of the errors inherent in the technique and uses as an example the measurement of waves breaking on a 1:30 plane sloping beach. Powell et al (1992) gives an account of the specific application to waves breaking on non-uniform beach slopes.

The first stage of PIV is to photograph the flow, which is seeded with tiny particles and illuminated with a pulsing laser light sheet. The idea is that the camera shutter is held open for several, say four, pulses of the light sheet so that on the photographic negative there will appear four multiple images of each of the seeding particles in the flow. Measuring the separation of the multiple images at any point will yield the velocity at that point, when coupled with the time separation of the light pulses. The seeding particles used in these experiments was Conifer Pollen which has a diameter of 50-70 μm and is almost exactly neutrally buoyant in water. Typical illumination pulse intervals are about 5ms, and the shutter speed of the camera is usually set at 1/60s for this type of flow.

The second stage of the technique is to analyse the developed negative. The analysis is carried out on an automated rig described in some detail in Greated et al (1992) and Quinn et al (1993). The process involves probing the negative on a regular grid (normally 1mm x 1mm) with the beam from a low power Helium-Neon laser. The multiple images of the seeding particles illuminated in the 1mm diameter laser beam cause an interference fringe pattern which is captured by a CCD camera and passed to a Personal Computer (PC). The separation and orientation of the fringes is calculated which gives the average displacement of the multiple images in the small interrogation area of the negative and hence the velocity vector at that point. This process is carried out for every point in the flow field and results in a 2-D velocity array.

Experiments

Beach slopes of 1:30 and 1:100 were used in this study. The experiments were carried out in the Universities of Edinburgh and Florence respectively.

Considering the 1:30 slope first, the experiments were conducted in a 10m wave flume with a SWL of 0.75 m. The wave maker is a hinged, absorbing paddle (Salter, 1982). Due to the restricted length of the flume a 1:30 beach does not reach the bottom of the tank. As we have a hinged paddle rather than a piston-type wave maker, the water depth cannot be altered significantly, so a ramp was installed from the bottom of the wave maker to the foot of the beach, 3m away. The water depth at the foot of the beach was 0.19m and as the waves were so small, ($T = 1$s, $H = 0.032$m) the set-up was deemed, if not ideal, then at least satisfactory. A section of the tank showing the beach and the PIV illumination system is shown in Figure 1.

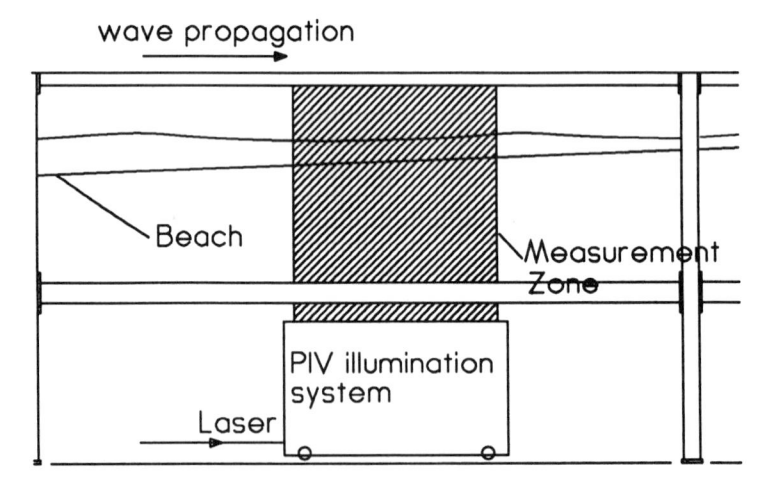

Figure 1: The Flume and PIV Illumination Section.

Regular waves with a period of 1s and a wave height of 0.032m were used. PIV measurements were made at five adjoining positions along the beach, each position being 0.6m in length. Four phases of the wave were recorded at each position.

As the 1:100 beach slope was too mild for the wave flume in Edinburgh the experiments on this slope were carried out in the 50m flume at The University of Florence, Italy. The laser and PIV illumination system were taken over to Florence, from Edinburgh, for these collaborative experiments. As the flume in Florence does not have a glass bottom a special section of the beach had to be built which allowed the PIV laser light sheet in through the side of the tank, below the level of the beach, and reflected it off an underwater mirror up through the beach and into the flow. This is shown in Figure 2.

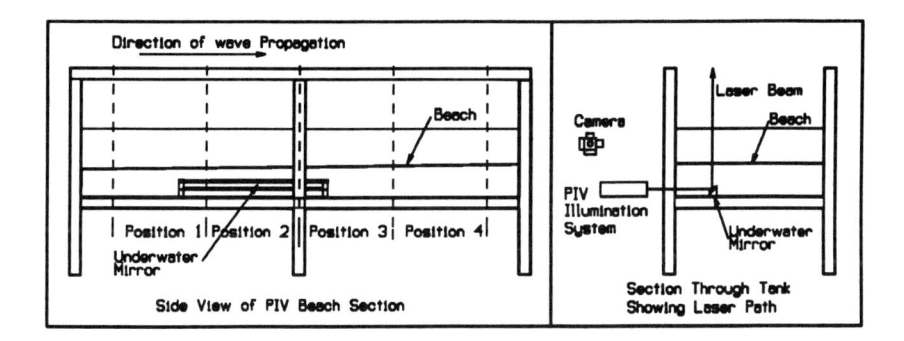

Figure 2: The PIV Beach Section in Florence.

Once again regular waves were used, this time with a period of 3s and a height of 0.1m. The waves were generated in a water depth of 0.42m and propagated about 25m up the beach before they broke. Eight positions were measured around the breaking zone and four phases of the wave were measured at each position.

The wave flume in Florence is fitted with a piston-type wave maker which does have an absorbing system, however, for this experimental set-up with an extremely mild beach slope and no structure to provide a significant reflection, the wave repeatability, as measured by wave-gauge analysis, was worse with the absorbing system activated. For this reason the absorbing system was switched off for these experiments. In order to minimise the effect of the unrepeatability of the waves the first PIV measurement was taken 40s after the start of the wave maker and the subsequent phases were taken from successive waves. The tank was allowed to settle for about 5 to 10 minutes between runs.

Velocity Profile Analysis

Returning to the main aim of the paper, to examine the equation providing a vertical profile of velocity from the depth-averaged velocity component given by the Boussinesq and Serre models. All that is required to calculate the vertical profile of the horizontal velocity component, $u(z)$, from Equation 1, is a spatial distribution of the depth-averaged horizontal velocity. This can easily be calculated from PIV measurements. If one considers Figure 3, which shows the velocity field of the crest phase of a 1s wave on a 1:30 slope, the depth-averaged velocity can be calculated for each column of vectors. Figure 4 shows the resulting spatial distribution of the depth-averaged velocity.

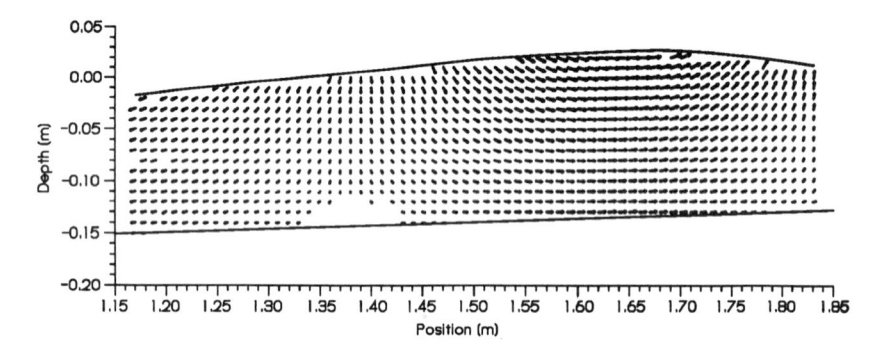

Figure 3: Velocity Vector map for a 1s wave on a 1:30 slope.

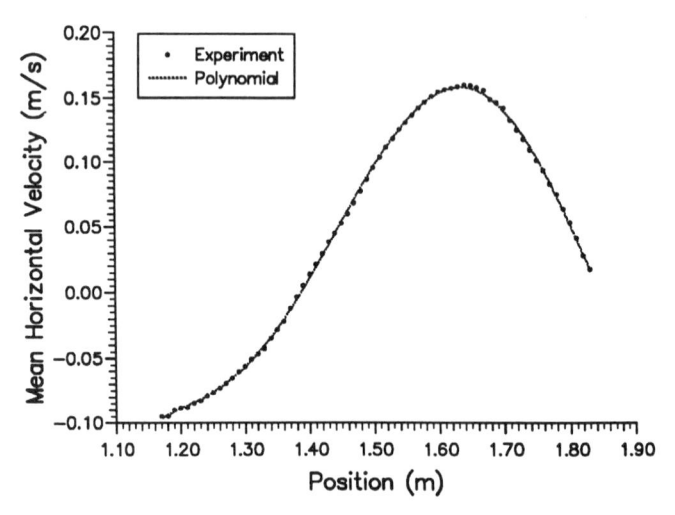

Figure 4: Depth-averaged velocity distribution for a 1s wave on a 1:30 slope.

In Figure 4 the points are the depth-average horizontal velocity components calculated from the vector map (Figure 3). The curve is a least squares polynomial and has been used to calculate $\bar{u}_{xx}$ and $(h\bar{u})_{xx}$. It was not possible to calculate these derivatives from the actual values of $\bar{u}$ because the standard deviation in calculating the means was of the same order of magnitude as the difference between the mean values at adjacent positions. This resulted in the second derivatives being wildly inaccurate. The curve has been used to smooth out this essentially statistical variation, and as the fit is so good it is not thought to introduce any significant errors.

One can now look at the parabolic velocity profiles calculated from the experimentally obtained values of $\bar{u}$, $\bar{u}_{xx}$ and $(h\bar{u})_{xx}$ together with the actual measured profiles. Figure 8 shows this at 5cm steps along the wave.

There are several things to notice from Figure 8: firstly the agreement in general is quite good and particularly so near the bed. For $x = 1.20$cm the poor agreement is due to an error introduced by trying to fit the polynomial approximation, used to calculate $\bar{u}_{xx}$, near the edges of the data. There is a tendency for the parabolic profile to exceed the measured values in the near-surface region of the crest. This effect is even more noticeable further up the beach where the wave is more non-linear. This is shown in Figure 9 for a 3s wave approaching breaking on a 1:100 slope, the vectorplot of which is shown in Figure 5. The same effect also occurs on the 1:30 slope.

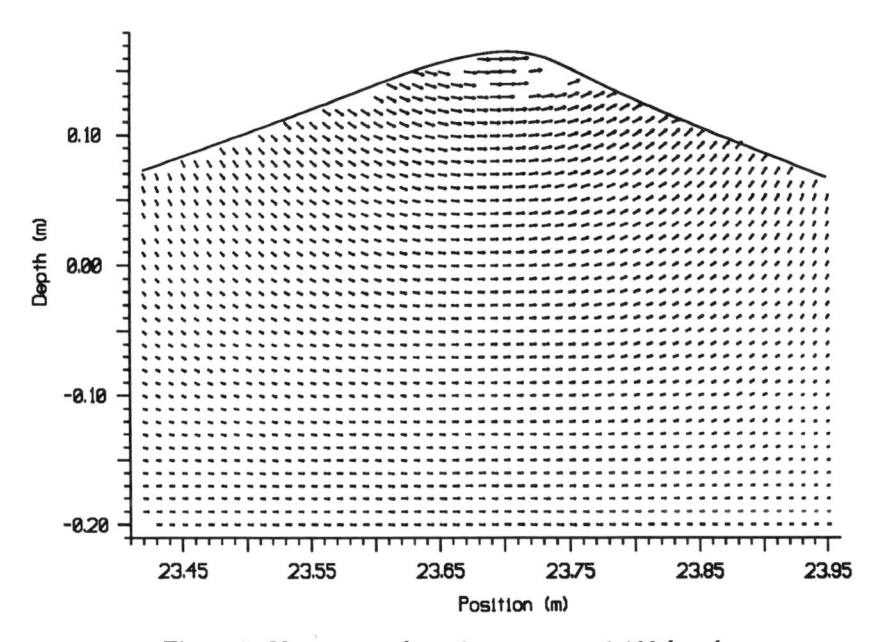

Figure 5: Vector map for a 3s wave on a 1:100 beach

The next step is to see how well the Boussinesq and Serre models can predict the depth-averaged horizontal velocities. The models were given the same wave input parameters, the bottom slope and initial depth for the 1:30 beach. Figure 6 shows the comparison of the distribution of the depth-averaged horizontal velocity given by the Boussinesq and Serre models with the measured values from the vector map in Figure 3.

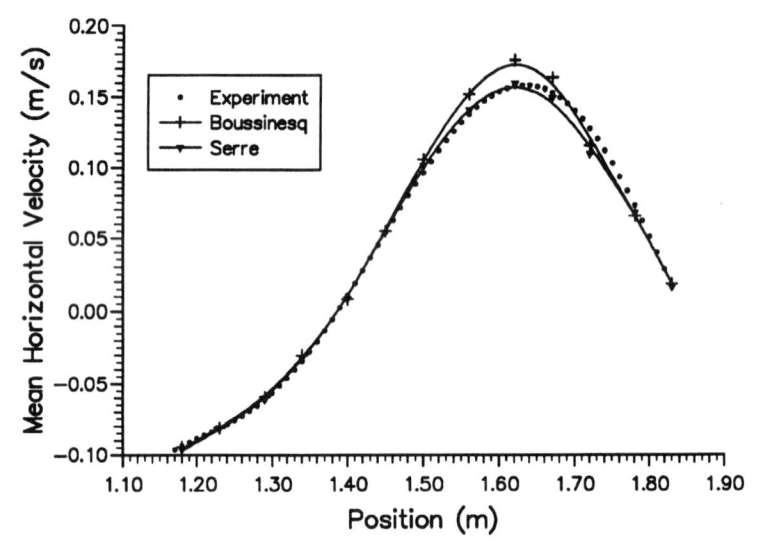

Figure 6: Comparison with the Boussinesq and Serre Models on a 1:30 Beach

In order to calculate the velocity profiles from the models' predictions a least squares polynomial was used, once again. This is also shown in Figure 6. The velocity profiles, calculated in the same way as before are now shown in Figure 10.

Doing the same for a position further up the beach we get the comparison shown in Figure 7 and the velocity profiles shown in Figure 11.

Conclusions

PIV results have been used to test the equation used by the Boussinesq and Serre models to provide a profile of velocity up through the wave.

In general the comparison of a parabolic profile calculated from experimental values agreed fairly well with the measured profiles. The agreement was particularly good near the bed.

There was a consistent overestimate of the measured velocity by the parabolic profile in the near-surface region of the crest of the wave, which appears to get worse as the wave steepens. The degree to which this discrepancy increases with non-linearity has yet to be quantified. One can briefly envisage one of the problems; Equation 1 is derived assuming the same range of validity as the

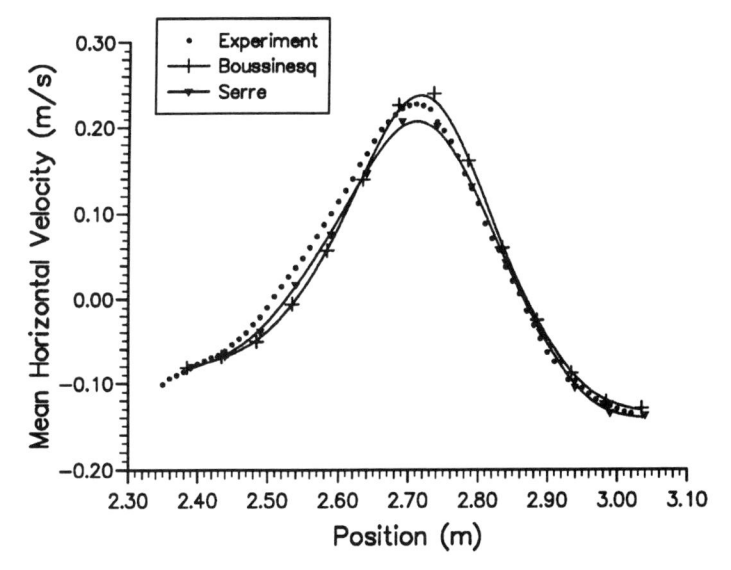

[Figure 7: Comparison with the Boussinesq and Serre Models on a 1:30 Beach]

Boussinesq and Serre models ie., they include terms of order $(kh)^2$, but they ignore those of order $(a/h)(kh)^2$, where a is the wave amplitude; however, as one approaches breaking $a/h = O(1)$ and so the neglected terms are of the same order of magnitude as the included ones.

The comparison of the Boussinesq and Serre model predictions of the depth-averaged horizontal velocity with the measured values shows an overestimate of the peak velocity by the Boussinesq model and a very close estimate for the Serre model at the first position shown. In the second position, further up the beach, the Boussinesq still overestimates the peak value but the Serre now underestimates it. In both positions the models provided a slightly sharper mean velocity distribution than that given by the measurements.

The velocity profiles calculated from the Boussinesq and Serre model predictions did not show particularly good agreement. This is due to the difference between the predicted and measured spatial distributions of the depth-averaged horizontal velocity, and manifests itself by a shift along the x-axis of the models' profiles with respect to the experiment's. The shape of the curves predicted by the models is similar to that of the experimental profiles, but it still tends to overestimate the curvature of the parabola leading to an overestimate of the near-surface velocity in the crest region. This indicates the sensitivity of the variable $\bar{u}_{xx}$ because small differences in the predicted and measured distributions of $\bar{u}$ can lead to large differences in the parabolic profile. Only the near-bed velocities were adequately modelled on a consistent basis.

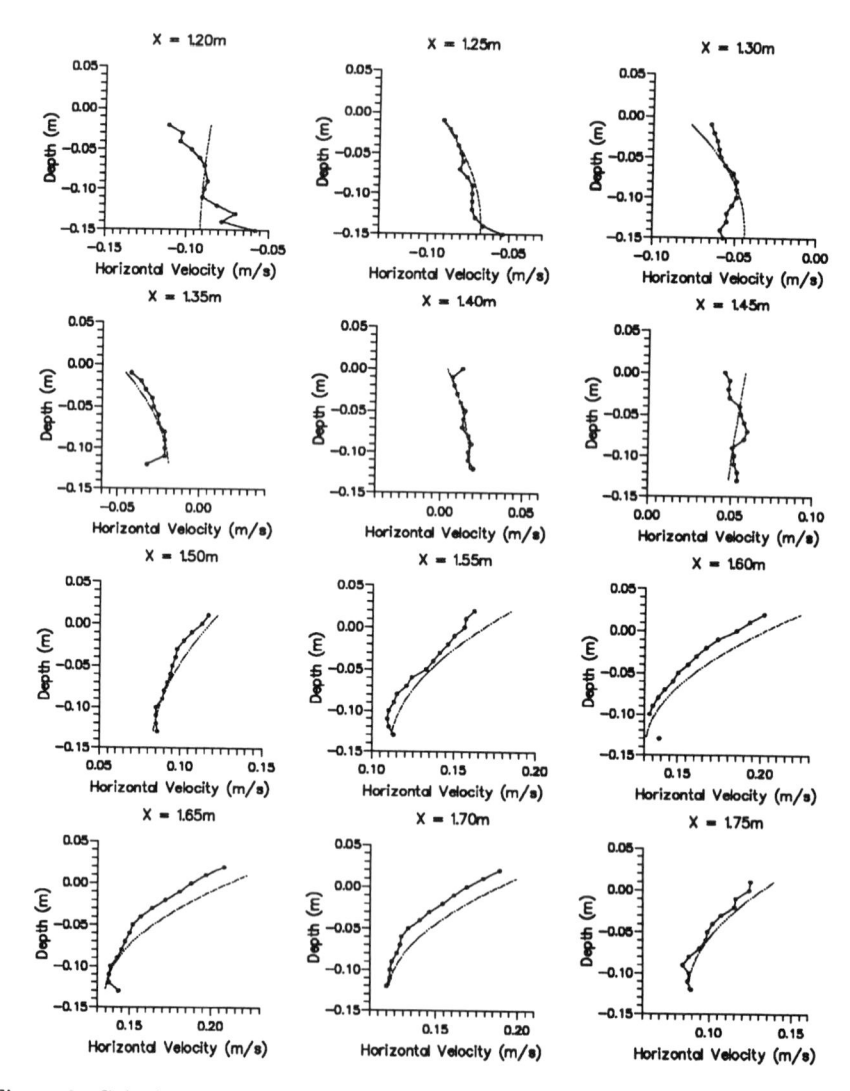

Figure 8: Calculated and Measured Velocity Profiles at Different Horizontal Positions (x) in the 1s wave on a 1:30 Beach.

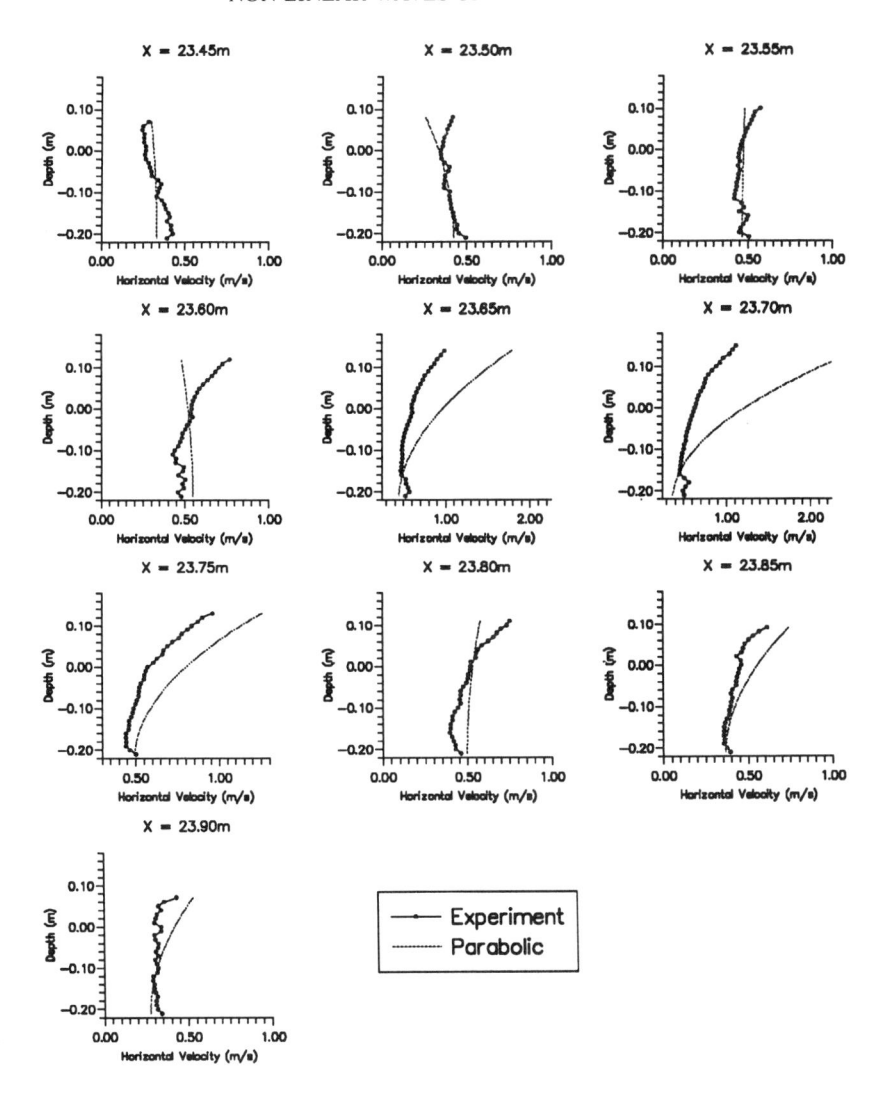

Figure 9: Calculated and Measured Velocity Profiles at Different Horizontal Positions (x) in the 3s wave on a 1:100 Beach.

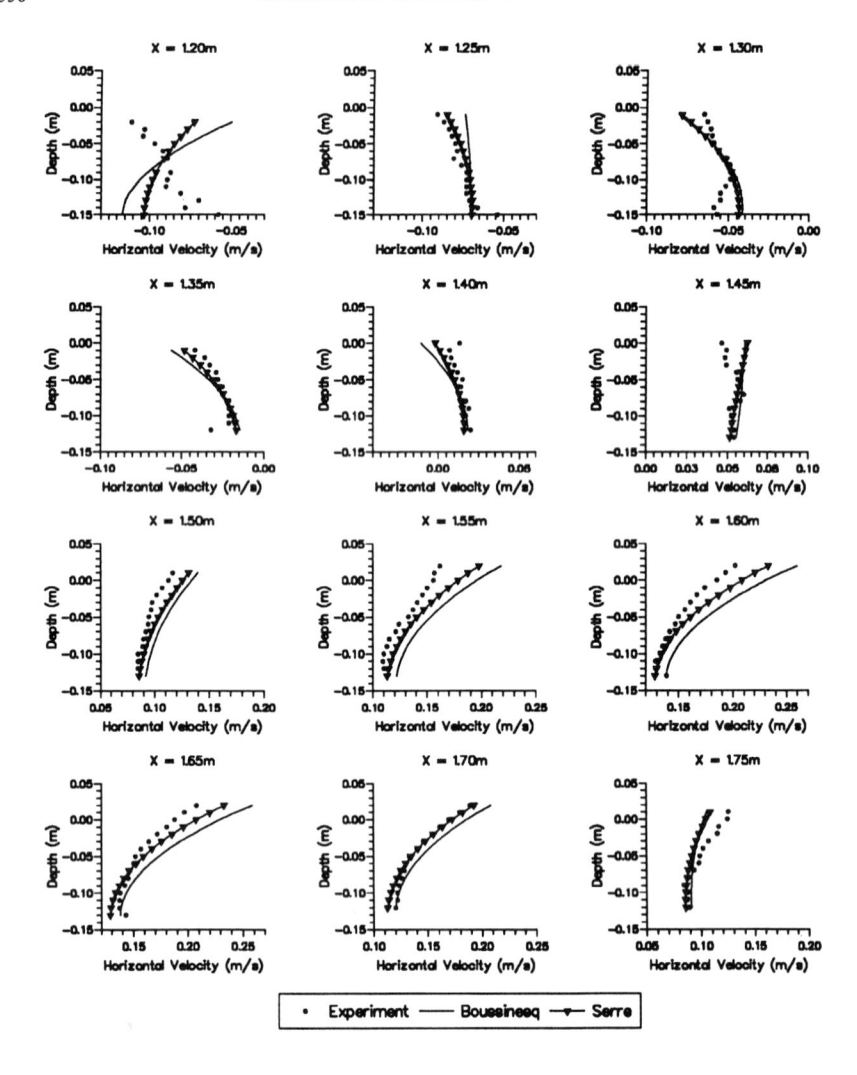

Figure 10: Modelled and Measured Velocity Profiles at Different Horizontal Positions (x) in the 1s wave on a 1:30 Beach.

NON-LINEAR WAVES ON MILD SLOPES 551

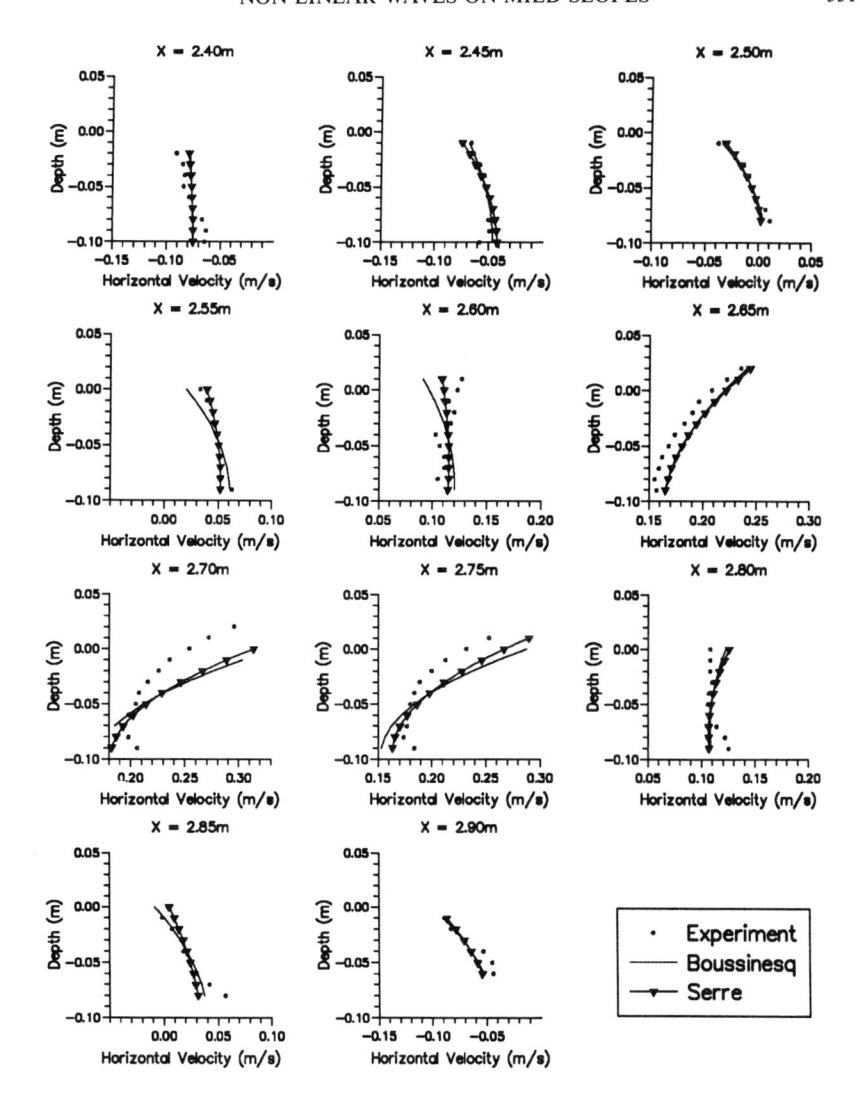

Figure 11: Modelled and Measured Velocity Profiles at Different Horizontal Positions (x) in the 1s wave on a 1:30 Beach.

Acknowledgements

This work was undertaken as part of the MAST G8-M Coastal Morphodynamics research programme. It was funded by the Commission of the European Communities Directorate General for Science, Research and Development under contract N°. EC MAST-II OCT 92 0027; their support is greatly appreciated.

The authors also wish to extend their gratitude to the technicians, Frank Morris, Mauro Gioli and Muzio Mascherini, who played an essential role in conducting the experiments in Italy with such a tight time limit.

References

Adrian, R.J. (1991) *Particle Imaging Techniques for Experimental Fluid Dynamics.* Ann. Rev. Fluid Mechanics, 23:261-304.

Brocchini, M., Drago, M. & Iovenitti, L., (1992) *The modelling of short waves in Shallow Waters. Comparison of numerical models based on Boussinesq and Serre equations.* Proc. 23rd Int. Conf. Coastal Eng. 4:76-88.

Dingemans, M.W. (1994a) *Boussinesq Approximations* Proc. EC MAST G8-M Workshop, Gregynog, Wales. (Abstract in Depth)

Dingemans, M.W. (1994b) *Water Wave Propagation over Uneven Bottoms.* World Scientific, Singapore. (To be published)

Greated, C.A., Skyner, D.J. and Bruce, T., (1992) *Particle Image Velocimetry (PIV) in the Coastal Engineering Laboratory.* Proc. 23rd Int. Conf. Coastal Eng. 15:212-225.

Powell, K.A., Quinn, P.A. and Greated, C.A., (1992) *Shingle Beach Profiles and Wave Kinematics.* Proc. 23rd Int. Conf. Coastal Eng. 181:2358-2369.

Quinn, P.A., Skyner, D.J., Gray, C., Greated, C.A. and Easson, W.J., (1993) *A Critical Analysis of the Particle Image Velocimetry Technique as applied to Water Waves.* In *Flow Visualization and Image Analysis*, Ed. F.T.M. Nieuwstadt. Kluwer Academic Publishers, Dordrecht, The Netherlands.

Salter, S.H., (1982) *Absorbing Wave Makers and Wide Tanks.* Proc. Conf. Directional Wave Spectra Applications, ASCE, 185-200.

CHAPTER 41

The Concept of Residence Time for the Description of Wave Run-Up, Wave Set-up and Wave Run-Down

Holger Schüttrumpf[1], Hendrik Bergmann[1], Hans-Henning Dette[2]

Abstract

Wave run-up, wave set-up and wave run-down are determined by using the residence time in the form of a duration curve. This method introduced by FÜHRBÖTER and WITTE (1991) considers the totality of a run-up series without considering individual run-up events separately. By using the residence time concept, wave run-up is not any more a timeless event but can be described as a time dependent variable. The results obtained by using this concept are in good agreement with previous studies based on other methods for defining wave run-up, wave set-up and wave run-down. The residence time concept has been used for regular and random waves. This concept is particularly useful for random waves, because there is no need to count the number of waves loading a dike.

1. Introduction

In the past different methods were used to define wave run-up, wave set-up and wave run-down in the swash zone. Furthermore these methods didn't consider the time the slope is covered by water during an individual run-up event. This led to the concept of the residence time developed by FÜHRBÖTER and WITTE (1991) which allows to define wave run-up, wave set-up and wave run-down. FÜHRBÖTER and WITTE verified this concept for a 1:6 slope and for wave run-up. But the concept of the residence time also allows to obtain results for wave

---

[1] Assistant Scientist Dipl.-Ing., Department of Hydrodynamics and Coastal Engineering, Leichtweiss-Institute for Hydraulics, Technical University of Braunschweig, Germany

[2] Academ. Director Dr.-Ing.

set-up and wave run-down. Therefore, the concept of the residence time has been verified with a more extensive amount of data obtained from tests in the Large Wave Flume of Hannover for different slope steepnesses as well as for composite slopes. First results are presented in this paper.

Results for shock pressures obtained by the same tests are reported by GRÜNE and BERGMANN (1994).

2. Experimental Set-Up and Test Conditions

Results of wave run-up tests in the Large Wave Flume (GWK) in Hannover, Germany, for dikes and similar structures are presented. These tests were carried out for uniform slopes with steepnesses of 1:6 and 1:12 as well as for composite slopes with a steepness of 1:3 for the lower slope and with a steepness of 1:6 for the upper slope (Fig. 1). A slope steepness of 1:6 was selected because it corresponds to the steepest slope recommended for seadikes in Germany (EAK, 1993). The composite slopes can be characterized by the elevation of the front edge d_k and the water depth d:

1.) position of the front edge d_k=3.3m, water depth d=4.0m
2.) position of the front edge d_k=3.3m, water depth d=5.0m
3.) position of the front edge d_k=4.5m, water depth d=4.0m
4.) position of the front edge d_k=4.5m, water depth d=5.0m

This yields ratios of d_k/d=0.825, 0.660, 1.125 and 0.900 for the composite slopes. For the two uniform slopes the water depth was d=5.0m. The slopes in the wave flume were composed of an asphalt concrete layer covering a sand core. For these experiments a wave run-up gauge was used (GRÜNE, 1982). All the tests were carried out using PIERSON-MOSKOWITZ and JONSWAP wave spectra covering a range between H_S/L_0=0.001 and H_S/L_0=0.031 (H_S = significant wave height, L_0 = deep water wave length).

3. The Concept of the Residence Time

Wave run-up and wave run-down occur in the swash zone (Fig. 2). The highest point represents the maximum wave run-up and the deepest point the maximum wave run-down. The elevation of the mean water level (MWL) over still water level (SWL) at the dike is defined as the maximum landward wave set-up. These three processes are described below using the residence time concept.

The wave run-up on a slope can be given as a function of time. The run-up height can be found by the difference between the highest run-up level on the slope and the mean water level (MWL) using common methods like a crest to crest, a trough to trough or an upcrossing method. The wave run-up is treated as a time dependent event. The time during which a step on the slope is covered by water provides the loading time and the time of overtopping can be derived.

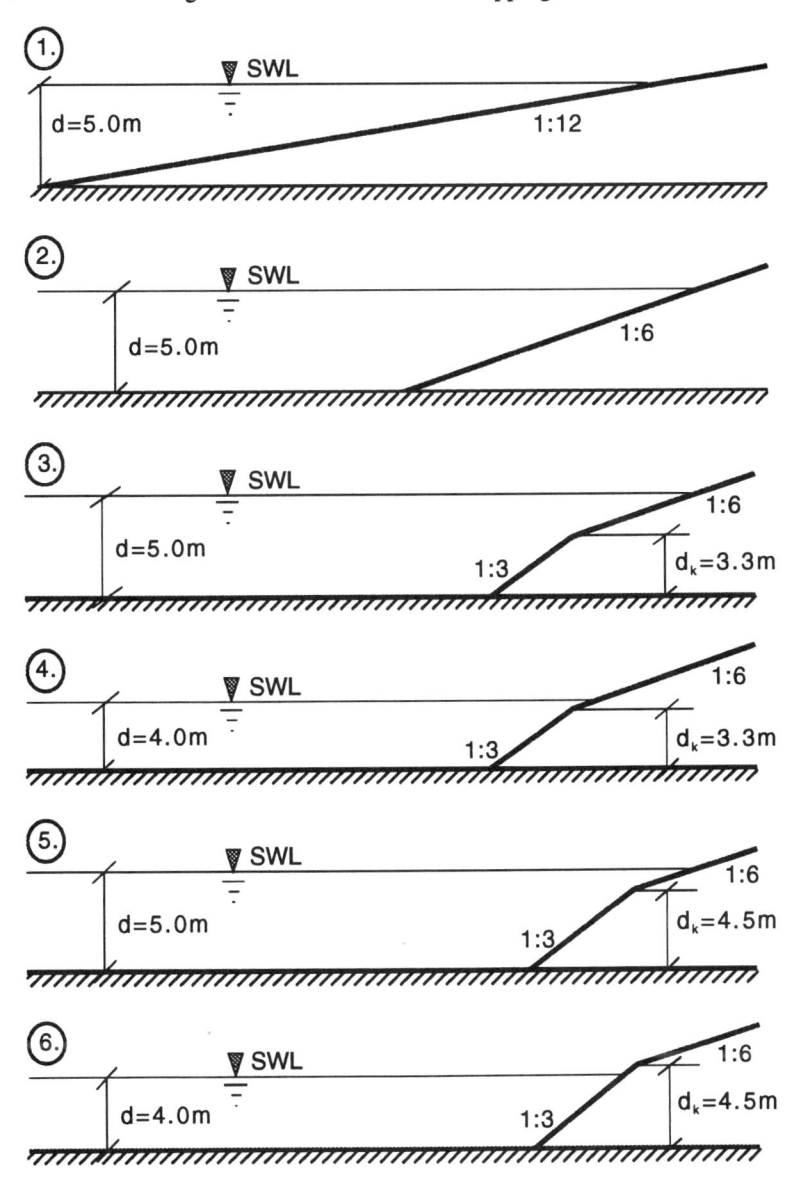

Fig. 1:   Slope Configuration of the Model Dikes

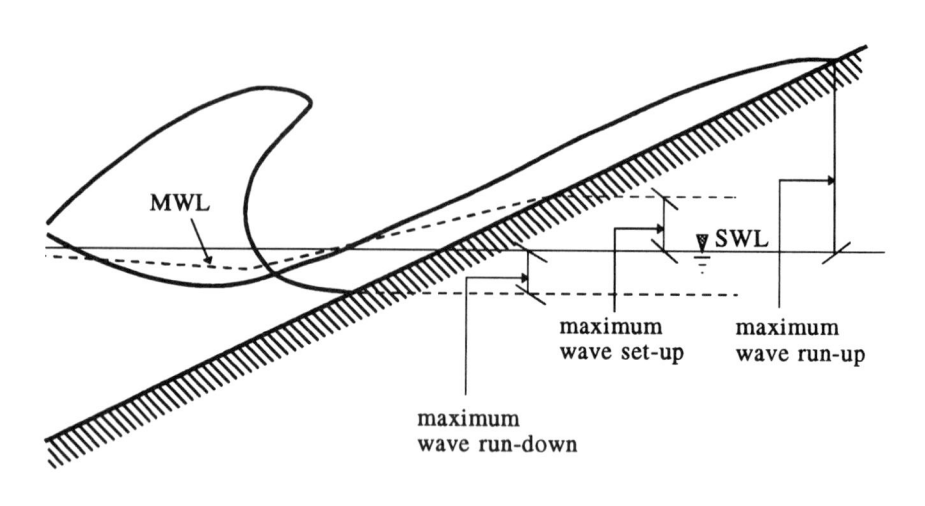

Fig. 2: Swash Zone - Definition Sketch -

The concept of residence time is illustrated by Fig. 3. In the left part the wave run-up is plotted versus time. The maximal wave run-up and the maximal wave run-down can easily be found. The time $\Delta t(r)_i$ during which a fixed step on the slope is covered by water can be calculated for each wave run-up i. The sum of the individual times $\Delta t(r)_i$ leads to the percental residence time D_r for each step r on the slope. The right part of Fig. 3 shows wave run-up as a function of the residence time D_r obtained by the residence time concept.

The residence time D_r can be calculated by the following equation:

$$D_r = \frac{1}{t} \left(\sum_{i=1}^{i=n} \Delta t(r)_i \right) \cdot 100 \quad [\%] \tag{1}$$

t = total duration of the record

The residence time is limited by the extreme values D=0% and D=100%.

D=0% corresponds to the highest run-up level r(D=0%) and D=100% to the deepest run-down level r(D=100%). The points r(D=0%) and r(D=100%) are shown on Fig. 3.

In analogy to the 2% run-up exceedance level $R_{u2\%}$ (2% of the run-up events exceed this run-up height by an upcrossing method) mainly used by other authors (WASSING, 1942, FÜHRBÖTER and WITTE, 1989, VAN DER MEER and JANSSEN, 1994) a r(D=2%) run-up height is defined. The run-up level r(D=2%) is exceeded within 2% of the total duration t. The wave run-down height is indicated by r(D=98%), exceeded within 98% of the total duration. Between these two extreme values wave set-up is defined as the wave-induced vertical elevation of the mean water level (MWL) above still water level (SWL) (GOURLAY, 1992). The value r(D=50%) fits well with this definition because the method of the residence time is highly time dependent.

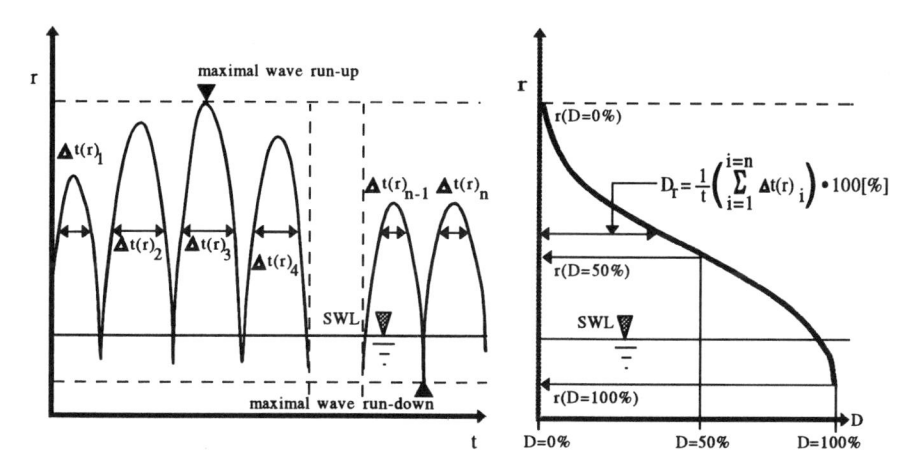

Fig. 3:   Definition of the Residence Time

The difference between r(D=2%) and r(D=98%) yields the loaded area on the slope:

$$r_{AB} = r(D=2\%) - r(D=98\%) \qquad (2)$$

An example for a duration curve is shown in Fig. 4 for a composite slope with a front edge at d_K=3.3m, a water depth d=4.0m, a wave height H_S=0.75m and a wave period T_p=8.0s. Wave run-up r(D=2%), wave run-down r(D=98%) and wave set-up r(D=50%) are also indicated.

4. Wave Run-Up, Wave Run-Down and Wave Set-Up Described by the Residence Time Concept

Wave run-up, wave run-down and wave set-up are dependent on the wave parameters (wave height H_S and wave period T_p) and the slope angle α. Using the wave steepness H_S/L_0 (H_S = significant wave height in front of the structure, L_0 = deep water wave period corresponding to T_p) and the slope angle α, the breaking wave can be described by the surf similarity parameter ξ_0 (BATTJES, 1974):

$$\xi_0 = \frac{\tan \alpha}{\sqrt{H_S / L_0}} \qquad (3)$$

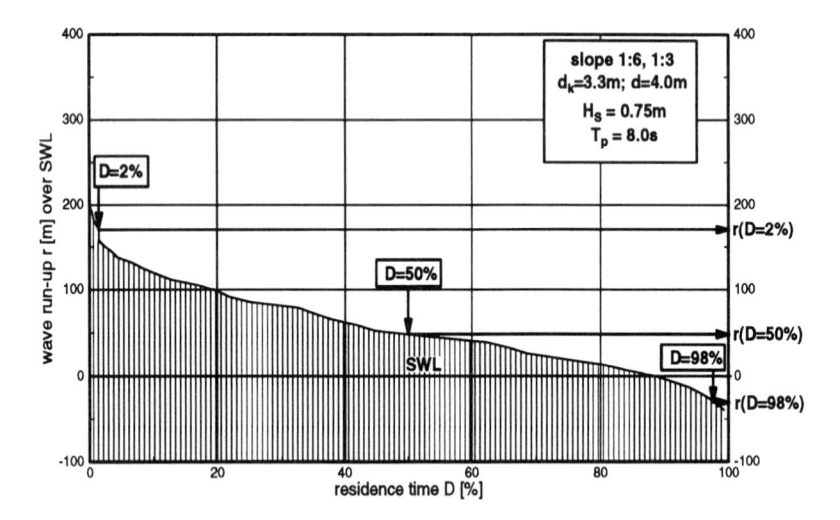

Fig. 4: The Residence Time as a Duration Curve for a Test on a Composite Slope (H_S=0.75m, T_p=8.0s)

The equivalent slope angle α' for the composite slopes is calculated using SAVILLE's method (SAVILLE, 1958) which has been verified by several authors (e.g. MAYER and KRIEBEL, 1994). This leads to an equivalent surf similarity parameter ξ_{eq} for composite slopes.

$$\xi_{eq} = \frac{\tan \alpha'}{\sqrt{H_S / L_0}} \qquad (4)$$

Relative wave run-up $r(D=2\%)/H_S$, relative wave run-down $r(D=98\%)/H_S$ and relative wave set-up $r(D=50\%)/H_S$ can be related to the surf similarity parameter ξ_{eq} by using the following expression:

$$\frac{r(D)}{H_S} = A + C \cdot \xi_{eq}^{E} \qquad (5)$$

A, C, E = coefficients

Fig. 5 shows a synopsis of the results for relative wave run-up, relative wave run-down and relative wave set-up versus surf similarity parameter ξ_{eq}.

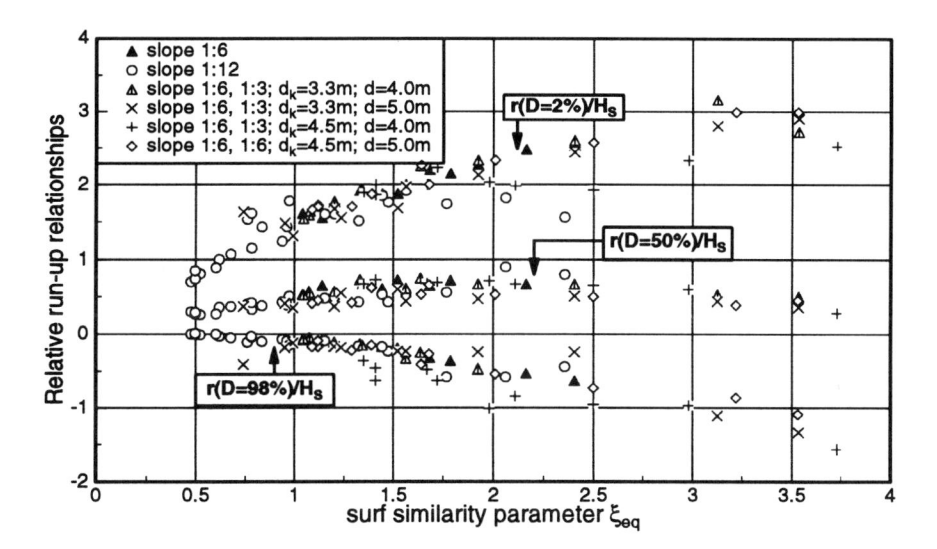

Fig. 5: Wave Run-up, Wave Set-up and Wave Run-down for Smooth Uniform and Composite Slopes

Wave Run-up

A large number of contributions to wave run-up have been published in the past. In most of them the recommended design formulas are extensions of the wave run-up formula proposed by HUNT (1959). Extensive studies were carried out in the Netherlands by VAN DER MEER and JANSSEN (1994) for uniform and composite slopes who suggested the following formula:

$$\frac{R_{u2\%}}{H_S} = 1.5 \, \xi_{eq} \qquad (6)$$

with a maximum relative run-up height of 3.0.

By using the concept of residence time wave run-up becomes smaller due to the slightly different definition of wave run-up. Wave run-up can be calculated by

$$\frac{r(D=2\%)}{H_S} = 1.44\ \xi_{eq} \qquad (7)$$

for small surf similarity parameters $\xi_{eq}<1.0$ (Fig. 6).

For $\xi_{eq}>1.0$ relative wave run-up can be expressed by the following formula:

$$\frac{r(D=2\%)}{H_S} = 1.44 + 0.75\ (\xi_{eq}-1.0) \qquad (8)$$

It can be seen that equations (7) and (8) lead to smaller wave run-up heights than equation (6) by VAN DER MEER and JANSSEN (1994). Therefore the ratio $r(D=2\%)/R_{u2\%}$ is plotted against surf similarity parameter ξ_{eq} in Fig. 7. The average deviation is about 15% with an increasing difference for higher surf similarity parameters ($\xi_{eq}>1.0$) and a smaller difference for smaller surf similarity parameters ($\xi_{eq}<1.0$).

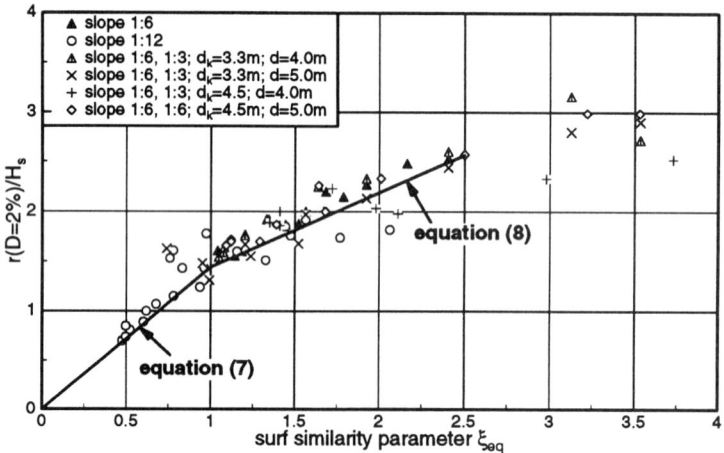

Fig. 6: Wave Run-up on Smooth Slopes

Wave Set-up

For wave set-up in the swash zone most of the known studies were performed for very smooth slopes (in GOURLAY, 1992) between $\tan\alpha=0.022$ and $\tan\alpha=0.100$.

[Figure 7: Ratio r(D=2%)/R_{u2%} versus surf similarity parameter ξ_{eq}]

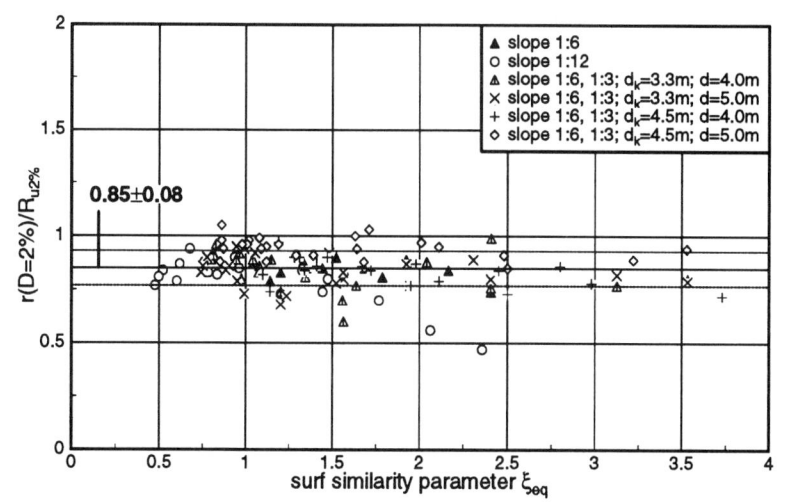

Fig. 7: Ratio $r(D=2\%)/R_{u2\%}$ versus surf similarity parameter ξ_{eq}

GOURLAY (1992) presented the following formulas for maximum set-up η_m based on field data:

$$(\eta_m - \eta_b)/H_b = 0.31 \pm 0.05 \quad \text{for } 0.083 < \tan\alpha < 0.1 \quad (9)$$

and

$$(\eta_m - \eta_b)/H_b = 0.14 \pm 0.03 \quad \text{for } 0.022 < \tan\alpha < 0.04 \quad (10)$$

η_b = Wave set-down at the breaking point

Previous studies based on field measurements by YANAGISHIMA and KATOH (1990) for a beach slope of 1:60 yield the following relationship between maximum wave set-up η_m, significant deep water wave height H_{0S} and surf similarity parameter ξ_0:

$$\frac{\eta_m}{H_S} = 0.267 \, \xi_0^{0.4} \quad (11)$$

In the present paper data for relatively steep slopes between $\tan\alpha=0.083$ (1:12) and $\tan\alpha=0.333$ (1:3) are presented. Fig. 8 shows a relation between the relative wave set-up $r(D=50\%)/H_S$ and the surf similarity parameter ξ_{eq}.

For $0.5<\xi_{eq}<2.5$ the relative wave set-up can be expressed by the following formula:

$$\frac{r(D=50\%)}{H_S} = 0.45\ \xi_{eq}^{0.56} \qquad (12)$$

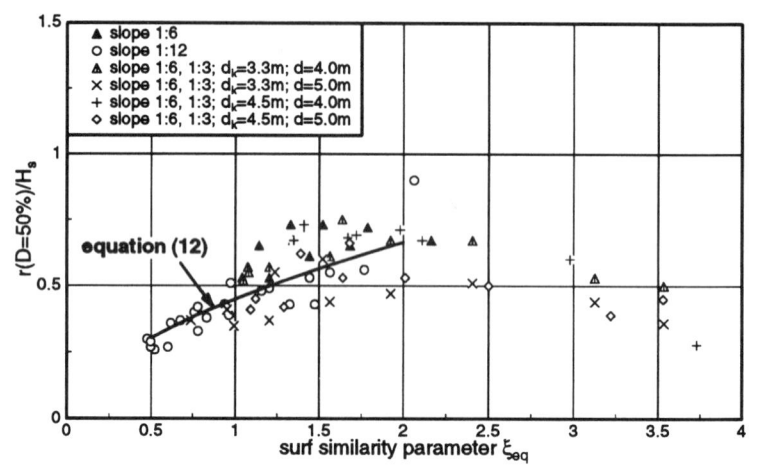

<u>Fig. 8:</u>   Wave Set-up on Smooth Slopes

Wave Run-down

Only very few investigations on wave run-down have yet been performed. VAN DER MEER and BRETELER (1990) presented large-scale model results for wave run-down r_d on a 1:3 smooth slope and found the following relationship for similarity parameters ξ_0 in the range between 2.0 and 4.3:

$$\frac{r_d}{H} = 0.1\ \xi_0^2 - \xi_0 + 0.5 \qquad (13)$$

The concept of residence time can also be used for wave run-down. The relationship between $r(D=98\%)/H_S$ and ξ_{eq} is shown in Fig. 9.

For $0.5<\xi_{eq}<2.5$ the following formula can be established:

$$\frac{r(D=98\%)}{H_S} = -0.1\ \xi_{eq}^{2.21} \qquad (14)$$

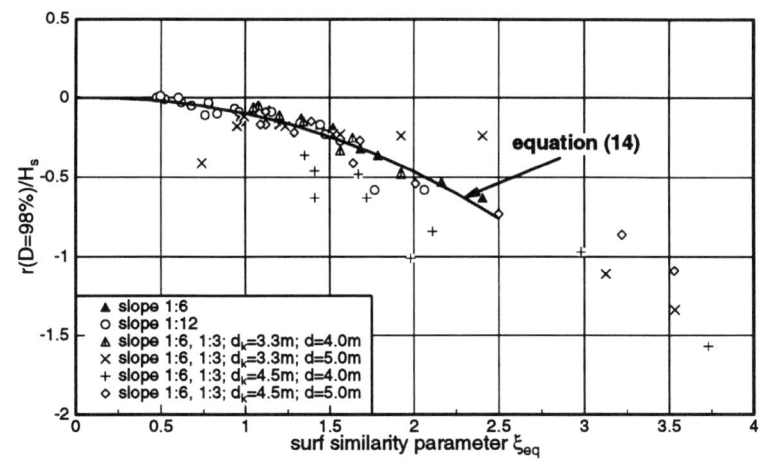

Fig. 9: Wave Run-down on Smooth Slopes

Due to a different definition of wave run-down by using the concept of residence time the obtained results are likewise smaller than the results by VAN DER MEER and BRETELER (1990).

5. Concluding Remarks

Wave run-up, wave run-down and wave set-up have been determined by using the concept of residence time in form of a duration curve. This method is highly time dependent. First results obtained by the concept of residence time have been compared to standard formulas, showing that this concept leads to:

- smaller wave run-up heights
- heigher wave set-up heights
- smaller wave run-down heights

Acknowledgements

This study is a part of the basic research Project A1 on sea dikes initiated by Prof. Führböter within the Coastal Engineering Research Unit "SFB 205". The support by the "Deutsche Forschungsgemeinschaft" (DFG), Bonn, is gratefully acknowledged.

6. References

BATTJES, J.A. (1974). Surf Similarity. Proc. 14th Coastal Eng. Conf.. Copenhagen

EAK (1993). Empfehlungen Arbeitsausschuß Küsteningenieurwesen. Die Küste, No. 45

FÜHRBÖTER, A., SPARBOOM, U. and WITTE, H.-H. (1989). Großer Wellenkanal Hannover: Versuchsergebnisse über den Wellenauflauf auf glatten und rauhen Deichböschungen mit der Neigung 1:6. Die Küste. No. 50.

FÜHRBÖTER, A. and WITTE, H.-H. (1991). Die dynamische Verweilzeit und ihre Beziehungen zum Wellenauflauf und -ablauf an einer Deichböschung 1:n=1:6. Die Küste. No. 52.

GOURLAY, M.R. (1992). Wave set-up, wave run-up and beach water table: Interaction between surf zone hydraulics and groundwater hydraulics. Coastal Engineering, No. 17.

GRÜNE, J. (1982). Wave Run-up caused by Natural Storm Surge Waves. Proc. 18th Coastal Eng. Conf., Kapstadt

GRÜNE, J. and BERGMANN, H. (1994). Wave loads on seadykes with composite slopes and berms. Proc. 24th Coastal Eng. Conf., Kobe

HUNT, I.A. (1959). Design for Seawalls and Breakwaters. Proc. ASCE, Vol. 85, No. WW3

MAYER, R.H. and KRIEBEL, D.L. (1994). Wave run-up on composite-slope and concave beaches. Proc. 24th International Conference on Coastal Engineering. Kobe, Japan

SAVILLE, T. (1958). Wave run-up on composite slopes. Proc. 6th Conf. on Coastal Eng., Gainesville

VAN DER MEER, J.W. and BRETELER, M.K. (1990). Measurements and computation of wave induced velocities on a smooth slope. Proc. 22nd Conf. on Coastal Eng.. Delft

VAN DER MEER, J.W. and STAM, C.-J. M. (1992). Wave runup on smooth and rock slopes of Coastal Structures. Journal of Waterway, Port, Coastal and Ocean Engineering, Vol. 118, No. 5

VAN DER MEER, J.W. and JANSSEN, J.P.F.M. (1994). Wave run-up and wave overtopping at dikes and revetments. Publications Delft Hydraulics. No. 485

WASSING, F. (1957). Model investigations of wave run-up on dikes carried out in the Netherlands during the last twenty years. Proc. 6th ICCE. Gainesville, Florida

YANAGISHIMA, S. and KATOH, K. (1990). Field observation on wave set-up near the shoreline. Proc. 22nd Conf. on Coastal Eng.. Delft

CHAPTER 42

Bottom Shear Stresses under Random Waves
with a Current superimposed.

Richard R.Simons[1]
Tony J.Grass[2]
Wameidh M.Saleh[3]
Mehrdad M.Tehrani[3]

Abstract

This paper describes experiments performed with the UCL shear plate device to make direct measurements of the bottom shear stress vector under the action of combined waves and currents. The three corresponding velocity components have also been recorded simultaneously, enabling the results to be expressed in terms of a friction factor. The scope of the work extends that of earlier tests on regular waves, and includes three sequences of random waves propagating above a fixed rough bed in still water and over two orthogonal currents.

The results show that, for the range of conditions considered, the addition of an orthogonal current has no discernable effect on the amplitude of the shear stress time series or on the friction factors used to characterize the complete sequence. If a single friction factor is used to describe the shear stress throughout a sequence of random waves, and if that factor is calculated from the RMS of the shear stress during the sequence scaled on an equivalent regular wave with a bottom orbital velocity amplitude of $u_{rms} \cdot \sqrt{2}$, then the results agree well with earlier observations from tests on regular waves, and f_w can be predicted from standard formulae.

---

[1]Senior Lecturer, Civil & Environmental Engineering Dept, UCL

[2]Professor, Civil & Environmental Engineering Dept, UCL.

[3]Research Fellow, Civil & Environmental Engineering Dept,
   University College London, Gower Street, London WC1E 6BT.

Introduction

The prediction of shear stresses at the sea bed is an important requirement for coastal engineers wishing to estimate wave energy loss, current strength, and sediment transport around our coastline. These stresses are applied both by wave action and by currents, so any non-linear interaction between the two scales of motion in a combined wave-current flow makes the prediction of bottom shear stress a non-trivial task.

There are over thirty different theories that have been developed taking into account non-linear interaction when predicting bed shear stresses under combined wave-current conditions. Many of these have been reviewed by Soulsby et al. (1993) in a paper that also brought together existing data sets against which the behaviour of the theories could be tested. But whereas this paper concentrated on monochromatic wave conditions, other researchers have tried to address the problem of currents interacting with a **random** wave sequence. For instance, Lee (1987) proposed a linearization technique for bed shear stress in order to predict bottom frictional dissipation under irregular waves; and, in the light of a recent field study under random wave conditions, Black and Gorman (1994) suggested a modified form of the Christoffersen and Jonsson model. Zhao and Anastasiou (1993) have presented a more rigorous approach to the problem, deriving a theory based on the regular wave model of O'Connor and Yoo (1988) - itself built on the work of Bijker (1966). Equations were derived separately for wave-dominated and current-dominated conditions as well as for the general case, and comparisons were made with published data. Ockenden and Soulsby (1994) have considered the same problem but with the specific requirement of estimating sediment transport rates under irregular waves and currents. They put forward a simple solution based on an equivalent regular wave ($u_b = \sqrt{2}.u_{rms}$, T_p = peak spectral period), and accounted for non-linear effects by adopting a parameterised version of the desired wave-current model as presented in Soulsby et al. (1993). And, finally, Madsen (1994) has developed a modified version of the Grant and Madsen (1979) model to be applied when irregular waves propagate over a current.

While there is increasing attention being paid to the theoretical prediction of bed friction under irregular waves on a current, experimental results from tests under combined wave-current conditions in general relate to monochromatic waves only, and relatively little work has been done under the practical case of random waves on a turbulent current. One of the reasons for this is the difficulty of making direct measurements of bed friction generated by unsteady flows, and shear stresses have often had to be deduced from modified velocity profiles or energy principles. However, Simons et al. (1992) have reported direct measurements of bottom shear stress using a novel shear plate device deployed in a large wave basin when regular waves are propagated orthogonally across a turbulent current, and the present paper extends that work to include sequences of random long-crested waves propagating across the same currents.

Wave Basin

The tests were performed in a wave basin measuring 20m by 18m, designed for a water depth of 1.5 m but with a raised central test area, 9 m by 6 m, over which the still water depth was reduced to 700 mm. This plateau area was coated with a fixed layer of sand (nominal diameter of 2 mm) to produce a uniform rough boundary with a Nikuradse roughness, k, of 1.5 mm - see Simons et al. (1992).

Ten ram-type wave generators were mounted along one wall of the basin. Each ram could be operated under independent control to produce waves with periods between 1 s and 3 s, and with heights up to 300 mm. The other three walls supported permeable beach units 2.5 m long round the perimeter of the basin. This beach system was constructed of synthetic "hairlock" sheets on a rigid frame, with a slope of 15^0 down to half-depth but with a vertical permeable face below that level. To avoid reflections of the relatively shallow water waves from this area, the 15^0 slope on the wall facing the wave generators was continued down with shingle "fill" to meet the raised bed in the centre of the basin.

Currents were introduced through a set of gate valves under the beaches in one of the side walls, flow being removed through a corresponding set of openings in the other side. The current strength was controlled by adjusting the speed of a pump which circulated water through a 2-compartment channel round the perimeter of the basin.

Instrumentation

For successful completion of the work, it was important that direct observations could be made of the shear stresses exerted on the bed of the basin by both currents and waves. A shear plate device had been developed in a preceding project exactly for this purpose, employing a circular "active" element supported on 4 thin columns and operating in sway mode in response to the instantaneous force vector. The system was described by Simons et al. (1992), and the same instrument was used in the present tests.

To determine the velocity field in the three-dimensional flow created by intersecting waves and currents, measurements were made using an ultrasonic current meter (UCM) capable of yielding three velocity components simultaneously. The transmitters on this instrument "pinged" at 100Hz, giving a response time of 1/30s and a resolution of 1mm/s in a range up to 1 m/s. The UCM was used to record the flow field immediately above the wave boundary layer (to correlate with the shear plate measurements), and up to the water surface. However, its size (with a measuring volume 15 mm in diameter) meant that it was unable to provide detailed information within the relatively thin wave boundary layer.

Table 1: Observed waves and current test conditions

Run Code:	Tz (s.)	Hsig (cm)	$U_{50}m$ curr. cm/s	U_{rms} wave cm/s	τ_m curr. N/m^2	τ_{rms} wave N/m^2	a/k	f_w
CU2			11.3	1.7	0.04	0.01		
CU1			20.1	2.7	0.08	0.02		
WR1P	1.28	14.5	-0.4	7.2	0.00	0.30	13.8	0.058
WR1CC	1.28	14.4	10.5	7.4	0.03	0.32	14.2	0.058
WR1C	1.29	14.5	19.5	7.8	0.08	0.35	15.1	0.058
WR2P	1.49	14.2	0.7	8.9	0.00	0.36	19.9	0.045
WR2CC	1.48	14.4	10.8	9.2	0.03	0.36	20.4	0.043
WR2C	1.50	14.7	19.7	9.6	0.08	0.38	21.6	0.041
WR3P	1.29	17.2	-0.8	8.9	0.00	0.37	17.2	0.047
WR3CC	1.29	17.2	10.6	9.3	0.04	0.37	18.0	0.043
WR3C	1.31	17.5	18.9	9.5	0.09	0.42	18.7	0.047

The water surface elevation was monitored over a 2m by 2m area centred above the shear plate using a square array of 16 resistance-type wave monitors. Data were sampled at approximately 100Hz on 22 channels for test runs lasting approximately 6 minutes.

Test Conditions

Three different "random" wave sequences were used (coded WR1, WR2 and WR3), all generated from a Jonswap target spectrum with peak periods of 1.3s and 1.5s - **fig.1**. Spectra observed at the four wave probes immediately surrounding the shear plate (**fig.2**) show that there were significant reflections in the basin at discrete frequencies, but these did not hinder the main objective of the research which was to correlate observed bottom shear stresses to the velocities immediately above the bed.

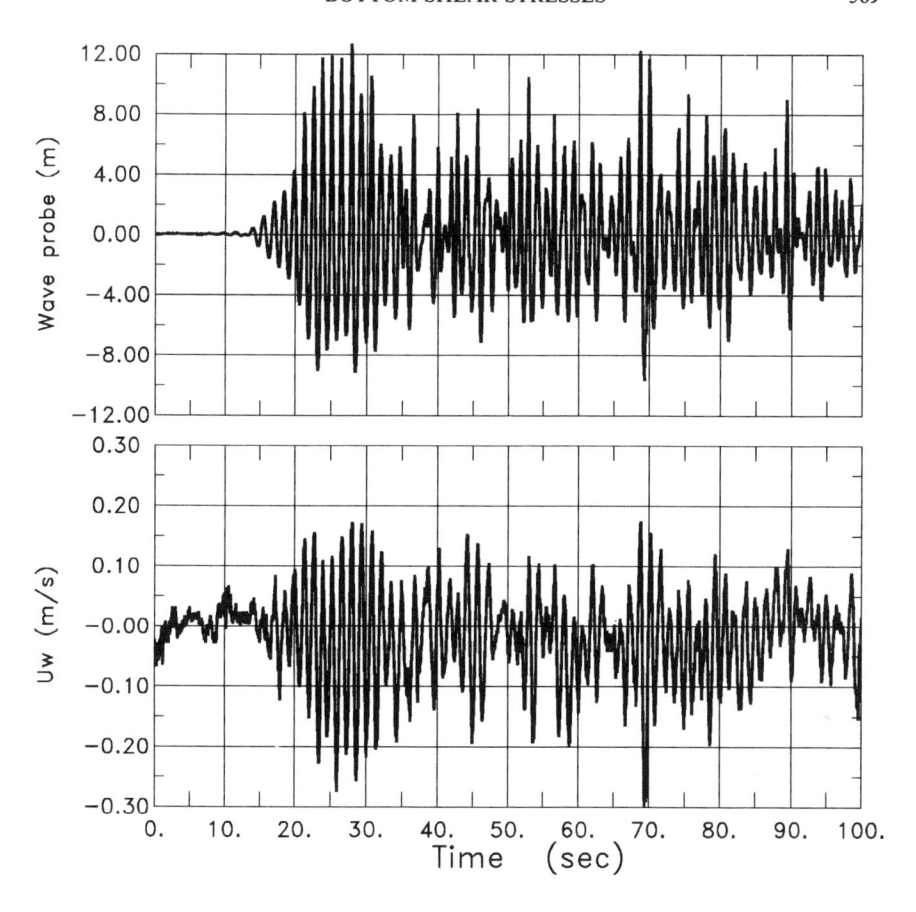

Figure 1 Sample of random wave sequence WR1 with strong current added.

Each wave spectral sequence was tested under three flow conditions: firstly through initially still water, secondly over a current with a mean velocity of 0.1 m/s, and finally over a current with a mean velocity of 0.2 m/s. The two currents were also tested without waves. Test conditions are listed in **Table 1**.

The wave sequences were reproducable, so that each test could be repeated with the UCM positioned at different heights above the bed. In this way velocity data were obtained at 18 positions through the flow depth for each test condition, while at the same time recording bottom stresses and wave surface elevations as a check on repeatability. The wave signal generator was not directly synchronised with the data recording system, so comparison between time-series from different tests required a small manual time-shift.

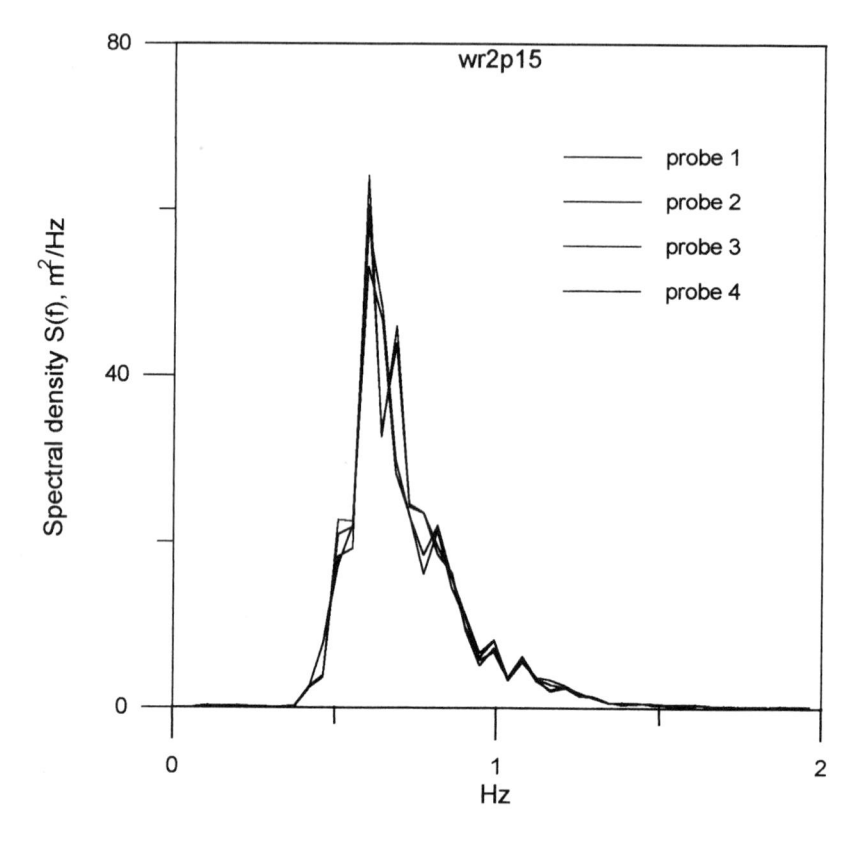

[Figure: Spectral density S(f) in m²/Hz vs Hz, titled "wr2p15", showing overlaid spectra from probe 1, probe 2, probe 3, and probe 4, with a peak around 0.5 Hz reaching approximately 60 m²/Hz.]

Figure 2 Spectra measured at 4 wave probes close to the shear plate.

The repeatability of the wave sequences was confirmed by the ease with which it was possible to overlay the time-series from tests with waves alone and those with orthogonal currents superimposed - as in **fig.3**.

Analysis

Because of the characteristics of the shear plate, it was necessary in the initial analysis of observed "shear stresses" to correct for the wave-induced pressure on the edge of the active plate. Under regular wave conditions it is possible (assuming an appropriate wave theory) to infer the pressure field across the bed of the basin from the fluid acceleration at a single point just outside the wave boundary layer above the centre of the plate. This procedure becomes more questionable under

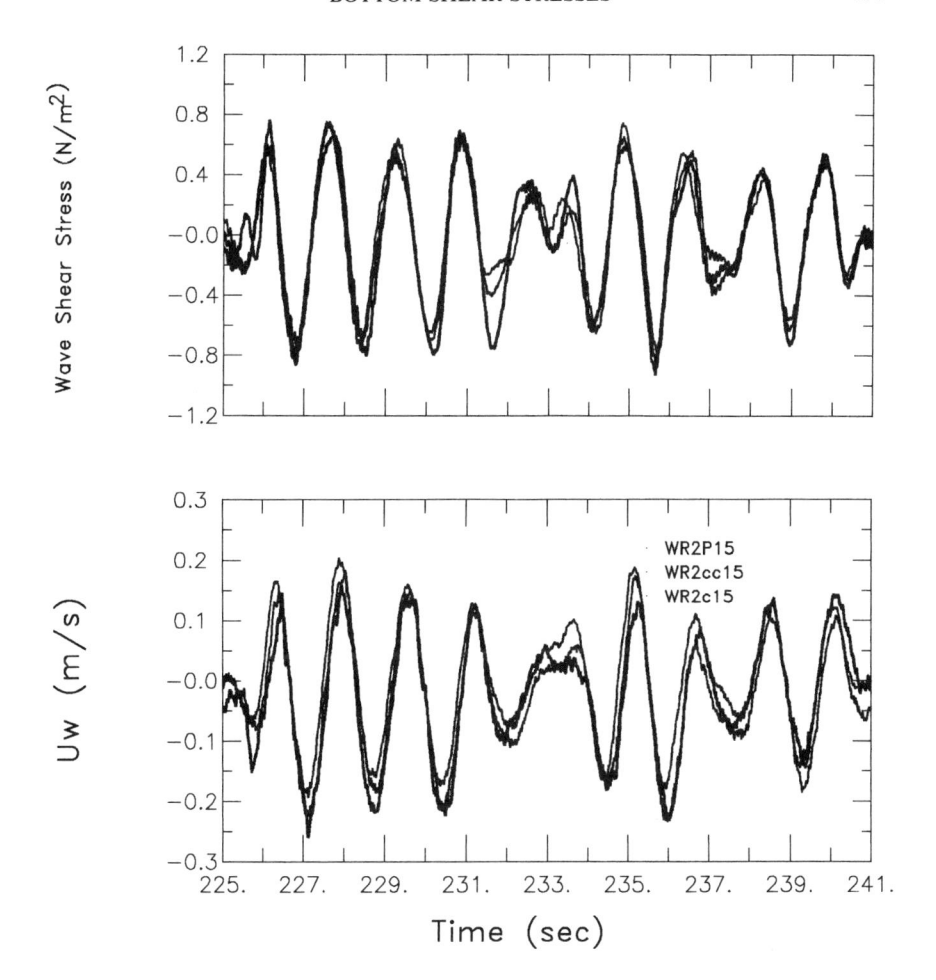

Figure 3 Wave-induced velocities and shear stresses for random waves:
a) in still water, b) with weak current, c) with stronger current.

irregular wave conditions, and in the present tests the calculation was simplified by taking the pressure gradient to be constant across the whole plate area.

It was also possible to apply a further correction to account for the pressure/inertia force being applied to the sand grain roughness attached to the plate. Although this force is a real effect, acting to generate sediment transport and dissipate wave energy, its consequences are far greater in laboratory scale experiments (with the ratio of wave length to grain diameter approximately 10^3) than in the field (ratio greater than 10^5), and might distort the interpretation of bottom forces. However, it was largely this force component that gave the "shear stress" a significant phase lead over the near-bed velocity.

Analysis of the random wave tests was performed both wave-by-wave, comparing maximum and minimum shear stresses and velocities during each wave crest and wave trough, and also in terms of the RMS of these quantities for the complete wave sequence (neglecting the first 20s start-up period), measured under the three different current conditions.

Results

Table 1 summarises results from the 11 test runs considered in the present paper, recording the changes in wave characteristics induced by the addition of currents, and vice versa. The effect of adding random waves to the mean flow was less significant than in the earlier tests using regular waves, but from the mean velocity profiles **(fig.4)** it was still possible to discern a reduction in velocity in the outer flow, with a corresponding increase closer to the bed. Experimental procedures meant that measurements at different heights above the bed for the same current flow were carried out on different days, so the scatter in mean velocity data can probably be attributed to the difficulty in resetting the current. The scale of the alteration in mean velocity profile can be judged from the negligible change in mean bed shear stress in the current direction sensed by the shear plate when the waves were superimposed.

The friction factor f_w and a/k ratio were both calculated from the root mean square properties of the full random wave sequences. The equivalent regular wave used in the normalisation procedure was chosen to have the period of the spectral peak T_p and an amplitude yielding the same RMS variation as that recorded in the random wave sequences:

$$f_w = \frac{\tau_{rms}}{\frac{1}{2}\rho(u_{rms}\sqrt{2})^2} \quad : \quad \frac{a}{k} = \frac{u_{rms}\sqrt{2}.T_p}{2\pi k}$$

Fig.5 shows these values of friction factor in comparison with previous data and theories for predicting wave-induced friction. The agreement is very good, and suggests that the use of $u_{rms}\sqrt{2}$ as the scaling velocity is an appropriate choice. It is also apparent that the data from all 9 tests lie very closely clustered and thus the addition of the two currents has made no significant difference to the wave-induced shear stress. Such a lack of enhancement is in agreement with the earlier results using regular waves (Simons et al. 1992) and with the work of Arnskov et al. (1993).

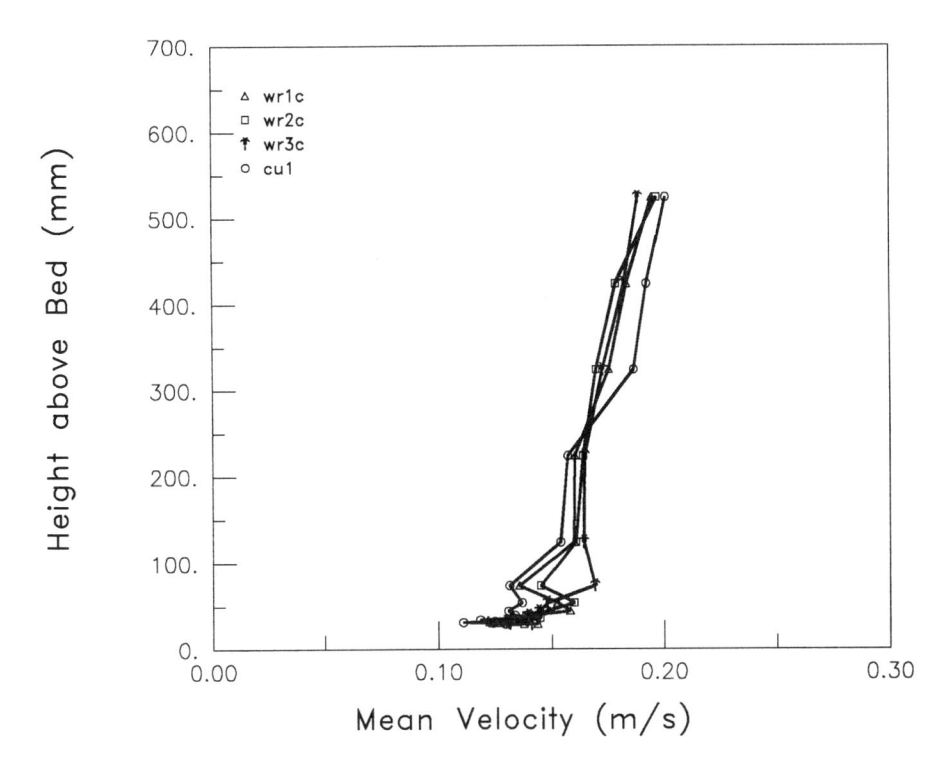

Figure 4 Mean velocity profiles for current alone and with random waves added.

While it is sometimes helpful to be able to characterise a random sea in terms of regular wave parameters, it is more important in assessing sediment transport to be able to predict particular "events" when the wave-induced shear stress exceeds that needed to initiate movement of the seabed material. To give an insight into the wave-by-wave behaviour of random waves, each of the present sequences of velocities and shear stresses was considered as independent half-cycles (between velocity zero-crossings). Friction factors were calculated from the half-cycle amplitude of shear stress (between consecutive maxima and minima) and the corresponding amplitude of wave-induced velocity, and a typical data sample is shown in **fig.6**. This method avoided the anomalous results caused by long waves if absolute maximum and minimum values were used in the calculations, when significant shear stresses could apparently be induced by infinitesimally small velocity fluctuations. It was also decided to ignore all values for waves with an amplitude less than 1.5 mm, as this was not felt to be within the measuring accuracy of the UCM.

Fig.7 shows the friction factors produced from one of the random wave sequences,

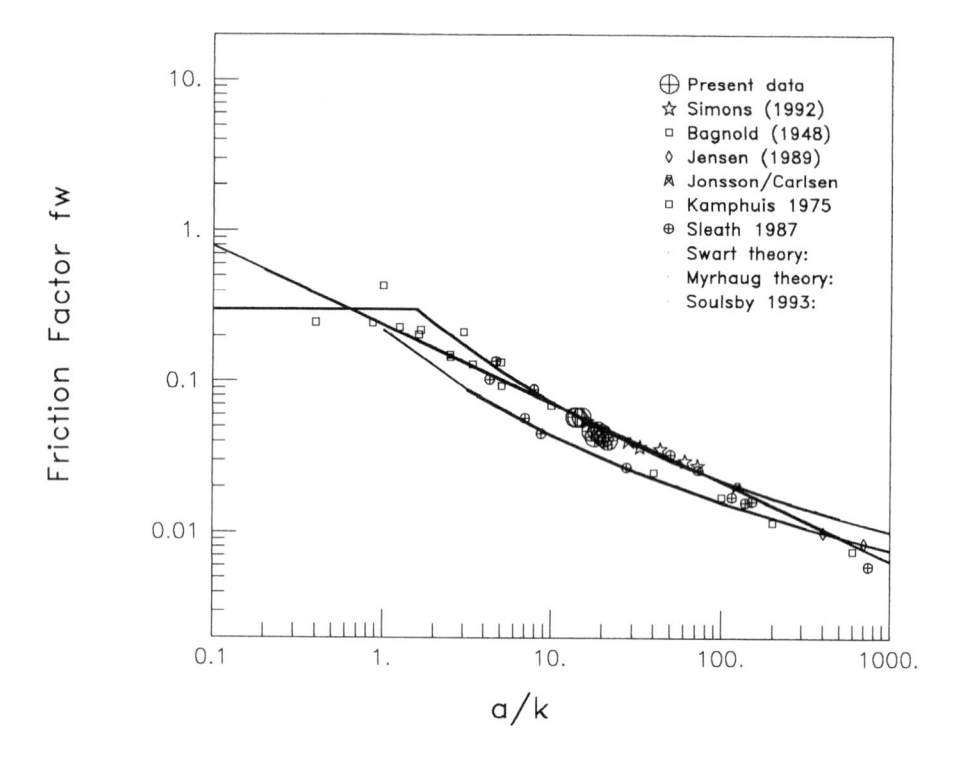

Figure 5 Variation of friction factor with relative bed excursion for random waves with/without current superimposed: (uses equivalent wave velocity $u_{rms}\sqrt{2}$)

firstly propagating through still water, then with the weaker current superimposed, and finally with the stronger current flowing. The first thing to note is that for low a/k, the friction factors are greater than predicted by theories for fully rough turbulent flow, but that as a/k increases, so the friction factor falls back in line with those formulae - Soulsby et al. (1993) for instance. While the underprediction of f_w at low a/k is at first sight worrying, it can almost certainly be attributed to the relatively low oscillatory Reynolds numbers [ua/υ] associated with these waves, when viscous effects are to be expected. In fact, the trend line formed by the scattered data lies parallel to (and almost exactly a factor of 2 above) the prediction for completely laminar flow over a smooth bed when the friction factor is given as:

$$f_w = \frac{2}{\sqrt{(Re)}} = \frac{2}{\sqrt{(ua/\nu)}}$$

wr1p15

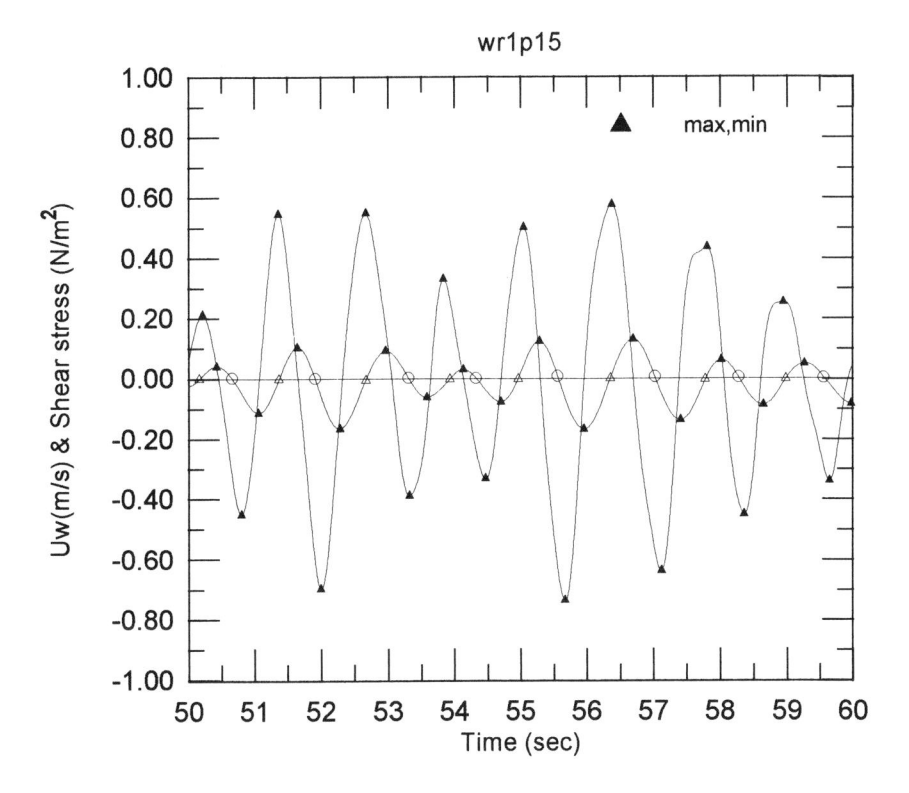

Figure 6 Sample of shear stress and velocity data used to identify friction factors.

Much of the data lies in the rough transitional flow regime, and these values are not out of line with the observations of Kamphuis (1975) for similar conditions.

The main aim of the project was to identify what effect the addition of an orthogonal current would have on the bottom friction, and it is clear from fig.7 that all three sets of data have very similar distributions, implying that the current has caused no obvious change. The same lack of sensitivity to the addition of an orthogonal current was also found for the other two random wave sequences, and confirms the results discussed above for the friction factors based on an equivalent regular wave representative of the complete test run, namely, that the current has little effect.

Although these results indicate that the presence of longshore currents can be ignored when considering wave energy dissipation in the onshore direction, it

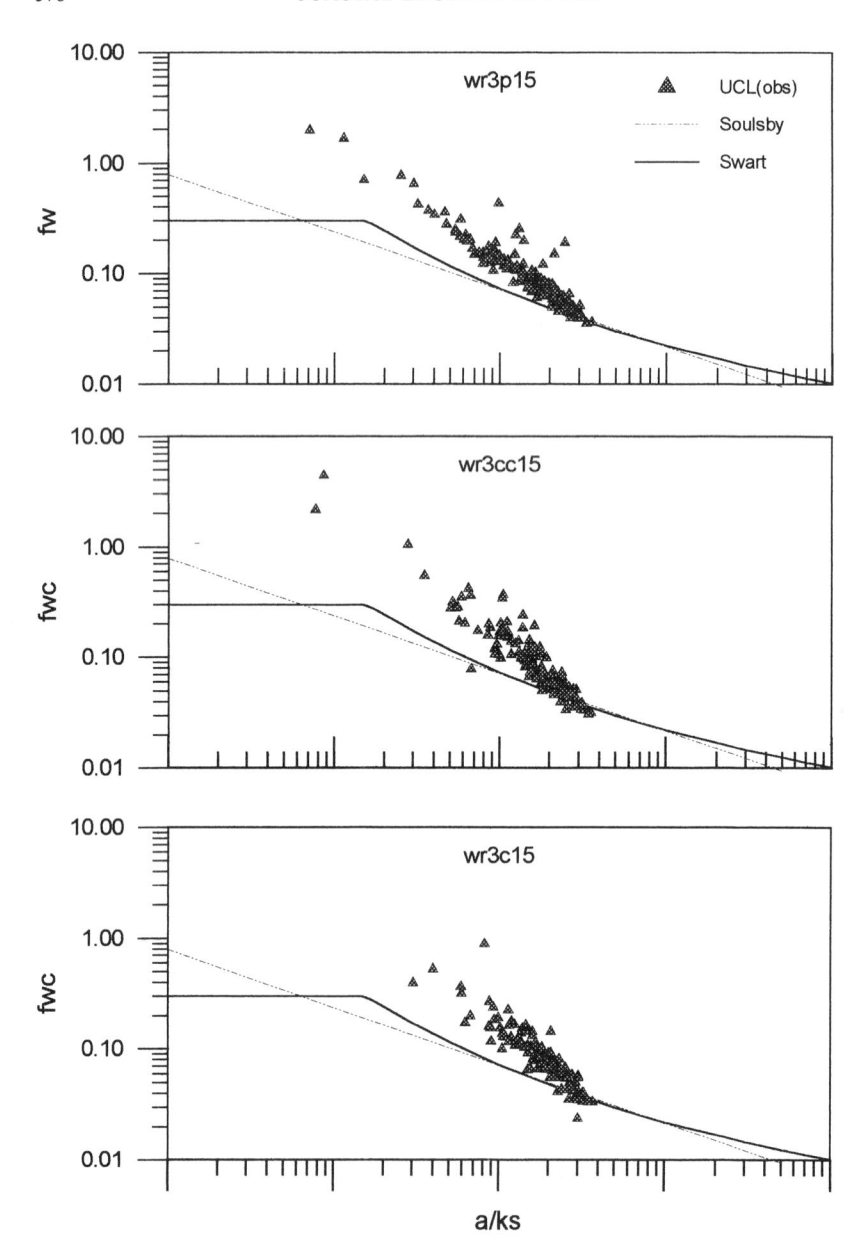

Figure 7 Friction factors for each half-cycle during random wave sequences: a) in still water, b) with weak current, c) with stronger current added.

should be noted that Lodahl et al. (1994) have recently reported results from tests over a smooth boundary. These suggest no increase in oscillatory shear stress when weak currents are added, but that friction factors do increase with the addition of currents at Reynolds numbers very much greater than those reported here. It is not unreasonable to expect that the interaction between a fully turbulent wave boundary layer and a turbulent steady current boundary layer is likely to be qualitatively different from the corresponding interaction at the low Reynolds numbers prevailing in the present tests.

Summary

Tests have being performed with 3 different sequences of Jonswap spectrum random waves, and direct measurements have been made of the bottom shear stresses with the UCL shear plate.

Friction factors have been calculated to represent the complete test sequence, using the RMS of the shear stress. When scaled on an equivalent regular wave with bed orbital velocity $u = u_{rms} \sqrt{2}$ and period $= T_p$, the results correspond to standard predictions for fully rough turbulent wave-induced motion.

Friction factors have also been calculated for each half-cycle during the tests. These reflect the transitional regime in which the data lie, at low a/k indicating higher values than predicted for fully rough turbulent oscillatory flow, but giving good agreement at higher a/k, where the flow approaches the fully turbulent regime.

For the three random wave spectra tested, the addition of a current was found to have little effect either on the wave-induced velocities near the bed or on the oscillatory shear stresses and friction factors.

Acknowledgements

This work was undertaken as part of the G8 Coastal Morphodynamics research programme. It was funded jointly by the Science and Engineering Research Council, and by the Commission of the European Communities Directorate General for Science, Research and Development under contract no. MAS2-CT-92-0027. The authors also acknowledge their gratitude to H.R.Wallingford for the use of their facilities.

References

Arnskov, M.M., Fredsoe, J. and Sumer, B.M. (1993) Bed shear stress measurements over a smooth bed in three-dimensional wave-current motion. Coastal Engineering, 20, pp.277-316.

Bijker, E.W. (1966) The increase of bed shear in a current due to wave motion. Proc. 10th Int. Conf. on Coastal Engng., Vol.1, pp.746-765.

Black, K.P. and Gorman, R.M. (1994) Observation of wave dynamics influenced by wave-current enhanced bed friction on an exposed coast. Victorian Institute of Marine Sciences, St Andrews Place, Melbourne, Australia.

Grant, W.D. and Madsen, O.S. (1979) Combined wave and current interaction with a rough bottom. J.Geophys. Res., 84 (C4), pp.1797-1808.

Kamphuis, J.W. 1975 Friction factors under oscillatory waves. J. Waterway Harbours Coastal Engng. Div., ASCE, 101, pp.135-144.

Lee, D-Y. (1987) Bottom frictional dissipation if irregular waves. Ocean Engng. 9, (1,2), pp.39-44.

Lodahl, C., Sumer, B.M. and Fredsoe J. 1994 Experiments on co-directional waves and currents at large Reynolds numbers. Proc. 24th Int. Conf. on Coastal Engng., Kobe, Japan, Oct.1994. ASCE, paper 82.

Madsen, O.S. (1994) Spectral wave-current bottom boundary layer flow. Proc. 24th Int. Conf. on Coastal Engng., Kobe, Japan, Oct.1994. ASCE, paper 83.

O'Connor, B.A. and Yoo, D. (1988) Mean bed friction calculation of combined wave-current flow. Coastal Engineering, 12, pp. 1-21.

Ockenden, M.C. and Soulsby, R.L. 1994 Sediment transport by currents plus irregular waves. HR Wallingford Report SR 376, February 1994.

Simons, R.R., Grass, A.J. and Mansour-Tehrani, M. (1992) Bottom shear stresses in the boundary layer under waves and currents at right angles. Proc. 23nd Int. Conf. on Coastal Engng, Venice, Italy, Ch.45, pp.604-617

Soulsby, R.L., Hamm, L., Klopman, G., Myrhaug, D., Simons, R.R., and Thomas, G.P. (1993) Wave-Current Interaction within and outside the bottom boundary layer. Coastal Engineering, 21, pp.41-69.

Zhao, Y and Anastasiou, K. (1993) Bottom friction effects in the combined flow field of random waves and currents. Coastal Engineering, 19, pp.223-243.

CHAPTER 43

EFFECTS FROM DIRECTIONALITY AND SPECTRAL BANDWIDTH ON NON-LINEAR SPATIAL MODULATIONS OF DEEP-WATER SURFACE GRAVITY WAVE TRAINS

Carl Trygve Stansberg [1]

Abstract

The non-linear nature of the largest waves in random wave trains is studied. Experimental wave elevation records from measurements in a large wave basin are analysed. Longcrested as well as shortcrested wave conditions, with various spectral shapes, are included. The spatial development of spectral and statistical properties of the wave trains are analysed. The results indicate that closer than 10-15 wavelengths, observed deviations from linear theory may be explained by 2nd order effects. Further away, higher-order modulational instabilities may lead to a stronger increase of extreme waves. This is especially observed in narrow-banded long-crested waves, while observations in short-crested sea show very little content of these space-dependent modulational effects.

1. Introduction.

Extreme waves observed in random wave trains on deep water have in many cases exceeded predicted levels based on linear wave theory and Rayleigh statistics. Examples on this may be found in laboratory experiments (Stansberg 1991, 1992) as well as in full scale records (Sand et. al. 1990, Kjeldsen 1990, Jonathan et. al. 1994, and others). Second-order random wave theory (Longuet-Higgins 1963) explains a significant part of the deviations from linear theory (Marthinsen and Winterstein 1991, Stansberg 1993, 1994, Vinje and Haver 1994). Some of the observed effects, however, need other explanations. Such results have been published in e.g. Stansberg (1992), where experiments with long-crested waves in a large wave basin were reported. Those experiments seemed to indicate that laboratory generated random wave trains travelling more than 10 - 15 wavelengths

---

[1] Senior Principal Research Engineer, Norwegian Marine Technology Research Institute (MARINTEK), P.O.Box 4125 Valentinlyst, N-7002 Trondheim, Norway.

from the wavemaker are modulated due to higher-order non-linear wave-wave couplings. These may then lead to extreme waves clearly higher than estimates from linear as well as from second-order theory. A possible connection between these observations and the well-known "Benjamin-Feir-modulations" in regular waves (Benjamin and Feir 1967, Lake et. al. 1977, Lo and Mei 1985, and others) was discussed on basis of the experiments. Numerical simulations on related problems have shown a similar connection (Wang et. al. 1992, Yasuda et. al. 1992, Yasuda and Mori 1994).

The significance of such higher-order modulational instabilities in random waves is, however, still not quite clear. Although their existence have been more or less documented, it is a question how stable they are in time and space, and how they depend on various parameters such as the spectral shape and the wave directionality. Previous works (see e.g. the experiments by Su et. al. 1982) have indicated, however, that there is a connection between non-linear modulations and wave directionality, and the simulations by Yasuda and Mori (1994) show some dependence on spectral bandwidth. On this background, new systematic and controlled experiments have been carried out with random waves in MARINTEKs Ocean Basin. Long-crested (unidirectional) as well as short-crested (directional) waves were run, with various spectral bandwidths. The present paper describes a preliminary, empirical analysis study based on the measurements. Thus some main results are shown and interpreted in light of the relationships discussed above. The presentation starts with some simple, but illustrating examples in bichromatic waves, and this is then followed by the main part dealing with random waves.

2. Tests with bi-chromatic waves

2.1. Test facility with set-up.

The tests were carried out in MARINTEKs 50mx80m Ocean Basin, which is equipped with a longcrested wavemaker along one full short side, and with a multiflap directional wavemaker along 63m of one long side. Rigid sloped beaches are installed at the opposite sides of the basin. The depth of the adjustable bottom was set to 4m, which corresponds to deep water for the actual waves considered. Wave elevation was measured with wave staffss at a number of locations at different distances from the wavemakers. In the present study, records taken at 10m meters intervals in the actual wave directions are reported.

2.2. Directional effects in bichromatic waves.

As shown in previous experiments (e.g. Stansberg 1992), non-linear wave modulations are rather rapidly developing in long-crested bichromatic wave trains with the 2 frequencies quite close to each other. These modulations, also identified through spatially developing side-bands in the frequency domain, give rise to asymmetry in the wave groups. This develops into groups with extreme individual waves that may become significantly higher than in the initial wave trains. In the new experiments, similar tests were run with longcrested as well as with shortcrested (bi-directional) waves. Examples from the results are shown in fig. 1, including time series samples taken at 3 different distances from the wavemakers. As seen from this, the strong modulations in the longcrested case (which confirms the previous results referred to above) almost disappear completely in the shortcrested case. This is a strong support of the idea that the directionality is an important parameter in the study of non-linear wave modulations.

As a check on the influence from the basin geometry, additional long-crested wave tests were run from the multiflap wavemaker, as well as from the longcrested wavemaker. The characteristic nature of the modulations in fig. 1a was observed in both cases.

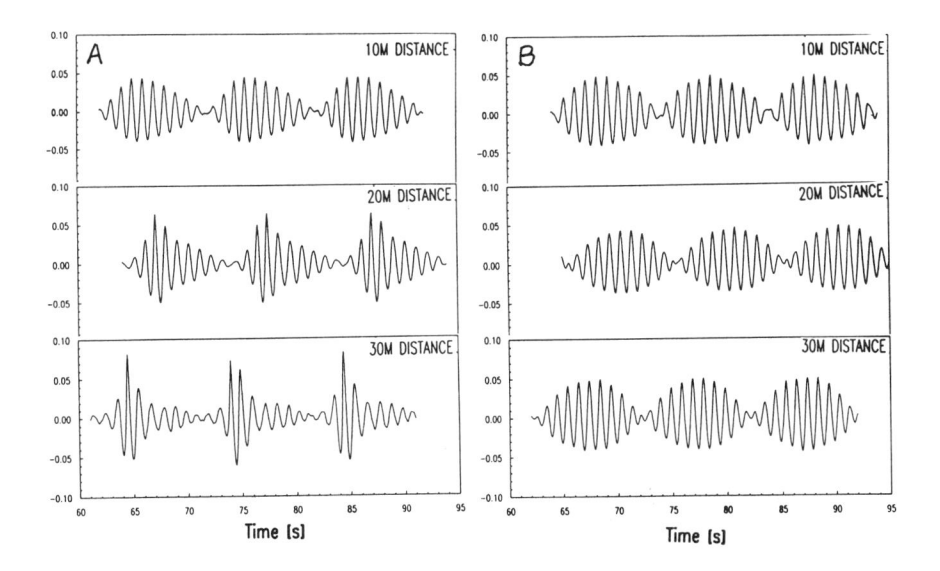

Fig. 1. Wave elevation from bichromatic test records.
A = long-crested. B = short-crested (bi-directional).

3. Tests with random waves.

All wave records presented below are from measurements in the MARINTEK Ocean Basin. For a brief description of the facility and the set-up, see section 2.1 above.

3.1. An example.

The random wave elevation sample records in fig. 2a illustrate how a long-crested wave group develops throughout space, finally resulting in a very large wave event. Similarly, fig. 2b shows how a group propagates in a directional sea. A basic difference is observed between the 2 cases: The long-crested group continues over a long distance, while the short-crested one exists only over a limited area. It seems reasonable to state that the spatial development of non-linear modulations in such wave groups must be different in the 2 cases. Intuitively, one would expect stronger modulational effects in the longcrested case, where these non-linear effects may grow over many wavelengths, than in the shortcrested case, where the wave groups disappear more quickly. (This can be seen in relation to a comment in Taylor and Haagsma (1994), where a simple argument indicates that these wave-wave interactions are less significant in spread sea than in unidirectional sea). Thus the study of non-linear extreme waves is closely connected to the study of non-linear wave group evolution, as indicated in Wang et. al. (1994).

3.2. Brief description of the tests and procedures.

A systematic test program with a large number of different longcrested and shortcrested wave conditions was run. Shortcrested waves were run from the multiflap wavemaker, while longcrested waves were basically run from the other, longcrested wavemaker. Since they generate waves in different directions in thebasin, some longcrested waves were also run from the multiflap wavemaker, as a check on possible noise effects from the basin geometry etc. Here we report mainly on tests with initial spectral peak periods Tp at 1.0 seconds, which corresponds to relatively short waves in this large basin. This is chosen in order to let the waves propagate a large number of wavelengths across the measuring area, for the study of spatially developing non-linear modulations.

The wave elevation was simultaneously recorded at a number of locations in the basin, as described in 2.1 above. Each random wave record includes about 1500 zero-crossing waves, in non-repeating wave trains. This is considered sufficient for the study of non-linear wave statistics. The following presentation focuses on the spatial development of spectral, grouping and statistical properties of the wave records. The dependence on the travelled distance (in number of wavelengths), directionality, spectral bandwidth and wave steepness is emphasized. Wave groups are analysed by the Hilbert envelope technique, where the square wave envelope represents the continous "group signal" (Stansberg 1983).

3.3. Presentation of results.

Figs. 3 & 4 show examples on the spatial development of the elevation power spectra. 3 different long-crested wave conditions are presented in fig. 3, while 2 short-crested conditions are presented in fig. 4. The short-crested conditions are

DIRECTIONALITY AND SPECTRAL BANDWIDTH EFFECTS 583

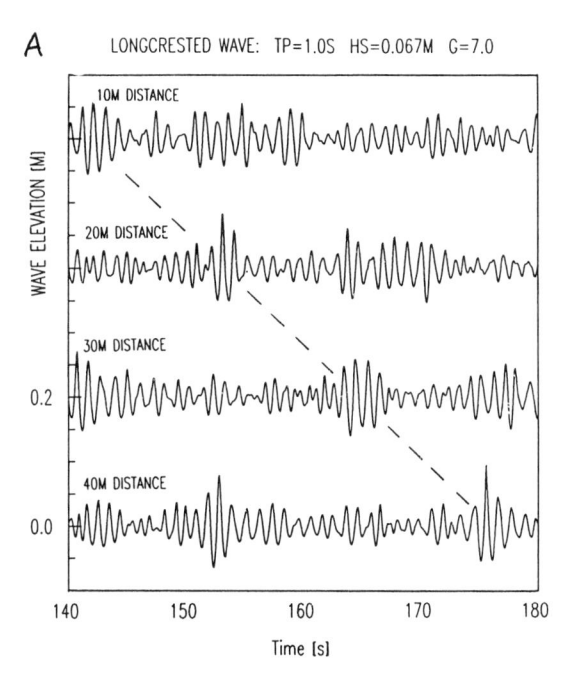

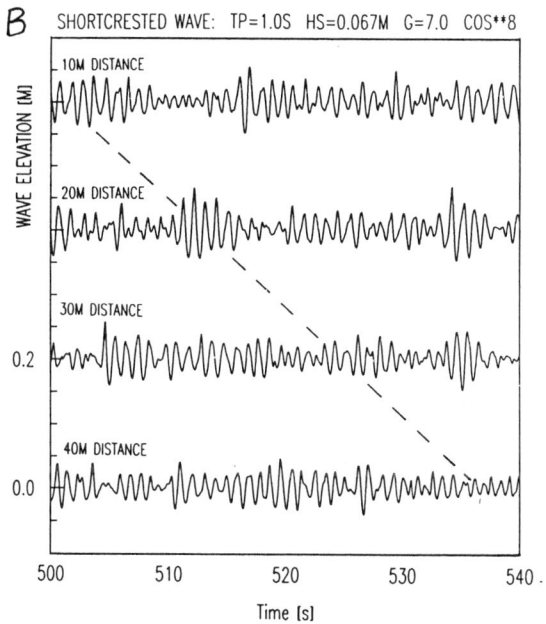

Fig. 2. Sample records illustrating wave group development.

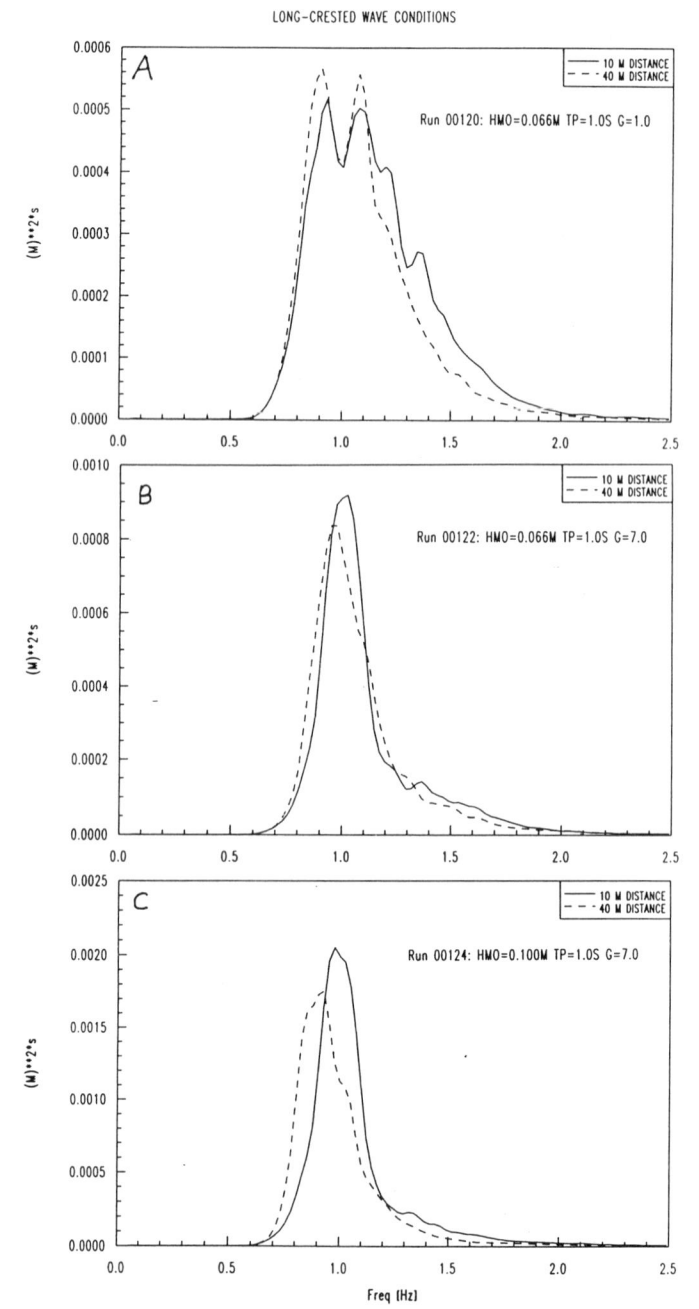

Fig. 3. Power spectra, 3 longcrested wave conditions, each at 2 locations (10 m and 40 m).

comparable, in scalar spectral parameters, to the 2 narrow-banded long-crested spectra. Each plot includes spectra measured both at 6 and 25 wavelengths distances (corresponding to 10 and 40 meters).

With location 10 m, nearest to the wavemaker, defined as the reference signal, amplitude amplification factors at location (40 m) are calculated for the same 5 sea states as in figs. 3 & 4. The results are presented in fig. 5, with 1 plot for longcrested, and 1 plot for short-crested waves. This factor is simply estimated as the square root of the ration between the corresponding spectra.

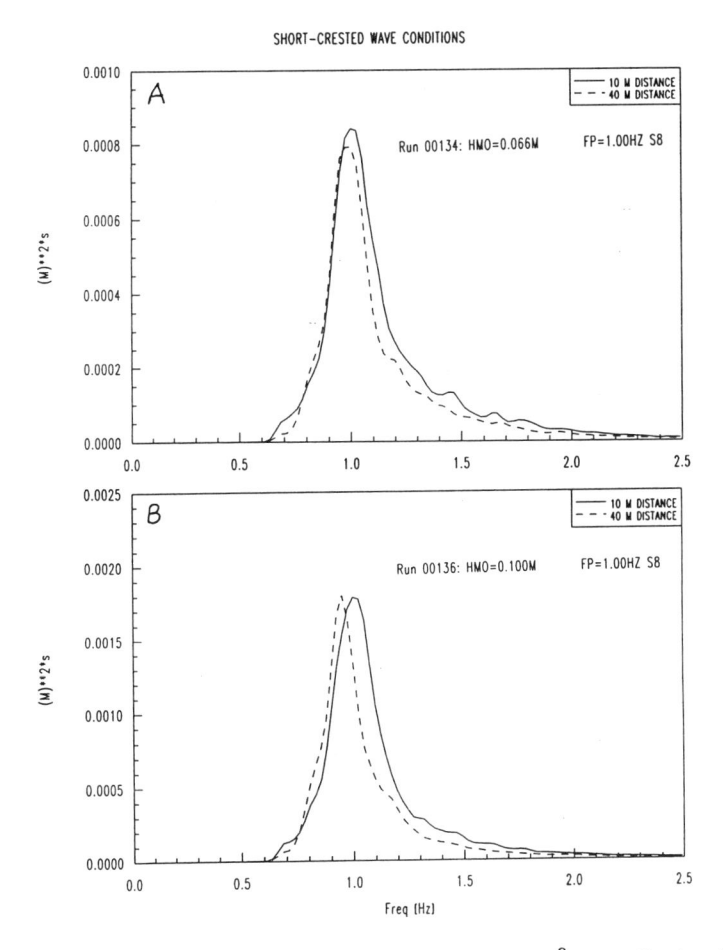

Fig. 4. As fig. 3, but for 2 shortcrested conditions. ($\cos^8$ - distributions).

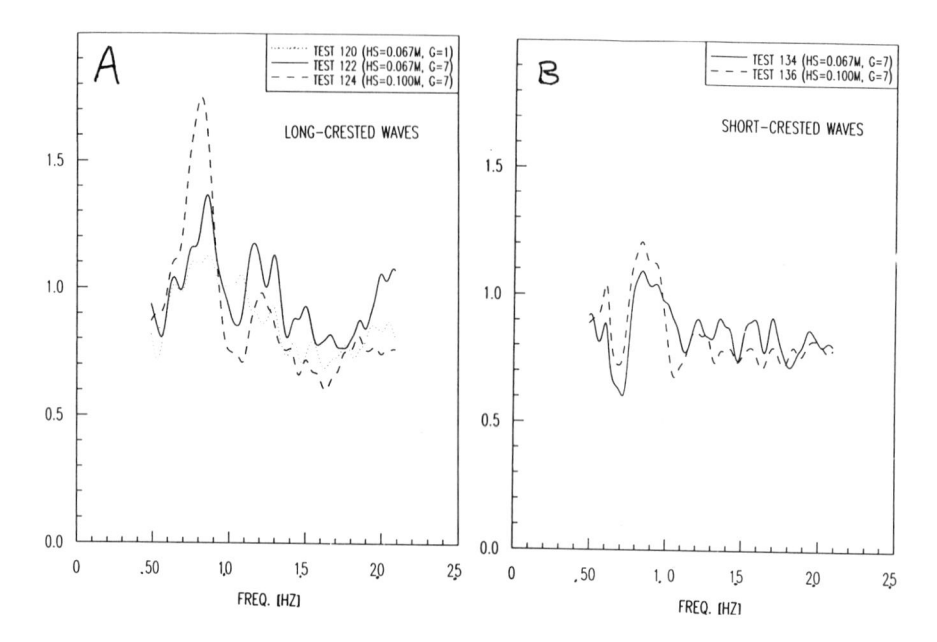

Fig. 5. Amplitude amplification from 10 m to 40 m distance.

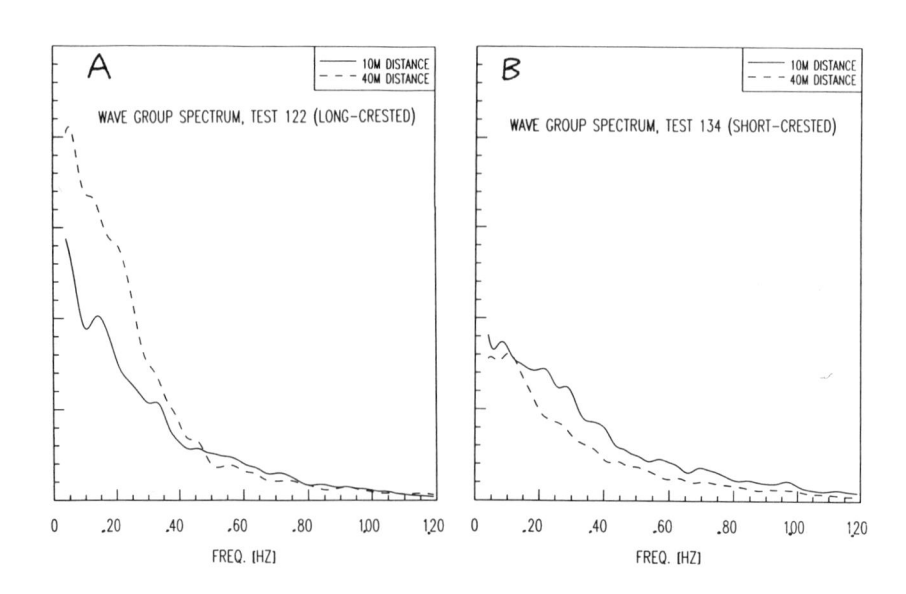

Fig. 6. Wave group spectra, for conditions in figs. 3B (long-crested) and 4A (short-crested).

The most pronounced modulations and non-Gaussian statistics are observed in the narrow-banded long-crested test case 122 (figure 3 B). For this case, wave group spectra and crest and waveheight distributions are presented in figs. 6A and 7, respectively. For comparison, the same type of results for the corresponding short-crested case are plotted in figs. 6B and 8.

For a concentrated illustration of the basic non-linear statistical properties of the recorded data, and their relations to other relevant parameters, the statistical skewness and kurtosis of some of the records are shown in fig. 9. The skewness is based on 3rd order statistical moment of the time series, and is calculated as:

$$\gamma_1 = \frac{1}{N\sigma^3} \sum_{i=1}^{N} (x_i - \overline{x})^3 \qquad (1)$$

where
x_i = sample value of time history
$\overline{x}$ = mean value of x_i
σ = standard deviation of x_i
N = number of time history samples.

The kurtosis is based on the 4th order statistical moment, and is calculated as:

$$\gamma_2 = \frac{1}{N\sigma^4} \sum_{i=1}^{N} (x_i - \overline{x})^4 - 3 \qquad (2)$$

(with this definition also called "the excess of kurtosis")

Both these parameters are expected to be zero for Gaussian records (linear waves), but they are always subject to some statistical scatter. With the long records available here, the scatter is believed not to be a problem. The skewness is an indicator of 2nd order contents in the wave signal (Vinje and Haver 1994), while the kurtosis indicates the presence of 3rd and higher order contents. Thus the skewness gives information on <u>crest</u> heights, while the kurtosis indicates possible non-linear <u>wave</u> height statistics. The skewness values are plotted for all records of the total experiment, as a function of the typical steepness of the record. ($S = H_s / L_p$, with $L_p = (g / 2\pi) \cdot T_p^2$, T_p = spectral peak period). The kurtosis values, however, are here plotted only for one set of Hs and Tp ($H_s = 0.067$m, $T_p = 1.0$s), as a function of the normalized travelled distance D_r (number of propagated wavelengths). ($D_r = D / L_p$). This is done in order to highlight observed relationships.

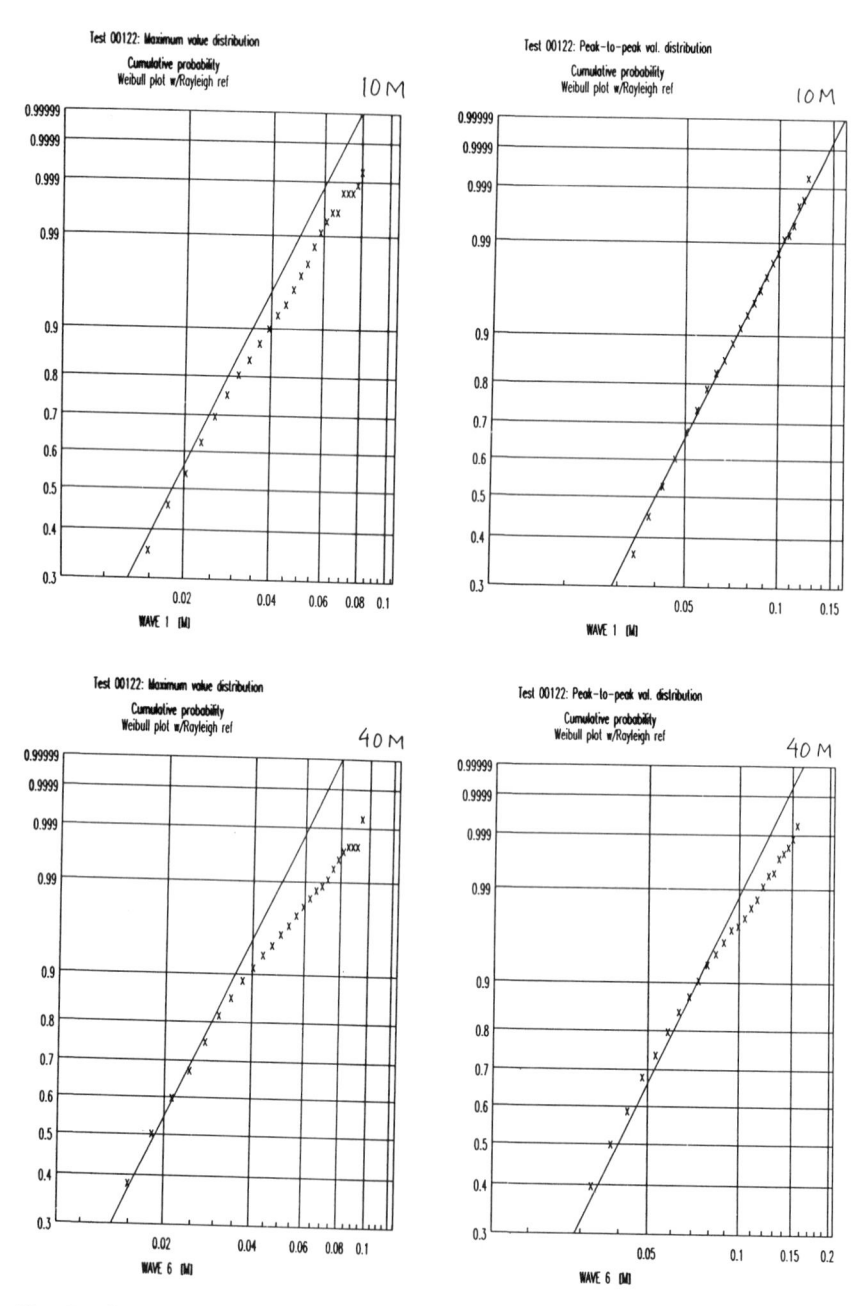

Fig. 7. Crest and wave height distributions, same conditions as in fig. 6A (long-crested).

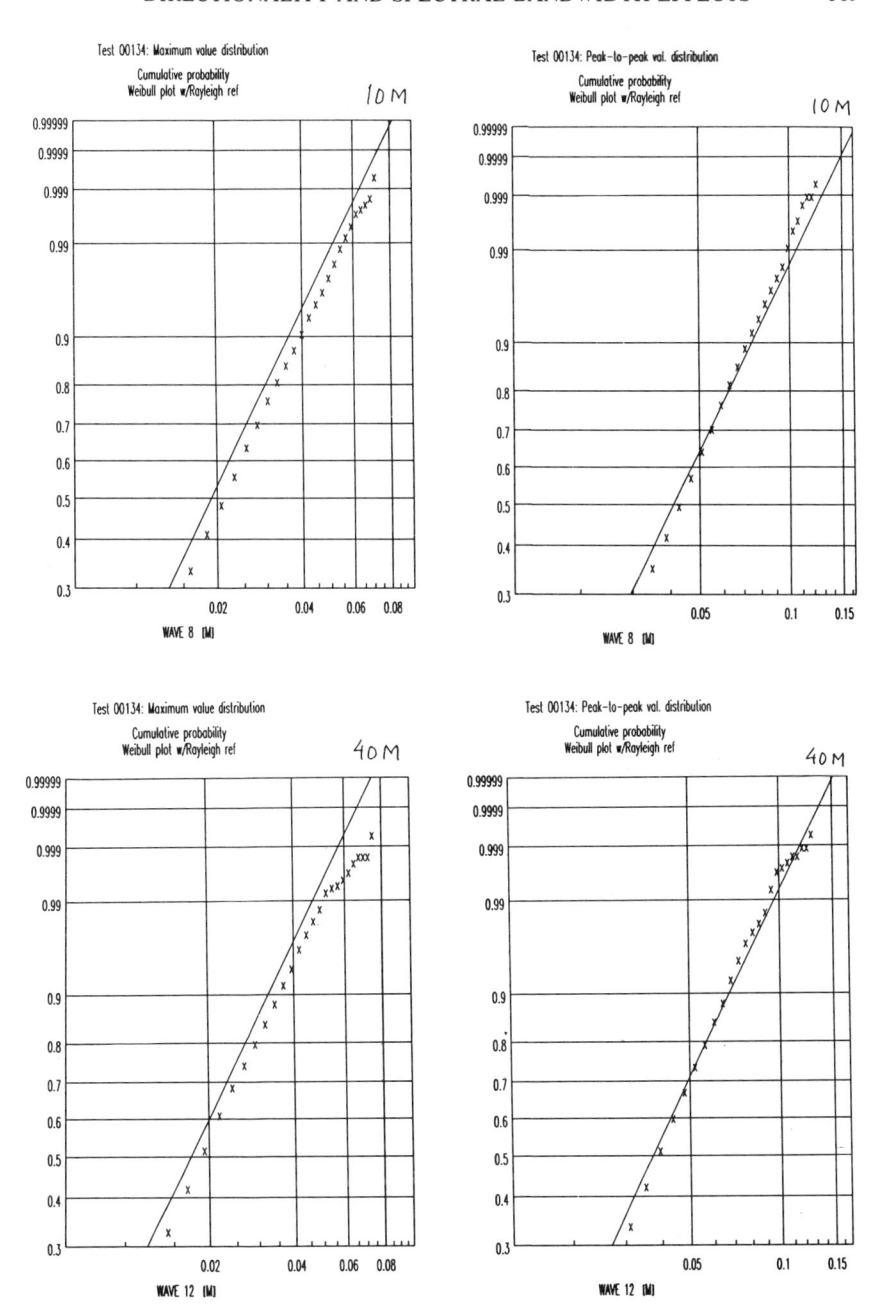

DIRECTIONALITY AND SPECTRAL BANDWIDTH EFFECTS 589

Fig. 8. As fig. 7, but same conditions as in fig. 6B (short-crested).

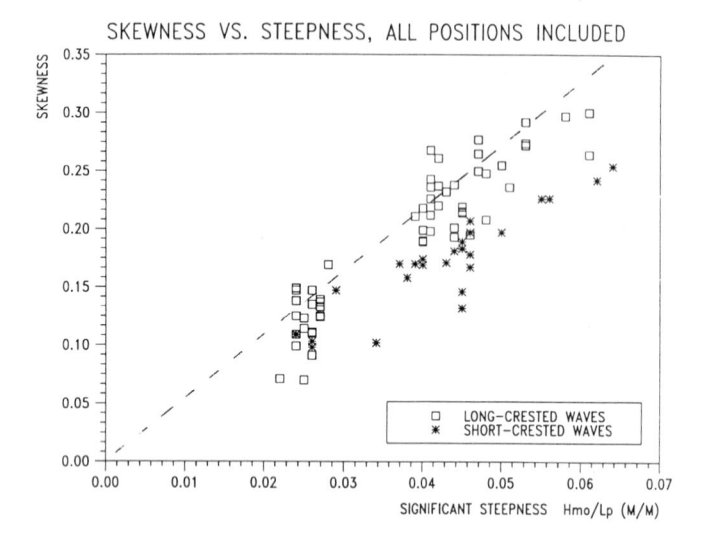

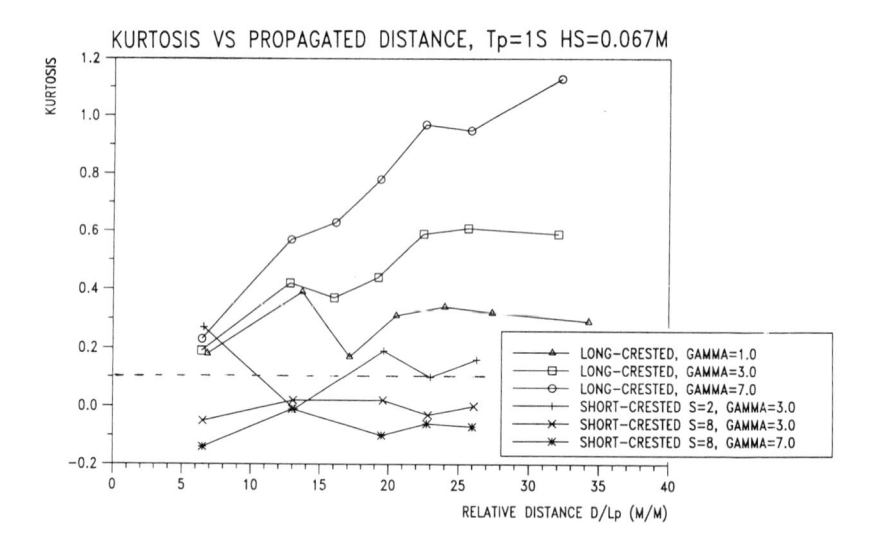

Fig. 9. Skewness vs. steepness, and excess of kurtosis vs. normalized travelling distance. Dotted line: estimated from 2nd order theory (Vinje & Haver 1994).

4. Discussion of results.

The power spectrum plots in figs. 3 & 4 show some changes in the spectra with increasing travelling distance. This is especially seen when the wave steepness is very high. In general, the energy in the high-frequency part of the spectrum is somewhat reduced due to dissipation. (At very high frequencies, however, the energy content seems to be more stable, probably due to bound harmonics). Energy shift is also observed from the peak frequency region to the surrounding upper and lower frequencies, particularly in longcrested waves. This is especially pronounced in the very steep sea condition in which case a net transfer to lower frequencies is also taking place together with some loss of energy (to some extent connected with dissipation due to breaking).

The energy transfer is further illustrated by the amplification plots in fig. 5. Of particular interest here is the apparent growth of side bands around the initial peak frequency, mainly observed in narrow-banded longcrested waves. This observation indicates a similarity with the side bands due to non-linear modulational effects in monochromatic and bichromatic waves discussed earlier. Therefore it fits well together with the observed increase in the far-distance wave group spectrum in fig. 6, and also with the crest height statistics deviating systematically from the Rayleigh model (fig. 7). While the distribution in fig. 7, at 10 m, (with extreme crests exceeding the Rayleigh estimate by 15-20 %) can be explained by 2nd order effects, the distributions at 40 m must be explained by higher-order effects. Crest levels up to 40% higher than the Rayleigh prediction are observed in the latter case.

The non-Gaussian statistics observed in the amplitude distribution plots are also reckognized in the skewness and kurtosis plots. The plot of skewness shows that it depends more or less only on the steepness of the sea state, and less on other parameters such as directionality, propagation etc. The theoretical curve from Vinje and Haver (1994), based on 2nd order theory, fits reasonably well. (A slight skewness reduction is observed in directional sea). We interpret the result in the way that the 2nd order contribution is quite stable, regardless of other circumstances. The kurtosis, however, varies significantly with travelling distance, wave directionality and with the spectral width.

A clear increase in non-linear modulation effects with narrower spectral width is observed from the measurements. We interpret this, qualitatively, as a consequence of modulation periods possibly around 4 - 6 times the wave periods. It is reasonable to assume that this will affect long wave groups (narrow spectra) stronger than it will affect short wave groups (broader spectra). (Side-band instabilities may more easily become "smeared out" by disturbing neighbouring frequencies).

As already commented, the above non-linear effects are only to a limited extent observed in short-crested waves in the present measurements. This is an important observation, which confirms the conclusion from the bichromatic test examples in fig. 1. Thus it seems reasonable to conclude from this that extreme wave events due to non-linear modulations are most pronounced in longcrested waves. It appears to be a possible connection between these effects and the direction of the local phase gradient relative to that of the energy gradient. For more firm conclusions to be drawn on this, more analysis work is to be made, possibly including comparisons to theoretical/numerical models.

5. Acknowledgement

This work has been financially supported by the Royal Norwegian Research Council.

6. References.

Benjamin, T.B. & Feir, J., (1967), "The Disintegration of Wave Trains on Deep Water", J. Fluid Mech., Vol. 27, pp. 417-430.

Jonathan, P., Taylor, P.H. & Tromans, P. S., (1994), "Storm Waves in the Northern North Sea", Proceedings, Vol. 2, the 7th BOSS Conference, MIT Cambridge, USA (Published by Pergamon / Elsevier, Oxford, UK), pp. 481-494.

Kjeldsen, S.P. (1990), "Breaking Waves", in Water Wave Kinematics, ed. A. Tørum & O.T. Gudmestad, Kluwer Academic, Dordrecht, the Netherlands, pp. 453-473.

Lake, B.M., Yuen, H.C., Rungaldier, H. & Ferguson, W.E., (1977), "Non-linear Deep-water Waves: Theory and Experiment. Part 2: Evolution of a Continuous Wave Train", J.Fluid Mech., Vol. 83, pp. 49-74.

Lo, E. & Mei, C.C., (1985), "A Numerical Study of Water-wave Modulation Based on a Higher-order Nonlinear Scrödinger Equation", J. Fluid Mech., Vol. 150, pp. 395-416.

Longuet-Higgins, M.S. (1963), "The Effect of Non-Linearities on Statistical Distributions in the Theory of Sea Waves", J. Fluid Mech. Vol. 17, pp. 459-480.

Marthinsen, T. and Winterstein, S. (1992), "On the Skewness of Random Surface Waves", Proceedings, Vol. III, the 2nd ISOPE Conference, San Francisco, USA, pp. 472-478.

Sand, S.E. et al. (1990), "Freak Wave Kinematics", in Water Wave Kinematics, ed. A. Tørum & O.T. Gudmestad, Kluwer Academic, Dordrecht, the Netherlands, pp. 535-549.

Stansberg, C.T. (1983), "Statistical Analysis of Slow Drift Responses", Journal of Energy Resources Technology, Vol. 105, pp. 310-317.

Stansberg, C.T. (1991), "Extreme Wave Asymmetry in Full Scale and Model Scale Experimental Wave Trains", Proceedings, the 10th OMAE Conference, Stavanger, Norway, pp. 215-222.

Stansberg, C.T., (1992), "On Spectral Instabilities and Development of Non-linearities in Deep-water Propagating Wave Trains", Proceedings, Vol. 1, 23rd International Conference on Coastal Engineering, Venice, Italy, pp. 658-671.

Stansberg, C.T., (1993), "Second-Order Numerical Reconstruction of Laboratory Generated Random Waves", Proceedings, Vol. II, the 12th OMAE Conference, Glasgow, Scotland, pp. 143-151.

Stansberg, C.T., (1994), "Second-Order Effects in Random Wave Modelling", Proceedings, Vol. II, the International Symposium, "Wave-Physical and Numerical Modelling", the University of British Columbia, Vancouver, Canada, pp. 793-802.

Su, M.-Y., Bergin, M., Marler, P. & Myrick, R., (1982), "Experiments on nonlinear instabilities and evolution of steep gravity-wave trains", J. Fluid Mech., Vol. 124, pp. 45-72.

Taylor, P.H. & Haagsma, I.J., (1994), "Focusing of Steep Wave Group on Deep Water", Proceedings Vol. II, International Symposium on "Waves-Physical and Numerical Modelling", University of British Columbia, Vancouver, Canada, pp. 862-870.

Vinje, T. & Haver, S. (1994), "On the Non-Gaussian Structure of Ocean-Waves", Proceedings, Vol. 2, the 7th BOSS Conference, MIT Cambridge USA, (Published by Pergamon / Elsevier, Oxford, UK), pp. 453-480.

Wang, P., Yao, Y. and Tulin, M.P., (1993), "Wave Group Evolution, Wave Deformation, and Breaking: Simulations Using LONGTANK, a Numerical Wave Tank", Proceedings, Vol. III, the 3rd ISOPE Conference, Singapore.

Yasuda, T., Mori, N. & Ito, K., (1992), "Freak Waves in Unidirectional Wave Trains and their Properties", Proceedings, Vol. 1, 23rd International Conference on Coastal Engineering, Venice, Italy, pp. 751-764.

Yasuda, T. & Mori, N. (1994), "High Order Nonlinear Effects on Deep-Water Random Wave Trains", Proceedings, Vol. II, International Symposium on "Waves-Physical and Numerical Modelling", University of British Columbia, Vancouver, Canada, pp. 823-832.

CHAPTER 44

SHEAR STRESSES AND MEAN FLOW IN SHOALING AND BREAKING WAVES

Marcel J.F. Stive[1)] and Huib J. De Vriend[2)]

ABSTRACT

We investigate the vertical, wave averaged distributions of shear stresses and Eulerian flow in normally incident, shoaling and breaking waves. It is found that shear stresses are solely due to wave amplitude variations, which can be caused by shoaling, boundary layer dissipation and/or breaking wave dissipation. The resulting shear stress and mean flow distributions for these cases are derived, and compared with earlier work.

The attractive, now frequently used modelling choice of specifying a shear stress at the mean surface level is discussed in the context of the constituent equations and related boundary conditions and constraints. A derivation of the shear stress at the mean surface level is given both by using the momentum balance and energy balance equations, which is shown to lead to the same result, if the effects of a changing roller are incorporated correctly[3)].

Finally, matching solutions for the shoaling and breaking wave cases between the boundary layer and the middle layer for the shear stresses and the wave averaged flow are derived.

INTRODUCTION

The intentions of the present paper are to present a conceptual view on the constituent equations and related boundary conditions and constraints for the vertical distributions of wave averaged shear stresses and Eulerian flow for the cases of

---

1) Netherlands Centre for Coastal Research, Delft University of Technology, Faculty of Civil Engineering, PO Box 5048, 2600 GA Delft, The Netherlands; and Delft Hydraulics, PO Box 152, 8300 AD Emmeloord, The Netherlands (Internet: marcel.stive@wldelft.nl)
2) University of Twente, Section Civil Engineering and Management, PO Box 217, 7500 AE Enschede, The Netherlands (Internet: h.j.devriend@sms.utwente.nl)
3) See Appendix to this paper: "Mean surface shear stress due to a changing roller" by Rolf Deigaard

shoaling and breaking waves, including boundary layer dissipation effects. While the relevant physical phenomena are indicated, it is not intended to suggest optimal closure hypotheses for all the phenomena, but rather to indicate their physical consequences in a transparent way. Examples are the eddy viscosity assumptions and the roller effects, for which simplified choices are made, which are, to a certain degree, not essential but only made for simplicity and transparency of the concept.

It is found that the sloping bottom does not introduce additional terms in the shear stress distributions found by Deigaard and Fredsoe (1989). The separate cases of no-breaking and breaking are discussed.

Furthermore, it is shown how a general expression for the mean Eulerian flow may be derived, which incorporates the effects of sloping bottom and wave amplitude variations, either due to boundary layer dissipation, shoaling or breaking. In the special cases of horizontal bottom and no-breaking and of sloping bottom and no-breaking, the expression reduces to Longuet-Higgins' conduction solution (1953) and to Bijker et al's shoaling solution (1974), respectively.

CONSTITUENT EQUATIONS AND CONDITIONS

If we neglect the effects of advective acceleration for the time-mean flow and assume a wave-averaged eddy viscosity approximation for the Reynolds' stresses, the local momentum equation most commonly used to solve the wave averaged Eulerian flow reads:

$$\frac{\partial}{\partial z} \nu_t \frac{\partial U}{\partial z} = g\overline{\eta}_x + (\overline{u^2})_x - (\overline{w^2})_x + (\overline{uw})_z \tag{1}$$

where:

U   is the wave averaged Eulerian flow,
ν_t is the eddy viscosity,
η_x is the mean water level set-up, and
u,w are the horizontal and vertical wave-orbital velocities.

This equation is assumed to be valid in the full vertical domain, except for the region near the free surface. Somewhat heuristicly, we assume the equation to be valid unto mean water level, since a wave averaged situation is considered, and we let the effects of the near surface layer (such as the roller) be effectuated in a shearstress acting at the mean water level.

Let us introduce the common assumption that the wave terms (orbital velocity moments) can be derived independently from the mean flow, i.e. we assume that there exists a sufficiently accurate wave theory to describe the wave terms which neglects the wave-current interaction. This is of course a simplification, and we may state that a true break-through here would be achieved if we could tackle the problem with a Lagrangian approach in which waves and currents would be considered simultaneously, which at the same time would allow us to deal consistently with the near surface layer (NSL). However, until such an approach is developed we rely on the experience that the suggested approach is shown to yield sufficiently accurate results for the moment.

Let us further assume that the turbulence viscosity is constant over depth. Note that this is only for the clarity of the argumentation. The rationale which follows would also allow for a depth-varying viscosity, as long as it is time invariant and does not depend on the undertow solution itself.

The above implies that we have one equation to solve for the undertow and the mean water level set-up. Integrating the equation twice yields:

$$\nu_t U = \frac{1}{2} g \overline{\eta}_x z'^2 + \int_0^{z'} \int_0^{z'} \overline{(u^2)}_x dz dz - \int_0^{z'} \int_0^{z'} \overline{(w^2)}_x dz dz + \int_0^{z'} \overline{(uw)} dz + C_1 z' + C_2 \qquad (2)$$

This expression contains three unknowns, viz. the two constants of integration, C_1 and C_2 remaining in this expression, and the set-up gradient. We therefore need, in addition to this equation, three boundary conditions and/or constraints:

(1) no-slip condition at the bottom ($U = 0$), from which we can find C_2;
(2) shear stress condition at the transition from the middle layer (ML) to the NSL (τ_t given), from which we can find C_1;
(3) mass balance constraint (total mass flux in the lower layers balances that in the NSL), from which we can derive the set-up gradient.

The essential information which we need here is related to the NSL. In fact, condition (2) is based on the assumption that we know the shear stress at the lower end of the NSL from the momentum balance for this layer. Similarly, when formulating the constraint (3), we assume the total mass flux in the NSL to be known.

This route is suggested by Stive and De Vriend (1987) and also followed by Deigaard et al. (1991). The former derive a formal expression through a third depth integration, while the latter use an iteration procedure which sees to it that a set-up gradient is created such that the correct depth-average mass flux is created. The result should be the same.

Alternatively, one could use the depth-averaged horizontal mass and momentum equations, with the wave-induced radiation stresses and mass fluxes properly modelled, instead of giving τ_t and imposing constraint (3). Note also that imposing constraint (3) is not correct in 3-D situations, where the mass flux in the NSL is not necessarily compensated in the lower parts of the same water column (cf. De Vriend and Kitou, 1990). In fact, later we shall use the depth-averaged momentum equation, with the usual expressions for the radiation stresses, to derive an expression for τ_t. Note that in either case we face the problem of describing the NSL, via τ_t and the mass flux, or via the wave-related terms in the depth-averaged mass and momentum equations.

DISTRIBUTION OF ORBITAL VELOCITY MOMENTS AND SHEAR STRESSES ON A SLOPING BOTTOM IN THE MIDDLE LAYER

In order to derive the shear stress distribution the orbital velocities outside and a little away from the bottom boundary layer need to be known. The idea is to use these results to look at the shear stress distribution in the middle layer for the cases of sloping bottom and boundary layer dissipation and of sloping bottom and breaking wave dissipation. Furtheron, we will look at these cases with boundary layer effects on the shear stress distribution included.

Since we are interested in the case of spatially varying waves on a sloping bottom, we must at least rely on a non-uniform depth approximation. This has been done by De Vriend and Kitou (1990) and recently also by Rivero and Arcilla (1994) for the shallow water approximation to shoaling waves. The result as obtained for the orbital velocity moments reads:

$$\overline{u^2} = \frac{1}{2} A^2 + O\left(A^2 \frac{\lambda^2}{L^2}\right)$$

$$\overline{w^2} = O\left(A^2 \frac{h^2}{L^2}\right) \tag{3}$$

$$\overline{uw} = \frac{1}{2} A^2 h_x - \frac{1}{4} (A^2)_x z'$$

where

λ = wave length,
L = scale of horizontal variations, and
A = $(a\omega)/(kh)$.

Based on these results we can investigate our two cases.

(1) The case of a sloping bottom and boundary layer dissipation

Using the above results the shear stress becomes

$$\tau(z')/\rho = v_t U_z = \overline{g\eta_x} z' + \frac{1}{4} (A^2)_x z' + \frac{1}{2} A^2 h_x + C_1 \tag{4}$$

Because of the absence of dissipation in the NSL there is no shear stress at the mean water level, sothat

$$\tau(z')/\rho = \overline{g\eta_x}(z' - d_m) + \frac{1}{4} (A^2)_x (z' - d_m) \tag{5}$$

In fact, no shear stresses will exist in the whole of the middle layer for (Deigaard and Fredsoe, 1989):

$$\rho g \overline{\eta_x} d_m = -\frac{1}{4} \rho (A^2)_x d_m = -\frac{1}{2} E_x \tag{6}$$

which complies with the fact that no shear stresses can be maintained in the middle layer due to the existence of irrotational flow. However, we will show later that due to the constraint of depth averaged zero mean flow, the existence of an additional mean water level gradient is required.

(2) The case of a sloping bottom and wave breaking dissipation

Again using the above results the shear stress in this case becomes:

$$\tau(z')/\rho = g\overline{\eta_x}(z' - d_m) + \frac{1}{4}(A^2)_x(z' - d_m) + \tau_t \tag{7}$$

Now a shear stress at mean water level exists due to the presence of the roller. The mean water level gradient again follows from the mean flow constraint.

DERIVATION OF THE SHEAR STRESS AT MEAN WATER LEVEL

Here derivations of the shear stress at mean water level are presented, and we show that the result is consistent between using either momentum flux or energy flux considerations.

From the above Equation (7) we find for the bottom shear stress:

$$\tau_b = -\rho g d_m \overline{\eta_x} + \tau_t - \frac{1}{4}\rho(A^2)_x d_m \tag{8}$$

Since we consider wave breaking dissipation only, the mean bottom shear stress τ_b should be zero.

The shear stress τ_t may be resolved between the τ_b equation and the depth mean horizontal momentum balance equation. In order to do this we need to introduce an expression for the radiation stress which at least needs to be extended with the roller effect. Following Svendsen (1984), Deigaard and Fredsoe (1989) suggest the shallow water approximation

$$S_{xx} = S_{xx,p} + S_{xx,u} = \frac{1}{2}E + \left[E + \rho \frac{Rc}{T}\right] \tag{9}$$

where R is the roller area.

The above implies that the set of equations available to resolve the shear stress τ_t reads:

$$\frac{dS_{xx}}{dx} + \rho g d_m \overline{\eta}_x = 0 \tag{10}$$

$$\frac{dS_{xx}}{dx} = \frac{3}{2} E_x + \frac{\rho}{T} (Rc)_x \tag{11}$$

$$\tau_b = -\rho g \overline{\eta}_x d_m + \tau_t - \frac{1}{4} \rho (A^2)_x d_m = -\rho g \overline{\eta}_x d_m + \tau_t - \frac{1}{2} E_x = 0 \tag{12}$$

which yields:

$$\tau_t = -E_x - \frac{\rho}{T}(Rc)_x \tag{13}$$

and

$$\tau(z') = -E_x \left(1 + \frac{d_m - z'}{2d_m}\right) - \rho g \overline{\eta}_x (d_m - z') - \frac{\rho}{T}(Rc)_x \tag{14}$$

These equations are equivalent to equations (56) and (58) of Deigaard and Fredsøe (1989), when introducing the relation

$$E_x = \frac{1}{c}(E_f)_x = -\frac{D}{c} \tag{15}$$

where E_f is the energy flux and D is the dissipation due to wave breaking.

In slightly different notation Equation (13) reads:

$$\tau_t = -\frac{1}{c}(E_f)_x - (2E_r)_x \tag{16}$$

where E_f is the energy flux, now using the expression of kinetic roller energy E_r (introduced by Svendsen, 1984, and used by Nairn et al., 1990):

$$2E_r = S_{xx,roller} = \rho Rc^2/L = \rho Rc/T \tag{17}$$

The presence of the roller leads to the additional term, which is due to the nonnegligible velocity contribution to the momentum flux, since the velocities are of magnitude c. The contribution due to the roller enhanced pressure is probably not only small, but may also be already included in $(E_f)_x$ since this property is quantified using the surface elevation variance.

One may, however, observe that the shear stress at NSL also appears in the energy balance equation, which according to Nairn et al. (1990) reads:

$$(E_f)_x + (E_r c)_x + \tau_t c = 0 \qquad (18)$$

where they have introduced Svendsen's (1984) suggestion for the energy flux due to the roller and the result given in Deigaard and Fredsoe (1989) that the dissipation D is due to the work done by the shear stress due to the roller acting on the fluid right below it. Again, the roller related contribution is due to the mean transfer of kinetic energy (proportional to c^2) with the roller velocity (equal to c).

Rearranging the last equation for τ_t yields

$$\tau_t = -\frac{1}{c}(E_f)_x - \frac{1}{c}(E_r c)_x \qquad (19)$$

which indicates that when we accept that the spatial variations in c are small relative to those in E_r we are faced with a factor 2 difference. Note that here also we have assumed that the horizontal component of the shear stress is negligibly different from the shear stress along the roller-wave interface.

It appears that the apparent inconsistency is caused by the complicated situation that occurs when the volume of the roller is changing in the wave propagation direction (see Appendix). Beside the shear layer between the roller and the wave there is a net transfer of water from the wave to the roller. If the roller is losing water (dR/dx < 0) the horizontal momentum transfer from the roller to the wave is not only that due to the shear layer, but also the momentum of the water leaving the roller. If the roller gains water (dR/dx > 0), the water leaving the wave has negligible horizontal velocity and does not change the momentum of the water remaining in the wave. In both cases, however, there is an additional energy dissipation of

$$\rho \frac{1}{2} c^2 \left| \frac{dR}{dt} \right| \qquad (20)$$

The corrections that these considerations give to the NSL shear stress and to the energy dissipation rate were derived in Deigaard (1993), and it appears that the energy balance equations in both cases yield:

$$(E_f)_x + (2E_r c)_x + \tau_t c = 0 \qquad (21)$$

which removes the discrepancy and implies a correction to Nairn et al. (1990) and Stive and De Vriend (1994).

MATCHING SOLUTION BETWEEN BOUNDARY LAYER AND MIDDLE LAYER

Where in the foregoing we have neglected the bottom boundary layer effects on the time- and depth-averaged momentum equation since it is a second-order effect in the overall momentum balance, we present in this section a matching solution to derive the mean Eulerian flow over the vertical, with accuracy near the bottom. Due to the existence of the bottom boundary layer the near-bottom horizontal orbital velocity may be shown to include a phase difference compared to the orbital velocity in the middle layer (cf. Longuet-Higgins, 1953). Because of the sloping bottom and/or due to wave breaking a variation of the wave amplitudes results which may be shown to yield the following matched result for the horizontal orbital velocity (cf. Bijker et al., 1974):

$$u = A(x)[\cos\chi - e^{-\phi}(\chi - \phi)] \tag{22}$$

where

$$\phi = z'/\delta$$
$$\delta^2 = 2v_t/\omega$$
$$\chi = \omega t - \psi$$
$$k = d\psi/dx.$$

By using continuity

$$w_z = -u_x \tag{23}$$

and depth integration

$$w = -\int_0^{z'} u_x \, dz' \tag{24}$$

we may derive the following expressions for the time-averaged values of the wave terms needed in Equation (1):

$$\overline{u^2} = \frac{1}{2}A^2[1 + e^{-2\phi} - 2e^{-\phi}\cos\phi] \tag{25}$$

$$\overline{uw} = A^2 k\delta \left[-\frac{1}{4} + \frac{\phi}{2}e^{-\phi}\sin\phi - \frac{e^{-\phi}}{4} + \frac{e^{-\phi}}{2}\cos\phi \right] -$$
$$- \left(\frac{A^2}{2}\right)_x \frac{\delta}{2} \left[-\frac{1}{2} + \phi + e^{-\phi}\cos\phi - \phi e^{-\phi}\cos\phi - \frac{e^{-2\phi}}{2} \right] \tag{26}$$

These equations were derived by Bijker et al. (1974) for the shoaling wave case, but they are equally valid for the case of wave breaking, since in the shoaling case it is the amplitude variation only that impacts on these terms. The difference being solely that during shoaling the amplitudes increase, while due to breaking they decrease.

Using the above expressions for the wave-averaged orbital terms we may resolve Equation (2) and the set of three boundary conditions and constraints to yield:

$$v_t U = \left(g\overline{\eta_x} + \frac{1}{4}(A^2)_x\right)\left[\frac{1}{2}z^2 - z'd_m\right] + \frac{\tau_t}{\rho}[z'] +$$

$$+ (A^2)_x \frac{\delta^2}{4}\left[2e^{-\phi}\sin\phi + \frac{1}{2}e^{-\phi}\cos\phi + \frac{1}{2}\phi e^{-\phi}(\sin\phi - \cos\phi) + \frac{1}{4}e^{-2\phi} - \frac{3}{4}\right] + \quad (27)$$

$$+ A^2 k \frac{\delta^2}{2}\left[-\frac{1}{2}e^{-\phi}\phi(\sin\phi + \cos\phi) - e^{-\phi}\cos\phi + \frac{1}{2}e^{-\phi}\sin\phi + \frac{1}{4}e^{-2\phi} + \frac{3}{4}\right]$$

Note that for transparency we have assumed an eddy viscosity invariant over depth, which is not necessary: it only yields an attractive, analytical expression over the total depth, which allows us to point out the respective approximations for the cases derived before. Longuet-Higgins' solution (1953) contains the fourth term of Equation (19) only. In the shoaling case of Bijker et al. (1974) we have

$$g\overline{\eta_x} = -\frac{1}{4}(A^2)_x \text{ and} \quad (28)$$

$$\tau_t = 0$$

so that only the last two terms of Equation (27) result. For breaking waves on a sloping bottom we need all four terms. Note that in all cases a small mean water level gradient may be necessary to comply with a depth-averaged zero Eulerian flow.

Here, we refrain from a deeper analysis of the surfzone situation with strong, breaking induced wave amplitude variations. Clearly, the mean water level gradient and the NSL shear stress (as represented by the first two terms in Equation (27)) will exert a major influence on the mean Eulerian flow distribution. However, we may expect that especially near the bottom the boundary layer term and the amplitude variation term will show their significance. An insight into their qualitative influence is given below.

For the shoaling wave case the mean Eulerian flow distribution reads:

$$U = \frac{A^2 k}{\omega} f'(\phi) + \frac{A}{\omega}\frac{dA}{dx} g'(\phi) \quad (29)$$

where the vertical form functions are given by

$$f'(\phi) = \left[-\frac{1}{2}e^{-\phi}\phi(\sin\phi + \cos\phi) - e^{-\phi}\cos\phi + \frac{1}{2}e^{-\phi}\sin\phi + \frac{1}{4}e^{-2\phi} + \frac{3}{4}\right] \quad (30)$$

$$g'(\phi) = \left[2e^{-\phi}\sin\phi + \frac{1}{2}e^{-\phi}\cos\phi + \frac{1}{2}\phi e^{-\phi}(\sin\phi - \cos\phi) + \frac{1}{4}e^{-2\phi} - \frac{3}{4}\right] \quad (31)$$

Their behaviour is graphically illustrated in Figure 1, which teaches us that close to the bottom the amplitude variation effect strengthens the streaming effect in the shoaling zone, but weakens it in the breaker zone. Whether this actually occurs is depending on the relative magnitude of these terms, which may be estimated as follows. At the edge of the boundary layer we find:

$$U_\infty = \frac{3}{4}\frac{A^2k}{\omega} - \frac{3}{4}\frac{A}{\omega}\frac{dA}{dx}$$

$$= \frac{3}{4}\frac{A^2k}{\omega}\left[1 - \frac{1}{Ak}\frac{dA}{dh}\frac{dh}{dx}\right] \quad (32)$$

Using linear shoaling it may be shown (Bijker et al., 1974) that $1/(Ak)\, dA/dh = O(1)$ for $kh \geq 1$, sothat only in the breaker zone and with relative steep slopes the amplitude variation effect becomes important.

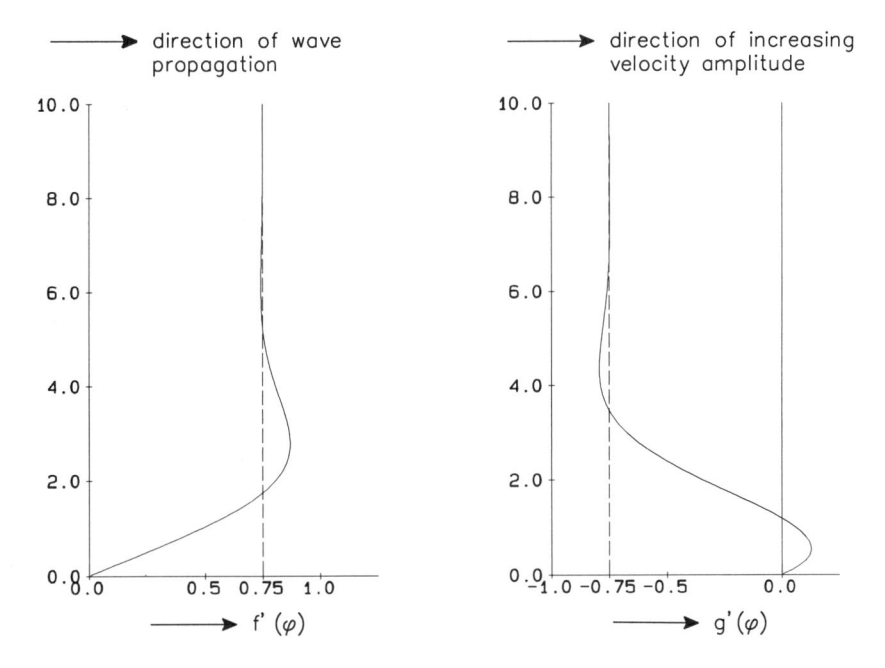

Figure 1 Form functions of streaming effect $(f'(\varphi))$ and amplitude variation effect $(g'(\varphi))$

CONCLUSIONS AND DISCUSSION

A conceptual view is presented on the constituent equations and related boundary conditions and constraints for the vertical distributions of wave averaged shear stresses and Eulerian flow for the cases of shoaling and breaking waves, including boundary layer dissipation effects. We have refrained from suggesting optimal closure hypotheses for the relevant processes, but we have tried to indicate their physical consequences in a transparent way. It is found that shear stresses are solely due to wave amplitude variations, which can be caused by shoaling, boundary layer dissipation and/or breaking wave dissipation. The resulting shear stress and mean flow distributions for these cases are derived, and compared with earlier work.

The specification of the shear stress at the mean surface level is discussed in the context of the constituent equations and related boundary conditions and constraints. A derivation of the shear stress at the mean surface level is given both by using the momentum balance and energy balance equations, which is shown to lead to the same result, if the effects of a changing roller are incorporated correctly (see Appendix). Finally, matching solutions for the shoaling and breaking wave cases between the boundary layer and the middle layer for the shear stresses and the wave averaged flow are derived.

We conclude by noting that the foregoing considerations concern the 2-D wave-induced water motion in a vertical plane parallel to the direction of wave propagation. The currents are weak and boundary-layer processes are primarily wave-induced. The wave orbital motion and the mean current higher up in the vertical can be treated separately, the former as inviscid, the latter as viscous flow. Advection plays no significant role in the mean current field.

Although this situation must occur now and then in nature, it rather concerns a wave flume. The common situation on a natural beach is more complicated, because

- the wave field is directionally spread, so whatever definition is chosen for "the" direction of wave propagation, there are always wave components in other directions;
- the waves are not monochromatic, there is always energy in the low-frequency bands (edge waves, surfbeat, shear waves, etc.);
- there can be strong currents, tidal, wave-driven or otherwise, and not necessarily in the alongshore direction; in general, these currents make a non-zero angle with the direction of wave propagation;
- strong spatial gradients of the bed topography and wave or current parameters can occur in any direction.

As a consequence, one would have to abandon a number of simplifying assumptions which underlie the present 2-D vertical (2-DV) model. One of these for instance is that in 3-D situations the mass flux in the NSL is not necessarily compensated in the lower parts of the same water column (cf. De Vriend and Kitou, 1990).

ACKNOWLEDGEMENT

A large part of this paper is based on work in the "G8 Coastal Morphodynamics" research programme, which is funded partly by the Commission of the European Communities in the framework of the Marine Science and Technology Programme (MAST), under contract no. MAS2-CT92-0027. The work is co-sponsored by Delft Hydraulics, in the framework of the Netherlands Centre for Coastal Research.

REFERENCES

Bijker, E.W., Kalkwijk, J.P.Th. and Pieters, T. (1974). Mass transport in gravity waves on a sloping bottom. Proc. 14th Conf on Coastal Eng, ASCE, pp 447-465

Deigaard, R. (1993). A note on the three-dimensional shear stress distribution in a surf zone. Coastal Engineering, 20, 157-171.

Deigaard, R. (1994). Mean surface shear stress due to a changing roller. Appendix to this paper.

Deigaard, R. and Fredsøe, J. (1989). Shear Stress Distribution in Dissipative Water Waves. Coastal Engineering, 13, 357-378.

Deigaard, R., Justesen, P. and Fredsoe, J. (1991). Modelling of undertow by a one-equation turbulence model. Coastal Engineering, 15, 431-458.

De Vriend H.J. and Kitou, N. (1990). Incorporation of wave effects in a 3D hydrostatic mean current model. Proc. 22nd Coastal Eng. Conf., ASCE, 1005-1018.

De Vriend, H.J. and Stive, M.J.F. (1987). Quasi-3D Modelling of Nearshore Currents. Coastal Engineering, 11, 565-601.

Longuet-Higgins, M.S. (1953). Mass transport in water waves. Philos. Trans. R. Soc., Ser. A, 245, 535-581.

Nairn, R.B., Roelvink, J.A. and Southgate, H.N. (1990). Transition zone width and implications for modelling surfzone hydrodynamics. Proc. 22nd Conf on Coastal Eng, ASCE, pp 68-81.

Rivero, F.J. and S.-Arcilla, A. (1994). On the vertical distribution of $<uw>$. Abstract MAST Overall Workshop, Gregynog, Wales.

Stive, M.J.F. and De Vriend, H.J. (1987). Quasi-3D nearshore current modelling: wave-induced secondary currents. Proc. Special Conf. on Coastal Hydrodynamics, ASCE, 356-370.

Stive, M.J.F. and De Vriend, H.J. (1994). Verification of the shear stress boundary conditions in the undertow problem. Proceedings Coastal Dynamics '94 (in press).

Svendsen, I.A. (1984). Wave Heights and Set-up in a Surf Zone. Coastal Engineering, 8, 303-330.

APPENDIX

MEAN SURFACE SHEAR STRESS DUE TO A CHANGING ROLLER

Rolf Deigaard

Institute of Hydrodynamics and Hydraulic Engineering, Technical University of Denmark, Building 115, DK-2800 Lyngby, Denmark

The apparent inconsistency, mentioned in Section 4 of the above paper, is caused by the complicated situation that arises when the volume of the roller is changing in the propagation direction. In this case the formulation of the energy dissipation and of the surface shear stress will have to be modified. In the following, the notation of the paper is used.

Due to the existence of a shear layer between the roller and the wave a shear stress (averaged over a wave length) τ_t exists which is assumed to apply at the mean water level. In addition there is a net transfer of water from the wave to the roller of:

$$\frac{1}{L}\frac{dR}{dt} = \frac{c}{L}\frac{dR}{dx} = \frac{1}{T}\frac{dR}{dx} \tag{A.1}$$

We are interested in the horizontal momentum transferred from the roller to the wave. If the roller is losing water ($dR/dx < 0$) this momentum is not just τ_t, but also the momentum of the water leaving the roller, which has a horizontal velocity of c. If the roller gains water ($dR/dx > 0$), the water leaving the wave has a negligible horizontal velocity, and it does not change the momentum of the water remaining in the wave.

If there is a net transfer of water to or from the roller this water will be mixed with the water in the roller ($dR/dx > 0$) or in the wave ($dR/dx < 0$). In both situations there is an additional energy dissipation, just as energy is lost when a rain drop hits the wind screen of a fast car, or if the car loses a drop of water on the road. The additional energy dissipation is:

$$\rho \frac{1}{2} c^2 \left|\frac{dR}{dt}\right| \tag{A.2}$$

The force between the roller and wave is split into a (symmetric) shear stress and a contribution from the net transfer of water makes the analysis complex. In Deigaard (1993) the shear stress in the shear layer was modelled as a Reynolds' stress with an exchange of water between the roller and the wave. A variation in the roller volume is then represented by a difference between the rate of fluid moving up and down.

With the modified energy dissipation and momentum transfer to the wave the analysis can proceed for the two cases, a decreasing roller and a growing roller:

Case 1: $\dfrac{dR}{dx} < 0$

Dissipation rate:

$$D = \tau_t c - \rho \frac{1}{2} \frac{c^2}{L} \frac{dR}{dt} = \tau_t c - \rho \frac{1}{2} c^2 \frac{c}{L} \frac{dR}{dx} = \tau_t c - \frac{\rho}{T} \frac{c^2}{2} \frac{dR}{dx} \qquad (A.3)$$

Energy dissipation expressed as a gradient in the energy flux:

$$D = -\frac{dE_f}{dx} - \frac{\rho}{T} \frac{d(\tfrac{1}{2} Rc^2)}{(dx)} \qquad (A.4)$$

Horizontal momentum transferred to the wave surface (averaged over a wave length):

$$\tau_s = \tau_t - \rho \frac{c}{L} \frac{dR}{dt} = \tau_t - \rho \frac{c}{L} c \frac{dR}{dx} =$$

$$\tau_t - \rho \frac{c}{T} \frac{dR}{dx} = \left(\frac{D}{c} + \frac{\rho}{T} \frac{c}{2} \frac{dR}{dx} \right) - \rho \frac{c}{T} \frac{dR}{dx} =$$

$$-\frac{1}{c} \frac{dE_f}{dx} - \frac{\rho}{2Tc} \frac{d(Rc^2)}{dx} - \frac{\rho}{2T} \frac{cdR}{dx} = \qquad (A.5)$$

$$-\frac{1}{c} \frac{dE_f}{dx} - \frac{\rho}{2T} \frac{dRc}{dx} - \frac{\rho}{2T} \frac{Rdc}{dx} - \frac{\rho}{2T} \frac{cdR}{dx} =$$

$$-\frac{1}{c} \frac{dE_f}{dx} - \frac{\rho}{T} \frac{dRc}{dx} = -\frac{1}{c} \frac{dE_f}{dx} - \frac{d(2E_r)}{dx}$$

Case 2: $\dfrac{dR}{dx} > 0$

Dissipation rate:

$$D = \tau_t c + \rho \frac{1}{2} \frac{c^2}{L} \frac{dR}{dt} = \tau_t c + \frac{\rho}{T} \frac{c^2}{2} \frac{dR}{dx} \qquad (A.6)$$

Horizontal momentum transferred to the wave surface:

$$\tau_s = \tau_t = \frac{1}{c}D - \frac{\rho}{T}\frac{c}{2}\frac{dR}{dx} =$$

$$-\frac{1}{c}\frac{dE_f}{dx} - \frac{\rho}{2Tc}\frac{d(Rc^2)}{dx} - \frac{\rho}{2T}\frac{cdR}{dx} = \qquad (A.7)$$

$$-\frac{1}{c}\frac{dE_f}{dx} - \frac{\rho}{T}\frac{dRc}{dx} = -\frac{1}{c}\frac{dE_f}{dx} - \frac{d(2E_r)}{dx}$$

As appears from the above derivations, an additional energy dissipation occurs due to a growing as well as a decreasing roller volume which causes additional work done, and consequently increases the shear stress, expressed by τ_s, which we have to apply at the mean water level. This removes the apparent inconsistency as noted in Section 4.

CHAPTER 45

Prediction of the maximum wave on the coral flat

Dede. M. Sulaiman[1], Shigeaki Tsutsui[2], Hiroshi Yoshioka[3], Takao Yamashita[3],

Shinichi Oshiro[4], and Yoshito Tsuchiya[5], M. ASCE

Abstract

Beach erosion has become a major problem at coral-sand beaches in Bali, Indonesia, due to the rapid changes in natural environment and utilization of coastal areas. An observation of waves was carried out at Sanur beach in Bali to know what waves are effective for beach processes. The representative waves are predicted by use of the mild-slope and KdV equations with evaluation of the coefficient of bottom friction. Empirical formulae of wave energy dissipation are proposed for taking into account of wave height damping due to breaking on the coral flat. The maximum wave on the coral flat, occurring in the reforming region of waves after breaking near the reef edge, can be decided by the theoretical results of the highest wave of permanent-type on water of uniform depth. Sanur beach has observed swells with a period longer than 15 sec, being incident onto the coral flat without breaking. This maximum wave of non-breaking progressive-type is, therefore, evaluated by the experimental wave height criteria.

INTRODUCTION

Indonesia, one of the largest archipelagos, is located in the tropics and is consisted of about 13,7000 islands. Among them the total area of islands surrounded with the coral reef, such as Bali, is about 1,900,000 km². Many rivers in the south parts of Bali island, as shown in Figure 1, supply sand sediment sources into the beach. The length of coastline of Bali is about 430 km, and main coastal engineering problems are beach erosion, river mouth closing, and tidal flood (Syamsudin, 1993). Especially in Sanur, Kuta, and Nusa Dua

---

[1] Researcher, Research Institute for Water Resources Development, Ministry of Public Works, Bandung, 40135, Indonesia.
[2] Professor, Dept. of Civil Engineering and Architecture, Faculty of Engineering, Univ. of the Ryukyus, Okinawa, 903-01, Japan.
[3] Instructor, Disaster Prevention Research Institute, Kyoto Univ., Uji, 611, Japan.
[4] Graduate Student, Civil Engineering and Architecture, Faculty of Engineering, Univ. of the Ryukyus, Okinawa, 903-01, Japan.
[5] Professor Emeritus, Kyoto Univ., and Professor, Meijo Univ., Nagoya, 468, Japan.

coral-sand beaches, located nearby the southern peninsular of coral limestone forming beautiful steep sea cliffs, beach erosion has become a major problem for which effective countermeasures are needed. There should be interrelation between coastal processes concerned with beach erosion in these beaches and changes in natural environment due to utilization of coastal areas. Clarification of the relation is very useful for other coral-sand beaches to predict what beach processes will occur in the future.

The successful control of coral-sand beach processes requires the knowledge of prediction of the maximum wave on the coral flat, i.e., what waves are the most effective for beach change. As the first step, therefore, an observation of waves, as joint investigation between the Research Institute for Water Resources Development and the Disaster Prevention Research Institute, Kyoto University, has been carried out at Sanur beach. This paper examines changes in wave heights and periods by using wave data, and predicts wave transformation in the coral reef beach, both by the mild-slope and KdV equations, and the maximum wave on the coral flat, where the effects of wave breaking and bottom friction are introduced.

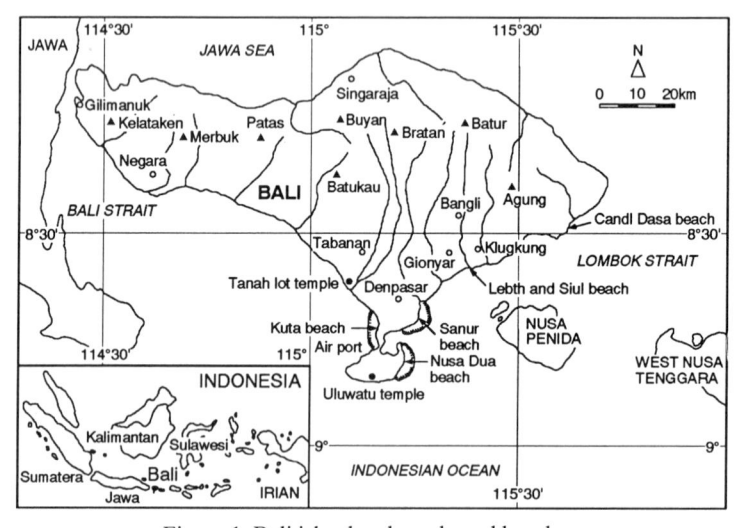

Figure 1. Bali island and coral-sand beaches.

COASTAL ENVIRONMENTS

Field measurements

(1) Bathymetry near Sanur coral reef beach and wave measuring stations

As seen in Figure 2(1), Sanur beach is about 7 km long having coral flats, 500 - 800 m wide. The coral reef beach profile is of slope-type, as seen in Figure 2(3), where the water depth increases gently in the outer reef with the bottom slopes of 1/20 - 1/50. This geometric feature is different from that in the Okinawa islands, Japan, where the step and barrier-type coral reefs are formed.

The linear array of five wave gauges was set up perpendicular to the reef edge, as shown in Figure 2(2) by the circles. The offshore station was set up at the location of 20 m water depth for measuring incident waves, and the others (St.1, 2, 3, and 4) were installed on the coral flat at intervals of 100 m from the reef edge. The offshore wave gauge is supersonic-type and those on the coral flat are of capacitance-type.

(1) Bathymetry near Sanur and Nusa Dua beaches.

(2) Wave array installed at Sanur beach, shown by circles.

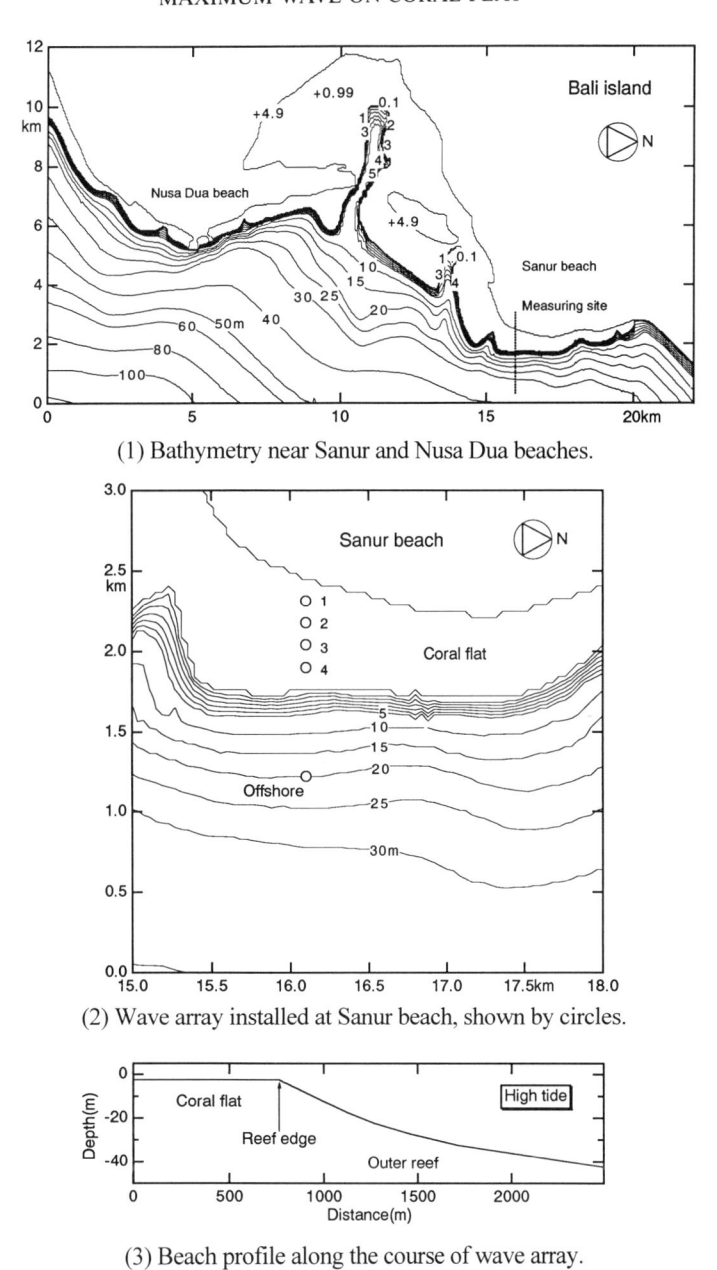

(3) Beach profile along the course of wave array.

Figure 2. Bathymetry around Sanur beach in Bali and the wave probe linear array.

(2) The period of observation

The seasonal variations of incident wave characteristics at Sanur coral reef beach, shown in Figure 3, indicate significance of the waves in the dry season for the beach process. The wave direction is almost unchanged throughout the year and the maximum wave heights and the corresponding wave periods (H_{max}, T_{max}) in the dry season (June - August) are greater than those in the wet season (December - February). The observation was therefore carried out in the period of July 16 - 19, 1992 under the usual sea conditions. After preliminary measurements on July 16, three runs of measurements were put into operation from 11:00 to 24:00 on July 17, from 9:00 on July 18 to 1:00 on July 19, and from 9:00 to 17:00 on July 19. The sample size of the data used is 20 minutes for all measuring stations, and the data were obtained at intervals of two hours for the offshore station and one hour for the other stations on the coral flat. At the offshore station, in addition, two-directional (NS and EW) water particle velocities were measured for a period of 20 minutes in every two hours, by using the sonic-type current meter.

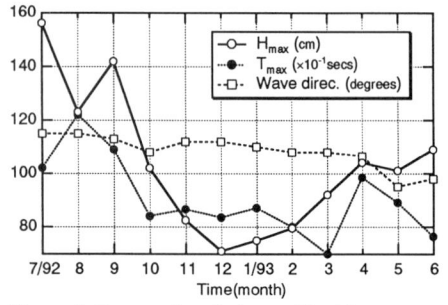

Figure 3. Seasonal variations of incident wave characteristics at Sanur beach.

(3) Wave direction

Figure 4 shows velocity vectors at the offshore measuring station, which were estimated from the two-directional (NS and EW) components of water particle velocities at high (12:00, 24:00) and low (18:00) tides. In the expression, the velocity components are normalized by each maximum value. The predominant directions of currents are ESE and WNW, indicating that offshore waves are incident from the direction of ESE. According to the results of numerical calculation of wave rays (Syamsudin, 1993), waves propagating in the Indonesian Ocean from the direction of SE are coming onto the beach, nearly perpendicular to the reef edge. Judging from the bathymetry of the beach in Figure 2(2) and the velocity vectors in Figure 4, offshore waves can be assumed to be incident normally to the reef edge in the period of the present observation.

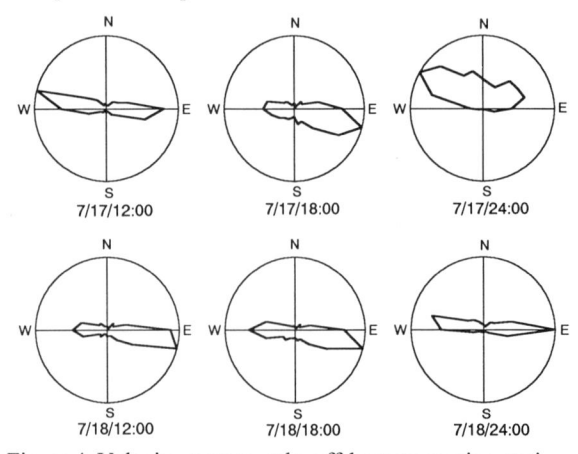

Figure 4. Velocity vectors at the offshore measuring station.

Wave characteristics

(1) Power spectra and wave energy dissipation

According to the tidal record measured at Benoa harbor nearby Sanur beach, the highest tidal level of water rises up to +2.3 m and +2.5 m in the period of July 17 - 18, as shown in Figure 5 by thin solid lines. The water depth at low tide is extremely shallow so that there appear some places dried up. The significant wave heights and periods ($H_{1/3}, T_{1/3}$) measured in the reef vary in phase with the tide level, as shown in Figure 5. Waves can enter onto the coral reef at high tidal phase of several hours, but the wave heights decrease gradually due to bottom friction with wave propagation toward the shore. Waves in the reef are, therefore, subjected to the water depth on the coral flat and incident waves at high tide are effective for coral-sand beach processes.

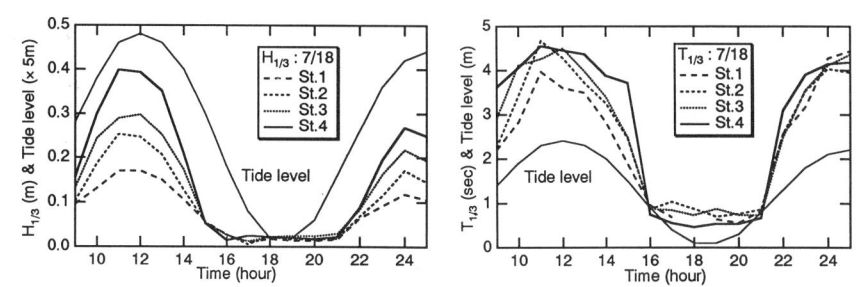

Figure 5. Significant wave heights (left) and periods (right), varying with the tide level on the coral flat (July 18).

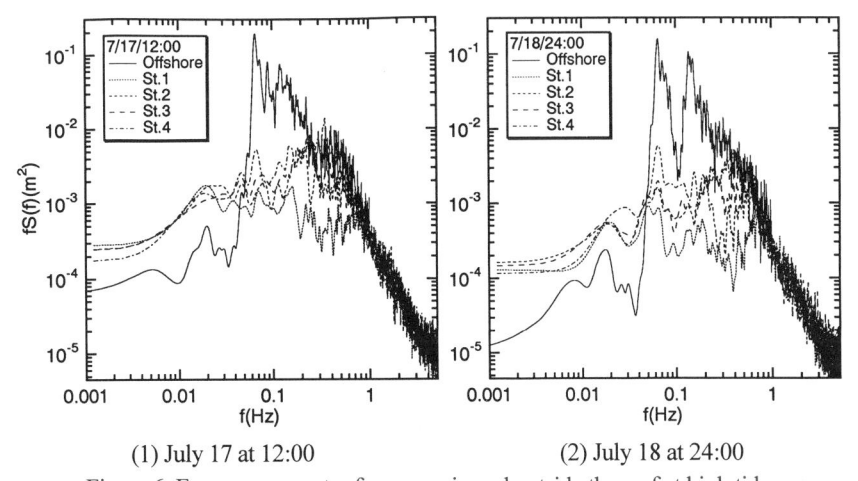

(1) July 17 at 12:00   (2) July 18 at 24:00

Figure 6. Frequency spectra for waves in and outside the reef at high tide.

Figure 6 shows typical examples of frequency spectra at high tide on July 17 - 18. In the ordinates, power spectra $S(f)$ are multiplied by the frequency f to obtain dominant peak frequencies clearly (Kai, 1985). The sampling time is 0.5 sec for offshore data and 0.1 sec for data on the coral flat.

Power spectra of offshore waves (solid lines) have several dominant peaks, showing that the waves include swells with a period of about 15 sec and wind waves with periods less than 8 sec. As can be seen in spectra at the stations 1 to 4 on the coral flat, high energy components with periods of 3 - 20 sec in offshore waves decrease suddenly by breaking near the reef edge. Long waves less than 0.05 Hz are then generated clearly on the reef. These long waves might be consisted of two components; incident swells onto the coral flat without breaking near the reef edge and waves generated by nonlinear interaction of incident and reflected waves on the reef, resulting in wave setup (Goda, 1975; Hino et al., 1989). The former can be confirmed in the following discussion on Figure 7. The power spectral shape on the coral flat becomes flat and the spectral width seems to be wider than that of offshore ones. Note, however, that the energy of swells with a period of about 15 sec, though decrease by breaking, is still predominant in the reef, as clearly shown in Figure 6(2). At the lowest tidal phase, on the other hand, the water depth becomes very shallow, and there exist strong offshore-wards currents from the reef edge, as shown in Figure 4. A few waves thus coming into the reef flat, most waves measured on the coral flat are, as shown in Figure 5, wind waves with periods less than 1 sec, of which the energies are very small.

If the frequency region of power spectra at high tide is divided into three regions; long waves ($f < 0.04$), swells ($0.04 < f < 0.1$), and wind waves ($f > 0.1$), respectively, each wave energy component E_i changes in and outside the reef due to wave breaking and bottom friction, as shown in Figure 7. The abscissa indicates the measuring stations and the ordinate does the energies relative to the total energy of incident waves, E_T. It is noted that the energy of long waves with periods greater than 25 sec, having small partition rate of the wave energy, are unchanged both in and outside the reef. The result shows that the coral reef in the beach is not effective for dissipating long period waves.

(2) Significant waves

At high tidal phase the significant wave heights and periods ($H_{1/3}, T_{1/3}$) change in and outside the reef, i.e., in the wave propagation direction, as shown in Figure 8. The abscissa shows the measuring stations and the subscript 0 denotes quantities of

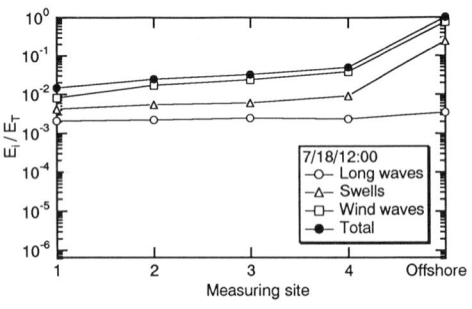

Figure 7. Wave energy dissipation on Sanur coral flat due to wave breaking and bottom friction at high tide (July 17 at 12:00).

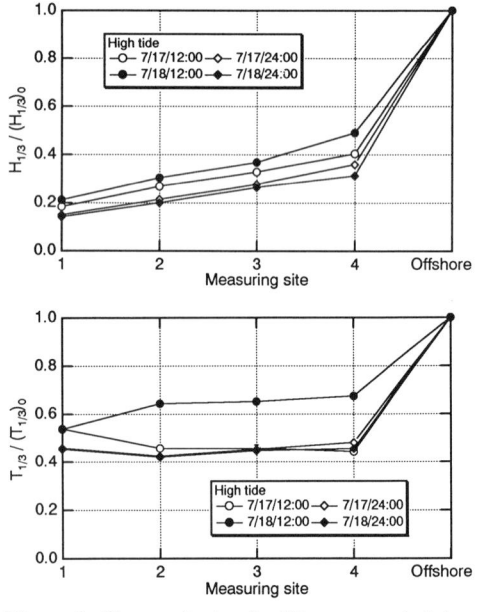

Figure 8. Changes in the significant wave heights (upper) and periods (lower) at high tide.

the offshore incident waves. After entering onto the coral flat, the significant wave heights decrease suddenly by breaking nearby the reef edge, and furthermore decrease gradually due to bottom friction on the coral flat. Similarly, the significant wave periods change suddenly near the reef edge, but they remain nearly constant on the coral flat. The main reasons of decrease in wave period are due to short-period waves generated in the course of wave breaking, and, probably, fission of long-period waves into solitons taking places on the coral flat. The prediction of wave properties should become taking into account of the change in wave period caused both by wave breaking and fission.

PREDICTION OF REPRESENTATIVE WAVE HEIGHTS

Model equations

The mild-slope (MS) and KdV equations are applied herein to predict representative wave heights for irregular waves on the coral flat. The MS equation is useful for estimation of transformation of waves propagating on the step-type reef with bathymetric discontinuity (Tsutsui & Zamami, 1993), as shown schematically on the left in Figure 9. Coral reefs of step-type are formed in the Okinawa islands, but those of slope-type are formed in Sanur beach, as shown schematically in the figure. As the offshore-side bottom slope changes gradually from 1/20 to 1/50, the KdV equation can be applied, but not for the step-type reef. Notice that all the physical quantities in this section are written in dimensionless form by using the representative quantities, the length being the nominal water depth h_0, the time $\sqrt{h_0/g}$, the velocity $\sqrt{gh_0}$, and the acceleration of gravity g.

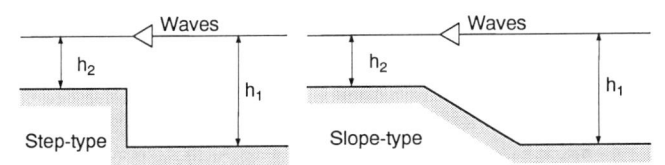

Figure 9. Step and slope-type reefs.

(1) The MS equation
The MS equation with the effects of wave energy dissipation due to wave breaking and bottom friction is given in the form (Dalrymple et al., 1984)

$$\nabla(cc_g\nabla\eta) + \left[(c_g/c)\sigma^2 - i\sigma f_d\right]\eta = 0 \qquad (1)$$

where $\nabla \equiv (\partial/\partial x, \partial/\partial y)$ is the horizontal differential operator, (x, y) are the horizontal coordinates, c is the wave velocity, c_g the group velocity, η the water surface displacement, $\sigma = 2\pi/T$ the frequency, T the wave period, and f_d the dimensionless coefficient of wave energy dissipation due to breaking and bottom friction. In the present study, after transformation into the two-dimensional form (Tsutsui, 1991), the MS equation (1) is calculated numerically by the finite element method.

(2) The KdV equation
The KdV equation with the effects of wave energy dissipation due to wave breaking and bottom friction is given in the form (Tsutsui, 1986; Yasuda & Nishio, 1990)

$$\eta_x + \frac{3}{2}h^{-3/2}\eta\eta_\xi + \frac{1}{6}h^{1/2}\eta_{\xi\xi\xi} + \frac{1}{4}h^{-1}h_x\eta = \frac{1}{2}\kappa h^{-3/2}\eta_{\xi\xi} - \frac{1}{2}sh^{-2}\eta|\eta| \qquad (2)$$

with $\xi = \int h^{-1/2}dx - t \qquad (3)$

where h is the local still water depth, κ the dimensionless coefficient of wave energy dissipation, s the coefficient of bottom friction, and the subscripts denote partial derivatives. Note that the wave energy dissipation is estimated by two terms on the right-hand side of Eq.(2). The term of diffusion, the first term, plays an important role in evaluating the effect of wave height damping due to breaking, and the second term does due to bottom friction. The equation (2) can be calculated numerically by the finite difference method (Tsutsui, 1986).

The sudden decrease in wave periods due to fission of long-period waves into solitons just after wave incidence upon the reef can be confirmed numerically, by making use of the KdV equation without wave breaking and bottom friction. Figure 10 shows an example for a gradually varying reef shown in the bottom of the figure, where U_r is the Ursell number, $\varepsilon \equiv h_2/h_1$ the ratio of the shallower water depth h_2 on the reef to the offshore water depth h_1, H is the wave height, and the subscript i stands for quantities of an incident wave. The wave height and period vary periodically with respect to wave propagation because of nonlinear interaction of solitons. Just after the fission, its period decreases suddenly to a value less than half of the incident wave period. As shown in Figure 8, the value nearly equals those obtained from observation. In the course of wave propagation the wave period becomes longest in the phase of reformation of waves, at which the wave height decreases a little.

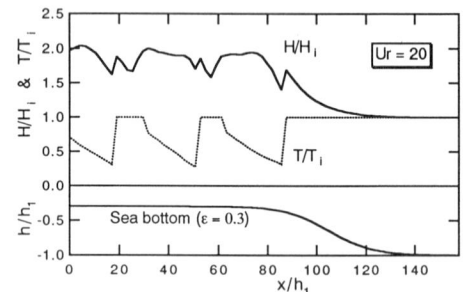

Figure 10. Changes in the wave height and period on the reef due to fission of a wave into solitons.

Coefficient of energy dissipation

In order to use Eqs.(1) and (2) for calculating wave properties on the coral reef, the coefficient of wave energy dissipation due to breaking and bottom friction must be evaluated. The equation of wave energy conservation is given by

$$\frac{d}{d\xi}(Ec_g) = -f_d E \qquad (4)$$

where E is the total wave energy and ξ is the horizontal coordinate in the direction of wave propagation. We use the expression for the coefficient of wave energy dissipation, f_d, introduced by Izumiya & Horikawa (1984), considering bottom friction, under the assumption of homogeneity of turbulence in the breaker zone. As the total water depth including wave setup is used in the expression, using linear wave theory, it can be modified (Yamashita et al., 1990; Tsutsui & Zamami, 1993) as

$$f_d = \left[\frac{1}{2} C_f^* + \frac{\beta_0}{8} \sqrt{\left(\frac{H}{h}\right)^2 - \left(\frac{H}{h}\right)_s^2} \right] \frac{H}{h} \frac{c_g}{h} \qquad (5)$$

in their notations, where $(H/h)_s$ is the minimum wave height relative to the water depth in the wave damping region, C_f^* the coefficient of bottom friction, and β_0 the dimensionless coefficient related with wave energy dissipation due to breaking.

Based on the experimental estimation of the coefficient β_0 for the step-type reef, the wave height distribution in the breaker zone can be estimated correctly (Tsutsui & Zamami,

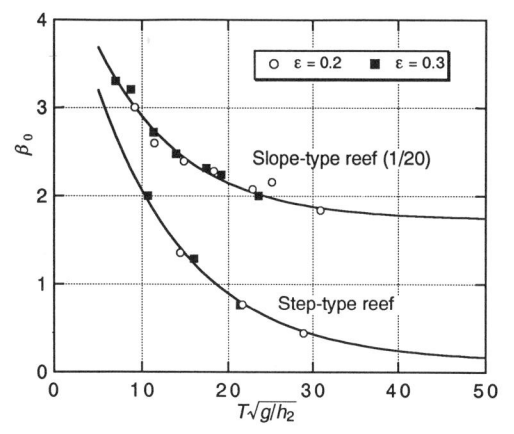

Figure 11. Coefficients β_0 for the slope and step-type reefs.

1993). Similarly, we can determine the coefficient β_0 on the slope-type reef, as shown in Figure 11, where the offshore slope is 1/20 and the abscissa indicates the dimensionless wave period $T\sqrt{g/h_2}$ on the coral flat. In the limiting state of $T\sqrt{g/h_2} \to \infty$, the value of β_0 on the slope-type reef can be approximated as $\beta_0 \approx \sqrt{3}$ (Appendix), by referring to the breaking model for the solitary wave (Le Méhauté, 1963). The empirical formulae of the coefficient β_0 are, for the slope-type reef,

$$\beta_0 = 3.27\exp\left[-0.1027T\sqrt{g/h_2}\right] + 1.73, \quad T\sqrt{g/h_2} > 5 \tag{6}$$

and for the step-type reef,

$$\beta_0 = 4.48\exp\left[-0.0921T\sqrt{g/h_2}\right] + 0.12, \quad T\sqrt{g/h_2} > 5 \tag{7}$$

Figure 12 shows typical examples of experimental wave height distributions for periodic waves on the slope-type reef, comparing with numerical results by the MS and KdV equations without bottom friction. The abscissa is taken as the distance from the reef edge and the ordinate denotes the dimensionless wave height. In the MS equation the coefficient of wave energy dissipation f_d with β_0 given by Eq.(6) is used in estimation of wave damping, whereas the coefficient of energy dissipation in the KdV equation is evaluated by

$$\kappa = f_d/4 \tag{8}$$

to express the effect of wave breaking by the term of diffusion. Generally, for shorter period waves the wave height distributions by the MS equation agree well with experimental ones, whereas

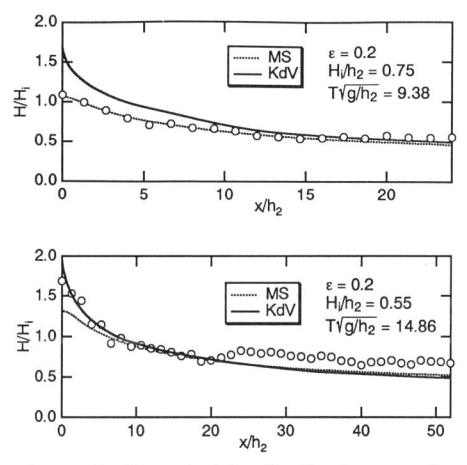

Figure 12. Wave height distributions on the slope-type reef.

the KdV equation gives good estimation of the wave height distributions for longer period waves. This reason is that, for longer period waves, the breaking wave heights at the reef edge estimated by KdV equation are agree fairly with experimental ones, but the MS equation shows to underestimate them. Therefore, the discrepancy between the calculated and experimental values of relative wave height near breaking point becomes remarkably. However, in a region far from the reef edge, both the MS and KdV equations are applicable for wave height estimation after wave breaking.

Wave height prediction on Sanur coral flat

As mentioned previously, wave setup takes place on the coral flat due to wave breaking. Hereafter, the amount of wave setup Δh is assumed to be 30 cm according to the results of

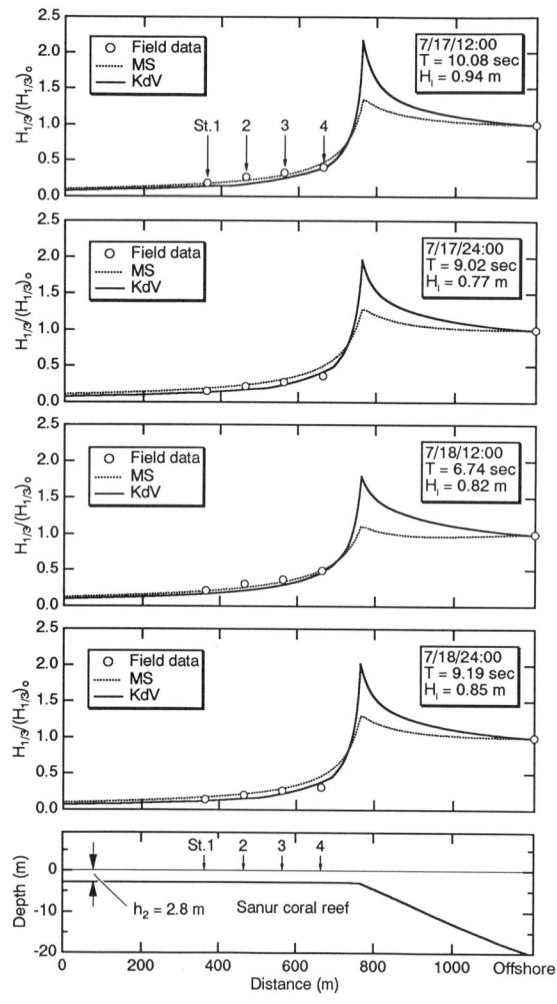

Figure 13. Wave height prediction on Sanur coral flat with the MS and KdV equations.

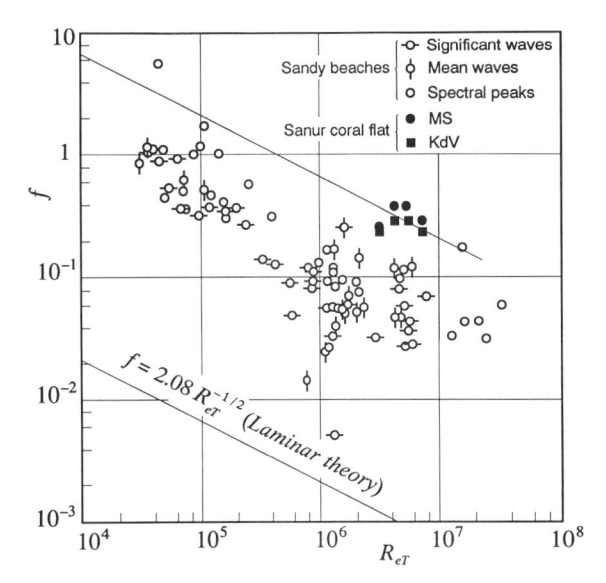

Figure 14. Coefficient of bottom friction on Sanur coral flat.

Goda (1975) and Hino et al. (1989).

When the significant waves on the coral flat at high tide were estimated by the MS and KdV equations, taking into account only of wave damping due to breaking, the values of wave heights were greater than those of observation. The result indicates the significant role of bottom friction to the wave energy dissipation on the coral flat. However, the value of the coefficient of bottom friction in the coral flat is unknown. Therefore, the coefficient of bottom friction was evaluated such that wave heights estimated coincide with the observed data. Figure 13 shows the results of prediction of significant wave heights on the coral flat at the highest tidal level of water, where the distance is taken offshore-ward from the shoreline. Discrepancy in numerical estimations by the MS and KdV equations (solid and dotted lines) becomes remarkably near the breaking point, i.e., the reef edge. The relative distances between the measuring station 4 and the reef edge are within $35 < x/h_2 < 40$, showing that all the measuring stations are located in the region where the damping of wave heights due to breaking is not effective, as shown in Figure 12. The MS and KdV equations, therefore, predict well the wave height distributions observed.

On the bottom friction, as seen in the data for sandy beaches shown by open symbols in Figure 14 (e.g., Iwagaki & Kakimuna, 1962), there exists a linear relation between the coefficient of bottom friction f and the wave Reynolds number $R_{eT} = u_{bmax}^2 T/\nu$, where $u_{bmax} = \sigma H/2\sinh kh$, k is the wave number, and ν is the kinematic viscosity. The values of bottom friction on Sanur coral flat are plotted by the solid symbols (●, ■). The coral flat composed of unevenness is covered by sea glasses so densely that the bottom friction is evaluated greater than those on sandy beaches. The values of bottom friction on the coral flat are not enough to design the relation, but may follow the linear relation shown in the figure by the solid line.

THE MAXIMUM WAVE ON THE CORAL FLAT

We consider two kinds of the maximum waves on the coral flat in relation to the rate of wave breaking near the reef edge. Waves of small wave heights can propagate on the coral flat without breaking but the waves may break after propagation of some distances when they become large. The first kind of the maximum wave is then for progressive waves propagating onto the coral flat without breaking. Figure 7 has already indicated the incident possibility of long-period waves into Sanur beach. Due to the low breaking rate of irregular waves at the reef edge, the maximum wave arriving at the shore is therefore subjected to this breaking criterion for progressive waves on the reef. As the results of experiments for periodic waves on the step-type reef (Tsutsui & Zamami, 1993), this breaking criterion is given by

$$\frac{H_b}{h_2} = \frac{c_0}{1 + c_1 \zeta + c_2 \zeta^2}, \quad \zeta \equiv \frac{2\pi h_2}{L_0} \tag{9}$$

with

$$\left. \begin{array}{l} c_0 = 2.45\,\varepsilon, \quad \text{for } \varepsilon < 0.2 \\ c_0 = 0.4\dfrac{1 + 1.34\sqrt{\varepsilon - 0.17}}{1 + 0.1\sqrt{\varepsilon - 0.17}}, \quad \text{for } \varepsilon \geq 0.2 \\ c_1 = 1 - 0.6\,\varepsilon, \quad c_2 = 0.02 \end{array} \right\} \tag{10}$$

On the other hand, when incident waves are relatively large to the water depth, they break very close to the reef edge on the step-type reef, forming partial standing waves in the offshore deeper region of the reef, but they break near the reef edge on the slope-type reef. In this case of high wave breaking rates of irregular waves, the second kind of the maximum wave arriving at the shore is subjected to the waves that are reformed after strong wave breaking.

As the results of experiments for irregular waves on the step-type reef, Figure 15 shows examples of the relationship between the wave height relative to the water depth on the coral flat, H/h_2, and the shallowness h_2/L_0, where all the individual waves measured in the reforming region after wave damping due to breaking are plotted by the open circle. In the parameters, L_0 is the wave length in deep water for the individual wave and H_0/L_0 is the wave steepness in deep water for the significant waves. In the upper figure (1), the significant wave height is smaller than the critical wave height shown by the solid line, Eqs.(9) and (10), and it seems that the rate of wave breaking is low. This solid line indicates the maximum wave, which shows the envelope of the data.

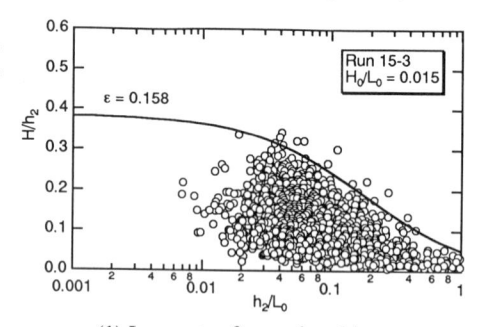

(1) Low rate of wave breaking.

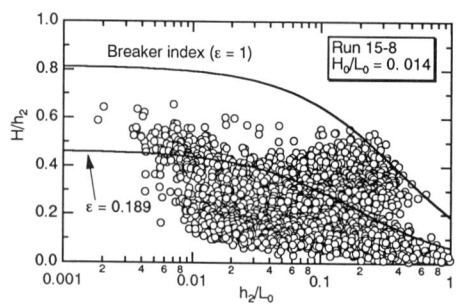

(2) High rate of wave breaking.

Figure 15. Maximum waves on the step-type reef according to the rate of wave breaking.

The result in the case of high wave breaking rate is shown in the lower figure (2), where the significant wave height is larger than the critical wave height shown by the lower solid line, and most waves break at the reef edge. The upper solid line with $\varepsilon = 1$ indicates the breaker index for the flat sea bottom (Goda, 1970), an approximation of the highest wave of permanent-type on water of uniform depth (Yamada and Shiotani, 1968), and it becomes the envelope of all the experimental data. Therefore, the maximum wave after strong wave breaking near the reef edge is determined by the breaker index for the flat sea bottom.

Similarly, the relation of the relative wave height H/h_2 on Sanur coral flat and the shallowness h_2/L_0 is shown in Figure 16, where all the individual waves measured at the station 4 near the highest tidal phase are plotted by the open circle. The breaker type near the reef edge is for progressive waves because Sanur beach is of slope-type reef with the offshore slopes of 1/20 - 1/50. Therefore, we can use Eqs.(9) and (10) as the breaking criterion, when the water depth ratio $\varepsilon = h_2/h_1$ is given as the ratio of the water depth h_2 on the coral flat to the offshore water depth h_1, where the shoaling coefficient given by the small amplitude wave theory takes unity, i.e., $h_1/L_0 = 0.057$. The solid lines in Figure 16 show the results. As $h_2/L_0 = 0.017 - 0.038$ and $H/h_2 = 0.303 - 0.346$ for the significant waves, the wave breaking rate is low. Therefore, the curves in Figure 15 indicate the empirical maximum wave on Sanur coral flat with envelopes for the field data. Furthermore, it has been confirmed by laboratory tests that the maximum wave on Sanur coral flat at the high wave breaking rate is also evaluated by the breaker index for the flat sea bottom, as well as for the step-type reef.

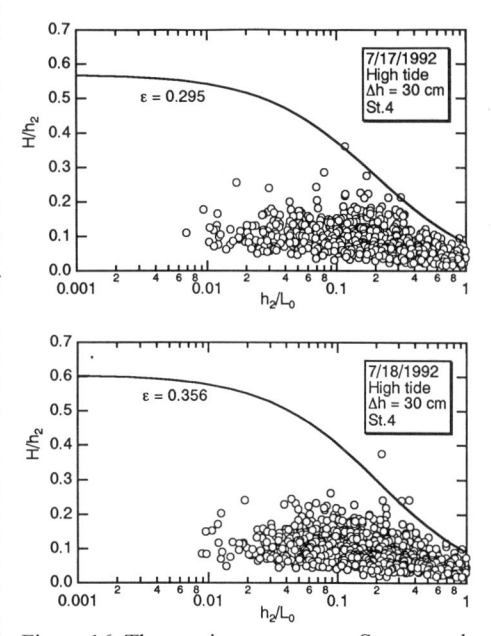

Figure 16. The maximum wave on Sanur coral flat.

CONCLUSIONS

Main conclusions of the present analysis of wave properties and the maximum wave based on the experimental and field data at Sanur coral beach are summarized as:
(1) In Sanur beach, waves in the dry season (June - August) are predominant for the beach process, and they are incident perpendicular to the reef edge. Swells with a period of about 15 sec are most effective for beach change.
(2) The MS and KdV equations with the effects of wave energy dissipation can predict the wave height distribution on the coral flat, when the coefficients of wave breaking and bottom friction are evaluated.
(3) In the case of the low wave breaking rate, the experimental wave height criteria for progressive waves, Eq.(9) and (10), give the maximum wave arriving at the shore without breaking. Contrarily, after full breaking of incident waves near the reef edge, the maximum wave on coral flats can be determined by the breaker index for flat sea bottom.

Acknowledgments

The present research, part of international joint investigation between the Research Institute for Water Resources Development and Disaster Prevention Research Institute, Kyoto University, associated with IDNDR, was supported by the Ministry of Education, Science and Culture of Japan. We wish to thank many individuals who have jointed the investigation, in particular, Dr. Badruddin Machbub, Director, Dr. Abdul R. Syamsudin, Chief, Water Resources division, Research Institute for Water Resources Development, Mr. T. Onuki, Nippon-Koei Co. Ltd., and Mr. K. Sakai, Toa Corporation, who kindly provided their supports and cooperation in carrying out this investigation.

Appendix: The coefficient β_0 for the slope-type reef as $T\sqrt{g/h_2} \to \infty$

If the small amplitude wave theory is applied in Eqs.(4) and (5), we have

$$\frac{d}{d\xi}(H^2 c_g) = -\frac{\beta_0}{8}\left[\left(\frac{H}{h}\right)^2 - \left(\frac{H}{h}\right)_s^2\right]\frac{H^3}{h^2} c_g \tag{A.1}$$

where the effect of bottom friction is neglected. Dividing by $H^2 c_g$ and integrating with respect to ξ within a small increment $\Delta\xi = \xi_2 - \xi_1$ give

$$\frac{H_2}{H_1} = \left(\frac{c_{g1}}{c_{g2}}\right)^{1/2} \exp\left[-\frac{\beta_0}{16}\sqrt{1-(H_s/H_b)^2}\left(\frac{H_1}{\bar{h}}\right)^2 \frac{\Delta\xi}{\bar{h}}\right] \tag{A.2}$$

where the subscripts 1 and 2 denote quantities at the locations ξ_1 and ξ_2, respectively, and $\bar{h}$ is the mean water depth within the increment $\Delta\xi$.

By introducing the bore model, Le Méhauté (1963) evaluated the energy dissipation rate of solitary wave as a spilling breaker as;

$$\frac{H_2}{H_1} = \frac{h_1}{h_2}\left(\frac{c_1}{c_2}\right)^{2/3} \exp\left[-\frac{\sqrt{3}}{16}\left(\frac{H_1}{\bar{h}}\right)^{5/2} B \frac{\Delta\xi}{\bar{h}}\right] \tag{A.3}$$

with $B = \dfrac{[(1 + H/h) - \beta(1 + \beta H/h)]}{(1 + H/h)(1 + \beta H/h)}(1 - \beta)^3$ (A.4)

where β is the dimensionless parameter that defines the range of surface disturbance due to wave breaking and then $0 \leq \beta \leq 1$. The condition that $\beta = 1$ corresponds to non-breaking ($B = 0$), and $\beta = 0$ to full breaking ($B = 1$). The coefficient B is then call the breaking coefficient. In case of non-breaking, Eq.(A.2) gives the so-called Green's law for long period waves as : $H_2/H_1 \approx (h_1/h_2)^{1/4}$, and Eq.(A.3) gives $H_2/H_1 \approx (h_1/h_2)^{4/3}$. This difference is due to the validity of using the solitary wave theory in analyzing wave motion on a slope. However, Eq.(A.3) is applicable for waves on the coral flat because the coefficients of the exponential terms in Eqs.(A.2) and (A.3) can be approximated as unity. Equation (A.2) must be asymptotic to Eq.(A.3) as $T\sqrt{g/h_2} \to \infty$. Comparing the coefficients in the exponential damping terms gives

$$\beta_0 = \sqrt{3}\sqrt{H_1/\bar{h}}\, D, \qquad D = B / \sqrt{1 - (H_s/H_b)^2} \tag{A.5}$$

It is easy to show that the effect of sea bottom slope on the parameter D in Eq.(A.5) through the wave height is very small, by using the breaker index (Goda, 1970) and assuming the breaking point to be the reef edge. Furthermore, waves on slopes steeper than 1/50 break completely (Le Méhauté, 1963). We can thus approximate as $\beta \approx 0$ and $D \approx 1$. Consequently, the coefficient β_0 for longer period waves depends only on the breaking wave height $(H/h)_b$, i.e.,

$$\beta_0 = \sqrt{3}\sqrt{(H/h)_b} \tag{A.6}$$

Since $\sqrt{(H/h)_b} = 1.01$ for solitary wave propagating on the gentle slope of 1/20, the coefficient β_0 can be approximated by $\beta_0 \approx \sqrt{3}$ when $T\sqrt{g/h_2} \to \infty$. Note that the breaking wave height takes values within $0.91 < \sqrt{(H/h)_b} < 1.17$ for the sea bottom conditions varying from the uniform water depth to the slope of 1/10.

References

Dalrymple, R. A., J. T. Kirby and P. A. Hwang (1984): Wave diffraction due to areas of energy dissipation, Jour. Waterway, Port, Coastal, and Ocean Eng., ASCE, Vol.110, pp.67-79.

Goda, Y. (1970): A synthesis of breaker indices, Proc. JSCE, Vol.180, pp.39-49 (in Japanese).

Goda, Y. (1975): Irregular wave deformation in the surf zone, Coastal Eng. in Japan, , JSCE, Vol.18, pp.13-26.

Hino, M., E. Nakaza, and S. Yogi (1989): Bore-like surf beat in reef coasts, Proc. Coastal Eng., JSCE, Vol.36, pp.75-79 (in Japanese).

Iwagaki, Y., T. Kakinuma (1962): On the coefficient of bottom friction at Akita coast, Proc. 9th Japanese Conf. on Coastal Eng., JSCE, pp.81-84 (in Japanese).

Izumiya, T. and K. Horikawa (1984): Wave energy equation applicable in and outside the surf zone, Coastal Eng. in Japan, Vol.27, pp.119-137.

Kai, K. (1985): Spectrum climatology of the surface winds in Japan, Part I: The 40-60 day fluctuations, Jour. Meteorological Soc. Japan, Vol.63, pp.873-882.

Le Méhauté, D. Sc. (1963): On non-saturated breakers and the wave run-up, Proc. 8th Conf. on Coastal Eng., ASCE, pp.77-92.

Syamsudin, A. R. (1993): Beach Erosion in Coral Reef Beaches and its Control, Doctoral Dissertation at Kyoto University, 221p.

Tsutsui, S (1986): Wind effects on the evolution of solitons, Proc. 33rd Japanese Conf. on Coastal Eng., JSCE, pp.61-65 (in Japanese).

Tsutsui, S. (1991): A wave prediction model in reef coasts, Bull. Fac. of Eng., Univ. of the Ryukyus, No.42, pp.35-43 (in Japanese).

Tsutsui, S. and K. Zamami (1993): Jump condition of energy flux at the line of bathymetric discontinuity and wave breaking on the reef flat, Coastal Eng. in Japan, JSCE, Vol.36, pp.155-175.

Yamada, H. and T. Shiotani (1968): On the highest water waves of permanent type, Bull. Disaster Prevention Research Inst., Kyoto Univ., Vol.18, Pt.2, pp.1-22.

Yamashita, T., Y. Tsuchiya, M. Matsuyama and T. Suzuki (1990): Numerical calculation of linear wave propagation in the coastal zone, Bull. Disaster Prevention Research Inst., Kyoto Univ., Vol.40, pp.15-40.

Yasuda, T. and Y. Nishio (1990): Model for random waves propagating in surf zones, Proc. Coastal Eng., JSCE, Vol.37, pp.66-70 (in Japanese).

CHAPTER 46

Development of a Submerged Doppler-Type Directional Wave Meter

Tomotsuka Takayama[1], Noriaki Hashimoto[2], Toshihiko Nagai[3]
Tomoharu Takahashi[4], Hiroshi Sasaki[5], and Yoshiki Ito[6]

Abstract

This paper presents a newly developed type of submerged directional wave meter having the capability to measure water surface elevation and multiple current velocity components. These components are determined from Doppler frequency shifts using the complex covariance method. The developed system was demonstrated to be capable of successfully obtaining directional wave data, as well as adequately estimating directional wave spectra.

1. Introduction

Many types of wave observation devices have been developed over the years for various scientific and engineering applications. Most, however, were designed to measure only wave height and period, and consequently, a practical instrument incorporating features for satisfactorily measuring directional seas has not yet been obtained.

The most commonly used wave meter in Japan is a submerged ultrasonic-type wave gage that can be installed at a water depth of up to about 50 m. A submerged ultrasonic-type current meter developed for measuring directional seas, however, is limited to 30-m-deep water due to a decay of water particle velocity caused by waves. Obviously then, there is a need to develop a combined measuring system that can operate at 50 m.

---
[1] Director, Hydraulic Engineering Div., Port and Harbour Research Institute (PHRI),
  1-1 Nagase 3-chome, Yokosuka 239, Japan
[2] Chief, Ocean Energy Utilization Lab., Hydraulic Engineering Div., PHRI.
[3] Chief, Marine Observation Lab., Marine Hydrodynamics Div., PHRI.
[4] Managing Director, Japan Marine Surveyors Association,
  14-12 Kodenma-cho, Nihonbashi, Chuo-ku, Tokyo 103, Japan
[5] Former manager, Measuring and Control System Div., Kaijyo Corporation,
  3-1-5 Sakae-cho, Hamura-shi, Tokyo 205, Japan
[6] Manager, Research and Development Div., Kaijyo Corporation.

This led to the present paper which describes a new device, named the submerged Doppler-type directional wave meter, that utilizes the Doppler effect of ultrasonic waves in water for measuring directional seas in coastal areas having a water depth of about 50 m.

2. Submerged Doppler-type directional wave meter

2.1 System description

Figure 1 shows a circuit block diagram of the measurement system of the submerged Doppler-type directional wave meter. This system is comprised of a submerged transducer, measurement section, and computation unit. The measurement section consists of transceiver and signal processing circuits. The former circuit connects the transducer to the latter one which converts the signals to water surface elevation and current velocities after digitizing received analog quantities. The computation unit statistically analyzes wave data and computes the resultant directional wave spectra.

To measure the water surface elevation, the transducer radiates vertically-upward-directed ultrasonic waves (Z-axis, **Fig. 1**) and then receives the waves reflected off the water surface. The water surface elevation is estimated using the wave propagation time. By also radiating ultrasonic waves upward in four directions separated by an inclination angle α, multiple current velocities are measured by receiving the waves scattered back from selected layers of water. The water particle velocity of each layer is estimated from the frequency shift caused by the Doppler

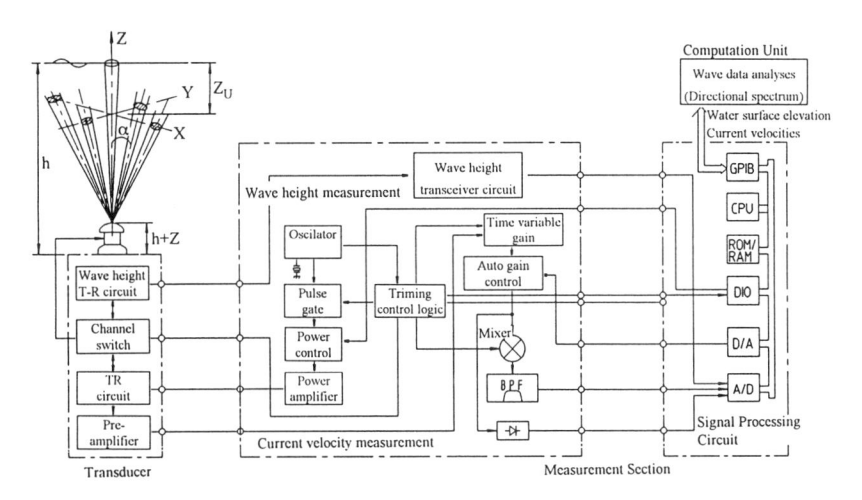

Fig. 1 Diagram of the measurement system of the submerged Doppler-type directional wave meter.

Table 1 Specifications of the Doppler-type directional wave meter.

Specification	Current Velocity	Wave Height
Mesurement system	Ultrasonic Doppler method	Ultrasonic wave propagation time
Operating frequency	500 kHz	200 kHz
Beam width	1	30
Mesurement channel	Orthogonal 4 channels	Vertical 1 channel
Beam tilt	300	—
Maximum output power	300 W	—
Mesurement range	0 to ± 5 m/s	0 ~ 30 m
Range resolution	2.3 m	—
Transmission interval	165 ms	165 ms
Mesurement period	900 ms	495 ms

effect between the radiated and scattered waves. Use of a time sharing system enables all these measurements to be successively carried out at very short time intervals. Specifications of the new wave meter are summarized in **Table 1**.

2.2 Doppler frequency estimation methods

Two different techniques exist for conducting Doppler frequency analysis: (1) use of analog signals, such as the phase lock loop (PLL) or analog filter bank methods; and (2) use of digital signals, such as the fast Fourier transformation (FFT) and complex covariance (CC) methods. Of these, real-time processing is normally performed with the FFT or CC method due to rapid developments in microcomputers and peripheral equipment.

The FFT method has an advantage in that even at a low S/N ratio it can analyze the scattered wave signal to estimate the spectral density over the entire frequency band. At a higher S/N ratio, however, the CC method is more advantageous because computing the spectral moment is easier and the frequency resolution is better.

Figure 2 shows a flow chart of the CC method's computational procedure, where the the input frequency signals $\cos(\omega_{in} t + \varepsilon)$ are first damped by the local oscillation frequency signals ($\cos\omega_0 t$ and $\sin\omega_0 t$) having a phase difference of 90°. Next, the first order moment μ_1 (average frequency) of the input signal is computed from the complex covariance function $R(\tau)$ obtained from the outputs $X_r(t_n)$ and $X_i(t_n)$. More specifically, the frequency components of $\omega_{in} + \omega_0$ and $\omega_{in} - \omega_0$ are obtained by mixing the input signals $\cos(\omega_{in} t + \varepsilon)$ with the cos and sin components of the local oscillation signals having frequency ω_0. The difference component $\omega_{in} - \omega_0$

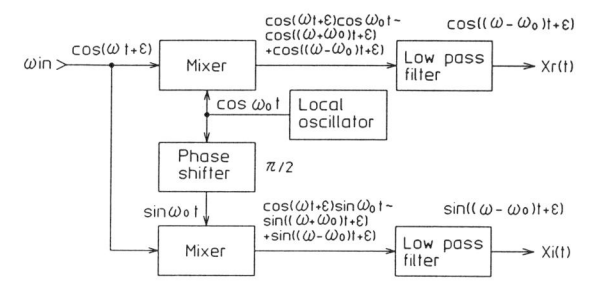

Fig. 2 Analog method for generating complex signals.

is then sent to a low pass filter (LPF) and A/D converter where it digitized. From $X_r(t_n)$ and $X_i(t_n)$, the complex values $Z(t_n)$ are obtained by $Z(t_n) = X_r(t_n) + iX_i(t_n)$, followed by computing μ_1 using

$$R(\tau) = \frac{1}{N}\sum_{n=1}^{N-1} Z(t_n)Z^*(t_n+\tau) \qquad (1)$$

$$\mu_1 = \frac{1}{2\pi\tau}\tan^{-1}\frac{\mathrm{Im}[R(\tau)]}{\mathrm{Re}[R(\tau)]}, \qquad (2)$$

where $Z^*(t_n)$ is the complex conjugate of $Z(t_n)$.

This procedure can also be performed by the simple homodyne complex covariance (SHCC) method which simplifies the computation (Ito et al., 1989). **Figure 3** shows a flow chart of this method's data processing procedure. Since the received signal is digitized, the sampling frequency f_s is easily synchronized with the transmitting wave which is four times larger than the transmitting frequency f_0. The reference signals for the homodyne detection can then be simulated as

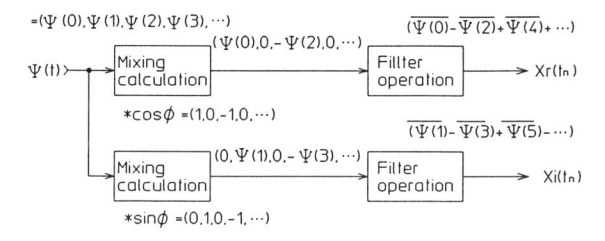

Fig. 3 Data processing for the SHCC method.

$$\cos\phi = (1,0,-1,0,1,0,-1,\cdots)$$
$$\sin\phi = (0,1,0,-1,0,1,0,\cdots).$$
(3)

Therefore, the outputs of homodyne circuit can be approximated by

$$\psi(t)\cos\phi = \psi(0),0,-\psi(2),0,\psi(4),0,\cdots$$
$$\psi(t)\sin\phi = 0,\psi(1),0,-\psi(3),0,\psi(5),\cdots,$$
(4)

where $\psi(t)$ is the received signal. If we utilize a LPF, by making block averages (16 points) of these series, the complex beat signal series, $X_r(t)$ and $X_i(t)$, can be obtained. Then $R(\tau)$ is subsequently determined by Eq. (1) and the Doppler or beat frequency μ_1 can be estimated as per the CC method (Eq. (2)).

When background noise is approximately stationary during a pulse repetition period, noise compensation can be accomplished using

$$\mu_1 = \frac{1}{2\pi\tau}\tan^{-1}\frac{\mathrm{Im}[R(\tau)]-\mathrm{Im}[R_n(\tau)]}{\mathrm{Re}[R(\tau)]-\mathrm{Re}[R_n(\tau)]},$$
(5)

where $R_n(\tau)$ indicates the complex covariance of the noise process.

Field experimental data shows this method estimates the current velocity at a S/N ratio corresponding to actual sea conditions with a mean error of about 2 to 6 cm/s and standard deviation of about 1 to 2 cm/s.

2.3 Analysis of directional spectrum

The extended maximum likelihood method (EMLM) developed by Isobe et al. (1984) was employed to estimate the directional spectrum because of its high versatility and relatively high accuracy.

Using polar coordinates (**Fig. 4**), the water particle velocity U in the direction r with coordinates (α,β,r) is obtained by

$$U(\alpha,\beta,r,h,z_0;\omega,\theta) = H(\alpha,\beta,r,h,z_0;\omega,\theta)\eta(\alpha,\beta,r;\omega,\theta),$$
(6)

where η is the water surface elevation and H the transfer function between U and η. H can be expressed using linear wave theory as follows:

$$H(\alpha,\beta,r,h,z_0;\omega,\theta)$$

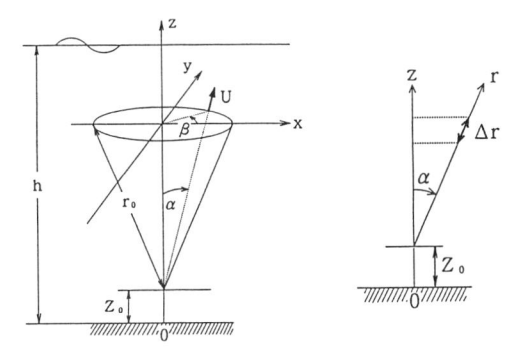

Fig. 4 Definition of polar coordinates.

$$= \frac{\omega}{\sinh kh}\Big[\cosh\{k(r\cos\alpha + z_0)\}$$
$$\times \sin\alpha \cos(\theta - \beta) \qquad (7)$$
$$-i\sinh\{k(r\cos\alpha + z_0)\}\cos\alpha\Big],$$

where h, k, ω, θ, and z_0 respectively represent the water depth, wave number, angular frequency, wave propagation direction, and height at which the meter is installed above the seabed.

When the water surface elevation η_0 above the origin $(0,0,0)$ is used as a basis, the transfer function between U and η_0 is obtained by multiplying Eq. (7) by $\exp[i\{kr\sin\alpha\cos(\theta-\beta) - \omega\Delta t\}]$, i.e.,

$$H_0(\alpha,\beta,r,h,z_0;\omega,\theta)$$
$$= \frac{\omega\exp(-i\omega\Delta t)}{\sinh kh}\Big[\cosh\{k(r\cos\alpha + z_0)\}$$
$$\times \sin\alpha\cos(\theta-\beta) - i\sinh\{k(r\cos\alpha + z_0)\} \qquad (8)$$
$$\times\cos\alpha\Big]\exp\{ikr\sin\alpha\cos(\theta-\beta)\},$$

where Δt is the time lag between the measurement of each velocity component and that of η_0.

The current velocity component detected by a Doppler-type wave meter is not the water particle velocity at a specified location, but instead an average velocity in a volume of known width Δr. Therefore, by disregarding the width around the r axis and integrating Eq. (8) in terms of r, the transfer function between U in the direction r with coordinates (α, β, r) and η_0 above the origin $(0,0,0)$ is represented as:

$$\begin{aligned}\overline{H}&(\alpha,\beta,r_0,\Delta r, z_0;\omega,\theta) \\ &= \frac{1}{\Delta r}\int_{r_0-\Delta r/2}^{r_0+\Delta r/2} H_0(\alpha,\beta,r,h,z_0;\omega,\theta)dr \\ &= \frac{-i\omega\exp(-i\omega\Delta t)}{\Delta r k \sinh kh}\Big[\cosh\{k(r\cos\alpha+z_0)\} \\ &\quad \times \exp\{ikr\sin\alpha\cos(\theta-\beta)\}\Big]_{r_0-\Delta r/2}^{r_0+\Delta r/2}.\end{aligned} \quad (9)$$

Employing Eq. (9) leads to the estimation equation of the directional spectrum by the EMLM, i.e.,

$$S(f,\theta) = \frac{\kappa}{H^{*t}\Phi^{-1}H}, \quad (10)$$

where H is the matrix comprised of the transfer functions given by Eq. (9), H^{*t} the complex conjugate of the transpose matrix of H, Φ^{-1} the inverse matrix of matrix Φ consisting of the cross-power spectra between each quantity, and κ is a proportionality constant used to normalize the energy of the directional spectrum.

3. Field observations

Field observations were carried out off the entrance of Kamaishi Bay (**Fig. 5**) after installing the transducer 0.95 m above the seabed at a water depth of 35 m. The

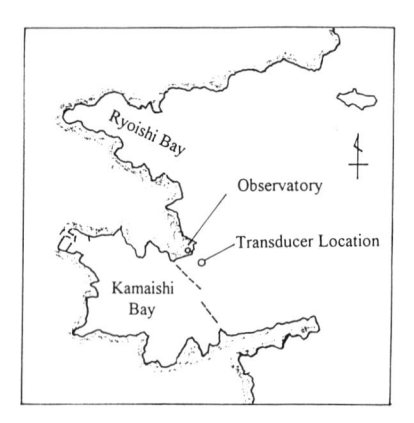

Fig. 5 Field observation location.

Table 2 Sea conditions during the field observations.

Obs. No.	Date	Runs	Wave height (m)	Wave period (s)	Water depth (m)	Depth of scattered layer Zu(m)
1	Aug.'90	12	0.69~1.51	9.8~10.6	34.51~35.6	3.56 − 9.90
2	Nov.'90	44	0.32~1.00	6.3~13.3	34.94~36.12	8.99 − 10.17
3	Dec.'90	20	1.16~1.88	9.3~12.7	35.08~36.32	9.13 − 10.37
4	Mar.'91	44	0.76~1.84	5.6~10.6	35.56~36.46	9.01 − 10.51

measured signals were transmitted via a submarine cable to a land-based observatory (Miyako Port Construction Office, 2nd District Port Construction Bureau, Ministry of Transport). The water surface elevation (Z) and four current velocity components (X+, X-, Y+, Y-) were recorded on an optical magnetic disk at a 0.99-s sampling interval. **Table 2** summarizes the sea conditions during the field observations.

The current velocities were measured 25 m above the transducer, with the thickness Δr of the scattering layers being about 2.3 m. α was set at 30° (**Fig. 1**). We assumed a constant sound velocity of 1500 m/s for the computation.

3.1 Validity of the transfer function

The validity of the Eq. (9) transfer function cannot be directly verified using the observed data. However, by linear wave theory, the value $\left\{\text{Re}(\bar{H})\right\}^2 + 2\left\{\text{Im}(\bar{H})\right\}^2$ can be approximated by dividing the sum of the power spectrum of water particle

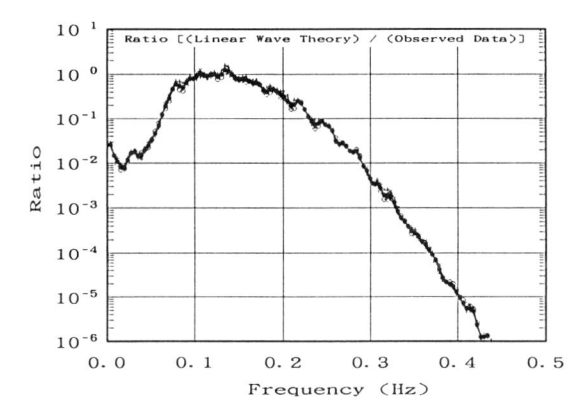

Fig. 6 Validity of transfer function

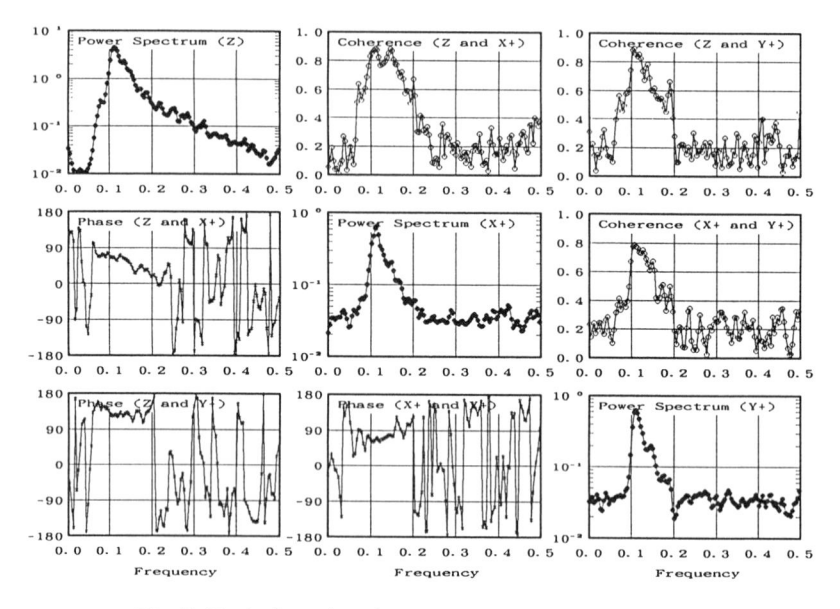

Fig. 7 Typical results of cross-power spectra estimations.

velocities in the X- and Y-directions by the power spectrum of the water surface elevation. **Figure 6** shows this ratio, which is near 1.0 from 0.08 to 0.15 Hz; hence, Eq. (9) can be considered as being valid.

3.2 Estimation of directional spectrum

Typical cross-power spectra estimated from the observed data are shown in **Fig. 7**, where the coherence between each wave quantity is high near the peak frequency of the power spectra and the phase angle shows continuous, smooth fluctuations from 0.06 to 0.2 Hz. These results indicate the field observation cross-power spectra are reasonable.

Figure 8 shows typical directional spectra estimated by the EMLM, where various combinations of quantities are indicated since the EMLM enables using a data set consisting of three or more quantities. The entrance of Kamaishi Bay faces east, and all these results clearly show that waves come from the appropriate direction.

In addition, a reasonable directional spectrum was estimated in 99 out of 122 observations. Of the 23 observations from which adquate results were not estimated, 16 had water surface elevation contaminated by the noise, while the remaining 7 contained erroneous measurements.

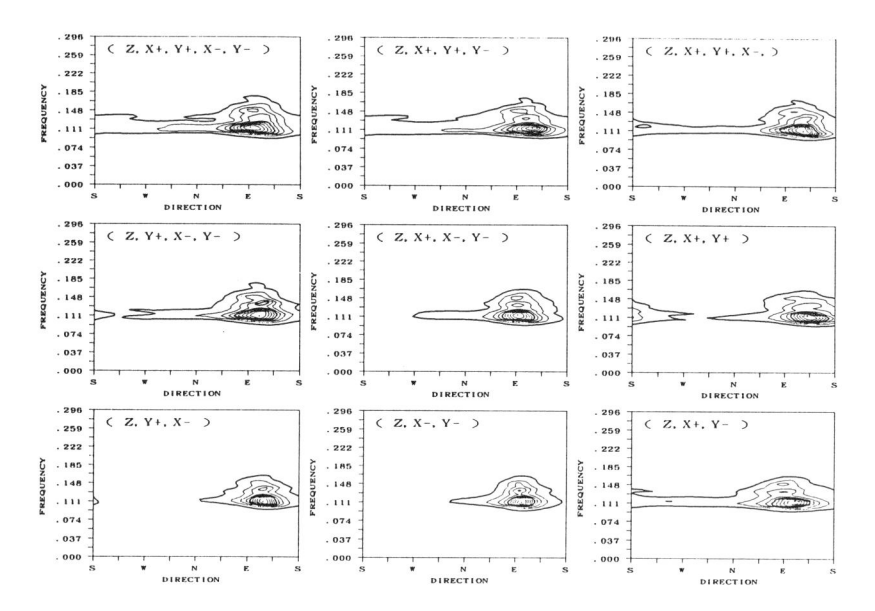

Fig. 8 Typical directional sprecta estimated using various combinations of quantities.

In cases where reasonable results were obtained, for z_U from 9 to 10 m, a measurement was obtained even when the significant wave height was only 35 cm at a significant wave period of about 10 s. Furthermore, at a significant wave period of about 6 s, waves with a significant wave height of 1 m were measured. It should be realized that the highest significant wave height for these observations was 1.85 m, which is not a severe sea condition. Under more severe sea conditions, however, measurement is believed to be possible by increasing z_U.

4. Conclusions

Knowledge of directional spectra is essential for clarifying various coastal engineering problems. However, difficulties in measuring directional seas have led to insufficient information being available. Since the submerged Doppler-type directional wave meter can be employed to measure directional seas with only one set of units and provide an accurate estimation of directional spectrum. It is considered to be the best measuring system for investigation and research on directional seas. In fact, in the near future, Japan's Nationwide Ocean Wave information network for Ports and HArbourS (NOWPHAS) (Nagai et al., 1993) will install this instrument at all their coastal wave observation stations.

Acknowledgments

This research was jointly carried out by the Port and Harbour Research Institute, Ministry of Transport, and Japan Marine Surveyors Association under the direction of the Committee for the Development of a Submerged Doppler-Type Directional Wave Meter (Chairman, Professor Y. Goda, Yokohama National University) organized by the Japan Marine Surveyors Association. The analyzed field wave data were contributed by the Second Port Construction Bureau, Ministry of Transport. Sincere gratidude is extended to all those involved in this study.

References

Isobe, M., K. Kondo and K. Horikawa (1984): Extension of MLM for estimating directional wave spectrum, Proc. Symp. on Description and Modeling of Directional Seas, Paper No. A−6, 15 p.

Ito, Y., Y. Kobori, M. Horiguchi, M. Takehisa and Y. Mituta (1989): Development of Wind Profiling Sodar, J. Atmos. Oceanic Tech., Vol. 6, pp.779−784.

Nagai, T., K. Sugahara, N. Hashimoto and T. Asai (1993): 20-year statistics of the Nationwide Ocean Wave information network for Ports and HArbourS (NOWPHAS 1970−1989), Tech. Note of Port and Harbour Res. Inst., No. 744, 247 p.

Nagai, T., K. Sugahara, N. Hashimoto and T. Asai (1993): Annual Report on the Nationwide Ocean Wave information network for Ports and HArbourS (NOWPHAS 1991), Tech. Note of Port and Harbour Res. Inst., No. 745, 304 p.

CHAPTER 47

Bragg Scattering of Waves over Porous Rippled Bed

Hajime Mase[1] and Ken Takeba[2]

ABSTRACT: A time-dependent and a time-independent wave equations are developed for waves propagating over porous rippled beds taking account of the effects of porous medium. The mean water depth and the thickness of porous layer are assumed to be slowly varying compared to the wavelength of surface gravity waves, and the spatial scale of ripples is assumed to be the same as the wavelength of surface waves. By using the time-independent equation, the Bragg scattering is examined in one-dimensional case. The results show that the reflected and transmitted waves become smaller than those in the case of impermeable rigid rippled bed due to energy dissipation in porous medium.

Introduction

Davies and Heathershaw (1984) studied the reflection from sinusoidal undulation over a horizontal bottom and derived a solution of reflection coefficient. Their experimental results showed a resonant Bragg reflection at the condition where the wavelength of the bottom undulation is one half the wavelength of the surface wave as predicted by their theory. Mei (1985) and Naciri and Mei (1988) developed theories of wave evolution at and close to the resonant condition by shore-parallel sinusoidal bars and two-dimensional doubly sinusoidal undulations over a slowly varying topography. For more realistic natural topography, Kirby (1986) derived a general wave equation which extends the mild slope equation of Berkhoff (1972). These existing theories don't take account of the effects of seabed permeability.

---

[1] Assoc. Prof., Dept. of Civil Eng., Kyoto Univ., Kyoto, 606, Japan.
[2] Grad. Student, Dept. of Civil Eng., Kyoto Univ., Kyoto, 606, Japan.

The strong reflection of incident waves due to the Bragg reflection results in rough offshore sea. Since it is desirable to reduce the transmitted and reflected waves, artificial porous ripples or bars would be more convenient.

In this study, a time-dependent wave equation is developed, by extending the theory of Kirby (1986), for waves propagating over permeable rippled beds in order to take into account the effects of porous medium. Some numerical calculations are carried out to show the effects of seabed permeability on wave transformations or on the Bragg scattering by ripples in one-dimensional case.

Derivation of Wave Equation over Porous Rippled Bed

The coordinate system and main quantities are shown in Fig.1. The actual depth, $h'(x)$, is divided into the rapidly varying small amplitude undulation, $\delta(x)$, and the slowly varying mean water depth, $h(x)$:

$$h'(x) = h(x) - \delta(x) \qquad (1)$$

where $x = (x, y)$. Thickness of the porous layer, $h'_s(x)$, is expressed as

$$h'_s(x) = h_s(x) + \delta(x) \qquad (2)$$

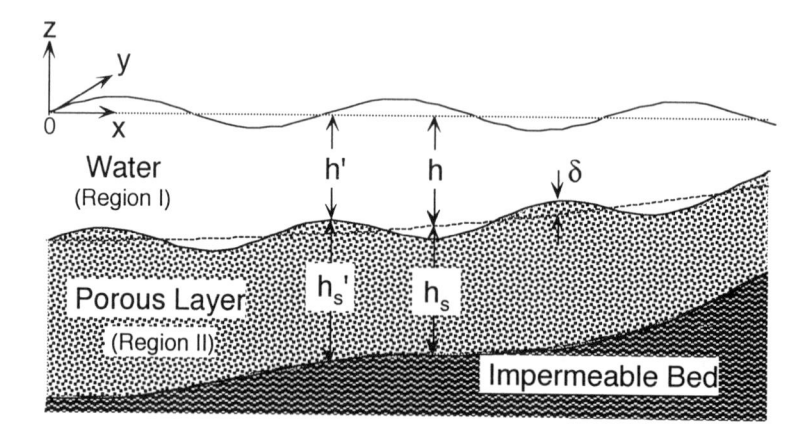

Fig.1 Definition of variables

where $h_s(x)$ is the slowly varying mean thickness. The bottom beneath the porous layer is assumed to be impermeable and rigid.

The horizontal scales of changes of $h(x)$, $h_s(x)$ and $\delta(x)$ are

$$O\left(\frac{\nabla_h h}{kh}\right) \approx O(k\delta) \ll 1 \tag{3}$$

$$O\left(\frac{\nabla_h(h+h_s)}{kh}\right) \approx O(k\delta) \ll 1 \tag{4}$$

and

$$O\left(\frac{\nabla_h \delta}{k\delta}\right) \approx O(1) \tag{5}$$

where ∇_h is the gradient operator as $(\partial/\partial x, \partial/\partial y)$, and k is the wavenumber.

The analytical domain is divided into two regions: the region (I) is the fluid domain above the porous layer; the region (II) is the porous layer, as shown in Fig.1.

In the region (I), the irrotational motion of incompressible and inviscid fluid is described by a velocity potential, ϕ, as follows:

$$\nabla_h^2 \phi + \phi_{zz} = 0 \; ; \; -h \leq z \leq 0 \tag{6}$$

$$\phi_{tt} + g\phi_z = 0 \; ; \; z = 0 \tag{7}$$

$$\phi_z = -\nabla_h h \cdot \nabla_h \phi + \nabla_h \cdot (\delta \nabla_h \phi) + w^{(I)} \; ; \; z = -h \tag{8}$$

Eq.(6) is the Laplace equation, Eq.(7) is the free surface boundary condition combined dynamic and kinematic boundary conditions, and Eq.(8) is the bottom boundary condition expanded about $z=-h$ to the order of $O(k\delta)$, where t is the time, g is the acceleration of the gravity, and $w^{(I)}$ is the discharge velocity at the interface between the region (I) and (II). The pressure, $p^{(I)}$, is given by

$$p^{(I)} = -\rho(\phi_t + gz) \; ; \; -h \leq z \leq 0 \tag{9}$$

where ρ is the density of the fluid.

In the region (II), after Sollit and Cross (1972) and Madsen (1974), the unsteady motion of the fluid in the porous medium is described by a continuity equation

$$\nabla \cdot \boldsymbol{u} = 0 \tag{10}$$

and by a momentum equation

$$\frac{\tau}{n}\frac{\partial \boldsymbol{u}}{\partial t} = -\frac{1}{\rho}\nabla\left(p^{(II)} + \rho g z\right) - f\frac{\omega}{n}\boldsymbol{u} \tag{11}$$

where $\boldsymbol{u}$ is the discharge velocity vector, ∇ is the gradient operator vector as $(\partial/\partial x, \partial/\partial y, \partial/\partial z)$, n is the porosity, τ is the inertia coefficient, f is the linearized friction factor, $p^{(II)}$ is the pressure, ω is the angular frequency. Assuming the irrotational motion of the fluid and introducing a discharge velocity potential, φ, we can rewrite Eqs.(10) and (11) as

$$\nabla_h^2 \varphi + \varphi_{zz} = 0 \quad ; \quad -(h+h_s) \leq z \leq -h \tag{12}$$

$$p^{(II)} = -\rho\left(\frac{\tau}{n}\varphi_t + g z + f\frac{\omega}{n}\varphi\right) \quad ; \quad -(h+h_s) \leq z \leq -h \tag{13}$$

The boundary condition at the upper face of the porous layer is

$$\varphi_z = -\nabla_h h \cdot \nabla_h \varphi + \nabla_h \cdot (\delta \nabla_h \varphi) + w^{(II)} \quad ; \quad z = -h \tag{14}$$

and the boundary condition at the bottom of the porous layer is

$$\varphi_z = -\nabla_h (h+h_s) \cdot \nabla_h \varphi \quad ; \quad z = -(h+h_s) \tag{15}$$

where $w^{(II)}$ is the vertical discharge velocity at the interface between the region (I) and (II).

At the interface, the pressure and the vertical discharge velocity should be continuous:

$$p^{(I)} = p^{(II)} \quad ; \quad z = -h \tag{16}$$

$$w^{(I)} = w^{(II)} \quad ; \quad z = -h \tag{17}$$

The solutions of velocity potentials, ϕ and φ, may be expressed as

$$\phi(\boldsymbol{x},z,t) = f^{(I)}(\boldsymbol{x},z)\ \tilde{\phi}(\boldsymbol{x},t) + (\text{non-propagating modes}) \tag{18}$$

$$\varphi(\boldsymbol{x},z,t) = f^{(II)}(\boldsymbol{x},z)\ \tilde{\varphi}(\boldsymbol{x},t) + (\text{non-propagating modes}) \tag{19}$$

Under the condition of horizontal bottom ($\nabla_h h = \nabla_h h_s = 0$), we can obtain the vertical distribution functions of $f^{(I)}$ and $f^{(II)}$ as follows:

$$f^{(I)} = \frac{1}{D}\{\cosh kh_s \cosh k(h+z) + \gamma \sinh kh_s \sinh k(h+z)\} \qquad (20)$$

$$f^{(II)} = \frac{1}{D}\gamma \cosh k(h+h_s+z) \qquad (21)$$

where

$$D = \cosh kh_s \cosh kh \left(1 + \gamma \tanh kh_s \tanh kh\right) \qquad (22)$$

$$\gamma = n/(\tau + if) \qquad (23)$$

The dispersion relation is given by

$$\omega^2 = gk \frac{\tanh kh + \gamma \tanh kh_s}{1 + \gamma \tanh kh \tanh kh_s} \qquad (24)$$

For the case of mild slope bottom, the velocity potentials could be described by Eqs.(18) and (19) with Eqs.(20) and (21). Substituting Eqs.(18) and (19) into the matching condition of pressure yields

$$\tilde{\phi} = \tilde{\varphi} \qquad (25)$$

Following Smith and Sprinks (1975) and Kirby (1986), we employ Green's second identity to the propagating component of ϕ and $f^{(I)}$:

$$\int_{-h}^{0} f^{(I)} \phi_{zz} \, dz - \int_{-h}^{0} \phi f_{zz}^{(I)} \, dz = \left[f^{(I)} \phi_z - \phi f_z^{(I)} \right]_{-h}^{0} \qquad (26)$$

Integrating the above equation yields

$$-\nabla_h \cdot \int_{-h}^{0} \nabla_h \tilde{\phi} f^{(I)2} \, dz - \int_{-h}^{0} k^2 \tilde{\phi} f^{(I)2} \, dz$$

$$= -\frac{1}{g}\left(\tilde{\phi}_{tt} f^{(I)2}\right)\bigg|_{0} - \frac{\omega^2}{g}\left(\tilde{\phi} f^{(I)2}\right)\bigg|_{0} - \nabla_h \cdot \left(\delta \nabla_h \tilde{\phi}\right) f^{(I)2}\bigg|_{-h}$$

$$- w^{(I)} f^{(I)}\bigg|_{-h} + \tilde{\phi} f^{(I)} f_z^{(I)}\bigg|_{-h} + \text{high order terms} \qquad (27)$$

Green's second identity to the propagating component of φ and $f^{(II)}$ is described by

$$\int_{-(h+h_s)}^{-h} f^{(II)} \varphi_{zz} dz - \int_{-(h+h_s)}^{-h} \varphi f_{zz}^{(II)} dz = \left[f^{(II)} \varphi_z - \varphi f_z^{(II)} \right]_{-(h+h_s)}^{-h} \quad (28)$$

and the integration yields

$$-\nabla_h \cdot \int_{-(h+h_s)}^{-h} \nabla_h \tilde{\phi} f^{(II)^2} dz - \int_{-(h+h_s)}^{-h} k^2 \tilde{\phi} f^{(II)^2} dz$$

$$= \nabla_h \cdot \left(\delta \nabla_h \tilde{\phi} \right) f^{(II)^2} \Big|_{-h} + w^{(II)} f^{(II)} \Big|_{-h}$$

$$- \tilde{\phi} f^{(II)} f_z^{(II)} \Big|_{-h} + \text{high order terms} \quad (29)$$

Eliminating $w^{(I)}$ and $w^{(II)}$ (actually $w^{(I)} = w^{(II)}$) from Eqs.(27) and (29) and using $\phi = \tilde{\phi}$, we obtain

$$\frac{1}{g}\left(\tilde{\phi}_{tt} + \omega^2 \tilde{\phi} \right) - \nabla_h \cdot \left(\alpha \nabla_h \tilde{\phi} \right) - k^2 \alpha \tilde{\phi} + \frac{\cosh^2 kh_s}{D^2}(1-\gamma)\nabla_h \cdot \left(\delta \nabla_h \tilde{\phi} \right) = 0 \quad (30)$$

where

$$\alpha = p + q/\gamma \quad (31)$$

$$p = \int_{-h}^{0} f^{(I)^2} dz = \frac{1}{4kD^2} \{ \cosh^2 kh_s \sinh 2kh(1 + 2kh/\sinh 2kh)$$
$$+ \gamma \sinh 2kh_s (\cosh 2kh - 1) + \gamma^2 \sinh^2 kh_s \sinh 2kh$$
$$\times (1 - 2kh/\sinh 2kh) \} \quad (32)$$

$$q = \int_{-(h+h_s)}^{-h} f^{(II)^2} dz = \frac{1}{4kD^2} \{ \gamma^2 \sinh 2kh_s \sinh 2kh(1 + 2kh_s/\sinh 2kh_s) \} \quad (33)$$

Eq.(30) is the time-dependent wave equation. Factoring the time out of $\tilde{\phi}$ as

$$\tilde{\phi} = \hat{\phi} e^{-i\omega t} \quad (34)$$

we can transform Eq.(30) into

$$\nabla_h \cdot \left(\alpha \nabla_h \hat{\phi} \right) + \alpha k^2 \hat{\phi} - \frac{\cosh^2 kh_s}{D^2}(1-\gamma)\nabla_h \cdot \left(\delta \nabla_h \hat{\phi} \right) = 0 \quad (35)$$

The effects of the porous medium are taken into account through the complex wavenumber k given by Eq.(24) and the complex coefficients of α and γ.

Relation to Existing Theories

Case of $h_S = 0$ and $\delta = 0$

In the case that there are not porous layer and rapid undulation, Eq.(35) reduces to the mild slope equations derived by Berkhoff (1972):

$$\nabla_h \cdot \left(CC_g \nabla_h \hat{\phi}\right) + k^2 CC_g \hat{\phi} = 0 \tag{36}$$

Case of $h_S = 0$ and $\delta \neq 0$

In the case that there is not porous layer and there exists rapid undulation, Eq.(35) reduces to the general wave equation over rippled bed derived by Kirby (1986):

$$\nabla_h \cdot \left(CC_g \nabla_h \hat{\phi}\right) + k^2 CC_g \hat{\phi} - \frac{g}{\cosh^2 kh} \nabla_h \cdot \left(\delta \nabla_h \hat{\phi}\right) = 0 \tag{37}$$

Case of very small permeability of porous layer

The case of very small permeability is treated as a mathematical limit of very small porosity ($n \to 0$) and very large friction factor ($f \to \infty$). In this condition, Eq.(35) reduces to the general wave equation expressed as Eq.(37).

Case of very large permeability of porous layer

Since $n \to 1$ and $f \to 0$, so $\gamma \to 1$. The resultant wave equation becomes the same as the mild slope equation expressed as Eq.(36) where the phase velocity C and the group velocity C_g are defined by using the water depth of $h+h_S$.

Numerical Calculations of Bragg Scattering

Boundary condition

At the seaward boundary condition, the following condition has to be satisfied:

$$\hat{\phi}_x = -ik(\hat{\phi} - 2\hat{\phi}_I) \tag{38}$$

where $\hat{\phi}_I$ is the incident wave potential amplitude. At the downstream boundary condition, the following transmitted condition was needed:

$$\hat{\phi}_x = ik\hat{\phi} \tag{39}$$

Numerical conditions

Numerical conditions followed the experimental ones of Davies and Heathershaw (1984). The water depth h and the thickness of porous layer h_S were constant, and the undulation δ was given by

$$\delta = D\sin(\lambda x) \; ; \; 0 \leq x \leq ml \tag{40}$$

where m, λ, l, and D are the number, the wavenumber, the wavelength and the amplitude of the ripples, respectively. Two cases of Case 1 ($m = 10$ and $D/h = 0.16$) and Case 2 ($m = 4$ and $D/h = 0.32$) were employed. Actual values of D, l and h_S in the experiments of Davies and Heathershaw (1984) were 5 cm, 1.0 m and 0 m, respectively. Here we changed h_S from 0 m to 0.2 m to examine the effects of permeability on the Bragg scattering.

Calculated results and discussion

Figure 2 shows the spatial distributions of wave amplitudes of wave period 1.0 s and 1.3 s, where the solid line corresponds to the case of impermeable rigid bottom, the dotted and the dash-dotted lines correspond to the case of porous bottom of $f = 1$ and $f = 10$, respectively, with $\tau = 1.0$, $n = 0.4$ and $h_S = 0.2$ m of Case 1. In Fig.2 (a), the reflection coefficient is less than 0.1; and, in Fig.2(b), the Bragg reflection condition is nearly satisfied, and the variations of the amplitudes are remarkable. It is seen from Fig.2 that when the bottom is permeable, the amplitudes and their variations become small.

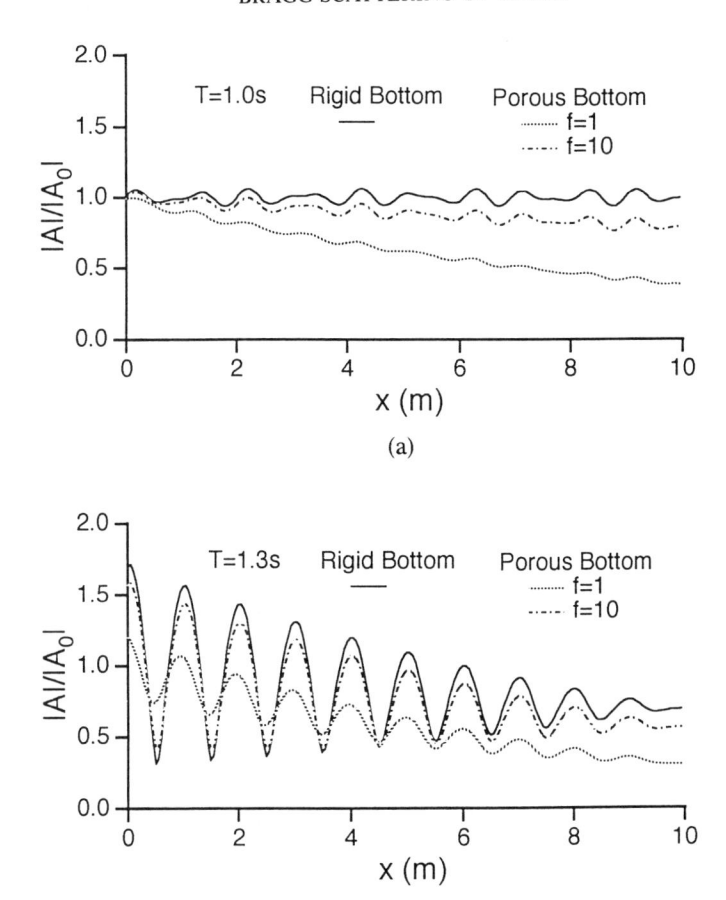

**Fig.2 Spatial distributions of wave amplitudes:
(a) non-resonant condition; (b) resonant condition**

Figure 3 shows the reflection coefficient, R, and the transmission coefficient, T, against the ratio of wavenumbers, $2k/\lambda$, where the k is taken as the real part for the case of permeable ripples. In the figures, the linearized friction factor was changed by 10, 5, and 1, keeping $\tau = 1.0$, $n = 0.4$ and $h_S = 0.2$ m. The calculated results for impermeable rigid bottom, shown by the solid lines, were obtained by setting $n = 0$. In the range of $1 \le f \le 10$, the reflection and transmission

(a)

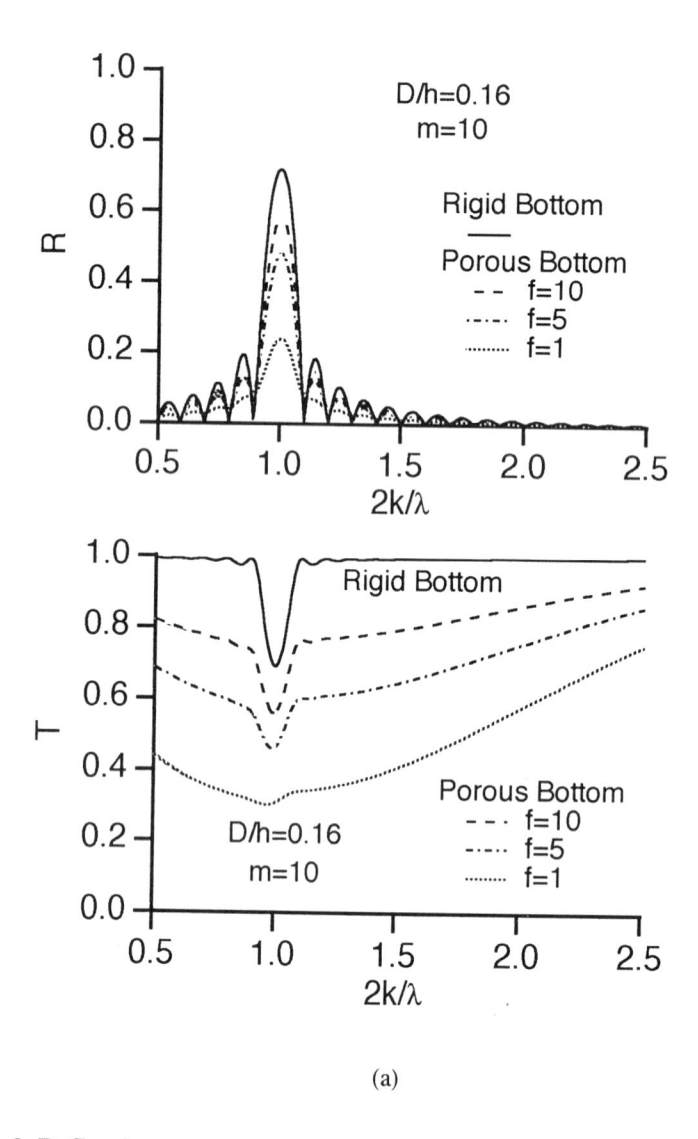

Fig.3 Reflection and transmission coefficients: (a) Case 1; (b) Case 2

(b)

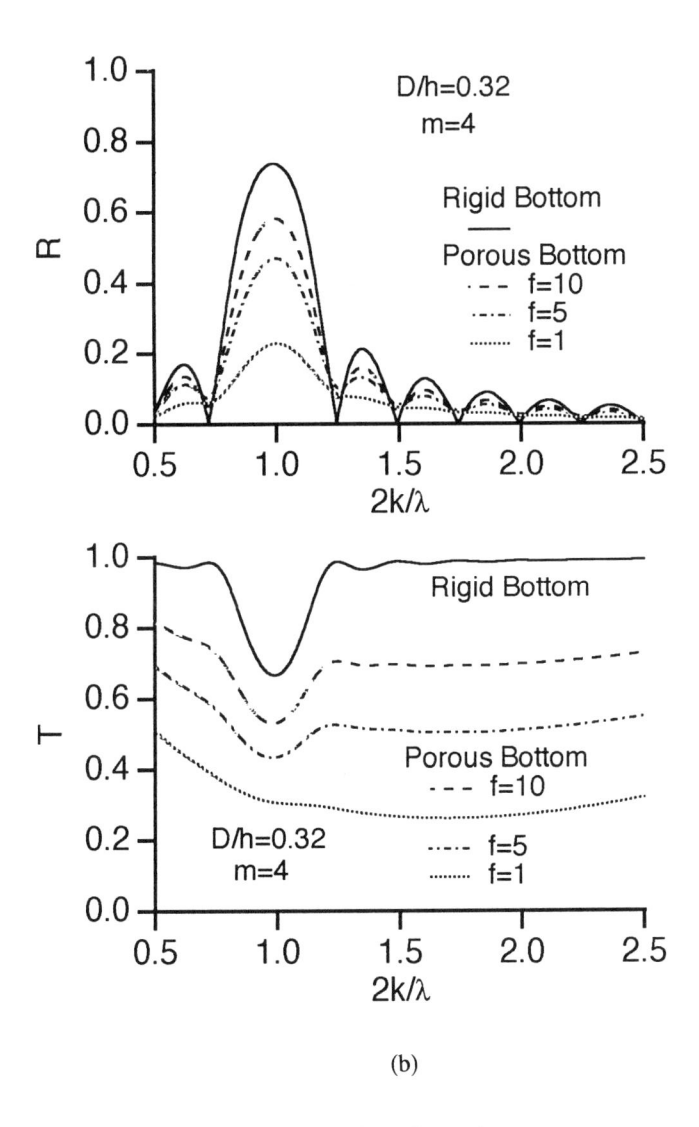

Fig.3 (Continued)

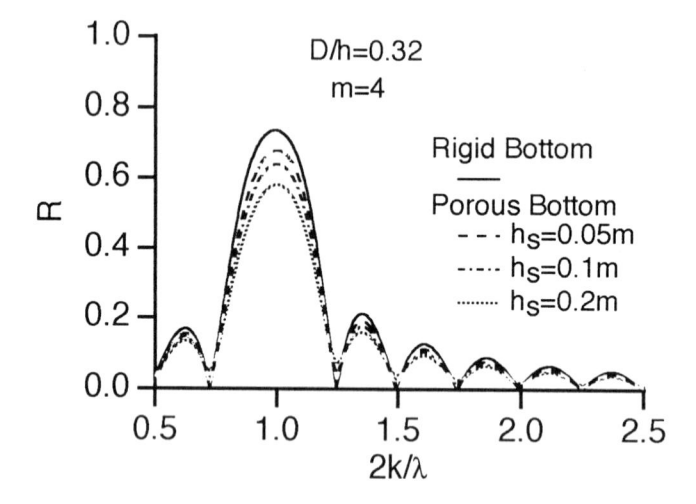

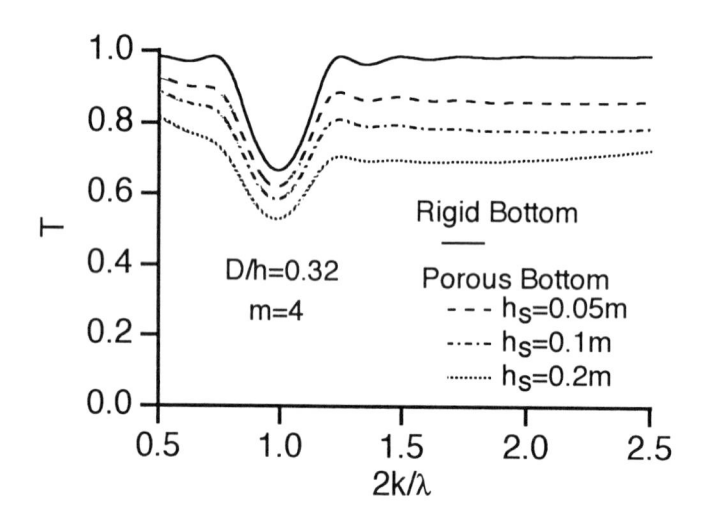

Fig.4 Effect of thickness of porous layer on reflection and transmission

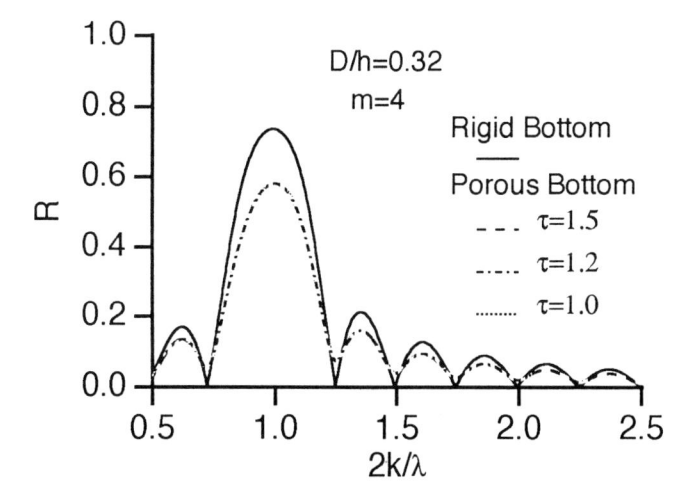

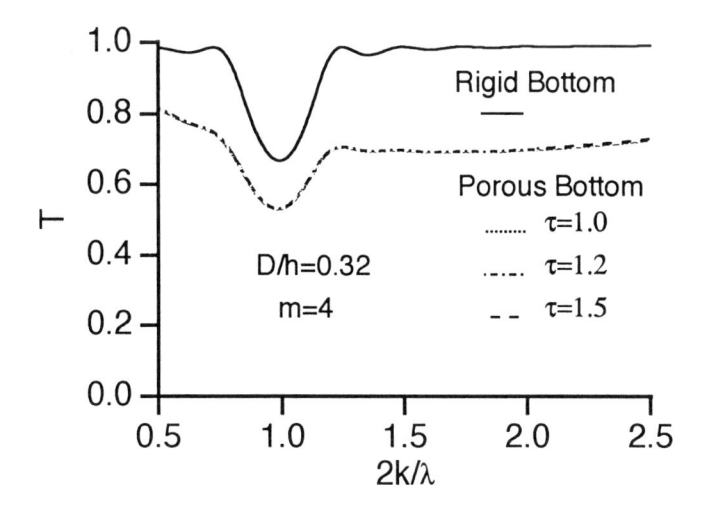

Fig.5 Effect of inertia coefficient of porous layer on reflection and transmission

coefficients become small with decrease in the linearized friction factor.

The effect of thickness of the porous layer on the Bragg scattering is shown in Fig.4, by changing h_S, with $\tau = 1.0$, $n = 0.4$ and $f = 10$. The reflection and transmission coefficients become small with increase in the porous layer thickness, as easily expected.

Figure 5 shows the effect of the inertia coefficient by changing τ, with $f = 10$, $n = 0.4$ and $h_S = 0.2$ m. It is seen from this figure that there is little effect of τ.

It should be noted that though the parameters of n, f, and τ were changed independently for the convenience in the calculations so far, these parameters are dependent each other. The determination of the parameters is difficult for arbitrary porous medium.

Conclusions

In order to deal with wave transformations over a permeable seabed with rapidly varying undulations, we developed the time-dependent and time-independent wave equations taking account of the effects of porous medium.

In the case that there are not porous layer and rapid undulation, the time-independent equation reduces to the mild slope equation derived by Berkhoff (1972). In the case that there is not porous layer and there exists rapid undulation, the time-independent equation reduces to the wave equation over rippled bed derived by Kirby (1986). When the permeability is very small, the time-independent equation reduces to the Kirby's equation. On the other hand, when the permeability is very large, the time-independent equation reduces to the Berkhoff's equation where the water depth is defined by a sum of the water depth and the thickness of porous layer.

Numerical examples of the Bragg scattering were shown in one-dimensional case. The reflected and transmitted waves became small due to the permeability of seabed bottom. The reflection and transmission coefficients were influenced by the friction factor, the thickness of porous layer, and the porosity; however, there was little effect of inertia coefficient.

Acknowledgments

This study was supported financially by the Maeda Memorial Engineering Foundation. The authors greatly appreciate Dr. Takashi Izumiya (assoc. Prof., Dept. Civil Eng., Niigata Univ., Japan) for showing the computer program to solve the dispersion relation.

References

Berkhoff, J.C.W. (1972). Computation of combined refraction-diffraction, Proc. 13rd Int. Conf. Coastal Eng., ASCE, pp.471-490.

Davies, A.G. and Heathershaw, A.D. (1984). Surface-wave propagation over sinusoidally varying topography, Jour. Fluid Mech., Vol.144, pp.419-443.

Kirby, J.T. (1986). A general wave equation for waves over rippled beds, Jour. Fluid Mech., Vol.162, pp.171-186.

Madsen, O.S. (1974). Wave transmission through porous structures, Proc. ASCE, Vol.100, No.WW3, pp.169-188.

Mei, C.C. (1985). Resonant reflection of surface waves by periodic sand-bars, Jour. Fluid Mech., Vol.152, pp.315-335.

Naciri, M. and Mei, C.C. (1988). Bragg scattering of water waves by a doubly periodic seabed, Jour. Fluid Mech., Vol.192, pp.51-74.

Smith, R. and Sprinks, T. (1975). Scattering of surface waves by a conical island, Jour. Fluid Mech., Vol.72, pp.373-384.

Sollit, C.K. and Cross, R.H. (1972). Wave transmission through permeable breakwaters, Proc. 13rd Int. Conf. Coastal Eng., ASCE, pp.1827-1846.

CHAPTER 48

Non-Reflective Multi-Directional Wave Generation by Source Method

Masahiro Tanaka[1], Takumi Ohyama[1], Tetsushi Kiyokawa[1], and Kazuo Nadaoka[2]

Abstract

A new wave-generation system, composed of a source and an absorber, is proposed to realize a non-reflective wave field in a laboratory basin as an advanced alternative to conventional methods. Theoretical studies based on the linear potential theory were carried out to investigate the characteristics of wave fields generated by the proposed method. Performance of the new system was compared to that of conventional *serpent-type* wave-makers. The proposed system was founded to provide a significantly larger effective area for multi-directional waves than does the serpent-type wave makers.

Experimental studies using a two-dimensional wave tank were also conducted to verify the performance and practicality of the new system. The results showed that the new devised equipment can produce a nearly non-reflective wave field and can generate waves efficiently even in deep water.

1. Introduction

Serpent-type wave-maker has been developed to generate multi-directional irregular waves, and has been introduced in many laboratories (Biesel, 1954; Gillbert, 1976; Takayama, 1982). However, this wave-generation system has not yet solved two serious problems. First, re-reflected waves from wave-generating paddles and side walls of a tank disturb the wave field around a model after certain time passed.

---

[1] Institute of Technology, Shimizu Co., 3-4-17, Etchujima, Koto-ku, Tokyo 135, Japan
[2] Tokyo Institute of Technology University, 2-12-1, Ohokayama, Meguro-ku, Tokyo 152, Japan

The second problem is that the effective wave area decreases according to increase of the incidence angle of waves.

Funk·Miles (1987), Isaacson (1989) and Dalrymple (1989) have proposed a directional wave-making method considering side-wall reflection to expand an effective wave field. Although this method can slightly widen the effective wave field by controlling reflection waves at side wall, proportion of ineffective wave area in the tank is still large.

On the other hand, Hirakuchi et al. (1991) and Ikeya et al. (1992) have proposed an active directional wave absorption method. In this method, the reflection waves are absorbed by controlling the amplitudes and phases of each paddle corresponding to the heights, phases and directions of the reflection waves. Since this method requires measurement and analysis of a significant amount of waves to control the wave-generating system in an instant time, it is quite difficult to apply to an actual system.

We have devised a non-reflective multi-directional wave-generating system that consists of a distributed wave making source and a wave absorber without using wave-generating paddles.

2. Concept of the New System

The wave-generation system consists of a circular wave-making source and an external wave absorber arranged in a circular tank as shown in Fig. 1. The idea of the proposed method is founded on the principle of two-dimensional numerical wave-generating model (Brorsen·Larsen, 1987; Ohyama et al., 1991). The two-dimensional numerical model employs an effective non-reflective open boundary treatment so that it can be applied to arbitrary wave fields including nonlinear random waves.

The proposed system generates waves by pumping water in and out without using paddles, and the heights and direction of waves are controlled by the intensity and phase of the source. Since this source method does not involve wave paddles unlike conventional wave-makers, it is expected that the reflected waves from model objects permeate through the source without re-reflection, and are absorbed by the external absorber. In addition, since the wave-making source is arranged in a circular shape, an effective wave field is independent of wave directions, whereas the performance of the conventional *serpent-type* wave-makers is significantly direction-dependent.

For practical application, the wave-making source can be composed of a series of small units mounted in a concentric semicircle. It would require careful provision to minimize the reflection from the source. A practical equipment and its performance of wave generation are shown in Section 4.

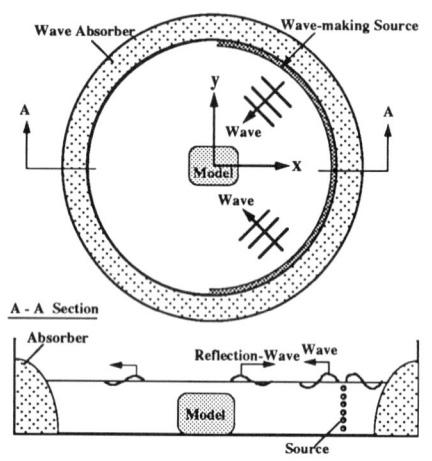

Fig. 1 Concept of non-reflective multi-directional wave generation by source method

3. Theoretical Analysis

3.1 Theoretical Formulation

Multi-directional irregular waves are composed of waves with various periods, heights and directions. Since the performance of waves generated by the source method is independent of wave direction, it can be evaluated by using single-directional regular waves.

The coordinate system and definition sketch used in this study are shown in Fig. 2. With the assumption of incompressible, inviscid fluid, and irrotational flow, the fluid motion can be described with a velocity potential, $\phi(r, \theta, z)e^{-i\omega t}$. The potential satisfies the following Poisson's equation within the fluid domain and which is subject to boundary conditions at the free surface and the tank bottom.

$$\nabla^2 \phi = U^*(\theta, z)\delta(r - A) \qquad (1)$$

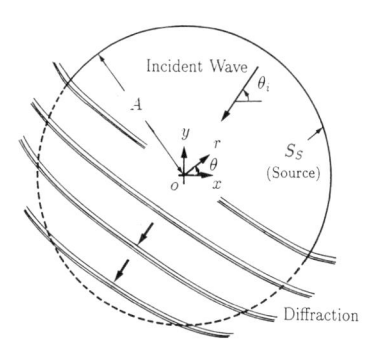

Fig. 2 Definition sketch of theoretical model

Where, U^* is the source intensity of formulated by Brorsen et al. (1987), δ: Dirac's Delta function, and A: distance (radius) between the center of a tank and a wave making source. Applying the Green's function method to Eq. 1, the velocity potential, $\phi(P)$, at an arbitrary point in a basin, $P(r_p, \theta_p, z_p)$, is expressed as

$$\phi(P) = -\frac{1}{4}\int_{S_S} U^* G\, ds \tag{2}$$

in which, S_S denotes the surface of wave-making source, and G: the Green's function derived by John (1950) as shown in Eq. 3 and 4.

$$G = \frac{2\pi i(\kappa^2 - v^2)}{h(\kappa^2 - v^2) + v}\cosh k(h+z_p)$$
$$\cdot \cosh k(h+z) H_0^{(1)}(\kappa R_p) + \sum_{n=1}^{\infty} \frac{4(\kappa_n^2 + v^2)}{h(\kappa_n^2 + v^2) - v}$$
$$\cdot \cos \kappa_n(h+z_p)\cos \kappa_n(h+z) K_0(\kappa_n R_p) \tag{3}$$

$$R_p = \sqrt{(r_p\cos\theta_p - r\cos\theta)^2 + (r_p\sin\theta_p - r\sin\theta)^2} \tag{4}$$

Where, k: wave number, k_n: eigenvalue, $H_0^{(1)}$: Hankel function, K_0: Bessel function, $v = \omega^2/g$, g: gravity acceleration.

Since waves generated by a source progress to both sides of the source, the intensity of the source can be given as the following equation (Brorsen·Larsen, 1987).

$$U^*(\theta,z) = 2\frac{\partial \phi_i}{\partial r}\bigg|_{r=A} \qquad (\theta_i - \frac{\pi}{2} \leq \theta \leq \theta_i + \frac{\pi}{2}) \qquad (5)$$

Substituting the potential ϕ_i to Eq. 5, the intensity U^* can be expressed as shown in Eq. 6.

$$U^*(\theta,z) = -ik\cos(\theta - \theta_i)\frac{gH_i}{\omega}\frac{\cosh k(h+z)}{\cosh kh}e^{-ikA\cos(\theta-\theta_i)}$$
$$(\theta_i - \frac{\pi}{2} \leq \theta \leq \theta_i + \frac{\pi}{2}) \qquad (6)$$

Substituting Eq. 3 and 6 into Eq. 2, analytical form of the potential $\phi(P)$ at an arbitrary point P in a tank can be obtained as follows:

$$\phi(P) = \frac{gH_i}{2\omega}\varphi(r_p, \theta_p)\frac{\cosh k(h+z_p)}{\cosh kh} \qquad (7)$$

$$\varphi = -\frac{kA}{2}\int_{\theta_i - \frac{\pi}{2}}^{\theta_i + \frac{\pi}{2}}\cos(\theta - \theta_i)\, e^{-ikA\cos(\theta-\theta_i)}H_0^{(1)}(k\hat{R}_p)\,d\theta \qquad (8)$$

$$\hat{R}_p = \sqrt{(r_p\cos\theta_p - A\cos\theta)^2 + (r_p\sin\theta_p - A\sin\theta)^2} \qquad (9)$$

3.2 Definition of Performance of Wave-generation

Performance of a wave generating system is generally evaluated with areal extent of effective wave-field. The effective area is defined as an area, in which errors of wave heights and directions from an intended wave are smaller than allowable tolerances, as shown in Fig. 3. The wave height and direction errors, ΔH and $\Delta \theta_w$, are described as shown in Eq. 10 and 11.

$$|\Delta H|/H_i = |H - H_i|/H_i \qquad (10)$$

$$|\Delta \theta_w| = |\theta_w - \theta_i| \qquad (11)$$

By defining the extent of the effective wave-field as a ratio of the effective zone's radius to that of basin, a/A, its dependence on incident wave-direction can be eliminated.

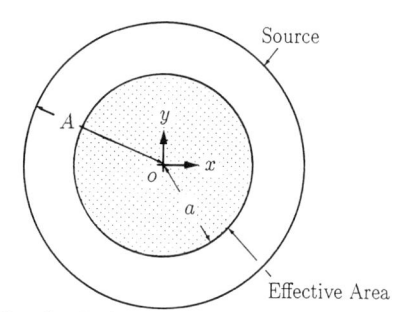

Fig. 3 Definition of effective wave area

3.3 Performance of wave-generation by the source method

Characteristics of wave field generated by the source method were examined using a theoretical model (see Fig. 1). In this calculation, non-reflection waves from the wave-making source and the absorber behind the source were assumed.

Wave-heights distributions normalized with an incident wave height is shown in Fig. 4 (a). Where, a tank radius is 3 times of the wave-length ($A/L = 3.0$). Fig. 4 (b) and (c) describe vectors of energy flux and distributions of wave-direction errors at $A/L = 3.0$, respectively. It can be seen in Fig. 4 (a) that the relative wave-height errors, $|(H - H_i)|/H_i$, are less than 0.05 in the wide area of the tank. The wave-direction errors are less than 5 degrees in the almost all the area as shown in Fig.4 (c). These results indicate that the proposed source method can provide the effective wave field in the wide area of the tank.

Distributions of normalized wave-height and wave-direction errors at $A/L = 1.0$ are shown in Fig. 5 (a) and (b), respectively. The magnitudes of both errors for this case are much bigger than those for the case of $A/L = 3.0$. This difference indicates that A/L is a significant parameter in the design of the proposed system.

Fig. 6 (a) shows the dependence of the a/A on the parameter A/L, when the error tolerance $(|\Delta H|/H_i)_{max}$ is varied from 0.05 to 0.2 with $|\Delta \theta_w|_{max}$ fixed at 4 degrees. Fig. 6 (b) also shows characteristic of a/A versus A/L corresponding to variation of wave-direction error, $|\Delta \theta_w|_{max}$, when $(|\Delta H|/H_i)_{max}$ is fixed at 0.1. It is seen that the effective area generally increases with A/L. The effective wave area can be evaluated using the results shown in Fig. 6 (a) and (b). For example, assuming allowable tolerance to be $(|\Delta H|/H_I)_{max} = 0.1$ and $\Delta \theta_w = 4$ degrees, when the tank radius is set at 2.4 times of incident wave-length, an area of about 80% radius in the tank is provided as the effective wave field.

(a) Wave heights distributions

(b) Vectors of energy flux

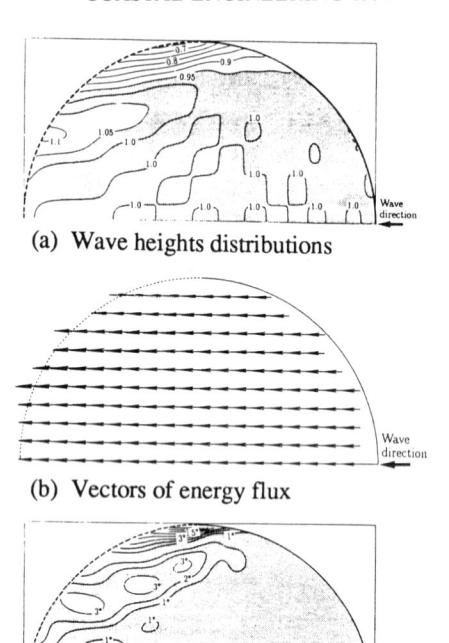

(c) Distributions of wave direction error $\Delta\theta_w$

Fig. 4 Wave field generated by the source method ($A/L = 3.0$)

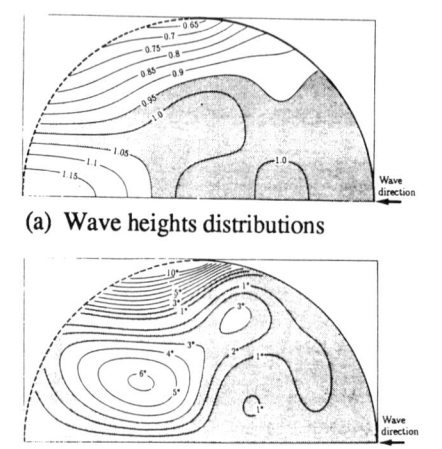

(a) Wave heights distributions

(b) Distributions of wave direction error $\Delta\theta_w$

Fig. 5 Wave field generated by the source method ($A/L = 1.0$)

(a) Effective wave field with $|\Delta H/H_I|_{max}$

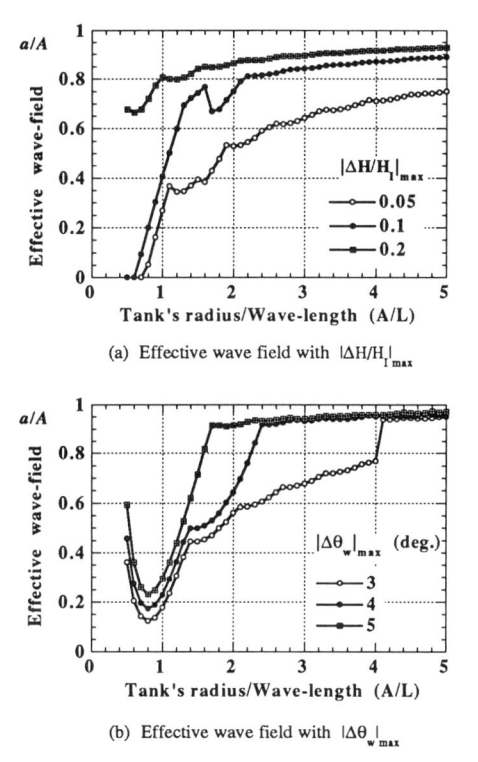

(b) Effective wave field with $|\Delta \theta_w|_{max}$

Fig. 6 Effective wave field by the source method

3.4 Comparison with serpent-type generator

Wave fields generated by a serpent-type wave-maker was simulated using a theoretical model (Dalrymple, 1989) to compare with performance of the proposed source method. In this calculation, the serpent-type generator was controlled so as to widen the effective wave field by using the side-wall-reflection of the tank.

Fig. 7 (a) and (b) show computed results of water-elevation distribution and vectors of energy flux of waves generated by a serpent-type. Where, the ratio of the tank width to wave length, B/L, and the incidence angle of wave, θ_i, are fixed at 6.0 and 40 degrees, respectively. In this case, the serpent-type wave generator was controlled so as to provide the best performance at x=0.

Although the serpent-type generator produces uniform wave field in the central area of the tank, it causes standing wave and diffraction waves at the corners and their

(a) Water elevation distributions   (b) Vectors of energy flux

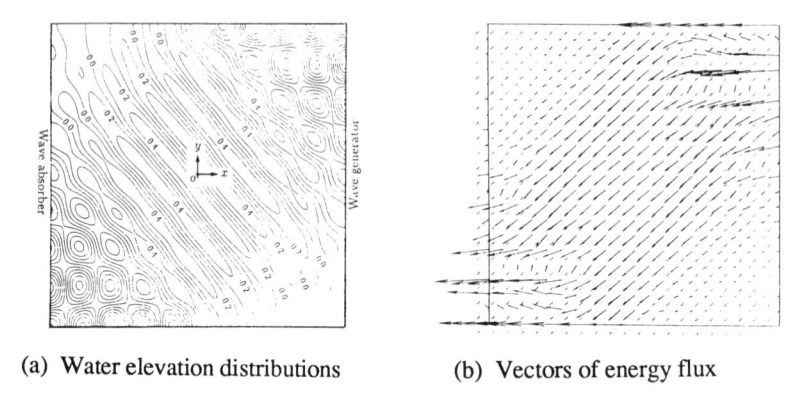

Fig. 7 Wave field generated by a serpent-type ($B/L = 6.0$)

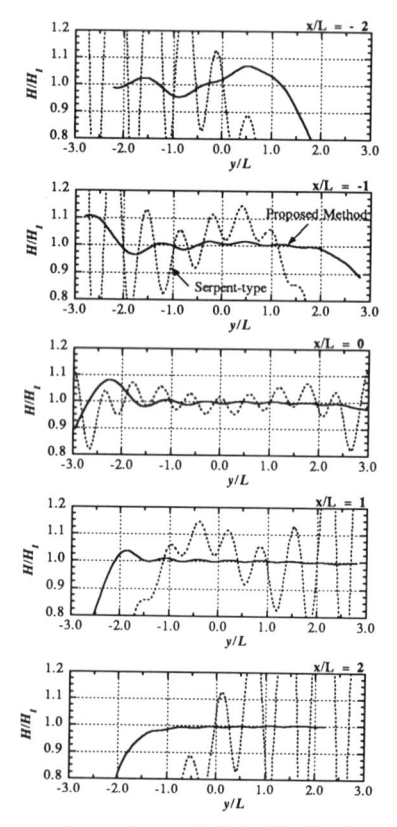

Fig. 8 Comparison of wave height distributions between for the source method and a serpent-type

perimeters. It can be seen in these areas that the magnitudes and directions of energy flux are different from those of the intended waves. As the results, the serpent-type generator provides large ineffective wave field in the tank. The effective wave-field of the serpent-type depends on incident wave directions, decreasing with increase of the incidence angle of waves.

In Fig. 8, normarized wave-height distributions at cross-sections of $x/L = -2, -1, 0, 1$ and 2 generated by a *serpent-type* are compared with those generated by the source method using the results shown in Fig. 4 (a) and 7 (a). The results show that the source method generates waves much more uniformly than does the *serpent-type* not only at $x/L = 0.0$, where the serpent-type gives the best performance, but also at $x/L = \pm 1.0$ and ± 2.0. It is expected, therefore, that the proposed source method can provide significantly larger effective fields for multi-directional waves compared to those provided by *serpent-type* wave makers.

4. Application of the source method

4.1 A devised equipment

We have carried out preliminary examinations of various devices to evaluate practicability of a wave-making source. As a result, a device, which supplies and discharges water at tank bottom, has proved to be the most practical source. The principle behind wave-generation by this device is similar to that by a non-reflective wave-making system devised by Goda (1965): it generates a vertically oscillating flow at channel bottom by using a paddle as shown in Fig. 9. The Goda's method was not of practical use because of inefficiency for deep water wave-generating. In addition, the method requires elaborate work to keep nearly water-tightness in an interstice between the pit's walls and the wave paddle.

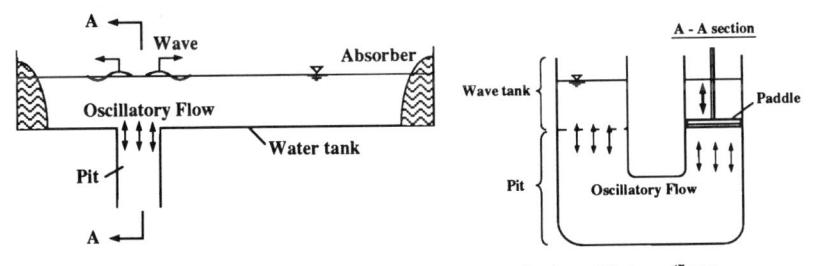

Fig. 9 Non-reflective wave generator with vertical oscillatory flow

To solve these problems, authors have made over this method into a practical equipment, which consists of pumps, valves, and an adjustable pit, as shown in Fig. 10. Vertical oscillating flow by using the pumps, valves and adjustable pit instead of using a wave-paddle: the pump P_I supplies water into the pit, and the pump P_O discharges water from the pit. Wave heights and periods are controlled by using two valves, V_I and V_O. The pit's height and width, d and B, can be adjusted to improve the efficiency of wave generation in deep water (see Fig. 10).

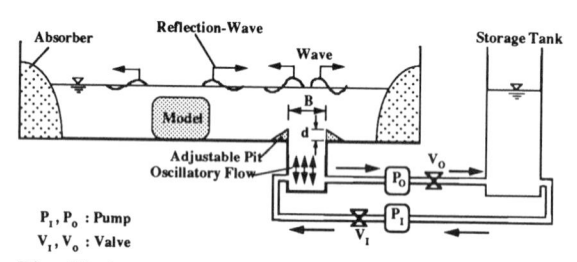

Fig. 10 A practical equipment of the source method

4.2 Performance of wave-generation by the equipment

It is important for the equipment to produce wave field with negligible reflection from the pit and to improve efficiency of wave generation in deep water. To evaluate the effects of the pit's width and height on characteristics of the reflection waves and the efficiency, experiments were carried out using a two-dimensional wave tank. The coefficient of reflection waves caused by the pit is calculated using Eq. 12 (Goda et al., 1964).

$$K_R = \frac{H_{max} - H_{min}}{H_{max} + H_{min}} \qquad (12)$$

Where, H_{max} and H_{min} are heights of standing wave in front of a vertical wall apart from the pit by N and $(N+1/2)$ times of a half wave length, respectively (N: an integer). On the other hand, the wave generation efficiency is evaluated using a ratio of a wave height to oscillatory flow amplitude in the pit. Calculations were also conducted to estimate the performance of the wave-generating pit by Green function method. As theoretical formulation about the wave generation pit has been described by Tanaka et al. (1993), only a definition sketch of the theoretical model is shown in Fig. 11.

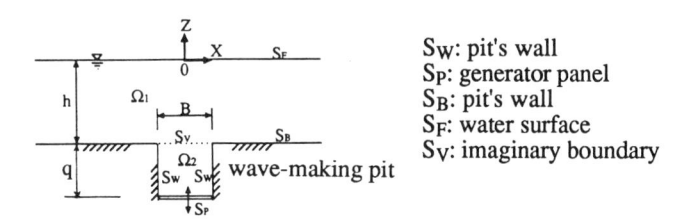

Fig. 11 Definition sketch of a theoretical model of the wave making pit

S_W: pit's wall
S_P: generator panel
S_B: pit's wall
S_F: water surface
S_V: imaginary boundary

Fig. 12 (a) shows response of reflection coefficients, K_R, to parameter, h/L, as B/h was varied from 1.2 to 3.2. In this case, the top of the pit is set at the same level of the tank bottom. Fig. 12 (b) also shows the dependence of wave generation efficiency, $H/2e$, on h/L upon the same conditions with Fig. 12 (a). Where, h is the water depth of the tank, L: the wave length, B: the width of the wave-making pit, and $2e$: an amplitude of the oscillatory flow in the pit. In the figures, solid lines denote the computed results. Although the experimental and the numerical results agree on the characteristics of $H/2e$, both results show large discrepancies in the characteristics of K_R. Since the detailed reason for the discrepancies has not been made clear, the characteristics of K_R for the new equipment are evaluated using the experimental results in this paper.

Although the increase in B/h improves the wave-generating efficiency of the equipment, it enlarges the reflection waves from the pit. Therefore, B/h should be set so that the equipment can achieve low reflection wave over a wide range of wave periods: K_R is 0.15 or less for $B/h = 1.5$, and further decreases to around 0.1 for $B/h = 1.2$.

In Fig. 13 (a), the effect of the pit height on the efficiency of wave generation is shown when d/h was varied 0, 0.305 and 0.432. Where, $B/h' = 1.2$, $B = 30$ cm, d: the pit height, h': depth between water surface and the top of the pit ($h = d+h'$). The wave-generation efficiency with $d/h = 0.305, 0.432$ is significantly improved in the range of $h/L > 0.3$ compared with that with $d/h = 0$. These results indicate that the adjustment of the pit height is effective for the improvement of the wave generation efficiency. In addition, the former maximum efficiencies are greater than the latter one.

Fig. 13 (b) shows characteristics of the reflection waves caused by the pit. The experimental conditions are the same with those of Fig. 13 (a). This equipment can achieve low reflection over a wide range of wave periods even when d/h is set at

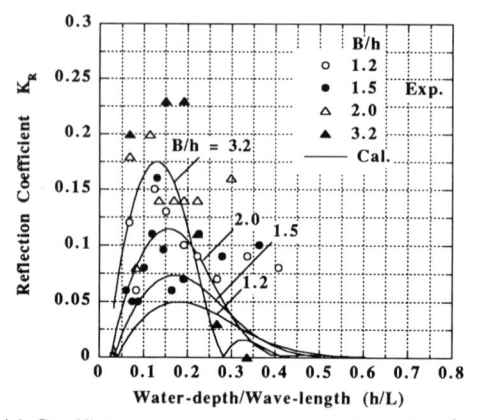

Fig. 12 The effects of pit width, B, on performance of wave generation ($d/h = 0.0$)

(a) Coefficients of reflection waves from the pit

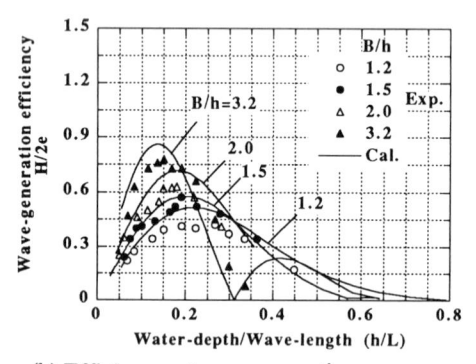

(b) Efficiency of wave generation

0.305 or 0.432. The results of Fig. 13 shows that the adjustable pit can improve the efficiency of the wave generation without increasing reflection waves.

From the experimental results, the proposed equipment can generate wave with negligible re-reflection from the wave-generating pit, and can efficiently make deep water waves.

It is expected that the equipment would provide nearly non-reflective wave field for multi-directional wave-generation by arranging a series of the units in a concentric semicircle in circle tank.

(a) Efficiency of wave generation

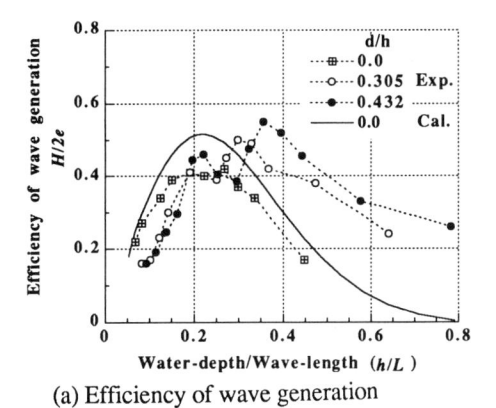

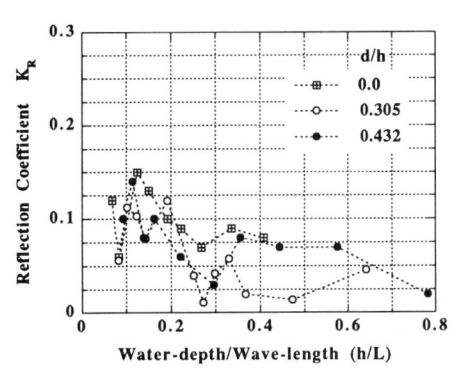

(b) Coefficients of reflection waves from the pit

Fig. 13 The effects of pit height, d, on performance of wave generation ($B/h' = 1.2$)

5. Conclusions

Wave-generation by source method, which does not depend on wave-making paddles, has been proposed to produce non-reflective multi-directional waves in a laboratory basin. The characteristics of wave fields generated by the source method were evaluated using a theoretical model. A practical equipment with a wave-making pit at channel bottom was devised to realize the source method, and the performance of the wave-generation was also examined. Major findings of the present study are summarized as follows:

1) The wave generation by the source method provides significantly larger effective area for multi-directional waves than does the *serpent-type* wave makers.

2) The equipment with a wave-making pit at a channel bottom can produce wave field with negligible reflection by adjusting the width and position of the pit, and can also efficiently generate waves in deep water.

References

Biesel, F. (1954): Wave machines, Proc. 1st Conf. on Ships and Waves, pp.121-125.

Brorsen, M. and J. Larsen (1987): Source generation of nonlinear gravity waves with the boundary integral equation method, Coastal Eng., Vol. 11, pp. 93 - 113.

Dalrymple, R. A. (1989): Directional wavemaker theory with side wall reflection, Jour. Hydraulic Research, Vol. 27, No. 1, pp 23-34.

Funke, E. R. and M. D. Miles (1987): Multi-directional wave generation with corner reflectors, National Research Council Canada, Hydraulic Laboratory, Tech. Rept., No. TR-HY-021.

Gilbert, G. (1976): Generation oblique waves, Hydraulics Research Station, Wallingford, England, Notes 18, pp. 3-4.

Goda, Y. and T. Kikura (1964): The generation of water waves with a vertically oscillating flow at a channel bottom, Report of Port and Harbor Tech. Res. Inst., No.9, pp. 1-24.

Hirakuchi, H., Y. Kajima, H. Tanaka and T. Ishii (1991): Characteristics of absorbing directional wavemaker, Proc. 38th Japanese Conf. on Coastal Eng., pp. 121-125 (in Japanese).

Ikeya, T., Y. Akiyama and K. Imai (1992): Active Directional Wave Absorption Theory, Proc. 39th Japanese Conf. on Coastal Eng., pp. 81-85 (in Japanese).

Isaacson, M. (1989): Prediction of directional waves due to a segmented wave generator, Proc. of 23rd Congress of the Int. Asso. Hydraulic Res. Vol. C, pp. 435-442.

John, F. (1950): On the motion of floating bodies II, Comm. Pure and Appl. Math., Vol. 3, pp. 45 - 101.

Ohyama, T. and K. Nadaoka (1991): Development of "Numerical Wave Tank" in nonlinear and irregular wave field with non-reflecting boundaries, Jour. of Hydraulic, Coastal and Environmental Eng., JSCE, Vol.429 II-15, pp. 77-86.

Takayama, T. (1982): Theoretical properties of oblique waves generated by serpent-type wave-maker, Report of Port and Harbour Research Inst., Vol. 21, No.2, pp. 3-48.

Tanaka, M., T. Ohyama and T. Kiyokawa (1993): Performance of non-reflective wave generator, Proc. 40th Japanese Conf. on Coastal Eng., pp. 41-45 (in Japanese).

CHAPTER 49

The Growth of Wind Waves in Shallow Water

L.A. Verhagen and I.R. Young
Dep. of. Civil Engineering, University College, University of New South Wales,
Canberra, Australia

Abstract
The field experiment at Lake George has brought valuable data that was not yet available. The analyses done to the data has shown new insights into shallow water waves. Already from the first analyses we can derive some interesting things. The one-dimensional shallow water spectra seem to have a clear peak enhancement and tail that is proportional to $f^{-4.5}$. The directional spread with two-dimensional spectra shows at the higher frequencies (twice the peak frequency) the waves move symmetrically at an angle with the wind. More analysis needs to be done to fit different kind of spectra and to find a function for the directional spread.

Introduction
Although most coastal engineering activity is in regions of shallow water, there is little know about the growth of shallow water waves. Many numerical models attempt to incorporate the growth with relationships which are mostly derived from deep water data. The lack of reliable field data in shallow water is a major problem. To overcome this lack of data, a large field experiment is being conducted to study the growth of shallow water waves. The experiment has been set in Lake George (25 km x 12 km) (fig 1) just north of Canberra (Australia). The lake has a flat bed with an approximate depth of 2 meters. There are no high obstacles around the lake to influence the wind significantly.

Experiment
Eight wave measuring stations have been established along the north-south axis of the lake (fig 1). Data from these stations are sent by telemetry to a station on the shore where they are logged. The data, after analysis, yield one-dimensional spectra. At station 6, located in the centre of the lake, there is also a cluster of 7 wave measuring gauges to determine the

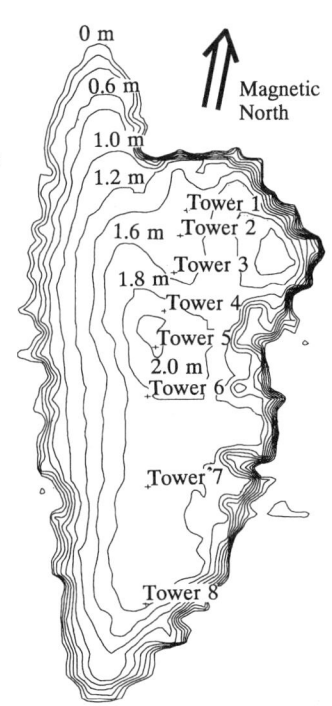

Figure 1
The bathometry of lake George and the position of the wave measuring towers. Note the platform is at tower 6.

directional spread. The wind speed and direction are measured at 5 locations along the lake. At various sites other meteorological data such as air and water temperature, air humidity, and solar radiation are also measured.

Seven wave measuring stations are nearly identical. They are space-frame towers, with a wave measuring device mounted on the side (fig 2). The towers are moored to the lake bed by 4 guy wires which were anchored into the bed. The wave measuring devices are Zwarts poles (Zwarts 1974). The Zwarts poles consist of an aluminium outer (55 mm) and inner pole (20 mm) which determine the electrical properties. These wave measuring devices have the advantage that the

Figure 2
Details of the tower. From left to right; the aerial to send back the data by telemetry. The box with batteries the radio and the electronics box for the ZWARTS pole, just in front the anemometer mast and the hydrometer, on top the solar cell, and on the back side the ZWARTS pole.

they do not require much maintenance. Other wave measuring devices like resistance gauges, capacitance gauges or waverider buoys needed too high level of maintenance or are outside the frequency range of the waves on Lake George. The poles were calibrated both dynamically and statically in the lab. The data were send to a base station on the shore via a telemetry link. The radio was powered by two solar cells on the top of the tower (fig 2). At five of the towers, there were 10 meter high masts with an anemometer on top (fig 3).

Figure 3
A complete view of one of the towers with a 10 meter anemometer mast.

At two towers the air humidity and temperature (fig 2) were measured together with the water temperature at three levels (0.5, 1, 1.5 meters for the lake bed).

Wind Analysis
The wind was measured at 5 stations (stations 2,4,6,7,8) on the lake. However there was only a very short time period that all 5 anemometers were working. The anemometer on the platform (station 6) worked most of the time. All anemometers

were mounted on 10 meter high mast above the water surface. The anemometer on the platform (station 6) was a "Young" anemometer and the anemometers on the towers were "VDO" anemometers.

A major problem is knowing the exact wind speed along the fetch. This is a serious problem when there is a change of roughness length of the surface (e.g. from land to water). Taylor (1984) gives some guidelines. The basic principle is that the boundary layer adjusts it self after a certain length. The flow will have an internal boundary layer with a depth of δ_i. Outside this internal boundary layer the wind velocity is the as equation 1.

$$U_z = \frac{U_*}{\kappa} \ln\left(\frac{z}{z_0}\right) \qquad (1)$$

In which Z is the elevation at which U_z is measured. κ is the von Karman's constant ($\kappa = 0.4$), Z_0 is the roughness length of the surface and U_* is the friction velocity. The internal boundary layer grows according to

$$\delta_i = 0.75 \, Z_0 \left(\frac{x}{Z_0}\right)^{0.8} \qquad (2)$$

Where x is the distance along the fetch and Z_0 is the roughness length at x. Inside the boundary δ_i layer the wind speed at elevation Z is

$$U_z = \frac{\ln\left(\dfrac{Z}{Z_0}\right) \ln\left(\dfrac{\delta_i}{Z_{0u}}\right)}{\ln\left(\dfrac{\delta_i}{Z_0}\right) \ln\left(\dfrac{Z}{Z_{0u}}\right)} U_{0u} \qquad (3)$$

With Z_{0u} being the roughness length of the upwind boundary, U_{0u} the upwind wind velocity and U_z the wind speed at elevation Z. The wind is usually only measured at a point or a few points along the fetch. As indicated in equations 2 & 3 the wind speed varies after a change of roughness length. Since the wave growth is dependent on the wind speed it is wise to calculate an average wind speed for the fetch. One could also argue that even a mean of the wind speed along the fetch is not good enough and a more sophisticated average should be taken.

We tested the Taylors (1984) method with the data set when all five anemometers were working. The results from the theory and the measured wind speeds were nearly the same. Therefore the anemometer at station 6 (the platform) was used to calculated the wind speed along the north south fetch for each event.

There is a large debate about which wind speed to use, in the making the wave height, wave frequency, fetch and water depth dimensionless. In general the wind speed is measured at 10 meters above the surface level. If the wind speed is measured at a different level it can be converted to U_{10} using the logarithmic form of the boundary layer (see equation 1). There are several other wind speeds to make the wave height, wave frequency, fetch and water depth dimensionless. They are usually based on either the friction velocity like Wu (1980) describes, the inverse wave age (U/c_p) like Donelan (1985) argues, or a combination of both like HEXOS (Smith et al. 1992). The wind speed used here was the wind speed measured at 10 meters above the surface, because the other did not give a clear advantage and they were not measured directly.

Growth Curves

The fetch limited growth curves for shallow water have two limits. One limit is the JONSWAP (Hasselmann et al. 1973) deep water growth, and the other is the maximum wave height by a finite depth and wind speed. Vincent in his first paper (1985) suggested that the relation between dimensionless water depth and dimensionless wave height,

$$\frac{H_s g}{u^2} = \left(\frac{d g}{u^2}\right)^{0.5} \quad (4)$$

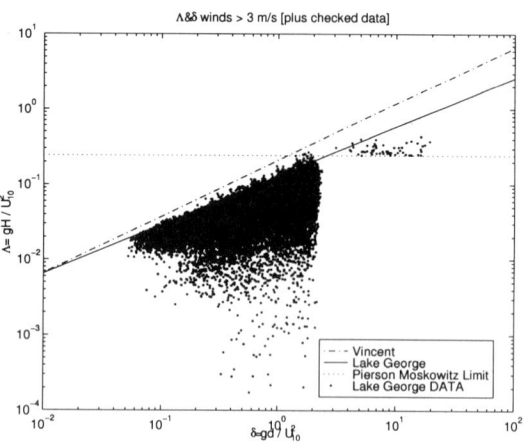

Where H_s is the significant waveheight, d the waterdepth, g the gravitational constant and u the wind speed at 10

Figure 4
Lake George Data. Dimensionless waterdepth is set out against dimensionless waveheigth. Note the spreading in the wave data with wind speeds greater than 3 m/s is caused by the fact that not all the wave are fully developed.

meters above the surface. Later Vincent and Hughes (1985) had taken measurements from Lake Okeechobee (Florida, USA) to determine the maximum dimensionless wave height. From these data and from an analytic procedure, they suggested that the relation between dimensionless water depth and dimensionless wave height was

$$\frac{H_s g}{u^2} = \left(\frac{dg}{u^2}\right)^{0.75} \tag{5}$$

Data from Lake George was used to test this. All the data from Lake George with wind speeds greater than 3 m/s were used together with selected data where the wind speed was smaller than 3 m/s. These last data were selected if the wind speed and direction were constant 8 hours before and 3 hours after the event. The Lake George data tend to indicate a less steep slope than what Vincent and Hughes suggested (fig 4). If one projects the Lake Okeechobee data on a graph similar to the Lake George data (fig 5) one can see that the Lake Okeechobee has the same tendency. Purely on empirical grounds the following seems to be more appropriate.

$$\frac{H_s g}{u^2} = \left(\frac{dg}{u^2}\right)^{0.65} \tag{6}$$

The upper limit for the growth in deep water is the Pierson Moskowitz limit (1964). Pierson Moskowitz (1964) proposed, integrating the spectrum resulting in a limit

$$\frac{H_s g}{u^2} = 0.2433 \tag{7}$$

Vincent and Hughes (1985) do not have a limit because they had not found evidence for this. Data for this limit is indeed very hard to find because of the long fetch and time necessary for the full development of the waves. This only happens during very low wind speeds. Because of swell and other wind waves generated elsewhere it is very hard to measure the growth of smaller waves in ocean environments. However in a lake this possible. During the Lake George experiment there were a number of these events. These events were specially selected because of the tendency of the wind varying in both speed and direction at low wind speeds. The events selected with wind speeds lower than 3 m/s had to a have a constant wind speed and wind direction, during a period of 8 hours before and 3 hours after. In figure 4 the Pierson Moskowitz limit with the selected data is shown.

It is believed that the lake sides have a small influence on the observed growth rates. Attempts to measure this effect using three different numerical models have been unsuccessful. The models [WAM (WAMDI 1988), HISWA (Holthuijsen et al.1988), ADFA1 (Young 1988)] produced conflicting results and all significantly deviate from the observed data. Therefore no correction factors have been applied to the data to compensate for the sides of the lake.

In general the data are consistent with the commonly used CERC (SPM 1984) growth curves, but at small values of dimensionless fetch tend to yield lower dimensionless energy. This could be due to the effects of the change in surface roughness length between land and water. Because of this and because of uncertainty in determining the exact location of the upwind boundary, the fetch of the first station is set equal to the corresponding JONSWAP (Hasselmann et al. 1973) result for the observed energy. From the data new curves were calculated (fig 6 and fig 6).

Figure 5
The same graph as with the Lake George data, except with the Lake Okeechobee data. The Lake George line looks like a better estimate.

Figure 6
Two growth curves for dimensionless Energy set out against dimensionless fetch. The curves are calculated from the points with the same dimensionless depth. The most upwind point is set on the deep water growth (JONSWAP, Hasselmann et al. 1973)

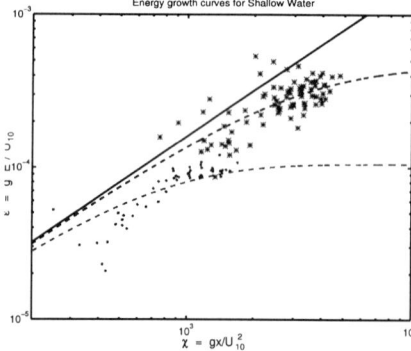

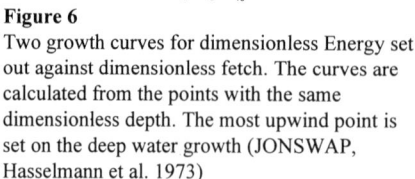

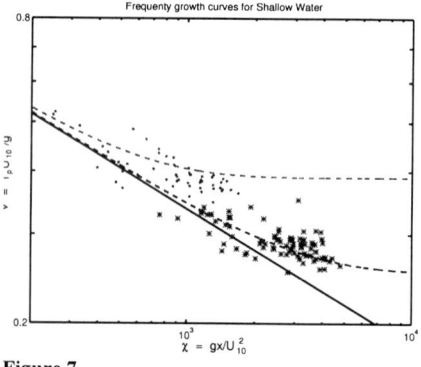

Figure 7
Two growth curves for dimensionless frequency set out against dimensionless fetch. The curves are calculated from the points with the same dimensionless depth. The most upwind point is set on the deep water growth (JONSWAP, Hasselmann et al. 1973)

One-Dimensional Spectra

The one-dimensional spectrum has some of the characteristics of deep water spectra. The data used in the analyses of the spectra along the lake is only during north and south wind events with wind speeds greater than 3 m/s. In figure 8 the spectral peak moves toward lower frequencies as the fetch increases. In figure 8 one can see a small but very distinct peak enhancement. The spectra decay with $f^{-4.5}$ from the peak frequency (f_p) until approximately 3 f_p, and at f^{-5} at higher frequencies (fig 9). The decay of f^{-5} however could be caused by a doppler shifting of the high frequency waves "riding" on the back off lower frequency waves. More analyses of these one-dimensional spectra are under way, including the differential growth between the wave measuring stations.

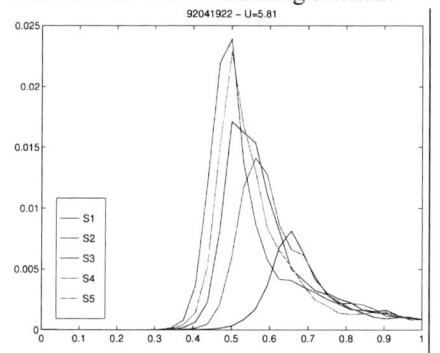

Figure 8
Different stages of the spectra during growth. The spectral peak moves toward the lower frequencies as the fetch (or growth) increases. The peak enhancement can easily be seen. The frequency is set out against the Energy

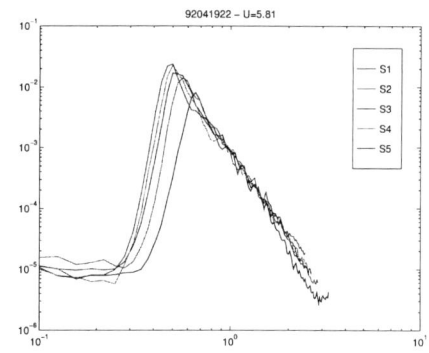

Figure 9
This is a loglog plot of the different stages of the spectral growth. The Energy decays in the tail of the spectrum is related to $f^{-4.5}$. At higher frequencies the Energy decay is related to f^{-5}. The frequency is set out against the Energy

Two-Dimensional Spectra

At the platform (station 6) the directional spectra were measured by an array of 7 Zwarts poles. The array could be operated by modem via the base station from the University College. The array was formed like a Mercedes star (Young 1995). For the analyses the Maximum Likelihood Method (MLM) was used. This method has the advantage over the Maximum Entropy Method (MEM) that it does not produce false peaks. The disadvantage is that the MLM produces a broader spectrum and has trouble analysing bi-modal seas. Lake George however, does not have any bi-model seas.

Due to the large number of wave measuring gauges in the array, the directional resolution is much higher than pitch-roll buoys or p, u, v meters. In figure 10 the normalised directional spreading function at ½f_p, f_p, 2f_p and 3f_p is shown. The spectrum at f_p is quite narrow and aligned with the wind direction. However at 2f_p

and $3f_p$ the spectrum is quite broad and develops a bidirectional mode, with more energy propagating at an angle to the wind than in the wind direction.

Conclusions

The first analyses look promising. However more work needs to be done in analysing all the data. We are trying to fit different types of spectra like the JONSWAP and TMA spectra to see which fit better and with which parameters. With the Two-Dimensional Spectra we are trying to fit a spreading function that is dimensionless frequency depended.

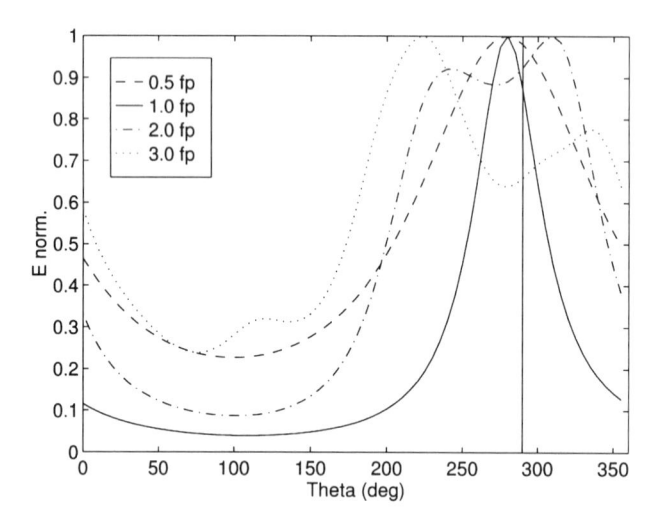

Figure 10
Cuts of the directional spectra at $½f_p$, f_p, $2f_p$ and $3f_p$. The spectrum has one peak at frequency smaller than $2f_p$. The waves are travelling in the direction of the wind (thick line). However at frequencies greater than $2f_p$ there is splitting of the directional spread. This means that the waves are travelling at an angel with the wind.

References

Donelan M.A., Hamilton J. and Hui W.H. 1985. Directional Spectra of Wind-Generated Waves. Phi. Trans. R. Soc. Lond. A 315, pp. 509-563

Hasselmann K., Barnett T.P., Bouws E., Carlson H., Cartwright D.E., Enke K., Ewing J.A., Gienapp H., Hasselmann D.E., Kruseman P., Meerburg A., Müller P., Olbers D.J., Richter K., Sell W., Walden H. 1973. Measurements of Wind-Wave Growth and Swell Decay during the Joint North Sea Wave Project (JONSWAP), Deutsches Hydrographisches Institut, Hamburg

Holthuijsen L.H., Booij N. and Hebers T.H.C. 1988 "A prediction model for stationary short-crested waves in shallow water with ambient currents", Coastal Engineering, 13, 23-54

Pierson W.J. and Moskowitz L. 1964. A proposed Spectral Form for Fully Developed Windseas Based on Similarity Theory of S.A. Kitaigorodskii, Journal of Geophysical Research, Vol. 69, pp. 5181-5190

Shore Protection Manual 1984. US Army Engineer Waterway Experiment Station, Coastal Engineering Research Center (CERC), US Government Printing Office, Washington, D.C., 4th edition, 2 Volumes

Smith D.S., Anderson J.R., Oost W.A., Kraan C., Maat N., DeCosmo J., Katsaros K.B., Davidson K.L., Bumke K., Hasse L., Chadwick H.M. 1992. Sea Surface Wind Stress and Drag Coefficients, The HEXOS Results, Boundary-Layer Meteorology, Vol. 60, pp. 109-142

Taylor P.A. and Lee R.J. 1984. Simple Guidelines for Estimating Wind Speed Variations due to Small Scale Topographic Features, Climatological Bulletin, Vol. 18, No. 2, pp. 3-32

Vincent C.L., Hughes S.T. 1985. Wind Wave Growth in Shallow Water, Journal of Waterway, Port, Coastal and Ocean Engineering, Vol. 111, No. 4, pp. 765-770

Wamdi Group 1988. The WAM Model- A Third Generation Ocean Wave Prediction Model, Journal of Physical Oceanography, Vol. 18, pp. 1775-1810

Wu J. 1980. Wind-Stress Coefficients over Sea Surface near Neutral Conditions-A Revisit, Journal of Physical Oceanography, Vol. 10 Part 1, pp 727-740

Young I.R.,1988, A shallow water spectral wave model, Journal of Geophysical Research, Vol. 93,C5, pp. 5113-5129

Young I.R., 1995, On the Measurement of Directional Wave Spectra, In press, Applied Ocean Research

Zwarts C.M.G. 1974. Transmission Line Wave Height Transducer, Proceedings International Symposium on Ocean Wave Measurement and Analysis, New Orleans, ASCE, Vol. 1, pp. 605-620

CHAPTER 50

Estimation of Typhoon-Generated Maximum Wave Height along the Pacific Coast of Japan Based on Wave Hindcasting

Masataka Yamaguchi[1] and Yoshio Hatada[2]

Abstract

Wave hindcastings in the West Pacific Ocean and at selected locations along the Pacific coast of Japan by the use of two wave prediction models are conducted for 118 typhoons which occurred from 1940 to 1991 to estimate maximum wave height at each point during a typhoon. Combination with maximum wave height data in the concerned area for 125 typhoons from 1934 to 1983 obtained by Yamaguchi et al.(1987a, 1987c) results in data sets of maximum wave heights caused by 243 typhoons from 1934 to 1991. Spatial distribution of the most extreme wave height and wave height for return period of 100 years, and their variation along the Pacific coast of Japan are re-examined on the basis of statistical analysis of the data and compared with the Yamaguchi et al. results (1987a, 1987c), in which case there is little change in the distribution except for the Western Kyushu coast. In addition, maximum wave height along the coast of the Japan Sea is evaluated in a similar manner.

1. Introduction

As the coastal areas of Japan facing the Pacific Ocean and the East China Sea have been heavily damaged by attacks of huge waves generated by powerful typhoons, the reasonable estimation of the typhoon-generated extreme waves around the Japanese coast is essential to planning and design of coastal protection systems for mitigation

---
1 Prof., Dept. of Civil and Ocean Eng., Ehime Univ., Bunkyocho 3, Matsuyama 790, Ehime Pref., Japan
2 Research Assistant, Dept. of Civil and Ocean Eng., Ehime Univ., Bunkyocho 3, Matsuyama 790, Ehime Pref., Japan

of wave-induced coastal hazards. The extreme waves expected in 50 or 100 years in the concerned sea area or location are usually evaluated based on statistical analysis of wave data hindcasted for many severe storms over a long term period of more than 30 years rather than measured wave data, because acquirement of measured data which is acceptable in quality and quantity is difficult.

Yamaguchi et al.(1987a, 1987c) estimated the spatial distribution of maximum significant wave height generated by typhoons in the West Pacific Ocean and around the Pacific coast of Japan, based on wave hindcasting for 125 typhoons in the past 50 years using two kinds of spectral wave prediction models. But re-estimation of probable extreme waves around the Pacific coast of Japan is required, because in recent years abnormally high waves and the consequent severe coastal hazards have been brought about by successive attacks of powerful typhoons such as Typhoons 8712 and 9119.

In this paper, wave hindcastings in the West Pacific Ocean and at selected locations along the Pacific coast of Japan by the use of two wave prediction models are conducted for 118 typhoons which occurred from 1940 to 1991 to estimate maximum wave height at each point during a typhoon. Combination with maximum wave height data in the concerned area for 125 typhoons from 1934 to 1983 obtained by Yamaguchi et al.(1987a, 1987c) results in data sets of maximum wave heights caused by 243 typhoons from 1934 to 1991. Spatial distribution of the most extreme wave height and wave height for return period of 100 years, and their variation along the Pacific coast of Japan are re-examined on the basis of statistical analysis of the data. In addition the most extreme wave height along the coast of the Japan Sea is also evaluated based on wave hindcastings for intense storms with use of the wave prediction models and their statistical analysis.

2. Outline of Wave Hindcast System

(1) Wind hindcast model
   A parametric typhoon model is used for the estimation of typhoon-generated winds. The model computes the spatial distribution of wind speed and wind direction in a typhoon by composing axisymmetyrical gradient wind components and wind components related to the movement of a typhoon. The exponential function is assumed for the pressure distribution in a typhoon. The model parameters such as central pressure, position of typhoon center, radius to maximum winds and wind inflow angle are given every 6 hours, and wind field every 1 hour is estimated

through a linear interpolation of the typhoon parameters. Correction factor to wind speed at the height of 10 m is 0.60.

(2) Wave hindcast model
The wave prediction models developed by the authors are used in the wave hindcasting. The first model referred to as the grid model(Yamaguchi et al., 1987a) is a coupled discrete model belonging to the second generation model which solves the energy balance equation on a regular grid. The grid with a spacing of 80 km is used for the estimation of spatial distribution of deep water extreme waves in the West Pacific Ocean shown in Fig. 1.

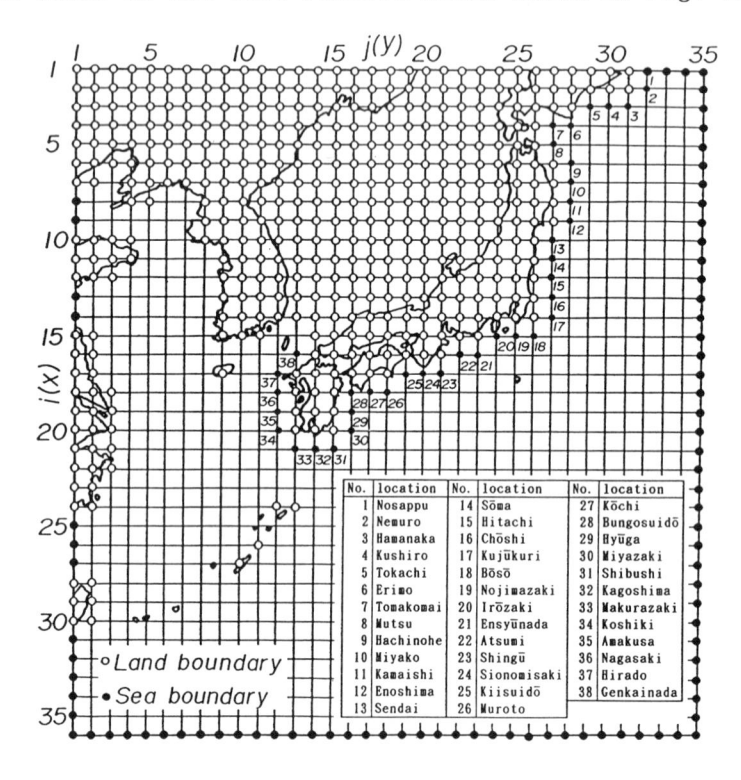

Fig. 1 Coarse grid used in wave hindcasting with grid model.

The second model referred to as the point model (Yamaguchi et al., 1987b) is a decoupled propagation model classified into a category of the first generation model in which the evolution of directional spectrum is traced along a wave ray for individual spectral component

TYPHOON-GENERATED MAXIMUM WAVE HEIGHT 677

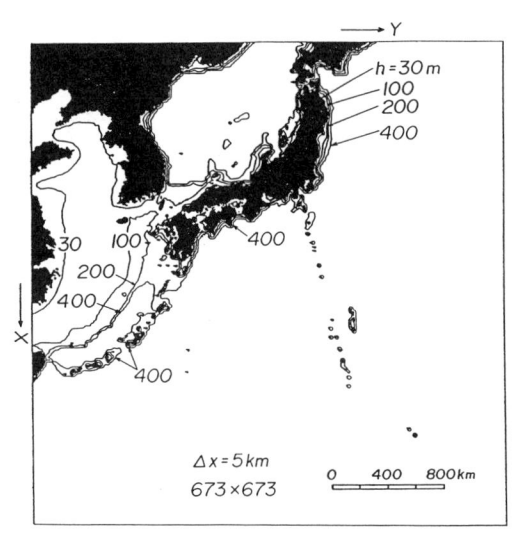

Fig. 2 Computation area used in wave hindcasting with point model and contour plot of water depth.

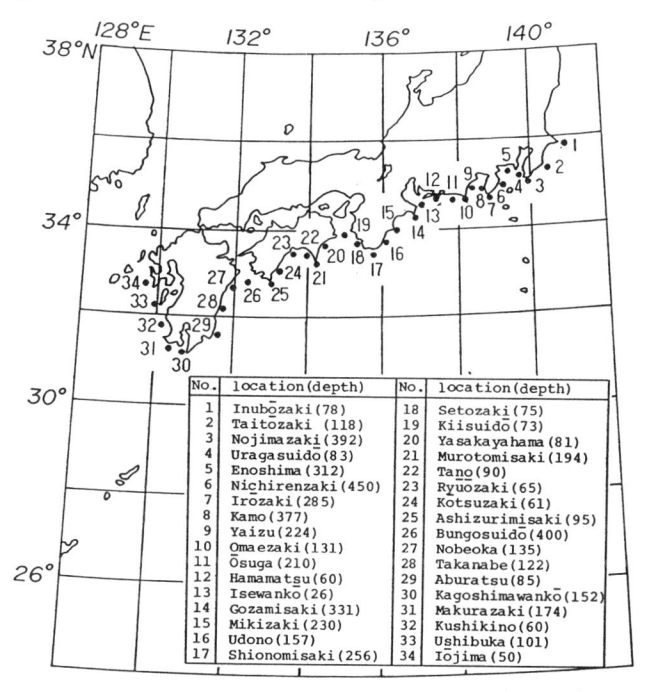

No.	location(depth)	No.	location(depth)
1	Inubōzaki(78)	18	Setozaki(75)
2	Taitōzaki(118)	19	Kiisuidō(73)
3	Nojimazaki(392)	20	Yasakayahama(81)
4	Uragasuidō(83)	21	Murotomisaki(194)
5	Enoshima(312)	22	Tano(90)
6	Nichirenzaki(450)	23	Ryuozaki(65)
7	Irōzaki(285)	24	Kotsuzaki(61)
8	Kamo(377)	25	Ashizurimisaki(95)
9	Yaizu(224)	26	Bungosuidō(400)
10	Omaezaki(131)	27	Nobeoka(135)
11	Ōsuga(210)	28	Takanabe(122)
12	Hamamatsu(60)	29	Aburatsu(85)
13	Isewanko(26)	30	Kagoshimawankō(152)
14	Gozamisaki(331)	31	Makurazaki(174)
15	Mikizaki(230)	32	Kushikino(60)
16	Udono(157)	33	Ushibuka(101)
17	Shionomisaki(256)	34	Iojima(50)

Fig. 3 Locations selected for wave hindcasting with point model.

focusing on a single position. The high topographical resolution grid with a spacing of 5 km is used in the point model to estimate the effect of shallow water on extreme waves in the nearshore region, as indicated in Fig. 2. Fig. 3 shows location of 34 selected points where wave hindcastings are conducted along the coastal region of Western Japan.

In hindcasting, 20 frequency components from 0.045 to 1.0 Hz and 19 directional components from 0 to 360 degrees are used in the grid model and 23 frequency components from 0.035 to 0.5 Hz and 37 directional components from 0 to 360 degrees are used in the point model. At the land boundary the directional spectrum is set at zero, and at the open boundary the directional spectrum computed from the product of frequency spectrum based on the Ross model(Ross, 1976) and an angular distribution is given in both models.

3. Typhoons Selected for Wave Hindcasting

In this study, 118 intense typhoons are newly selected for wave hindcasting by surveying the published reports. They include annual reports of wave data with ocean buoys deployed around Japan by the Japan Meteorological Agency, annual reports of wave data acquired with sonic-type wave gauge around the coastal area of Japan by the Harbor Construction Bureau, Ministry of Transport, a report of Typhoon Summary of Japan 1940-1970 edited by the Japan Meteorological Association and a report of Tropical Cyclone Tracks in the Western North Pacific 1951-1990 edited by the Japan Meteorological Agency. Extremes of typhoon-generated wave height are estimated based on the results of wave hindcasting for the 243 typhoons from 1934 to 1991, in which case wave hindcastings for the 125 typhoons from 1934 to 1983 have already been carried out by Yamaguchi et al.(1987a, 1987c).

Fig. 4 shows tracks of the 243 typhoons with those of the 125 typhoons used in wave hindcasting by Yamaguchi et al.(1987a, 1987c). Tracks of the 243 typhoons cover a wide area of the West Pacific Ocean far more densely than tracks of the 125 typhoons. Yearly consecutive typhoon data is available only for the period of 42 years from 1950 to 1991, because the typhoon data preceding the period are limited to huge typhoons which caused severe damages. Data of annual maximum wave height used in extreme wave statistics are those estimated for the 221 typhoons which occurred over the period of 42 years.

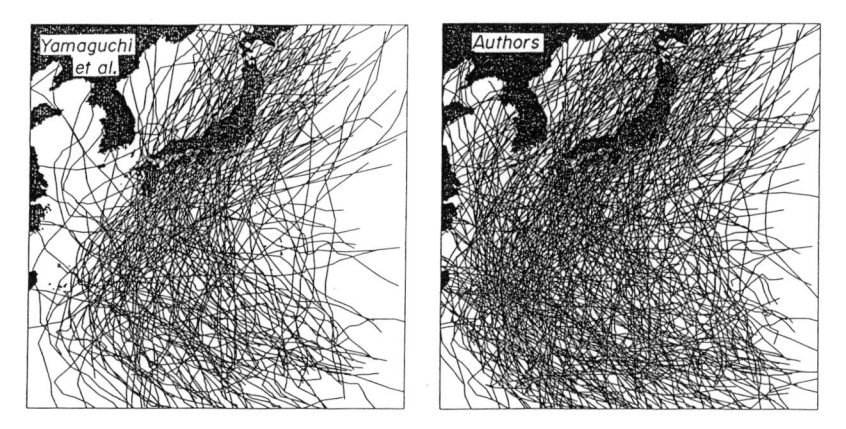

Fig. 4 Tracks of typhoons used in wave hindcasting.

4. Estimation of Maximum Wave Height

(1) Spatial distribution of maximum wave height

Superimposing maximum wave height obtained from hindcasted wave heights every 1 hour for the 243 typhoons at a point produces the most extreme wave height from 1934 to 1991 at the point. Fig. 5 shows spatial distribution of the most extreme wave height estimated by the wave hindcasting using the grid model of Yamaguchi et al. (1987a) and the authors' revised results. As the typhoon-generated extreme wave height at a point is strongly dependent on tracks of powerful typhoons which passed over near the point, it varies remarkably from place to

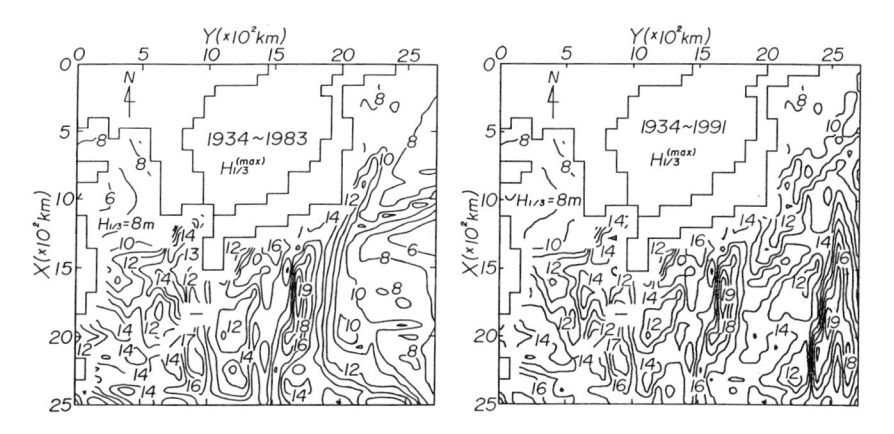

Fig. 5 Spatial distribution of the most extreme wave height.

place. Differences between both the results are seen at the southeast and southwest parts of the computation area and the western part of Kyushu Island. Especially, maximum wave height along the western part of Kyushu Island is replaced from 12-13m to 14 m associated with Typhoons 8712 and 9119.

(2) Alongshore distribution of the most extreme waves
Fig. 6 illustrates alongshore distributions of the most extreme wave height and frequency of occurrence of waves more than 10 m high estimated from wave data hindcasted with the grid model and observed data. Name and number of the location indicated along the horizontal axis in the figure are given in Fig. 1. In a qualitative sense, alongshore distribution of extreme wave height looks similar, but in a quantitative sense, hindcasted results give much higher estimation than observed results except for the northern part of the Japanese coast such as Sendai or Tomakomai where the most extreme waves are frequently generated by strong monsoon winds in winter. The reasons are that the wave model does not take into account the shallow water effects, that the grid system of poor topographical resolution is used in the wave hindcasting and that the period of wave observation is too short compared to the period of wave hindcasting. The most extreme wave height estimated by Yamaguchi et al.(1987a) is updated 2 m near Nagasaki by Typhoons 8712 and 9119, and updated 1 m near Chōshi by Typhoon 8913. Maximum of the most extreme wave height exceeding 15 m is observed at the sea area near Shionomisaki.

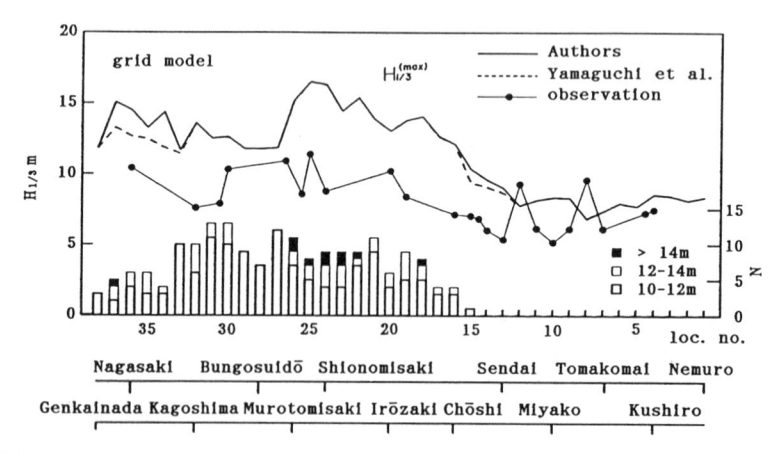

Fig. 6 Variation around the Pacific coast of Japan of the most extreme wave height and frequency of occurrence of waves more than 10 m high estimated from wave data hindcasted with grid model and observed wave data.

In addition, Fig. 7 shows alongshore distributions of the most extreme wave height estimated from wave hindcasting with the point model. Name of location corresponding to the figure along the abscissa is summarized in Fig. 3. The most extreme wave height and occurrence rate of high waves estimated with the point model are relatively lower than that with the grid model and varies significantly from location to location, because the point model can take into account the topographical effect in much more detail than the grid model.

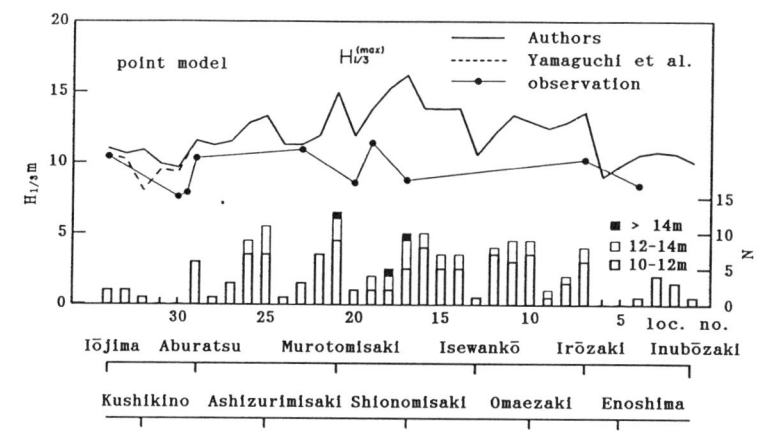

Fig. 7 Variation around the Pacific coast of Western Japan of the most extreme wave height and frequency of occurrence of waves more than 10 m high estimated from wave data hindcasted with point model and observed wave data.

Alongshore distributions of the most extreme wave height estimated by both the grid model and the point model and observed wave data are compared in Fig. 8. Notation of the abscissa is the same as Fig. 6. Extreme wave height estimated by the point model gives a consistently smaller value than that of the grid model and varies remarkably from location to location, as the point model can take into account the effect of water depth variation and topography by usage of fine computation grid. The most extreme wave height estimated with the point model shows a closer value to the observed extreme wave height than the result with the grid model. As mentioned before, the most extreme wave height observed near Sendai and Tomakomai exceeds the estimate with the grid model. This suggests that the most extreme wave height in northern Japan is caused by not only typhoons but also by monsoons.

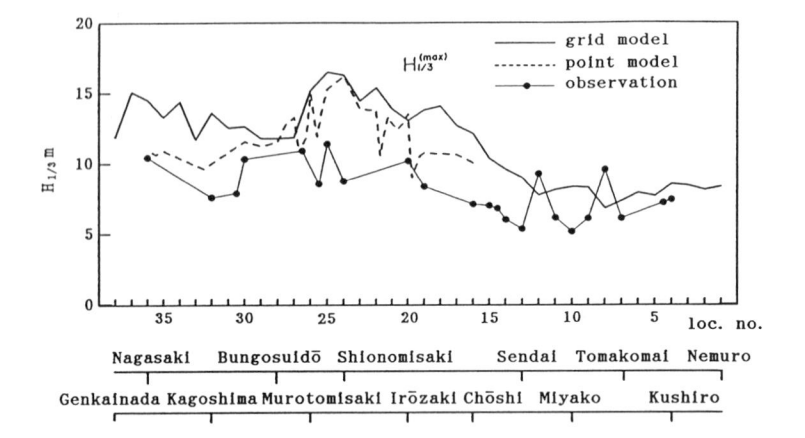

Fig. 8 Variation around the Pacific coast of Japan of the most extreme wave height estimated from wave data hindcasted with both grid model and point model and observed wave data.

(3) Wave height for return period of 100 years
Yearly variations of the annual maximum wave heights at Irōzaki estimated with the grid model and point model and the corresponding observed data are shown in Fig. 9. Both hindcasted results agree relatively well with the observed data, because Irōzaki is exposed directly to open sea and water depth of the installed wave gauge of 50 m is enough to measure deep water waves. Maximum wave height varies remarkably from year to year due to strong annual change of powerful typhoons attacking the Pacific coast of Japan, in which cases the severest sea state occurred in the nineteen-fifties and the nineteen-sixties.

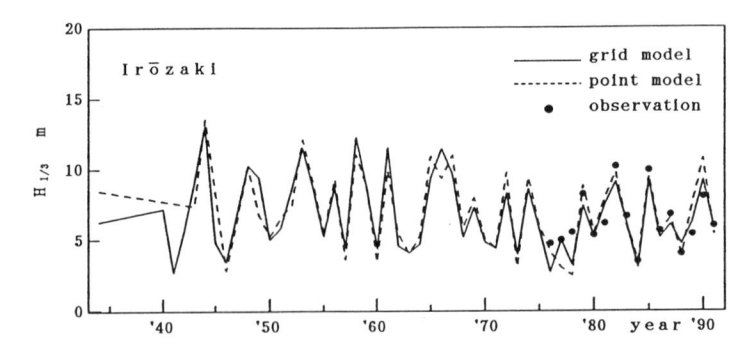

Fig. 9 Yearly variation of annual maximum wave height at Irōzaki.

The log-normal distribution with three parameters is fitted to the data of annual maximum wave height for the period of 42 years from 1950 to 1991, in which the estimation of parameters is due to the moment method. Fig. 10 shows an example of the fitting. At Irōzaki, wave heights with a return period of 100 years are 13.8 m and 13.5 m for the data hindcasted with the grid model and point model, and 11.9 m for the observed data. Observed data give a smaller estimation of extreme wave height, because the observation period is short compared to the wave hindcast period and the observed waves are affected by bottom topography in rough sea condition even if the wave gauge is installed at the point of 50 m deep.

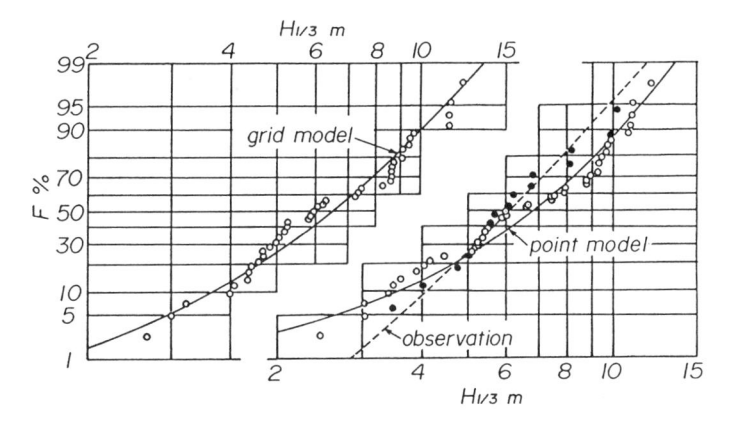

Fig. 10 Fitting of log-normal distribution to annual maximum wave height data at Irōzaki.

Fig. 11 describes the spatial distribution of wave height for a return period of 100 years over the West Pacific Ocean and the East China Sea. The spatial distribution is estimated by the fitting of log-normal distribution with three parameters to the annual maximum wave height data obtained from wave hindcastings by use of the grid model for the period from 1950 to 1991. Being similar to the spatial distribution of the most extreme wave height in this area, return wave height over 19 m is observed in the offshore sea area and return wave height becomes smaller around the Japanese coast, especially in the northern part because of the decay in typhoon intensity.

Fig. 12 illustrates alongshore distributions of wave height for a return period of 100 years estimated from wave data hindcasted with the grid model and observed wave data. Name of the location corresponding to the number along abscissa in the figure is given in Fig. 1. Except near Nagasaki, the present results are similar to

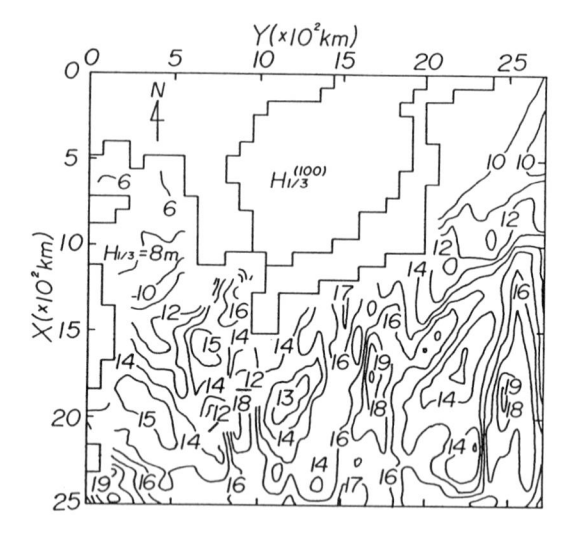

Fig. 11 Spatial distribution of wave height for return period of 100 years.

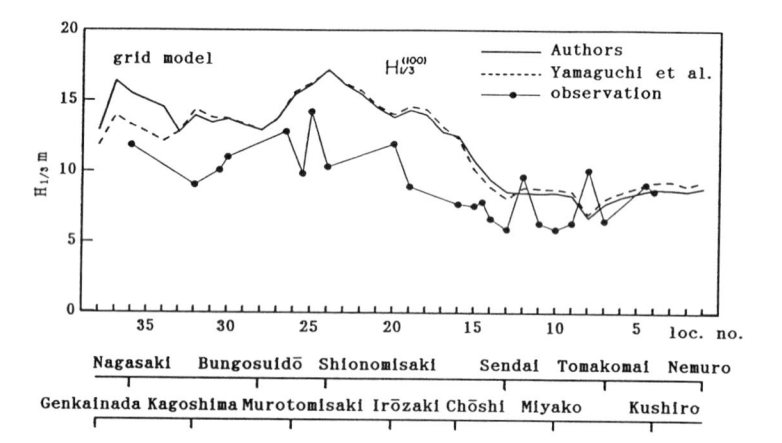

Fig. 12 Variation around the Pacific coast of Japan of wave height for return period of 100 years estimated from wave data hindcasted with grid model and observed data.

the results of Yamaguchi et al.(1987a), despite the fact that the number of annual maximum wave height data and cases of typhoon for wave hindcasting increased from 34 to 42 and from 107 to 221 respectively.

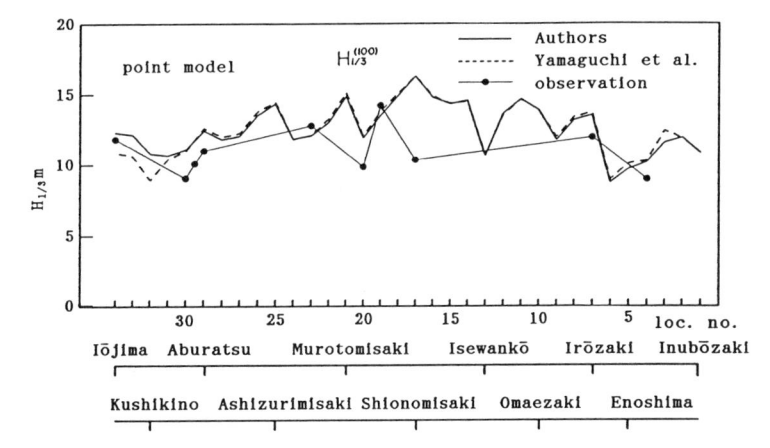

[Figure: plot of $H_{1/3}$ (m) vs loc. no., showing point model results for $H_{1/3}^{(100)}$ from Authors, Yamaguchi et al., and observation, along the Pacific coast from Iōjima/Kushikino to Inubōzaki/Enoshima.]

Fig. 13 Variation around the Pacific coast of Western Japan of wave height for return period of 100 years estimated from wave data hindcasted with point model and observed wave data.

Fig. 13 shows alongshore distributions of wave height for a return period of 100 years estimated with the point model by the authors and Yamaguchi et al. (1987c), which includes the corresponding results based on observed data. Name of numbered location is indicated in Fig. 3. Hindcasted results produce smaller estimates at Ryūōzaki and Kiisuido than observed results. But, the reliability of the observed results is relatively low, because the number of annual maximum wave height data used in extreme analysis is less than 10 at these points.

Alongshore distributions of wave height for a return period of 100 years evaluated with three kinds of wave data are given in Fig. 14. Return wave height estimated from the results hindcasted with the grid model varies from 9 m to 17 m and is greater than the other results because of poor topographical resolution of the grid system used in wave hindcasting. But at the open coastal locations, both models produce comparative results. Return wave height based on the point model using the grid system with high topographical resolution varies greatly from location to location, which is in closer agreement with that estimated with observed data. It should be noted that at most locations, the return wave height evaluated with observed data takes a lower value than that with hindcasted data, because the observation period is far shorter compared to the hindcast period.

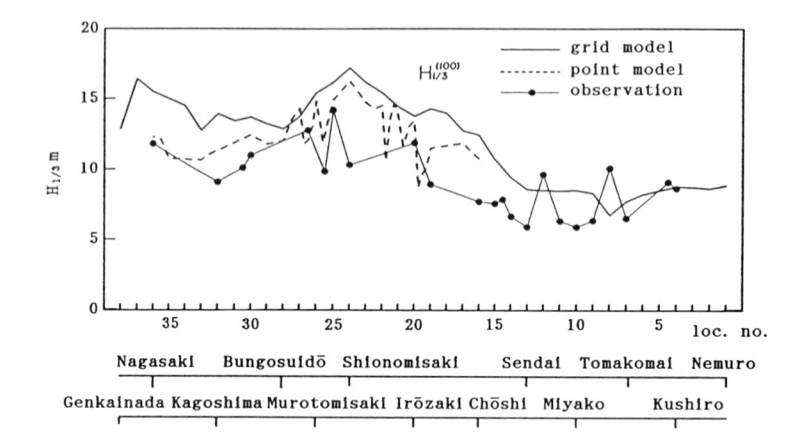

Fig. 14 Variation around the Pacific coast of Japan of wave height for return period of 100 years estimated with both grid model and point model and observed wave data.

5. Estimation of Maximum Wave Height along the Coast of the Japan Sea

Finally, maximum wave height along the coast of the Japan Sea is briefly discussed to contrast the characteristics of the maximum wave height along the Pacific coast. To evaluate maximum wave height along the coast of the Japan Sea, wave hindcasting for 143 storms which occurred from 1962 to 1991 is carried out with the same spectral wave prediction models as mentioned before. Wind distributions are estimated by weather map analysis, as high waves in the Japan Sea are predominantly generated by winter monsoons. The weather map analysis consists of a spline interpolation onto regular grid of irregularly-distributed atmospheric pressure data and an application of the Bijvoet wind model(1957).

Fig. 15 shows alongshore distribution of the most extreme wave height estimated from the wave data hindcasted for intense meteorological disturbances over 30 years from 1962 to 1991 with both the grid model and the point model and the observed wave data of shorter period. Usually, annual maximum wave height along the Japan Sea coast does not change so greatly as that along the Pacific coast, because in most cases it is generated by winter monsoons with weak yearly variation and the Japan Sea is a semi-closed and limited basin. The most extreme wave height along the coast of the Japan Sea ranges from 6 m to 10 m and is about 5 m lower than that along the southern Pacific coast shown in Fig. 8 due to the limited sea area.

Sheltering effect of waves by Sado Island can be observed from Naoetsu to Sakata by comparing the estimates with the grid model and point model. Grid system with a grid distance of 40 km used in the grid model is too coarse to resolve the presence of Sado Island and consequently the grid model overestimates the wave conditions in the coastal sea area of Hokuriku district behind Sado Island.

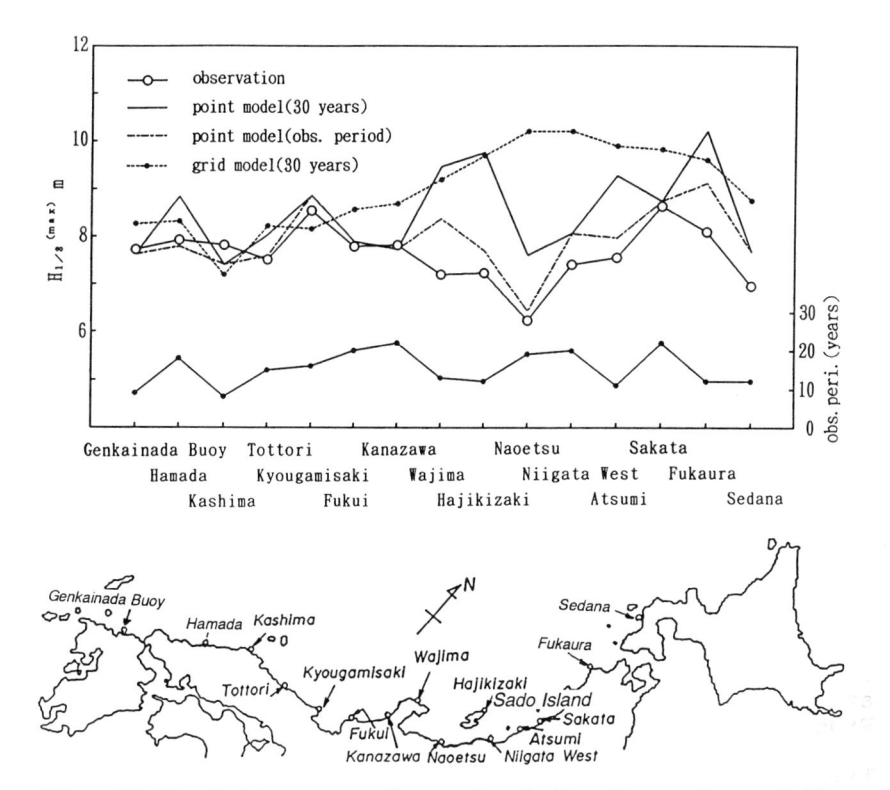

Fig. 15 Variation around coast of the Japan Sea of the most extreme wave height estimated from wave data hindcasted with both grid model and point model and observed wave data.

6. Conclusions

The most extreme waves expected in 100 years in the West Pacific Ocean and around the Japanese coast were evaluated based on statistical analysis of the wave data hindcasted with the use of two kinds of wave prediction models for the 243 typhoons which occurred from 1934 to 1991. It is deduced that maximum wave height around the Japanese coast exceeds 15 m. The most extreme wave heights along the Japan Sea coast were also estimated in similar manner. The result is that the wave height ranges from 6 m to 10 m which is about 5 m lower than that along the Pacific coast for the reasons of a limited sea area with restricted fetch and monsoon-dominated wind conditions in the sea area.

7. Acknowledgements

A part of this study was accomplished with the support of the Science Research Fund (No. 06680431) of the Ministry of Education, for which the authors express their appreciation. Thanks are also due to Mr. M. Ofuku, Technical Officer of Civil and Ocean Engineering, Ehime University and Mr. H. Satoh, former graduate student of Civil and Ocean Engineering, Ehime University for their sincere assistance during the study.

References

Bijvoet, H.C.(1957): A new overlay for the determination of the surface wind over sea from surface weather charts, KNMI, Mededelingen en Verhandelingen, Vol. 71, pp. 1-35.

Ross, D.B.(1976): A simplified model for forecasting hurricane generated waves, Bull. American Meteorol. Soc., 113.

Yamaguchi, M. et al.(1987a): Estimation of alongshore distribution of typhoon-generated maximum wave height around the Japanese coast based on wave hindcasting, Proc. JSCE, No. 381/II-7, pp. 131-140 (in Japanese).

Yamaguchi, M. et al.(1987b): A shallow water prediction model at a single location and its applicability, Proc. JSCE, No. 381/II-7, pp. 151-160 (in Japanese).

Yamaguchi, M. et al.(1987c): Alongshore distribution of extremes of typhoon-generated waves around the Pacific coast of Western Japan on the basis of wave hindcasting, Proc. JSCE, No. 387/II-8, pp.281-290 (in Japanese).

CHAPTER 51

RUN-UP OF IRREGULAR WAVES ON GENTLY SLOPING BEACH

YOSHIMICHI YAMAMOTO[1], KATSUTOSHI TANIMOTO[2], KARUNARATHNA G. HARSHINIE[3]

ABSTRACT
Irregular wave run-up on beaches has been studied on the basis of laboratory and field data, which confirmed that long-term wave run-up corresponding to surf beat (or infragravity waves) appears in case of a sea bottom slope gentler than about 1/20. Moreover analytical and numerical models to calculate surf beat caused by wave groups are investigated, and empirical and numerical models to predict the long period wave run-up are proposed.

1. INTRODUCTION
Various patterns of wave run-up on beaches due to irregular incident waves with surf beat are observed in laboratories and fields. Figure 1 shows schematic patterns of time profiles of run-up height for incident wave groups.
1) Pattern-(a): When a sea bottom slope is steep, the state that individual run-up heights almost correspond to individual incident waves is dominant. This pattern is called "predominant incident wave type".
2) Pattern-(b): When the bottom slope is gentler than that of pattern-(a), because the width of the surf zone with swash zone is wider than that of pattern-(a), the disappearance of small waves by rundown, and the capturing and overtaking of waves predominates in the swash zone. This pattern is called "intermediate type".
3) Pattern-(c): When the bottom slope is sufficiently gentle, because the width of the surf zone is wide enough, short period waves are almost eliminated by wave breaking, and the reflection coefficient of short period waves becomes very small. Thus the run-up of long period waves (surf beat or infragravity waves) is dominant. This

---
[1] Dr.-Eng., Coastal Engineer, INA Corporation, 1-44-10 Sekiguchi, Bunkyo-ku, Tokyo, 112, Japan.
[2] Dr.-Eng., Professor, Department of Civil & Environmental Engineering, The University of Saitama, Urawa, Japan.
[3] Graduate Student, Doctoral Course, Ditto.

(a) Predominant incident wave type (slope = 1/5)

(b) Intermediate type (slope = 1/10)

(c) Predominant long period wave type (slope = 1/20)

Fig.1 Type of wave run-up profiles.

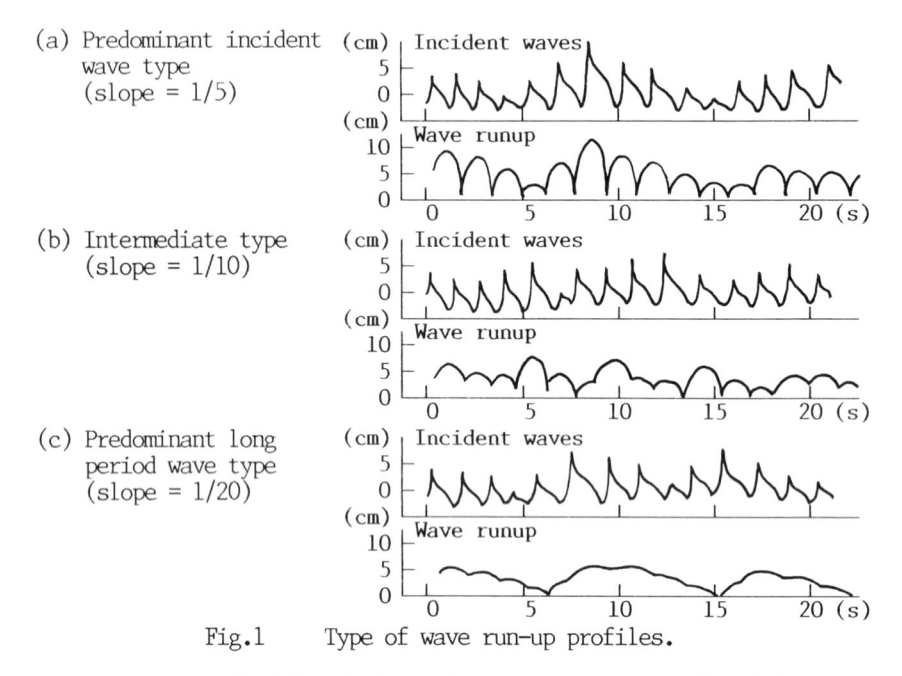

pattern is called "predominant long period wave type" and the periods of these correspond to those of wave groups.

Mase and Iwagaki (1984) showed that the number of run-up waves becomes smaller as the surf similarity parameter, ξ [= the bottom slope / (the wave steepness)$^{1/2}$], becomes smaller. Since the distribution width of bottom slopes is wider than that of the square root of wave steepness, and the groupiness of incident waves can be expressed by the total run length, it is possible to arrange available data by using the mean bottom slope in the surf zone i and the mean total run length j_{2m}. Figure 2 shows the relation between i and j_{2m} for different types of run-up (Yamamoto and Tanimoto, 1994 : by experimental and

○ : Predominant incident wave type.

◐ : Intermediate type.

● : Predominant long period wave type.

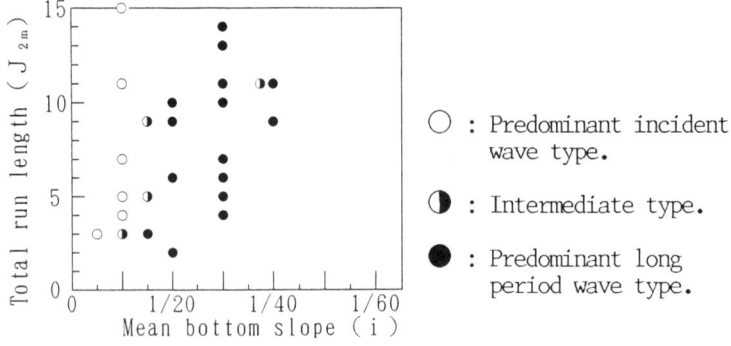

Fig.2 Relation between i and j_{2m} for different run-up types.

field data). As shown in Fig.2, it can be stated that the predominant incident wave type appears mostly when i is steeper than about 1/10, while the predominant long period wave type appears when i is gentler than 1/20. As reasons why data at the slope $i=1/37$ is peculiar, unsuitable estimation of the slope i and disregard of wave steepness are considered.

In this paper, the predominant long period wave run-up type is investigated. First, the generation mechanism of two-dimensional surf beat is discussed using analytical and numerical models. Then, empirical and numerical models to predict the run-up height are proposed.

2. GENERATION MECHANISM OF SURF BEAT

2.1 Synopsis of Generation Theories

Various theories have been presented concerning the mechanism of generation of surf beat. Since there is a strong correlation between surf beat and run length of offshore waves, the contribution of the long period waves due to wave groupiness under two-dimensional conditions is considered to be very significant.

Two theories have been presented on the generation of long period waves due to wave groupiness.
a) Bound Long Wave (BLW) Theory

The BLW theory was presented by Longuet-Higgins and Stewart(1962). According to this theory, long period waves, whose velocity is constrained by the group velocity, are generated by the periodic variation of the mean water level due to wave groupiness outside the breaker zone. Since the velocity of these waves becomes equal to the velocity of long waves near the breaking point, it is possible to consider that the long period waves propagate landwards as free long waves after breaking of incident waves.
b) Breakpoint-Forced Long Wave (BFLW) Theory

The BFLW theory was presented by Symonds et al.(1982). According to this theory, long period waves are generated by the periodic variation of the mean water level in the breaker zone, caused by the periodic movement of the breakpoint due to wave groupiness. Because the discrepancy between calculated values by their model and measured values was not negligible, Nakamura and Katoh (1992) was later modified model of Symonds et al.

A numerical model embracing both these theories has recently been developed by Goda (1990) and List (1992a,b). The numerical results given by List indicate that, in the case of a uniformly sloping sea bottom, BLW will predominate when the period ratio of the long period waves to incident waves falls below around 10, while BFLW will predominate when the above ratio is greater than 10.

The characteristic of surf beat in surf zone is investigated in the following.

2.2 Treatment of Irregular Waves with Wave Groupiness

In an attempt to simplify the treatment of irregular waves, the wave height H_{off} at the offshore boundary is expressed as follows :

$$H_{off} = H_{offm} - a_H \times \sin(2\pi t / T_{Lm}) \quad (1)$$

where, H_{offm} is the mean wave height, a_H is the amplitude of the wave height fluctuation, T_{Lm} is the mean period of surf beat.

By Substituting Eq.(1) into the theoretical equation for BLW given by Longuet-Higgins and Stewart, the wave height of BLW can be expressed as follows :

$$\frac{H_L}{H_{offm}} \fallingdotseq \frac{1}{2} \left[\frac{(2c_g/c) - 1/2}{1 - (c_g^2/gh)} \right] \frac{a_H}{h} \quad (2)$$

where, c_g is the group velocity, c is the wave velocity, g is the acceleration due to gravity, and h is the water depth.

By assuming that the period distribution of irregular waves is narrow, the distribution of H_L corresponds to the distribution of a_H. Thus the incident wave height corresponding to the statistical value of the wave height of surf beat is given by Eq.(1) with specified a_H.

The amplitude a_H corresponding to the significant wave height of surf beat can be expressed (Nakamura and Katoh, 1992) as follows :

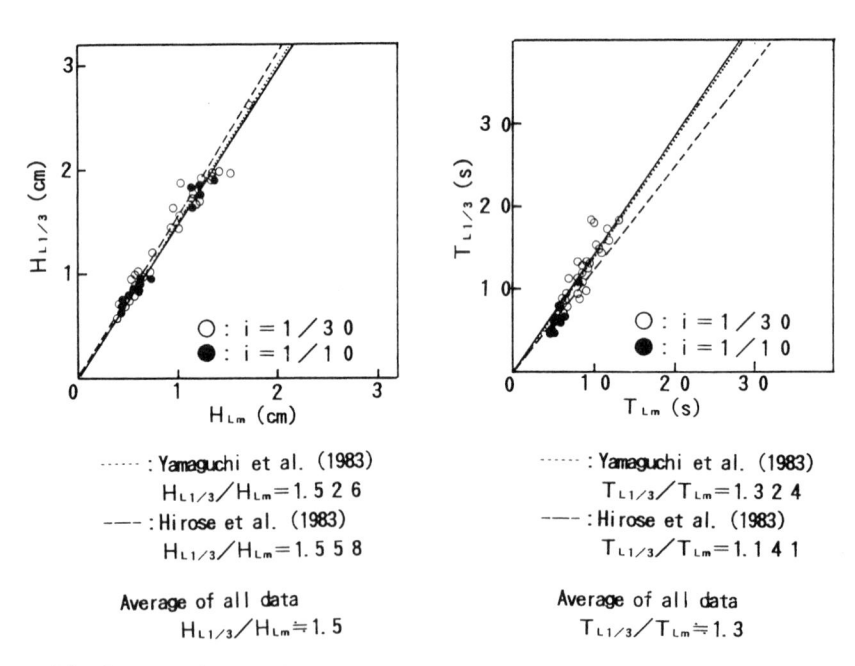

$$a_H \fallingdotseq 0.4714 \, H_{off 1/3} \quad (3)$$

······ : Yamaguchi et al. (1983)
$H_{L1/3}/H_{Lm} = 1.526$
——— : Hirose et al. (1983)
$H_{L1/3}/H_{Lm} = 1.558$

Average of all data
$H_{L1/3}/H_{Lm} \fallingdotseq 1.5$

······ : Yamaguchi et al. (1983)
$T_{L1/3}/T_{Lm} = 1.324$
——— : Hirose et al. (1983)
$T_{L1/3}/T_{Lm} = 1.141$

Average of all data
$T_{L1/3}/T_{Lm} \fallingdotseq 1.3$

Fig.3 Relation between mean and significant wave heights, and that of wave periods on surf beat respectively (Yamamoto and Tanimoto, 1994).

where, $H_{off1/3}$ is the significant wave height.

Moreover, by experimental and field data as shown in Fig.3, the relationship between the mean wave height and the significant wave height of surf beat (H_{Lm}, $H_{L1/3}$) is expressed as follows :

$$H_{Lm} \fallingdotseq H_{L1/3} / 1.5 \qquad (4)$$

Thus the amplitude a_H corresponding to the mean wave height of surf beat can be expressed as follows :

$$a_H \fallingdotseq 0.4714\, H_{off1/3} / 1.5 \fallingdotseq 0.50\, H_{offm} \qquad (5)$$

2.3 Investigation by Analytical Model

Nakamura and Katoh have shown that it is possible to provide a fairly accurate representation of the surf beat in the breaker zone by considering the time lag of breaking accompanying the propagation of waves according to theory of Symonds et al.

The on-offshore distribution of wave heights of surf beat obtained by their model is indicated by the broken lines in Fig.4. In the top three graphs, the calculated values are compared with field data taken at Hazaki Beach (Ibaraki Prefecture in Japan) by Katoh et al. (1991). The bottom two graphs show a comparison between experimental data (at the University of Saitama) and calculated results. In all cases, the surf beat to incident wave period ratio is around 10:1.

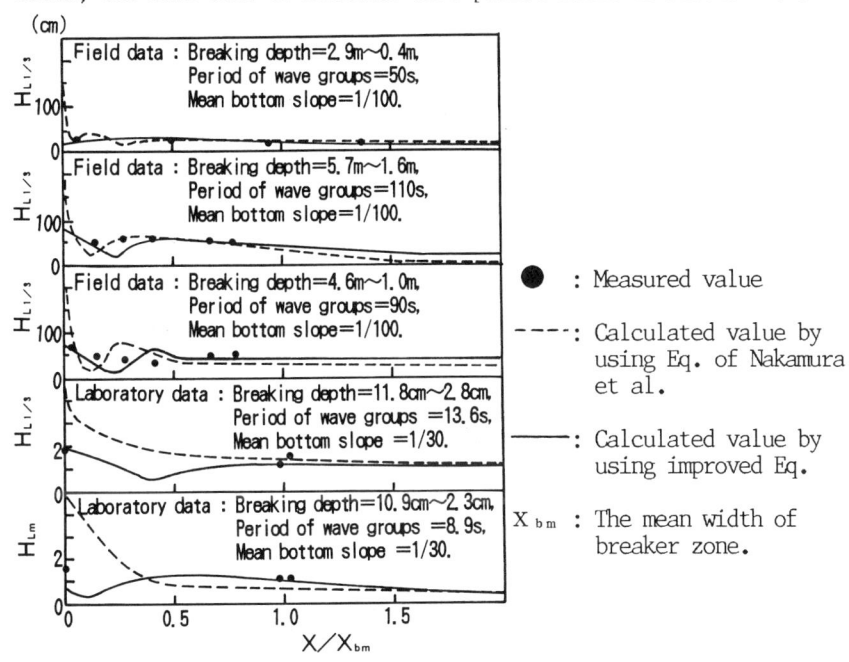

Fig.4   Comparison between observed data and predicted curve.

As can be understood from Fig.4, however, the results obtained by method of Nakamura et al. are abnormally large values near the shoreline. The following improvements were made here, as this study is also concerned also with wave heights at the shoreline.

One of the reasons for the abnormally large values near the shoreline could be the absence of considerations for the water level rise in the values used for the water depth in the basic equation proposed by Katoh et al. An improvement can be made for this by solving the following basic equation :

$$\left.\begin{array}{l}\dfrac{\partial U}{\partial t}+g\dfrac{\partial \zeta}{\partial x}=-\dfrac{1}{\rho D}\dfrac{\partial S_{xx}}{\partial x} \\ \dfrac{\partial \zeta}{\partial t}+\dfrac{\partial (DU)}{\partial x}=0, \quad D=h+\zeta_o\end{array}\right\} \quad (6)$$

where, U and ζ are the horizontal velocity and the water level of long period waves respectively, t is the time, x is the horizontal coordinate whose origin is the point of intersection between the bottom slope and the still water level, ρ is seawater density, D is the water depth including mean water level rise (ζ_o), and S_{xx} is radiation stress.

The details of the solution are described in the appendix.

The variables relating to BFLW in the breaker zone include the minimum and the maximum widths of the breaker zone (X_{b1} and X_{b2}), the period of the surf beat (T_L), sea bottom slope in the surf zone (i), and the wave height - water depth ratio (2γ) in Eqs.(A.2), (A.5) and (A.6) in the appendix. γ can be ignored as it shows little variation on a gently sloping sea bottom. T_L can be related to the period of short waves (T) using the expression $T_{L1/3} \propto T_{1/3}$ given by Nakamura and Katoh (1992). Furthermore, $L_o = (g/2\pi) T^2$; hence $T_L^2 \propto L_o$. There is a proportionality between X_b and H_o. Thus H_o/L_o and i can be used as approximate indices for the

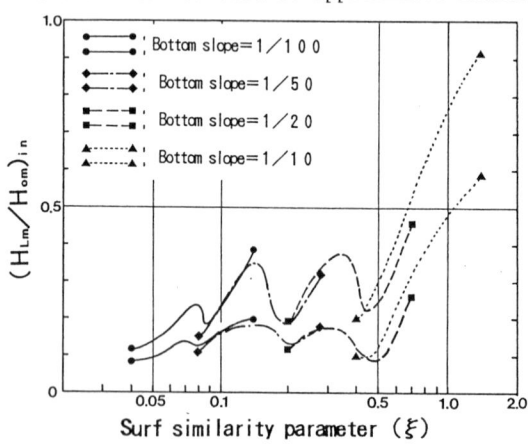

The subscript "in" stands for mean values inside the breaker zone.

Fig.5   Relation between $\xi\ [=i/(H_o/L_o)^{1/2}]$ and H_{Lm}/H_{om}.

determination of BFLW in the breaker zone.

In Fig.5, the mean wave heights of surf beat (H_{Lm}) are rendered dimensionless and averaged out within the breaker zone, and then arranged by the surf similarity parameter ξ. The values for the cases falling within the range 0.01 < wave steepness < 0.06 have been selected, and ranges in which these values fall are shown by double curves.

2.4 Investigation by Numerical Model

Since real surf beat composed of BFLW and BLW, List's numerical model is improved by taking the change of depth due to the long period waves into account [$D = h + \zeta$ is used instead of $D = h + \zeta_o$ of Eq. (6)]. In this model, incident waves propagate with group velocity. Radiation stress is calculated by the small amplitude wave theory. Breaking points are determined by using Goda's breaker index (1975). BFLW is calculated by giving the radiation stress only in the breaking zone.

The example of calculated wave heights given in Fig.6 are for the BFLW on a beach with a bottom slope of 1/40, caused by wave groups with a mean wave height of 0.8m at a water depth of about 8m, a period of 8.0s and a long wave period of 61.6s. The solid lines in the figure indicate the results obtained using the modified version of the model of Symonds et al., while the dotted lines in the figure indicate the results obtained with the numerical model. This example indicates that the numerical model will give values for BFLW which are more or less the same as those obtained by the theory of Symonds et al. Since the numerical model does not allow the water level to fall below the ground level, the discrepancy between calculated values of both models becomes great near the shoreline.

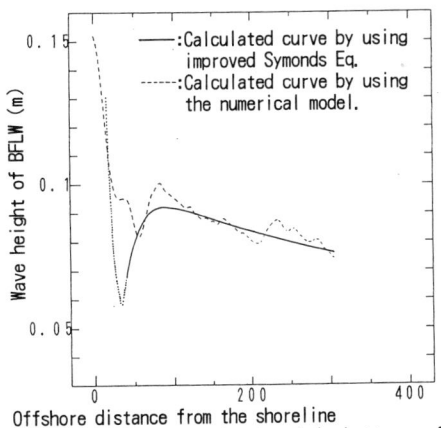

The same conditions as List's experiment are used.

$H_{off} = 0.8 - 0.4 \times \sin(2\pi t / 61.6)$,
$T = 8.0$, $T_L = 61.6$,
(units in m and s).

Fig.6   Wave height of BFLW (i=1/40).

Given in Fig.7 are the calculated results for the waves generated on a beach with a relatively steep slope at points shallower than 0.8m and a gentle slope of around 1/70 at points further offshore. Incident wave groups are taken to have mean offshore wave height 0.36m, period 10.9s and long wave period 70s. Since the shallow area where the group velocity is close to the phase velocity of long waves extends widely outside the surf zone, BLW can develop sufficiently.

The same condition as List's experiment are used.
$H_o = 0.36 - 0.09\cos(2\pi t/70)$,
$T = 10.9$, $T_L = 70$,
(units in m and s).

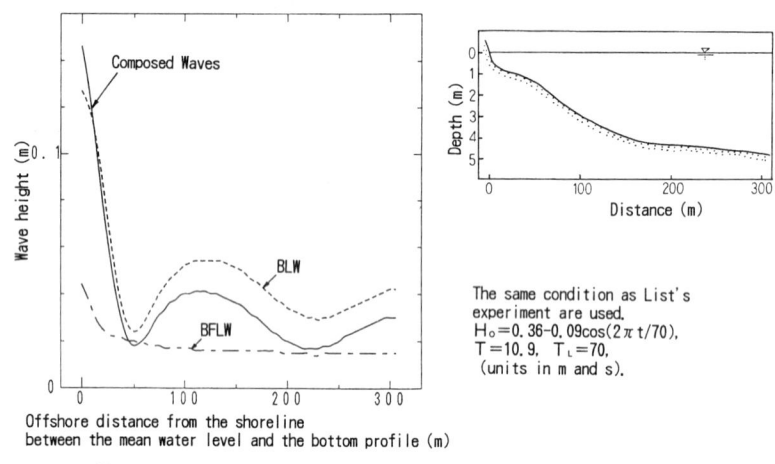

Fig.7 Wave height of surf beat (a concave type coast).

If this assumption is correct, one can expect BFLW to predominate on a beach with opposite slope characteristics. Given in Fig.8 are the calculated results for the waves generated on a beach with a horizontal reef with a water depth of about 2m extending for approximately 500m from the shoreline. Offshore incident waves in this case are considered to have a mean wave height of 2.88m, period of 13.8s, and

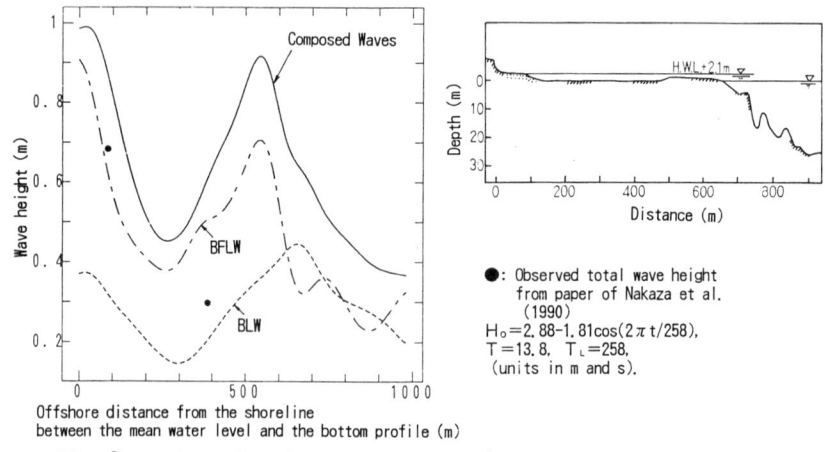

● : Observed total wave height from paper of Nakaza et al. (1990)
$H_o = 2.88 - 1.81\cos(2\pi t/258)$,
$T = 13.8$, $T_L = 258$,
(units in m and s).

Fig.8 Wave height of surf beat (a convex type coast).

3. EMPIRICAL AND NUMERICAL MODELS OF WAVE RUNUP

3.1 Empirical Method

long wave period of 258s. After comparison of calculated results with actual field data, it can be stated that the numerical model has sufficient accuracy.

By improving Goda's equation (1975) with field and experimental data, the significant wave height of surf beat can be expressed by the following relation (Yamamoto and Tanimoto, 1994):

$$\left. \begin{array}{l} \dfrac{H_{L1/3}}{H_{o1/3}} \fallingdotseq \dfrac{0.066\, i^{1/6}}{[(H_{o1/3}/L_{o1/3}) \times (1+h/H_{o1/3})]^{1/2}} \\[2mm] \dfrac{H_{Lm}}{H_{om}} \fallingdotseq \dfrac{0.067\, i^{1/6}}{[(H_{om}/L_{om}) \times (1+h/H_{om})]^{1/2}} \\[2mm] (1/10 \geq i \geq 1/70) \end{array} \right\} \quad (7)$$

where, $L_{o1/3}$ and L_{om} is significant, mean wavelengths in deep-water respectively, i is the mean bottom slope in the surf zone, and h is the water depth below the still water level.

Assuming the run-up oscillation of surf beat is a parabolic motion, the wave run-up length along the slope can be expressed as follows:

$$y_r \fallingdotseq C_1\, U_s\, t - C_2\, g\, i t^2 / 2 \qquad (8)$$

where, U_s is a maximum velocity at shore line [$\fallingdotseq C_3\, (gH_{Ls})^{1/2}$], C_1, C_2 and C_3 are empirical coefficients, and H_{Ls} is the wave height of surf beat at the shoreline.

Substitution of Eq.(7) into Eq.(8), and the determination of coefficients C_1, C_2 and C_3 by using field and laboratory data, the period and the run-up height of the surf beat can be expressed by the following equations (Yamamoto and Tanimoto, 1994):

$$\left. \begin{array}{l} (T_{Lm})_{in}^2 \fallingdotseq 288.0^2\, (H_{Lm}/g)_{in} \\ (T_{L1/3})_{in}^2 \fallingdotseq 305.7^2\, (H_{L1/3}/g)_{in} \\ (1/20 > i \geq 1/60 \text{ and } H_{om}/L_{om} \geq 0.015) \end{array} \right\} \quad (9)$$

$$\left. \begin{array}{l} R_{Lm}/H_{om} \fallingdotseq 1.52\, i/(H_{om}/L_{om})^{1/2} \\ R_{L1/3}/H_{o1/3} \fallingdotseq 1.50\, i/(H_{o1/3}/L_{o1/3})^{1/2} \\ (1/20 \geq i \geq 1/60) \end{array} \right\} \quad (10)$$

where, T_{Lm}, $T_{L1/3}$ are the mean, significant periods of the surf beat respectively, R_{Lm} and $R_{L1/3}$ are the mean, significant run-up heights of the surf beat respectively, and the subscript "in" stands for the mean value inside the surf zone.

Substitution of Eq.(7) and $T_{1/3} \fallingdotseq 3.86(H_{1/3})^{1/2}$ (units in s and m, Bretshneider, 1954) into Eq.(9), and by assuming $H_{o1/3}/h \sim 1$ in the surf zone, the same relationship that Nakamura and Katoh (1992) obtained by analysis of field data can be derived ($T_{L1/3} \propto T_{1/3}$).

The agreement between R_{Lm} calculated by Eq.(10) and experimental

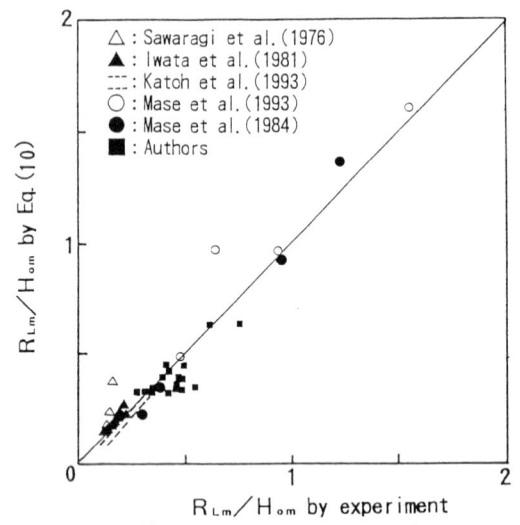

Fig.9　Comparison of R_{Lm} calculated by Eq.(10) with measured data.

and field data are good as shown in Fig.9.

3.2　Numerical Model

Considering the convection and friction terms, the basic equations averaged over the wave period are expressed as follows :

$$\left.\begin{array}{l}\dfrac{\partial \zeta}{\partial t}+\dfrac{\partial (DU)}{\partial x}=0 \\ \dfrac{\partial (DU)}{\partial t}+\dfrac{\partial}{\partial x}(\dfrac{(DU)^2}{D})+gD\dfrac{\partial \zeta}{\partial x} \\ =-\dfrac{1}{\rho}\dfrac{\partial S_{60}}{\partial x}-\dfrac{f}{D^2}(DU)|DU| \\ D=h+\zeta\end{array}\right\} \quad (11)$$

where, S_{60} is the 60% value of the radiation stress [Mase et al. (1986) states that the equation by the small amplitude wave theory gives excessively large values for the radiation stress and recommends the use of values corresponding to 60% of the obtained value.], and f is the mean bottom friction coefficient, which is obtained by using the following equation based on Freeman and LeMehaute's wave run-up height equation (1964) :

$$f \simeq \left[\dfrac{(1+C_v)(1+2C_v)}{C_L}-1\right] \dot{\iota} \ C_v{}^2 \quad (12)$$

where, C_v is the wave verocity - particle velocity ratio, and C_L is the loss between potential energy of wave run-up and kinetic energy of waves on a shoreline. According to Yamamoto et al.(1994), these parameters for sand beach are expressed as follows :

$$\left. \begin{array}{l} C_V \fallingdotseq 24.2\, i^{19/30}\, (H_{om}/L_{om})^{1/4} \\ C_L \fallingdotseq 2221\, i^{22/15}\, (H_{om}/L_{om})^{1/4} \\ \qquad (i \leqq 1/30) \end{array} \right\} \qquad (13)$$

Some calculations on the run-up of the surf beat have been conducted by using the numerical model based on Eq.(11). An example [the incident wave height is $0.8 - 0.4 \times \sin(2\pi t / 61.6)$ (units in m and s), the wave period is 8.0s, the bottom slope is 1/40] is shown in Fig.10. A relatively close agreement is found between the result obtained by the numerical model and that by the empirical Eq.(10).

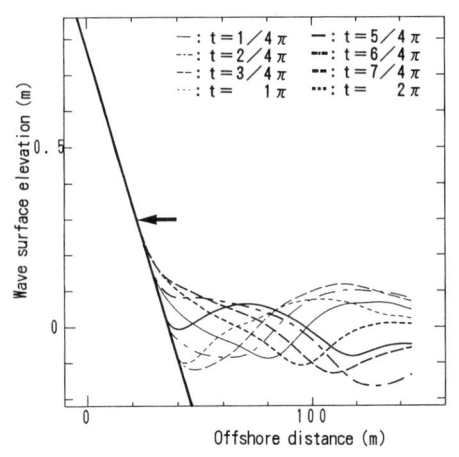

Fig.10　An example of long period wave runup.
〔The arrow shows R_{Lm} calculated by Eq.(10)〕

A comparison between wave run-up heights computed by the numerical model for beaches with bottom slopes ranging from 1/30 to 1/60 and those obtained by empirical Eq.(10) is shown in Fig.11. Relatively close agreement is found between the model and Eq.(10).

4. CONCLUSIONS
1) Patterns of irregular wave run-up can be classified into three types : predominant incident wave type, intermediate type and predominant long period wave type. The predominant long period wave type appears when the mean bottom slope in the surf zone is gentler than 1/20.
2) Model of Symonds et al. and List's model were modified to allow more accurate calculation of surf beat near the shoreline. The tests conducted with the numerical model show that BLW can develop when the shallow area having a group velocity close to the phase velocity of long waves extends widely outside the surf zone, on the other hand, BFLW can develop when a wide shallow area exists within

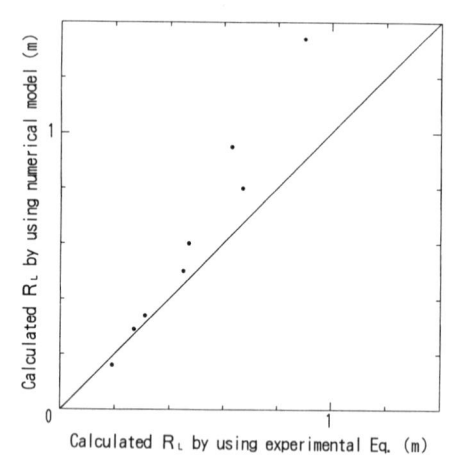

Fig.11 Relation between wave runup height using the numerical model and those using the empirical Eq.(10).

the surf zone such as a reef.
3) Run-up height of long period waves caused by wave groupiness in incident irregular waves can be predicted by both empirical and numerical models.

ACKNOWLEGEMENT

The authors are grateful to Dr. Kiyoshi Horikawa and Mr. Hiroaki Nakamura of the University of Saitama for making useful suggestions.

REFERENCES

(1) Bretschneider, C. L.(1954). Generation of wind waves over a shallow bottom. U.S. Army Corps of Engrs., Beach Erosion Board Tech. Memo., No.51, 24pp.
(2) Freeman, J. C. and B. LeMehaute(1964). Wave breakers on a beach and surges on a dry bed. Proc. ASCE, Vol.90, No. HY2, pp.187~216.
(3) Goda, Y.(1975). Deforation of irregular waves due to depth-controlled wave breaking. Report of the Port and Harbour Research Institute, Vol.14, No.3, pp.59~106 (in Japanese).
(4) Goda, Y., T. Isayama and S. satoh(1990). Field observation and experiment on development mechanism of long period waves inside surf zone. Proc. 37th Japanese Conf. Coastal Eng., pp.96~100 (in Japanese).
(5) Iwata, K., T. Sawaragi and W. Nobuta(1981). wave run-up height and period of irregular waves on a gently sloping beach. Proc. 28th Japanese Conf. Coastal Eng., pp.330~334 (in Japanese).
(6) Katoh, K., S. Nakamura and N. Ikeda(1991). Estimation of infragravity waves in consideration of wave groups. Report of the Port and Harbour Research Institute, Vol.30, No.1, pp.137~163 (in Japanese).

(7) Katoh, K. and S. Yanagishima(1993). Beach erosion in a storm due to infragravity waves. Report of the Port and Harbour Research Institute, Vol.31, No.5, pp.73~102.
(8) List, J. H.(1992a). A model for two-dimensional surf beat. J.G.R., Vol.97, No.C4, pp.5623~5635.
(9) List, J. H.(1992b). Breakpoint-forced and bound long waves in the nearshore. Proc. 23rd Coastal Eng. Conf., ASCE, pp.860~873.
(10) Longuet-Higgins, M. S. and R. W. Stewart(1962). Radiation stress and mass transport in gravity waves, with application to "Surf Beat". J.F.M., Vol.13, pp.481~504.
(11) Mase, H. and Y. Iwagaki(1984). Run-up of random waves on gentle slopes. Proc. 19th Coastal Eng. Conf., ASCE, pp.593~609.
(12) Mase, H., A. Matsumoto and Y. Iwagaki(1986). Calculation model of random wave transformation in shallow water. Proc. JSCE, No.375, II-6, pp.221~230 (in Japanese).
(13) Mase, H. and N. Kobayashi(1993). Low-frequency swash oscillation. Proc. JSCE, No.461, II-22, pp.49~57 (in Japanese).
(14) Nakamura, S. and K. Katoh(1992). Generation of infragravity waves in breaking process of wave groups. Proc. 23rd Coastal Eng. Conf., ASCE, pp.990~1003.
(15) Nakaza, E., S. Tsukayama and M. Hino(1990). Bore-like surf beat on reef coasts. Proc. 22nd Coastal Eng. Conf., ASCE, pp.743~756.
(16) Sawaragi, T., K. Iwata and A. Morino(1976). wave run-up characteristic on a gently sloping beach. Proc. 23rd Japanese Conf. Coastal Eng., pp.164~169 (in Japanese).
(17) Symonds, G., D. A. Huntley and A. J. Bowen (1982). Two-dimensional surf beat : long wave generation by a time-varying breakpoint. J.G.R. Vol.87, No.C1, pp.492~498.
(18) Yamamoto, Y and K. Tanimoto(1994). A study on long period run-up waves caused by irregular incident waves. Proc. JSCE, No.503, II-29, pp.109~118 (in Japanese).
(19) Yamamoto, Y., K. Tanimoto and H. Nakamura(1994). Run-up of long period waves caused by wave groups. Proc. 49th Annu. Conf. JSCE, II, pp.644~645 (in Japanese).

APPENDIX : SOLUTION WITH CONSIDERATIONS FOR MEAN WATER LEVEL RISE

The radiation stress in Eq.(6) may be expressed as follows in terms of its relationship to the amplitude [$A_{mp} = \gamma (x i + \zeta_0)$] of the incident waves :

$$S_{xx} = (3/4) \rho g (A_{mp})^2 \quad (A.1)$$

Eq.(6) is then made dimensionless by using the following dimensionless parameters :

$$\left. \begin{array}{l} A = \dfrac{a}{X_{bm}}, \quad X = \dfrac{x}{X_{bm}}, \quad \tau = t \dfrac{2\pi}{T_L}, \quad X_{b1} = \dfrac{X_{b1}}{X_{bm}}, \quad X_{b2} = \dfrac{X_{b2}}{X_{bm}} \\ U(X,\tau) = \dfrac{3}{2} \gamma^2 \dfrac{2\pi}{T_L} X u, \quad Z(X,\tau) = -\dfrac{\zeta(x,t)}{1.5 \gamma^2 X_{bm} i} \end{array} \right\} $$

$$(A.2)$$

where, a is $(X_{b1} - X_{b2})/2$, X_{bm} is $(X_{b1}+X_{b2})/2$, γ is (wave height - water depth ratio at breaking limit) $/2$, and i is bottom slope.

The dimensionless horizontal flow velocity U is then eliminated from the dimensionless form of the basic equation, and the differential equation for the dimensionless water level Z is obtained. If the term for the radiation stress in this differential equation is expressed by Fourier series, the following equation will be given:

$$X' \frac{\partial^2 Z}{\partial \tau^2} - \frac{dX''}{dX} \frac{\partial Z}{\partial X} - X'' \frac{\partial^2 Z}{\partial X^2}$$
$$= \frac{d}{dX} [X'' \frac{dX''}{dX} (a_0 + 2\sum_{n=1}^{\infty} a_n \cos(n\tau) + 2\sum_{n=1}^{\infty} b_n \sin(n\tau))] \quad (A.3)$$

where, $X' = (\frac{2\pi}{T_L})^2 \frac{X}{gi}$, $X'' = X + \frac{3}{2} \gamma^2 Z_0(X)$,

Z_0 is dimensionless mean water level rise, and a_0, a_n, b_n is Fourier coefficients (n: term number) - The methods for calculation of these coefficients are given in Nakamura and Katoh (1992).

The solution to Eq.(A.3) may be expressed as follows.

$$Z(X,\tau) = Z_0(X) + \sum_{n=1}^{\infty} Z_n(X,\tau) \quad (A.4)$$

Z_0 can easily be obtained from the dimensionless equation of motion, as shown in the first expression in Eq.(A.5) below. For areas further offshore than X_{b1}, however, the values given by Symonds et al. for Z_0 are used without alteration, assuming that the effects of the mean water level rise on the water depth here are negligibly small.

Z_n can be obtained in a similar manner to Symonds et al. by means of approximated expression, $dX''/dX \approx 1$ (found to be a valid approximation in investigations conducted by substituting actual physical values) during the calculation of the particular solutions between X_{b1} and X_{b2}.

a) From shore-edge to X_{b1}

$$\left. \begin{aligned} Z_0(X) &= (1 - X + 1.5\gamma^2 A)/(1+1.5\gamma^2) \\ Z_n(X,\tau) &= -(I_{bJ} + I_{aN}) J_0(z) \cos(n\tau) \\ &\quad - I_{bN} J_0(z) \sin(n\tau) \\ z &= 2n(2\pi/T_L)[X_{bm}/(g\,i)]^{1/2} \\ &\quad \times (1+1.5\gamma^2)(X+1.5\gamma^2 Z_0)^{1/2} \end{aligned} \right\} \quad (A.5)$$

Here, $J_0(z)$ is zero-degree Bessel function.

b) From X_{b1} to X_{b2}

$$\left. \begin{aligned} Z_0(X) &\approx \{(1-X)\cos^{-1}[(X-1)/A] \\ &\quad - [A^2 - (X-1)^2]^{1/2}\}/\pi \\ Z_n(X,\tau) &= [-(I_{bJ} + I_{aN}) J_0(z) \\ &\quad + C_n N_0(z) + \eta_{pa}] \cos(n\tau) \end{aligned} \right|$$

$$\left.\begin{array}{l} \quad + [-I_{bN} J_0(z) + \eta_{pb}] \sin(n\tau) \\ z = 2n(2\pi/T_L)[X_{bm}/(g\ i)]^{1/2} \\ \quad \times \{1 + 1.5\gamma^2 \cos^{-1}((X-1)/A)/\pi\} \\ \quad \times (X + 1.5\gamma^2 Z_0)^{1/2} \\ \eta_{pa} \doteq 2\pi[\int_{x_M}^{x} X_{an} N_0(z) dX\ J_0(z) \\ \quad - \int_{x_M}^{x} X_{an} J_0(z) dX\ N_0(z)] \\ \eta_{pb} \doteq 2\pi[\int_{x_M}^{x} X_{bn} N_0(z) dX\ J_0(z) \\ \quad - \int_{x_M}^{x} X_{bn} J_0(z) dX\ N_0(z)] \\ X_{an} = d[(X + 1.5\gamma^2 Z_0) a_n]/dX \\ X_{bn} = d[(X + 1.5\gamma^2 Z_0) b_n]/dX \end{array}\right\} \quad (A.6)$$

Here, $C_n = 0$ (onshore), $C_n = I_{aJ}$ (offshore), and $N_0(z)$ is zero-degree Neumann function.

c) Offshore from X_{b2}

$$\left.\begin{array}{l} Z_0(X) = 0 \\ Z_n(X,\tau) = -I_{bJ} J_0(z) \cos(n\tau) \\ \quad - I_{bJ} N_0(z) \sin(n\tau) \\ z = 2n(2\pi/T_L)[X_{bm}/(g\ i)]^{1/2}\ X^{1/2} \end{array}\right\} \quad (A.7)$$

I_{aJ}, I_{aN}, I_{bJ}, and I_{bN} in Eqs.(A.5) to (A.7) can be obtained as follows using z, X_{an}, and X_{bn} in Eq.(A.6):

$$\left.\begin{array}{l} I_{aJ} \doteq 2\pi \int_{x_{b1}}^{x_{b2}} X_{an}\ J_0(z)\ dX \\ I_{aN} \doteq 2\pi \int_{x_{b1}}^{x_{b2}} X_{an}\ N_0(z)\ dX \\ I_{bJ} \doteq 2\pi \int_{x_{b1}}^{x_{b2}} X_{bn}\ J_0(z)\ dX \\ I_{bN} \doteq 2\pi \int_{x_{b1}}^{x_{b2}} X_{bn}\ N_0(z)\ dX \end{array}\right\} \quad (A.8)$$

It should be noted here that partial integration was used for the integration of the above equations.

CHAPTER 52

Soliton-Mode Wavemaker Theory and System for Coastal Waves

Takashi Yasuda[1], Seirou Shinoda[2] and Takeshi Hattori[3]

Abstract

A theory of wavemaker system for coastal waves (nonlinear random waves in very shallow water) having the desired statistics is developed under the assumption that they are treated as random trains of the KdV solitons and further its applicability is examined experimentally. This wavemaker system is shown usefull to generate accurately nonlinear waves having the desired water surface profiles of which nonlinearity is specified by the Ursell number $U_r \geq 15$ for regular waves (cnoidal waves) and $U_r \geq 8$ for random ones (coastal waves).

Introduction

Waves in shallow water come to behave like trains of independent individual waves with the shoaling water. In shallow to very shallow water, particle-like property of each individual wave is more strengthened because of the pronounced nonlinear effect, so that its time sequence becomes increasingly important to coastal engineering problems, such as dynamic response of coastal

---
[1]M. ASCE, Prof., Dept. of Civil Eng., Gifu Univ., 1-1 Yanagido, Gifu 501-11, Japan
[2]Assoc. Prof. of Institute for River Basin Environmental Studies, Gifu Univ., 1-1 Yanagido, Gifu 501-11, Japan
[3]Civil Engineer, Penta-Ocean Construction Co. LTD., 2-2-8 Koraku, Bunkyo-ku, Tokyo 112, Japan

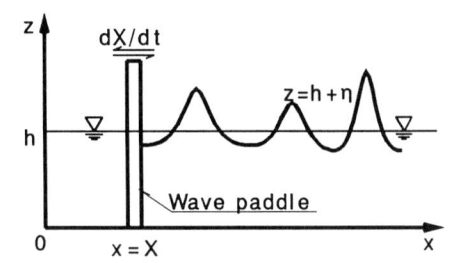

Figure 1. Definition of coordinate system and symbols

structures, nearshore current system and so on. It hence is required in experimental works to generate the wave train having the desired sequence of wave crest. However, usual methods based on the Biesel-Suquet wavemaker theory and the envelope theory assuming a narrow-banded Gaussian process cannot be used to generate strongly nonlinear random waves in very shallow water (called here as coastal waves).

Authors(Yasuda et. al., 1987) already showed that coastal waves can be treated as a random train of solitons and their kinematics can be described using the asymptotic multi-soliton solution of the KdV equation. Hence, treating individual waves defined by the zero-crossing method as solitons and using a stochastic model (Shinoda et al., 1992) of solitons, we can generate straightforwardly coastal waves having the desired sequence of individual waves as a train of solitons.

In this study, we suggest a nonlinear wavemaker theory for the coastal waves and then develop a soliton-mode wavemaker system to generate them in a wave tank. Further, the performance of the system is examined by generating cnoidal waves (treated as uniform trains of solitons) and coastal waves (as random trains of solitons) having the desired sequence.

Wavemaker Theory

The boundary value problem for the wavemaker in a wave tank follows directly from that for two-dimensional waves propagating in an incompressible and irrotational fluid. For the coordinate depicted in **Figure 1**, it is required to solve the following equations,

$$\phi_{xx} + \phi_{zz} = 0, \tag{1}$$

$$\eta_t + \eta_x \phi_x - \phi_z = 0|_{z=h+\eta}, \tag{2}$$

$$\phi_t + (\phi_x^2 + \phi_z^2)/2 + g\eta = 0|_{z=h+\eta}, \tag{3}$$

$$\phi_z = 0|_{z=0}, \tag{4}$$

$$dX/dt = \phi_x|_{x=X}, \tag{5}$$

where ϕ denotes the velocity potencial, η the water surface elevation, g the acceleration of gravity, h the mean water depth, and X the displacement of a piston type paddle.

Integrating eq.(1) from $z = 0$ to $z = h + \eta$ and substituting eqs.(2) and (4) into the resultant equation, we can obtain the following equation,

$$\int_0^{h+\eta} \phi_x dz = -\int \eta_t dx + \text{const.} \tag{6}$$

Since eq.(5) can not be satisfied for the arbitrary z-coordinate as long as piston-type wave paddle is used, the vertically integrated equation instead of eq.(5) is employed as the boundary condition with regard to the depth-averaged velocity by following Goring and Raichlen's approach(1980). Integration of eq.(5) from $z = 0$ to $z = h + \eta$ yields to

$$\int_0^{h+\eta} \frac{dX}{dt} dz = \int_0^{h+\eta} \phi_x dz \bigg|_{x=X}. \tag{7}$$

Substituting eq.(6) into eq.(7), we obtain the following relation satisfying eqs.(1), (2), (4) and (5).

$$\int_0^{h+\eta} \frac{dX}{dt} dz = -\int \eta_t dx + \text{const.} \bigg|_{x=X}. \tag{8}$$

Since the value of η_t becomes zero in the still water and the wave paddle is also at rest, the integral constant in eq.(8) is regarded as zero. Further, since dX/dt is independent of z in the present piston-type wavemaker system, we can derive the following equation for the paddle motion to generate the wave profile η from eq.(8).

$$\frac{dX}{dt} = \frac{-1}{h+\eta} \int \eta_t dx \bigg|_{x=X}. \tag{9}$$

Here, the following dimensionless variables with asterisk are defined for convenience of numerical computations.

$$\phi^* = \phi/\sqrt{gh^3},\ \eta^* = \eta/h,\ X^* = X/h,\ x^* = x/h,\ z^* = z/h,\ t^* = t\sqrt{g/h} \tag{10}$$

However, the asterisks are omitted for simplicity hereinafter, so that eq.(9) is rewritten as

$$\frac{dX}{dt} = \frac{-1}{1+\eta} \int \eta_t dx \bigg|_{x=X}. \tag{11}$$

SOLITON-MODE WAVEMAKER THEORY

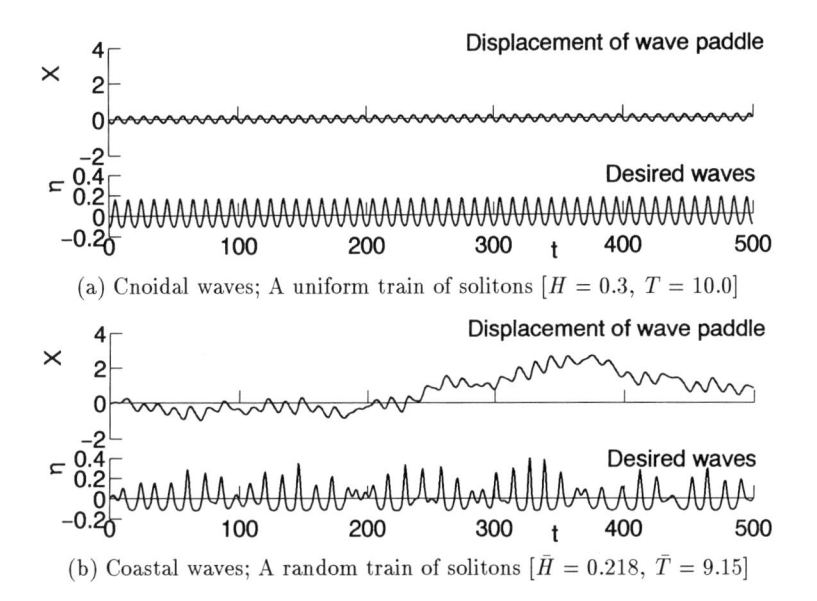

(a) Cnoidal waves; A uniform train of solitons [$H = 0.3$, $T = 10.0$]

(b) Coastal waves; A random train of solitons [$\bar{H} = 0.218$, $\bar{T} = 9.15$]

Figure 2. Wave profiles of a soliton train having the desired sequence and the displacement of wave paddle to generate it; H is the dimensionless wave height corresponding to the amplitude of solitons, T the dimensionless wave period corresponding to the distance of adjoining solitons, $\bar{H}$ the dimensionless mean wave height, and $\bar{T}$ the dimensionless mean wave period

If waves to be generated are trains of solitons and satisfy eqs.(1)~(4) in the order of the KdV equation, η can be represented as

$$\eta(x,t) = \sum_{j=1}^{N} \eta_j(x,t) \tag{12}$$

$$\left. \begin{array}{l} \eta_j(x,t) = A_j \operatorname{sech}^2 \frac{\sqrt{3A_j}}{2} \{x - c_j(t - \delta_j)\} - \frac{\eta_0}{N}, \\ c_j = 1 + \frac{A_j}{2} - \frac{3}{2}\eta_0, \quad \eta_0 = 4\sqrt{\frac{A_j}{3}} \frac{c_j T_o}{h} \end{array} \right\} \tag{13}$$

where A_j is the amplitude of the j-th soliton non-dimensionalized with h, N the number of solitons, T_o the observation period and δ_j the phase constant of the j-th soliton.

As a result, the problem to solve eqs.(1)~(5) can be finally reduced to that to solve the following nonlinear ordinary differential equation with regard to the

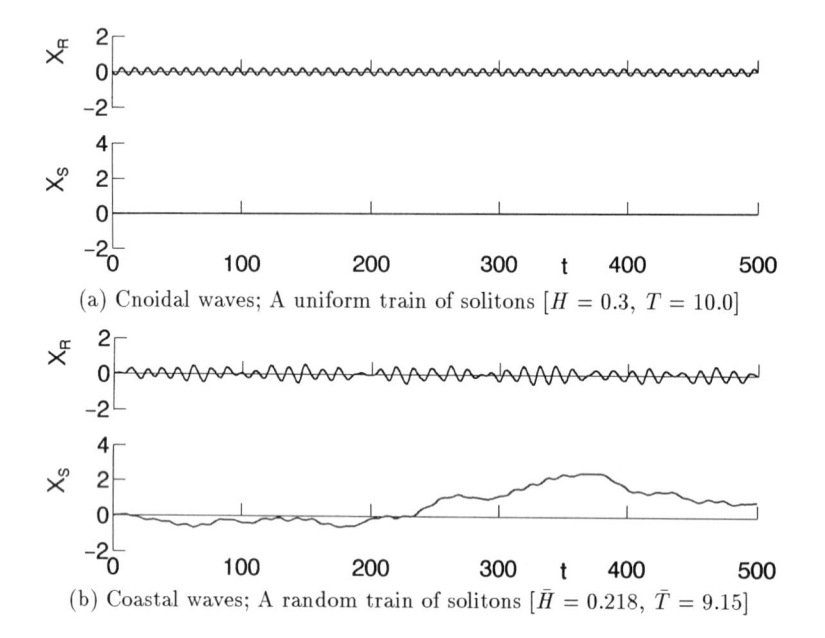

(a) Cnoidal waves; A uniform train of solitons [$H = 0.3$, $T = 10.0$]

(b) Coastal waves; A random train of solitons [$\bar{H} = 0.218$, $\bar{T} = 9.15$]

Figure 3. Decomposed displacement of the wave paddle; time history of X_R and X_S

displacement X, by combining eqs.(11) and (12),

$$dX/dt = \sum_{j=1}^{N} c_j \eta_j(X, t) / \left\{ 1 + \sum_{j=1}^{N} \eta_j(X, t) \right\}. \tag{14}$$

Thus, eq.(14) is solved numerically for given values of η_j and c_j ($j = 1, \ldots, N$), so that the displacement X of the wave paddle required to generate the desired waves is uniquely determined. Hence, if the waves to be generated are represented as a train of solitons described by eq.(12) and the ensembles of the soliton parameters A_j and δ_j governing them are substituted into eq.(14) through eqs.(12) and (13), the displacement X required to regenerated the waves in a wave tank can be easily known by solving eq.(14).

Wavemaker System

Wavemaker system to be used here is required to enable the wave paddle to follow accurately the displacement X given by eq.(14). Hence, in order to investigate the displacement X required to generate cnoidal waves (uniform trains of solitons) with $H = 0.3$ and $T = 10.0$ and coastal waves (random

Photo 1. A rotary actuator and cylinder actuator attached to the wavemaker

trains of solitons) with $\bar{H} = 0.218$ and $\bar{T} = 9.15$, time histories of X are shown in **Figure 2** together with their surface profiles. Here, H and T denote the dimensionless wave height and period, respectively, and the over-bar means their mean values. While the displacement for cnoidal waves is periodic and stays within a stroke less than the mean water depth, that for coastal waves is aperiodic and accompanies a slow drift oscillation with a large amplitude exceeding two times of the water depth. Thus, a wavemaker system having a long stroke wave paddle is required to generate coastal waves. However, the motion of wave paddle seems to be a rapid movement with short stroke riding on a slow drift with long stroke, although the stroke of the paddle exceeds two or more times distance of water depth. By aiming at this point, the displacement $X(t)$ was decomposed into a rapid but short displacement $X_R(t)$ and a slow but long one $X_S(t)$ as shown in **Figure 3**, and thus a double mode wavemaker having a high-speed driving part with short stroke and a low-speed driving part with long stroke was newly developed.

The double-mode wavemaker has two types of movement corresponding to X_R and X_S. **Photo 1** shows the high-speed driving part with a rotary

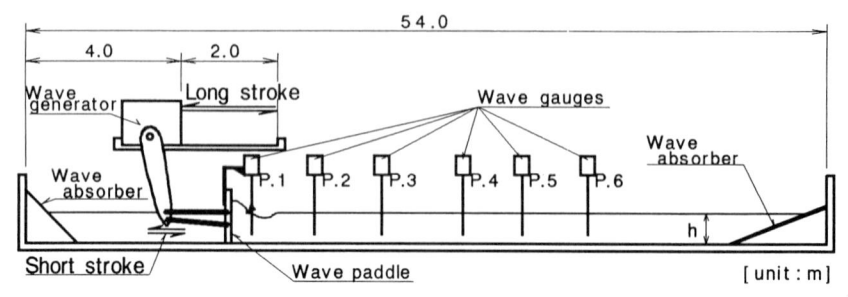

Figure 4. Sketch of the wave tank used here and location of wave gauges installed

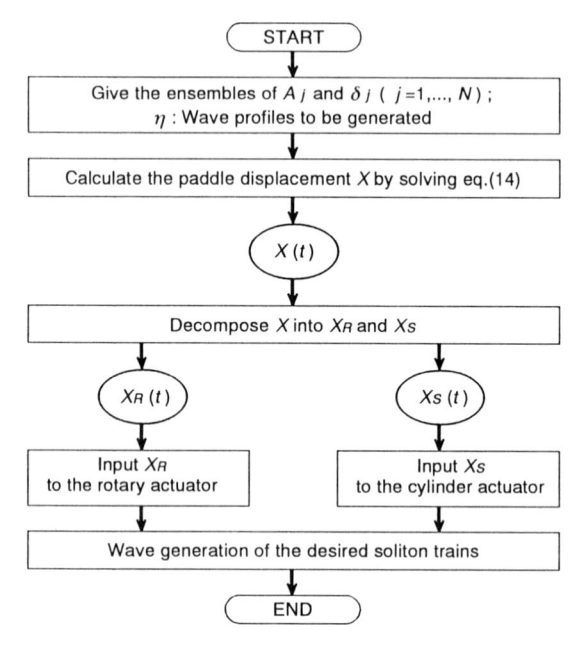

Figure 5. Flow chart of the procedure to generate the soliton trains as desired

actuator and the low-speed driving part with a cylinder actuator attached to the wavemaker, and **Figure 4** illustrates its schematic sketch. The rotary actuator enables short stroke but high speed motion corresponding to X_R and has a performance of the maximum speed of 60cm/s and stroke of 30cm. The cylinder actuator enables slow but long stroke motion corresponding to X_S and has a performance of the maximum speed of 7.3cm/s and stroke of 2m.

As a result, the procedure using the double-mode wavemaker to generate the desired waves is summarized as follows: The ensemble of A_j and δ_j for the

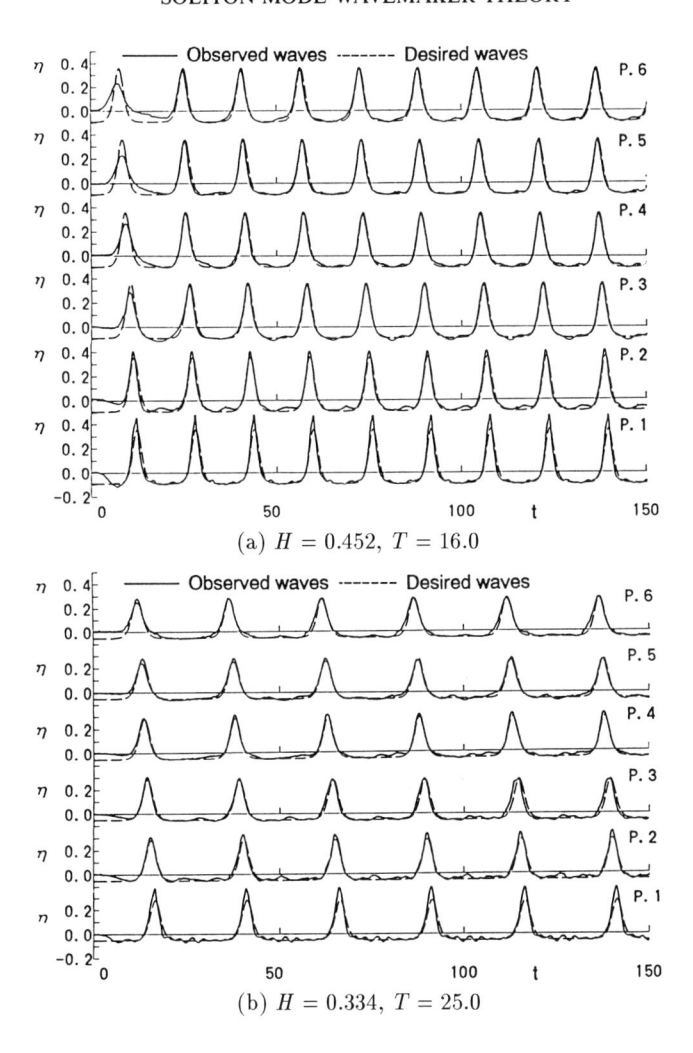

Figure 6. Comparisons for a train of uniform solitons between the desired wave profiles and the observed ones

soliton trains to be generated are given to eq.(14) and then the displacement X required to generate them is determined by solving eq.(14). Further, the displacement X is decomposed into X_R and X_S and the values of X_R and X_S at each time step are given to the rotary actuator and cylinder one, respectively, as input signal. Thus, the wave paddle follows the composed motion of X_R and

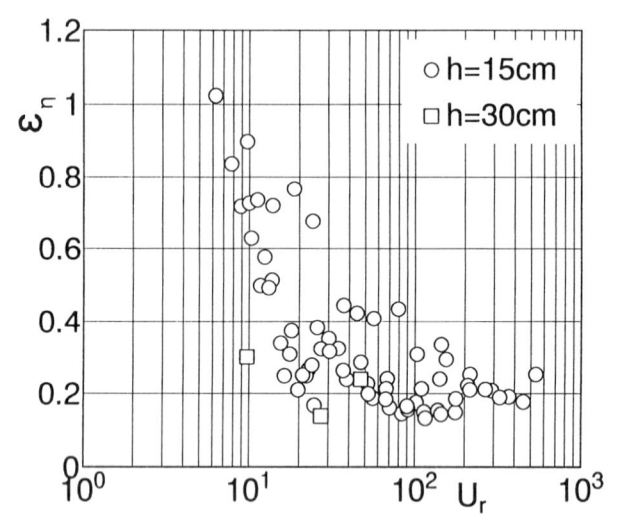

Figure 7. Relationships for cnoidal waves between the Ursell number U_r and the error index ε_η

X_S and produce the soliton trains as desired. Flow chart of the procedure explained above is illustrated in **Figure 5**.

Performance Test

Experiments using the wave tank and gauges shown in **Figure 4** were undertaken to examine the performance of the double-mode wavemaker system. Six wave gauges were installed at six locations from P.1 to P.6 in the tank. Distance between each wave gauges is evenly about 2.5m.

Cnoidal waves (Uniform trains of solitons)

Since cnoidal waves can be represented as a uniform train of solitons having constant amplitude and period, they can be generated in a wave tank by producing a uniform soliton train having a given constant amplitude and period in order. Water surface elevations of generated trains of solitons can be regarded as cnoidal waves as shown in **Figure 6**, where the comparisons between the measured temporal water surface elevations of the waves generated as a train of solitons and theoretical ones of the cnoidal waves to be generated are made This result further demonstrates that the soliton is the elementary excitation of waves in very shallow water and therefore the desired cnoidal waves can be generated as train of solitons.

However, the soliton can not exist in shallow to deep water and becomes

(a) $H = 0.097$, $T = 10.0$, $\varepsilon_\eta = 0.897$

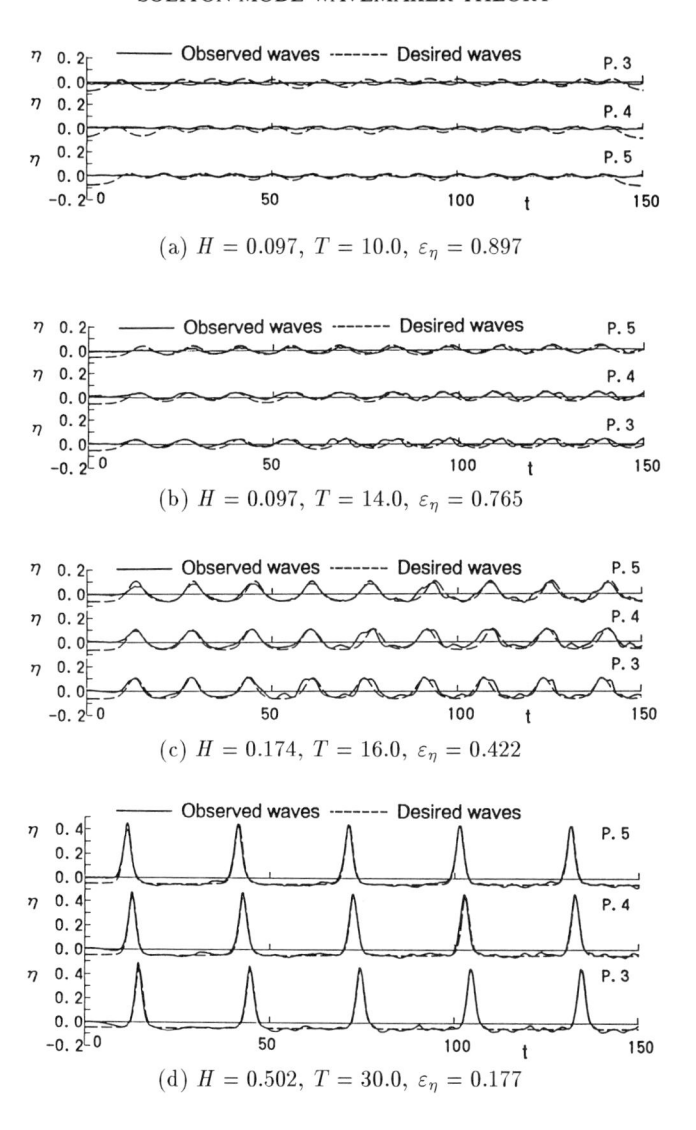

(b) $H = 0.097$, $T = 14.0$, $\varepsilon_\eta = 0.765$

(c) $H = 0.174$, $T = 16.0$, $\varepsilon_\eta = 0.422$

(d) $H = 0.502$, $T = 30.0$, $\varepsilon_\eta = 0.177$

Figure 8. Comparisons for a train of uniform solitons between the desired wave profiles and the observed ones having a typical value of ε_η

physically meaningless wave there. The present soliton-mode wavemaker can not generate waves in shallow to deep water and therefore its applicable region should be made clear since there is the applicable limit to the wavemaker. In

Table 1. Experimental code numbers and conditions for coastal waves

Code No.	$\bar{H}$	$\bar{T}$	U_r	Skewness
R01	0.218	9.15	23.95	1.136
R02	0.321	10.25	40.44	0.760
R03	0.102	10.12	13.06	1.308
R04	0.207	12.21	38.21	1.305
R05	0.297	11.80	55.16	1.612

order to clarify quantitatively the applicable limit, an error index ε_η is defined as

$$\varepsilon_\eta = \sqrt{\int_0^{T_o}(\eta_{obs}-\eta_{des})^2 dt \Big/ \int_0^{T_o}(\eta_{des})^2 dt}, \tag{15}$$

where η_{obs} denotes the water surface elevation of the observed waves in the wave tank and η_{des} that of the desired waves given to eq.(12) as input data. Relationships between the Ursell number U_r of generated waves and the error index ε_η are shown in **Figure 7**. The values of ε_η increase with the decreasing of the value of U_r and particularly is amplified rapidly from the region where the value of U_r begins to be less than about 15. Moreover, the difference of surface profile between η_{obs} and η_{des} can be approximately ignored as shown in **Figure 8** if the value of ε_η is less than 0.8. It hence could be stated that the present wavemaker system can generate cnoidal waves with $U_r \geq 15$ as desired.

Coastal waves (Random trains of solitons)

Intensive experiments generating various trains of solitons were undertaken to examine how to present wavemaker can accurately regenerate coastal waves having various statistics as desired. Representative cases of them are shown in **Table 1**, in which their code numbers and wave statistics are indicated. R01 and R02 denote the coastal waves observed on Ogata coast facing the Japan sea, and R03, R04 and R05 indicate the coastal waves simulated as random trains of solitons by using a digital simulation method (Shinoda et al., 1992). $\bar{H}$ and $\bar{T}$ are their mean wave heights and periods respectively. The values of U_r and Skewness show that all the waves to be regenerated have pronounced nonlinearity and can not be generated with usual methods based on the Biesel-Suquet wavemaker theory.

Figure 9 shows the comparisons of wave profiles of R01 and R02 between the observed waves in field and the regenerated waves in the wave tank. Both the wave profiles are in almost complete agreement and demonstrate that the

(a) Code No. R01

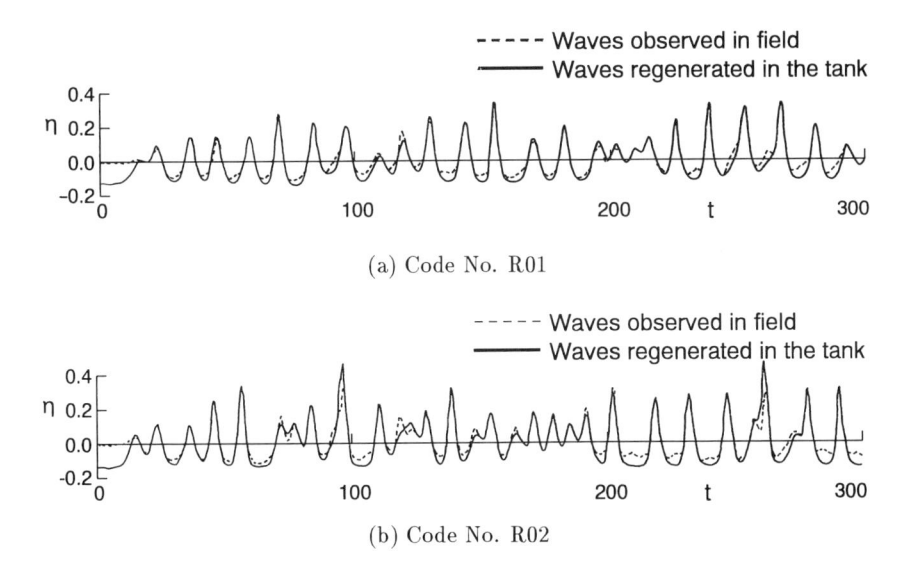

(b) Code No. R02

Figure 9. Comparison of wave profiles between waves observed in field and regenerated ones in the wave tank

present wavemaker can regenerate coastal waves having the given statistics including time sequence of wave crest as desired. It should be particularly noted that coastal waves in field can be almost completely regenerated in the tank even if they partially break.

Further, **Figure 10** shows the comparisons of wave profiles of R03, R04 and R05 between the digitally simulated waves using a stochastic mode of solitons and the regenerated waves in the tank. Although the surface profiles of the regenerated waves get to deviate slightly from those of the simulated waves as the waves propagate from P.2 toward P.3 and further P.6, both the profiles keep a good agreement and shows that the present wavemaker can generate the coastal waves having arbitrary wave statistics by making random trains of solitons in order only if the value of the statistics are given.

In order to make clear the applicable limit of the present wavemaker system to coastal waves, the relations between the Ursell number U_r of coastal waves to be generated and the error index ε_η for the regenerated waves in the tank are examined and shown in **Figure 11**. The value of ε_η is less than 0.8 almost independently of the Ursell number. Since the values of ε_η in all the cases shown in **Figures 9** and **10** are at most 0.644, the generated waves in the tank could

(a) Code No. R03

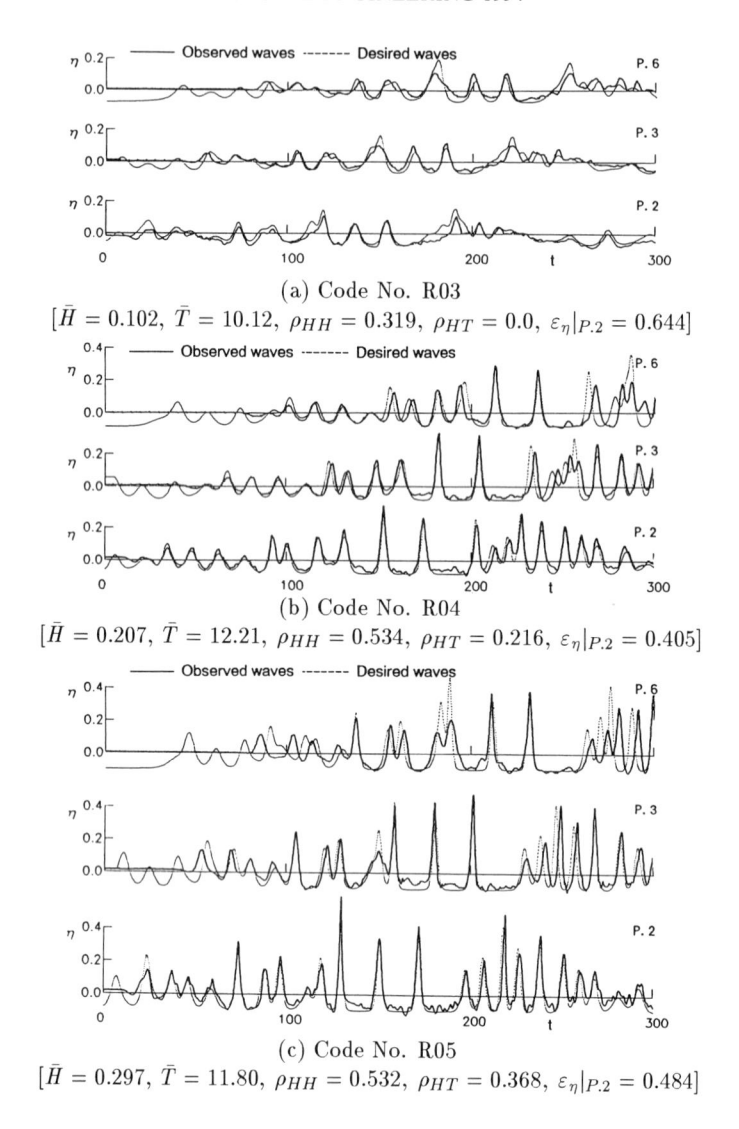

$[\bar{H} = 0.102,\ \bar{T} = 10.12,\ \rho_{HH} = 0.319,\ \rho_{HT} = 0.0,\ \varepsilon_\eta|_{P.2} = 0.644]$

(b) Code No. R04

$[\bar{H} = 0.207,\ \bar{T} = 12.21,\ \rho_{HH} = 0.534,\ \rho_{HT} = 0.216,\ \varepsilon_\eta|_{P.2} = 0.405]$

(c) Code No. R05

$[\bar{H} = 0.297,\ \bar{T} = 11.80,\ \rho_{HH} = 0.532,\ \rho_{HT} = 0.368,\ \varepsilon_\eta|_{P.2} = 0.484]$

Figure 10. Comparisons between the desired wave profiles for a random train of solitons and those measured in the wave tank

be regarded as almost desired if the value of ε_η is less than 0.8, that is, the value of U_r is almost more than 8. It hence could be stated that the present wavemaker can generate the coastal waves having given statistics as desired if their Ursell numbers are more than 8.

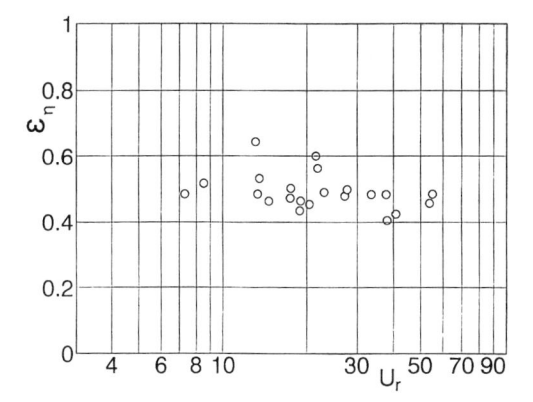

Figure 11. Relationships for coastal waves between the Ursell number U_r and the error index ε_η at the measurement point P.2

Conclusions

A wavemaker theory to generate nonlinear waves in very shallow water as trains of solitons was developed by solving mathematically wavemaker problem in the order of the KdV equation. The paddle motion given by the theory showed that the displacement required to generate them as desired exceeds two times or more of the water depth and a wavemaker having long stroke paddle is necessary. Hence, a new type wavemaker was thought out and developed under the cooperation of ISEYA Manufacturing Co. in order to solve this problem, by noticing that the displacement can be separated into two parts; the one is rapid but short stroke part and the other is slow but long stroke part. This wavemaker can generate the desired soliton trains in order by equipping a double-mode movement having a rotary actuator enabling short but rapid motion and a cylinder actuator enabling slow but long stroke motion was thought out and newly developed. Experimental examination verifies that the system is useful to generate strongly nonlinear waves having the desired water surface profiles of which nonlinearity is specified by the Ursell number $U_r \geq 15$ for cnoidal waves and $U_r \geq 8$ for coastal waves. Thus, the double-mode movement is essential to the soliton-mode wavemaker to generate the desired coastal waves.

Hence, it could be stated that the present wavemaker system can almost completely generate both the cnoidal and coastal waves having not only the desired statistics including sequence of wave crest but also the desired wave profiles, if the input signal to the wave paddle is given by the present wavemaker theory and given to the soliton-mode wavemaker as input signal and further the waves to be generated are within the aforementioned applicable region of this system.

The authors wish to thank Iseya Manufacturing Co. for their developing and manufacturing the present double-mode wavemaker. Further, the authors are grateful for the supports by the Kajima Foundation's Research Grant and the Grant-in-Aid for Scientific Research (No. 06750542), The Ministry of Education, Science and Culture.

References

Goring, D. and F. Raichlen (1980). The generation of long waves in the laboratory. *Proc. 17th Conf. on Coast. Eng., Sydney, A.S.C.E.*, pp. 763–783.

Shinoda, S., T. Yasuda, T. Hattori, and Y. Tsuchiya (1992). A stochastic model of random waves in shallow water. *Proc. of 6th IAHR Int. Symp. on Stochastic Hydraurics*, pp. 329–336.

Yasuda, T., S. Shinoda, and R. A. Darlymple (1987). Soliton mode representation for kinematics of shallow water swell. *Coastal Hydrodynamics, ASCE, R. A. Darlymple ed.*, pp. 751–764.

CHAPTER 53

On a Method for Estimating Reflection Coefficient in Short-Crested Random Seas

Hiromune Yokoki[*]
Masahiko Isobe[†]
Akira Watanabe[†]

Abstract

A method for estimating the directional spectrum as well as the reflection coefficient in a incident and reflected wave field is developed for practical uses. In the method, the directional spectrum is assumed to be expressed by a circular normal distribution which includes several parameters, Then the parameters are estimated by the maximum likelihood method. The validity of the method is verified by numerical simulation.

1 Introduction

The randomness of sea waves has recently become accounted for in the design of coastal and ocean structures. Directional spectra have often been used to describe multi-directional random sea waves. However, to evaluate the reflection coefficient of structures, a theory for uni-directional random waves has usually been applied with a slight modification. This is because a theory for multi-directional random waves has not yet been established for practical uses.

The purpose of this paper is to derive a method to estimate the directional spectrum as well as the reflection coefficient of structures for multi-directional random waves and to examine its validity by applying it to simulated data.

In various methods to estimate the directional spectrum, the maximum likelihood method (MLM) has a high resolution (Capon, 1969). Isobe and Kondo

---
[*]Research Associate, Dept. of Civil Eng., Univ. of Tokyo, Bunkyo-ku, Tokyo, 113 Japan
[†]Professor, ditto

(1984) proposed the modified maximum likelihood method (MMLM) to estimate the directional spectrum in a combined incident and reflected wave field, taking into account the fact that there is no phase difference between the incident and reflected waves at the reflective wall.

Recently for practical purposes, Isobe (1990) proposed a method to estimate the directional spectrum of a standard form in which the spectrum is expressed in terms of a few parameters. Yokoki et al. (1992) extended this method in order to estimate the directional spectrum in an incident and reflected wave field.

In this study, we modify the method to improve the numerical efficiency by assuming a circular normal distribution for the directional function.

2 Theory

2.1 Formulation of cross power spectrum

2.1.1 Definition of cross power spectrum

In a monochromatic wave field which consists of incident and reflected waves, the water surface fluctuation, $\eta(\boldsymbol{x},t)$, at the position, $\boldsymbol{x}$, are represented by Eq. (1):

$$\eta(\boldsymbol{x},t) = A(\boldsymbol{k},\sigma)\{\cos(\boldsymbol{k}\boldsymbol{x} - \sigma t + \epsilon) + r\cos(\boldsymbol{k}_r\boldsymbol{x} - \sigma t + \epsilon)\} \tag{1}$$

The definitions of the variables in the above equation are given in **Table 1**. We integrate Eq. (1) with respect to $\boldsymbol{k}$ and σ, and get the expression of $\eta(\boldsymbol{x},t)$ for multi-directional random waves as Eq. (2):

$$\eta(\boldsymbol{x},t) = \int_{-\infty}^{\infty}\int_{\boldsymbol{k}} A(d\boldsymbol{k},d\sigma)$$
$$\times \{\exp[i(\boldsymbol{k}\boldsymbol{x} - \sigma t + \epsilon)] + r\exp[i(\boldsymbol{k}_r\boldsymbol{x} - \sigma t + \epsilon)]\} \tag{2}$$

where complex variables are introduced to represent the amplitude and phase. From Eq. (2), the directional spectrum (wavenumber-frequency spectrum), $S(\boldsymbol{k},\sigma)$, is defined as Eq. (3):

$$S(\boldsymbol{k},\sigma)d\boldsymbol{k}d\sigma = <A^*(d\boldsymbol{k},d\sigma)A(d\boldsymbol{k},d\sigma)> \tag{3}$$

Table 1: Definitions of variables in Eq. 1

$A(\boldsymbol{k},\sigma)$	Amplitude (Eq. (1)),
$A(d\boldsymbol{k},d\sigma)$	Complex amplitude (Eq. (2))
$\boldsymbol{k}$	Wave number vector of incident waves
$\boldsymbol{k}_r$	Wave number vector of reflected waves
σ	Angular frequency
ϵ	Phase

ESTIMATING REFLECTION COEFFICIENT

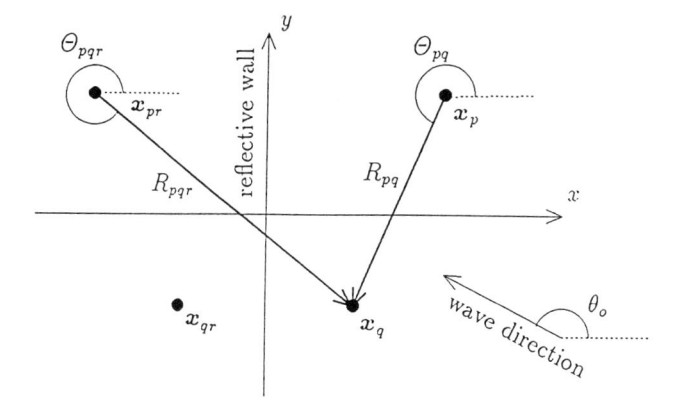

Figure 1: Definitions of variables

where $<\ >$ represents the ensemble average and A^* the complex conjugate of A. Then, the cross power spectrum, $\Phi_{pq}(\sigma)$, can be defined by Eq. (4) (Horikawa, 1988, Isobe and Kondo, 1984):

$$\Phi_{pq}(\sigma) = \int_k S(\mathbf{k}, \sigma) \\ \times \{\exp(i\mathbf{k}\mathbf{x}_p) + r\exp(i\mathbf{k}\mathbf{x}_{pr})\}\{\exp(-i\mathbf{k}\mathbf{x}_q) + r\exp(-i\mathbf{k}\mathbf{x}_{qr})\}\,d\mathbf{k} \quad (4)$$

in which the variables are defined in **Fig. 1**.

We rewrite Eq. (4) by using the transformation of variables as follows. The definitions of variables are also indicated in **Fig. 1**.

$$\left.\begin{array}{rl} \mathbf{k} & = (k\cos\theta,\ k\sin\theta) \\ \mathbf{x}_p - \mathbf{x}_q & = (R_{pq}\cos\Theta_{pq},\ R_{pq}\sin\Theta_{pq}) \\ \mathbf{x}_{pr} - \mathbf{x}_{qr} & = (R_{pq}\cos(\pi-\Theta_{pq}),\ R_{pq}\sin(\pi-\Theta_{pq})) \\ \mathbf{x}_{pr} - \mathbf{x}_q & = (R_{pqr}\cos\Theta_{pqr},\ R_{pqr}\sin\Theta_{pqr}) \\ \mathbf{x}_p - \mathbf{x}_{qr} & = (R_{pqr}\cos(\pi-\Theta_{pqr}),R_{pqr}\sin(\pi-\Theta_{pqr})) \end{array}\right\} \quad (5)$$

Then we get the expression of the cross power spectrum, $\Phi_{pq}(f)$:

$$\Phi_{pq}(f) = \int_0^{2\pi} S(f, \theta) \\ \times [\exp\{ikR_{pq}\cos(\theta - \Theta_{pq})\} + r^2\exp\{ikR_{pq}\cos(\theta - \pi + \Theta_{pq})\} \\ + r\exp\{ikR_{pqr}\cos(\theta - \Theta_{pqr})\} + r\exp\{ikR_{pqr}\cos(\theta - \pi + \Theta_{pqr})\}]\,d\theta \quad (6)$$

where f represents the frequency connected with the wave number vector, $\mathbf{k}$, by the dispersion relation, and r the reflection coefficient of the reflective wall (shown in **Fig. 1**).

2.1.2 Directional spectrum function

Three standard functions have been proposed to express the directional spectrum except the Fourier series.

The first one is the function proposed by Mitsuyasu et al. (1975) shown as:

$$G(\theta|f) \equiv \frac{S(f,\theta)}{P(f)} \propto \cos^{2s}\left(\frac{\theta - \theta_o}{2}\right) \tag{7}$$

where $P(f)$ indicates the frequency spectrum, s the degree of directional concentration, and θ_o the peak wave direction. Yokoki et al. (1992) used this standard directional function. However, it took fairly much time to calculate the cross power spectrum, Φ_{pq}.

The second one is the square of the hyperbolic secant function, which is proposed by Donelan et al. (1985):

$$G(\theta|f) \propto \text{sech}^2 \beta(\theta - \theta_o) \tag{8}$$

where β indicates the degree of directional concentration.

The last one is the circular normal distribution function proposed by Borgman (1969):

$$G(\theta|f) \propto \exp\{a\cos(\theta - \theta_o)\} \tag{9}$$

where a indicates the degree of directional concentration.

In the present study, we assume the directional spectrum, $S(f,\theta)$, to be expressed by using the circular normal distribution function, because we can obtain the cross power spectrum analytically to some extent and the numerical calculation becomes faster. Therefore, the directional spectrum, $S(f,\theta)$, is expressed by Eq. (10):

$$S(f,\theta) = P(f)\frac{1}{2\pi I_o(a)}\exp\{a\cos(\theta - \theta_o)\} \tag{10}$$

where I_o is the modified Bessel function.

2.1.3 Formulation of cross power spectrum

To rewrite Eq. (6) in a simpler form, we define a new function φ as follows:

$$\varphi(a, \theta_o, R, \Theta|f)$$
$$= \frac{1}{2\pi I_o(a)}\int_0^{2\pi} \exp\{-ikR\cos(\theta - \Theta)\}\exp\{a\cos(\theta - \theta_o)\}d\theta \tag{11}$$

By using the above definition, Eq. (6) can be written as Eq. (12):

$$\Phi_{pq}(f) = \left\{\begin{array}{l} \varphi(\ a,\theta_o,R_{pq},\ \Theta_{pq}\ \ |f) \\ + r^2\varphi(\ a,\theta_o,R_{pq},\ \pi - \Theta_{pq}\ |f) \\ + r\varphi(\ a,\theta_o,R_{pqr},\Theta_{pqr}\ \ |f) \\ + r\varphi(\ a,\theta_o,R_{pqr},\pi - \Theta_{pqr}|f) \end{array}\right\} \times (1 + \delta_{pq}\varepsilon_p)P(f) \tag{12}$$

Table 2: Directional spectrum parameters

a	Degree of directional concentration
θ_0	Peak wave direction
r	Reflection coefficient
$P(f)$	Frequency or power spectrum
ε_p	Ratio of the noise component to the power (at the p-th point; $p = 1 \sim 3$)

where ε_p indicates the ratio of the noise component to the power, $\Phi_{pp}(f)$. We finally get the expression of the cross power spectrum in terms of the seven parameters which are summarized in **Table** 2 and called the directional spectrum parameters in this paper.

The integral on the right-hand side of Eq. (11) can be expressed by using the integral expression of the Bessel function and the modified Bessel function (*e.g.* Abramowitz and Stegun, 1972):

$$\varphi(a, \theta_o, R, \Theta | f) = J_o(kR)$$
$$+ 2 \sum_{m=1}^{\infty} (-1)^m \cos(2m\beta) J_{2m}(kR) \frac{I_{2m}(a)}{I_0(a)}$$
$$+ 2i \sum_{m=0}^{\infty} (-1)^m \cos[(2m+1)\beta] J_{2m+1}(kR) \frac{I_{2m+1}(a)}{I_0(a)} \qquad (13)$$

In the present study, we use Eqs. (12) and (13) to calculate the cross power spectrum.

2.2 Maximum likelihood method

2.2.1 Definition of likelihood

The maximum likelihood method is used to get the most probable values of the directional spectrum parameters. The likelihood, L, is defined by Isobe (1990):

$$L(A^{[j]}; \Phi) = \left\{ p(A^{[1]}) \times p(A^{[2]}) \times \cdots \times p(A^{[j]}) \times \cdots \times p(A^{[J]}) \right\}^{1/J}$$
$$= \frac{1}{(2\pi \Delta f)^M |\Phi|} \exp\left(-\sum_{p=1}^{M} \sum_{q=1}^{M} \Phi_{pq}^{-1} \widehat{\Phi}_{qp} \right) \qquad (14)$$

where $p(A^{[j]})$ is a joint probability density function of the Fourier coefficients, $A^{[j]}$, of the time series data, Δf the frequency interval, and $|\Phi|$ the determinant of the matrix, Φ_{pq}. The quantity $\widehat{\Phi}_{qp}$ which is represented by Eq. (15) corresponds to the power spectrum ($q = p$) or the cross spectrum ($q \neq p$) with a use of rectanglar filter.

$$\hat{\Phi}_{qp} = \frac{1}{2J\Delta f}\sum_{j=1}^{J}\overline{A}_{q}^{[j]}A_{p}^{[j]} \tag{15}$$

where $\overline{A}$ denotes the complex conjugate of A.

2.2.2 The most probable values of the parameters

In this Section, we show the procedure to estimate the directional spectrum parameters including the reflection coefficient by using the likelihood defined above.

The maximum likelihood method implies that the most probable values of λ_i are the solutions of the algebraic equation:

$$\frac{\partial L}{\partial \lambda_i} = \sum_{j=1}^{M}\sum_{l=1}^{M}\frac{\partial L}{\partial \Phi_{jl}}\frac{\partial \Phi_{jl}}{\partial \lambda_i} = 0 \tag{16}$$

From Eq. (14), Eq. (17) is obtained.

$$\begin{aligned}\frac{\partial L}{\partial \Phi_{jl}} &= \frac{\partial}{\partial \Phi_{jl}}\left[\frac{1}{(2\pi\Delta f)^M|\Phi|}\exp\left(-\sum_{p=1}^{M}\sum_{q=1}^{M}\Phi_{pq}^{-1}\hat{\Phi}_{qp}\right)\right] \\ &= -\frac{1}{(2\pi\Delta f)^M|\Phi|^2}\exp\left(-\sum_{p=1}^{M}\sum_{q=1}^{M}\Phi_{pq}^{-1}\hat{\Phi}_{qp}\right)\frac{\partial|\Phi|}{\partial \Phi_{jl}} \\ &\quad + \frac{1}{(2\pi\Delta f)^M|\Phi|}\exp\left(-\sum_{p=1}^{M}\sum_{q=1}^{M}\Phi_{pq}^{-1}\hat{\Phi}_{qp}\right) \\ &\quad \times \left(-\sum_{p=1}^{M}\sum_{q=1}^{M}\frac{\partial \Phi_{pq}^{-1}}{\partial \Phi_{jl}}\hat{\Phi}_{qp}\right) \\ &= -L \times \left(\frac{1}{|\Phi|}\frac{\partial|\Phi|}{\partial \Phi_{jl}}\right) + L \times \left(-\sum_{p=1}^{M}\sum_{q=1}^{M}\frac{\partial \Phi_{pq}^{-1}}{\partial \Phi_{jl}}\hat{\Phi}_{qp}\right) \end{aligned} \tag{17}$$

Also, the following relations are obtained from the theorem of the matrices.

$$\frac{\partial|\Phi|}{\partial \Phi_{jl}} = |\Phi|\Phi_{lj}^{-1} \tag{18}$$

$$\frac{\partial \Phi_{pq}^{-1}}{\partial \Phi_{jl}} = -\Phi_{lq}^{-1}\Phi_{pj}^{-1} \tag{19}$$

By using Eqs. (18) and (19), Eq. (17) is rewritten as follows:

$$\frac{\partial L}{\partial \Phi_{jl}} = L \times \left\{-\Phi_{lj}^{-1} + \sum_{p=1}^{M}\sum_{q=1}^{M}\Phi_{lq}^{-1}\hat{\Phi}_{qp}\Phi_{pj}^{-1}\right\} \tag{20}$$

Substituting Eq. (20) into Eq. (16) and considering that $L \neq 0$, we obtain

$$\sum_{j=1}^{M}\sum_{l=1}^{M}\left\{-\Phi_{lj}^{-1}+\sum_{p=1}^{M}\sum_{q=1}^{M}\Phi_{lq}^{-1}\widehat{\Phi}_{qp}\Phi_{pj}^{-1}\right\}\frac{\partial\Phi_{jl}}{\partial\lambda_{i}}=0 \qquad (21)$$

The directional spectrum parameters, λ_i, which satisfy Eq. (21) for all i ($i = 1 \sim 7$) are the most probable values. Then the directional spectrum parameters, including the reflection coefficient, are estimated.

The solutions, λ_i, of Eq. (21) are obtained numerically by using the Newton-Raphson method. The left-hand side of Eq. (21) is first defined as a function of the directional spectrum parameters:

$$f_i(\lambda_{i'}) = \sum_{j=1}^{M}\sum_{l=1}^{M}\left\{-\Phi_{lj}^{-1}+\sum_{p=1}^{M}\sum_{q=1}^{M}\Phi_{lq}^{-1}\widehat{\Phi}_{qp}\Phi_{pj}^{-1}\right\}\frac{\partial\Phi_{jl}}{\partial\lambda_{i}} \qquad (22)$$

In the Newton-Raphson method, the values of $\lambda_i^{(j+1)}$ at the $(j+1)$-th iteration of the calculation is expressed in terms of the previous values, $\lambda_i^{(j)}$, as follows.

$$\lambda_i^{(j+1)} = \lambda_i^{(j)} - \left[\sum_{i'=1}^{I}\left[\frac{\partial f_i}{\partial \lambda_{i'}}\right]^{-1} f_{i'}\right]_{\lambda_i=\lambda_i^{(j)}} \qquad (23)$$

where $\partial f_i/\partial \lambda_{i'}$ is expressed by Eq. (24):

$$\begin{aligned}\frac{\partial f_i}{\partial \lambda_{i'}} =& \sum_{j=1}^{M}\sum_{l=1}^{M}\left\{-\Phi_{lj}^{-1}+\sum_{p=1}^{M}\sum_{q=1}^{M}\Phi_{lq}^{-1}\widehat{\Phi}_{qp}\Phi_{pj}^{-1}\right\}\frac{\partial^2 \Phi_{jl}}{\partial \lambda_{i'}\partial \lambda_i}\\ &-\sum_{j'=1}^{M}\sum_{l'=1}^{M}\sum_{j=1}^{M}\sum_{l=1}^{M}\frac{\partial \Phi_{j'l'}}{\partial \lambda_{i'}}\frac{\partial \Phi_{jl}}{\partial \lambda_i}\\ &\times\left[-\Phi_{l'j}^{-1}\Phi_{lj'}^{-1}+\left\{\begin{array}{l}\Phi_{lj'}^{-1}\sum_{p=1}^{M}\sum_{q=1}^{M}\Phi_{l'q}^{-1}\widehat{\Phi}_{qp}\Phi_{pj}^{-1}\\ +\Phi_{l'j}^{-1}\sum_{p=1}^{M}\sum_{q=1}^{M}\Phi_{lq}^{-1}\widehat{\Phi}_{qp}\Phi_{pj'}^{-1}\end{array}\right\}\right]\end{aligned} \qquad (24)$$

In calculations of the present method, it sometimes happened that we obtained the parameters, $\lambda_i^{(j+1)}$, which made the value of the likelihood, L, smaller than the preceding parameters, $\lambda_i^{(j)}$. This was because the present procedure sometimes finds the minimum likelihood in stead of the maximum likelihood. Therefore, we put the additional procedure that we replace $\Delta\lambda_i$ ($= \lambda_i^{(j+1)} - \lambda_i^{(j)}$) with $-\Delta\lambda_i$ if $(\partial L/\partial \lambda_i)\Delta\lambda_i < 0$. With this procedure, we can obtain converged solutions of the paramters of Eq. (21) from various initial values of the parameters.

3 Numerical calculation

3.1 Simulated cross power spectrum

To verify the validity of the present method, we applied it to simulated data.

Table 3: Values of the directional parameters

Case No.	f(Hz)	s	$\theta_o(°)$	$P(f)$ (m^2s)	r	ε_1	ε_2	ε_3
PA01	0.1	5.0	120	1.0	0.1	0.1	0.1	0.1
PA02	0.1	10.0	120	1.0	0.1	0.1	0.1	0.1
PA03	0.1	5.0	120	1.0	0.5	0.1	0.1	0.1
PA04	0.1	5.0	120	1.0	0.1	0.3	0.3	0.3
PA05	0.1	5.0	120	1.0	0.1	0.1	0.2	0.3
PA06	0.2	5.0	120	1.0	0.1	0.1	0.1	0.1

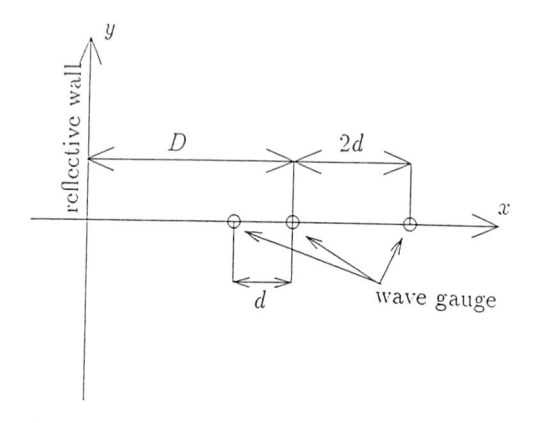

Figure 2: Arrangement of wave gauges

First, we created sets of $\hat{\Phi}_{pq}$ by using of the directional spectrum proposed by Mitsuyasu et al. (1975) for each set of directional spectrum parameters given in **Table 3**.

The arrangement of wave gauges used to calculate $\hat{\Phi}_{pq}$ is shown in **Fig. 2**. In the figure, we vary the distance between the reflective wall and the wave gauges, D, and that between the wave gauges, d, as shown in **Table 4**. We finally calculated $\hat{\Phi}_{pq}$ for each combination of the arrangement of wave gauges and the set of the directional spectrum parameters. Then we applied the present method to each set of the cross power spectra, $\hat{\Phi}_{pq}$.

3.2 Results of estimation

Figure 3 shows contour maps of likelihood in terms of the directional spectrum parameters, a, ε_p, r and θ_o. In these figures, the directional spectrum is calculated

Table 4: Arrangement of wave gauges

Case No.	D(m)	d(m)
WG01	20	5
WG03	40	5
WG05	40	15
WG07	60	5
WG09	60	15

from the set of the parameters, PA01, shown in **Table** 3. WG01, WG03 and WG05 indicates the arrangement of wave gauges as shown in **Table** 4.

From these figures, we can see that the likelihood becomes maximum at the points which correspond to the given values of the directional spectrum parameters.

Table 5 shows the results of estimations for all combinations of the directional spectrum parameters and the arrangements of wave gauges. For these directional spectrum parameters, estimated parameters agree quite well with the true values, as shown in **Table** 3.

However, for the directional spectrum parameters, PA06, the agreement is not good. This is because the distance between the wave gauges are too long compared to the wavelength.

4 Conclusion

By using a circular normal distribution as the directional function, a parametric method for estimating the directional spectrum as well as the reflection coefficient in an incident and reflected wave field is proposed. From the results of numerical simulations, the method is proved to be valid as long as the arrangement of wave gauges is appropriately designed.

References

[1] Abramowitz, M. and I. A. Stegun, (ed.) (1972): Handbook of Mathematical Functions with Formulas, Graphs, and Mathematical Tables, Dover Pub. Inc., pp. 355–433.

[2] Borgman, L. E. (1969): Directional spectra models for design use for surface waves, Tech. Rep. HEL 1-12, Univ. of Calif. 56p.

[3] Capon, J. (1969): High-resolution frequency-wavenumber spectrum analysis, Proc. IEEE, Vol. 57, No. 8, pp. 1408–1418.

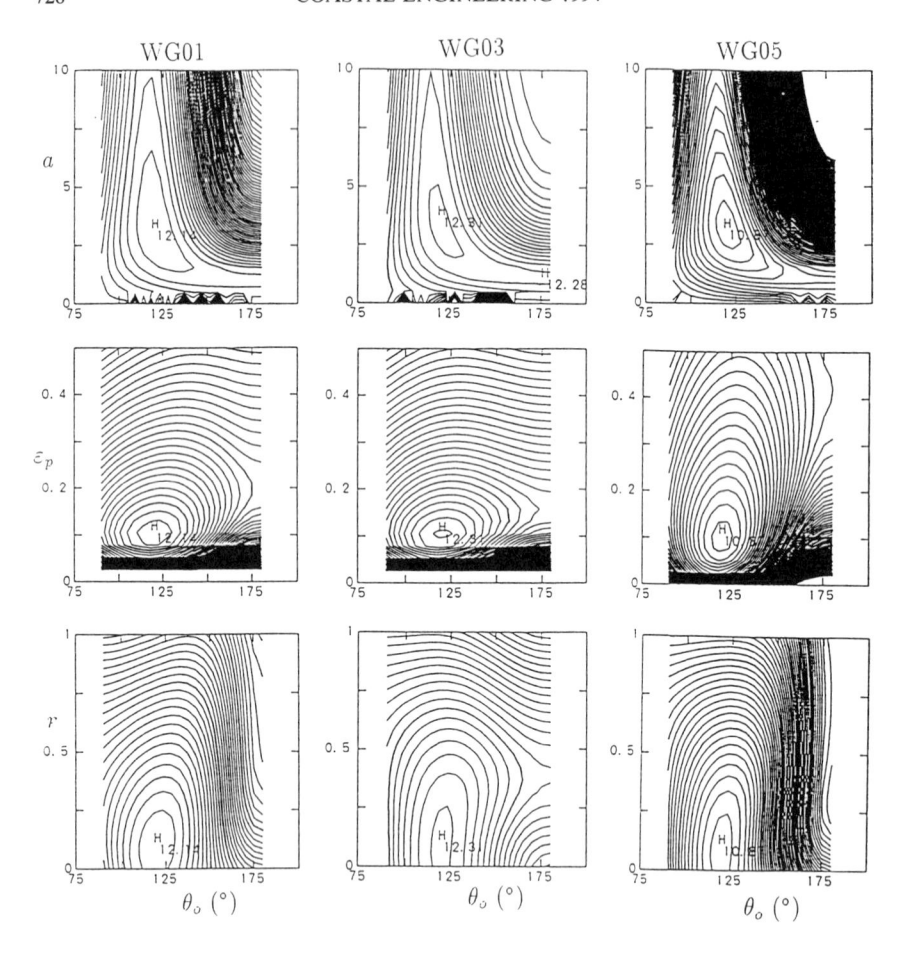

Figure 3: Contour map of the likelihood

Table 5: Results of estimation

Case No.	Case No.	f(Hz)	a	$\theta_o(°)$	$P(f)$ (m^{2}s)	r	ε_1	ε_2	ε_3
PA01	WG01	0.1	3.22	120	1.00	0.098	0.10	0.099	0.10
	WG03	0.1	3.22	120	1.00	0.10	0.099	0.10	0.099
	WG05	0.1	3.01	121	1.01	0.11	0.096	0.088	0.096
	WG07	0.1	3.55	121	0.99	0.13	0.10	0.098	0.10
	WG09	0.1	3.08	120	1.01	0.081	0.095	0.091	0.10
PA02	WG01	0.1	5.65	120	1.00	0.098	0.10	0.10	0.099
	WG03	0.1	5.74	120	1.00	0.10	0.098	0.10	0.097
	WG05	0.1	5.58	120	1.00	0.092	0.10	0.096	0.096
	WG07	0.1	5.97	121	0.99	0.12	0.10	0.099	0.099
	WG09	0.1	5.62	120	1.00	0.094	0.099	0.098	0.098
PA03	WG01	0.1	3.04	120	1.01	0.48	0.10	0.099	0.10
	WG03	0.1	2.81	120	1.02	0.46	0.099	0.10	0.097
	WG05	0.1	2.82	119	1.03	0.45	0.11	0.097	0.093
	WG07	0.1	3.64	121	0.99	0.50	0.10	0.099	0.099
	WG09	0.1	2.75	118	1.04	0.44	0.095	0.098	0.10
PA04	WG01	0.1	3.13	121	1.00	0.095	0.30	0.30	0.30
	WG03	0.1	3.72	120	1.00	0.12	0.29	0.30	0.29
	WG05	0.1	3.07	120	1.01	0.10	0.29	0.28	0.29
	WG07	0.1	3.35	121	0.99	0.11	0.30	0.30	0.30
	WG09	0.1	3.05	120	1.01	0.085	0.30	0.29	0.30
PA05	WG01	0.1	3.07	120	1.01	0.096	0.099	0.20	0.30
	WG03	0.1	3.74	120	1.01	0.12	0.097	0.20	0.29
	WG05	0.1	3.08	120	1.01	0.11	0.096	0.19	0.29
	WG07	0.1	3.37	121	0.99	0.11	0.10	0.20	0.30
	WG09	0.1	3.07	120	1.01	0.081	0.096	0.19	0.30
PA06	WG01	0.2	3.17	122	0.98	0.19	0.13	0.063	0.11
	WG03	0.2	3.17	121	1.00	0.12	0.10	0.091	0.10
	WG05	0.2	2.89	120	1.02	0.10	0.071	0.068	0.077
	WG07	0.2	3.12	120	1.01	0.082	0.10	0.095	0.10
	WG09	0.2	2.81	120	1.04	0.12	0.056	0.048	0.060

[4] Donelan, M. A., J. Hamilton and W. H. Hui (1985): Directional spectra of wind-generated waves, Phil. Trans. Roy. Soc. Lond., Ser. A, vol. 315, pp. 509–562.

[5] Horikawa, K. (ed.) (1988): Nearshore Dynamics and Coastal Processes, Univ. Tokyo Press, Part. V, pp. 407–411.

[6] Isobe, M. and K. Kondo (1984): Method for estimating directional wave spectrum in incident and reflected wave field, Proc. 19th Int. Conf. on Coastal Eng., pp. 467–483.

[7] Isobe, M. (1990): Estimation of directional spectrum expressed in standard form, Proc. 22nd Int. Conf. on Coastal Eng., pp. 647–660.

[8] Mitsuyasu, H., F. Tasai, T. Suhara, S. Mizuno, M. Ohkusu, T. Honda and K. Rikiishi (1975): Observations of directional spectrum of ocean waves using a cloverleaf buoy, Jour. of Physical Oceanography, Vol. 5, pp. 750–760.

[9] Yokoki, H., M. Isobe and A. Watanabe (1992): A method for estimating reflection coefficient in short-crested random seas, Proc. 23rd Int. Conf. on Coastal Eng., pp. 765–776.

CHAPTER 54

The Directional Wave Spectrum in the Bohai Sea*

Yuxiu Yu[1] and Shuxue Liu[2]

Abstract

The directional spectrum is one of the basic characteristics of sea waves. The observations of directional spectrum of sea waves are successfully conducted at platform Bohai 8 during 1991 and 1992 using a wave gage array and it is the first time in China. Based on the field data, the directional spectrum which depends on the wave growth is given in this paper. Before observations, the effects of the figure of gage array, the distance between the gages and the platform itself on the measured results and the precision of some methods for estimating the directional spectrum were investigated and compared with the methods of numerical simulations and model tests of multi-directional irregular waves. This ensures the quality of the observations and estimations of the directional spectrum.

Introduction

One of the main practical problem for research of the sea wave theory and its application is to seek the directional spectrum because it is the key for studying the generating mechanism of the wind waves and the wave deformation as it propagate into shallow water. It is also the important reference for the design of maritime structures. But there is less research work on the directional spectrum due to the complex of the sea wave phenomena and the difficulty of the observation and the analysis of directional spectrum.

At present, the main method for obtaining directional spectrum is observations in field. The commonly used direct measurement methods may be of three types. Early in 1954, Barber(1954) had used wave gage array to measure the directional

---

* The project is supported by National Natural Science Foundation of China.
1. Professor, State Key Laboratory of Coastal and Offshore Engineering, Dalian University of Technology, Dalian 116024, China.
2. Assistant Researcher, State Key Laboratory of Coastal and Offshore Engineering, Dalian University of Technology, Dalian 116024, China.

spectrum and gave proposals concerning the wave gage number and the distance between gages for a linear array. After that, some researchers studied several types of gage array with numerical and physical simulation. Longuet-Higgins et al.(1963) used pitch-and-roll buoy to measure the water surface elevation and the orthogonal components of wave surface slopes, and estimated the directional spectrum with Fourier series method. Mitsuyasu et al.(1975) designed a cloverleaf buoy which can measure three components of surface curvature besides the water surface elevation and two components of wave slopes. Therefore, the directional resolution of buoy measurement was enhanced. Also, he gave a formula of directional spectrum which was called Mitusuyasu type spectrum according to the measurement results in the sea of Japan. The another commonly used method is to measure the two horizontal components of orbital velocities by a two-axis current meter and the wave surface which was first proposed by Nagata(1964). This method is simpler and has a higher directional resolution. Besides, the remote sensing such as the stereo wave observation and the microwave technique is also used to measure directional spectrum.

The observations of directional spectrum in China were carried out later. After the ENDECO 956 buoy was introduced from U.S.A. in 1982, it is used to measure the waves for general engineering design. Because there are some disputes about their measured directional spectrum, so less systematic study has been done on it. The Ocean University of Qingdao tried observing directional spectrum with stereophotogrametry several years ago. In this paper, the observation and estimation methods of directional spectrum is mainly studied. Based on the analysis of existing observations of the directional spectrum, and the existing facilities(equipments) and the condition of experiment, wave gage array method is used to measure the directional spectrum for two times. The advance research works were conducted in laboratory and then the available methods of measurement and analysis were set up. Based on the observation results, the directional spectrum of the wind-waves is given in the paper.

Measurement Program

Observation method

At present, there are several measurement and analysis methods of directional spectrum. But their results have a big difference each other. So the measurement and estimation of directional spectrum themselves are an important research problem and they are studied with the numerical simulation and the laboratory experiments(Yu and Liu, 1992, 1993). Considering the present conditions of the equipments and laboratory in China, the wave gage array method which is more reliable and easily carried out on platform is selected to measure the directional waves.

Type and dimension of array

Platform Bohai-8 is almost situated in the center area of the Bohai Sea(39°09'N, 119°42'E). The water depth d is 27.0 m. It consists of a life platform and a working platform and they are linked with a bridge of 50m long. The orientation of the bridge is west-east. The number of wave gages must be greater than 3. 5 wave gages of BCS vertical line type (made by the Institute of Oceanology, Academia Sinica) are used in the observation and they can be arranged in many types. The effects of the array type including linear array, T-type array and pentagon array and the spacing of wave gages on the directional resolution are investigated through numerical simulation and model tests in 3-dimensional basin respectively(Yu and Liu, 1992). The results show that the linear array is only suitable for the waves which have a narrow directional spreading and whose main direction has a large angle with the axis of the array and the gage number must be greater than 5. Both T-type and pentagon array have higher resolutions and the imposing restrictions on gage spacing are not so strict. Considering the situation of the Platform Bohai 8, T-type array is selected.

The wave gage spacing in an array has a big effect on the directional resolution. The results of the numerical and physical simulation show that the minimum spacing should be less than the half of the measured wave length. Considering the irregularity of the sea waves, the ratio between the minimum spacing and the wave length corresponding to the peak frequency of the spectrum can be about equal to 0.3. The vector distance of each pair of gage should be distributed as uniformly as possible in a range as wide as possible and no pair of wave gages should have the same vector distance. According to above guidelines and the real situation of the platform, the T-type array is arranged as Fig.1, in which the minimum distance between wave gages is 4.9m and the maximum value is 21.2m. It can be available for the waves with the period larger than 2.5s.

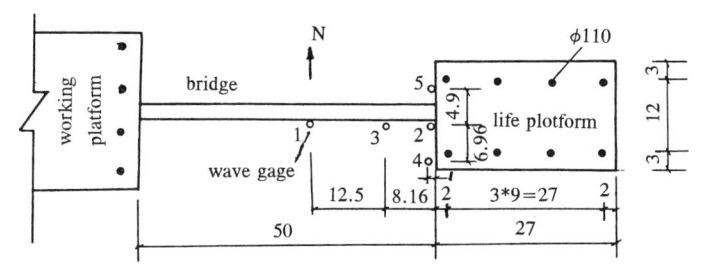

Fig.1 The arrangement of the wave gage array

Effect of the platform on the results

The life platform is supported by 8 steel cylinders and there are some horizontal and inclined cylinders between them. To check the effect of the platform on the waves from all directions, the model tests with the multi-directional irregular waves are carried out in the basin of the State Key Laboratory of Coastal and

Offshore Engineering in Dalian University of Technology, the model scale is 1:40. The test results show that the effects of the platform structure on the directional spreading are small.

The Observations and Data Analysis

Situation of the measurement

The T-type array consisted of 5 gages is arranged as Fig.1. The wave data are acquired simultaneously and automatically with computer for 1200 seconds every hour. The time interval is 0.25s. At the circumstance of strong wind, the continuous wave data are acquired by artificial control. The velocity and direction of wind are recorded simultaneously by wind-velocity-direction meter. At the same time, according to the request of the Specification of Ocean Inspection, the sea state, the wave pattern and the wave direction are observed visually in the daytime. During October 20 to November 4, 1992, 271 runs of 4 wind histories are recorded. There are 144 runs whose significant wave height are larger than 0.5m. The measured maximum wind velocity is 21m/s and the maximum significant wave height is 2.58m.

The Selection of the data

The Bohai sea is basically a closed inner sea and can not be affected by the ocean swell. The directional spectrum formed by a single wind is emphatically investigated in this paper and the wave data are selected according to following principles: (1) The frequency spectra have only one peak, (2) the frequency spectra measured by the 5 gages are almost identical each other and (3) the directional spreading function also basically has one peak and the directional range of the wave energy is narrower. So the 51 runs of data (Table 1) are analyzed. In the table, U is the wind velocity at 10m elevation above the sea level. Cp is the wave velocity corresponding to the peak frequency. $H_{1/3}$, $T_{H1/3}$ and Ls are the significant wave height, period and length respectively. Because U is used in most of the present formula of the directional spectrum, 32 runs of wave data with the angle between the wind direction and the main wave direction being less than 45^0 are used as the observation results being compared with the formulas.

Analysis method of directional spectrum

There are several methods for analyzing directional spectrum. For selecting the optimum method, the single direction per frequency method(Yu and Liu 1991) is used to simulate the multi-directional waves on computer and in 3-dimensional basin and then the waves are measured by a T-type array including 5 wave gages. Afterwards, the directional spreading is analyzed by the parameterized method, the direct Fourier transform method, the extension maximum likelihood method and the Bayesian approach and their results are compared with the input one. It is

Table.1 The main parameter of the selected data

Date	Runs	Time	U/Cp	$H_{1/3}/L_s$	$H_{1/3}$	$T_{H1/3}$
Oct. 22	8	11:00~19:00	0.76~1.17	0.0282~0.0374	0.90~1.45	4.50~5.07
Oct. 23	8	3:00~17:00	1.01~1.32	0.0296~0.0366	0.93~1.23	4.06~4.92
Oct. 24	2	9:00~10:00	1.20~1.44	0.0286~0.0302	0.63~0.72	3.64~4.02
Oct. 25	2	7:00~10:00	0.38~0.49	0.0159~0.0202	0.50~0.66	4.49~4.58
Oct. 26	1	13:00	0.55	0.0157	0.55	4.74
Oct. 28	7	7:00~22:00	0.76~1.31	0.0235~0.0335	0.60~1.36	3.92~5.26
Oct. 29	3	5:00~13:00	0.50~0.84	0.0264~0.0298	1.01~1.07	4.66~5.10
Oct. 30	2	19:00~20:00	1.03~1.35	0.0390~0.0397	0.86~0.87	3.75~3.76
Oct. 31	12	4:00~23:00	0.96~2.52	0.0247~0.0447	0.63~2.58	3.83~6.53
Nov. 1	6	0:00~5:00	0.68~1.14	0.0249~0.0346	1.48~2.12	5.98~6.27
Sum	51		0.38~2.52	0.0157~0.0447	0.50~2.58	3.66~6.40

shown(Yu and Liu 1993) that the Bayesian approach(Hashimoto, et. al., 1987) has the highest directional resolution and small effect by the wave gage number(greater than 4 only), the distance between wave gages and the width of the directional spreading. So the Bayesian approach is used.

The Frequency Spectrum of the Measured Waves

There are many types of frequency spectrum. The commonly used one is JONSWAP spectrum. In the Technical Specification for Port Engineering of China, the wind-wave frequency spectrum proposed by Wen et al(1989) is adopted. To the equilibrium ranges at high frequency of wave spectra, the slope, p of -5 is adopted by the former spectrum and the latter one use $p=-(2-0.626Hs/d)$(where d is the water depth) i.e., $|p|<4$. Figure.2 show the measured results of the slopes and the mean value can be taken as $p=-4.125$.

For comparing each other, the frequency spectra are nondimensionalized. Comparing the above two spectra with the measured average spectrum shows that the shape of the JONSWAP spectrum is wider and that of the Wen's spectrum is narrower. The spectrum shape proposed by Hong and Yong(1980) is consisted of two parts and can fit the measured one easily(Fig.3), i.e.

$$s(f)=0.117H_s^2T_p(f/f_p)^{-4.125}\exp[-1.171(f/f_p)^{-11}][1+3.893(f/f_p)^{-11}] \quad (1)$$

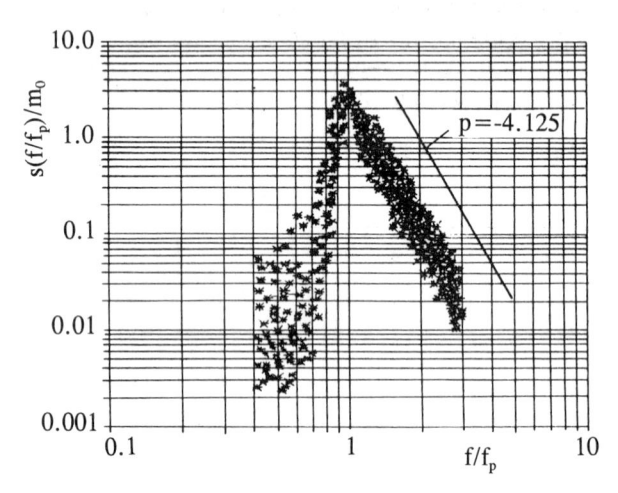

Fig.2 The slope of the equilibrium ranges of the frequency spectrum

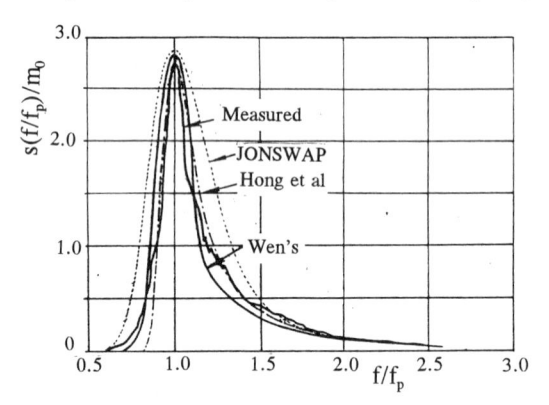

Fig.3 Comparation of the measured frequency spectrum with the fitted one

The Directional Spectrum of Sea Waves

The main aim of the observations is to obtain the directional spectrum of sea waves. The directional spectrum is generally expressed as the product of the frequency spectrum and the directional spreading function $G(f,\theta)$, i.e.,

$$S(f,\theta)=S(f)\cdot G(f,\theta)$$
$$\int_{-\pi}^{\pi} G(f,\theta)d\theta =1 \qquad (2)$$

The measured directional spreading functions of the 51 runs data are obtained by the Bayesian approach and compared with 5 existing typical models.

The present used directional spectrum model

Equation (2) is used in most models, but the spreading function in each model is different.

1. Simple empirical formula. Assuming the directional spreading is independent of wave frequency.

$$G(f,\theta)=G(\theta)=C(n)\cos^{2n}\theta \qquad (3)$$

The coefficient C(n) can be determined by Eq.(2). When n=1, $C(n)=1/\pi$,; n=2, $C(n)=8/3\pi$.

2. Mitsuyasu-type spreading. Longuet-Higgins et al(1963) expressed the directional spreading function as

$$G(f,\theta)=G_0(s)\cos^{2s}\frac{\theta-\theta_0}{2}$$
$$G_0(s)=\frac{1}{\pi}2^{2s-1}\frac{\Gamma^2(s+1)}{\Gamma(2s+1)} \qquad (4)$$

where θ_0 is the main direction of waves, s is the spreading parameter. According to the data measured by a cloverleaf buoy, Mitsuyasu et al(1975) gave the relationship between s and frequency and wind velocity as following

$$s=s_{max}(f/f_p)^5 \qquad f\leq f_p$$
$$s=s_{max}(f/f_p)^{-2.5} \qquad f>f_p$$
$$s_{max}=11.5(2\pi f_p U/g)^{-2.5}=11.5(C_p/U)^{2.5} \qquad (5)$$

where f_p is the peak frequency of the spectrum and Goda(1985) proposed that f_p can be estimated by $f_p=1/(1.05T_{H1/3})$. When $f=f_p$, $s=s_{max}$, it means that the spreading at the peak frequency is the narrowest. In Eq.(5), Cp/U is the wave age and the less is Cp/U, the younger is the waves. The mean value of Cp/U is around 1.0 for Mitsuyasu's data and it corresponds to fully developed sea waves.

3. Hasselmann directional spreading. Hasselmann et al(1980) used Eq.(4) for

spreading function but gave the following parameters according to the data of JONSWAP:

$$s = s_{max}(f/f_p)^{\mu} \tag{6}$$

When $f/f_p \geq 1.0$
$$\begin{cases} s_{max} = 9.77 \pm 0.43 \\ \mu = -(2.33 \pm 0.06) - (1.4 \pm 0.45)(\dfrac{U}{C_p} - 1.17) \end{cases} \tag{7}$$

When $f/f_p < 1.0$
$$\begin{cases} s_{max} = 6.97 \pm 0.83 \\ \mu = 4.06 \pm 0.22 \end{cases} \tag{8}$$

This model is similar to the Mitsuyasu-type spreading, but s_{max} and the power are different and only at the high frequency side, the spreading function is dependent on the wind velocity. In Eq.(7), U/Cp should be greater than 1.0 because of their measured wave data $U/C_p = 1.0 \sim 1.8$.

4. Donelan's spreading function. Donelan et al(1985) measured the directional spectrum systematically using an array consisted of 14 wave gages at the Lake Ontario in Canada. U/Cp of the data is $0.83 \sim 4.6$. The directional spreading function was given as

$$G(f,\theta) = \frac{1}{2}\beta sech^2 \beta\theta \tag{9}$$

$$\left. \begin{array}{l} \beta = 2.61(f/f_p)^{1.3}; \quad 0.56 \leq f/f_p \leq 0.95 \\ \beta = 2.28(f/f_p)^{-1.3}; \quad 0.95 < f/f_p < 1.6 \\ \beta = 1.24 \quad\quad\quad\quad\quad others\ f/f_p \end{array} \right\} \tag{10}$$

This model does not include the parameter which expresses the wave growing. When $f/f_p = 0.95$, $\beta_{max} = 2.44$ and the spreading is the narrowest. When $f/f_p < 0.56$ and $f/f_p \geq 1.6$, the spreading is a constant.

5. Wen's directional spectrum. Wen et al obtained the directional spectrum as follows from the frequency spectrum of each wave propagating direction θ which is derived analytically.

$$\left. \begin{array}{l} \tilde{F}(\tilde{\omega},\theta) = \dfrac{k_1 k_3}{k_2} P cos^n \theta\, \tilde{\omega}_\theta^{-p_\theta} exp[-\dfrac{p_\theta}{q_\theta}(\tilde{\omega}_\theta^{-q_\theta} - 1)]; \quad \tilde{\omega} \leq \tilde{\omega}_L \\ \tilde{F}(\tilde{\omega},\theta) = \dfrac{\tilde{F}(\omega_L,\theta)}{\tilde{\omega}^4}\tilde{\omega}_L^4 ; \quad\quad\quad\quad\quad\quad\quad\quad\quad\quad\quad \tilde{\omega} > \tilde{\omega}_L \end{array} \right\} \tag{11}$$

where $\tilde{F}(\tilde{\omega},\theta) = \dfrac{\omega_p F(\omega,\theta)}{m_0}$; $\tilde{\omega}_\theta = \dfrac{1}{k_2 cos^{n_2}\theta}\tilde{\omega}$, $\tilde{\omega} = \omega/\omega_p$, $\tilde{\omega}_L = \dfrac{\omega_L}{\omega_p}$, $\omega_L = 2.38 P^{-0.406}$

$$P = 1.69 \frac{U}{C_p} = 1.69 \frac{\omega_p U}{g} \tag{12}$$

P express the peakedness of the frequency spectrum and also the wave growing. Other parameters are related to P.

The selection of the spreading model

Figure 4 is an example of the measured directional spectrum and directional spreading function. It is shown that the main directions of the component waves corresponding to different frequency are close to each other. The wave energy are distributed in a narrow direction range and most directional spreading functions are symmetric approximately. The spreading is the narrowest near the peak frequency. These are basically the same as all above models.

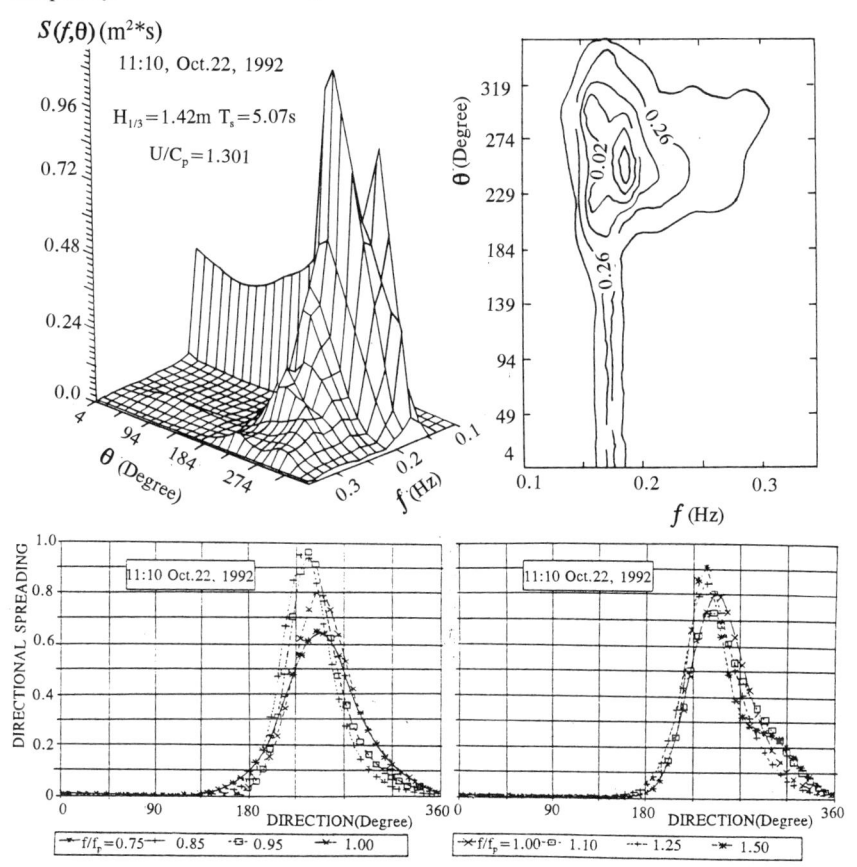

Fig.4 Example of the measured directional spectrum

To check the adaptability of 5 models to the measured results, these results are divided into 2 groups according to that $U/C_p \geq 1.0$ or $U/C_p < 1.0$. The measured results are compared with above 5 models according to the directional spreading curves, the directional cumulative spreading curves(refer to Eqs(13)~(15)), the maximum of the directional spreading function and the standard deviation of the spreading respectively. The results show that the Donelan's model conforms to the measured spreading best and the Wen's model takes the second place, but Donelan's model can not consider the effect of the wave growing and Wen's model is too complex. On the other hand, in Mitsuyasu model, the spreading parameter s is dependent on the condition of wave growing and this model has been commonly used in the recent years. So the Mitsuyasu model is used to fit the measured spreading and the Donelan's model will be modified in this paper.

Fitting the measured spreading with Mitsuyasu's Model

Let the cumulative curve of the directional spreading, $G(f,\theta)$ be

$$F(f,\theta) = \int_{-\pi}^{\theta} G(f,\theta)d\theta \qquad (13)$$

then the deviations of two directional spreading functions G_X and G_Y, and their cumulative curves are as followings respectively

$$\Delta G_{xy}(f) = \{\int_{-\pi}^{\pi} [G_x(f,\theta) - G_y(f,\theta)]^2 d\theta\}^{1/2} \qquad (14)$$

$$\Delta F_{xy}(f) = \{\int_{-\pi}^{\pi} [F_x(f,\theta) - F_y(f,\theta)]^2 d\theta\}^{1/2} \qquad (15)$$

When the Mitsuyasu's model X is used to fit the measured spreading, Y, let $\Delta G_{xy}(f)$ and $\Delta F_{xy}(f)$ take the minimum, the parameter s for a given frequency can be calculated and the values of s are varied with f/f_p as shown in Fig.5. In this figure, the Mitsuyasu's spreading of $U/C_p = 0.7$, 1.0 and 1.3 are also given. Figure 5 shows that when $f/fp > 1.0$, the power -2.5 basically conforms to the average measured condition, but when $f/f_p \leq 1.0$ the power 5 is obviously too large and 2.5 may be suitable. s takes the maximum value at $f/f_p = 1.0$. So the parameter s can be determined by

$$\left. \begin{array}{l} s = s_{max}(f/f_p)^{2.5}, \quad f \leq f_p \\ s = s_{max}(f/f_p)^{-2.5}, \quad f > f_p \end{array} \right\} \qquad (16)$$

Concerning s_{max}, the measured results show that s_{max} is tending to decrease with U/C_p increasing, but their relationship is not apparent. The growing state of the stable wind-generated waves can be described by U/C_p, but in fact the wind velocity and wind direction are usually changeable and the wave growth much lags

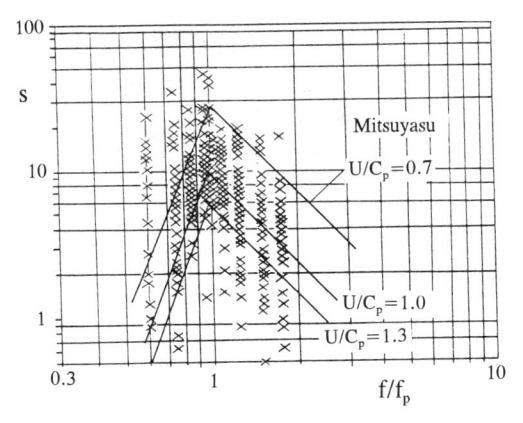

Fig.5 Relation between s and f/f_p

behind the wind changing. Furthermore, it is very difficult to determine the wind velocity corresponding to the design waves for engineering design. So Goda(1985) proposed that $s_{max}=10$, 25 and 75 for wind-waves, swell with short decay and swell with long decay distance respectively, but no observation data yet certify it. In another respect, Sverdrup and Munk had obtained the famous relationship between wave steepness (H/L) and wave age (C/U). When C/U>0.4, wave steepness decreases with wave age increasing. The similar results are obtained by this measurement(Fig.6). So the wave steepness is used to express the growing state of the waves, it is convenient for engineering application, and s_{max} can be determined as follows(Fig.7).

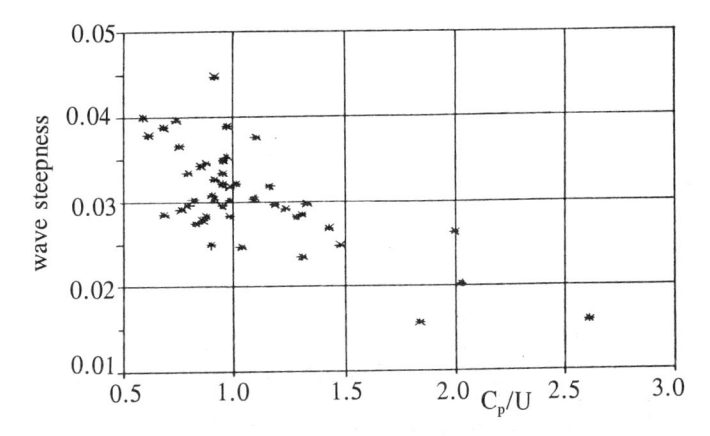

Fig.6 Relation between the significant wave steepness and C_p/U

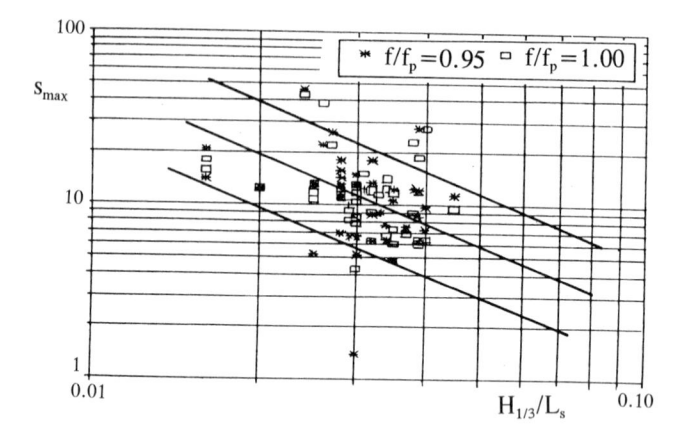

Fig.7 Relation between s_{max} and $H_{1/3}/L_s$

$$s_{max} = 0.13(H_{1/3}/L_S)^{-1.28} \quad (Mean)$$
$$s_{max} = 0.26(H_{1/3}/L_S)^{-1.28} \quad (Upper) \qquad (17)$$
$$s_{max} = 0.065(H_{1/3}/L_S)^{-1.28} \quad (Lowwer)$$

The spreading functions with s calculated by Eqs(16) and (17) are compared with the measured. It is shown that they are close each other.

Wang(1992) obtained similar results from the wave data in severe seas during storms that occurred offshore California.

$$s_{max} = 0.11(H_S/L_P)^{-1.28} \quad (Mean)$$
$$s_{max} = 0.20(H_S/L_P)^{-1.28} \quad (Upper) \qquad (18)$$
$$H_S = 4.0\sqrt{m_0}$$

Fitting with Donelan model

Similar to the above, let deviations $\Delta G_{xy}(f)$ and $\Delta F_{xy}(f)$ between measured spreading, X and that of Donelan's model, Y take the minimum, the variation of β with f/f_p is shown in Fig.8. Also the relationship between β in Eq.(9) and s in Eq.(4) is shown in Fig.9 Figure 8 shows that Donelan model is basically identical to the measured. But the value of β in Eq.(9) is in the large side. Moreover, for considering the effect of wave growing, β can be determined by Eq.(17) and Fig.9.

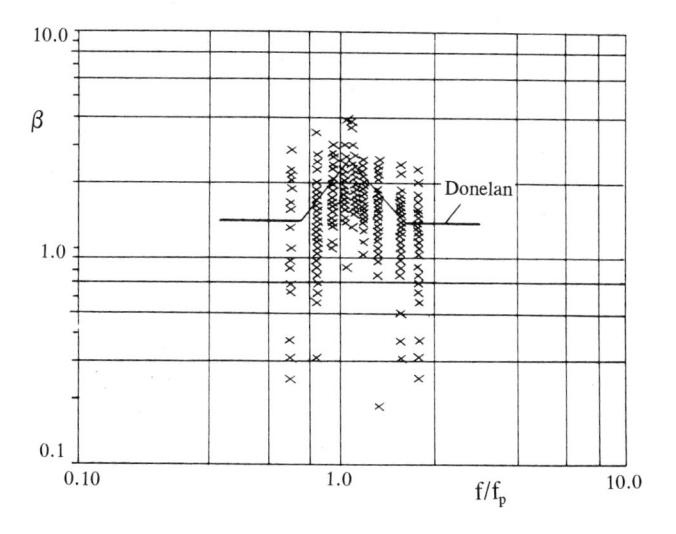

Fig.8 Relation between β and f/f_p

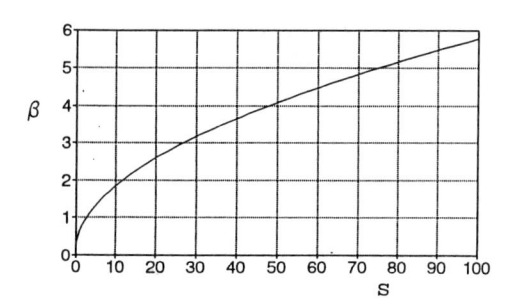

Fig.9 Relation between β and s

Conclusion

1. The observation and estimation methods of directional spectrum are investigated in this paper by combining the numerical simulation and the laboratory experiments with the field observations. The results show that a reliable directional spreading can be obtained using a 5-gage array to measure the waves and the Bayesian approach for the directional spectrum analysis. This method is more suitable for measuring waves in laboratory or on a field platform.

2. The results of the spectral analysis show that the power of the equilibrium range of frequency spectra can be selected as $p = -4.125$. The frequency spectrum can be expressed as Eq.(1).

3. The wave steepness is used to describe the growing state of wind-waves and the relationship between the directional spreading parameter s_{max} and the wave steepness is given in this paper. It is convenient for engineering applications.

4. Using Mitsuyasu model to fit the measured results, the directional spreading function in the Bohai Sea is obtained preliminarily as Eqs(4), (16) and (17). The variation of β in Donelan model with wave steepness is also given in this paper.

5. The process of sea wave growing is very complex and the measured wave data are affected by many factors. In this paper, even if only 51 runs of wave data which generated by a single wind in the center part of the Bohai Sea are used, the datum points obtained from analysis are rather scattered. For obtaining the directional spectrum at coastal and offshore area of China, much more observations with gage array or other methods whose directional resolution is high are needed.

Acknowledgments---The work described herein is a part of the project supported by both National Natural Science Foundation of China and the Ministry of Communications. The observations of sea waves were carried out by Mr. X. Yu, J. Xie and B. Fan, First Institute of Oceanography. The authors sincerely wish to express their appreciation to Mrs. X. Wu, Mr. N. Chu, G. Lu, G. Li and the people worked on the platform Bohai 8 for their great helps to this study.

References

Barber, N.F., 1954, Finding the direction of travel of sea waves, Nature, Vol.174, 1048~1050.
Donelan M. A., et al, 1985, Directional spectra of wind-generated waves, Phil. Trans. R. Soc. Lond, A315, 509~562.
Goda Y., 1985, Random seas and design of maritime structures, University of Tokyo Press.
Hashimoto N. et al., 1987, Estimation of directional spectrum using the Bayesian approach and its application to field data analysis, Report of the Port and Habour Research Institute, Japan, Vol.12, 58~100.
Hasselmann K et al, 1980, Directional wave spectra observed during JONSWAP 1973, J. Phys. Oceanogr., 10(8), 1264~1280.
Hong Guangwen and Yang Zhengyi, 1980, The wind-wave frequency spectrum and its statistical properties in shallow water, J. of Nanjing Hydraulics Research Institute(in Chinese), No.2, 68~86.
Longuet-Higgins, M.S., D.E. Cartwright and N. D. Smith, 1963, Observation of directional spectrum of sea waves using the motion of floating buoy, Proc. Conf. Ocean Wave Spectra, Prentice-Hall Inc., 111~132.
Mitsuyasu, H. et al, 1975, Observations of directional spectrum of ocean wave using a cloverleaf buoy, J. Phys. Oceanogr., 5(4), 750~760.
Nagata, Y., 1964, The statistical properties of orbital wave motions and their

application for the measurement of directional wave spectra, J. of the Oceanographic Society on Japan, 19(4), 169~191.

Wang David Wei-Chi, 1992, Estimation of wave directional spreading in severe seas, Proc. Second Int. Offshore and Polar Engineering Conf., Vol.3, 146~153.

Wen, Shengcheng et al., 1989, Improved form of wind-wave frequency spectrum, Acta Oceanologica Sinica, 8, 467~483.

Yu Yuxiu, S. X. Liu and L. Li, 1991, Numerical simulation of multi-directional random waves, Proc. 1st Int. Offshore and Polar Engineering Conf., Edinburgh, United Kingdom, 26~32.

Yu Yuxiu and S. X. Liu, 1992, Study on measurement methods of multi-directional waves by wave gage array, Proc. 2nd Int. Offshore and Polar Engineering Conf., San Francisco, U.S.A, 179~185.

Yu Yuxiu and S. X. Liu, 1993, The Analyzing methods of directional spectrum, Proc. 7th China Coastal Engineering Conf.(in Chinese).

CHAPTER 55

IRREGULAR WAVES OVER AN ELLIPTIC SHOAL

Xiping Yu[1] *and* Hiroyoshi Togashi[2]

ABSTRACT

A numerical model for the transformation of narrow-banded irregular waves over gradually varying bottom topography is presented. The model is based on the mild slope wave equation for component waves. Perturbation of the mild slope wave equation with respect to the deviation of the angular frequency of any component wave from that of a principal wave, which is a small quantity for waves of narrow-banded spectra, is carried out. The mild slope wave equation, which depends on the frequency of the component wave, can thus be replaced by the perturbation equations in terms of the principal wave parameters. The finite element method is considered for numerical solutions of the perturbation equations. Since the matrix of the linear algebraic finite element equations depends on neither the component wave properties nor the order of the perturbation, numerical solution of an irregular wave field can be efficiently obtained. The model is applied to the computation of the wave motion over an elliptic shoal. The computed wave height distribution shows satisfactory agreement with the available experimental data.

INTRODUCTION

The mild slope wave equation (Berkhoff, 1972) has been established as an effective model for describing the combined refraction and diffraction of small amplitude waves in nearshore zone. In spite of its widely recognized validity,

---

[1]Assoc. Prof., Dept. Civ. Eng., Nagasaki Univ., Bunkyo-Machi 1-14, Nagasaki City, Nagasaki 852, Japan
[2]Prof., ditto

application of this model in practice has, however, not always been easy. One of the reasons is the considerable computational efforts necessitated to solve the elliptic partial differential equation if the domain of interest has a dimension of over several wavelengths. The situation becomes even more critical if the wave irregularity is assumed to be of primary importance and, therefore, a large number of component waves must be dealt with independently. The research efforts on developing effective numerical methods so that the mild slope wave equation can be applied to the problems with large domain have led to many distinguishable achievements in the past decade. On improving the approach to the wave irregularity, however, less progress has been made.

Real sea waves are always irregular. However, most engineering analyses of the nearshore wave motion had long been based on representation of the real sea by monochromatic waves, usually the significant waves. The inaccuracy of such representation has been pointed out by many researchers who compared the results by the representative wave method with those by other more accurate methods (see, *e.g.*, Goda, 1985). Since the wave transformation processes are always frequency dependent, it can not be expected that the significant wave parameters of an irregular wave field are even close to the wave parameters following the transformation of the significant wave. This is particularly true in a region where waves undergo significant refraction and diffraction.

The most direct approach to the wave irregularity may be the superposition method. This classical method is based on decomposing irregular waves into monochromatic components with different frequencies. By applying a regular wave theory to each of these component waves and reassembling the solutions, the irregular wave field can be computed. As long as the wave is of small amplitude, or, is linear, the superposition method is authoritative. However, a large number of component waves must be considered to ensure the accuracy of results. Since the component waves are numerically independent, considerable computational efforts are necessary.

A rather different approach to the superposition is the energy method originally proposed by Karlsson (1969). This method is based on a governing equation in terms of the energy spectrum, which is generally expressed by the product of a wave height related parameter with a distribution function describing the spread of the wave energy with respect to the frequency and the directional angle. Once the distribution function is assumed to be invariant or to be in a known form in the domain of interest, the wave height can be accordingly solved from the governing energy equation. This method is direct but can not be widely applied because the spectrum is in fact part of the solution of an irregular wave field and it is not always appropriate to assume its invariance or foreseeability. In particular, if there is a permeable breakwater in the domain of interest, because of the frequency selectivity of the breakwater to reflection and transmission, the wave spectrum, at least, in the vicinity of the breakwater may undergo significant transformation. Any presumption on its form under this circumstance may lead to mistakes.

The present study is to provide a rather different approach to the irregular wave motion in nearshore zone. The method is essentially superposition but the computational effort involved is equivalent to that required by the representative wave method. In the following sections, we first describe the governing equation for the component wave and perturb the equation with respect to the deviation of the angular frequency of the component wave to that of the principal wave. Then, we illustrate the finite element method for solutions of the perturbation equations. Finally, we apply our numerical model to the wave motion over an elliptic shoal and compare the computational results with experimental data.

THEORY

The Basic Equation

For the component wave, with a small amplitude and an arbitrary angular frequency, over a gradually varying bottom topography, we employ the mild slope wave equation to describe its motion. Denote the water surface elevation caused by the wave motion by $\tilde{\eta} = \eta(x,y)e^{-i\sigma t}$, where σ is the angular frequency and $\eta(x,y)$ is called the complex amplitude of the component wave (the modulus of η denotes the usual wave amplitude and the argument of η represents the relative phase of the water surface elevation). The governing equation for η can then be written as (Berkhoff, 1972)

$$\nabla \cdot (CC_g \nabla \eta) + k^2 CC_g \eta = 0 \qquad (1)$$

where ∇ is the horizontal gradient operator, k the wavenumber, C the wave celerity and C_g the group velocity. Eq. (1) is for waves without any dissipation. If the dissipation effect is not negligible, we may have to introduce a factor $\mu = 1 + i\xi$ in the equation so that

$$\nabla \cdot (CC_g \nabla \eta) + \mu^2 k^2 CC_g \eta = 0 \qquad (2)$$

Eq. (2) is slightly different from a previous equation proposed by Dalrymple et al. (1984). It is expected that the parameter ξ, that is, the imaginary part of the factor μ, can be more closely related to the conventional energy decaying factor

Φ_D (see, e.g., Horikawa, 1989) so that the dissipation effect can be readily evaluated. The relation between ξ and Φ_D is clear if we consider a progressive wave in the positive x direction over a constant water depth. Under this circumstance, the energy decaying factor Φ_D is

$$\Phi_D = -\frac{1}{EC_g}\frac{d(EC_g)}{dx} = -\frac{1}{E}\frac{dE}{dx} \tag{3}$$

since C_g, which depends on the wave frequency and the water depth, is constant. E in (3) is the average wave energy, which can be expressed by

$$E = \frac{1}{8}\rho g H^2 \tag{4}$$

for small amplitude waves, where H is the wave height, ρ the fluid density and g the gravitational acceleration. Inserting (4) into (3) gives

$$\Phi_D = -\frac{2}{H}\frac{dH}{dx} \tag{5}$$

On the other hand, for a unidirectional wave, Eq. (2) reduces to

$$\frac{d^2\eta}{dx^2} + \mu^2 k^2 \eta = 0 \tag{6}$$

Eq. (6) has two independent solutions, representing the progressive decaying waves in the positive and negative x directions, respectively. The wave in the positive x direction can be expressed by

$$\eta = \eta|_{x_0} e^{-\xi k(x-x_0)} e^{ik(x-x_0)} \tag{7}$$

where x_0 denotes a reference point. Eq. (7) gives the variation of the wave height H as follows:

$$H = H|_{x_0} e^{-\xi k(x-x_0)} \tag{8}$$

Inserting (8) into (5), we readily obtain

$$\xi = \frac{1}{2k}\Phi_D \tag{9}$$

Perturbation of the Basic Equation

We introduce a principal angular frequency and denote it by $\bar{\sigma}$. $\bar{\sigma}$ may, but not necessarily, be defined as the peak angular frequency of the incident wave spectrum. With the principal angular frequency, the angular frequency σ of any component wave can be expressed as

$$\sigma = \bar{\sigma}(1+\epsilon) \tag{10}$$

where $\epsilon = (\sigma - \bar{\sigma})/\bar{\sigma}$. As long as we consider only narrow-banded waves, ϵ is a small quantity and all the frequency dependent variables may then be expanded into power series of ϵ at their principal values. In particular, for the surface elevation η, the wavenumber k and the product of the wave celerity C with the group velocity C_g, we have the following perturbation expressions:

$$\eta = \eta^{(0)} + \epsilon \eta^{(1)} + \epsilon^2 \eta^{(2)} + \cdots \tag{11}$$

$$k = \bar{k}(1 + \epsilon \alpha^{(1)} + \epsilon^2 \alpha^{(2)} + \cdots) \tag{12}$$

$$CC_g = \bar{C}\bar{C}_g(1 + \epsilon \beta^{(1)} + \epsilon^2 \beta^{(2)} + \cdots) \tag{13}$$

where

$$\alpha^{(1)} = \frac{\bar{\sigma}}{\bar{k}}\overline{\frac{dk}{d\sigma}} = \frac{1}{\bar{n}} \tag{14}$$

$$\alpha^{(2)} = \frac{1}{2}\frac{\bar{\sigma}^2}{\bar{k}}\overline{\frac{d^2k}{d\sigma^2}} = -\frac{1}{2\bar{n}^3}[(2\bar{n}-1)\bar{m} - \bar{n}^2] \tag{15}$$

$$\beta^{(1)} = \frac{\bar{\sigma}}{\bar{C}\bar{C}_g}\overline{\frac{dCC_g}{d\sigma}} = \frac{1}{\bar{n}^2}[(2\bar{n}-1)\bar{m} - \bar{n}] \tag{16}$$

$$\beta^{(2)} = \frac{1}{2}\frac{\bar{\sigma}^2}{\bar{C}\bar{C}_g}\overline{\frac{d^2CC_g}{d\sigma^2}}$$
$$= \frac{1}{2\bar{n}^4}[(2\bar{n}-1)\bar{m}^2 + 3\bar{n}(\bar{n}-1)(2\bar{n}-1)\bar{m} - \bar{n}^2(5\bar{n}-4)] \tag{17}$$

and

$$\bar{n} = \frac{1}{2}\left(1 + \frac{2\bar{k}h}{\sinh 2\bar{k}h}\right) \tag{18}$$

$$\bar{m} = \bar{n} + \frac{1}{2}\left(1 - \frac{2\bar{k}h}{\tanh 2\bar{k}h}\right) \tag{19}$$

The bars are used to denote principal values. It is obvious that $\alpha^{(1)}$, $\alpha^{(2)}$, $\beta^{(1)}$ and $\beta^{(2)}$ are all single functions of $\bar{k}h$ and, consequently, the principal wave effect parameter $\bar{\sigma}^2 h/g$ if the following dispersion relation for the principal wave is taken into consideration:

$$\frac{\bar{\sigma}^2 h}{g} = \bar{k}h \tanh \bar{k}h \tag{20}$$

IRREGULAR WAVES OVER ELLIPTIC SHOAL

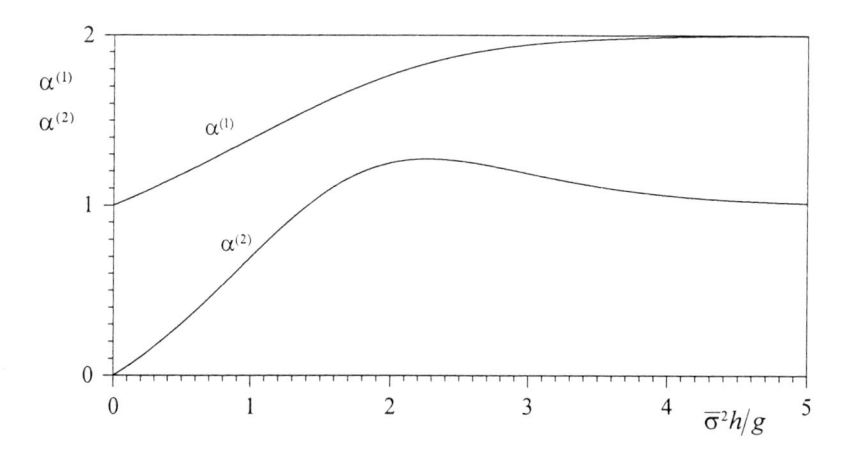

Fig. 1: Variation of $\alpha^{(1)}$ and $\alpha^{(2)}$ versus $\bar{\sigma}^2 h/g$

In **Figs. 1** and **2** we show these functional relations for $0 \leq \bar{\sigma}^2 h/g \leq 5$. It can be noted that when the principal wave tends to be a long wave, that is, $\bar{\sigma}^2 h/g$ tends to zero, $\alpha^{(1)}$ tends to 1 while $\alpha^{(2)}$, $\beta^{(1)}$ and $\beta^{(2)}$ all tend to zero. On the other hand, when the principal wave becomes a deep water wave, or, when $\bar{\sigma}^2 h/g$ tends to infinity, $\alpha^{(1)}$ tends to 2, $\alpha^{(2)}$ tends to 1, $\beta^{(1)}$ tends to -2 and $\beta^{(2)}$ tends to 3.

Substituting (11), (12) and (13) into the mild slope wave equation (2) and collecting all the terms for each order of ϵ, we obtain

$$\nabla \cdot (\bar{C}\bar{C}_g \nabla \eta^{(0)}) + \mu^2 \bar{k}^2 \bar{C}\bar{C}_g \eta^{(0)} = 0 \tag{21}$$

$$\nabla \cdot \left[\bar{C}\bar{C}_g \nabla \eta^{(1)} + \beta^{(1)} \bar{C}\bar{C}_g \nabla \eta^{(0)}\right] + \mu^2 \bar{k}^2 \bar{C}\bar{C}_g \left[\eta^{(1)} + \beta^{(1)} \eta^{(0)} + 2\alpha^{(1)} \eta^{(0)}\right] = 0 \tag{22}$$

$$\nabla \cdot \left[\bar{C}\bar{C}_g \nabla \eta^{(2)} + \beta^{(1)} \bar{C}\bar{C}_g \nabla \eta^{(1)} + \beta^{(2)} \bar{C}\bar{C}_g \nabla \eta^{(0)}\right] + \mu^2 \bar{k}^2 \bar{C}\bar{C}_g \left[\eta^{(2)} + \beta^{(1)} \eta^{(1)} \right.$$
$$\left. + 2\alpha^{(1)} \eta^{(1)} + \beta^{(2)} \eta^{(0)} + 2\beta^{(1)} \alpha^{(1)} \eta^{(0)} + \alpha^{(1)^2} \eta^{(0)} + 2\alpha^{(2)} \eta^{(0)}\right] = 0 \tag{23}$$

......

where μ is treated as frequency independent. By considering (21) in (22) and, (21) and (22) in (23), Eqs. (22) and (23) may be simplified to give

$$\nabla \cdot (\bar{C}\bar{C}_g \nabla \eta^{(1)}) + \mu^2 \bar{k}^2 \bar{C}\bar{C}_g \eta^{(1)} + \bar{C}\bar{C}_g \nabla \beta^{(1)} \cdot \nabla \eta^{(0)} + 2\mu^2 \bar{k}^2 \bar{C}\bar{C}_g \alpha^{(1)} \eta^{(0)} = 0 \tag{24}$$

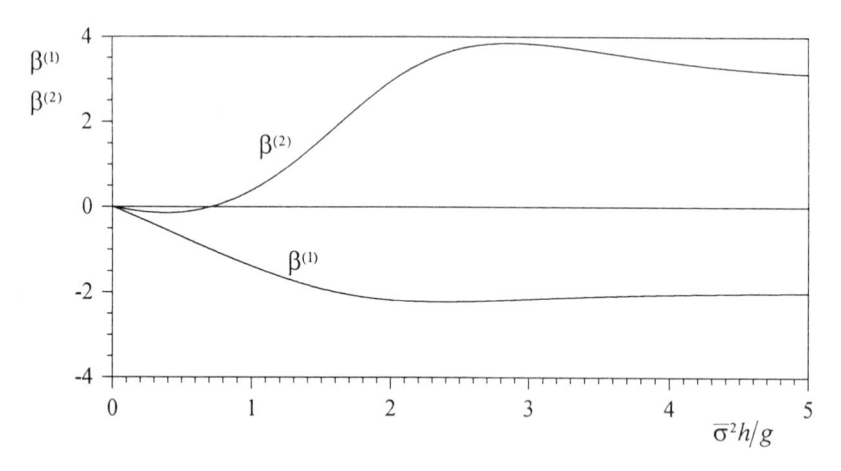

Fig. 2: Variation of $\beta^{(1)}$ and $\beta^{(2)}$ versus $\bar{\sigma}^2 h/g$

$$\nabla \cdot (\bar{C}\bar{C}_g \nabla \eta^{(2)}) + \mu^2 \bar{k}^2 \bar{C}\bar{C}_g \eta^{(2)} + \bar{C}\bar{C}_g \left[\nabla \beta^{(1)} \cdot \nabla \eta^{(1)} - \beta^{(1)} \nabla \beta^{(1)} \cdot \nabla \eta^{(0)}\right.$$
$$\left. + \nabla \beta^{(2)} \cdot \nabla \eta^{(0)}\right] + 2\mu^2 \bar{k}^2 \bar{C}\bar{C}_g \left[2\alpha^{(1)} \eta^{(1)} + \alpha^{(1)} \alpha^{(1)} \eta^{(0)} + 2\alpha^{(2)} \eta^{(0)}\right] = 0 \quad (25)$$

Eqs. (21), (24) and (25) may be used to solve $\eta^{(0)}$, $\eta^{(1)}$, $\eta^{(2)}$ and, therefore, η approximately. It is obvious that the zeroth order equation (21) describes the motion of the principal wave. This implies that the representative wave method is the leading order approximation of the present approach. Eqs. (24) and (25) govern the higher order modifications. The higher order equations all include a source term which depends on the lower order solutions. We also note that the equations for all orders can be expressed in a unified form as follows:

$$\nabla \cdot (\bar{C}\bar{C}_g \nabla \eta^{(m)}) + \mu^2 \bar{k}^2 \bar{C}\bar{C}_g \eta^{(m)} + q^{(m)} = 0 \quad (m = 0, 1, 2, \cdots) \quad (26)$$

where

$$q^{(0)} = 0 \quad (27)$$

$$q^{(1)} = \bar{C}\bar{C}_g \nabla \beta^{(1)} \cdot \nabla \eta^{(0)} + 2\mu^2 \bar{k}^2 \bar{C}\bar{C}_g \alpha^{(1)} \eta^{(0)} \quad (28)$$

$$q^{(2)} = \bar{C}\bar{C}_g \left[\nabla \beta^{(1)} \cdot \nabla \eta^{(1)} - \beta^{(1)} \nabla \beta^{(1)} \cdot \nabla \eta^{(0)} + \nabla \beta^{(2)} \cdot \nabla \eta^{(0)}\right]$$
$$+ 2\mu^2 \bar{k}^2 \bar{C}\bar{C}_g \left[2\alpha^{(1)} \eta^{(1)} + \alpha^{(1)} \alpha^{(1)} \eta^{(0)} + 2\alpha^{(2)} \eta^{(0)}\right] \quad (29)$$

......

Boundary Conditions

We consider two kinds of boundaries. The first kind is where the free surface oscillation is given by

$$\eta = ae^{i\theta} \tag{30}$$

where a is the amplitude and θ the phase. For each order, we require

$$\eta^{(0)} = ae^{i\theta} \tag{31}$$

$$\eta^{(m)} = 0 \quad (m \geq 1) \tag{32}$$

The second kind of boundary condition is assumed to be expressed in the following form:

$$\frac{\partial \eta}{\partial n} - i\lambda k\eta + i\nu k = 0 \tag{33}$$

where λ and ν are constants related to the physical situation (Yu et al., 1992). In particular, an impermeable boundary is represented by $\lambda = 0$ and $\nu = 0$.

Substituting (11) and (12) into (33) and collecting all the terms for each order of ϵ, we obtain

$$\frac{\partial \eta^{(0)}}{\partial n} - i\lambda \bar{k}\eta^{(0)} + i\nu \bar{k} = 0 \tag{34}$$

$$\frac{\partial \eta^{(1)}}{\partial n} - i\lambda \bar{k}\eta^{(1)} - i\lambda \alpha^{(1)}\bar{k}\eta^{(0)} + i\nu \alpha^{(1)}\bar{k} = 0 \tag{35}$$

$$\frac{\partial \eta^{(2)}}{\partial n} - i\lambda \bar{k}\eta^{(2)} - i\lambda \bar{k}(\alpha^{(1)}\eta^{(1)} + \alpha^{(2)}\eta^{(0)}) + i\nu \alpha^{(2)}\bar{k} = 0 \tag{36}$$

. . . . . .

The above equations have the following unified form:

$$\frac{\partial \eta^{(m)}}{\partial n} - i\lambda \bar{k}\eta^{(m)} + p^{(m)} = 0 \quad (m = 0, 1, 2, \cdots) \tag{37}$$

where

$$p^{(0)} = i\nu \bar{k} \tag{38}$$

$$p^{(1)} = -i\lambda \alpha^{(1)}\bar{k}\eta^{(0)} + i\nu \alpha^{(1)}\bar{k} \tag{39}$$

$$p^{(2)} = -i\lambda \bar{k}(\alpha^{(1)}\eta^{(1)} + \alpha^{(2)}\eta^{(0)}) + i\nu \alpha^{(2)}\bar{k} = 0 \tag{40}$$

. . . . . . \hfill (41)

FINITE ELEMENT METHOD

It is known that the solution of the elliptic equation (26) when subjected to the boundary condition (37) stagnates the following functional in terms of $\eta^{(m)}$:

$$\Pi = \int_\Omega \left[\frac{1}{2}\bar{C}\bar{C}_g \nabla\eta^{(m)} \cdot \nabla\eta^{(m)} - \frac{1}{2}\mu^2\bar{k}^2\bar{C}\bar{C}_g\eta^{(m)}\eta^{(m)} - q^{(m)}\eta^{(m)}\right] d\Omega$$
$$+ \int_{\Gamma_2}\left[-\frac{1}{2}i\lambda\bar{k}\bar{C}\bar{C}_g\eta^{(m)}\eta^{(m)} + p^{(m)}\eta^{(m)}\right] d\Gamma \qquad (42)$$

where Ω is the domain of interest and Γ_2 the part of the boundary of Ω where (37) must be satisfied. For a finite element solution of $\eta^{(m)}$ which stagnates Π in some approximate sense, we discretize the domain Ω into elements and let all the elements be related to each other through the nodes located on the common boundaries of the elements. Denote

$$\Pi^e = \int_{\Omega^e} \left[\frac{1}{2}\bar{C}\bar{C}_g \nabla\eta^{(m)} \cdot \nabla\eta^{(m)} - \frac{1}{2}\mu^2\bar{k}^2\bar{C}\bar{C}_g\eta^{(m)}\eta^{(m)} - q^{(m)}\eta^{(m)}\right] d\Omega$$
$$+ \int_{\Gamma^e}\left[-\frac{1}{2}i\lambda\bar{k}\bar{C}\bar{C}_g\eta^{(m)}\eta^{(m)} + p^{(m)}\eta^{(m)}\right] d\Gamma \qquad (43)$$

where Ω^e is an element and Γ^e its boundary, $\lambda = 0$ and $\nu = 0$ if $\Gamma^e \not\subset \Gamma_2$. Eq. (42) can then be written as

$$\Pi = \sum_e \Pi^e \qquad (44)$$

We introduce the primed indices $1', 2', \cdots, N'$ in the anti-clockwise fashion in each element for locally numbering the nodes related to the element and assume the global nodal numbers of $1', 2', \cdots, N'$ to be $n^e_{1'}, n^e_{2'}, \cdots, n^e_{N'}$, respectively. As interpolation functions $L^e_{i'}(x,y)$ ($i' = 1', 2', \cdots, N'$) are defined in each element, any function $F(x,y)$ in Ω^e can be approximated in terms of its nodal values $F_{1'}$, $F_{2'}$, $\cdots$, $F_{N'}$ as

$$F(x,y) = L^e_{i'} F_{i'} \qquad (45)$$

where the summation convention is implicit. Therefore, Π^e can be partially evaluated so as to give

$$\Pi^e = \frac{1}{2}K^e_{i'j'}\eta^{(m)}_{i'}\eta^{(m)}_{j'} + f^{(m)e}_{j'}\eta^{(m)}_{j'} \qquad (46)$$

where $K^e_{i'j'}$ is a $N' \times N'$ matrix depending on the interpolation functions as well as the local features of the principal wave. With the following matrix

$$T^e_{i'i} = \begin{cases} 1 & \text{when } n_{i'} = i \\ 0 & \text{when } n_{i'} \neq i \end{cases} \quad (i' = 1', 2', \cdots, N' \text{ and } i = 1, 2, \cdots N) \qquad (47)$$

where N is the total number of nodes, we have

$$\eta_{i'}^{(m)} = T_{i'i}^e \eta_i^{(m)} \tag{48}$$

Eq. (46) can then be expressed by

$$\Pi^e = \frac{1}{2}\left(K_{i'j'}^e T_{i'i}^e T_{j'j}^e\right)\eta_i^{(m)}\eta_j^{(m)} + \left(f_{j'}^{(m)e} T_{j'j}^e\right)\eta_j^{(m)} \tag{49}$$

Therefore,

$$\Pi = \frac{1}{2}\left(\sum_e K_{i'j'}^e T_{i'i}^e T_{j'j}^e\right)\eta_i^{(m)}\eta_j^{(m)} + \left(\sum_e f_{j'}^{(m)e} T_{j'j}^e\right)\eta_j^{(m)} \tag{50}$$

$$= \frac{1}{2}K_{ij}\eta_i^{(m)}\eta_j^{(m)} + F_j^{(m)}\eta_j^{(m)} \tag{51}$$

where

$$K_{ij} = \sum_e K_{i'j'}^e T_{i'i}^e T_{j'j}^e \quad \text{and} \quad F_j^{(m)} = \sum_e f_{j'}^{(m)e} T_{j'j}^e \tag{52}$$

From the necessary condition for Π to be stagnated:

$$\frac{\delta \Pi}{\delta \eta_j^{(m)}} = 0 \tag{53}$$

we obtain the following linear algebraic equations:

$$K_{ij}\eta_i^{(m)} + F_j^{(m)} = 0 \tag{54}$$

since the matrix K_{ij} is symmetric. When modified so that the forced boundary condition is satisfied, Eq. (54) gives the nodal values of $\eta^{(m)}$. It may be noted that the matrix K_{ij} is independent of the order of perturbation. This implies that, if the LU decomposition method is utilized to solve the finite element equations, we need to carry out the decomposition only once for solving all the perturbation equations. The computational efforts involved in our numerical model are, therefore, equivalent to those required by the representative wave method.

WAVE MOTION OVER AN ELLIPTIC SHOAL

We apply our numerical model to the study of the unidirectional and narrow-banded irregular wave transformation over an elliptic shoal, a problem which has been investigated by Vincent and Briggs (1989) and by Panchang et al. (1990) with different methods. The topography and the incident wave conditions in our study are made identical to the U4 case of Vincent and Briggs (1989) and Panchang et al. (1990) so that our numerical results may be verified. The computational domain is sketched in **Fig. 3**, where the shoal is centered at $x = 0$ and $y = 0$ and the perimeter of the shoal is described by

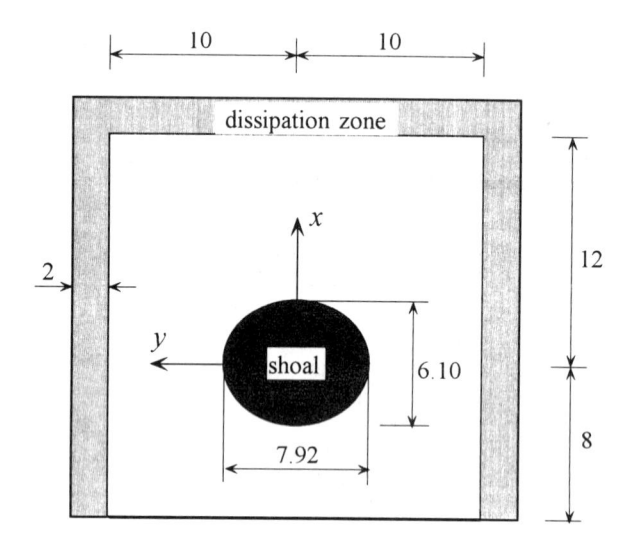

Fig. 3: The computational domain (all numbers are measured in meters)

$$\left(\frac{x}{3.05}\right)^2 + \left(\frac{y}{3.96}\right)^2 = 1 \tag{55}$$

The water depth is

$$h(x,y) = 0.9144 - 0.7620 \left\{ 1 - \left(\frac{x}{3.81}\right)^2 - \left(\frac{y}{4.95}\right)^2 \right\}^{\frac{1}{2}} \quad (\text{m}) \tag{56}$$

over the shoal and is 0.4572m in the rest of the domain.

The incident wave is assumed to have the following σ-spectrum:

$$S(\sigma) = \phi\psi g^2 \sigma \exp\left\{ -1.25\left(\frac{\bar{\sigma}}{\sigma}\right)^4 + \ln\gamma \exp\left[-\frac{(\sigma-\bar{\sigma})^2}{2\chi\bar{\sigma}^2}\right] \right\} \tag{57}$$

where S is so defined that the energy associated with the component waves of the angular frequency between σ and $\sigma + \Delta\sigma$ is $E = \rho g S(\sigma)\Delta\sigma$; the depth-effect parameter ϕ is evaluated through

$$\phi = \begin{cases} 0.5\nu^2 & \text{for } \nu < 1 \\ 1 - 0.5(2-\nu)^2 & \text{for } 1 \le \nu \le 2 \\ 1 & \text{for } \nu > 2 \end{cases} \tag{58}$$

where $\nu = \sigma(h/g)^{1/2}$; the shape factor χ is

$$\chi = \begin{cases} 0.07 & \text{for } \sigma < \bar{\sigma} \\ 0.09 & \text{for } \sigma \ge \bar{\sigma} \end{cases} \tag{59}$$

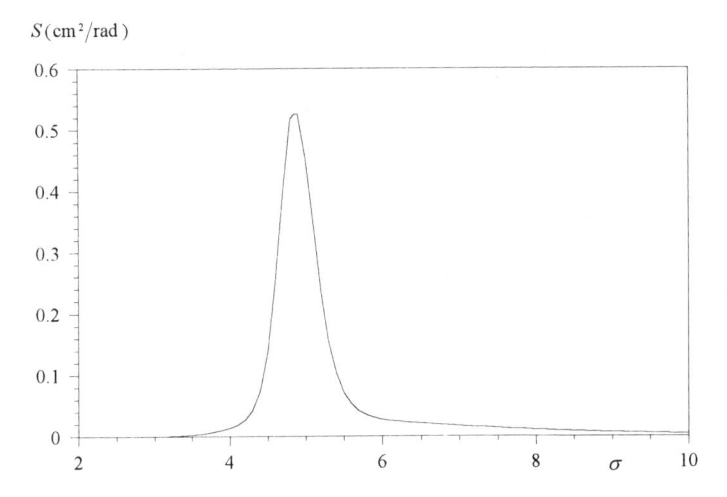

Fig. 4: The incident wave spectrum

γ is the peak enhancement factor and ψ the Phillips constant. For the present case where $\gamma = 20$, $\psi = 0.00047$ and the peak angular frequency $\bar{\sigma} = 4.833$, the spectrum is demonstrated in **Fig. 4**. As it can be noted, most of the incident wave energy is banded between $\sigma = 4$ and 6. It is then reasonable to represent the spectral incident wave by the superposition of the component waves with the following discrete angular frequencies:

$$\sigma_n = 4.0 + n\Delta\sigma \qquad (n = 0, 1, \cdots, 50) \tag{60}$$

where $\Delta\sigma = 0.04$. Following Longuet-Higgins (1957) we have

$$\tilde{\eta} = \sum_{n=0}^{50} a_n e^{i\theta_n} e^{i\sigma_n t} \tag{61}$$

where $a_n = \sqrt{2S(\sigma_n)\Delta\sigma}$ is the amplitude of the nth component wave and θ_n are random values with a uniform distribution between 0 and 2π. The forced boundary condition for the nth component wave in our problem can then be specified as

$$\eta = a_n e^{i\theta_n} \tag{62}$$

The lateral and downwave boundaries in our problem should be totally transmissive, that is, the boundary conditions should be specified so that the outgoing waves are totally absorbed by the boundaries. This requirement can be approximately met through the following numerical installation. We place a two-meter

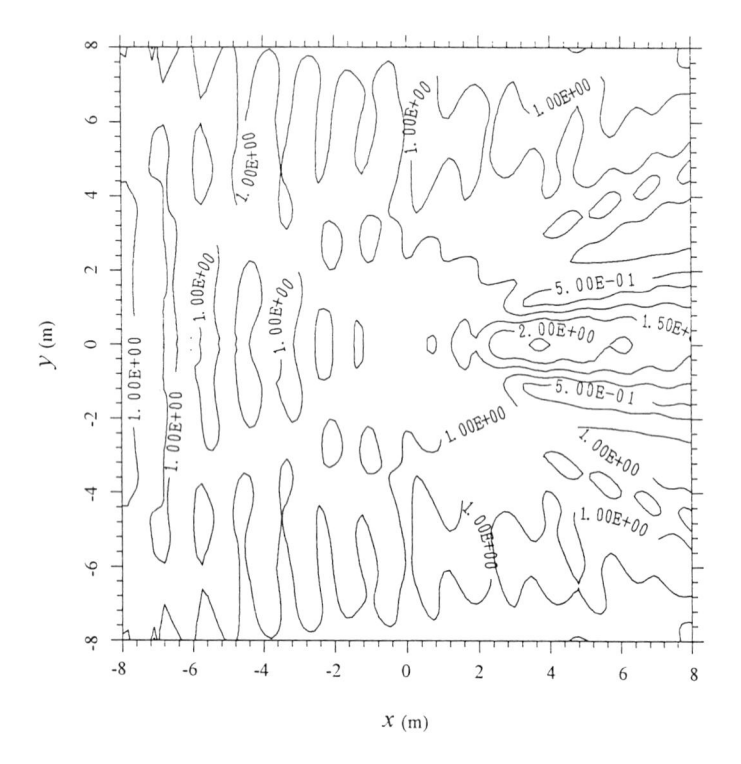

Fig. 5: Computed wave height distribution

dissipative layer with $\mu = 1 + 0.5i$ along the transmissive boundaries. The function of the dissipative layer is as same as the wave absorber in a physical model. The artificial boundaries at $y = \pm 12$m are assumed to be impermeable and, therefore, the boundary conditions there are expressed by (33) with $\lambda = 0$ and $\nu = 0$. The boundary at $x = 14$m is required to be non-reflective to the principal wave. The boundary condition can then be expressed by (33) with $\lambda = 1$ and $\nu = 0$.

In the computation, the domain is discretized into triangular elements with a dimension equivalent to one-fifteenth of the principal wavelength over the flat bottom. Linear interpolation functions are employed. **Fig. 5** shows the resulted distribution of the significant wave height (normalized by the significant incident wave height H_0). In **Fig. 6** we compare the computed wave height, at $x = 2.28m$, with the experimental data obtained by Vincent and Briggs (1989) and by Panchang et al. (1990). The agreement between the numerical solution

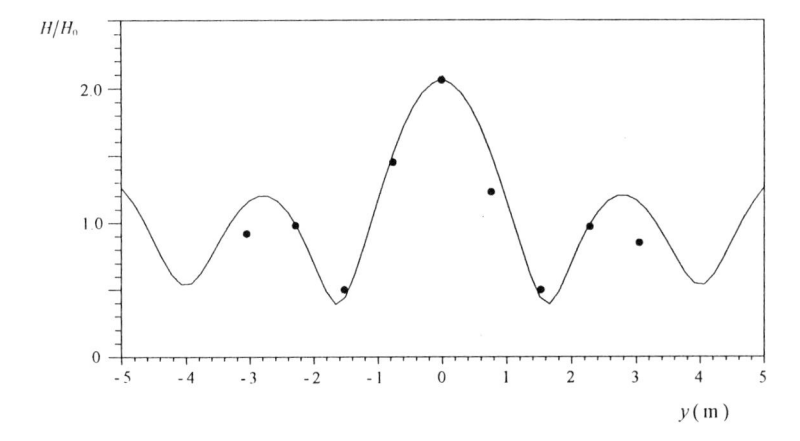

Fig. 6: Comparison of the computed wave height with experimental data

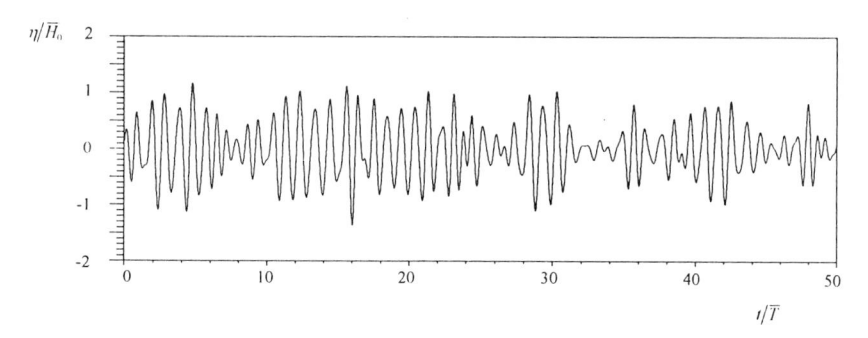

Fig. 7: Water surface elevation over the top of the shoal

and laboratory measurement is shown fairly satisfactory. In **Fig. 7** we plot the assembled irregular wave profile (normalized by the mean incident wave height $\bar{H}_0$) over the top of the shoal.

CONCLUSIONS

We presented a numerical model for the analysis of narrow-banded irregular wave transformation over gradually varying bottom topography. The model is based on the mild slope wave equation for component waves. By regular perturbation, the mild slope wave equation, which depends on the frequency of the component wave, leads to the perturbation equations in terms of the principal wave parameters. The finite element method has been suggested for numerical solutions of the perturbation equations. Since the matrix of the linear algebraic finite element equations depends on neither the component wave properties nor the order of the perturbation, the computational efforts involved in the present model is equivalent to those required by the representative wave method. The model has been applied to the computation of the wave motion over an elliptic shoal. Satisfactory agreement between the numerical solution and experimental data has been obtained.

REFERENCES

Berkhoff, J. W. C. (1972). Computation of combined refraction-diffraction. *Proc. 13th Conf. on Coast. Engrg.*, 471–490.

Dalrymple, R. A., Kirby, J. T. and Hwang, P. A. (1984). Wave diffraction due to areas of energy dissipation, *J. Wtrwy., Port, Coast. and Oc. Engrg.*, ASCE, 110(1), 67–79.

Goda, Y. (1985). *Random Seas and Design of Maritime Structures.* University of Tokyo Press, 323p.

Horikawa, K. (1989). *Nearshore Dynamics and Coastal Processes—Theory, Measurement and Predictive Model*, University of Tokyo Press, 522p.

Karlsson, T. (1969). Refraction of continuous ocean wave spectra, *Proc. Amer. Soc. Civ. Engrs.*, 95(WW4), 437-448.

Longuet-Higgins, M. S. (1957). The statistical analysis of a random, moving surface, *Phil. Trans. Roy. Soc. London*, 249(A), 321-387.

Panchang V. G., Wei, G., Pearce, B. R. and Briggs, M. J. (1990). Numerical simulation of irregular waves propagation over shoal, *J. Wtrwy., Port, Coast. and Oc. Engrg.*, ASCE, 116(3), 324–340.

Vincent, C. L. and Briggs, M. J. (1989). Refraction and diffraction of irregular waves over a mound. *J. Wtrwy., Port, Coast. and Oc. Engrg.*, ASCE, 115(2), 269–284.

Yu, X., Isobe, M. and Watanabe, A. (1992). Finite element solution of wave field around structures in nearshore zone, *Cost. Engrg. in Japan*, JSCE, 35(1), 21-33.

CHAPTER 56

PERFORMANCE OF A SPECTRAL WIND-WAVE MODEL IN SHALLOW WATER

Gerbrant Ph. van Vledder [1], John G. de Ronde [2] and Marcel J.F. Stive [1,3]

Abstract

A full spectral third-generation wave prediction model has been extended with formulations for surf breaking and nonlinear triad interactions and a first assessment of its performance against shallow water wave data has been examined. The formulation for wave energy loss by surf-breaking in shallow water is based on the expression of Battjes and Janssen (1978), which has heuristically been modified to predict the energy loss per spectral component. The source term for nonlinear triad interactions was taken from Abreu et al. (1992). Results of the extended model have been compared against laboratory and field data. The results of the model computations indicate that surf breaking and triad interactions are important processes in the coastal zone. Surf breaking is mainly responsible for the decay of wave energy, whereas triad interactions are mainly responsible for changes in the mean wave period. The applicability of the Abreu formulation is limited and needs further attention.

Introduction

The modelling of wind waves in shallow water is important for many coastal engineering applications in the nearshore zone. Especially the prediction of both the significant wave height and the mean wave period is still difficult with the presently available wave process formulations. A major problem is that most of the interactions between waves, bottom and currents are nonlinear and poorly understood. This is particularly true for directionally spread random waves in areas with a varying bottom topography.

---

1 Delft Hydraulics, PO Box 152, 8300 AD Emmeloord, The Netherlands.
2 National Institute for Coastal and Marine Management, PO Box 20907, 2500 EX, The Hague, The Netherlands.
3 Netherlands Centre for Coastal Research, Delft University of Technology, Faculty of Civil Engineering, PO Box 5048, 2600 GA, Delft, The Netherlands.

A number of model classes exists to compute wave conditions in the coastal zone (cf. Hamm et al., 1993). Commonly used models are the spectral and probabilistic models because they are relatively efficient to use, partly because they neglect diffraction effects. Such a model is also the topic of this paper. More advanced models solve Boussinesq type equations in the time domain. An advantage of such models is that they are able to model processes in a more attractive, more physical way, and also the interactions between the different processes. A disadvantage, however, is that they are rather time-consuming in comparison with spectral models.

Depending on the dimensionless water depth kd, different physical effects are important. In deep water ($kd \gg 1$) the waves are mainly influenced by three physical processes: wave growth by wind, dissipation of energy by white-capping and non-linear quadruplet wave-wave interactions. In water of intermediate depth ($kd \approx 1$) additional effects become important such as bottom friction and depth- and current refraction. In shallow water ($kd \ll 1$) also the effects of surf breaking, triad interactions and the effect of waves on currents become noticeable.

The concept of modelling the wave field in terms of the wave spectrum was introduced by Gelci et al. (1957). Since then, many spectral wave models have been developed which are usually classified in terms of their generation, which has mainly do to with the treatment of the nonlinear quadruplet wave-wave interactions and the degrees of freedom of the spectral representation of the wave field.

The first generation of spectral wave models described the evolution of the wave field in terms of parameterized spectra using simple rules, with that implicitly incorporating the effects of nonlinear wave-wave interactions. Spectral models of the second generation incorporated some effects of nonlinear interactions, but they still put limitations to the spectral shape. Only by the development of the discrete interaction approximation for nonlinear quadruplet interactions (Hasselmann et al., 1985) it became possible to develop models of the third generation. Such models explicitly compute all physical effects and they do not impose limitations to the spectral shape. The first full spectral wave model has been developed by the WAM group (WAMDI, 1988). The concept of the WAM model has been extended by Tolman (1991) to account for the effect of instationary current and water level fields. The WAM model can be applied in areas with deep or intermediate water depths, but not in shallow water because it lacks descriptions for typical shallow water processes such as surf breaking and nonlinear triad interactions.

A second generation model for shallow water was described by Holthuijsen et al. (1989). This model, which is parametric in frequency and discrete in direction, includes a formulation for surf breaking but not one for triad interactions. The formulation for wave breaking has been verified against laboratory experiments by Dingemans et al. (1986). Since this model uses one characteristic frequency per direction sector it is not always able to predict the change of the mean wave period in areas with a varying bottom topography. The only way to properly predict the change of the mean wave period with spectral models is by using a full spectral third-generation wave model which includes all physical processes affecting the waves in shallow water.

The above physical processes have various effects on the mean wave period in shallow water. The main effect of bottom friction is that it reduces wave energy in the lower frequencies and they will decrease the mean wave period, whereas quadruplet wave-wave interactions increase the mean wave period (cf. Young and van Vledder, 1993). Little is known about the spectral modelling of source terms for energy dissipation in shallow water due to breaking waves. The well-known Battjes-Janssen (1978) model predicts the rate of change of the total amount of wave energy, but no information is given about its spectral distribution. Recent experiments by Beji and Battjes (1993), however, indicate that the wave breaking process does not change the shape of the spectrum. Instead, nonlinear triad wave-wave interactions change the spectral shape by the generation of both lower and higher harmonic components. Recently a model was presented by Abreu et al. (1992) which models these triad wave-wave interactions in the spectral domain. Although this model is based on inconsistent assumptions, it is nonetheless considered as a first step in the development of a spectral source term for triad wave-wave interactions.

The purpose of this paper is to further extend the concept of spectral wave modelling in shallow water by introducing the newly available modelling techniques for energy dissipation by breaking waves and nonlinear triad wave-wave interactions, and to give a first assessment of these processes against laboratory and field measurements. In addition, the relative importance of various physical processes in shallow water is assessed.

The present study has been carried out within the framework of the HYDRA project, which is aimed at the determination of hydraulic boundary conditions along the Dutch coast. For the present study DELFT HYDRAULICS's third-generation wave prediction model PHIDIAS has been used.

2   The PHIDIAS wave model

The third-generation wave model PHIDIAS has been developed by DELFT HYDRAULICS for application on oceanic, shelf-sea and coastal zone scales, and for application in deep, intermediate depth and shallow water. In addition, it has successfully been implemented in a real-time data assimilation system. The PHIDIAS model can be applied in areas with a spatially-varying bottom topography and time-dependent current fields.

The PHIDIAS model is based on a spectral description of the sea surface in terms of wave action density, which is a convenient description of the wave field in the presence of currents. Wave action density is considered in the PHIDIAS model as a function of the location x and y, wave number k and direction θ. The PHIDIAS model solves the time-dependent action balance equation (cf. Hasselmann et al., 1973):

$$\frac{\partial N}{\partial t} + \frac{\partial}{\partial x}(\dot{x}N) + \frac{\partial}{\partial y}(\dot{y}N) + \frac{\partial}{\partial k}(\dot{k}N) + \frac{\partial}{\partial \theta}(\dot{\theta}N) = S \quad (1)$$

in which $N = N(x,y,k,\theta;t)$ is the action density defined as the energy density divided by the relative frequency σ. The dot-terms in Eq. (1) are the spectral velocities which follow from linear wave theory (cf. Mei, 1983):

$$\dot{x} = c_{g,x} + U_x \quad (2)$$

$$\dot{y} = c_{g,y} + U_y \quad (3)$$

$$\dot{k} = -\left[\frac{\partial \sigma}{\partial d}\frac{\partial d}{\partial s} + \vec{k}\cdot\frac{\partial \vec{U}}{\partial s}\right] \quad (4)$$

$$\dot{\theta} = -\frac{1}{k}\left[\frac{\partial \sigma}{\partial d}\frac{\partial d}{\partial n} + \vec{k}\cdot\frac{\partial \vec{U}}{\partial n}\right] \quad (5)$$

where d is the local water depth, $c_{g,x}$ and $c_{g,y}$ the x- and y-components of the group velocity, U_x and U_y the velocity components of the ambient flow field, and $\vec{s}$ and $\vec{n}$ the components of unit vectors in the direction of and the direction perpendicular to the direction θ of a wave component.

The source term S on the right-hand side of equation (1) contains expressions of all physical processes that affect the action density of a spectral wave component. For deep water applications the source term S contains state-of-the art expressions for wave growth by the action of the wind (Snyder et al., 1981), dissipation by white-capping (Komen et al., 1984) and nonlinear quadruplet interactions computed with the discrete interaction approximation of Hasselmann et al. (1985). In intermediate water depth applications, the above deep water source terms are scaled as described in WAMDI (1988) and supplemented with state-of-the art formulations for bottom friction (Hasselmann et al, 1973; Madsen et al., 1988). For shallow water, the PHIDIAS model also uses source terms for surf breaking and triad wave-wave interactions. The expressions for surf breaking and triad interactions in a full spectral wave model are recently developed and need some explanation.

A method incorporating the dissipation of wave energy by breaking waves was given by Young (1988). In this method a limit is imposed on the total wave energy and the excess of wave energy is removed from the lowest energy containing waves. This is a rather coarse method because it only removes energy from the lowest frequencies and because it is not formulated in terms of a source term, i.e. a dissipation rate.

A successful model for computing the energy dissipation in random waves by wave breaking in shallow water was given by Battjes and Janssen (1978). This dissipation model is formulated in terms of the rate of change of the total wave energy E_{tot}:

$$\frac{\partial E_{tot}}{\partial t} \approx -\frac{1}{4} \alpha f_p Q_b H_{max}^2 \tag{6}$$

where α is a factor of about 1, f_p the peak frequency, Q_b a measure for the fraction of breaking waves and H_{max} a maximum wave height. The parameter Q_b is computed from:

$$\frac{1 - Q_b}{\ln(Q_b)} = \left\{\frac{H_{rms}}{H_{max}}\right\}^2 \tag{7}$$

in which H_{rms} is the root mean square wave height, which can be computed from the total wave variance σ^2 according to:

$$H_{rms} = \sqrt{8}\,\sigma \tag{8}$$

In the Battjes-Janssen model the maximum wave height is computed according to a combined steepness and depth-limited breaking criterion (Battjes and Stive, 1985):

$$H_{max} = \frac{\gamma_1}{k_p} \tanh\left(\frac{\gamma_2}{\gamma_1} k_p h\right) \tag{9}$$

in which γ_1 and γ_2 are coefficients and where k_p is the peak wave number. In this expression the coefficient γ_1 controls the breaking on wave steepness and γ_2 the depth-limited wave breaking. For application in a full spectral wave model three adaptions are needed of the original Battjes-Janssen model. Firstly, an assumption has to be made about the spectral distribution of wave energy by breaking waves and secondly, the criterion for computing H_{max} should be adapted to avoid the double counting of breaking on wave steepness in the presence of a white-capping dissipation source term, and thirdly to replace the peak frequency f_p with a more stable measure of a representative frequency, for instance by the mean frequency f_{m01}.

The most simple method of distributing the energy dissipation by breaking waves over the spectrum is to assume that this dissipation rate is in proportion to the energy density of each spectral component:

$$S_{brk} = -\frac{1}{4} \alpha f_{m01} Q_b H_{max}^2 \frac{E(k,\theta)}{E_{tot}} \tag{10}$$

This method seems to be supported by laboratory experiments of Beji and Battjes (1993). The criterion for computing the maximum wave height is simplified to:

$$H_{max} = \gamma_2 h \tag{11}$$

An advantage of expression (11) is that it becomes negligible for deep water. The above formulation has also been included in a recently developed spectral wave model for the coastal zone, developed by Delft University of Technology (Holthuijsen et al., 1993).

The triad interactions are computed by the method proposed by Abreu et al. (1992) on the basis of their equation (34). Their model contains the parameter kd_{lim} which limits the range in which interactions between triads can take place. Based on theoretical arguments they set this limit at $kd_{lim} = \pi/10$, but on the basis of a comparison against field measurements they suggest that this limit should be close to 1. Triad interactions conserve energy and they do not directly affect the significant wave height.

Performance studies

A number of studies has been performed to compare the extended PHIDIAS wave model in shallow water conditions against two sets of laboratory and one set of field data. The laboratory data were collected in wave flumes, using mechanically-generated uni-directional random waves. Field data were obtained from the Egmond site along the coast of the Netherlands for the case of a double bar system. The primary objective of these studies was to investigate the performance of the wave model with respect to the prediction of both the significant wave height and mean wave period. For the field experiment also the relative importance of the modelled physical processes was examined.

The first set of laboratory data were collected by Battjes and Janssen (1978). For the present study results of their experiment numbers 13 and 15 were used, corresponding to wave propagation over an underwater bar. The incident wave conditions and parameter settings are summarized in Table 1. In the Battjes-Janssen experiments no attention was paid to changes in wave periods, and related results will not be shown here.

Experiment number	H_{rms0} (m)	bar-depth (m)	γ_1	γ_2	f_p (Hz)
13	0.113	0.267	0.88	0.75	0.497
15	0.154	0.120	0.88	0.75	0.530

The results obtained with the PHIDIAS model are shown in Fig. 1. The computed data for the significant wave height show good agreement with the measured data.

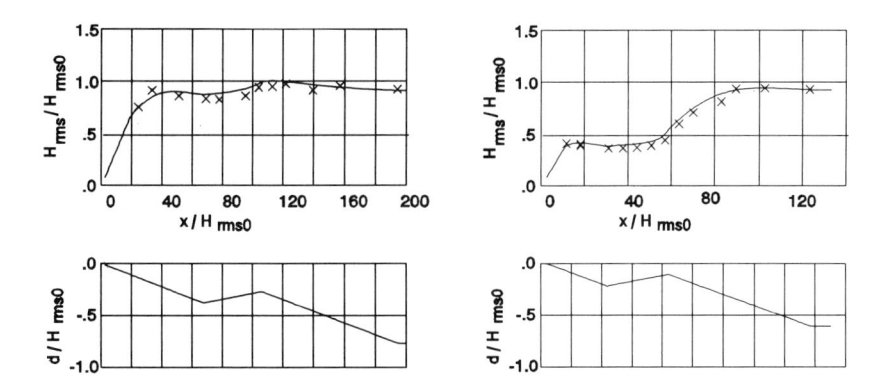

Figure 1: Observed (crosses) and computed (solid line) variation of the root mean square wave height H_{rms} and water depth d, normalized with the deep water rms wave height H_{rms0}, for the cases 13 (left panel) and 15 (right panel) of Battjes and Janssen (1978).

The second set of laboratory wave data were collected in the Schelde flume of DELFT HYDRAULICS in the framework of the MAST program of the European Community. One of these experiments consisted of random wave propagation over an underwater bar. Detailed spectral measurements were performed at 26 locations along the wave flume.

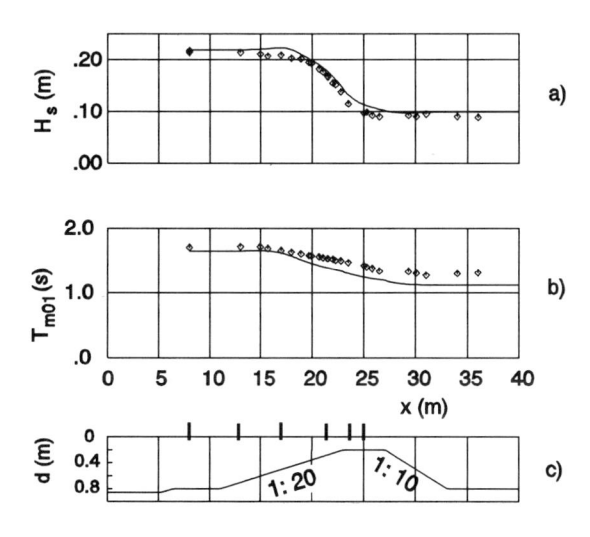

Figure 2: Observed (diamonds) and computed (solid line) variation of significant wave height (H_s) and mean wave period (T_{m01}) in Schelde flume experiment. The tick marks in panel c) refer to the locations for which frequency spectra are shown in Fig. 3.

The input wave conditions comprised a JONSWAP spectrum with a peak period $T_p = 1.77$ s, a significant wave height $H_s = 0.22$ m and a peak enhancement factor $\gamma = 3.3$. After a number of trial runs with different parameter settings for surf breaking and the triad interactions, good agreement was found for the settings $kd_{\lim} = 0.75$ and $\gamma_2 = 0.75$. The cross-section of this flume and the results for the change in significant wave height and the mean wave period (T_{m01}) are shown in Fig. 2. The observed and computed frequency spectra for six locations are shown in Fig. 3.

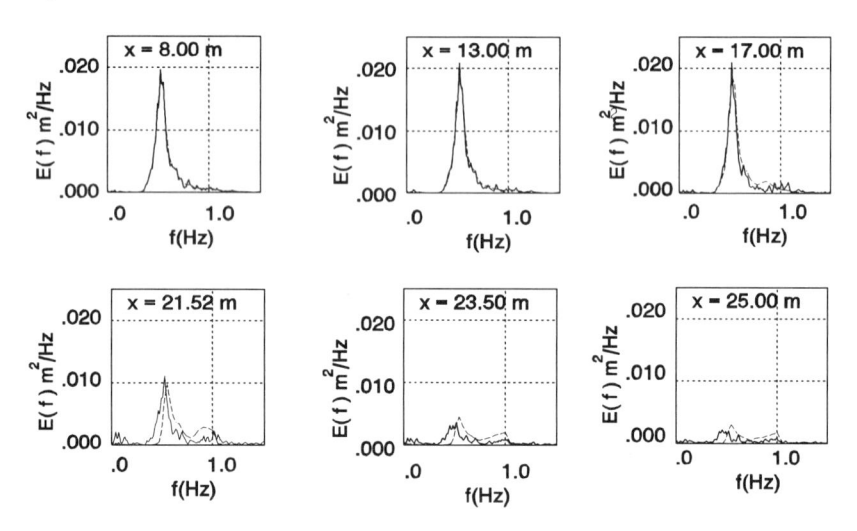

Figure 3: Observed (solid line) and computed (dashed line) frequency spectra in Schelde flume experiment at 6 locations.

The results of the computations show good agreement, not only for the significant wave height but also for the change in the mean wave period. As can be seen in Fig. 3, the change in the mean wave period is due to the generation of a second spectral peak in the vicinity of twice the peak frequency. The generation of this second peak is basically due to the effect of triad interactions. This was demonstrated by performing a computation in which the effect of triad interactions was omitted. Results of other computations (not shown here) with different parameters setting for the triad interactions indicate that the generation of the second spectral peak is controlled by the choice of the upper limit $kd_{\lim}$. In the case of a high value, the second spectral peak showed excessive growth, such that this peak became larger than the first spectral peak.

Field data were collected for a site along the coast of the Netherlands near the town of Egmond. The bottom profile consists of almost parallel bottom contours with a double bar system. Wave data were collected at four locations along a ray protruding into the sea. The bottom profile of this ray and the locations of wave measurement instruments are shown in Fig. 4. The outer measurement system consisted of a WAVEC, a directional wave buoy. Closer to the coast three wave poles were used which collected time series of the surface elevation. Directional information was not obtained with the wave poles. Unfortunately, no reliable wave data were obtained with wave pole 2.

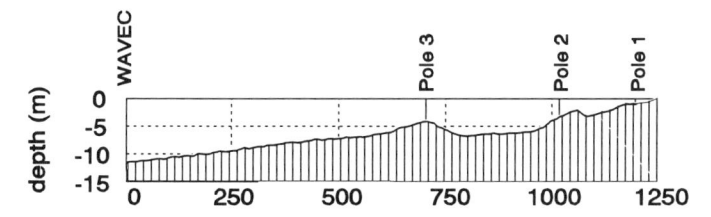

Figure 4: Bottom profile and location of measuring instruments along the Egmond ray.

The wave characteristics observed with the WAVEC buoy were transformed into two-dimensional wave spectra that were used as the boundary conditions for wave computations with the PHIDIAS model. Per frequency a directional distribution was reconstructed on the basis of the mean wave direction and directional spreading according to the $\cos^{2s}(\theta/2)$-model. Wind and water level information was obtained from nearby coastal stations.

For a number of situations wave model computations have been performed with the PHIDIAS model with the objectives of predicting the changes in the spectral shape and of studying the relative importance of the various physical processes in the coastal zone.

Results are presented of one computation for the situation on October 16, 1992, 3 hours, the wind speed was 5 m/s, blowing to the shore. The variation of the significant wave height, mean wave period, incident wave direction and directional spreading are shown in Fig. 5. This figure clearly shows the effect of the underwater bars on the above-mentioned wave parameters. As expected, energy dissipation takes place on the bars and in the area closer to the coast. The spatial variation of the incident wave direction and directional spreading resemble the bottom profile. The waves turn towards the coast and the directional spreading becomes smaller as the water becomes shallower, two effects which are both in agreement with the theoretical expectation.

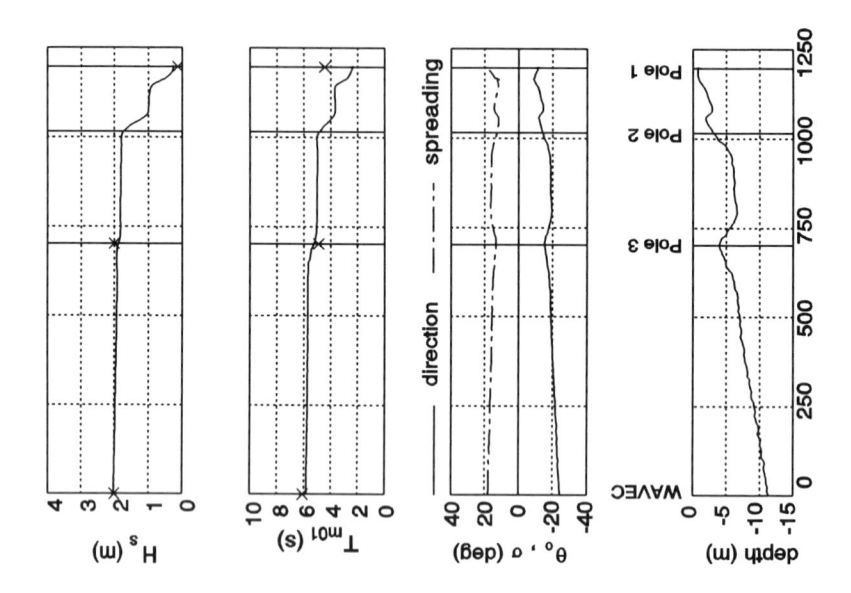

Figure 5: Observed (crosses) and computed (lines) variation of significant wave height (H_s), mean wave period (T_{m01}), mean wave direction (θ_0) and directional spreading (σ) along the Egmond ray on October 16, 1992, 3 hours.

A comparison between the computational results with the measurements shows good agreement for the significant wave height, but not for the mean wave period. As can be seen in Fig. 6, the change of the mean wave period is due to the growth of a strong second spectral peak which reduces the mean wave period. Different parameter settings for the triad interactions were tried to obtain better agreement with the observations. However, this was not possible. The origin of the observed second spectral peak at wave pole 3 could not be predicted with the present implementation of the triad interactions. A closer look at the computational results revealed that this second peak could not be generated by the effect of wind. Clearly, the Abreu model is not capable of handling this situation.

The results of the computations for the Egmond site were also analyzed with respect to the strength of the various source terms, representing the various physical processes. The strength of a source term is defined as the integral over all spectral bins. For the nonlinear interaction source terms (quadruplets and triads) the integral over the absolute values was taken because these processes conserve energy. For the same case as above, the spatial variation of the strength of each source term is shown in Fig. 7. For the present case, the source terms for surf breaking and triads interactions are dominant over all other source terms. It can be seen that the strength of all source terms is influenced by the local water depth. The wind input source function increases in strength because the waves are slowed down on the bar systems such that the relative wind speed increases. All dissipation source terms become stronger (more negative) over the bars, which is an effect that can also be seen in the strength of the nonlinear interaction source terms.

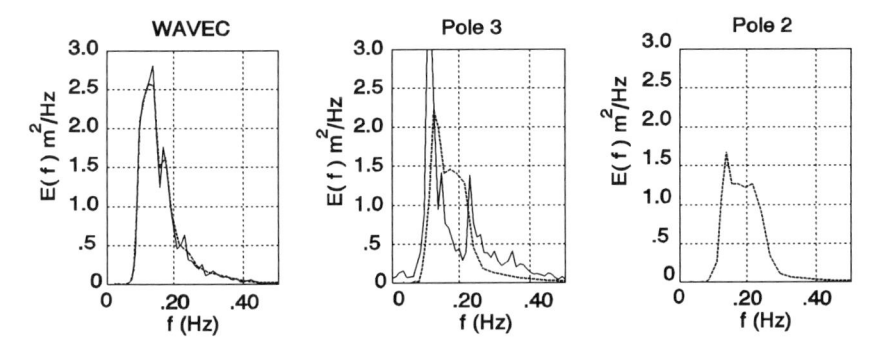

Figure 6: Observed (solid line) and computed (dashed line) frequency spectra along Egmond ray on October 16, 1992, 3 hours.

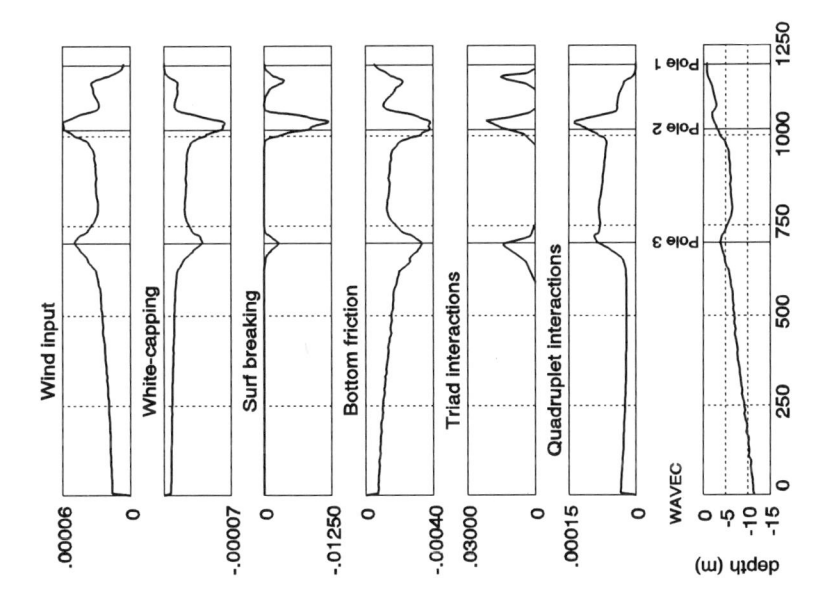

Figure 7: Computed spatial variation of the magnitude of the various physical processes for the Egmond computation of October 16, 1992, 3 hours.

Discussion

The computational results obtained with the extended PHIDIAS wave model show good agreement with measurements regarding the prediction of the significant wave height in situations were surf breaking plays a dominant role. This is not surprising since the source term for surf breaking is based on a well-tested dissipation model (Battjes and Janssen, 1978; Battjes and Stive, 1985). The spectral distribution of dissipation by surf breaking in proportion to the existing energy density was straightforward and seems to be supported by measurements, although a theoretical basis is still missing.

In the coastal zone the change in the mean wave period is mainly due to the effect of triad interactions, especially if waves propagate over an underwater bar. In such a case relatively large changes occur over short distances. The results of the computations for the Schelde flume indicate that the inclusion of a source term for triad interactions is essential for computing a change in the mean wave period. The results for the Egmond site show that it was not possible to find a proper parameter setting for the triad interactions such that the change of the mean wave period would be predicted correctly. This implies that the spectral method of Abreu et al. (1992) for computing the triad interactions is incomplete.

The problem with the Abreu method is that it is based on inconsistent assumptions (Elgar et al., 1993). It is based on the non-dispersive, shallow water equations and a natural asymptotic closure for directionally spread, non-dispersive waves. A consequence is that only exact resonance is taken into account. Moreover, since the waves are assumed to be frequency non-dispersive, only triads containing waves travelling in the same direction are considered resonant. One of the results is that too much wave energy is transferred to higher frequencies resulting in an excessive growth of secondary spectral peaks. This was found to occur on a sloping beach. In the Schelde flume, however, just enough high frequency energy was produced on the bar to obtain a good prediction of the mean wave period. In the deeper water behind the bar the effect of the triad interactions became negligible.

The nonlinear interactions between quadruplets and triads conserve energy. They do not affect the total amount of wave energy, but only the spectral shape. In shallow water triad interactions are much stronger than quadruplet interactions, whereas in deep water the latter process is more dominant.

The present results provide quantitative information about the relative importance of the various physical processes in the nearshore zone, see also Battjes (1994). Such knowledge can be useful in the preparation of a computational run, e.g. by omitting the relative costly computation of the nonlinear quadruplet interactions.

Conclusions and future work

The present study was aimed at obtaining a better understanding of modelling waves in shallow water. To that end a full spectral third-generation wave model was extended with formulations for surf breaking and triad interactions, and compared against shallow water wave data. The results of this study led to the following conclusions:

1 In the coastal zone surf breaking is dominant over the physical effects of wave growth, bottom friction and white-capping dissipation, and triad interactions are dominant over quadruplet interactions.
2 The spectral distribution of wave energy dissipation by surf breaking does not affect the spectral shape.
3 The inclusion of a source term for triad interactions is necessary for the prediction of changes in the mean wave period, especially if waves propagate over an underwater bar.
4 The method of Abreu et al. (1992) for the computation of triad interactions in a wave spectrum is based on inconsistent assumptions. An improved formulation for these interaction is needed, possibly by the inclusion of the proper frequency dispersion characteristics.

Acknowledgements

The authors thank Yasser Eldeberky of Delft University of Technology and Maarten Dingemans of DELFT HYDRAULICS for discussions on triad interactions.

References

Abreu, M., A. Lazzara and E. Thornton, 1992: Nonlinear transformation of directional wave spectra in shallow water, J. Geophys. Res., Vol. 97. No. C10, 15579-15589.

Battjes, J.A., 1994: Shallow water wave modelling, Proc. Int. Symp.: Waves - physical and numerical modelling, Vancouver, 1-23.

Battjes, J.A., and J.P.F.M. Janssen, 1978: Energy loss and set-up due to breaking of random waves. Proc. 16th Int. Conf. on Coastal Eng., 569-587.

Battjes, J.A. and M.J.F. Stive, 1985: Calibration and verification of a dissipation model for random breaking waves. J. Geophys. Res., Vol. 90, C5, 9159-9167.

Beji, S., and J.A. Battjes, 1993: Experimental investigation of wave propagation over a bar, Coastal Engineering, Vol. 19, 151-162.

Dingemans, M.W., M.J.F. Stive, J. Bosma, H.J. de Vriend and J.A. Vogel, 1988: Directional nearshore wave propagation and induced currents. Proc. 20th Int. Conf. on Coastal Eng., 1092-1106.

Elgar, S., R.T. Guza and M.H. Freilich, 1993: Observations of nonlinear interactions in directionally spread shoaling surface gravity waves. J. Geophys. Res., Vol. 98, No. C11, 20299-20305.

Gelci, R., H. Cazalé and J. Vassal, 1957: Prévision de la houle. La méthode des densités spectroangulaires. Bulletin d'information du commité central océanographie et d'étude des côtes, Vol. 9, 416-435.

Hamm, L., P.A. Madsen and D.H. Peregrine, 1993: Wave transformation in the nearshore zone: a review, Coastal Engineering, Vol. 21, 5-39.

Hasselmann, K., T.P. Barnett, E. Bouws, H. Karlson, D.E. Cartwright, K. Enke, J.A. Ewing, H. Gienapp, D.E. Hasselmann, P. Kruseman, A. Meerburg, P. Müller, D.J. Olbers, K. Richter, W. Sell and H. Walden, 1973: Measurements of wind-wave growth and swell decay during the Joint North Sea Wave Project (JONSWAP), Erganzungsheft zur Deutschen Hydrographischen Zeitschrift, 12.

Hasselmann, S., K. Hasselmann, J.H. Allender and T.P. Barnett, 1985: Computations and parameterizations of the nonlinear energy transfer in a gravity wave-spectrum. Part II: Parameterizations of the nonlinear transfer integral for application in wave models. J. Phys. Oceanogr., Vol. 15, 1378-1391.

Holthuijsen, L.H, N.Booij and T.H.C. Herbers, 1989: A prediction model for stationary, short-crested waves in shallow water with ambient currents, Coastal Engineering, Vol. 13, 23-54.

Holthuijsen, L.H., N. Booij and R.C. Ris, 1993: A spectral model for the coastal zone. Proc. Int. Conf. WAVES'93, New Orleans, 630-641.

Komen, G.J., S. Hasselmann and K. Hasselmann, 1984: On the existence of a fully developed wind-sea spectrum. J. Phys. Oceanogr., Vol. 14, No. 8, 1271-1285.

Madsen, O.S., Y.-K. Poon and H.C. Graber, 1988: Spectral wave attenuation by bottom friction: theory. Proc. 21st Int. Conf. on Coastal Eng., 492-504.

Mei, C.C., 1983: The applied dynamics of ocean surface waves, Wiley, New York.

Snyder, R.L., F.W. Dobson, J.A. Elliott and R.B. Long, 1981: Array measurements of atmospheric pressure fluctuations above surface gravity waves. J. of Fluid Mech., Vol. 102, 1-59.

Tolman, H.L.,1991: A third-generation model for wind waves on slowly varying, unsteady, and inhomogeneous depth and currents, J. Phys. Oceanogr., Vol. 21, No. 6, 782-797.

WAMDI, 1988: The WAM Model - A third generation ocean wave prediction model, J. Phys. Oceanogr., Vol. 18, 1775-1810.

Young, I.R., 1988: A shallow water spectral model. J. of Geophys. Res., Vol. 93, No. C5, 5113-5129.

Young, I.R., and G.Ph. van Vledder, 1993: A review of the central role of nonlinear interactions in wind-wave evolution. Phil. Trans. Roy. Soc. London, Vol. 342, 505-524.

PART II

Long Period Waves, Storm Surges and Wave Groups

CHAPTER 57

THE GENERATION OF LOW-FREQUENCY WAVES BY A SINGLE WAVE GROUP INCIDENT ON A BEACH

G. Watson [1] T.C.D. Barnes [2] and D.H. Peregrine [3]

Abstract

The generation of a single low-frequency wave (LFW) pulse by a single group of waves incident on a beach is investigated by means of laboratory experiments and a numerical model. This simplified case allows the LFW to be measured in isolation, after the incident group has passed and before there is any reflection from the wavemaker. A beach consisting of two different slopes (1:100 and 1:20) was used, and runs were made with the water level on each slope. The results were simulated using a composite numerical model, with Boussinesq equations in the deeper water and nonlinear shallow water equations in the surf zone. For some calculations, a friction term was included. For the 1:20 slope, the outgoing LFW is well predicted even without the friction term. With a 1:100 slope, a friction factor of 0.01 gave a good result, in this case reducing the amplitude of the outgoing LFW by a factor of about 2 compared with the frictionless result. The nondimensional equations show that the friction term is insignificant if the beach slope is large compared with the friction factor. Runs of the surf zone part of the model show the outgoing LFW to be correlated with the swash motion. Its amplitude is largest if the duration of the wave group is similar to the swash period of the largest wave in the group. The model also showed a slightly stronger than linear dependence of LFW amplitude on incident wave amplitude.

---

[1]Research Associate, University of Bristol, Mathematics Department, University Walk, Bristol BS8 1TW, United Kingdom. *Current address:* Disaster Prevention Research Institute, Kyoto University, Gokasho, Uji, Kyoto 611, Japan. (*gary.watson@bristol.ac.uk*)
[2]Research Assistant, University of Bristol. (*tim.barnes@bristol.ac.uk*)
[3]Professor of Applied Mathematics, University of Bristol. (*d.h.peregrine@bristol.ac.uk*)

1. Introduction

Low-frequency waves (LFW) with periods typically 5 to 15 times those of wind-generated waves are formed when wind waves meet beaches. Hamm, Madsen & Peregrine (1993) give a general review of the phenomenon, whilst Herbers et al. (1992) give a good review of recent field measurements. Recent work on the theory of LFW generation has been published by Cox et al. (1992), List (1992a,b), Roelvink et al. (1992), Roelvink (1993) and Schäffer (1993). Low-frequency swash oscillations, with bichromatic waves, have also been investigated by Mase (1994).

In this work, our aim was to throw some light on the question of how such waves are generated, by investigating the simplest possible case of LFW generation in as much detail as possible. Watson & Peregrine (1992) investigated the generation of low-frequency waves in the surf zone using a numerical model based on the nonlinear shallow-water equations. A single group of waves was used to illustrate the process whereby a group forces a variable set-up near the shoreline, which then travels offshore as a single LF pulse. This process has now been investigated more quantitatively, by means of laboratory experiments and a wider range of numerical computations. The range of validity of the model has also been extended by coupling the surf zone model to a Boussinesq model in the deeper water.

The reason for choosing a single group of waves rather than continuous bichromatic waves (such as those used by Kostense, 1984) was that this allows the outgoing LFW to be measured in isolation. By the time it reaches the deeper water, the incident wave group has already passed by. It is also measured before reflecting off the paddle. With continuous waves such reflections will contaminate the results unless they are mechanically removed or taken into account in the data analysis. Both of these are difficult to achieve with any degree of confidence.

Details of the experiments, and discussion of one run showing the generation of an outgoing LF pulse, are given in Section 2. A more detailed account of the experiments will be available in the Ph.D. thesis of T. Barnes. The numerical model is described briefly in Section 3 and its results are compared with the measurements, showing the importance of friction effects on the shallower beach slope. Section 4 discusses the influence of friction in more detail. Section 5 illustrates the correlation between swash motions and the outgoing LFW, using results from various other runs of the surf zone part of the model. The importance of the relative timescales of group period and swash period is pointed out. Discussion and conclusions are given in Section 6.

2. Experimental Results

Description of Experiments

Experiments were performed in a 50 m long wave flume at Hydraulics

Research Ltd. in Wallingford, England. The flume is 1 m wide. It is equipped with a piston-type wavemaker with two paddles, controlled by a PC. A concrete beach with two slopes, 1:100 and 1:20, was installed in the flume as shown in Figure 1. Runs were performed using two different water depths of 0.35 m and 0.20 m, so that the undisturbed shoreline would lie on a different slope in each case.

An array of resistance-type wave probes was used, and Figure 1 shows their layout for the deeper water runs. Probe 1 was located near the paddle in order to measure the generated wave signal. Probes 2 and 3 were in the deeper water. Probes 4–11 were on the 1:100 slope and Probe 12 was on the 1:20 slope.

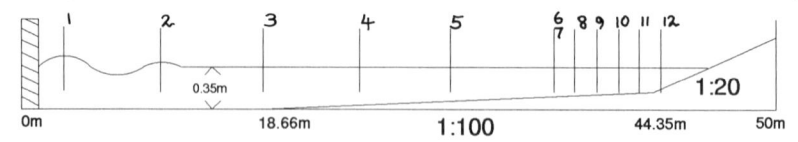

Figure 1: Cross-section of the HR wave flume, showing beach and positions of wave probes (not to scale). A current meter was located at Probe 6.

Since a number of numerical runs were to be performed using just the surf zone model, particular attention was paid to measurements at a point where all but the smallest waves had already broken. In the case illustrated, this was near Probes 6 and 7, (depth 15.5 cm). Two probes were used at this point in order to check for consistency across the tank.

Along-tank velocity data were also collected at this point, using an ultrasonic current meter. The sensor head was placed at about half the water depth, the results having been found not to be very sensitive to the precise placement of the probe. At this location, the incident and outgoing signals are not separated in time, but can be approximately separated by calculating the Riemann invariants. This is done by combining the surface elevation and velocity signals as explained in Section 3.

Wavemaker Signal

Initially, single groups consisting of modulated sine waves were used. However, these waves acquired large second harmonics as they propagated along the tank, with each wave effectively splitting into two. To solve this problem, groups consisting of solitary wave solutions to the Korteweg-deVries equation were used instead. For the shallow water regime of these experiments, this is more appropriate than Stokes theory, which is better in deeper water. Stokes theory requires the Ursell parameter $U = ka_0/(kh)^3$, to be small, whereas in these experiments it was of order 1. In the formula for U, k is the wavenumber, a_0 the fundamental amplitude and h the undisturbed water depth.

The wavemaker signal was designed to produce surface elevation time series of the form

$$\eta = \sum_i \eta_i \text{sech}^2 \sqrt{\frac{3\eta_i}{4h_i^3}} c_i(t - t_i) - (a + a_s)\sin \omega t \qquad 0 \leq t \leq \pi/\omega \quad (1)$$

and zero outside that time interval. The wave times, t_i, were equally spaced. The wave amplitudes, η_i, were sinusoidally modulated by the function $\eta_i = A \sin \omega t_i$. The water depth for each wave, h_i, must be adjusted for the additional sine function: $h_i = h_0 - (a + a_s)\sin \omega t_i$. The wave speed c_i is $\sqrt{g(h_i + \eta_i)}$. Velocity was estimated from the shallow-water approximation $u = \sqrt{g/h_0}\,\eta$ and then integrated to give the wavemaker displacement signal.

The stroke of the paddle is not great enough to make a succession of solitary waves, each with its own net displacement. For this reason, the sine function was added. This has the same duration as the group and corresponds to a set-down beneath it. This means that there is little net mass flux in the incident wave groups.

Results

Surface elevations from a typical run in the deeper water (1:20 slope) are shown as the thick lines in Figure 2. All of the runs yielded results which were qualitatively similar. Numerical predictions from the model described below are included for comparison. The shape of the initial wave group, consisting of five waves, is seen in the trace from Probe 1. The form of the solitary waves is preserved quite well along the constant depth section, except for the development of a small dispersive tail (1–3). As the wave group travels into shallower water, the waves steepen and at Probe 6/7 all except the first are breaking. Near the shoreline (11 & 12) a LFW is seen to develop consisting of a peak, apparently like wave set up, followed by a trough. At the same time the short waves dissipate their energy and get smaller, so that the LFW is more prominent. This wave then propagates offshore, decreasing in amplitude as it does so. At Probe 5 it is separated in time from the incident group. Further offshore it can be identified quite clearly. The moving peak and trough are marked with arrows. It then reflects off the wavemaker (also marked). At Probe 1, the amplitude appears to be twice as big, due to the superposition of the reflection.

The shape and amplitude of the outgoing LFW can be seen more clearly if the vertical scale of the plots is expanded. This is shown for Probes 2–5 in Figure 3.

3. Composite Numerical Model

A coupled numerical model, based on the nonlinear shallow-water equations (NLSWE) in the surf zone (Watson and Peregrine, 1992) and the Boussinesq equations in deeper water, is able to simulate the phenomenon.

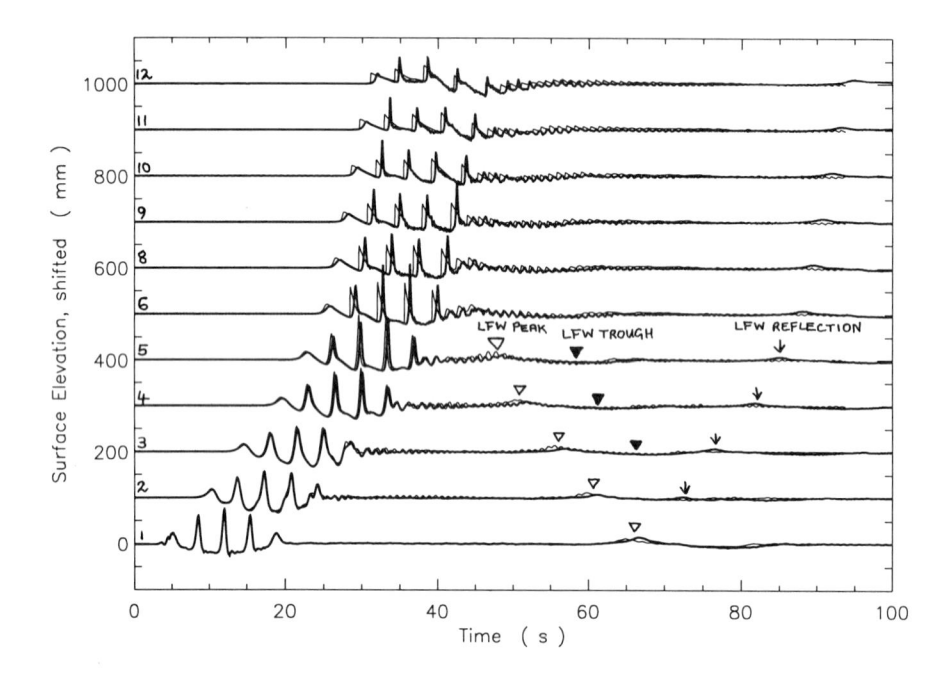

Figure 2: *Thick line:* Surface elevation time series at each probe (except 7) for a group of five solitary waves. *Thin line:* Numerical prediction.

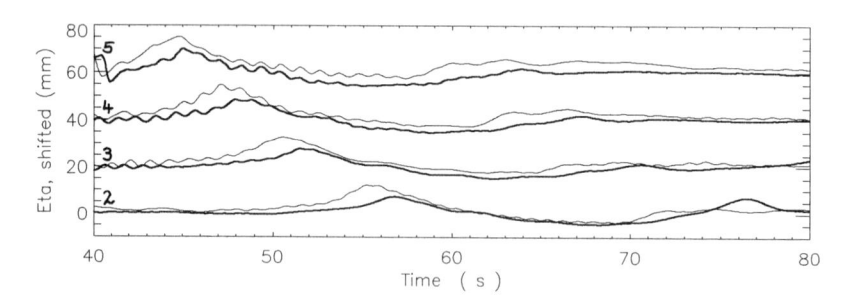

Figure 3: Expanded view of the outgoing LFW in Figure 2, for Probes 2–5. *Thick line:* Measurements. *Thin line:* Numerical prediction.

Our primary interest here is in the mechanisms by which LFW are generated. LFW have their greatest amplitudes in the surf zone, and the most important components of the generation process occur there. It is therefore important to use a model that includes all the important physics in the surf zone. The nonlinear shallow-water equations for the conservation of mass and momentum, with a friction term, were used:

$$d_t + (ud)_x = 0 \tag{2}$$

$$u_t + uu_x + g\eta_x + \frac{1}{2} f \frac{u \mid u \mid}{d} = 0 \tag{3}$$

where $\eta = d - h$ is the surface elevation, d is the total water depth, h the undisturbed water depth, u the flow velocity, g gravity and f an empirical friction coefficient.

These equations are able to represent spilling breakers as travelling hydraulic jumps or bores, which manifest themselves as discontinuities in the solution. Some information on these equations without the friction term, and the numerical method used to solve them has been given previously (Watson & Peregrine, 1992; Watson, Peregrine & Toro, 1992). Treatment of the moving shoreline boundary condition is discussed in the latter paper.

The friction term in the above equations is the simplest that is conventionally used to represent friction. Typical values for f are of order 10^{-2}. For the present study it was necessary to modify the numerical procedure slightly to allow for this term. This was done by first solving the frictionless equations at timestep n as before, a procedure which is second-order accurate. After this a simple first-order forward difference step was applied at each grid point, altering the velocity u_n by an amount

$$-\frac{1}{2} f \frac{u_n \mid u_n \mid}{d_n} \Delta t \tag{4}$$

Δt must be small enough for this to be reasonably accurate. The inclusion of friction in this manner destroys the second-order accuracy of the scheme. However, since the empirical friction term is rather approximate in any case, the degradation of numerical accuracy here is not important.

In deeper water, the Boussinesq equations are appropriate because they include dispersive terms. The Boussinesq equations in the form due to Peregrine (1967) were used:

$$d_t + (ud)_x = 0 \tag{5}$$

$$u_t + uu_x + g\eta_x = \frac{1}{2}h(hu_t)_{xx} - \frac{1}{6}h^2 u_{xxt} \tag{6}$$

Recently, Dingemans has shown that in some circumstances, particularly on a barred beach, it is important to include further dispersive terms. See Dingemans (1995) for a full discussion. However, in our case it will be seen that the above equations give sufficiently good results. The friction term was not included in these equations, since it becomes small in the deeper water.

The equations were solved numerically using a finite difference scheme due to Peregrine (1967).

Matching the two models together

The models were matched at the join using a characteristics boundary condition. In the frictionless shallow-water equations, the wave signals propagating in each direction are given by the Riemann invariants $R^+ = 2c + u$ and $R^- = 2c - u$, where $c = \sqrt{gd}$ and d is the total water depth. This is also approximately true for the Boussinesq equations, which for waves of sufficiently gentle slope approximate to the shallow-water equations. Thus, an almost non-reflecting join can be made by taking R^+ from the last point of the Boussinesq section and feeding it into the first point of the NLSWE section, and taking R^- from the first point of the NLSWE section and feeding it into the last point of the Boussinesq section.

This being done, a number of runs were made in order to test the sensitivity of results to the position of the join. It was found that the join could be moved a significant distance without much change in the output, as long as it was somewhere in the vicinity of the break point. The run reported here was done with the join at the location of Probe 6 in Figure 1.

The Seaward Boundary Condition

At the offshore end of the computational domain, a boundary condition similar to that at the matching point was applied. This used the characteristic equations to allow outgoing waves to pass out with no reflection. The outgoing R^- was found from data immediately inside the domain, and the surface elevation was forced to be equal to that measured by Probe 1. This was sufficient information to define the incident R^+. Although the reflections from the paddle that were present in the wave flume were not reproduced, these are not of any interest.

Comparison of Numerical and Experimental Results

The numerical results for the data in Figures 2 and 3 are shown as thin lines in those figures. The main feature of interest, i.e. the amplitude, shape and propagation of the outgoing LF pulse, is reasonably well predicted — although the amplitude is a little too large and the timing is not precise. Some properties of the short waves are not reproduced very well as the group travels shoreward. Their amplitude is underpredicted, the sharp wave crest is flattened out, and there are small timing errors. Note however that these details do not appear to have a significant effect on the LFW generation process.

The success of this model in predicting the measured LFW shapes and amplitudes is illustrated in more detail using data from Probe 6. Results from two runs are shown. One has the water level set so that the shoreline is on the 1:20 slope, the other with the shoreline on the 1:100 slope.

GENERATION OF LOW-FREQUENCY WAVES 783

In both cases the model was run using the measured water depth and velocity at Probe 6, to construct a time series of the Riemann invariant $R^+ = 2\sqrt{gd} + u$. The outgoing LFW was then examined by computing the other invariant $R^- = 2\sqrt{gd} - u$. This is plotted using dashed lines in Figures 4 and 5 for the two runs, together with the numerical result. Both cases show the LFW quite clearly. There are spikes in the data every time an incident wave passes the probe. This effect is due to the fact that the shallow-water equations on which the Riemann invariant analysis is based do not accurately describe the details of the crests of steep waves, breakers, or bores.

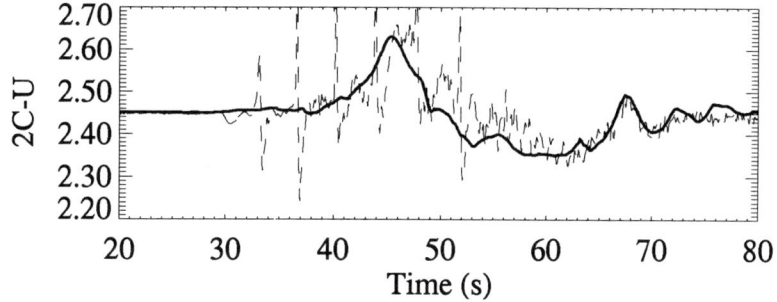

Figure 4: Outgoing signal at seaward boundary in deep water case (composite slope with shoreline on the 1:20 slope). Data (broken), frictionless model (solid).

In the deeper water case (1:20, Figure 4), the agreement between model and data is quite good, except for precise details of the wave shape. In this case, the model gave almost the same result with or without friction. The frictional result is not plotted.

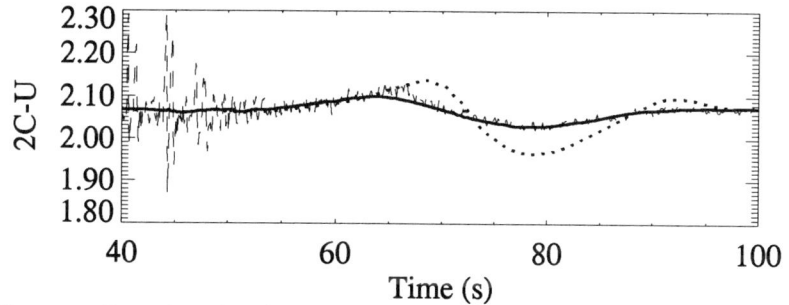

Figure 5: Outgoing signal at seaward boundary in shallow water case (1:100 slope). Data (broken), frictionless model (dotted), frictional model (solid).

In the shallower case (1:100, Figure 5), agreement in the frictionless case is not so good (dotted). The LFW is overpredicted by a factor of about 2. The shape of the pulse is also different, in that the peak occurs about 5 s late. However when the friction term is included, with the value $f = 0.01$, the result is much better — as shown by the solid curve. The effect of friction is discussed in more detail in the following section.

4. The Effect of Friction

The frictional drag force acting between the bottom and the water has an increasingly strong effect as the depth decreases. This is simply because the friction force is primarily determined by the near-bottom velocity of the water, and in the shallow-water approximation it acts on a mass of water that is proportional to the depth. It will thus have its strongest effect in the swash zone, where the water is shallowest. Since this is where LFW processes are particularly important, it is to be expected that friction may have some effect on LFW generation.

For a plane beach of slope α extending fom the shoreline to the offshore boundary, the relative importance of friction effects may be seen from the following scaling argument. Eqs. 2 & 3 may be written using the following non-dimensional variables:

$$x' = x/x_1, \quad d' = d/h_1, \quad \eta' = \eta/h_1, \quad u' = u/u_1, \quad t' = t/t_1$$

where the scaling variables are x_1 (the distance between the offshore boundary and the undisturbed shoreline), h_1 (the undisturbed water depth at the offshore boundary), $u_1 = \sqrt{gh_1}$ and $t_1 = x_1/u_1$. Dropping primes, the equations become:

$$d_t + (ud)_x = 0 \tag{7}$$

$$u_t + uu_x + \eta_x + \frac{1}{2}\frac{f}{\alpha}\frac{u|u|}{d} = 0 \tag{8}$$

where α is the slope h_1/x_1.

For breaking waves, u^2/d is of order 1, so the friction term is of order f/α. Thus friction can be expected to have a negligible effect if $f \ll \alpha$, a noticeable effect if $f \sim \alpha$ and to dominate everywhere if $f \gg \alpha$. In the two cases under consideration, with $f = 0.01$, f/α takes the values 0.2 and 1.0. In the former case friction had little effect, whereas in the latter the outgoing LFW amplitude was reduced by a factor of about 2.

These conclusions were further confirmed by model runs made using a range of values of f. The same incident wave group as in Watson & Peregrine (1992) was used, and the height of the outgoing LF pulse was determined by taking the difference between maximum and minimum in R^- at the offshore boundary. Six runs with values of f/α ranging from 0.0 to 2.0 were performed. The results are plotted in Figure 6.

The figure shows a strong dependence of outgoing LFW amplitude on the scaled friction factor f/α. With no friction ($f/\alpha = 0$), the amplitude has its maximum value, but this is reduced quite rapidly as f/α is increased, so that for $f/\alpha = 0.2$ the LFW amplitude is 78% of its maximum value. As f/α is increased, the amplitude continues to decrease, although more slowly. for $f/\alpha = 1.0$, it is 47% of its maximum value. This is roughly consistent with the results in Figures 4 and 5.

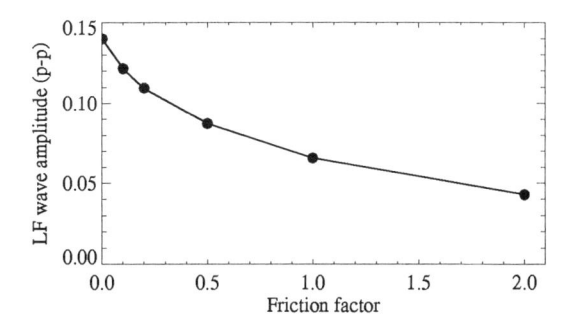

Figure 6: The dependence of outgoing LFW amplitude on scaled friction factor, f/α.

This means that the agreement which was obtained in the 1:100 case with $f = 0.01$ is open to the suggestion that it may have been merely coincidental. However, this value was chosen from experience as being one that is typically used.

5. The Importance of Swash Motions

During the time that the wave group is in the swash zone, there is a complex interaction between the waves in the group and the swash motion from previous waves. This affects the amplitude and shape of the LFW that finally emerges. The nature of the swash, and the properties of the LFW, depend on the relative values of the various timescales in the problem. Rather than attempt to understand the swash motion in detail, this effect was investigated in a more empirical manner by performing a number of runs of the model with different timescales and hence different swash regimes.

The timing of the incident group is determined by two parameters: the wave period, τ, and the number of waves in the group, N. The total duration of the group is then $T = N\tau$. There is a third timescale in the problem, namely the natural period of swash on the beach. Let us call this t_s. This depends on the amplitude of the incident waves. In the frictionless case (considered here), the swash motion approximates fairly closely to free motion under gravity on the sloping beach. Thus, the larger the incident waves, the greater the initial velocity of the uprush and the longer the swash period t_s.

A number of runs of the model were performed with different wave periods, different numbers of waves and and different swash periods. These were based on variations about a control case, summarized in Figure 7.

The top panel of this figure shows a perspective view of the space-time plot of surface elevation. The input is an idealized group of five sinusoidally modulated sawtooth-shaped waves, which are intended to represent waves that are already breaking. The peak and trough values have been set so that there is no net mass flux in each wave. The curved wave trajectories show how each wave slows down as it approaches the shore, and the wave heights can be seen to diminish in the process. The wave group can be seen to force a mass of water up the beach face on the timescale of the wave group. The seaward-propagating LFW pulse that this forcing generates is just about visible to the right of the wave group in the plot.

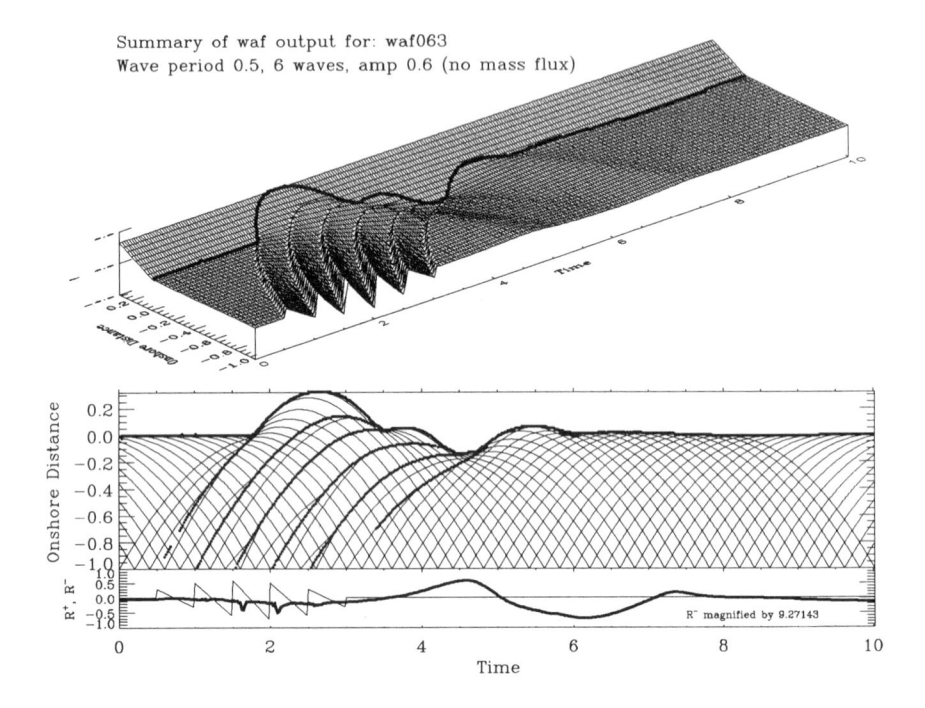

Figure 7: Result from the control case (see text).

The next panel shows shallow-water equation characteristics and bore trajectories (indicated by black dots). The shoreline motion is also shown in plan view. Beneath this, the incident and outgoing signals at the seaward boundary of the model are plotted. These are computed from the Riemann invariants R^+ and R^- as explained in Section 3. Note that the outgoing signal, which shows the shape of the LFW pulse, has been magnified by the indicated factor in order to make it more clearly visible. In this case it is about 10 times smaller than the incident signal. The main thing to notice from this plot is that the shape of the LFW pulse is very similar to the shape of the runup, especially after the runup has been averaged over the short-wave oscillations.

The generality of this observation was tested by performing a variety of model runs with different timescales. These were then used to investigate the effect of wave group timing on the LFW amplitude. Runs were performed with a range of values of τ and then with a range of values of N, varying the wave group duration T in two different ways. Both sets span approximately the same range of T-values for each wave amplitude, with t_s roughly central in each case. The runs were repeated with two different amplitudes for the largest wave in each group: 0.1 and 0.6 in dimensionless units. This provided two different values of the swash period. Separate runs using just one incident bore showed that for a wave of height 0.1 above still water level, the swash period was $t_s = 1.4$; whereas for a height of 0.6 it was $t_s = 2.9$.

The results of these runs are summarised in Figure 8, with runup plotted next to the LFW signal (at the offshore boundary) for each run. These curves have been smoothed a little to remove discretization effects in the runup, and spikes (due to incident bores) in the LFW signals. Note the clear correlation between the shapes of each pair of signals, with the LFW signal occurring somewhat later than the runup signal. It is significant that this is true despite the runup signals being quite varied in nature.

The runup is plotted so that whatever the range, the vertical extent of the plot is the same. The LFW plots all have the same relative scaling. Ratios of LFW elevation amplitude to incident wave amplitude were in the range 0.015–0.07 (a factor of 4.7). However, ratios of LFW elevation amplitude to runup amplitude were in the narrower range 0.035–0.125 (a factor of 3.6).

These two observations indicate the reflected LFW is better correleted with runup than with the incident wave envelope. This is consistent with the idea that the time-varying set-up within the wave group is manifested as swash when it is near the shore, and then propagates offshore as the outgoing LFW.

The amplitude of the outgoing LFW was found to depend in a consistent way on the relative values of wave group duration T and swash period t_s. This is shown in Figure 9. Here the LFW height, defined as the difference between the maximum and minimum in the LFW pulse, is plotted against T for each run. Four sets of results are presented. Figure 9(a) contains the results for amplitude 0.1, and Figure 9(b) for amplitude 0.6. The thick lines are runs with variable N, and the thin lines for runs with variable τ.

The main feature to note is that each of these curves has a maximum close to the swash period for a single wave of the respective amplitude. This period, t_s, is marked with an arrow on each plot. There are notable differences between the variable-N and variable-τ series, and at present the explanation for this is not clear. However, it is suggested that the presence of a peak in each curve can be explained as a quasi-resonance between the wave group forcing and the natural swash motion of the water on the beach face. If the wave group period is similar to the swash period ($T \approx t_s$), then a sequence of waves can carry water up the beach following, reinforcing and being carried by the initial swash motion. Thus, a large (wave-averaged) swash motion develops.

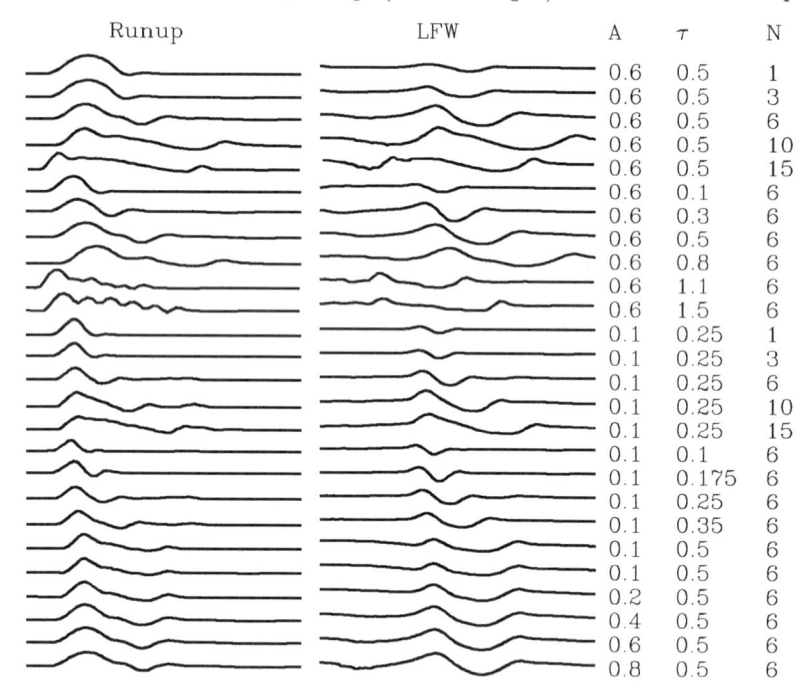

Runup	LFW	A	τ	N
		0.6	0.5	1
		0.6	0.5	3
		0.6	0.5	6
		0.6	0.5	10
		0.6	0.5	15
		0.6	0.1	6
		0.6	0.3	6
		0.6	0.5	6
		0.6	0.8	6
		0.6	1.1	6
		0.6	1.5	6
		0.1	0.25	1
		0.1	0.25	3
		0.1	0.25	6
		0.1	0.25	10
		0.1	0.25	15
		0.1	0.1	6
		0.1	0.175	6
		0.1	0.25	6
		0.1	0.35	6
		0.1	0.5	6
		0.1	0.5	6
		0.2	0.5	6
		0.4	0.5	6
		0.6	0.5	6
		0.8	0.5	6

Figure 8: Swash motion and LFW signal for various runs of the model.

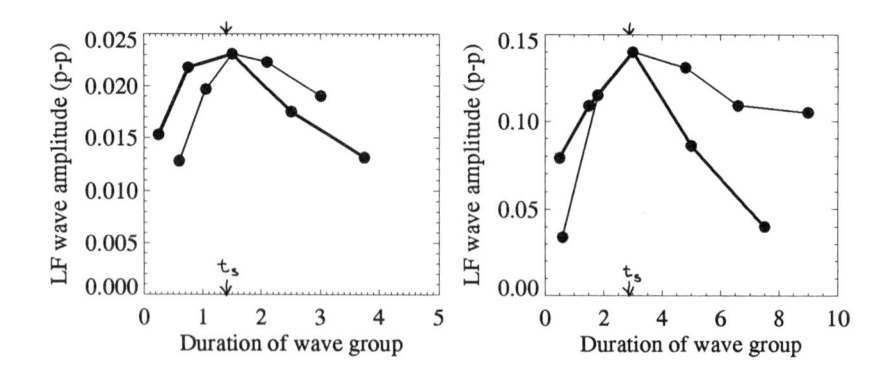

Figure 9: Dependence of LFW amplitude on wave group period T. (a) Amplitude 0.1. (b) Amplitude 0.6.

If the wave group period is longer than this swash period ($T \gg t_s$), then succeeding waves tend to meet swash already coming down the beach, and this opposing effect restricts the development of a large swash. In the extreme case of uniform waves, the swash zone becomes very narrow. For $T \ll t_s$, there is a strong fall-off of LFW amplitude as T decreases. Figure 8 shows that in these cases, the period of the LF motion does not decrease beyond t_s. These shorter groups have less momentum and energy, so can drive less and less fluid up the beach as T decreases.

With continuously modulated waves, rather than a single group, the number of wave groups or fraction of a wave group that is within the surf zone at any one time is expected to be relevant. This may be represented by a group-based surf-similarity parameter, $G = T/t_b$, where T is the duration of the group and t_b is the time it takes the largest wave to reach the shore after breaking. G is small for a wide surf zone and large for a narrow one. If G is small, several groups may simultaneously be generating LFW. There will then be substantial interference, which is likely to be destructive. If it is large, there will only be one or two waves in the surf zone at once: little interaction can occur and LFW generation is expected to be minimal. The strongest LFW generation is expected to occur when G is of order one. This idea remains to be investigated more thoroughly.

6. Discussion and Conclusions

The experiments clearly show the generation of LFW, and the numerical modelling successfully predicts their form and amplitude, even though the finer details of wave breaking are not included. It is clear that friction is important on the gentler slopes, on a laboratory scale, and reduces LFW amplitudes.

In interpreting the numerical and laboratory experiments it appears that the swash zone, and in particular the period of swash from the largest wave of a group, is an important feature. It is not entirely clear however, whether this is an indication of set-up generated in the approach to the shore line or a process centred close to and in the swash zone. In addition, for the practical case where there may be continuous wave modulation we suggest a group surf-similarity parameter based on the size of the surf zone relative to individual groups should be important.

The numerical experiments also indicate a slightly stronger than linear dependence of LFW amplitude on incident wave amplitude, in accord with field observations (Herbers et al., 1992). However, the question as to what LFW are incident on the surf zone, has yet to be resolved. The behaviour of wave groups with differing set-down has been investigated, but is not reported here. The development and decoupling of incident bound waves, as they enter shallow water where the Stokes theory becomes invalid and wave crests behave more like individual solitary waves needs to be determined. This is currently under investigation using a fully nonlinear potential flow solver.

Acknowledgements

This work was funded by the Commission of the European Communities, DG XII, under Contract No. MAS2-CT92-0027 (MAST-II G8-M Coastal Morphodynamics). The authors are grateful to the staff of HR Wallingford for providing facilities and support for the experiments. T. Barnes was funded by an EPSRC CASE research studentship with HR Wallingford. Dr. E.F. Toro of Cranfield Institute of Technology provided the core of the numerical algorithm used for solving the nonlinear shallow water equations.

References

Cox, D.T., N. Kobayashi and A. Wurjanto (1992). Irregular wave transformation processes in surf and swash zones. *Proc. 23rd Int. Conf. Coastal Eng.*, 156–169.

Dingemans, M.W. (1995). *Water wave propagation over uneven bottoms*. World Scientific, in press.

Hamm, L., P.A. Madsen and D.H. Peregrine (1993). Wave transformation in the nearshore zone: a review. *Coastal Eng.*, **21**, 5–39.

Herbers, T.H.C., S. Elgar, R.T. Guza & W.C. O'Reilly (1992). Infragravity-frequency (0.005–0.05 Hz) motions on the shelf. *Proc. 23rd Int. Conf. Coastal Eng.*, 846–859.

Kostense, J.K. (1984). Measurements of surf beat and set-down beneath wave groups. *Proc. 19th Int. Conf. Coastal Eng.*, 724–740.

List, J.H. (1992a). Breakpoint-forced and bound long waves in the nearshore: a model comparison. *Proc. 23rd Int. Conf. Coastal Eng.*, 860–873.

List, J.H. (1992b). A model for two-dimensional surf beat. *J. Geophys. Res.*, **97**, 5623–5635.

Mase, H. (1994). Uprush-backrush interaction dominated and long wave dominated swash oscillations. *Proc. Int. Symp.: Waves — Physical and Numerical Modelling*, UBC Vancouver, August 21–24 1994, eds. M. Isaacson and M. Quirk, 316–325.

Peregrine, D.H. (1967). Long waves on a beach. *J. Fluid Mech.* **27**, 815–827.

Roelvink, J.A., H.A.H. Petit and J.K. Kostense (1992). Verification of a one-dimensional surf-beat model against laboratory data. *Proc. 23rd Int. Conf. Coastal Eng.*, 960–973.

Roelvink, J.A. (1993). Surf beat and its effect on cross-shore profiles. *Ph.D. Thesis*, Technical University of Delft, 116 pp.

Schäffer, H.A. (1993). Infragravity waves induced by short-wave groups. *J. Fluid Mech.* **247**, 551–588.

Watson, G. and Peregrine, D.H. (1992). Low frequency waves in the surf zone, *Proc. 23rd Int. Conf. Coastal Eng.*, 818–831.

Watson, G., Peregrine, D.H. and Toro, E.F. (1992). Numerical solution of the shallow-water equations on a beach using the weighted average flux method. In *Computational Fluid Dynamics '92*, Volume 1, C. Hirsch et al. (Eds.), 495–502.

CHAPTER 58

Influence of Long Waves on Ship Motions in a Lagoon Harbour

Volker Barthel[1] and Etienne Mansard[2]

ABSTRACT

A physical model study of a vessel moored inside a lagoon harbour was undertaken in order to optimize a breakwater layout that best reduces the penetration of long wave energy inside the lagoon. The longer the breakwater at the entrance: the smaller the surge motion inside the lagoon. A combination of a short breakwater at the entrance channel and a pair of groynes in front and in the rear of the vessel also led to an efficient reduction of the surge motion. However, cost estimates and evaluation of vessel manoeuvrability are required to make an optimal choice of the breakwater layout. The use of a deterministic approach in wave generation led to a quicker testing procedure.

INTRODUCTION

In the design process of a harbour, its location, size and infrastructure are usually determined by economic factors such as the magnitude and structure of the expected marine traffic, the hinterland, the amount of throughput cargo and many other parameters. A preliminary lay-out of basins, berths, access channels, jetties, breakwaters and other coastal structures is then optimized with the help of numerical and/or physical models of tidal motions, currents, waves and sediment transport. However, the fine-tuning of the design such as the length and orientation of breakwaters can be carried out only with the help of a ship mooring study. This is generally done in a physical model, although various promising numerical approaches have been presented in literature (see for instance Sand and Jensen, 1990).

---

[1] Federal Admin. of Waterways and Navigation, Directorate North Kiel, Germany.
[2] Institute for Marine Dynamics, National Research Council of Canada, Ottawa, Canada.

This paper describes a physical model study of a 35,000 tdw bulk carrier moored in a lagoon harbour. The objective of the study was to evaluate the magnitude of vessel response induced by irregular waves. It was therefore, focussed on the propagation of irregular waves through an access channel, past a breakwater, into a lagoon harbour. The modification of the irregular wave train and its group-bound long wave components, during the shoaling process and the effect of various breakwater layouts on their penetration into the lagoon harbour were evaluated by measuring the motions of the moored vessel with highly sensitive instrumentation.

EXPERIMENTAL SET-UP

Figure 1a is a sketch of the lay-out of the lagoon harbour used in this study. It consisted of an offshore section of 25m depth, a 1:100 sloped bathymetry and a channel at the 15m contour cutting into this slope. At the end of the 1250m access channel, a circular lagoon with a radius of 280m accommodated a trestle structure berth and a ship moored to it. Permeable gravel beaches were installed on either side of the entrance channel to absorb incident wave energy. Slopes of the harbour basin were at 1:10. Since a model bulkcarrier of 35,000 tdw was readily available at a scale of 1:70, the same scale was chosen for the entire investigation. The hydrodynamic similarity of the vessel had been established in previous tests, including the definition of CG, GM, roll period and pitch gyradius. The model ship was ballasted for a 100% load condition.

In order to obtain the highest possible accuracy in vessel response, mooring lines and fenders of the model were designed to reproduce non-linear characteristics of the desired prototype lines and fenders including hysteresis. Forces were measured by high-resolution load cells, while the six-degrees-of-freedom motions of the model were monitored using an optical tracking system. Details of the instrumentation and mooring line simulators can be found in (Barthel, et al., 1989; Laurich, 1989). The model was moored with 12 polypropylene mooring lines arranged in a pattern which had been previously optimized (Kubo and Barthel, 1992, Barthel et al, 1994). Figure 1b shows a photograph of the model set-up.

Water surface elevations were measured in the model basin using capacitance type wave gauges. Their locations are shown in Fig. 1a. The array of gauge 1 - 5 was used to determine reflections coming from the model layout. Simulation of waves in the model basin was carried out using the wave generation package developed at the Hydraulics laboratory of the National Research Council Canada. This package contains several algorithms for simulation of uni and multidirectional waves.

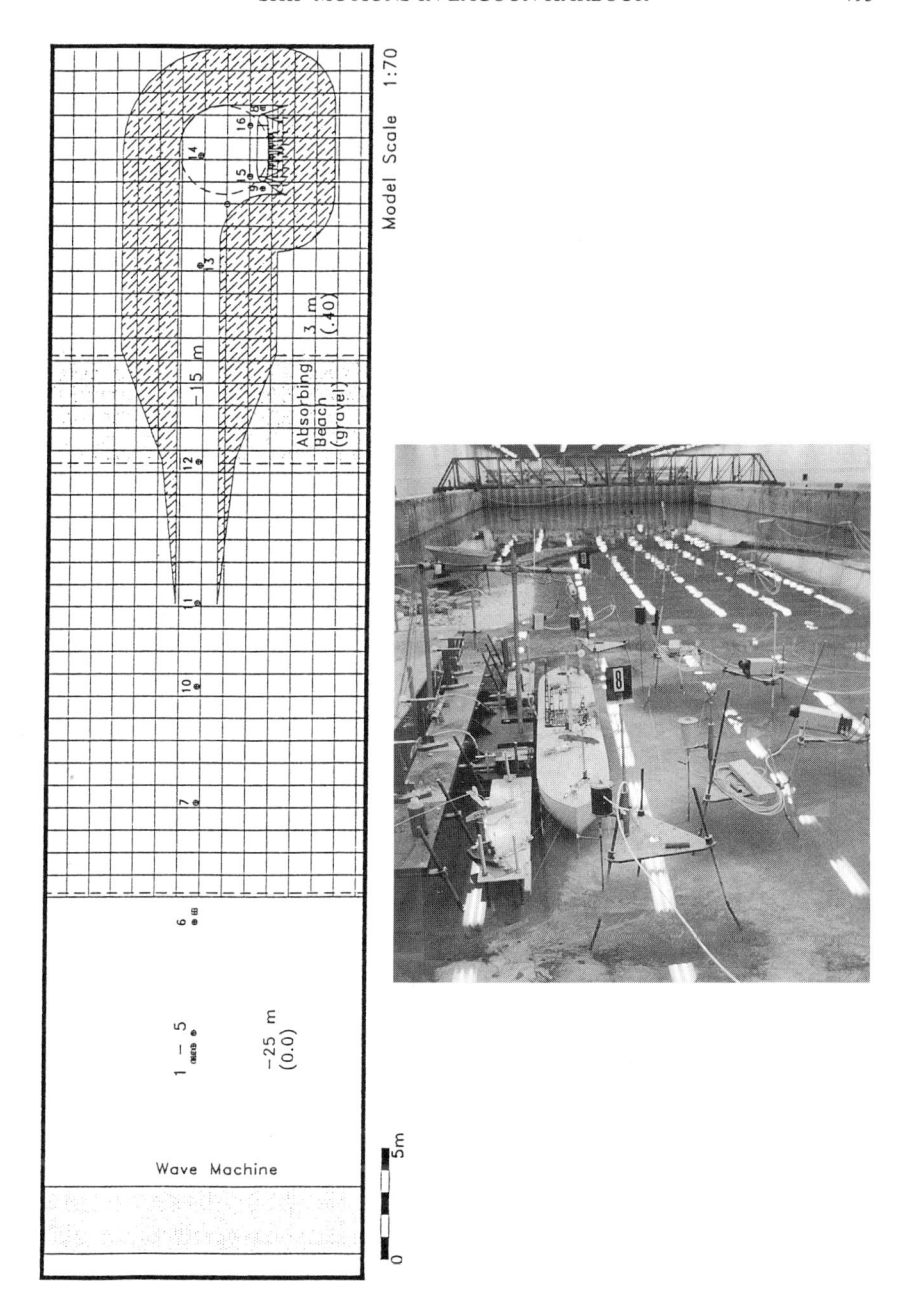

Figure 1. Layout of the Experimental Set-up : a) Sketch indicating the position of gauges; b) Photograph showing the vessel in place.

PROPAGATION OF LONG WAVES INTO THE LAGOON

The understanding of the physical processes involved in the propagation of waves on a complex bathymetry and layout such as the one used in this study is limited. However, it can be expected that, as the waves propagate the shoaling of wave groups and their associated group-bound long waves will induce the breaking of the first order (short) waves and cause their energy dissipation. The long waves can then be expected to detach themselves from the group and continue to travel at free wave velocity to be either reflected off the beach or to propagate along the beach at an angle (set-up / edge waves). An experimental description of this process was attempted by (Mansard and Barthel, 1984). The present situation is however more complicated. From what could be observed during tests, the following happens:

During the propagation of waves from deep to shallow water, the shoaling process induces breaking of the first order waves and releases the energy of the group bound long waves that continue then to propagate into the lagoon. These long waves, although hardly visible to the observer because of their small amplitudes, can move the vessel to its maximum limit. In this particular set-up, the surge motion and the loads on the mooring lines which restrained the surge motion, were most sensitive to long waves.

The main objective of this study was therefore to optimize a breakwater lay-out that will reduce the penetration of the long wave energy and the resulting ship motions.

SHIP MOORING TESTS

Experimental Procedure

Many laboratories in the world use the stochastic approach in ship mooring tests by subjecting the model to a very long time series of water surface elevation, in order to ensure the inclusion of worst combination of waves that will cause the maximum motions of the vessels, as well as maximum loads on the mooring lines. In this particular study, a deterministic approach was adopted by simulating short time series of pre-defined surface elevation, in order to save testing time and cost of model investigations. This use of short time series was also expected to minimize any potential build-up of long wave energy in the model set-up.

The time series used in this study were 2.4 minutes long in model scale and it corresponded to 20 minutes in prototype duration. This length was also equivalent to 12 periods of 100s long waves that were found to excite the ship substantially in an earlier study (see Barthel et al, 1994). In order to select time series that would be appropriate for this deterministic approach, the methodology described below was used.

Synthesis and selection of time series

From a JONSWAP spectrum, time series of water surface elevation were first synthesized by the commonly used random phase spectrum method. This method pairs a given amplitude spectrum with a phase spectrum generated by random numbers. By varying the random numbers, different realizations of wave records, with varying time domain characteristics can easily be simulated.

Since it is believed that narrower spectra result in higher degree of wave grouping, JONSWAP spectra with two different values of γ (i.e. $\gamma=1$ and $\gamma=7$) were used in this synthesis. (According to Sand (1982), higher degree of grouping results in larger amplitude of group-bound long waves that are responsible for the excitation of the horizontal motions of the vessel). From each target spectrum, ten realizations of time series were synthesized and reproduced in the model. The response of the vessel was then measured for all the twenty realizations of the time series described above. Figure 2 presents the results obtained by this procedure.

The RMS values (i.e. standard deviation) of the long waves measured in the lagoon, in the proximity of the vessel are shown in Figure 2a for the various realizations of time series. These long waves were derived by low-pass filtering the water surface elevation with a frequency cut-off of 0.03 Hz (full scale). Figure 2b illustrates the RMS values of the corresponding surge motions.

The results show that narrower spectra ($\gamma=7$) results in higher long wave content and consequently larger surge motions. Amongst the ten realizations of the time series under each γ value, the 8th time series generally leads to the largest response of the vessel. This time series will be denoted in this paper as TRN8. It is also noticeable that the 4th realization (TRN4) generally provides the smallest response. The above findings were also confirmed by analyzing the sway motion and the loads on the mooring lines.

Figure 3 shows the time series of water surface elevations corresponding to TRN4 and TRN8 under the two γ values of JONSWAP. The corresponding SIWEH functions which illustrate their grouping pattern are also illustrated along with the appropriate values of Groupiness Factors (see Funke and Mansard, 1979 for the definitions of SIWEH and Groupiness Factor).

Since the length of each of these time series was only 2.4 minutes in model scale, it was decided to use all the four time series (i.e. TRN4 of $\gamma=1$ and $\gamma=7$ and TRN8 of $\gamma=1$ and $\gamma=7$) for the optimization of the breakwater layout, in order to validate the resulting solutions under four different sea states.

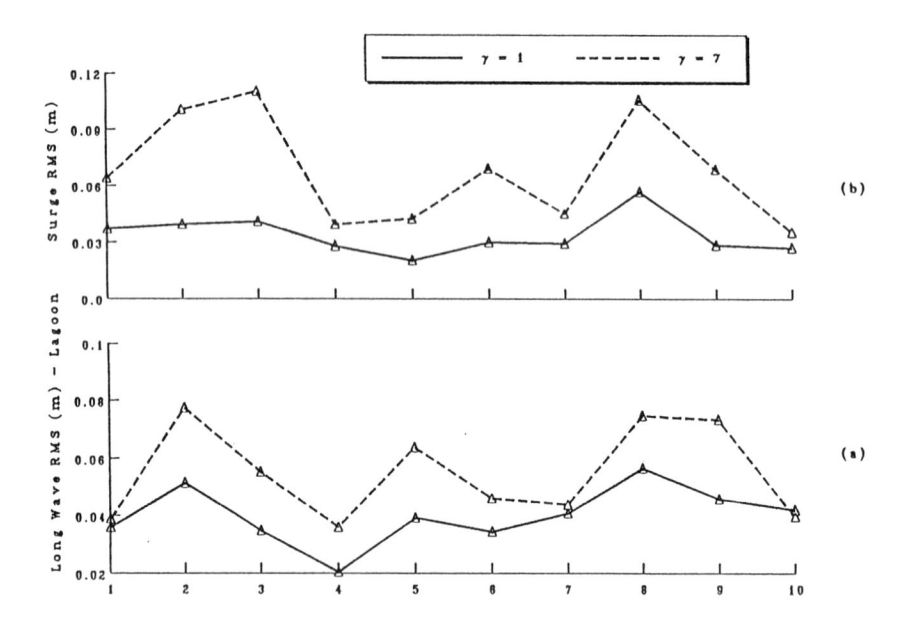

Figure 2. RMS values of long waves and surge motion inside the lagoon harbour for the ten realizations of time series.

LAYOUT OF BREAKWATERS

Several breakwater configurations were evaluated in this study during the optimization procedure. They are, as can be seen in Figure 4: a long version of the breakwater (BWL), a medium length structure (BWM) and a short breakwater with (BWD) and without (BWS) a second barrier. In addition several versions of groynes (BWSG1, BWSG2 and BWSGD) upstream and downstream of the moored vessel were also tested in conjunction with the short breakwater BWS.

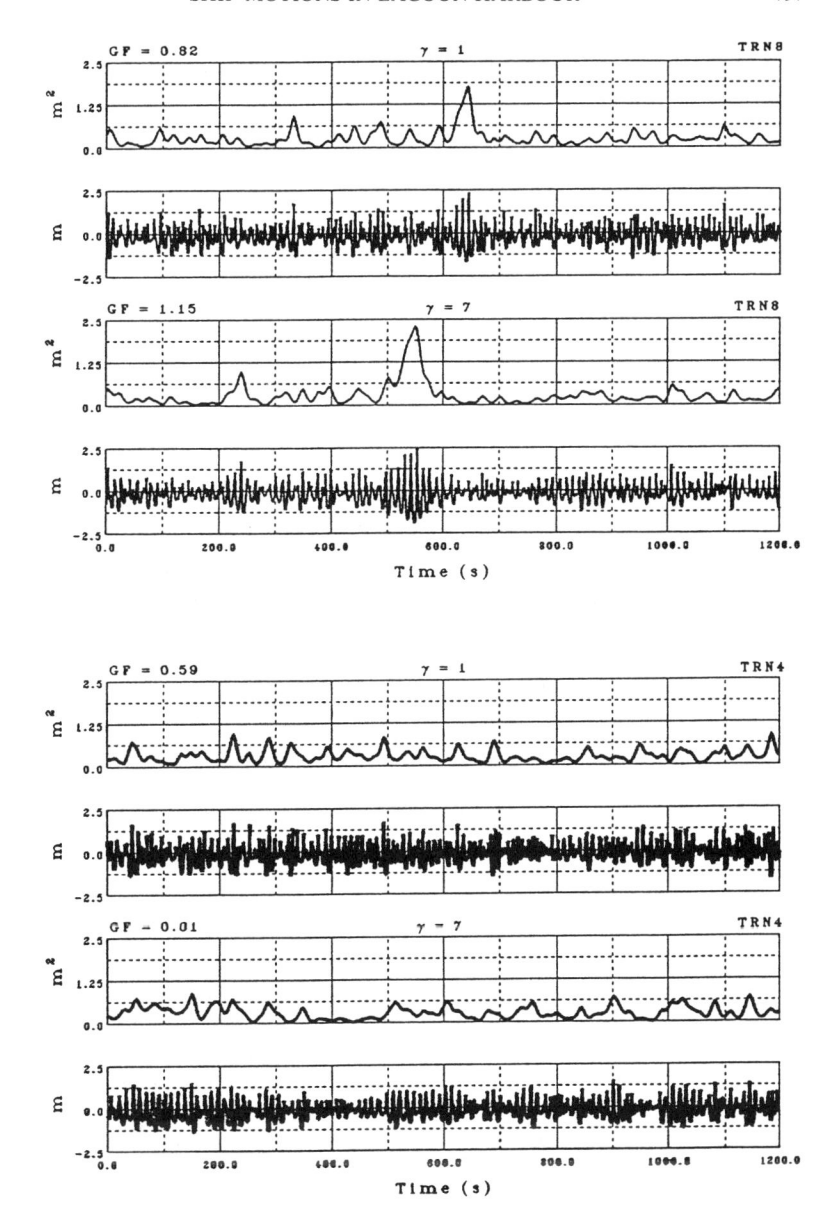

Figure 3. Time Series and SIWEH of the Selected Time Series

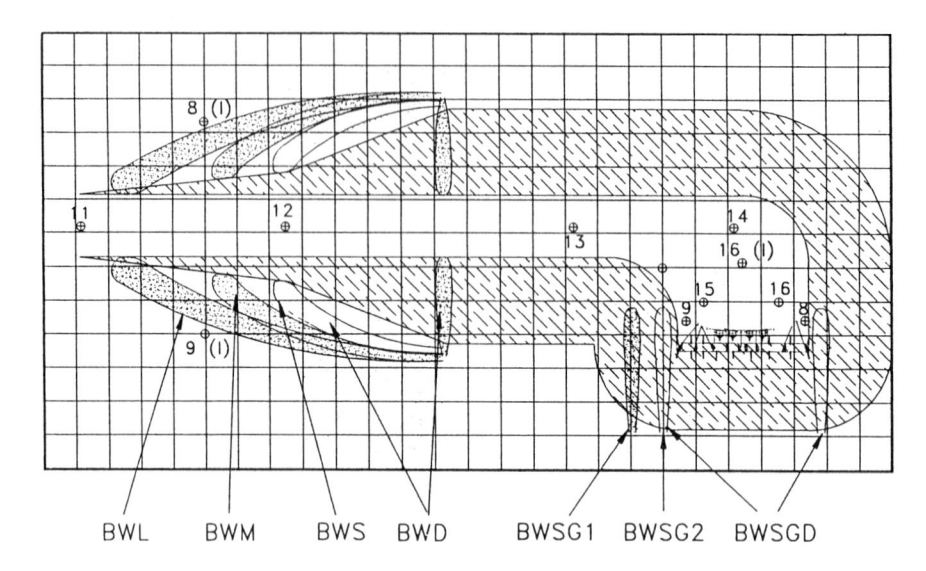

Figure 4: Sketch of the various breakwater options tested

The obvious questions that were expected to be answered by installing these breakwater configurations were the following:
Can a breakwater help reduce the penetration of long wave energy?
Is there an optimal layout which ensures an efficient reduction of energy with smaller capital costs?

PEFORMANCE OF BREAKWATER LAYOUTS

Influence of Short Breakwater (BWS)

In terms of engineering practice and as an obvious economical measure for reducing the wave excitation inside the harbour, the short breakwater seemed to be a reasonable solution. This breakwater extended from the high water line at the beach to the edge of the sloped channel. Its performance is illustrated in Figure 5 by comparing it with results obtained without any breakwater (i.e. NBW). In Figure 5a, a comparison of the spectra and the time series measured at the offshore region during the No and Short breakwater tests, are presented to illustrate the similarity in the inputs. Also shown in this Figure is the standard deviation of the long wave energy measured at different locations of the experimental set up (see Figure 1 for the locations).

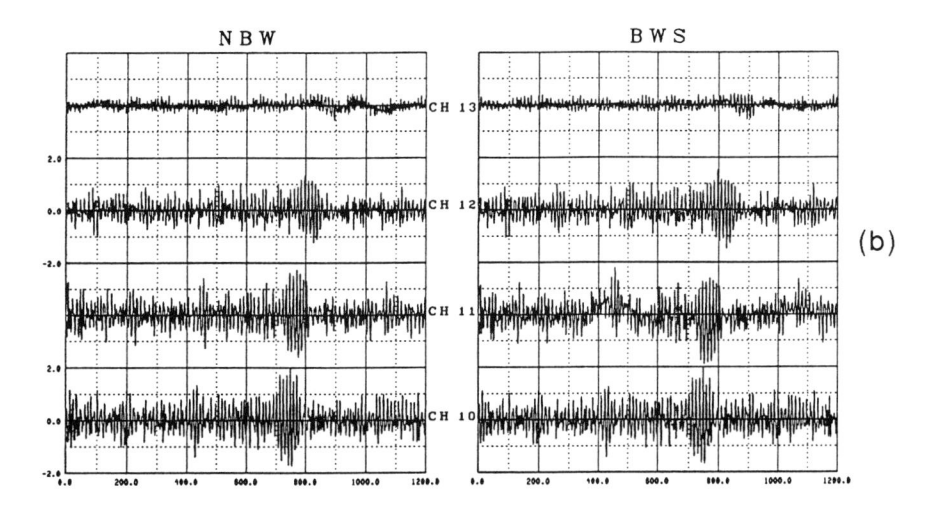

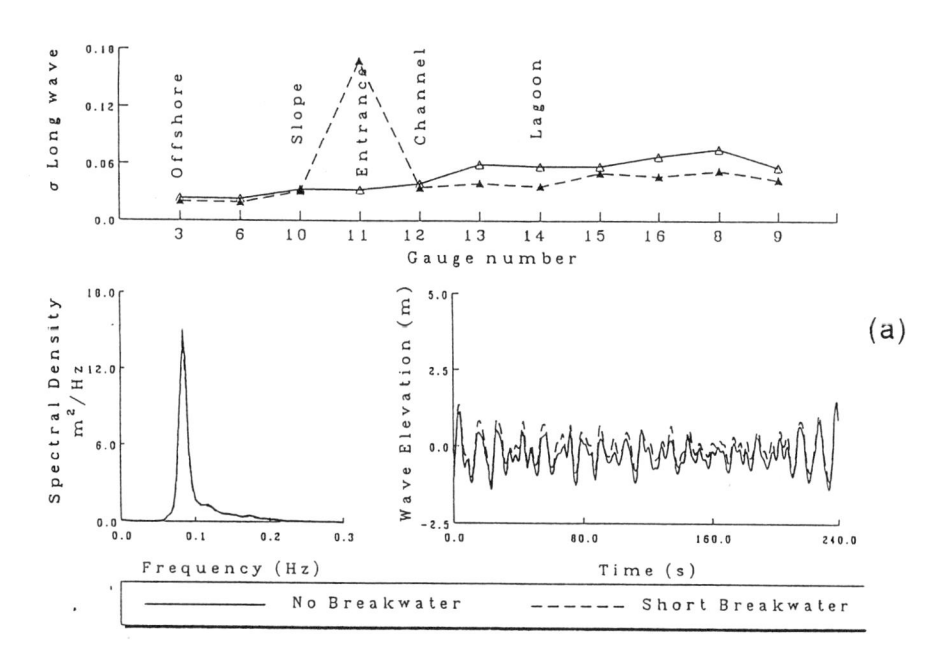

Figure 5. Comparison of results with and without short breakwater

It is interesting to note in Figure 5a that the long wave energy prevailing in the lagoon is substantially reduced by the presence of the short breakwater while at the entrance of the channel, the presence of short breakwater induces a sharp rise in the energy of the long waves. On examining the time series presented in Figure 5b, it appears that this sharp rise may be due to the reflective nature of the breakwater. In spite of this, the net long wave energy that penetrates the channel is shown to be smaller, as the value at gauge 12 illustrates. This reduction leads subsequently to small amplitudes of long waves inside the lagoon. A similar physical process was found when testing with other sea states.

Figure 6 provides an overall view of the results for the four pre-selected sea states, by presenting the spectral densities of long waves measured in the proximity of the vessel by the gauge 8 and the resulting surge spectra. The legends presented in this Figure correspond to the following situations:

GAM7_TRN8 correspond to $\gamma=7$ and TRN8; NBW and BWS correspond to No and Short Breakwater situations.

As can be seen in Figure 6a, the shape of the long wave spectra encountered near the vessel remains nearly the same under the two situations of breakwater. The sea state with the largest degree of grouping exhibits the most dramatic influence of the breakwater, possibly due to severe breaking induced by the largest wave group.

Tests using Bichromatic waves

In order to confirm the good performance of the short breakwater a test series was also undertaken using bichromatic waves. Bichromatic waves are waves composed of two primary frequency frequency components f_1 and f_2, and a long wave component with a frequency (f_1-f_2). In an earlier study of this lagoon harbour, Barthel et al (1994), determined the critical frequency at which the ship responded significantly, by running several combinations of (f_1-f_2). The particular bichromatic wave which created this critical frequency (i.e. 100s prototype) was also used in this study to assess the performance of the short breakwater. Figure 7 illustrates the results obtained for three different amplitudes of the long waves characterized by the gain factor. The influence of the short breakwater causing an effective reduction in the long wave amplitude and in the surge motion can easily be appreciated.

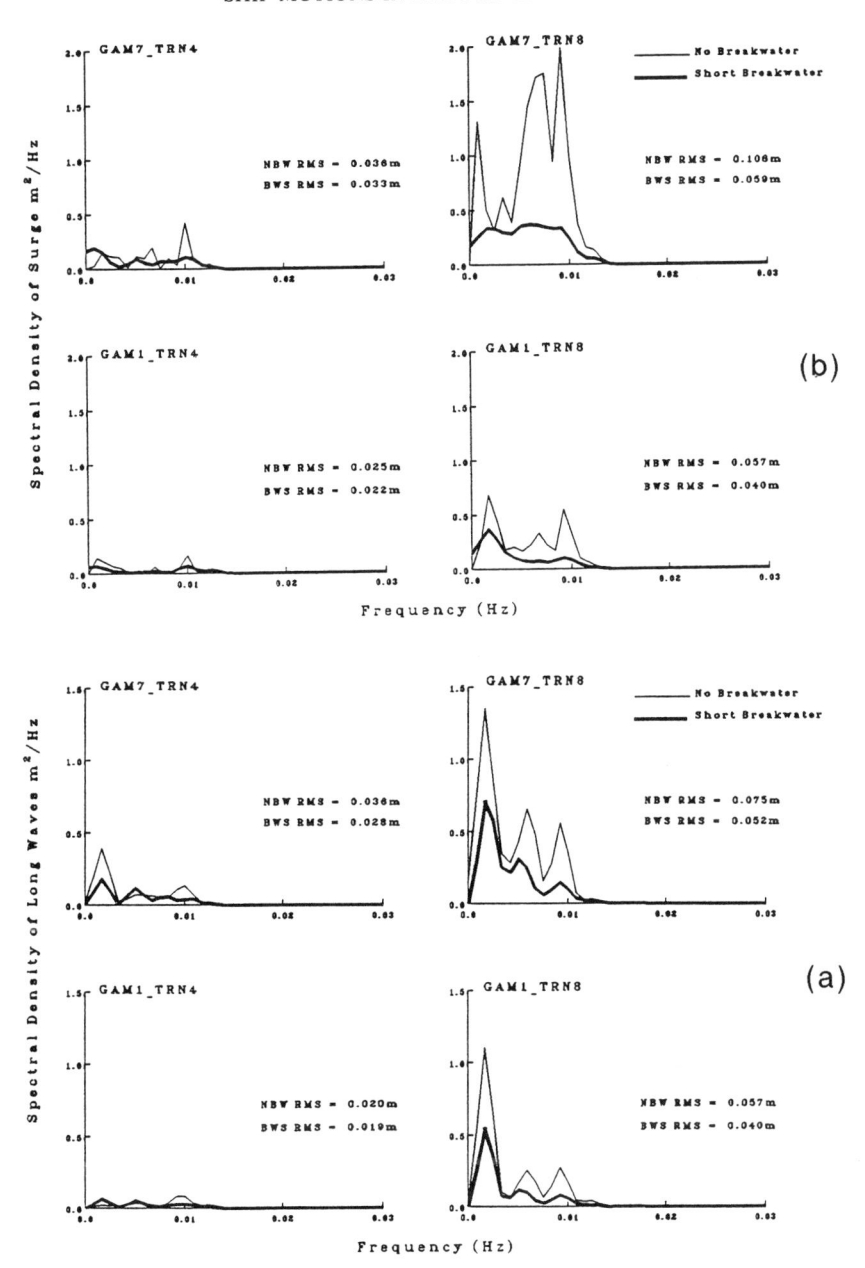

Figure 6 : Spectra of long waves and surge motions under No and Short breakwater situations.

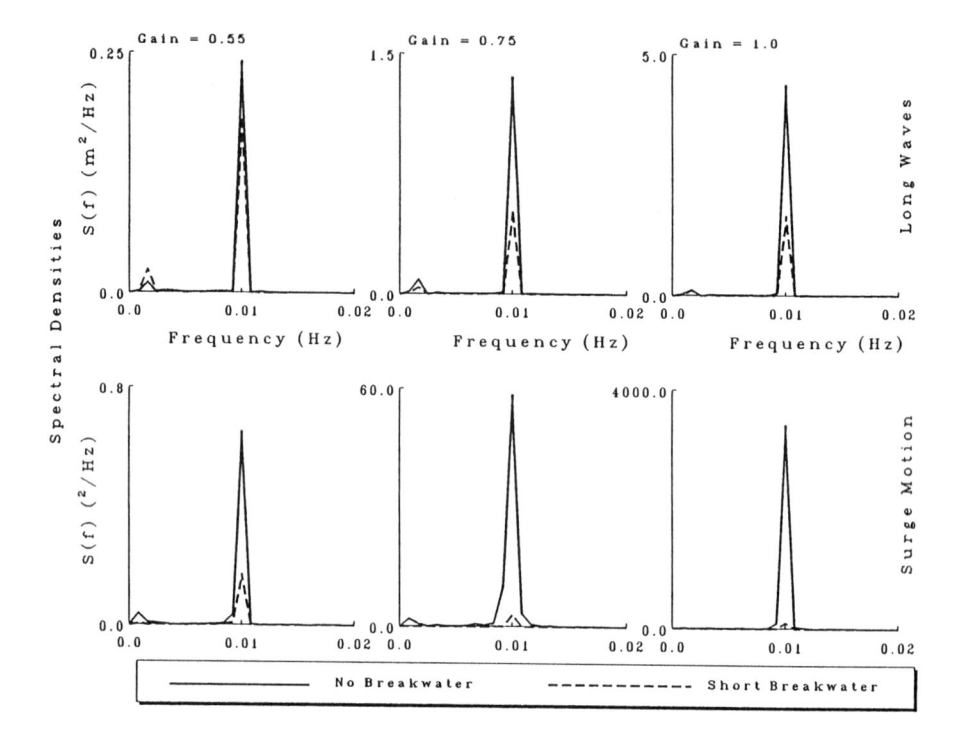

Figure 7. Test results under bichromatic waves

Performance of Other Breakwater Options

A limited series of tests with longer breakwaters (BWM AND BWL) and with the entrance barrier (BWD) showed that they can reduce the ship motions further. However, their results have to be analysed in conjunction with the cost of the structures and with the ship manoeuvrability studies for vessel approach.

Influence of Groynes.

It was speculated that although the short breakwater is effective in reducing the penetration of the long wave energy inside the lagoon, the groynes close to the structure can effectively cut-off the influence of orbital current associated with the long waves and thereby reduce the motions of the vessel and the loads on the mooring lines. Although the installation of groynes will restrict the manoeuvrability of the vessel, it is believed that tug assistance will be needed in a closed environment such as this lagoon. Hence the trade-off between a longer berthing procedure and a safer mooring system had to be explored. Tests were therefore initiated with one groyne in front of the moored vessel (BWSG1 and BWSG2) and pursued with a combination of one in front and another in the rear (BWSGD). Note that these groynes were evaluated with the short breakwater in place.

As expected, the combination of groynes in front and in the rear generally provided the best protection.

Test Series with Grouped wave trains

In order to validate this optimization process under different severities of sea states, a grouped wave train, synthesized using the SIWEH concept was used. A JONSWAP spectrum with γ =3.3 and a peak period T_p = 12 s was chosen for this simulation. The SIWEH function adopted for this purpose had a Groupiness Factor of 0.9 and a peak period of 120s (see Funke and Mansard 1979 for details on this synthesis process). Six different values of H_{m0} were used in this test series. The results were then used to assess the relationship that may exist between long wave energy and vessel response. Figure 8 presents the RMS values of the surge measured under different breakwater configurations as a function of the long wave energy measured at gauge 10, just before the entrance of the channel. Although the model was subjected to its maximum response, these results clearly illustrate the the good performance of the BWSGD option (short breakwater with two groynes). Note that for this particular test series, other options such as BWL, BWD and BWSGD1 were not unfortunately explored.

CONCLUSIONS

A physical model of a ship moored inside a lagoon harbour under the attack of irregular waves with different grouping properties permitted to make the following observations:

- Sophisticated simulation and wave generation techniques including non-linear properties of mooring lines and fenders are required to achieve the necessary accuracy for ship mooring studies;

- Instead of reproducing long time series of water surface elevation, a deterministic approach of using short realizations of pre-selected time series was adopted successfully in the optimization of the breakwater layout; This approach helped reduce the testing time;

- While inside the lagoon first-order waves were hardly visible, long waves of low amplitudes induced significant motions of the vessel;

- The arrangement of breakwaters at the entrance channel enhanced the dissipation of the long wave energy and reduced the ship motions; The longer the breakwater: the greater the reduction in surge motion.

- Substantial reduction of the surge motion achieved by the short breakwater option was well illustrated by the test series;

- The construction of groynes interrupting the orbital currents associated with long waves is another option to reduce the surge motion further;

- Although efficient breakwater lay-outs have been established in terms of vessel motions inside the lagoon during this experimental program, manoeuvrability of the vessel in the presence of reflections in front of the main breakwater and in the presence of groynes inside the lagoon was not part of this study and therefore has to be explored.

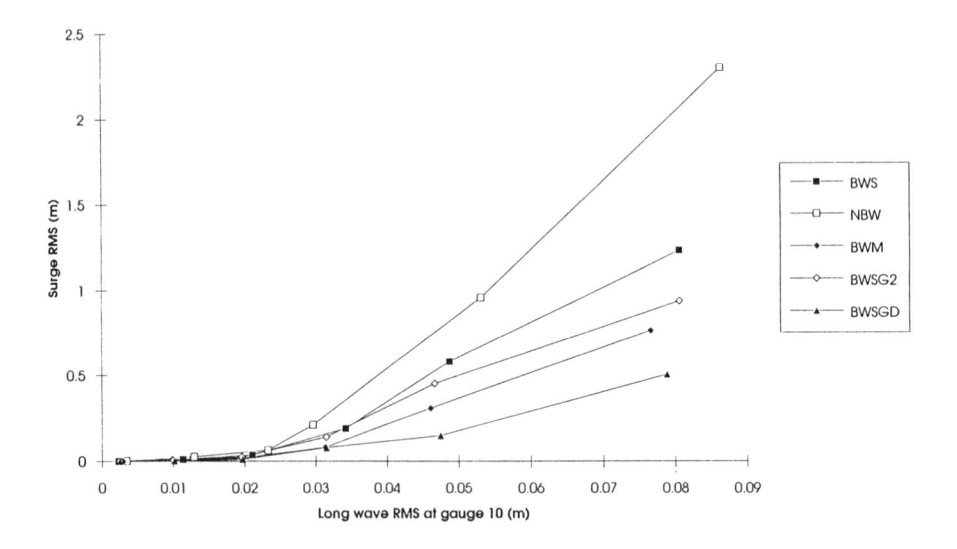

Figure 8 : RMS values of the surge motions as a function of the long wave RMS measured at gauge 10, under different breakwater options.

ACKNOWLEDGEMENTS

The authors acknowledge their sincere thanks to D.D. Macdonald for his technical assistance. The large number of tests, their analysis and presentation would not have been possible without his help and patience.

REFERENCES

Barthel, V., Mansard, E.P.D. and Macdonald, D.D., 1994, "Penetration of Long waves into a Lagoon Harbour and Resulting Ship Motions", Proc. HYDRO-PORT'94, Port and Harbour Research Institute, Yokosuka, Japan.

Barthel, V., Funke, E.R. and Laurich P.H., 1989, "Simulation of a Moored Ship in Waves", Instrumentation Workshop of the IAHR. Burlington, Canada.

Funke, E.R. and Mansard, E.P.D., 1979, "On the Synthesis of Realistic Sea States in a Laboratory Flume", Technical Report, National Research Council of Canada, LTR-HY-66.

Kubo, M. and Barthel, V., 1992, "Some Considerations How to Reduce the Motions of Ships Moored at an Open Berth", The Journal of Japan Institute of Navigation, Vol. 87, 91/92.

Laurich, P.H., 1989, "Instrumentation Used at the Hydraulics Laboratory of the National Research Council Canada", Proc. IAHR Workshop on Instrumentation, Burlington, Ontario, Canada.

Mansard, E.P.D. and Barthel, V., 1984, "Shoaling Properties of Bounded Long Waves", Proc 19th International Conference on Coastal Eng. Houston, TX, USA: 798-814.

Sand, S.E., 1982, "Wave Grouping Described by Bounded Long Waves", Journal of Ocean Engg., Vol. 9, No. 6.

Sand, S.E. and Jensen, O.J., 1990, "Integration of Marine Simulation in Harbour Design", Proc. 22nd ICCE, Delft, The Netherlands.

CHAPTER 59

Resonant Forcing of Harbors by Infragravity Waves

Gordon S. Harkins, A. M. ASCE and Michael J. Briggs[1] M. ASCE

Abstract

An extensive study of Barbers Point Harbor, including field wave gaging and numerical and physical modeling, was conducted to analyze the harbors response to infragravity wave energy. Infragravity waves have been defined as waves whose periods are greater than 25 sec. Field data collected over a 4-year period were used to calibrate the numerical model while eight extreme events were simulated in the physical model. Agreement between the numerical model, physical model, and the prototype data is good. The importance of spectral shape in the frequency domain also was analyzed by comparing the results from broad and narrow spectra.

Introduction

Frequency spectra of ocean waves consist of wind waves whose periods are less than 25 sec and longer period infragravity waves whose periods are greater than 25 sec. Although infragravity waves are rarely seen by the casual beach goer, they are important for coastal processes, including harbor seiching. Harbors with sides on the order of 500 m in length and depths on the order of 10 m are subject to harbor oscillation on the order of 1 minute and longer. Barbers Point Harbor on the Island of Oahu, Hawaii, is a prime example of a harbor that is subject to infragravity harbor resonance.

Sea and swell wave periods contribute the majority of energy in the frequency spectra. Sea waves are locally generated wind waves whose peak period is usually less than 10 sec. Sea waves also are characterized by a broad

---

[1] Research Hydraulic Engineer, USAE Waterways Experiment Station, Coastal Engineering Research Center, 3909 Halls Ferry Rd., Vicksburg, MS 39180.

spectrum both in direction and frequency. Swell waves, on the other hand, are generated far from the area of interest and have peak periods usually greater than 10 sec. These waves, unlike sea waves, tend to exhibit wave groupiness and have both narrow directional and frequency spectra.

Infragravity waves contribute the remaining energy in the frequency spectrum. Infragravity energy can be divided into bound and free wave energy. Bound or forced infragravity waves are nonlinearly coupled to wave groups, traveling at the group velocity of wind waves, and phase locked to sea and swell waves (Longuet-Higgins 1962). Free infragravity waves are further subdivided into leaky and edge waves.

The importance of frequency spread in the generation of infragravity energy has been widely reported from field data and theoretical analysis. Sands (1982) showed theoretically that when peakedness of the frequency spectrum increased, so did the amplitude of long-period waves generated.

This study is unique in that field wave gaging occurred prior to physical and numerical model studies. Numerical and physical models of Barbers Point Harbor were constructed to evaluate the resonant response of six different harbor expansion configurations; however, only results for the existing harbor layout will be discussed in this paper. Eight field wave cases were simulated in the physical model and compared to numerical and prototype results. To evaluate the importance of frequency spread on the generation of long-period waves, 18 empirical seas were analyzed in which the width of the frequency spectra was varied.

Description of Harbor

Barbers Point Harbor is located on the southwest coastline of Oahu, Hawaii (Figure 1). The harbor complex presently consists of an entrance channel, deep-draft harbor, barge basin, and a resort marina (often referred to as West Beach Marina). The parallel entrance channel is 140 m wide, 945 m long, and 13 m deep (MLLW). The deep-draft harbor basin is 11.6 m deep, 670 m wide, and 610 m long, covering an area of 0.37 sq km. Rubble-mound wave absorbers line approximately 1,400 linear meters of the inner shoreline of the harbor basin. The barge basin, located just seaward of the harbor on the south side of the entrance channel, is poorly sheltered from incident wave energy. It is 67 m by 400 m and is 7 m deep. The private West Beach Marina was built to the west of the deep-draft harbor. It shares the same entrance channel, is 4.6 m deep, and covers approximately 0.08 sq km. The marina was designed to accommodate 350 to 500 pleasure boats.

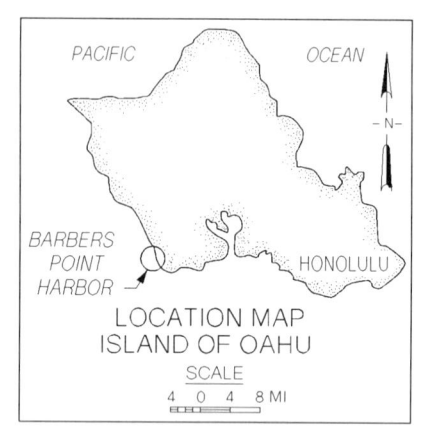

Figure 1. Project location

Prototype Measurements of Waves

Prototype measurements of waves were made in Barbers Point Harbor between July 1986 and March 1990 as part of the Monitoring Completed Coastal Projects (MCCP) Program and the Coastal Data Information Program (CDIP). These programs provide a network of real-time wave gages and are jointly sponsored by the Corps of Engineers, California Department of Boating and Waterways, and Scripps Institution of Oceanography (SIO). Figure 2 shows selected sites in the main harbor, entrance channel, and nearshore region. Bottom-mounted pressure gages were used to minimize interference with navigation (Briggs et al. 1994).

A four-gage S_{xy} array was used offshore to measure incident directional spectra conditions in 8.4 m of water. Individual gages were used elsewhere to measure frequency spectra. Other gages included the offshore (Of) and onshore (On) gages, both located shoreward of the S_{xy} gage. Channel entrance (Ce) and channel mid-point (Cm) gages were located in the entrance channel, where navigation conditions were a consideration. Finally, a gage was located in the south (Sc) corner of the harbor to measure anticipated maximum amplification factors.

A sampling scheme that collected both wind waves (energy) and long period waves (surge) was designed. Initially, energy and surge were obtained from separate records collected by each sensor: 1,024 samples at 1.0 Hz for the energy, and 2,048 samples at 0.125 Hz for the surge. After January 1989, a system upgrade permitted a single record of 4.6 hr at 0.5 Hz (8,192 samples) inside the harbor, or 1.0 Hz (16,384 samples) outside the harbor, to be collected by each sensor.

Sampling interval was controlled by varying the call-up schedule in the software. The standard interval was every 6 hr in summer and every 3 hr in winter. A threshold routine was built into the system that automatically switched the interval back to 3 hr if significant wave height exceeded 1 m offshore, or 30 cm in the harbor. On the 3 hr schedule, an enhanced sampling scheme provided a continuous record.

The S_{xy}, Ce, Cm, and Sc gages were installed in July 1986. The S_{xy} gage experienced two major data gaps from cable failures when vessels pulling barges snagged the cable with their tow bridles. This problem was eliminated by moving the shore station to the navigation aid and re-routing the cable away from the entrance channel. Data from the second position of the S_{xy} (S_{xy2} in Figure 2) were believed to be more reliable because this gage was further from the edges of the

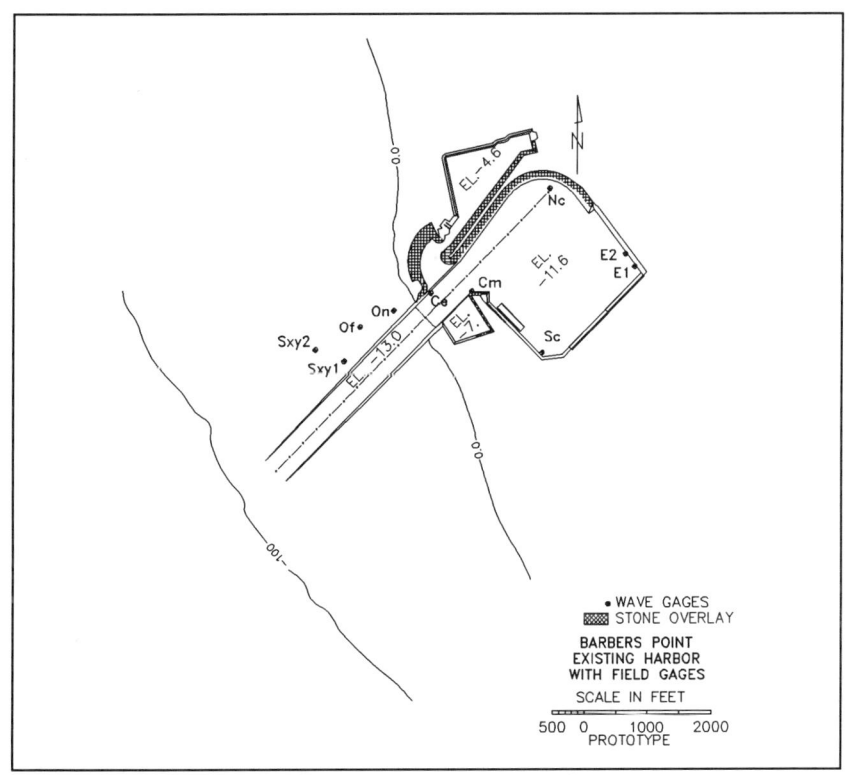

Figure 2. Prototype gage locations

entrance channel and any refractive effects which might have influenced the first position of the array. Construction in the harbor caused longer gaps in the Sc gage. In January 1989, additional sensors were installed in the north (Nc) and east corners to improve spatial resolution. The east corner gages are labeled E_1 and E_2 to differentiate the two different locations. At this time, the entire system was upgraded to the longer sampling scheme.

Numerical Model Description

The numerical harbor wave-response model, HARBD, was used to estimate wave oscillations in Barbers Point Harbor. HARBD is a steady-state finite element model which calculates linear wave oscillations in harbors of arbitrary configuration and variable bathymetry. The effects of bottom friction and boundary absorption (reflection) are included. Bottom friction is assumed to be proportional to flow velocity with a phase difference. Boundary reflection is based on a formulation similar to the impedance condition in acoustics and is expressed in terms of the wave number (i.e wavelength) and reflection coefficient of the boundary. The model uses a hybrid element solution method which involves the combination of analytical and finite element numerical solutions to determine the response of a harbor to an arbitrary forcing function.

Numerical Model Formulation

In model formulation for arbitrary depth water waves (i.e., shallow, intermediate, and deepwater waves), the water domain is divided into near and semi-infinite far regions. The near region includes the harbor and all marine structures and bathymetry of interest and is bounded by an artificial 180 deg semicircular boundary offshore of the harbor entrance. The far region is an infinite semicircular ring shape bounded by the 180 deg semicircular boundary of the near region and the coastline. The semi-infinite far region extends to infinity in all directions and is assumed to have a constant water depth and no bottom friction (Chen and Houston 1987). The finite near region, which contains the area of interest, is subdivided into a mesh of nonoverlapping triangular shaped elements. The length of side of each element is determined from the desired grid resolution and design wave parameters. The water depth and bottom friction coefficient are specified at the centroid of each element, and a reflection coefficient is assigned to each element along the solid, near region boundaries. The model requires wave period and direction as input. The solution consists of an amplification factor (i.e., the ratio of local wave height to incident wave height) and a corresponding phase angle for each grid point in the near region. Phase angle represents the difference in phase between the grid point and the incident wave. Contour plots of the amplification factors and corresponding phase angles are used to determine oscillation patterns occurring throughout the harbor.

The governing partial differential equation is derived through application of linear wave theory to the continuity and momentum equations. All dependent variables are assumed to be periodic in time with angular frequency, ω. These steps yield the generalized Helmholtz equation (Chen 1986).

The HARBD model is intended to simulate waves which can be adequately described by the governing generalized Helmholtz equation. Therefore, HARBD does not simulate nonlinear processes such as wave breaking, wave transmission and overtopping of structures, entrance losses, steep bathymetric gradients, and wave-wave and wave-current interaction. Fortunately, these limitations are not dominant for many harbors and HARBD can be applied with some degree of confidence. Since nonlinear processes naturally occur in the prototype, consideration of these effects must be taken in interpretation of results.

A hybrid element method is used to solve the boundary value problem. In this solution, a conventional finite element approximation is used in the finite near region, while an analytical solution with unknown coefficients is used to describe the semi-infinite far region. Conditions in the near and far regions must be matched along the artificial semicircular boundary. This requirement is met by HARBD routines which automatically match the solutions, using the stationarity of a functional, to a series of Hankel Functions which give the solution for the semi-infinite far region (Farrar and Chen 1987). The hybrid element numerical techniques used in the formulation are discussed in greater detail in Chen and Mei (1974).

Numerical Model Calibration

The numerical model grid was designed with a grid resolution, the length of each element, equal to approximately one-sixth of the local wavelength, based on linear wave theory using a wave period of 10 sec and the localized water depth. After the grid was generated, individual monochromatic waves with periods from 45 to 24,576 sec were run at increments of 0.00004069 Hz (1/24,576 sec) and the results were compared to the prototype data.

The HARBD numerical model has two free parameters which can be adjusted to match prototype data: bottom friction and reflection coefficients. Other nonlinear processes such as dissipation at the sidewalls and entrance are not included in this model. The boundaries for these long-period waves were felt to be nearly perfectly reflecting. The bottom friction coefficients, however, should be a function of the type of bottom material as a function of the wave period and corresponding wavelength. Therefore, the bottom friction coefficients were varied to calibrate the model predictions to the measured prototype values at each frequency peak or mode.

The HARBD model computes a standing wave for a given frequency. For a low frequency, or very long wavelength, the entire harbor responds as if it were a reflecting wall. A standing wave against a reflecting wall has a height of twice the incident wave. Therefore, the low-frequency wave height amplitudes predicted by HARBD for input frequencies between 0.000122 (8,196 sec) and 0.001343 (745 sec) were divided by two. Only the Helmholtz mode (or pumping mode of the harbor) was affected by this criteria because the wavelength of this wave encompasses the entire domain of the harbor and outer region to the S_{xy} gage.

The model was then tested at 0.00004069 Hz frequency increments with varying bottom friction coefficients. Resulting wave height amplifications from each test were compared with prototype measurements to investigate the reduction of wave energy due to the increase of bottom friction. This procedure was repeated until an accurate match of wave height amplification between the model predictions and prototype measurements was possible.

Physical Model Design

An undistorted, three-dimensional model of Barbers Point Harbor was constructed at a model-to-prototype scale $L_r = 1:75$, in accordance with Froude scaling laws. The nearshore area extends to the 30.5 m MLLW contour and includes approximately 1,065 m on either side of the entrance channel. Total area of the model was over 1,000 m^2.

Waves were generated with the directional spectral wave generator (DSWG) which can produce directional seas at multiple periods. The (DSWG) is an electronically controlled, electromechanical system, designed and built by MTS Systems Corporation, Minneapolis, MN. It is 90 ft long and consists of 60 paddles, each 1.5 ft wide and 2.5 ft high. Each wave paddle is independently driven at its joint by an electric motor operating in piston mode. This configuration, along with flexible plastic plate seals between the paddles, produces a smoother, cleaner wave form (Outlaw and Briggs 1986, Harkins 1991).

Physical Model Wave Conditions

Eight field events and eighteen empirical unidirectional spectral wave climates were generated. The eight field events were chosen from prototype wave data to obtain the largest wave height and a representative range of wave periods and direction within model constraints. All eight wave conditions represent rare events because of their large wave heights.

Directional spectra were recorded at the S_{xy} gage for the eight field events chosen. A control signal file for the 60-paddle DSWG was generated by reproducing 30 frequency bins from 0.01 - 0.3 Hz (prototype units) at 2.5° directional bins. Random phase was applied to the control signal generation and thus the long-period waves were not bound to the shorter period waves at the wavemaker. An iterative process was used until a suitable control signal spectrum was obtained (Briggs et al. 1994).

The entrance channel is aligned approximately S45°W with the principal wave direction of the eight cases distributed around this direction. Table 1 shows the wave parameters for the simulated field wave cases.

Table 1
Simulated Target Wave Conditions

No.	Peak Period sec	Significant Wave Ht ft	Average Direction deg	Spread deg
1	12.6	7.1	80	15
2	7.7	9.8	38	17
3	8.3	7.1	45	19
4	10.0	7.4	63	16
5	9.1	10.2	58	7
6	16.7	6.5	43	14
7	16.7	8.2	43	9
8	14.2	7.1	45	9

The eighteen empirical cases consisted of three prototype wave periods (7.69, 11.11, and 16.67 sec), three principal directions (perpendicular to the shoreline and 25 and 30 deg on either side of the orthogonal) and two spectral peakedness parameters ($\gamma=3.3$ a broad spectrum, and $\gamma=7.0$ a narrow spectrum).

Long-Wave Harbor Response

Okihiro (1993) postulates that both bound and free infragravity waves are the forcing function for harbor resonance at Barbers Point Harbor. Infragravity waves are long-period waves in the range of 25 to 200 sec on the Pacific coast. Infragravity wave heights are much smaller than wind-wave heights, typically only 10 percent as large. Bound infragravity waves are non-linearly forced by and coupled to wave groups. Bound long waves appear to be the controlling mechanism when swell energy outside the harbor is large (Bowers 1977, Mei and Agnon 1989, Wu and Liu 1990). For this condition, it may be possible to predict harbor resonance given the wind-wave spectrum outside the harbor. Also, they found that wind-wave energy present at swell frequencies produces more bound wave energy than the equivalent amount of energy in sea frequencies.

Recent research (Okihiro and Seymour 1992, Elgar et al. 1992, Herbers et al. 1992, Bowers 1993) indicates that free long waves, in the form of leaky or edge waves, are important and may contribute the bulk of infragravity energy in depths corresponding to the S_{xy} location. Leaky waves are generated in shallow water and reflected or radiated seaward to the open ocean. Edge waves are generated and radiated seaward like leaky waves but become trapped on the continental shelf due to reflection and refraction and propagate in the longshore direction. Bound waves may even be a source of free infragravity waves in shallow water. The discontinuity of bound infragravity waves across the harbor mouth may nonlinearly generate free infragravity waves. These free waves would then have energy comparable to bound long waves from outside the harbor.

Outside Barbers Point Harbor, Okihiro and Seymour (1992) found a nearshore coupling between infragravity and wind-wave energy, with a larger infragravity wave height for swell conditions than for higher frequency sea waves. Inside the harbor, they found that infragravity wave heights were highly correlated with infragravity wave heights measured outside the harbor. Furthermore, infragravity wave heights increased as swell energy increased outside the harbor.

Prototype Analysis Methods

The amplification factor A(f), which is a function of frequency, was used to compare the long-period wave height outside the harbor with the long-period wave energy inside the harbor. A(f) is defined as

$$A(f) = \frac{G_{yy}(f)}{G_{xx}(f)} \tag{1}$$

where G_{xx} and G_{yy} are the input and output auto-spectral density functions. The S_{xy} gage value is used as the input and the harbor gages values are used as the output. Estimates of the auto-spectral density functions $\hat{G}_{xx}(f)$ and $\hat{G}_{yy}(f)$ were obtained for each data record by breaking the 4.6-hr-long time series into 2.3-hr records and ensemble averaging the two raw spectral density functions. Estimates of the amplification factor $\tilde{A}(f)$ were calculated from a linear regression on $\hat{G}_{xx}(f)$ and $\hat{G}_{yy}(f)$ from all the records as shown in Figure 3 (Lillycrop et al. 1993).

Physical Model Analysis Methods

In the physical model, a slightly different analysis method called transfer function estimates was utilized. The transfer function is defined as

$$|\hat{H}(f)| = \frac{|\hat{G}_{xy}(f)|}{\hat{G}_{xx}(f)} \tag{2}$$

where $\hat{G}_{xy}(f)$ is the cross-spectral estimate between input x and output y channels and $\hat{G}_{xx}(f)$ is the auto-spectral estimate for the input x channel. The auto-spectral estimate is just the frequency spectrum for the S_{xy2} gage for each wave case. Cross-spectral estimates are similar to auto-spectral estimates except that both input S_{xy2} and output harbor gages are used in the calculation. One advantage of the cross-spectral analysis over the auto-spectral analysis is that the estimate is not as easily biased by noise in the input or output signal (Briggs 1981). For the transfer function for the south gage (Sc), the cross-spectral estimate contains information from both the S_{xy2} and Sc gage. A single line was plotted on Figure 3 by averaging the transfer function for the eight wave cases. This increased the statistical confidence of the results.

Numerical Model Analysis Methods

To analyze long period harbor response in the numerical model, individual monochromatic waves with an amplitude of one were input with direction perpendicular to the bottom contours. The increment between wave frequency was 0.000041 Hz (1/24,576 sec) for wave periods between 45 and 24,576.1 sec. Since the numerical model was tested at three times the frequency of the analyzed prototype measurements, the results were averaged over wave periods one increment above and one increment below the prototype frequencies, analogous to band averaging. This was done so that the numerical frequencies matched those of the prototype. The single line shown in Figure 3 is thus made up of numerous runs. Since the input wave has unit amplitude, the amplification factor is simply the value of the wave amplitude at a particular location.

Model Comparison

Frequency response of prototype and physical and numerical models is shown in Figure 3. As can be seen, there is good agreement between the three. Values greater than one represent harbor resonance, since there is more energy at that particular frequency inside the harbor than outside. One might note that the physical model does not replicate the longest period resonant mode. This is because limited duration experimental runs were conducted and these long-period modes were not extractable from the data record.

Since the numerical model was tuned to existing conditions of the prototype, one would expect the results to be very similar. The reason for calibrating the numerical model was to investigate alternative harbor layouts.

The prototype results inherently show the effects of both free and bound infragravity wave energy, since both components are present in nature. The physical model was run under extreme wave events and in nature bound waves would be the principal component in the infragravity spectra for these wave conditions. To avoid reflections of wind waves off the sidewalls, wave absorbers were placed along the perimeter of the model. This also dampened the leaky wave energy released at the breaker zone which then would have to reflect off the side walls before propagating into the area of interest. Infragravity energy found in the physical model was generated from nonlinear interactions of the wind wave spectra.

Waves generated in the numerical model are analogous to free leaky waves. The numerical model is linear and is not capable of modelling nonlinear mechanisms important for the generation of infragravity wave energy.

Independent of the forcing mechanism of infragravity waves, the harbor response is a function of harbor geometry. The numerical and physical model

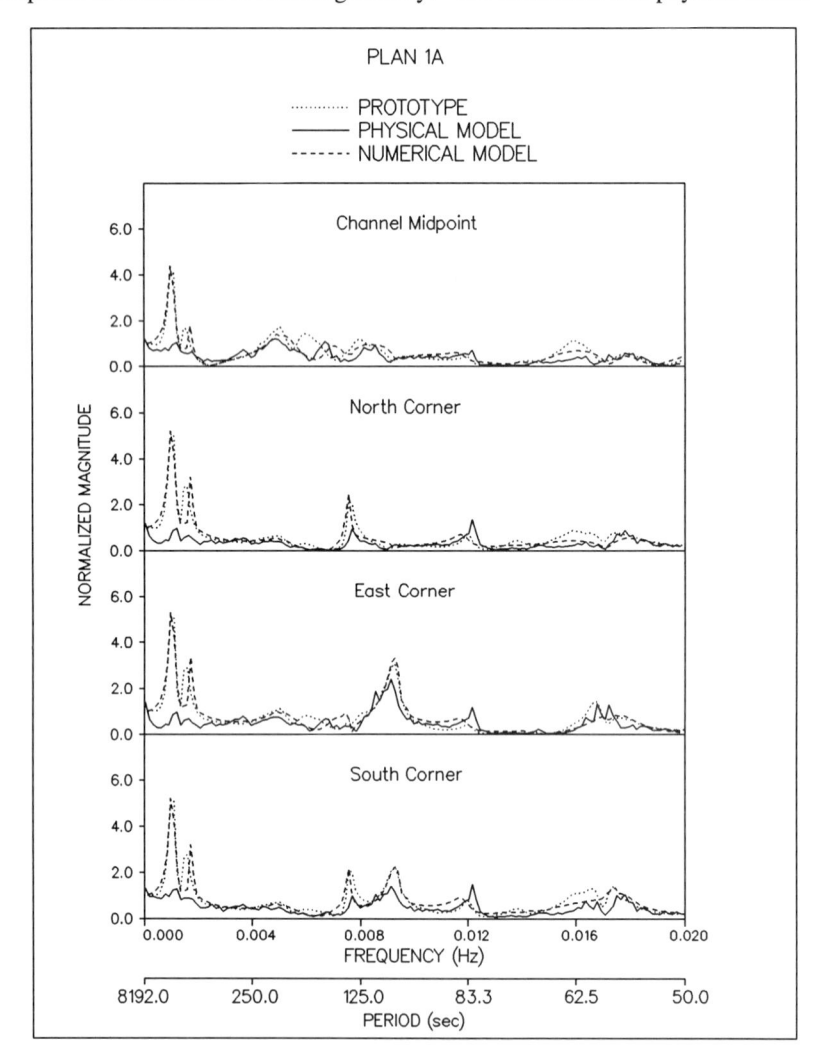

Figure 3. Transfer functions for existing harbor

accurately predict the harbor response.

Importance of Frequency Spread

The transfer functions shown in Figure 3 for the four corners show amplification or reduction of long-period waves between the S_{xy2} gage and the four corner gages. To calculate the amplitude of the long-period wave at a particular period inside the harbor, the amplitude of that wave outside the harbor must be known. In nature, there are always some low-amplitude long-period waves present, and in most cases these low levels do not interfere with ship mooring or navigation. One aspect that does appear to be important in the growth of infragravity wave energy is the peakedness of the spectra in the frequency domain.

The JONSWAP spectrum proposed by Hasselman (1973) is characterized by a parameter called the peak enhancement factor, γ, which controls the peakedness of the frequency spectra (Goda 1985). A peakedness parameter $\gamma = 3.3$ (broad spectrum) was found to be the average for the results of the joint wave observation program and is characteristic of sea waves. Waves that have travelled long distances (on the order of 9,000 km) exhibited narrow banded spectrum on the order of $\gamma = 8-9$ (Goda 1983).

To ascertain the importance of the peakedness parameter on generation of infragravity energy, seven gages were located along the 30-m depth contour (offshore gage array) and seven gages were located along the 8.5-m contour (nearshore gage array). The eighteen empirical wave cases then were run with two peakedness parameters. The zeroth moment m_0 or sum of the energy spectra between 25 and 660 sec was used to evaluate the growth of infragravity wave energy. By averaging results between the nine broad-band spectra for the fourteen different wave gages, there was a 183 percent increase in the infragravity wave energy between 30m and 8.5m. The nine narrow band spectra cases showed a growth of 224 percent between the offshore and nearshore gage arrays.

A comparison between the spectral response of two wave cases with identical parameters, except the peakedness parameter, is shown in Figure 4. The general trend shows an increase in the infragravity wave energy for the narrow band spectra.

Summary

Results from the physical and numerical models were compared against prototype data. Harbor response of the physical and numerical models showed good agreement with field data. Slightly different analysis techniques were used by each but the final results produced by each technique are comparable.

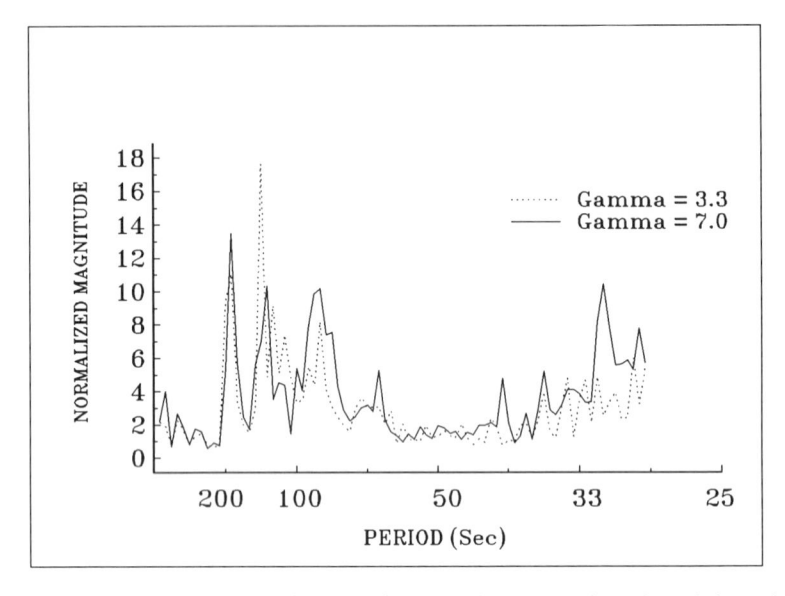

Figure 4. Comparison of infragravity growth between broad $\gamma=3.3$, and narrow $\gamma=7.0$ spectra.

Comparisons between the physical model, numerical model and field data show the resonant response of the harbor. The amplitude of long period waves in the harbor is a function of wave amplitude outside the harbor. One aspect that is important for the evolution of infragravity wave energy outside the harbor is the spectral shape of the wind wave energy in the frequency domain. The growth of infragravity wave energy was evaluated between 30m and 8.5m water depth. Empirical wave spectra with two peakedness parameters were analyzed. The physical model showed that more infragravity wave energy evolved from the narrow banded spectra then evolved from the broad banded spectra.

Acknowledgements

The authors wish to acknowledge Headquarters, U.S. Army Corps of Engineers, U.S. Army Engineer Division, Pacific Ocean (POD), and Harbors Division, Department of Transportation (HDOT), State of Hawaii for authorizing publication of this paper. It was prepared as part of a joint effort among the Civil Works Research and Development Program, Monitoring Completed Coastal Projects Program, POD, and DOT. The author would also like to thank the following individuals for their assistance and participation in this project: Messrs. Stan Boc, Dennis Markle, Edward Thompson, Ernie Smith, Frank Sargent, David Daily, Hugh Acuff, Larry Barnes and Mses. Debra R. Green and Linda Lillycrop.

Bibliography

Bowers, E.C. (1977). "Harbor Resonance Due to Set-Down Beneath Wave Groups," J. Fluid Mech., 79, pp 71-92.

Bowers, E.C. (1993). "Low Frequency Waves in Intermediate Depth," Proceedings of 23rd International Conference on Coastal Engineering (ICCE).

Briggs, M. J. (1981). "Multichannel Maximum Entropy Method of Spectral Analysis Applied to Offshore Structures," M.S. thesis, Massachusetts Institute of Technology.

Briggs, M. J., Lillycrop, L. S., Harkins, G. S., Thompson, E. F., and Green, D. R. (1994). "Physical and Numerical Model Studies of Barbers Point Harbor, Oahu, Hawaii," Technical Report CERC-94-14, U.S. Army Engineer Waterways Experiment Station, Vicksburg, MS.

Chen, H. S. (1986). "Effects of Bottom Friction and Boundary Absorption on Water Wave Scattering," Applied Ocean Research, Vol 8, No. 2, pp 99-104.

Chen, H. S. and Houston, J. R. (1987). "Calculation of Water Oscillation in Coastal Harbors: HARBS and HARBD User's Manual," Instruction Report CERC-87-2, U.S. Army Engineer Waterways Experiment Station, Vicksburg, MS.

Chen, H. S., and Mei, C. C. (1974). "Oscillations and Wave Forces in an Offshore Harbor,"Report No. 190, Department of Civil Engineering, Massachusetts Institute of Technology, Cambridge, MA.

Elgar, S., Herbers, T. C., Okihiro, M., Oltman-Shay, J., and Guza, R. T., (1992) "Observations of Infragravity Waves," Journal of Geophysical Research, Vol 97, No. C10, 15573-15577.

Farrar, P. D., and Chen, H. S. (1987). "Wave Response of the Proposed Harbor at Agat, Guam: Numerical Model Investigation," Technical Report CERC-97-4, US Army Engineer Waterways Experiment Station, Vicksburg, Miss.

Goda, Y. (1985) *Random Seas and Design of Maritime Structures.* University of Tokyo Press, Toyko, Japan.

Goda, Y. "Analysis of Wave Grouping and Spectra of Long-Travelled Swell", Rept. Port and Harbour Res. Inst., Vol 299, 1983, pp. 3-41.

Harkins, G. S. (1991). "Sensitivity Analysis For Multi-Element Wavemakers," Thesis presented to the University of Delaware in partial fulfillment of the requirements of the Degree of Masters of Applied Science.

Hasselman, K. et al. Measurement of Wind Wave-Growth and Swell Decay During the Joint North Sea Wave Project (JONSWAP), Deutsche Hydr. Zeit, Reiche A (8°), No. 12, 1973.

Herbers, T. C., Elgar, S., Guza, R. T., and O'Reilly, W. C. (1992). "Infra-gravity-frequency (0.005-0.05 Hz) Motions on the Shelf," Proceedings of 23rd International Conference on Coastal Engineering (ICCE).

Lillycrop, L. S., Briggs, M. J., Harkins, G. S., Boc, S. J., and Okihiro, M. S. (1993b). "Barbers Point Harbor, Oahu, Hawaii monitoring study," Technical Report CERC-93-18, U.S. Army Engineer Waterways Experiment Station, Vicksburg, MS.

Longuet-Higgins, M. S. (1962). Resonant Interaction Between Two Trains of Gravity Waves, J. Fluid Mech., Vol 12, pp 321-332.

Mei, C.C., and Agnon, Y. (1989). "Long-Period Oscillations in a Harbor Induced by Incident Short Waves," J. Fluid Mech., Vol 208, pp 595-608.

Okihiro, M., (1993). "Seiche in a Small Harbor," Dissertation presented to the Scripps Institution of Oceanography, University of California, San Diego, in partial fulfillment of the requirements of the Degree of Doctor of Philosophy in Oceanography.

Okihiro, M., and Seymour, R. J. (1992). "Barbers Point Harbor Resonance Study," Scripps Institute of Oceanography, (unpublished manuscript), pp 1-26.

Outlaw, D. G., and Briggs, M. J. (1986). "Directional Irregular Wave Generator Design for Shallow Wave Basins," 21st American Towing Tank Conference, August 7, Washington, D.C., pp 1-6.

Sands, S. E., (1982) Long Waves in Directional Seas, Coastal Engineering, 6, 195-208.

Wu, J.-K., and Liu, P.L.-F. (1990). "Harbor Excitations by Incident Wave Groups," J. Fluid Mech., Vol 217, pp 595-613.

CHAPTER 60

Numerical Simulation of the 1992 Flores Tsunami in Indonesia
: Discussion on large runup heights in the northeastern Flores Island

Fumihiko Imamura[1], Tomoyuki Takahashi[2] and Nobuo Shuto[3]

ABSTRACT - Tsunami source model is disscused by numerical analysis for the 1992 Flores Island, Indonesia. Computed results with the composite fault model with two different slip values show good agreements with the measured run-up heights in the northeastern part of Flores Island, except for those in the southern shore of Hading Bay and at Riangkroko. The landslides in the southern part of Hading Bay could generate local tsunamis of more than 10 m, which could be the reason of discrepancy between the measured and computed one. The circular-arc slip model proposed in this study for wave generations due to landslides shows better results than the subsidence model. It is, however, difficult to reproduce the tsunami run-up height of 26.2 m at Riangkroko, which was extraordinarily high compared to other places. Two villages located on the southern coast of Babi Island, back side of the tsunami source region, received severe damages. The simulation model shows that the reflected wave along the northeastern shore of Flores Island, accompanying a high hydraulic pressure, could be the main cause of huge damages in the southern coast of Babi Island.

INTRODUCTION

On 12 December 1992, a earthquake of 7.5 Ms and accompanying destructive tsunami struck the northeastern coast of Flores Island (Figure 1). 1,713 casualties and 2,126 injured, half of which were due to tsunami, were reported [*Tsuji et al.*, 1993]. An extremely large tsunami run-up height of 26.2 m was measured at Riangkroko in the northeastern peninsula of Flores Island. There are only two examples in this century --- the 1933 Sanriku earthquake tsunami and the 1993 Hokkaido nansei-oki earthquake tsunami --- in which more than a 20 m tsunami run-up height has been observed.

An International Tsunami Survey Team (ITST) consisting of engineers and scientists from Indonesia, Japan, U.K., Korea and U.S. conducted a field survey

----

[1] Assoc.Prof., School of Civil Eng., Asian Inst. of Tech., G.P.O.Box 2754, Bangkok 10501, Thailand.

[2] Res. Assoc., Disaster Control Research Center, Tohoku Univ., Aoba, Sendai 980, Japan.

[3] Prof., Disaster Control Research Center, Tohoku Univ., Aoba, Sendai 980, Japan.

along the northeastern coast of Flores Island and smaller offshore islands, and measured tsunami run-up heights as shown in Figure 2.

In this paper, two of problems ITST pointed out are selected because of their important role in preventing tsunami disasters in the future. First is that run-up heights at Riangkroko and at Uepadung and Waibalan in Hading bay of 17.3 - 26.2 m and more than 10 m, respectively, are surprisingly larger than those on the western part of Flores Island[Gonzalez et al., 1993] (Figure 2). Second is the damage at two villages in Bali Island, which are located on the back side of tsunami source where usually damage due to tsunami is not large.

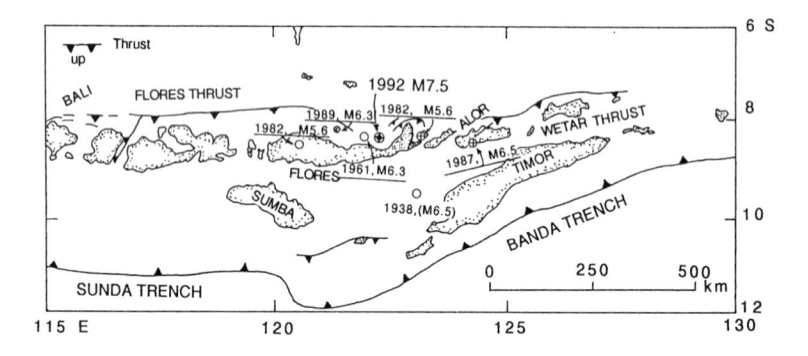

Figure 1 Map of tectonic and epicenter of earthquake in Flores Island which is located in the back arc of eastern Sunda and western Banda thrusts.

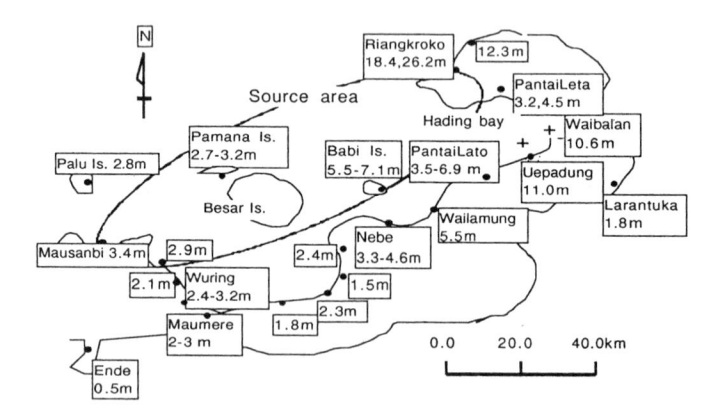

Figure 2 Measured tsunami run-up heights in meter in the eastern Flores Island and the tsunami source [Tsuji et al., 1993]. A general trend of increasing run-up height from west to east in this region is significant. The symbols of cross indicate the location of landslide in the Hading Bay.

TSUNAMI SOURCE MODEL

Our initial tsunami model uses one fault model, Model-A in Table 1: the quick Harvard CMT solution [event file M121292Y]. Taking into account the tectonics in this region, that is thrust in the back arc [*Hamilton, 1988*], a shallow dip thrust fault as the source with (strike,dip, slip) = (61°,32°,64°) is selected. According to the distribution of the aftershocks determined by the United State Geological Survey (USGS), the fault plane is estimated to be 100km long and 50km wide. Thus an average dislocation of 3.2 m can be determined by using a rigidity of 4.0×10^{11} dyne/cm^2.

Table 1 Fault parameter of the 1992 Flores earthquake

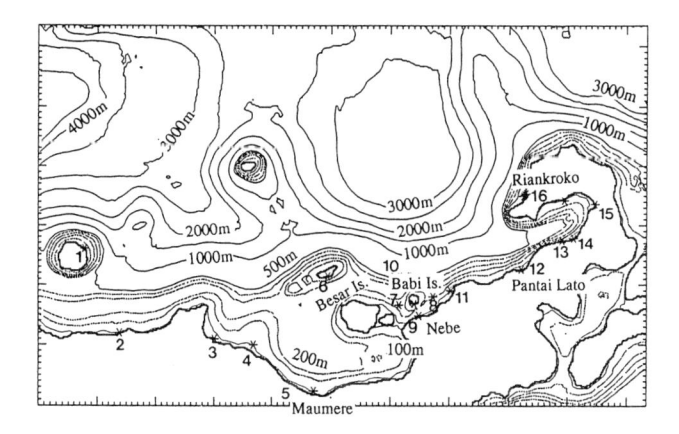

	Mo ($\times 10^{27}$dyn-cm)	Depth (km)	Strike (deg)	Dip (deg)	Slip (deg)	Length (km)	Width (km)	Dislocation (m)
Model-A	6.4	15	61	32	64	100	50	3.2
Model-B								
(East)	1.6	15	61	32	64	50	25	9.6
(West)	4.8	15	61	32	64	50	25	3.2
Model-C								
(East)	1.6	3	61	32	64	50	25	9.6
(West)	4.8	3	61	32	64	50	25	3.2

Figure 3 Computational region and geometry, countor of water depth, in the northeastern Flores Island. Spatial grid size of 300 m is used.

A numerical simulation of tsunami generation and propagation, the TUNAMI - N1 code [*Shuto et al.*, 1990 and *Imamura et al.*, 1992], was carried out using this fault model. The area for computation is shown in Figure 3. The lack of detailed topographical data in shallow region and on the land makes it difficult to carry out run-up simulation. Therefore the computed maximum levels along the coastline are directly compared with the measured run-up heights. The computational condition is summarized in Table 2.

Table 2 Computational condition

Governing Equation	Shallow water theory (nonlinear theory)
Spatial grid siz	300 m
Time step	1 sec
B.C. of coastal line	Perfect reflection condition (vertical wall)
Reproduction time	1 hour

The computed results with Model-A are smaller than the measured run-up heights in the eastern part, suggesting that the slip value on the eastern side of the fault might have been larger than that on the western side. After several trials, a composite fault model with different slip values : 3.2 and 9.6 m was, therefore, proposed [*Imamura & Kikuchi*, 1994]. In this model, the seismic moment of 6.4×10^{27} dyne-cm is kept and both segments have the same fault area of 50 km long and 25 km wide (Figure 4). The crustal movements estimated by the composite fault model with a depth of 3 km coincides with the crustal deformation measured by *Tsuji et al.* (1993). Computed results using composite model, Model-B and C, are compared to the measured run-up heights in Figure 5, which shows that the numerical model with the composite fault model reproduces well the distribution of the measured tsunami heights along the northeastern coastline of Flores Island, except for the southern shore of Hading Bay and Riangkroko.

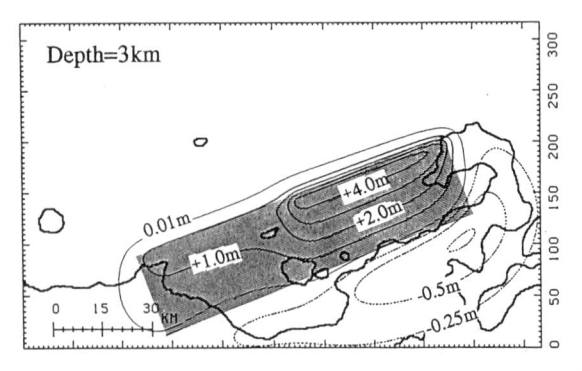

Figure 4 Composite fault model [Imamura & Kikuchi, 1994] and its vertical displacement of ground. Moment release in the eastern part is larger than that in the

western part.

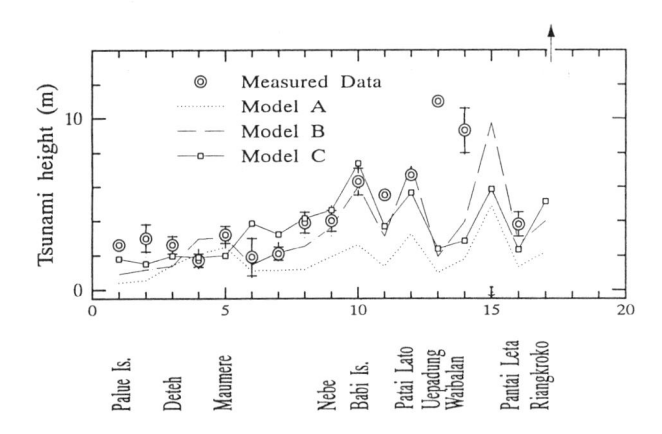

Figure 5 Comparison between the measured run-up heights and the computed maximum water level along the northeastern coast of Flores Island. Computed result show a good agreement with the measured one except for Uepadung, Waibalan and Riangkroko.

TSUNAMI IN HADING BAY AND RIANGKROKO

Landslide in Hading Bay

As shown in Figure 5, the measured run-up heights of 10.6 m at Waibalan and 7.6 - 11.0 m at Uepadung in the southern shore of Hading Bay are much higher than those at other places in this bay such as Pantai Lato and Pantai Leta, and the computed results with composite fault model. It is possible that abnormal run-up heights were generated by not only earthquake but another geological agents. ITST noticed large landslides along the southern shore of Hading Bay [*Yeh et al.*, 1993] and reported that the landslide shear planes are almost vertical, forming a steep and vertical cliff along the present shoreline and its area is approximately 150 m wide and 2 km long. The large run-up heights observed are limited to the area of subaqueous slumps, suggesting a possibility that they were caused by landslide-generated waves rather than tectonic tsunami waves.

Wave Generation Model due to Landslide

Until now several models for wave generation due to landslides, ranging from the inflow volume method and unit-width discharge method [*Ming & Wang*, 1993] to two-dimensional boxes dropped vertically at the end of a semi infinite channel [*Noda*, 1970] have been proposed. Yet, all these methods cannot be applied directly to the present case, because higher tsunami heights were observed not on the opposite side the landslide area but along the landslide and its surrounding area. In addition, the observed run-up were higher than the level of the top of the landslide. In the present study two models, subsidence and circular-arc slip model (Figure 6), are proposed [Gica, 1994], because the cavity formed by a sudden collapse could generate a wave propagating toward the coast. In the discussion for wave generation due to landslides, the run-up modeling is used. Numerical condition that the averaged depth, 20 m, and height of land, 8 m, is estimated from the field survey, but the radius of circular-arc slip, 30 m, and property of soil are assumed here because there are no data about them. Assuming the same maximum vertical displacement in both models, the computed results showed that the wave run-up height, 3.9 m, of the subsidence model is half that of the circular-arc slip model, 8.2 m. The reason is that the toe of the landslide mass in the circular-arc slip model increases the sea bottom, which causes the disturbances of water surface propagating toward the coastal line. The maximum run-up heights of the circular-arc slip model is 8.2 m and almost same as that of the ground level. The above result suggests that the circular-arc slip model is more suitable to reproduce the tsunami of more than 10 m in height. Unfortunately the lack of field data such as landslide scale under the sea, and soil properties, does not allow us to compare the computed values with the measured in Hading Bay.

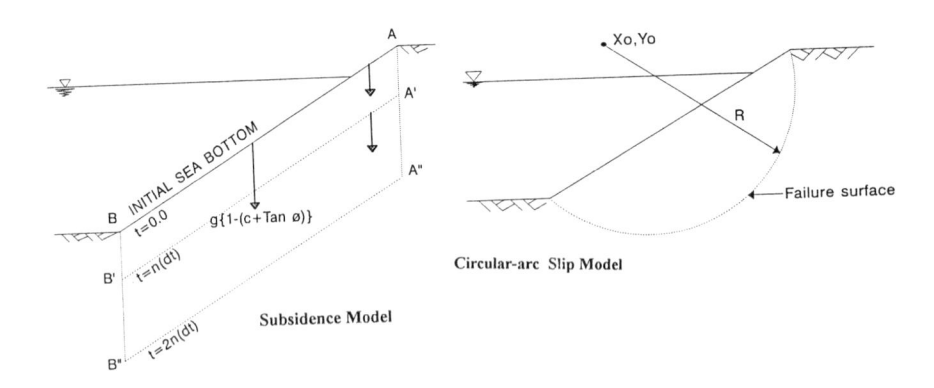

Figure 6 Subsidence (left) and circular-arc slip (right) models for the wave generation due to landslides. The deformation sea bottom is calculated every steps based upon two models. c; soil cohesion, ϕ; soil friction angle, g: gravitaional acceleration.

Riangkroko

An extremely large tsunami run-up was measured at the small village of Riangkroko (Figure 7), where 137 people lost their lives (population 406 prior to the event). The village is located at the mouth of a small river, the Nipar River, with its northwestern side facing the Flores Sea. Offshore from the village there is a remarkable steep slope of sea bottom, which is similar to the case of Okushiri Island in the 1994 Hokkaido Nansei Oki earthquake tsunami. The maximum run-up height measured at Riangkroko was 26 m and the average of heights based on four different tsunami marks was 19.6 m, indicating that the magnitude of the tsunami run-up is probably not an isolated local phenomenon like wave splash-up. Figure 5 shows that the computed results are much smaller than the measured at Riangkroko. The ratio of the measured to the computed values is more than 5, which is considerably large. The reason of such a large run-up height and discrepancy to the computed results should be explained, however no significant geological phenonema such sa landslide at Hading bay was observed on the land. A propagation process on the steep slope of sea bottom could cause the significant amplification of wave, or another geophysical phenomena such as submarine landslide could generate wave. For further discussions, a field investigation on this area and a numerical analysis with the more detailed topography at the offshore and on the land are required.

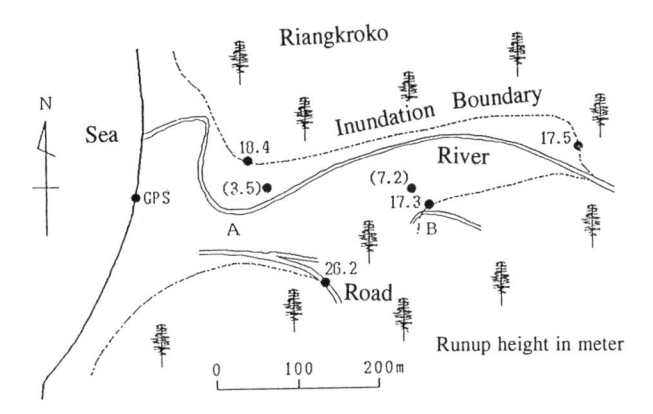

Figure 7 Inundation area and tsunami run-up at Riangkroko. All houses were completely washed away by tsunami. Numerals in pareuthese are the inundated tsunami heights.

H UGE DAMAGE IN BABI ISLAND

Babi Island

Babi Island (Figure 8) received significant damage with 263 casualties for 1,100 residences, and a complete loss of all houses. Measured tsunami run-up heights are 5.6 m in the Christian village and 3.6 m in the Moslem village. A maximum of 7.2 m was measured on the cliff in the western side of the island. Although the run-up heights are not as high as those at other damaged places, the damage due to tsunamis was quite severe.

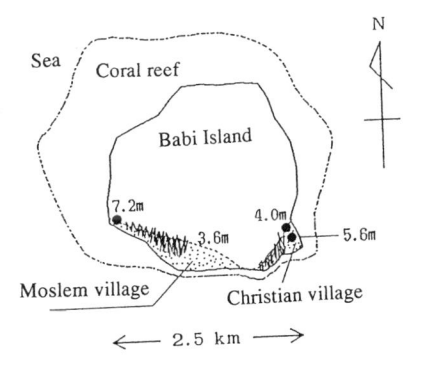

Figure 8 Map of Babi Island.

Babi Island is about 5 km offshore from Flores. It has a conical shape with a 351 m summit elevation and approximately 2 km in diameter. The northern shore faces the Flores Sea and has a wide coral reef, while the southern shore where the two villages were located has a much narrower reef. Even when strong wind waves and swells of the Flores Sea attack from the north, wave conditions on the southern side are usually calm, because most incoming wave energy is dissipated on the wide coral reef on the northern side [*Yeh et al, 1993*]. However, southern part of the island severely received damages due to tsunami current rather than large wave.

Reflected Waves and Hydraulic Pressure

In order to explain why the southern coast of Babi Island suffered huge damages due to the tsunami, a numerical simulation with the same condition shown in Table 2 was carried out specifically focusing on Babi Island. Figure 9 shows the water elevation contours and the current vector distributions around this island from 4 minutes to 10 minutes after tsunami generation. Figure 10 shows the time histories of water level, velocity, and hydraulic pressure in front of the Moslem village. The hydraulic pressure is defined as follows [*Aida, 1977*];

[Hydraulic Pressure (m³/s²)] =
 [Tsunami Inundated Height (m)] x [Current Velocity (m/s)]²   (1)

It is noticed that current velocity makes effect on hydraulics pressure more than tsunami height.

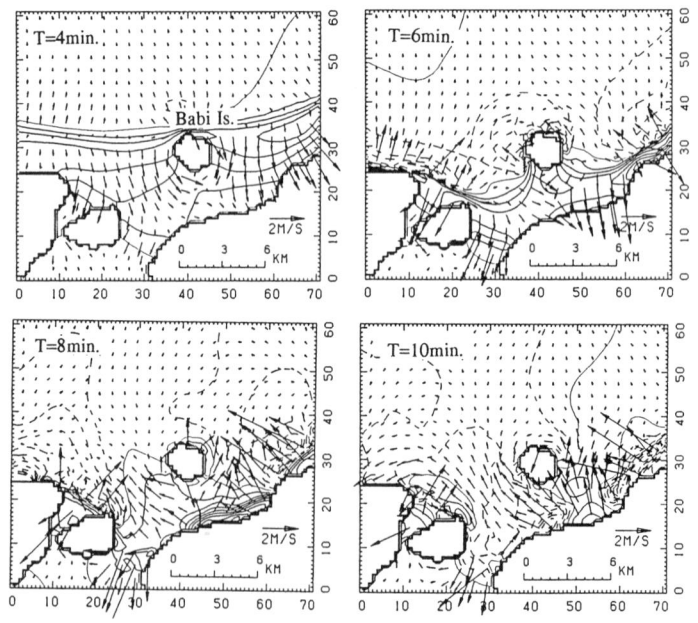

Figure 9 Water elevation contours with intervals of 1 m and current vectors around Babi Island. The first wave was reflected along the northern shore of Flores Island at 8 minutes after earthquake and attacked the southern coast of Babi Island.

Hatori's(1984) empirical result of the relationship between damage on houses and hydraulic properties showed that the hydraulic pressure is the most suitable parameter to estimate the degree of house damages. Figures 9 and 10 show that the first wave attacked the island from the north direction with not a high hydraulic pressure, but the same wave was reflected off the coast of Flores Island and again attacked Babi Island on the southern part accompanied by a high hydraulic pressure. This result is consistent with the eyewitnesses at this island. In addition, the island is located at the nodal point of the standing wave generated along the coast of Flores Island, suggesting that the wave height is small whereas the current is large.

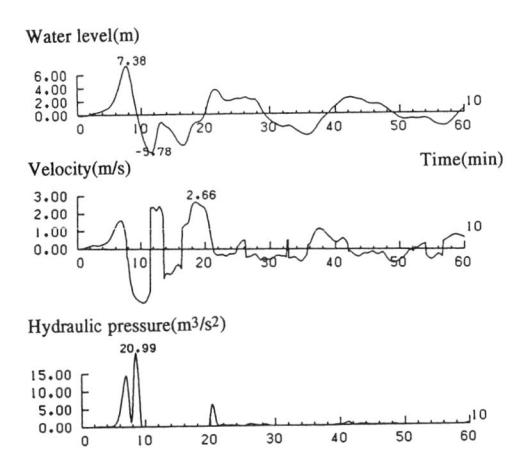

Figure 10 Time histories of water level, velocity and hydraulic pressure. The positive values of velocity compoment indicate that tsunami propagate from the northern side to the southern one.

Measured and Computed Hydraulic Pressure

We could estimate hydraulic pressure and velocity at Wuring near Maumere by measuring the tsunami traces on the wall of the Mosque [*Matsutomi*, 1993; *Imamura et al.*, 1993] (see Figure 11). Here almost all of the wooden houses were destroyed due to the tsunami. Applying the Bernoulli theorem by assuming energy conservation between the front and back of the Mosque, we calculated the velocity to be 2.7 - 3.6 m/s and the hydraulic pressure to be 6.2 - 15.2 m^3/s^2. This estimation coincides with Hatori's(1984) criterion that the hydraulic pressure over 5-9 m^3/s^2 corresponds to damage of over 50 %. The result of the present simulation, a velocity of 2.38 m/s and a hydraulic pressure of 7.70 m^3/s^2, agreed with the estimated data well, which supports the accuracy and reliability of this simulation model. In terms of numerical results, the places recorded the hydraulic pressure over 10 m^3/s^2 are Babi Island, Waibalan, Pantai Lato and Riangkroko. Surprisingly the hydraulic pressure at Riangkroko is over 30

m^3/s^2.

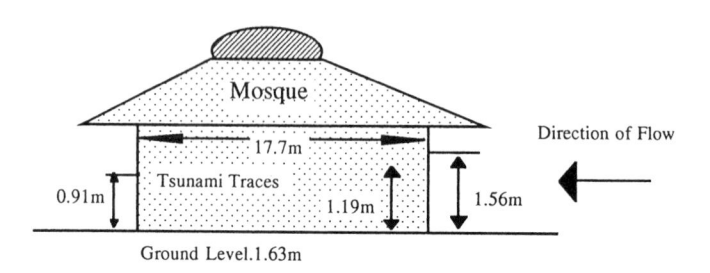

Figure 11 Tsunami traces on the mosque at Wuring, which are used to estimated the current velocity and the hydraulic pressure.

CONCLUSIONS

The landslide found in the southern shore of Hading bay could generate local waves which are much higher than those obtained by the tsunami simulation with only crustal movement. A new model for wave generation due to landslide, circular-arc slip model, is proposed. This model might reproduce higher run-up heights along the coastline than the subsidence model. The tsunami run-up height at Riangkroko was extraordinarily high compared to other places. The wave propagation process on a steep slope of sea bottom as well as the geological agents such as submarine landslide could be related to this data. Numerical model shows that the reflected wave along the northeastern shore of Flores Island is the main cause of huge damages in southern Babi Island. The computed hydraulic pressure and velocity at Wuring is well corresponding to the measured one.

ACKNOWLEDGEMENTS
The study was partially supported by a Grant-in-Aid No.04306024 for Cooperative Research (A) and No.06044015 for International Scientific Research Program from the Ministry of Education, Science and Culture, Japan.

REFERENCES
Aida,I.(1977), *Numerical experiment for the tsunami inundation - in the case of Suzaki and Usa in Kochi prefecture,* Bull. Earthq. Res. Inst. Univ. Tokyo, 52, 441-460 (in Japanese).
Gica, E.(1994), *A study on the 1992 Flores Indonesia earthquake tsunami ; Numerical model on the wave generation due to landslide,* Master thesis in Asian

Institute of Technology, 67p..
Gonzalez,F., Sutisna,S., Hadi,P.,Bernard,E., and Winnarso,P.(1993), *Some observations related to the Flores Island earthquake and tsunami*, Proc. Int. Tsunami Symp. in Wakayama, 789-801.
Hamilton,W.B.(1988), *Plate tectonics and Island arcs*, Geological Soc.Am.Bull., 100,1503-1527.
Hatori,T.(1984), *On the damage to house due to Tsunamis*, Bull. Earthq. Res. Inst. Univ. Tokyo, 59, 422-439.
Imamura,F., Shuto,N., Ide,S.,Yoshida,Y. and Abe,Ka(1992),*Estimate of the tsunami source of the 1992 Nicaraguan earthquake from tsunami data*, Geophys. Res.Let., 20 , 1515-1518.
Imamura,F., Matsutomi,H., Tsuji,Y., Matsuyama,M., Kawata,Y. and Takahashi,T. (1993), *Field survey of the 1992 Indonesia Flores Tsunami and its analysis*, Proc. of Coastal Engng in Japan, 40, 181-185 (in japanese).
Imamura,F. and Kikuchi,M (1994),*Moment release of the 1992 Flores Island earthquake inferred from tsunami and teleseismic data*, Sci. Tsunami Hazards, Accepted.
Matsutomi,H.(1993),*Tsunami and damage in the northeast part of Flores Island*, Monthly Ocean, 25 , 756-761. (in Japanese).
Ming,D. and Wang,D.(1993), *Studies on waves generated by landslide*, Proc.XXV congress of IAHR, Tech.Session c, 1-8.
Noda,E.K.,(1970), *Water waves generated by landslide,* J.Waterways, Harbors and Coastal Eng Div., ASCE, 96, 835-855.
Shuto,N, Goto,C. and Imamura,F.(1990), *Numerical simulation as a means of warning for near-field tsunami*, Coastal Engng in Japan, 33, 2, pp.173-193.
Tsuji.Y., Imamura,F., Kawata,Y., Matsutomi,H., Takeo,M.,Hakuno,M.,J.Shibuya, Matsuyama,M. and Takahashi,T.(1993), *The 1992 Indonesia Flores earthquake tsunami*, Monthly Ocean, 25 , 735-744,(in Japanese).
Yeh,H, Imamura,F.,Synolakis,C., Tsuji,Y., Liu,P. and Shi,S. (1993), *The Flores Island tsunamis*, EOS, Trans. Am. Geophys. Union, 74, 33, 371-373.

CHAPTER 61

A Comparative Evaluation of Wave Grouping Measures

E.P.D. Mansard[1] and S.E. Sand[2]

ABSTRACT

This paper describes the limitations associated with some of the wave grouping measures that are currently available in the literature. A brief review of the new concepts that have been proposed recently to overcome these limitations is also included in it.

Extensive analysis of prototype waves is carried out worldwide to establish a relationship between the degree of grouping in waves and their variance spectral density. But some of these analyses have been unable to identify any relationship because of the statistical variability inherent to records of finite length. A brief discussion of this variability is included in this paper.

INTRODUCTION

The occurrence of wave groups and their importance are well recognized by the engineering community. During the last fifteen years, many studies have illustrated the relevance of wave groups to the design of maritime structures. Much research is therefore being carried out in this field and their main objectives can be summarized as follows:

(i) to develop an adequate measure of the degree of grouping;
(ii) to establish a statistical model for the wave groups; and
(iii) to achieve a better understanding of the effect of wave groups on structural response.

Several measures have been proposed to quantify the degree of grouping in a sea state, but none of them seems fully adequate for practical use. Some of the limitations of these measures are reviewed in this paper, using numerical simulation.

---

[1]Senior Research Officer, Coastal Engineering Program, IMD, NRC.
[2]Manager, Marine Simulation Department, Danish Maritime Institute, Lyngby, Denmark.

In order to establish statistical models for wave groups, many researchers have advanced the concept that all information about wave groups is contained within the primary spectrum. This implies that, once the shape of the primary spectrum is known, the information about wave groups can easily be deduced. A brief evaluation of this hypothesis is carried out in the later sections.

Attempts are made to achieve a better understanding of the complex physics involved in the interaction of wave groups with coastal structures in order to relate the degree of wave grouping to group-induced response of structures. Some of the techniques used for this purpose are also discussed in this paper.

BRIEF REVIEW OF THE WAVE GROUPING MEASURES

Amongst the large number of concepts that have been proposed to describe wave grouping, the following are some which have been explored in detail.
   i)   Statistics of Run Lengths;
   ii)   Wave Grouping from the Square of the Water Surface Elevation;
   iii) Correlogram;
   iv)   Phase Spectrum; and
   v)   Concepts Based on Hydrodynamics of Wave Motion.

Amongst these concepts, only the first two are used extensively. This paper will focus on them, nevertheless indicating relevant bibliography for the remaining three.

Statistics of Run Lengths

The Run Length is defined as the number of successive waves with wave heights exceeding a specified threshold; the threshold can be the average, the median or the significant wave height of a sea state. The average run length (j_1) within a wave record is then used as the measure of the degree of grouping. The total run is analogous to a wave group period, and is defined as the number of waves contained in the interval which commences with the first threshold exceedance of one run and ends with the first exceedance of the next run. The average value of the total run within a wave record is usually denoted as j_3.

Kimura (1980) provides the necessary expressions for predicting the mean and the standard deviation of the run lengths and of the total runs, as a function of a correlation parameter κ derived from wave records. This parameter is related to the correlation coefficient between two successive heights, ρ_{HH}, and can be calculated from the zero crossing-analysis of waves. But if the time series has to be used anyway for determining this correlation coefficient, the run length statistics can be easily inferred at the same time. It appears therefore that there is no distinct advantage in using Kimura's theory except to compare his predictions with observed values in field data. Hence researchers such as Battjes and van Vledder (1984) promote the use of the correlation parameter κ derived from the spectral density of the sea state as follows:

$$\kappa_f^2 = \frac{1}{m_0^2} \left\{ \left[\int_0^\infty S(f) \cdot \cos(2\pi f \tau) df \right]^2 + \left[\int_0^\infty S(f) \cdot \sin(2\pi f \tau) df \right]^2 \right\} , \quad (1)$$

where: m_0 is the zeroth moment of the spectral density function $S(f)$ and τ is the mean wave period.

If the run length statistics can be satisfactorily predicted from this spectrum-derived correlation parameter κ_f, it can be concluded that most information regarding wave groups is contained in the spectrum itself.

Wave Grouping From the Square of the Water Surface Elevation

Several measures have been proposed to quantify the degree of grouping present in a sea state based on the concept of the square of the water surface elevation. A brief description of some of these measures is given below.

Smoothed instantaneous wave energy history (SIWEH)

One of the well known measures for characterizing the degree of grouping in a sea state is the Groupiness Factor based on the SIWEH (Funke and Mansard, 1979). The SIWEH is a time series of the low frequency part of the square of the elevation, $\eta_\ell^2(t)$. It is computed by squaring the instantaneous water surface elevation and then smoothing it by a Bartlett low-pass filter of size $2 \times T_p$ (where T_p is the peak period of the spectrum). The Groupiness Factor, GF, which is a measure of the degree of grouping, is defined as the standard deviation of SIWEH normalized with respect to its mean value.

Hilbert transform of wave record

A technique which involves no low-pass filtering to compute the low frequency part of the square of the water surface elevation is the Hilbert transform technique, proposed by Bitner-Gregersen and Gran (1983) and extensively used by Medina & Hudspeth (1988, 1990).

If the water surface elevation and its conjugate are expressed as $\eta(t)$ and $\hat{\eta}(t)$ respectively, it can be shown that they are the Hilbert transform of each other. From these two functions, the envelope of the water surface elevtion, $A(t)$, the wave height function, $H(t)$, and the low frequency part of the square of the water surface elevation, $\eta_\ell^2(t)$, can be computed easily (see Medina et al., 1990 and Medina & Hudspeth, 1990).

A concept similar to the SIWEH Groupiness Factor can also be used in this case to define the degree of grouping. Medina et al. (1990) propose the use of GF defined as the standard deviation of the square of the wave height function normalized by $8m_0$ (m_0 is the zerth moment of the sea state spectrum). This approach of computing GF is obviously superior to SIWEH analysis since it involves no low-pass filtering. However, it has some limitations which will be discussed later in this paper.

Expected wave grouping spectrum

For a given sea state, the expected spectrum of the low frequency part of the square of the water surface elevation (known also as spectrum of wave groups) can be computed easily from the auto-correlation of the primary spectrum, without having to deal with the time series of water surface elevation and its Hilbert transform (see Pinkster, 1984). Figure 1 shows three JONSWAP spectra with different peak enhancement factors, γ, and their corresponding wave grouping spectra. The cut-off frequency used in the low pass filtering of SIWEH is also indicated in Figure 1b for easy reference.

It should be pointed out that individual realization of finite length wave records, simulated from these JONSWAP spectra, can result in grouping spectra that differ from the expected functions illustrated in Figure 1.

Correlogram

The concept of using the correlogram (i.e. auto-correlation function) to describe wave grouping was attempted by Rye and Lervik (1981) and Sobey & Read (1984) without much success. More recently, Medina and Hudspeth (1990) showed that the correlation parameter κ used in the prediction of run lengths can be derived from the auto-correlation function of the square of the wave height function.

Phase Spectrum

Funke and Mansard (1981) and Sobey and Read (1984) speculate that phase spectra of sea state may contain some information on the degree of grouping. In fact, Funke and Mansard (1979) showed that, by just changing the phase spectrum, wave trains with different degrees of grouping can be simulated from a given variance spectral density, without however establishing any specific relationship between the degree of grouping and the phase spectrum. Goda (1983) analyzed some swell records for the probability distribution of phases, and concluded that the phases were indeed uniformly distributed between $-\pi$ and π but not completely random. Nevertheless, the deviation from randomness was only slight.

Concepts Based on Hydrodynamics of Wave Motions

All the wave grouping concepts presented so far are based only on statistical wave properties such as wave energy spectrum or wave energy envelope. But other approaches, based on the hydrodynamics of wave motions, have also been explored in the literature.

Mase and Iwagaki (1986a and b) carried out investigations using modulational instability theory to describe the wave grouping process. They suggest that a wind wave field could be considered as a modulated non-linear wave train with a single carrier wave. The other approach used to describe wave grouping is the envelope soliton model. Yasuda et al. (1986) postulate that all wave groups observed in waves behave as envelope solitons and can be defined by the multi-envelope soliton solutions of the plural non-linear Schrödinger equations. More research is required in this field for practical applications.

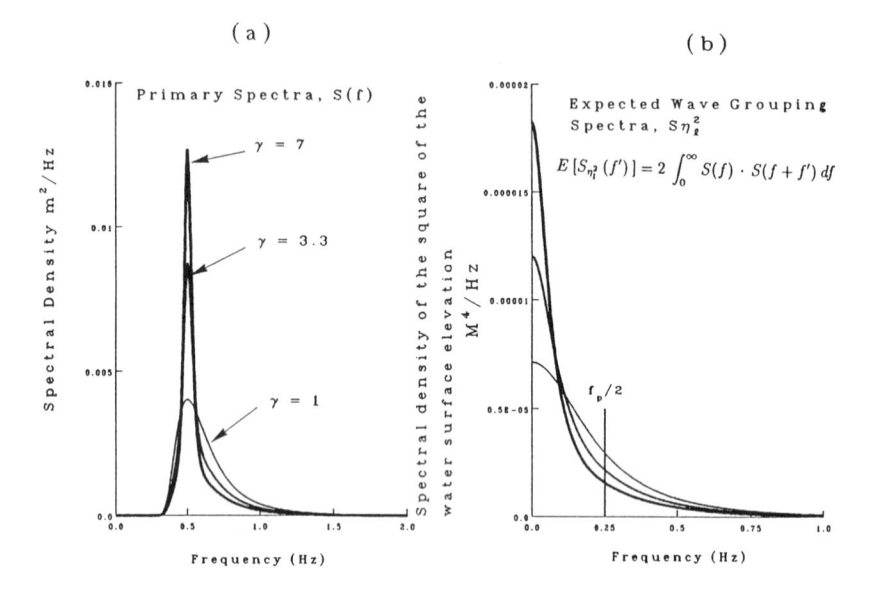

Figure 1. Expected spectra of wave groups for JONSWAP sea states

ANALYSIS OF WAVE GROUPING AND THEIR STATISTICAL VARIABILITY

As indicated above, only the run length statistics and Groupiness Factor are extensively used for quantifying the degree of grouping in a sea state. Hence the detailed evaluation presented below discusses only these two measures.

Many prototype waves have been analyzed in different parts of the world, in order to relate the degree of grouping to the spectral width of the sea state, or to the correlation parameter κ, using the measures of run length and Groupiness Factor. However, some of these analyses have not been too successful because of the various limitations associated with the two wave grouping measures. In order to get a better understanding of these limitations, some investigations were undertaken in this study, using numerical simulation.

Statistical Variability of Wave Grouping Measures

It can be shown that, for records of finite length, it is not easy to establish relationship between the measures of grouping such as GF and j_1 and spectral width parameters such as 'Q_p' or correlation parameter κ. The statistical variability associated with sea states of finite record length obscures any relationship that might exist between these parameters. To illustrate this point, the following investigations were undertaken.

JONSWAP spectra with three different peak enhancement factors $\gamma = 1$, 3.3 and 7 were first formulated. From each one of these three spectra, 200 different realizations of wave trains ($N_R = 200$) were simulated numerically using the random phase spectrum method. This random phase spectrum method is a technique that pairs a given spectral density with a randomly selected phase spectrum. By varying the random numbers used in this phase spectrum, wave records with different time domain characteristics, and therefore with different Groupiness Factors and run lengths can be generated (Funke et al, 1988). The record length of each of these wave trains, T_R, was chosen arbitrarily to be 200s (model scale). At a scale of 1:36, this would represent a 20 minute record in prototype. The peak frequency of the spectrum was made equal to 0.55 Hz, and thus the total number of waves contained in each of these time series was about 120 to 150 waves.

The synthesized time series were subjected to wave grouping analysis, and the 200 values of Groupiness Factor obtained from each one of these spectra, were then subjected to statistical analysis. For all three JONSWAP spectra, the estimated Groupiness Factors exhibited a large variability. It was therefore considered relevant to determine if there is an optimum record length that would ensure a distinct relationship between GF and γ. For this purpose, the numerical simulation described earlier was extended to include several other record lengths during the synthesis process. For instance, instead of simulating 200 time series each 200 s long, 100 time series each 400 s long were simulated and then subjected to wave grouping analysis. In a similar fashion, the record lengths were increased to 800 s, 1600 s, 3333 s, 5000 s and 6667 s respectively. The corresponding number of time series used in the analysis were therefore equal to 50, 25, 12, 8 and 6 respectively. The results of all these analyses are summarized in Figure 2a for the three types of spectra under consideration. This figure shows the mean values of the Groupiness Factors, their maxima and minima and the mean values $\mp$ one standard deviation obtained from the statistical analysis. Figure 2b illustrates the run length statistics obtained from similar numerical simulations. These statistics correspond to the average run length, $\bar{j}_1$, exceeding the significant wave height. Two conclusions are evident from these figures 2a and 2b:

Mean values of GF(i.e $\overline{\overline{GF}}$) and $\bar{j}_1$ (average of $\bar{j}_i$) exhibit an increase with narrower spectra. The statistical variability of GF and j_1 decrease as the record length increases. However, even with a record length of 111.12 minutes, there is a small variability in the values of GF and $\bar{j}_i$. For instance, the standard deviation of GF which is estimated to be 2 to 3% of the mean value, for record length $T_R = 55.56$ minutes, does not seem to decrease with more increase in the record length. If the scale factor of 1:36, used earlier to represent 200 s of model time to full scale 20 minutes, is adopted here, the record length of 55.56 minute would correspond to 5.56 hours prototype.

Since natural wave records are often of short duration, it is believed that this figure would be useful for a judicious interpretation of wave group statistics.

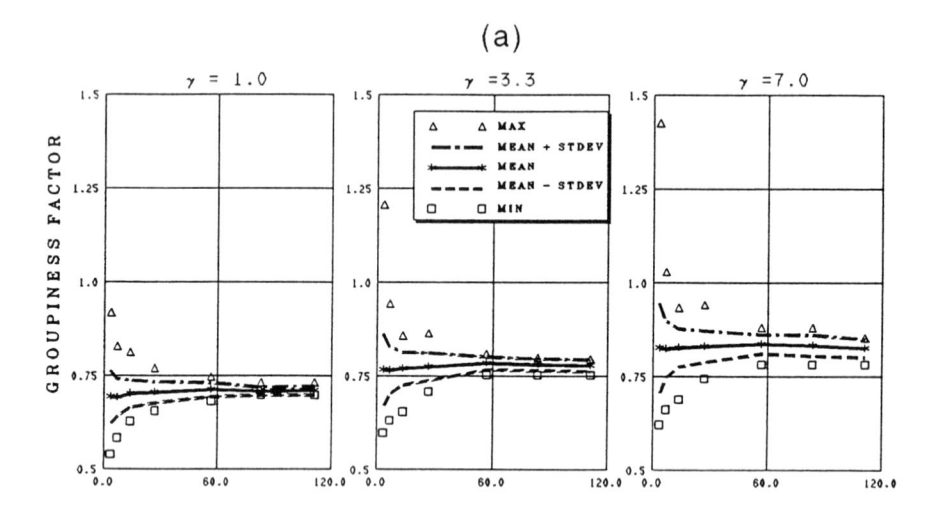

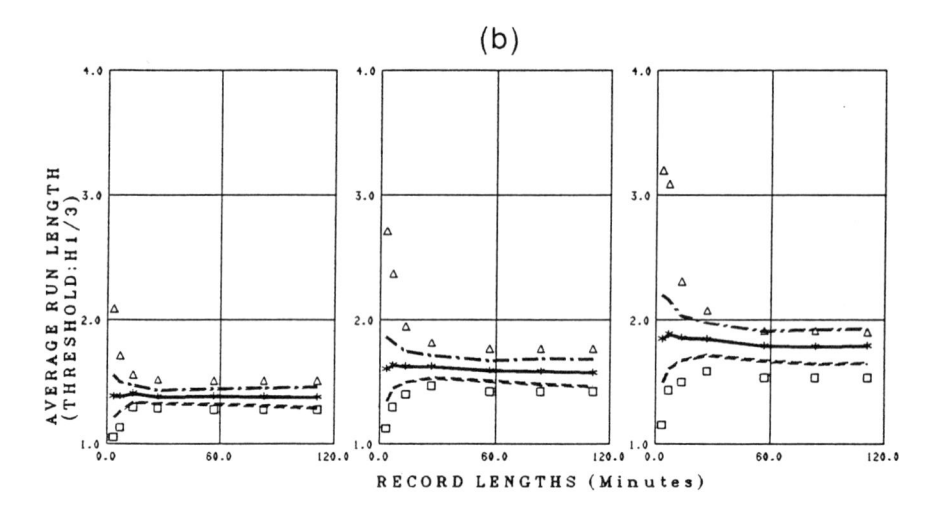

Figure 2. Statistical variability in Groupiness Factors and Run Length statistics.

COMPARISON BETWEEN VARIOUS WAVE GROUPING MEASURES

In spite of the statistical variabilities described above, GF and run length statistics are being commonly used to provide prototype information on wave groups, for purposes of model testing in laboratory basins. But these two measures do not characterize the degree of grouping in the same fashion, i.e. a large Groupiness Factor does not necessarily mean a long run length.

A discussion on the lack of relationship that exists between these two measures and on the advantages of the Hilbert Transform are given below.

Comparison Between Run Length Statistics and Groupiness Factors Derived from SIWEH

The measures of GF and average run length $\overline{j_i}$, are based on two different approaches: The Groupiness Factor is based on the square of the water surface elevation while the run length is based on the count of successive waves which exceed a certain threshold. A large GF implies the existence of many distinct energy packets of large waves but a long run length does not say much about the wave heights contained in the group except that they exceed the given threshold. It is therefore possible for a wave train with a small GF to have long run lengths and vice versa. A relationship can however be found between the average period of SIWEH and the total run (see Goda, 1983).

Between these two measures, GF is perhaps the parameter which is more commonly used in numerical and physical model studies, since simulation techniques exist to control the value of GF without having to change the spectral characteristics of the sea state. A similar technique is not available for a control of run lengths. Furthermore, the GF provides a measure of the low frequency energy contained in a sea state and therefore could be related to structural response, such as the slow drift oscillations of floating structures, seiches in harbours etc. The run length statistics do not readily provide any indication on the low frequency energy but they may be more appealing for the study of fixed structures.

Comparison Between the Concepts of SIWEH and Hilbert Transform

The main criticism that the SIWEH based Groupiness Factor is subjected to is the necessity to perform filtering operations and the arbitrary choice of the filter width. Because of this, researchers such as Medina and Hudspeth (1988, 1990) have been promoting the concept of Hilbert transform to compute $\hat{\eta}_t$ without any low pass filtering. Figure 3 illustrates the difference between SIWEH and Hilbert Transform approaches. The computation used for obtaining these results was carried out as follows: first a 200s long time series of water surface elevation was synthesized from the JONSWAP spectrum, shown in Figure 3a as the primary spectrum. This time series was then subjected to wave grouping analysis using the SIWEH and Hilbert transform analysis. The time series of the low frequency part of the square of the water surface elevations, η_t^2, derived from these two analyses were then subjected to spectral analysis. It can be seen from Figure 3a that the spectrum of wave groups derived from SIWEH goes nearly to zero at $f_p/2$ (i.e.

0.25 Hz), while the spectrum from the Hilbert Transform analysis overlaps the primary spectrum. Since the Groupiness Factor is a global measure of the variance contained in the entire low frequency part of the square of the water surface elevation, its value derived from the results of Hilbert transform is larger than the one provided by SIWEH analysis (see Figure 3b). However, neither of these values is directly applicable to predict the structural response of floating structures for the reason indicated below.

The response of a floating structure depends only on the variance contained within the range of its frequency response. Hence the Groupiness Factor which characterizes the total variance of η_t^2 is not suitable for the prediction of structural response. To overcome this problem, Mansard and Sand (1992) suggest the use of some alternative concepts.

Care must also be exercized in using the concept of Groupiness Factors for evaluating the sensitivity of fixed structures to the degree of grouping. By using the SIWEH concept, Vidal et al. (1995) and Mansard et al (1994) show that there is a correlation between the degree of grouping defined by the Groupiness Factor and the statistics of large wave heights. (This correlation can be expected, to a certain extent, since the Groupiness Factors are defined by the standard deviation of instantaneous wave energy history in the time domain). Unlike floating structures whose frequency response of horizontal oscillations are in the range of wave group frequency, stability of fixed structures such as breakwaters is highly sensitive to the statistics of large wave heights. Hence higher damage in breakwaters with larger values of GF does not necessarily mean that the degree of grouping is the main cause of damage: it could be induced by the large wave heights associated with high GF. Hence, when evaluating sensitivity of fixed structures to wave grouping, the time series that are selected for testing purposes should exhibit comparable wave height statistics, but different values of GF (see Mansard et al., 1994). Researchers such as Johnson et al. (1978) and Galand and Manoha (1991), who illustrated the influence of wave grouping on breakwater stability, did not specifically evoke the aspects related to wave height statistics, possibly because their intent was to illustrate the unsuitability of qualifying a sea state solely by its frequency domain characteristics. Medina et al. (1990) have recently proposed a new concept which can overcome the difficulty described above. This concept is included below in the discussion on new concepts.

NEW CONCEPTS OF WAVE GROUPING MEASURES

Improved Predictor of Run Length Statistics

In recent years, the prediction of run length statistics from the spectrum-derived correlation parameter, κ_f, has undergone substantial advances. Earlier investigations using the spectrum-derived correlation parameter (see Equation 1) resulted in the underprediction of run lengths since the correlation coefficient between successive wave heights computed by zero-crossing analysis was larger than the value derived using κ_f. The main reason for this discrepancy is in the implicit assumption of the narrow-banded process used in defining κ_f. Recently van Vledder using the investigations of Tayfun (1990) on broad-banded spectrum, proposed a modified expression for the spectrum-derived correlation coefficient between successive wave heights $\rho_{HH,f}$ that correlates well with the estimation made in the time domain, $\rho_{HH,t}$. The modified expression is:

$$\rho_{HH,f,new} = \frac{\rho_{HH,f}(\frac{1}{2}\hat{T}) + 2\rho_{HH,f}(\hat{T}) + \rho_{HH,f}(\frac{3}{2}\hat{T})}{2 + 2\rho_{HH,f}(\frac{1}{2}\hat{T})} \quad (2)$$

where :

$$\hat{T} = T_{m02}(1 - \frac{1}{2}v^2) \quad (3)$$

T_{m02} is the average period based on the second spectral moment and v is the spectral width parameter proposed by Longuet-Higgins (1975). The relationship between $\rho_{HH,f}$ and the κ_f parameter given in Equation 1 can be found in van Vledder (1992).

Motion Equivalent Groupiness Factor

Mansard and Sand (1992) claim that a large Groupiness Factor, derived either from SIWEH or from the Hilbert transform, does not necessarily lead to higher structural response for all test structures. It is possible that the large variance of η_ℓ^2 which results in high value of GF, may be outside the range of frequency response of the structures. Mansard and Sand (1992) suggest therefore a new concept of Motion Equivalent Groupiness Factor, which provides an integral measure of the degree of grouping in waves and the frequency response of test structures. As an example of this concept, a new expression was developed relating the surge motion of a simple floating structure to the degree of grouping in waves.

Groupiness Factor Distribution Function

Often, the frequency response of structures may be difficult to formulate and will vary from one structure to another. To account for this, Mansard and Sand (1992) suggest the use of a Groupiness Factor Distribution Function, defined by the following equation:

$$GFDF(f_c) = \frac{1}{m_0} * \sqrt{\int_0^{f_c} s_{\eta_\ell}^2(f) \cdot df} \quad \text{for} \quad 0 < f_c < \infty \quad (4)$$

Instead of using a single value of GF based on the total variance of the low frequency part of the square of the water surface elevation, it is proposed here to take into account the distribution of this variance over different frequency ranges. (Hilbert transform technique is recommended for this application). Although this concept does not provide a unique parameter to characterize the degree of grouping, it is believed that it would find wider application because of its suitability to any arbitrary frequency response.

Envelope Exceedance Coefficient

Recently, Medina et al (1990) advanced a new groupiness measure called Envelope Exceedance Coefficient. It is based on the concept of the wave height function computed using the Hilbert transform technique and can be expressed as follows :

$$\alpha = \frac{\alpha'}{E(\alpha')}, \quad \alpha' = \frac{1}{N} \sum_{n=1}^{N} \left[\frac{H(n\Delta t) - H_{1/10}}{H_{1/10}}\right]^2 \cdot \delta(n) \quad (5)$$

$$\delta(n) = 1 \quad \text{if } H(n\Delta t) > H_{1/10}, \quad \delta(n) = 0 \quad \text{if } H(n\Delta t) < H_{1/10}$$

where:

$H_{1/10}$ = 1.27 Hm_0 (Hm_0 is the estimate of significant wave height derived from the spectrum); and N is the number of data points in the wave height function sampled at Δt intervals.

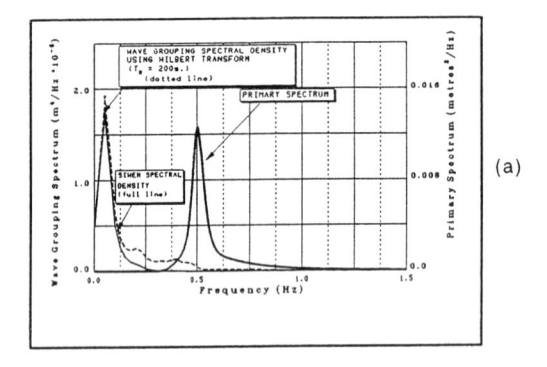

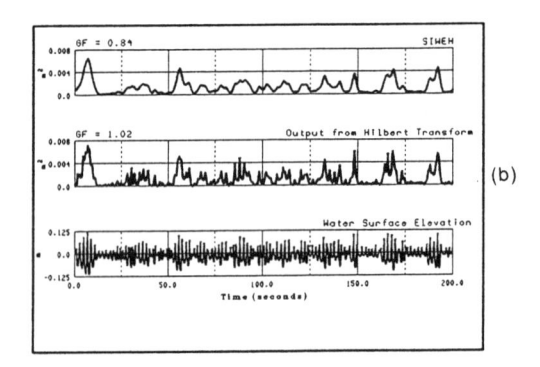

Figure 3. Comparison between SIWEH and Hilbert Transform concepts.

The main assumption used in this concept is that $H_{1/10}$ is the relevant parameter for breakwater stability as suggested in the Shore Protection Manual of 1984. However, this concept can easily be adapted to account for other wave height parameters that may be considered more relevant to structural response. Its main limitation is that it does not lend itself easily to the control of wave group frequency for evaluating the response of floating structures. The SIWEH concept may still be the most useful technique for this particular application, because it gives the user the flexibility of varying the peak frequency and the spectral width of wave grouping spectrum.

As the envelope exceedance coefficient promises to be a useful tool for testing of fixed structures, it was considered relevant to provide a comparative evaluation of this concept with other well known measures through numerical simulation described below.

Comparison of Envelope Exceedance Coefficient with Groupiness Factor and Run Lengths by Numerical Simulation

From a JONSWAP spectrum characterized by its γ value, 100 different realizations of time records were synthesized, using the well known random phase spectrum method. Each wave record, having approximately 650 waves, was then analyzed to provide the values of Envelope Exceedance Coefficient, Groupiness Factors and Run Lengths. For the value of $H_{1/10}$, it was decided to use, as Medina et al. (1990) did, the relationship $H_{1/10} = 1.27\ Hm_0$. In the run length analysis, the threshold was chosen to be $H_{1/10}$. Two Groupiness Factors, one based on SIWEH and another on the expression $\sigma[H^2(t)]/8m_0$ proposed by Medina et al. (1990) were computed.

The above simulation and analysis were carried out for three different values of peak enhancement factor $\gamma = 1$, 3.3 and 7. In order to establish the correlation that exists between the exceedance coefficient α and other measures, α values were sorted in the ascending order making sure that their corresponding values of Groupiness Factors and run lengths were also properly re-arranged. A linear polynomial fit was then applied to the various combinations of data sets.

The complete results of these simulations are presented in Figure 4 and they lead to the following conclusions:

The Groupiness Factors, GF, derived from SIWEH increase with γ values. The reasons for this increase can be easily inferred from Figure 1. The goodness of fit between α and GF in the order of 70%.

The GF values from the wave height function are not sensitive to the spectral width. When $\alpha=1$, the GF value is equal to 1. There is a distinct relationship between these two parameters with a goodness of fit in the order of 78%. However, they are not interchangeable.

There is no correlation between α and run length (goodness of fit less than 10%). The average run length increases with narrower spectral width.

CONCLUSIONS

All the concepts of wave grouping measures proposed in the literature have some limitations in terms of their practical applications.

Statistical variability of groupiness measures increases as the record length decreases. This statistical variability does not permit an easy evaluation of the relationship that exists between wave grouping and spectral width of prototype sea states.

Expected values of wave grouping measures can be deduced from the spectral information of the primary sea state.

The two commonly used wave measures, Groupiness Factor and run length, do not characterize a given sea state in the same fashion.

The Hilbert transform technique of computing the square of the water surface elevation, eliminates the necessity of smoothing it with a Bartlett filter, but needs some improvements in terms of its application to predict structural response of floating structures.

The three new concepts, Envelope Exceedance Coefficient, Motion Equivalent Groupiness Factor, Groupiness Factor Distribution Function, proposed in the literature, provide promising improvements to the existing concepts and deserve further research in terms of their broader applicability.

REFERENCES

Battjes, J.A. and van Vledder, G.Ph. 1984. Verification of Kimura's Theory for Wave Group Statistics. Proc. 19th Int. Conf. on Coast. Engg., Houston, USA.

Bitner-Gregersen, E.M. and Gran, S. 1983. Local Properties of Sea Waves Derived from a Wave Record. Applied Ocean Research, Vol. 5, No. 4.

Funke, E.R. and Mansard, E.P.D. 1979. On the Synthesis of Realistic Sea States in a Laboratory Flume. Tech. Report, Hydraulics Laboratory, National Research Council of Canada LTR-HY-66.

Funke, E.R. and Mansard E.P.D. 1981. On the Meaning of Phase Spectra in the Fourier Transform of Random Waves. Proc. Int. Conf. Hydrodynamics in Ocean Eng., Trondheim, Norway.

Funke, E.R., Mansard, E.P.D. and Dai, G. 1988. Realizable Wave Parameters in a Laboratory Flume. Proc. 21st Int. Conf. on Coast. Engg.,Malaga, Spain.

Galand, J.C. and Manoha, B. 1991. Influence of Wave Grouping on the Stability of Rubblemound Breakwaters, Proc. XXIV IAHR Congress, Vol. B, Madrid, Spain.

Goda, Y. 1983. Analysis of Wave Grouping and Spectra of Long Travelled Swell. Rept of the Port and Harbour Res. Inst. Japan, Vol. 22, No. 1.

EVALUATION OF WAVE GROUPING MEASURES 845

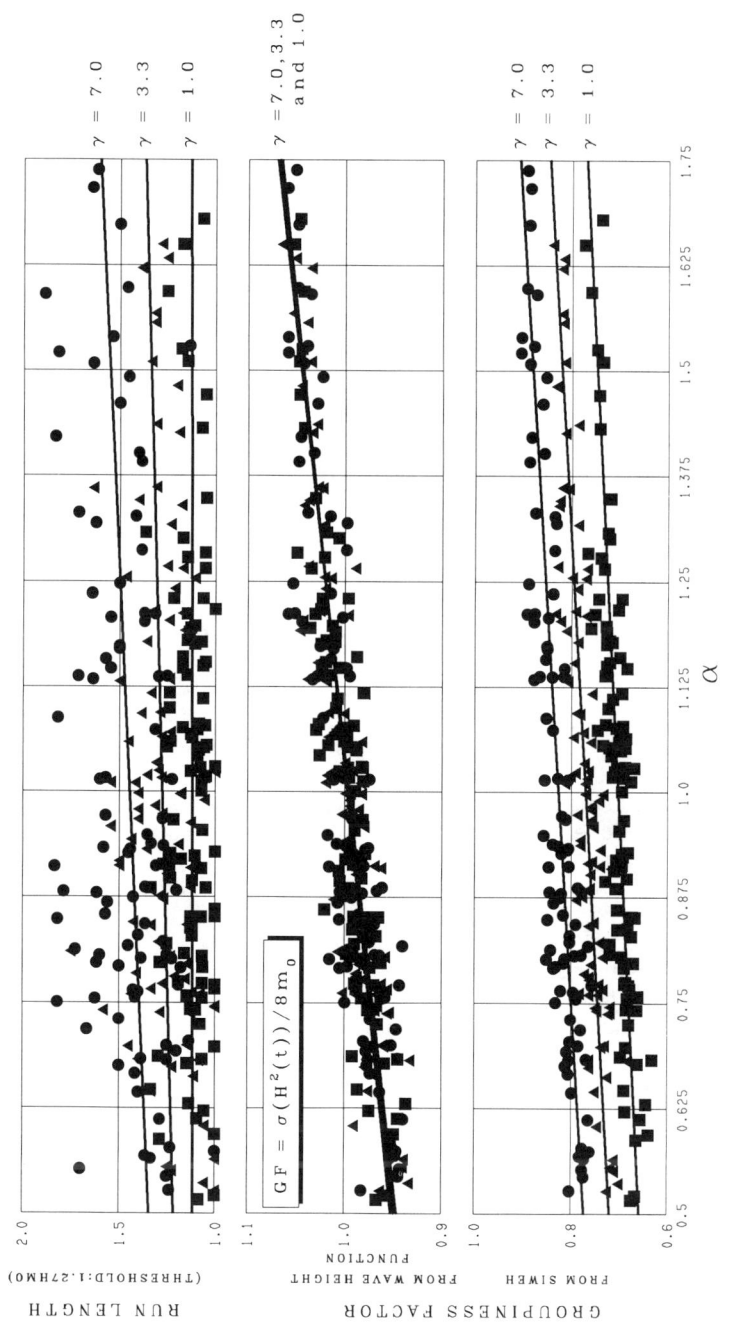

Figure 4. Comparison of four different wave grouping measures

Johnson, R.R., Mansard, E.P.D. and Ploeg, J. 1978. Effects of Wave Grouping on Breakwater Stability. Proc. 16th Int. Conf. on Coast. Engg.,Hamburg, Germany.

Kimura, A. 1980. Statistical Properties of Random Wave Groups. Proc. 17th Int. Conf. Coast. Eng., Sydney, Australia.

Longuet-Higgins, M.S. 1975. On the Distribution of Wave Periods and Amplitudes of Sea Waves. J. of Geophysical Research, Vol. 86 (5).

Mansard, E.P.D., and Sand, S.E. 1992. Towards Improved Concepts for the Description of Wave Grouping. Proc. OMAE'92 Conference Calgary, Canada.

Mansard, E.P.D., Miles, M.D. and Nwogu, O.G. (1994). Laboratory Simulation of Waves: An Outlineof Wave Generation and Analysis Procedures, Proc. Int. Symp. on Waves-Physical and Numerical Modelling, Vancouver, Canada.

Mase, H. and Iwagaki, Y. 1986a. Wave Group Analysis of Natural Wind Waves Based on Modulational Instability Theory. Coastal Eng., Vol. 10, pp. 341-354.

Mase, H. and Iwagaki, Y. 1986b. Wave Group Property of Wind Waves from Modulational Instability. Coastal Engineering in Japan, pp. 565-577.

Medina, J.R., Fassardi, C. and Hudspeth, R.T. 1990. Effects of Wave Groups on the Stability of Rubble-Mound Breakwaters. Proc. 22nd Int. Conf. on Coastal Engineering, Delft, The Netherlands.

Medina, J.R. and Hudspeth, R.T. 1988. Analysis of Wave Groups in Random Fields, Proc., Ocean Structural Dynamics Symposium '88, Corvallis, Oregon

Medina, J.R. and Hudspeth, R.T. 1990. A Review of the Analyses of Wave Groups. Coastal Engineering, Vol. 14, pp. 515-542.

Pinkster, J.A. 1984. Numerical Modelling of Directional Seas. Symp. on Description and Modelling of Directional Seas. Copenhagen, Denmark.

Rye, H. and Lervik, E. 1981. Wave Grouping Studies by Means of Correlation Techniques. Int. Symp. on Hydrodynamic in Ocean Eng., Trondheim, Norway.

Sobey, R.J. and Read W.W. 1984. Wave Groups in the Frequency and Time Domains. Proc. 19th Int. Conf. in Coast. Eng. Houston, USA.

Tayfun, M.A. 1990. Distribution of Large Wave Heights, ASCE J. Waterway, Port, Coastal and Ocean Engineering, Vol 116, No.6.

Vidal, C., Losada, M., and E.P.D. Mansard, 1995. A Suitable Wave Height Parameter for Stability Formulae of Rubble-Mound Breakwaters, to be published in the Mar/Apr issue of ASCE J. on Waterway, Port, Harbour and Coast. Engg.

Van Vledder, G.P. 1992. Statistics of Wave Group Parameters, Proc. 23rd Int. Conf. on Coastal Engineering, Venice, Italy.

Yasuda, T., Nakashima, N.K. and Tsuchiya, Y. 1986. Grouping of Waves and Their Expression on Asymptotic Envelope Solution Modes. Coast. Engg in Japan, pp. 864-876.

CHAPTER 62

Relationship of a moored vessel in a harbour and a long wave caused by wave groups

Toshihiko Nagai [1] Noriaki Hashimoto [1] Tadashi Asai [1] Isao Tobiki [2]
Kazunori Ito [3] Takao Toue [3] Akio Kobayashi [3] Takao Shibata [4]

Abstract

Loading operations failed at a berth in Sendai New Harbour in 1992. The vessels continuously moved along the berth, in spite of small wave heights and weak winds, which were much less than the criteria for loading operation at the berth. To clarify the cause of the failures, the authors analyzed the observed wave data. The phenomenon related to the time characteristics of wave groupiness. The vessels' motions were caused by the long waves that were bounded by wave groups. Mean Wave Group Period ($\overline{T_g}$) proposed here, can explain the phenomenon.

1. Introduction

The harbour tranquility should be evaluated not only by the wave height at a berth but by the wave period, wind condition, the scale of vessels and so on.

An experienced berth master decides when the loading operation should start considering wave heights, wave periods and wind conditions. But it is frequently reported that due to the continuous motion of a moored vessel, the vessel has to be released, or at worst the mooring devices get damaged, even when the wave heights are small and the winds are weak. Since these failures of the loading operations reduce the rate of effective working days and the safety of working, it is very important to reveal the cause of the failure.

With the observed wave data from the *NOWPHAS* (Nationwide Ocean Wave information network for Ports and HArbourS,Nagai et al,1994) and the situation

---

[1] Port and Harbour Research Institute,Ministry of Transport,1-1,Nagase 3-chome,Yokosuka,Kanagawa,239,Japan

[2] Second District Port Construction Brureau,Ministry of Transport, 5-57,Kitanaka-dori,Naka-ku,Yokohama,231,Japan

[3] Taisei Corporation,Technology Research Center,344-1,Nasemachi,Totsuka-ku,Yokohama,245,Japan

[4] Tohoku Oil Corporation, LTD,1-1,Minato 5-chome,Miyagino-ku,Sendai,985,Japan

of the failure, we investigated the wave characteristics and the situation, and clarified the cause of the operation failure and proposed a new parameter for the harbour tranquility in this report.

2. Location and condition of the field study

Sendai New Harbour is located in Miyagi prefecture in Japan as shown in Figure 1. The harbour faces the Pacific Ocean, and it is sheltered by the outer breakwaters and an offshore breakwater that is under extension work. The yearly averaged wave direction is in between SE and SSE. The berth studied here is No. 1 berth owned by Tohoku Oil Company.

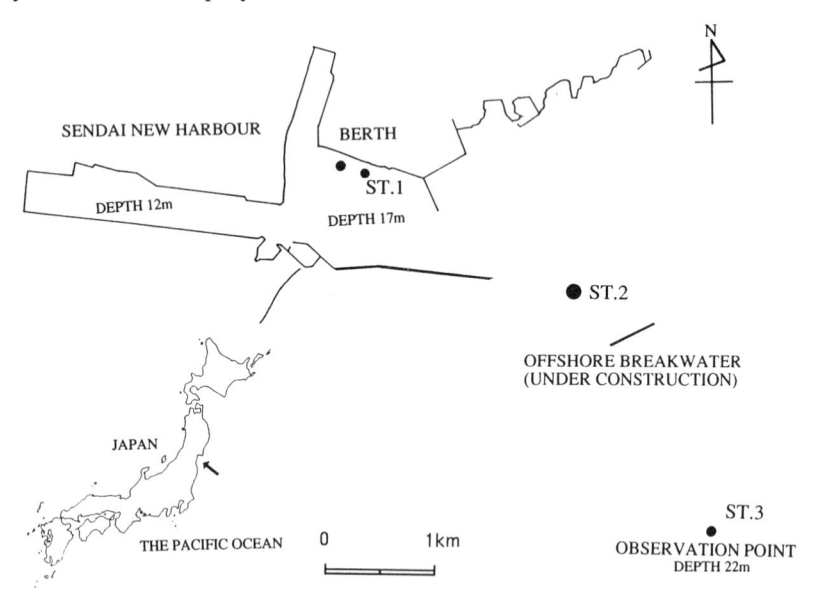

Figure 1 Location of Sendai New Harbour

Wave conditions are measured at three points. One(ST.1) is located at the east side of the berth where the water depth is 17 m, operated by the Tohoku Oil Company. The second(ST.2) is located behind the breakwater, where the water depth is 18 m. The third(ST.3) is located at 2.4 km offshore from the harbour mouth where the water depth is 22m, one of the *NOWPHAS* offshore observation stations operated by the Second District Port Construction Bureau of the Ministry of Transport. Wave heights are measured by a ultrasonic-type wave gage (USW), and the horizontal currents are measured by an ultrasonic-type current meter (CWD). The sampling interval is 0.5 s, and the data are recorded for 20 minute in every two hours due to *NOWPHAS* standard. The winds are measured at the height of 20m above the land of the berth.

The conditions of the loading operations are recorded for every operations. The duration time in which a vessel is moored at the berth and waiting time at offshore is also recorded. The beginning time of the loading operation is equal to the time that the berth master judges if the loading operation is possible. At first, the data of ST.1 and ST.3 are analyzed from June to October in 1990 and 1991 and the two cases in which the operation failed in 1992 are also analysed. Futhermore the data of ST.1, ST.2 and ST.3 are analyzed from January to February in 1994.

3. Criteria for loading operation

Start time of loading operation entirely depends on the judgment of the berth master. The berth master generally decides the time considering the wave condition, weather, vessel scale, conditions of the mooring devices, duration time for the loading and so on. It is, however, true that the judgment is quite experiential.

Table1 Wave heights and periods at the berth

YEAR	MONTH	DAY	TIME	OFFSHORE WAVES H1/3 m	T1/3 s	θ	VESSEL ×10,000 DWT	WAVES AT THE BERTH H1/3 m	T1/3 s
90	6	30	12	0.76	5.98	SE	—	—	—
90	7	7	10	0.7	7.51	SE	—	0.26	4.39
90	7	8	14	0.61	6.19	SEE	—	0.15	4.28
90	7	15	10	0.74	7.57	SE	8.1	0.24	5.5
90	7	17	12	0.76	6.90	SSE	—	0.67	4.46
90	7	18	18	0.79	7.27	SSE	—	0.2	4.04
90	8	8	16	1.06	7.46	SEE	5.4	0.16	6.93
90	8	14	12	0.48	8.36	SE	—	0.09	5.08
90	8	15	18	1.2	5.75	SSE	18.6	0.13	5.55
90	8	24	10	0.94	8.27	SSE	6.5	0.11	6.57
90	8	25	14	0.45	6.74	SSE	—	0.09	4.08
90	8	26	14	0.41	7.20	SSE	—	0.09	4.71
90	9	2	10	0.65	6.59	SE	5.7	0.18	4.33
91	6	7	12	0.73	8.97	SE	6.1	—	—
91	6	17	16	0.89	7.66	SE	—	0.26	5.65
91	6	19	10	0.93	7.11	SSE	—	0.29	5.57
91	6	20	16	0.82	7.55	SE	—	0.22	5.34
91	6	25	10	0.97	7.43	SEE	—	0.23	4.78
91	7	14	12	0.51	6.34	SEE	—	0.41	5.67
91	7	16	10	0.5	7.22	SE	—	0.44	4.23
91	8	3	10	0.5	7.61	SE	7.3	0.19	3.85
91	8	4	14	0.64	4.83	SE	6.2	0.23	3.45
91	8	10	12	0.59	7.77	SEE	—	0.15	3.46
91	8	24	10	1.19	8.07	SE	—	0.34	4.61
91	8	25	18	0.98	9.29	SE	6.1	0.2	5.96
91	8	27	12	0.54	8.42	SE	—	0.15	3.84
91	8	29	14	0.68	6.01	SSE	6.1	0.18	4.24
91	9	4	10	0.55	8.92	SE	21.2	0.14	4.61
91	9	6	12	0.47	8.48	SEE	6.2	0.16	3.43
91	9	18	10	0.78	8.06	SEE	—	0.22	4.75
91	10	3	12	0.6	7.63	S	9.6	—	—
91	10	6	12	0.47	7.48	SE	6.1	—	—
92	10	22	10	2.54	12.3	SEE	22	0.26	5.78
92	11	24	12	1.56	12.67	SE	22	0.22	5.05

θ : WAVE DIRECTION
— : MISSING OF DATA

The wave heights and periods both at the berth and at the offshore in the beginning of the loading operation are listed in Table 1. In the table, a solid line (—) denotes the missing of data. Two rows from the bottom of the table show the data for failed operations. Wave heights and periods at the berth are smaller than those of offshore as expected. The criteria of the loading operation can be estimated from the table.

The maximum wave height is about 0.5m, and the maximum wave period is about 6 s for the judgment of the berth master. The table also shows that the judgment of the berth master is generally quite reasonable. It should be noted that the wave heights are small, and periods are shorts in the failed operation.

4. Investigation of Fail-Operation
4.1 Condition of Fail-Operation

Fail-Operation is defined here as a phenomenon that the loading is not well operated or failed in spite of small wave heights and winds. As the antonym, Normal-Operation is also used. The examples of the Fail-Operation were found on Oct. the 22nd and Nov. the 10th in 1992. The former is defined as case 1 and the later is case 2. The offshore wave heights in Fail-Operations are larger than those of Normal-Operation in 1990 and 1991. But the wave heights at the berth are much less than the criteria for the loading operation as mentioned before due to the extension of the offshore breakwater in 1992.

According to the berth master, the situations in Fail-Operations are as follows;

case 1
> When 220,000 DWT oil tanker was moored at the berth, the tanker began to move along the berth with a surging motion. The amplitude of motion was about 1.5m. When the amplitude become 1.0 m after 4.5 hours, the loading finally could be started.

case 2
> Immediately after being moored at the berth, the vessel continuously moved along the berth also with a surging motion for 24 hours. The vessel was finally had to be released without loading operation.

In both cases, the vessel was moored at the berth with mooring ropes. When the tensile force of mooring ropes achieved the breaking force of the winch, the mooring rope was released automatically, and it was rewinded adequately.

4.2 Offshore wave condition in Fail-Operations

Figure 2 shows that the time series of wave profiles of case1 and case 3, where case 3 is of Normal-Operation on Sep. 4th in 1991. The type of the vessels for each cases are all the same. The wave heights of case 1 are bigger than that of case 3. But the wave height of case 1 at the berth was 0.26m ,which is small enough for the loading operation.

The wave direction of case 1 was *SEE* and that of case 3 was *SE*. The offshore breakwater is quite effective to these wave directions. Figure 3 indicates a typical wave directional spectra of case 1. The wave directional spectrum was calculated by the maximum entropy method (*MEP*) developed by Hashimoto et al (1985). The

(a) case 1 Fail-Operation

(b) case 3 Normal-Operation

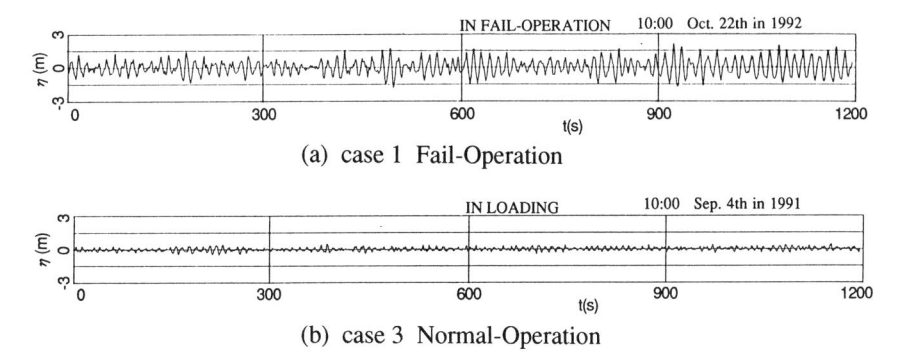

Figure 2 Wave profiles of offshore wave

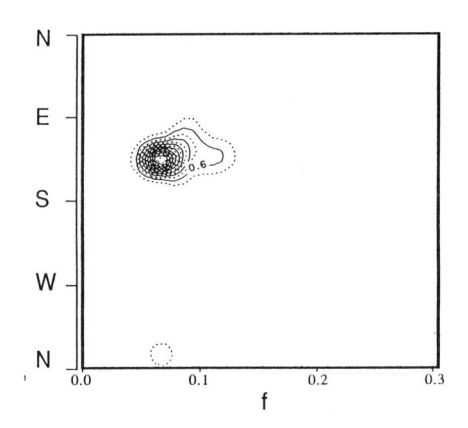

Figure.3 Directional spectrum of observed data

directional spectrum was narrow, and it had one peak. The directional spectra of case 1, case 2 and case 3 were of the same forms.

The wave trains of case 1 and case 2 are characterized by remarkable wave groups. On the contrary, the clear wave group can not be seen in the wave trains of case 3. The characteristics of the wave group are discussed in the next chapter.

5. Characteristics of Wave group
(1) Evaluation of Wave Groupiness

The characteristics of the wave groups are usually evaluated by Groupiness factor(GF ; Funke and Mansard(1980)), mean run length($\overline{j1}$), and envelop correlation parameter(κ ; Battjes and Vledder(1984)). Figure 4 shows the time history of GF and $\overline{j1}$ during the Fail-Operations. GF and $\overline{j1}$ of case 3 are also shown for comparison. In Figure 4, the horizontal axis is a elapsed time (hour) from the start of the operation, and the vertical axis is the value of GF and $\overline{j1}$. GF and $\overline{j1}$ do not have any difference between the Fail-Operation and the Normal-Operation. As shown in Figure 2, the wave trains of case 1 and 2 have much stronger wave groupiness than case 3. Strong wave groupiness means that the amplitude of the envelop wave is large, and the modulation of high waves is gradual / smooth.

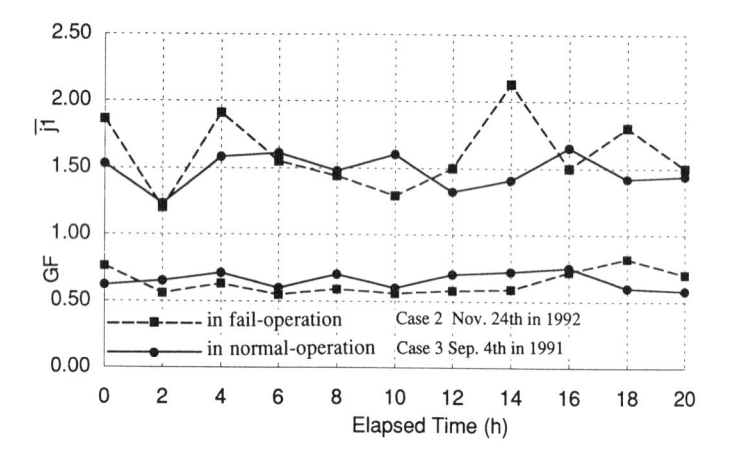

Figure 4 Time series of GF and $\overline{j1}$

GF is defined by covarience of $SIWEH$ (Smoothed Instantaneous of Wave Energy History). $SIWEH$ expresses the energy distribution of wave train, and is similar to the envelop. It is clear that GF can not express the sequence of heigh waves and magnitude of wave height because of the difinition. Therefore, if the sequences of two wave trains are same in the shape but different in the amplitude, GF gives a same value for both wave trains. Therefore, GF may have large value for the wave train without remarkable wave groups, if the sequence of high waves do not change gradually and the wave train has isolated high waves.

On the other hand, since $\overline{j1}$ is defined as the number of high waves that exceed the threshold value, $\overline{j1}$ expresses the time characteristics of wave groupiness. But, since the unit of $\overline{j1}$ is number, if the sequences of two wave trains are the same in the shape but different in the period, $\overline{j1}$ gives the same value for both wave trains. And also, the similar discussion of GF can be made for $\overline{j1}$. In the calculation of $\overline{j1}$, the threshold value must be known to define the run length. Since the threshold value is usually taken as a statistic value of wave train such as the significant wave height, $\overline{j1}$ can not express the difference of the magnitude of wave heights between Fail and Normal-Operation.

Figure 5 shows the time history of κ. In the figure, a white circle denotes when

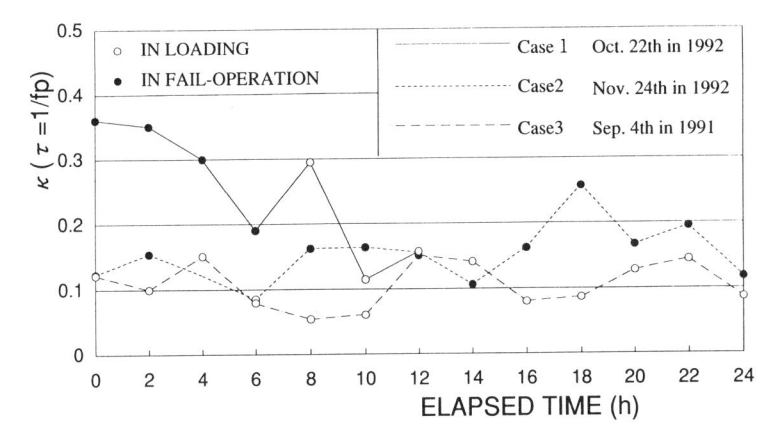

Figure 5 The comparison of κ

the loading was possible, and a black circle denotes when indicates the loading was not possible. κ in Fail-Operation, illustrated by a solid line, indicates that κ decreases with the elapsed time. When the moored ship was continuously moving, κ was about 0.35, but when the loading was possible, κ was about 0.15. The κ values can express the difference between Normal and Fail Operation, in contract with the GF and $\overline{j1}$ values. Because κ is actually auto-correlation of the amplitudes of the wave envelop with a time lag, τ. i.e., κ includes a time characteristics of wave groups. But the difference between Fail-Operation of case 2 and Normal-Operation is not clear. κ is directly calculated by a frequency spectra form based on the narrow band linear theory. If frequency spectrum form is not narrow enough, κ has some error in the calculation.

As shown above, Fail-Operation has a relationship to wave groups. Especially, the time characteristic of the wave groups is important. Since the common parameters such as GF, $\overline{j1}$ and κ can not express the difference of two Operations, a new parameter must be introduced to explain the differences.

(2) A new parameter for wave groupiness

Time characteristics of wave groups are an important factor to explain the Fail-Operation. One of the time characteristics of wave groups is a time interval of wave groups. Mean wave Group Period, $\overline{T_g}$, is defined as the mean interval of each wave groups. T_g is equivalent to the total run. T_g, however, has a dimension of time. T_g is calculated by eq.(1) and also illustrated in Figure 6. (Of course, we could apply SIWEH to the method that define the wave group periods.)

$$T_g = \sum_{i=1}^{N} T_i \tag{1}$$

where, T_i is the zero-crossing wave period, N is the number of T_i, where N is equal to a total run($j2$).

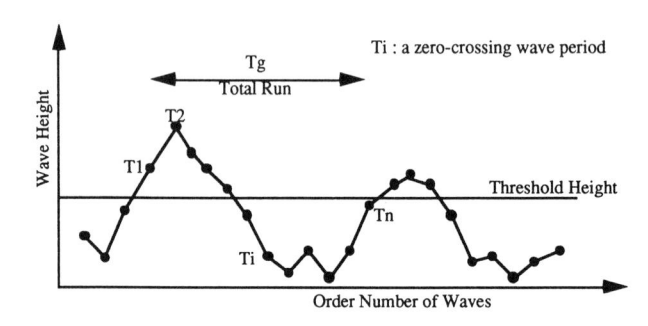

Figure 6 Definition sketch of T_g

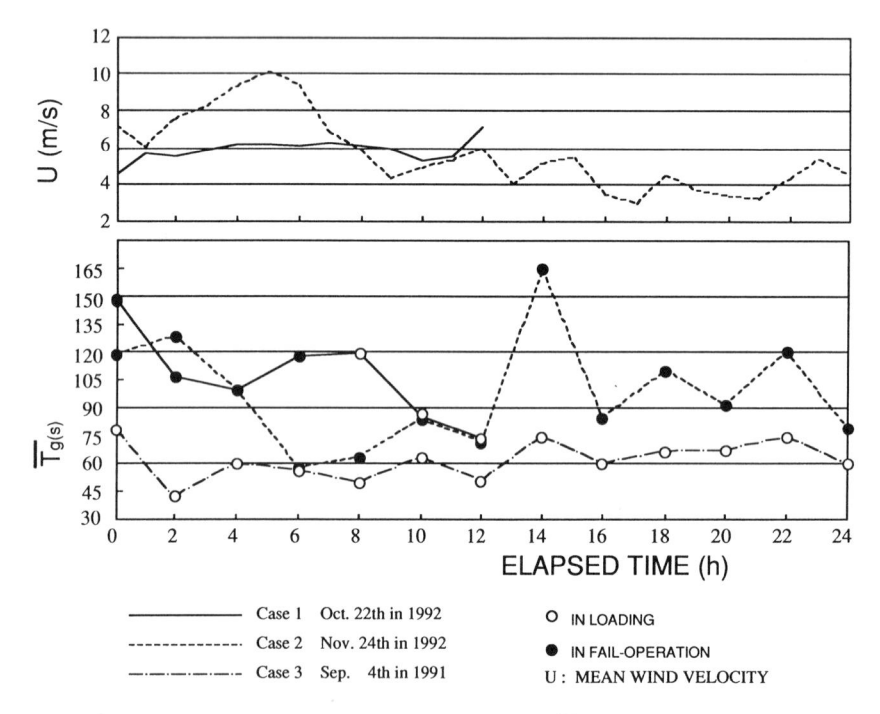

Case 1   Oct. 22th in 1992
Case 2   Nov. 24th in 1992
Case 3   Sep. 4th in 1991

○ IN LOADING
● IN FAIL-OPERATION
U : MEAN WIND VELOCITY

Figure 7 Time series of $\overline{T_g}$

Figure 7 shows $\overline{T_g}$ for case 1, case 2 and case 3 with the wind velocity data. In case 1, $\overline{T_g}$ ranges from 105s to 150s, when the loading was not possible. But, when the loading was possible, $\overline{T_g}$ was around 75s. $\overline{T_g}$ in case 2 was also long when the loading was not possible. On the contrary, $\overline{T_g}$ in case 3 was always around 60s.

In case 2, there is the time when the loading is not possible in spite of the relatively short $\overline{T_g}$ (from 6 to 12 of the elapsed time in the figure). This is because of the strong wind. The shorter $\overline{T_g}$ is caused by the growth of the wind waves. But, since the moored vessel at the berth was moved due to the wind, the loading was not possible.

It is clear that when the loading was not possible, $\overline{T_g}$ was long, on the other hand, when the loading was possible, $\overline{T_g}$ was short. Therefore, $\overline{T_g}$ is an useful parameter for the harbour tranquility when we consider the loading operation.

5.2 Physical meanings of $\overline{T_g}$

It is well known that the wave set-down occurs when wave groups propagate. Assuming that temporal wave trains are homogeneous in space and wave trains propagate unidirectional, the wave set-down can be calculated after Longuet-Higgins (1962) by eq.(2) and eq.(3).

$$\overline{\zeta} = -S_x/\rho(gh - c_g^2) \tag{2}$$

$$S_x = 0.5\rho g a^2 c_g (2c_g/c - 0.5) \tag{3}$$

where, h is the water depth, ρ is the specific gravity of water, g is the gravity acceleration, a is a half of zero-crossing wave height, c is the phase velocity, c_g is the group velocity, and S_x is the radiation stress. The radiation stress S_x is defined by each zero-crossing wave heights and periods, and c and c_g are defined by zero-crossing wave periods.

Figure 8 shows the wave set-down calculated from the time series of wave heights in case1. The amplitude of the wave set-down corresponds to the strength of the wave groups.

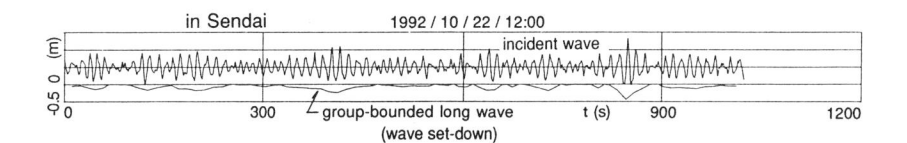

Figure 8   Group-bounded long wave (wave set-down)

Figure 9 indicates the frequency spectrum of the incident wave and the calculated wave set-down. The figure shows the variation of spectrum from Fail-Operations to Normal-Operations. The variation of the power in the low frequency correspond to the events of Fail or Normal-Operation. When the power in the low frequency is large, the loading is not possible. On the contrary, when the power is not large, the loading is possible. Thus, the power in the low frequency is a dominant factor to control the loading operation.

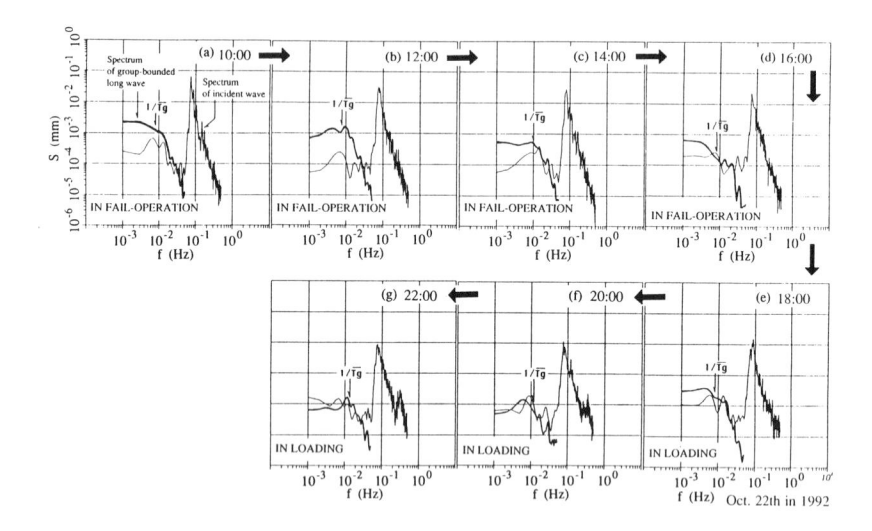

Figure.9 The frequency spectra of incident waves and group-bounded long wave

When the peak frequency in the low frequency of the incident waves is clear, e.g. (b), (c) and (g) in the Figure 9, the peak frequency of the wave set-down and $1/\overline{T}_g$ are nearly equal to the peak frequency of the incident waves. Therefore, it may be concluded that $1/T_g$ is the peak frequency of the long waves in the incident waves and the peak frequency of the wave-set down.

In Fail-Operation from (a) to (c), we can recognize the difference between the spectrum of wave set-down and one of the incident waves. This is because in the low frequency region, the incident waves are consist of the group bounded long waves and the progressive long waves, and since the observation time of wave data is only 20 min., it is not long enough to calculate the group-bounded waves.

6. Propagation of bounded long wave

It is clear now that the incident waves in Fail-Operation has strong wave groupiness, and the long waves bounded by the wave group are important to explain the difference of the wave groupiness between in Fail and Normal Operation. The discussion above, however, are limited in the offshore waves.

The berth was sheltered by the offshore breakwater, thus, the wave height was small enough for the loading operation. We can guess that the groupiness of the incident waves would disappear by the existence of the offshore breakwater, but, the group-bounded long waves can be free waves and can propagate into the harbour.

As shown in Figure 1, a new field observation point(ST.2) was set in 1994, we observed wave condition at three points. Figure 10 indicates the variations of spectra from ST.1 to ST.3.

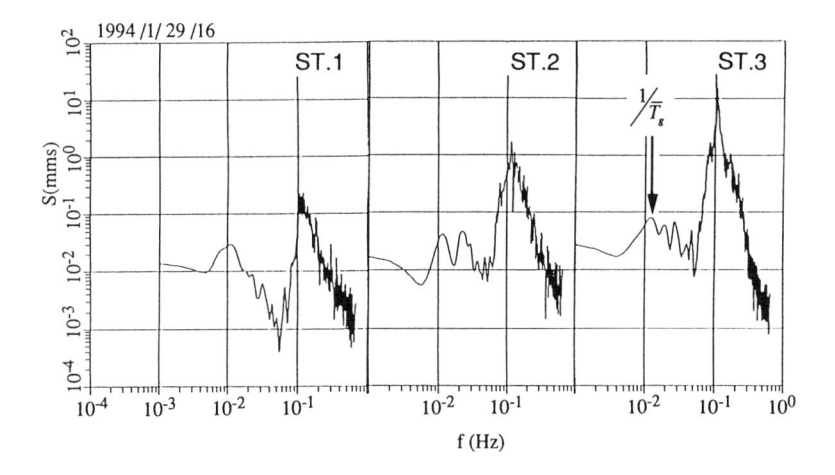

Figure 10 Variation of spectrum as propagation

The power of high frequency region which has the peek frequency around 0.1 Hz decrease as their propagation . On the other hand, the power of low frequency region does not denote remarkable dissipation between ST.1, ST.2 and ST.3. And, the peek frequency of low frequency region is corespond to $1/\overline{T_g}$ between ST.1, ST.2 and ST.3. Figure 11 shows comparison of statistical values between ST.2(behind offshore breakwater) and ST.3(at offshore). $\hat{\eta}$ rms is the root mean square value of low frequency components that are less than 0.05 Hz. The significant wave heights at ST.2 are much smaller than those of ST.3 because of the offshore breakwater. GF and $\overline{T_g}$ at ST.2 are also smaller than those of ST.3. $\hat{\eta}$ rms at ST.2 does not dissipate so much in spite that the waves pass the offshore breakwater. Thus, from figures 10 and 11, we may conclude that the group bounded long wave at offshore exchanges to the progressive long wave behind the offshore breakwater, and, the progressive long wave propagate into the harbour.

7. Relationship between harbour oscillation and motion of vessel

It is well known that the slow drift oscillation of vessels are caused by the low frequency modes of waves (Pinkstar(1974)). The motion of the vessel in Fail-Operation might be the slow drift oscillation. Unfortunately, the time series of wave data at the berth in Fail-Operation was not obtained. Thus, the relationship of the waves at the berth and the motion of the moored vessel can not be directly investigated. But, we can guess two causes of a moored vessel's continuous motion(slow drift oscillation) . One is that a free long wave directly attacked the moored vessel. The other is that a free long wave excites harbour oscillation.

We simulated the harbour oscillation by J.J.Lee's method(1971). Figure 12 shows

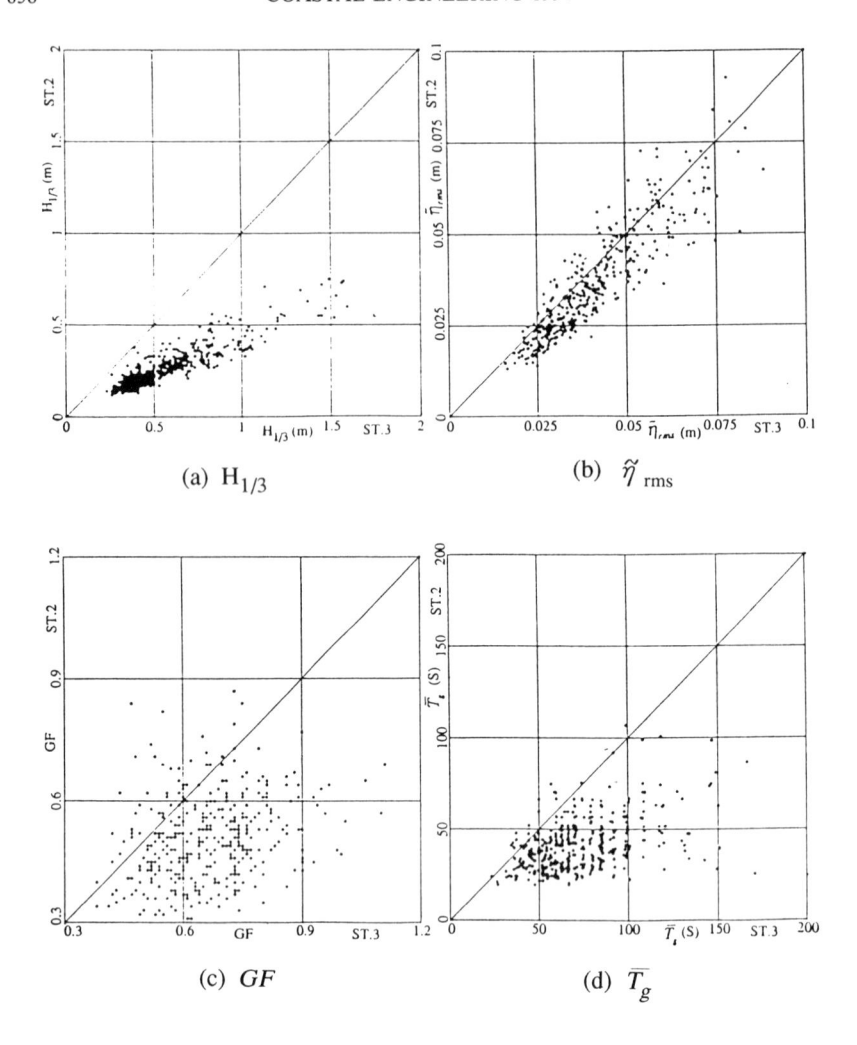

Figure 11 Comparison of statistic values between ST.2 and ST.3

the amplification factor distribution in the harbour. The period of incident wave is equal to $\overline{T}_g$ (100s). The position of the berth is near the node point, and the direction of the motion is surging direction. Therefore, the result of simulation implies that the cause of the slow drift oscillation is the harbour oscillation.

If the periods of the long waves were very close to the natural frequency of surging motion of the vessels, the slow drift oscillation would be excited.

A natural frequency of the surging motion of the vessel (220,000DWT) in Fail-

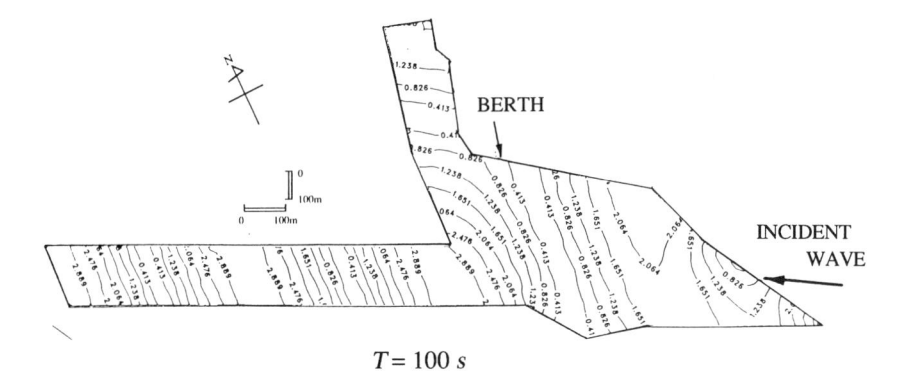

[Figure: contour plot with BERTH and INCIDENT WAVE labels]

$T = 100$ s

Figure 12 Amplification factor of numerical result

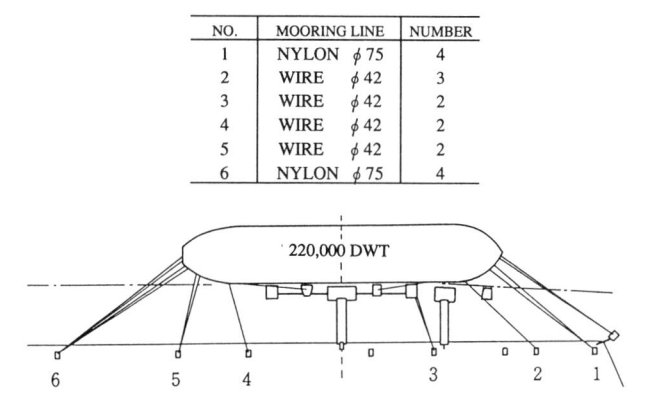

NO.	MOORING LINE	NUMBER
1	NYLON ϕ 75	4
2	WIRE ϕ 42	3
3	WIRE ϕ 42	2
4	WIRE ϕ 42	2
5	WIRE ϕ 42	2
6	NYLON ϕ 75	4

Figure 13 Mooring condition of numerical model

Operation are estimated to 82 s for the mooring system shown in Figure 13. In the estimation, it is assumed that the tensile force of each mooring rope is 3 tf, that is constant through the operation. The restoring force in the direction of the berth is also assumed to be the same as the value obtained from the steady analysis. Furthermore the added mass is not considered. As shown in 4.1, the actual operation of mooring is not steady, and the real tensile force of each mooring rope might be weaker than 3 tf. Therefore the true natural frequency must be longer than 82 s, i.e., the natural period could be close to the period of the group bounded long wave or $\overline{T_g}$.

8. Conclusion

In spite of small waves and weak winds, there was a event that the loading operation failed(Fail-Operation). By analyzing the cause of Fail-Operation, it is found that the group bounded long waves cause Fail-Operation. This means that the information related to the wave group is required for the loading operation in harbors. Wave group period, $\overline{T_g}$, proposed here can be a useful information for the harbour tranquility in controlling the loading operation. $\overline{T_g}$ represents a period of long wave that is bounded by wave groups. A long wave bounded by wave groups at offshore changes to a free long wave behind offshore breakwater, and a free long wave which propagates into a harbour excites a harbour oscillation. When the period of this long wave is close to natural frequency of vessels, and energy of the long wave is large in the harbour, fail-operation would occur.

ACKNOWLEDGEMENTS

The *NOWPHAS* is operated by Ports and Harbours Bureau and its associated agencies including District Port Construction Bureaus and PHRI(Port and Harbour Research Institute) of the Ministry of Transport. Authors must express sincere gratitude to all the persons working for the *NOWPHAS* operations, especially to the engineers responsible for the offshore wave observation of the Shiogama Port Construction Office of the Second District Port Construction Bureau. The authors are also express gratitude to engineers, Oil Receiving and Shipping Sect.,Tohoku Oil Co. for their cooperation.

REFERENCES

Battjes,J.A. and Van G.Ph.Vledder(1984) : Verification of Kimura's theory for wave group statistics, Proc.19th Conf.,ASCE,pp.642~648
Funke,E.R.and E.P.D.Mansard (1980) : On the synthesis of realistic sea states,Proc. 17th Coastal Eng. Conf., ASCE,pp.2974~2991
Hashimoto,N. and K.Kobune(1985): Estimation of directional spectra from the maximum entropy principle,Rept. of P.H.R.I,vol.26,No.5,pp.57~100.(in Japanese),or Kobune,K.and N.Hashimoto(1986): Estimation of directional spectra from the maximum entropy principle, Proc. 5th Inter. OMAE, symp., Tokyo,pp.80~85
Lee.J.J.(1971) : Wave-induced oscillations in harbours arbitrary geometry,J.Fluid Mech., Vol.45, No.2 ,pp.375~394
Longet-Higgins,M.S. and Stewart,R.W.(1962) : Radiation stress and mass transport in gravity waves,with application to 'surf beats',J.Fluid Mech., Vol.13,pp.481~504
Nagai, T, K.Sugahara,N.Hashimoto,T.Asai, S.Higashiyama and K.Toda (1994):Introduction of Japanease NOWPHAS System and its Recent Topics,Proce. of Hydro-Port' 94,pp.83 ~92
Pinkstar,J.A.(1974) : Low frequency phenomena associated with vessels moored at sea, Soc. of Petroleum Engineers of AIME,SEP Paper No. 4837

CHAPTER 63

Coherent Structure of Tidal Turbulence in a Rotating System of Osaka-Bay

Tsukasa Nishimura[1], Tomonao Kobayashi[1], Goichi Furuta[1]

Abstract

Coherent structures dominating in the tidal turbulence field inside the Osaka-Bay were detected by flow visualization through satellite image sensors. To obtain the hydrodynamical characteristics, scale down model experiments were done with the Froude/Rossby similarity law. The tidal residual circulations were reproduced on the ensemble mean turbulence field, which is made very similar to each instantaneous turbulence owing to the robustness of the coherent structure interlocked inside a closed vessel. The earth rotation effects were verified to dominate the turbulence processes in this closed vessel, and a hysteresis due to the moon aging was revealed.

Introduction

Hydrodynamical characteristics of turbulence structures interlocked inside the tidal flow field in Osaka-Bay was investigated by employing the coherent structure concept, satellite-based flow visualization and scale down model experiments performed in a rotating system. Figure 1(a) shows an example of the visualized flow pattern, in which the high gain visible data obtained with Mos-1/MESSR is enhanced to snapshot an instantaneous turbulence field at 1028, 23 December 1987.

The Osaka-Bay is an inland sea basin located in the middle part of Japan-Islands, which is enclosed by some populous and industrialized areas (Figure 1(b)). It is about 50km in size and about 50m in average depth, and is therefore 1,000 in the ratio of horizontal and vertical scales. The water quality of this bay, however, is not so seriously polluted owing to the tidal exchange through the two narrow channels connecting to adjoing sea areas; Akashi-Strait to the Harima-Sea and Tomogashima-Strait to the Pacific-Ocean. Further, tidal jets injected through these straits accelerate the transport process inside the bay feeding with its turbulence components.

---
[1]Dept. of Civil Engineering, Science University of Tokyo, Noda-City, 278, Japan

862 COASTAL ENGINEERING 1994

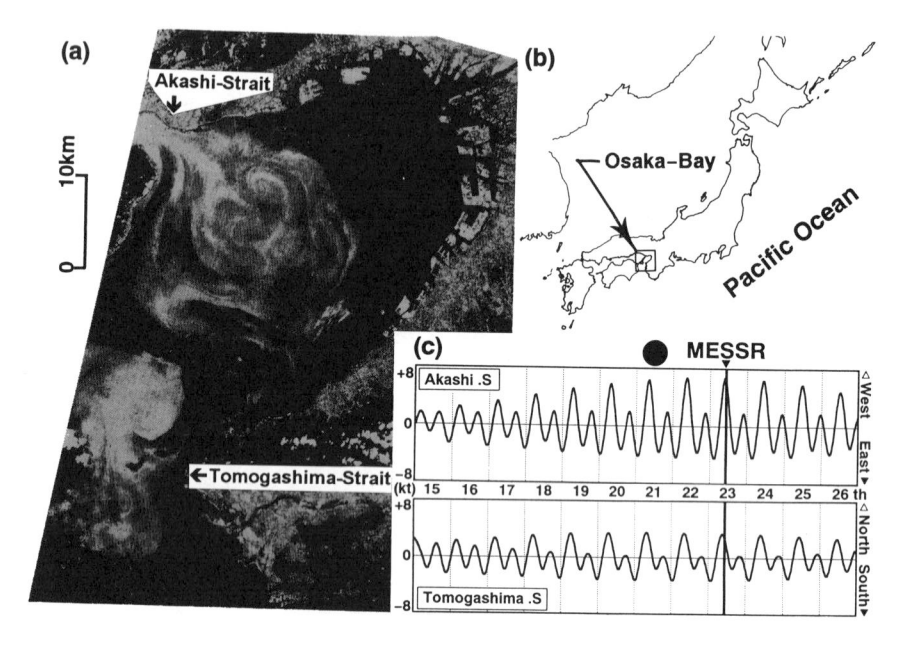

Figure 1. Tidal turbulence in the Osaka-Bay. **(a)** Flow visualization with Mos-1/MESSR high gain visible data at 1028, 23 December 1987. **(b)** Location of the site. **(c)** Tidal change showing with the running speed at Akashi and Tomogashima-Straits in 15-26 December 1987 (Source: forecasted value by the Maritime Safety Agency of Japan).

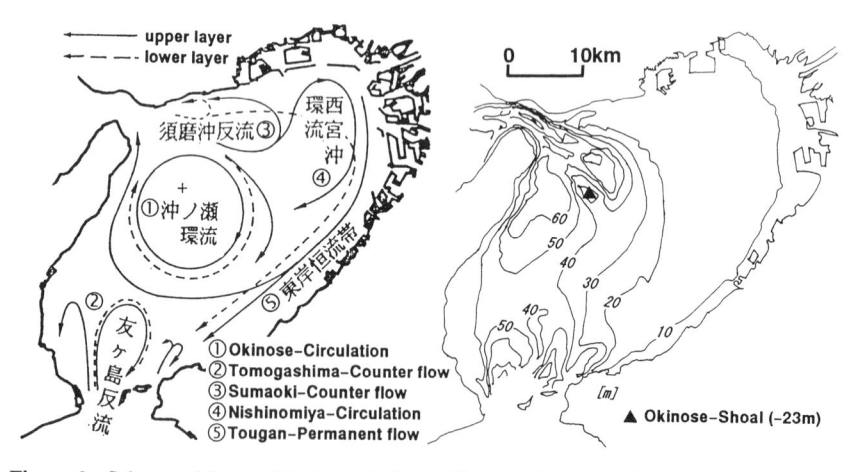

① Okinose-Circulation
② Tomogashima-Counter flow
③ Sumaoki-Counter flow
④ Nishinomiya-Circulation
⑤ Tougan-Permanent flow

Figure 2. Schema of tidal residual circulations in the Osaka-Bay. (by Fujiwara, 1989)

Figure 3. Topography of the rotating closed vessel of the Osaka-Bay.

Flow visualization of coherent structure

On the MESSR image, we detect a coherent structure in the tidal turbulence, which is fed by the tidal jet and is interlocked inside this closed sea basin. Figure 1(c) shows the tidal change indicated with running speed at the two straits. When the MESSR data were received, north tide of about 3 knots in speed is flowing into the bay through the Tomogashima-Strait. On the satellite image, we identify a dipole of a cyclone and an anticyclone is now growing through the inverse cascade of turbulent eddies, which are injected riding on the northward tidal jet. Near the Akashi-Strait, we find another dipole. Through the strait, however, the west tide is discharging out of the bay with about 6 knots in speed. Here, we identify the dipole with that formed by the east tides in the past, which transported the turbulent eddies into the bay. The most impressive is that the dipoles are not independent, but interconnect to organize a coherent structure interlocked in the whole basin. The cyclones and anticyclones are rotating so smoothly just like the gears interlocked in a machine gear box.

Morphology and scales of coherent structure

Tide curves forecasted at the two straits are the time series index of the input of turbulence energy into the Osaka-Bay. The dominant frequency components are those of semidiurnal, diurnal and of half a month due to the moon aging. The coherent structure is equipped with three stages of morphology. The first one is the 3-dimensional turbulent eddies, which are injected into the bay riding on the tidal jets. The time and space scales are minutes and 10 meters, respectively. The second stage relates to the 2-dimensional dipoles near the straits, which is formed through the inverse cascade of injected turbulece. The scales are hours and kilometers. The third stage is the 2-dimensional coherent structure filling the whole bay, which is formed through the inverse cascade of 2-dimensional dipoles. The scales are days and 50km.

Tidal residual circulation

The coherent structure hadn't been so explicitly recognized until the satellite based flow visualization was realized. From surface surveys performed in the past, however, we can extract some onsite information implicitely indicating the physics of turbulence. Figure 2 shows the schema of tidal residual circulations compiled by Fujiwara (1989). The most dominant one is the anticyclonic circulation near the Akashi-Strait, the Okinose-Circulation named after a shallow shoal locating there.

Comparing it with the MESSR image, we find a geometrical similarity. The key of this similarity is in the topographical boundary of this closed vessel (Figure 3). If the vessel interlocked tight the coherent structure, then we would gain ensemble mean turbulence quite similar to each sample of instantaneous turbulence. Here, we identify the tidal residual circulations with the ensemble mean of coherent structure dominating in this rotating closed vessel. To observe the turbulence processes, however, the satellite survey has some disadvantage; such a large period of repetitive observation and so frequent cloud covering against the sequential observation.

Hydrodynamical experiments in a rotating system

Some laboratory experiments were performed by employing the scale down model of prototype turbulence. Figure 4 outlines the experimental arrangements and methods. The 3-dimensional model of Osaka-Bay was set on a turntable of 2 meter in diameter. Horizontal scaling ratio is 1/50,000, which is determined by the diameter of table. Vertical scaling ratio is set as 1/1,000 to make the surface tension effects negligible. With the Froude's similarity law, the scaling ratio for the time is calculated at 1/1,581. Employing the Rossby's similairty law, the rotation period of the turntable is set as 98 seconds. Under these conditions, topographical β-effect is satisfied.

The tide was generated by two plunger-type tide generators installed outsides of the two straits, which are computer-controlled to reproduce the forecasted tide curves at the two straits. The flow field is visualized with floating fine particles of aluminum powder, and the movement was recorded sequentially by a VTR camera.

Daily-mean velocity and vorticity fields

The sea surface velocity vector distribution was estimated in every hour by applying the pattern matching method to sequential VTR image frames sampled with a certain time interval. Through the time averaging, daily-mean velocity fields were obtained. Taking a spatial differential, the daily-mean vorticity fields were calculated.

Figure 5 shows the daily-mean velocity and vorticity fields in the model basin corresponding to the MESSR image obtained on 23 December 1987. On the velocity field, we detect the tidal residual circulations quite similar to the prototype ones shown in Figure 2 compiled by Fujiwara (1989). The vorticity field looks much like the prototype eddy patterns interpreted on the MESSR image shown in Figure 1.

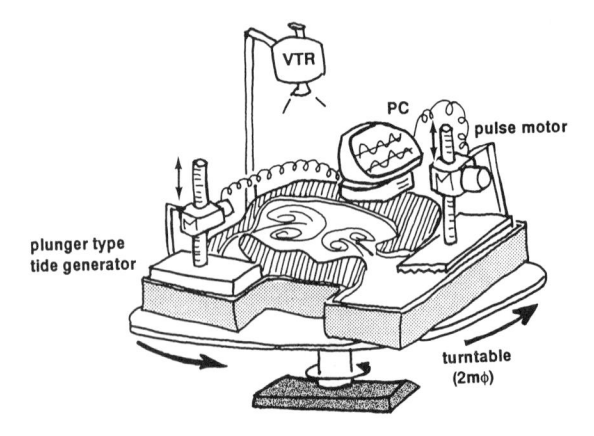

Figure 4. Experimental apparatus and methods for the hydrodynamical experiments on the tidal turbulence of Osaka-Bay. The diameter of turntable is 2 meter.

Earth rotation effects on turbulence

The present model is characterized by the Rossby's similarity law. For the experimental limitation, the model is one order smaller in size, and one order larger in the distortion ratio than the existing models. Table 1 indicates the comparison of some parameters for the present model to those for existing two model basins, which had been also driven for the physical simulation of the tidal flow field in Osaka-Bay. These two models are those designed with the Froude's similarity law alone. The sizes were set one order larger than the present model in order to simulate the bottom friction, but the earth rotation effects on the turbulence were assumed negligible.

For the comparison, the earth rotation effects on the tidal turbulence inside Osaka-Bay was estimated here using the present model. Figure 6 shows one of the experimental results, which were obtained on the stationary table. We see that this Froude model fails to simulate the prototype coherent structure. The vorticity fed through the Akashi-Strait diffuse into the whole basin, and the bay is dominated by a basin scale anticyclonic tidal residual circulation. In the traditional design of physical models, the earth rotation effects were usually neglected for the sea basin of less than 100km in size. There, the bottom friction was estimated as the dominant parameter, and the Reynolds number was made as close as possible to the prototype to take the similarity in the turbulence mixing. Refering to the experiments on geophysical coherent structure, however, the turbulent eddies is made of higher coherence when it is in a rotating system (Nihoul, 1989). Comparing the turbulent fields in Figures 5 and 6, it is shown that the earth rotation effects should be put the first priority in the physical modelling of tidal turbulence in Osaka-Bay of 50km in size.

Table 1 Scale down model experiments and parameters.

	horizontal scale	vertical scale	distortion ratio	Reynolds number	Froude's similarity	Rossby's similarity
Ishikawa(1979)	1/2,000	1/159	12.5	1×10^5	+	
Imamoto(1988)	1/5,000	1/500	10.0	3×10^4	+	
Present model	1/50,000	1/1,000	50.0	1×10^4	+	+

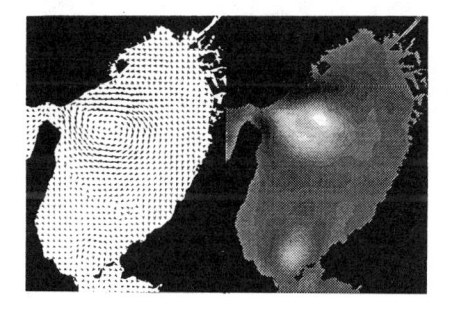

Figure 5. Daily-mean velocity and vorticity fields in the Froude/Rossby model.

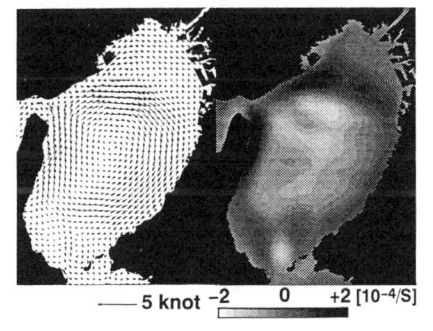

Figure 6. Daily-mean velocity and vorticity fields in the Froude model.

Hysteresis with moon aging

An impressive phenomenon in this Froude/Rossby model is the hysteresis detected in the process of coherent structures. Figure 7(a) shows the daily-mean turbulence fields, which were gained at some representative phases of moon aging; neap tide, flood tide and transient phases. Figure 7(b) shows the modelled tide curve at the Akashi-Strait, which corresponds to the prototype in the December 1987, when the MESSR image shown in Figure 1(a) was obtained.

In the case (A) of the neap tide, little trace of the coherent structure is found. In the cases (C)–(E) at the flood tide, the coherent structure is formed, which is similar to the MESSR image and to the schema of tidal residual circulation. By the detailed inspection, we see that the coherence of eddy structure and the similarity to the prototype are both made higher as the moon ages.

The remarkable difference is observed in the transient phases of the moon aging. Comparison between the daily-mean turbulence fields in the cases (B) and (F) denotes a hysteresis in the turbulence processes due to the aging of moon. In the transient phase (B) from neap tide to flood tide, we scarecely find any coherent structures. In the transient phase (F) from flood tide to neap tide, we find a coherent structure, which was established by the preceding flood tides. Although the strength of tide is same, the coherent structure in the bay is much more robust in the decreasing phase of the tide than in the increasing phase. The time scale of this hysteresis phenomenon is estimated at days, which is comparable to the earth rotation.

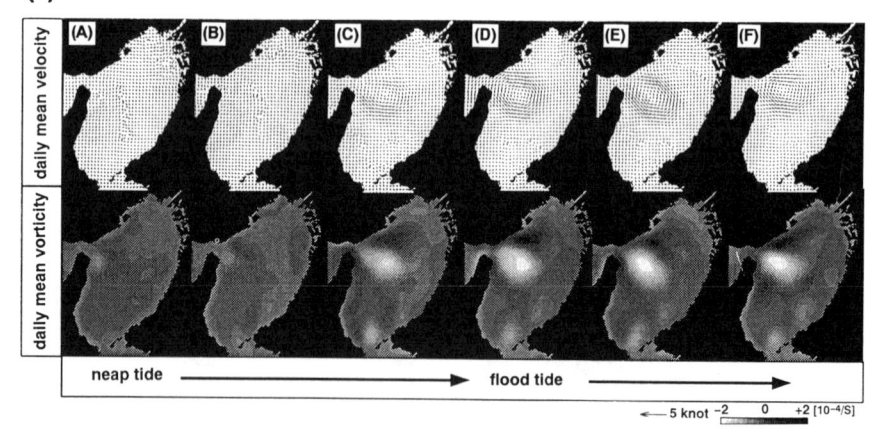

Figure 7. Hysteresis in the process of coherent structure in the Froude/Rossby model of Osaka-Bay.
(a) Daily-mean turbulence fields.
(b) Tide curve at Akashi-Straits.

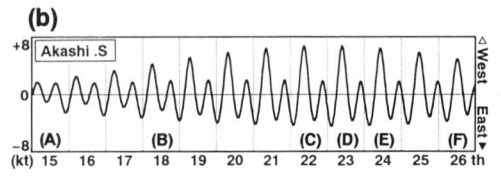

Satellite based verificaion of the hysteresis

For the physical reasoning of the hysteresis phenomenon, we should remark the three stages of the coherent structure and two stages of inverse cascade processes. Figure 8 shows a Landsat/TM image of the neap tide flow field inside the Osaka-Bay. Although the dipoles are formed near the two straits, the coherent structure in the third stage is not yet organized. In this case, the second stage of inverse cascade fails to proceed, and we can scarecely expect the tidal residual circulations.

Figure 9 shows a Landsat/MSS image in the final phase of flood tide. On the image, we can interpret three dipoles, which were produced through the inverse cascade from the 3-dimensional turbulence injected by the eastward tidal jets of preceding flood tides. They are labelled as A, B and C respectively on the Landsat/MSS image and on the tide curve at the Akashi-Strait. These dipoles are riding on the anticyclonic Okinose-Circulation formed by the preceding flood tides.

Here, we notice that the central part of the set of dipoles is composed of three anticyclones, which would strengthen the Okinose-Circulation feeding it with the anticyclonic vorticity. In the decreasing phase from flood tide to neap tide, such an inverse cascade proceeds effectively, and the tidal residual circulation is maintained to gain the robustness. In the increasing phase, on the other hand, this process is scaresely expected because of the absence of existing Okinose-Circulation. This difference in the second stage inverse cascade cause the hysteresis in the coherent structure with moon aging.

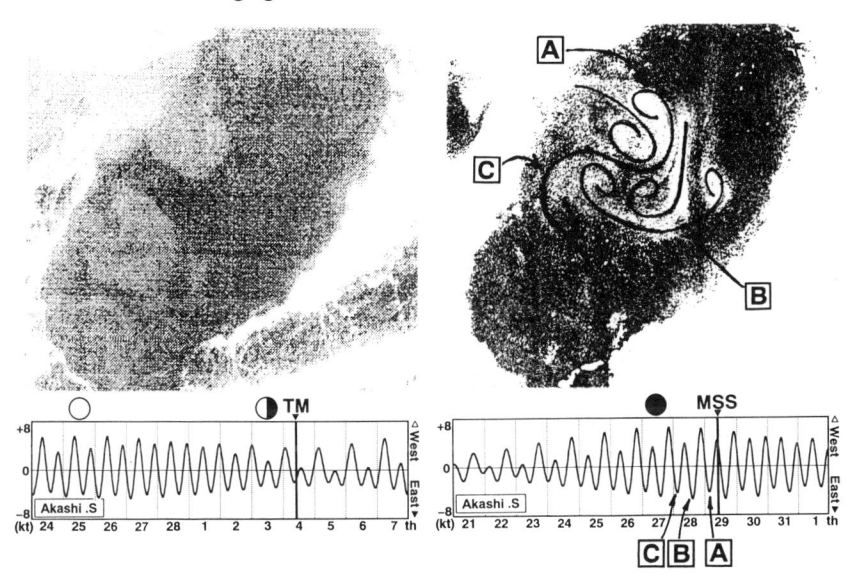

Figure 8. Landsat/TM image of the neap tide flow field on 4 March 1986.

Figure 9. Landsat/MSS image of the flood tide flow field on 29 August 1984.

Inverse cascade of 3-dimensional tidal jets

Based on past research, it is well known that a pulse-like 3 dimensional jet injected into a vessel forms a compact dipole when it is in a rotating system (Nihoul, 1987). Figure 10 shows an experimental verification of the earth rotation effects on the inverse cascade of 3-dimensional tidal jet. To abstract the process, the topography of Osaka-Bay is simplified to a rectangular basin with a single strait and a flat bottom as shown in Figure 10(b). This 2-dimensional model is intended to compare with the Froude model of tidal residual circulation by Yanagi (1976) with 5 times larger size. The tide generator was driven to generate a sinusoidal tide curve at the strait.

Figure 10(a) shows the daily-mean velocity and vorticity fields. In the Froude model (A), the flow field similar to Yanagi(1976) was gained, in which the Okinose-Circulation was formed inside the whole bay similar to the 3-dimensional model shown in Figure 6. In the Froude/Rossby model (B), it was made compact near the strait like the 3-dimensional model shown in Figure 5. For the physical model of the first stage inverse cascade, however, this model is oversimplified. As the topography was made 2-dimensional from the first, the turbulence injected into the bay is composed of 2-dimensional line vortices, which is well known to inverse cascade easily in nature. In order to simulate the inverse cascade from 3-dimensional tidal jets,

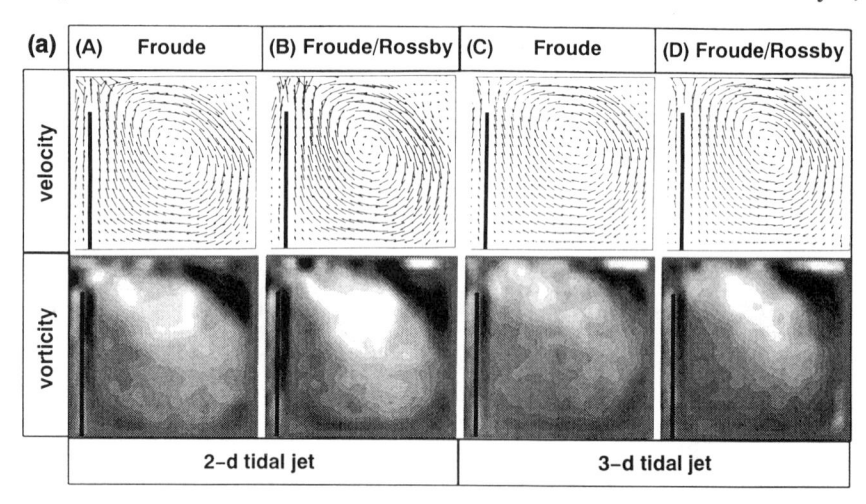

Figure 10. Experimental verification of earth rotation effects on the inverse cascade of 3-dimensional tidal jets. (a) Time-mean turbulence field in a simplified model. (b) Simplified modelling of Osaka-Bay referring to the Froude model of tidal residual circulation by Yanagi (1976).

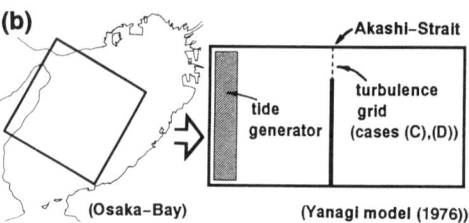

the last two experiments were done installing a turbulence grid at the Akashi-Strait. In the Froude model (C), the inverse cascade didn't proceed so efficiently as in the case of 2-dimensional tidal jets. In the Froude/Rossby model (D), a more compact Okinose-circulation is formed near the strait.

Earth rotation, roughness and 2-dimensionality of turbulence

In the numerical modelling of tidal flow fields, the earth rotation effects are usually formulated with the Coriolis term. On the simplified physical model, however, the eddy viscosity is also influenced in the rotating system significantly. Then, some turbulence characteristics are expected only from the physical model, however, it is under an experimental limitation from the size of available turntable. As shown in Table 1, the distortion ratio of Froude/Rossby model is 5 times higher than the representative Froude model used by Ishikawa(1979), which has been practically applied to the assessments of physical influence due to reclamations. The higher the distortion rate is set up, it enhances more the 3-dimensionality of the turbulence in the model, and the first stage inverse cascade is relatively underestimated. Figure 11 shows the experimental estimation of the balance of hydrodynamical effects of earth rotation and 2-dimensionality of turbulence, in which the daily-mean turbulence on 23 December 1987 are compared. Case (A) shows the model of 50 in the distortion ratio, which has been examined here in detail. In the Froude/Rossby model, the prototype coherent structure was well simulated excepting some deviations found in the location

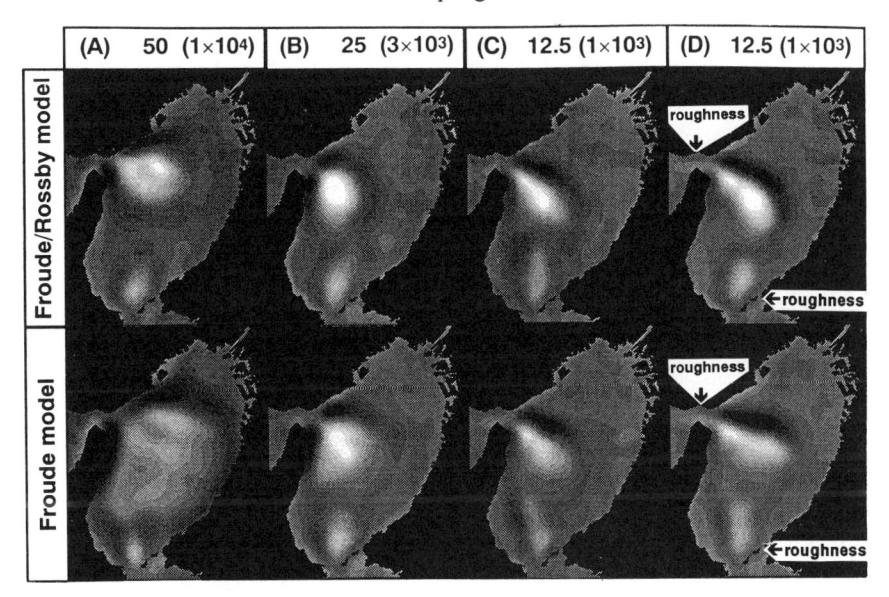

Figure 11. Experimental estimation of the balance of hydrodynamical effects of earth rotation and 2-dimensionality of turbulence. The parameters are distortion ratio (Reynolds number) of the 3-dimensional model. The representative length and speed are the mean depth and maximum speed at the Akashi-Strait.

of the Okinose-Circulation. Case (B) shows the results in a model basin of 25 in distortion ratio. Owing to the improvement in the 2-dimensionality of tidal jet, the similarity of Froude/Rossby model is made higher. Here, we also notice that the difference due to the earth rotation is made smaller. In the case (C), the distortion ratio was set at 12.5, same to the representative Froude model for the practical use. As the 2-dimensionality of tidal jet is improved further, the Okinose-Circulation is made positioned nearer to the Okinose-Shoal in the Froude/Rossby model. In the same time, the difference caused by the earth rotation effects are made much smaller.

Here, we should mention that the Reynolds number of the third model is two order less than that of the representative Froude model, and four order less than in the prototype Osaka-Bay. The turbulence energy injected into the model basin is underestimated in this model. To simulate the first stage inverse cascade of tidal jet in the last Case (D), some artificial roughness were attached at the narrowest part of the straits of Akashi and Tomogashima. On the Froude/Rossby model, the Okinose-Circulation is made positioned exactly overhead the Okinose-Shoal. On the Froude model, however, the reproducibility is made less than in the case without roughness, as the first stage inverse cascade of 3-dimensional turbulence scarcely proceeds.

Concluding remarks

From the viewpoint of a turbulence interlocked inside a rotating closed vessel, the physics of the tidal flow field in Osaka-Bay was examined. Through the flow visualization by Mos-1/MESSR, a coherent structure very similar to the tidal residual circulation was detected, which is interlocked tight inside the bay. The morphology and scales of the coherent structure are revealed. To discuss the turbulence processes, laboratory experiments were done using a hydrodynamical model of 1/50,000 in scaling ratio with the Froude/Rossby similarity law. A hysteresis in the turbulence processes was detected, and was verified through the hydrodynamical interpretation of Landsat images. Based on a simplified model experiments, it was shown that the earth rotation accelerates the inverse cascade of 3-dimensional tidal jet. Using four types of scale down models with some difference in distortion ratio and surface roughness, the hydrodynamical balance of earth rotation, bottom friction and 2-dimensionality of turbulence were estimated experimentally.

References

Fujiwara, T., Higo, T., Takahashi, Y.: On the residual circulations, tidal flows and eddies, Proc. of Coastal engineering , JSCE, Vol.36, pp.209-213, 1989(in Japanese)

Imamoto, H., Otoshi, K.: Hydrodynamical model experiments on the tidal flow in Osaka-Bay. Proc. of 43th meeting JSCE, pp.778-779, 1988(in Japanese)

Ishikawa, M., Kumagai, M., Nishimura, H., Fujiwara, T., Hayakawa, N.: Hydrodynamical model experiments on the tidal residual circulation under the stratified condition. Proc. of Oceanogr. Soc. Japan, pp.41-42, 1979(in Japanese)

Nihoule, C.J., Jamart, M.(Editor): Mesoscale/Synoptic coherent structures in geophysical turbulence. ELSEVIER OCEANOGRAPHY SERIES 50, 1989

Yanagi, T.. Fundamental study on the tidal residual circulation-I, J. Oceanogr. Soc. Japan, Vol.32, pp.199-208, 1976

CHAPTER 64

Development of a Partially Three-Dimensional Model for Ship Motion in a Harbor with Arbitrary Bathymetry

Takumi Ohyama [1] and Mitsuru Tsuchida [2]

Abstract

A numerical method has been developed for the analysis of ship motions in a harbor with arbitrary bathymetry. A BEM-based 3-D model, applied partially to a near-field surrounding a ship, is combined with a FEM-based 2-D model, utilized in the remainder of harbor domain. This combination may achieve an efficient computation of the ship motions with taking into account of wave deformation in a harbor. Preliminary examinations have been performed to investigate appropriate location of a matching boundary where these two models are coupled. It is found that, for reliable prediction, $(2 \sim 3)h$ (h: water depth) is required for the distance between the matching boundary and a body. The numerical results of added mass and damping coefficients for a rectangular floating body in a rectangular basin are then compared with those obtained from a conventional numerical model. Favorable agreement between the results verifies the present numerical method. Ship motions in a harbor with slowly varying depth are also demonstrated.

---

[1]Centre for Water Research, The University of Western Australia, Nedlands, WA6009, AUSTRALIA (on leave from Institute of Technology, Shimizu Corporation, Etchujima 3-4-17, Koto-ku, Tokyo 135, JAPAN)

[2]Institute of Technology, Shimizu Corporation, Etchujima 3-4-17, Koto-ku, Tokyo 135, JAPAN

1. Introduction

Precise prediction of ship motions in a harbor is essential to reliable harbor design. Coastal structures such as breakwaters and bottom topography in a harbor cause incoming waves to be diffracted and refracted before their reaching a floating body. In addition to such deformation of incoming waves, diffracted and radiated waves propagating from the body may be re-reflected by breakwaters and come back to the body again. Therefore, in general, hydrodynamic forces acting on a floating body in a harbor significantly differ from those in the case of open sea.

Under such circumstances, it may be necessary that the wave-ship interaction problem is solved simultaneously with the wave deformation in a harbor. In spite of this, corresponding numerical models, taking into account of coastal structures and bottom topography, have rarely been provided for predicting ship motions in a harbor. Oortmerssen (1976) developed a numerical method to calculate wave-induced motions of a ship moored at a straight quay, whereas Sawaragi and Kubo (1982) applied a two-dimensional boundary element method (2-D BEM) to the case of a rectangular floating body in a rectangular harbor. These basic studies may indicate essential influences of harbor boundaries on the ship motions. However, since these approaches have utilized the principle of mirror image, their applications are limited to harbors with straight boundaries.

In the light of this, Sawaragi et al. (1989) proposed a numerical method applicable to harbors of an arbitrary horizontal configuration. In their approach, a 3-D BEM model using a Green's function derived by John (1950) is applied only to the near-field around a floating body, and is combined with a 2-D BEM model utilized in the remainder of harbor domain. Although the basic idea of this "partially three-dimensional model" may address more practical situations as compared to the aforementioned methods, its application is still restricted to the case of constant depth in a harbor.

In this connection, the present study attempts to develop an alternative "partially three-dimensional model" for more general situations with a slowly varying bottom. A finite element method (FEM) based on the mild-slope equation (Berkhoff, 1972) is employed as a 2-D model in a harbor domain excluding the near-field around a floating body. Taking into account of continuities of fluid mass and momentum, this 2-D FEM model is coupled with a 3-D BEM model in the vicinity of the body.

The basic theory and the numerical formulation are described in Section 2, where the 3-D BEM and 2-D FEM models are solved simultaneously with continuity conditions imposed on a matching boundary. Since the 2-D model is based on the mild-slope equation, this model is applicable only to the domain where evanescent modes are negligible. In this connection, Section 3 investigates an appropriate location of the matching boundary in a vertical two-dimensional plane. In Section 4, in order to verify the present numerical model, comparisons are made with a conventional model (Sawaragi and Kubo, 1982) for radiation-force coefficients of a rectangular floating body in a rectangular basin. Lastly, a numerical example is given in Section 5 for ship motions in a harbor with an inclined bottom.

2. Partially Three-dimensional Numerical Model

Governing equation and boundary conditions

Let us consider wave diffraction and radiation by a floating body in a harbor with arbitrary bathymetry. Assuming an irrotational small-amplitude motion of an incompressible and inviscid fluid, the fluid motion can be expressed by using the linear potential theory. The velocity potential for the wave-body interaction may be separated into the *propagation mode* and the *evanescent modes*, where the latter diminishes exponentially with distance away from the body. Therefore, in the present model, the harbor domain considered is subdivided into two regions, Ω_1 and Ω_2, as shown in **Fig. 1**. The former region Ω_1 denotes a small near-field around the floating body where the evanescent modes are significant, whereas the latter Ω_2 represents a whole domain in a harbor excluding Ω_1. It is assumed that the matching boundary, S_C, between Ω_1 and Ω_2 is located sufficiently apart from the body and, hence, only the propagation mode is predominant in Ω_2.

The velocity potential, $\Phi(x, y, z, t)$ can be expressed by the combination of incident, diffraction and radiation potentials:

$$\Phi(x, y, z, t) = \phi_7(x, y, z)e^{-i\sigma t} + \sum_{l=1}^{6} \frac{d}{dt}(D_l e^{-i\sigma t})\phi_l(x, y, z), \tag{1}$$

where (x, y, z) represents Cartesian coordinates (see **Fig. 1**), t is the time, σ is the angular frequency of incident waves, $l = 1, 2, \cdots, 6$ correspond to the surge, sway, heave, roll, pitch and yaw body motions, respectively, ϕ_7 is the sum of incident and diffraction potentials, and ϕ_l and D_l ($l = 1 \sim 6$) are the radiation potential and the complex amplitude of the l-th directional body motion.

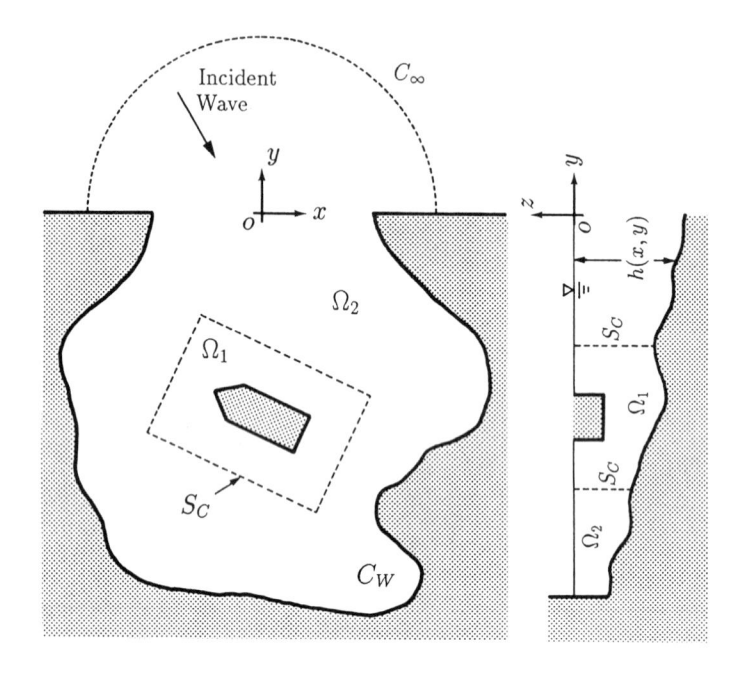

Fig. 1 Schematic diagram of partially three-dimensional model.

The governing equation and the boundary conditions for ϕ_l ($l = 1 \sim 7$) are

$$\frac{\partial^2 \phi_l^{(i)}}{\partial x^2} + \frac{\partial^2 \phi_l^{(i)}}{\partial y^2} + \frac{\partial^2 \phi_l^{(i)}}{\partial z^2} = 0, \quad (l = 1 \sim 7; \text{ in } \Omega_1 \, [i = 1] \text{ and } \Omega_2 \, [i = 2]), \quad (2)$$

$$\frac{\partial \phi_l^{(i)}}{\partial z} - \frac{\sigma^2}{g}\phi_l^{(i)} = 0, \quad (l = 1 \sim 7, \, i = 1, 2; \text{ on } S_F), \quad (3)$$

$$\frac{\partial \phi_l^{(i)}}{\partial n} = 0, \quad (l = 1 \sim 7, \, i = 1, 2; \text{ on } S_B), \quad (4)$$

$$\frac{\partial \phi_l^{(1)}}{\partial n} = v_l, \quad (l = 1 \sim 7, \text{ on } S_V), \quad (5)$$

$$\begin{aligned}
&v_1 = n_x, \quad v_2 = n_y, \quad v_3 = n_z, \\
&v_4 = (y - y_G)n_z - (z - z_G)n_y, \\
&v_5 = (z - z_G)n_x - (x - x_G)n_z, \\
&v_6 = (x - x_G)n_y - (y - y_G)n_x, \\
&v_7 = 0,
\end{aligned} \qquad (6)$$

where $\phi_l^{(1)}$ and $\phi_l^{(2)}$ denote the velocity potentials in Ω_1 and Ω_2, respectively, S_F is the free surface, S_B is the seabed, S_V is the submerged body surface, n is the outward normal on each surface, (n_x, n_y, n_z) are the $x-$, $y-$ and $z-$components of the outward unit normal on S_V, and (x_G, y_G, z_G) is the gravitational center of the body.

On the other hand, partial wave reflection is considered along the harbor boundary C_W. Although Isaacson and Qu (1990) proposed a corresponding boundary condition including the effects of wave direction and phase shift, the present study utilizes the following simple condition:

$$\frac{\partial \phi_l^{(2)}}{\partial n} = \frac{i\sigma}{C} \frac{1 - K_{RW}}{1 + K_{RW}} \phi_l^{(2)}, \quad (l = 1 \sim 7, \text{ on } C_W), \qquad (7)$$

where C is the wave celerity and K_{RW} represents the reflection coefficient imposed.

3-D BEM model in Ω_1

Applying Green's theorem to the fluid domain Ω_1, the Laplace's equation Eq. (2) is transformed into the following integral equation:

$$\alpha(P)\phi_l^{(1)}(P) + \int_S \left\{ \phi_l^{(1)} \frac{\partial G}{\partial n} - \frac{\partial \phi_l^{(1)}}{\partial n} G \right\} ds = 0, \quad (l = 1 \sim 7), \qquad (8)$$

where S represents the closed boundary surface containing Ω_1, P denotes an arbitrary position in Ω_1, and G is a Green's function. The coefficient $\alpha(P)$ is 2π if P is on S, and is 4π in other cases.

Using a Green's function derived by John (1950) and substituting the boundary conditions, Eqs. (3), (4) and (5), into Eq. (8), the following boundary integral equations can be obtained:

$$\alpha(P)\phi_l^{(1)}(P) + \int_{S_C} \left\{ \phi_l^{(1)} \frac{\partial G}{\partial n} - \frac{\partial \phi_l^{(1)}}{\partial n} G \right\} ds + \int_{S_V \cup S_B} \phi_l^{(1)} \frac{\partial G}{\partial n} ds = \int_{S_V} v_l G \, ds,$$

$$(l = 1 \sim 7). \qquad (9)$$

If the water depth is constant in Ω_1, the integral on S_B involved in Eq. (9) can be eliminated since $\partial G/\partial n = 0$ (on S_B).

2-D FEM model in Ω_2

In Ω_2, sufficiently far from the body, we may consider only the propagation mode in the fluid motion over a slowly varying seabed. This allows us to utilize the mild-slope equation (Berkhoff, 1972; Smith and Sprinks, 1975; Lozano and Meyer, 1976).

Thus, the velocity potential ϕ_l in Ω_2 is approximated by the form

$$\phi_l^{(2)}(x, y, z) = \varphi_l^{(2)}(x, y) \frac{\cosh k(h+z)}{\cosh kh}, \quad (l = 1 \sim 7), \tag{10}$$

where $k(x, y)$ and $h(x, y)$ are the wave number and the water depth, respectively. The corresponding mild-slope equation, derived from Eq. (2), is given as

$$\nabla \cdot (CC_G \nabla \varphi_l^{(2)}) + k^2 CC_G \varphi_l^{(2)} = 0, \quad (l = 1 \sim 7, \text{ in } \Omega_2), \tag{11}$$

where $\nabla \equiv (\partial/\partial x, \partial/\partial y)$ and C_G is the group velocity.

According to Chen and Mei (1975), a variational approach is employed in a FEM-based formulation to solve Eq. (11). Along the open boundary, denoted by C_∞ in **Fig. 1**, the finite elements are coupled to the *superelement* which satisfies the radiation condition analytically. The exterior region outside C_∞ is assumed to have a constant water depth.

The variational function for the governing equation [Eq. (11)], the boundary condition on C_W [Eq. (7)] and the radiation condition is given as

$$J_l = (J_1)_l + (J_2)_l + (J_3)_l + (J_4)_l, \quad (l = 1 \sim 7), \tag{12}$$

$$\begin{aligned}
(J_1)_l &= \int_{\Omega_2} \frac{1}{2} \left\{ CC_G (\nabla \varphi_l^{(2)})^2 - \frac{C_G}{C} \sigma^2 \left(\varphi_l^{(2)}\right)^2 \right\} d\Omega, \\
(J_2)_l &= \int_{C_\infty} CC_G \left\{ \left(\frac{1}{2}\overline{\varphi_l} - \varphi_l^{(2)} + \frac{1}{2}\delta_{l7}\varphi_0 \right) \frac{\partial \overline{\varphi_l}}{\partial n} - \frac{1}{2}\delta_{l7}(\overline{\varphi_l} - \varphi_0) \frac{\partial \varphi_0}{\partial n} \right\} dc, \\
(J_3)_l &= -\int_{C_W} \frac{1}{2} i\sigma C_G \frac{1-K_{RW}}{1+K_{RW}} \left(\varphi_l^{(2)}\right)^2 dc, \\
(J_4)_l &= -\int_{C_C} CC_G \frac{\partial \varphi_l^{(2)}}{\partial n} \varphi_l^{(2)} dc,
\end{aligned} \tag{13}$$

where δ_{ij} is the Kronecker's delta, φ_0 is the velocity potential of incident waves, C_C is the intersection of the free surface and the matching boundary S_C, and $\overline{\varphi_l}$ denotes the exterior analytical solution (Chen and Mei, 1975).

Discretizing Eq. (12) by using linear triangular elements and applying the variational principle, we finally obtain a set of simultaneous equations for $\varphi_l^{(2)}$ in Ω_2 including C_C and $\partial \varphi_l^{(2)}/\partial n$ on C_C.

Continuity conditions on the matching boundary

The velocity potential, ϕ_l, and its normal derivative, $\partial \phi_l / \partial n$, must be continuous across the matching boundary S_C:

$$\phi_l^{(1)} = \phi_l^{(2)}, \quad \frac{\partial \phi_l^{(1)}}{\partial n} = \frac{\partial \phi_l^{(2)}}{\partial n}, \quad (l = 1 \sim 7, \text{ on } S_C). \tag{14}$$

Since the magnitude of evanescent modes is assumed negligible on S_C, these equations are rewritten by

$$\left. \begin{array}{l} \phi_l^{(1)} = \varphi_l^{(2)} \dfrac{\cosh k(h+z)}{\cosh kh}, \\[6pt] \dfrac{\partial \phi_l^{(1)}}{\partial n} = \dfrac{\partial \varphi_l^{(2)}}{\partial n} \dfrac{\cosh k(h+z)}{\cosh kh} + \varphi_l^{(2)} \dfrac{\partial}{\partial n}\left(\dfrac{\cosh k(h+z)}{\cosh kh}\right), \end{array} \right\} (l = 1 \sim 7, \text{ on } S_C). \tag{15}$$

Substituting Eq. (15) into Eq. (9), we obtain

$$\alpha(P)\phi_l^{(1)}(P) + \int_{C_C} \left\{ \varphi_l^{(2)}(I_A - I_B) - \frac{\partial \varphi_l^{(2)}}{\partial n} I_C \right\} dc + \int_{S_V \cup S_B} \phi_l^{(1)} \frac{\partial G}{\partial n} ds,$$

$$= \int_{S_V} v_l G \, ds \quad (l = 1 \sim 7), \tag{16}$$

where

$$\left. \begin{array}{l} I_A = \displaystyle\int_{-h}^{0} \dfrac{\cosh k(h+z)}{\cosh kh} \dfrac{\partial G}{\partial n} dz, \\[8pt] I_B = \displaystyle\int_{-h}^{0} \dfrac{\partial}{\partial n}\left\{\dfrac{\cosh k(h+z)}{\cosh kh}\right\} G \, dz, \\[8pt] I_C = \displaystyle\int_{-h}^{0} \dfrac{\cosh k(h+z)}{\cosh kh} G \, dz. \end{array} \right\} \tag{17}$$

Equation (16) is discretized into a finite number of facets on S_V and S_B, and of line elements on C_C. In the resultant discretized equation, the position of the control point P is set at the center of each element on S_V, S_B and C_C. This leads to $(N_V + N_B + N_C)$ linear algebraic equations involving $\phi_l^{(1)}$ (on S_V, S_B), $\varphi_l^{(2)}$ (on C_C) and $\partial \varphi_l^{(2)}/\partial n$ (on C_C) as unknown variables, where N_V, N_B and N_C are the number of elements on S_V, S_B and C_C, respectively. These equations are solved simultaneously with those obtained from the 2-D FEM model so as to determine the velocity potential in the both domains.

3. Location of the matching boundary

Prior to computing ship motions in a harbor, preliminary examinations are conducted to investigate the effect of the location of the matching boundary. As described in the previous section, the matching boundary must be set far enough away from the body so as to satisfy the assumption of negligible magnitude of evanescent modes.

For this examination, we consider a wave-diffraction problem in a vertical plane. In general, the horizontal variation of the magnitude of evanescent modes may be represented by

$$A_m e^{-k_m X}, (m = 1, 2, \cdots),$$

where X denotes the horizontal distance from the origin of wave scattering, k_m is the m-th eigen value ($k_m \tan k_m h = -\sigma^2/g$), and A_m is a constant corresponding to the magnitude of the m-th mode at $X = 0$. Since $e^{-k_1 X} \geq e^{-k_2 X} \geq \cdots$ ($k_1 < k_2 < \cdots$), we may examine only the first mode, $A_1 e^{-k_1 X}$. The magnitude ratio of the first mode, normalized by A_1, is then defined as

$$p(X) \equiv e^{-k_1 X}. \tag{18}$$

As an example, **Fig. 2** shows the computed reflection coefficient of a submerged rectangular shelf, K_R, for different values of p. In the figure, the abscissa denotes the normalized wave frequency, the solid line is the corresponding analytical solution obtained from an eigen-function expansion method, and S_R and S_T are the matching boundaries. The combination of a 1-D FEM model with a 2-D BEM model was employed for these computations.

As seen in the result for $p = 0.1$, a large value of p (small X) causes an apparent deviation from the analytical solution. In this case, the magnitude of evanescent modes may be still significant at the matching boundaries. While a slight discrepancy is observed for $p = 0.05$, the numerical results for $p = 0.01$ and 0.005 agree well with the linear theory in the entire frequency range examined, indicating that the assumption of negligible evanescent modes becomes appropriate.

In **Fig. 3**, the normalized distance, X/h_0 (h_0: the water depth), is plotted for different values of p. It is found that X/h_0 corresponding to a constant p is insensitive to the wave frequency. The results shown in **Figs. 2** and **3** may conclude that the distance X, required for reliable prediction, is $2 \sim 3$ times as long as the water depth.

SHIP MOTION WITH ARBITRARY BATHYMETRY

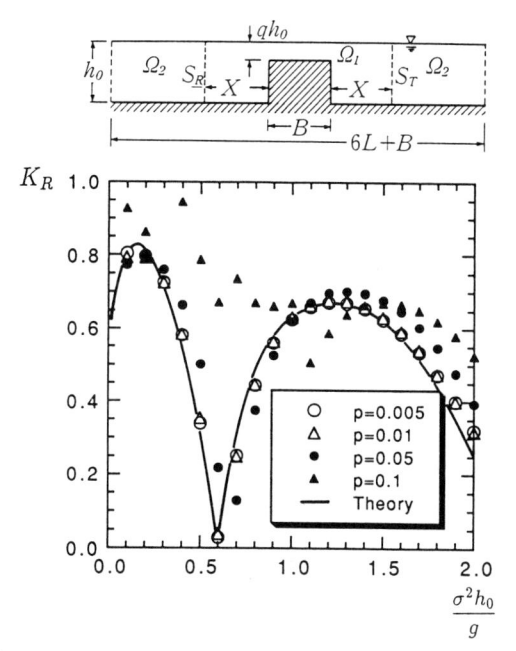

Fig. 2 Variation in computed reflection coefficients of a submerged rectangular shelf, with the location of matching boundaries.

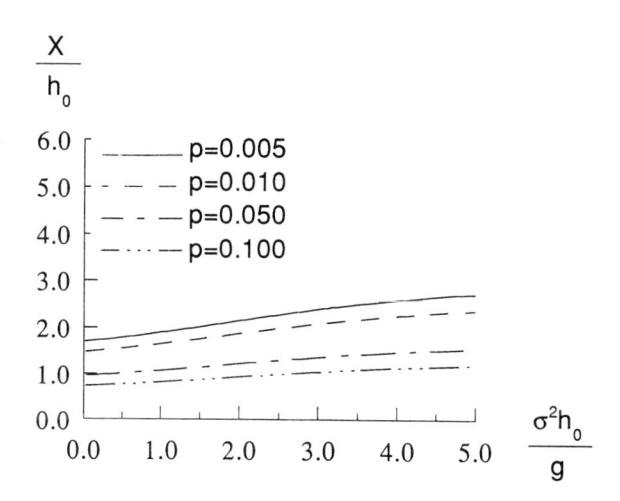

Fig. 3 Relation between the normalized distance X/h_0 and the parameter p.

For 3-D cases, the distribution of evanescent modes in a horizontal plane may be expressed as

$$B_{nm} K_n(k_m R) \cos n\theta, \quad (n = 0, 1, \cdots, m = 1, 2, \cdots),$$

where (R, θ) is local polar coordinates, B_{nm} is a constant and K_n is the modified Bessel function of the $n-$th order. Although the examinations were performed only in a vertical 2-D domain, the results obtained here may also be valid for 3-D cases since $K_n(k_m R)$ can be represented by using an exponential function $e^{-k_m R}$ for large R.

4. Comparison with a conventional numerical model

Sawaragi and Kubo (1982) computed hydrodynamic forces on a rectangular floating body in a rectangular basin by using a 2-D BEM model. Although the application of their model is limited to a rectangular harbor with a constant depth, the comparison with their results may confirm the validity of the present numerical model.

The numerical results for added mass and damping coefficients in the sway motion, M_{22} and N_{22}, are given in **Fig. 4**, where $\triangle$ and L_s are the weight and the length of the floating body. In the figure, the configurations of the basin and the floating body are also illustrated in a horizontal plane. The body's submergence and the water depth are $0.2m$ and $0.5m$, respectively, and the harbor boundaries are fully reflective ($K_{RW} = 1$).

As shown in this figure, predominant peaks of the hydrodynamic forces emerge at certain frequencies corresponding to harbor resonance. This peak phenomenon may suggest an importance of surrounding-boundary effects in predicting ship motions in a harbor. A favorable agreement between the results obtained from the present model and those from the 2-D BEM model (Sawaragi and Kubo, 1982) indicates the reliability of the present numerical model.

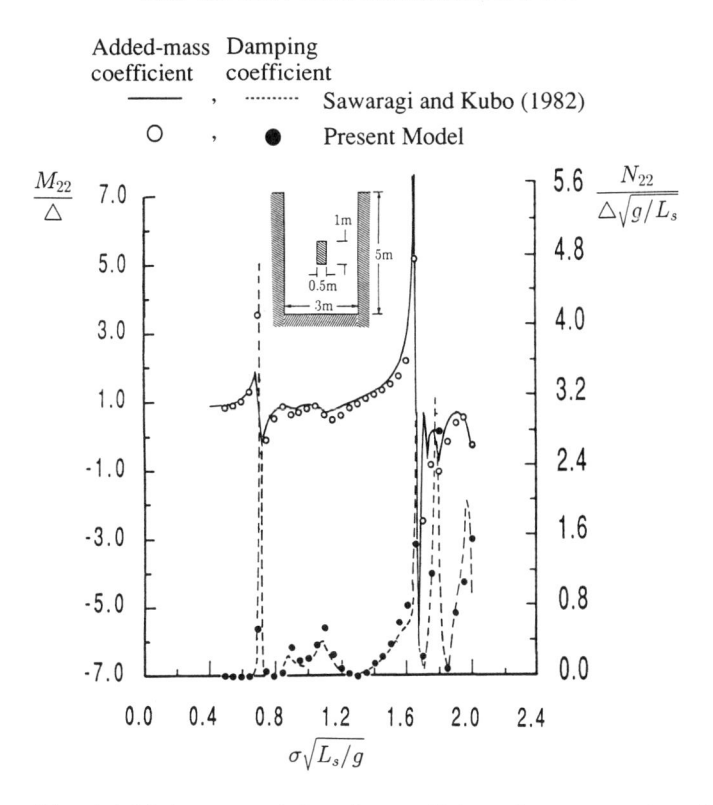

Fig. 4 Added-mass and damping coefficients in sway motion of a rectangular floating body in a rectangular basin.

5. Numerical Example

In addition to the influence of harbor boundaries, seabed topography may be one of the important factors for precise prediction of ship motions in a harbor. However, as mentioned in Introduction, most conventional approaches applied to this problem have assumed a constant water depth in a harbor. In this section, therefore, a numerical example is given for the case of a harbor with an inclined bottom.

The configurations of a floating body and a harbor examined are illustrated in **Fig. 5**. For reference, computations were also conducted for a flat-bottom case, for which the water depth is denoted by a broken line in this figure. The propagation direction of the incident waves is 30 degrees oblique to the harbor mouth.

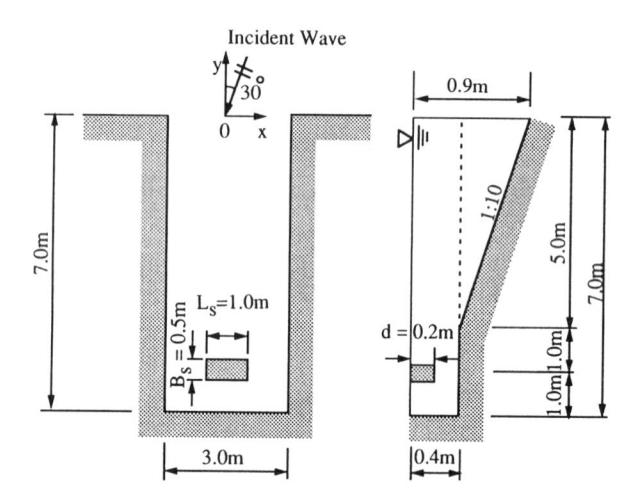

Fig. 5 Configurations of a floating body and a harbor examined.

The resultant frequency responses of ship motions are shown in **Fig. 6**, where the solid and dotted lines represent the numerical results for the flat bottom and the inclined bottom, respectively, and H_0 is the incident wave height. Although resonant peaks emerge in the both numerical results, the corresponding peak frequencies are different between the cases. In particular, the computed responses for the inclined-bottom case shows distinctive peaks at $\sigma^2 L_s/g = 1.8$, which are not observed in the result for the flat-bottom case. This difference is attributed primarily to the variation in natural frequencies of the harbor. The results given in **Fig. 6**, therefore, may indicate an importance of a depth-variation effect on the ship motions in a harbor.

(a) Surge

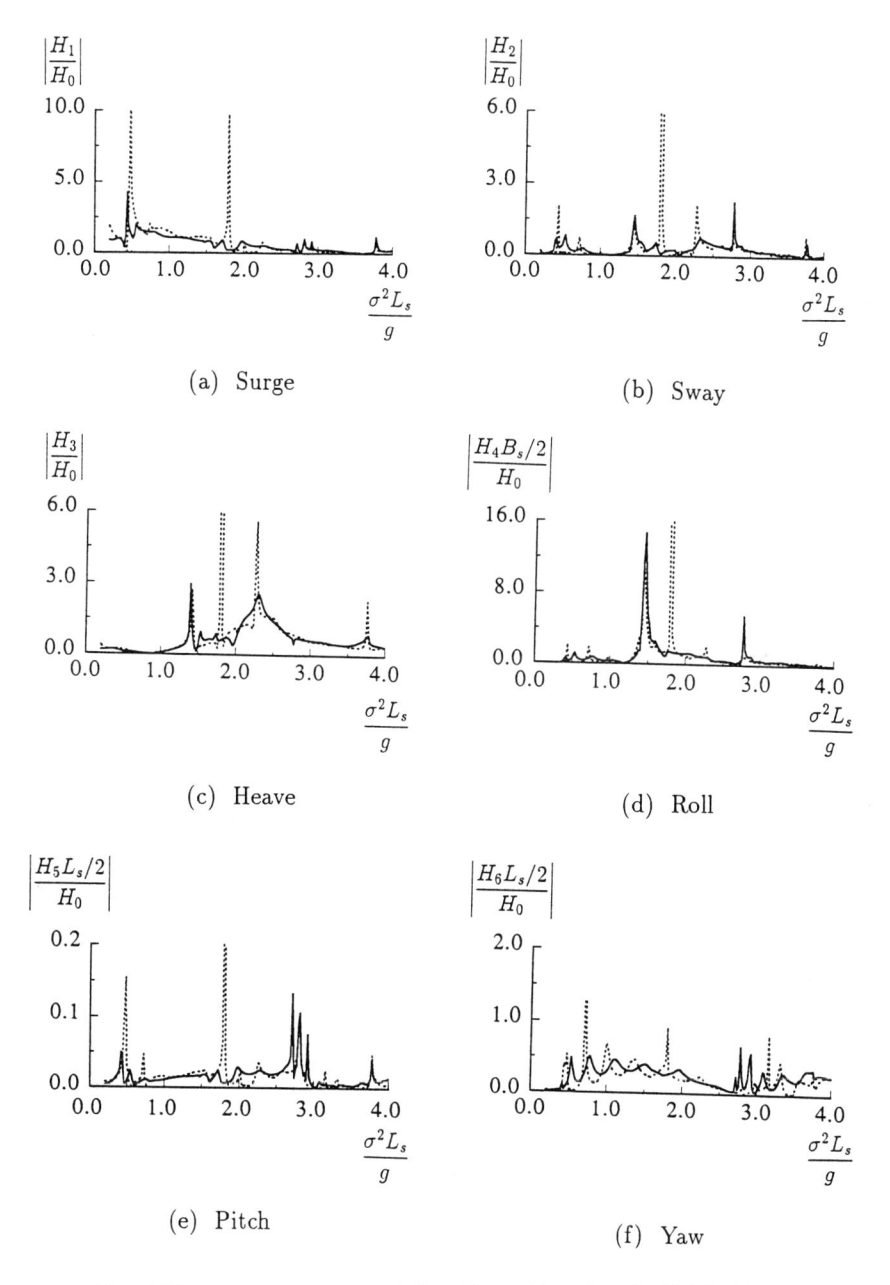

(b) Sway

(c) Heave

(d) Roll

(e) Pitch

(f) Yaw

Fig. 6 Frequency responses of the ship motions for the flat-bottom case (────) and the inclined-bottom case (········).

6. Conclusions

A numerical method, composed of a BEM-based 3-D model and a FEM-based 2-D model, has been developed for the analysis of ship motions in a harbor with arbitrary bathymetry. This combination of the two different models may achieve efficient computation with taking into account of wave deformation in a harbor. Since the mild-slope equation is employed in a horizontal 2-D domain in a harbor, the present method can be applied to more general cases with varying water depth as compared to a conventional "partially 3-D model" (Sawaragi et al., 1989).

Basic examinations have been performed to investigate appropriate location of a matching boundary where the two models are coupled. The results show that, for reliable prediction, the distance between the matching boundary and a body is required to be $2 \sim 3$ times as long as the water depth. This practically satisfies a basic assumption that the magnitude of evanescent modes is negligible at the matching boundary.

The numerical results of radiation-force coefficients for a rectangular floating body in a rectangular basin are then compared with those obtained from a conventional method (Sawaragi and Kubo, 1982). Favorable agreement between the results verifies the present numerical method.

Lastly, ship motions in a harbor with an inclined bottom are demonstrated. The comparison between the numerical results for the flat-bottom and inclined-bottom cases indicates that the bottom topography may cause a variation in natural frequencies of harbor, which significantly influence the resultant ship motions.

References

Berkhoff, J. C. W.(1972). "Computation of combined refraction-diffraction," *Proc. 13th Coastal Eng. Conf., ASCE*, 471-490.

Chen, H. S. and C. C. Mei (1975). "Hybrid-element method for water waves," *Proc. the Modelling Techniques Conf.*, **1**, 63-81.

Isaacson, M. and S. Qu (1990). "Waves in a harbour with partially reflecting boundaries," *Coastal Eng.*, **14**, 193-214.

John, F. (1950). "On the motion of floating bodies II," *Comm. Pure and Appl. Math.*, **3**, 45-101.

Lozano, C. J. and R. E. Meyer (1976). "Leakage and response of waves trapped round islands," *Phys. Fluids*, **19**, 1075-1088.

Oortmerssen, G. V. (1976). "The motions of a moored ship in waves," N.S.M.B. publication No.510, 1-138.

Sawaragi, T. and M. Kubo (1982). "The motion of a ship in a harbor basin," *Proc. 18th Int. Conf. Coastal Eng., ASCE*, 2743-2762.

Sawaragi, T., S. Aoki and S. Hamamoto (1989). "Analysis of hydrodynamic forces due to waves acting on a ship in a harbor of arbitrary geometry," *Proc. 8th Int. Conf. Offshore Mech.and Arctic Eng., ASME*, **2**, 117-123.

Smith, J. D. and T. Sprinks (1975). "Scattering of surface waves by a conical island," *J. Fluid Mech.*, **72**, 373-384.

CHAPTER 65

THE MEASURED AND COMPUTED HOKKAIDO NANSEI-OKI EARTHQUAKE TSUNAMI OF 1993

Tomoyuki Takahashi[1], Nobuo Shuto[1]
Fumihiko Imamura[2] and Hideo Matsutomi[3]

Abstract

The linear long-wave theory is used in the first stage to find the tsunami initial profile. The perfect reflection is assumed at the land boundary. On comparing the computed with the measured overall runup distribution, the best fault model is found among 24 models. It consists of three sub-faults dipping westward. It satisfies the measured land subsidence and overall distribution of runup heights in Okushiri Island.

Then, the shallow-water theory with bottom friction is used to simulate runup. The runup condition is used as the land boundary condition. The arrival time and runup heights agree well with the measured.

Introduction

At 22:17, July 12th, 1993, an earthquake of Ms=7.8 occurred off the southwestern coast of Hokkaido, in the Japan Sea. A giant tsunami was generated. It hit the island of Okushiri about 5 minutes after the earthquake, and claimed more than 200 lives. A big fire, caused by the tsunami, burned houses which were saved from the direct attack of the tsunami in Aonae, a small town at the southern end of Okushiri Island. The area of concern is shown in Fig.1 (Hokkaido Tsunami Survey Group, 1993).

Since the tsunami hit at night, its details could not be obtained from witnesses. The most reliable data would be the tsunami traces. Many parties were dispatched to

---

[1]Disaster Control Research Center, Tohoku Univ., Sendai 980-77, Japan
[2]School of Civil Eng., Asian Inst. of Tech., G.P.O.Box 2754, Bangkok 10501
[3]Dept. of Civil Eng., Akita Univ., Akita 010, Japan

measure runup heights and collect such data as tide records. The authors also organized a survey team of 19 persons, and sent it to the site two days after the disaster. At the same time, the authors simulated the tsunami with the Harvard CMT solution. On exchanging results by faxes and phone calls between the survey team and the simulation team every morning and night, places and density of measurement points were determined. The authors considered that if there was a difference between the measured and the computed, this difference was really important to distinguish the special feature of the tsunami.

In the second section, tsunami data are shown and the special feature of this tsunami which should be satisfied by the simulation is summarized. In the third section, the method of simulation is briefly given. In the fourth section, initial profiles

Fig.1 Location of the area studied.

obtained from seismic data are shown and their defect are discussed. In the fifth section, a comparison is made between the measured tsunami runups and the simulated with the Harvard CMT solution. In the sixth section, a model of two sub-faults, DCRC-2, is proposed and examined. In the seventh section, a model of three sub-faults, DCRC-17a, is introduced. Among 24 models examined by the authors, the DCRC-17a model is considered the best at present. In the same section, the problems which requires a further study are also discussed.

Measured Runup Heights, Tide Records, Arrival Time and Ground Subsidence

Positions of aftershocks are shown in Fig.2 (Hokkaido Univ. et al, 1993). The vertical distribution of aftershocks in each section implies that it is very hard to determine the inclination angle of fault, which is closely related to the location of the crest of the initial tsunami profile. A reverse-slip fault dipping eastward (westward) makes the crest far from (near to) Hokkaido Island and results in a late (early) arrival of the tsunami.

Figure 3(a), (b) shows the measured runup heights in Okushiri Island and the mainland of Hokkaido. The highest runup of 31.7 m was obtained on the western shore of Okushiri Island directly facing the earthquake fault. It was found at the

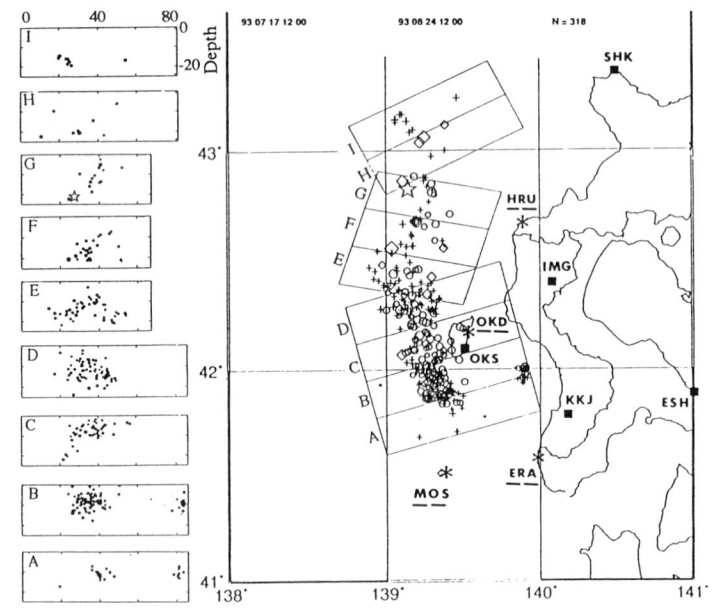

Fig.2 Distribution of aftershocks. Left figures are vertical distributions in the areas given by rectangles in the right figure.

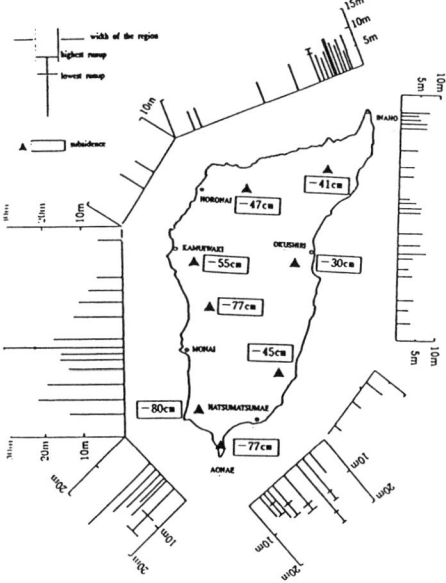

Fig.3(a) Measured runup heights in Okushiri island with the measured subsidence shown in rectangles.

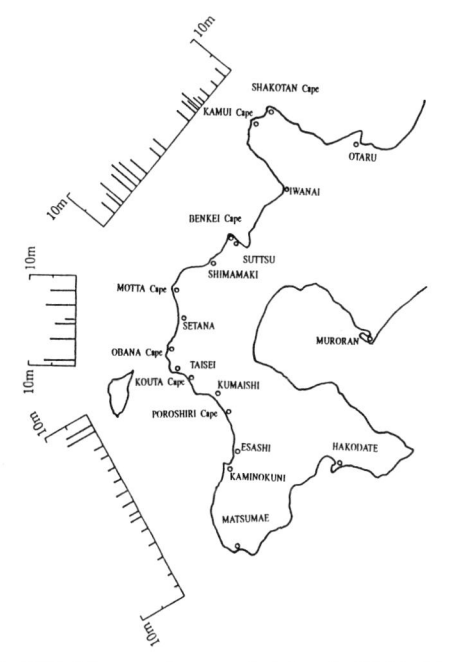

Fig.3(b) Measured runup heights in Hokkaido.

bottom of a tiny valley 50 m wide at its mouth on a pebble pocket beach 250 m long backed by cliffs. At the valley mouth and on the cliffs, runup heights were 22 m to 24 m. In order to simulate this high runup, very fine spatial grids (and, of course, very detailed maps) will be required.

Even on the eastern coast of Okushiri Island, high runup of 20 m was measured at Hamatsumae which is located in the sheltered area of Aonae Point. This may be a result of concentration due to refraction caused by the Okushiri Spur, a very wide shallow extending in the sea south of the point. In order to simulate the refraction effect, not only the detailed sea bottom contours but also the angle of tsunami incidence is important. This is closely related with the strike angle of the fault.

Along the eastern coast of Okushiri Island, runups show a wavy distribution, which implies the interaction of the tsunami entrapped around the island and the tsunami reflected from the mainland of Hokkaido.

Along the Hokkaido coast, tsunami runups are high from Taisei to Shimamaki and not high in the other region.

In the first column of Table 1, the arrival times collected from witnesses or determined from submerged clocks are summarized.

In Fig.3(a), the measured ground subsidence are also shown (Kumaki,Y. et al., 1993). Any fault model should yield the ground displacement coincident with them.

Method of Simulation

Two kinds of equations are used in the following simulation; the linear long-wave theory and the shallow-water theory with the bottom friction term included. The former is reduced to a set of difference equations, TUNAMI-N1, and the latter to TUNAMI-N2, by using the staggered leap-frog scheme. The details are given elsewhere (e.g., Shuto et al., 1986).

The area of simulation is from 138°30'E to 140°33'E and from 40°31'N to 43°18'N.

The initial tsunami profile is calculated with the Mansinha-Smylie method (1971) from fault parameters.

In simulations for modification and examination of fault models, TUNAMI-N1 is used with the perfect reflection condition at the land boundary. Overall agreements between the measured and computed are compared. The spatial grid is 450 m wide

Table 1 Arraival times

Aonae (west)	Aonae (east)	Esashi	Taisei	Setana	Sukki	Shima-maki	Iwanai
4-5	4-5	11	5	5	3	5	15

(unit:minute)

and the time step interval is 1 second.

TUNAMI-N2 is applied to the DCRC-17a model, selected as the best among 24 models. The spatial grids are varied; 450 m in Domain A, 150 m in Domain B and 50 m in Domain C. The runup condition is used at the land boundary,

Initial Profiles Proposed

Figure 4 compares tsunami initial profiles by different researchers and institutions. These profiles are obtained, based upon seismic data. Differences are resulted from the frequency range of seismic waves used in the analysis and from the locations of stations where the data were acquired.

With the accumulation of detailed and accurate seismic data, the situation became more and more complicated. No seismologists dare to determine what happened. Different from ordinary interplate earthquakes, no clear evidence was found about which plate is sinking below another. Almost a year later, many seismologists agree to think that the fault may dip westward.

Seismologists are now skeptical about the validity of vertical ground displacements which are calculated with such a method as the Mansinha-Smylie's. On contrary to the homogeneous displacement on the fault plane assumed in these methods, an actual fault motion is heterogeneous, and the resulted ground displacement is more complicated.

There is no other method and no other theory than the use of tsunami simulations, at present, to estimate the vertical displacement of sea bottom. One way is the tsunami inversion method introduced by Satake (1985), by which the initial profile is calculated back from tide records. The present paper uses another method, the trial-and-error repetition. We start the numerical tsunami simulation with an assumed initial profile, compare the simulated results with the measured runup heights to find the difference, modify the assumed initial profile, and repeat the procedure again.

When a tsunami model is determined, the number of faults is important. In the beginning, the authors used the Harvard CMT solution, which assumed a single fault. This model, however, could not simulate the tsunami well. Then, models of plural faults were examined. In this paper, one of two-fault models, DCRC-2, and one of three-fault models, DCRC-17a, are discussed.

Harvard CMT Solution

Immediately after the earthquake, seismic data seemed to support one of the Harvard CMT solution, the fault plane of which gently dipped eastward. Figure 5

(a)Harvard

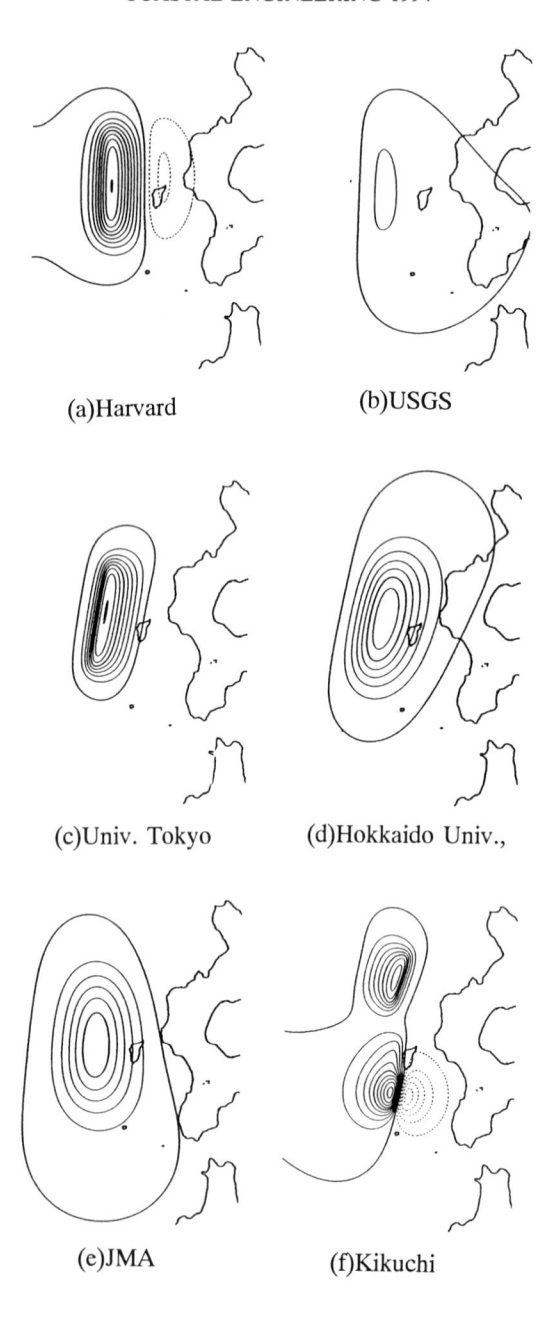

(b)USGS

(c)Univ. Tokyo

(d)Hokkaido Univ.,

(e)JMA

(f)Kikuchi

Fig.4 Proposed tsunami initial profiles. Solid lines are for upheaval and dotted lines for subsidence.

shows the vertical ground displacement (= tsunami initial profile) calculated for this solution. The crest is located at the western end of the tsunami source area.

With accumulation of seismic, geographical and tsunami data, differences of this model from the measured became non-negligible. Although this solution and the measured give the subsidence in Okushiri Island (Fig.6), the former gives the larger subsidence at the northern end of the island, while the latter reveals the larger subsidence of about 90 cm at the southern end.

The most important difference is found in the arrival time of tsunami. Every witnesses at Aonae confirm the fast arrival of the tsunami 4 to 5 minutes after the earthquake. The simulation, however, gives the arrival 7.5 minutes after the earthquake, as shown in Fig.7. In order to simulate this early arrival, the crest of tsunami initial profile should be located near the island; i.e., the fault should dip westward.

In Fig.8(a), (b), the computed runups are compared with the measured in Okushiri Island and Hokkaido. At Hamatsumae of Okushiri Island, the computed is too small. The same difference is found at Shimamaki of Hokkaido.

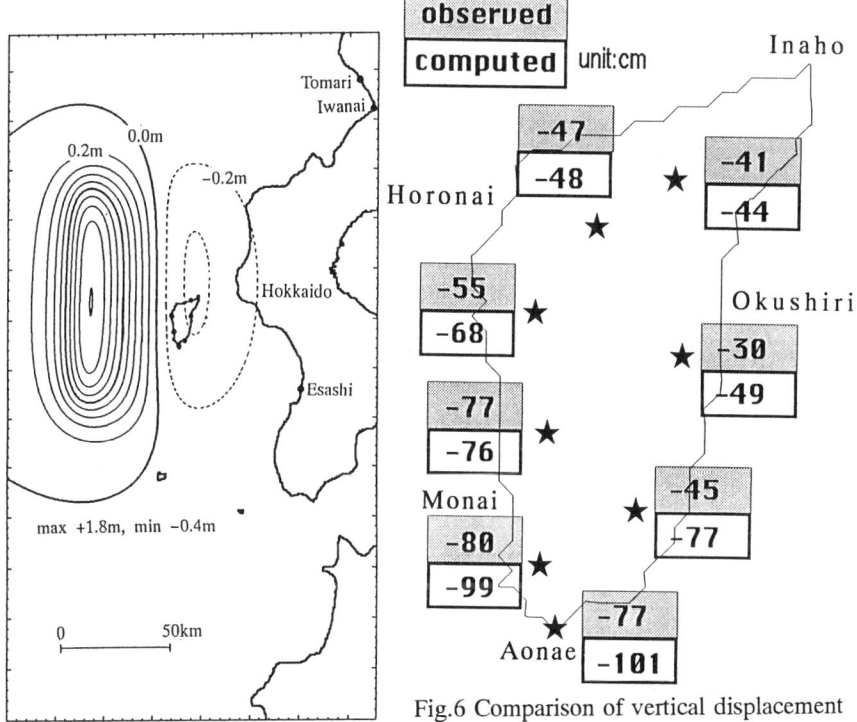

Fig.5 Tsunami initial profile computed for the Harvard CMT solution.

Fig.6 Comparison of vertical displacement in Okushiri Island between the measured and the computed with the Harvard CMT solution.

DCRC-2

Stimulated by the Kikuchi model (Fig.4 (f)), the authors assume plural faults. The first trial is DCRC-2, composed of two fault planes.

At the beginning of the modification, the authors take the following facts into consideration; aftershock distribution, ground subsidence in Okushiri Island and the witnessed arrival time of tsunami. The conclusion is that the faults must be reverse-slip faults inclining westward with a large dip angle.

Then, high runups at Hamatsumae is simulated. According to authors' assumption that these high runups are due to refraction, the strike direction of the south fault is adjusted because it governs the incident direction of the tsunami.

With these consideration, the initial profile for DCRC-2 is obtained as shown in Fig.9. Different from the Harvard CMT solution, there are two wave crests, both of which are located close to Okushiri Island. Contours are of more complicated shape.

The computed runups are compared with the measured in Fig.10. There are differences along the western coast. Along the south shore, the computed is much smaller than the measured, while the former is slightly smaller than the latter.

Fig.7 Arrival time computed with the Harvard CMT solution. Numerals are in minutes after the earthquake.

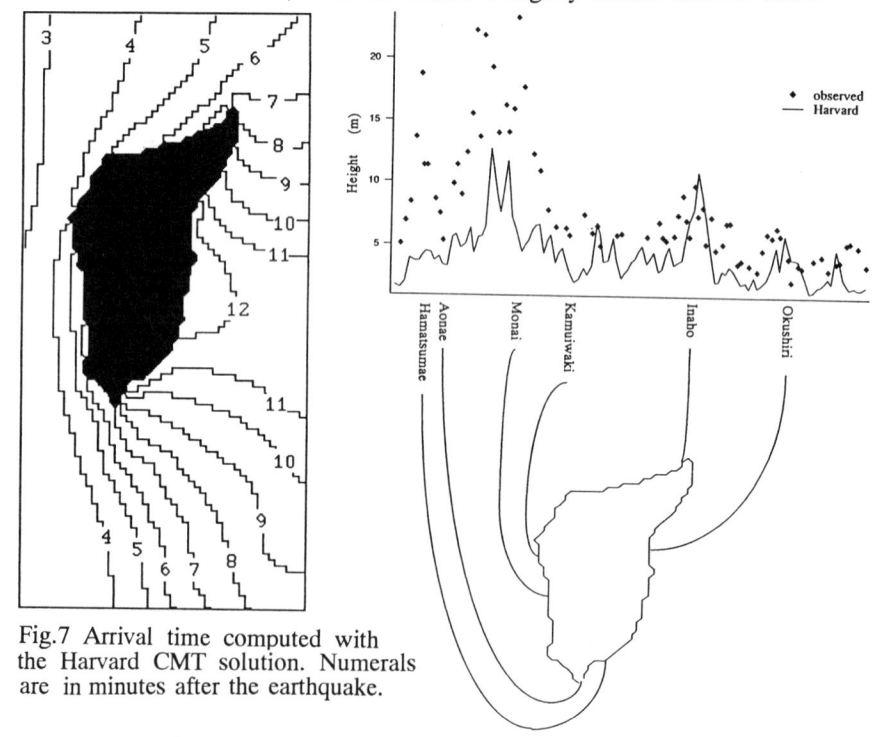

Fig.8(a) Comparison of runup heights between the computed with the Harvard CMT solution (solid line) and the measured (marks) along Okushiri Island.

EARTHQUAKE TSUNAMI OF 1993   895

Fig.8(b) Comparison of runup heights between the computed with the Harvard CMT solution (solid line) and the measured (marks) along Hokkaido.

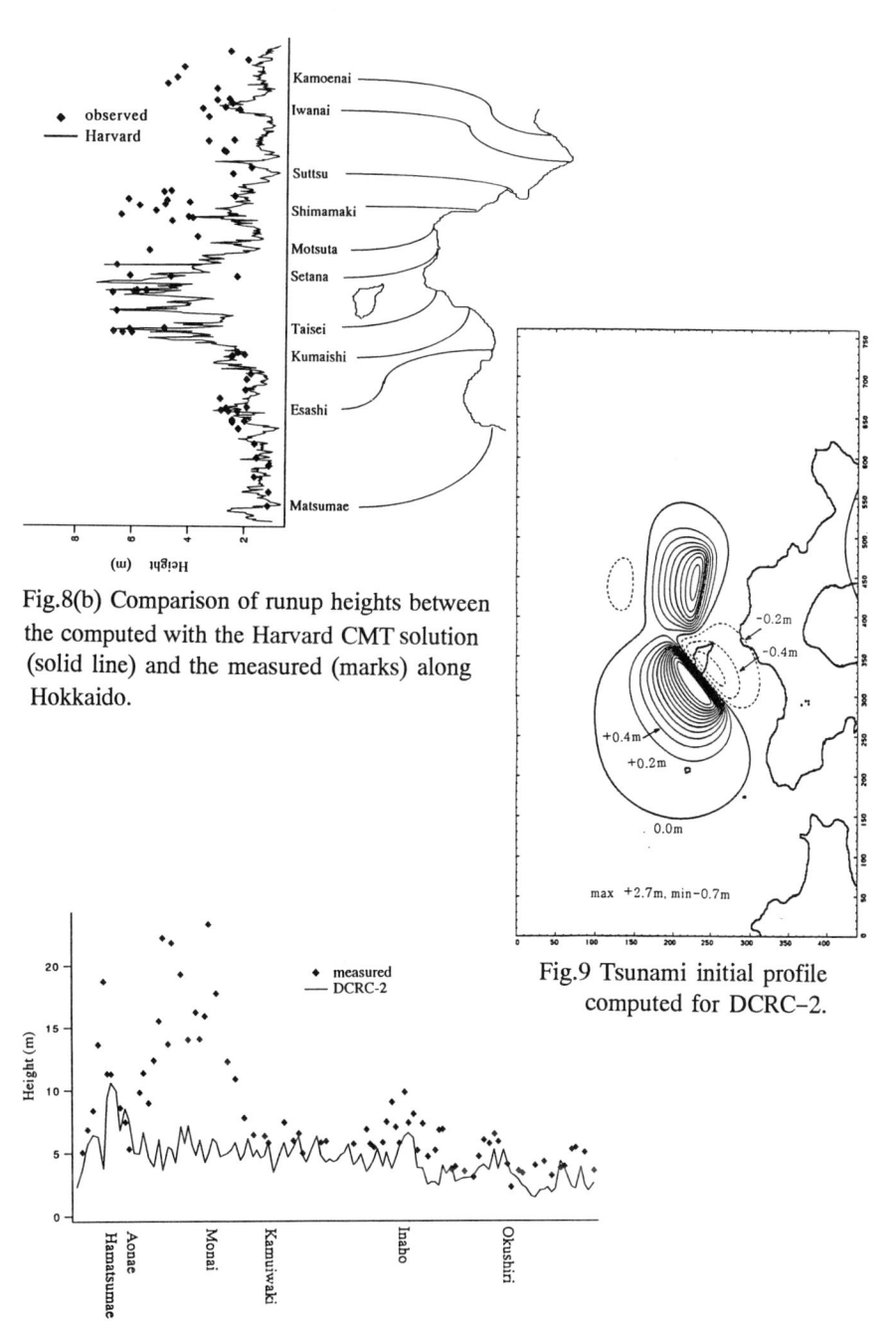

Fig.9 Tsunami initial profile computed for DCRC-2.

Fig.10 Comparison of runup heights between the computed with DCRC-2 (solid line) and the measured (marks) along Okushiri Island.

DCRC-17a

In order to solve the difference above mentioned, the authors divide the south fault into two faults with almost same strike directions. After several trials, DCRC-17a is found as the best. Its three fault planes are shown in Fig.11. The north fault is made longer than DCRC-2 in order to cover locations of aftershocks. Fault parameters are given in Table 2.

The initial profile of DCRC-17a is shown in Fig.12. The crest corresponding to the south sub-fault is 4.9 m high. Another crest is 2.2 m high, corresponding to the north sub-fault. The central fault does not yield any crest, because its slip is small.

The computed vertical displacement in Okushiri Island is compared with the measured in Fig.13. Compared to the Harvard CMT solution (see Fig.6), the agreement is much better.

Figure 14 shows the computed arrival time. The tsunami arrives at Aonae about five minutes after the earthquake, while it does 7.5 minutes in Fig.7 for the Harvard CMT solution.

Figure 15 compares runup height distributions around Okushiri Island. Overall agreement is fairly well, except for the neighborhood of Monai on the western coast, where the differences between the measured and the computed can only be explained by another simulation with very fine spatial grids as stated in the section of the measured runup heights. In order to evaluate the agreement, Aida's parameters, K and κ, the geometric mean of the ratio of the measured to the computed runup height and the corresponding variance, are used. For the whole island of Okushiri, K=1.048 and κ = 1.469. Distribution of K and κ for every coast is shown in Fig.16.

Figure 17 compares runup height distributions along Hokkaido. In the neighborhood of Esashi, the computed is higher than the measured. This indicates the need of a further modification. The initial tsunami height in the southern part of south fault should be made smaller than in the present model. In addition, a slight modification may be necessary to adjust the computed results near Suttsu and its neighborhood.

Table 2 Fault parameters of DCRC-17a.

Fault	$M_0(\times 10^{27})$	depth	Strike(°)	Dip(°)	Slip(°)	Length	Wide	dislocation
North	3.85	10km	188	35	80	90km	25km	5.71m
Central	0.56	5km	175	60	105	30km	25km	2.50m
South	2.21	5km	163	60	105	24.5km	25km	12.00m
Total	6.62							

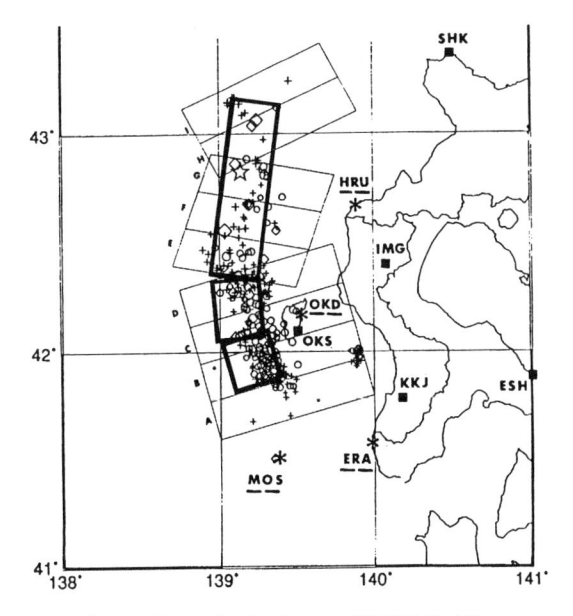

Fig.11 Three fault planes of DCRC-17a.

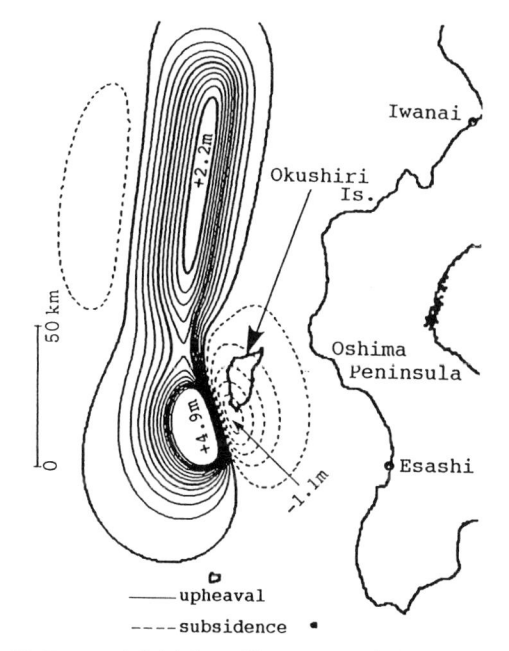

Fig.12 Tsunami initial profile computed for DCRC-17a.

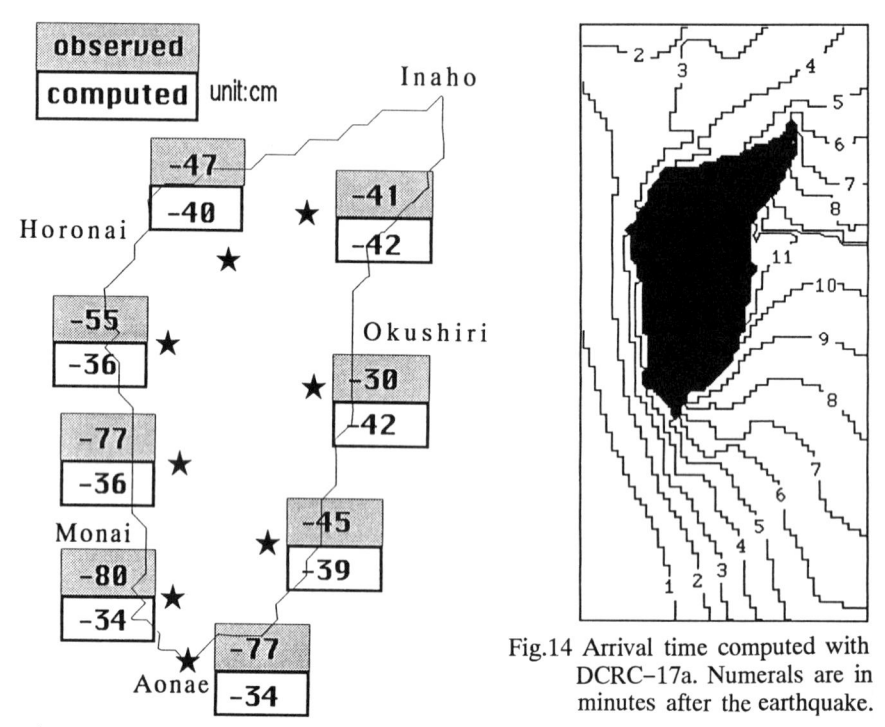

Fig.13 Comparison of vertical displacement between the computed with DCRC-17a and the measured in Okushiri Island.

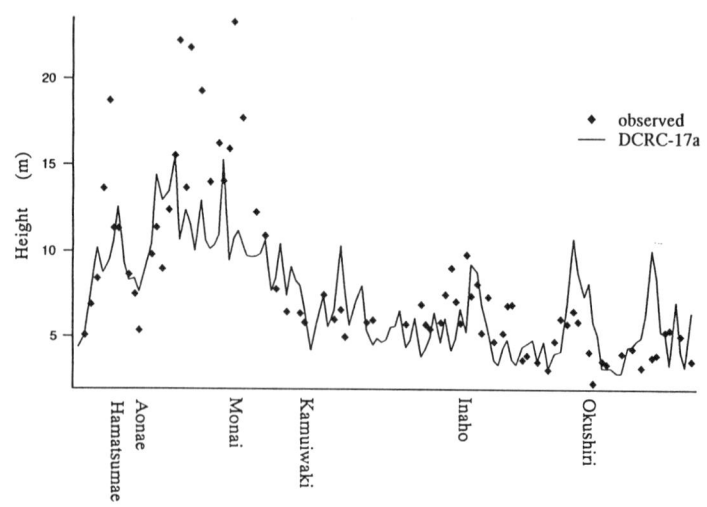

Fig.14 Arrival time computed with DCRC-17a. Numerals are in minutes after the earthquake.

Fig.15 Comparison of runup heights between the computed with DCRC-17a (solid line) and the measured (marks) along Okushiri Island.

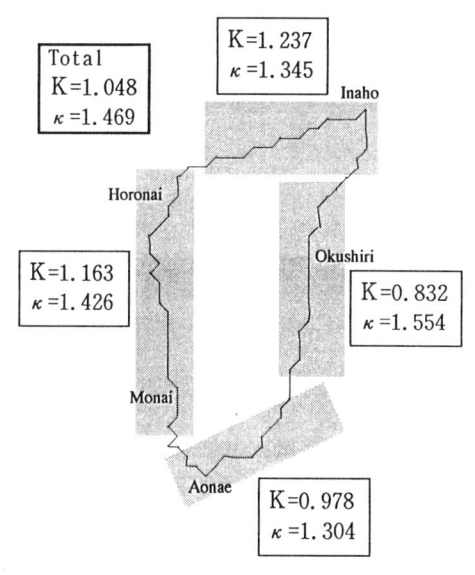

Fig.16 Distribution of K and κ for DCRC-17a along Okushiri Island.

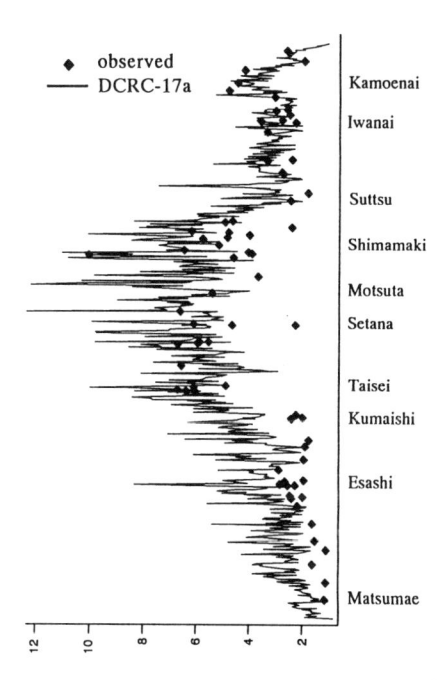

Fig.17 Comparison of runup heights between the computed with DCRC-17a (solid line) and the measured (marks) along Hokkaido.

Conclusions

In order to simulate the 1993 tsunami around Okushiri Island, the DCRC-17a model composed of three fault planes is obtained, by mostly adjusting the south and central faults. It satisfies the arrival time, runup distribution including high runups at Hamatsumae which is located in the sheltered area, and land subsidence.

In the present simulation, the maximum runup of 31.7 m in the neighborhood of Monai in Okushiri Island is not taken into consideration. A simulation with very fine spatial grids is required to simulate it.

Further adjustments of the south fault is necessary to explain the tsunami in the neighborhood of Esashi, Hokkaido. In addition, another adjustment is necessary for the northern part of the north fault. These adjustments may affect, to some extent, the wavy distribution of runup heights on the eastern coast of Okushiri Island, too.

References

Hokkaido Tsunami Survey Group:1993, Tsunami devastates Japanese coastal region, Eos, Transaction, AGU, 74, 417–432.

Mansinha, L. and Smylie, D.E.:1971, The displacement of the earthquake fault model, Bull. Seismol. Soc. America, 61, 1433–1400.

Satake, K.:1985, The mechanism of the 1983 Japan Sea earthquake as inferred from long period surface waves and tsunamis, Phys. Earth Planet. Interiors, 37, 249–260.

Satake, K. et al.:1988, Tide gauge response to tsunamis: Measurements at 40 tide gauge stations in Japan, J. Marine Res., 46, 557–571.

Shuto, N. et al.:1986, A study of numerical techniques on the tsunami propagation and run up, Science of Tsunami Hazard, 4, 111–124.

CHAPTER 66

Quasi-Three-Dimensional Model for Storm Surges and Its Verification

Takao Yamashita[1], Yoshito Tsuchiya, M.ASCE[2] and Hiroshi Yoshioka[1]

Abstract

Three-dimensional model for storm surges was developed, in which momentum equations first were solved in the vertical direction together with mass conservation and turbulence model, then mass conservation in the horizontally two-dimensional coordinates was solved by the finite difference method. Several fundamental tests concerning swing and wind-induced current, were performed to examine the model properties. Finally the hindcasting of the current profile in the Tanabe bay was conducted to test model applicability.

Introduction

Storm surge models have been developed as a two-dimensional model, which is called a vertical integrated model or one(single) level model. Because of Japan's recent need for waterfront development, significant environmental problems and coastal disasters will persist during accelerated future developments in the coastal ocean. It is desired to develop a more sophisticated numerical model for storm surges which predicts the profile of current, turbulence, and surge heights. A three dimensional model for tide and storm surges combined with a turbulence and wind-wave prediction model may become popular in coastal ocean assessment in the near future.

Heaps (1973) studied dynamic response of the Irish Sea to a stationary wind stress field by a finite difference model which solves the homogeneous hydrodynamic equations taken in linearlized form by transforming them to eliminate the vertical coordinate z. The system including the coordinates x, y, t, and z was solved by an explicit finite difference scheme. This model only considers homogeneous systems, and neglects the convection terms. A series of numerical experiments was also performed with the model to determine the steady state circulation of the Irish Sea system. Sundermann (1974) and Laevastu (1975) extended the method for three dimensional, multi-level, and multi-layer flow systems. Simons (1974 & 1975) studied the circulation in Lake Ontario under strong stresses by

---
[1] Instructor, Disaster Prevention Research Institute, Kyoto University, Uji, Kyoto, 611 Japan
[2] Professor Emeritus of Kyoto University & Professor, Meijyo University, Tenpaku-ku, Nagoya, 467 Japan

using a four layer model. Sundermann (1974) extended Hansen's (1962) basic two dimensional scheme to study three dimensional tidal circulation of the North Sea. This extension involves an assumption of homogeneous density distributions. The basic two-dimensional explicit scheme of Hansen (1962) also has been extended by Laevastu (1975) by using multi-layer modeling.

Another three-dimensional model which is the simple version of σ-coordinates was developed by Koutitas and O'Connor (1980), in which momentum equations together with mass conservation equation and turbulence model first are solved in the vertical direction. Then mass conservation in the horizontally two-dimensional coordinates is solved. This model elaborates on the vertical motion of fluid and reducing the matrix of difference equations.

Numerical prediction in shallow bays or estuaries must be focused to develop an effective three-dimensional model for environment assessment in the coastal ocean, since this is our most probable development space in the coming century. In this paper, a quasi-three-dimensional (Q3-D) model is developed based on Koutitas and O'Connor's (1980) modeling.

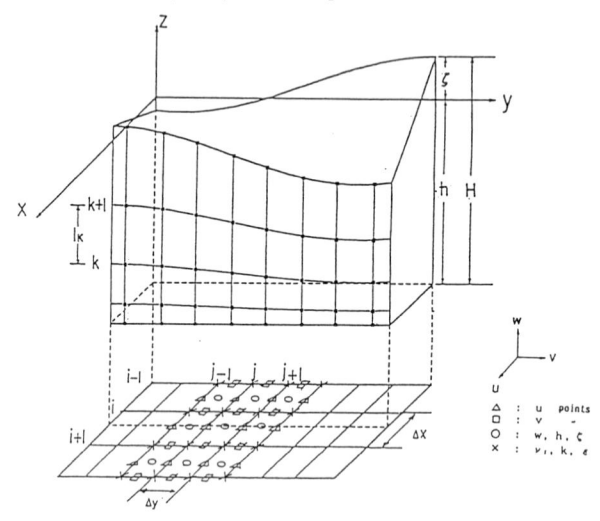

Figure 1 Coordinate system and definition of variables.

Basic Equations and Solution Method

The basic equations for the Q3-D model consist of the momentum equations of mean flow, the equation of continuity and the equations of turbulence model. The following additional relations for the vertical velocity w and the surface elevation ζ are used.

$$w(x,y,z) = -\frac{\partial}{\partial x}\int_{-h}^{z} u\,dz - \frac{\partial}{\partial y}\int_{-h}^{z} v\,dz \qquad (1)$$

$$\frac{\partial \zeta}{\partial t} + \frac{\partial (HU)}{\partial x} + \frac{\partial (HV)}{\partial y} = 0 \qquad (2)$$

which is derived from the mass continuity equation by integrating in the vertical direction. This equation satisfies total mass conservation in the horizontally two-dimensional system.

Vertical distribution of turbulence has to be specified to simulate the velocity distribution of wind-induced mean currents. Several turbulence models are now available for the Q3-D modeling: (1) The so-called zero equation model in which the vertical eddy viscosity ν_t is assumed. (2) Multi-equation models in which turbulence kinetic energy budget is described by the standard k-ε model. In order to get an efficient numerical solution to the three-dimensional system of equations, Koutitas and O'Connor (1980) employed a time fractional finite difference method in which advection and propagation are split. The computational domain is covered by a rectangular-grid system in the x-y plane, and the water depth is divided into a several layer-elements as shown in Figure 1. The finite element Galerkin method is applied to the momentum equations to obtain nodal velocities over the depth. For the depth-averaged continuity equation (2), the finite difference method is applied.

Applying an explicit scheme, the difference equation of the momentum equation in the x-direction is written as:

$$\frac{u^{n+1} - u^n}{\Delta t} = -u^n \frac{\partial u^n}{\partial x} - v^n \frac{\partial u^n}{\partial y} - w^n \frac{\partial u^n}{\partial z} - g \frac{\partial \zeta^{n+1/2}}{\partial x}$$

$$+ \frac{\partial}{\partial x}\left(\nu_T \frac{\partial u^n}{\partial x}\right) + \frac{\partial}{\partial y}\left(\nu_T \frac{\partial u^n}{\partial y}\right) + \frac{\partial}{\partial z}\left(\nu_T \frac{\partial u^{n+1}}{\partial z}\right) + f v^n \quad (3)$$

The y-momentum equation can be shown by the same manner.

Assuming fractional time step t^*, the momentum equations can be split up as; for the x-momentum equation,

$$\frac{u^*}{\Delta t} = \frac{u^n}{\Delta t} - u^n \frac{\partial u^n}{\partial x} - v^n \frac{\partial u^n}{\partial y} - w^n \frac{\partial u^n}{\partial z} - g \frac{\partial \zeta^{n+1/2}}{\partial x} \quad (4)$$

$$\frac{u^{n+1}}{\Delta t} - \frac{\partial}{\partial z}\left(\nu_T \frac{\partial u^{n+1}}{\partial z}\right) = \frac{u^*}{\Delta t} + \frac{\partial}{\partial x}\left(\nu_T \frac{\partial u^n}{\partial x}\right) + \frac{\partial}{\partial y}\left(\nu_T \frac{\partial u^n}{\partial y}\right) + f v^n \quad (5)$$

for the y-momentum equation,

$$\frac{v^*}{\Delta t} = \frac{v^n}{\Delta t} - u^n \frac{\partial v^n}{\partial x} - v^n \frac{\partial v^n}{\partial y} - w^n \frac{\partial v^n}{\partial z} - g \frac{\partial \zeta^{n+1/2}}{\partial y} \quad (6)$$

$$\frac{v^{n+1}}{\Delta t} - \frac{\partial}{\partial z}\left(\nu_T \frac{\partial v^{n+1}}{\partial z}\right) = \frac{v^*}{\Delta t} + \frac{\partial}{\partial x}\left(\nu_T \frac{\partial v^n}{\partial x}\right) + \frac{\partial}{\partial y}\left(\nu_T \frac{\partial v^n}{\partial y}\right) - f u^n \quad (7)$$

and the vertical velocity w is calculated by

$$w(x, y, z) = -\frac{\partial}{\partial x} \int_{l_k} u(x, y, z) dz - \frac{\partial}{\partial y} \int_{l_k} v(x, y, z) dz \quad (8)$$

The finite element Galerkin method is applied to Eqs. (4) to (7) to solve the nodal velocities over the depth. Substituting the approximated velocity $\hat{u}$ into the original momentum equation and integrating it between nodal points over a whole area, we get the following residual equation in the x-direction.

$$\int \tilde{u}\frac{u^*}{\Delta t}dz = \int \tilde{u}\frac{u^n}{\Delta t}dz$$

$$-\int \tilde{u}u^n\frac{\partial u^n}{\partial x}dz - \int \tilde{u}v^n\frac{\partial u^n}{\partial y}dz - \int \tilde{u}w^n\frac{\partial u^n}{\partial z}dz - \int \tilde{u}\frac{\partial \zeta^{n+1/2}}{\partial x}dz \qquad (9)$$

Applying this relation to Eqs. (4) to (7), a set of linear equations can be reduced to

$$\begin{array}{ll}\mathbf{A}u^* = \mathbf{a}^n, & \mathbf{B}u^{n+1} = \mathbf{b}^* \\ \mathbf{C}v^* = \mathbf{c}^n, & \mathbf{D}v^{n+1} = \mathbf{d}^*\end{array} \right\} \qquad (10)$$

where $\mathbf{A}$ is the 2×2 matrix, $\mathbf{B}$ the 2×4 matrix, $\mathbf{a}$ and $\mathbf{c}$ are two component vectors, and $\mathbf{b}$ and $\mathbf{d}$ are four component vectors.

The linear interpolation function is used as

$$u = N_k u_k + N_{k+1} u_{k+1} \qquad (11)$$

where

$$N_k = \frac{1}{l_k}(z_{k+1} - z), \qquad N_{k+1} = \frac{1}{l_k}(z - z_k) \qquad (12)$$

where l_k is the element length defined by $l_k = z_{k+1} - z_k$.

The three-dimensional current field is obtained by computing these equations which can be solved by the Gaussian elimination method. When velocities $u(x,y,z)$ and $v(x,y,z)$ are calculated in the time step n, the corresponding vertical velocity $w(x,y,z)$ is computed by

$$w(x,y,z) = -\int_{l_k} u(x,y,z)dz - \int_{l_k} v(x,y,z)dz \qquad (13)$$

The surface elevation ζ is calculated by

$$\frac{\zeta^{n+1/2} - \zeta^{n-1/2}}{\Delta t} = -\frac{\partial(HU)^n}{\partial x} - \frac{\partial(HV)^n}{\partial y} \qquad (14)$$

In the above-mentioned computation processes, the kinematic eddy viscosity ν_t in the time-space domain has to be specified. Three types of turbulence models are incorporated into the numerical model for Q3-D storm surge prediction:

(1) For the zero-equation turbulence model, ν_t is assumed as:

$$\nu_t = \text{constant} \quad \text{or} \quad \nu_t = l^2\left|\frac{\partial u}{\partial z}\right| \qquad (15)$$

(2) For the one-equation turbulence model, ν_t is calculated as:

$$\nu_t = \sqrt{k}l \qquad (16)$$

where l is the eddy length scale and k is calculated by the k-equation.
(3) For the two-equation turbulence model, ν_t is calculated as:

$$\nu_t = C_D \frac{k^2}{\varepsilon} \tag{17}$$

where k and ε are ontained by the k-ε model. Details of the k-ε model are shown below.

The k equation is discretized by the Chrank-Nicholson scheme as:

$$\frac{k^{n+1}}{\Delta t} + \frac{1}{2}\left(w^n \frac{\partial k^{n+1}}{\partial z}\right) - \frac{1}{2}\left(\frac{\partial}{\partial z}\left(\frac{\nu_T}{\sigma_k}\frac{\partial k^{n+1}}{\partial z}\right)\right)$$

$$= \frac{k^n}{\Delta t} - \frac{1}{2}\left(w^n \frac{\partial k^n}{\partial z}\right) + \frac{1}{2}\left(\frac{\partial}{\partial z}\left(\frac{\nu_T}{\sigma_k}\frac{\partial k^n}{\partial z}\right)\right) + P^n - \varepsilon^n \tag{18}$$

The finite element formulation is made by multiplying the weighting function, $\tilde{k}$, to this equation and integrating it over the depth as:

$$\int \tilde{k} \frac{k^{n+1}}{\Delta t} dz + \int \tilde{k}\left\{\frac{1}{2}w^n \frac{\partial k^{n+1}}{\partial z}\right\} dz - \int \tilde{k}\left\{\frac{1}{2}\frac{\partial}{\partial z}\left(\frac{\nu_T}{\sigma_k}\frac{\partial k^{n+1}}{\partial z}\right)\right\} dz = \int \tilde{k}\frac{k^n}{\Delta t} dz$$

$$- \int \tilde{k}\left\{\frac{1}{2}w^n \frac{\partial k^n}{\partial z}\right\} dz + \int \tilde{k}\left\{\frac{1}{2}\frac{\partial}{\partial z}\left(\frac{\nu_T}{\sigma_k}\frac{\partial k^n}{\partial z}\right)\right\} dz + \int \tilde{k}P^n dz - \int \tilde{k}\varepsilon^n dz (19)$$

The linear shape function, Eq. (20), is assumed in the finite element formulation.

$$N_k = \frac{1}{l_k}(z_{k+1} - z), \quad N_{k+1} = \frac{1}{l_k}(z - z_k) \tag{20}$$

where k is the nodal number, l_k the element length, and z_k the position of k in the z axis.

The point of z is interpolated by,

$$z = N_k z_k + N_{k+1} z_{k+1} \tag{21}$$

Applying the Galerkin method, the integrations of each term in the equation are shown respectively as:

Equations for ε are derived in the same munner, and are solved numerically togther with the k equations by the Gaussian elimination method.

Fundamental Tests of Q3-D Model

Model parameters in the k-ε equations will now be discussed along with several fundamental tests in the rectangular flume (i.e. swing and wind-induced current tests).

Boundary conditions at the bottom and surface: For time-averaging turbulence models, such as the k-ε and one equation models, the boundary conditions at the bottom and surface have to be carefully discussed because of its strong shear flow characteristics. Usual modeling method in this area apply the knowledge called the universal law of the wall, which depicts the velocity distribution of shear flow by a logarithmic law given by

$$\frac{u}{u_\tau} = \frac{1}{\kappa}\ln\frac{Eu_\tau y}{\nu} \qquad (22)$$

where u_τ is the friction velocity, κ the Kármán constant, E the roughness parameter, y the distance from the wall, and ν the kinematic viscosity of the fluid. When we assume a hydraulically smooth wall, the roughness parameter may be evaluated by $E = 9$ in the range of $30 < u_\tau y/\nu < 100$. This concept is useful in the modeling of mean flow field in the boundary layer. Equation (22) is used to develop a quasi-three dimensional model for simulation of storm surges.

On the other hand, the boundary condition of the turbulence model is posed by the assumption of balancing production and dissipation, which results in the following relation

$$\frac{k}{u_\tau^2} = \frac{1}{\sqrt{c_\mu}}, \qquad \varepsilon = \frac{u_\tau^3}{\kappa y} \qquad (23)$$

for the k-ε model.

To compute the mean flow in the interior region, we have to specify the velocity at the interface between the interior region and boundary layer with the relation of shear stresses and eddy viscosity given respectively by

$$\left(\nu_t \frac{\partial u}{\partial z}\right)_{z=\zeta} = \tau_s \qquad \text{at the surface} \qquad (24)$$

$$\left(\nu_t \frac{\partial u}{\partial z}\right)_{z=-h} = \tau_b \qquad \text{at the bottom} \qquad (25)$$

where τ_s and τ_b are the shear stresses at the surface and bottom of the sea respectively.

When we specify the shear stresses (τ_s and τ_b), and turbulence properties (k and ε), it is possible to get the boundary condition for mean flow field by velocity gradient. The following relation is used to evaluate the eddy viscosity, ν_t.

$$\nu_t = c_\mu \frac{k^2}{\varepsilon} \qquad (26)$$

Dirichlet's problem for the disecrate system brings a difficulty in determining the velocity gradient without dependence on grid size. Fine mesh makes the model extremely expensive in the three-dimensional case. To overcome these problems the point where the boundary condition is posed is fixed in this investigation. At the surface layer the point of Δz_s=1.35m and Δz_b=0.1m at the bottom are set. This means that the distance from the bottom or surface, y, is fixed by Δz_s or Δz_b in the wall law relation of Eq.(22) or Eq.(23). In the case of the zero-equation turbulence model we do not have any information about turbulence near the boundary, therefore we need some assumptions in evaluating the eddy viscosity ν_t or shear stresses τ_b and/or τ_s. If the two-equation turbulence model is used, we do not need any assumption. We do need much CPU time and information of turbulence at the open boundary. This is usually difficult to get.

For zero-equation modeling, the following text elaborates on the way to determine the bottom shear stresses from the numerical results of swing and wind-induced current tests. For this purpose, both the quasi-three dimensional simulation of k-ε model version and the horizontally two-dimensional simulation are also conducted under the same computational condition. These results are used as indexes to determine the bottom shear stresses at the fixed point Δz_b by comparison of surface elevations. Moreover, a simulation of flow combined tides and wind-induced currents is performed by the model of zero-equation turbulent version.

Swing test Free oscillation of water in the closed flume may be simulated fairly well by the so-called single level model (horizontally two-dimensional model). By comparing the water surface oscillations computed by the single level model and those by the Q3-D model, the bottom shear stress of Q3-D model can be evaluated. As previously stated, this is one of the parameters which has to be determined in the Q3-D model. A closed flume which is used for the swing test has a dimensions of uniform water depth of 20 m deep, L=10,000 m in length and B=3,000 m in width. The horizontal spacing of the grid system is Δx=1000 m and the number of the vertical nodes is 35. Time increment is fixed by Δt=10 sec. The initial surface profile is given by $\zeta(x) = 0.0001x - 0.5$. For the zero-equation version of Q3-D model, the vertical distribution of eddy viscosity is assumed by the following equation which is obtained by fitting the mean velocity distribution with Baines and Knapp's experiment.

$$\nu_t = 0.01 + \kappa C_\mu \left(1 - 3.6\frac{z}{h}\right)\frac{z}{h} \qquad (27)$$

This assumption of eddy viscosity distribution will be further discussed later by comparing zero and two equation models in $\nu_t(z)$.

Surface elevation at the right end is shown in Figure 2 together with the results computed by the leap-frog scheme(dotted line). Bottom friction is neglected in the computation by leap-frog scheme. The computed propagation speed of the disturbance on the uniform depth of 20 m is 1428 sec, which is equivalent to the long wave celerity. It can be recognized that almost uniform velocity distribution of the mean current is computed in the case of swing test with no bottom friction.

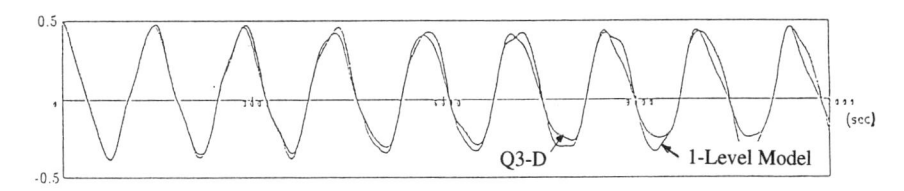

Figure 2 Comparison of water surface oscillation at the flume computed by Q3-D model (solid line) and 1-Level Model (dotted line).

Wind-induced current test The three-dimensional model is of course more effective than the horizontally two-dimensional in the computation of wind induced current which is characterized by the strong shear near the surface. In the case of a closed basin, a return flow is observed near the bottom.

A wind-induced flow test in the closed flume is conducted and the vertical distribution of eddy viscosity as well as mean flow is discussed together with surface elevation property. This is calibrated by a comparison with theoretical formulation of wind set-up using Eq. (28).

$$\zeta_{max} = \frac{3\rho_a C_D L W^2}{4 g D \rho_w} \qquad (28)$$

where ζ_{max} is the maximum water surface elevation, C_D the drag coefficient at the surface, L length of the flume, W the wind velocity, D the total depth, ρ_a the density of air, and ρ_w the density of sea water.

As mentioned before, shearing stresses both at the surface and bottom are not uniquely determined in the Q3-D model of the zero-equation version of turbulence. They depend on the spacing of nodes just inside the interface because an approximation accuracy of the mean flow gradient depends on node spacing. When the number of vertical node points is fixed, shearing stresses are dependent upon the total depth. Figure 3 depicts this effect, in which four different results in the maximum surface elevation under the same condition are compared. Changing the still water depth in the range of 5 to 40 m, the maximum wind set-up is computed under the condition of uniform wind speed of 20 m/sec in the same flume as swing test. Gradually increasing wind speed from 0 to 20 m/sec for 3000 sec is assumed in the test to reduce time to be steady state which means small tank oscillation is expected. In the figure, Q3-D CONST indicates the results computed by using the uniform distribution of eddy viscosity in the Q3-D model of zero-equation turbulence version, Q3-D QUARD indicates those expressed by the distribution resulting from Eq. (27), 1LEVEL MODEL indicates those developed by the horizontally two dimensional model.

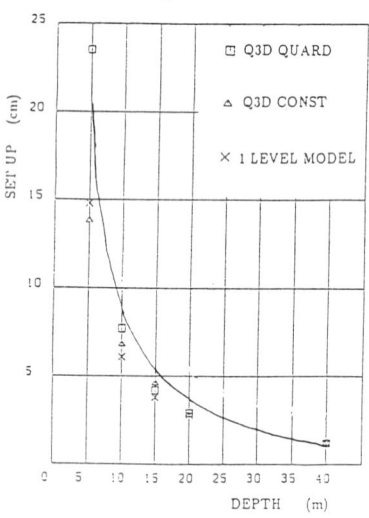

Figure 3 Relation between the water depth and the computed set-up.

It can be recognized that the wind set-up computed by Q3-D QUARD is larger than that computed by other models in the region shallower than 10 to 5 m. This means that it may result in over-estimate of the shearing stress at the surface, and a lower-estimate at the bottom in the very shallow region, because the their real distribution may become uniform as the depth becomes shallower. The difference between 1LEVEL MODEL and the solid line comes from the difference of out-put points. The out-put point of numerical computation is just inside of the end-wall.

Vertical distributions of the mean flow and eddy viscosity at the center of the flume in steady state is shown in Figure 4 where the column is the depth nondimensionarized by the still water depth, while the abscissas indicates the horizontal velocity normalized by friction velocity. White circles in the figure indicate the experimental values of Baines-Knapp showing the null-velocity point at 0.7 in nondimensional depth.

(a) Constant distribution of eddy viscosity

(b) Parabolic distribution of eddy viscosity

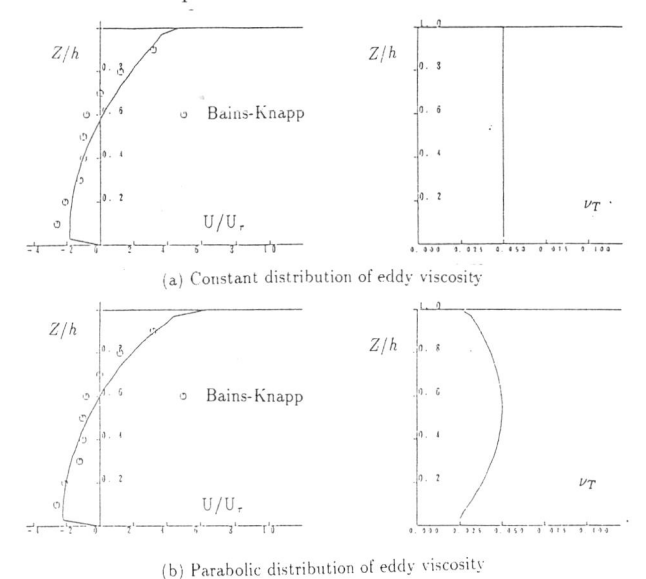

Figure 4 Vertical distribution of the mean flow and eddy viscosity at the center of the flume in steady state by Q3-D model of zero-equation turbulence version.

On the other hand, Figure 5 also shows the vertical distribution of both mean flow velocity and eddy viscosity which were computed in terms of the Q3-D model of k-ε version whose boundary conditions are given by Eq. (23) for k and Eq. (29) for ε.

$$\varepsilon_s = \frac{k_s^{\frac{3}{2}}}{\kappa L_s}, \quad \varepsilon_b = \frac{k_b^{\frac{3}{2}}}{\kappa L_b} \tag{29}$$

where the length scale parameters are assumed as $L_s = 100\Delta z_s$ and $L_b = \Delta z_b$ by comparison of mean flow velocity distribution with Baines and Knapp's experiments.

From these figures it is found that the null-velocity point of the vertical distribution of mean flow is located around the nondimensional depth of 0.6 which is slightly lower than that by experiment. Moreover the vertical distribution of computed eddy viscosity by the k-ε model is similar to that of a parabola,

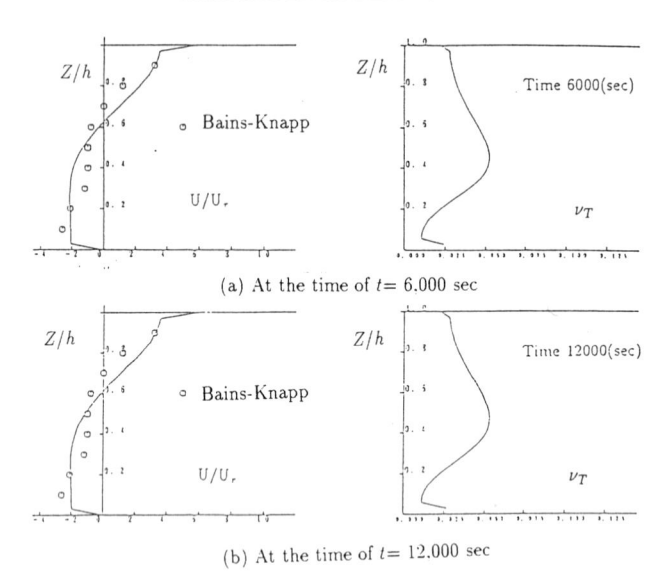

Figure 5 Vertical distributions of the mean flow and eddy viscosity at the center of the flume in steady state by Q3-D model of k-ε version.

Eq. (27), which is assumed for the model of zero-equation version. The computed eddy viscosity, however, skews downward with its maximumm at 0.4 in the nondimensional depth.

Application to wind-induced currents in Tanabe Bay

Observations of vertical velocity distribution in terms of an acoustic Doppler current profiler (ADCP) have been conducted since 1989 in Tanabe Bay by the Shirahama Oceanographic Observatory, Disaster Prevention Research Institute, Kyoto University. Tide and wind vector on the sea surface are also observed at the Oceanographical observation tower whose location is indicated in the map of Tanabe Bay (Figure 6) together with ADCP's location and depth contours.

As this data is very useful for calibration of three-dimensional numerical model, simulation tests of mean flow in Tanabe Bay are carried out to compare with observation in the current profile. Because the model assumes a well-mixed situation in the sea, the observed profile of mean current of horizontal component in winter season is selected for comparison. Figure 7 shows the time changes in wind vector(upper), tide(middle) and current profile(bottom) which are used for model calibration here. Current vector in this figure is defined by NS-WE system in which the northward component is plotted upward and the westward current is leftward. A typical feature of this data is the strong wind-induced currents whose direction is ESE and the return flow to the direction of W.

The sea bottom topography is reproduced from the chart (No. 74) with the grid system of $\Delta x = 100$ m. Total point number is 89×110 and the origin is $33°$ 44' 40" north latitude and $135°$ 17'00" east longitude. For actual computation, 200m horizontal grid size, 19-point vertical nodal points, and 5sec time increment

MODEL FOR STORM SURGES 911

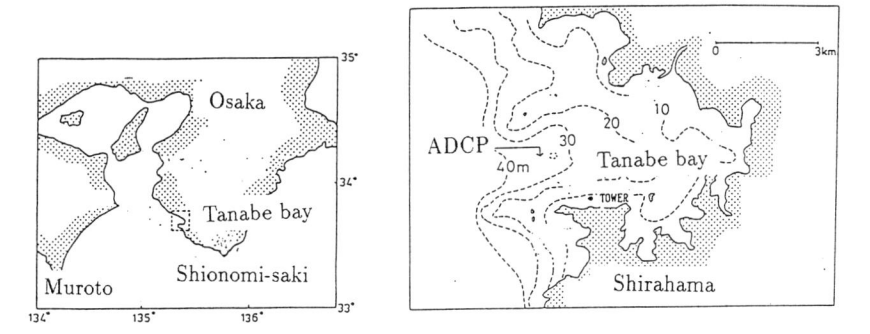

Figure 6 Location of the Tanabe Bay and the oceanographical observation tower together with ADCP's observation point.

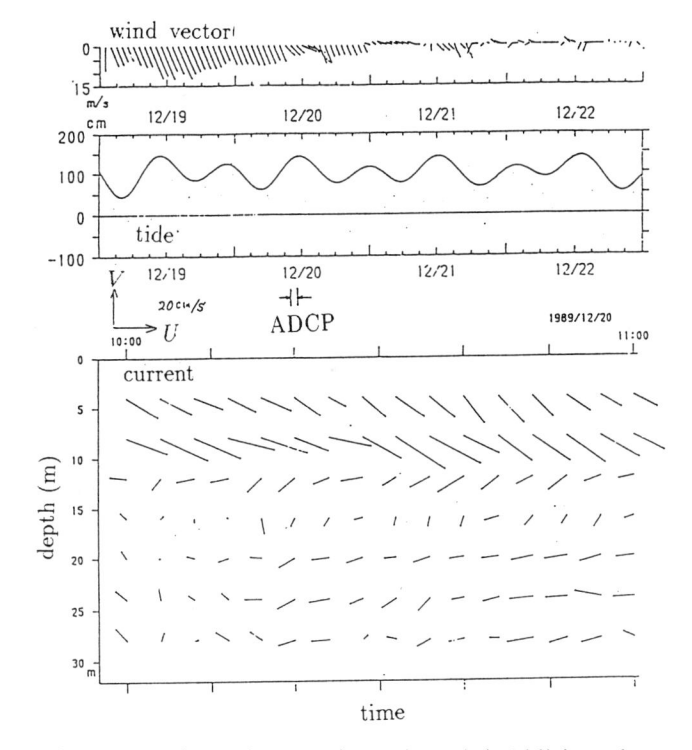

Figure 7 The data of wind vector(upper), tide(middle) and current profile(bottom), observed on December 12, 1989.

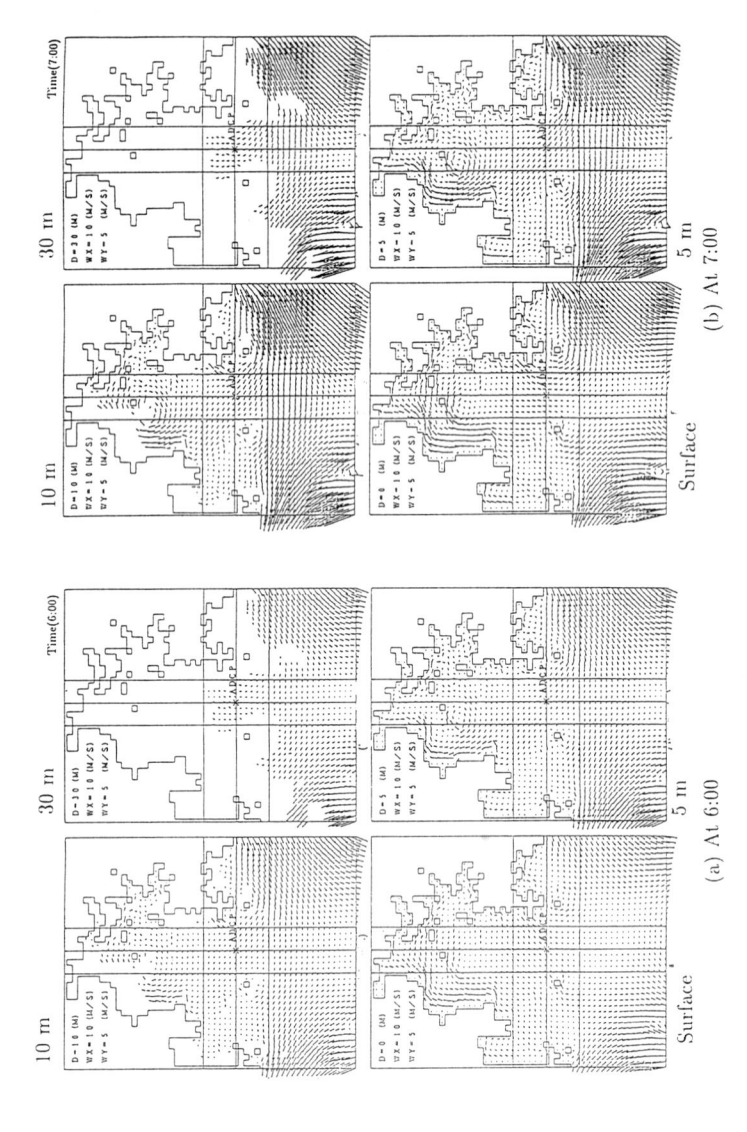

Figure 8 Computed mean flow vectors in Tanabe Bay at the surface level, 5m level, 10m level and 30m level by Q3-D model of zero-equation version.

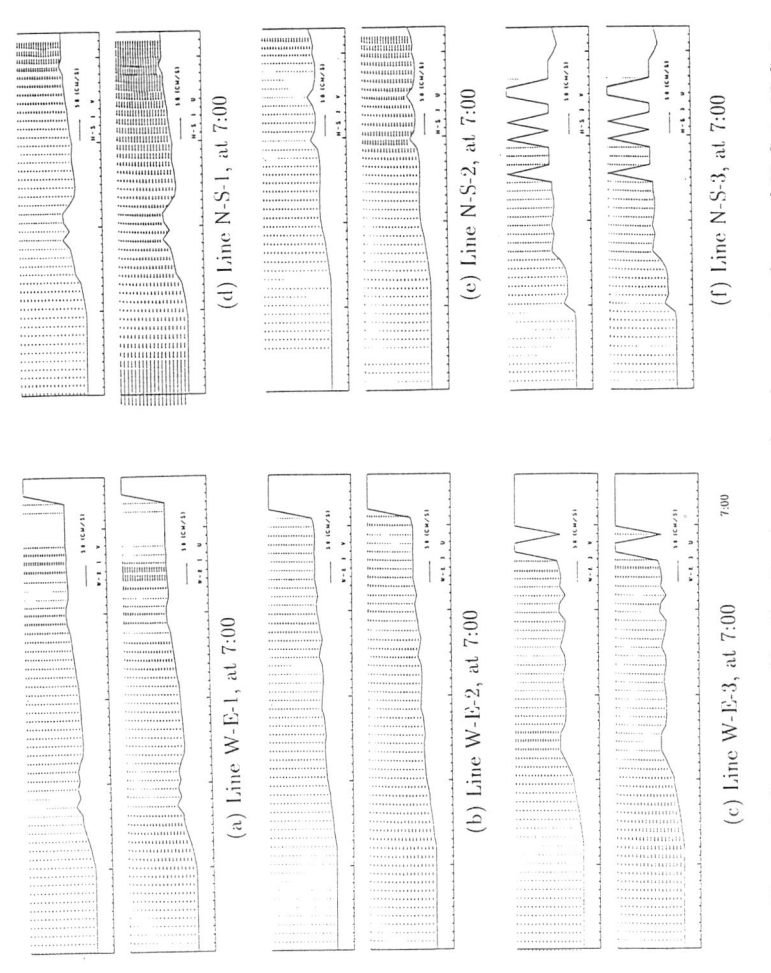

Figure 9 Current profile distribution on the observation lines of N-S-1, N-S-2, N-S-3, W-E-1, W-E-2 and W-E-3.

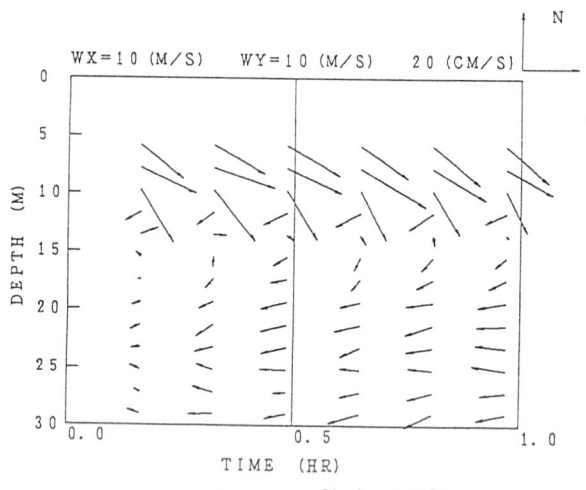

(a) Observed velocity profile by ADCP

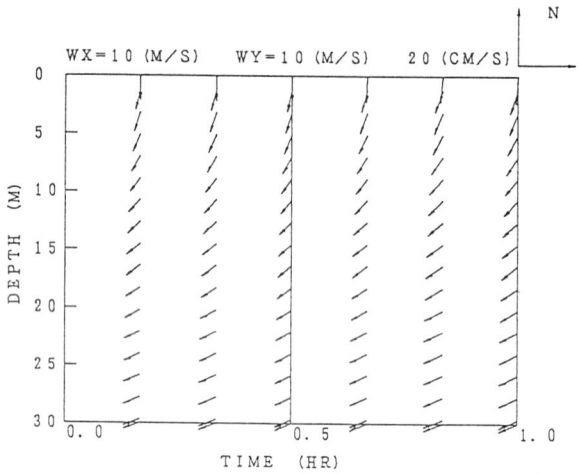

(b) Computed velocity profile by Q-3D model

Figure 10 Comparison between observation and simulation of current profile. Observation of 2 min and 10 min averaging and computation output of 10 min interval.

are employed. Simulation period is 5:00-11:00, 20 December, 1989 which covers ADCP observation, 10:00-11:00. Wind condition is assumed to be constant NW wind whose velocity is 10 m/sec.

A sample of the computed mean flow vectors at the surface level, 5m level, 10m level and 30m level are shown in Figure 8. The vertical velocity distribution on six lines of N-S-1, N-S-2, N-S-3, W-E-1, W-E-2 and W-E-3 are shown in Figure 9. Figure 10 is the comparison between observation and simulation, in which observation of 2 min and 10 min averaging and computation output of 10 min interval are shown. The strong wind-induced flow is not simulated in the numerical model, however, the return flow (probably the tidal current) is well simulated both in its magnitude and direction. One possibility which may cause this discrepancy may be the open boundary condition. The other may be some trouble in the ADCP observation. Further discussion about this will be needed in terms of continuous efforts in executing simulations and ADCP observations.

Conclusions

A three-dimensional model for storm surges was developed and its fundamental tests were carried out. The obtained main results in this paper are shown below.
(1) From the swing test, propagation speed of the initial disturbance was confirmed to be in good agreement with that of free long waves.
(2) From the wind-induced current test in the two-dimensional flume, it was found by fitting the mean velocity profile with the Baines and Knapp's experiment that the eddy viscosity ν_t was around 0.01-0.02 m^2/s for Q3-D model of zero-equation version. Moreover, the profile of eddy viscosity computed by the k-ε model was similar to that assumed the parabolic profile in the zero-equation version.
(3) The simulation of the flow fields in the Tanabe Bay was performed, in which both tides and wind-induced currents were taken into consideration. The velocity profile observation with ADCP also was conducted in the bay. From the comparison of velocity profile between observations and computations, it was found that the strong surface wind-induced flow was not simulated, however the return flow was well simulated both in its magnitude and direction.

References

Hansen, W.(1962), Hydrodynamical methods applied to oceanographic problems, Mitt. Inst. Meereskd., University of Hamburg, Humburg.

Heaps, N.S.(1973), Three-dimensional model of the Irish Sea, Geophys. J. R. Astron. Soc., Vol.35, pp.99-120.

Koutitas, C. and B.A. O'Connor(1980), Modeling three-dimensional wind-induced flow, Proc. ASCE, Jour. Hydraulics Div., Vol.106, HY11, pp.1843-1865.

Laevastu, T.(1975), Multilayer hydrodynamical-numerical models, Proc. Symp. Modelling Tech., ASCE, p.1010.

Simons, T.J.(1974), Verification of numerical models of Lake Ontario: Part I. Circulation in spring and early summer, Jour. Phys. Oceanogr., Vol.4, p.507.

Simons, T.J.(1975), Verification of numerical models of Lake Ontario: Part II. Stratified circulation and temperature changes, Jour. Phys. Oceanogr., Vol.5, pp.98-110.

Sundermann, J.(1974), A three-dimensional model of a homogeneous estuary, Proc. 14th ICCE, ACSE, pp.2337-2390.

PART III

Coastal Structures

CHAPTER 67

Analysis of practical rubble mounds

Allsop N.W.H.[1], Jones R.J.[2], Besley P.[2] & Franco C.[3]

ABSTRACT

This paper describes results from an unusual research project completed under Topic 3R2 of the European Union's MAST research project G6-S Coastal Structures. In this project, data were collected from the major European hydraulics laboratories on the hydraulic and structural responses of example rubble mound breakwaters and sea walls that each laboratory had previously studied in wave flume or wave basin tests. The main responses considered here are:
- a) Main armour stability, given by measurements of armour movement and/or displacement under wave action.
- b) Wave overtopping, described by the number of waves passing over the structure crest, or by the mean overtopping discharge;

The paper describes some of the analysis of armour stability and hydraulic performance of these structures, and explores the potential to develop general conclusions from ad hoc studies. This paper develops some of the analysis described initially within the G6-S project by Allsop & Franco (1992), but also re-considers and revises some of the early analysis and initial conclusions.

1. OUTLINE OF THE STUDY

Within the MAST I project G6-S Coastal Structures, work under Topic 3R addressed the performance of rubble mound breakwaters. Such structures may be used to protect harbours, cooling water intakes or outfalls, and related areas of coastal development, against wave action. Rubble mound breakwaters are formed by constructing the inner part of the mound, termed the core, from quarried rock. The core is protected against erosion by armour layers, supported by filter or under-layers. The size of the armour is closely related to the height of the design waves. Such structures may include a crown wall, a number of armour and underlayers on the seaward and lee faces. They are usually designed with a number of different levels from foundation and toe layers to crest armouring (Fig 1).

---

[1] Manager Coastal Structures, HR Wallingford, Howbery Park, Wallingford, UK; Professor (associate) University of Sheffield

[2] Engineer, Coastal Structures, HR Wallingford, Howbery Park, Wallingford, UK.

[3] Institute of Hydraulic Engineering, University of Rome

Rubble mounds are used to reduce levels of wave activity by limiting wave overtopping or transmission, and/or to protect against erosion. The degree of wave reduction needed, and hence the hydraulic responses required, depend on the requirements of the harbour or coastal development. The structural design of the breakwater must ensure that it can serve its stated purpose over its full design life, and that damage to the structure is therefore kept below accepted limits.

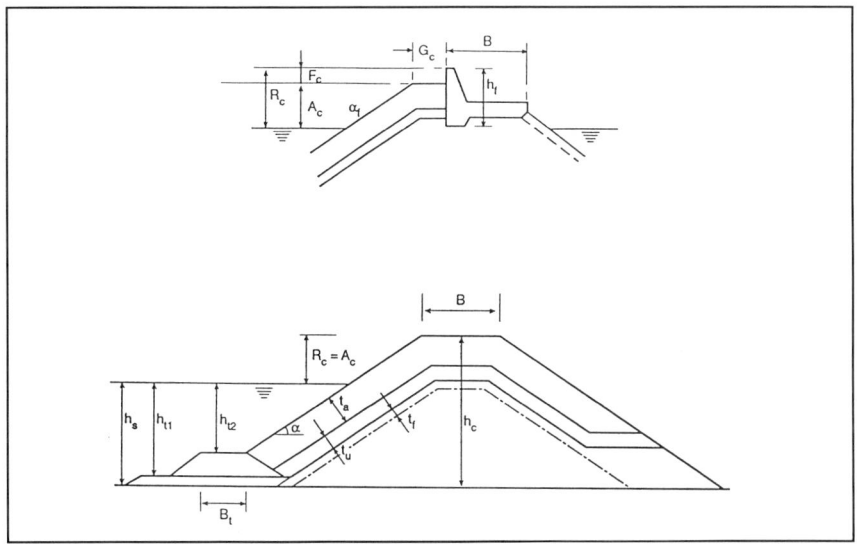

Figure 1   **Rubble mound breakwaters, main geometrical parameters**

The main methods used in the design of rubble mound breakwaters are based on empirical formulae, supported by results from hydraulic model tests. Such methods are therefore derived for simplified structure sections tested under normal wave attack. Very few methods address the stability of structures incorporating complex or "non-standard" details; under oblique wave attack; at and around the outer breakwater end or roundhead; or at junctions with dis-similar construction.

These types of structures are used worldwide wherever quarried rock is available in adequate sizes and quality for construction of coastal structures. Many such structures have however suffered significant damage, with an apparent peak between about 1977 and 1988. In analysing these failures, and difficulties with similar structures, it has become clear that analysis and design methods have been insufficiently reliable. A programme of studies were therefore proposed to the European Union. The programme that was contracted was somewhat restricted, but included key elements of the original proposal.

1.1 Work in G6-S Coastal Structures

The MAST I research project "G6-S Coastal Structures" addressed techniques available for the analysis and design of coastal and harbour structures such as sea walls, revetments, and breakwaters. The research covered three technical topic areas:
       Topic 1.   Wave action on and in coastal structures;

Topic 2. Wave impact loading on vertical structures;
Topic 3B/R. Berm and rubble mound breakwaters.

Within this project, studies under Topic 3R addressed the stability and hydraulic performance of rubble mound breakwaters and sea walls using analysis of previous model test data, described here, and by new model tests described by Galland (1994).

The main objective of the desk study described here was to provide information on the stability and performance of rubble mound breakwaters, where possible at singular points. These include: roundheads; junctions; bends; toe and rock berms. The study was based on the collection of data from study reports from the major hydraulic laboratories in Europe. It was agreed that HR Wallingford would design the study approach; that each laboratory would be responsible for the collation of their own test results; data; and that HR collect together and analyse the results. The objectives of the project were to:
   a) Identify whether data on responses from practical studies could be used to draw general design guidance;
   b) Collect data from many different institutes, and retain in consistent form for analysis;
   c) Contrast data from ad hoc studies with predictions made using methods based on idealised structures;
   d) Identify new or modified design methods based on these data;
   e) Identify gaps in present design information or methods;
   f) Provide justification for further research on rubble mound breakwaters.

The approach taken was firstly to identify the main breakwater failure modes and the principal parameters influencing failure, and to develop standardised parameter definitions and notation. These were used to design a database spreadsheet to hold model test data on breakwater structures and responses. Test results were then collected from each partner in the project. In parallel with this work, the prediction methods for the main design parameters were summarised, see Allsop (1993) and these were then used to devise the analysis strategy.

2 DATA COLLECTED

The data analysis was based on responses measured previously in site specific studies conducted by the institutes. The data used were therefore confined to those aspects of structure performance of concern to the designers of the particular structures, and were limited to those combinations of wave conditions and water levels for which the tests had been conducted.

The principal responses recorded were:
   a) Toe armour stability, given by measurements of toe armour movement and/or displacement;
   b) Main armour stability, again by movement and/or displacement, often using the damage parameters S or $N_{d\%}$.
   c) Wave overtopping, described by the number of waves passing the crest, $N_{wo\%}$, or by the mean overtopping discharge, Q;

Each set of results were combined by response types, and then compared with the simple design methods to test their use, and/or to identify whether new prediction methods could be derived. The data collected in this study were too great to handle as a single spreadsheet. During the analysis process, the larger data sets were split into smaller and increasingly more specific sets, mirrored by the sections within this chapter. Typical

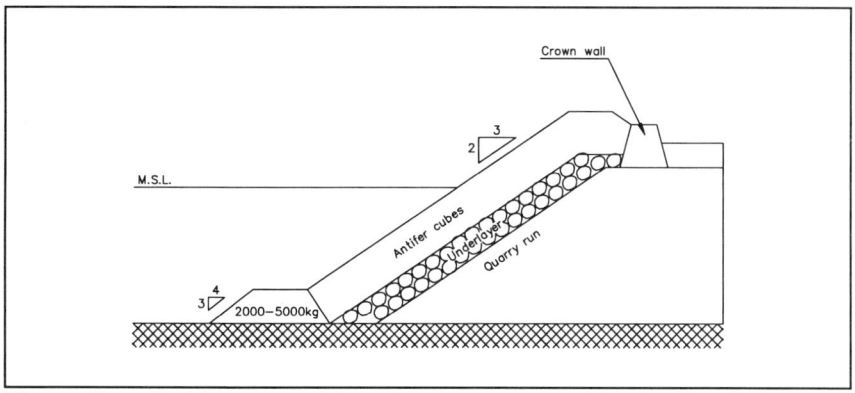

Figure 2 Rubble mound breakwater armoured by Antifer Cubes

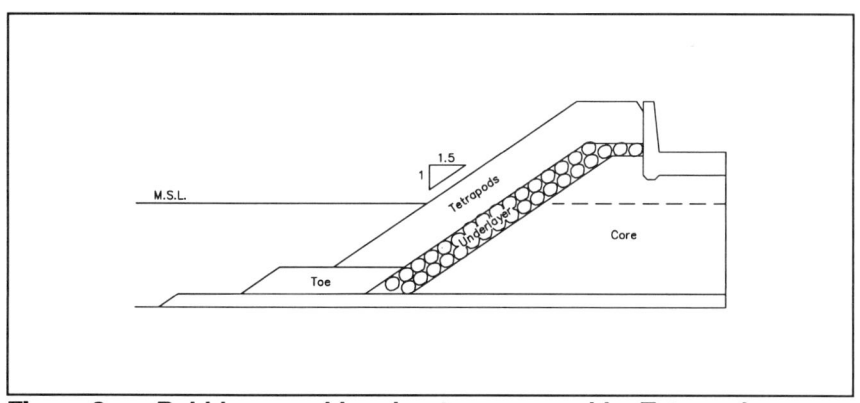

Figure 3 Rubble mound breakwater armoured by Tetrapods

cross-sections through the structure are shown in Figures 2-5. The influences on overtopping and armour damage of armour type, cross-section geometry, and plan configuration have been treated separately. This paper describes only the analysis of armour movement and wave overtopping.

3 ANALYSIS OF ARMOUR MOVEMENTS

Design methods for rubble mound armour layers focus principally on the calculation of the median armour unit mass, M_{50}, or the nominal median stone diameter, D_{n50}, for given levels of armour damage. In most instances damage is defined in terms of erosion area A_e or number (%) of armour units displaced, $N_{d\%}$. Damage may also be described by N_{od} referring to the number of unit displaced related to a width along the breakwater of $1.0D_n$. For a Tetrapod, D_n is 0.65 D where D is the height of the Tetrapod; and for Accropode D_n is 0.7 D. The definition of units displaced N_{od} may be compared with the damage parameter S. Generally S is about 2 N_{od}, but the relationship differs for different armour units laid at different porosities:

For Cubes	$S = 1.8 N_{od} + 0.4$	(1a)
Tetrapods and Accropode	$S = 2 N_{od} + 1$	(1b)

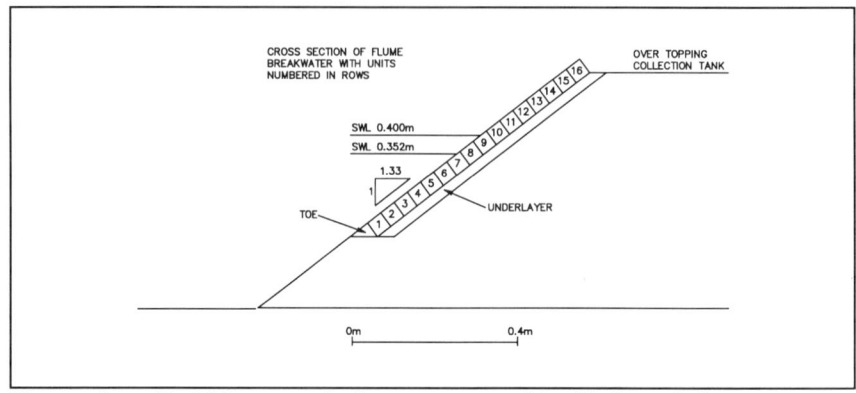

Figure 4 Rubble mound slope armoured by Hollow Cubes

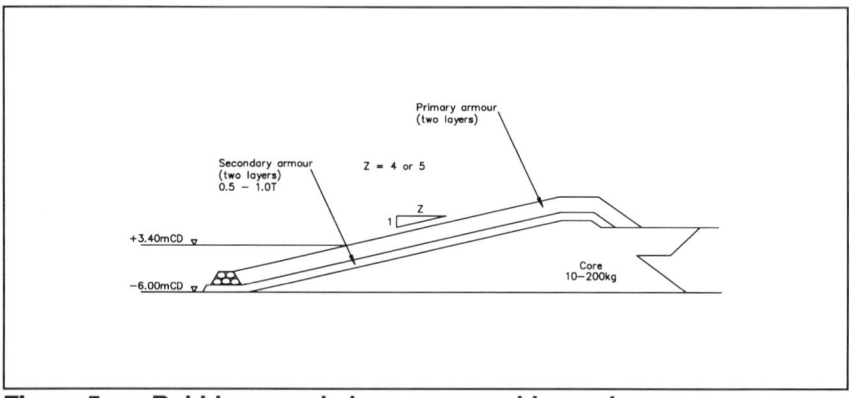

Figure 5 Rubble mound slope armoured by rock

The initial analysis of armour response presented by Allsop & Franco (1992) sought to identify the effects of wave obliquity and trunk versus roundhead on armour damage. Data from tests on structures armoured with Rocks, Tetrapods and Cubes formed the main body of the analysis. The data was collected from diverse sources, and much effort was expended to try to harmonise values of the input parameters for this analysis. Damage was presented in the database as displacement in % of units related to a certain area, or as the damage parameter $S=A_e/D_{n50}^2$ in which A_e is the area eroded around SWL. When the level of displacement was given, this was often divided into classes in relation to the nominal diameter D_{n50} of the armour unit:

N_{D1}: units displaced less than $0.5D_{n50}$
N_{D2}: units displaced more than $0.5D_{n50}$ and less than $1D_{n50}$
N_{D3}: units displaced more than $1D_{n50}$

Sometimes the number of rocking units were also given. Comparisons with other data sets and with predictions methods demanded that damage be presented in a consistent way, and in this analysis, damage was always defined by S. Unfortunately, values of damage have not often been expressed as S in the past, so an alternative approach was needed. In the initial analysis, a simple method was suggested to relate damage in these various classes to S using:

$$S = 0.8 (0.25N_{D1} + 0.75N_{D2} + 1.0N_{D3}) \tag{2}$$

Some of the data sets contain information on very small movement, perhaps as small as $0.1D_{n50}$, but these were not included in this anlysis. In many cases the collaborating laboratories themselves combined categories N_{D2} and N_{D3}.

A damage formula developed by Van der Meer (1988) from the Hudson formula was used to compare damage data with values of Hudson's stability coefficient, K_D:

$$H_s/\Delta D_{n50} = a (K_D \cot\alpha)^{1/3} S^b \tag{3}$$

where for rock armour   a=0.67, b=0.16
and for Tetrapods and Cubes   a=0.69, b=0.14

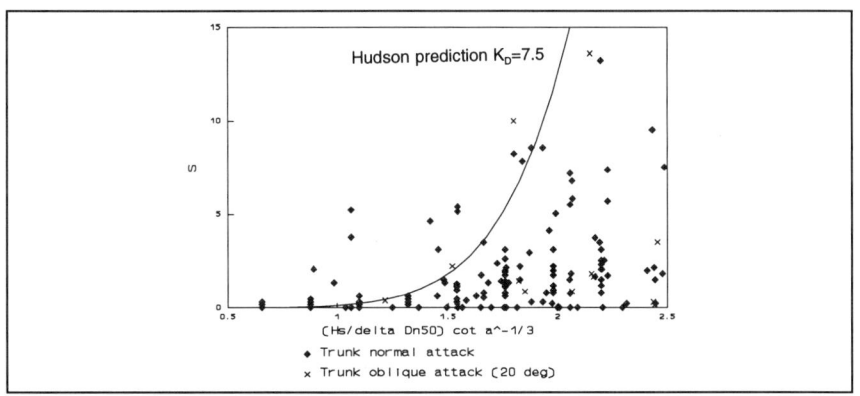

Figure 6   Stability of cube armour, analysis includes movement in categories N_{D2} and N_{D3}

The stability of cube armour is illustrated in Figure 6, which shows the damage S against $H_s/\Delta D_{n50} \cot\alpha^{-0.333}$. The plot shows a great deal of scatter with many data points above the prediction line using $K_D=7.5$. Analysing measurements in category N_{D3} only, it was clear that relatively few small movements had been recorded, and the damage vaules calculated for N_{D3} only lie in very similar positions.

Designers of breakwaters using concrete armour units have been more rigorous in requiring more information on small movements in recent years, particularly for slender units. The use of these small categories in calculating an equivalent value of S less certain than suggested in eqn (2). It is probable that the use of $N_d\%$ alone to calculate S will under-estimate the damage, but the simple method used by Allsop & Franco (1992) to include the influence of small movements appears to lead to significant over-estimates of damage compared to existing prediction methods. This is particularly so where cumulative damage is estimated by summing damage in individual tests. This problem was illustrated when considering damage to Tetrapod armoured slopes.

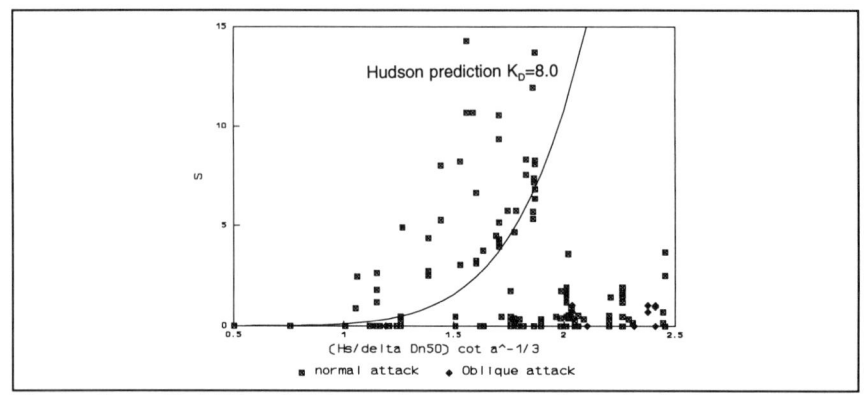

Figure 7 Stability of Tetrapod armour, analysis includes movement in Categories N_{D2} and N_{D3}

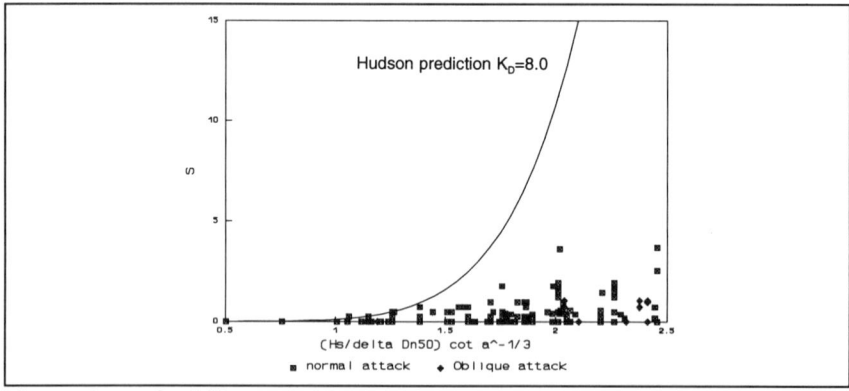

Figure 8 Stability of Tetrapod armour, analysis only includes movement in category N_{D3}

Damage to Tetrapods was also analysed, with damage categories N_{D2} and N_{D3} used to calculate S, plotted in Figure 7. Again the damage appears to be much greater than predicted by $K_D=8.0$. The affect of plotting points in category N_{D3} only is shown in Figure 8, which shows that the contribution of the smaller movements had a substantial influence on the comparison. Now all the data point lie below the prediction curve.

This exercise leads to a rather disappointing, but not altogether surprising conclusion. Unless the damage definitions used in design formulae precisely reflect those used by design engineers in site specific studies, a correlation of data from these studies with simple formulae may lead to considerable uncertainty, if not confusion. This is however to be expected, as the sophisticated designer and experienced modeller should be expected to use more sophisticated descriptions of structure response than those appropriate for simple design formulae.

4 ANALYSIS OF OVERTOPPING

Wave overtopping may be described by the number or percentage of waves passing over the crest expressed as N_{wo}; or by the mean overtopping discharge per unit length, Q. The data returned seldom identified both responses, so analysis had to concentrate on the two sets of data separately. Most recent research has been concentrated on the prediction of the mean overtopping discharge Q, so most of this section will address this response. Some initial work was however completed on analysing the data returns that only gave the number of waves overtopping, expressed as N_{wo}.

4.1 Number of waves overtopping, N_{wo}.

The test data examined in this study were limited to structures under normal wave attack. Structures were constructed with crown wall elevations equal to or below the front armour crest level. Four sets of data for which the number of waves overtopping had been recorded were analysed.

Example results for slopes armoured with Cubes and Tetrapods under waves of constant steepness of s_m=0.030 were analysed by plotting $\ln(N_{wo\%}/100)$ against R^{*2} as derived by Owen (1980, 1982). The scatter of the data on $N_{wo\%}$ was wide, even when restricted to a single sea steepness. Agreement between measured and predicted values were not good for the Cube structure. A better agreement was found for the Tetrapod armoured structure. The methods used to predict the number of overtopping waves is described by Allsop and Franco (1992). It was concluded that there was little to be gained by extending the analysis. These uncertainties confirmed that the mean overtopping discharge Q gives a more reliable description of overtopping of such structures.

4.2 Mean overtopping discharge, $\overline{Q}$.

The main aim of this analysis was to examine the influence on overtopping discharges of singular points such as crown wall element geometries, armour crest levels, and slope configurations. The overtopping performance of structures armoured with rock, Antifer cubes, Tetrapods, and high-porosity Hollow Cubes, were examined.

The mean wave overtopping discharge depends on freeboard R_c, H_s and T_m. The prediction method developed by Owen (1980) relates dimensionless parameters Q* and R* by an exponential equation with a roughness coefficient, r, and coefficients A and B for each slope angle:

$$Q^* = A \exp(-B R^* / r) \qquad (4)$$

Where $Q^*=\overline{Q}/(gT_m H_s)$ and $R^*=R_c/T_m(gH_s)^{0.5}$, and for smooth slopes, r = 1.0, and values of A & B have been derived for slopes from 1:1.0 to 1:4.0:

Slope	A	B
1:1.0	0.0079	20.1
1:1.5	0.0102	20.1
1:2.0	0.0125	22.1
1:3.0	0.0163	31.9
1:4.0	0.0192	47.0

Table 1 Values of A and B for smooth slopes, r=1

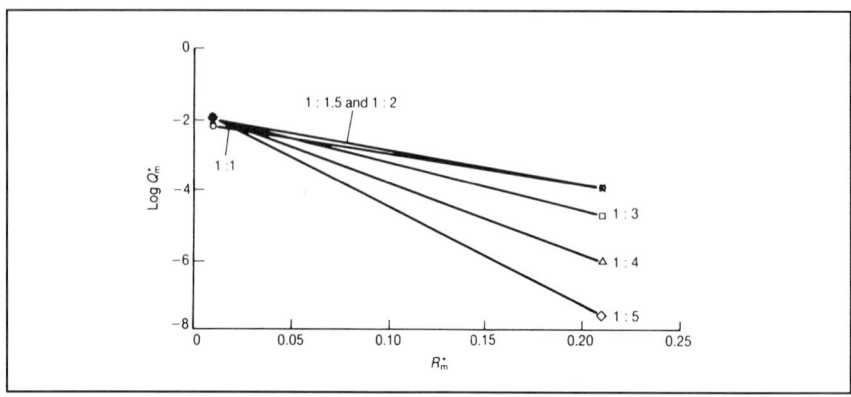

Figure 9 Overtopping of smooth straight slopes (Owen 1980)

A design graph such as that shown in Figure 9 can be compiled by plotting Q* as a function of R* and using the constants A and B described in Table 1. For structures with small relative freeboards and/or large wave heights, the regression lines come together at one point, indicating that the slope angle, and relative roughness are no longer effective in controlling the overtopping discharge. The discharge characteristics for slopes 1:1, 1:1.15 and 1:2 are very similar, but overtopping reduces significantly for slope angles less than 1:2.

Owen's method was developed initially from results for smooth slopes only, but the use of the roughness factor, r, allowed its extrapolation to study the overtopping performance of rough, and even armoured slopes. Since 1980, various researchers have explored alternative prediction methods for armoured slopes, see Bradbury & Allsop (1988) and Aminti & Franco (1988), but no new method has proved any more reliable. The advantage of Owen's method is its simplicity, and the ready availability of data to support particular coefficient values. Three alternative approaches have therefore been developed:
- a) Use Owen's method and coefficients A and B with r derived from tests with the correct slope geometry;
- b) Use Owen's general equation, but with new values of A and B derived for similar cross section, and r = 1.0;
- c) Develop alternative equation, with new coefficients for that section.

The overtopping performance of armoured breakwater structures with and without crown walls have been studied. Owen (1980, 1982) showed that in relatively shallow water berms or beaches in front of a structure will reduce overtopping. The toe design of the structures vary somewhat, but it is likely that these differences will have little bearing on the overtopping discharge during the deep water test analysed here. Although the crest detail of the various structures was not identical, the crown wall level was generally equal or below the armour crest, so it might be expected that the variations of crest detail would have had little affect on the overtopping discharge.

Measurements show that there is a good relationship between Q* and R* for all the structures studied. Data for Tetrapod and Antifer cube armoured structures are shown in Figures 10 and 11 respectively. The data presented here show that for the rough armour

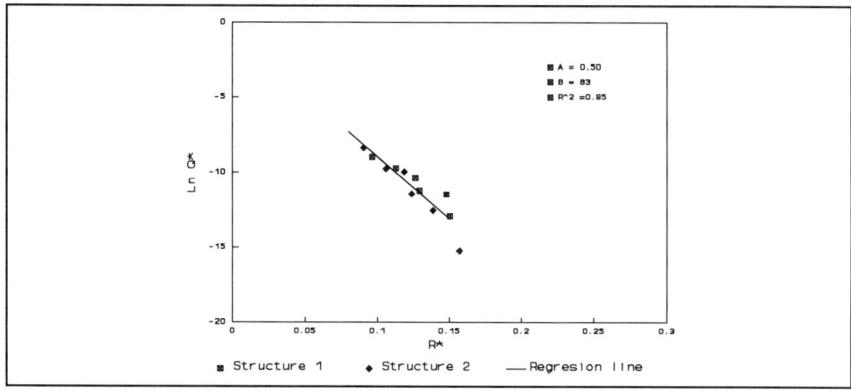

Figure 10 Overtopping of Antifer Cube armoured structure

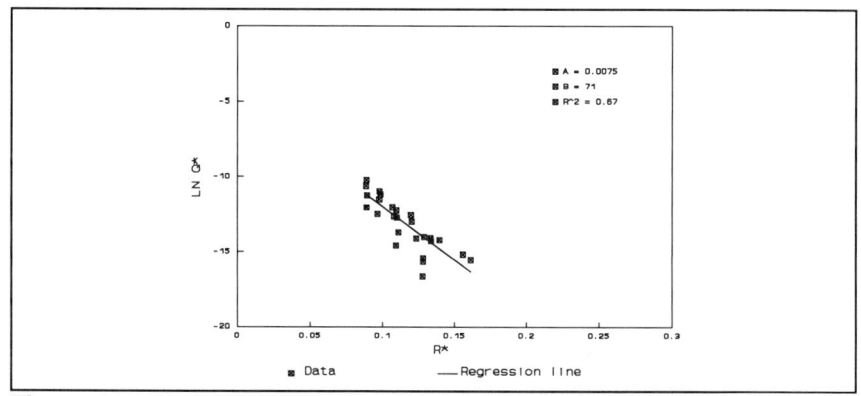

Figure 11 Overtopping of Tetrapod armoured structure

structures analysed the overtopping performance is better described by modifying the coefficients A and B. The overtopping data for rough structures shows that regression lines passing through data points have varying steepness. This change in steepness is due to the increased turbulence and friction caused by the 'rough armour'. The regression lines cross the R* axis at different points depending on the armour type. This is not taken into account in the earlier theory as the coefficient A remains constant for a constant slope angle. This suggests that the hydraulic responses cannot be represented correctly by the roughness coefficient alone.

For armoured slopes, it is therefore suggested that the original Owen formula equation (4) should be used for overtopping, but that the coefficients A and B should be changed depending on the armour type and structure slope. The original Owen method using values of the roughness coefficient is not as accurate as using regression lines for site specific data. The simple Owen method is however very quick and easy to use where little site specific data is available.

The values of the coefficients A and B for rough slopes analysed during the study are tabulated below. The regression lines of the Tetrapod and Antifer cube structures are shown in Figures 10 and 11.

Structure	Slope	A	B
Cob units	1:1.33	0.00839	46.5
Shed units	1:1.33	0.00268	29.9
Antifer Cubes	1:1.5	0.496	82.7
Efficient units	1:1.5	0.016	60.0
Tetrapod	1:1.5	0.0075	71.0

Table 2   **Values of A and B for armoured structures, r=1**

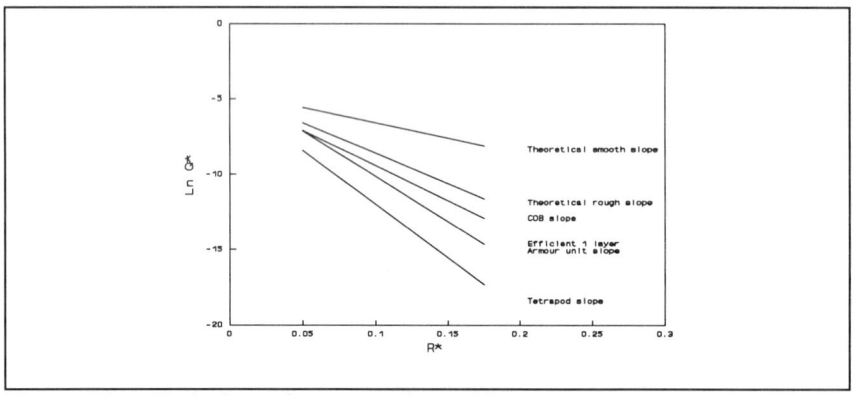

Figure 12   **Variation of overtopping with armour type**

The overtopping performance of 1:1.5 sloping structures armoured with Tetrapod, Antifer cube and efficient single layer armour units are compared in Figure 12. For similar armour units and structure designs it may be expected that for a given R* the overtopping discharge would be equal. However, Figure 12 shows that as the porosity of the armour units increases, overtopping discharge decreases. The two layer Tetrapod system performs significantly better than the single layer structures. The armour efficiency increases as the relative freeboard increases.

The effect of armour layout is shown in Figure 13. This figure describes the overtopping performance of a single layer and a double layer hollow cube armour system. The armour units had a porosity of about 60%, and were placed to a tight patter on a slope of 1:1.3. The overtopping is reduced for the two layer structure, but not as effectively as on the much thicker Tetrapod armour.

So far only simple structures have been considered. A similar method can be used to assess structures with berms or crest detail. Two test series have been carried out on two layer rock structures and the overtopping performance is shown in Figure 14. The data at the crest of the 1:4 slope shows a slightly lower overtopping discharge compared with the Owen theoretical rough slope. The 1:5 slope crest data shows a small increase

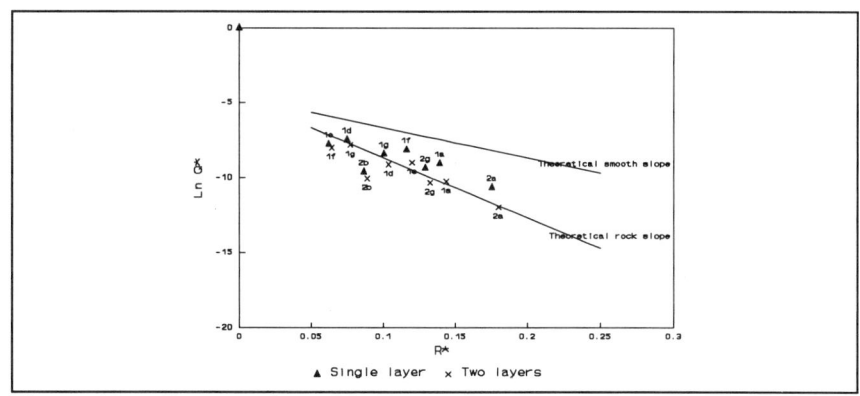

Figure 13   Overtopping of hollow cube armoured structure

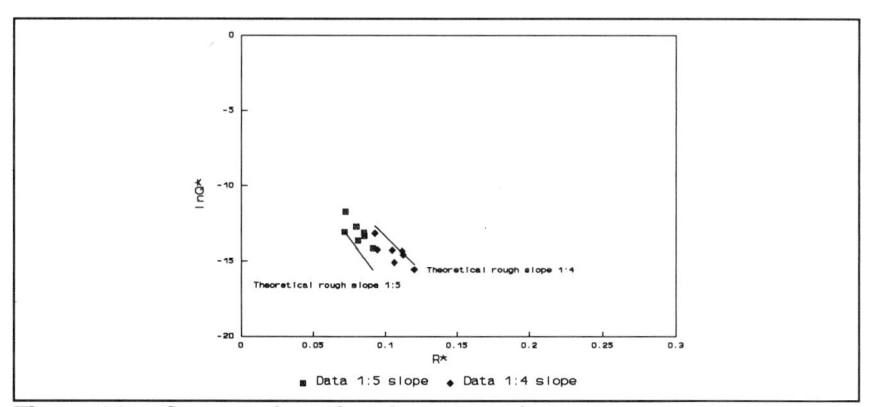

Figure 14   Overtopping of rock armoured structure

in overtopping discharge compared with the Owen theoretical rock slope r=0.5. Within the accuracy of the measurement, the prediction lines show a relatively good fit for the rock armour slopes with shallow slope angles.

The data shown in Figure 14 suggests that the roughness coefficient may be affected by the slope of the structure and the wave conditions. A value of r may be calculated for the measured discharges using Owen's value of A and B for a given slope angle. When the r value is plotted against the Iribarren number I_{rm}, the value of r decreases as I_{rm} increases, this relationship is shown in Figure 15. The result is consistent with the conclusions described earlier noting that both A and B values need to be modified when investigating the overtopping of permeable seawalls when using the Owen formula for structures armoured with 'rough' material.

For the same wave conditions overtopping discharges 10m behind the crest were about one-tenth of those measured at the crest. The crest detail of the 1:4 slope structure was

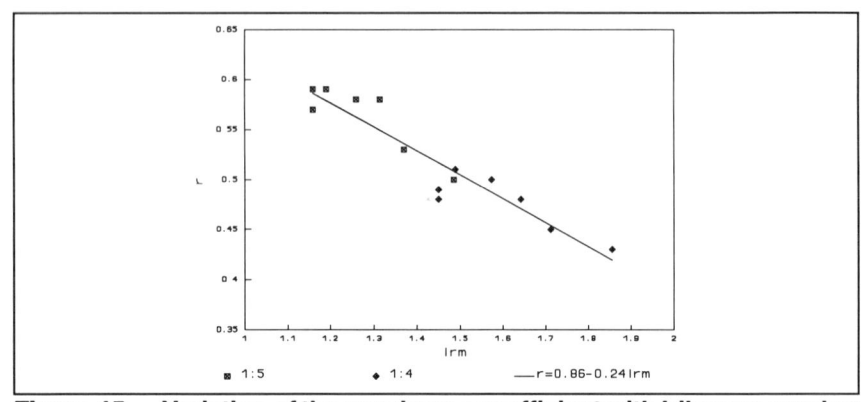

Figure 15 Variation of the roughness coefficient with Iribarren number for rock armoured structures

altered to include a wave return wall. The armour crest freeboard A_c = 6.8m and the wave wall freeboard R_c = 8.8m. The performance of this structure can be compared with the structure with no wave wall, $R_c = A_c$ = 8.8m. Although the inclusion of a wave wall shows a higher overtopping discharge compared with the full length armour slope, the wave wall design does have it's advantages. The number of units and the volume of underlayer material required to build the structure is reduced and the structure may therefore be cheaper to build.

Where values of A and B cannot be calculated using site specific data, the analysis has shown that the original Owen formula with values of A and B for various slopes can be used with a roughness coefficient appropriate for the armour concerned. Values of the roughness coefficient for various armour units are given in Table 3.

Armour type	r
Rock	0.5-0.6
Hollow cubes	0.5
Dolos	0.4
Stabits	0.35
Tetrapods	0.3

Table 3 Recommended values of r for armoured structures using A and B values given in Table 1

5 CONCLUSIONS

<u>Overall</u> The spreadsheet database worked adequately, but needed plans and sections to convey important information.

<u>Armour on trunks and roundheads</u> Analysis of damage was complex. Initial comparisons show wide scatter, with many tests showing little damage when prediction methods suggest severe damage. Cumulative damage is not given by design methods.

Overtopping Few studies recorded both $N_{wo\%}$ and Q. Selected studies gave data on overtopping allowing new coefficients to be derived. The original Owen formula should be used for rough slopes, but both the coefficients A and B must be changed depending on the armour type and structure slope. The original Owen method is not as accurate as using regression lines for site specific data, but is simple to use where little site specific data is available.

ACKNOWLEDGEMENTS

This research study was part-supported under the European Union's MAST research programme. Additional support was provided by Delft Hydraulics, Danish Hydraulics Institute, Laboratoire National d'Hydraulique, CEPYC in Madrid, and HR Wallingford. The model test results analysed in this report were supplied by the partner laboratories of G6-S. Their work was not limited to supplying the original data returns, but in supplying supplementary information and advice during the analysis period.

The analysis of test data was conducted by Claudio Franco on research attachment to HR Wallingford from the Hydraulics Department of the University of Rome, assisted the Coastal Structure Section at HR Wallingford. The preparation of this paper was supported by HR Wallingford and the University of Sheffield.

REFERENCES

Allsop N.W.H. (1993) "Formulae for rubble mound breakwater failure modes" Final report of PIANC, PTC II, Report of Sub-grou A of Working group 12, PIANC, Brussels, April 1993

Allsop NWH & Franco C (1992) "MAST G6-S Coastal Structures Topic 3R: Performance of rubble mound breakwaters singular points" Paper 3.12 to G6-S Final Overall Workshop, Lisbon, November 1992

Aminti P.L. & Franco L. (1988) "Wave overtopping on rubble mound breakwaters". Proc 21st ICCE, Malaga, ASCE, June 1988.

Bradbury A.P. & Allsop N.W.H. (1988) "Hydraulic effects of breakwater crown walls" Proc Conference Breakwaters '88, ICE, Eastbourne, May 1988

CERC (1984) Coastal Engineering Research Centre. "Shore Protection Manual", Vols I-II, US Gov Printing Off, Washington, 4th edition 1984

CIRIA (1991) Simm J.D. (Ed) "Manual on the use of rock in coastal and shoreline engineering" CIRIA Special Publication 83 / CUR Report 154, November 1991

Galland (1994) "Rubble mound breakwaters stability under oblique waves: an experimental study" 24th ICCE, Kobe, ASCE, October 1994

Owen M.W. (1980) "Design of seawalls allowing for wave overtopping" Report EX 924, HR Wallingford, June 1980.

Owen M.W. (1982) "Overtopping of sea defences" Proc. Conf. Hydraulic Modelling of Civil Engineering Structures, BHRA, Coventry, September 1982.

Van der Meer JW. (1988) "Rock slopes and gravel beaches under wave attack", PhD thesis Delft University of Technology, April 1988.

CHAPTER 68

Friction and clamping forces in wave loaded placed block revetments

Adam Bezuijen[*]

Abstract

A calculation method is presented to determine the strength of a block revetment against wave loading. The critical loading on such a revetment is present during wave run-down when still relatively high pore pressures exist in the filter layer and the blocks can be pushed out of the revetment. The strength of placed block revetments against this loading is not only composed of the weight of the blocks, but also of the friction and clamping forces. Friction and clamping forces are incorporated in a calculation method, using a simplified model that takes into account the horizontal and vertical forces separately. Example calculations show the forces that can be expected in such a cover layer.

1 Introduction

The placed block revetment is a commonly used type of revetment in the Netherlands to protect the dikes around the estuaries against wave attack. A typical cross-section of such a revetment is shown in Figure 1. In a long term research project the failure mechanisms for this type of revetment were investigated, calculation models were developed to simulate the loading on the revetment (Bezuijen et al, 1990; Burger et al, 1990; Bezuijen & Klein Breteler, 1992). Furthermore the strength of the dike itself after damage to the revetment was studied (Rigter, 1994). Results of this research have been used for the design of revetments, see Bezuijen et al. (1988) for an example.

The principal loading on such a revetment is different from the loading on for example a rip-rap revetment. Damage to a placed block revetment is not caused by severe water motion on the revetment, but by the pore pressures present in the filter layer. At wave run-down the water level in the filter layer is considerably higher than on the cover layer. Furthermore flow in the filter layer underneath the next incoming wave causes high pore pressures in the filter layer. The combination of a high water level and high pore pressures in the filter layer leads to an uplift pressure underneath the blocks just in front of

---
[*]Delft Geotechnics, P.O. Box 69, 2600 AB Delft, The Netherlands

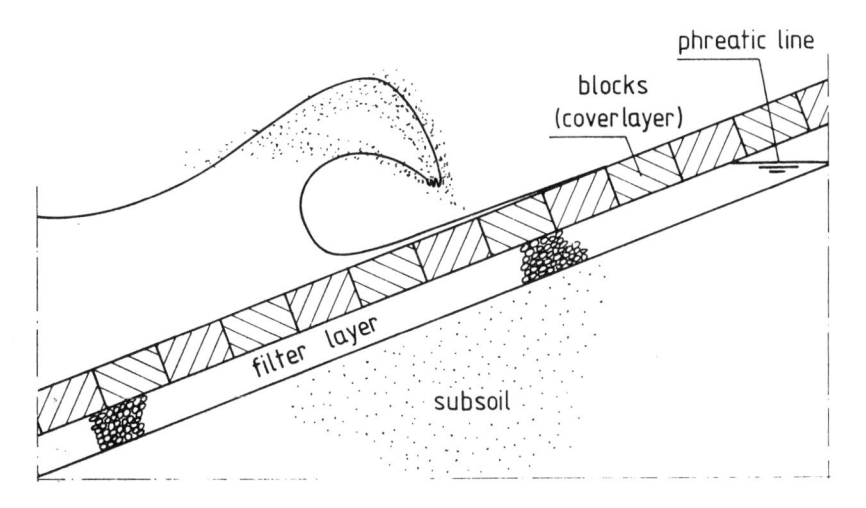

Figure 1: Typical cross-section of a placed block revetment with wave loading.

the next incoming wave, the situation as sketched in Figure 1. The strength of the revetment must be large enough to resist this uplift pressure.

Field measurements and large scale model tests have shown that the weight of a single block in a placed block revetment is often not sufficient to resist the maximum uplift pressure that is recorded underneath the blocks during severe wave attack. The strength of the revetment against lifting of blocks however, is also determined by the clamping and friction forces between the blocks. Only a small area of the revetment is loaded with the maximum uplift pressure and due to the clamping and friction forces the maximum load can be distributed over a larger area of the block revetment. Some calculation models, see for example Burger et all (1990), include the influence of friction between the blocks, and include the influence of block movement and inertia (Townson, 1988 and Bezuijen et all, 1990). However, these models are not capable to explain the high pull out forces, necessary to lift one block out of the revetment, as measured during field tests (Stoutjesdijk et all, 1992). The reason is that the influence of clamping was not considered. If the friction forces between two blocks exceed a certain limit, then one moving block causes rotation of adjacent blocks. Since the possibilities of rotation of blocks are rather limited in a placed block revetment, the result of such a rotation will be a force in the plane of the slope, see Figure 2, leading to higher friction forces and in this way increasing the strength of the revetment considerably.

2 Theory, strength of revetment

During wave attack a high load will be present on a part of the block revetment. How this area is calculated will be dealt with in the next section. This area can

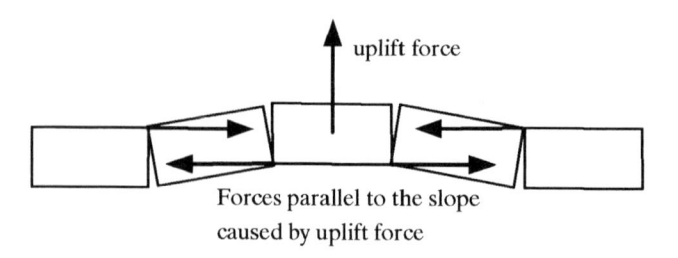

Figure 2: Sketch how forces develop in a row of blocks when one block is lifted out of the revetment.

cover several rows a and columns (the blocks on the same level in a horizontal line along the revetment are called a row in this paper and the blocks to be seen in a cross-section are called a column). When uplift force exceeds the weight of the blocks, there will be some, very small, movement causing the development of friction and clamping forces. Whether or not this situation is unstable depends on the friction and clamping forces between the blocks. The influence of the friction forces also depends on the area of blocks that is lifted. If this is a large area, than the influence of the friction forces becomes less, because the friction forces only act on the boundaries of the loaded area.

It appeared from the field tests that the clamping forces can be very high. The question is, what conditions are necessary for clamping forces to develop. There is no need for a detailed calculation of deformations, because it is known that for blocks of 0.5x0.5x0.25 m^3 with a self weight of 85 kg under water, in case of good clamping a lift force of 900 kg leads to a vertical displacement of less than 25 mm (Stoutjesdijk, 1992). Furthermore such a calculation is only possible with detailed knowledge of the flexibility of the joints, the width of the joints and the exact dimensions of all blocks.

Therefore the calculation model described in this paper only takes into account the forces acting on the blocks. The deformations are disregarded. Another simplification is that either forces in a horizontal row of blocks are calculated, or in a column along the slope from top to the lower end of the revetment. This means that, as another simplification, the situation of blocks shifted half a block on each row is not taken into account.

When blocks in a row are loaded with a lift force higher than the weight of a block, as shown in Figure 2, some movement will occur. This movement is assumed to be very small, but the contact between the block and the subsoil will be lost. Stability is only possible if the total lift force, caused by the excess pore pressure in the filter layer, can be compensated by the total weight of the blocks and the number of blocks without contact with the subsoil will be larger than the number of blocks with a lift force exceeding the weight of the block.

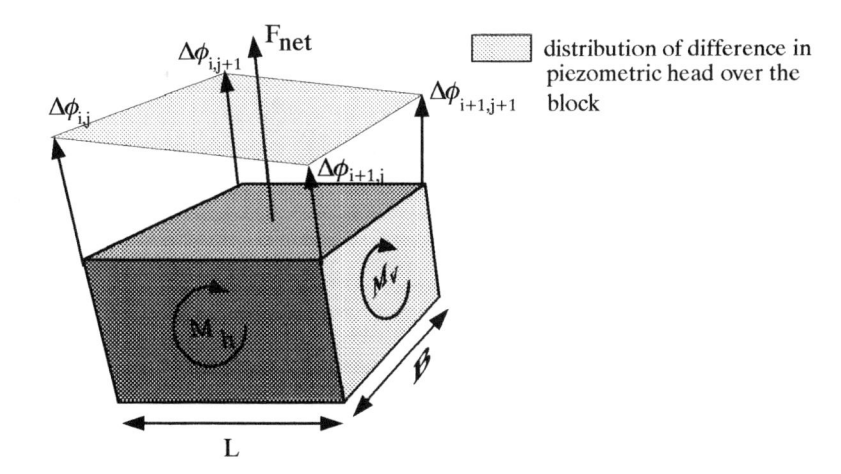

Figure 3: Definition sketch Forces and momentum on a block.

The calculation method assumes that the piezometric head is known on the corners of the block and that there is a linear distribution of the piezometric head over the block, see Figure 3. This leads to the following relation for the force on the block and the horizontal and vertical momentum:

$$F_{net} = \rho_w g BL \frac{\Delta\phi_{i,j} + \Delta\phi_{i+1,j} + \Delta\phi_{i,j+1} + \Delta\phi_{i+1,i+1}}{4} - (\rho_b - \rho_w) g DBL \cos\alpha \quad (1)$$

$$M_v = \frac{1}{12}\rho_w g LB^2 \frac{\Delta\phi_{i,j} + \Delta\phi_{i+1,j} - \Delta\phi_{i,j+1} - \Delta\phi_{i+1,j+1}}{2} \quad (2)$$

$$M_h = \frac{1}{12}\rho_w g BL^2 \frac{\Delta\phi_{i,j} + \Delta\phi_{i,j+1} - \Delta\phi_{i+1,j} - \Delta\phi_{i+1,j+1}}{2} \quad (3)$$

Where F_{net} is the "net" force on the block. If this force is larger than zero the block will be unstable in a situation without friction or clamping. M_v and M_h is the momentum in vertical and horizontal direction respectively, see also Figure 3. D is the thickness of the block and L and B the length and width of the block, B is in the direction of the slope, see Figure 4. $\Delta\phi_{i,j}$ is the difference in piezometric head over the block, expressed in metres water. Finally ρ_b and ρ_w is the density of the blocks and water respectively. With equation (1) it can be calculated whether or not the uplift pressure is higher than the weight of the block. If it is higher, then stability can only be obtained by friction and clamping forces.

Whether or not clamping will occur depends on the friction coefficient between the blocks and the initially available pre-stress. In a row the pre-stress is most important. In a column the forces along the slope are composed of

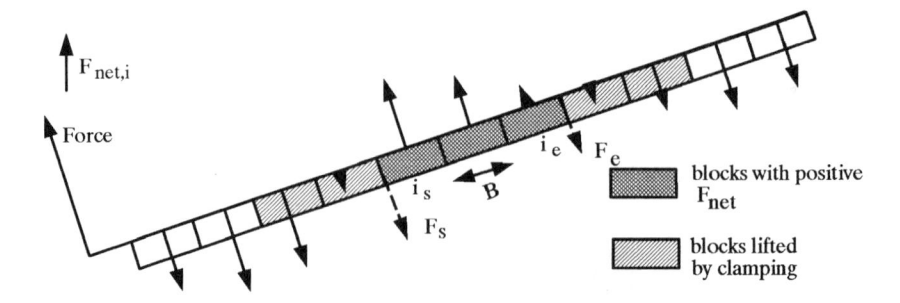

Figure 4: Example of the distribution of forces over a column of blocks for the determination of F_s and F_e. Explanation F_s and F_e see text.

the weight component along the slope of some blocks above the most loaded row. During storm conditions some blocks above the most loaded row will slide downwards when the friction coefficient with the subsoil is exceeded. With a given friction coefficient and load distribution on the blocks, it can be calculated whether the blocks will slide along each other or that clamping will occur.

As is shown in Figure 2, block movement with clamping leads to forces in the plane of the cover layer. In a horizontal row these forces can become very large. However, in a column these forces are limited by the weight of the blocks above the loaded block and some friction between the blocks and the subsoil. When the calculated force parallel to the slope becomes too large, the blocks will be pushed up along the slope and failure will occur. To calculate the friction forces it is essential to know the horizontal force that develops between the blocks. In the situation of Figure 2 the clamping force is a function of the loading and this situation will be calculated. Looking at a number of blocks in a column, see Figure 4, equation (1) can be used to determine the number of blocks with an uplift pressure higher than the weight of the blocks and the total force that is not compensated by the weight of the blocks that are "lifted". This force has to be compensated by the blocks next to lifted blocks. This means that due to friction there must be two forces next to the "lifted" blocks together as large as the sum of the net uplift forces but with the direction opposite to the uplift forces, see Figure 4. "Lifted" is put between quotes because the blocks are not really lifted but only the contact with the subsoil is lost (grain stress is zero).

Assuming that the force in the plane of the blocks is constant, which is nearly correct because the forces due to the weight are in most cases much smaller than the forces due to clamping, the forces F_s and F_e can be calculated using the momentum equation. Using the point of application of the force F_s as the point from which the momentum is calculated, the momentum equation reads:

$$(i_e - i_s + 1)BF_e - \sum_{i=i_s}^{i_e}(F_{net,i}(i - i_s + 0.5)B + M_{v,i}) = 0 \qquad (4)$$

A comparable equation can be derived for F_s. According to "action is reaction" the forces F_s and F_e at the end of the blocks with an uplift pressure higher than corresponding to the weight of the block will lead to an equal force with opposite sign in the block next to it. With these forces the total number of blocks that start to move can be calculated. A block will loose contact with the subsoil as long as:

$$\sum_{i=i_e}^{i} F_{net,i} - F_e > 0 \qquad (5)$$

for $i_n > i_e$. Again a comparable equation can be derived for F_s. F_s or F_e will be the largest vertical force in the revetment near a joint. Let's assume F_s is the largest, then the clamping force (F_{vm}) in the cover layer will be approximately:

$$F_{vm} = \frac{B}{D} F_s \qquad (6)$$

In the computer program this is calculated more accurately taking into account the momentum (M_v) and net force (F_{net}) on each block. This force parallel to the slope can be compared with the maximum possible force in the column, the weight component of the blocks along the slope and the friction of these blocks with the subsoil. If this force is exceeded, then the blocks will be pushed upward and failure will occur.

The same calculation principle can be used to calculate the force in a row. However, no minimum force in the revetment row before clamping starts can be calculated. This depends strongly on the way the blocks are placed. There can be loose blocks in a row without horizontal friction and consequently no clamping. Furthermore it is not possible to determine a maximum clamping force. In principle this force can be very large.

3 Loading on revetment

Calculation of the clamping forces is only possible if the load distribution on a placed block revetment is known. This load is the uplift pressure over the blocks. The computer program STEEN3D (Bezuijen, 1992) calculates uplift pressure over an area of the block revetment. This finite difference program takes into account the turbulent flow conditions that will mostly occur in the filter layer and through the cover layer. To perform a calculation, the wave pressure distribution on the revetment at a certain moment, the geometry and the permeabilities of cover layer and filter layer must be known. An example of such a wave distribution, for regular waves measured in a wave basin, is presented in Figure 5. The resulting calculated distribution of the uplift pressure is presented in Figure 6.

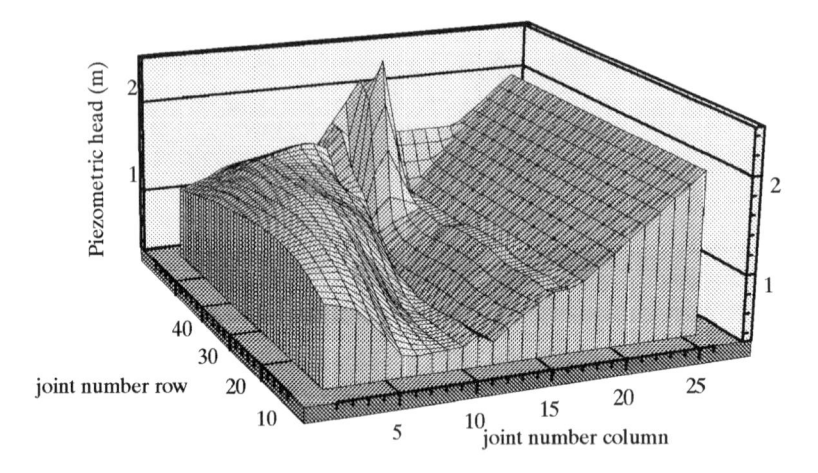

Figure 5: Example of measured distribution of piezometric head on the slope. Oblique wave with an angle of incidence of 30° from the line perpendicular to the slope. Wave height is 1.2 m, wave period 1.4 s.

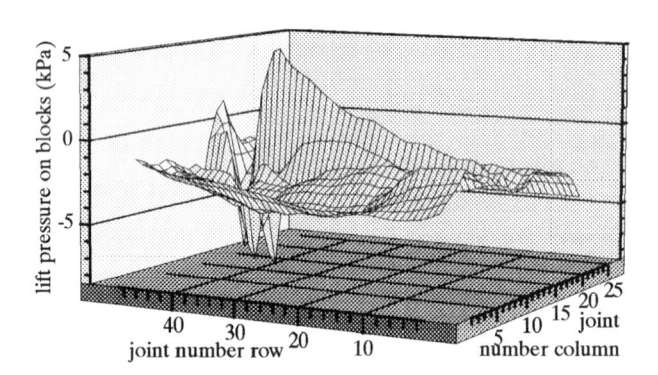

Figure 6: Calculated uplift pressure distribution for a wave loaded block revetment. For positive loading the revetment is potentially unstable.

In this figure the pressure corresponding to the weight of the blocks is subtracted from the calculated pressure, which means that a pressure higher than 0 is a potentially unstable situation. The calculation is performed for blocks of 0.5x0.5x0.2 m^3, a wave height of 1.2 m and a leakage length of 1.1 m. The leakage length (Λ) is defined as $\Lambda = \sqrt{bDk/k'}$, where b is the thickness of the filter layer, D the thickness of the blocks, k the permeability of the filter layer and k' the permeability of the cover layer. Comparing Figure 5 and Figure 6 it appears that the calculated uplift pressures are at maximum before the wave impact.

4 Computer model BLOKKEN

The computer model BLOKKEN, made to evaluate the equations presented in section 2, uses a pressure distribution as shown in Figure 6 to calculate the force on individual blocks. For each row and column the blocks are selected with a lift force higher than the weight of the block and it is checked whether or not the friction forces between the blocks are large enough to cause initial movement of the blocks next to the blocks with the high lift force. The horizontal force in the the revetment is calculated as explained in section 2. From Figure 6 it appeared, that the pressure distribution in horizontal direction differs from the loading in vertical direction. In horizontal direction there is only one row that is really loaded (the 'crest' in the figure). Looking at different columns the loading is comparable. This means that the stability of the maximum loaded row is increased by the friction forces. If on the other hand a column is loaded to failure, then the neighbouring columns will have a comparable loading and the loading can only be resisted by the strength of the column itself. The program takes this into account by assuming a friction force between the different rows but not between the columns. The program is written as a post-processing program on the results of the STEEN3D program. This has the advantage that for a calculated pressure distribution the reaction of the blocks can be calculated for different weights of the blocks or different friction coefficients. Figure 7 is a result of a calculation with the model along a horizontal row. This calculation is performed for the row with the maximum loading from Figure 6. In this case the friction with other rows is assumed to be zero. The figure shows clearly that the blocks with a positive lift force are stabilized by blocks with a negative lift force. For all of these blocks the grain stress with the filter layer will be zero. Note the high clamping force parallel to the cover layer of the revetment necessary to obtain a stable situation, 11.55 kN, approximately 10 times the weight of a single block. The situation shown in this figure can only exist if the initial friction force between the blocks is large enough to make sure that one moving block will cause movement of other blocks in the row. In Figure 7 the loading on a row is presented. However, the result of a calculation presents the situation for the blocks in an area of the revetment, see the Figures 8 and 9 to be explained in the next section.

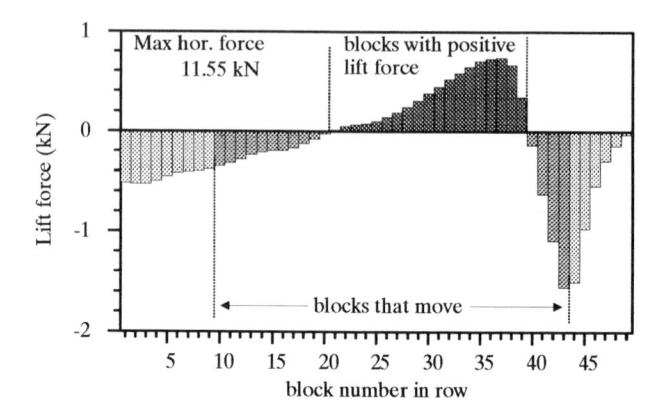

Figure 7: loading on blocks and lifted blocks.

5 Example calculations

Calculations will be presented for two situations. The first example is a typical revetment in an estuary. Calculations were run to show the influence of the leakage length on the loading on the revetment. The second example shows the influence of a non permeable transition in a revetment on the loading and strength. These These calculations were performed as a part of a study for a revetment in the southern part of Terschelling, one of the Dutch Wadden islands. Calculations were made to advise on the stability of this revetment. For these calculations it was not possible to simulate the revetment exactly, because wave pressure measurements were not available this location.

For the first example, calculations were performed for a leakage length (Λ) of 1.1 and 0.84 m. The parameters used in the calculations are shown in Table 1 and the results in the Figures 8 and 9. These figures show the loading on the blocks. They show only a part of the revetment area for which the calculation was run, the part which the highest loading. The area for which the calculation was run had 26 rows and 49 columns. The stability of the columns was investigated. Different grey values show the influence of the loading on the blocks, see Figure 8. It appears that there are three areas with a high loading. Comparing this figure with Figure 6 it appears that the largest area with high loading on the blocks is present during wave run-down, as could be expected from Figure 6, but that there are also two small area's near the wave impact. The black blocks indicate that for those blocks the force parallel to the slope, caused by the weight component of the column of blocks above, is not high enough. This means that the column above these blocks will be pushed upwards and failure will occur. However, this depends on the situation. This calculation was run assuming that the blocks, that are loaded to a degree that

		1st example	Terschelling
slope		1:4	1:4
block length (L)	(m)	0.5	0.3
block width (B)	(m)	0.5	0.22
block thickness (D)	(m)	0.2	0.18
density blocks (ρ_b)	(kg/m^3)	2350	2900
friction coeff. block-block (wb)	(-)	0.4	0.3
friction coeff. block-filter (wo)	(-)	0.4	0.3
blocks with pos. lift force			
Λ=1.1m		65	
Λ=0.84m		28	
Λ=0.52m middle of revetment			26
Λ=0.52m near transition			20

Table 1: Parameters used in calculations with BLOKKEN. In the calculations the wave pressure distribution as shown in figure 5 is used.

they slide downwards, have a contribution to the strength. Blocks higher on the revetment do not contribute, because there is a gap between the blocks that have slided downwards and those that have not. This gap prevents friction and clamping. Such gaps will not exist in a well maintained revetment and in that case all blocks above the most loaded block will contribute to the force parallel to the slope. The calculation was also run for a well maintained revetment and in that case the black blocks disappear, which means that the revetment is stable. The calculations also show that the leakage length has a large influence on the loading on the blocks. A reduction in the leakage length from 1.1 to 0.84 m reduces the number of blocks with a force higher than the weight of the block from 65 to 28. The revetment on Terschelling is made with granite blocks of, on average, 0.3*0.22*0.18 m^3 with a higher density and a lower friction coefficient than concrete, see Table 1. Relatively large joints are present between the blocks, leading to a permeable cover layer and as a consequence to a short leakage length (Bezuijen et al, 1990). An impermeable transition is planned on the top of the revetment. At this height a shoulder is planned in the dike profile and on this shoulder a road will be built. During design conditions this road will be well below the water line and severe wave attack can be expected on the transition. The calculated maximum wave loading on various sections of the dike varied between 0.6 and 1.3 m. Calculations were run for a wave height of 1.2 m, again with an incident wave angle of 30°. When the maximum loading was exerted on the middle of the revetment it appeared that the uplift force exceeds the weight of the block for 20 blocks. These blocks were found on 3 different rows. Friction between the blocks appeared enough to prevent failure. However, when the model was run for the situation with the

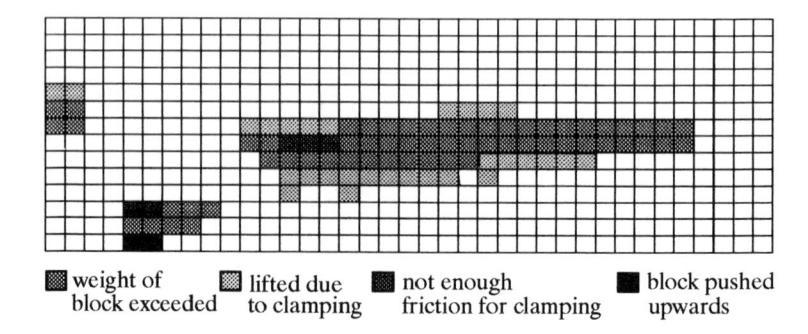

■ weight of   ▨ lifted due   ■ not enough   ■ block pushed
  block exceeded to clamping   friction for clamping   upwards

Figure 8: Calculated loading on a revetment. Each square represents a block. $\Lambda = 1.1$ m, wave loading see Figure 5

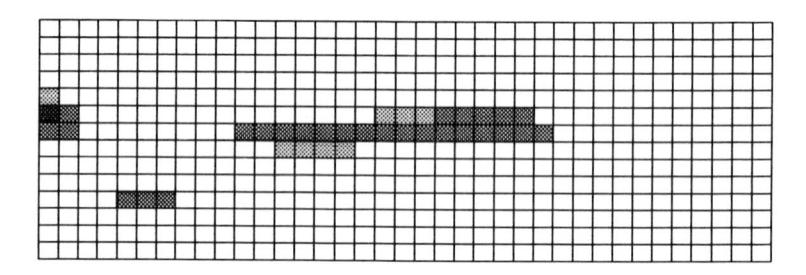

Figure 9: Calculated loading on a revetment. Each square represents a block. $\Lambda = 0.84$ m, wave loading see Figure 5.

maximum loading near the transition, then the number of blocks with an uplift force higher than its weight increased to 26 and all these blocks were found in the top row of the revetment near the transition. According to the calculation this top row will not be stable for this loading condition. From these results it was concluded that the transition will be a weak point in the revetment. The loading is increased on what can be called the weakest row in the revetment, because there are no overlaying blocks that contribute by their wheight to the force parallel to the revetment. It was therefore decided to strengthen the transition by applying bitumen between the blocks and in the filter layer up to 0.6 m below the transition.

6 Conclusions

The strength of a placed block revetment is not determined by the weight of the blocks only. The interaction between the blocks is of great importance. Calculations with the computer model BLOKKEN showed that, for a given friction coefficient, a horizontal pre-stress in the revetment can greatly increase the strength. This pre-stress can be obtained by putting gravel over the revetment with a comparable diameter as the joints. This will decrease the permeability and in this way increase the loading during wave attack. However, the increase in strength will more than compensate this and a more stable revetment will be obtained. For this aspect the model presents a theoretical base for a general practice in the Netherlands. The results of this model are qualitatively in agreement with results of large scale model test performed for this type of revetments (Burger, 1985). The applicability of the model as presented in this paper is limited by:

- the still limited quantitave knowledge of the forces parallel to the slope in a row before it is "lifted". Therefore the possible contribution of clamping in a row to the stability cannot be calculated.

- the limited number of wave pressure registrations on a slope area. These registrations are only available for regular waves. Most registrations were made in a wave flume for a wave loading perpendicular to the slope. However, the results show that the obliqueness of the waves has a large influence on the number of blocks that are "lifted" and this will certainly also influences the stability.

The method as presented is already useful to investigate the influence of transitions on the loading and the number of blocks loaded.

The model predicts a decrease in strength for blocks higher on the slope. This means that for a constant wave loading the likhood of failure increases when the water level increases.

7 References

Bezuijen A., Laustrup C. and Wouters J.(1988), Design of block revetments with physical and numerical models. *Proc. 21th Int. Conf. on Coastal Engineering, Malaga*

Bezuijen A., Klein Breteler M. and Burger A.M. (1990) Placed block revetments *Chapter in 'Coastal Protection' Pilarczyk (ed.) Balkema, Rotterdam ISBN 90 6191 1273*

Bezuijen A. and Klein Breteler M. (1992) Oblique wave attack on block revetments *Proc. 23rd ICCE, Venice*

Burger A.M. (1985), Strength dams along the Eastern Scheldt under concentrated wave loading (in Dutch), *Delft Hydraulics, Project M2036*

Burger A.M. et all. (1990) Analytical design method for relatively closed block revetments. *ASCE Journal Waterway Port, Coastal and Ocean engineering, Vol 115, pp 525-544*

Rigter B.P, Klein Breteler M and Kruse G.A.(1994) Strength of sea dikes is limited (in Dutch). *Land + Water No 5*

Stoutjesdijk T.P., B.P. Rigter and A. Bezuijen (1992) Field measurements on placed block revetments *Proc. 23rd ICCE, Venice*

Townson, J.M. (1988) The simulated motion of a loose revetment block *Journal of Hydraulic research, vol. 26, no. 2, pp 225-242.*

CHAPTER 69

A FIELD EXPERIMENT ON THE INTERACTIONS WAVES-REFLECTING WALL

Paolo Boccotti

Department of Fluid Mechanics and Offshore Engineering
University of Reggio Calabria
Via E. Cuzzocrea, 48 - 89100 Reggio Calabria - Italy
Fax. +39 965 875220

Abstract

A special reflecting wall 12 m long and 2.1 m high was built off the beach at Reggio Calabria, and 30 wave gauges were assembled before the wall and were connected to an electronic station on land. It was possible to observe the reflection of wind waves generated by a very stable wind over a fetch of 10 Km. The experiment aimed to verify the general closed solution for the wave group mechanics (Boccotti, 1988, 1989), for the special case of the wave reflection.

1 Introduction

Starting on 1990, a few small scale field experiments were executed off the beach at Reggio Calabria on the eastern coast of the Straits of Messina. The original aim was to verify the closed solution for the mechanics of the highest 3-D wave groups in the wind generated sea states (Boccotti, 1988, 1989).
The tide excursion in the area is very small and a local wind often remains constant from the North West for several consecutive days. After two days of NW wind, the Southerly swells vanish and the sea states in front of Reggio Calabria consist of pure wind waves with significant height typically within 0.2 and 0.4 m and dominant period within 1.8 and 2.5 s.
Because of the small wave size and of the very small tide excursion, it is possible to operate like in a big wave tank. Moreover, the water is exceptionally clear because of the Strait current which flows for about one half an hour in 12 hours. The clearness of the water enables

underwater works of high precision.

The first experiment (May, 1990) was concerned with the progressive waves on deep water. An array of nine wave gauges and nine pressure transducers supported by vertical iron piles provided space-time information on waves generated over a fetch of approximately 10 Km. It was confirmed that the general 3-D configuration of the extreme wave groups was consistent with the theoretical predictions (Boccotti et al., 1993-a).

Given that the closed solution for the mechanics of the highest wave groups holds (to the Stokes first order) for an arbitrary shape of the solid boundary, it particularly predicts the reflection of a 3-D wave group by a wall. Thus, a second experiment was executed on May, 1991 with the main purpose to verify whether the reflection of the 3-D wave groups was consistent with the theoretical prediction. Here a few results of this verification are shown. Previously, the experiment's data were used by Boccotti et al. (1993.b) to test a few predictions on the variations of wave energy and band-width with distance from the wall.

2 The closed solution for the mechanics of the wave groups on the sea surface (Boccotti, 1988, 1989)

Let us consider a random wind-generated sea state assumed Gaussian (Longuet-Higgins, 1963 and Phillips, 1966). The extreme wave events have been shown by the writer to occur in a well defined way that can be specified in terms of the autocovariance function. The theory admits that the random wave field is generally non-homogeneous: it can be homogeneous like in an open sea or non-homogeneous like before a reflecting wall.

Specifically, if the extreme wave crest occurs at $\underline{x}_o = (x_o, y_o)$ at time t_o, with a crest-to-trough height of H, the mean surface configuration in space and time is given by

$$\eta_G(\underline{x}_o + \underline{X}, t_o + T) = \frac{H}{2}\left\{\frac{\Psi(\underline{X}, T) - \Psi(\underline{X}, T - T^*)}{\Psi(\underline{0}, 0) - \Psi(\underline{0}, T^*)}\right\} \quad (2.1)$$

where Ψ is the autocovariance function of the surface displacement of the random wave field

$$\Psi(\underline{X}, T) = <\eta(\underline{x}_o, t)\eta(\underline{x}_o + \underline{X}, t + T)> \quad (2.2)$$

and T^* is the abscissa of the absolute minimum of the autocovariance function $\Psi(\underline{0}, T)$, which is assumed to exist and to be the first minimum after T=0. Superimposed on the deterministic form (2.1) is of course the random noise of the residual wave field, but when $H/\sigma(\underline{x}_o)$ is large ($\sigma(\underline{x}_o)$ being the r.m.s. wave elevation at location $\underline{x}_0$), the variations in the actual sea surface configuration surrounding $\underline{x}_o, t_o$ are small compared with η_G itself.

Associated with the configuration (2.1) is a distribution of velocity

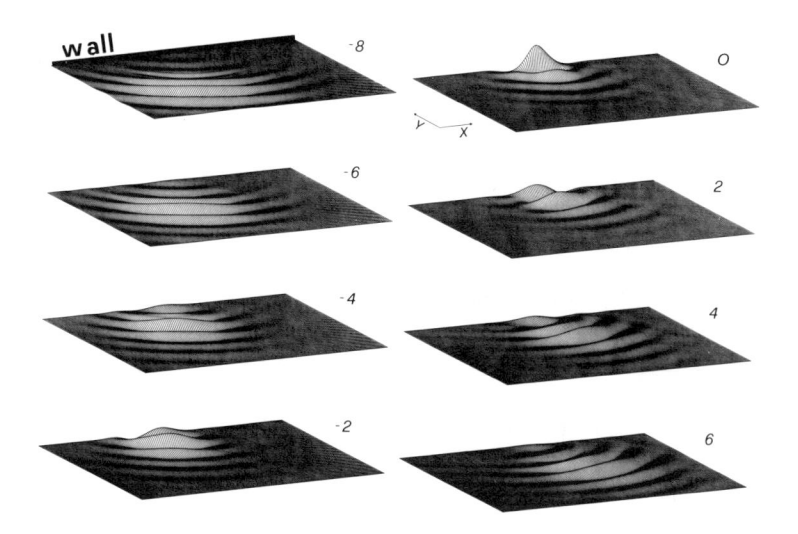

*Fig.1 Occurrence of a wave of given very large height at a reflecting
wall. The wall is along the upper x-parallel side of the framed rectangle,
and the interval between two consecutive configurations is of 2 Td (Td
being the dominant period of the random sea state). The wave group
impacts the wall, at the apex of the development stage (minimum width
of the wave front and maximum height of the central wave), it is
reflected mirrorwise and goes back seaward.*

potential in the water, which to the lowest order in a Stokes expansion
is given by

$$\phi_G(\underline{x}_o + \underline{X}, z, t_o + T) = \frac{H}{2} \left\{ \frac{\Phi(\underline{X}, z, T) - \Phi(\underline{X}, z, T - T^*)}{\Psi(\underline{0}, 0) - \Psi(\underline{0}, T^*)} \right\} \quad (2.3)$$

where

$$\Phi(\underline{X}, z, T) = <\eta(\underline{x}_o, t)\phi(\underline{x}_o + \underline{X}, z, t + T)>. \quad (2.4)$$

The theory holds for arbitrary solid boundary conditions, and it
can be formally proved that deterministic functions η_G and ϕ_G satisfy
the Stokes equations to the first order as well as an arbitrary set
of solid boundary conditions, if random functions η and ϕ satisfy those
equations and boundary conditions. Note that the hypothesis that $H/\sigma(\underline{x}_o)$
is large is not necessarily inconsistent with the use of the lowest
order (linear) terms in a Stokes expansion, provided H remains small
with respect to the wave length and the water depth.

The covariances (2.2) and (2.4) can be readily obtained from the
directional frequency spectrum of the random sea state. In the case
that the random sea state interacts with some obstacle, e.g. with a

reflecting wall, the covariances can be obtained from the directional frequency spectrum of the wave field that there would be if the obstacle was not there (see Appendices A and B).

From equation (2.1) we find that a wave of given very large height H at a location x_o in an open sea occurs because *a well defined wave group* transits that location, when it is at the apex of its development stage. A few 3-D pictures of the wave group can be seen in the papers of Boccotti (1988) and (1989) and of Boccotti et al. (1993-a). The basic phenomena that occur during the course of evolution of the group are not dependent on the detailed shape of the wave spectra, though the shape of the group does vary somewhat. Specifically, the wave group has always a development stage in which the height of its central wave grows to a maximum and the width of the wave front reduces to a minimum. Then a decay stage follows with the opposite features. The individual waves have a propagation speed greater than the envelope so that each wave crest is born at the tail of the group, grows to a maximum when it reaches the central position, and then dies at the head of the group.

From equation (2.1) we find also that a wave of given very large height at a location x_o of a reflecting wall, occurs because a wave group impacts the wall, when it is at the apex of the development stage -see Figure 1-. Then we find that a very high wave at a location x_o far from the wall occurs because of the collision of two wave groups -the first one approaching the wall, and the second one going back seaward after having been reflected.

Finally, equation (2.1) shows that a very high wave at a location x_o behind a wall (we mean "very high with respect to the mean wave height at this location") occurs because a wave group targets on the wall-end, and, after the impact with the wall, one half of the wave front goes on and penetrates into the shadow cone. The relevant 3-D pictures can be seen in the papers of Boccotti (1988).

3 The experiment of May 1991 off the beach at Reggio-Calabria

Eq.(2.1) holds not only for the surface waves but also for the pressure head waves at some fixed depth. In this case, η is to be intended as the fluctuating pressure head at the fixed depth. This property is a consequence of the fact that, to the Stokes first order, the fluctuating pressure head at some fixed depth represents a stationary Gaussian process of time like the surface displacement. Therefore, aiming to test eq.(2.1), the decision was taken to deal with the pressure head waves at a fixed depth, given that the fluctuating pressure head can be measured by transducers of high precision and low cost.

A set of pressure transducers was assembled before (seaward) a special upright reflecting wall 12 m long and 2.1 m high with a rubble mound .2 m high. The structure consisted of a steel truss whose stability was ensured by a dead weight of pig iron discs, and the reflecting plane was formed by aluminium panels with a thickness of .05 m.

The transducers were supported by three horizontal beams .6 m below

INTERACTIONS WAVES-REFLECTING WALL 949

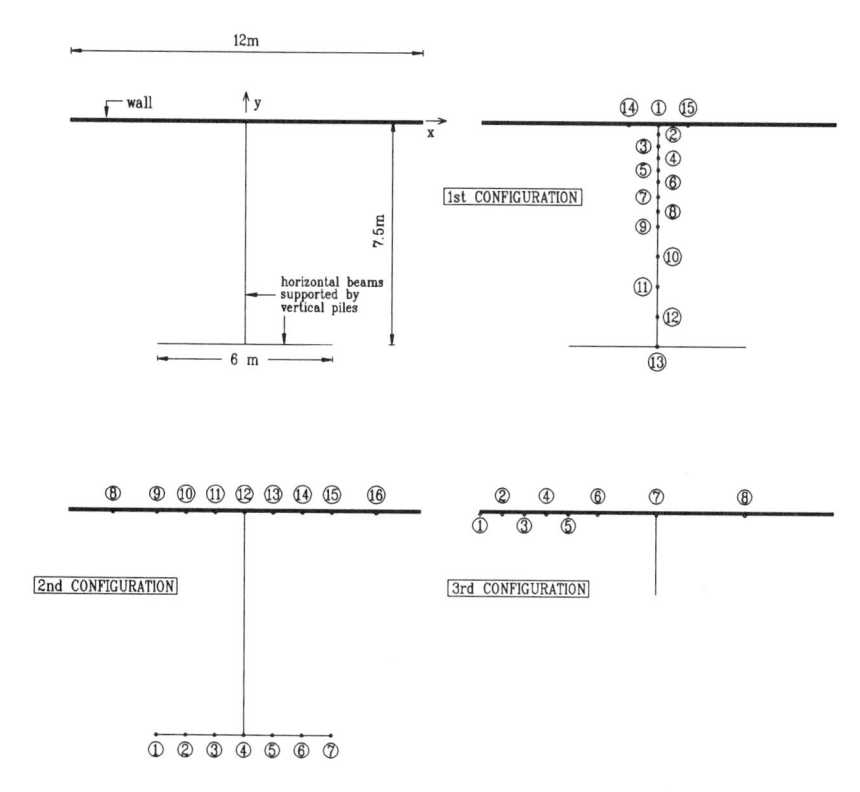

Fig.2 Plan view of the pressure transducers before the wall.

the mean water level -see Figure 2-. The average bottom depth in the area covered by the gauges was of 2.0 m. Both the horizontal beams and the supporting piles were designed to get an high degree of stiffness as well as a small section, and in particular each upright pile of the diameter of .05 m was stiffened by four small steel cables. The pressure transducers were connected by submarine cables to an A/D converter unit in an onshore building and the sampling rate was 10 hz. Besides the pressure transducers, an ultrasonic wave probe far from the wall provided information on the undisturbed waves.

Three different configurations of the gauges were assembled during the experiment -see Figure 2-. A set of 9 records was obtained with the first configuration, and sets of 27 and of 16 records were obtained respectively with the second and with the third configuration. The total number of the records was then of 52, each of 9 minutes and containing 250 to 300 dominant waves. The significant height H_s ranged within .17 and .42m and the dominant period T_d within 1.9 and 2.6s.

4 A few results of the experiment

The covariances can be found directly from the measurements by cross-correlation of the time series obtained at the discrete measurement locations and if an extreme wave of the pressure head, with a crest-to-trough height of H, is encountered at one such location, the time history of the expected pressure head configuration at this and the other locations can be obtained from (2.1).

With $\underline{x}_o$ taken as location 1 at the wall-center (first configuration of the gauges), the vectors $\underline{X}$ were specified by the relative locations of the other gauges. The time series data of record 17 provided measured auto-covariances as a function of T for the various gauge locations and these were used *without smoothing* on the right hand side of equation (2.1) to estimate $\eta_c(t_o+T)$ at these locations in an extreme wave. The results are shown in Figure 3.

The covariances were found also from a theoretical spectrum. The classic JONSWAP frequency spectrum (Hasselmann et al., 1973) and the spreading direction function of Mitsuyasu et al. (1975) were assumed for the two dimensional spectrum of the wave field that there would be if the wall was not there. The spreading direction parameter n_0 of Mitsuyasu et al. was taken equal to 20, that is the suggested value for the conditions of our experiment: fetch 10000 m, wind velocity 7÷8 m/s. The dominant wave period and the dominant wave direction were given the values of record 17. The dominant period was of 1.9 s. The direction was estimated accurately since the front of wave A at the traverse of locations 14-1-15 (at the wall) proves to be nearly straight. The relative phases indicated an angle of incidence of 13°.

The substitution of the theoretical covariances in equation (2.1) leads to Figure 4. Like Figure 3, this shows the expected waves in the time domain, at the various gauge locations, if a wave of given very large height H occurs at location 1 at the wall-center, with the first configuration of the gauges. The likeness of the two figures is amazing -figure 3 was obtained from the time series data, while figure 4 was obtained from a theoretical spectrum!

In the figures, A represents the wave of given very large height H at location 1, B, C and D are the waves immediately before this one, and B', C', D' are the waves immediately after A. At the wall, A occupies the envelope center.

The wave height exhibits a local minimum (node) at location 5, because the waves approaching the wall overlap the reflected waves in phase opposition. Then, a local maximum of the wave height (antinode) takes place at location 8, where the waves overlap themselves in phase coherence. The dominant wave length L_d was of nearly 6 m, so that the node was at $\frac{1}{4}L_d$ from the wall, and the antinode was at $\frac{1}{2}L_d$ from the wall. At the antinode, two waves of the same height occupy the envelope center, the first one is the overlap of wave A approaching the wall and of preceding wave B going back after having been reflected; the second one is the overlap of wave A going back seaward and of succeeding wave B' approaching the wall. At greater distances from the wall, nodes and antinodes tend to disappear because the central waves of the group

FROM CROSS-CORRELATION OF THE TIME SERIES DATA

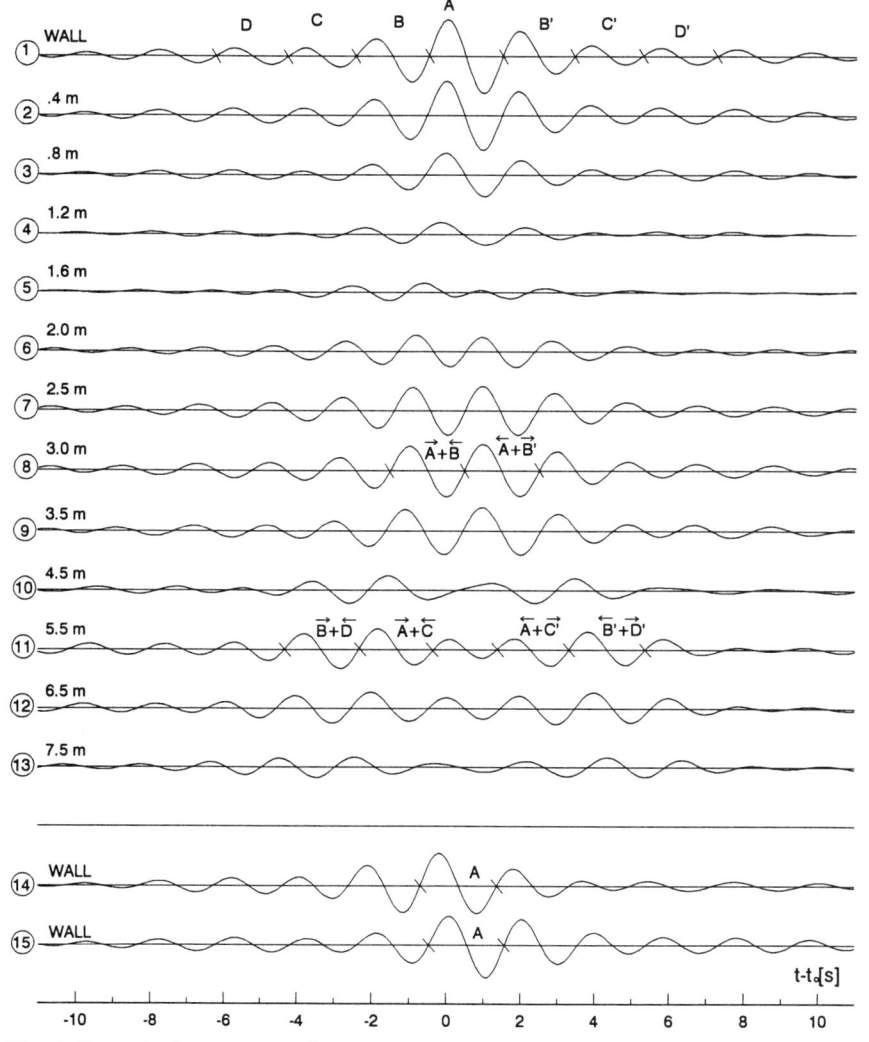

Fig.3 Expected waves at the various gauge locations if a wave of given very large height H occurs at location 1 (wall-center) with the first configuration of the gauges. Calculation from equation (2.1), directly with the time series data of record 17, <u>without smoothing</u>.

FROM THE JONSWAP SPECTRUM

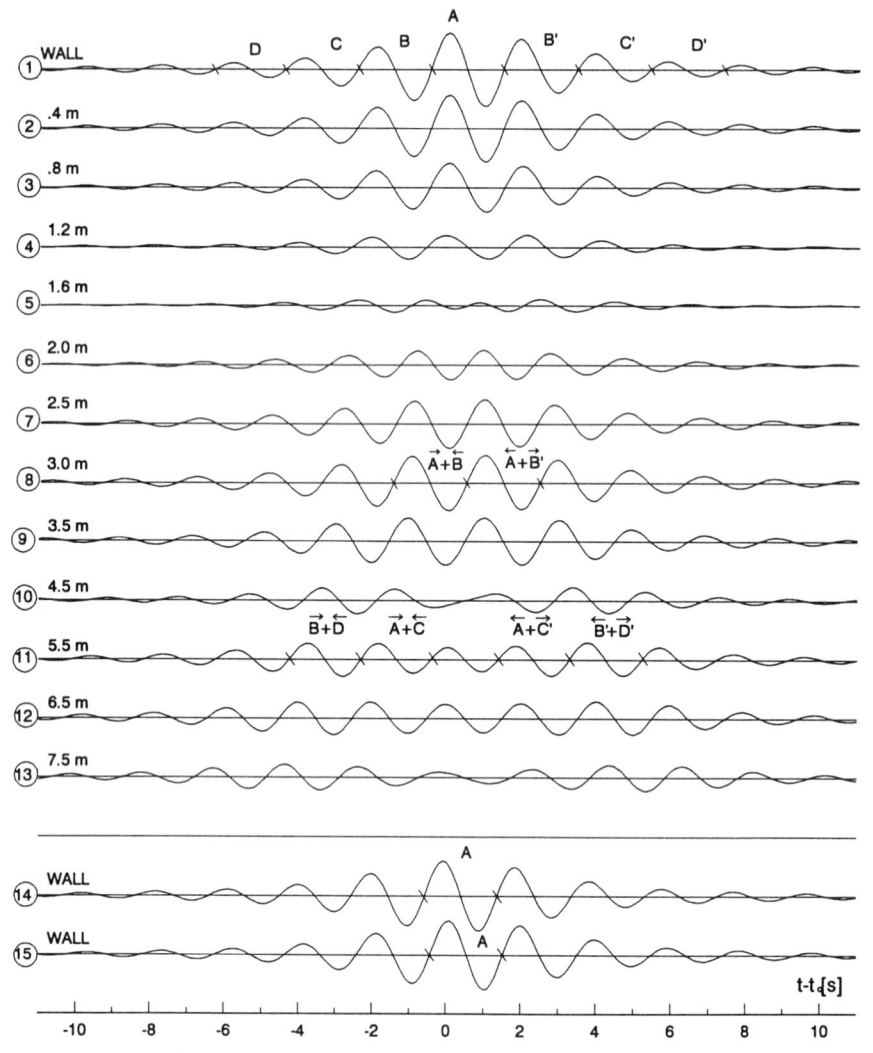

Fig.4 Expected waves at the various gauge locations if a wave of given very large height H occurs at location 1 (wall-center) with the first configuration of the gauges. Calculation from equation (2.1), with the JONSWAP spectrum and the spreading direction function of Mitsuyasu et al.(1975).

FROM CROSS-CORRELATION OF THE TIME SERIES DATA

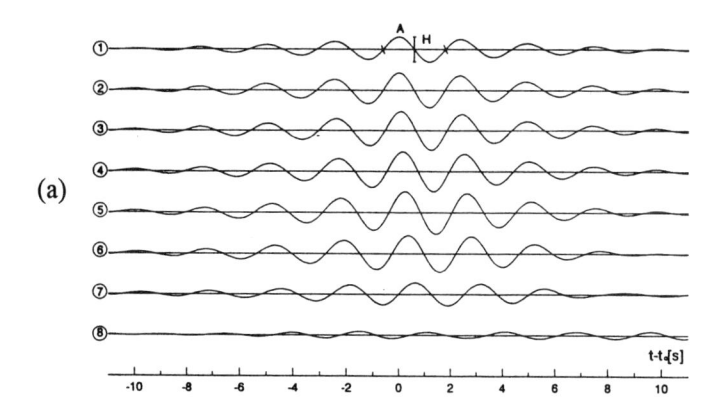

(a)

FROM THE JONSWAP SPECTRUM

(b)

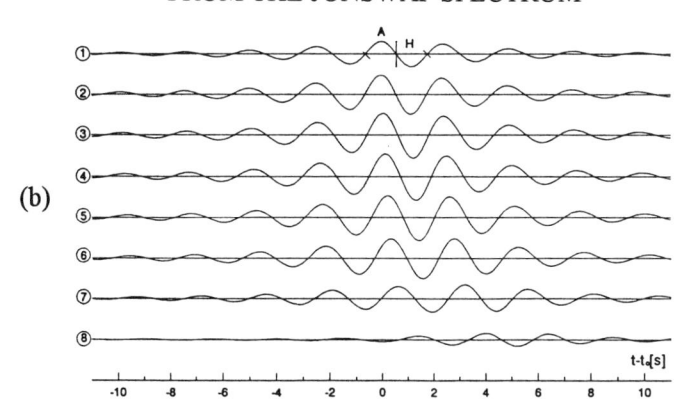

Fig.5 Expected waves at the various gauge locations if a wave of given very large height H occurs at location 1 (wall-end) with the third configuration of the gauges. Calculation from eq.(2.1). Panel (a): covariances obtained by cross-correlation of the time series data of record 30. Panel (b): covariances obtained with the JONSWAP spectrum and the spreading direction function of Mitsuyasu et al.(1975).

do not overlap themselves, and it is possible to distinguish the wave group approaching the wall from the wave group going back seaward. In particular, at location 13, which is the most remote from the wall, we see the wave group approaching the wall, then a short calm, then the wave group going back.

Fig.5 shows the expected time histories at the locations of the *third gauge configuration* if a wave of fixed very large height H occurs at the wall-end (H is intended to be very large with respect to the mean wave height at the wall-end). The calculation was made by means of eq.(2.1). The covariances used for Fig.5.a were obtained by cross-correlation of the time series of record 30, while the covariances used for Fig.5.b were obtained from a theoretical spectrum (see Appendix B). The general form of the theoretical spectrum was that used for Fig.4, and the dominant wave period and wave direction were given the values of record 30. The dominant period was 2.5 s and the angle of the dominant wave direction with the wall-orthogonal was 22°.

A very high wave at the wall-end, with a very high probability, occurs because the front center of a wave group targets on the wall-end. For this reason, the wave height along the wall, on the one hand has to decrease starting on the wall-end because of the increasing distance from the front center, and on the other hand it has to increase because of the raising produced by the wall. The result is that a local maximum of the wave height takes place at some distance from the wall-end. This observation should permit to understand the overall configurations of figures 5.a and 5.b. These configurations are strongly similar to each other, and particularly we can see, in both two the figures, that the local maximum of the wave height falls within locations 4 and 5, at nearly $\frac{1}{3}$ wave length from the wall-end.

Figs.3 and 5.a prove that the wind wave field has something like a "genetic code" showing what essentially happens if a very high wave occurs at any fixed location. Given that x_0 can be anyone of the gauge locations, we drew a number of pictures like Figs.3 and 5.a from each record, and the expected time histories were always smooth and consistent with one another as in the figures shown in this paper. Although the records from single realizations of very high waves were more irregular than the expected profiles, the essential features of these profiles were still evident and provided good support for the theoretical connection. This topic is dealt with in a forthcoming paper giving also a more simple formal proof of the theory.

Appendix A. Autocovariance of the wind waves being subject to reflection
A.a Surface waves

According to the theory of the sea waves to the Stokes first order (Longuet-Higgins, 1963), the surface displacement of a progressive wave field is

$$\eta(x,y,t) = \sum_{i=1}^{N} a_i \cos(k_i x \sin\theta_i + k_i y \cos\theta_i - \omega_i t + \epsilon_i), \quad k_i \tanh(k_i d) = \omega_i^2/g, \quad (A.1)$$

where it is assumed that number N is very large, phases ϵ_i are distributed purely at random in $(0, 2\pi)$, frequencies ω_i are different from one another. Then it is assumed that a_i, ω_i and θ_i (this being the angle of wave direction and y-axis) are such as to form a directional frequency spectrum

$$S(\omega,\theta)\delta\omega\,\delta\theta = \sum_i \tfrac{1}{2}a_i^2 \text{ for } i \text{ such that } \omega < \omega_i < \omega + \delta\omega \text{ and } \theta < \theta_i < \theta + \delta\theta. \quad (A.2)$$

If a vertical reflecting wall is put along the x-axis (line y=0), random wave field (A.1) takes on the form (Boccotti, 1988)

$$\eta(x,y,t) = 2\sum_{i=1}^{N} a_i \cos(k_i x \sin\theta_i - \omega_i t + \epsilon_i)\cos(k_i y \cos\theta_i). \quad (A.3)$$

From this equation of the surface displacement, and definition (2.2) of the autocovariance, we have

$$\Psi(\underline{X},T) = 4 \sum_{i=1}^{N} \sum_{j=1}^{N} a_i a_j \cos(k_i y_o \cos\theta_i)\cos[k_j(y_o+Y)\cos\theta_j] \cdot$$

$$\cdot <\cos(k_i x_o \sin\theta_i - \omega_i t + \epsilon_i)\cos(k_j x_o \sin\theta_j - \omega_j t + \epsilon_j + k_j X \sin\theta_j - \omega_j T)>, \quad (A.4)$$

where

$$\underline{x_o} = (x_o, y_o), \quad \underline{X} = (X,Y), \quad (A.5)$$

and the angle brackets denote an average with respect to time t. Eq. (A.4) is reduced to

$$\Psi(\underline{X},T) = 4\sum_{i=1}^{N} \frac{1}{2} a_i^2 \cos(k_i X \sin\theta_i - \omega_i T)\cos(k_i y_o \cos\theta_i)\cos[k_i(y_o+Y)\cos\theta_i], \quad (A.6)$$

and with the definition (A.2) of $S(\omega,\theta)$

$$\Psi(\underline{X},T) = 4\int_0^\infty \int_0^{2\pi} S(\omega,\theta)\cos(kX\sin\theta - \omega T)\cos(ky_o\cos\theta) \cdot$$

$$\cdot \cos[k(y_o+Y)\cos\theta]d\theta\, d\omega. \quad (A.7)$$

Thus, if we specify function $S(\omega,\theta)$ and location $\underline{x_o}$ we can evaluate the autocovariance. Note that eq. (A.7) depends on y_0 but does not depend on x_0. This is a consequence of the fact that the wall is assumed to be very long, so that the autocovariance generally changes with distance $|y_0|$ from the wall, but does not change with position x_0 along the wall.

A.b Pressure head waves

In this section $\eta(x,y,t)$ is intended to be the fluctuating pressure head at some fixed depth z_o. Before a long reflecting wall, this is given by

$$\eta(x,y,t) = 2\sum_{i=1}^{N} a_i \frac{\cosh k_i(d+z_o)}{\cosh k_i d}\cos(k_i x \sin\theta_i - \omega_i t + \epsilon_i)\cos(k_i y \cos\theta_i), \quad (A.8)$$

so that the autocovariance (2.2) turns out to be

$$\Psi(\underline{X},T) = 4\int_0^\infty \int_0^{2\pi} S(\omega,\theta)\frac{\cosh^2 k(d+z_o)}{\cosh^2 kd}\cos(kX\sin\theta - \omega T)\cos(ky_o\cos\theta) \cdot$$

$$\cdot \cos[k(y_o+Y)\cos\theta]d\theta d\omega. \quad (A.9)$$

This equation was used to obtain Fig.4.

Appendix B. Autocovariance of the wind waves being subject to diffraction

Near the wall-end, a more appropriate theory is that of the semi-infinite breakwater. If a semi-infinite breakwater is put along the x-axis, random wave field (A.1) takes on the form

$$\eta(r,\alpha,t) = \sum_{i=1}^{N} a_i[F(r,\alpha;\omega_i,\theta_i)\cos(\omega_i t + \epsilon_i) + G(r,\alpha;\omega_i,\theta_i)\sin(\omega_i t + \epsilon_i)], \quad (B.1)$$

where r, α are polar coordinates with origin at the breakwater-end ($\alpha = 0$ at the protected wall, $\alpha = 2\pi$ at the wave beaten wall) and $F(r,\alpha;\omega,\theta)$ and $G(r,\alpha;\omega,\theta)$ are the functions of Penny and Price (1952) for a wave whose angular frequency is ω and whose direction makes an angle θ with the wall-orthogonal. If we define

$$\underline{x}_o = (r_o, \alpha_o) \quad , \quad \underline{x}_o + \underline{X} = (r,\alpha), \tag{B.2}$$

from general eq.(2.2) of the autocovariance we have

$$\Psi(\underline{X},T) = \sum_{i=1}^{N} \sum_{j=1}^{N} a_i a_j < [F_i(r_o,\alpha_o)\cos(\omega_i t + \epsilon_i) + G_i(r_o,\alpha_o)\sin(\omega_i t + \epsilon_i)] \cdot$$

$$\cdot [F_j(r,\alpha)\cos(\omega_j t + \epsilon_j + \omega_j T) + G_j(r,\alpha)\sin(\omega_j t + \epsilon_j + \omega_j T)] >, \tag{B.3}$$

where, for simplicity, we have used the compact notation

$$F_i(r,\alpha) = F(r,\alpha;\omega_i,\theta_i). \tag{B.4}$$

Eq.(B.3) is reduced to

$$\Psi(\underline{X},T) = \sum_{i=1}^{N} \frac{1}{2} a_i^2 \{[F_i(r_o,\alpha_o)F_i(r,\alpha) + G_i(r_o,\alpha_o)G_i(r,\alpha)]\cos\omega_i T +$$

$$+ [F_i(r_o,\alpha_o)G_i(r,\alpha) - G_i(r_o,\alpha_o)F_i(r,\alpha)]\sin\omega_i T\}, \tag{B.5}$$

and from the definition (A.2) of $S(\omega,\theta)$

$$\Psi(\underline{X},T) = \int_0^\infty \int_0^{2\pi} S(\omega,\theta)\{[F(r_o,\alpha_o;\omega,\theta)F(r,\alpha;\omega,\theta) +$$

$$+ G(r_o,\alpha_o;\omega,\theta)G(r,\alpha;\omega,\theta)]\cos\omega T + [F(r_o,\alpha_o;\omega,\theta)G(r,\alpha;\omega,\theta) +$$

$$- G(r_o,\alpha_o;\omega,\theta)F(r,\alpha;\omega,\theta)]\sin\omega T\}d\theta d\omega. \tag{B.6}$$

This is the autocovariance of the surface waves. Then, autocovariance of the pressure head waves at some fixed depth z_o has one more factor in the integral. This is

$$\cosh^2 k(d + z_o)/\cosh^2 kd.$$

Eq.(B.6) with this additional factor was used to obtain Fig.5.b.

References

Boccotti P., 1988, Refraction, reflection and diffraction of irregular gravity waves, Excerpta of the Italian Contribution to the Field of Hydraulic Engng., vol.3, Libreria Progetto Publ., Padova.
Boccotti P., 1989, On mechanics of irregular gravity waves, Atti Acc. Naz. Lincei, Memorie (Trans. Nat. Acc. Lincei), VIII, 19.
Boccotti P. et al., 1993-a, A field experiment on the mechanics of irregular gravity waves, J. Fluid Mech., 252.
Boccotti P. et al., 1993-b, An experiment at sea on the reflection of the wind waves, Ocean Engng., 20.
Hasselmann K. et al., 1973, Measurements of wind wave growth and swell decay during the Joint North Sea Wave Project (JONSWAP), Deut. Hydrogr. Zeit., A-8.
Longuet-Higgins M.S., 1963, The effects of non-linearities on statistical distributions in the theory of sea waves, J. Fluid Mech., 17.
Mitsuyasu H. et al., 1975, Observation of directional spectrum of ocean

waves using a clover-leaf buoy, J. Phys. Oceanography, 5.
Penny W.G. and Price A.T., 1952, The diffraction theory of sea waves by breakwaters, Phil. Trans. Roy. Soc., A-244.
Phillips O.M., 1967, The theory of wind generated waves, Adv. in Hydroscience, 4.

Presented at ICCE'94, Kobe, Japan, October 1994

CHAPTER 70

The application of load-cell technique in the study of armour unit responses to impact loads

Hans F. Burcharth[1]   Zhou Liu[2]

Abstract

The slender, complex types of armour units, such as Tetrapods and Dolosse are widely used for rubble mound breakwaters. Many of the recent failures of such structures were caused by unforeseen early breakage of the units, thus revealing an inbalance between the strength (structural integrity) of the units and the hydraulic stability (resistance to displacements) of the armour layers. Breakage is caused by stresses from static, pulsating and impact loads. Impact load generated stresses are difficult to investigate due to non-linear scaling laws. The paper describes a method by which impact loads on slender armour units can be studied by load-cell technique. Moreover, the paper presents Dolos design diagrams for the prediction of both breakage and hydraulic stability.

Introduction

The slender complex types of armour units, such as Tetrapods and Dolosse are widely used. Breakage of the armour units has caused many of the recent breakwater failures. Thus there is a need for studying stresses in the units.

Due to the stochastic nature of the wave loads, the complex shape of the armour units and their random placement, the problem cannot be dealt with on a deterministic basis, but must be handled as a probabilistic problem.

Consequently, a very large number of situations must be investigated. This can be performed at reasonable costs only by small scale experiments. Stresses in small scale armour units are studied by the use of load-cells inserted in the units. Burcharth et al.(1992) presented design diagrams for structural integrity based on stress exceedence probability, which, however, do not express the proportion of the units that will break. The present paper presents a new set of diagrams for the prediction of the amount of breakage.

---

[1]Prof. of Marine Civil Engineering, Aalborg University, Denmark.
[2]Research Engineer, Ph.D. Aalborg University, Denmark.

Load-cell technique involves a number of complications. The installation of the load-cell makes the material properties of the unit different from those of the homogeneous prototype units and consequently the impact responses, which depend on the elastic behaviours of the bodies, cannot be directly reproduced in model tests. Besides this, the responses of the instrumented units could involve dynamic amplification effects. Moreover, the ultra short duration of solid body impact loads and the wave slamming necessitates high frequency sampling which results in data storage capacity problem. The frequencies of the impact stresses are in the order of 800-1500 Hz for the applied model units.

The paper first discusses the scale law for the impact stress in the armour units and presents results of impact calibration of the load-cell instrumented Tetrapods and Dolosse. The paper then presents the model test results on impact stresses of Dolosse and finally presents the design diagrams which incorporate both the hydraulic stability and the structural integrity of Dolos armour layers. The diagrammes are different from the earlier ones presented by the authors in that they contain information on the proportion of the units that will break, instead of the stress exceedence probability.

Duration of impacts

When two solid bodies collide the impact force and the related stresses will depend on the duration of the impact, i.e. the time of contact, τ. Due to the non-linear material properties of concrete and to the complex shape of slender armour units it is not possible to establish a formula by which τ can be quantified. However, it is sufficient for the present research to formulate a qualitative expression for τ. In the following are discussed two realistic models for estimation of τ. It is shown that for geometrically similar systems and constant Poisson's ratio it is reasonable to assume

$$\tau \sim L\sqrt{\frac{\rho_A}{E_A}} \qquad (1)$$

where $\sim$ means proportional to.

Case 1. Impacting blunt bodies of identical linear elastic material.

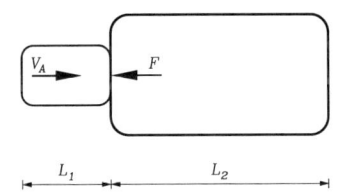

L_1 and L_2 are proportional to the characteristic length L of the system.

It is assumed that the impact generates mainly one-dimensional compression longitudinal shock waves which travel with the rod wave speed, $C = \sqrt{E_A/\rho_A}$, the distances L_1 and L_2 to the free edges, where they are reflected as tension waves. The two bodies will loose contact at the first return of a tension wave to the impact surface. Consequently, because $L_1 < L_2$

$$\tau = \frac{2L_1}{C} = \frac{2L_1}{\sqrt{\frac{E_A}{\rho_A}}} \quad \text{or} \quad \tau \sim L\sqrt{\frac{\rho_A}{E_A}}$$

Case 2. Slender body impacted by blunt body of identical linear elastic material.

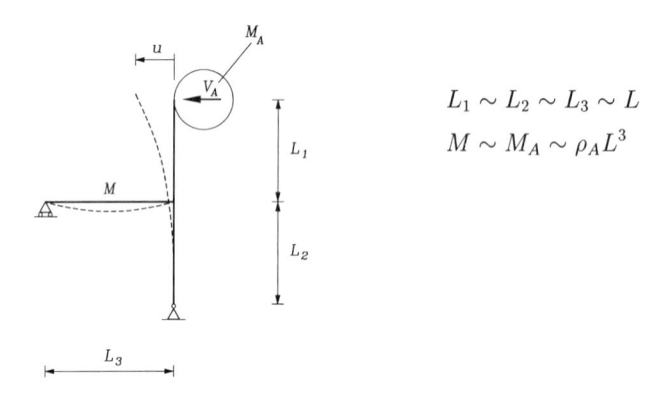

$$L_1 \sim L_2 \sim L_3 \sim L$$
$$M \sim M_A \sim \rho_A L^3$$

The impacting blunt body of mass M_A hits the slender structure of mass $M \sim M_A$ with impact velocity V_A by which a vibration mainly caused by bending and shear is initiated. It is assumed that the maximum value of τ corresponds to contact between the two bodies during approximately one half period T of the first mode of vibration for the slender body.

If it is assumed that the slender structure has a linear response corresponding to transverse impacts on free and simply supported beams then the system corresponds in principle to a mass-spring system with spring stiffness

$$k \sim \frac{E_A I}{L^3} \tag{2}$$

where $I \sim L^4$ is the moment of inertia.

The deflection time defined as one half period of the first mode of vibration is

$$\tau \simeq \frac{T}{2} \simeq \sqrt{\frac{M_A + M_o}{k}} \tag{3}$$

where $M_o \sim M \sim M_A \sim \rho_A L^3$ is the modal mass of the slender body.

From eqs (2) and (3) is then obtained eq (1). This conclusion was already presented in Burcharth (1984).

Scaling law for impact stresses of armour units

Case 1. Scale law in case of free fall impinging body

Geometrical similarity and constant coefficient of restitution are assumed

$$V_A \sim \sqrt{2gL}$$
$$M_A \sim \rho_A L^3 \qquad (4)$$

The momentum equation reads

$$F\tau = M_A \Delta V_A \sim M_A V_A \qquad (5)$$

where τ is the duration of the impact and ΔV is the velocity difference of the impinging body before and after the collision. $\Delta V \sim V_A$ is due to the assumed constant coefficient of restitution.

Inserting eqs (1) and (4) in eq (5) yields

$$F \sim \frac{\rho_A L^3 (gL)^{0.5} E_A^{0.5}}{L \rho_A^{0.5}} = \rho_A^{0.5} E_A^{0.5} g^{0.5} L^{2.5}$$

Introducing $\lambda = \dfrac{\text{model value}}{\text{prototype value}}$ we obtain

$$\lambda_{\sigma_{Impact}} = (\lambda_{\rho_A} \lambda_{E_A} \lambda_L \lambda_g)^{0.5} \qquad (6)$$

Case 2. Scale law of impinging body affected only by flow forces

V_A is found from Newton's equation

$$F_W = M_A \frac{dV_A}{dt} \qquad (7)$$

$$V_A = \frac{F_W}{M_A} t = \frac{F_W}{\rho_A L^3} t \qquad (8)$$

where F_w is the flow force on the impinging body.

By the use of eqs (1), (5) and (8) is obtained

$$F \sim \frac{\rho_A L^3 F_W t E_A^{0.5}}{\rho_A L^3 L \rho_A^{0.5}} = \rho_A^{-0.5} L^{-1} F_W t E_A^{0.5}$$

$$\lambda_{\sigma Impact} = \lambda_{\rho_A}^{-0.5} \lambda_L^{-3} \lambda_{F_W} \lambda_t \lambda_{E_A}^{0.5} \qquad (9)$$

Because in the Froude model, $\lambda_{F_W} = \lambda_{\rho_W} \lambda_L^3$ and $\lambda_t = \lambda_L^{0.5}$, then

$$\lambda_{\sigma Impact} = \lambda_{\rho_A}^{-0.5} \lambda_{\rho_W} \lambda_{E_A}^{0.5} \lambda_L^{0.5} \qquad (10)$$

The variation in F_w due to viscous effects is neglected. This, however, introduces some unknown bias, the size of which depends on the Reynolds number range.

Case 3. Collision between impinging water (slamming) and a solid body.

The air-cushioning effect is neglected because it is unlikely that air-pockets will be entrapped due to the limited size and rounded shape of the elements.

$$V_W \sim \sqrt{g L}$$
$$M_W \sim \rho_W L^3 \qquad (11)$$

τ is assumed given by (1) because the solid body stress wave is reflected from a free surface of the armour unit long time before reflection from a free surface of the wave (travel distance $\simeq H_s >$ dimension of armour unit; shock wave speed is smaller in water than in concrete) and the deflection time will be shorter than the transverse time of the elastic wave in the water.

From the momentum equation

$$F \tau = M_W \Delta V_W \sim M_W V_W \qquad (12)$$

and eqs (1) and (8) is obtained

$$F \sim \frac{\rho_W L^3 g^{0.5} L^{0.5} E_A^{0.5}}{L \rho_A^{0.5}} = \rho_A^{-0.5} \rho_W E_A^{0.5} L^{2.5} g^{0.5}$$

and consequently

$$\lambda_{\sigma Impact} = \lambda_{\rho_A}^{-0.5} \lambda_{\rho_W} \lambda_{E_A}^{0.5} \lambda_L^{0.5} \lambda_g^{0.5} \qquad (13)$$

The difference between the above three scaling laws eqs (6), (10) and (13) is related to the scales of the densities only, because generally

$$\lambda_{\rho_A} \neq \lambda_{\rho_A}^{-0.5} \lambda_{\rho_W} \qquad (14)$$

As long as the model is made of approximately the same concrete as the prototype, eq (6) can be chosen as the scaling law for the impact stresses, as it introduces less than 1% error for $0.97 \leq \lambda_{\rho_A} \geq 1.00$ and $0.98 \leq \lambda_{\rho_w} \geq 1.00$.

Apparent elasticity of the units with load-cell

The scaling law for the impact stresses of armour units is related to the elasticity of the material. Unfortunately, the insertion of the load-cell destroys the homogeneity of the material. This means that the impact stresses recorded in the small scale model tests cannot be scaled up to prototypes by the use of eq (6) valid only for homogeneous materials. Fig.1 shows the 200 g Dolos and 280 g Tetrapod with the load-cells.

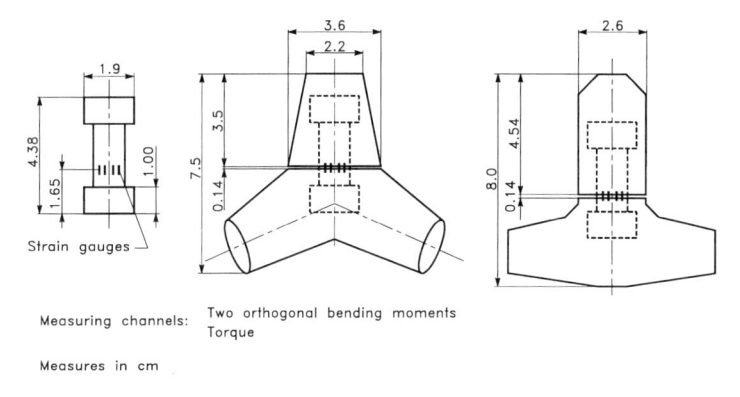

Fig.1. 200 g Dolos and 280 g Tetrapod with the load-cells.

However, by comparison of small scale impact test results for Dolosse and Tetrapods with results of the similar large scale impact tests (Burcharth, 1980, Bürger et al. 1990), it is possible to obtain an apparent elasticity for the small scale units. The apparent elasticity is then used for the interpretation of the impact signals recorded in the hydraulic flume tests.

The impact calibration results of the small scale Dolosse with load-cell have been published in Burcharth et al.(1990). The results of impact calibration of the Tetrapods are given in Fig.2. For the applied pendulum test set-up the reference is made to Bürger et al.(1990).

A way of checking the apparent elasticity is to compare the impact duration of the small load-cell mounted units with those of the various large size units, cf. eq (1). Fig.3 shows the ratio of dimensionless stress of various sizes of Dolosse using the apparent elasticity of the 200 g Dolos. Even though there is a big scatter, it can be seen that most ratios are around the value of 1, thus confirming the value of the apparent elasticity.

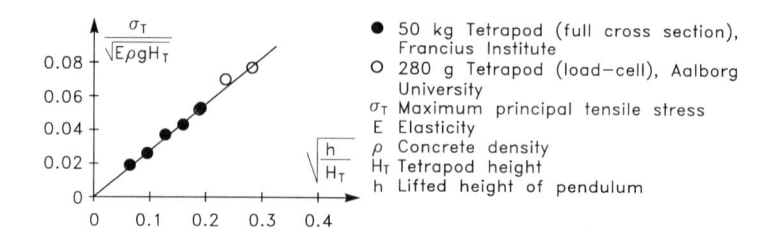

Fig. 2. Comparison of pendulum test results of the large scale Tetrapod with surface mounted strain gauges and the small scale Tetrapod with load-cell. Apparent elasticity $E = 4799\ MPa$.

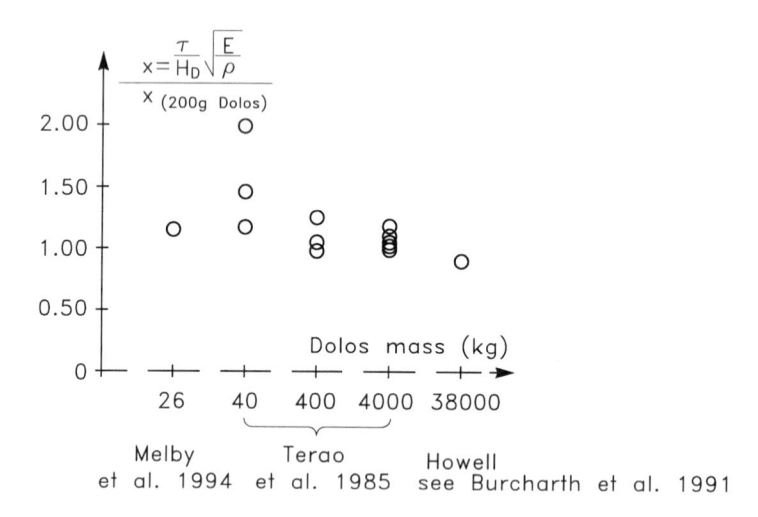

Fig. 3. Ratios of the dimensionless impact duration of large scale Dolosse against 200 g Dolos with load-cell. Apparent elasticity $E = 3500\ MPa$.

Sampling frequency

The ultra short duration of solid body impact loads and wave slamming requires a very high sampling frequency. The following analysis gives the underestimation of the stress corresponding to a certain sampling frequency.

Suppose the stress signal is recorded at frequency f_s and the stress signal is sinusoidal with the maximum stress σ_p and the frequency f.

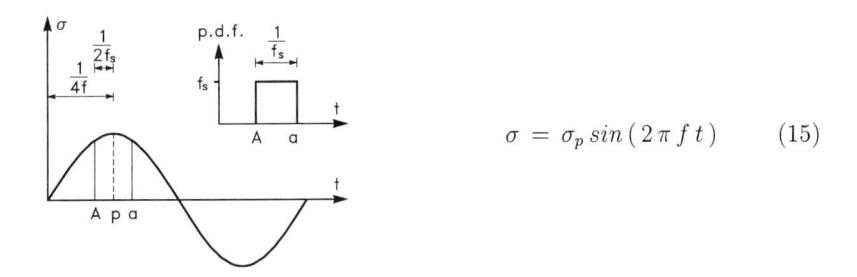

Fig. 4. *Sinusoidal stress signal.*

$$\sigma = \sigma_p \sin(2\pi f t) \qquad (15)$$

The most unfavourable case is when the two adjacent sampling points, A and a, are symmetrically located around the center of the peak p. For this case the sampled maximum stress σ_A is

$$\sigma_A = \sigma_p \sin(2\pi f t_A) = \sigma_p \sin\left(2\pi f \left(\frac{1}{4f} - \frac{1}{2f_s}\right)\right) \qquad (16)$$

and *the maximum relative error* is

$$\frac{\sigma_p - \sigma_A}{\sigma_p} = 1 - \sin\left(\pi\left(\frac{1}{2} - \frac{f}{f_s}\right)\right) \qquad (17)$$

On the other hand, if the sampling points are uniformally distributed along the length (A-a), the average of the sampled maximum stress is

$$\overline{\sigma} = \int_{t_A}^{t_a} \sigma_p \sin(2\pi f t) f_s \, dt$$

$$= \frac{\sigma_p}{2\pi} \frac{f_s}{f} \left(\cos\pi\left(\frac{1}{2} - \frac{f}{f_s}\right) - \cos\pi\left(\frac{1}{2} + \frac{f}{f_s}\right)\right) \qquad (18)$$

and *the average relative error* is

$$\frac{\sigma_p - \overline{\sigma}}{\sigma_p} = 1 - \frac{1}{2\pi} \frac{f_s}{f} \left(\cos\pi\left(\frac{1}{2} - \frac{f}{f_s}\right) - \cos\pi\left(\frac{1}{2} + \frac{f}{f_s}\right)\right) \qquad (19)$$

The maximum relative error and the average relative error are depicted in Fig.5. However, the actual impact signals are not sinusoidal, cf. Fig.6. In order to check the influence of the sampling frequency a series of Dolos pendulum tests with different sampling frequencies have been performed. The results are depicted in Fig.6. It can be seen that the sinusoidal results hold also for the actual impact signal when the offset for $f_s = 10000\ Hz$ is considered.

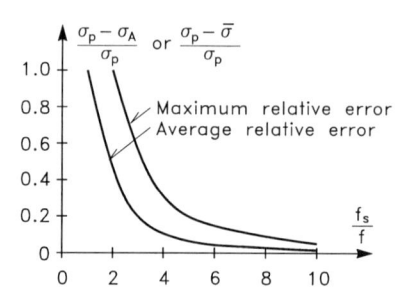

Fig. 5. Maximum relative error and average relative error due to the limited sample frequency.

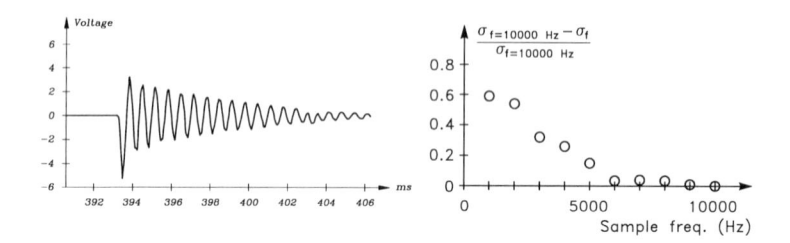

Fig. 6. Example of the impact signals of the 200 g load-cell instrumented concrete Dolosse ($f = 1500\ Hz$) and the relative error of the Dolos pendulum test results as function of the sampling frequency.

In the Dolos hydraulic model test, the applied sampling frequency is $f_s = 6000\ Hz$ and the damped natural frequency of the instrumented Dolosse $f = 1500\ Hz$. On average the sampled maximum impact stress is underestimated by 10% due to the limited sampling frequency. Therefore, in the data processing all sampled maximum impact stresses were increased by 10%.

Check for the dynamic amplification by wave slamming

It is well-known that resonance occurs when the frequency of the load is close to the natural frequency of the system. The installation of the load-cell into the model Dolos makes its natural frequency smaller, cf. eq (1). In order to check if the reduced natural frequency of the Dolosse is close to the wave slamming frequency, and hence introduces dynamic amplification, the frequency of the wave slamming on the Dolos armour layer was recorded by a pressure transducer installed in the stem of the Dolos. The pressure transducer did in all tests face the breaking waves. The results are given in Fig.7, showing the highest frequency of the wave slamming on the Dolos armour to be 330 Hz, far away from the natural frequency of 1500 Hz for the Dolosse with the load-cells. Consequently, no dynamic amplification are present in the model tests.

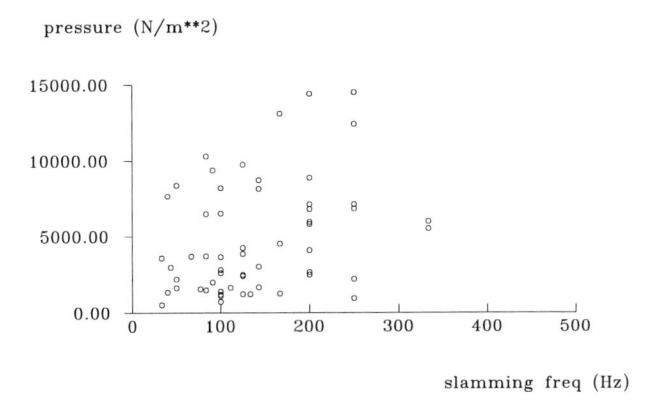

Fig. 7. Recorded frequencies of the wave slamming on the Dolos armour units.

Description of the experiments

A 1 : 1.5 slope armoured with 200 g concrete Dolosse of waist ratios 0.325, 0.37 and 0.42 were exposed to irregular waves in a wave flume with a foreshore slope of 1 : 20. Fig. 8 shows the set-up of the model and the cross section of the breakwater. The hydraulic stability formula of Dolos armour layer is given by (Burcharth et al. 1992)

$$N_s = \frac{H_s}{\Delta D_n} = (47 - 72\,r)\,\varphi_{n=2} D^{1/3} N_z^{-0.1}$$

$$= (17 - 26\,r)\,\varphi_{n=2}^{2/3} N_{od}^{1/3} N_z^{-0.1} \quad (20)$$

where H_s significant wave height in front of breakwater
Δ $\left(\rho_{concrete}/\rho_{water}\right) - 1$, ρ is the mass density
D_n length of cube with the same volume as Dolosse
r Dolos waist ratio
$\varphi_{n=2}$ packing density
D relative number of units within levels SWL $\pm$ 6.5 Dn displaced one Dolos height h, or more (e.g. for 2% displacement insert $D = 0.02$)
N_{od} number of displaced units within a width of one equivalent cubic length D_n.
N_z number of waves. For $N_z \geq 3000$ use $N_z = 3000$.

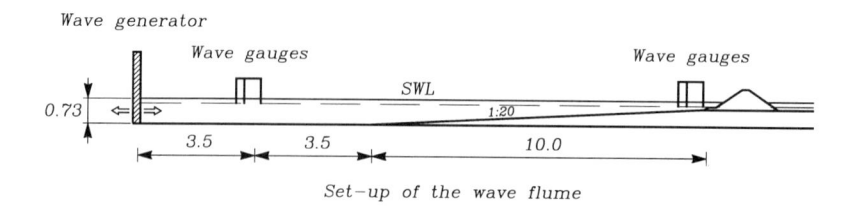

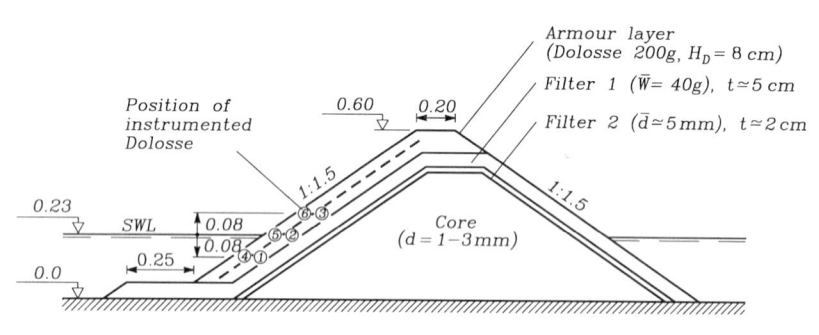

Fig. 8. Set-up of the wave flume and the cross section of the breakwater.

Distribution of stresses over the slope

The distribution of σ_T over the slope is of interest in order to identify the potential areas for armour breakage.

Fig. 9 shows typical distributions given by the 2% exceedence values of σ_T for each of the six instrumented Dolos positions for 10 t and 50 t Dolos of waist ratios 0.325 and 0.42 exposed to wave action levels, $N_S = \dfrac{H_{m_o}^t}{\Delta D_n} = 0.9, 1.8$ and 2.6.

The following conclusions can be drawn from the analyses of a large number of distributions of maximum σ_T over the slope:

- The contribution of the impact stress to the maximum principal tensile stress σ_T is small for $N_S \leq 2.0$.

- The contribution from the impact stress to σ_T is small in the bottom layer.

- The contribution from the impact stress to σ_T is very significant in the top layer.

- Breakage will in most cases start in the top layer in the zone just below SWL. This zone is more vulnerable to breakage than the zone above SWL.

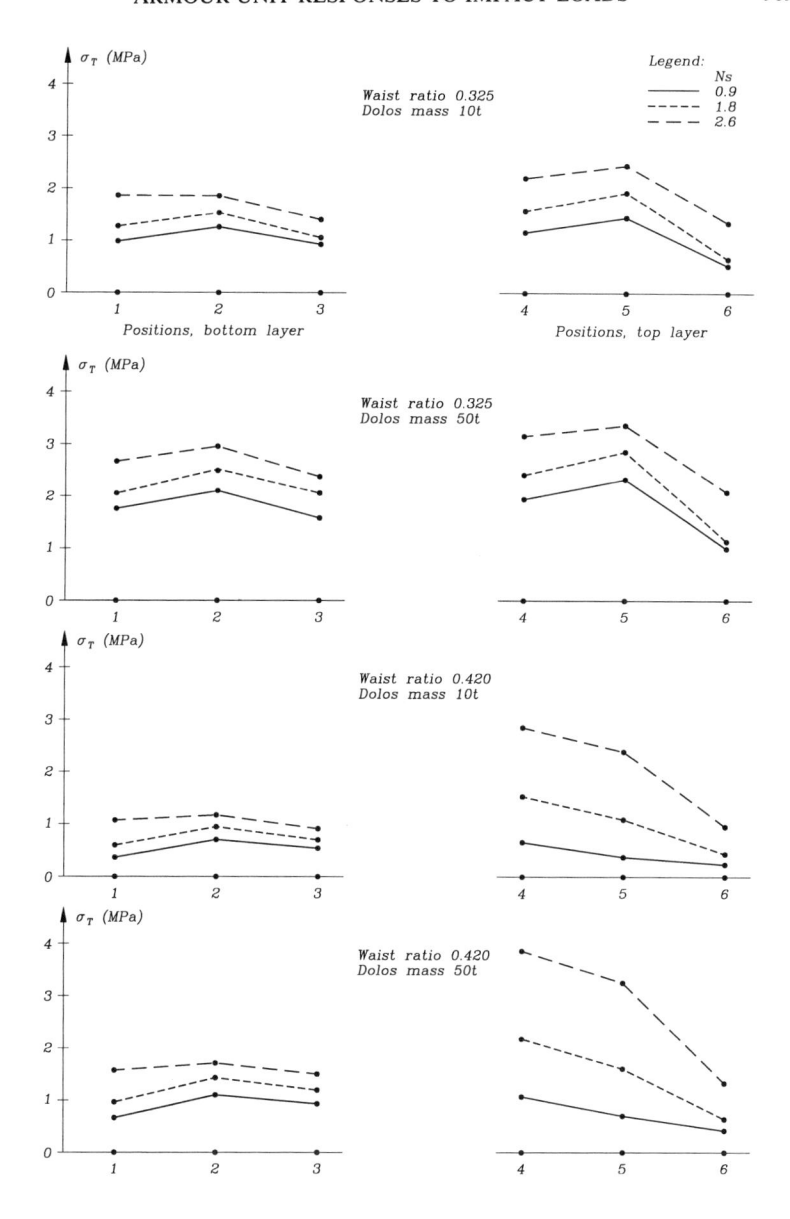

Fig. 9. Distribution of σ_T over the slope, Dolos positions 1 to 6, cf. Fig. 8. 2% stress exceedence probability level.

Design diagramme

In the analysis of breakage it is only the maximum value of σ_T in each instrumented Dolos within a test run which is of interest. Repeated short test runs of 100-300 waves were used because most movements take place in the beginning of each test.

The authors presented Dolos design diagrammes in ICCE'92 based on stress exceedence probability, which do not give exactly the proportion of the units that will break. A reanalysis was performed in which the maximum stress of each load-cell instrumented Dolos within each test run was compared with the strength of concrete in order to obtain the relative number of units that will break.

The results are given in the design diagrams, one of which is shown in Fig.10. For the complete set of the design diagrams reference is made to Burcharth (1993).

The concrete tensile strength in the diagrams is the one corresponding to static load. However, the diagrams take into account the dynamic amplification of the strength when impacts are involved.

The design diagrams have been checked against observed behaviour of prototype Dolos breakwaters and good agreement was found, cf. Table 1.

Table 1. *Observed and predicted damage of some Dolos breakwaters*

	Crescent City USA	Richards Bay SA	Sines POR
H_s (m)	10.7 [1]	5 [2]	9 [3]
slope	1:4	1:2	1:1.5
Dolos mass (ton)	38	20	42
Waist ratio	0.32	0.33	0.35
Dolos packing density	0.85	1	0.83
Concrete density (kg/m^3)	2500	2350	2400
Elasticity (MPa) [4]	30,000	30,000	30,000
Tensile strength (MPa) [4]	3	3	3
Reported displacement	7.3%		
Reported breakage	19.7%		
Reported displacement+breakage	26.8%	4%	collapse
Predicted displacement	3.6%	0.6%	3.6 %
Predicted breakage	> 10%	5%	> 10%

(1) depth limited in front of breakwater (2) in front of breakwater
(3) offshore ≈ in front of breakwater (4) estimated values

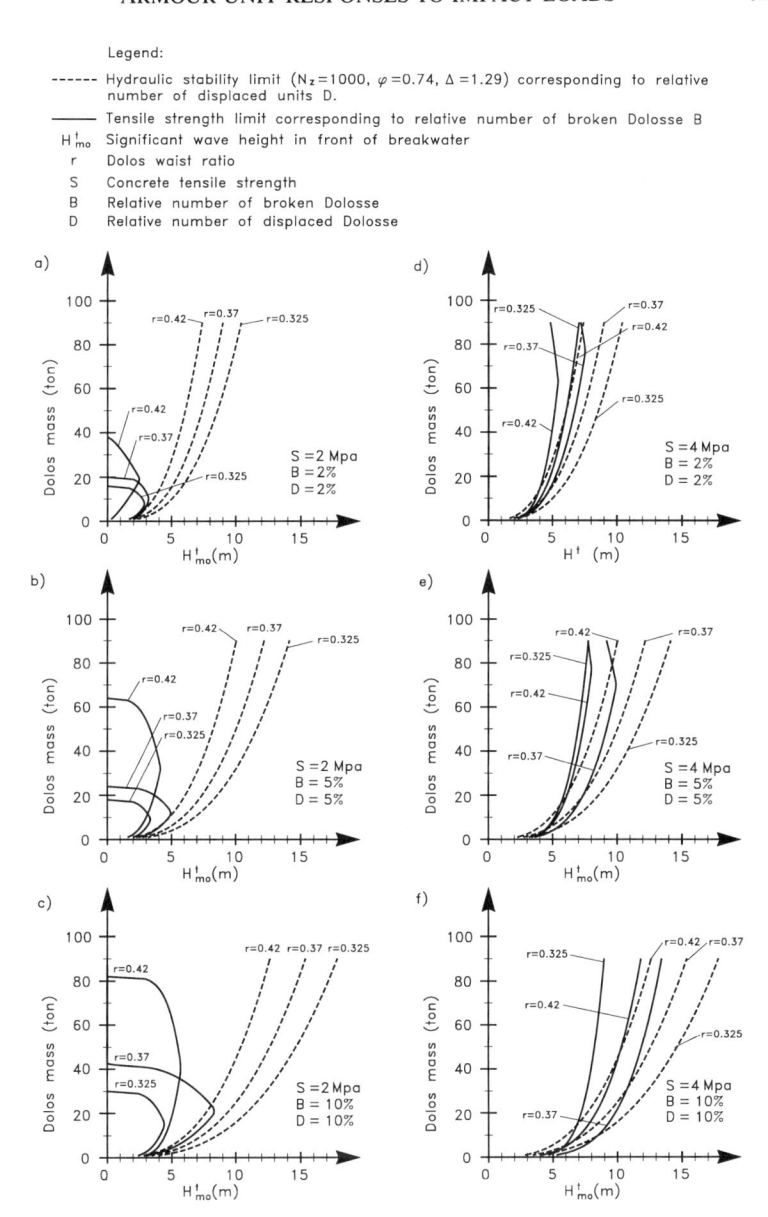

Fig. 10. Design diagrams for structural integrity and hydraulic stability of Dolos armour. Reference area SWL $\pm 6.5 D_n$.

Acknowledgements

The load-cells applied in this study were kindly made available by CERC, USA. The project was partly sponsored by the Commission of the European Community under the Marine Science And Technology Research Program (MAST II), and partly sponsored by the Danish Technical Research Council under the MARIN TEKNIK Research Program.

References

Burcharth, H.F. (1981). Full-scale dynamic testing of Dolosse to destruction. Coastal Engineering, 4 (1981).

Burcharth, H.F. (1984). Fatigue in breakwater armour units. Proceedings of the 19th International Conference on Coastal Engineering, Houston, Texas, 1984.

Burcharth, H.F. and Liu, Z.(1990). A general discussion of problems related to the determination of concrete armour unit stresses inclusive specific results related to static and dynamic stresses in Dolosse. Proc. Seminar on Structural Response of Armour Units. CERC, Vicksburg, MISS, USA.

Burcharth, H.F., Howell, G.L. and Liu, Z.(1991). On the determination of concrete armour unit stresses including specific results related to Dolosse. Coastal Engineering, 15 (1991), pp 107-165.

Burcharth, H.F. and Liu, Z. (1992). Design of Dolos armour units. Proceedings of the 23rd International Conference on Coastal Engineering. Venice. Italy.

Burcharth, H.F. (1993). Structural integrity and hydraulic stability of Dolos armour layers. Series Paper 9, published by the Department of Civil Engineering, Aalborg University, Denmark, 1993

Bürger, W.W., Oumeraci, H., Partenscky, H.W. (1990). Impact strain investigations on tetrapods results of dry and hydraulic tests. ASCE Proc. Seminar Stresses in Concrete Armor Units.

Melby, J.A. and Turk, G.F. (1994). Scale and modeling effects in concrete armor experiments. Coastal Dynamics'94, Barcelona, Spain.

Terao, T., Terauchi, K., Ushida, S., Shiraishi, N., Kobayashi, K. and Gahara, H. (1985). Prototype testing of Dolosse to destruction. Proc. Workshop on Measurement and Analysis of Structural Response in Concrete Armor Units. U.S. Army Corps of Engineers, CERC, WES, Vicksburg, MISS., USA.

CHAPTER 71

Steep wave diffraction by a submerged cylinder

J. T. Aquije Chacaltana & A. F. Teles da Silva

Abstract

The steep wave diffraction by a fixed circular submerged cylinder is numerically simmulated as a completely nonlinear time domain evolution from an initial condition. Attention is focused on two aspects of the problem. The first is a study of the perturbation introduced by the cylinder on the wave field; the second is a study of the hydrodynamic forces on the cylinder due to the waves.

Introduction

We study the diffraction of steep waves on a fixed, submerged circular cylinder in deep water; being particularly interested in the hydrodynamic forces induced by the waves on the cylinder and the disturbances produced by the cylinder on the wave field.

The problem of a cylinder held fixed beneath waves field was first studied by Dean (1948) in the context of gentle waves in deep water; Dean finds, to the linear approximation, that incoming waves are not reflected and the only disturbance produced by the cylinder on the waves is an uniform phase delay. Ogilvie (1963) extend the method of Ursell(1950) to study some related problems of wave–body interaction and shows, in the linear approximation for the problem of a restrained circular cylinder, that the hydrodynamic force components oscillate in quadrature with the wave period, have the same amplitude and a phase difference of $\frac{\pi}{2}$; a second order,

---

Inst. de Matemática & Lab. Mec. Fluidos.
Address :
PEM – COPPE / UFRJ, CP 68503, CEP 21945–970, R.J.–BRAZIL

steady vertical force component is calculated from the first order potential, but the steady horizontal force component is found to be zero. Mehlum (1980) gives a first order potential solution for a restrained cylinder under waves which is very easy to compute. Salter et al. (1976) measures forces on a submerged circular cylinder and finds that mean horizontal forces are quite small and come to be negative for steeper waves. Chaplin (1984) makes experiments with a submerged cylinder under waves finding no wave reflection, even for the highest waves; it is also found that the phase lag is smaller than predicted by the linear theory for Keulegan Carpenter, K_c, numbers smaller than 2. Vada (1987) calculates the second order oscillatory forces on a cylinder of arbitrary cross section; for the case of circular cylinders and K_c numbers smaller than results agree well with Chaplin's (1984). Palm (1991), working with a high order perturbation scheme, shows that there is no reflection at the order m and frequency $m\omega$, where m is an integer and ω is the frequency, for monochromatic incident waves. Bichromatic waves are shown to have a null second order reflection coefficient; this result is shown to hold for higher order terms. Liu, Dommermuth & Yue (1992) working with a time domain, higher order perturbation scheme calculate, numerically, the wave diffraction by a circular cylinder, giving results for the steady force components on the cylinder, and the transmission coeficients.

Mathematical Formulation

We assume that the flow is incompressible and irrotational being thus described by a velocity potential ϕ. The free surface is described parametrically by $\mathbf{R}(\xi,t) = [X(\xi,t), Y(\xi,t)]$ to allow for overturning. The velocity potential satisfy the nonlinear boundary value problem:

$$\nabla^2 \phi = 0 \tag{1}$$

$$\phi_t + \frac{1}{2}(\nabla\phi)^2 + gY = 0 \tag{2}$$

$$\frac{D\mathbf{R}}{Dt} = \nabla\phi \tag{3}$$

$$\frac{\partial \phi}{\partial n} = 0 \tag{4}$$

(1) is to be valid inside the fluid region, (2) and (3) on the free surface and (4) along the cylinder's surface. At some initial time, e. g. $t = 0$, values for **R** and $\phi|_\mathbf{R}$ must be prescribed; we set a steep nonlinear steady wave, impulsively, above a submerged cylinder in deep–water, see figure 1. The steep steady wave, was computed through a numerical code described in Teles da Silva & Peregrine (1988). The nonsteady evolution of the initial condition is computed by a Boundary Integral code which is an extension of the one developed by Dold & Peregrine (1984) to compute the evolution of surface waves on water of uniform finite depth; in the original scheme the free surface is mapped on a closed curve and the horizontal bed into a circle which is surrounded by the mapped free surface; the mapped free surface is then reflected about the bed to ensure impermeability; in the present case the transformed free surface is reflected about the transformed cylinder contour, instead of about the bed, Peregrine (1989).

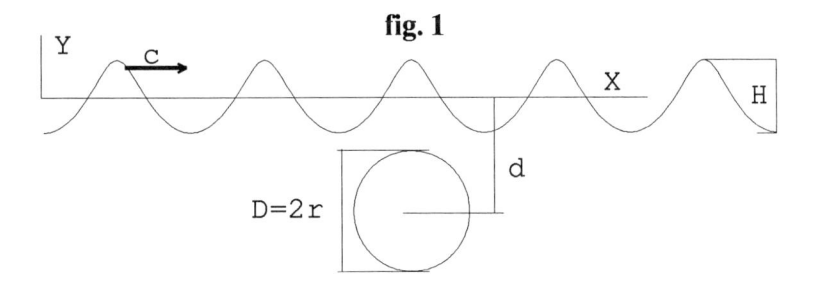

fig. 1

Results

Lengths have been scaled by $\frac{1}{k}$, where k is the wave number, accelerations by gravity g, and time by $(gk)^{-\frac{1}{2}}$.

The solutions of the problem depend on three parameters which are: i) wave-height H, ii) cylinder radius R, iii) the depth of the cylinder d. The potential flow modelling, given in the last section, poses strong limitations on the variation of these parameters: first, The Keulegan-Carpenter numbers, K_c, used as measure of the vortex-shedding, must be kept low, smaller than 1.5 or 2.0 according to the depth of the cylinder, Chaplin (1984); second, wave breaking stops the computations; and third, because of the imaging, the cylinder cannot be uncovered.

Most of the computations were made on a Sun Sparc Station 2. The simmulation of five wave crests moving over the submerged cylinder, during five wave periods, with twenty numerical points per wave takes 9.9 CPU seconds.

Figure 2 presents the time evolution of an initial wave, for $r = \frac{4}{3}$, $d = \frac{5}{3}$, which are the data used by Dean in his 1948 paper; the wave must be small, $H = 0.1$, for the cylinder being near the mean level, higher waves would break or uncover it; the centre of the cylinder, is at the start, located below the sixth crest at the position $x = \pi$; the wave length is $\frac{1}{11}\pi$; the numerical simmulation is made during a time of 11 wave periods.

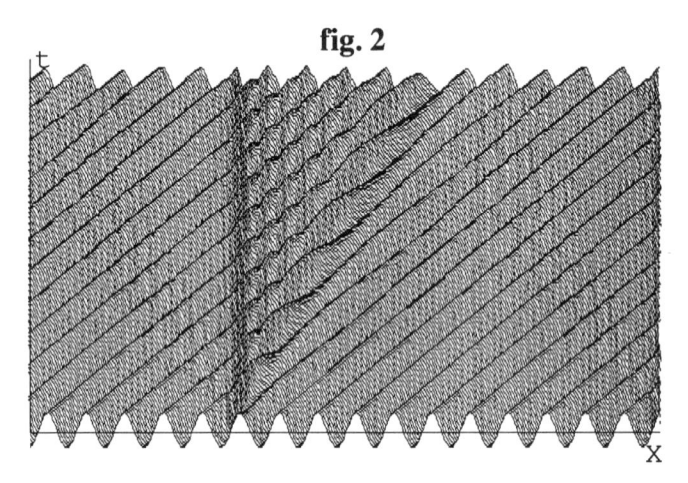

fig. 2

Figure 2 reveals some important aspects of the wave diffraction. With the exception of a faint disturbance starting above the cylinder at time zero, propagating upstream, due to the impulsive start of the numerical experiment, no other disturbances move upstream of the cylinder. This appears to be a nonlinear confirmation of Dean, and Palm's results about the absence of wave reflection in deep water. Following the lines of the crests in the space time diagram, Figure 2, we see that waves lose their steady shape as they pass a region of perturbation shaped like a 'V'; putting a ruler over a crest line we see that as the waves pass the cylinder the crests are over this line meaning that they are delayed; it is also possible to see for some crests, those that start nearer the cylinder, with the ruler, that after they leave the 'V' region of perturbation they get back to the original line, meaning that once outside the 'V' crests recover the delay. Another feature, in Figure 2, is that the wavelength of the perturbation on the 'V' appears to get longer towards its downstream edge as it would be for a wave group; with a ruler it is possible to estimate the slope of the 'V's edge and hence the velocity of propagation of the perturbations; these estimated velocities agree with the group velocity of small deep water waves with a wave-length which is a half of the wave length of the incident waves. The fact that the perturbed region be defined by a second harmonic of the transmitted wave suggests an analysis of the harmonics of the incident and the perturbed wave.

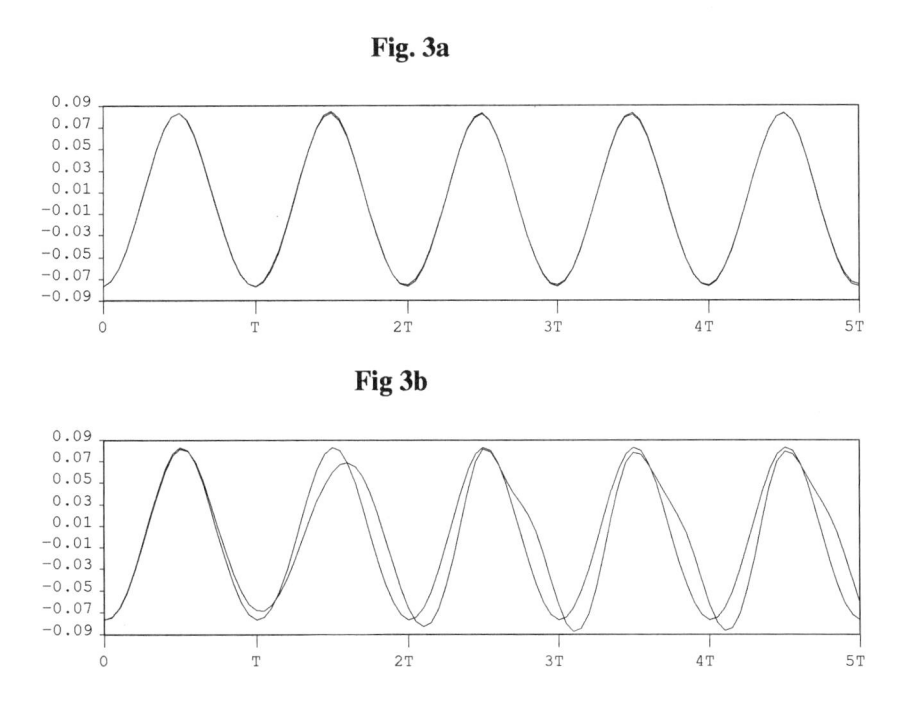

Fig. 3a

Fig 3b

In order to have an idea of the relative magnitude of wave harmonics in simmulations, the time history of the elevation of 2 points on the free surface have been recorded for some cases; the points have been placed at a horizontal distance of eight times the radius of the cylinder, one downstream and the other upstream. A typical case is shown in Figure 3 where, for an incident wave of heigth $H = 0.16$ and a cylinder of radius $r = 0.4$ submerged at the depth of $d = 0.8$, the time histories of the upstream, Figure 3a, and downstream, Figure 3b, surface elevation are depictured; in both cases, for comparison, the time history for the original steady wave has been included. In Figure 3a the time history of the upstream free surface elevation is undiscernible from that of the steady wave supporting the conclusion that waves are not reflected; the some does not happens for the upstream elevation where from the third wave period a conspicuous second harmonic appears in the perturbed wave. To see it more clearly we decomposed the time histories in its Fourier modes in the same way as for the forces in equations 5 and 6 below. The results are shown in Table 1 where the most important features, when we compare the upstream elevation with the elevation of the incident steady wave, appear to be the permanence of the first harmonic and the great increase of the higher harmonics; among the higher harmonics the most important is the second which in fact shapes the 'V'.

Table 1

Harmonic	Steady Wave	Upstream Wave	Downstream Wave
a_0	0.000010	-0.000106	0.000200
r_1	0.079812	0.078808	0.078772
r_2	0.003213	0.003252	0.016048
r_3	0.000213	0.000219	0.003312
r_4	0.000029	0.000003	0.001447
r_5	0.000022	0.000008	0.000195
r_6	0.000004	0.000006	0.000167

A remarkble fact in Figure 2 is that outside the perturbed 'V' region the waves recover their original shape and steadyness with no traces of the interaction with the cylinder although the interaction is nonlinear; this is the same for the steeper waves provided they do not break. Comparing the position of the crests going past the cylinder with the position they should have in the absence of the

STEEP WAVE DIFFRACTION

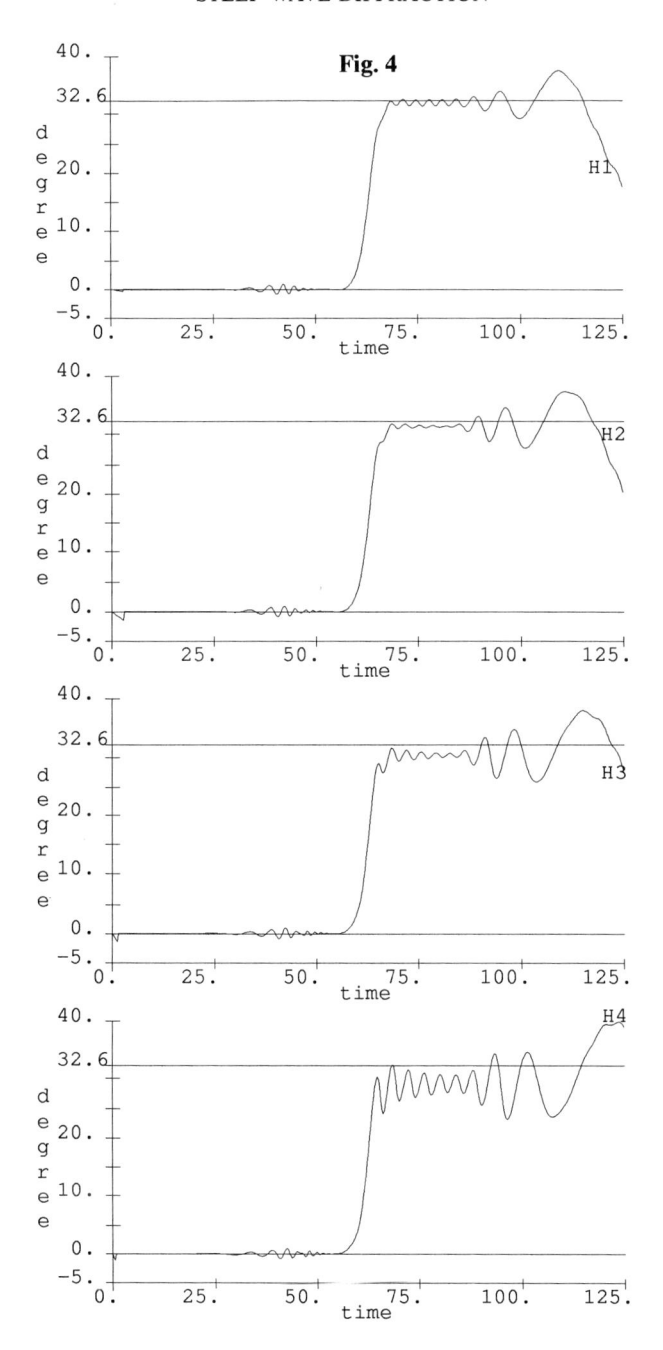

Fig. 4

cylinder, we find the phase lag. Figure 4 shows the time evolution for the phase lag, vertical axis in degrees, for four different waves; namely, wave 1: $H = 0.1$, $T = 6.275$, $c = 1.001$, wave 2: $H = 0.2$, $T = 6.252$, $c = 1.005$, wave 3: $H = 0.3$, $T = 6.213$, $c = 1.011$, wave 4: $H = 0.4$, $T = 6.1588$, $c = 1.0202$; H, T and c are respectively the nondimensional wave-height, period and phase speed. In all this experiments $r = 2\pi$ and $\frac{d}{r} = 1.2$; for this same radius and submergence Mehlum (1980) calculates a phase lag of 32.6 degrees. Note that this is nearly exact for wave 1, but, the phase lag tend to decay with wave-height. A bigger wave with $H = 0.5$ would break. In all the above cases, in given time, the phase lag returns to zero as the tails of the first two waves suggest; a permanent phase lag is an asymptotic result

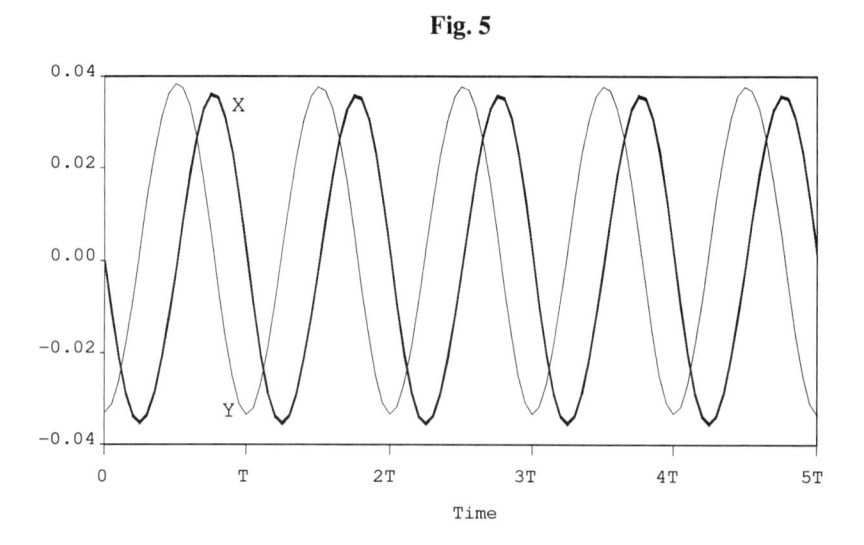

Fig. 5

A typical time history for the X and Y force components is shown in figure 5 for a wave of heigth $H = 0.5$; the figure shows a remarkble periodicity from a few periods after the begining of the experiment. For small waves the linear theory predicts that the X and Y, horizontal and vertical, force components have the same amplitudes, oscillate in quadrature with the wave period and there is a phase difference of $\frac{1}{2}\pi$ between them. We observed these results to hold for smaller waves; however for the higher waves we find the phase difference between the X and Y force components to become smaller, by as much as 10%, than $\frac{\pi}{2}$. The X and Y force components have been, both, decomposed in Fourier modes:

$$\frac{a_0}{2} + \sum(a_n \cos\theta_n t + b_n \sin\theta_n t)$$

where $\theta_n = \frac{2n\pi}{T}$, T being the wave period; or simmilarly as

$$\frac{a_0}{2} + \sum r_n \cos[\theta_n t - \delta_n] \qquad (5)$$

$$r_n = (a_n^2 + b_n^2)^{\frac{1}{2}} \qquad (6)$$

where δ_n is a phase shift. Results for fixed parameters 'R' and 'd' and a set of four waves with increasing wave height are displayed in table 2. We find for all waves a negative horizontal drift, $\frac{a_0}{2}$, which increases with the wave-height.

Despite the fact that the amplitudes of the X and Y force components are the same, irrespective of wave-height, the amplitudes of the Fourier modes of the same frequency increasingly differ for the higher waves. The vertical drift, $\frac{a_0}{2}$, increases with wave height reaching, for the last wave $H = 0.7$, an amplitude greater than any other Fourier mode, except the first.

As should be expected, the force harmonics for the smaller waves behave as in a perturbation expansion in the small parameter $\epsilon = H$, which is the dimensional wave height scaled by $\frac{1}{k}$; representing either the vertical or horizontal force components as:

$$f_1 \epsilon + f_2 \epsilon^2 f_3 \epsilon^3 + \ldots$$

where the first term in the expansion represents the first harmonic of the vertical or horizontal force component, the second term represents the second harmonic and so on. Observe in Table 2 that: the first harmonic for $H = 0.3$ is roughly three times the first one for $H = 0.1$; the second harmonic for $H = 0.3$ is roughly nine times the second one for $H = 0.1$; the steady components for $H = 0.3$

are roughly nine times the steady components for $H = 0.1$. For the waves with $H = 0.5$ and $H = 0.7$ these relations hold approximately only for the first harmonic. This provides a measure of the range of validity of the linear and weakly nonlinear results and also a way to estimate the force components of waves from the force components of particular wave. This is not new but it is very interesting to know that it holds for the first harmonics of a wave with $H = 0.7$ that is over 80% the height of the highest mathematically, not physically, possible wave.

Table 2

Wave/cylinder	Harmonic	X-component	Y-component
H=0.1	a_0	-0.000002	0.000289
T=6.2753	r_1	0.007285	0.007285
c=1.0012	r_2	0.000029	0.000029
D=0.5,d=1.	r_3	0.000001	0.000001
Kc=0.23	r_4	0.000000	0.000001
H=0.3	a_0	-0.000021	0.002489
T=6.2129	r_1	0.021710	0.021704
c=1.0113	r_2	0.000310	0.000313
D=0.5,d=1.	r_3	0.000003	0.000004
Kc=0.69	r_4	0.000003	0.000003
H=0.5	a_0	-0.000111	0.006417
T=6.0899	r_1	0.035615	0.035549
c=1.0317	r_2	0.001097	0.001102
D=0.5,d=1.	r_3	0.000035	0.000028
Kc=1.08	r_4	0.000005	0.000008
H=0.7	a_0	-0.000200	0.010207
T=5.9107	r_1	0.044369	0.044097
c=1.0630	r_2	0.002526	0.002534
D=0.5,d=1.	r_3	0.000192	0.000183
Kc=1.56	r_4	0.000037	0.000019

Conclusions

A numerical scheme for the time domain simmulation for the nonlinear steep wave diffraction by a submerged cylinder has been successfully implemented. Results have been checked, with very good agreement, with analytic, semi-analytic and experimental results given in the literature, namely Ogilvie (1963), Mehlum (1980), Vada (1987) and Chaplin (1984). In these comparisons an important point appears: many of the analytic and semi-analytic results presented in the literature are for waves that actually break; specially these for small cylinder depth where the diffraction effects are enhanced; by one side when waves break the flow ceases to be a potential flow and results should not be valid, but by another side Chaplin finds with experiments that, in his case E experiment, the mean vertical force agrees well with Ogilvie's results and the total force is smaller than that predicted by Ogilvie's results. This may suggest that approximate results frequently provide a good prediction of phenomena and sometimes give an upper bound for quantities.

Despite the limitations inherent to potential theory, and the computational costs in a time domain simmulations, which preclude their use for a thorough description of the phenomenon, some important aspects of the problem can be studied and complemented with frequency domain and experimental investigation; since computational costs are comparatively very cheap for frequency domain calculations and potential modelling is not a problem with experiments.

With respect to the disturbances on the waves due to the presence of the cylinder we have two main aspects: the phase-lag on the transmitted wave and the transmission and reflection coefficients. With respect to the phase lag, we find that the linear potential given by Mehlum (1980) gives cheap and accurate results for smaller waves and for the higher ones these results provide an upper bound. With respect to reflection our results support the linear prediction of no reflection; regarding the transmitted wave we find that the presence of the cylinder greatly enhances the higher harmonics; but to this stage there is not yet a satisfactory description of the dependence of this phenomenon on the parameters; neither have we yet atempted to quantify it with the help of transmission coefficients.

Regarding the forces on the cylinder due to waves an important result is that for the very steep waves the horizontal and vertical steady force components become evident and the vertical steady component come to be at the same order of magnitude as the first harmonic. Also important is the increase of the magnitude of the higher harmonics which is inherent to the very nonlinear character of steep waves; this increase in magnitude of the harmonics is followed by a loss of symmetry between the horizontal and vertical components.

Acknowledgements

The authors acknowledge support from the Conselho Nacional de Desenvolvimento Científico e Tecnológico, CNPq, for this research.

Bibliography

AQUIJE, J. T. 1993 Non–linear diffraction of water waves by a submerged body, M.Sc. degree, Prog. Eng. Mech. COPPE/UFRJ

CHAPLIN, J. R. 1984 Nonlinear forces on a horizontal cylinder beneath waves, *J. Fluid Mech.* **147**, 449 – 464.

DEAN, W.R. 1948 On the reflection of surface waves by a submerged cylinder, *Camb. Phil. Soc.* **44**, 483 – 491.

DOLD, J.W. & PEREGRINE, D.H. 1984 Steep unsteady water waves. An efficient computational scheme, *Proc. 19th ICCE*, Houston, Texas, **1**, 995 – 967.

LIU, Y., DOMMERMUTH, D. G. & YUE, D. K.P. 1992 A high–order spectral method for nonlinear wave–body interactions, *J. Fluid Mech.* **245**, 115 – 136.

MEHLUM, E. 1980 A circular cylinder in water waves, *App. Ocean Res.* **2**, 171 – 177.

OGILVIE, T. F. 1963 Fristand second order forces on a cylinder submerged under a free surface, *J. Fluid Mech.* **16**, 451 – 472.

PALM, E. 1991 Non–linear wave reflection from a submerged circular cylinder *J. Fluid Mech.* **233**, 49 – 63.

PEREGRINE, D. H. 1989 private communication

SALTER, S. H., JEFFREY, D.C. & TAYLOR, J.R.M 1976 *The Naval architect*, Jan. 1976, pp.21 – 24.

TELES DA SILVA, A. F. & PEREGRINE, D. H. 1988 Steep, steady finite depht const. vort. water waves, *J. Fluid Mech.* **195**, 281 – 302.

VADA, T. 1987 A numerical solution of the second order wave-diffraction problem for a submerged cylinder of arbitrary shape, *J. Fluid Mech.* **174**, 23 – 37.

URSELL, F. 1950 Surface waves on deep water in the presence of a submerged circular cylinder, I, *Proc. Camb. Phil. Soc.* **46**, 141 – 152.

CHAPTER 72

Wave Stresses on Rubble-Mound Armour

Andrew Cornett[*] and Etienne Mansard[*]

Abstract

In this study, physical experiments are used to investigate the relationship between fluid velocities on the surface of two rubble-mounds and the shear stresses τ and normal stresses σ acting on the surface layer of rock armour. Results presented herein indicate that the peak magnitude of slope-parallel hydrodynamic forces on rubble-mound armour located below the still waterline can be reasonably well estimated using wave friction factors originally developed for rough turbulent oscillatory flow over impermeable horizontal beds. Stresses on rubble-mound armour are also compared to the prevailing surf similarity (Iribarren) parameter. The largest stresses are found to result from waves that form collapsing breakers.

1 Introduction

Fluid flows on a rubble-mound under wave attack are highly variable in space and time. The spatial and temporal distribution of kinematics depends in a complex manner on the character of the incident waves, the type of wave breaking, and properties of the structure such as slope, roughness and permeability. Not surprisingly, the normal and shear stresses induced by such kinematics are also temporally and spatially variable.

Kobayashi et. al. (1990a, 1990b, 1990c) describe and present results from a numerical model of wave interaction with rough permeable slopes that includes an analysis of the stability of armour stones. In this model, the hydrodynamic forces acting on armour stones in the surface layer are separated into drag, inertia and lift forces that are calculated in terms of the fluid velocity and acceleration parallel to, and above, the surface of the structure. In this formulation, the forces on armour stones are independent of the kinematics normal to the surface of the rubble-mound.

Tørum (1994) measured regular wave loads on a single irregularly shaped armour stone with a mass of $0.152\,kg$ located $10.3\,cm$ below the still waterline on the

---

[*]National Research Council of Canada, Institute for Marine Dynamics, Building M32, Ottawa, Ontario, Canada K1A 0R6

surface of a reshaped berm breakwater. Measurements of fluid kinematics close to the stone were used to model the slope-parallel force component as a summation of drag and inertia forces calculated from slope-parallel kinematics. The slope-parallel force was found to be drag dominated, such that the force peaks occurred in phase with the peaks of slope-parallel velocity. The slope-normal force was modelled using a similar drag and inertia formulation, augmented by an additional term to represent lift force. In this case, drag and inertia forces were calculated from slope-normal kinematics, while the lift force was computed from the slope-parallel velocity. The peaks of the slope-normal force led those of the slope-parallel force, and were significantly smaller in magnitude. Tørum concludes that the slope-normal force is not dominated by lift, and could not be adequately modelled by the assumed augmented formulation of the Morison equation.

Shear stresses on an impermeable horizontal bed under oscillatory flow are commonly expressed in terms of a wave friction factor f_w, defined by

$$f_w = \frac{2\tau_{\max}}{\rho u_{\max}^2} \tag{1}$$

where $\tau_{\max}$ is the maximum shear stress at the bed, and $u_{\max}$ is the maximum orbital velocity just outside the boundary layer. Riedel et. al. (1972) and Kamphuis (1975) report results on f_w obtained with a shear plate in an oscillating water tunnel as functions of the maximum amplitude Reynolds number $Re = u_{\max} a/\nu$ and the relative roughness a/k_s where ν is the kinematic velocity, a is the amplitude of water particle orbits just outside the boundary layer, k_s is the Nikuradse sand grain roughness, given by $k_s \simeq 2D_{90} \simeq 2.5 D_{n50}$, where D_{n50} is the nominal diameter of particles on the bed. Different flow regimes were delineated, corresponding to laminar, smooth turbulent, and rough turbulent flow. For rough turbulent flow, the wave friction factor was found to be independent of Re and could be well represented by the simple expression

$$f_w = \frac{2}{5}\left(\frac{a}{k_s}\right)^{-3/4} \quad \text{for } \frac{a}{k_s} \leq 100. \tag{2}$$

2 Experiments

Experiments to measure the wave stresses on rubble-mound armour were performed in the 'Coastal Wave Basin' of the National Research Council of Canada at Ottawa. Figure 1 shows a sketch of the lay-out for these experiments. Three separate test channels, each $0.65\,m$ wide, were constructed near the centre of a $14\,m$ wide wave flume. The entrances to each channel were calibrated to ensure very similar wave conditions. Two dimensional sections of a rubble-mound breakwater were constructed in two of the three test channels. The third test channel was left open to record incident wave conditions using an array of capacitance wave probes. Stresses on the armour layer due to waves were measured in the central test channel using a pair of instrumented armour panels, described below

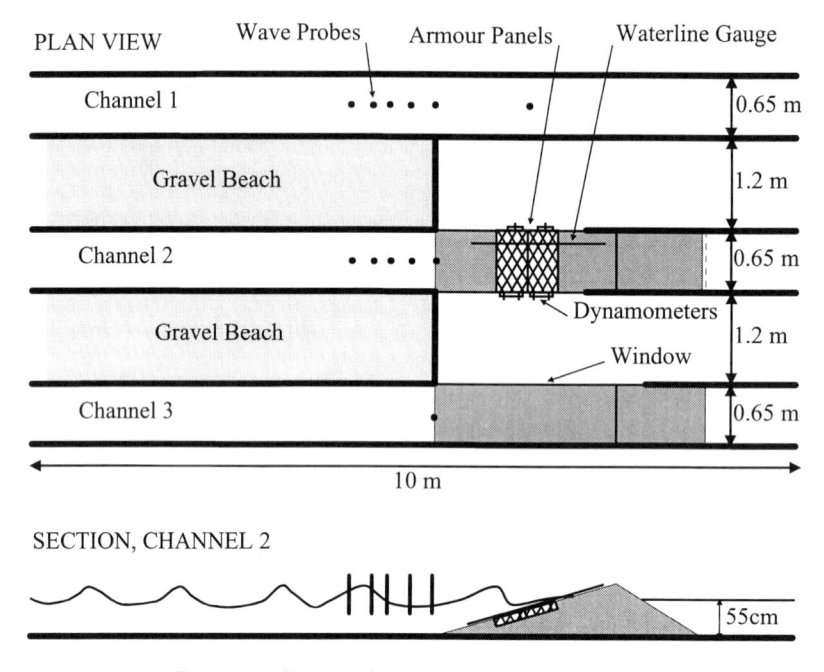

Figure 1: Sketch of the experiment lay-out.

in more detail. The third test channel was used to monitor damage to an identical rubble-mound constructed entirely from loose stones. The end of the wave flume was lined with a porous gravel beach that limited wave reflections to 5 % or less. Similar absorbing beaches were installed between the three test channels. This set-up allowed simultaneous measurement of the incident waves, stresses on armour stones, and armour damage, relatively uncontaminated by spurious wave reflections.

Results for two different structures (test series 4 and 5) are reported here. Pertinent characteristics are summarized in Table 1. Cross-sections are sketched in Figures 2 and 3. The armour for both structures consists of finely graded, angular, granite rock with nominal diameter $D_{n50} = (M_{50}/\rho_a)^{1/3} = 4.2\,cm$ placed in two layers. The series 4 rubble-mound features a sea-ward slope of $\cot\alpha = 1.75$, a permeable core with $D_{n50} = 2.6\,cm$, $D_{85}/D_{15} = 1.3$ and no filter layer. The series 5 structure features a milder slope of $\cot\alpha = 3$, a thin filter layer with $D_{n50} = 1.9\,cm$, $D_{85}/D_{15} = 1.3$ over an impermeable core.

2.1 Armour panels

Each instrumented armour panel consists of 50 individual, irregularly shaped, aluminum model rocks, bonded together by spot-welds into a rigid, porous, rect-

	Series 4	Series 5
water depth (cm)	55	55
slope, $cot(\alpha)$	1.75	3
core	permeable	impermeable
filter thickness (cm)	-	4
filter D_{n50} (cm)	-	1.9
filter D_{85}/D_{15}	-	1.3
armour thickness (cm)	8	8
armour $D_{n50}(cm)$	4.2	4.2
armour D_{85}/D_{15}	1.1	1.1
panel 1, centroid elev. (cm)	43.5	41.5
panel 2, centroid elev. (cm)	31.5	33.5

Table 1: Characteristics of the series 4 and 5 rubble-mounds.

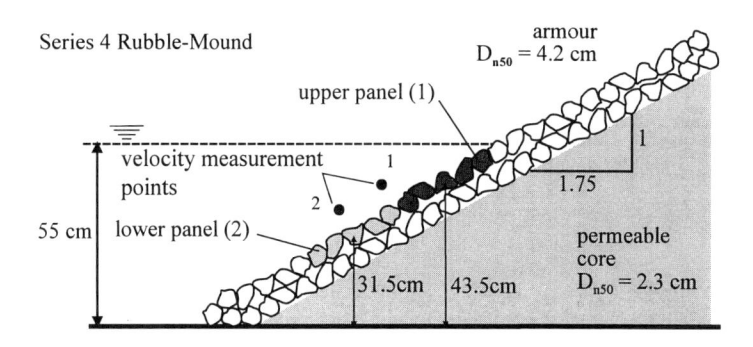

Figure 2: Cross-section sketch of the series 4 rubble-mound.

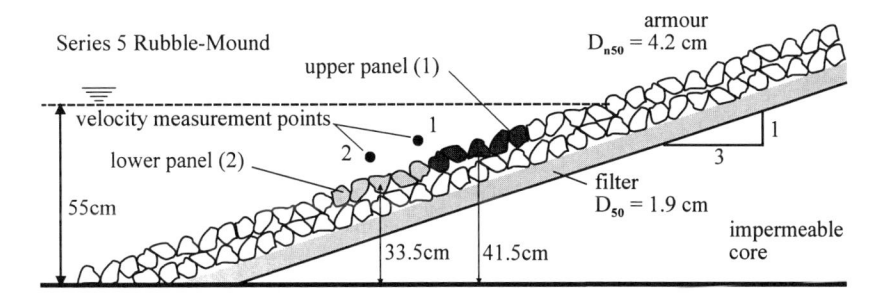

Figure 3: Cross-section sketch of the series 5 rubble-mound.

angular mat of armour stones in a single layer with approximate overall dimensions 64 cm by 23 cm by 5 cm. Each panel represents a 0.15 m^2 rectangular patch of armour stones. The panels were installed on the outer surface of the rubble-mound test sections, just below the still waterline, as sketched in Figures 2 and 3. The upper and lower panels are denoted as panels 1 and 2, respectively. The elevations of the panel centroids are included in Table 1. Each panel was connected at three points to a custom-built, five degree-of-freedom dynamometer located outside the walls of the test channel. Armour stones immediately surrounding the panels were glued in place so that the panels were isolated from adjacent materials by a thin ($\sim 2\,mm$) gap that followed their irregular shape. Away from the armour panels, wire mesh was placed over the surface of the rubble-mound to restrain the motion of loose armour stones. This set-up was adopted as a trade-off between the need to isolate the panels from the surrounding breakwater materials, and the desire to minimize distortions in the modelling of the rubble-mound.

The total fluid force $\mathbf{F}_f$ acting on an armour panel can be written $\mathbf{F}_f = \mathbf{F} + \mathbf{F}_b$ where $\mathbf{F}$ is the hydrodynamic force and $\mathbf{F}_b$ is the buoyancy force. For a panel that remains fully submerged, the buoyancy force is constant and equals the mass of fluid displaced by the panel and some dynamometer hardware. In this case, isolation of the hydrodynamic force is relatively straightforward. When the submergence of the panel varies with time, the magnitude of the time varying buoyancy force $F_b(t)$ can be estimated from the instantaneous elevation of the waterline on the surface of the rubble-mound $\eta_s(t)$ as

$$F_b(t) = |\mathbf{F}_b(t)| = \begin{cases} F_{b,\max} & \text{for } \eta_s(t) > \eta_{s,0} + l\sin\alpha\,, \\ F_{b,\max}\left(\frac{\eta_s(t)-\eta_{s,0}}{l\sin\alpha}\right) & \text{for } \eta_{s,0} \geq \eta_s(t) \leq \eta_{s,0} + l\sin\alpha\,, \\ 0 & \text{for } \eta_s(t) < \eta_{s,0}\,, \end{cases} \quad (3)$$

where $F_{b,\max}$ is the buoyancy force when fully submerged, $\eta_{s,0}$ is the largest value of η_s for which the panel remains entirely dry, and $l\sin\alpha$ is the effective vertical distance between the lowest and highest parts of the panel. Appropriate values of l and $\eta_{s,0}$ for each panel were determined by experiment.

Both panels were located entirely below the still waterline. The lower panel remained entirely submerged during all wave conditions, while the upper panel became partially submerged during attack by larger waves. When necessary, (3) was applied to estimate the time-varying buoyancy force in order to isolate of the hydrodynamic force acting on the upper panel.

2.2 Waves, waterline motion and kinematics

The rubble-mounds were exposed to a variety of regular and irregular waves; however, results in regular waves alone are reported here. Incident wave characteristics were computed by zero-crossing analysis of the water surface elevation $\eta(t)$ recorded in the side channel at a location corresponding to the toe of the rubble-mounds. The regular waves ranged in height from 10 cm to 22 cm at periods of 1.5, 2.0 and 3.0 s. For each wave condition, reported values were averaged over 100 wave cycles.

The vertical motion of the waterline $\eta_s(t)$ on the surface of the instrumented rubble-mound was recorded using a capacitance wire wave gauge inclined parallel to, and located approximately 1 cm above, the surface of the structure.

Fluid velocities were measured using a pair of bi-directional electromagnetic velocimeters located 4 cm above the upper surface of the instrumented armour panels at the positions shown in Figures 2 and 3. These locations are believed to be outside the boundary layer. Sleath (1992) gives the approximate displacement thickness δ of an oscillatory boundary layer as $\delta \simeq 0.5 f_w a$. For rough turbulent flow, (2) may be re-arranged to give $a = 0.295 k_s f_w^{-4/3}$, whence the displacement thickness can be estimated by

$$\delta \simeq 0.147 k_s f_w^{-1/3} \ . \tag{4}$$

With $k_s = 2.5 D_{n50} = 0.105\,m$, and $f_w = 0.15$, this equation gives $\delta \approx 0.029\,m$, which suggests that the positions of velocity measurement are outside the boundary layer. Velocity components parallel and normal to the surface of the rubble-mound are denoted by u and w, respectively, where u is positive up-slope and w is positive away from the structure.

Water particle orbits are estimated by integration of the Eulerian velocity signal. In particular, the water particle displacement parallel to the face of the structure at time t_1 is given by

$$s(t_1) = \int_0^{t_1} u(t)\,dt \ . \tag{5}$$

3 Shear and Normal Stresses

The armour panels provided a steady, repeatable measure of the fluid forces exerted on a patch of surface layer armour stones due to wave attack. The hydrodynamic component of the fluid force was isolated by compensating the measured force for the buoyancy of the armour panel. The hydrodynamic force was separated into orthogonal components F_P and F_N, acting parallel and perpendicular to the surface of the rubble-mound. F_P was defined positive up-slope, while F_N was defined positive away from the structure. These hydrodynamic force components can be expressed as a shear stress τ and a normal stress σ, defined by

$$\tau(t) = \frac{F_P(t)}{A}, \quad \sigma(t) = \frac{F_N(t)}{A} \tag{6}$$

where $A = 0.15\,m^2$ is the surface area occupied by a panel. The surface area can be expressed in terms of the number of armour stones N_a, and the porosity of the armour n, as $A = N_a D_{n50}^2 / (1-n)$. The average hydrodynamic force components acting on a single armour stone located within the patch are denoted by f_P and

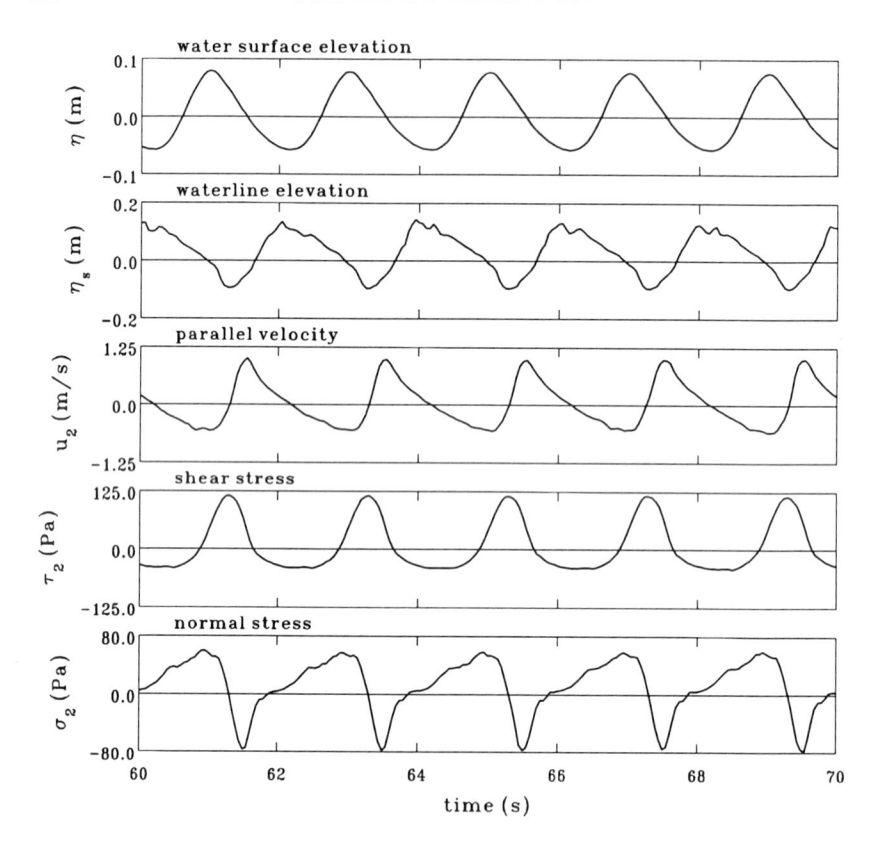

Figure 4: $\eta(t)$, $\eta_s(t)$, $u_2(t)$, $\tau_2(t)$, $\sigma_2(t)$ on the series 5 rubble-mound ($H = 14\,cm$, $T = 2\,s$ and $\xi = 2.2$).

f_N, and can be estimated in terms of the shear and normal stresses as

$$f_P(t) = \frac{D_{n50}^2}{1-n} \cdot \tau(t) \ , \quad f_N(t) = \frac{D_{n50}^2}{1-n} \cdot \sigma(t) \ . \tag{7}$$

The shear and normal stresses defined by (6) include contributions from all hydrodynamic forcing mechanisms, including drag, inertia and lift forces.

Short segments of $\eta(t)$, $\eta_s(t)$, $u_2(t)$, $\tau_2(t)$ and $\sigma_2(t)$ from a series 5 regular wave test with $H = 14\,cm$ and $T = 2\,s$ are presented in Figure 4. The velocity signal features an asymmetric 'saw-tooth' shape that indicates large positive (upslope) accelerations immediately prior to the maximum up-slope velocity. Negative (down-slope) accelerations are less intense, but prevail for a longer duration during each flow cycle. The shear stress time series features sharp positive (upslope) peaks in approximate phase with the large positive accelerations, which suggests that accelerations are a dominant forcing mechanism. Negative shear stress peaks are significantly smaller and more broad, which parallels the character of the negative fluid accelerations. In these wave conditions, the maxima and minima of the normal stress coincide with the negative and positive velocity peaks, respectively. Significant seepage flows normal to the surface of the rubble-mound, associated with drainage of the permeable outer layers, were observed during the tests, even for structures with an impermeable core. These slope-normal flows are an important factor contributing to the normal stresses on the surface layer of armour.

Individual flow cycles on the rubble-mound were defined to start and finish at the times of maximum runup. During each flow cycle, the maximum and minimum values of velocity, fluid particle displacement, shear stress, and normal stress were obtained and then averaged over 100 regular waves to give representative quantities, which are denoted by the subscripts $_{min}$ and $_{max}$; i.e. u_{min}, u_{max}, etc. Peak-to-peak values are denoted by subscript $_{pp}$, and are defined as the positive difference between the maximum and minimum value; i.e. $u_{pp} = u_{max} - u_{min}$.

For a given wave height incident to a given rubble-mound structure, the wave period influences the type of wave breaking that prevails. The effect of wave period on the type of wave breaking, and on the stability of rock armour, is commonly quantified in terms of the surf similarity (or Iribarren) parameter $\xi = \tan\alpha\sqrt{gT^2/(2\pi H)}$. Battjes (1974) observed that the character of wave breaking on a slope depends on ξ, such that collapsing breakers prevail for $\xi \sim 3$, while plunging breakers prevail for $\xi < 2$, and surging breakers prevail for $\xi > 4$.

Van der Meer (1988) presented separate design formulae for the stability of rubble-mound armour under plunging and surging wave attack. His equations suggest that minimum stability occurs under collapsing breakers and that wave period has a strong influence on stability for plunging waves ($\xi < 2$), but has much less influence on stability for surging waves ($\xi > 4$). Van der Meer's equations also describe a dependence between armour stability and the permeability of a rubble-mound, such that stability is enhanced on more permeable structures.

Figure 5 shows peak-to-peak values of **shear** stress as a function of surf similarity for the upper and lower panels of the series 4 and 5 rubble-mounds. Here, τ_{pp} has been made non-dimensional by the factor $\rho g H$, which represents the pressure under a static column of water with height H. Several observations can be made.

- Panel 1 (the upper panel) always experiences greater shear stresses. This result is consistent with the numerical simulations of Kobayashi et. al. (1990c) which suggest that armour damage is most likely to occur at an elevation approximately $0.75H$ below the still waterline, and that the likelihood of damage decreases below and above this elevation.

- On both panels, the normalized peak-to-peak shear stress is maximized for $\xi \sim 2.5$. For a given wave height, the shear stress exerted on a patch of armour stones is greatest for waves with periods such that $\xi \sim 2.5$. These

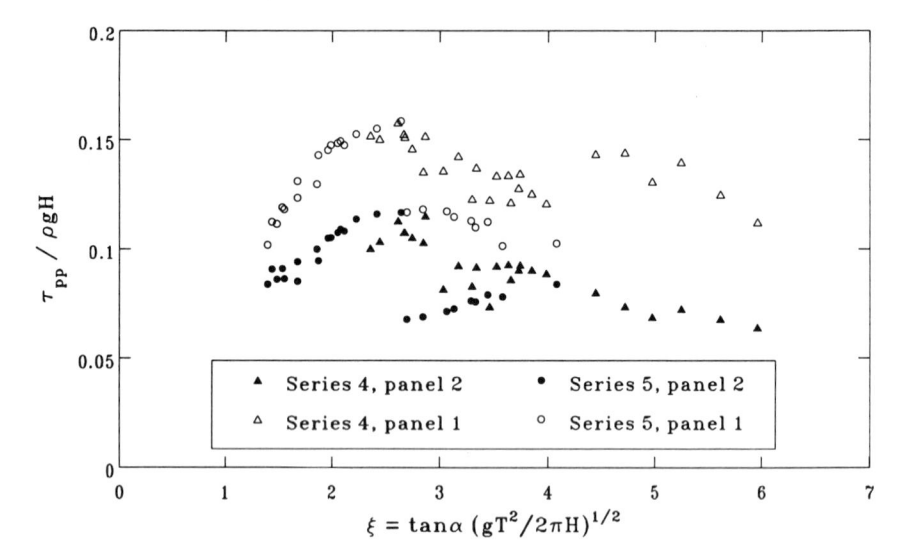

Figure 5: Non-dimensional peak-to-peak shear stress.

conditions are close to those associated with collapsing breakers. This result is consistent with idea that stability is minimized for collapsing breakers, and supports the form of stability design equation proposed by van der Meer (1988).

- Surf similarity has a strong effect on the normalized peak-to-peak shear stress in plunging waves ($\xi < 2$), but has relatively little influence on the shear stress in surging waves ($\xi > 4$). This result is also consistent with the design equations of van der Meer (1988), and suggests that plunging and surging breakers represent distinctly different processes.

- Tørum (1994) computes a shear stress of $\tau = 216\,Pa$ for a specific regular wave with $H = 0.2\,m$ and $T = 1.8\,s$. In non-dimensional terms, this is equivalent to $\tau/\rho g H = 216/(9810 \cdot 0.2) = 0.11$, which is in general agreement with the shear stresses reported here.

A similar plot of non-dimensional maximum **normal** stress $\sigma_{max}/\rho g H$ on the lower and upper panels of the series 4 and 5 rubble-mounds is presented in Figure 6. Positive normal stresses act to lift armour stones out of the surface layer and thus are critical to the stability and initial motion of armour units. In general, the variation in maximum normal stress measured on these two structures in various wave periods are less well described by ξ alone. In spite of this, the following observations can be made.

- Normal stresses are greater on the upper panel. Again, this supports the result that armour stones just below the still waterline are more susceptible

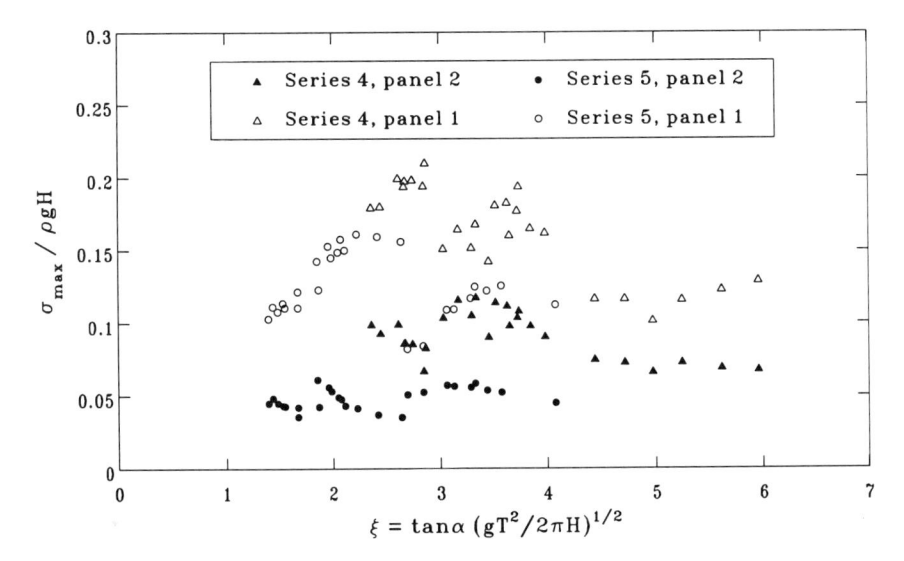

Figure 6: Non-dimensional maximum normal stress.

to damage than those lower down the rubble-mound. The normal stress maxima measured on the upper panel generally exceed the peak-to-peak shear stress in the same wave condition.

- For the same value of ξ, larger normal stresses are exerted on the armour of the steeper, more permeable series 4 rubble-mound.

- Surf similarity has a significant effect on the normal stress exerted on the upper panel, but has only a small influence on the normal stress acting on the lower panel.

4 Friction Factors for Rubble-Mound Armour

Flows on the surface of a rubble-mound under wave attack exhibit similarities and differences compared to wave driven flows on a horizontal seabed. While both flows are fundamentally oscillatory, flows on a rubble-mound tend to exhibit proportionately larger high-frequency content that produces 'saw-tooth' velocity fluctuations and larger accelerations. Flows on a rubble-mound also vary greatly with position. Near the toe, flows are similar to those on a horizontal seabed at similar depth, however, approaching the still waterline, velocities and accelerations are significantly amplified. Above the point of minimum rundown, the surface of the rubble-mound is only intermittently submerged. In this zone, the depth of flow can vary greatly throughout a wave cycle. When the water level outside the structure exceeds the level of the internal phreatic surface, water flows into the

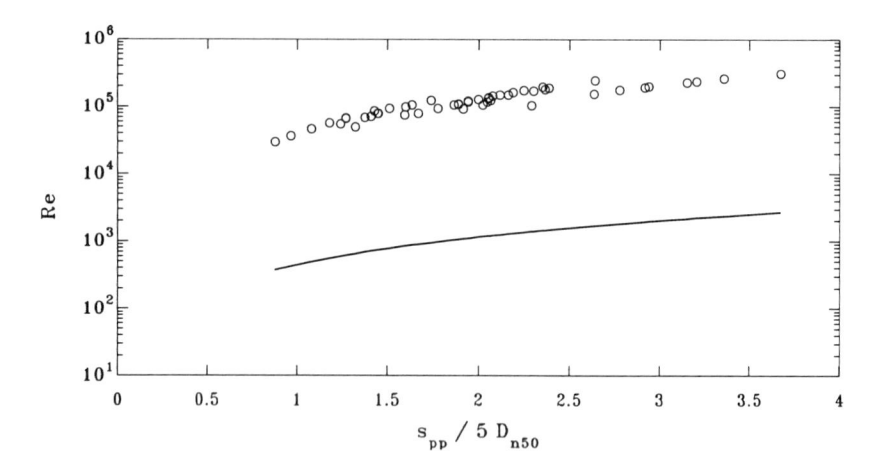

Figure 7: Orbital amplitude Reynolds number at velocimeter 2.

permeable outer layers. As the external water level recedes below the internal phreatic surface, water flows out of the permeable outer layers. Infiltration flow generally occurs above the mean water level during uprush while seepage flow is concentrated below the mean water level during downrush.

Kamphuis (1975) suggests the following criterion for the lower limit of rough turbulent flow:

$$Re = \frac{u_{\max}a}{\nu} \geq 200\frac{a}{k_s}\sqrt{\frac{f_w}{2}} \quad \text{for rough turbulent flow.} \quad (8)$$

Substituting (2) for f_w yields

$$Re \geq 447\left(\frac{a}{k_s}\right)^{1.375} \quad \text{for } \frac{a}{k_s} \leq 100. \quad (9)$$

An equivalent orbital amplitude Reynolds number for the surface of a rubble-mound can be written

$$Re_{pp} = \frac{\frac{u_{pp}}{2}\frac{s_{pp}}{2}}{\nu} = \frac{u_{pp}s_{pp}}{4\nu} \quad . \quad (10)$$

The relative roughness of the flow on a rubble-mound can be written

$$\frac{s_{pp}/2}{2.5D_{n50}} = \frac{s_{pp}}{5D_{n50}} \quad (11)$$

which is equivalent to the quantity a/k_s. Figure 7 shows values of Re_{pp} at velocimeter 2, in which the solid line indicates the threshold criterion for rough turbulent flow given by (9). The data span the range of $s_{pp}/5D_{n50}$ (relative roughness) from 0.8 to 3.7 and indicate that rough turbulent flow prevails for all test conditions.

Measurements of the shear stress on panel 2 and the kinematics at position 2 have been used to construct friction factors for rubble-mound armour. Separate

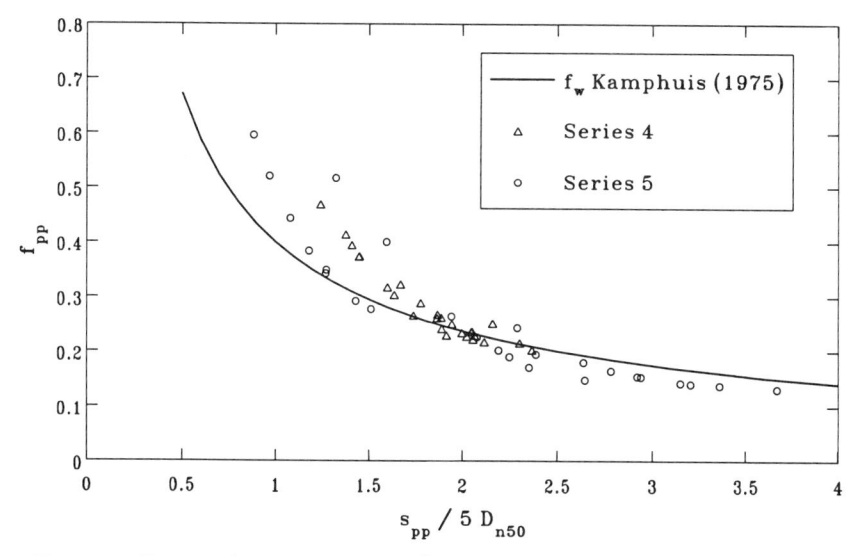

Figure 8: Friction factors on panel 2 for the series 4 and 5 rubble-mounds.

friction factors are computed for the uprush and downrush portions of the flow cycle according to

$$f_u = \frac{2\tau_{max}}{\rho u_{max}^2} \quad \text{for uprush}$$

$$f_d = \frac{-2\tau_{min}}{\rho u_{min}^2} \quad \text{for downrush.}$$

(12)

A friction factor for the complete flow cycle is defined in terms of the peak-to-peak velocity u_{pp} and peak-to-peak shear stress τ_{pp} as

$$f_{pp} = \frac{2\left(\tau_{pp}/2\right)}{\rho\left(u_{pp}/2\right)^2} = \frac{4\tau_{pp}}{\rho u_{pp}^2} \quad \text{for the complete flow cycle.} \quad (13)$$

Friction factors for the complete flow cycle f_{pp} measured on panel 2 during regular wave attack of the series 4 and series 5 rubble-mounds are favorably compared in Figure 8 to estimates of f_w from (2). Data from the two different rubble-mounds show good agreement, however, there is evidence of a stronger variation with relative roughness than predicted by (2). Overall, these results suggest that the relation between friction factor and relative roughness on a rubble-mound under wave attack is similar to that for a rough, impermeable, horizontal bed under oscillatory flow. Moreover, these results suggest that wave friction factors might be used to estimate the shear stresses acting on rubble-mound armour exposed to wave attack.

Friction factors for the uprush and downrush portions of the flow cycle are presented in Figures 9 and 10. Most of the data for the uprush friction factor suggest that $f_u < f_w$, while most of the data for the downrush friction factor

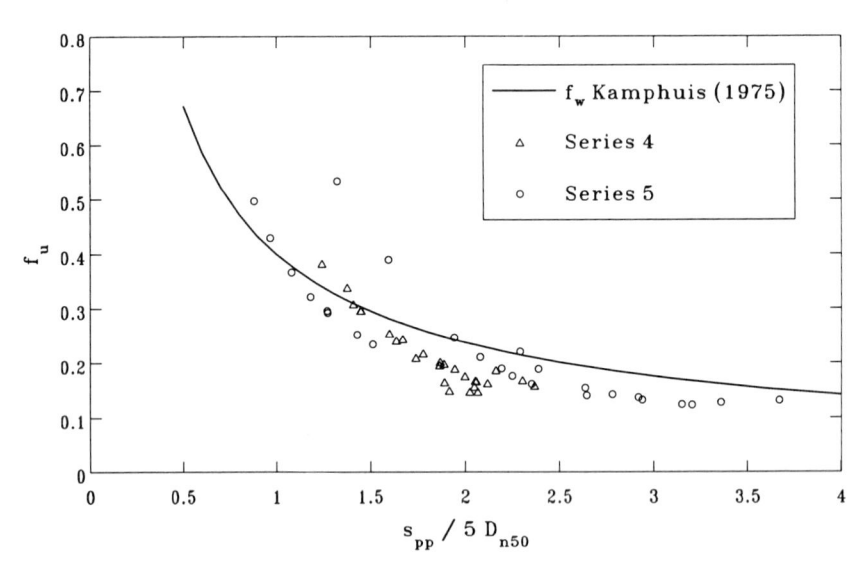

Figure 9: Uprush friction factors on panel 2 for the series 4 and 5 rubble-mounds.

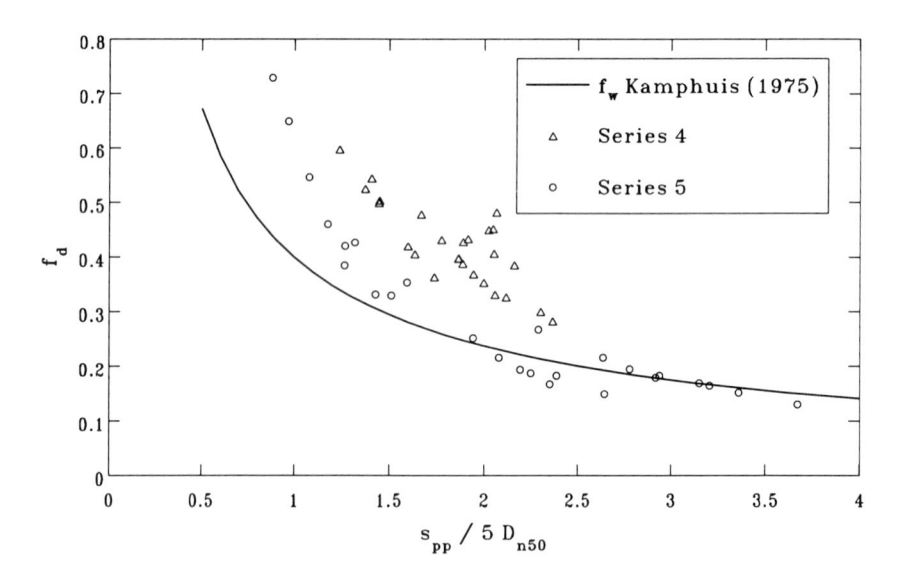

Figure 10: Downrush friction factors on panel 2 for the series 4 and 5 rubble-mounds.

suggest that $f_d > f_w$. In other words, the peak shear stress acting down-slope is generally larger than that predicted by (2), while the peak shear stress acting up-slope is generally smaller. The trend is particularly evident for the steeper, more permeable, series 4 rubble-mound. This behavior is not entirely surprising, considering the distorted, asymmetric, quasi-oscillatory flow cycles that prevail on the surface of a rubble-mound (see Figure 4), in contrast to the more symmetric flow cycles that prevail over horizontal beds under waves.

Kobayashi et. al. (1990a, 1990b, 1990c) assumed a constant friction factor of $f_w = 0.3$ in numerical simulations of regular and irregular waves on conventional rubble-mounds. Tørum and van Gent (1992) used a constant wave friction factor $f_w = 0.15$ in their numerical simulations of regular wave interaction with a permeable berm breakwater, while Tørum (1994) used $f_w = 0.3$. These values are in general agreement with the results reported here. However, the present results confirm that the friction factor for rubble-mound armour is a variable quantity that, for rough turbulent flow, depends on the relative roughness in a manner that is similar to that for oscillatory flow over a rough, impermeable, horizontal bed, and can be reasonably well predicted by (2).

5 Conclusions

Measurements of the shear and normal stresses acting on rubble-mound armour under regular wave attack have been presented. The stresses were measured on a pair of rigid porous armour panels, each supported by a five degree-of-freedom force dynamometer, and installed to minimize distortions to internal flows within the test structures. Results from two different structures have been considered.

- Greater shear and normal stresses were measured just below the still water-line than lower down the rubble-mound.

- Shear and normal stresses are maximized in waves for which $\xi \sim 2.5$, suggesting that collapsing breakers exert greater forces on armour stones and are therefore most critical to the their stability.

- The influence of wave period on shear stress is strong for plunging waves ($\xi < 2$), and relatively weak for surging waves ($\xi > 4$).

- The shear stresses reported here are in general agreement with the force measurements on a single armour stone reported by Tørum (1994).

- In some cases, normal stresses were measured that exceed the shear stresses.

- Greater normal stresses were recorded on the steeper, more permeable, series 4 rubble-mound.

- Friction factors for rubble-mound armour under regular wave attack are in general agreement with the wave friction factors f_w for rough turbulent flow computed from (2).

- Friction factors for uprush are generally less than f_w, while friction factors for downrush are generally greater. This behavior is related to the asymmetric, quasi-oscillatory nature of surface flows on a rubble-mound.

6 Acknowledgment

The authors gratefully acknowledge the contributions to this work made by Mr. E. Funke, the staff at M32 and the support of the National Research Council.

References

[1] Battjes, J.A. 1974. Surf Similarity. *Proc. 14th Int. Conf. Coastal Eng.* Copenhagen, Denmark. Chapter 26.

[2] Kamphuis, J.W. 1975. Friction Factor under Oscillatory Waves. *J. Waterways, Harbors and Coastal Eng.* ASCE, Vol. 101 (WW2), pp. 135-144.

[3] Kobayashi, N., Wurjanto, A. and Cox, D. 1990a. Irregular Waves on Rough Permeable Slopes. *J. Coastal Research.* Special Issue No. 7, pp. 167-184.

[4] Kobayashi, N., Wurjanto, A. and Cox, D. 1990b. Numerical Model for Waves on Rough Permeable Slopes. *J. Coastal Research.* Special Issue No. 7, pp. 149-166.

[5] Kobayashi, N., Wurjanto, A. and Cox, D. 1990c. Rock Slopes Under Irregular Wave Attack. *Proc. 22nd Int. Conf. Coastal Eng.* pp. 1306-1319.

[6] Nielsen, P. 1992. *Coastal Bottom Boundary Layers and Sediment Transport.* World Scientific, London.

[7] Riedel, H.P., Kamphuis, J.W. and Brebner, A. 1972. Measurement of Bed Shear Stress under Waves. *Proc. 13th Int. Conf. Coastal Eng.* Vancouver, Canada. pp. 587-603.

[8] Sleath, J.F.A. 1984. *Sea Bed Mechanics.* John Wiley & Sons, Toronto.

[9] Tørum, A. and van Gent, M. 1992. Water Particle Velocities on a Berm Breakwater. *Proc. 23rd Int. Conf. Coastal Eng.* pp. 1651-1665.

[10] Tørum, A. 1994. Wave-Induced Forces on Armour Unit on Berm Breakwaters. *J. Waterway, Port, Coastal and Ocean Eng.* ASCE, Vol. 120, No. 3, pp. 251-268.

[11] Van der Meer, J.W. 1988. Rock Slopes and Gravel Beaches under Wave Attack. Doctoral Thesis, Delft Univ. of Tech. Delft, The Netherlands.

CHAPTER 73

DAMAGE ANALYSIS FOR RUBBLE-MOUND BREAKWATERS

Michael H. Davies
Etienne P.D. Mansard, and
Andrew M. Cornett[1]

Abstract

Storm damage to rubble-mound structures can range from the piece-wise removal of individual armour stones to the large-scale sliding of the entire armour layer. This paper presents an overview of qualitative and quantitative descriptions of various types of armour layer damage due to wave action. This work is based on a series of large-scale physical models of revetment stability. The discussion includes methods for measuring the change in profile and the use of the normalized area eroded, S. A new parameter, d_c is introduced as a measure of the minimum depth of cover remaining on the structure. An analysis method is presented which predicts the armour layer's factor of safety against sliding.

Introduction

Coastal structures are often designed to survive severe storms with no significant damage while undergoing tolerable damage during more extreme events. For such designs it is essential that the damage be both predictable and repairable - the failure mode should be ductile not brittle. Physical hydraulic models are a key tool in the design of structures where damage levels need to be determined. In addition to accurately reproducing the wave loading on the structure and the physical characteristics of the structure it is important to be able to accurately measure and quantify the damage which occurs.

Damage analysis

The two common measures for rubble-mound damage due to wave action are visual assessment and profiling. Visual assessment includes the monitoring of the development of holes in the armour, observation of the exposure of under-layers and counting the number of armour units displaced. Profile analysis includes measurement of the eroded profile and calculation of the eroded cross-sectional area, A.

---

[1] Coastal Engg. M-32, Institute for Marine Dynamics, National Research Council Canada, Montreal Rd, Ottawa, Ontario, K1A 0R6.

At low damage levels the breakwater damage can be readily estimated by visually counting the number of rocks displaced. This becomes difficult however, when more than, say, 50 stones are displaced or when stones which have been moved slightly by one storm condition are subsequently moved further downslope by a subsequent storm sequence. Image processing allows visual techniques to be used at higher damage levels. The size of the damaged area can be measured using photogrammetry and digital image processing. Also, the movement of individual stones can be tracked using image processing software. The limitations on these techniques are that it is often necessary to drain the basin prior to image-capture to allow measurement below the still water line and that image quality can restrict accuracy. Furthermore, most techniques allow an estimation of only the damaged surface area, not the depth of penetration of the damage. This can sometimes be overcome by colour coding the various layers of revetment.

For damage quantification based on stone count, the 1984 Shore Protection Manual (CERC, 1984) defines damage as the percentage of stones removed from within the active zone. The active zone is defined as extending from the middle of the breakwater crest down to one zero-damage wave height below the still water level.

$$\text{Damage (\%)} = \frac{\text{Number of stones removed}}{\text{Number of stones within active zone}} \times 100 \qquad (1)$$

This has the limitation that it depends on visual stone counts (which can be unreliable at higher damage levels) and that the number of stones within the active zone is very dependent upon structure geometry.

Several mechanical and electro-mechanical profiling methods exist for the determination of breakwater profile. All have their limitations. The more sophisticated profilers which operate on the principles of electrical resistivity or light or acoustic beams to locate the profile generally require the flume to be either fully drained or fully flooded so that measurements are made in either the dry or in the wet - measurement through the air-water interface is difficult. The need to flood or drain the basin between each test sequence can be very time consuming. Measurement methods which use mechanical contact with the rocks to locate the profile can readily cross the air-water interface but often provide lower resolution and sparser data.

Measurement of the cross-sectional profile is often the fastest and most reliable means of damage assessment and is used in most hydraulics laboratories. Hughes (1993) reviews some of the visual and profiling methods available for damage analysis. Van der Meer (1988) reports on the use of a mechanical profiler for measuring damage to rubble-mound break-waters. Latham et al. (1988) report on the use of high resolution mechanical profiling methods for breakwater damage analysis. Medina et al. (1994) reports on the use of mechanical profiles for damage analysis.

If profiles of the damaged cross-section are obtained, the damage can be represented by the mean cross-sectional area removed from the profile, A. In this analysis the eroded quantity, A is determined by integration of the difference between the damaged and initial profiles. The most widely accepted method for interpreting the damage data from profiling is the damage index, S as adopted by van der Meer. The damage index,

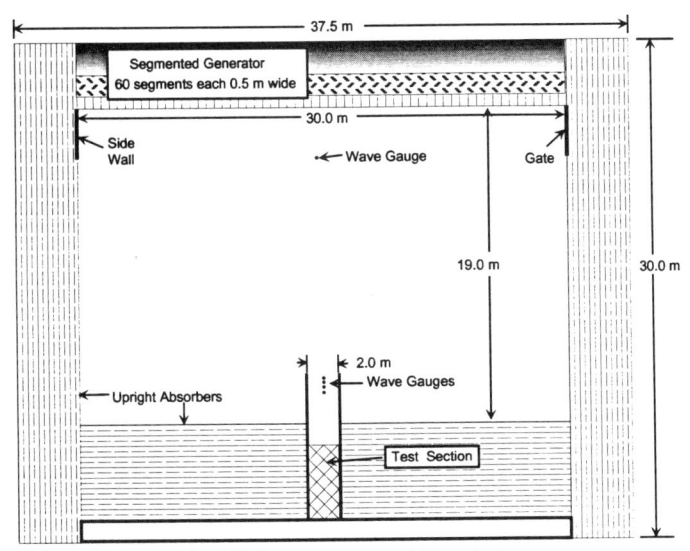

Figure 1 Test configuration for NRC revetment stability study.

S is defined as:

$$S = \frac{A}{D_{n50}^2} \quad \text{where} \quad D_{n50} = \sqrt[3]{\frac{M_{50}}{\rho_s}} \tag{2}$$

S corresponds to the number of cubic stones of size D_{n50} removed from the average profile line. M_{50} is the median of the cumulative stone mass distribution and ρ_s is the density of stone (typically 2,650 kg/m³).

For comparison between counted stone damage and profile analysis a visual estimate of the damage, S_{visual} can be derived as follows:

$$S_{visual} = \frac{\text{\# stones removed}}{(\text{Structure width}/D_{n50})(1-n)} \tag{3}$$

Here n is the placement density determined by profiling the armour and underlayers and from knowing the total mass of rock placed.

NRC Experiments on breakwater damage

NRC has recently completed a series of large-scale 2-D experiments on revetment stability for the Société d'énergie de la Baie James. This study examined the existing revetment along with proposed repair schemes for some of the earth dams and dykes of the La Grande hydroelectric complex near James Bay in northern Québec. Some details of this model study are reviewed in Mansard et al (1994). This test series was performed in the NRC Multi-directional Wave Basin at scales ranging from 1:10 to 1:15

depending on the structure being tested. Figure 1 shows the experimental setup used for these tests. The wave climates used in the tests were calibrated prior to installation of the structures in the test section to allow determination of the time-domain characteristics of the incident wave climate in the absence of reflections from the structure. The results from some of the testing performed in this study will be used in this paper to illustrate various damage analysis methods.

The NRC breakwater profiler (shown schematically in Figure 2) uses a carriage-mounted mechanical pivoting arm to track the breakwater profile. The arm is connected to a rotational potentiometer which measures the angle of declination, θ. The horizontal location of the carriage, R is measured using a potentiometer connected to the cable system which drives the carriage. The contact point between the rotating arm and the armour layer is a wheel of diameter roughly equal to the mean stone size, D_{50}. Thus each profile is averaged mechanically so that gaps in the armour layer smaller than D_{50} are ignored. Output from the two potentiometers is sampled at 20Hz using the NRC Gedap data acquisition and analysis system. In processing, the data files, R(t) and θ(t) are reduced to a single file containing z(x) using the relationships:

$$x = R - L\cos(\theta) \quad \text{and}$$
$$z = \text{(Elevation of beam)} - L\sin(\theta) \quad \text{where L is the profile rod length.} \tag{4}$$

Typically nine evenly-spaced profile lines are measured. Analysis can be performed on individual profile lines to obtain details on the spatial distribution of damage or the measured profile lines can first be averaged then integrated. The cross-sectional area of material eroded from the armour layer is determined by integration using Simpson's method, $A_i(x)$:

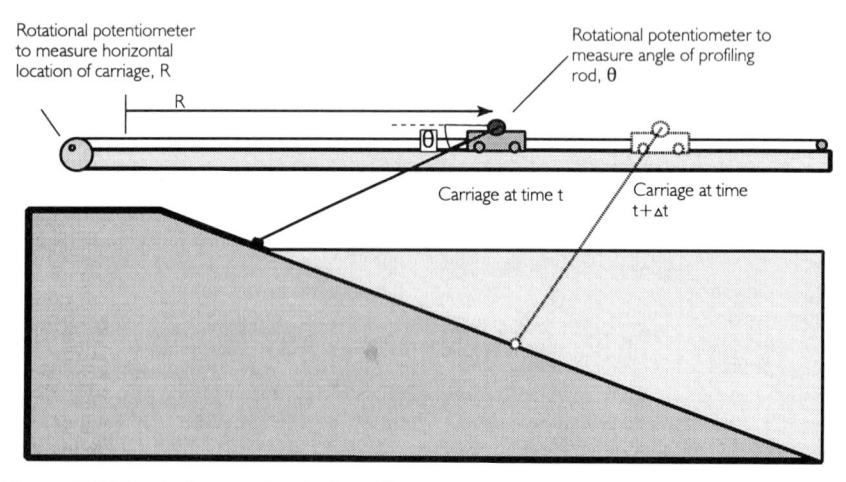

$$A_i(x) = \int_{x_0}^{x} [z_i(x) - z_0(x)] dx \tag{5}$$

Figure 2 NRC electro-mechanical profiler.

DAMAGE TO RUBBLE-MOUND

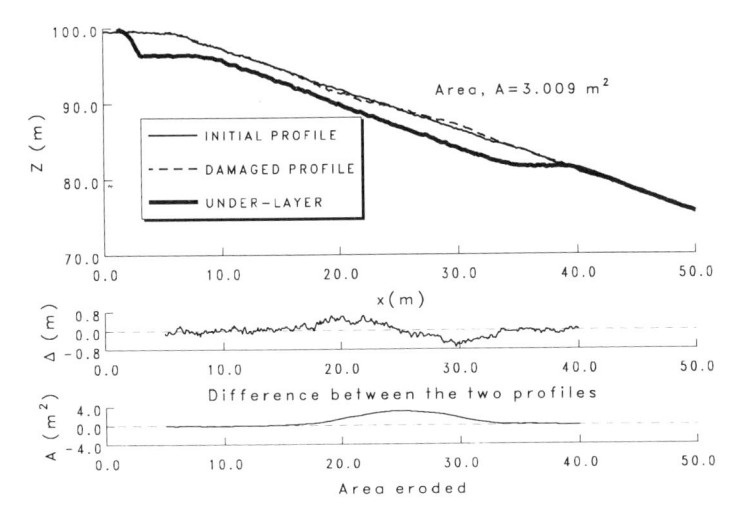

Figure 3 Typical average profile analysis.

where $A_i(x)$ is the indefinite area integral and z_i and z_o are the damaged and original (undamaged) profile lines, respectively, x_o is the lower limit of integration (typically the top of the slope). S is determined as the maximum value of $A_i(x)$ normalized by the square of the nominal diameter.

$$S_i = \frac{\max[A_i(x)]}{D_{n50}^2} \qquad (6)$$

Figure 3 shows the results of average profile analysis - the upper plot shows the original and damaged profiles ($z_o(x)$ and $z_i(x)$), the middle plot shows the vertical difference between the two profiles, $\Delta(x)$ and the bottom plot shows the indefinite integral, $A_i(x)$.

When sparse data is collected (e.g. foot method) curve fitting is often used to define a function describing the eroded area, A(x) (e.g Medina et al, 1994 using a sinusoidal smoothing function). Previously integration had been performed only on the measured erosion by first removing the negative portion of the $\Delta(x)$ curve in an attempt to increase the accuracy of the determination of S (e.g. Vidal et al, 1992). This method then requires accurate delineation of the range of integration since random noise in the $\Delta(x)$ signal produces a positive bias in the integration. Recent refinements to NRC's profiling and data analysis methods have made such pre-processing of the data unnecessary.

While S analysis does give a reasonable measure of the average cross-sectional area of material eroded from the profile, it does not necessarily give sufficient information regarding the severity of damage. As illustrated in Figure 4 a value of S=3 could represent a deep hole in the armour layer near the water line, or equally it could represent shallow damage spread over a wider area. The significance of S depends strongly on stone size and armour layer thickness. When the damage index, S is applied to berm-type revetments the absolute value of S loses its physical significance although the stabilization of damage (attainment of an equilibrium profile) is characterized by a constant value of S.

Analysis of Conventional Armour Layers

Van der Meer suggests using a value of S=2 for the initiation of damage (zero-damage condition of Shore Protection Manual - 0 to 5% damage). Medina et al. suggest that lower values of S perhaps better describes the initiation of damage. Based on a series of 40 experiments on revetment stability performed at the National Research Council, it has been seen that S=2 provides a good estimate of the start of damage. Lower damage levels have been seen to exhibit greater experimental variability.

Figure 5 illustrates a repeatability test, two realizations of the exact same test, this figure shows that agreement between the two experiments improves as S increases above 2. Figure 6 shows a comparison between the values of S obtained from profile analysis and the estimates of S based on visual stone counts. The good agreement obtained between these two methods illustrates the accuracy of the profiling method. Once the stone count gets above 60 to 80 stones the reliability of the visual method drops - it becomes difficult to visually monitor the movement of so many stones.

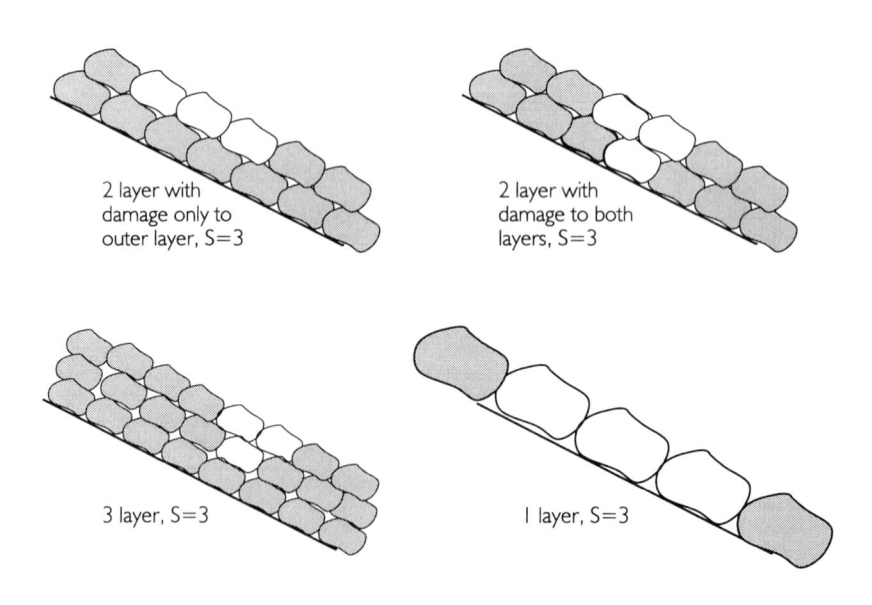

Figure 4 Various possible damage configurations with same S.

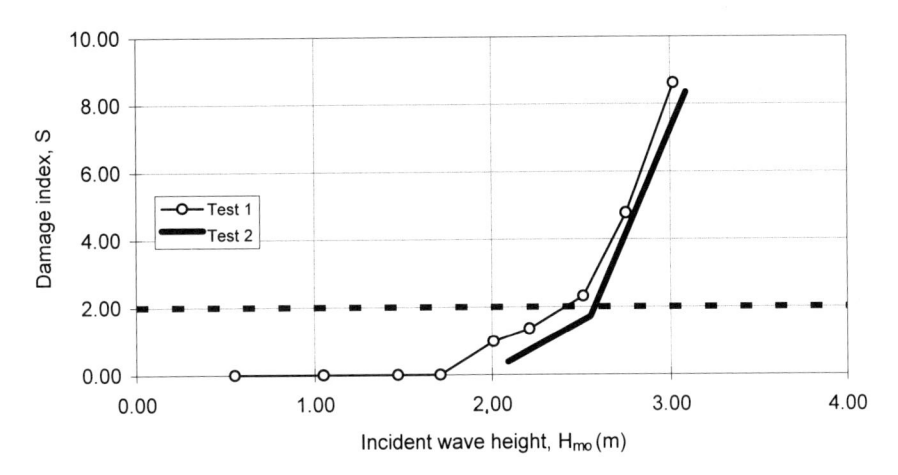

Figure 5 Repeatability test for typical revetment.

For berm breakwaters, the absolute value of S loses its direct relevance to damage (since many stones can move in the process of re-shaping to a stable profile). However, the shape of the damage evolution curves (S vs time, t or S vs number of waves, N) does serve as a useful indication of performance. A stable berm should tend

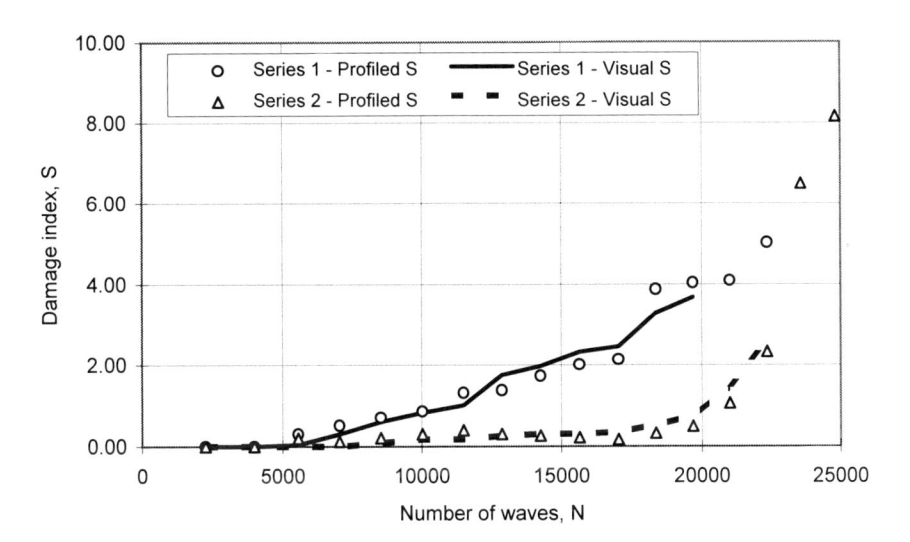

Figure 6 Comparison between S and S_{visual} for two typical revetment tests.

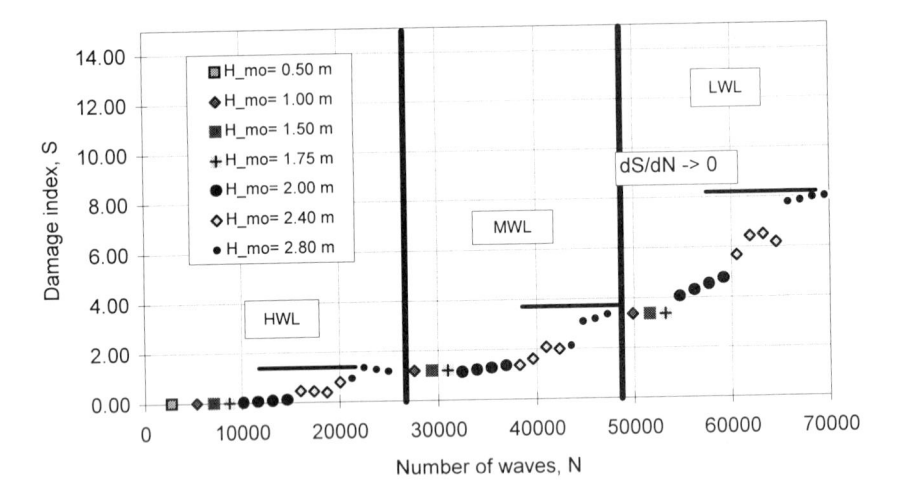

Figure 7 S vs N for a berm tested and found stable (dS/dN=0) at three water levels.

towards dS/dN=0 under design conditions. This pattern is illustrated in Figure 7 where, during waves of H_{mo}=3.0 m, the berm stabilizes and dS/dN approaches zero.

Depth of cover

The purpose of the outer revetment on a breakwater is to prevent the wave action from attacking the finer underlayers. Since failure of a breakwater is primarily caused by removal of the armour layer resulting in exposure and subsequent rapid removal of the finer underlayers it is reasonable to quantify the damage in terms of the existence of holes in the armour layer. For such analysis the quantity of material removed is inconsequential, it is the protection remaining which matters. This can be quantified as the depth of cover remaining on the structure, d_c. By measurement of the profile of the underlayer during construction and taking the difference between measured profiles of the revetment during testing and the reference underlayer profile one can calculate the thickness of the revetment, and the minimum thickness of the revetment, d_c. For most analyses the average minimum depth of cover is used. For example, in a series of two-dimensional flume experiments, nine profiles were taken over the 2 m width of the test section, the average profile was then subtracted from the average underlayer profile and the minimum cover thickness, d_c was computed. The resulting measured average minimum depth of cover can be compared to D_{n50}, such that when d_c drops below D_{n50} it can be inferred that there is some exposure of underlayer across the width of the test section. Should d_c drop to zero, the damage is severe enough that significant damage to the underlayer is likely to have occurred. Figure 8 shows typical depth of cover analysis for a revetment scheme. The left hand plot shows the profile (initial, damaged and under-layer), the armour layer thickness for both the initial and damaged conditions is shown in the right hand plot. The cover thickness is calculated as the slope-normal distance between the armour layer profile and the under-layer. The depth of cover, d_c is taken as the minimum of the cover thickness over the range of interest. As shown in the plot, d_c for the initial conditions occurs at a different location than d_c for the damaged conditions.

When applied to berm-style revetments, the performance criteria is typically that the depth of cover, d_c should not drop below a certain level under design storm conditions. Figure 9 shows a d_c vs N plot for a berm-style structure (again tested at three water levels).

Sliding failure

Other than the conventional unravelling of coastal structures due to the removal of individual stones by wave action, revetment failure can also occur by the large-scale sliding of the entire protection layer. This geotechnical failure mode has been discussed by other researchers but without an analytical framework within which to analyze such failures, the literature on this subject remains mostly anecdotal. In several of the revetment stability tests performed at NRC large-scale sliding of the armour layer was observed. Often this occurred with minimal prior damage to the structure. Several such failures were observed for steep sloped (1.6:1 and 1.7:1) structures. None were observed for structures with milder slopes (1.8:1 and 2.2:1). Figure 10 shows the measured profile of a revetment after sliding occurred near the mean water level (MWL). The profile of the slope after a sliding failure looks similar to other wave-eroded profiles - the key difference being the rapid rate of damage. This damage occurred during a single wave group (5 or 6 individual waves) and involved the downslope movement of the armour layer 'en masse'. Profile analysis is of little use in interpreting sliding - the main tools are visual observation and geotechnical analysis of the slope's factor of safety against sliding.

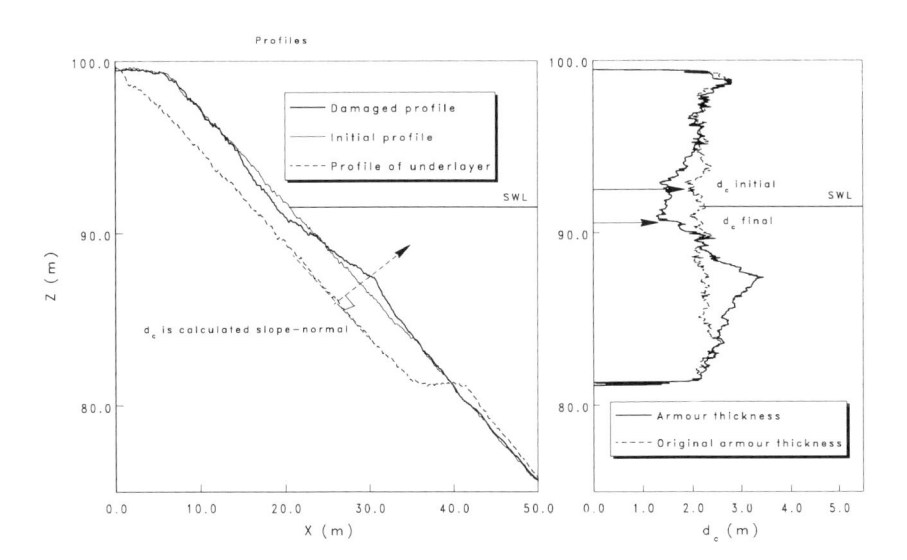

Figure 8 Typical depth of cover analysis.

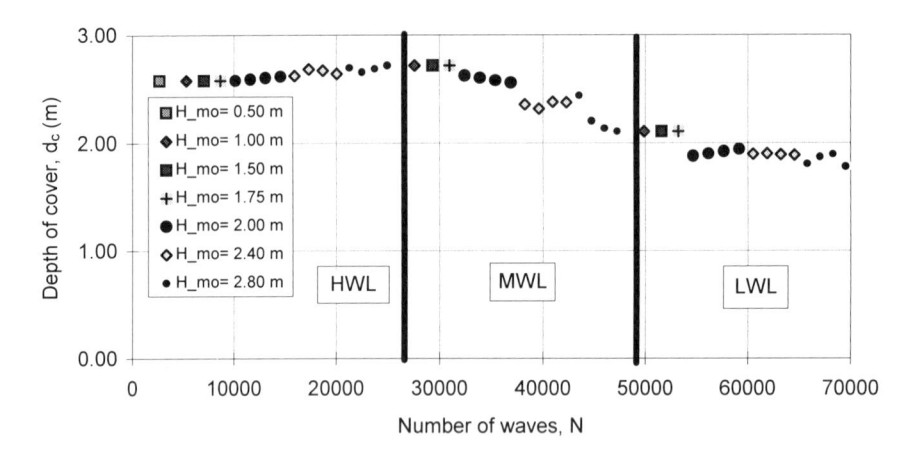

Figure 9 Depth of cover, d_c vs N for a berm.

A separate series of NRC experiments measuring the force exerted on breakwater revetments by wave action provides us with a reliable tool for estimating the wave loading on the revetment (Cornett and Mansard, 1994). These wave forces are then integrated into a limiting equilibrium translational sliding model. This uses conventional geotechnical stability analysis in conjunction with estimates of wave-induced lift and drag forces to evaluate the factor of safety of the slope against large-scale sliding failure. Comparisons between the sliding model results and breakwater testing have verified the ability of this method to predict the likelihood of sliding (Davies et al., 1994). This failure mode is most predominant for very steep breakwater slopes - 1:1.6 up to 1:1.3. Much of the literature on sliding of armour layers is related to the sliding which can occur once wave attack has created a hole in the armour layer, thereby providing an unsupported protection layer above the hole. This analysis is valid for such problems but is of more interest in the case of steep slopes where large-scale sliding is shown to occur without prior removal of the armour layer.

Block model for sliding

Under the action of gravity alone, a slope of cohesionless soil will fail through translatory sliding. Under wave action, application of the driving force is limited to the zone around the still water line. In this case, the failure will be a block translational slide.

For the static problem, the planar transitional slide predominates, and the factor of safety, FS (in the absence of external loadings) is simply the ratio of the tangent of the mobilised angle of shear resistance at failure, ϕ' to the tangent of the slope angle, θ.

$$FS = \frac{\tan \phi'}{\tan \theta} \qquad (7)$$

The factor of safety is reduced when additional forces are exerted by wave action. The downslope shearing force due to wave action, F_{shear} can be estimated using the wave friction factor of Kamphuis (1975). The applicability of this type of analysis for the steep and very rough slopes of rubble-mound breakwaters has recently been verified at NRC by Cornett. In a series of hydraulic model tests, the wave forces exerted on breakwater armour have been measured and compared to the friction factor, f_w calculated using wave orbital velocity measurements just above the measuring section (see Cornett's figure 8 these proceedings, 1995). It is assumed that the wave-induced shear stress is applied uniformly over the region from the still water level down to H (measured vertically). This corresponds approximately to the range over which force measurements were made by Cornett.

Applying solitary wave theory to the breakwater slope, we can estimate the peak orbital velocity, u_b and the corresponding orbital amplitude, a_b as follows:

$$u_b = 0.61\sqrt{gd_b} \quad \text{and} \quad a_b = \frac{u_b T}{2\pi} \qquad (8)$$

here u_b is determined at an elevation equal to 1/2 the breaking depth of the wave (i.e. at $z=d_b/2$, where $d_b=1.28\,H$).

Applying the friction factor relationship of Kamphuis:

$$f_w = 0.4 \left(\frac{k_s}{a_b}\right)^{0.75} \quad \text{where } k_s = 2D_{90} \qquad (9)$$

This allows estimation of the shear stress acting on the surface, τ:

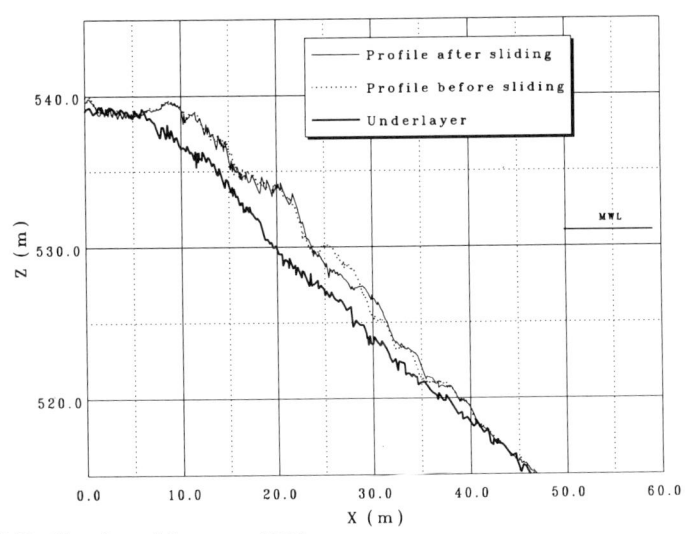

Figure 10 Profile after sliding near MWL.

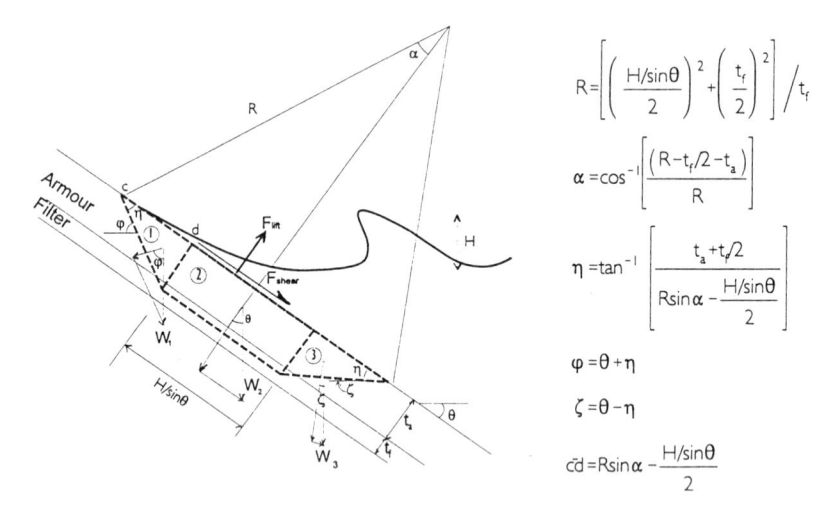

$$R = \left[\left(\frac{H/\sin\theta}{2}\right)^2 + \left(\frac{t_f}{2}\right)^2\right]/t_f$$

$$\alpha = \cos^{-1}\left[\frac{(R - t_f/2 - t_a)}{R}\right]$$

$$\eta = \tan^{-1}\left[\frac{t_a + t_f/2}{R\sin\alpha - \frac{H/\sin\theta}{2}}\right]$$

$$\varphi = \theta + \eta$$

$$\zeta = \theta - \eta$$

$$\bar{cd} = R\sin\alpha - \frac{H/\sin\theta}{2}$$

Figure 11 Block model for sliding failure.

$$\tau = \frac{f_w \rho u_b^2}{2} \qquad (10)$$

Assuming that this shear stress acts over a vertical range of H, the total shear force (per unit width of structure), F_{shear} becomes:

$$F_{shear} = \tau \frac{H}{\sin\theta} \qquad (11)$$

The measurements by Cornett integrate the effects of internal fluid velocities within the filter and armour layers, internal pore pressures and lift forces on the armour layer due to velocity gradients through the measurement of the lift force on the armour panel, measured perpendicular to the slope. Generally, these measurements indicate that $F_{lift} \approx F_{shear}$. From observations of the tests at NRC, we have observed that sliding failures occurring near the still water level have a failure plane passing through the filter layer over a distance approximately equal to the wave height, H. In the present analysis, we divide the failure zone into three blocks as shown in Figure 11. As described in Davies et al. a failure zone is assumed which passes through the filter zone over an along-slope distance of $H/\sin\theta$ and passes through the armour zone at angle η. This geometry is defined by assuming the failure zone follows a 'slip-circle' which passes though the filter zone over an along-slope distance of $H/\sin\theta$.

Using this assumed failure model, we can analyze the resisting and driving forces along the three failure planes (at the bases of blocks 1, 2 and 3).

Forces acting on Block 1:

Weight: $W_1 = 1/2 \text{ base} \times \text{height} \times \gamma'$
$= \gamma'/2 \; \bar{c}d \; (t_a + t_f/2)$
Driving force along failure plane: $F_{D1} = W_1 \sin\varphi$ (12)
Resisting force: $F_{R1} = W_1 \cos\varphi_1 \tan\varphi'$

Forces acting on Block 2:

Weight: $W_2 = \dfrac{H\sin\theta}{2} (t_a + t_f/2) \gamma'$

Driving force along failure plane: $F_{D2} = W_2 \sin\theta + F_{shear}$

where $F_{shear} = \tau_{waves} \dfrac{H\sin\theta}{2}$ (13)

Resisting force: $F_{R2} = (W_2 \cos\theta - F_{lift}) \tan\varphi'$

Forces acting on Block 3:

Weight: $W_3 = \gamma'/2 \; \bar{c}d \; (t_a + t_f/2)$
Driving force along failure plane: $F_{D3} = W_3 \sin\zeta$ (14)
Resisting force: $F_{R3} = W_3 \cos\zeta \tan\varphi'$

The overall stability of the failure zone is calculated as the ratio of the sum of all resisting forces to the sum of all driving forces:

$$\text{Factor of safety, FS} = \dfrac{\sum F_{Resisting}}{\sum F_{Driving}}$$

$$FS = \dfrac{F_{R1} + F_{R2} + F_{R3}}{F_{D1} + F_{D2} + F_{D3}}$$ (15)

Using the above translational block sliding analysis, we can estimate the factor of safety of the revetment against failure.

Calculations have been made of the factor of safety against sliding for a slope of 1:1.7, an armour thickness of $t_a=1.3$m, a filter layer thickness of $t_f=0.6$m and a placement density of $\gamma=1850$kg/m^3. In this analysis, the shear strength of the rockfill is taken to be $\varphi'=42.8$. This is based on the average of a series of tilting tests, where a cross-section of the dyke core, filter and armour layers were prepared in a rigid box, the box was then slowly lifted and the angle at which the slope failed was measured. This value corresponds well with reported natural angles of repose from field observations.

The translational block sliding analysis has been used to estimate the factor of safety against sliding for a given wave height for various layer geometries. In this analysis we have used $H_{1/100}$ (roughly equal to $1.8H_s$) as being representative of the larger waves in

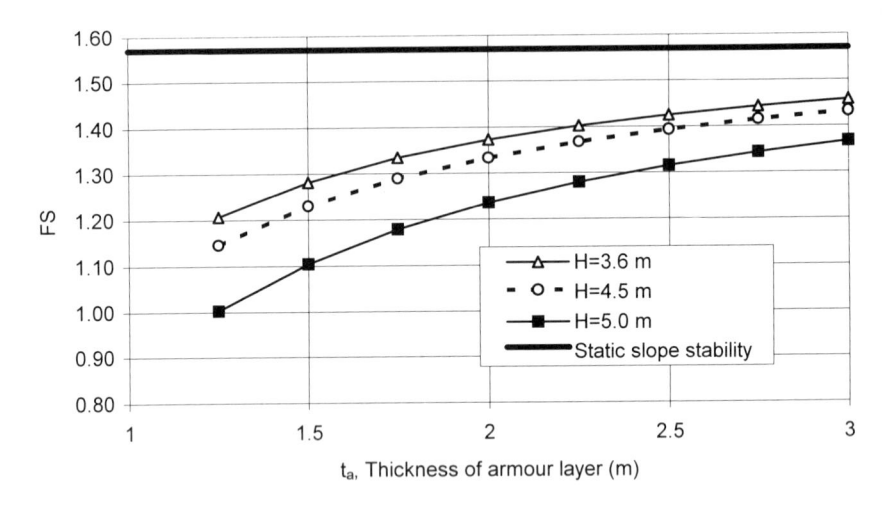

Figure 12 Effect of armour layer thickness on factor of safety against sliding.

the time series. Figure 12 shows how the factor of safety varies with slope geometry. This analysis suggests that layer thickness has a strong influence on sliding stability (assuming that the preferred failure plane is through the filter layer).

This analysis suggests that single layer revetment is particularly prone to sliding failures and that the thickness of the armour layer is a significant factor in the sliding process. Thicker armour layers have a greater resistance to sliding failures which pass through the filter layer. Preliminary application of this translational block sliding analysis to observed sliding failures agrees well with observations, indicating that layer thickness and slope are the key factors affecting sliding potential. Detailed analysis of individual failures requires a less simplified model of the block sliding process. Recent work by Altae (1994) shows that a non-linear, stress and consolidation finite element model can be applied to this revetment stability problem. This work further supports the existence of a potential sliding failure zone in the filter layer just below the still water level. Work on these analysis methods is continuing.

Conclusions

For profile analysis, reliable and efficient techniques have been developed using electro-mechanical profiling resulting in excellent agreement between S measured by profiling and S_{visual} measured from stone counting. There are inherent limitations in S as an analysis tool for breakwater damage due to the variability in the significance of S with structure type. Depth of cover, d_c has been shown to be a useful quantification of the damage related directly to the main failure criterion - the exposure of underlayers. This type of analysis is valid regardless of structure type.

Sliding has been seen to be a second failure mechanism for steep sloped structures. Visual observation, coupled with evaluation of the slope's factor of safety against sliding in the presence of wave action is the best tool for damage analysis when sliding occurs. Block sliding analysis illustrates the importance of layer thickness in providing resistance to sliding. The translational block sliding model provides an analytic framework for interpreting the sliding potential of armour layers.

Acknowledgements

The authors would like to thank Herman Claes, David Watson and Rod Girard without whose assistance and attention to detail this work would not have been possible. The input and advice of many of the staff of the Hydraulics and Geology and Soil Mechanics Departments of SEBJ are also gratefully acknowledged.

References

Altae, A. (1994). Feasibility study for numerical analysis of revetment stability under wave loading. Internal report by Anna Geodynamics Inc. to NRC (Ref. 93-1466/6297).

Cornett, A.M. and Mansard, E.P.D. (1994) Wave stresses on rubble-mound armour. Proc. 24th Int. Coastal Engg. Conf. ASCE, New York, NY.

Davies, M.H., Mansard, E.P.D. and Parkinson, F.E. (1994) Études sur modèles de la stabilité des ouvrages TA13 et KA3. NRC Controlled Technical Report IECE-CEP-CTR-001. Report prepared for Société d'énergie de la Baie James (in french).

Hughes, S.A. (1993) Physical models and laboratory techniques in coastal engineering. Advanced Series on Ocean Engg. - Vol. 7.World Scientific Publishing, Singapore, 568 pp.

Kamphuis, J.W. (1975) Friction factor under oscillatory waves. J. Waterways, Harbours and Coastal Engg. ASCE Vol. 101 (WW2). pp. 135-144.

Latham, J.P., Poole, A.B., and Mannion, M.B. (1988) Developments in the analysis of armour layer profile data. Design of Breakwaters - Proc. of the Conference Breakwaters '88, ICE, Thomas Telford Ltd., London, UK, pp. 343-361.

Medina, J.R., Hudspeth, R.T., and Fassardi, C. (1994) Breakwater armour damage due to wave groups. J. of Waterway, Port, Coastal and Ocean Engg., Vol. 120, No. 2, March/April 1994. pp. 179-198.

Coastal Engineering Research Centre (1984) Shore Protection Manual. Coastal Engineering Research Center, U.S. Army Waterways Experiment Station, U.S. Govt. Printing Office, Washington, DC, 1984.

van der Meer, J. (1988). Rock slopes and gravel beaches under wave attack. Ph.D. thesis, Delft Technical University, April. 152 pp.

Vidal, C., Losada, M.A., Medina, R., Mansard, E.P.D. and Gomez-Pina, G. (1992). A universal analysis for the stability of both low-crested and submerged breakwaters. Proc. 23rd Int. Coastal Engg. Conf. ASCE, New York, NY.

CHAPTER 74

NUMERICALLY MODELING PERSONNEL DANGER ON A PROMENADE
BREAKWATER DUE TO OVERTOPPING WAVES

Kimihiko Endoh[1] and Shigeo Takahashi[2]

ABSTRACT

Prototype experiments were carried out to quantify when personnel on top of a promenade breakwater will lose their balance due to overtopping waves. The danger of being carried out into the sea was also investigated using model experiments. Based on our results, we developed an empirical formula for calculating the wave height at which personnel danger occurs.

1. INTRODUCTION

Public access to breakwater areas is usually prohibited in Japan due to safety reasons, yet many people nevertheless enter these areas to enjoy the comfortable sea environment. The Japanese Ministry of Transport (MOT) has recently developed a new type of breakwater, named the "Promenade Breakwater," which serves a dual purpose of protecting a harbor from storm waves while also providing the public with amenity areas. **Figure 1** shows a photograph of a promenade breakwater constructed in Wakayama Port.

Because Japanese breakwaters are typically the low-crown type, wave overtopping sometimes occurs, and therefore, it is essential for the design of a promenade breakwater to consider maintaining personnel safety. Based on this important concern, the Port and Harbour Research Institute (PHRI) initiated research to investigate the types of danger a person may be subjected to while on a top of a breakwater.[1,2]

---

1 Port and Harbor Engineering Section, Civil Engineering Research Institute, Hokkaido Development Agency, 1-3 Hiragishi, Sapporo 062, Japan
Tel:+81-11-841-1111(283)   Fax:+81-11-842-9169
2 Chief of Maritime Structures Lab., Hydraulic Div., Port and Harbour Research Institute, Ministry of Transport, 3-1-1 Nagase, Yokosuka 239, Japan
Tel:+81-468-44-5011   Fax:+81-468-42-7846

Figure 1 Promenade breakwater constructed in Wakayama port

2. PERSONNEL DANGER WHILE ON A BREAKWATER

Personnel danger on a breakwater is closely related to the wave conditions around it. **Figure 2** shows four successive stages of danger that a person can be subjected to while on a breakwater. Here, we classify personnel danger into the following four stages:

- 1st stage: A splash occurs over the breakwater.
- 2nd stage: An overtopping wave occurs.
- 3rd stage : A person is knocked over by the overtopping wave.
- Final stage: A person is carried into the sea by overtopping flow.

When a wave splashes over a breakwater (1st stage), a person may feel scared although no substantial danger exists. The danger, however, is substantial at the 2nd stage, with the 3rd and final stages being extremely dangerous since a serious accident may inevitably happen.

Using experimental results, empirical models were developed to quantify the four stages of personnel danger. **Figure 3** shows a basic flow chart of how the models were employed in the calculations. Wave conditions in the 1st and 2nd stages can be calculated by the overtopping wave model (OWM), which is explained in detail in **Ref. 3**. The "loss of balance model" and "carry model" were developed to quantify the 3rd and final stages, respectively. The present study discusses the danger of each stage, being focused on the 3rd and final stages, and quantitatively describes these conditions using the wave height and wave crest height included in the models.

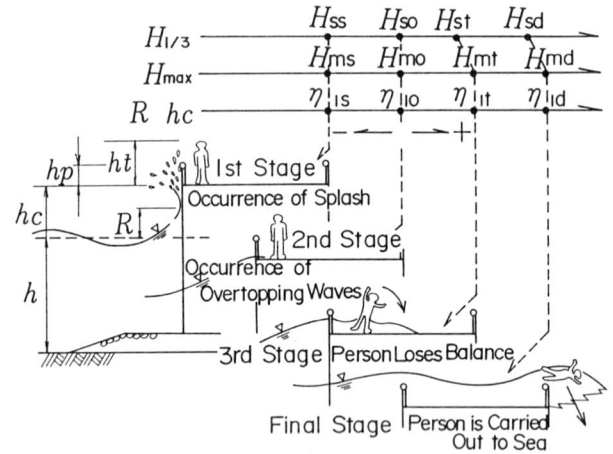

Figure 2 Four successive stages of personnel danger while on a breakwater

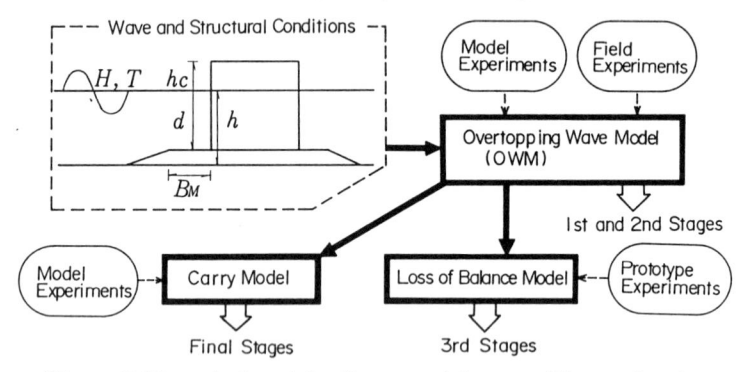

Figure 3 Numerical models of personnel danger while on a breakwater

3. PERSONNEL LOSS OF BALANCE IN OVERTOPPING FLOW
3.1 Prototype Experiments

A series of prototype experiments were conducted in a large current basin (50-m-long, 20-m-wide) to investigate the stability of a person under various flow conditions, i.e., we measured the current force on a person and observed personnel loss of balance.

Component load cells were used to measure the forces acting on three human bodies subjected to steady flow. Table 1 summarizes the physique of each person (sample A-C). The angle of the person's body against the current, θ, was varied (0, 45 and

Table 1 Physiques of personnel participating in the experiment

	Sample A	Sample B	Sample C
Height (cm)	183	164	174
Weight (kg)	73	65	64
Deviation from standard weight (%)	-2.3	12.8	-3.9
Length of inside leg (cm)	88	73	80
Waist (cm)	78	80	76

90°), as was the width between feet, *Lf* (0, 25 and 50 cm). Three different types of clothes were used.

Loss of the balance under various flows was observed with a high-speed video camera (200 frames/s). Two pairs of shoes with different type soles were used, having a coefficient of friction μs of 0.71 and 0.37. *Lf* was 25 cm and θ was 0, 90, and 180°.

Figure 4 shows the current force acting on a person's body in steady flow, where the x-axis indicates the current velocity and the y-axis indicates the force of current acting on the body. This force is proportional to the flow velocity squared, and can be expressed as a drag force:

$$F = \frac{w_0}{2g} C_D \cdot A \cdot U^2 \ . \tag{1}$$

The coefficient of the drag force, C_D, is dependent on several parameters, i.e., θ, *Lf*, and water depth, and can be expressed as

$$C_D = 1.1(1 - Lf/ht) \quad : \theta = 0°$$
$$1.1(1 + Lf/ht) \quad : \theta = 45, 90°, \tag{2}$$

where *A* is the projected area of the body against overtopping flow, *U* the current velocity during wave overtopping, w_0 the specific weight of sea water, and *g* the gravitational acceleration. Note that when $Lf \ne 0$, $C_D = 1.1$.

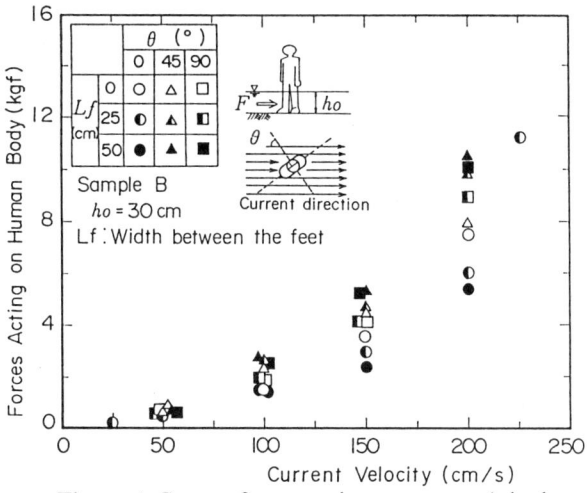

Figure 4 Current forces acting on a person's body

3.2 Loss of Balance Model

Figure 5 represents the loss of balance model which considers the two main types: "tumbling" and "slipping". In the tumbling type, a person is knocked over by the overtopping flow in the downstream direction. This type occurs when the moment produced by the current forces around the feet is greater than the moment produced by the person's weight, being expressed by

$$F \cdot h_G \geq W_0 \cdot l_G, \tag{3}$$

where h_G is the vertical distance from the floor to the point where the resultant force acts, W_0 is the weight of a human body in overtopping flow, and l_G is the horizontal distance between the center of the gravity and the fulcrum of the moment.

In the slipping type, a person is knocked over in the upstream direction, which occurs when the current force is greater than the friction force between the shoes and ground. This type of loss of balance is represented using

$$F \geq \mu s \cdot W_0, \tag{4}$$

where μs is the coefficient of friction between the shoes and the ground.

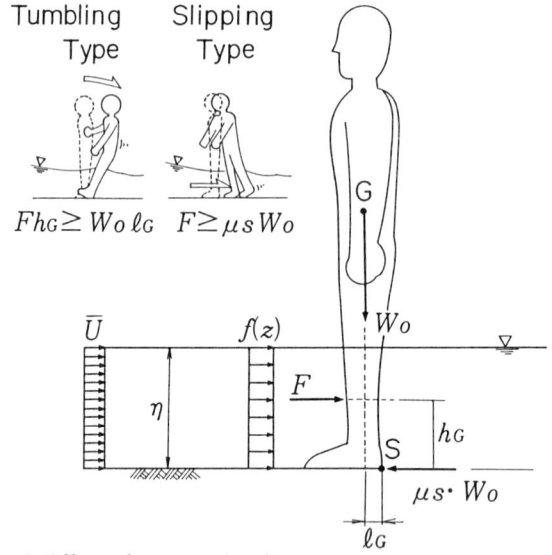

Figure 5 Effect of overtopping flow on the static balance of a person

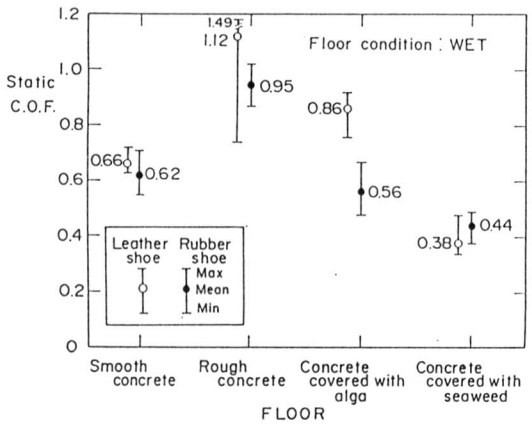

Figure 6 Coefficient of friction between shoe and wet concrete

Figure 6 shows measured field values of μs between the shoes (leather and rubber soles) and various types of wet concrete (smooth concrete, rough concrete, concrete covered with alga, and concrete covered with seaweed), where μs for the rough concrete is larger than that for smooth concrete. If the floor is covered with alga or seaweed, μs naturally decreases. In the loss of balance model, we used a μs value of 0.4 for covered concrete and 0.6 otherwise.

3.3 Loss of Balance Observation

Figure 7 shows analogue data of water depth, current velocity, and force acting on the human body, where the current velocity increases after the current generator starts. When the current velocity is 165 cm/s and the force acting on the human body is 14 kgf, a person is knocked over (slipping type loss of balance) as shown in **Fig. 8** (the flow direction is from left to right and the person faces upstream). μ_s between the shoes and ground was about 0.3.

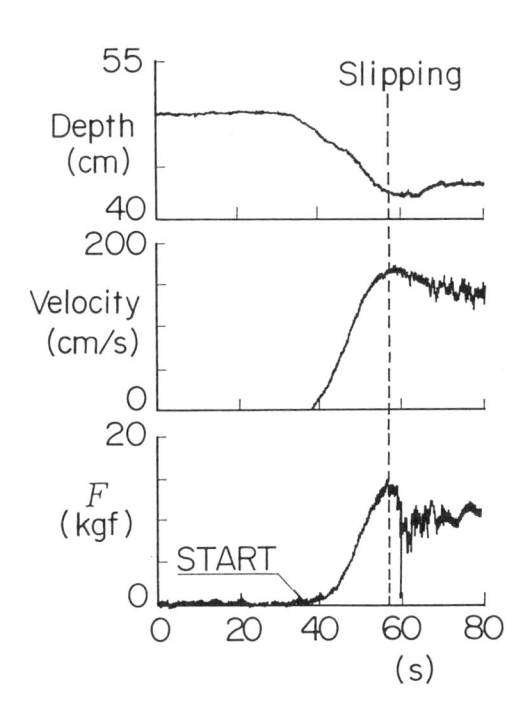

Figure 7 Measured analogue data during loss of balance experiment

Figure 8 Loss of balance by slipping

Figure 9 shows experimental results of a person's stability under various flow conditions. The x-axis is the water depth where the person is standing, while the y-axis is the current velocity there. The solid line indicates the calculated stable limit as determined by the slipping type loss of balance model. The good agreement between the calculated and experimental results is clearly apparent.

Figure 10 shows the critical water depth on a caisson at the seaward edge, η_{lt}, when a person is knocked over. η_{lt} is calculated by the loss of balance model in which we assumed that $\mu_s = 0.4$, the person is standing at the most dangerous location facing the seaward side ($\theta = 0°$), and the person's legs are spread to 22% of their height. Note that η_{lt} tends to increase as the person's height increases or as the body shape becomes more sleder. If a person is 152 cm in height and has a standard body shape, $\eta_{lt} = 50$ cm. In this condition, the tumbling type loss of balance occurs at the seaward edge of a caisson where the maximum current velocity is 0.9 m/s.

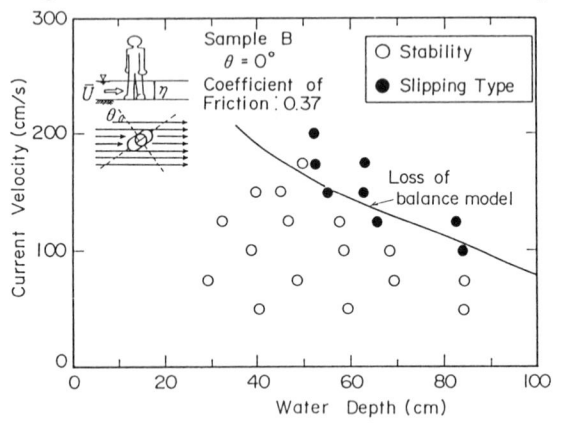

Figure 9 Verification of the loss of balance model

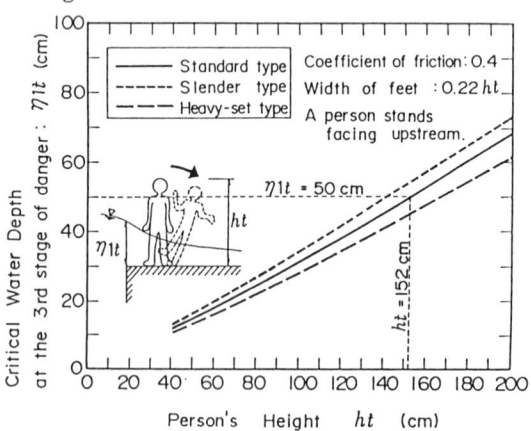

Figure 10 The relation between a person's height and wave crest elevation when loss of balance occurs

4. CARRIED INTO THE SEA BY OVERTOPPING WAVES
4.1 Outline of Model Experiments

A series of model experiments were conducted in a wave channel to investigate the danger of a person being carried into the sea by overtopping waves. Experiments involve measuring overtopping flow with handrails present and observation of human body movement.

We measured the motion of overtopping waves using a wave gauge, current meter, and high-speed video camera. Handrails were installed at the seaward and landward edge of a caisson. **Figure 11** shows model handrails having four opening ratios $\varepsilon = 0$, 0.24, 0.44, or 0.61.

The motion of a human body model in overtopping flow was also observed by a high-speed video camera under various wave conditions with different shaped handrails. The human body model has a cylindrical shape with a diameter of 2 cm and height of 7.6 cm. At a model scale of 1/20, the corresponding height is 152 cm. It is made from wood whose specific gravity is 0.8.

Figure 11 Diagram of model handrails

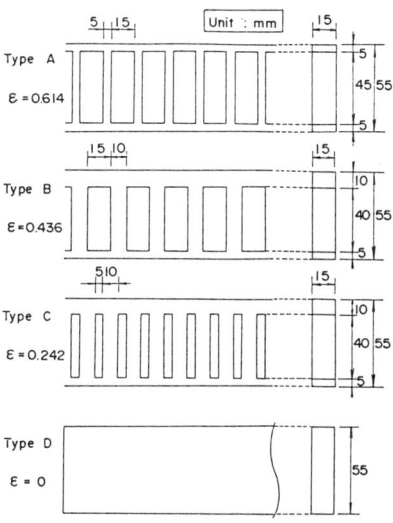

4.2 Handrail Effects on Overtopping Flow Motion

Figure 12 shows the seaward handrail's effect on water depth on a caisson during wave overtopping. The x-axis is the maximum water depth on a caisson at the seaward edge, η_1, and the y-axis is that 40 cm from the seaward edge, η ($x = 40$). When no handrail is installed, η ($x = 40$) is about 40% of η_1. If a handrail is installed at the seaward edge, the water depth behind it decreases in comparison to that if no handrail is present. Note that the water depth decreases as the opening ratio of the handrail decreases. The effect of a handrail can be quantitatively estimated as

$$\eta(x) = 0.4\{\eta_1 - hp(1 - \varepsilon_1)^2\}, \tag{4}$$

where hp is the height of the handrail and ε_1 is the opening ratio of the seaward handrail.

Figure 13 shows the distribution of the maximum flow velocity on a caisson near the landward handrail with an opening ratio of 0.61. The length of a vector indicates the flow velocity value, while the solid line indicates the maximum water depth on the caisson. Since this handrail has a high ε_2 value, the flow near it is not even disturbed. However, if the handrail's ε_2 is small, the water depth near it will significantly increase. This effect can be formulated as

$$\eta_2 = \min\{\eta_1, 0.4\eta_1 + (1 - \varepsilon_2)hp\}, \tag{5}$$

where ε_2 is the opening ratio of the landward handrail and η_2 is the maximum water depth at the landward edge.

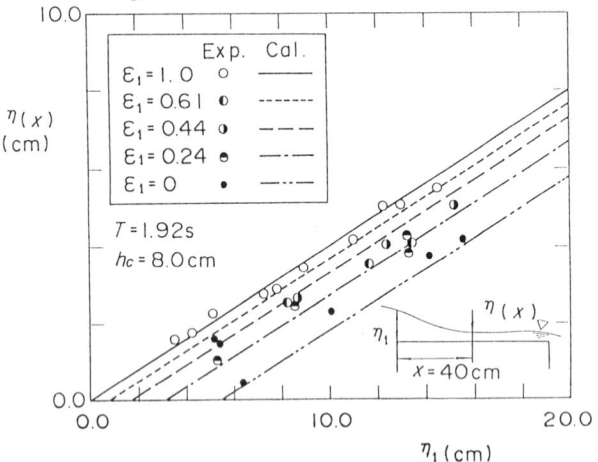

Figure 12 Maximum water depth behind the seaward handrail

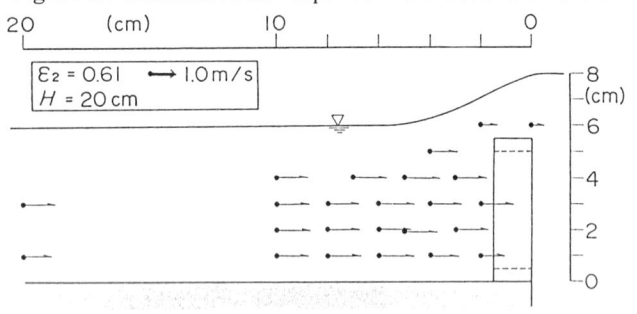

Figure 13 Distribution of current velocity near the landward handrail

4.3 Body Motion in Overtopping Flow

Figure 14 shows wave-induced movement of the human body model when it is placed near the landward handrail. Each diagram depicts the effect of ε_2 on model movement. When $\varepsilon = 0$, the higher handrail is considered to be an impermeable wall. Note that as ε_2 decreases, the model is lifted up; thereby indicating a strong relation exists between movement and the water depth near the handrail, i.e., the maximum water depth at the landward handrail strongly influences whether or not a person will be carried into the sea. The critical water depth at which the model was carried over the handrail into the sea was found to be about 17% of its height. A expression representing this danger is

$$\eta_{2cr} = 0.17 ht + hp, \tag{6}$$

where η_{2cr} is the critical water depth at the caisson's landward edge which will carry a person into the sea, and ht is a person's height.

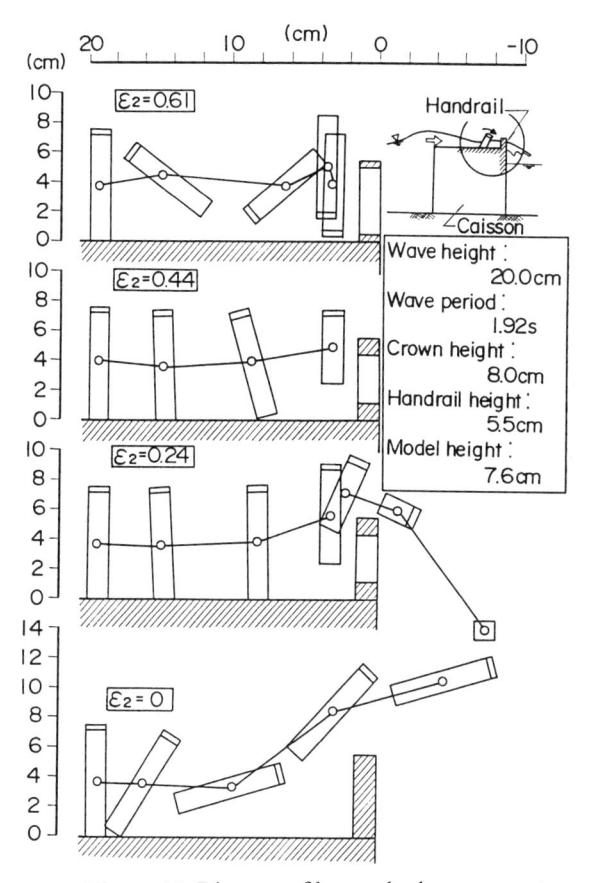

Figure 14 Diagram of human body movements in overtopping flow at various ε_2 values

4.4 Carry Model

The carry model can estimate the effect of the seaward and landward handrails on the maximum water depth using Eqs. (4) and (5), respectively, whereas Eq. (6) enables calculating the maximum water depth at the caisson's seaward edge which will carry a person into the sea.

Figure 15 shows the critical water depth (contour lines) at the caisson's seaward edge as determined by the carry model. The x- and y-axis respectively indicate the opening ratio of the seaward and landward handrail. It is assumed that a person is 152-cm tall and the height of the handrails is 110 cm. As shown, if no handrails are installed, the critical water level is only 0.7 m. However, if both handrails have an opening ratio of 0.7, this increases the critical water depth to 2.1 m. These results verify that handrails effectively prevent a person from being carried into the sea.

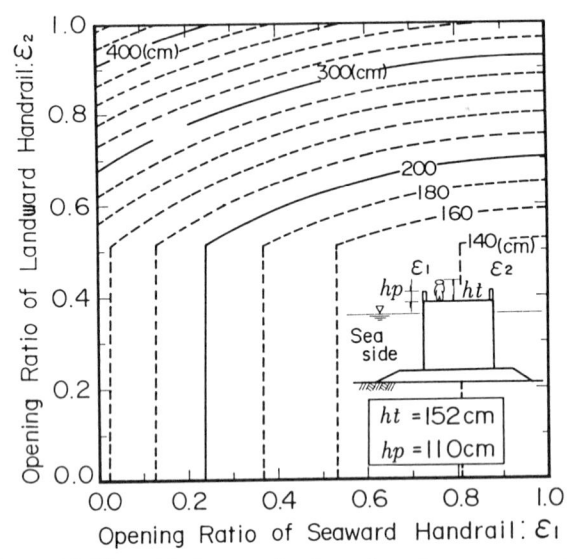

Figure 15 The critical water depth on the caissons seaward edge which carries into the sea

5. WAVE HEIGHT DURING EACH DANGER STAGE
5.1 Dangerous Wave Height Formula
As shown next, the wave height at each successive stage of danger can be formulated by the OWM.
5.1.1 Critical wave height of splash: *Hms*

$$Hms = \frac{-1 + \sqrt{1 + 2.8\alpha_1 hc/hm}}{2\alpha_1} \times hm \tag{7}$$

$$hm = \begin{array}{ll} d & : B_M/L \geq 0.16 \\ d + (h-d)\dfrac{0.16 - B_M/L}{0.05} & : 0.11 < B_M/L \leq 0.16 \\ h & : 0.11 < B_M/L \end{array} \tag{8}$$

where *hc* is the breakwater's crown height, and α_1 is a coefficient representing the breakwater superstructure, being 1.0 for a vertical breakwater and 0.5 for a slit-caisson breakwater or a composite breakwater with wave-dissipating blocks. Also, *d* is the water depth above the mound foundation, *h* the water depth above the sea bottom, B_M the width of the mound shoulder, and *L* the wave length at the depth of the breakwater.

5.1.2 Critical wave height at overtopping: *Hmo*

$$Hmo = \frac{-1 + \sqrt{1 + 4\alpha_1(hc + hp^*)/hm}}{2\alpha_1} \times hm \tag{9}$$

$$hp^* = 0 \quad : \varepsilon_1 \neq 0$$
$$hp \quad : \varepsilon_1 = 0 \quad (10)$$
where *hp* is the handrail height and ε_1 is the opening ratio of the seaward handrail

5.1.3 Critical wave height for being knocked over: *Hmt*

$$Hmt = \frac{2(hc^* + \eta_{1t})}{1 + \sqrt{1 + 4\alpha_1 hc^*/hm}} \quad (11)$$

$$hc^* = hc \quad : \varepsilon_1 \geq 0.4$$
$$hc + hp \quad : \varepsilon_1 < 0.4 \quad (12)$$

where η_{1t} is the maximum water depth at the caisson's seaward edge when a person is knocked over. In Eq. (11), η_{1t} is assumed as 0.5 (m), which is the critical value for a 152 cm tall person (average 12 years old person in Japan).

5.1.4 Critical wave height for being carried into the sea: *Hmd*

$$Hmd = \frac{2(hc + \eta_{1d})}{1 + \sqrt{1 + 4\alpha_1 hc^*/hm}} \quad (13)$$

where η_{1d} is the maximum water depth at the caisson's seaward edge when a person is carried into the sea (**Table 2**, **Figure 16**).

Table 2 η_{1d} against the types of handrail (*hp* = 1.1 m) for 152 cm tall person
(unit : m)

Seaward handrail Landward handrail	Fence type	Wall type	Chain type or nothing
Fence type	2.1	3.1	2
Wall type	1.5	2.5	1.4
Chain type or nothing	0.8	1.3	0.7

Fence type   Wall type   Chain type

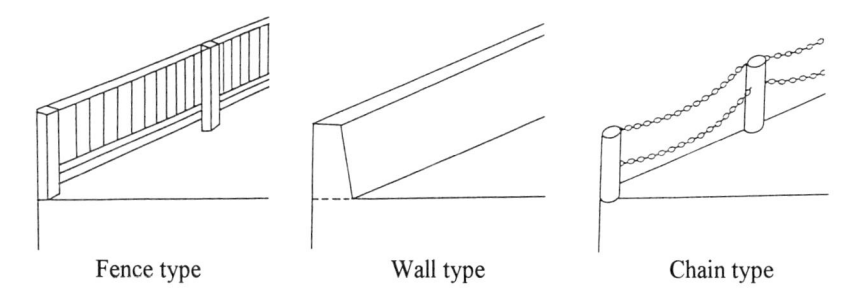

Figure 16 Handrail types

5.2 Example Calculation

Using the following conditions, an example will be shown calculation that determines the critical wave for causing danger at each danger stage;
$hc = 2$ m, $hp = 1.1$ m, $d = 8$ m, $h = 10$ m, $B_M = 5$ m, $\varepsilon_1 = \varepsilon_2 = 0.7$, $T = 6$ s.

Under these conditions, a splash occurs over the breakwater at a maximum wave height $H_{max} \geq 1.2$ m (*Hms*), while overtopping wave occur at $H_{max} = 1.7$ m (*Hmo*). A person on the breakwater is knocked over at $H_{max} \geq 2.1$ m (*Hmt*), and is carried over the handrail into the sea at $H_{max} \geq 3.5$ m (*Hmd*). If no handrails are installed, however, the final stage requires an H_{max} value of only 2.3 m.

A significant wave height for occurring dangers without breaking condition are assumed to be the H_{max} value divided by 1.8 for splash and overtopping danger (*Hss* and *Hso*), while the H_{max} value divided by 2.0 for 3rd and final stages of danger (*Hst* and *Hsd*), i.e., *Hss* = 0.7 m, *Hso* = 0.9 m, *Hst* = 1.1 m, and *Hsd* = 1.8 m under the same condition.

6. EFFECTS OF OVERTOPPING FLOW RATE ON PERSONNEL DANGER

Figure 17 shows the relationship between the significant wave height and the overtopping flow rate under the same conditions used in the example calculation. The overtopping flow rate is calculated using the OWM[4], which can evaluate this flow rate and the maximum overtopping flow rate for a regular wave. The overtopping flow rate for irregular waves is calculated based on the assumption that the Rayleigh distribution of wave heights holds. We assume the wave number is 1700, in which the maximum wave height is just two times the significant wave height.

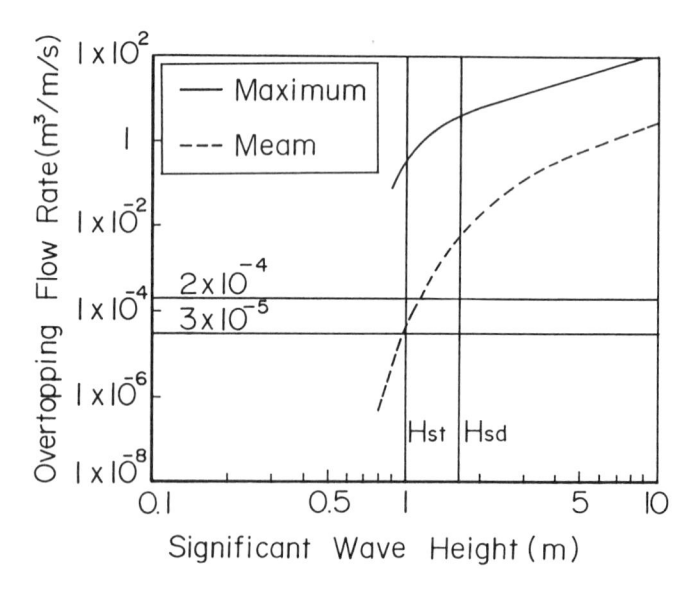

Figure 17 Overtopping flow rate when a person faces the several danger

Under these conditions and those used in the example calculations, the mean overtopping flow rate that knocks over a person is 4×10^{-5} m³/m/s, while at 6×10^{-3} m³/m/s the person is carried into the sea.

Fukuda et al. (1974) found that the wave overtopping flow rate to provide a probability of 50 and 90% personnel safety on a seawall is 2×10^{-4} and 3×10^{-5} m³/m/s, respectively. Note that our overtopping flow rate for being knocked over falls within these probabilities of safety.

7. CONCLUDING REMARKS

The types of overtopping wave-induced personnel dangers that can occur while standing on a breakwater were experimentally investigated. Our main conclusions are:
1) Based on prototype experiments, we developed a loss of balance model to calculate the critical water depth at a breakwater's seaward edge. If a person is 152-cm-tall and has standard body physique, the critical water depth is 0.5 m which causes a person to their balance.
2) The proposed carry model can calculate the critical water depth at the breakwater's seaward edge which will carry a person into the sea. This depth is dependent on the opening ratios of handrails installed at the breakwater's seaward and landward edge. If fence-type handrails having a 0.7 opening ratio are installed at the both edges, the critical water depth is 2.1 m for a 152-cm-tall person.
3) When no handrails are present, the calculated critical water depth which carries a person into the sea is only 0.7 m for a 152-cm-tall person, thus handrails are demonstrated to be a very effective measure for preventing a person from being carried into the sea by overtopping waves.
4) The proposed breakwater formula for evaluating the wave height at which personnel dangers will occur during successive stages of wave overtopping should be employed in the design of promenade breakwaters.

REFERENCES
1) S. Takahashi, K. Endoh and Z. Muro: Experimental study on people's safety against overtopping waves on breakwaters, Rept. of Port and Harbour Res. Inst., Vol. 31, No. 4, 1992, pp. 3 - 29. (in Japanese)
2) S. Takahashi et. al. : Experimental study on people's carriage into the sea caused by overtopping waves on breakwaters, Rept. of Port and Harbour Res. Inst., Vol. 33, No. 1, 1994, pp. 3 - 29. (in Japanese)
3) A. Hujii, S. Takahashi and K. Endoh: An investigation of the wave forces acting on breakwater handrails, 24th ICCE, 1994
4) S. Takahashi, K. Endoh and Z. Muro: Wave height and overtopping rate in relation to the risk of falling on promenade breakwater, Proceeding of Civil Engineering in the Ocean, Vol. 9, 1993, pp.295-300. (in Japanese)
5) Fukuda, N., T. Uno and I. Irie : Field observations of wave overtopping of wave absorbing revetment, Coastal Engineering in Japan, No.17.

CHAPTER 75

WAVE OVERTOPPING ON VERTICAL AND COMPOSITE BREAKWATERS

L. Franco[1], M. de Gerloni[2], J.W. van der Meer[3]

Abstract

After an extensive series of 2-D model tests on the overtopping response of various caisson breakwaters, general conceptual design formulae and graphs have been derived which relate the mean discharge with the relative freeboard. The influence of geometrical changes is described by reduction factors with reference to the pure vertical structure. A simple correlation has been made with the overtopping performance of sloping structures. Overtopping volumes per wave were also measured and fitted with a universal probability function; their effects on model persons and cars behind the crownwall were statistically evaluated, thus allowing an upgrading of the existing criteria for the admissible overtopping on breakwaters.

Introduction

Wave overtopping is one of the most important hydraulic responses of a breakwater, since it significantly affects its functional efficiency and to a minor extent even its structural safety (though the latter effect is often negligible for monolithic breakwaters). The overtopping discharge is in fact the main parameter for the design of shape and height of the breakwater crest.

However, little research work had been addressed to this subject in the past, since most attention had been paid to wave forces and breakwater stability. It may be

---

[1] Prof. Coastal Eng., DIIAR Politecnico di Milano, piazza Leonardo da Vinci 32, 20133 Milano, Italy.
[2] Manager Modelling Unity, ENEL SpA - CRIS, via Ornato 90/14, 20162 Milano, Italy.
[3] Dr., Delft Hydraulics, P.O.Box 152, 8300 AD Emmeloord, The Netherlands.

noted that these aspects are interrelated since the wall crest elevation influences the amount of wave force. Few overtopping studies are available for seawalls (Owen, 1980; Goda, 1985;) and more recently for rubble mound breakwaters (Jensen and Juhl, 1987; Aminti and Franco, 1988; Bradbury and Allsop, 1988; de Waal and van der Meer, 1992). Goda (1985) remains the main reference for the design of vertical and composite structures.

An extensive laboratory investigation on the overtopping performance of modern vertical-face breakwaters has been started in Milano since 1989, with random wave flume model testing. Preliminary results were presented by de Gerloni et al. (1989, 1991).

Additional model tests and a more detailed analysis of the test results have been carried out in 1993 with the support of the UE-MAST 2-MCS (Monolithic Coastal Structures) project funding. Moreover the results obtained by other European hydraulic laboratories from specific studies on similar structures have been incorporated in the analysis to enlarge the data set and improve validation. The first main results have been reported by Franco (1993, 1994) at the MAST workshops.

Experimental setup, test conditions and procedures

Model tests were carried out in the 43 m long, 1.5 m deep random wave flume of ENEL SpA - Center for Hydraulic and Structural Research (CRIS) laboratory in Milano. A special device was used for measuring the overtopping volumes: a tray suspended through a load cell to a supporting beam. The load cell signal reading after each overtopping wave allows the measurement of its individual volume; the number of overtopping waves, the total volume and hence the mean discharge in each test can then be easily computed.

The effects of each overtopping wave were analyzed by placing a few model cars and model persons along the center of the crown slab behind the wall, and by accurately observing the number of displacements and relative distance from the former position after each overtopping event (then repositioning the "targets").

The proper model scaling of the human behaviour (to 1:20) was assessed with a simplified full-scale test procedure: a large bucket and a plastic pipe were used to direct known amounts of water, from an elevation of 4.5 m and without notice, against both a volunteer (the first author of this paper) and a human-size plastic dummy. It was found that the dummy had to be ballasted up to 1500 N (twice the man weight) to have the same falling response of the man.

To improve the statistical validity rather long test durations were used with no less than 1000 waves. Peak periods (T_p) of JONSWAP spectra (bimodal spectra were also generated) varied between 7 and 13 s, significant wave heights (H_s) between 2.5 and 6 m with water depths/wave heights ratios (h/Hso) ranging between 3 to 5 (all figures are expressed in prototype terms for easier engineering interpretation). A total of about 250 tests with non-breaking wave conditions were performed.

Model breakwater configurations are shown in Fig. 1. They include traditional vertical-face caissons, perforated ones (14%, 25%, 40% porosity), shifted sloping parapets and a caisson with rubble mound protection (horizontally composite) with variable elevation and width of the homogeneous porous rock berm (S1 to S6 in Fig.1). All structures were designed for low overtopping conditions (i.e. high freeboard).

Additional results from model studies on similar structures designed in Italy and carried out by other European laboratories were included in the analysis, to enlarge the data set by covering a wider range of geometric and hydraulic conditions ($H_s=2\div8$m $T_p=6\div15$s h=$9\div18$ m). They were performed at Delft Hydraulics (DH) on vertical and shifted caissons and at Danish Hydraulic Institute (DHI) on perforated shifted-wall caissons. Further model test results from a research study on a simple vertical wall were supplied by CEPYC laboratory in Madrid. All these additional model test data typically only refer to the mean overtopping rate.

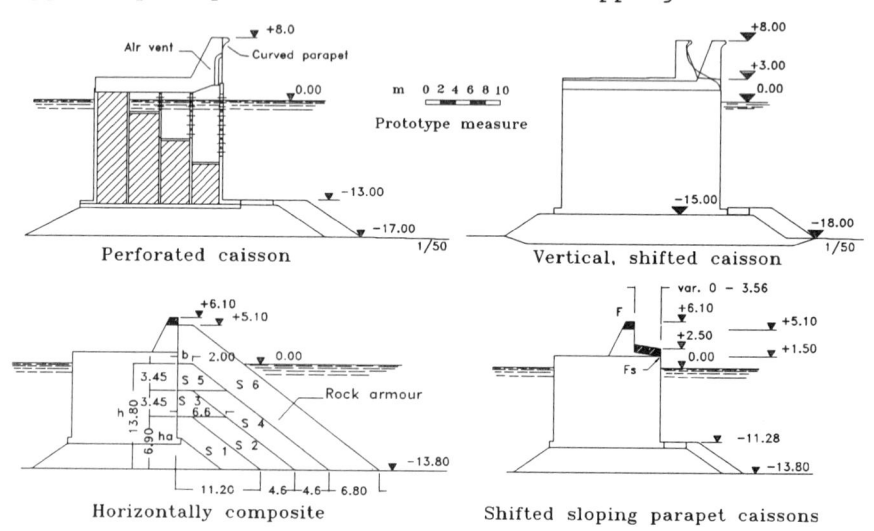

Fig 1 Model test sections of caisson breakwaters.

The influence of onshore winds on the overtopping effects was considered negligible particularly for large water flows above vertical walls. This assumption has been confirmed by a recent laboratory study (de Waal, 1994) which shows that the additional spray transport due to wind never exceeds 3 times the mean rate and is less than 1.4 times in typical deep water conditions similar to the tested ones. This increase is small compared to the much larger variability of overtopping with the wave height.

Admissible overtopping rates

The definition of tolerable limits for overtopping is still an open question, given the high irregularity of the phenomenon and the difficulty of measuring it and its consequences. Many factors, not only technical ones, should be taken into account to define the safety of the increasing number of breakwater users such as the psicology, age and clothing of a person surprised by an overtopping wave.

Still the current admissible rates (expressed in m^3/s per m length) are those proposed by the Japanese guidelines, based on impressions of experts observing prototype overtoppings (Fukuda et al. 1974; Goda, 1985). They are included in CIRIA/CUR-manual (1991), and in British Standards (1991) (Fig. 2). The lower limits of inconvenience to pedestrians may correspond to safe working conditions on the breakwater, while the upper limits of danger to personnel may correspond to safe ship stay at berth.

Obviously the overtopping criteria for design depend upon the function and degree of protection required, and upon the associated risk considerations, taking into account the joint probability of wave heigths and water levels. In fact relatively large overtopping might be allowed during extreme storms (structural design conditions) if transit on the breakwater is then prohibited (functional limit).

The structural safety of the breakwater typically demands less restrictive limits than the safety of its utilization (functional safety). The maximum admissible overtopping discharge for the structural safety of dikes and revetments are shown by Goda (1985). If the features of cast-in-situ concrete superstructures of modern caisson breakwaters are considered, the higher limit given for paved revetments (0.2 $m^3/s/m$) can be assumed.

As far as the functional safety is concerned (e.g. drainage behind seawalls), the figure of 0.01 $m^3/s/m$ is considered as the tolerable discharge for direct protection of densely populated coastal areas (Goda, 1985). Much lower discharges (0.0001 m^3/s/m) are given for the safe transit of vehicles, such as along a coastal highway. Few data are available on the critical overtopping discharges for the safety of various harbour operations and ship mooring on the breakwater rear side. A value of 0.00042 m^3/s/m in the 50 year design storm is proposed by Sigurdarson and Viggosson (1994) as criterion for damage to equipment and cargo on quay.

One of the aims of this model study was to assess better criteria in the case of caisson breakwaters. It was then believed that the overtopping volume per wave (V), being actually responsible for the damage on the breakwater crown, was a far better hydraulic parameter than the mean discharge for this analysis, when there is no need of land reclamation drainage.

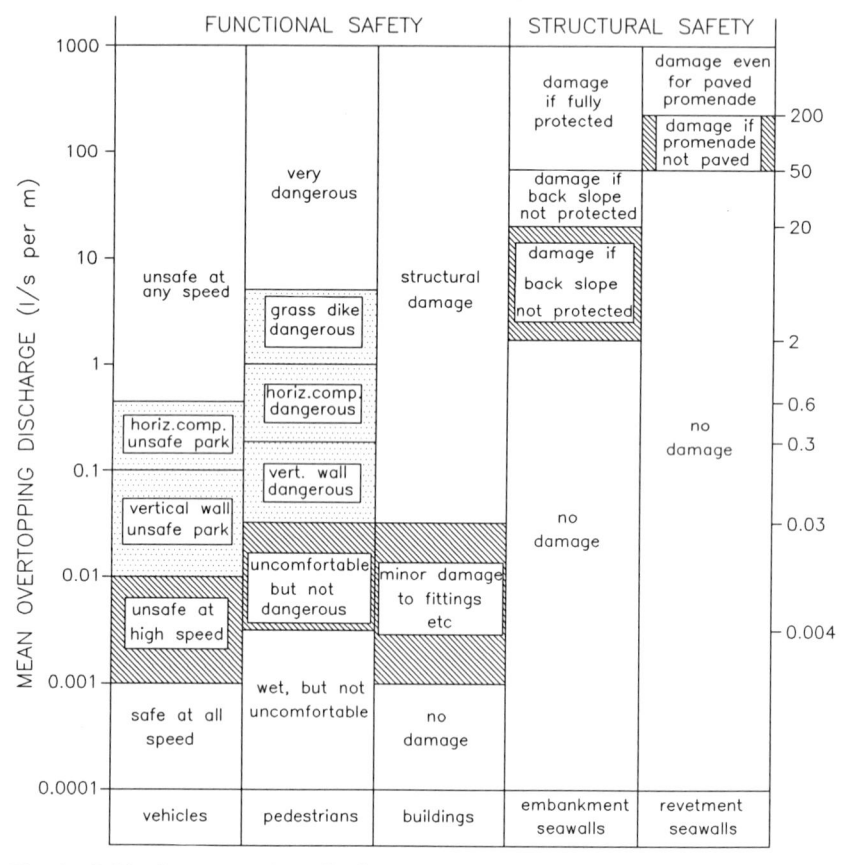

Fig. 2 Critical overtopping discharges of existing guidelines integrated with new safety bands (dotted) for transit on breakwaters.

For each breakwater configuration the individual overtopping volumes recorded in any tests were divided in classes of 0.1 m^3/m and the corresponding effects on model cars and pedestrians were statistically evaluated for each class. Some results obtained for pedestrians are shown in Fig. 3. It is interesting to observe that the effect is dependent on the structure geometry itself. The same overtopping volume is likely to be more dangerous if the breakwater is purely vertical than in the case of perforated or shifted-parapet caissons or horizontally composite ones. This is probably due to the different overflow mechanism which produces a more concentrated and fast water jet falling down from the crest of a vertical wall in comparison with a slower, more aerated, horizontal flow over a sloping structure. It was also observed that pedestrians are slightly more "stable" than vehicles undre the same overtopping wave.

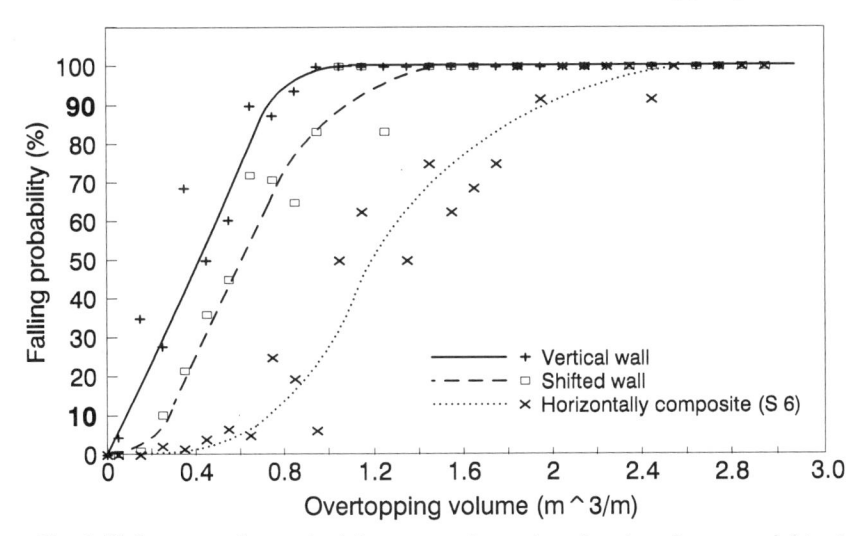

Fig. 3 Risk curves for pedestrians on caisson breakwaters from model tests.

From Fig. 3 it can be seen that the "critical bands" of overtopping volume (being dangerous above the upper limit and safe below the lower one) lie between 0.2 and 2.0 m^3/m (but a concentrated jet of 0.05 m^3/m on the upper body can be enough to make a person fall down as shown by the full scale calibration tests)

These results may be translated in terms of traditional mean rates for easier comparison with the previous criteria, by using the graph of Fig. 4, where the correlation between the mean discharge and the maximum (one in 1000 waves) volume is shown for some structural configurations. The critical band volumes within the range of 10% to 90% falling probability in Fig. 3 are assumed as $V_{max}=V_{0.1\%}$ in a conservative approach to enter in Fig. 4.

The new proposed critical discharges, which are at least 10 times higher than the previous ones, are illustrated in Fig. 2, which also includes the results of recent large scale tests (Smith et al., 1994) showing that the dangerous discharge for a man standing on a dike is in the range $(1\div10)\cdot10^{-3}$ m^3/s/m. From Fig. 4 it can be observed that this discharge actually corresponds to maximum volumes of $0.5\div2$ m^3/m while in the same range of V_{max} the average discharge over a vertical wall is $(0.1\div0.5)\cdot10^{-3}$ m^3/s/m. The ratio V_{max}/q can in fact vary between 100, for large mean discharges and percentage of overtopping waves, and 10,000, for small mean discharges and for vertical structures.

Thus it is confirmed that the significant parameter for the breakwater functional safety is the overtopping volume rather than the mean discharge. A relationship exists between the two parameters but it varies with the structure geometry and wave conditions.

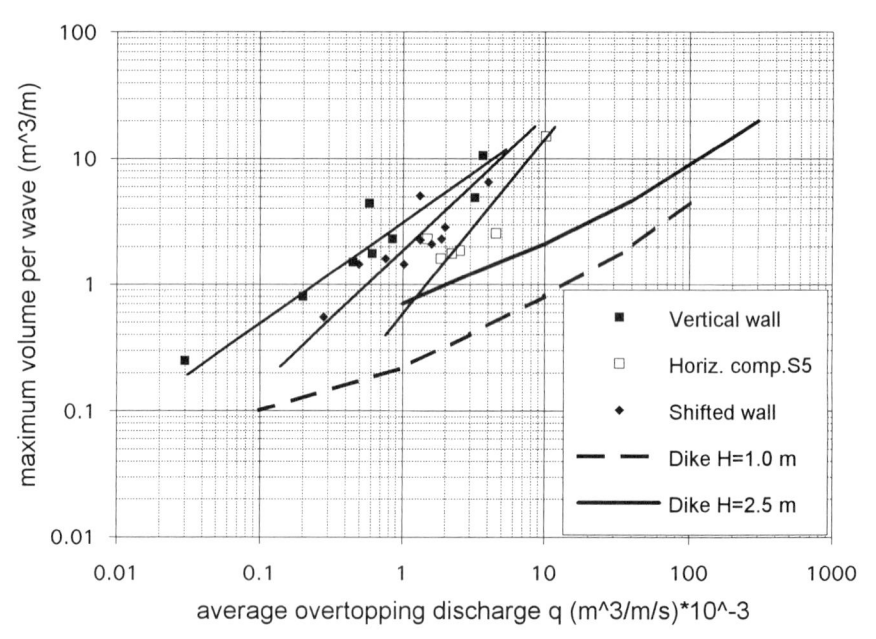

Fig. 4 Relation between mean discharge and maximum overtopping volume.

Probability distribution of individual overtopping waves

Overtopping events occur unevenly both in time and amount, often just a few waves overtopping among the thousands. The measurement of the individual overtopping volumes carried out during the model tests allowed the definition of their probability distribution. Considering the tests with a number of overtopping waves $N_{ow} > 30$ (for obvious reasons of statistical significance) the exceedance probability of each overtopping volume P_v was calculated and a 3-parameter Weibull distribution function gave best fit to the data as shown in the example of Fig. 5:

$$P_v = 1 - \frac{i}{N_w + 1} = C \cdot \exp\left(-\frac{V}{A}\right)^B = \exp\left(-\frac{V-C}{A}\right)^B \quad (1)$$

where N_w is the number of waves in the test, V is a volume in the ith rank and A,B,C are fitting constants.

It would also be possible to use in Fig. 5 on the horizontal axis the probability related to the number of overtopping waves (N_{ow}) instead of the incident waves (N_w). In that case the graph would start at 100% and C in eq. 1 would become 1.0.

It can be demonstrated with some mathematics that the scale parameter A can be defined by the equation:

$$A = 0.84 \cdot \overline{V} = \frac{0.84 \cdot T_m \cdot q}{\frac{N_{ow}}{N_w}} \quad (2)$$

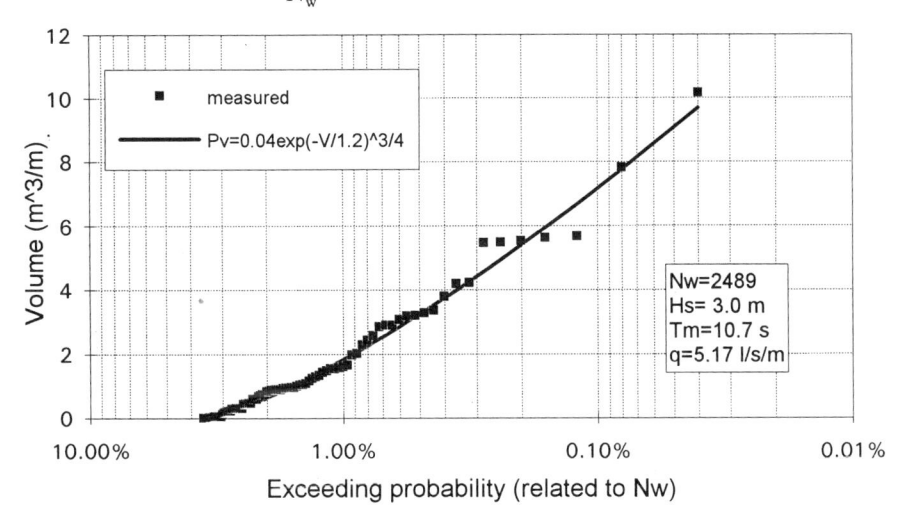

Fig. 5 Example of probability distribution of overtopping volume per wave: vertical wall caisson, test 21.

where T_m is the mean wave period, q is the average overtopping discharge and $\overline{V}$ is the average overtopping volume per wave. Hence the exceedance probability of a given volume is related to the mean discharge and to the overtopping probability. Even V_{max} can be related to the other two overtopping parameters from eq. 1 as:

$$V_{max} = A \cdot \ln(N_{ow})^{4/3} \qquad (3)$$

The shape parameter B was found to have a little variability around a mean value of 0.75, which is the same found by Van der Meer and Janssen (1994) for dikes. Then B=0.75 is assumed to be constant.

The "set-off" coefficient C, being the intercept with the x axis (Fig. 5), represents the percentage (probability) of overtopping waves (N_{ow}/N_w), which is assumed to be Rayleigh distributed, and can be expressed by the following equation (Fig. 6):

$$C = \frac{N_{ow}}{N_w} = \exp\left(-\frac{1}{k}\frac{R_c}{H_s}\right)^2 \qquad (4)$$

where best fitting gives k=0.91 for caisson breakwaters and $R_c/H_s=R$ is the relative freeboard, R_c being the wall crest height above the sea level.

The assumption that C represents the overtopping probability was confirmed experimentally as shown in Fig. 7, which presents the comparison between measured ratios $C=N_{ow}/N_w$ and the corresponding values calculated from data fitting in eq. 1; the experimental verification of eq. 2 for coefficient A is also shown.

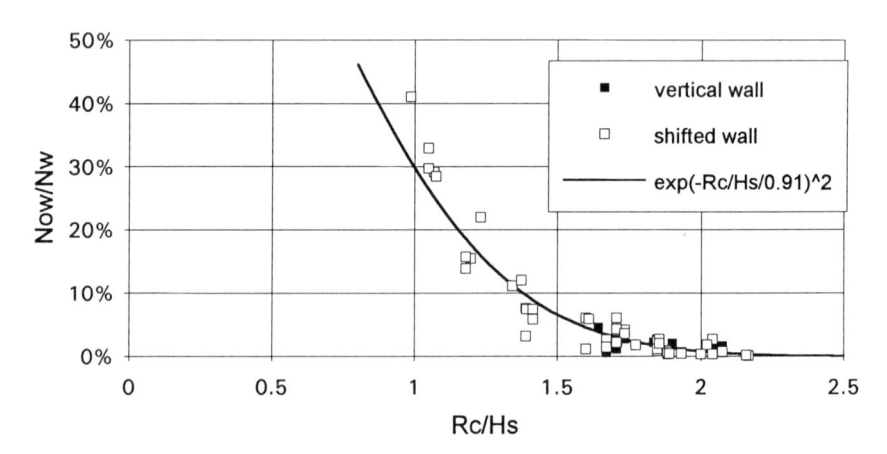

Fig. 6 Correlation between the percentage of overtopping waves and the relative freeboard.

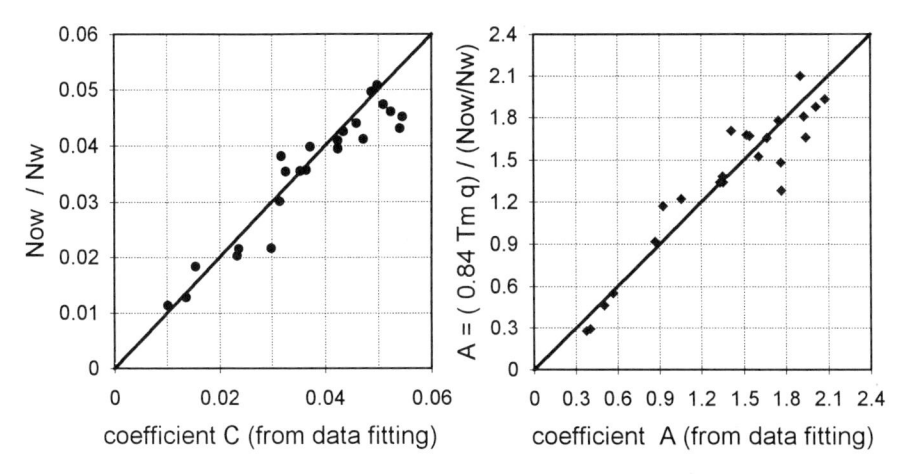

Fig. 7 Verification of functional relationship of coefficients C and A.

Conceptual design formulae

The method of analysis proposed by van der Meer and de Waal (1992) to derive a general design formula was applied to the tests results (restricted to a wave steepness range of 0.018-0.038) in terms of mean overtopping discharge allowing a direct comparison with the above admissible limits and an easy evaluation of the overtopping volumes per wave with eq. 1 and 2.

Consistent curves have been fitted with the least square method to the experimental data representing the dimensionless mean overtopping discharge $Q = q / \sqrt{g \cdot H_s^3}$ against the relative freeboard R_c/H_s, which is the most important parameter. Since an exponential relationship is assumed according to Owen (1980), the data should give a straight line on a log-linear plot:

$$Q = a \cdot \exp\left(-\frac{bR_c}{H_s}\right) \qquad (5)$$

It was found (Fig. 8) that for vertical-face breakwaters b=4.3 and a=0.192, which is close to the one found by van der Meer and Janssen (1994) for sloping structures (a=0.2); the value a=0.2 was then kept constant for the successive regressions with different geometries which generally showed a high correlation coefficient (Fig. 9). The physical interpretation of "a" is the dimensionless mean discharge when the freeboard is set at the mean water level. It may be observed in Fig. 9 that the overtopping rates predicted by eq. 5 are

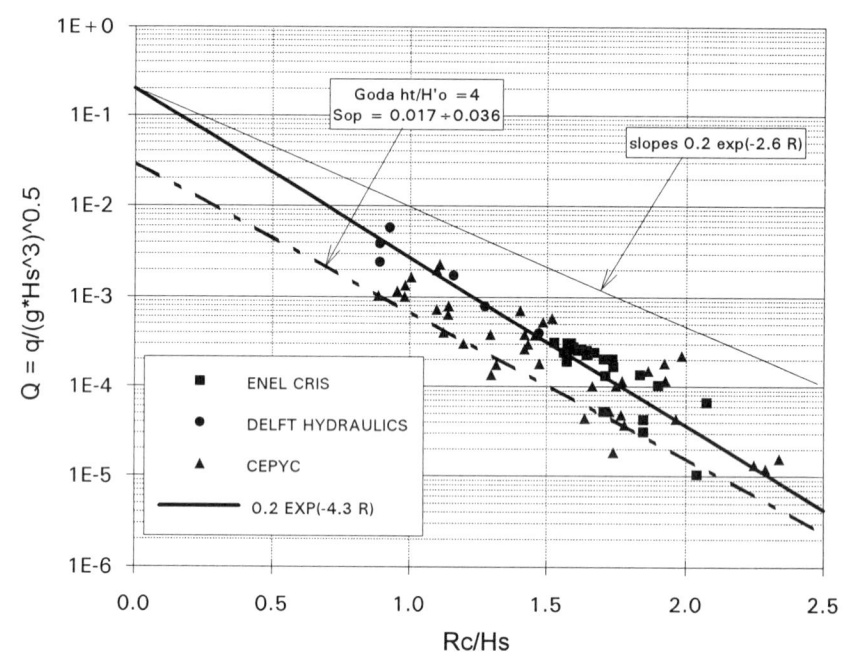

Fig. 8 Regression of wave overtopping data for vertical wall breakwaters.

slightly larger than those predicted by Goda (1985) for deep water vertical walls and with similar wave steepness range, besides the small differences in seabed slope and toe geometry. A similar underprediction of Goda's curves was also found by De Waal (1994).

Then the influence of structural modifications with reference to the vertical-face breakwater can be described by suitable freeboard reduction factors (γ), which are the ratios between the reference value b=4.3 and the various b coefficients fitted by eq. 5 as given in Fig. 9. Even the sloping structures (under non-breaking conditions) can be easily compared with the vertical ones considering the ratio γ_s=4.3/2.6=1.66, 2.6 being the fitting coefficient obtained by van der Meer and Janssen (1994) as also shown in Fig. 8. All the data can be plotted together (Fig. 10) after correction of the Rc/Hs values for each geometry with the corresponding γ, the general equation thus becoming:

$$Q = 0.2 \cdot \exp\left(-\frac{4.3}{\gamma}\frac{R_c}{H_s}\right) \tag{6}$$

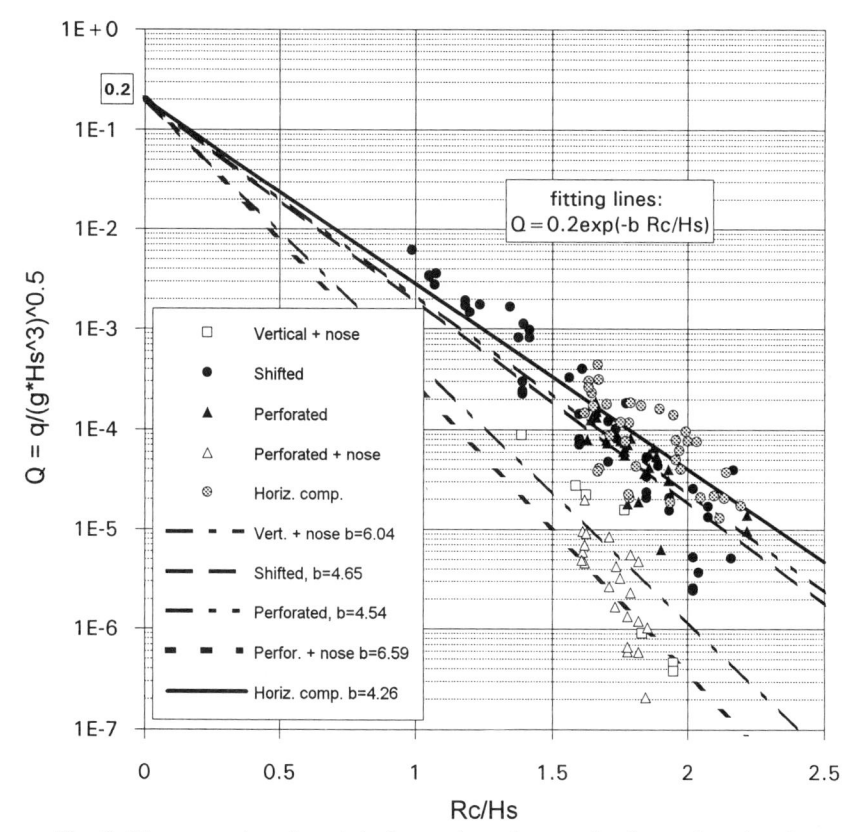

[Figure: plot with y-axis $Q = q/(g \cdot H_s^3)^{0.5}$ ranging 1E-7 to 1E+0, x-axis R_c/H_s from 0 to 2.5. Legend markers: Vertical + nose, Shifted, Perforated, Perforated + nose, Horiz. comp. Fitting lines: $Q = 0.2\exp(-b\, R_c/H_s)$ with Vert.+nose b=6.04, Shifted b=4.65, Perforated b=4.54, Perfor.+nose b=6.59, Horiz. comp. b=4.26.]

Fig. 9 Wave overtopping data for various types of caisson breakwaters.

which can be effectively used for the preliminary design of vertical breakwaters. The reliability of the formula 6 can be given by taking the coefficient 4.3 as a normally distributed stochastic variable with a standard deviation $\sigma=0.3$.

From the influence factors of the various caisson geometries, as compared to the plain vertical wall some useful engineering conclusions can be distilled:

- the greatest overtopping reduction can be achieved by introducing a recurved parapet (nose) at the crest of a vertical front wall: the corresponding $\gamma_n=0.7$ means a 30% crest elevation reduction to get the same overtopping rate; this may however be limited to relatively small discharges;
- for simply perforated or shifted caissons the freeboard saving is only 5÷10%;

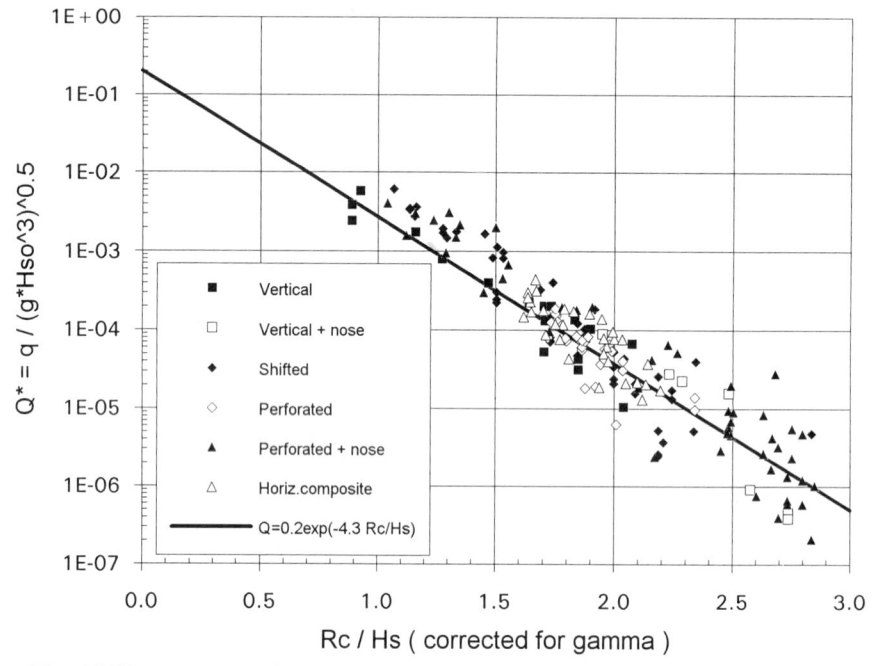

Fig. 10 Wave overtopping on vertical and composite breakwaters: conceptual design graph.

- if a nose is adopted at the crest of a perforated caisson, then the combined reduction factor can achieve 0.65, while its effect on a shifted parapet is negligible;
- the overtopping of horizontally composite breakwaters is influenced by porosity, slope, width and elevation of the mound. Overtopping increases if the armour crest is below or at mean sea level (max $\gamma_{ss}=1.15$).

Conclusions and further work

The results of an extensive 2D model test investigation on the overtopping performance of caisson breakwaters have been analysed to produce updated criteria for their functional safety and a new comprehensive conceptual design method.

The following remarks, valid for structures designed for relatively small overtopping, may be outlined:

- the proposed admissible overtopping discharges (q) for the safety of people and vehicles on breakwaters are

larger than those presently recommended in the manuals and are dependent on the structure geometry;
- the overtopping volume (V) is however a better parameter for allowable criteria;
- overtopping discharges on deepwater vertical walls are slightly larger than those predicted by Goda (1985) and quite smaller than those for an equivalent sloping structure (dike);
- the same exponential relationship between Q and Rc/Hs applies for both structure types (the reduction factor for vertical walls being 0.6), whereas very different and large ratios V_{max}/q can be observed;
- the combination of a perforated wall with a recurved crest (nose) on the front wall produces the largest overtopping reduction, whereas a rock protection in front of the caisson up to the sea level can increase overtopping;
- the probability distribution of overtopping volumes per wave is well defined by a Weibull distribution with a shape factor of 0.75 and a scale factor dependent on q and on the percentage of overtopping waves.

Further work is necessary to take into account the effects of wave obliquity and directional spreading. Actually a 3D caisson model study has just been carried out at Delft Hydraulics within the same European MAST-MCS project and results will be published soon. Additional analysis should also be performed to verify the influence of other less important structural and hydraulic parameters. The critical overtopping for the structural integrity of the caisson foundation system as well as for other facilities on and behind the breakwater should be also evaluated.

Finally, large scale model studies and prototype measurements of wave overtopping on real breakwaters are recommended to verify the present guidelines.

Acknowledgements

The experimental research study has been carried out with the financial support of ENEL-CRIS, MURST (Italian Ministry of University and Research) and the European Union within the MAST-MCS project contract 0047. The contribution of J.Juhl (DHI) and B.G. Madrigal (CEPYC) for supplying their flume data is also gratefully acknowledged.

References

Aminti P., Franco L. (1988), "Wave overtopping on rubble mound breakwaters." Proc. XXI ICCE, Malaga, ASCE, N.Y.

Bradbury A.P., Allsop N.W.H. (1988), "Hydraulic effects of breakwater crownwalls." Proc. ICE Conference BREAKWATERS'88, Eastbourne, T. Telford, London.

British Standard BS 6349: part 7 (1991), "Maritime Structures: Guide to the design and construction of breakwaters." London.

CIRIA/CUR (1991), "Manual on the use of rock in coastal and shoreline engineering." CIRIA S.P.83-CUR Rep. 154.

de Gerloni M., Franco L., Noli A, Rossi U. (1989), "Porto Torres industrial Port. Model tests on the new west breakwater." 2nd AIOM Congress, Napoli (in italian).

de Gerloni M., Franco L., Passoni G. (1991), "The safety of breakwaters against wave overtopping." Proc. ICE Conf. on Breakwaters and Coastal Structures, T. Telford, London.

de Waal J.P. (1994), "Wave overtopping of vertical coastal structures. Inlfluence of wave breaking and wind." Delft Hydraulics report H1635.

De Waal J.P. and van der Meer J.W. (1992), "Wave runup and overtopping on coastal structures." Proc. XXIII ICCE, Venice, ASCE, N.Y., Vol. 2, pp. 1758-1771.

Franco L. (1993), "Overtopping of vertical breakwaters: results of model tests and admissible overtopping rates." MAST2-MCS, 1^ workshop, Madrid.

Franco L. (1994), "Further results of hydraulic model tests on wave overtopping." MAST2-MCS, 2^workshop, Milano.

Fukuda N., Uno T. and Irie I. (1974), "Field observations of wave overtopping of wave absorbing revetment." Coastal Eng. in Japan, 17.

Goda Y. (1985), "Random seas and design of maritime structures." University of Tokyo Press.

Jensen O.J. and Juhl J. (1987), "Wave overtopping on breakwaters and sea dikes." Proc. II COPEDEC Conf., Beijing.

Owen M.W. (1980), "Design of seawalls allowing for wave overtopping." Rep. EX924, Hydraulics Research Wallingford.

Smith G.M., Klein Breteler M, Seijiffert J.W.W., van der Meer J.W.(1994), "Erosion and overtopping of a grass dyke: 1:1 scale model tests." proc. ICCE 94, Kobe.

Sigurdarson S, Viggosson G. (1994), " Berm breakwaters in Iceland, practical experiences", Proc. Hydro-port `94, PHRI, Yokosuka.

Van der Meer J.W., Janssen P.F.M. (1994). "Wave forces on inclined and vertical wall structures" to be published on ASCE, ed. Z.Demirbilek.

WAVE OVERTOPPING 1045

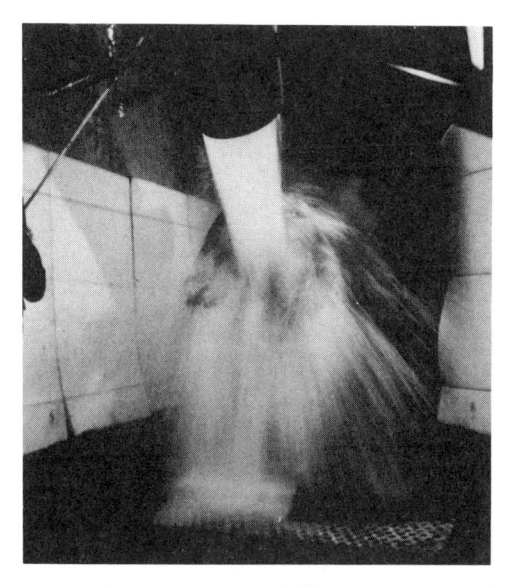

Photo 1 Assessing first author's stability under water jets; a ballasted manikin (on the left) is waiting for its turn.

Photo 1 Effects of wave overtopping on model cars and model persons on a caisson breakwater.

CHAPTER 76

AN INVESTIGATION OF THE WAVE FORCES ACTING ON BREAKWATER HANDRAILS

Atsushi HUJII[1], Shigeo TAKAHASHI[2] and Kimihiko ENDOH[3]

ABSTRACT
We carried out model experiments to quantitatively elucidate the motion of overtopping waves on a caisson breakwater, with results leading to the development of an empirical model that can evaluate the maximum current velocity and depth of such waves on a breakwater. In addition, we developed a formula for calculating the wave forces acting on breakwater handrails, then confirmed its validity using field experiments.

1. INTRODUCTION

Seawalls and breakwaters are ideal locations from which people can comfortably enjoy a seafront environment. In Japan, such structures allowing public access are called promenade seawalls or breakwaters. Because such facilities are a dangerous place for personnel, thorough safety measures must be in place before opening them to the public. Handrails installed on seawalls and breakwaters are a basic safety feature, being present on almost all of Japan's promenade facilities.

Many facilities, however, have been damaged by high overtopping waves, thereby creating serious problems for facility safety management. **Figure 1** shows a photograph of damaged promenade breakwater handrails. Obviously then, overtopping wave forces are a major consideration in handrail design, yet only a limited amount of associated quantitative information is available. In addition to determining the wave forces acting on handrails, the motion of overtopping waves must also be clarified.

Figure 1 Photograph of damaged breakwater handrails

1 Niigata Investigation and Design Office, the First District Port Construction Bureau, MOT, 1-332 Hakusan-ura, Niigata 951, Japan
2 Maritime Structures Laboratory, Port and Harbour Research Institute, MOT
3 Port and Harbor Engineering Section, Civil Engineering Research Institute, HDA

This led to the present study which uses laboratory model experiments to (i) examine the characteristics of overtopping waves above a breakwater and (ii) develop a formula for obtaining wave forces acting against a handrail. We also examine the characteristics of wave forces on handrails, then verify the proposed formula using field experiments.

2. MOTION OF OVERTOPPING WAVES ON A CAISSON
2. 1 Experiment Outline

To examine the basic characteristics of overtopping wave motion on a composite breakwater, and to develop a model of the motion, we conducted a laboratory model experiment using relatively low wave heights. During this experiment (Experiment A), a 163-m-long, 1-m-wide, and 1.5-m-deep wave flume was employed to measure water depths and current velocities above a model caisson.[1]

We also conducted another laboratory experiment using relatively high wave heights over other types of model breakwaters; hence enabling the development of an empirical formula. During this experiment (Experiment B), a 38-m-long, 1-m-wide, and 1.5-m-deep wave flume was employed to measure water depths and current velocities above composite breakwaters, breakwaters covered with wave-dissipating blocks, and slit-caisson breakwaters. Wave forces acting on model handrails were also measured.[2]

This paper primarily reports on the fundamental motion characteristics of overtopping waves (Experiment A).

2. 2 Overtopping Wave Motion

Figure 2 shows overtopping wave motion above a caisson at $hc/H = 0.293$, $H/L = 0.051$, and $\Delta t = 0.05$, being obtained using a high-speed video camera that records 200 frames/s. Note that after the water level in front of the breakwater rises and the wave grows higher than the crown, its shape gradually shifts toward the crown's landward edge. The overtopping wave then crashes on the breakwater, being transformed into one-directional landward flow. We classify these phenomena into the following two stages:
1) "Green wave stage": the phenomena occurring from the wave's run-up above the crown to its crash onto it.
2) "Overtopping flow stage": the phenomena occurring when the crashed overtopping wave is transformed into fast, landward-directed flow.

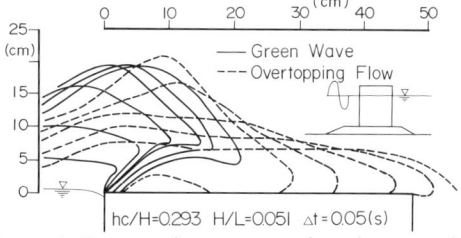

Figure 2 Overtopping wave motion above a caisson

3. OVERTOPPING WAVE MODEL

Figure 3 shows a diagram representing the "overtopping wave model (OWM)," which we apply to empirically evaluate the maximum water depth and current velocity which occurs above a caisson during overtopping. Input data consists of wave and breakwater structural conditions, as well as water depth in front of the caisson.

The green wave stage is characterized by the overtopping wave front having a parabolic trajectory. The distance between the caisson's seaward edge and the crash point of the wave front is called the "green wave range," and is represented by l_3. The value of l_3 is determined by the velocity of the rising wave crest (wave crest rise velocity) above the caisson's seaward edge and the wave velocity. The maximum water depth and current velocity are considered to be constant within the green wave range.

In the overtopping flow stage, the maximum water depth above the crown changes over time, i.e., it decreases within a certain distance from the seaward edge, called the "accelerated flow range," then tends to be almost constant, called the "constant flow range." The maximum current velocity in the former range increases as the wave approaches the landward side, becoming nearly uniform in the latter range. In the model, η_1 represents the maximum water depth at the top of the caisson's seaward edge, while l_1 represents the horizontal distance over which the water depth changes.

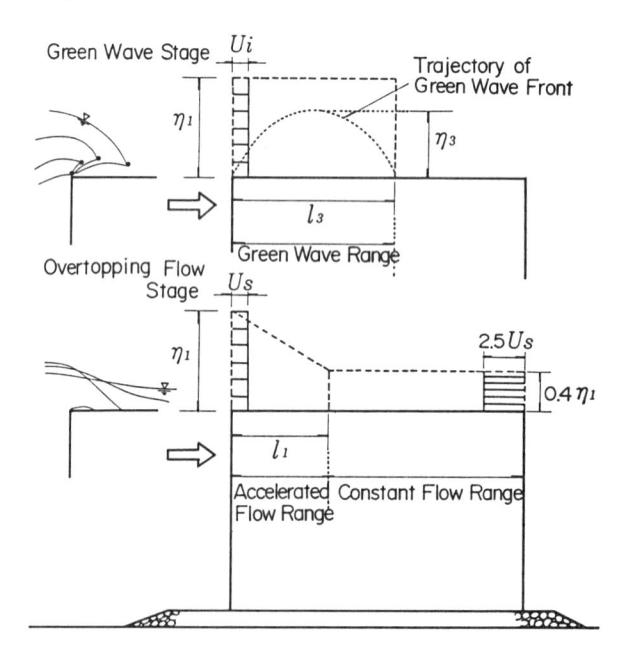

Figure 3 Overtopping wave model

3.1 Green Wave
3.1.1 Wave Crest Rise Velocity

The rise velocity of the overtopping wave crest at the moment it runs up above the crown, $V\text{sf}$, is equal to the vertical velocity of the wave's surface profile. $V\text{sf}$ can be evaluated using a second-order approximation formula for finite amplitude standing waves. If the breakwater superstructure is the composite type, the effects of its rubble mound must be considered because they differ depending on wave length L and mound shape. It is reasonable to consider that the mound has a significant effect on waves when the mound shoulder width, B_M, is relatively large in comparison with L, and also that the mound has only a slight effect when B_M is relatively small. Thus, use of the ratio B_M/L enabled us to develop a functional formula for obtaining the representative water depth hm, which subsequently determines overtopping wave motion, i.e.,

$$hm = \begin{cases} d & : B_M/L \geq 0.16 \\ d + (h-d)\dfrac{0.16 - B_M/L}{0.05} & : 0.11 < B_M/L \leq 0.16 \\ h & : B_M/L < 0.11 \end{cases} \qquad (1)$$

where h is the natural water depth in front of the breakwater and d is the water depth above the rubble mound.

Figure 4 shows the relationship between B_M/L and $V\text{sf}$ using Eq. (1), where the experimental and calculated values clearly indicate good agreement.

Overtopping waves move landward as soon as they run up over the crown, and at this moment, the wave crest horizontal velocity at hm has a proportional relation with wave velocity Cm, i.e., it is about 30% smaller that Cm regardless of crown height. Thus, if we consider the motion of the front to be in free-fall, having come over the caisson's seaward edge at $V\text{sf}$ with a horizontal velocity of $0.3Cm$, then a functional formula can be developed to determine the trajectory of the green wave front, with l_3 being subsequently expressed as

$$l_3 = 0.6 Cm \cdot V\text{sf}/g, \qquad (2)$$

where g is gravitational acceleration.

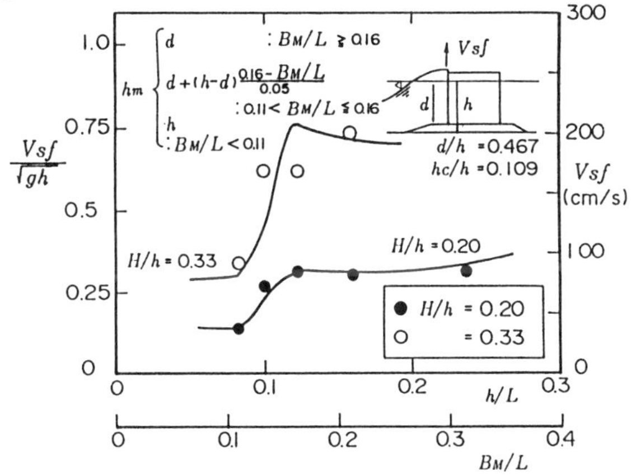

Figure 4 Relation between B_M/L and $V\text{sf}$

3.1.2 Maximum Water Depth and Current Velocity

The distribution of the maximum water depth in the green wave stage is considered to be constant within the green wave range, being equal to the overtopping flow stage's maximum water level η_1 at the caisson's seaward edge. Experiment results enable the corresponding maximum current velocity, Ui, to be expressed as a function of Cm and wave height H, i.e.,

$$Ui = 0.8Cm(1.67H/hm - 0.67)^{1/3} \quad : H/hm \geq 0.4 \quad (3)$$
$$0 \quad : H/hm < 0.4$$

3.2 Overtopping Flow
3.2.1 Maximum Crest Height over The Caisson's Seaward Edge

Goda et al.[3] developed the following formula for determining the wave crest height ratio K for standing waves, being the ratio of the run-up height above the still water level in front of a caisson, R, to the wave height, H:

$$K = \min\left\{\left[1.0 + a(H/h) + b(H/h)^2/Ksb\right], c\right\}, \quad (4)$$

where Ksb is the coefficient expressing nonlinear shoaling effects and breaker decay, while $(a, b, c) = (1.0, 0.8, 10.0)$ for vertical walls. K is 1.0 for small amplitude standing waves and exceeds 1.0 as their finite amplitude increases. The second term in the brackets of Eq. (4) represents the effect in which K increases in proportion to H/h under the condition that no breakers exist, while the third term represents a green wave generated in front of the caisson by breakers. When breaking waves hit the caisson, overtopping waves run up high above it and produce splashing. The quantity of splashed water, however, is relatively small; thus we neglected the third term.

In general, breakwaters can handle large quantities of water from overtopping waves, and because wave overtopping reduces the value of the reflection coefficient, K tends to become constant and is not proportional to H per Eq. (4). We therefore included the effect of overtopping waves using the following formula:

$$K = 1 + \alpha_1 H/hm \quad : H/hm < \frac{-1 + \sqrt{1 + 4\alpha_1 hc/hm}}{2\alpha_1}$$
$$\frac{1 + \sqrt{1 + 4\alpha_1 hc^*/hm}}{2} \quad : H/hm \geq \frac{-1 + \sqrt{1 + 4\alpha_1 hc/hm}}{2\alpha_1} \quad (5)$$

$$hc^* = hc \frac{H/hm}{2H/hm - \frac{-1+\sqrt{1+4\alpha_1 hc/hm}}{2\alpha_1}} . \quad (6)$$

The resultant K values enable calculating the maximum wave crest run-up height at the caisson's seaward edge, i.e.,

$$\eta_1 = K \cdot H - hc , \quad (7)$$

where hc is the breakwater crown height and α_1 is a correction factor dependent on the breakwater structure. For a composite breakwater, $\alpha_1 = 1.0$; whereas $\alpha_1 = 0.5$ for a breakwater covered with wave-dissipating blocks or a slit-caisson breakwater.

Figure 5 shows the experimental (when overtopping occurred) and calculated (Eq. (5)) K values for Experiment A, where the experimental values ranged from 0.9 to 1.1 when $hc/h = 0.109$, and from 1.0 to 1.2 when $hc/h = 0.207$: results that confirm a higher crown height will increase K. In addition, when overtopping waves occurred, the values calculated using Eq. (5) tended to be constant and independent of changes in wave height, being in good agreement with the experimental values.

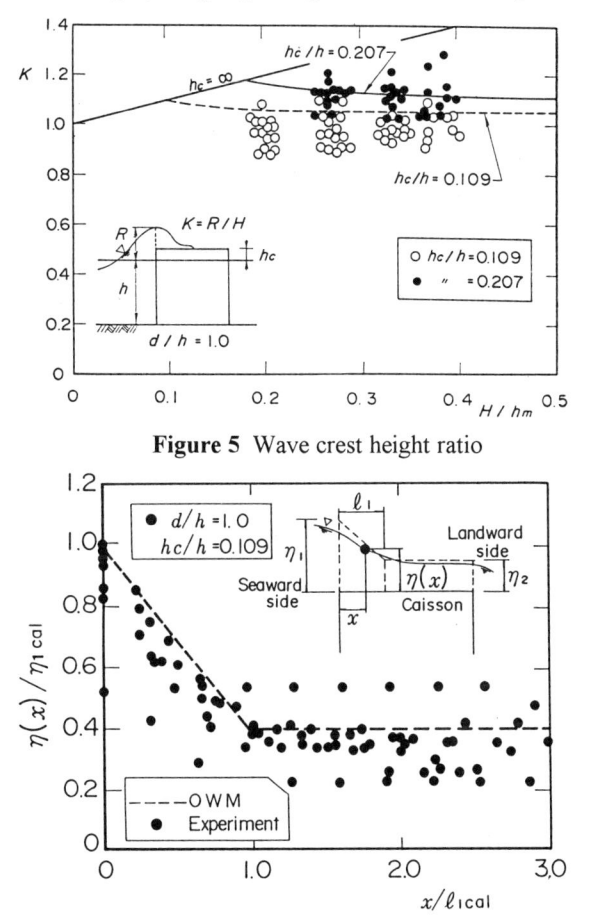

Figure 5 Wave crest height ratio

Figure 6 Distribution of maximum water depth above a caisson

3.2.2 Maximum Water Depth

The accelerated flow range l_1 can be expressed as a function of hc, maximum water depth η_1, and wave velocity Cm. The maximum water depth in the constant flow range is known to be about 0.4 times smaller than η_1 at the caisson's seaward edge, and therefore, the maximum water level at an arbitrary crown point $\eta(x)$ can be calculated using the values of η_1 and l_1 occurring during overtopping flow stage, i.e.,

$$\eta(x) = \begin{array}{ll} \dfrac{l_1 - 0.6x}{l_1}\eta_1 & :x < l_1 \\ 0.4\eta_1 & :x \geq l_1 \end{array} \qquad (8)$$

Figure 6 shows the distribution of experimental and calculated maximum water levels above the crown of a vertical breakwater with $hc = 8.2$ cm ($hc/h = 0.109$), in which the water depth and distances from the caisson's seaward edge are respectively non-dimensionalized by η_1 and l_1. Note that the experimental values are scattered, and also that when $x/l_1 = 0$, η/η_1 ranges from 0.8 to 1.0, becoming 0.25 to 0.55 in the constant flow range. Since the experimental values are roughly dispersed around the calculated ones, this indicates good agreement.

3.2.3 Maximum Current Velocity

Since our Experiments A clarified that the current velocity in the overtopping flow stage has a relationship with water depth, and also that their maximum values have only a slight phase difference, this indicates the maximum current velocity $Us(x)$ can be determined by dividing the maximum overtopping wave quantity q_{max} by $\eta(x)$ (Eq.(8)). Based on a formula modeling steady flow in a dam, we used tests to examine the flow coefficient C_1, which enables calculation of the overtopping wave quantities. Consequently, a functional formula was developed to obtain q_{max}, i.e.,

$$q_{max} = \begin{cases} (0.68 + 1.10 \cdot H/hm)C_1\eta_1^{3/2} & : H/hm < 0.4 \\ \left(0.8 + \dfrac{0.32}{(10H/hm - 4)^2 + 1}\right)C_1\eta_1^{3/2} & : H/hm \geq 0.4 \end{cases} \qquad (9)$$

where $C_1 = 1.61$ (m$^{0.5}$/s).

Figure 7 shows the distribution of the maximum current velocity above the caisson for $d/h = 1.0$ and $hc/h = 0.109$. The x-axis is the distance from crown's seaward edge, x, divided by l_1, while the y-axis is the maximum current velocity at each point, $Us(x)$, divided by the maximum current velocity in the constant flow range, $Uscal(\infty)$. When x/l_1 is less than 1.0, i.e., within the accelerated flow range, the experimental current velocity increases as x increases. It is clear that in the constant flow range, the value of $Us(x)/Uscal(\infty)$ tends to be constant, ranging from 0.9 to 1.15. Note the rough agreement between experimental and calculated values.

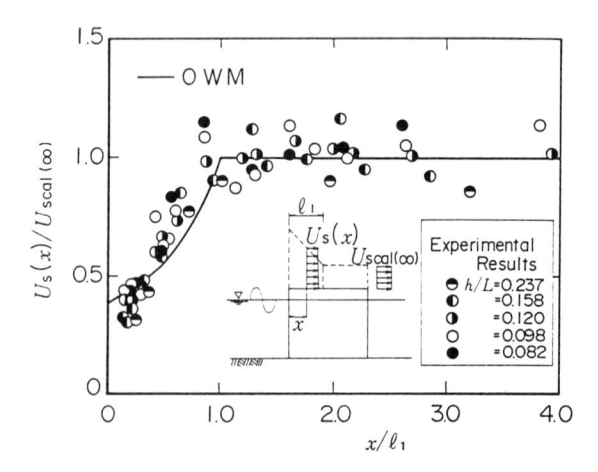

Figure 7 Distribution of maximum current velocity above a caisson

4. FORMULA FOR WAVE FORCE ACTING ON HANDRAILS

Overtopping wave forces acting on column-shaped handrail members may include: (i) an impulsive wave force that occurs when an overtopping wave hits, and (ii) a drag force that is dependent on the current velocity of overtopping waves. Because these members have relatively small diameters, the impulsive component was assumed to be negligible, and therefore, only the drag force was considered. Neglecting the effect of impulsive pressure was confirmed to be appropriate using field experiments. The drag force can be expressed as

$$F = \frac{w_0}{2g} C_D \cdot A \cdot U_{max}^2, \qquad (10)$$

where A is the horizontally projected area of a submerged handrail at the maximum water depth, C_D is the drag coefficient, and U_{max} is the maximum current velocity estimated to occur at the handrail installation point. We used the largest values of U_{max} observed in the green wave and overtopping flow stages.

Figure 8 shows the ratio of experimental to calculated wave forces, F_{exp}/F_{cal}, acting on a handrail. Wave forces acting on handrail members installed on the caisson's seaward and landward edge are indicated. Since the handrail was made of cylindrical column-shaped members, we assumed $C_D = 1.0$. At the seaward edge, the maximum current velocity in the green wave stage is larger than that in the overtopping flow stage, and consequently, the wave forces at this edge were calculated using the maximum velocity Ui in the green wave stage. Since the landward edge is in the constant flow range outside the green wave range, we used the maximum current velocity Us. At the seaward edge handrail, F_{exp}/F_{cal} ranged from 0.7 to 1.2, with good agreement being present between experimental and calculated values. At the landward edge handrails, F_{exp}/F_{cal} was slightly less (0.4 to 0.7), which indicates that calculated values are comparatively slightly larger. These experimental results confirm that Eq.(10) is suitable if $C_D = 1.0$ for cylindrical column-shaped members.

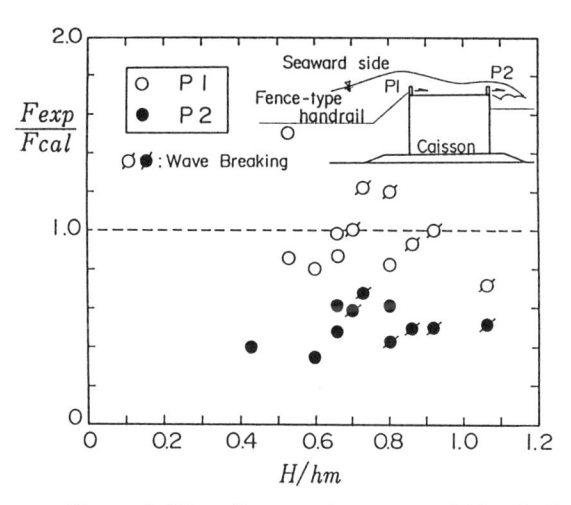

Figure 8 Wave forces acting on a model handrail

At a high Reynolds number, Re, however, the C_D value is lower for this type of handrail. Consequently, since Re values are higher in the field than in experiments, a possibility exists that C_D is actually less than assumed. It should be realized that the characteristics of impulsive wave forces could not be determined because the experimental response characteristics of handrail members did not agree with field results, possibly being due to measuring the wave forces with strain gages. This problem led to using field experiments to examine the characteristics of wave forces and impulsive wave forces (Section 5).

5. FIELD EXPERIMENTS
5. 1 Background
5.1.1 Location of Field Experiments

Field experiments were conducted from November 1991 to March 1992 above the Second North Breakwater of Sakata Port in northeastern Japan. **Figure 9** shows plane and cross-section views of the field site. Although this front-line breakwater is installed in deep water ($\approx$ 16 m), its crown height is relatively low (4.5 m) because it was under construction during the tests. The upper seaward edge of the caisson has a 45° slope. An observation house was built about 2 km away from the test breakwater to record measured data transmitted through optical fiber cables.

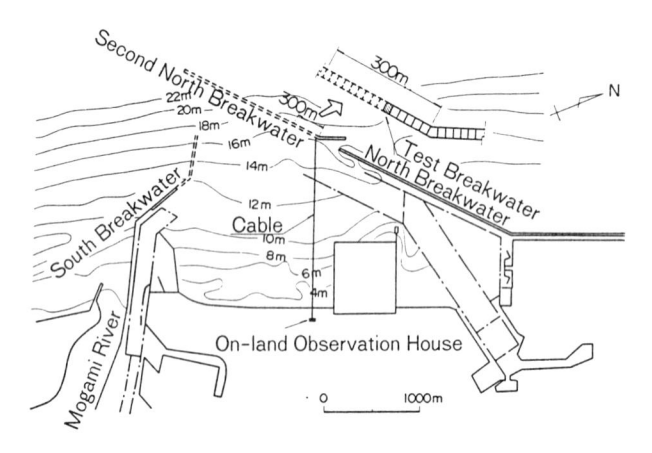

Figure 9 (a) Plane view of Sakata Port

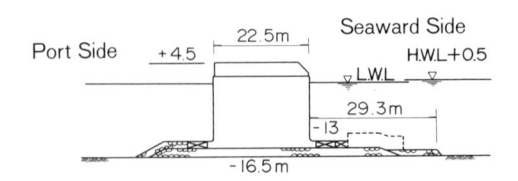

Figure 9 (b) Cross-section view of test breakwater

Table 1 Stanchion shapes

Name	Shape	Diameter (mm)	Thickness (mm)	Allowable wave height (m)
A1	Round steel pipe	216.3	8.2	--------
B1-4	Round steel pipe	101.6	4.2	--------
C1-4	Round steel pipe	48.6	3.2	10
D1	Square steel pipe	100	4.5	--------
E1	Square steel pipe	75	4.5	--------
a	Round steel pipe	27.7	2.3	4.6
b	Round steel pipe	76.3	3.2	14.8
c	Round steel pipe	139.8	4.5	--------
d	Round steel bar	13	---	4
e	Round steel bar	25	---	7.7
f	Round steel bar	38	---	12.3
g	Round steel bar	48	---	--------
F1	Fence-type	60.5	3.2	4
F2	Chain-type	114.3	6	--------

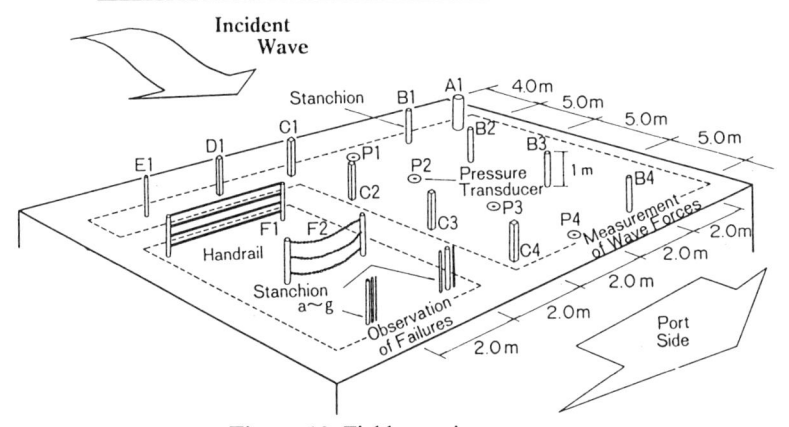

Figure 10 Field experiment setup

5.1.2 Experiment Outline

A wave force measurement test and failure test were carried out in the field. For the wave force test, five kinds of test stanchions (1-m-tall, round or square steel pipes) were installed on the breakwater crown (**Table 1**), and we measured the strain generated at their bottoms. **Figure 10** shows the employed stanchion setup. Stanchions of each type (*A1-E1*) were installed 4 m away from the seaward edge. Stanchions *B* and *C* were also installed 9, 14, and 19 m away from the seaward edge.

For the failure test, we installed conventionally used fence- and chain-type handrails (*F1* and *F2*), round steel pipes (*a-c*), and round steel bars (*d-g*) in order to observe tilt angles and other failure conditions that can occur after a major storm (**Table 1, Fig. 10**). Wave pressure, current velocity, and wave height were obtained corresponding to when the bottom of a stanchion reaches allowable stress. The fence-type handrail (*F1*) had an allowable wave height of only 4.0 m, which indicates that it will fail at comparatively low wave heights.

To examine the motion characteristics of overtopping waves above the crown, we installed four pressure transducers (*P1-4*) at various locations on the crown. Water depth during an overtopping wave was determined by removing the impulsive component from vertical pressure, i.e., only the static pressure component was used to obtain water depth. We calculated the current velocities during overtopping waves using phase differences in the pressure profiles.

5.1.3 Analysis Method

By assuming the wave pressure acting on the test stanchions is equally distributed, we could calculate strain-produced pressure. Peak pressures were then compared with observed maximum wave heights. These heights were measured about 250 m away from the breakwater along its normal line; hence some errors may arise when evaluating the wave pressure characteristics. Wave direction data was obtained about 7 km directly seaward of the breakwater, being a location where good correlation is present.

5.2 Wave Forces Acting on Handrails
5.2.1 Typical Strain and Water Pressure Profiles

Figure 11 shows typical profiles of the strain which occurred at the bottom of the test stanchions and those of water pressure in response to a maximum wave height of 7.4 m (hc/H_{max} = 0.50) (January 24, 1992, 8:14-8:34a.m.). Strain profiles for stanchions *A1*, *B1-4*, *C1-2*, and *D1* are shown with water pressure profiles from *P1-4*. The maximum water depth calculated from the water pressure profiles is 3.2 m at *P1*, the most seaward point, and 1.2 m at *P4*, the most landward point, which indicates that the water depth exceeded the height of the stanchions over the entire the crown area. The average current velocity determined from the phase difference of these maximum water levels is 9.1 m/s between *P1* and *P2*, 11.1 m/s between *P2* and *P3*, and 12.5 m/s between *P3* and *P4*, showing a slight increase toward the landward edge.

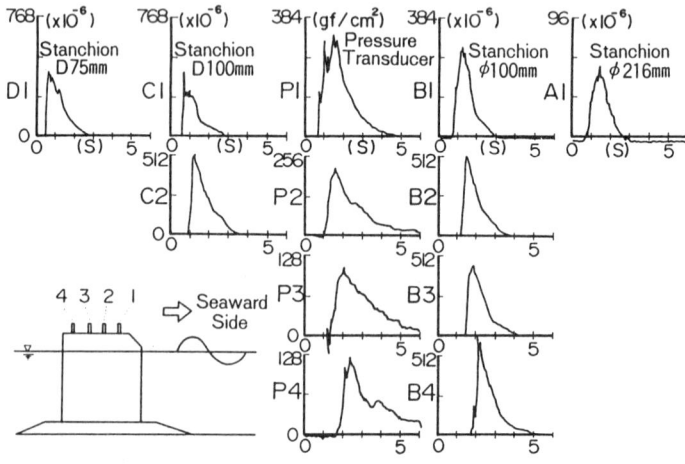

Figure 11 Strain and pressure profiles

As can be seen in *C1*'s profile, strain in the stanchion includes both impulsive and drag force components. The impulsive component seems to occur when a portion of an overtopping wave smashes into the stanchions. Its pressure profile is a strong, sharp peak acting over a relatively short time. In comparison, the drag force component is dependent on current velocity, displaying a relatively mild peak acting over a long time. With the exception of *C1*, the profiles of *C2*, *A1*, and *B4* indicate the presence of an impulsive component, although it is much smaller than the drag force.

Regarding the drag force component of strain, similar to current velocity, strain increased toward the landward edge, e.g., the non-dimensional wave pressure, p/w_oH, at *B1-4* was respectively 0.28, 0.47, 0.46, and 0.57.

The stanchion shape also affected the wave force acting against it. If we compare the drag force component of p/w_oH at the most seaward stanchions, the cylindrical column-shaped ones (*A1* and *B1*) had values of 0.2 and 0.28, whereas those of the square-column ones (*C1* and *D1*) were twice as large at 0.42 and 0.45. This result suggests a difference occurs in the drag coefficient.

5.2.2 Wave Pressure Acting on Stanchions

Figure 12 shows the wave pressure acting on stanchion *B1* in the form of the non-dimensional wave pressure p/w_oH, where peak values of the drag force component and of any impulsive component are indicated. Although the data is scattered due to the effects of different tide levels and wave directions, at a maximum wave height of 6 m or greater, p/w_oH ranges from 0.25 to 0.4 for the drag force component and from 0.3 to 0.5 for the impulsive component. Similarly, the drag force component of the pressure ranges from 0.2 to 0.6 for the cylindrical column stanchions and from 0.3 to 1.0 for the square column stanchions, though the degree of scattering varies with position.

Since impulsive wave pressures vary with position depending on the overtopping wave conditions, as well as showing great variations in scale, peak impulsive pressures were compared with peak drag pressures as shown in **Fig. 13**. The x-axis indicates drag pressures *Ps*, while the y-axis indicates *Pn*, the peak values

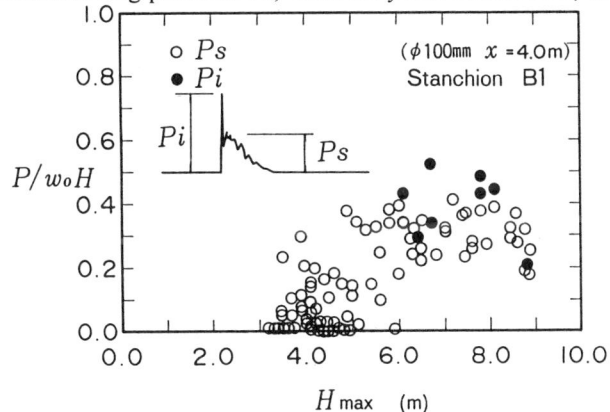

Figure 12 Non-dimensional wave pressure acting on stanchion B1

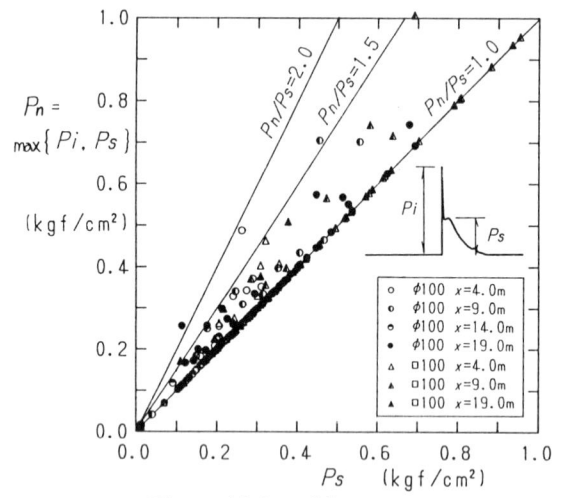

Figure 13 Impulsive pressure

of Ps and impulsive pressure Pi. When Pn/Ps equal to 1.0, the peak pressures are caused by the drag force component, whereas when greater than 1.0, they are caused by the impulsive component. Although the maximum Pn/Ps value is around 2.0, it generally is less than 1.5; hence, impulsive pressures do not generally exceed the drag pressures by much. In addition, the impulsive pressures measured at the stanchions were less frequent and smaller than those measured by pressure transducers located the same distance away from the seaward edge. The relatively small impulsive pressures observed on stanchions may be due to their thin columnar shape or a dynamic response at their natural frequencies (1.5-7 Hz). Measured impulsive wave pressures are expected to be conservative in comparison to those on actual handrails, because actual handrails have the same quality and are fabricated from same material as the test stanchions, and also their natural frequencies are slightly less due to no transverse members being present. Therefore, neglecting impulsive wave pressures during handrail design, as is done in the conventional breakwater design method, is considered acceptable.

5. 3 Comparison with Wave Force Formula
5.3.1 Comparison Between Measured and Calculated Wave Forces

Figure 14 compares measured and calculated wave forces, where the x-axis indicates the non-dimensional value of l_1 divided by the distance x away from the caisson's seaward edge, and the y-axis indicates the ratio of the measured wave force Fm to the calculated one Fc. For the measured wave force, only the drag force component was included, and for the calculated force, a C_D value of 1.0 and 2.0 was used for the cylindrical and square stanchions, respectively. As shown, in a relatively seaward area where x/l_1 is less than 1.0, the measured forces for the cylindrical and square columnar stanchions are only 40-90% of their calculated forces (ave.≈60%). On the other hand, in a relatively landward area where x/l_1 is greater than 1.0, the

F_m/F_c values are close to 1.0 on the average, though they are widely scattered from 0.3 to 1.5. These results confirm the calculated values are roughly suitable.

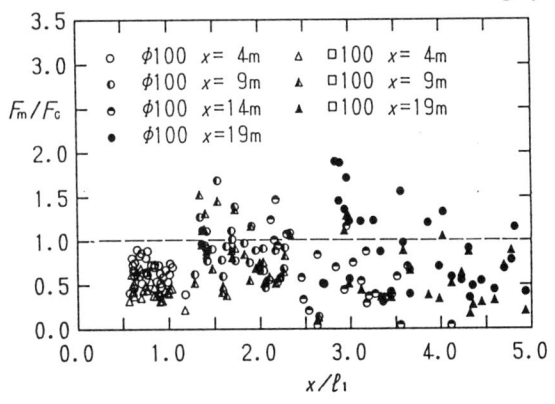

Figure 14 Comparison of measured and calculated forces acting on stanchions

5.3.2 Reproducing Failure Conditions Using Calculated Stress

Structures that tilted or completely fell over during the experiment are: stanchion E used in the wave force measurement; and stanchions a, d, e, and handrail $F1$ used in the failure test. Since the allowable wave heights for all these damaged members is 10 m or less (**Table 1**), these failures could easily be predicted by wave observation results. The strongest storm during the experiment produced a maximum wave height of 10.5 m (Dec. 28-30, 1991), and stanchion e was tilted. Stanchions a, and d and handrail $F1$ failed in a storm on Nov. 19-21, 1991.

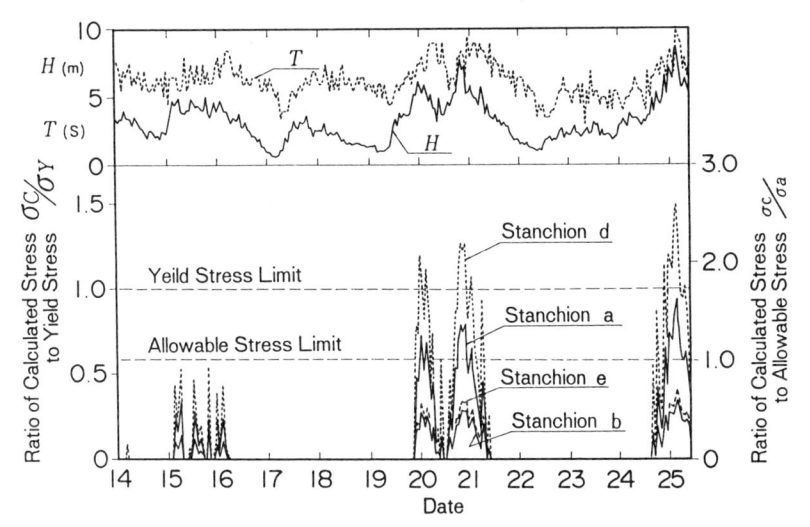

Figure 15 Comparison of calculated and allowable stresses at the bottom of stanchions

Figure 15 shows wave conditions and calculated stresses at the bottom of several stanchions from Nov. 14-25, 1991. We examined damaged stanchions a and d and undamaged stanchions b and e. The wave condition data was used to determine calculated stress σc, which was non-dimensionalized by dividing with either allowable stress σa or yield stress σ_Y.

When wave forces act on a stanchion, the stresses generated in it are generally the greatest at the bottom. When the stresses there exceed allowable stresses, i.e., $\sigma c/\sigma a > 1.0$, a greater possibilities exists for tilting or other failures to occur. From Nov. 14-25, two major storms produced maximum wave heights of 5.0 m greater, with one wave having a maximum height of 8.8 m. During storms before Nov. 19, stanchions a and d both had $\sigma c/\sigma a < 1.0$, and no failure took place. However, when their maximum $\sigma c/\sigma a$ ratios increased to 1.4 and 2.1 during the Nov. 19-22 storms, both suffered failure. In addition, their maximum value of $\sigma c/\sigma_Y$ increased to 1.5 and 0.75. In contrast, both these ratios for stanchions b and e were less than 1.0, and they were not damaged. For the fence-type handrail FI, which failed during same period, $\sigma c/\sigma_Y > 1.5$. These results indicate that the stress calculated using Eq. (10) is accurate enough to be used to the practical design.

6. CONCLUDING REMARKS

We have shown that the motion of an overtopping wave above a caisson can be classified two distinct stages: the "green wave" and "overtopping flow" stages. The fundamental characteristics of each stage were also classified, which enabled developing a model of the motion. Our resultant model can effectively determine maximum water depth and the maximum current velocity above a caisson.

Through the field experiments using handrails, it was possible to understand the effects of impulsive wave forces and the practical wave force characteristics in the region of high Reynolds numbers. It also was possible to estimate the wave force acting on a breakwater handrail as a drag force and to verify the appropriateness of the estimation through wave force measurement and failure tests.

ACKNOWLEDGMENTS

Sincere gratitude is extended to Dr. Tomotsuka Takayama, Director of the Port and Harbour Research Institute for his valuable advice and encouragement.

REFERENCES

1) Takahashi, S., Endoh, K., and Muro, Z. : Experimental study on motions and forces of overtopping waves, Rept. of Port and Harbour Res. Inst. Vol.31, No.1, 1992, pp.3-50. (in Japanese)
2) Sugawara, K., et. al. : Experimental study of wave forces acting on handrails on breakwaters, 39, pp.716-720. (in Japanese)
3) Goda, Y., et. al. : Laboratory investigation on the overtopping rate of seawalls by irregular waves, Rept. of Port and Harbour Res. Inst. Vol.14, No.4, 1975, pp.3-45. (in Japanese)

CHAPTER 77

Rubble Mound Breakwater Stability Under Oblique Waves :
An Experimental Study

J.-C. Galland[1]

Abstract

A series of model tests was carried out in a wave basin at LNH to quantify the effect of long-crested, oblique waves on rubble mound breakwaters. Four types of armouring units (quarry stone, Antifer cube, tetrapod and ACCROPODE®) were tested, under six angles of wave attack (0°, 15°, 30°, 45°, 60°, and 75°). Overtopping, toe berm and main armour stability were studied as functions of wave obliquity. A method is proposed to reduced the results obtained under oblique waves to those obtained under normal waves, allowing thus for the use of formulae derived under normal waves for the design of a breakwater under oblique wave attack.

Introduction

Scale model tests aimed to define guidance for the design of breakwaters are mainly conducted in wave flumes, because this is the easiest (and then the cheapest) way to vary the numerous structural parameters involved. Therefore, the effect of wave obliquity on the stability of such structures has been hardly investigated so far, and the estimation of the possible influence of that parameter is often based upon conjectures or derived from tests not directly related to breakwater stability (run-up measurements on smooth mild slopes mainly).

A few tests, or re-analysis of tests, have been found in the literature, which give some trends for the stability of rubble mound breakwaters under oblique waves.

Whillock (1977) made tests on a 1:2 slope armoured with dolosse under regular wave attack at a fixed period. Results of his tests showed a slight decrease in stability up to an angle of wave attack β of 60° and then, at $\beta = 75°$, quite a large increase in stability.

This trend for dolosse was also mentioned by Gravesen and Sorensen (1977) who, reviewing tests data with random waves, stated a slight decrease in stability when increasing β (with a minimum at $\beta = 45°$), although they did not noticed such a large increase for angles higher than 60°. For quarry stone, the same authors found

---

[1] EDF - Laboratoire National d'Hydraulique, 6 Quai Watier, 78400 Chatou, France.

that stability was not much affected for β ranging from 0° to 45°, but was then increased a lot at higher angles.

Van de Keeke (1969) performed tests with regular waves (at a fixed period) on rocks for 3 different slope angles (1:1.5, 1:2 and 1:3) and for β = 0°, 30°, 45°, 60° and 90°. He found exactly the same trends as the above mentioned ones.

Gamot (1969) reported tests on a breakwater armoured with tetrapod and stated that the armour stability was increased with increasing angle of incidence, this effect being noticeable as soon as β > 40°. He also mentioned that, once they are initiated, damage increase faster under oblique waves than under normal waves.

Finally, Markle (1989) conducted tests in which toe berm stability was investigated, some of these tests being performed at β = 45°. What he concluded, despite very few tests were conducted, is that there was no great difference in stability associated with different angles of wave attack. Some general trend of higher stability under oblique wave could however be seen, although not well defined.

As only very few papers exist in the literature, a series of model tests was carried out at LNH to quantify the effect of long-crested, oblique waves on the stability of rubble mound breakwaters. Advantage was taken from these tests to study also the influence of oblique waves on overtopping and toe stability. Four types of armouring units - quarry stone, Antifer cube, tetrapod (two layers units) and ACCROPODE® (one layer unit)- were tested, under six angles of wave attack β - 0° to 75°, each 15°. A comprehensive description of these tests is given in Galland (1993).

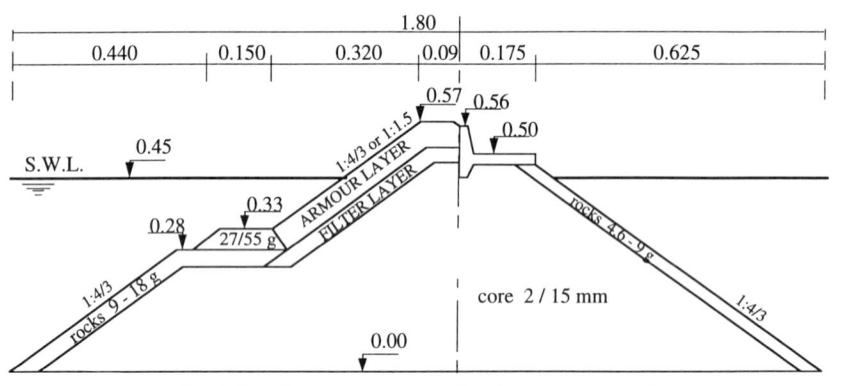

Fig. 1 Breakwater cross-section (measures in m).

Model design and layout

Although these tests do not refer to any field study, the cross-section of the breakwater was defined in such a way that it could represent an actual one. With reference to the maximum significant wave height to be tested (H_s = 0.135 m), the depth of the toe berm below the water level (h_t = 0.12 m) was chosen so that the toe was under strong influence of hydrodynamic forces and the crest elevation (h_c = 0.12 m) so that overtopping was allowed at the same time as damage should start (under normal waves). Doing this way also ensures that the armour layer (which is the main element under study) has a realistic number of units rows, and then a behaviour similar to the one that could occur in the field.

The breakwater consisted in four trunks, each one being 3 m long and armoured with one of the studied units. The total length of the breakwater, including two roundheads, was about 16 m. The rear slope and the crown wall blocks were artificially stabilised, in order to prevent destruction of the breakwater by rear slope degradation and/or crown wall tilting.

Breakwater cross-section

The cross-sections for concrete units were identical and a 1:4/3 slope was chosen (fig. 1). For quarry stones however, the slope was changed into 1:1.5, which is a more realistic value for that armouring unit.

The crown wall blocks, the core, the toe berm, the rear armour layer and the lower part of the front cover layer were identical for all armour units.

Characteristics of the units

Characteristics of the armouring units and toe rocks are given in table 1. The weights and mass density are mean values over at least 30 dry units. In the following, Δ is the relative mass density and D_n the nominal diameter of the unit considered.

Block	ρ_r (kg/m^3)	W (10^{-3} kg)	ΔD_n (10^{-2} m)
ACCROPODE®	2310	44.7	3.51
Antifer block	2400	48.6	3.82
Tetrapod	2540	61.6	4.48
Armour rock	2850	(W_{50}) 90.0	5.85
Toe rock	2500	(W_{50}) 38.0	3.75

Table 1. Characteristics of units.

Grading characteristics of the rocks used for that study are presented in table 2, together with their mass density.

Rock	W_{min} (10^{-3} kg)	W_{max} (10^{-3} kg)	W_{50} (10^{-3} kg)	W_{85}/W_{15}	ρ_r (kg/m^3)
Armour layer (main)	55.0	145.0	90.0	2.05	2850
Underlayer	1.8	9.3	5.0	1.83	2650
Toe berm	27.0	55.0	38.0	1.34	2500
Rear slope	2.8	10.0	6.6	1.57	2660
Cover-layer (lower part)	5.0	17.0	10.0	1.93	2630

Table 2. Characteristics of rocks.

Placement of armouring units was realised according to the procedure relative to each unit : for concrete units, this means that the prescribed mesh was respected and, for quarry stones, that they were tipped on the slope. A cover layer armoured with a given armouring unit was always built by the same person, to ensure a good reproducibility in the way of placing the units.

Test conditions

The programme included 6 series of tests, each one being defined by its angle of wave incidence (0°, 15°, 30°, 45°, 60° and 75°). The normal wave test was aimed to be the reference when analysing the results. To limit the number of investigated stability parameters, the breakwater was placed on a flat bottom (water depth h = 0.45 m) and, as one test consisted in 8 steps with increasing wave height, the peak wave period was tuned at each step so that the wave steepness remains a constant ($s_{0p} \cong 4\%$). The steps duration was adapted so that the total number of waves in each one was 2000. It was large enough a number to ensure both a suitable statistical distribution of waves and a stabilised damage evolution at the end of each step. Targeted wave characteristics are presented in table 3.

Step	1	2	3	4	5	6	7	8
H_s (m)	0.030	0.045	0.060	0.075	0.090	0.105	0.120	0.135
T_p (s)	0.71	0.87	1.00	1.12	1.22	1.32	1.41	1.50
L_{0p} (m)	0.78	1.18	1.56	1.96	2.32	2.72	3.10	3.51
Duration (')	22	26	30	34	37	40	42	45

Table 3. Targeted wave characteristics.

Test facilities

The tests were performed in a wave basin (fig. 2) which overall dimensions are 33 m x 28 m x 1 m, the maximum water depth being 0.45 m. This basin is fitted out with a hydraulic flap-type wave maker, which paddle is 17 m long, 0.85 m high and can move round the basin, allowing this way for a 180° rotation. The four walls of the basin were equipped with wave absorbers in order to avoid re-reflection.

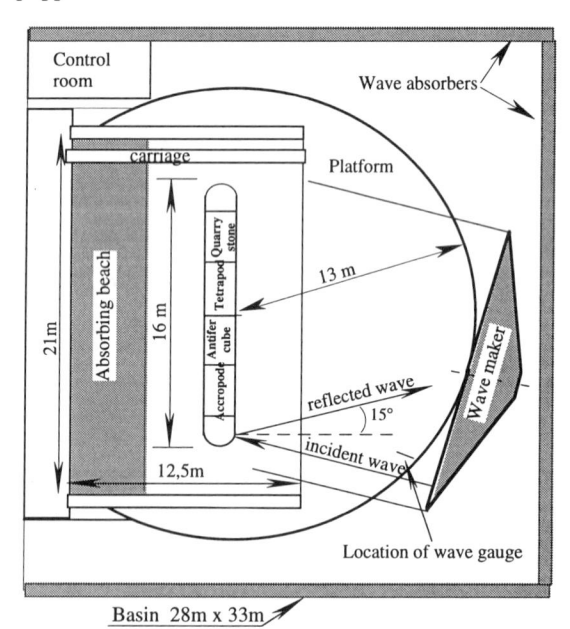

Fig. 2. Test set-up and location of surface elevation measurement for oblique wave attack (example, $\beta = 15°$).

Test procedures
Generation and measurement of waves

Random waves were generated according to a JONSWAP type spectrum, with $\gamma = 3.3$. The wave signals were calculated following the "Deterministic Spectral Amplitude" method, by adding 100 sinusoidal components. They covered the whole steps duration and were computed prior to the tests, which ensured a good reproducibility in the succession of waves during a given step for all tests.

Evaluation of the wave heights is of prime importance for such a study. Under normal waves, water surface elevation was measured with three gauges, which enables to separate incoming and reflected waves by mean of a spectral analysis. Incoming waves characteristics (peak period T_p and significant wave height H_s) were determined this way.

Under oblique waves however, it was not possible to separate incoming and reflected waves. It has then been decided to measure only a "global wave height" (incoming+reflected) at a location in the wave basin that was not optically under direct influence of reflection (cf. fig. 2).

Surface elevations were also measured in front of each test section to ensure that the wave field was homogenous. Wave measurements and their analysis were performed over the whole duration of each step, i.e. over 2000 waves.

Damage evaluation

Damage level to the armour layer D_a was determined by counting, at the end of each step, both blocks that were removed out of the cover-layer and blocks that were distinctly displaced. Armouring units were coloured and the armour layer consisted in a succession of horizontal coloured stripes, the width of which was two blocks (two median diameters for rock), so that displaced blocks were those displaced out of their coloured band, while still standing in the cover layer. All displaced or removed blocks were counted at the end of each step, so that *cumulated* damage was evaluated.

Cumulated damage to the toe berm D_t was determined in the same way, but by counting the number of removed blocks only (reshaping not considered).

Damage measurements were made on a 1 m wide section in the centre part of each trunk, called the test section in the following, in order to avoid side effects at the junction of two trunks when oblique waves are performed. Before counting, the water level was lowered to the toe mound. Damage levels are expressed as the percentage of displaced and removed units in the test section.

Overtopping measurement

The number of overtopping waves was recorded on a paper-recorder, over the whole duration of each step, at the centre of each trunk by use of a wave gauge placed on top of the structure. Calibration enabled to discriminate between green water and broken spray. Overtopping was expressed as the percentage of overtopping waves N_{ov}.

Experimental results

Wave field

The main problem here is the impossibility to separate incoming and reflected waves for the tests under oblique waves, which makes questionable the comparison between the results of these tests, and even more with those under normal waves. To clear up that point, all steps 7 were re-run with normal waves after all tests had been performed, ensuring very carefully that waves were measured in strictly identical conditions : same wave signal, same location of the wave gauges, same position for the wave guides as during the tests. The global wave height measured under oblique wave was then compared to the incident normal wave height. As their relative difference is expected to grow up with increasing wave height (the reflected wave becoming higher), the comparison between these two values for step 7 is a good evaluation of the accuracy of the global wave height as an estimate for the incident wave height under oblique waves. As this relative difference was lower than 6% for all tests, the global wave height as measured under oblique waves was thus considered as a reasonable approximation of the actual incident one.

Armour stability

The curves shown hereafter present the percentage of moved armour blocks D_a as a function of the non-dimensional significant wave height $H_s/\Delta D_n$, and are limited to $D_a < 20\%$, which is their most significant part.

Table 4 below gives the relative variation in wave height with respect to normal wave when increasing obliquity, corresponding to several given damage levels (start of damage, $D_a = 5\%$ and $D_a = 10\%$).

β (°)	15	30	45	60	75
		Antifer Cube			
$D_a < 0.1\%$	+ 56 %	+ 52 %	+ 54 %	+ 130 %	+ 140 %
$D_a = 5\%$	+17 %	+ 31 %	+ 31 %	-	-
$D_a = 10\%$	+ 4 %	+ 13 %	+ 18 %	-	-
		Tetrapod			
$D_a < 0.1\%$	+ 52 %	+ 50 %	+ 44 %	+ 48 %	+ 140 %
$D_a = 5\%$	+ 8 %	+4%	+ 5 %	+ 22 %	-
$D_a = 10\%$	- 3 %	- 4 %	+ 2 %	+ 16 %	-
		Rock			
$D_a < 0.1\%$	- 17 %	+ 65 %	+ 35 %	+ 49 %	+ 98 %
$D_a = 5\%$	0 %	+ 10 %	+ 6 %	+ 6 %	-
$D_a = 10\%$	- 4 %	+ 6 %	+ 15 %	+ 18 %	-
		ACCROPODE ®			
$D_a < 0.1\%$	- 23 %	+ 23 %	+ 38 %	+ 21 %	-
$D_a = 1\%$	+ 12 %	+ 19 %	+ 26 %	-	-

Table 4. Relative variation of $H_s/\Delta D_n$ with β, with respect to normal waves, for start of damage ($D_a < 0.1\%$), $D_a = 5\%$ and $D_a = 10\%$, or $D_a = 1\%$. (- indicate that the corresponding damage level was not reached).

Some trends can be seen in table 4 which indicate an increase in stability with increasing angle of wave incidence. Results are detailed below for each unit.

Antifer cube

Four trends can be observed from table 4 and fig. 3 :
- stability increases with increasing wave obliquity,
- start of damage is delayed under oblique waves : it corresponds to a wave height 50 % higher for β = 15°, 30° and 45° than under normal waves,

- damage, once initiated, increases faster under oblique waves than under normal waves : about two times faster for β = 15°, 30° and 45°,
- For β > 45°, the increase in stability is so high that nearly no damage occurs.

Tetrapod

Exactly the same trends (cf. table 4 and fig. 4) as for Antifer cube can be noted, although they are somewhat less pronounced and valid mainly for $D_a < 10\ \%$

These results (increasing stability with increasing obliquity, faster damage increase under oblique waves) are consistent with those reported by Gamot (1969).

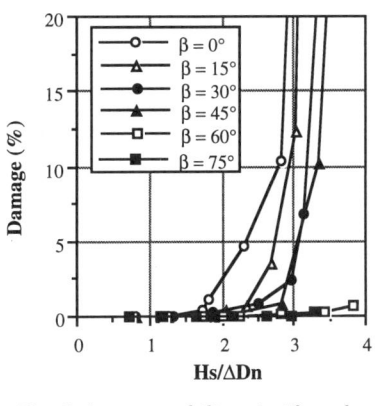

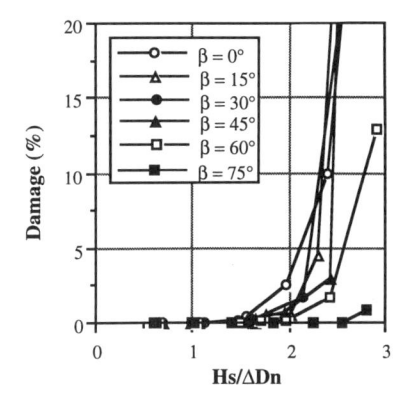

Fig. 3. Armour stability - Antifer cube

Fig. 4. Armour stability - Tetrapod

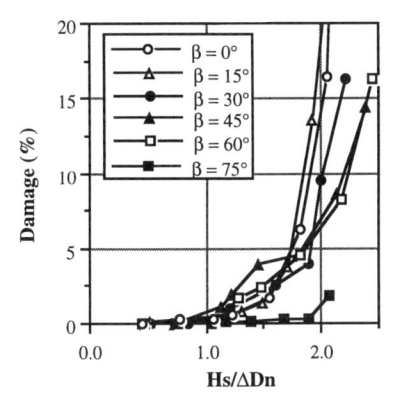

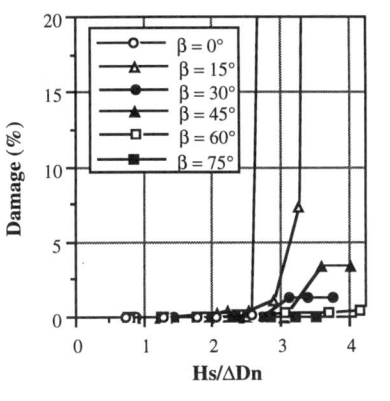

Fig. 5. Armour stability - Quarry stone

Fig. 6. Armour stability - ACCROPODE®

Quarry stone

From fig. 5 and table 4, start of damage seems to be slightly delayed under oblique waves, but quarry stone is seen to be not very sensitive to wave obliquity, at least at low damage levels (D < 5 %). For higher damage levels and β ≥ 30°, some trend is noticeable that could indicate an increasing stability for increasing angle of incidence. However, stability strongly increases at β = 75° only.

These points are in accordance with the results of previous works (Gravesen and Sorensen (1977), Van de Keeke (1969)), although the increase in stability was noticed as soon as $\beta \geq 60°$.

ACCROPODE®

Evolution of ACCROPODE® stability with increasing wave obliquity (fig. 6) is quite different from those observed till now. At $\beta = 15°$, the armour layer behaves similarly as under normal wave attack, with a very sudden failure (characteristic of a one layer unit) which has led to retain a zero-damage criteria for the design of breakwaters armoured with this unit. At higher angles, its behaviour however significantly differs : after some damage, units rearrange so that the armour is stable again and no more damage occur. This could be explained by the high interlocking of this one layer unit.

Toe berm stability

The curves shown below present the percentage of removed toe blocks D_t as a function of the non-dimensional significant wave height $H_s/\Delta D_n$.

A distinction can be made here between the results for Antifer cube and tetrapod on one hand, and quarry stone and ACCROPODE® on the other hand.

Toe berm at Antifer cube and Tetrapod armour layer

The same trends as for the corresponding armour unit can be made (fig. 7 and 8) :
- stability increases with increasing wave obliquity,
- start of damage is delayed under oblique waves : it corresponds to a wave height 45 % higher for $\beta = 15°$, 30° and 45° than under normal waves,
- damage, once initiated, increase faster under oblique waves than under normal waves : about 1.6 time faster for $\beta = 15°$, 30° and 45°,
- For $\beta = 75°$, the increase in stability is so high that nearly no damage occurs.

What could be add is that, provided that high damage levels to the toe are accepted, a 15° wave incidence could be more dangerous than normal wave.

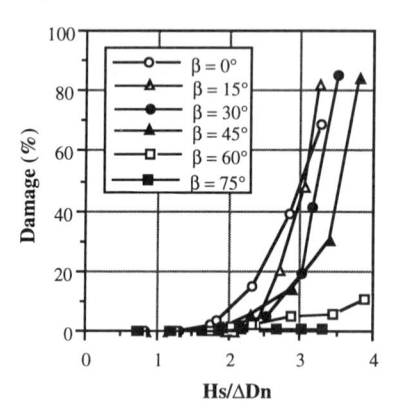

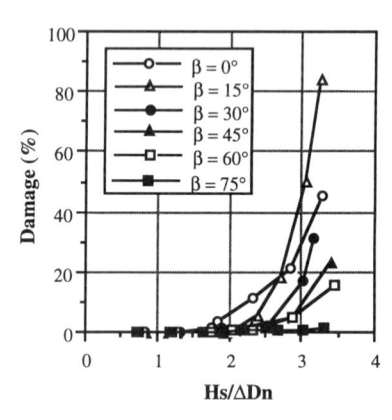

Fig. 7. Toe stability - Antifer cube   *Fig. 8. Toe stability - Tetrapod*

Toe berm at quarry stone and ACCROPODE armour layer

The behaviour of toe berm at quarry stone and ACCROPODE® armour layer (fig. 9 and 10) is rather atypical. Initiation of damage is not so much delayed under oblique waves, and the only marked trend is an increase in stability with increasing wave incidence for $\beta \geq 30°$. However for ACCROPODE®, the same trend as for Antifer cube and tetrapod can be noticed, that is that damage could possibly be higher at 15° than under normal waves.

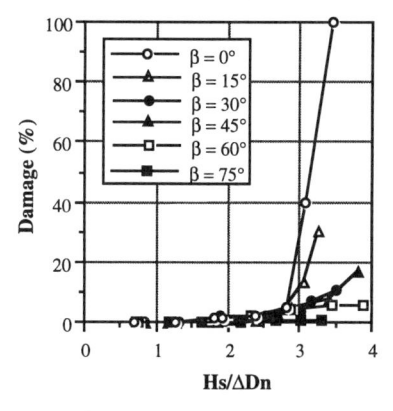

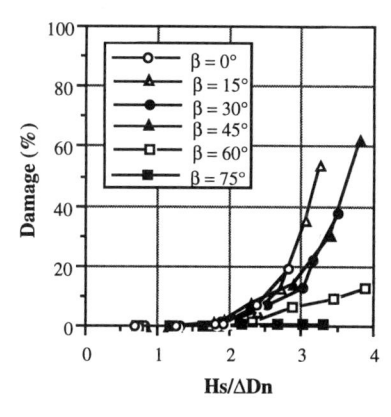

Fig. 9. Toe stability - Quarry stone   *Fig. 10. Toe stability - ACCROPODE®*

Out of the scope of this study but important to notice, is that very high damage levels to the toe berm were reached without endangering the stability of the armour layer.

Overtopping

The curves shown hereafter present the percentage of overtopping waves N_{ov} as a function of the significant wave height H_S.

For all the units studied here, the percentage of overtopping wave N_{ov} is seen to present the same features (fig. 11 to fig. 14):
 - overtopping is equivalent under normal waves and a 15° incidence,
 - overtopping then decreases with increasing wave obliquity,
 - at $\beta \geq 60°$, overtopping is reduced to nearly zero, except for quarry stone.

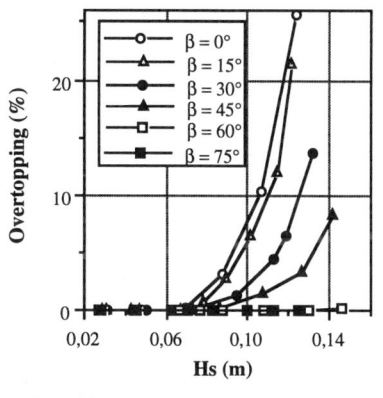

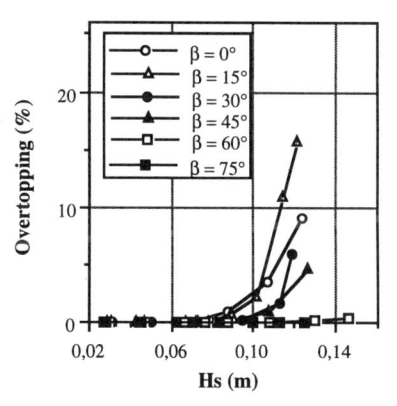

Fig. 11. Overtopping - Antifer cube   *Fig. 12. Overtopping - Tetrapod*

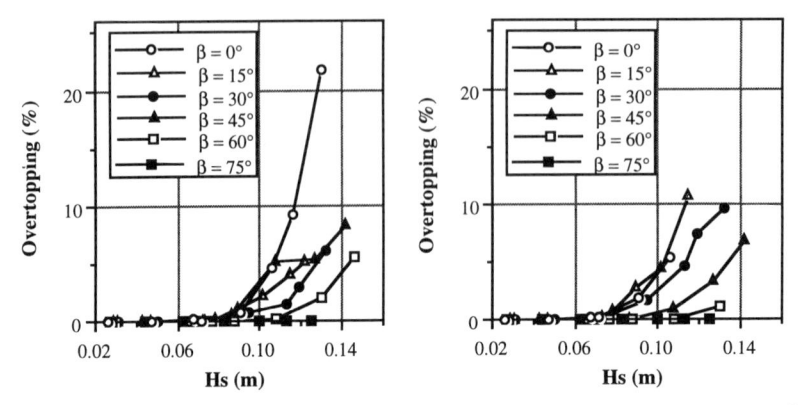

Fig. 13. Overtopping - Quarry stone   *Fig. 14. Overtopping - ACCROPODE®*

Discussion

The aim of this study was not to derive new formulae for the design of rubble mound breakwaters armour layer and toe berm or for the evaluation of overtopping rates, but to provide a better idea of the influence of wave obliquity and also to propose a way for taking into account that influence in existing formulae.

In this idea, and as the apparent slope of the breakwater turns from tgα under normal waves into tgα.cosβ under a β angle of wave attack, it is of interest to study all the above mentioned phenomena as functions of $H_s \cos\beta^x$ instead of functions of H_s only, $H_s \cos\beta^x$ being then an *equivalent normal wave height*.

The remaining problem is of course to determine the coefficient x : results are not numerous enough to allow for the use, from a scientific point of view, of a numerical "best fit" method and the evaluation of x has then to be derived by visual adjustment, being thus subject to some interpretation.

All the curves for armour, toe berm stability, and overtopping have been re-plotted against $H_s \cos\beta^x$, with the best estimates for x (fig. 15 and fig. 16). A summary of the values obtained for x is given in table 5 below.

Such an approach is not fully satisfactory : first because, in some cases, the adjustment is not very good and second because it can not represent the change in the slope of some of the curves that occurs under oblique wave. Nevertheless it could be very useful for the preliminary design of breakwaters.

	Antifer cube	Tetrapod	Quarry stone	ACCROPODE®
Armour stability	0.6	0.3	0.25	1
Toe stability	0.6	0.4	0.6	0.4
Overtopping	1	0.6	1/3	0.75

Table 5. Coefficient x for the equivalent normal wave height $H_s \cos\beta^x$.

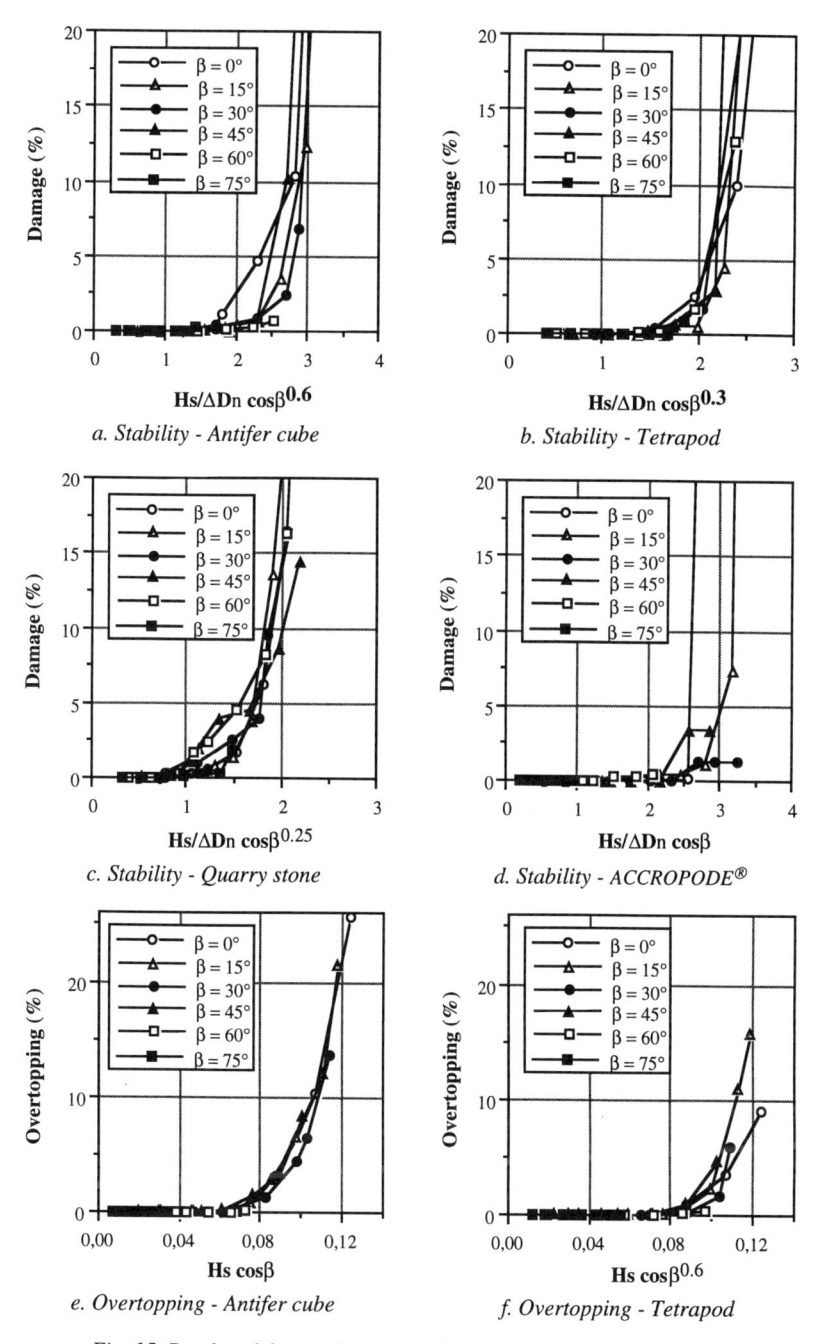

Fig. 15. Re-plot of the results against the equivalent normal wave height

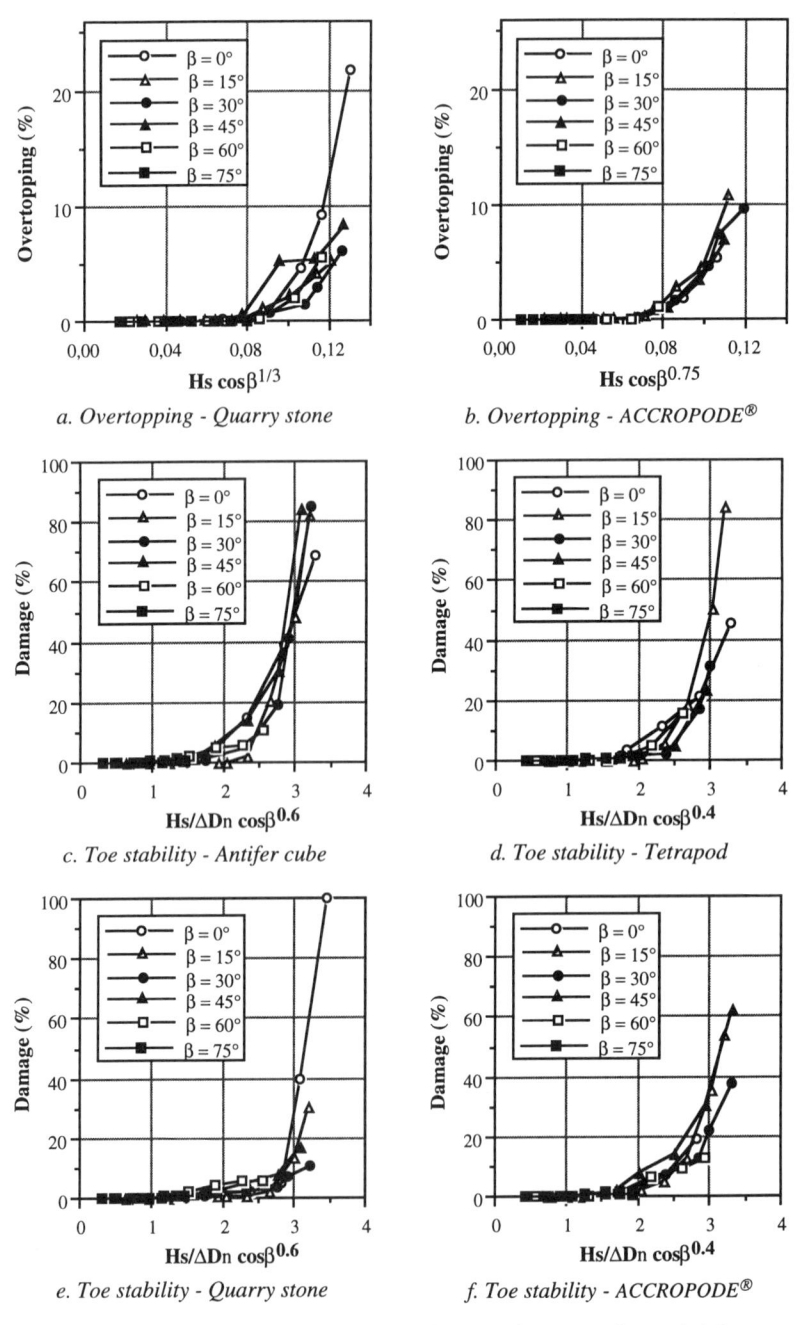

Fig. 16. Re-plot of the results against the equivalent normal wave height

As a general remark, it should be noted that the results obtained within this study do not support the vague belief in a lower stability at small angles of wave incidence (say β = 10°-30°). Such a belief is mainly based on run-up measurements, which have been shown to be maximum at β = 10°-30° on smooth, mild slopes (Tautenhain et al. (1982), CUR/CIRIA (1991)).

The numerical values derived within this study are to be taken with great care, as they result from a single test at each wave angle. Many authors (Jensen (1984), Burcharth et al. (1986), Galland and Manoha (1991) for instance) have indeed stated a large scatter in results of such tests, so that a well-defined characterisation of the influence of wave obliquity should be derived from series of identical tests.

But what is important here, is that the results are not drowned into the scatter but *do really define* the trends reported above ; only the actual values of the gains obtained under oblique waves are not known exactly.

Conclusions

Under the conditions tested :
- flat bottom (no shoaling effect),
- JONSWAP spectrum, γ = 3.3 (no influence of spectrum width or wave groupiness),
- constant wave steepness, $s_{0p} \cong 4\%$ (constant surf similarity parameter ξ_{op}),
- 2000 waves per step (no influence of storm duration),
- one series of tests per angle of wave incidence (scatter not considered),

the following conclusions can be drawn :

Armour stability

Concrete units
- armour stability increases with increasing angle of incidence,
- initiation of damage is delayed under oblique waves,
- once initiated, damage increases faster under oblique waves (Antifer cube, tetrapod).

Quarry stone
- armour stability is not much influenced by wave obliquity at damage levels lower than 5%,
- at higher damage levels and at incidence higher than 15°, armour stability slightly increases with increasing wave obliquity.

Overtopping

All units
- overtopping is equivalent under normal waves and at a 15° angle of wave incidence,
- then, it continuously decreases with increasing incidence.

Toe berm stability

Concrete units
- toe berm stability is equivalent, and can even be lower, at a 15° angle of wave attack and under normal waves,
- then it increases with wave obliquity.

Quarry stone
toe berm stability continuously increases with increasing obliquity.

Equivalent normal wave height

Results obtained at an angle of wave attack β and a wave height H_s are equivalent to those obtained under a normal wave height $H_s.\cos\beta^x$, x being a coefficient which depends on the phenomenon and the unit under consideration. This way, it is possible to use formulae derived under normal wave attack to take into account wave obliquity, for preliminary design of breakwaters.

However it is important to keep in mind that these conclusions result from a single series of tests at each wave incidence, and therefore do not take into account the scatter often reported. Trends presented in this paper are assessed because of the continuity in the evolution of the phenomena they represent, but numerical values should be taken just as estimates. Further testing is still required in order to derive reliable laws taking into account the influence of wave obliquity on the stability of rubble mound breakwaters.

Acknowledgement

This work was funded partly by the EC under contract n° MAST - 0032 - M (JR) and partly by the French Sea State Secretary (STCPMVN).

References

Burcharth, H. F. and Brejnegaard-Nielsen, T., 1986. The influence of waist thickness of dolosse on the stability of dolosse armour. I.C.C.E. Taiwan, vol. 2, chp. 130, pp. 1783-1796.

CUR / CIRIA, 1991. Manual on the use of rock in coastal and shoreline engineering. Published by CIRIA, London, UK and CUR, Gouda, The Netherlands.

Galland, J.-C. and Manoha, B., 1991. Influence of wave grouping on the stability of rubble mound breakwaters. Proc. XXIV IAHR Conf., Madrid.

Galland, J.-C., 1993. Rubble mound breakwaters stability under oblique waves. LNH Report HE-42/93-18, prepared for MAST G6-S Coastal Structures.

Gamot, J.-P., 1969. Stabilité des carapaces en tétrapodes de brise lame à talus. La Houille Blanche n° 2 (in French).

Gravesen, H. and Sorensen, T., 1977. Stability of rubble mound breakwaters. PIANC, 24th Int. Navigation Congress, Leningrad.

Jensen O. J., 1984. A monograph on rubble breakwaters. DHI, 209 p.

Markle, D. G., 1989. Stability of toe berm armour and toe buttressing stone on rubble mound breakwaters and jetties; Physical model investigation. CERC, Vicksburg, Technical Report REMR-CO-12.

Tautenhain, E., Kohlhase, S., Partensky, H.W., 1982. Wave run-up at sea dikes under oblique wave approach. I.C.C.E. Cape Town, vol. 1, chp. 50, pp. 804-810.

Van de Kreeke, J., 1969. Damage function of rubble mound breakwaters. J. Waterways and Harbours Division, Proc. ASCE, Vol. 95, n° WW3.

Whillock, A. F., 1977. Stability of dolos blocks under oblique wave attack. Hydraulics Research Station, Wallingford, report n° IT 159.

CHAPTER 78

WAVE LOADS ON SEADYKES WITH COMPOSITE SLOPES AND BERMS
by
Joachim Grüne[1] and Hendrik Bergmann[2]

Abstract

This paper deals with investigations on loads from wave-induced shock pressures on seadykes with composite slopes and berms. The investigations have been done with large scale laboratory tests, using different wave climate characteristics (regular waves, PM - spectra and field spectra). Two different types of dyke cross-sections have been investigated: one with an upper slope 1:6 and a berm in front, and the other one with composite slopes 1:3 in the lower part and 1:6 in the upper part. The presented results were discussed with respect to the influences of wave climate, of absolute waveheights and of geometrical conditions.

Introduction

Due to the increasing of storm surges and the supposed long-term rising of water levels at the coastlines of the North Sea, wave loads on seadykes have become again more important for savety analysis of excisting dykes. Air entrainment and wave climate characteristics under real sea state conditions play an important role both on shock pressure occurrence and on wave run-up process. Thus boundary effects and scale effects have to be minimized by using field data or large scale laboratory test data. The results presented in this paper were obtained from large scale laboratory tests in the "Large Wave Channel" (GWK) at Hannover, Germany, within an extensive research program on stability on sloping seadykes and revetments (supervision Prof. Führböter +), which had been supporting partly by the German Research Foundation (DFG) for more than 10 years. Previous results from this research program on wave loads on **uniform slopes** have been reported recently for example by Grüne (1992) for shock pressures and by Sparboom, Grüne, Haidekker & Grosche (1990) for wave run-ups.

---

[1] Dipl.-Ing., Senior researcher, Deputy Operation Manager of Joint Institution Large Wave Channel (GWK) of the University of Hannover and the Technical University of Braunschweig, Merkurstrasse 11, 30419 Hannover, Germany.

[2] Dipl.-Ing., Researcher, Hydrodynamic and Coastal Engineering Dept. of Leichtweiß Institute of the Technical University of Braunschweig.

First results from the recent investigations with **composite slopes and berms** on loads from shock pressure occurrence will be presented in this paper. First results on loads from the wave-run-ups measured synchronously within the same testseries are reported by Schüttrumpf, Bergmann & Dette (1994).

Test equipment and data recording

The investigations have been conducted in the Large Wave Channel (GWK) at Hannover (Grüne, Führböter, 1976), which is a joint institution both of the University of Hannover and the Technical University of Braunschweig.

Two different types of dyke cross-sections were installed, which are shown in figure 1. A uniform slope of 1:6, which had been used for previous investigations, in its lower part, firstly was replaced by a berm up to 3.3 m above bottom (upper cross-section in Fig.1), then secondly replaced by lower slopes of 1:3 (middle and lower cross-section in Fig.1). To get a wide range for the vertical distance Dc between stillwaterlevel SWL and slope junction, the replacing by lower slopes of the composite slope cross-sections has been done in two steps. In the first step the lower slope was built up to the same level as the berm (3.3 m above bottom), in the second step up to 4.5 m above bottom.

Waves were measured with wire gauges in front of the wave generator and in front of the dyke cross-sections. The wave induced shock pressures were measured with pressure cells on the slope surfaces in vertical distances of roughly 10 cm. The areas on the dyke surface with installed pressure cells are indicated in figure 1. For the testseries with the berm and the composed slopes with a lower slope up to 3.3 m above bottom pressure cells only were installed on the upper slope 1:6. Furthermore wave run-up was recorded with a step run-up gauge.

The testseries for each of the three dyke cross-sections in figure 1 were conducted with two water levels (4.0 m and 5.0 m above bottom). For each testseries with one of these both waterlevels three different types of wave characteristics were generated: Regular waves, PM - spectra (narrow banded) and field spectra (wide banded); roughly 15 single tests with each of these three wave climate types. The generated field spectra had been measured in front of similar dyke cross-sections at the german coast during high storm surges.

For data collecting two different modes have been used: With the first one all analog signals from pressures cells, waves gauges and wave-run-up gauges were continuously digitized with 100 Hz and then storaged by the main processing computer. With the second mode only higher individual shock pressure events were storaged additionally by a personal computer, using 200 ms long time intervals of the pressure signals as uncontinuous windows with a digitation rate of 1 kHz. The advantage of this sample method is the high digitation rate, but due to that the disadvantage is, that the sample is time-restricted, which leads to lossings of total

BERM

COMPOSED SLOPES

COMPOSED SLOPES

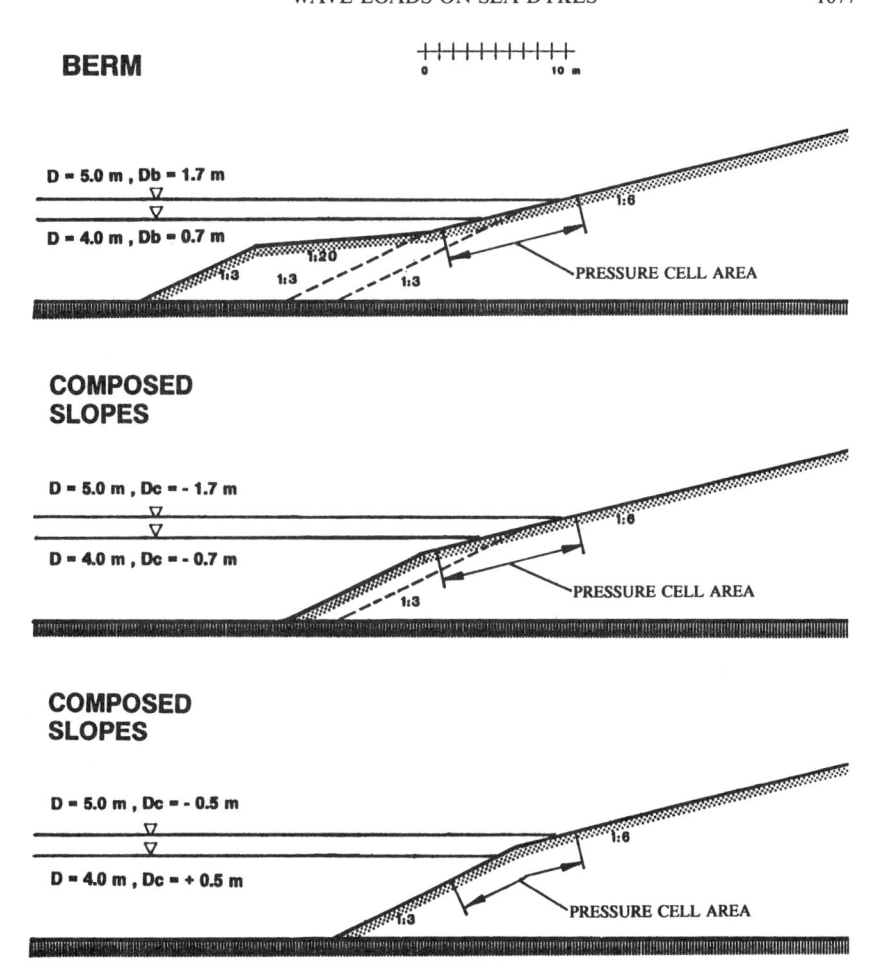

Fig. 1 Dyke cross-sections with berm and composed slopes, investigated in the LARGE WAVE CHANNEL (GWK)

events, especially if the threshold level is accomplished by high quasi-static pressures or by spikes from noises. Therefore a complete data analyzing of all testseries with all pressure cells was done from the 100 Hz digitized datafiles. It must be mentioned, that there were some reduced peak pressure values for higher shock pressure events compared with the 1 kHz-digitation-mode, depending on the peak pressure rising time. The reductions mostly were small, but some rare events were reduced up to approximately 30%. With respect to the stochastic behaviour the influence of partly reduced maximum peak pressure values on the general characteristics of shock pressure ocurrence seems to be small.

Analyzing of shock pressures, definitions

Shock pressures occuring on sloping dyke surfaces are damped more frequently compared to those on vertical walls. Furthermore, especially under real sea state conditions, partly they are mixed with pressure components from waves and wave run-ups. Thus it is often very difficult to distinguish the real shock pressures from the steep wavefronts or the wave run-up fronts, especially if one use normal speed records. In previous reports the author has described an detailed method of analyzing pressure-time histories from high-speed records, which allows a summerizing of the numberless different occuring shapes of pressure-time histories by anatomy parameters (Grüne 1988a, 1988b, 1992).

Using this parameterizing mode, it was possible to evaluate a generalized model of shock pressure occurrence with respect to deterministic and stochastic characteristics. The deterministic parts of the model are represented by local distributions of the anatomy parameters on dyke surface. The local distributions may be devided into different local ranges, which each of it represents a certain state during the breaking process.

For the stochastic part of the generalized model stands the superposition with the stochastic fluctuations of the anatomy parameters. Most data of the different anatomy parameters spread like stars at the sky. But nevertheless there are some clear tendencies and some envelope conditions. There are many possibilities of relating the parameters one with another, reasonable trends were found in dependence on the peak pressure values *Pmax*. In all cases the quality of correlation increase with increasing *Pmax*. Thus the higher peak pressures may be seen as indicators both for general shock pressure occurrence and for details of anatomy parameters.

The following definitions and notations have to be given with respect to the presentation of the results in this paper:

- The maximum pressure value (peak pressure) of an individual shock pressure event usually is defined as *Pmax* [10^4 Pa , (m waterhead)].
- The maximum value of all peak pressures *Pmax*, recorded during one single test is defined as *Max Pmax* and is either analyzed from the data of each pressure cell separately, when the results are presented as local distributions or analyzed from all pressure cells, when presented in dependence of the breakerindex ξ. All pressure data are related to waveheights *H* and are therefore dimensionless, using [m waterhead] for pressures and [m] for waveheights.
- The waves were generated with a closed loop system, to prevent rereflexion at the wave generator. Thus the generated waveheights have been used. A comparison with the waves, measured in front of the dyke shows sufficient agreements, especially for irregular waves. All waveheights *H* are defined as mean waveheights for regular waves and as significant waveheights *H* 1/3 for irregular waves.

WAVE LOADS ON SEA DYKES 1079

- The unit for the local distribution on the slope surface is definied as waveheight related vertical distance *Delta D/H* between pressure cell and stillwaterlevel *SWL* as shown in figure 2. The vertical distance between the slope junction and *SWL* is defined as *Dc*, as shown in figure 3. The waterdepth in front of the dyke is defined as *D* and the waterdepth on the berm as *Db*.
- The pressure data are either presented in dependence of the breakerindex

$$\xi = H * \tan \alpha / \sqrt{H/Lo}$$

with $H = Hm$, $T = Tm$ for regular waves and $H = H1/3$, $T = Tp$ for irregular waves or as local distributions. An example for a local distribution of the waveheight related maximum peak pressures *Max Pmax/H* is shown in figure 4. The lefthand plot shows for one single test the local distribution of *Max Pmax/H* for each pressure cell. In the righthand plot such local distributions from all tests within a testseries with constant boundary conditions (for this example: $D = 5.0$ m, $Dc = -0.5$ m, PM - spectra, $H = 1.10$ m) have been summerized by an envelope curve.

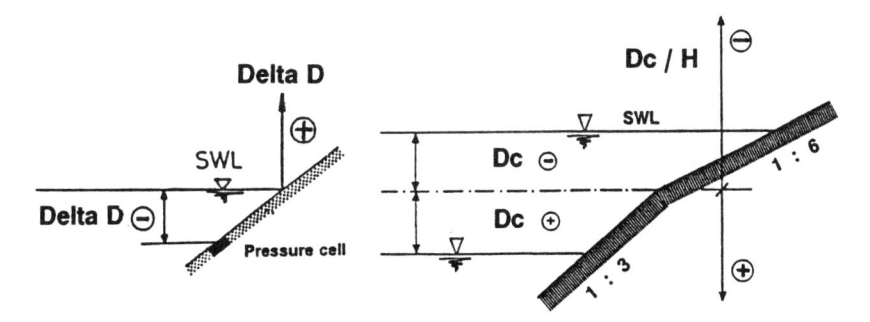

Fig. 2 Definition of *Delta D*   Fig. 3 Definition of *Dc*

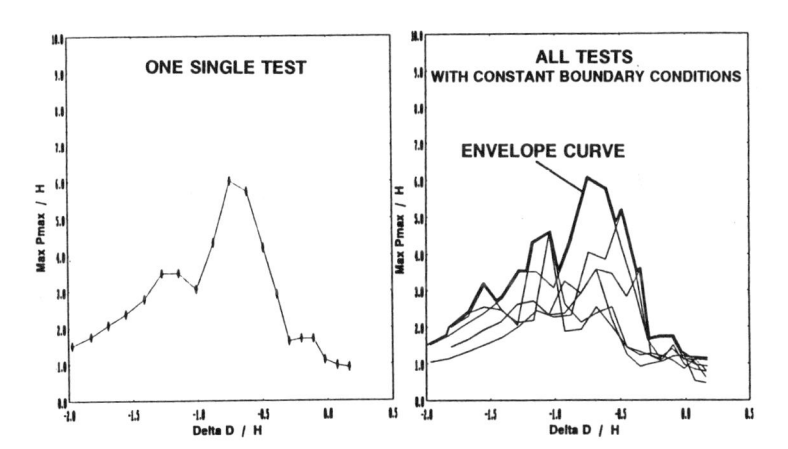

Fig. 4 Example for a local distribution of pressure data *Max Pmax/H*

Results for the dyke with a berm

Figure 5 shows the local distributions of the envelope curves for the testseries with regular waves. It is obvious, that for the testseries with a waterdepth of 5.0 m (lefthand plot) there is a clear trend of higher waveheight related pressures *Max Pmax/H* with decreasing absolute waveheight *H*, which may easily be explained by stronger breaking of higher waves on or in front of the berm due to restricted waterdepths *Db* on the berm. There is also a trend of shifting of the local maximum value area to smaller related waterdepths *Delta D/H* with increasing absolute waveheight *H*, which is also caused by the restricted waterdepth on the berm. A comparison with the corresponding results for the 4.0 m waterdepth in the left hand plot shows, that the trend to higher pressure values is only dominant for waveheights $H \leq 0.5$ m. With the waterdepth *Db* of 0.7 m all higher waves must be broken at least partly.

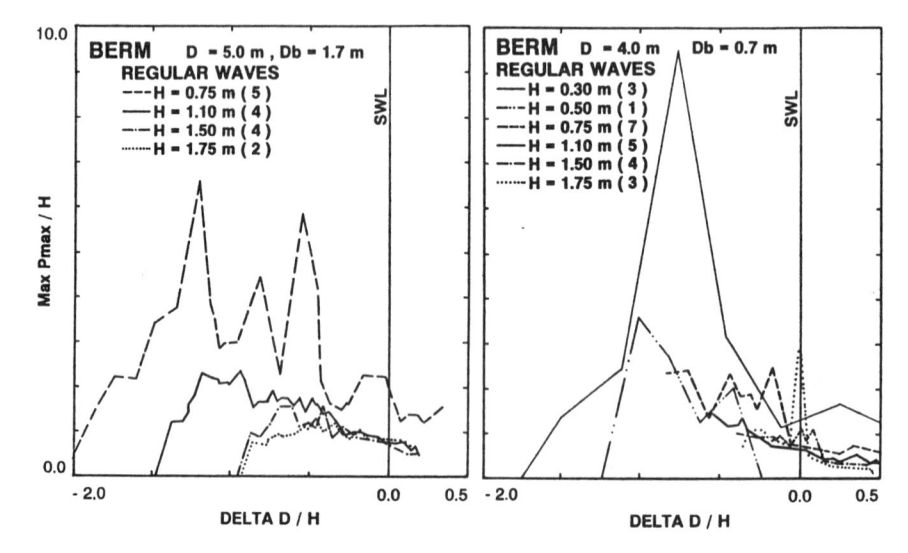

Fig. 5 Local distributions of *Max Pmax/H* for testseries with regular waves

The same trends were found for the results of the testseries with PM - spectra in figure 6, but the differences between the values *Max Pmax/H* of the different waveheights $H1/3$ are smaller. That is caused by the fact, that not all waves in the irregular wave train were broken compared to a regular wave train with same waveheight H as $H1/3$ in the irregular wave train. The same trend also is obvious in figure 7, where the data *Max Pmax/H* of all single tests with regular and irregular waves are plotted versus the breakerindex ξ.

The envelope curves of the local distribution for all tests with an absolute waveheight H ($H1/3$) of 1.10 m are plotted for the different wave characteristics in figure 8. The results indicate for all tests with regular waves smaller pressure values *Max Pmax/H* than for irregular waves. Similar results were found for other waveheights. The influence of wave characteristic also comes out in figure 9, where all test results are plotted versus the breakerindex ξ. Both the influence of wave characteristics and of restricted waterdepth can be seen distinctly.

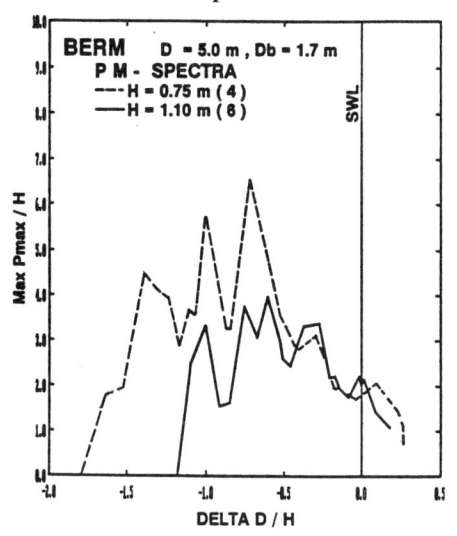

Fig. 6 Local distribution of *Max Pmax/H* for testseries with PM - spectra

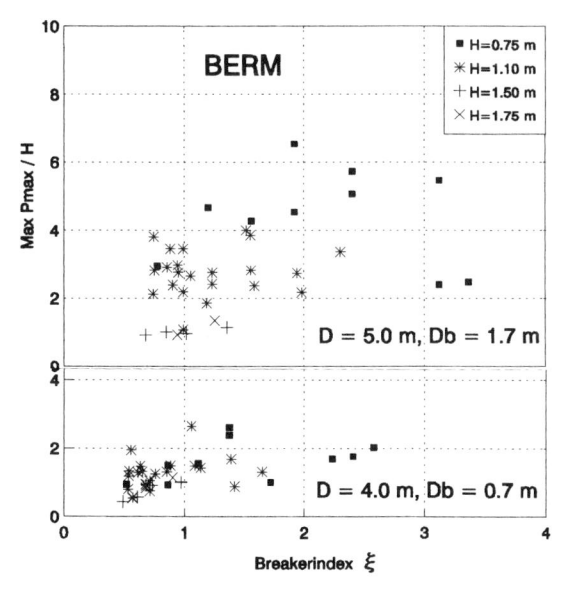

Fig. 7 *Max Pmax/H* versus breakerindex ξ for of regular and irregular waves

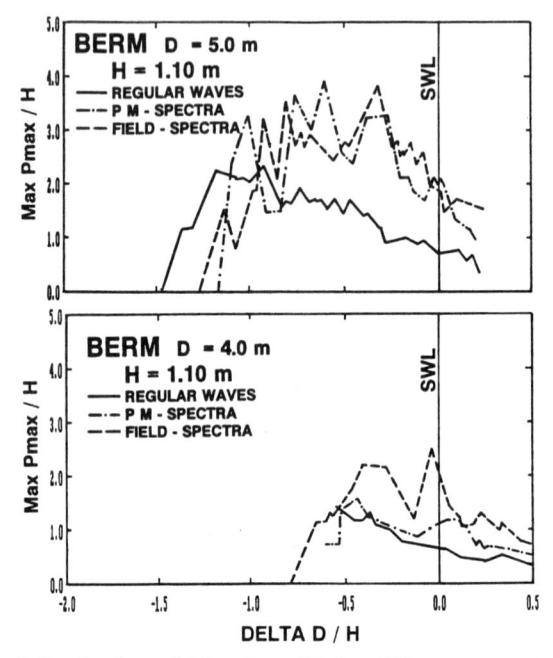

Fig. 8 Local distribution of *Max Pmax/H* for different wave characteristics

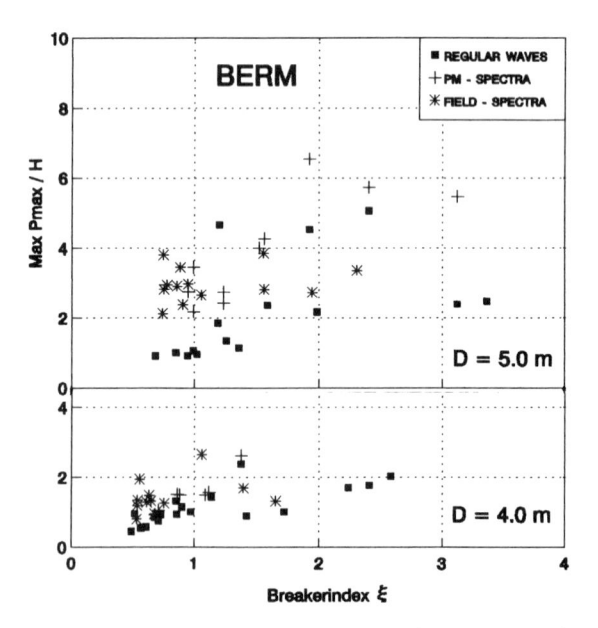

Fig. 9 *Max Pmax/H* versus breakerindex ξ for different wave characteristics

The mentioned trends lead to summerizing all test results as shown in figure 10, where the maximum peak pressures *Max Pmax/H* are plotted versus the waveheight related waterdepth *Db/H* on the berm. Both for regular and for irregular waves the same tendency comes out clearly, but with a different mean or upper value gradient. The corresponding maximum value *Max Pmax/H* for an uniform slope 1:6 is plotted in figure 10 as a dotted line. This value is a constant one due to much higher waterdepth and was estimated from previous field and large scale laboratory investigations.

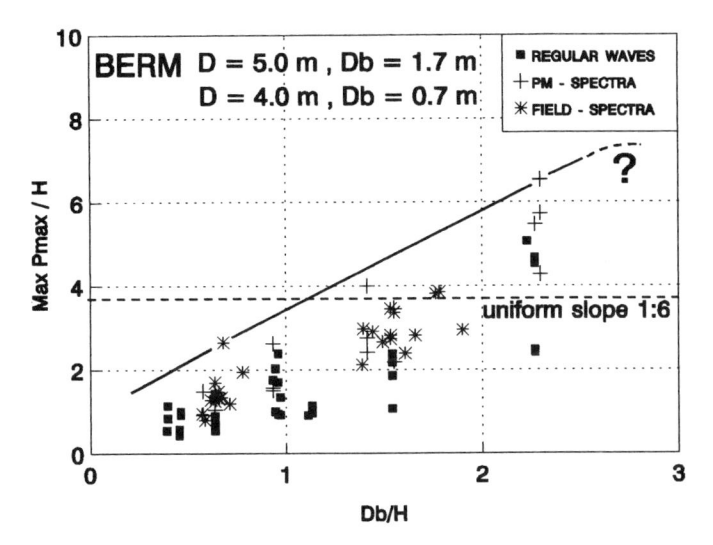

Fig. 10 *Max Pmax/H* versus waveheight related waterdepth *Db/H* on the berm

It is obvious, that compared to the uniform slope value for the berm, there are as well lower as higher pressure values, depending on the waveheight related waterdepth *Db/H* on the berm. Both effects may be explained either by the wave breaking effect in front of the berm (slope 1:3) for lower *Db/H* or by the wave shoaling effect on the 1:20 slope of the berm for higher *Db/H*. Further it can be expected, that the increasing effect due to shoaling is limited by higher related waterdepths and that the upper value curve goes finally back to the uniform slope 1:6 value. Finally it must be remarked, that decreasing pressure values on upper slope 1:6 due to wave breaking automaticly cause higher pressure values on the berm slopes 1:20 and 1:3, which could be recognized during the tests visually and acuostically.

Results for the dyke with composed slopes

The trends with respect to absolute waveheight and to wave characteristics differ for the results of the testseries with composite slopes between the testseries in dependence on the level of slope junction and stillwaterlevel. In contrast to the results for the berm the trend of higher peak pressures *Max Pmax/H* in dependence on the absolute waveheights *H* for the testseries, which is most influenced by the upper slope 1:6 ($D = 5.0$ m, $Dc = -1.7$ m), is negligible for regular waves and only small for irregular waves, respectively, as shown in figure 11.

If one compare the influences of the different wave characteristics on the results in figure 12, there is the same trend to lower pressure values *Max Pmax/H* for regular waves, which was found for the berm, but with a smaller magnitude.

The maximum values of all single tests of this testseries are plotted in dependence on the breakerindex ξ in figure 14a. There is a certain influence of the wave characteristics, but the influence of the absolute waveheights was found to be more or less negligible.

A different trend was found for the testseries, which is mostly influenced by the lower slope 1:3, as the junction of the slope is 0.5 m above *SWL* ($D = 4.0$ m, $Dc = +0.5$ m). The tendency on the influence of absolute waveheights in figure 13 changes to higher values *Max Pmax/H* with higher absolute waveheights, less clear for regular waves, but very clear for the PM - spectra.

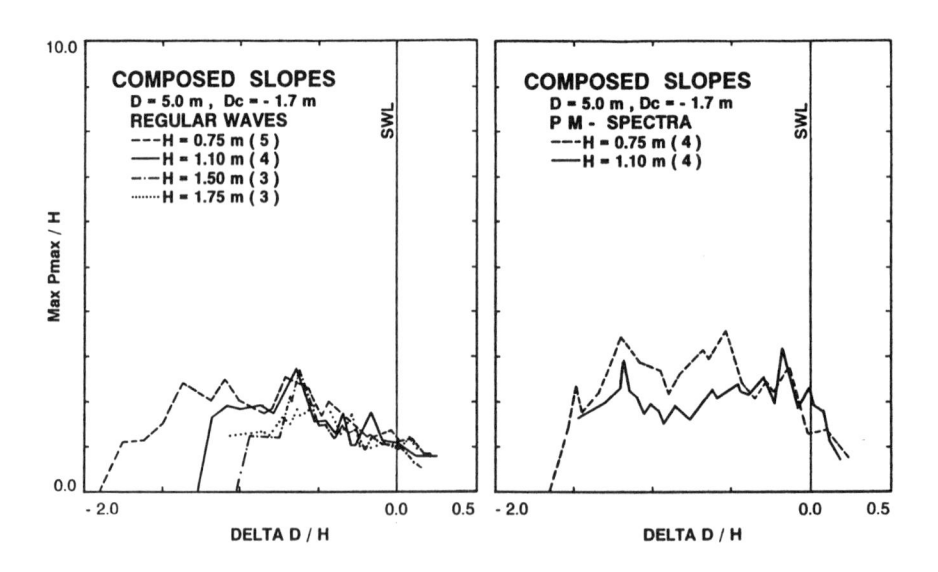

Fig. 11 Local distribution of *Max Pmax/H* for testseries with regular waves.

For the different wave characteristics in figure 15 still a trend was found to higher pressure values with irregular wave trains, which also comes out clearly with the plot of all results versus the breakerindex ξ in figure 14d.

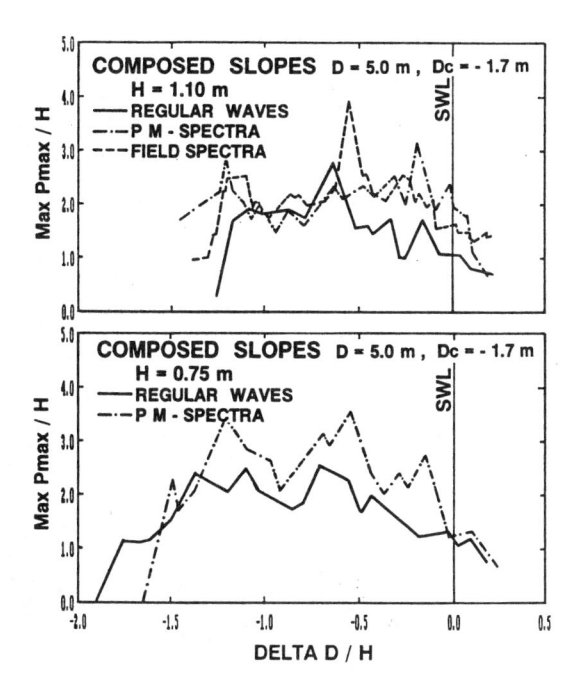

Fig. 12 Local distribution of *Max Pmax/H* for different wave characteristics

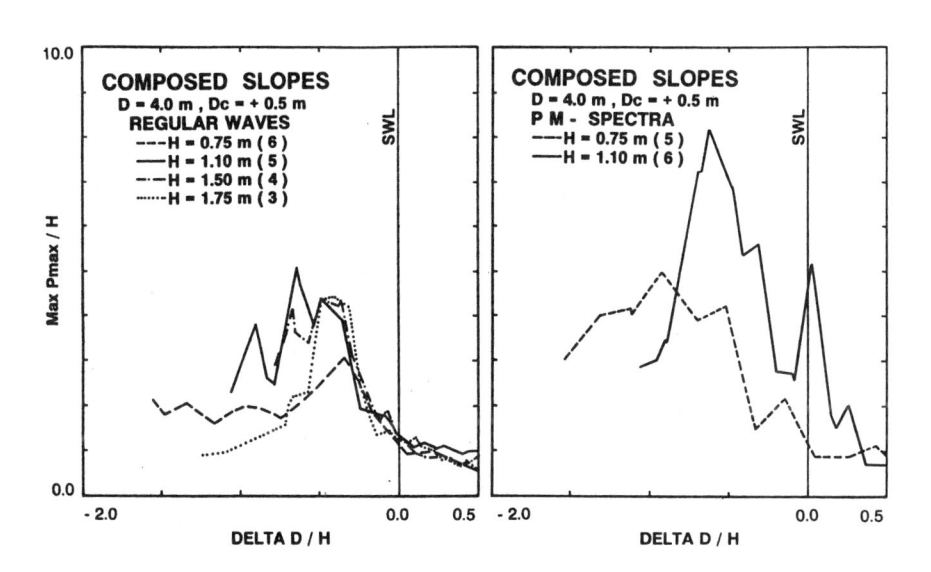

Fig. 13 Local distribution of *Max Pmax/H* for regular waves and PM - spectra

Fig. 14 a

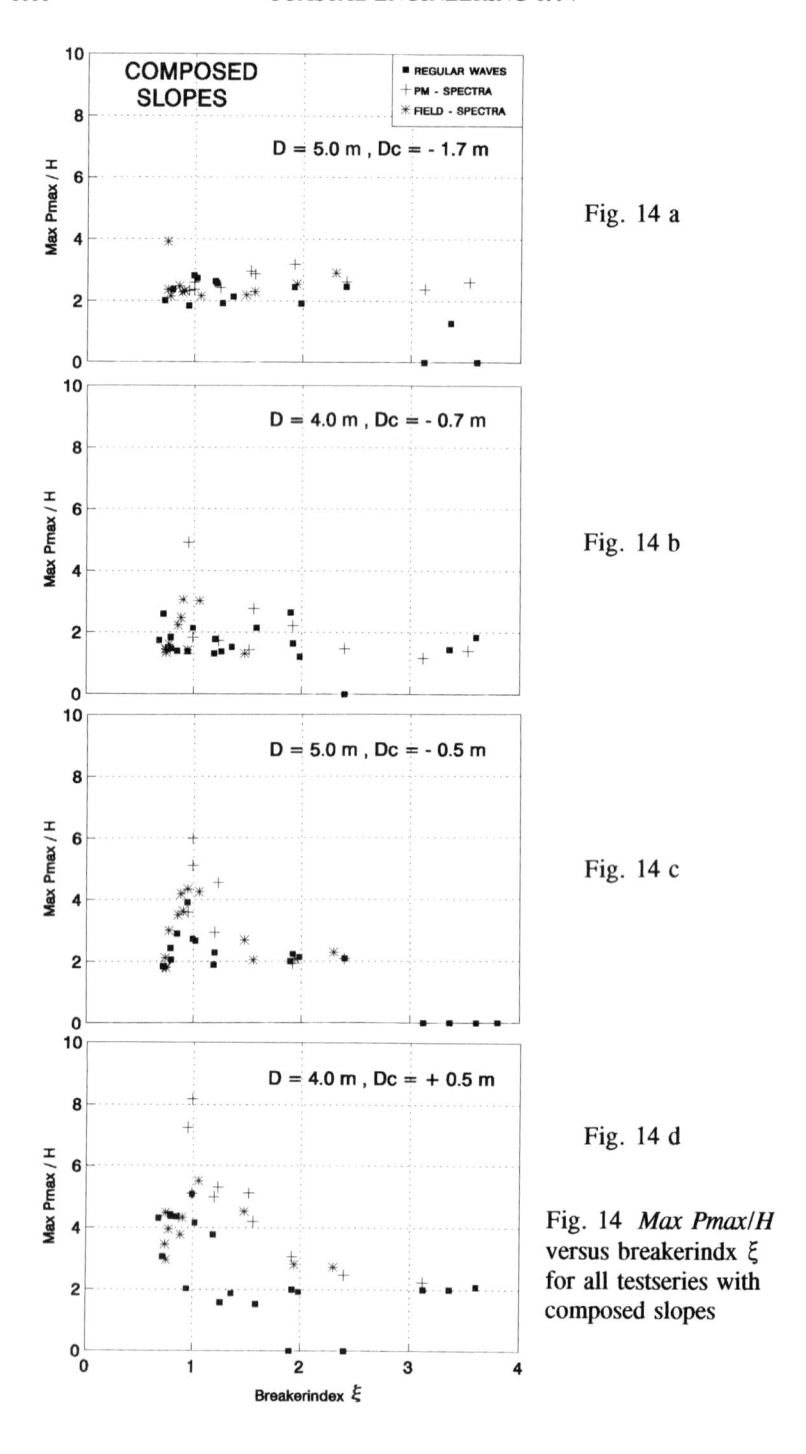

Fig. 14 b

Fig. 14 c

Fig. 14 d

Fig. 14 *Max Pmax/H* versus breakerindx ξ for all testseries with composed slopes

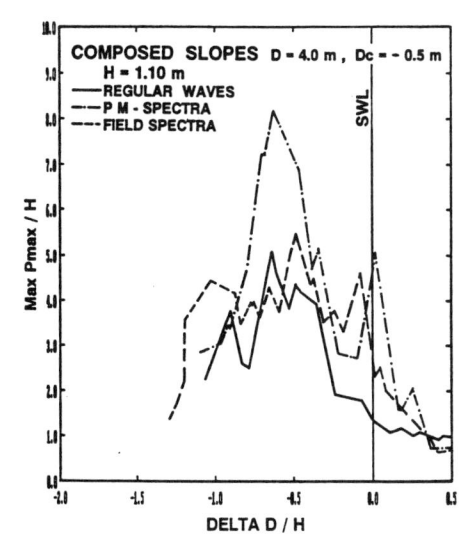

Fig. 15 Local distribution of *Max Pmax/H* for different wave characteristics

Similar results were found with the other both testseries ($D = 4.0$ m, $Dc = -0.7$ m; $D = 5.0$ m, $Dc = -0.5$ m): Firstly, no distinct trend for the dependence on absolute waveheight H for regular waves, but a small one for irregular waves. Secondly, the influence of the wave characteristics is evident as shown in figure 14b and 14c, where all test results are plotted versus the breakerindex ξ. Smaller pressure values *Max Pmax/H* were found for regular waves, as well as in all other testseries.

To summerize the pressure results from all 4 testseries with composed slopes, the pressure values *Max Pmax/H* of all single tests have been plotted versus the waveheight related vertical distance *Dc/H* between slope junction and stillwaterlevel *SWL* as shown in figure 16 for regular waves (upper part) and irregular waves (lower part). If the slope junction lies below *SWL*, the slope 1:6 seems to be dominant. If the slope junction lies above *SWL*, the lower slope 1:3 with higher pressure values is dominant. The transition range has a high gradient, if the slope junction is around till roughly half a waveheight below *SWL*. The eye-fitted upper value distributions are plotted as dotted lines in figure 16. A qualitative generalisation is given in figure 17. The asymetry of the influences from both slopes with respect to the distance *Dc/H* between slope junction and *SWL* is distinct.

Also peak pressure values smaller than for any of both slopes occur at *Dc/H* roughly half a waveheight below *SWL* from the testseries with $D = 4.0$ m, $Dc = -0.7$ m. From the local distributions of the results it was found, that these results may be partly effected by the restricted pressure cell area below *SWL*, because for this testseries pressures only have been measured on the upper slope 1:6.

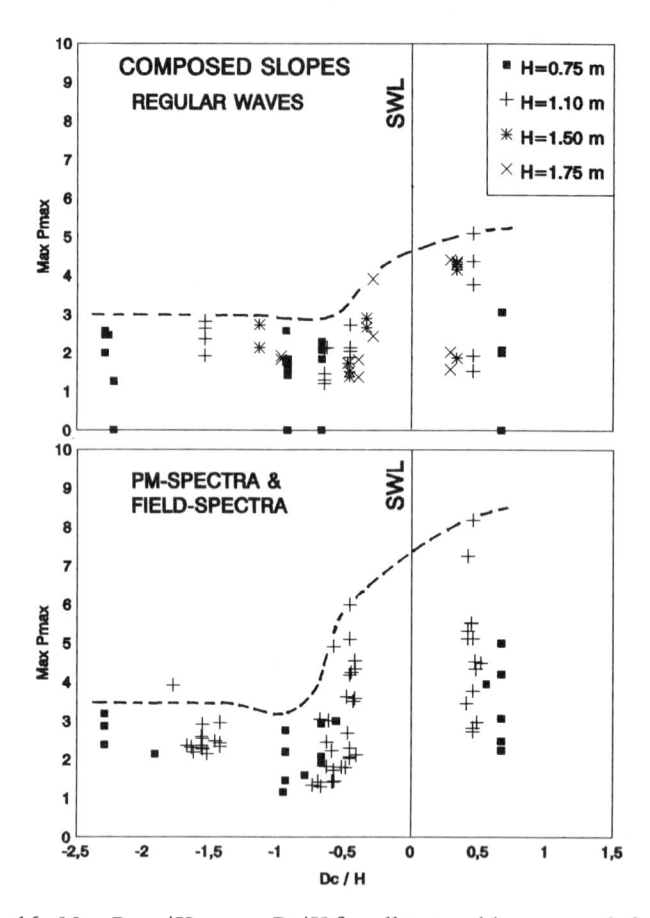

Fig. 16 *Max Pmax/H* versus *Dc/H* for all tests with composed slopes

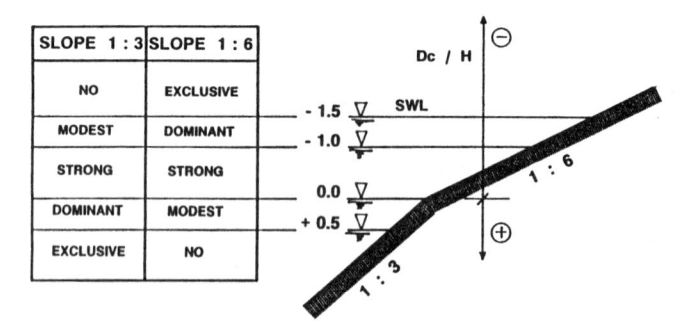

Fig. 17 Scheme of influence of slopes on shock pressure occurrence

Conclusion

From the results the following statements on the maximum peak pressure occurrence may be concluded:

- There are trends in dependence of the absolute waveheight, but in different ways:
a.) Where wave trains are broken at least partly, before they reach the slope (for instance at the berm), the waveheight related maximum pressure values $Max\ Pmax/H$ increase with decreasing absolute waveheights H.
b.) Where waves trains are broken completely by the slope (for instance at composed slopes), the pressure values $Max\ Pmax/H$ may increase with increasing absolute waveheights H.

- There are strong influences of the wave characteristics:
Regular waves always lead to smaller related pressure values $Max\ Pmax/H$, if regular waveheights H are compare with irregular waveheights $H1/3$. This confirmed previous results for the uniform slopes, that for comparison one should use mean waveheights both for regular and irregular waves. Nevertheless one should not use regular wave results for design of coastal protection works generally.

- Some special effects of geometrical conditions can be stated:
a.) For a slope with a berm in front, the peak pressures $Max\ Pmax/H$ may decrease due to breaking effect or increase due to shoaling effect depending on waterdepth on the berm.
b.) For composed slopes there is a transition range, which is asymetric with respect to the distance Dc between slope junction level and stillwaterlevel SWL.

References :

GRÜNE, J., FÜHRBÖTER, A. (1976). Large Wave Channel for "full-scale modelling" of wave dynamics in surf zones. Proc of Symp. on Modelling Techniques, San Francisco, pp. 82-100.
GRÜNE, J. (1988a). Anatomy of shock pressure (surface and sand core) induced by real sea state breaking waves. Proc. of the Intern. Symp. on Modelling Soil-Water-Structure Interactions (SOWAS 88), Delft, pp. 261-270.
GRÜNE, J. (1988b). Wave-induced shock pressures under real sea state conditions. Proc. of the 21st International Conference on Coastal Engineering (ICCE'88), Malaga, pp. 2340-2354.
GRÜNE, J. (1992). Loads on sloping seadykes and revetments from wave-induced shock pressures. Proc. of the 23rd International Conference on Coastel Engineering (ICCE '92), Venice, pp. 1175-1188.
SCHÜTTRUMPF, H., BERGMANN, H., DETTE, H.-H. (1994). The concept of residence time for the description of wave run-up, wave set-up and wave run-down. Proc. of the 24st International Conference on Coastel Engineering (ICCE'94), Kobe.

CHAPTER 79

COMPUTERISED METHODOLOGY TO MEASURE
RUBBLE MOUND BREAKWATER DAMAGE

Bruno CHILO' , Franco GUIDUCCI[1]

Abstract

The damage of rubble mound breakwater is a basic interest in the field of design and realisation of coastal works. The individuation of formulations to associate this element with the geometrical characteristics of structures and wave climate conditions is carried out using physical model tests. It is therefore necessary to implement an objective damage valuation method able to point out armour units movements on the layer from the settlement phase to the collapse.

A methodology able to permit an exact measurement of the external layer armour units movements and its damage has been carried out by ESTRAMED S.p.A.; it consists on a computerised analysis of armour units movements in order to give an objective valuation of the damage.

Such methodology is based on photographic images analysis of the armour taken from a fixed point. The procedure permits to determine the position of each armour unit on the images relative to the subsequential damage stages and the movement of each armour unit.

Introduction

The experiences relevant to the Port of Sines Main Breakwater, the improvement of computers capabilities, the wider use and the development of the physical modelling and the studies carried out in the last years improved the analysis of the rubble mound breakwater damage. Relations between rubble mound breakwater stability and incident waves and section characteristics were deduced.

---

[1] ESTRAMED S.p.A. - Via Campobello n°6 - 00040 POMEZIA -ROMA - ITALY.

The breakwater stability is usually referred to the eroded volume or to the percentage of units moved more than a fixed threshold.(A.I.P.C.N. 1992, British Standard, 1991).

The measurements of these parameters are extremely important to carry out investigations on the rubble mound breakwater damages and to grow research in this field.

The use of objective methodology to describe the phenomenon from the first movement up to the collapse is essentially.

Methodology purposes

At ESTRAMED S.p.A. a methodology that allows to measure the external layer armour units movements by a computerised procedure, was developed.

Aims of the methodology were the following:
- to get information on the small armour units movements;
- to obtain objective measurements of the layer damages;
- to analyse armour units movements in a time comparable with the stability test duration.

It is necessary to obtain information on the small armour units movements as the experience on the physical models to test the rubble mound breakwater stability indicates. Small armour units movements are very important because they allow the description of layer damages under the action of ordinary wave motions and to evaluate the entity of the first settlement of the breakwater and therefore, on experiments in which comparative or repetitive tests are required, to make a correct comparison of breakwaters behaviour under the action of extreme waves.

A precise knowledge of small and large movements of the units make possible to draw graphs of the damage on the basis of the incident waves characteristics. These graphs are useful tools for the designer in order to choose the best solution considering the entire structure behaviour, not only in severe damage conditions. Small armour units movements analysis is of interest for the comparison of different solutions even when no severe damage level is reachable, for example, during tests on structures where the wave attack is limited by bottom effects.

The use of a standard laboratory method to measure rouble mound breakwater damage guarantees the repetitivity of test results and permits to compare results obtained from a large number of experiences carried out by different laboratories, increasing in this way the base of available experimental data for further formulations.

To obtain analysis results of armour units movements in a time comparable with the tests duration allows to make strategy choices on the next tests to be carried out. This permits the optimization of the tests programme and therefore, a large number of damages stages can be easily examined obtaining more information compared to the methodologies often used to measure rubble mound breakwater stability.

Methodology description

The methodology consists of five subsequent activities:
- set up of rubble mound breakwater armour units;
- acquisition of breakwater images in the subsequent conditions of damage and their importation into the computer;
- images processing;
- check and eventual corrections of images processed;
- print out of the analysis results.

Armour units in nine different colours are employed to built the external layer of the armour so that two units of different colours are located between two units of the same colour and the individuation of the position of each armour unit on each images is facilitated.

The colour selection for armour units painting is very important in this methodology. Each colour must be detected as different from the others by the computer analysis. The achievement of this objective depends on armour units colouring, lighting sources, breakwater contour characteristics, photographic equipment and images filtration; a useful set of colours was selected by Estramed carrying out several tests.

Once placed the units , it is necessary to draw a white point on the sight view of each of them as shown in figure 1 where in the white window are highlighted the units with the same colour. The movement of this point on the sequential images will show the armour unit movement.

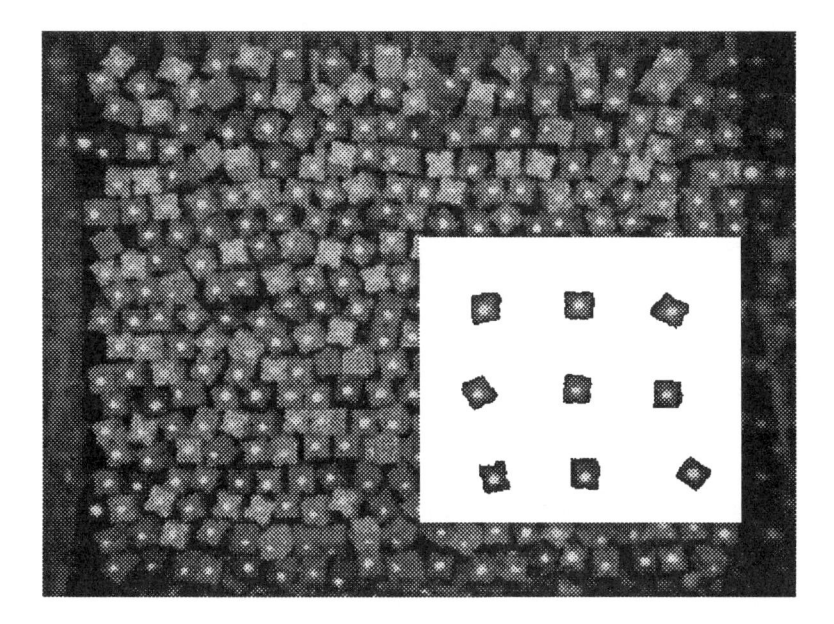

Figure 1 - Set up of rubble mound breakwater armour units

In order to permit the comparison of the different images, at the beginning of the experiments it is necessary to put three marks in the model so that the software can be able to compensate eventual shot point movements and to adopt one single coordinates system to locate each armour unit on each image.

The methodology permits to acquire rubble mound breakwater images in different damage stages and to import them into the computer. The camera station must be fixed and located in such a way that the shooting plane is parallel to the breakwater armour. Images have to be acquired without water in order to avoid reflection and refraction phenomena. The procedure uses a still video camera and a digitiser card with a resolution of 768x576 pixel to acquire and to import the images into a personal computer. Images must be taken at the end of the section construction and after each wave attack.

Once imported into the computer the images are analysed by a software that was set-up by Estramed. Such analysis individuates for each armour unit its colour and the coordinates of the white point located on it as pointed out in figure 2.

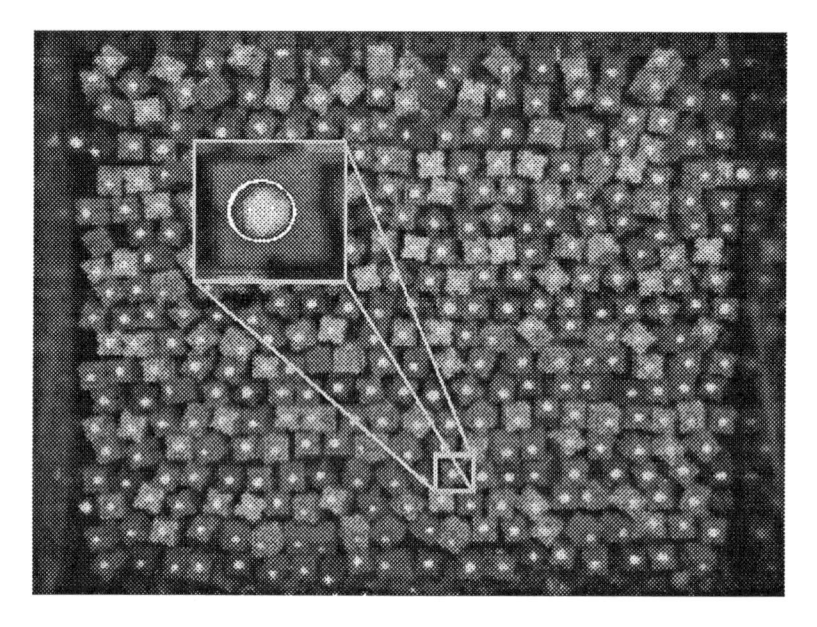

Figure 2 - Example of the analysis result

At the end of this phase, a further analysis is carried out on the files containing, for each image, the coordinates and the colours of the units in order to verify the number of armour units of each colour among all the images. If any inconsistencies arises (for example the number of armour units can be lower compared to the previous one in the case where one or more units rotate on themselves without originate unit barycentric movements, hiding the white evidence point) the operator, who disposes of an edit menu, points out and solves them.

Scope of the second analysis phase is to individuate the position of the same armour unit among all the different images. The procedure perfoms an analysis examining two files relevant to subsequent images: the file relevant to the less damaged condition is considered as reference for the computation of the units movements. The analysis is carried out taking into account separately armour units of the same colour. As first step, each armour unit of the first image is coupled with the nearest unit of the second image having the same colour in a range less than 0.5 times the characteristic length of the unit. Successively, the analysis is carried out on the not yet associated units increasing the range up to two and half times the characteristic length of the unit as shown in figure 3.

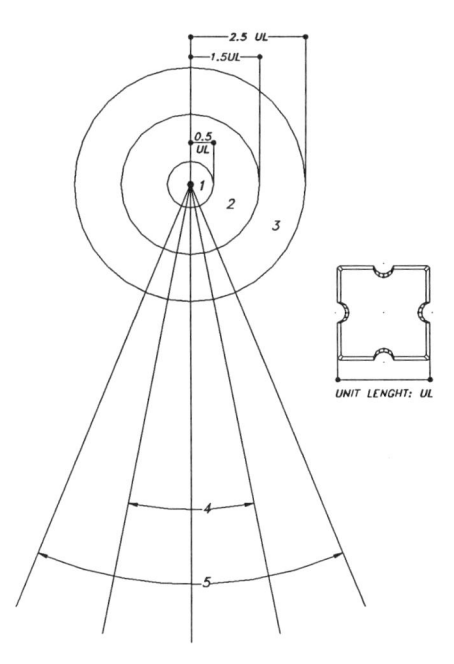

Figure 3 - Scheme of criteria to identify same unit in two successive images

The successive step on the not yet associated units is the coupling of the units present inside circular sectors (see figure 3) which centres are located on the units of the first image, which axis of symmetry is a vertical line and which angle is initially equal to 10° and is successively increased by step of 10° up to 180°. If two possible associations are individuated in the sector, the procedure chooses the one with the minimum angular value compared to the vertical.

Once the coupling of the units has been carried out, the results of the analysis are submitted to the operator for check and for eventual corrections.

The software shows a screen with four windows as shown in figure 4. An image is shown in the upper left window while the successive one appears in the upper right one. The lower right window shows the armour units movements and the lower left one displays an edit menu and the following information:

- number of units for each colour individuated in the two images;
- total number of units individuated in each image;

- number of reference marks individuated in each image.

If any inconsistencies arise the operator can correct the results using the edit menu in a proper way.

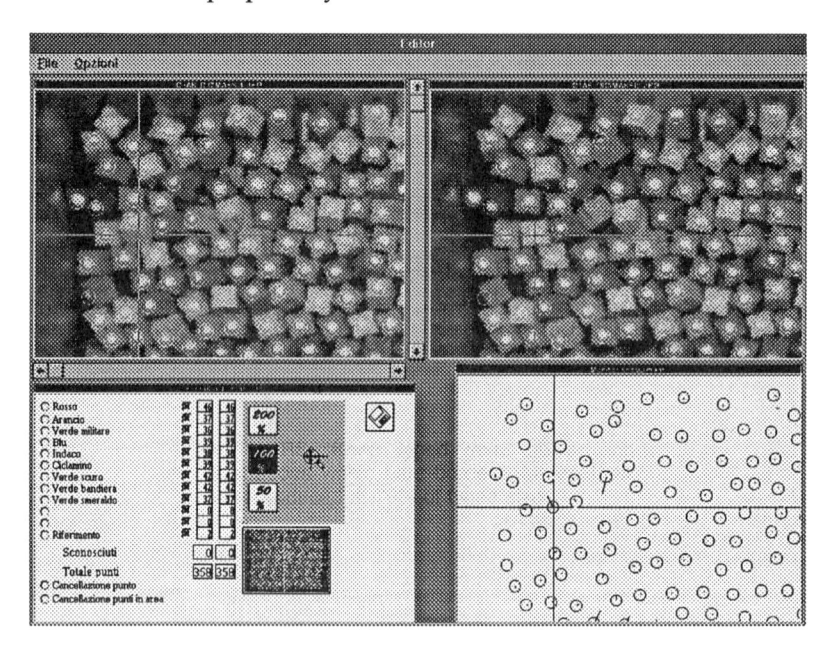

Figure 4 - An example of the results of the units movements analysis

A sequence of the results of the units movement analysis relevant to a test series with increasing waves is shown in figure 5.

The first image (upper left) of figure 5 was taken after the construction of the model and before the start of the tests. The second image (upper right) shows the units movements measured after a test with a very low significant wave height while the last image (lower right) shows the damage of the breakwater after a test characterised by a significant wave height higher than the design one.

COMPUTERISED METHODOLOGY 1097

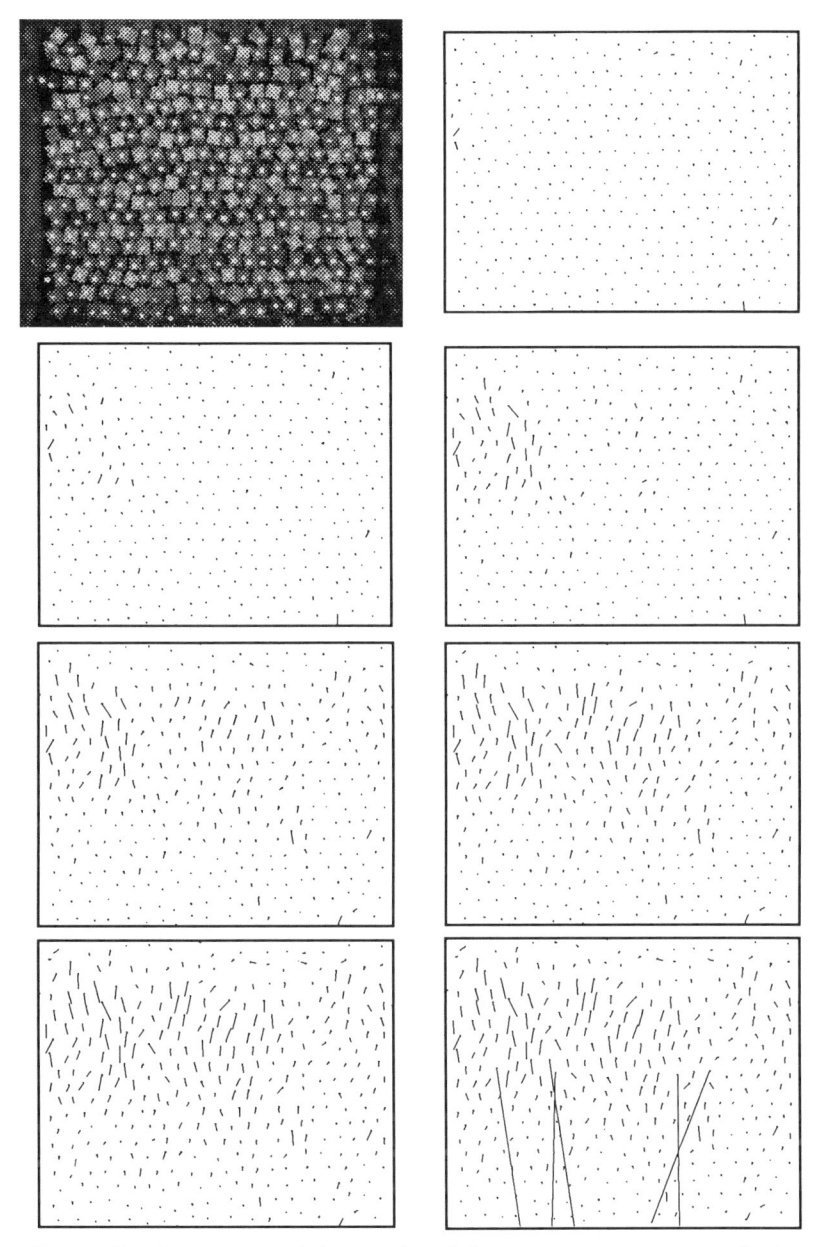

Figure 5 - A sequence of the results of the units movements analysis

The methodology allows to obtain a wide range of information about the units movements. Many information on the small movements of the units can be obtained, too.

A standard plot of the results of the analysis indicates for each test or wave height the percentage of armour units which movements were less than fixed values that are generally part of the characteristic unit length (B). In figure 6 a plot of the results relevant to the analysis of figure 5 is shown.

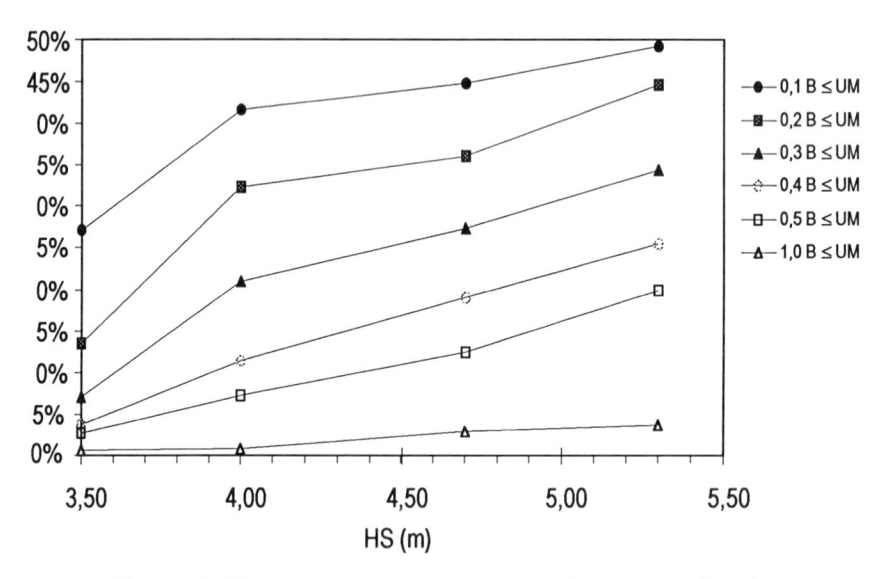

Figure 6 - Units movements versus significant wave height

The methodology has been tested comparing its results with those obtained using a traditional approach. The agreement was very good, at any rate, it must be noted that both methods find difficulties when an unexpected collapse of the whole breakwater happens (with more than 30% of armour units dislocated). This is not a real limit to the method application because, in such conditions the single armour unit movements doesn't give any essential indication on breakwater stability.

Fields of application of the methodology

The methodology has several fields of application in laboratory testing.

A first field is relevant to the two dimensional physical modelling. The methodology allows to carry out objective measurements which are essential for repetitive or comparative tests; to describe accurately the damage behaviour of the breakwater armour and to obtain results in a very short time. In the Estramed S.p.A. Laboratory the methodology has been employed in two dimensional tests to individuate also the movements of toe protection blocks and submerged crest blocks.

A second field is relevant to the three-dimensional physical modelling. Treating these models as plane images, the methodology cannot describe the armour units absolute movements but the parameters to asses a roundhead stability are various and not limited to the armour units movement. Therefore the computerised analysis gives a useful contribute that can be surpassed only using a very sophisticated three dimensional survey system.

Two-dimensional and three-dimensional applications of the methodology are currently used at Estramed S.p.A. Laboratory.

Conclusion

The computerised methodology allows a reliable and objective measurement of rubble mound breakwaters damage.

The analysis of a stability test is obtained at the end of each run in a shorter time than the test duration.

The use of this methodology permits the standardisation of test modalities and damage levels measurements.

The application in reality of this methodology, with some modifications could bring a remarkable increasing to rubble mound breakwaters know-how.

Acknowledgements

A special thank goes to Mr. Spartaco Coletta, to Mr. Pompeo Forte and to Mr. Danilo Marinucci for their contribution and cooperation.

Bibliography

Abecasis,F. - Pita,C.: Monitoring mound breakwater : the case of Sines - Coastal Engineering - 1992.

British Standards: "Maritime Structures" Part. 7 - Guide to the Design and Construction of Breakwaters, 1991

Burcharth, H.F.: Design innovations, including recent research contributions - Coastal structures and breakwaters - Institutions of Civil Engineers - London -November 1991.

PIANC : Analysis of rubble mound breakwaters - Report of Working Group n°12 - Supplement to PIANC Bulletin n° 78/79 - 1992.

Van der Meer,J.W. : Rock slopes and gravel beaches under wave attack - PhD Thesis - Delft University of Technology - Delft - 1988-

CHAPTER 80

WAVE BREAKING OVER PERMEABLE SUBMERGED BREAKWATERS

Masataro Hattori[1] and Hiroyuki Sakai[2]

Abstract

A laboratory experiment was carried out in a wave tank to examine significant features of wave breaking over permeable submerged breakwaters and to determine the breaker height and depth indices. Submerged trapezoidal-shaped breakwaters were place on a 1/20 steel slope, and permeability of the breakwaters was varied.

The experiment showed that the breaker height and depth change with the breakwater permeability governing the strength of return flow over the breakwater. Breaker height index was developed in terms of the integrated parameter $\xi_{ss}"$ proposed by Hara et al. (1992), which consists of the geometrical and structure properties of the breakwater as well as properties of incident waves. To determine the breaking position or depth, additional indices were also proposed. Validity of the computation scheme of the breaker height and position were confirmed by comparisons with the experiments.

INTRODUCTION

As is commonly known, permeable and submerged, low-crested rubble-mound, breakwaters force to break incoming steep waves and dissipate effectively the wave energy. Therefore, many breakwaters of this type have been built or planned at various locations to stabilize an eroded beach and to reduce damages of coastal and harbor structures due to severe wave actions.

---

[1] Professor, Department of Civil Engineering, Chuo University, Kasuga 1-13-27, Bunkyou-ku, Tokyo 112.

[2] Engineer, Urban Development Bureau, Yokohama City, Minato-cho 1-1, Naka-ku, Yokohama City 231, Kanagawa Prefecture.

To meet requirements of the design and construction of the structures, many theoretical and experimental studies are extensively making to deepen understanding of the physics of wave breaking due to the submerged breakwater. The wave breaking may be sufficient to provide the principal control over the wave motion and resulting phenomena around the submerged breakwater, especially behind the breakwater, such as wave transmission, wave set-up, and so on. It seems likely that lack of a sound understanding of wave breaking over the submerged breakwater acts as a bottleneck of the physical and numerical examinations on the hydraulic function of the submerged breakwater.

A series of laboratory experiments was conducted to examine breaking wave properties over permeable submerged breakwaters and to determine the empirical relationships or breaker indices developed from measurements.

The results from the study will be useful for modeling the nearshore wave field around the breakwater and for predicting the evolution of the protected beach profile.

EXPERIMENTAL EQUIPMENT AND MEASUREMENTS

Experimental

Figure 1 shows the general arrangement of equipment. Experiments were carried out in a 20-m-long, 0.30-m-wide, and 0.55-m-high glass-walled wave tank, containing a steel beach of 1 on 20 slope. Monochromatic waves were produced by a reflection-absorbed wave maker of flap-type, installed at one end of the tank.

Six model breakwaters of same geometry, as shown in Fig. 2 and composed of the gravel materials or armor concrete blocks, were placed on the slope. Size and

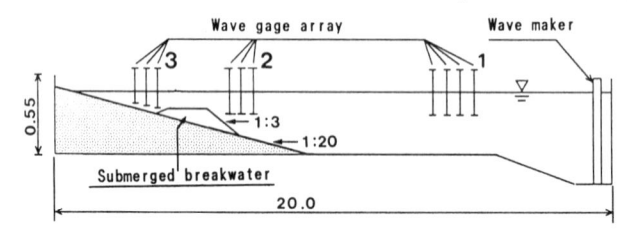

Fig. 1 General arrangement of equipment (units: m).

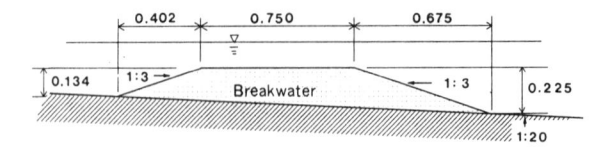

Fig. 2 Dimension of model breakwaters (units: m).

placing condition of the breakwater were determined based on actual breakwaters in Japan. The breakwaters had an upslope and downslope (tanθ) of 1:3, a 0.75m crest width (B), and a 0.225m height (d_s). The porosity (ε) were varied between 0.17 and 0.52 by changing the material size. In situ porosities were determined from the weight, volume and specific gravity of the materials used. Physical properties of the materials are listed in Table 1.

Table 1 Physical properties of materials used.

Porosity ε	Material Used	Material Size (cm)
0	Concrete	---
0.17	Gravel	1.7
0.25	Gravel	2.8
0.38	Gravel	4.0
0.50	Armor Block	7.5
0.52	Gravel	5.0

Measurements

As see in Fig. 1, water surface elevations were recorded at three wave gage arrays, which consist of either three or four capacitance-type gages. Wave array 1 was located at un uniform water depth section to resolve the incident and reflected wave heights using the method of Mizuguchi (1990). Wave array 2 was placed at the section of the seaward toe of structure slope to measure the incident wave height. Array 3 was set behind the breakwater to measure the transmitted wave height.

Wave data were collected for 2 minutes at 100 Hz, but only 10 successive waves were analyzed to obtain the wave height, mean surface elevation $\bar{\eta}$ and its variance η_{rms}. Wave profiles on and around the breakwater were recorded by a high-speed video of 200 frames per second. The wave properties were read from the still pictures, with aids of 1.0 cm square grid system attached on the sidewall glass. Incident wave parameters are listed as follows;

    Incident Wave Height   : H_I = 2.0 cm – 10.0 cm
    Incident Wave Period   : T = 0.8 s – 1.5 s
    Water Depth
       on Horizontal Bottom : h_I = 31.2 cm and 35.0 cm
       above Breakwater Crest : R = 3.75 cm and 7.50 cm

INCIPIENT BREAKING

A schematic of a wave profile at breaking on the structure seaward slope of breakwater is depicted in Fig. 3. H_B and h_B are the wave height and the water depth at breaking.

There are several ways of defining the break point. After careful inspection of the video pictures of breaking waves profile, we defined the breaking point as the point where the wave height is maximum. This definition ensured us stable judgement of the breaking point from video pictures. Breaker types observed over the submerged breakwater were basically the same to these over a plane sloping beach, spilling, plunging, and collapsing.

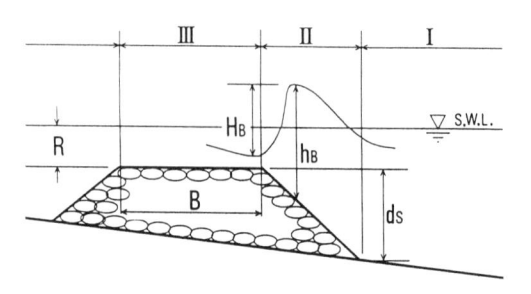

Fig. 3 Definition sketch of breaking wave and demarcation of breaker zone.

Breaking range of waves passing across the breakwater is demarcated into the three following zones, as shown in Fig. 3;

    Zone I : on the offshore sloping beach in front of the breakwater,
         II : on the structure seaward slope, and
        III : on the breakwater crest.

Thus we will examine characteristics of the wave breaking for each zone.

High-speed videos of the wave profile over the breakwater revealed that return flow over the crest and offshore slope of the breakwater influences the wave height and position at incipient breaking. In particular, a strong return flow over a impermeable submerged breakwater produces a variety of collapsing breaker, which the wave crest remains unbroken and the lower part of the front face breaks just like a hydraulic jump due to rapid flows (Smith and Kraus, 1990; Katano et al., 1992). The effects of the return flow on the breaker characteristics are weaken rapidly with decrease of the return flow strength, in other wards, with increase of the breakwater permeability.

BREAKER HEIGHT INDEX ON OFFSHORE SLOPING BEACH, ZONE I.

Due to the permeability and low crest of the submerged breakwater, wave reflection from the breakwater is low, but slightly higher than the plane sloping beach (Katano et al., 1992). As the result, experiments confirmed that the incipient breaker height in Zone I is predicted reasonably well by Goda's breaker index (1978), as given by Eq. (1), with a slight change of the value of empirical constant A from 0.17 to 0.15.

$$\frac{H_B}{L_0} = A \left[1-\exp\{-1.5\frac{\pi h_B}{L_0}(1+15\tan^{4/3}\beta)\} \right] \tag{1}$$

, in which H_B and h_B are the wave height and water depth at breaking, L_0 the wave length in deep water, and $\tan\beta$ the offshore beach slope. It is, therefore, considered that a steep wave approaching the breakwater is transformed principally by the shoaling effect caused by the plane slope of offshore beach.

BREAKER HEIGHT INDEX ON THE BREAKWATER, ZONES II AND III.

Integrated Parameter for Breaker Height.

Previous studies have been made on the wave breaking forced by low crested mound-type breakwaters as well as by barred and reef beaches. However, those have still failed to determine an appropriate breaker index as a function of the breakwater properties and the incoming wave parameters.

Hara et al. (1992) conducted extensive numerical experiments on the transformation of a solitary wave passing across an impermeable submerged breakwater of trapezoidal type, located on horizontal sea bottom. Based on regression analyses of the numerical computations, they propose a parameter for the breaker height index, ξ_s'', given by Eq. (2). It refers to a modified surf similarity parameter.

$$\xi_s'' = \left[\frac{B}{h_s} + \frac{(d_s/h_s)}{(3.5\tan\theta)^{0.2}} \right] \frac{(d_s/h_s)}{(H_s/h_s)^{0.4}} \tag{2}$$

, where H is the wave height, and subscript S denotes the quantity at the offshore toe of breakwater. The parameter comprises the physical properties and placing condition of the breakwater as well as incident wave properties. From this context, the parameter ξ_s'' is named the integrated parameter of breaker height.

Prior to employment of the integrated parameter, we will discuss briefly influences induced by differences in the experimental conditions of the present and Hara's studies, such as the character of incident wave, bottom profile, and permeability of the breakwater.

The experiments indicated that an incident periodic wave passing across the breakwater behaves almost like solitary wave. In addition, it was found that wave and current field in the vicinity of the breakwater display a very similar one produced in the numerical simulation, owing to the low wave reflection of the breakwater. As for the bottom slope, the wave shoaling on a plane slope was taken account by replacing the wave height in ξ_s'' with that at offshore toe of the breakwater.

Breaker Height Index as a Function of Integrated Parameter.

Complete analyses of the wave data on incipient breaking waves yield an empirical relation for the breaker height in terms of the integrate parameter ξ_s'', as given by Eq. (3).

$$\frac{H_B}{L_0} = A_B\left(\varepsilon, \frac{R}{h_s}\right)\left(\frac{2B}{5d_s}\right)^3 \left(\frac{h_s}{L_0}\right) \xi_s'' \qquad (3)$$

, where R is the water depth above the crest, and A_B is the empirical function and is written as

$$A_B\left(\varepsilon, \frac{R}{h_s}\right) = \left[1.0 - 0.12\left(\frac{R}{h_s}\right) - 0.6\varepsilon\right] \exp \varepsilon \qquad (4)$$

, representing the permeability effect of total breakwater system. However, the value of A_B varies very slightly between 1.0 and 1.2 within limits of the experiments.

Figures 4 and 5 show plots of the breaking wave steepness, H_B/L_0 versus the relative wave height at seaward toe of the breakwater, H_s/h_s, to illustrate the ability of Eq. (3). It is noticed that the breaker height index, Eq. (3), describes surprisingly well the measurements.

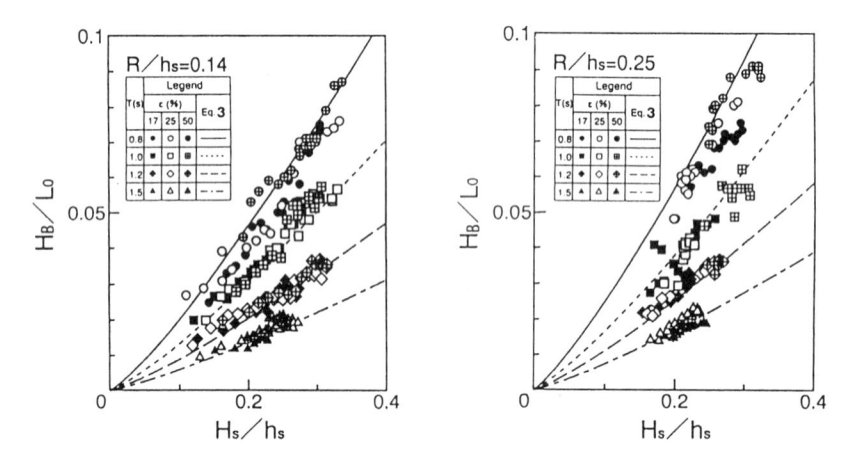

Fig. 4 Relation between H_B/L_0 and H_s/h_s.   Fig. 5 Relation between H_B/L_0 and H_s/h_s.

Calibration of Eq. (3) is also made by using wave data of Izumiya et al. (1989). They carried out experiments of the wave transformation due to a two-layered submerged breakwater having rather complicated structure, as shown in Fig. 6. Porosity of the main breakwater body is 0.20. Figure 6 is an example of the comparison of Eq. (3) with their measurements. Although there is some scatter to the measurements, Fig. 7 shows reasonable agreements between the measured and computed breaker heights for their whole data.

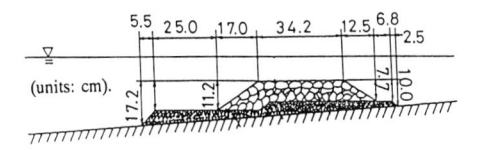

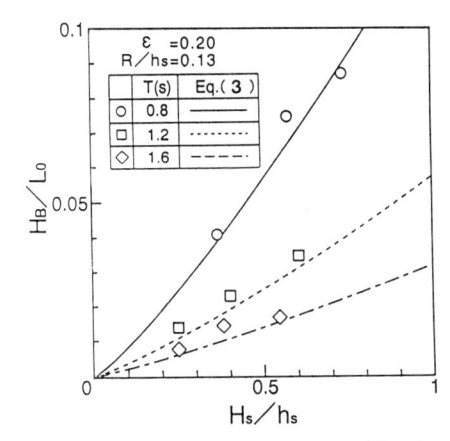

Fig. 6 Comparison of Eq. (3) with data of Izumiya et al. (1989).

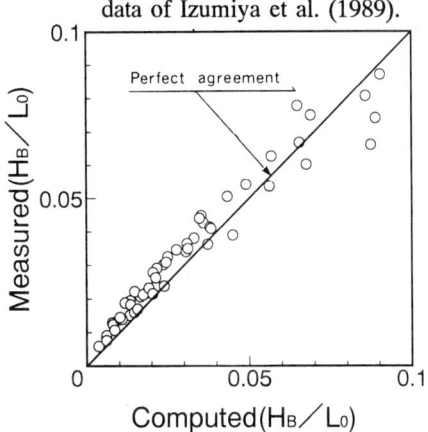

Fig. 7 Agreement of Eq. (3) with the whole data of Izumiya et al. (1989).

The agreements in Figs. (4) through (7) substantiate the validity and applicability of the proposed breaker height index Eq. (3). In addition to this, the integrated parameter ξ_s'' is likely to be very promising to other problems, such as wave breaking on barred and reef beaches. Using data of wave breaking over a triangle-shaped bar of Smith and Kraus (1990), breaker steepness H_B/L_0 is shown, in Fig. 8, as a function of the relative wave height at offshore toe of the bar. Equation (5), represented by lines in Fig. 8, agrees the data and follows their trend well.

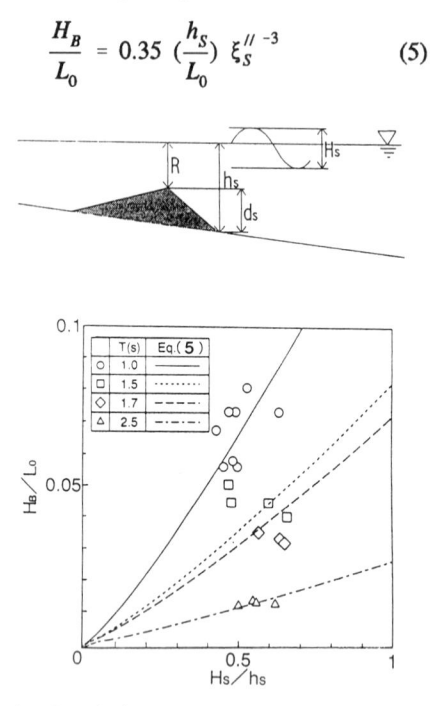

$$\frac{H_B}{L_0} = 0.35 \left(\frac{h_S}{L_0}\right) \xi_S''^{-3} \qquad (5)$$

Fig. 8 Relation of H_B/L_0 to H_S/h_S, illustrating the ability of ξ_s''. Wave breaking over triangle-shaped bar. (Smith and Kraus, 1992)

Breaking Position or Depth on the Breakwater

Breaker depth h_B can not be calculated directly from the breaker height index, because Eq. (3) does not include explicitly the breaker depth. It is, therefore, required an additional index completely to determine the breaking condition on the breakwater. To do this, at first, we will examine the wave breaking on the breakwater crest, in Zone III, where the still water depth is very shallow and constant.

Following Goda (1964), the relative wave height H_B/R is plotted, in Fig. 9, as

WAVE BREAKING 1109

a function of the relative wave crest height h_c/R to determine the breaking position on breakwater crest. h_c and R are the wave crest height and the still water depth above breakwater crest. Plus and cross marks represent measurements for nonbreaking wave, and various solid and open marks for breaking waves. η_c is the wave crest height from still water level. The lines of $\eta_c = 0.5H$ and H refer to relations for small amplitude wave and solitary wave, respectively. Broken line in Fig. 9 represents the bounds that incident waves pass across the breakwater without breaking.

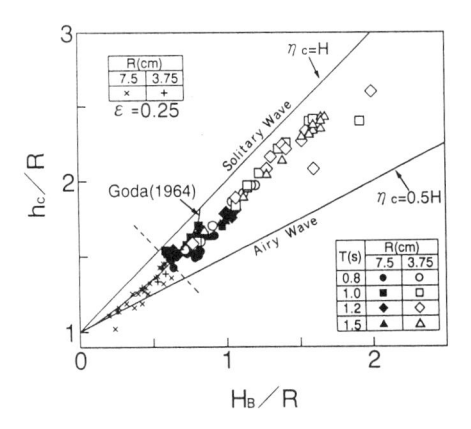

Fig. 9 h_c/R as a function of H_B/R for the incipient breaking on breakwater crest.

As noticed from Fig. 9, the breaker crest height approaches that of a solitary wave, as increasing the breaker height, $\eta_c \approx 0.75H_B$. This indicates that the wave breaking on breakwater crest depends clearly on the trough depth of preceding wave and not on the still water depth above the crest, R. As a result, wave breaking on the crest is of a depth-limited wave, mainly controlled by the vertical asymmetry of incoming wave profile. Based on the results, the conditions under which waves break on the breakwater crest are examined by relationships between H_B/R and ε. In Fig. 10, solid circle and triangle marks represent the lowest wave height when breaking occurs on the crest. And open

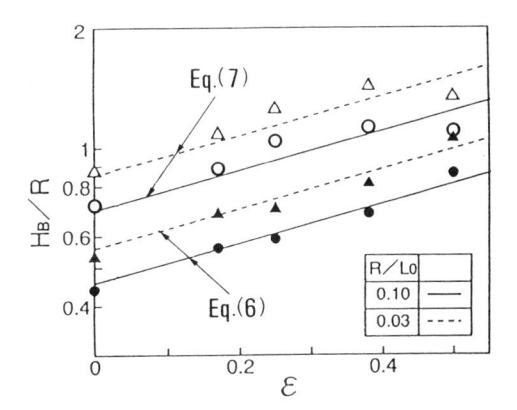

Fig. 10 Relation between H_B/L_0 and ε for determining the bounds of wave breaking on breaker crest.

ones show the breaker height at top of the structure slope. Two conditions of wave breaking on the breakwater crest can be determined as follows; (1) the lowest breaker height at the shoreward limit of wave breaking is given by,

$$\frac{H_B}{L_0} = (\frac{R}{L_0})^{\frac{6}{7}} \exp(2.7\varepsilon - 1.8) \qquad (6)$$

Therefore, Eq. (6) refers to the condition that if breaker height computed from Eq. (3) is lower than that from Eq. (6), the wave passes across the breakwater without breaking. And (2) the breaker height at top of the seaward structure slope is

$$\frac{H_B}{L_0} = (\frac{R}{L_0})^{\frac{6}{7}} \exp(1.2\varepsilon - 0.8) \qquad (7)$$

Visual analyses of the wave profile videos reveal that the breaker height changes exponentially with distance from top of the seaward breakwater crest, and that the shoreward limit of breaking position on the crest depends mostly on the breakwater porosity ε. Equation (8) is derived from the results with respect to the shoreward distance, x_{NB}, from top of the structure slope to the bound beyond which wave breaking does not occur.

$$\frac{x_{NB}}{B} = 0.5 - 0.75\varepsilon \qquad (8)$$

Equation (8) indicates that the breaking position displaces toward the top of structural slope with increasing the breakwater porosity, this results in reduction of the return flow strength over the crest. The wave breaking on breakwater crest depends basically on the permeability rather than roughness of the crest surface.

By taking account of Eqs. (6) and (7), an empirical equation (9) for estimating the position of wave breaking on the crest is determined:

$$\frac{x}{B} = \frac{1}{2} [\ln \{ (\frac{H_B}{L_0}) (\frac{R}{L_0})^{-\frac{6}{7}} \} - 1.2\varepsilon + 0.8] \qquad (9)$$

This relation refers to as the breaking position index.

Breaking Position or Depth in Zone II

As pointed out by the previous experimental and numerical studies (Rojanakamthorn et al.,1990), we also found experimentally that wave breaking on the seaward structural slope of the breakwater can be described reasonably well by a criterion, similar to Miche (1951) and Hamada (1951). Their criterion can be rewritten by using the dispersion relation of linear waves, as Eq. (10).

$$\frac{H_B}{L_0} = A'_m \tanh^2 (k_B h_B) \quad (10)$$

, in which k_B and h_B are the wave number and water depth at breaking. A_m' is the coefficient determined from the continuity condition of breaker height at top of the seaward structure slope (the boundary between Zones II and III). Using Eq. (7), A_m' is written as·

$$A'_m = [(\frac{R}{L_0})^{\frac{6}{7}} \exp(1.2\varepsilon - 0.8)] / \tanh^2 (k_R R) \quad (11)$$

, in which k_R is the wave number for the water depth above the crest. Figure 11 displays the range of the minimum and maximum values of A_m' as a function of the porosity ε, within the limits of the experiments. It is noticed from this figure that the value of A_m' in case of $\varepsilon = 0$ is almost the same to Miche's coefficient of $A_m'=0.142$ for a plane sloping beach.

To confirm applicability of the breaker height index Eq. (10), Fig. 12 shows predicted and measured values of H_B/L_0 as a function of $K_B h_B$. Equation (10) agrees with the measured breaker height on the seaward structure slope.

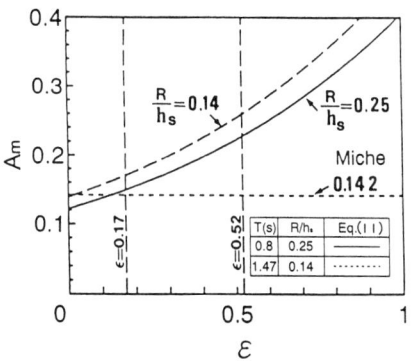

Fig. 11 Relation between A_m'' and ε.

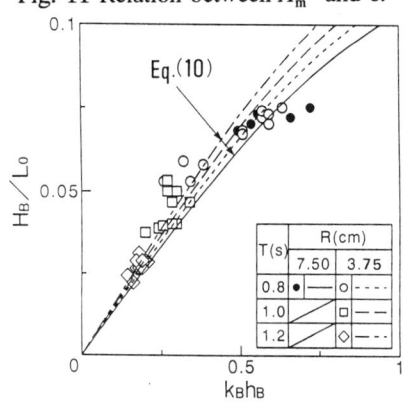

Fig. 12 H_B/L_0 as a function of $k_B h_B$, illustrating the validity of Eq. (10).

Comparisons Between Measured and Calculated Breaker Height over the Breakwater

In the two previous sections, the two additional breaker indices, Eqs. (9) and

(10), have been developed for the breaking zones II and III in order to determine the breaking position. Consequently, substitution of Eq. (3) into Eqs. (9) and (10) yields the distance of breaking position for Zone III and the breaking depth for Zone II, respectively.

Figures 13 and 14 are prepared to confirm this approach. Wave steepness H_B/L_0 at incipient wave breaking over permeable submerged breakwater is shown as a function of the relative distance, x/B, from top of the structure slope. It is noticed that synthetic trends generated by Eqs. (9) and (10) follow observed data trends very well. Discrepancies between predicted and observed values appear to result from variability inhering in the wave breaking.

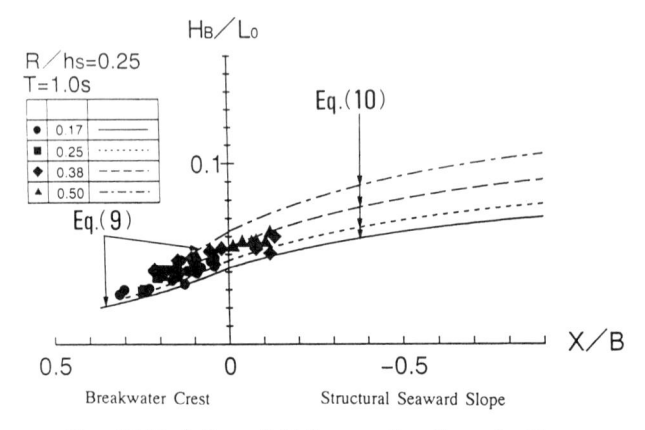

Fig. 13 Variation of H_B/L_0 as a function of x/B.

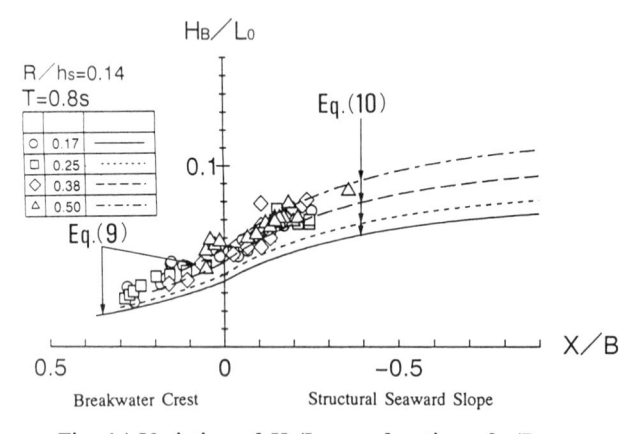

Fig. 14 Variation of H_B/L_0 as a function of x/B.

CONCLUDING REMARKS

A laboratory experiment was conducted to discuss and determine the conditions of incipient breaking over permeable submerged breakwaters. In this study, Breaking range of waves passing across the breakwater was divided into the three following zones; Zone I: on the offshore sloping beach, Zone II: on the seaward structure slope of the breakwater, and Zone III: on the breakwater crest. Height and position of the breaking wave were examined for each zone, and the breaker height and position indices were developed from measurements.

Main findings are summarized as follows;

(1) Permeability of the submerged breakwater plays an important role in reduction of the strength of return flow over the breakwater. The return flow changes the breaker height and breaking position as well as the breaker type.
(2) The integrated parameter ξ_s'' of Hara et al. (1992) expresses remarkably well synthetic trends of observed data of the wave breaking over the breakwater. The parameter is versatile; it can be applied to cases of wave breaking over barred and reef beaches.
(3) Breaker height and depth for Zone I are calculated from Goda's breaker index, Eq. (1), by changing value of the empirical constant to $A=0.15$. Breaker height index, Eq. (3), for Zones II and III is determined as a function of ξ_s''.
(4) Additional indices, Eqs. (9) and (10), are developed to estimate the breaking position in Zone III and breaker depth in Zone II, respectively.

In conclusion, the computation scheme of the breaker height and depth (position) on and around permeable submerged breakwaters is shown as follows;

Zone I: On the offshore sloping beach,
H_B and h_B, calculated from Eq. (1).

Zone II: On the seaward structure slope of the breakwater,

$$Eq.(1)|_{h_B=h_s} > \frac{H_B}{L_0}\bigg|_{Eq.(3)} \geq Eq.(7)$$

H_B and h_B, calculated from Eqs. (3) and (10).

Zone III: On the breakwater crest,

$$Eq.(7) > \frac{H_B}{L_o}\bigg|_{Eq.(3)} \geq Eq.(6)$$

H_B and x_B, calculated from Eqs. (3) and (9).

If wave height calculated from Eq. (3) is smaller than that given by Eq. (6), the wave will pass across the submerged breakwater without breaking.

ACKNOWLEDGEMENTS

We are grateful to Mr. K. Tachikawa, student of the Graduate School of Science and Engineering of Chuo University, for his help in the experiments and data analyses.

REFERENCES

Goda, Y. (1964): Wave forces on a vertical circular cylinder, Experiments and a proposed method of a wave force computation, Rept. of Port and Harbour Res. Inst., No. 8.

Goda, Y. (1970): A synthesis of breaker indices, Proc. Japan Soc. Civil Engrs., No. 180, pp. 39–49 (in Japanese).

Hamada, T. (1951): Breakers and beach erosion, Rept. Transportation Tech. Res. Inst., Ministry of Transportation, No. 1, 165 pp.

Hara, M., T. Yasuda, and Y. Sakakibara (1992): Characteristics of a solitary wave breaking caused by a submerged obstacle, Proc. 23rd ICCE, pp. 253–266.

Izumiya, T., H. Komata, and J. Mizukami (1990): Evaluation of friction factor and local reflection coefficient for a permeable structure, Proc. of Coast Eng., JSCE, Vol. 36, pp. (in Japanese).

Katano, A., M. Hattori, and S. Murakami (1992):,Proc. of Coastal Eng., JSCE, Vol. 39, pp. 646–650 (in Japanese).

Miche, R. (1951): Le pourvoir réfléchissant des ouvrages maritimes exposés à l'action de la houle, Annales Pont et Chaussees., 121ᵉ Annee, pp. 285–319.

Mizuguchi, M. (1990): Reflection from swash zone on natural beaches, Proc. 22nd ICCE, pp. 570–583.

Rojanakamthorn, S., M. Isobe, and A. Watanabe (1990): Modeling of wave transformation on submerged breakwater, Proc. 22nd ICCE, pp. 1060–1073.

Smith, E.R. and N.C. Kraus (1990): Laboratory study on Macro-features of wave breaking over bars and artificial reefs, Tech. Ret. 90-12, CERC, 155 p.

CHAPTER 81

WAVE FORCES ACTING ON A VORTEX EXCITED VIBRATING CYLINDER IN WAVES

Kenjirou Hayashi[1], Futoshi Higaki[2], Koji Fujima[3]
Toshiyuki Sigemura[4], M.ASCE and John R. Chaplin[5], M.ASCE

ABSTRACT

An experimental investigations into the wave forces acting on the vortex-excited vibrating vertical circular cylinder in regular waves have been performed with emphasis being placed on the amplification of the wave forces caused by the fluid-structure interaction. The cylinder is vibrating only in the transverse, cross-flow, direction by means of the restriction of the vibration of the in-line direction. The results indicate that the existence of amplification of the lift forces acting on the vortex-excited vibrating cylinder in comparison with the stationary cylinder is a function of the ratio of wave frequency f_w to the natural frequency of the cylinder in water f_{nw} and Keulegan-Carpenter number at still water level CKC. The in-line forces are also amplified in the range of CKC where large amplitude of oscillation in the transverse direction occurs.

INTRODUCTION

The wave forces acting on a small diameter offshore structure are usually resolved into two components. One, the inline force, acts in the direction of wave propagation and the other , the lift force or transverse force, acts in the transverse direction of it. The predominant frequency of lift force caused by vortex shedding is a multiple of that of the inline force. Therefore, the

---
[1] Assoc. Prof., Civ. Engrg. Dept., The National Defense Academy, 1-10-20 Hashirimizu Yokosuka, 239, Japan.
[2] Post graduate student, Civ. Engrg. Dept., N.D.A.
[3] Lecturer, Civ. Engrg. Dept., N.D.A.
[4] Professor, Civ. Engrg. Dept., N.D.A.
[5] Professor, Civ. Engrg. Dept., The City University, Northampton Square London, EC1V 0HB, England.

structure's dynamic response to the lift forces "Vortex-excited vibration" must be considered more significantly.

A great number of studies for the vortex-excited vibration of a circular cylinder in steady flow have been made. An important phenomenon of this vibration is that of "lock-in" between the frequency of the vortex shedding and the frequency of the vibrating cylinder. Under "lock-in" condition, large resonant vibration occurs and the lift forces acting on this vibrating cylinder are amplified by the fluid-structure interaction, Blevins (1977).

A similar phenomenon may occur under certain conditions if a flexible cylinder is placed in planar oscillatory flow or in waves. However, this has not been sufficiently understood and relatively little work has been carried out into this interesting problem in hparmnic flow and in waves.

The results for the amplification of the forces acting on a vortex-excited vibrating cylinder in harmonic flows have been reported by Sarpkaya and Rajabi (1979), Hayashi et al.(1990) and Sumer et al.(1994). Sarpkaya and Rajabi(1979) show that at perfect resonance, the lift forces acting on a vortex-excited vibrating cylinder in plannar oscillating flow are amplified nearly two times compared to that of rigidly mounted cylinder. Hayashi et al.(1990) show that the lift and in-line forces acting on a vortex-excited cylinder vibrating only in cross flow direction being resonant with the second harmonic component of the lift are amplified in the ranges of Keulegan-Carpenter number 4<KC<8 and 6<KC<16 respectively. The maximum increase in the lift force is about 200% at around KC=6 and that in the in-line force is about 100% at around KC=8. Sumer et al.(1994) have studied the influence of KC, reduced velocity V_r, amplitude of cylinder vibration and wall-proximity effect to the forces acting on a cylinder vibrating only in cross-flow direction. Their results show that the increase in the drag coefficient is about 50-200% for around KC=10 and nearly negligible for the range 60<KC and the increase in lift force is up to 200%.

The results for the amplification of the forces acting on a vortex-excited vibrating vertical cylinder in waves have been reported by Isaacson and Maull(1981), Zedan and Rajabi(1981), Angrilli and Cossalter(1982). They show that the increases of lift forces acting on a vortex-excited vertical cylinder are 70-290% for f_w/f_{nw}=1/2, CKC=10-12 and kd=1-3.9, 60% for f_w/f_{nw}=1/3, CKC=17.8, and kd=0.88, and 90% for f_w/f_{nw}=1/4, CKC=35.9, and kd=0.62, where f_w and f_{nw} are the incident wave frequency and the natural frequency of cylinder in water, k is the wave number, and d is the mean water depth.

Hayashi(1984) shows that the lift forces acting on

a vortex-excited vibrating vertical cylinder in waves are amplified in the range of 5<CKC<15 for $f_w/f_{nw}=1/2$ and kd=1.83 and in the range of 18<CKC<30 for $f_w/f_{nw}=1/3$ and kd=1.01. Maull and Kaye(1988) shows similarly that when f_w/f_{nw} is fixed at 1/2, the lift forces are amplified in the range of 5<CKC<12 for kd=1.32 and 1.73.

The forces acting on a partial parts of a flexibly mounted vertical cylinder in waves have been measured by Bearman(1988) and Borthwick and Herbert(1988). Force coefficients of in-line and lift forces from the flexibly mounted cylinder are found to be larger than those for the same cylinder but with a rigid mounting.

In the present paper, experimental investigations into the wave forces acting on the vortex-excited vibrating vertical cylinder in regular waves have been described with emphasis being placed on the amplification of the forces acting on the cylinder caused by the fluid-structure interaction.

EXPERIMENTS

Laboratory experiments were carried out in a wave flume 40m long, 0.8m wide, and 1m deep. The general arrangement of the test cylinder is shown in Figure 1. In order to give a high degree of rigidity to the test cylinder, both ends of it are connected to the core cylinder. The flange weights are attached to the core cylinder above the test cylinder to adjust the equivalent mass m_e of the test cylinder.

The support plate is attached to the holder flag at the bottom end of the test cylinder. Both end in the inline direction of the support plate are pivoted on the floor of the flume to prevent vibration in the inline direction. The upper end of the core cylinder, above the water level, is mounted with springs only in the transverse direction. Each spring is connected to a strain-gauged steel canti-

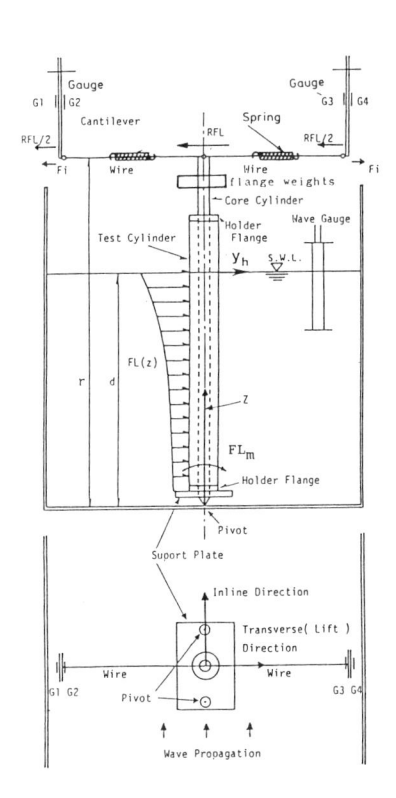

Fig. 1 General Arrangement of Test Cylinder

lever by wire. Two strain gauges were fixed on each cantilever to measure its bending moment produced by the force acting on the end of it through the springs and wires. Strain gauges were connected into a wheatstone bridge circuit in the bridge conditioner to produce the out put signal corresponding to the displacement of the top end of the core cylinder in the transverse direction.

In order to obtain the relationship between the output voltage signals from the bridge circuit and the top end displacements of the core cylinder, known loads are applied horizontally to the top end of the core cylinder by weights hung over a pulley.

Free vibration tests are performed in air and in still water. The damping factor, obtained from the logarithmic decrement, and the natural frequency of the test cylinder are measured by releasing its top end from an initial displacement and recording amplitude decay of the transient oscillation of the cylinder. The equivalent mass per unit length of the test cylinder in water, m_e, is calculated from the measurements of natural frequency and stiffness of test cylinder in still water, Hayashi(1984). Thus the value of m_e includes the mass of structure and the entrained fluid.

Two kinds of experiments, Tests-A and Tests-B, are carried out. The experimental conditions in Tests-A are shown in Table 1. Where $k(=2\pi/L$, L=wave length) is the wave number, f_{na} and f_{nw} are the natural frequencies of the test cylinder in air and in water, h_{ta} and h_{tw} are damping factor of the cylinder in air and in water, ρ is the density of water, and $\beta (=D^2 f_w/\nu$, ν =kinematic viscosity of water) is the viscous-frequency parameter.

The still water depth d is 40cm. The test cylinder, of outside diameter D=1.9cm and 73.7cm length, is at-

Table 1 Experimental Conditions in Tests-A

Case	D (cm)	d (cm)	f_w (Hz)	f_w/f_{nw}	kd ($2\pi \cdot d/L$)	CKC	β
V1	1.9	40	1.07	1/2	1.92	5.2-26.4	386
V2	1.9	40	0.7	1/3	1.03	5.6-38.0	253
V3	1.9	40	0.53	1/4	0.73	4.5-35.1	191

f_{na}=2.17Hz, f_{nw}=2.12Hz
h_{ta}=0.002, h_{tw}=0.005
Mass ratio : $m_e/(\rho_w D^2)$=14.9
Reduced damping : $K_s=2m_e(2\pi h_{ta}/(\rho_w D^2))$=0.37

S1	1.9	40	1.07	1.92	5.5-25.5	386
S2	1.9	40	0.70	1.03	2.8-39.9	253
S3	1.9	40	0.53	0.73	2.1-36.4	191

tached to the support plate(25x10x0.2cm) with the core cylinder of 0.9cm diameter and 78.5cm length.

In Case V1-V3, the relationship between the vortex-excited vibration of the test cylinder and the Keulegan-Carpenter number at still water level $CKC = U_{ms} \cdot T/D$, is measured. Where U_{ms} is the maximum horizontal water particle velocity at still water level and T is the wave period. In each of these Cases, the frequency ratio, f_w/f_{nw}, is fixed at around one value of the resonance frequency ratios, $f_w/f_{nw} = 1/2$, 1/3, and 1/4.

In Case S1, S2, and S3, the base bending moment FL_m(=the moments about the bottom of the test cylinder) due to the lift forces acting on the test cylinder rigidly mounted with strings replacing the springs in the transverse direction are measured in the similar waves used in the Case V1, Case V2 and Case V3.

In Tests-B, in-line and transverse forces acting on a partial part of the test cylinder are measured by using a force sleeve supported by two components small load cell installed in the test cylinder as shown in Figure 2. The force sleeve of outside diameter D=3cm and 3cm length is positioned 15.9 cm below still water level with water depth d=80cm. The test cylinder of outside diameter D=3cm and 98.5cm length is attached to the support plate(20x5x0.6cm) with the core cylinder of 1cm diameter and 104.6cm length. The conditions are shown in Table 2. The measurements are made for both cases of vortex-excited cylinder being resonant with the second to sixth harmonic components of lift forces($f_w/f_{nw} = 1/2 - 1/6$) and rigidly mounted cylinder for comparison.

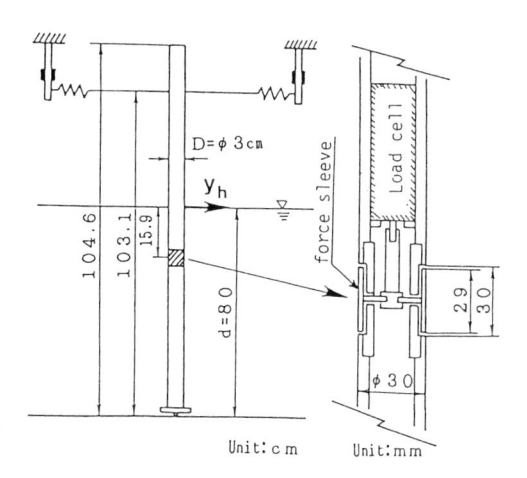

Fig. 2 Experimental Set-up (Force Sleeve and Load Cell)

Table 2 Experimental Conditions in Tests-B

Case	D (cm)	d (cm)	f_w (Hz)	f_w/f_{nw}	kd ($2\pi \cdot d/L$)	CKC	β
LV2	3.0	80	0.85	1/2	2.41	6.0-20.6	655
LV3	3.0	80	0.56	1/3	1.20	6.9-30.5	426
LV4	3.0	80	0.42	1/4	0.90	4.4-29.7	342
LV5	3.0	80	0.34	1/5	0.69	9.3-29.4	274
LV6	3.0	80	0.28	1/6	0.56	5.2-40.7	228

f_{na}=1.79Hz, f_{nw}=1.69Hz
h_{ta}=0.0035, h_{tw}=0.014
Mass ratio, : $m_e/(\rho_w D^2)$=8.34
Reduced damping : $K_s = 2m_e(2\pi h_{ta}/(\rho_w D^2))$=0.37

MODEL FOR VIBRATION

The definition sketch of the test cylinder, which illustrates the transverse response of the cylinder which is rod pivoted on the bottom of flume and supported by spring in the transverse direction, is shown in Figure 3. The dynamic response of the cylinder to the lift forces may be described by using the equation of motion as

$$M_{mt} \cdot \ddot{y}_h + C_{mt} \cdot \dot{y}_h + K_{mt} \cdot y_h = FL_m, \quad ---(1)$$

where $\ddot{y}_h$, $\dot{y}_h$ and y_h are the transverse displacement, velocity and acceleration of the cylinder at still water level, M_{mt}, C_{mt} and K_{mt} are the effective mass, damping and stiffness of system. M_{mt} includes the mass of cylinder and added mass in water. C_{mt} includes the structural damping and fluid damping. K_{mt} includes the stiffness due to the spring force, and the buoyancy and distributed weight when the cylinder is in a deflected position from the vertical. FL_m is the bending moments around the pivot which is produced by the lift force acting on the cylinder.
$M_{mt} \cdot \ddot{y}_h$, $C_{mt} \cdot \dot{y}_h$ and $K_{mt} \cdot y_h$ show the moments

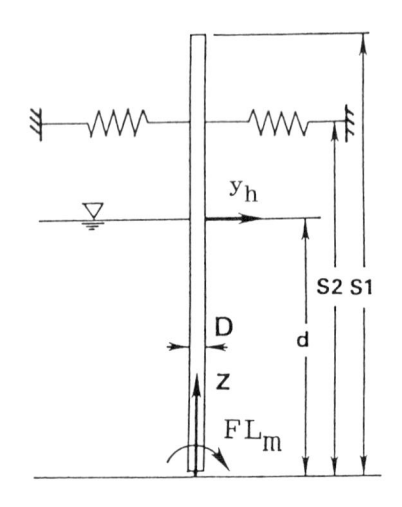

Fig. 3 Coordinate System

produced by the inertia forces, the damping forces and the stiffness of the structure respectively. Thus, Eq.(1) shows the equivalence of the moments taken about the pivot of the base, Hayashi(1984). FL_m may be expressed in a series form as

$$FL_m = FL_m(n)*\sin[2\pi \cdot n \cdot f_w \cdot t + \phi(n)], \quad ---(2)$$

where $FL_m(n)$, n=1,2,3---, is the nth frequency component of FL_m and $\phi(n)$ is the phase lag. The amplitude $Y_h(n)$ of response vibration y_h to $FL_m(n)$ may be given as the solution of Eq.(1) as

$$Y_h(n) = FL_m(n)/K_{mt} * [\{1-(n \cdot f_w/f_{nw})^2\}^2 + (2h_{tw} \cdot n \cdot f_w/f_{nw})^2]^{-1/2}$$

$$---(3)$$

, where h_{tw} is the total damping factor of the cylinder in water and may be expressed as the sum of the structural damping factor h_s and fluid damping factor h_f, Hayashi and Chaplin (1991).

RESULTS AND DISCUSSION FOR TESTS-A

Figure 4 shows the decay of free oscillations y_h of the test cylinder in air and in water, which is used in the experiments of Tests-A. Figure 5 shows the variation of the damping factor h_{tai} in air and h_{twi} in still water, d=40cm, with the non dimensional amplitude Y_{hi}/D. Where Y_{hi} is the amplitude of i-th oscillation of the cylinder y_h at still water level. The values of h_{tai} and h_{twi} for each amplitude of Y_{hi}/D are defined as

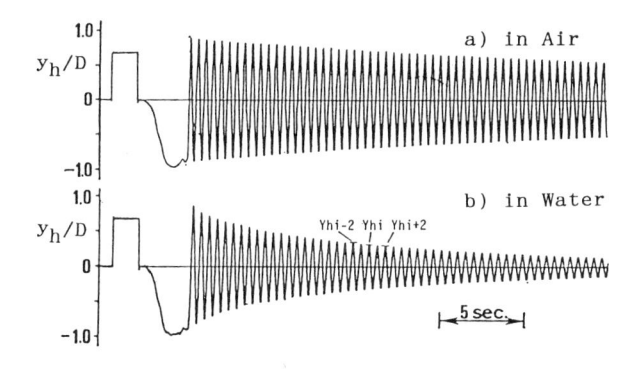

Fig. 4 Decay of Free Oscillation

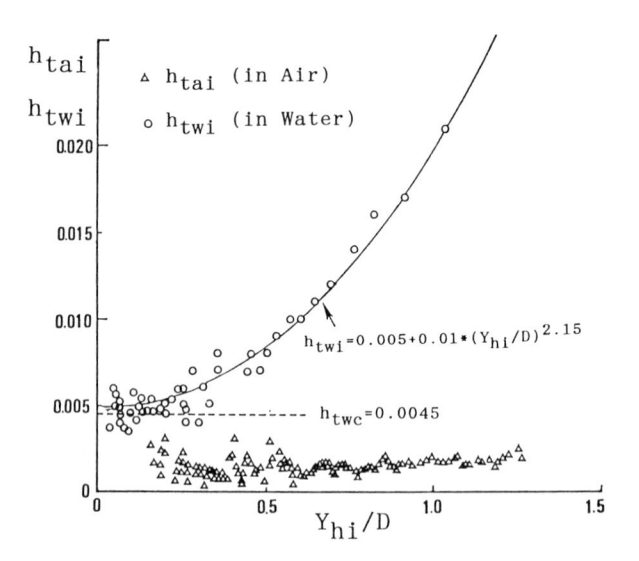

Fig. 5 Variation of Damping Factor with Y_h/D

$$h_{tai} \text{ or } h_{twi} = 1/(2\pi) * \{\ln(Y_{hi-2}/Y_{hi+2})\}/4, \quad ---(4)$$

where Y_{hi-2} and Y_{hi+2} are the amplitudes of the (i-2)th and (i+2)th periods respectively (see Figure 4).

The value of h_{tai}, which shows the structural damping factor, is nearly independent of the value of Y_{hi}/D. The constant value of $h_{ta}=0.002$ is written in Table 1. On the other hand, the value of h_{twi}, which is composed of structural damping and fluid damping, is independent of amplitude effect only for low values of Y_{hi}/D and it becomes amplitude dependent at higher value of Y_{hi}/D owing to the characteristics of the fluid damping. The constant value of h_{twi}, $Y_{hi}/D<0.3$, is nearly corresponding to the theoretical value of $h_{twc}=0.0045$ which is derived from the Stokes's theory, Stokes (1851), for the forces on a cylindrical pendulum bobs oscillating at low KC number, Hayashi and Chaplin (1991). The increase of h_{twi} in $Y_{hi}/D>0.3$ may be due to the appearance of vortex-sheddings. The variation of the h_{twi} with Y_{hi}/D is approximated by the regression equation as

$$h_{twi} = 0.005 + 0.01*(Y_{hi}/D)^{2.15}. \quad ---(5)$$

The variation of $Y_{hm}(n)/D$ and $Y_{hc}(n)/D$, n=2,3,4 are shown in Figure 6 (a),(b),(c) where $Y_{hm}(n)$ is the Fourier amplitude of measured displacement y_h and $Y_{hc}(n)$ is a calculated value obtained by substituting the measured moments $FL_m(n)$ acting on the rigidly mounted cylinder into Eq.(3). Here $FL_m(n)$, n=1-4, is the first four harmonics of FL_m, which are measured for the three values of kd=1.92, 1.03 and 0.73 in Case S1, S2 and S3. In calculation, Eq.(3) is coupled to Eq.(5) by means of $Y_h(n)=Y_h=Y_{hi}$.

The measured response $Y_{hm}(n)$, n=1, 2, 3, is above the predicted response $Y_{hc}(n)$ in the range of 8<CKC<15 for the case of $f_w/f_{nw}=1/2$, in the range of 12<CKC<23 for the case of $f_w/f_{nw}=1/3$, and 25<CKC for the case of $f_w/f_{nw}=1/4$ respectively. These phenomena may be due to the amplification of the lift acting on the vortex-excited vibrating cylinder caused by the fluid-structure interaction. Similar phenomena are obtained by Maull and Kaye (1988) for the case of kd=1.73 and 1.32.

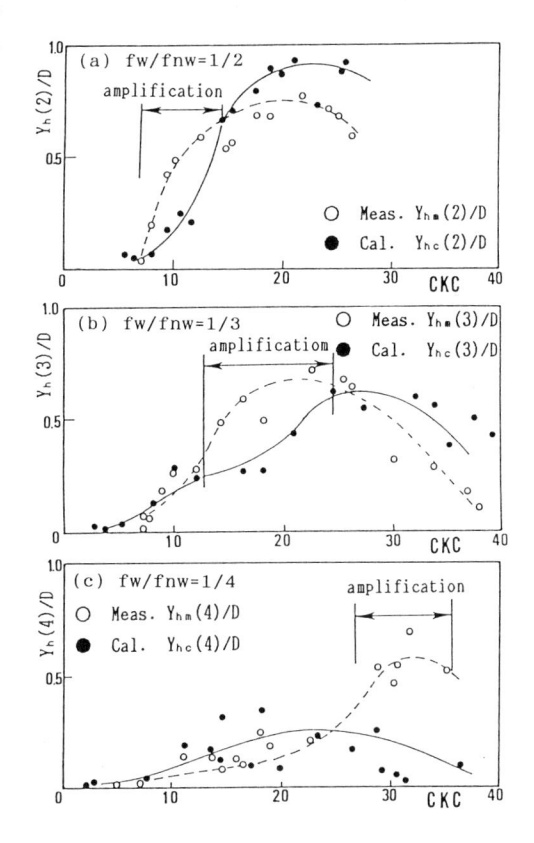

Fig. 6 $Y_{hm}(n)$ and $Y_{hc}(n)$

RESULTS AND DISCUSSION FOR TESTS-B

The traces for free decay of oscillations y_h and of the forces dF_Y in air and in water are shown in Figure 7 (a) and (b), where dF_Y is the transverse, Y-direction, forces acting on the unit length of the force sleeve, which is measured by the load cell in the test cylinder used in Tests-B. The relationship between these dF_Y and y_h/D are shown in Figure 7 (c). From this figure, dF_Y in air and in water may be expressed as

$$dF_Y, \text{(in air)} = dF_{YIa} = m_{sa}*d^2y_h/dt^2 = 2.37*y_h/D, \quad ---(6)$$

$$dF_Y, \text{(in water)} = dF_{YIw} = m_{sw}*d^2y_h/dt^2 = 5.17*y_h/D, \quad ---(7)$$

where dF_{YIa} and dF_{YIw} are inertia forces acting on the unit length of force sleeve in air and in water, and m_{sa} and m_{sw} are reduced mass per unit length of the force sleeve in air and in water.

The variations of η, y_h/D, dF_Y, dF_{Ya}, dF_{Yw}, dF_X with time for both cases of vortex-excited cylinder and rigidly mounted cylinder for comparison for the case of $f_w/f_{nw}=1/2$, CKC=12, kd=2.41 are shown in Figure 8 (a),(b) and (c). Here η is the water surface elevation. The Keulegan-Carpenter number at the level of force sleeve, 16cm below the still water level is LKC=7.5. dF_X and dF_Y are in-line and lift force acting on the unit length of force

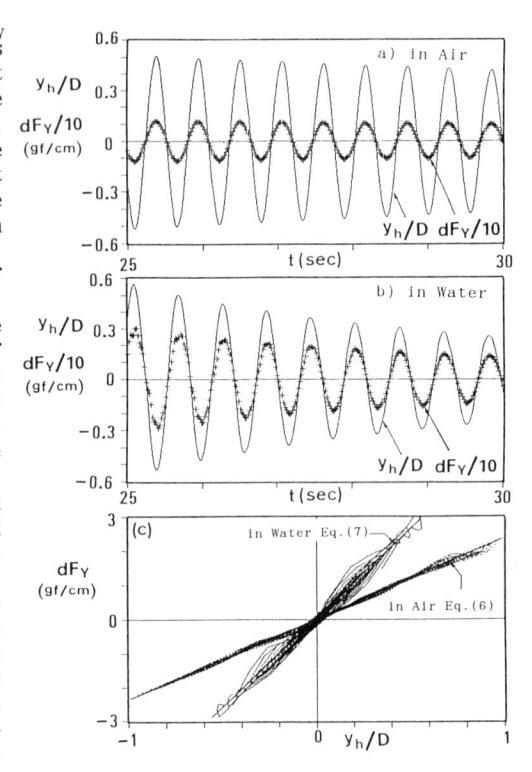

Fig. 7 Decay of Free oscillation

(A) Vibrating Cylinder (B) Rigid Cylinder

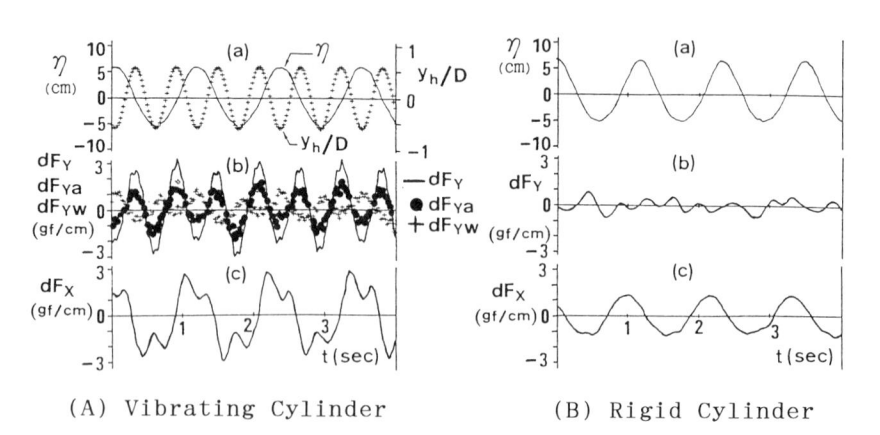

Fig. 8 η, y_h/D, dF_Y, dF_{Ya}, dF_{Yw}, dF_X versus time

WAVE FORCES ON VORTEX EXCITED VIBRATING CYLINDER   1125

sleeve, which are measured by the two components small load cell. dF_{Ya} and dF_{Yw} are defined as

$$dF_{Ya} = dF_Y - dF_{YIa} = dF_Y - 2.37 \cdot y_h/D, \quad ---(8)$$

$$dF_{Yw} = dF_Y - dF_{YIw} = dF_Y - 5.17 \cdot y_h/D. \quad ---(9)$$

dF_{Ya} shows the transverse fluid force acting on the unit length of the force sleeve. dF_{Ya} may be composed of the inertia force due to the added mass of water, the fluid damping force and the loading force per unit length of force sleeve. Thus dF_{Yw} shows the transverse force which is composed of the damping and the loading force per unit length of force sleeve in water. When the test cylinder is vibrating in resonance condition, f_w/f_{nw} = 1/2, 1/3, 1/4, 1/5, 1/6, dF_{YIa} and dF_{YIw} may be estimated by using Eq.(6) and Eq.(7).

dF_X shows the in-line fluid force acting on a unit length of force sleeve in water because the vibration of the in-line direction is restricted.

We can recognize that the forces dF_X, dF_{Ya} and dF_{Yw} acting on the

- Isaacson and Mall
  [fw/fnw=1/2 Ks=3.54]
- ▼ Zedan and Rajabi
  [fw/fnw=1/2 Ks=0.83]
- ■ Anglilli and Cossalter
  [fw/fnw=1/2 Ks=0.98]
  [fw/fnw=1/3 Ks=0.98]

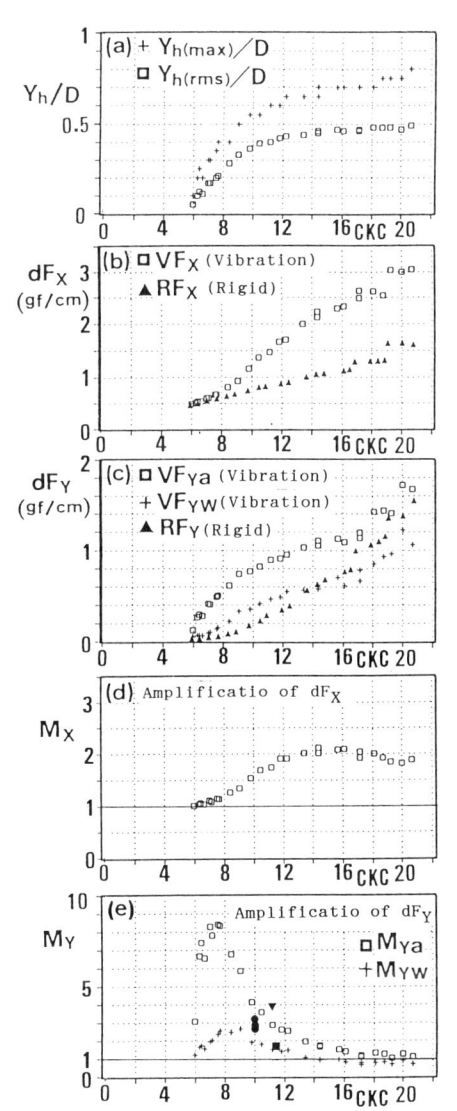

Fig. 9   Amplification versus CKC for fw/fnw=1/2

vibrating cylinder are large compare to the forces dF_X and dF_Y acting on the rigidly mounted cylinder.

Fig. 9 (a)-(e) show the variation of the non-dimensional cylinder vibration (Y_h/D), wave forces (dF_X, dF_Y), and amplification factor (M_X, M_Y) with CKC for the case of $f_w/f_{nw}=1/2$. Here $Y_{h(max)}$ and $Y_{h(rms)}$ are the maximum half-amplitude and the root mean square value of the cylinder displacement y_h at still water level. VF_X, VF_{Ya} and VF_{Yw} are the root mean square values of in-line and transverse forces, dF_X, dF_{Ya} and dF_{Yw}, acting on the force sleeve of vibrating cylinder. RF_X RF_{Ya} and RF_{Yw} are the root mean square values of those of the rigidly mounted cylinder. The amplification factors M_X, M_{Ya} and M_{Yw} are defined as

$$M_X = VF_X/RF_X, \qquad ---(10)$$

$$M_{Ya} = VF_{Ya}/RF_Y, \qquad ---(11)$$

$$M_{Yw} = VF_{Yw}/RF_Y. \qquad ---(12)$$

Figure 9 (d) and (e) shows that M_X increases with increasing Y_h/D and M_{Ya} is little affected with increasing Y_h/D. M_{Ya} is large in the range of 6<CKC<13 where Y_h/D increases rapidly with increasing CKC. This range of CKC, where the amplification of lift force is large, is nearly consistent with the results obtained for $f_w/f_{nw}=1/2$ in Tests-A, see Figure 6 (a). The amplification factor M_Y obtained by Isaacson and Maull(1981), Zedan and Rajabi(1981), and Anglilli and Cossalter(1982) for the case of $f_w/f_{nw}=1/2$ are plotted in Figure 9 (e) only for reference. It should be noted that their results are obtained for the total lift forces acting on the vertical cylinder in waves.

The variations of the cylinder vibration Y_h/D and the amplification factors with CKC for $f_w/f_{nw}= 1/3$, 1/4, and 1/6 are shown in Figure 10 (A),(B) and (C) respectively. We can also recognize that M_X increases with increasing Y_h/D, which is nearly consistent with the results obtained in the case of steady flow, Griffin et al.(1975) and Sumer et al. (1994), and M_{Ya} is a little affected with increasing Y_h/D. M_{Ya} is a function of f_w/f_{nw} and CKC. The amplification factor M_Y obtained for the total lift acting on the vertical cylinder by Anglilli and Cossalter (1982) for the case of $f_w/f_{nw}=1/3$ is also plotted in Figure 11.

The value of M_{Yw} is bigger than 1 in the range of 6<CKC<13 for $f_w/f_{nw}=1/2$. On the other hand, it is less than 1 for $f_w/f_{nw}=1/3$, 1/4 and 1/6. We need more consideration to the characteristics of M_{Yw}.

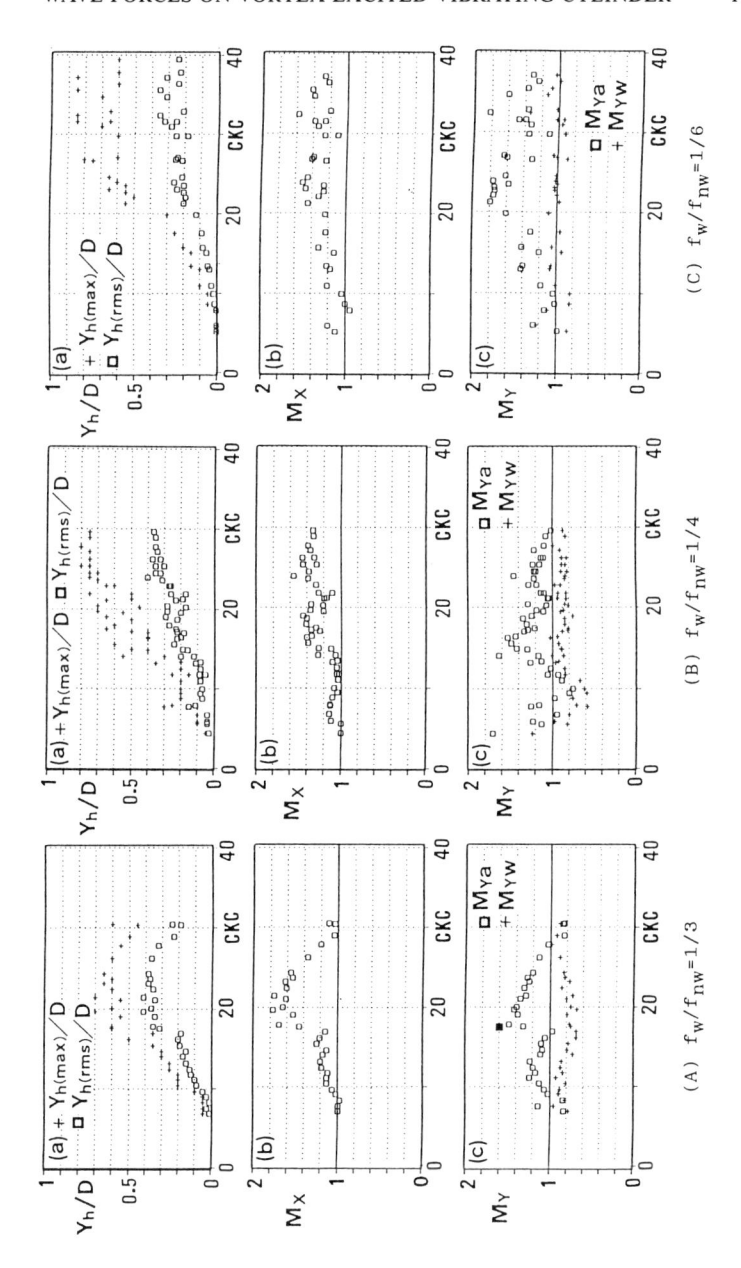

Fig. 10 Cylinder Vibration and Amplification Factor

CONCLUSIONS

The main conclusions obtained in this study are summarized as follows:

1). The existence of the amplification of lift force acting on the vortex-excited vibrating cylinder in comparison with the rigidly mounted cylinder is a function of the ratio of wave frequency f_w to the natural frequency f_{nw} of the cylinder in water and Keulegan-Carpenter number at still water level. When f_w/f_{nw} is fixed at about 2, large amplitude of transverse, cross-flow, vibration Y_h/D of cylinder occurs in the wide range of CKC, but the amplification of lift occurs only in the range of 6<CKC<13.

2). The in-line force acting on the partial part of the vortex-excited vibrating cylinder is amplified with increase of cross-flow vibration Y_h/D.

REFERENCES

Angrilli, F., and Cossalter, V. (1982). "Transverse oscillations of a vertical pile in waves." Journal of Fluids Engineering, Vol.104, pp.46-53.

Bearman, P. W. (1988). "Wave loading experiments on circular cylinders at large scale." Proc., Int. Conf. on Behavior of Offshore Structures BOSS 88, Trondheim, Norway, June, 2, pp.471-487.

Blevins, D. B. (1977). "Flow-induces vibration." Van Nostrand Reinhold, New York, N.Y.

Borthwick, A.G.L, and Herbert, D.M. (1988). "Loading and response of a small diameter flexibly mounted cylinder in waves." Journal of Fluids and Structures, Vol.2, pp.479-501.

Hayashi, K. (1984). "The non-linear vortex-excited vibration of the vertical cylinder in waves." Ph.D. Thesis, University of Liverpool.

Hayashi, K., Ogihara, Y., Fujima, K., and Shigemura, T. (1990). "Forces acting on a vortex-excited vibrating cylinder in planar oscillatory flow." Hydraulic Engineering Proceedings 1990 National Conference, HY. Div /ASCE, San Diego, Vol.3, pp.91-96.

Hayashi, K and Chaplin, J.R. (1991). "Damping of a vertical cylinder oscillating in still water." Proceedings of 1st International Offshore and Polar Engineering Conference, Edinburgh, UK., pp.346-353.

Isaacson M.Q. and Maull D.J. (1981). "Dynamic response of vertical piles." International Symposium on Hydrodynamics in Ocean Engineering, The Norwegian Institute of Technology, pp.887-904.

Maull, D.J. and Kaye, D. (1988). "Oscillations of a flexible cylinder in waves." Proceedings of International Conference on Behavior of Offshore Structures BOSS'88 , Trondheim, pp.535-549.
Sarpkaya, T. and Rajabi F. (1979). "Dynamic response of piles to vortex shedding in oscillating flows." Offshore Technology Conference in Houston, pp.2523-2528.
Stokes, G.G. (1851). "On the effect of the internal friction of fluids on the motion of pendulums." Trans. Camb. Phil. Soc. 9, pp.8-106.
Sumer, B.M., Fredsoe, J., Jensen, B.L., and Christiansen, N.(1994). "Forces on vibrating cylinder near wall in current and waves." Journal of Waterway, Port, Coastal and Ocean Engineering, Vol.120, No.3, pp.233-248.
Zedan, M.F. and Rajabi, F. (1981). "Lift forces on cylinders undergoing hydroelastic oscillations in waves and two dimensional harmonic flow." International Symposium on Hydrodynamics in Ocean Engineering, The Norwegian Institute of Technology, pp.239-262.

CHAPTER 82

Overtopping of sea walls under random waves

D M HERBERT[1], N W H ALLSOP[1] and M W OWEN[2]

Abstract

Over the last fifteen years a long running research programme has been undertaken at HR Wallingford to investigate the overtopping discharge performance of a wide range of sea walls. The research, which is funded by the Ministry of Agriculture, Fisheries and Food, is principally aimed at deriving methods to enable design engineers to determine the overtopping performance of a particular sea wall cross-section under a range of wave and water level conditions. The studies have generally used random wave physical model tests in order to collect data which can then be employed to derive empirical equations that describe the level of overtopping discharge.

1 Introduction

This paper describes recent research work at HR Wallingford, based on the results of physical model tests, aimed at quantifying the overtopping performance of recurved and vertical sea walls. The work is a continuation of a large research programme which has resulted in the derivation of empirical methods for assessing overtopping discharges on embankment sea walls.

2 Summary of previous work

Considerable stretches of the United Kingdom (UK) coastline are protected by a simple earth embankment, consisting of a sloping seaward face, a horizontal crest just a few metres wide and possibly a rear slope. These embankments are

---

[1] Coastal Group, HR Wallingford, Wallingford, OX10 8BA, UK

[2] Llanfair Caereinion, Welshpool, Powys, SY21 0DS, UK

particularly frequent in rural areas, where the seaward face is often protected either by grass or pitched stone. In the late 1970's the then Hydraulics Research Station carried out an extensive research programme to determine the overtopping discharge behaviour of embankment type sea walls, culminating in the production of design guidelines (Owen, 1984).

The design method established for embankment sea walls is based on a dimensionless discharge parameter, Q_*, and a dimensionless freeboard, R_*. These two parameters are defined below:-

$$Q_* = Q / (T_m \, g \, H_s) \tag{1}$$

$$R_* = R_c / (T_m \sqrt{gH_s}) \tag{2}$$

where Q is the mean discharge overtopping the crest of the sea wall,
T_m is the mean wave period,
H_s is the significant wave height,
g is acceleration due to gravity
and R_c is the sea wall freeboard (the height of the sea wall crest above still water level).

The dimensionless parameters are connected by the following exponential equation:-

$$Q_* = A \exp(-BR_*/r) \tag{3}$$

where r is a roughness coefficient
and A and B are empirically derived coefficients dependent upon the structure slope.

Typical values of these empirical coefficients vary from A=0.00794 and B=20.12 for a 1:1 slope to A=0.025 and B=65.2 for a 1:5 slope. Recommended values of the roughness coefficient vary from $r = 1.0$ for smooth impermeable slopes, $r = 0.85$-0.9 for turf, $r = 0.8$ for one layer of stone rubble on an impermeable base and $r = 0.5$-0.6 for two or more layers of rubble.

Further work on bermed sea walls (Owen, 1984) showed that equations (1) - (3) could also be applied to these type of structures by modifying the empirical coefficients A and B. This work illustrated that, in general, the most effective berm for reducing overtopping is located at or close to still water level.

Wave basin tests using long crested waves (Owen, 1984) indicated that overtopping can increase for angles of approach up to 30° off normal with the worst overtopping occurring at about 15° off normal. Under short crested seas (CIRIA, 1991) the overtopping discharge remains roughly constant for wave directions between 0° and 30° off normal before tailing off at larger angles.

Allsop and Bradbury (1988) completed model tests in which measurements were made of the overtopping discharge for vertically faced crown walls mounted on top of rock revetments or breakwaters. A change to the relationship given by Owen (1984) was suggested with the introduction of a new dimensionless freeboard parameter, F_*, defined as follows:-

$$F_* = R_c^2 / (H_s^2 \, g \, T_m) \tag{4}$$

The equation connecting the dimensionless discharge and freeboard was also modified:-

$$Q_* = A \, F_*^B \tag{5}$$

where A and B are coefficients dependent upon the geometry of the structure cross-section.

3 Recurved Walls

In many urban areas the traditional embankment type sea wall frequently incorporates a wave return wall at its crest. This wall can be located either at the top of the seaward slope, or else it can be sited a few metres back allowing the crest berm to be used as a promenade. A series of physical model tests were subsequently undertaken to measure the overtopping discharges of a range of recurved wave return walls for different sea wall slopes, water levels and wave conditions (Owen and Steele, 1991).

The model tests were carried out in a wave flume at a nominal geometric scale of 1:15. Smooth impermeable sea wall slopes of 1:2 and 1:4 were tested under a range of wave and water level conditions but always with a constant sea steepness (based on the mean deep water wave length) of s=0.045. Although wave return walls with a very wide range of profiles have been constructed at different locations around the UK coastline, only the basic profile originally suggested by Berkeley - Thorne and Roberts (1981) was used in this study. However, the distance between the top of the seaward slope and the foot of the wave recurve was varied throughout testing. Figure 1 illustrates the general configuration of the model tests.

Two options were investigated as a means of defining the effectiveness of wave return walls. The two alternative definitions were:

- the ratio of the measured overtopping discharge to the discharge which would have occurred if the return wall had been removed, and the seaward slope had been extended up to the same elevation as the top of the return wall.

- the ratio of the measured overtopping discharge to the discharge which would have occurred if the return wall had been absent.

The second definition was the most appropriate as it appeared to be a much more direct indicator of the performance of the return wall. During the course of the analysis it became clear that one factor governing the effectiveness of the return wall was the height of the wall relative to its position above the still water line. Accordingly the dimensionless height of the wave return wall was defined as:-

$$W_* = W_h/R_c \tag{6}$$

where W_h is the height of the wave return wall from its base to its top
and R_c is the freeboard between the top of the seaward slope (which is at an identical elevation to the base of the return wall) and the still water line.

Using the above definition of the effectiveness of the wave return wall, it is necessary to know the overtopping discharge which would have resulted during the tests if the wave return wall had been absent, for identical wave conditions, water levels and sea wall geometry. Measurements of these discharges were not made specifically for this study but used the results of the earlier research programme (Owen, 1984).

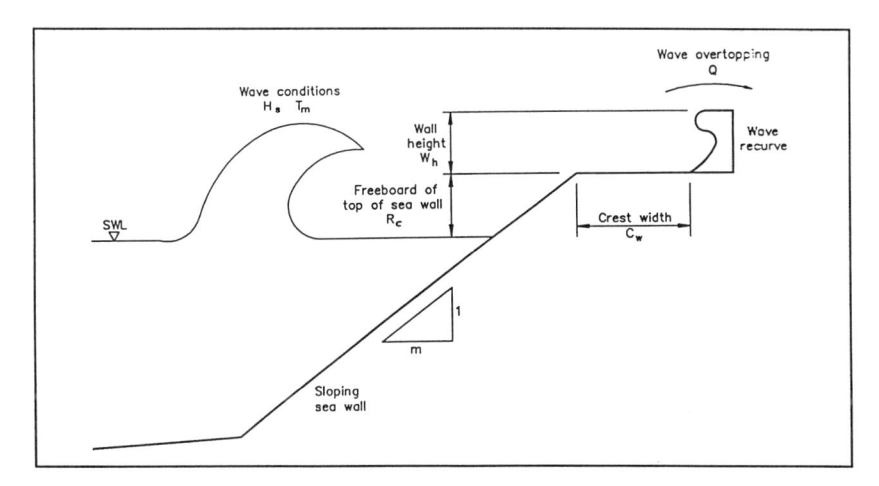

Figure 1   Definition of parameters, recurve wall

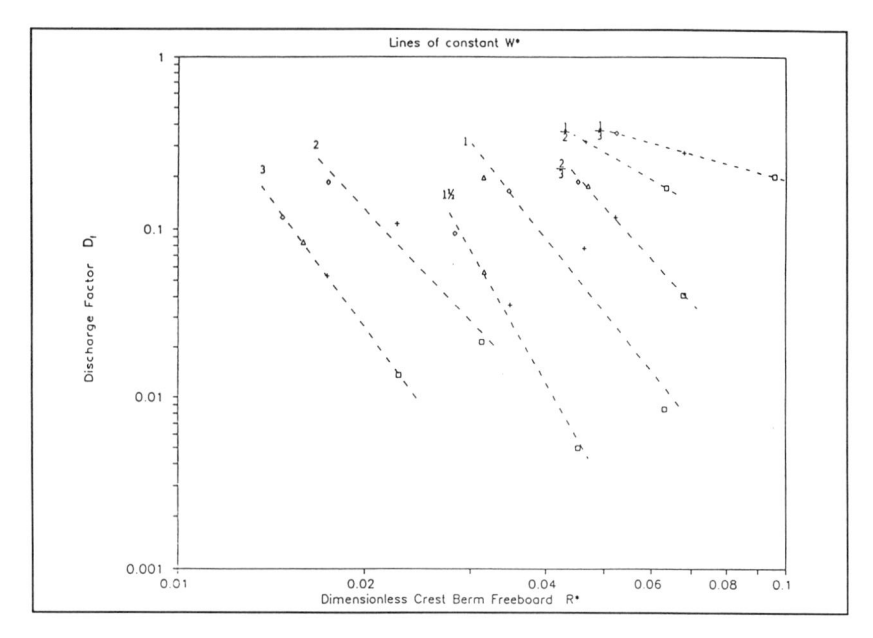

Figure 2   Recurve wall results, 1:2 slope, 0m crest width

For each test in the present study, the overtopping discharge to be expected without the wave return wall was calculated using equations (1) - (3). The measured discharge overtopping the wave return wall, expressed in dimensionless terms as Q_{*w}, could then be compared with the dimensionless discharge at the crest of the sea wall, Q_*, (ie the recurve has been removed) to give the discharge factor, D_f. Thus:-

$$Q_{*w} = Q_w / (T_m\, g\, H_s) \qquad (7)$$

$$D_f = Q_{*w} / Q_* \qquad (8)$$

where Q_w is the mean discharge overtopping the wave recurve.

In selecting a method of presenting the data, consideration was given to the way in which a designer could use the information. Figure 2 shows the form of presentation which was finally selected, in this case for a sea wall with a 1:2 slope and with the wave return wall placed directly at the top of the slope (ie the crest width, $C_w = 0$). In this graph the abscissa is the dimensionless crest berm freeboard, R_*, as defined in equation (2), which can be calculated from the actual freeboard and the wave height and period. Each line on the graph represents a constant value of the dimensionless wave return wall height, W_*, which can be determined from the wall height and the actual freeboard. Knowing the values of R∗ and W∗ allows a discharge factor D_f, to be established from Figure 2. Use of equation (3) to calculate the dimensionless discharge at the crest of the sea wall, Q∗, then enables Q_{*w} to be determined from equation (8). The mean discharge overtopping the wave recurve, Q_w, may be determined from equation (7).

The method outlined above allows the overtopping discharge of a sea wall with a recurve wall to be estimated provided that the crest width and sea wall slope are equal to one of those combinations tested. However, a single design graph would be preferable, together with some means of estimating the overtopping discharge for conditions not specifically tested. Given the scatter of results in Figure 2, and the fact that fewer than the ideal number of tests were completed for each structure cross-section, it was decided to investigate whether a standard slope could be fitted to all lines having the same dimensionless wall height.

All of the individual graphs were overlain and, using the 1:2 slope with a zero crest width as the baseline, the data sets displaced in the horizontal direction. With the appropriate displacements the individual W∗ data sets tended to collapse on to a straight line. An iterative procedure was used to find the displacements which, using the method of least squares to find the line of best fit, gave the highest overall coefficient of correlation for all the data sets. This overall coefficient of correlation was taken as the average of all the coefficients of correlation of all the data sets for different dimensionless wall heights.

The result of the analysis described above was a single design graph which is illustrated in Figure 3. In this figure the abscissa is the dimensionless adjusted crest berm freeboard, X∗, which is defined as follows:-

$$X_* = R_* A_f \qquad (9)$$

where A_f is an adjustment factor dependent on the structure cross-section.

Hence R∗ may be calculated from equation (2), whilst typical adjustment factors are given in Table 1. Thus a discharge factor can be obtained from Figure 3 and the overtopping discharge calculated as before (see equations (1), (7), and (8)).

The results of the model tests showed that recurved wave return walls can have a very dramatic effect on the overtopping discharges of sea walls. For some test conditions the discharge was reduced by almost three orders of magnitude compared to the expected situation without the return wall. Although some reduction would be obtained by simply raising the basic sea wall by the same amount as the height of the return wall, calculations indicated that only one order of magnitude reduction in overtopping could be expected. This point is well illustrated in Figure 4. For either a 1:2 or a 1:4 sea wall, the figure shows a plot of the overtopping discharge against the total height of the sea wall, for a

Sea wall slope	Crest width, C_w	$W_h/R_c \geq 2/3$ Adjustment factor, A_f	$W_h/R_c \leq 1/2$ Adjustment factor, A_f
1:2	0	1.00	1.00
1:2	4	1.07	1.34
1:2	8	1.10	1.38
1:4	0	1.27	1.27
1:4	4	1.22	1.53
1:4	8	1.33	1.67

Table 1   Adjustment factors

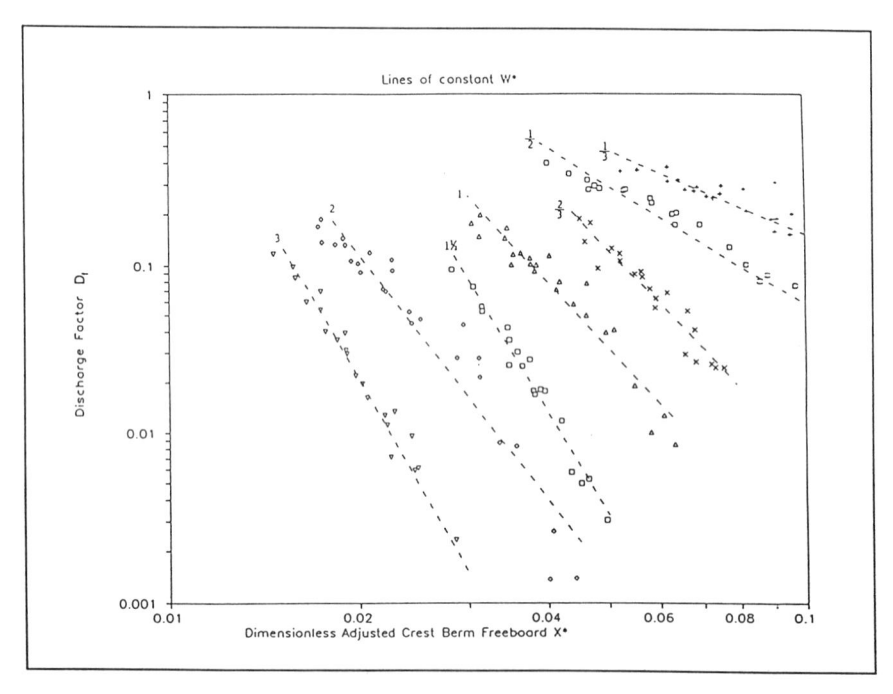

Figure 3   Design graph, recurve wall

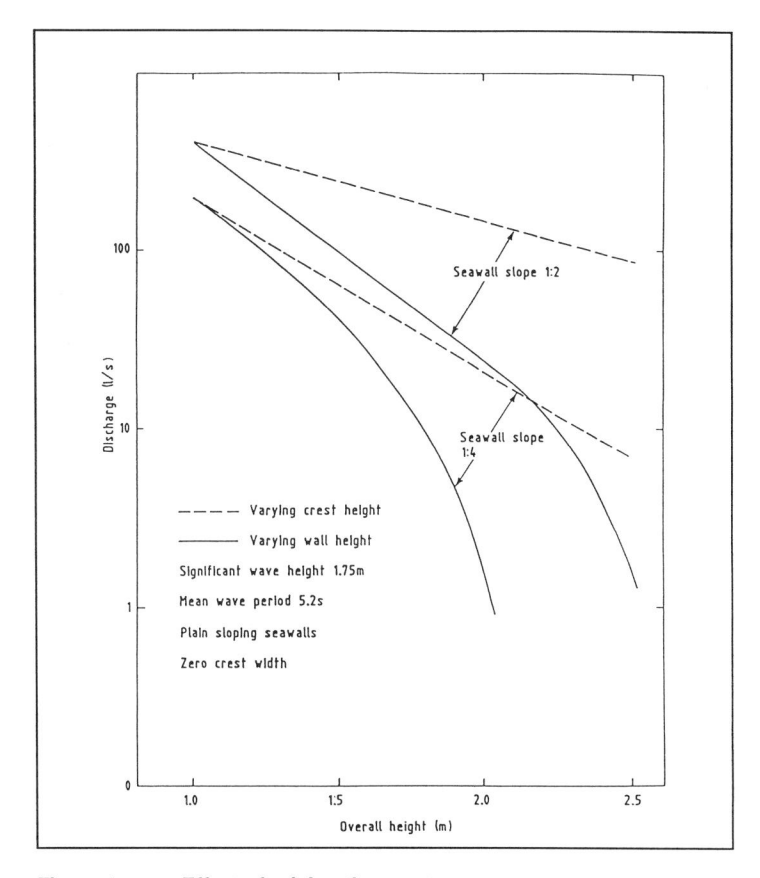

Figure 4   Effect of raising the crest

particular wave condition and water level. Starting from a crest elevation of 1.0m with no wave return wall, the solid lines show the reduction in discharge which is obtained by adding a return wall of gradually increasing height. The broken line shows the reduction obtained by raising the crest height, without any wave return. For a given total height of sea wall, the incorporation of a wave return wall greatly reduces the overtopping discharge compared to simply raising the crest.

The analysis outlined above was also applied to the data obtained by Allsop and Bradbury (1988) who measured the discharges overtopping a rock armoured slope topped with a crown wall. In all cases these tests were carried out for a seaward slope of 1:2 and a crest width equivalent to two rock diameters. The crown wall had a significantly less efficient profile than that proposed by Berkeley-Thorn and Roberts (1981) as used in the other recurved wall tests.

The results, which are illustrated in Figure 5, showed considerably more scatter than for the smooth impermeable slopes. This was thought to be due to the different degrees of energy absorption on the slope and of drainage into the crest for different wave and water level conditions. Also shown in Figure 5 are the discharge factors for the equivalent smooth impermeable slope with a crest width of $C_w = 4.0$ metres. The discharge factors for a return wall mounted on top of a rock slope are very much better (lower) than for a smooth slope despite the less effective recurve. The reduction in discharge factor must therefore be due to the effects of the rock slope.

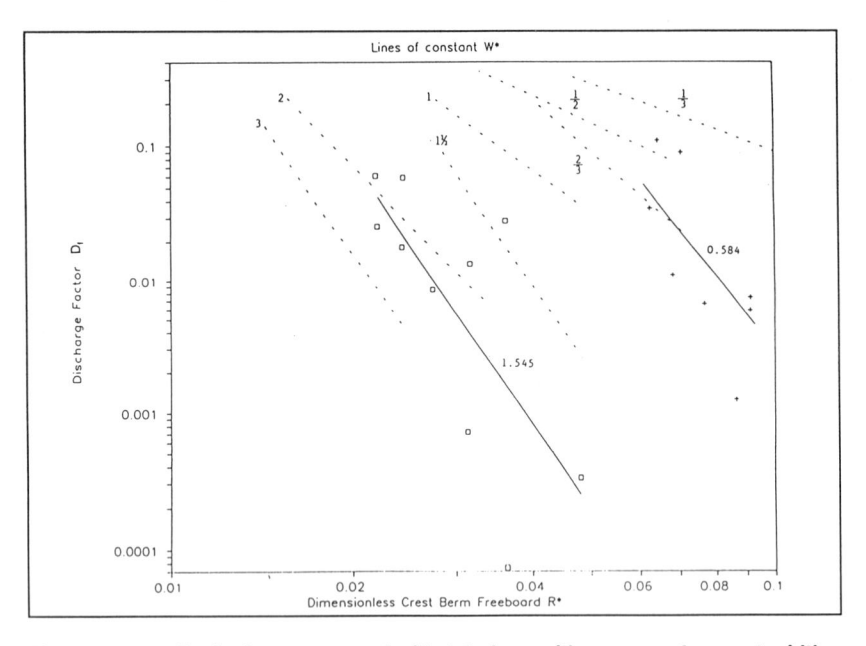

Figure 5   Rock slope compared with 1:2 slope with recurve, 4m crest width

The probable explanation for the lower discharge factors on the rock slope is as follows. As the wave runs up the slope and on to the crest, its forward progress is arrested by the return wall, increasing the depth of water on the crest. For an impermeable slope the remainder of the wave run-up to some extent rides over this cushion of water and a fraction overtops the wave return wall. On a permeable slope water reaching the return wall is able to drain away through the armour layer thereby limiting the depth of water at the crest. Hence wave run-up finds it more difficult to overtop a wave return wall on permeable than an impermeable slope.

Recent work has concentrated on assessing the efficiency of recurve walls under oblique wave attack. Physical model tests have been completed, at a scale of 1:25, using both short and long crested random waves with angles of wave attack ranging from 15° to 45° off normal. The preliminary analysis of the oblique data appears to indicate that discharge factors are larger, and hence the recurve is not as efficient, when compared to normal wave attack.

4 Vertical walls

Vertical or near vertical sea walls are common in urban areas and are often sited behind shingle or sandy beaches. Work detailing the overtopping performance of vertical walls had previously been completed by Goda (1975). Goda investigated approach slopes of 1:10 and 1:30 and offshore sea steepness of $s_{om} = 0.012, 0.017$ and 0.036. These conditions were considered to be unrepresentative of conditions around the UK coastline where the steepness of storm waves is greater and the bathymetry of approach is generally shallower.

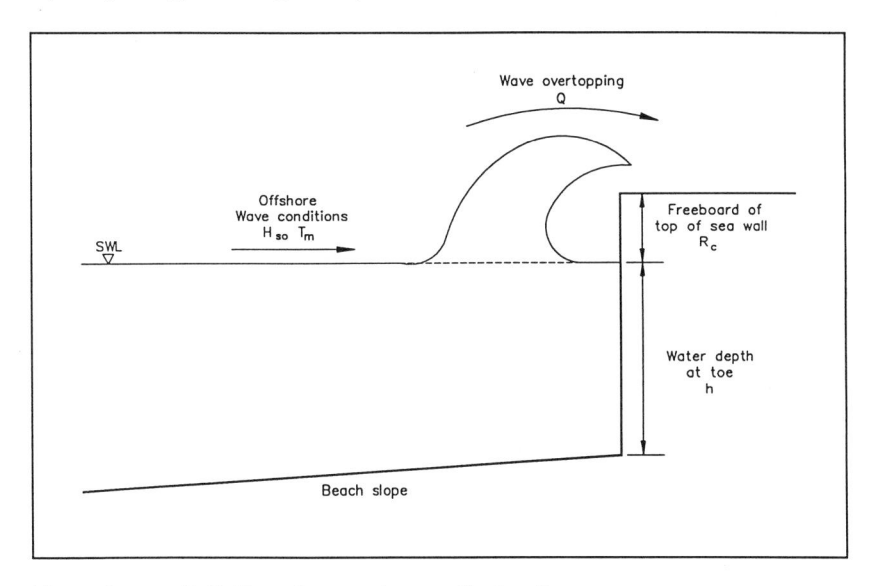

Figure 6   Definition of parameters, vertical wall

A series of physical model tests was therefore undertaken with the aim of confirming and extending the work of Goda so that it was more applicable to UK coasts (Herbert, 1993). Consequently three approach bathymetries of 1:10, 1:30 and 1:100 were used in the model with offshore sea steepness ranging from 0.017 - 0.060. Other parameters that were varied included the offshore wave height, the water depth at the toe of the sea wall and the freeboard of the wall. These parameters, which are illustrated in Figure 6, were varied to ensure that the model tests were completed in the zone of interest identified by Goda.

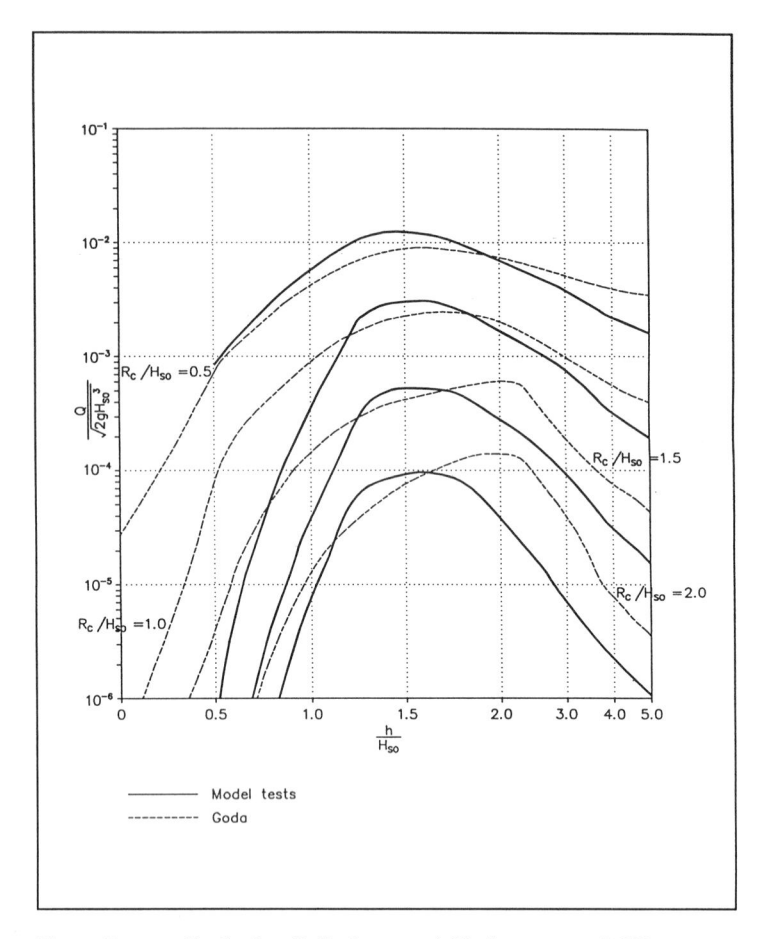

Figure 7 Vertical wall discharges, 1:30 slope, s_{om} = 0.036

The model data gave good agreement with the work of Goda. This is illustrated in Figure 7 where a dimensionless discharge, $Q^{\#}$, is plotted against a dimensionless water depth, h/H_{so}, where:-

$$Q^{\#} = Q / (2gH^3_{so})^{\frac{1}{2}} \qquad (10)$$

H_{so} is the offshore significant wave height
and h is the water depth at the toe of the structure.

Lines of constant values of the dimensionless freeboard, R_c/H_{so}, are illustrated on the graph where R_c is the height of the crest above still water level.

For a given dimensionless freeboard, maximum overtopping discharges occurred when $1.4 < h/H_{so} < 2.0$. These conditions correspond to waves breaking immediately seaward of the structure toe. The breaking waves often pass directly over the crest of the seawall. For $h/H_{so} < 1.4$ waves break before they reach the vertical wall. A considerable amount of energy is dissipated during breaking and hence overtopping is reduced. If $h/H_{so} \ll 1.4$ the waves break farther offshore and overtopping is further reduced. Conversely, for $h/H_{so} > 2.0$, waves are unbroken when they reach the structure and this also leads to a reduction in peak overtopping discharge. As waves travel into shallower water they steepen before breaking. When the water depth at the structure is large relative to H_{so} little or no shoaling occurs and overtopping is commensurately lower. For unbroken waves, as the h/H_{so} ratio increases the level of shoaling, and hence overtopping, is reduced. Eventually the effect of water depth at the structure and bed slope will become insignificant and overtopping will be a function of wave height and freeboard only. Therefore, for large vales of h/H_{so}, overtopping discharges will approach a constant value for a given dimensionless freeboard, R_c/H_{so}.

5 Further work

A fieldwork deployment, designed to measure overtopping discharges at prototype sites, has been completed on the North Wales coast. This work is being analysed and compared with results from physical model tests. Furthermore measured discharges are being compared with present guidance on allowable overtopping with particular reference to vehicular and pedestrian safety.

Work is presently being undertaken to assess the performance of a wide range of sea wall cross-sections under oblique wave attack. The structure cross-sections, which are being physical model tested using both long and short crested waves, include simply sloping sea walls with and without recurved walls, bermed sea walls and vertical walls.

A design manual is being planned which will describe and detail the overtopping performance of sea walls and the standards to which they should be designed. It is anticipated that draft copies of this manual will be available in the latter part of 1995.

6 Acknowledgements

The authors would like to thank the Ministry of Agriculture, Fisheries and Food for their continued support for the work described in this paper. The work of Mr A A J Steele, who performed the majority of the experimental work, is also gratefully acknowledged.

7 References

Allsop N W H and Bradbury A P. 'Hydraulic performance of breakwater crown walls'. HR Wallingford, Report No. SR146, March 1988.

Berkeley - Thorn, R and Roberts A C. 'Sea defence and coast protection works'. Published by Thomas Telford Ltd, 1981.

CIRIA. 'Manual of the use of rock in coastal and shoreline development'. CIRIA Special Publication 83/CUR Report No. 154, 1991.

Goda, Y, Kishira, Y, and Kamiyama, Y. 'Laboratory investigation on the overtopping rates of seawalls by irregular waves'. Ports and Harbour Research Institute, Vol 14, No. 4, pp 3-44, 1975.

Herbert D M. 'Wave overtopping of vertical walls'. HR Wallingford, Report No. SR316, February 1993.

Owen M W. 'Design of sea walls allowing for wave overtopping'. Hydraulics Research Station, Report No. EX924, June 1984.

Owen M W and Steele A A J. 'Effectiveness of recurved wave return walls'. HR Wallingford, Report No. SR261, February 1991.

CHAPTER 83

Stability of High-specific Gravity Armor Blocks

Masahiro Ito[1], Yuichi Iwagaki[1], Hiroshi Murakami[2]
Kenji Nemoto[3], Masato Yamamoto[3] and Minoru Hanzawa[3]

Abstract

The stability coefficient of the well-known the Hudson formula, K_D, has been established by many laboratory tests for various types of armor blocks with a range of specific gravity 2.16~2.47 representative of normal concrete. Design manuals for coastal structures do not describe the adoption of high-specific gravity armor blocks for use in field construction. To examine the effect of specific gravity on stability of Tetrapods, then a laboratory test was conducted by using Tetrapod models of different weights which are composed of five values of specific gravities. Temporal changes of damage ratio arranged test conditions that satisfy the Hudson formula do not coincide among each other, depending on wave period and specific gravity. The effect of the specific gravity on their stability is discussed. As a result, it is found that the relationship between the stability and the specific gravity is considerably affected by the surf similarity parameter and Reynolds number.

Introduction

The stable weight of armor units which are used in the cover layer of groins, breakwaters and wave absorbing works is evaluated by the Hudson formula (1959). The Hudson's stability coefficient of armor blocks made from a normal concrete with specific gravity of about 2.30, such as Tetrapod and Dolos, have been determined on the basis of laboratory tests. These tests have been performed using a wide range of experimental conditions for wave height, wave period, and structure slope, using regular waves. As for the use of the armor blocks on beaches and coast lines, there are the following problems and needs:

---

[1] Professor, Department of Civil Engineering, Meijo University, Shiogamaguchi 1-501, Tempaku-ku, Nagoya 468, JAPAN.
[2] Doctor Course, Department of Civil Engineering, Meijo University.
[3] Nippon Tetrapod Co. Ltd.

armor blocks, that is to say, the maximum weight of Tetrapod as a armor block is 784 kN (80 tf) at the present time, but this weight becomes unstable for wave heights over 9 m for seaward slopes of 1 on 4/3 of typical breakwaters.

(2) When large scale construction equipment such as floating cranes or derrick barges are difficult to move the place such a far detached-island, then small size and more stable armor blocks have to be used.

(3) Construction site with limited storage space for armor blocks.

(4) Armor blocks in a corner or head of breakwaters, groins or wave absorbing works receive the complex wave action due to the combined wave-flow induced due to waves. Then, thus parts a weak point in armor blocks and are easily broken.

(5) To hold the natural beauty of beaches, coastlines, and harbors, smaller and stable armor blocks are required.

For these problems and needs, high-specific gravity armor blocks are considered more suitable. These armor blocks are smaller than the same weight units made from normal concrete. They are more stable against waves. Also it is possible to reduce structure size by using these armor blocks.

A manufacturing method for producing high specific gravity armor blocks is made possible by mixing a heavy stone, or crushed or grained iron ore such as pyrite into the concrete. The "heavy stone" concrete reaches specific gravity of about 2.7~2.8. In the iron ore concrete mixture, the specific gravity becomes greater than 3.0. Such high-specific gravity armor blocks have been already put to a practical use in Japan.

Previous investigation of the high-specific gravity armor blocks, Zwamborn (1978) discussed the effect of specific gravity as the ratio of armor unit volume to wave height based on the past laboratory data that had been performed using rubble- stones of different specific gravity by Kydland and Sodefjed, and dolosse of different specific gravity by Gravesen and Sørensen. Zwanborn reported that by rearranging some test data with dolosse of different specific gravities (2.4, 2.75 and 3.05), the dimensionless wave height, $(V/H^3)\cot\theta$, based on block volume, V, wave height, H, and slope angle of armor layer, θ, is proportional to the specific gravity in water to -3.0, -3.0 and -4.0 power for 1, 2, and 5 per cent damage, respectively. But in the Hudson formula, its exponent value is constant (-3rd power) and well accepted. The authors (1992a, 1992b, 1994) have been carrying out the laboratory study using Tetrapod models of various sizes and specific gravities in order to investigate the effect of specific gravity on the stability for waves. On the basis of the tests results, this paper discusses the effect of specific gravity, surf similarity parameter, and Reynolds number as a function of the block-size and wave characteristic. Recently, Takeda et al. (1993) discussed the relationship between stability and specific gravity, based on the laboratory tests using high-specific gravity armor blocks. He pointed out that the wave period considerably affects the stability-specific gravity relationship.

Thus, investigations on the stability of high-specific gravity armor blocks have been not insufficient up until the present time. Design manuals for coastal structures do not describe the adoption of high-specific gravity armor blocks for use in field construction. The effect of the specific gravity on armor unit stability, therefore, must be further investigated.

Advantages of high-specific gravity armor blocks

The effect of the specific gravity on the resistance against wave action is possible to be calculated using the Hudson formula. Assuming that the Hudson's stability coefficient K_D is a constant for Tetrapod (8.3), the slope angle of armored structure is a constant, the unit weight of normal concrete w_c is 22.54 kN/m^3 (2.3 tf/m^3), and the unit weight of water w_c is 9.8 kN/m^3 (1 tf/m^3), then, we can obtain the ratio of the design wave height for high-specific gravity armor block, H/H_c to the change of the relative specific gravity against the normal concrete, S_r ($=w_r/w_w$), from the Hudson formula assuming that both blocks are the same size (l_B=const.). Where w_r, w_c and w_w are the unit weight of any specific gravity armor units, normal concrete, and water, respectively. The effect of S_r on H/H_c, the resistance against wave action is shown with a solid curve in Fig. 1. The results show the relative block weight W/W_c of high-specific gravity to normal concrete, which is also shown with a broken line curve in Fig. 1. Here H and H_c are the design wave height of any specific gravity and normal concrete, respectively. It can be seen from this figure that H in the armor blocks of specific gravity S_r=3.0 is possible to resist about 1.5 times wave height, H_c, for the normal concrete of the same size. However, this finding holds only when K_D-values are the constant though the specific gravity ranges.

Laboratory test

To examine the effect of specific gravity on stability of Tetrapod, models of different weight were used, composed of five different values of specific gravities (1.82, 2.30, 2.77, 3.40, 4.27). A total of 21 different Tetrapod models were used. The range of Tetrapod models used varied between a vertical height of l_B=3.29~16.6 cm and a weight of W=0.27~22.85 N (27.7~2329.6 gf). The sizes and weights of Tetrapod models used in the laboratory tests are listed in Table 1. Model weights on each row (A, B, C, D, and E) within this table are obtained for each specific gravity

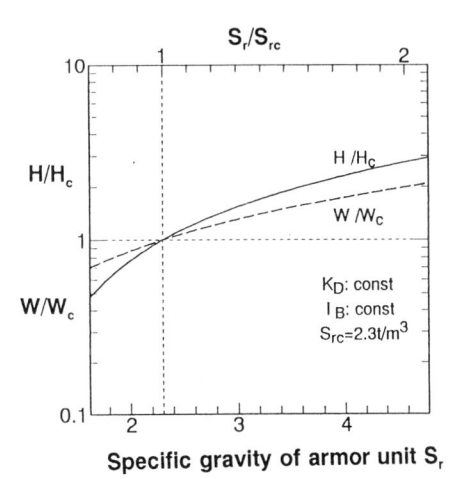

Fig. 1. Effect of specific gravity on stability of armor blocks for waves

Table 1. Specific gravities, sizes and weights of Tetrapod models

Marks of model	Class Nos. of specific gravity	5	4	3	2	1	Design Wave height at 1%-damege H(cm)
	W_r/W_w	1.82	2.30	2.77	3.40	4.27	
A	ℓ_n (cm)	7.16	4.52	3.29			8.6
	W (gf)	186.4	58.9	27.7			
B	ℓ_n (cm)	9.04	5.68	4.14			10.8
	W (gf)	372.7	117.8	55.4			
C	ℓ_n (cm)	12.24	7.72	5.68	4.14	3.08	14.7
	W (gf)	931.8	294.4	141.8	68.0	34.2	
D	ℓ_n (cm)	16.62	10.48	7.72	5.68	4.14	19.9
	W (gf)	2329.6	736.0	354.6	174.1	85.4	
E	ℓ_n (cm)		14.15	10.48	7.72	5.68	26.9
	W (gf)		1811.0	886.4	435.2	218.6	
F	ℓ_n (cm)		8.28				15.7
	W (gf)		368.0				

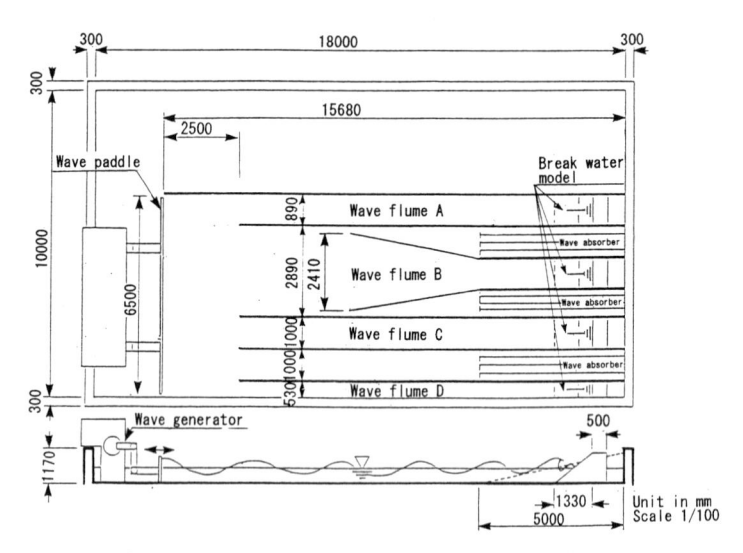

Fig. 2. Wave flumes to minimize the multiple wave reflection

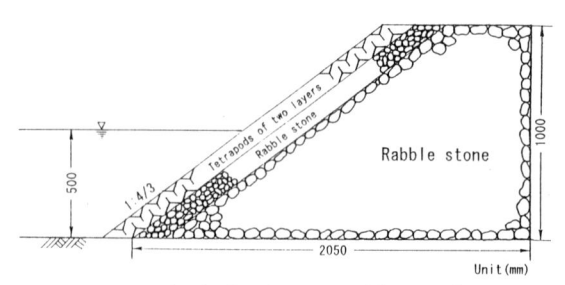

Fig. 3. Breakwater model armored with Tetrapods of two-layer

Table 2. Experimental conditions

Waves	Water depth (cm)		50, 60
	Wave height (cm)		5~40
	Period (sec)		1~3
	Wave steepness Ho/Lo		~0.14
Tetrapods	Specific gravity w_r/w_w		1.82, 2.30, 2.77 3.40, 4.27
	Weight (gf)		34.2~2329.0
	Vertical height (cm)		3.08~16.6.0
Breakwater	Front slope of cover layer		1:4/3
Slope of Sea-floor			horizon
Wave operation time (min) (Wave numbers t/T)			50~70 (80~2000)

by giving K_D=8.3 (in 0<Damage ratio≤1%), cotθ=4/3, w_w=9.8 kN/m³ (1 tf/m³), and given a constant Hudson formula design wave height. The design wave height of the models in each row of Table 1, becomes nearly equal, despite difference in specific gravity.

As shown in Fig. 2 the wave basin was divided into the seven flumes, each having a free water zone of 2.5 m in front of the wave paddle. Model breakwaters used in the stability test were constructed within every other flume. The control flumes had wave absorbing net-mats, which were set on a slope of 1 on 10 used to minimize the multiple wave reflection which takes place between wave paddle and breakwater model in normal flumes. In the laboratory test, regular waves were generated continuously on the Tetrapod model for a total of about 1,500 waves under no-overtopping conditions. The 1,500 waves are equivalent to prototype wave action of about 4 hours for a wave period of 10 sec. Wave flume B was constructed with a geometry to increase wave height. Each Tetrapod model was constructed randomly in two layers, on a slope of 1 on 4/3 as shown in Fig. 3. Laboratory test conditions are summarized in Table 2. The damage progression of Tetrapod models was observed continuously by a 8 mm Video-camera. The damage rate, defined as the ratio of the number of blocks that moved to the total number of armor blocks in the two layers, was measured with a display playing back the Video.

Test results

A temporal damage change of Tetrapod models attributed to the difference of specific gravity based on test data, is shown in Fig. 4. The damage to the Tetrapod models of each different specific gravities should be similar to each other. These Tetrapod models (A-3, A-4, A-5) are listed the same row of Table 1. The design wave height of these armor units is nearly equal. However, these damage curves differ due to specific gravity. As to a reason for the difference,

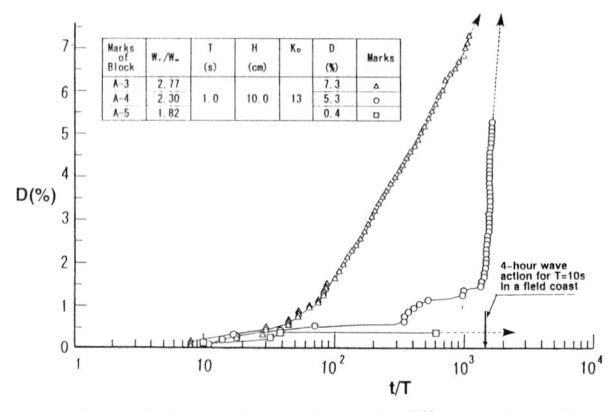

Fig. 4. Temporal damage changes due to the difference of specific gravity

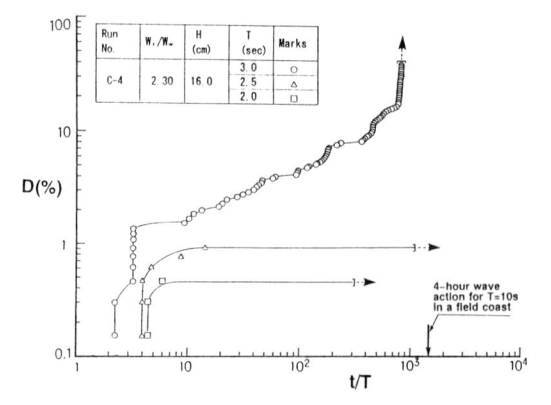

Fig. 5. Temporal damage changes due to the difference of wave period

Table 3. Damages of detached breakwaters, groins, and seawall with Tetrapods due to storm waves in Japan

Locations of port and harbour in Japan	W (tf)	T (s)	Damage ratio (%)	$H_{1/3}$ (m)	K_D	R_e	ξ	Remarks
Hamana	3	13	50	5.6	64.6	6.7E+06	4.3	Typhoon 7219
Hamada	20	13	50	8.0	23.6	1.6E+07	3.6	Depression, Jan. 1971
Simoda	8	12	30	5.0	14.4	9.3E+06	4.2	Typhoon 7010
Simoda	16	12	30	5.0	7.2	1.2E+07	4.2	Typhoon 7020
Nobeoka	13	14	50	4.0	4.7	9.6E+06	5.4	Typhoon 7123
Miyazaki	8	14	50	4.0	7.4	8.3E+06	5.4	Typhoon 7119
Miyazaki	5	14	50	4.0	11.8	7.1E+06	5.4	Typhoon 7119

considerations must be given to the fact that stability is affected by the specific gravity. Selecting identical wave heights from the test data in the same size Tetrapod model, a temporal damage change is shown for each wave period in Fig. 5. Damage curves, generally, must be similar to each other regardless of the difference in wave periods. Hudson's (1959) stability coefficient is independent of wave period. But from Fig. 5 it is recognized that stability becomes a function of the wave period.

From field data, we can investigate damage ratio, significant wave height, and block weight of Tetrapod. The technical report by Takeyama and Nakayama (1975) reported on damage of failure of coastal structures by large waves in Japan. These field data are listed in Table 3. The field data includes the effects of oblique incident waves, deflection waves, and scour in or around the toe of breakwater slope. The damage ratios shown in this table, then, are considerable large (range from 30% to 50%). Hudson's stability coefficient in this table is obtained by converting wave height to equivalent significant wave height. Regular waves were converted based on the equation, $H = 1.5 H_{1/3}$. This equation follows the equations proposed by Fan et al. (1983) and Tanimoto (1985) as

$H = 1.4 H_{1/3}$ (Fan et al.)   (1)
$H = (1.5 \pm 0.37) H_{1/3}$ (Tanimoto et al.)   (2)

The evaluated surf similarity parameter ξ and Reynolds number R_e are also listed in Table 3.

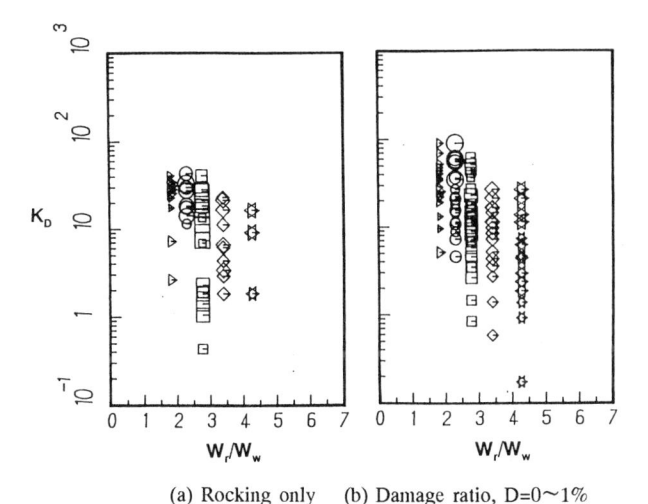

(a) Rocking only   (b) Damage ratio, $D = 0 \sim 1\%$

Fig. 6. The relationship between K_D-values and specific gravity in two types of damages

Effect of specific gravity on stability

The rocking only (no-damage) and the 0~1 % damage are selected from the laboratory test data. Fig. 6 shows these data which show the relationship between Hudson's stability coefficient, K_D, and the specific gravity in water, S_r-1. In this figure, the K_D-values scatter wildly for the difference in the specific gravity in water. The effect of the specific gravity on stability (rocking only and 0~1% damage) can not be clearly established from this figure. In the Hudson formula, the stability coefficient K_D should be constant because it should considerably depend on the shape of armor units. Therefore, the stability coefficient must be independent of the change of specific gravity. But K_D-values scatter wildly for the different S_r-1 values as shown in Fig. 6. Then the Hudson formula may not satisfy the test data.

Theoretical discussion

As above discussed, the effect of the specific gravity on the stability disagrees with the Hudson formula. As this reason it is considered that many factors affect Hudson's stability coefficient, including effects due to wave steepness, relative depth, Reynolds number, drag coefficient, acceleration of wave motion etc. as described by Hudson (1959).

Reviewing the Hudson formula, the stability coefficient K_D is expressed by;

$$K_D = \frac{w_r H^3}{W(S_r-1)^3 \cot\theta} \qquad (3)$$

where, W is the weight of armor unit, w_r is the unit weight of armor unit, H is the design wave height, S_r (=w_r/w_w) is the specific gravity of armor unit, w_r is the unit weight of armor unit, w_w is the unit weight of water, θ is the angle of structure slope measured from horizontal in degrees, and K_D is the stability coefficient varying primarily with the shape of the armor units, degree of interlocking, damage rate, etc.

The behaviors of armor units is affected by slope angle, wave breaking, wave runup, and rundown, and is sensitive to the surf similarity parameter ξ,

$$\xi = \frac{\tan\theta}{\sqrt{H/L_o}} \qquad (4)$$

where L_o is the wave length in deep water.

Hudson (1959), Dai and Camel (1969), and Sakakiyama and Kajima (1990) have pointed out that the stability of armor units is affected by the Reynolds number related the scale of laboratory tests. The Reynolds number is expressed using the velocity of wave motion and a characteristics length of the armor block as

$$R_e = \frac{\sqrt{gH}(W/w_r)^{1/3}}{\nu} \tag{5}$$

where ν is the kinematic viscosity, g is the acceleration due to gravity. Rewriting Eq. (3) using Eqs. (4) and (5), the stability coefficient K_D is expressed as follows:

$$K_D = \frac{R_e^2}{\xi^4(S_r-1)^3} C_R \tag{6}$$

where

$$C_R = \frac{L_o^2 \nu^2}{g}(\frac{w_r}{W})^{\frac{5}{3}} \tan^5\theta \tag{7}$$

From Eq. (6), it is shown that the stability coefficient K_D is proportional to the −3rd power of the specific gravity in water, the −4th power of the surf similarity parameter, and the 2nd power of the Reynolds number.

Stability and Surf similarity parameter

Fig. 7 shows the relationship between K_D-values and the surf similarity parameter for the test data and the field data listed Table 3. In this Figure, the K_D-

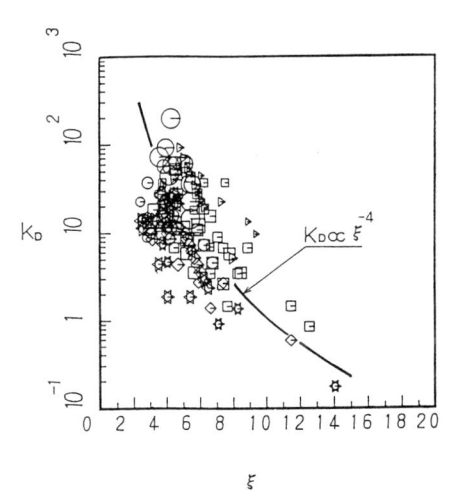

Fig. 7. The relationship between K_D-values and surf similarity parameter of test data and field data

(a) Specific gravity: 1.82

(b) Specific gravity: 2.30

(c) Specific gravity: 2.77

(d) Specific gravity: 3.40

(e) Specific gravity: 4.27

Fig. 8. The relationship between K_D-values and surf similarity parameter for each specific gravity

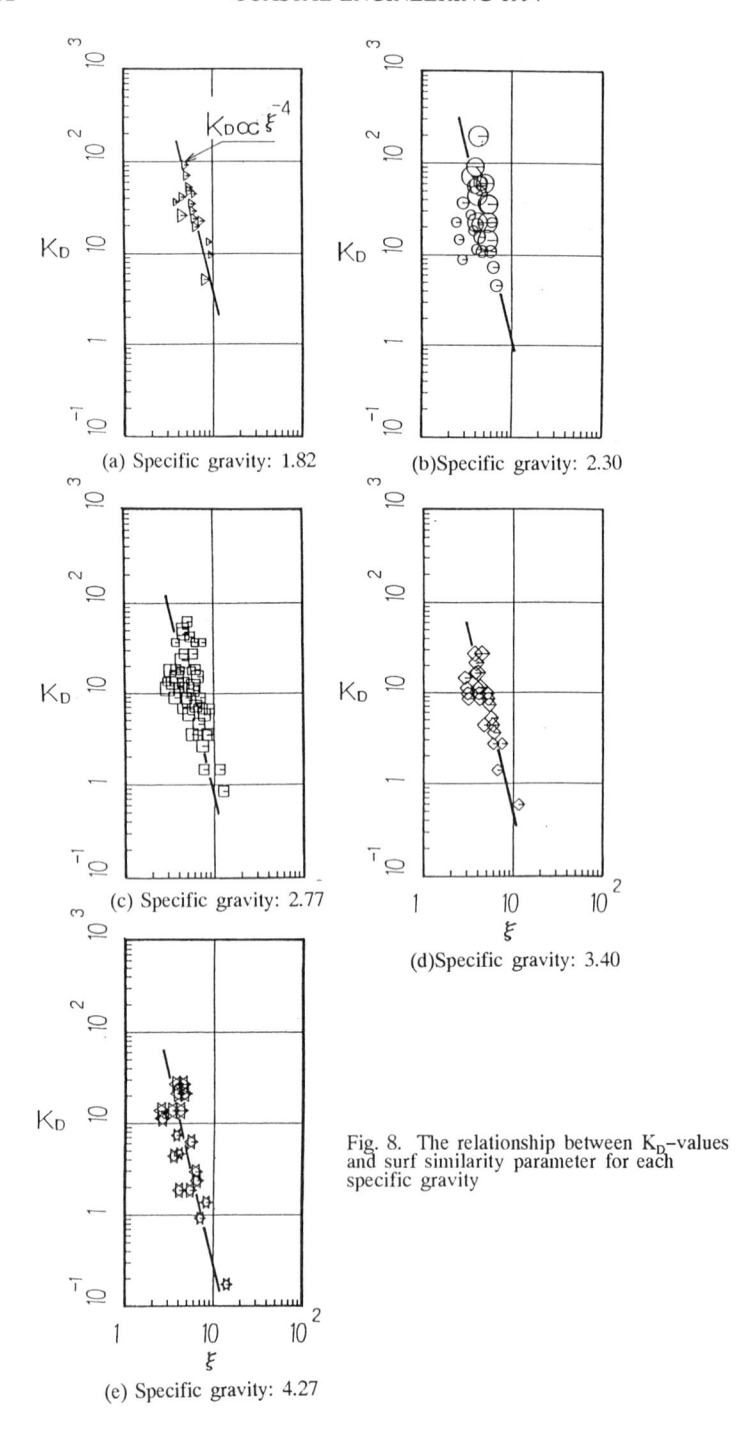

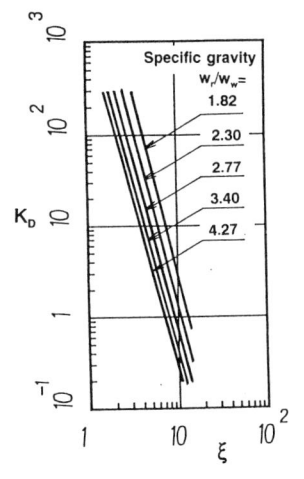

Fig. 9. The relationship between K_D-values and surf similarity parameter due to the difference of specific gravity

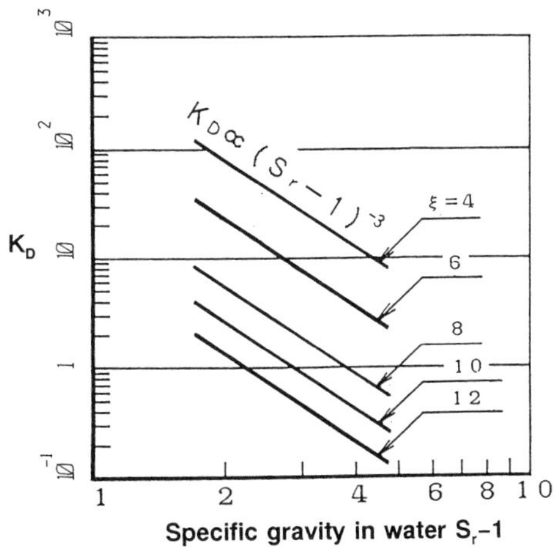

Fig. 10. The relationship between K_D-values and specific gravity in water due to the difference of surf similarity parameter

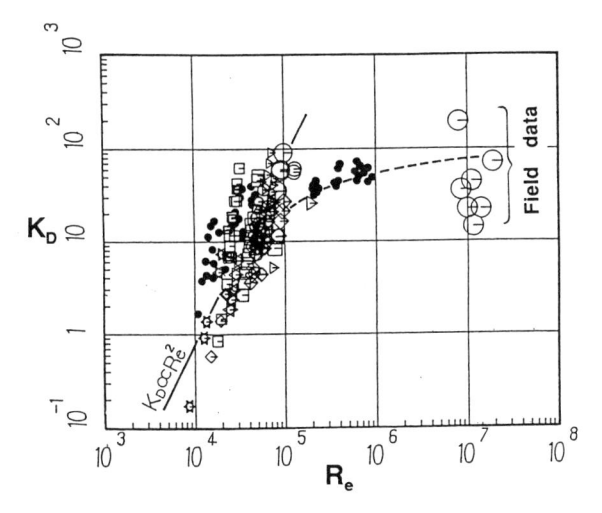

Fig. 11. Effect of Reynolds number on K_D-values

values agrees well with the solid curve expressing the relationship of $K_D \propto \xi^{-4}$ in Eq. (6). The relationship between the K_D-values and the surf similarity parameter for each specific gravity are shown in Fig. 8(a)~(e), respectively. From these figures, the relationship between the K_D-values and the surf similarity parameter agrees well the solid line indicating the relationship of $K_D \propto \xi^{-4}$. Fig. 9 shows the relation of ξ to K_D arranged together from Fig. 8(a) to Fig.8(e) using the specific gravity as a parameter. Fig. 10 shows the relationship between the stability coefficient and specific gravity in water. This figure was obtained by rearranging Fig. 9 taking the surf similarity parameter as a parameter. From Fig. 10, it can be seen that the stability coefficient decreases with increasing specific gravity in water dependent on $K_D \propto (S_r-1)^{-3}$, as indicated in Eq. (6). From Fig. 10, stability is affected by the surf similarity parameter, and increases with decreasing the surf similarity parameter.

Sawaragi et al. (1983) pointed out that Iribarren's number reaches a minimum value around the surf similarity parameter of $\xi \approx 3$ due to the occurrence of wave-armor unit resonance on the basis of the laboratory tests of rubble mound breakwaters. Losada (1990) discussed the effect of the surf similarity parameter on the stability function, $\varphi=1/(K_D\cot\theta)$, by rearranging laboratory test data for various armor units. Then he pointed out that the stability function increases inversely proportional to decreasing surf similarity parameter. These findings are contrary to authors results as above discussed. The discrepancy between the authors and Losada may cause the reason that the range of the surf similarity parameter used and the size of armor unit model is not in agreement.

Stability and Reynolds number

It can be considered from Eq. (6) that the stability coefficient K_D is affected by the Reynolds number. Then, the relationship between the Reynolds number and the K_D-values of the zero damage test data and the field data listed in Table 3, is plotted in Fig. 11. This figure is also plotted with the dark circle indicating Tetrapods only from the paper by Simada et al. (1986), who investigated the stability of Tetrapods, Dolosse, and Koken-blocks. Theses armor units are made from normal concrete ($w_c/w_w=2.30$). The armor blocks had a weight range of W=0.16~484 kN (16 gf~49.39 kgf), and were tested in the medium and large wave tanks. It is shown from Fig. 11 that the K_D-values increase rapidly with increasing Reynolds number, up to $R_e=10^5$ which corresponds to small and medium test scales in the range of Reynolds number $R_e=8\times10^3\sim10^5$. Over this value the K_D-values gradually reach a constant. The K_D-values fit well with the solid line indicating the relation of $K_D \propto R_e^2$. The K_D-values beyond $R_e=10^5$ deviates from the trend depicted by the solid line. Such changes in K_D-values may be caused by scale effects. This has been indicated by Dai and Camel (1969) and Shimada et al. (1986). A wider range of scales and numerous test data are needed to further the understanding of the affects of specific gravity and the surf similarity parameter as they related to K_D-values and the Reynolds number.

Conclusions

The effect of the specific gravity on the hydraulic stability of Tetrapod is examined on the basis of numerous laboratory test data. These data were obtained using the total of 21 different models composed five values of specific gravity;

1.82, 2.30, 2.77, 3.40, and 4.27. The discussion includes test and field data reported by another researchers. Conclusions drawn from the results obtained in this study are as follows:

(1) Hudson's formula does not include the effect of wave period. Then Hudson's stability coefficient must be independent of wave period. But this coefficient is considerably influenced by weave period.

(2) Trends are not seen for the relationship between K_D-values and specific gravity. But by arranging the relationship as a function of the surf similarity parameter, it is found that the K_D-values decrease with increasing surf similarity parameter, where $K_D \propto \xi^{-4}$.

(3) The relationship between the K_D-values and the specific gravity in water is shown by the relation, $K_D \propto (S_r - 1)^{-3}$, when the surf similarity parameter is taken into account. The K_D-values decrease with increasing specific gravity.

(4) K_D-values from the test data correspond well to the 2nd power of the Reynolds number within a range of Reynolds number between $R_e = 8 \times 10^{-3} \sim 10^5$. In $R_e \geq 10^5$, the K_D-values gradually reach a constant as Reynolds number increases. Therefore the stability of high-specific gravity armor blocks, is affected significantly by scale effects.

It should be remarked that the conclusions of (1), (2), and (3) are results that are discussed on the basis of the data of small and medium test scale included the scale effects.

Acknowledgement

The authors would like to thank Mr. George F. Turk, Research Hydraulic Engineer of U.S. Army Coastal Engineering Research Center for proofreading the draft paper. We also grateful to the undergraduate and graduate students of Department of Civil Engineering at Meijo University for their kind assistance in the laboratory test.

References

Bandtzaeg, A. (1978). "The effect of unit weights of rock and fluid on the stability of rubble mound breakwaters." *Proc. 16th Int Conf. on Coast. Engrg*, ASCE, 990–1021.

Dai, Y.B. and Kamel, A.M. (1969). "Scale effect tests for rubble–mound breakwaters hydraulic model investigation." Research Report H–69–2.

Fan, Q., Horikawa, K. and Watanabe, A. (1983). "Experimental study on the stability of a concrete block mound breakwater under irregular waves." *Proc. of 30th Coastal Eng. Conf.*, JSCE, 352–356, (in Japanese).

Gravesen, H. and Sørensen T. (1977). "Stability of rubble mound breakwaters." *Proc. 24th Int. Nav. Congress*, PINAC, Leningrad.

Hudson, R. Y. (1959). "Laboratory investigation of rubble–mound breakwaters." *J. Wrtway. and Harb. Div.*, ASCE, 93–118.

Ito, M., Iwagaki, Y., Yamada, T., Nemoto, K., Yamamoto, H. and Hanzawa, M. (1992a). "An influence of surf similarity parameter on stability of high–specific gravity armor units." *Proc. of Coastal Engineering*, JSCE, Vol.39, 666–670, (in Japanese).

Ito, M. Iwagaki, Y., Yamada, T., Nemoto, K., Yamamoto, H. and Hanzawa, M. (1992b). "On the effect of specific-gravity change on stability in armor units". *Proc. of Civil Engineering in Ocean*, JSCE, Vol.8, 75–80, (in Japanese).

Ito, M., Ogawa, K and Murakami, H. (1994). "An experimental study on the vibration of armor unit by wave action". *Proc. of Civil Engineering in Ocean*, JSCE, Vol.10, 171–175, (in Japanese).

Losada, M. A. (1990). "Recent Development in the Design of Mound Breakwaters". *Handbook of Coastal and Engineering*, Vol.1, Herbich, J.B. Ed., Gulf Publishing Co.

Sakakiyama, T. and Kajima, R. (1990). "Scale effect of wave force on armor units." *Proc. 22nd Int Conf. on Coast. Engrg.*, ASCE, 1716–1729.

Sawaragi, T., Iwata, K. and Ryu, C. (1983). "Consideration of the destruction mechanism of rubble mound breakwaters due to resonance phenomenon." *Proc. 8th Int. Harbor Congress*, 3.197–3.208.

Shimada, M., Fujimoto, T., Saito, S., Sakakiyama, T. and Hiraguchi, H. (1986). "Scale effect on stability of amour units." *Proc. of 33th Coastal Eng. Conf.*, JSCE, 442–445, (in Japanese).

Takeda, H, Yamamoto, Y., Sasajima, T., Kikuchi, S. and Mizuno, Y. (1993). "Stability of high-specific gravity artificial concrete blocks by irregular-wave test." *Proc. of Civil Engineering in Ocean*, Vol. 9, 313–318, (in Japanese).

Takeyama, H. and Nakayama, T. (1975). " Disasters of breakwaters by wave action." *Technical Note*, No.200, Port and Harbor Res. Inst. Ministry of Trans., Japan, (in Japanese).

Tanimoto, K., Haranaka, S. and Yamazaki, K. (1985). "Experimental study on the stability of wave dissipating concrete blocks against irregular waves." *Report of the Port and Harbor Res., Ins.*, Vol.24, No.2, 86–121, (in Japanese).

Zwamborn, J.A. (1978). "Effect of relative block density." *Proc. 16th Int. Conf. Coast. Engrg.*, ASCE, 2285–2304.

CHAPTER 84

Rock armoured beach control structures on steep beaches

Jones R.J.[1] & Allsop N.W.H.[2]

ABSTRACT

This paper presents the results of a research study to quantify the stability of rock armoured groynes on steep shingle beaches. The research was needed to identify why significant numbers of rock revetments and groynes have suffered greater damage than would be predicted by conventional design methods. The paper describes the design and execution of tests in a large wave basin, at notional scales of 1:10-20, on four types of typical beach control structures. The test structures were constructed on a model shingle beach of slope 1:7, and were subjected to random waves of steepnesses s_m=0.02 or 0.04.

The tests confirmed that armour damage may be substantially greater than predicted by existing methods, even on simple slopes under normal wave attack. The results of the damage analysis have been used to suggest modified coefficients to van der Meer's plunging wave formulae.

1. INTRODUCTION

The use of shingle beaches in a coastal defences increasingly requires control structures to maintain and retain the beach. A variety of structures are in use (Fig 1), including rock groynes and/or breakwaters, and revetments. The main structure types in use, not necessarily in the UK, may be summarised:

    a)   near-shore, detached, breakwaters;
    b)   low-crest or reef breakwaters;
    c)   submerged breakwaters or sills;
    d)   rock groynes, bastion or inclined;
    e)   rubble revetments.

Studies at Wallingford on the influence of the control structures on the beach response to wave action, reported by Coates (1994, suggest that rubble groynes are often the most cost-effective of these structures in controlling movements of shingle beaches, and significant research is underway to describe the effect of structure plan configuration and geometry on the plan re-shaping of the beach. As shingle beaches may become depleted without re-nourishment, or may be locally denuded by strong oblique wave attack, back-

---

[1]   Engineer Coastal Group, HR Wallingford, Howbery Park, Wallingford, UK

[2]   Professor (associate), Department of Civil Engineering, University of Sheffield; Manager Coastal Structures, HR Wallingford, Howbery Park, Wallingford, UK.

beach rubble revetments may also be required to form a stronger rear defence.

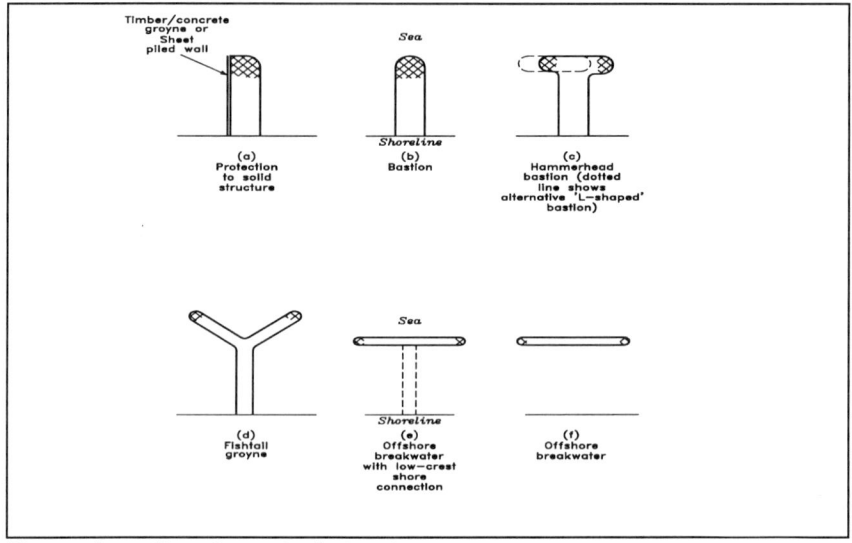

Figure 1 Beach control structures

The main parameter set by structural considerations on all armoured control structures is the median armour size. The armour slope angle chosen interacts with the armour size needed for a given stability level, but is also set by consideration of wave reflections and overtopping. Many other dimensions, and aspects of construction practice, relate closely to the unit armour size. If under-sized, rock armour will move excessively, leading to deterioration of the armour, and/or erosion of fill. Structural changes to the groyne will in turn lead to higher wave overtopping, and/or beach erosion. It is important therefore to ensure that any damage to the structure remains below acceptable limits. Various methods may be used to calculate the armour size, but are only fully valid for normal attack on simple cross-sections. Allsop et al (1995) present evidence where some structures have experienced more damage than expected by their designers, and/or deemed acceptable by their owners.

The problem of increased damage armour appears to have been most severe on steep shingle beaches, typically in the UK at slopes between 1:5 to 1:10, with the average near to 1:7. It is likely that such steep slopes have significant influences on the form and strength of wave breaking onto any structure constructed on the beach, thus modifying the waves from the form of deep or intermediate depth waves for which most design methods have been derived. Most of this paper therefore describes hydraulic model studies conducted to measure armour damage on beach control structures on steep beaches, and the analysis then conducted to devise appropriate design methods.

2 METHODS TO CALCULATE ARMOUR STABILITY

Design methods for rock armour focus on calculation of the median armour unit mass, M_{50}; or the nominal median diameter D_{n50} defined in terms of the median unit mass and rock density ρ_r: $D_{n50} = (M_{50}/\rho_r)^{1/3}$. The most common calculation methods are the Hudson formula given in the Shore Protection Manual by CERC (1984); or equations by Van der Meer (1988). Hudson developed a simple expression for the minimum armour weight for regular waves which may be written in terms of the median armour unit mass, M_{50}, and wave height, H:

$$M_{50} = \rho_r H^3 / (K_D \cot\alpha \, \Delta^3) \tag{1}$$

where ρ_r is the density of rock armour (Kg/m³); ρ_w is the density of (sea) water; Δ is the buoyant density of rock, = $(\rho_r/\rho_w)-1$; α is the slope of angle of the structure face; and K_D is a stability coefficient to take account of the other variables. Values of K_D corresponded to the wave height giving least stability in tests with regular waves on permeable cross-sections subject to little overtopping. Slight re-shaping of armour was expected, and values of K_D correspond to "no damage" where 0-5% of the armour was displaced.

An alternative method was derived by Van de Meer (1988) who included model data by Thompson & Shuttler at Wallingford, extended this by further tests at Delft, and derived new formulae for armour damage which include the effects of random waves, range of core / underlayer permeabilities, and distinguish between plunging and surging wave conditions respectively:

$$H_s/\Delta D_{n50} = 6.2 \, P^{0.18} \, (S/\sqrt{N_z})^{0.2} \, \xi_m^{-0.5} \tag{2a}$$

$$H_s/\Delta D_{n50} = 1.0 \, P^{-0.13} \, (S/\sqrt{N_z})^{0.2} \, \sqrt{\cot\alpha} \, \xi_m^P \tag{2b}$$

where the parameters not previously defined are:
- P   notional permeability factor
- S   design damage number = A_e/D_{n50}^2, and A_e is erosion area
- N_z number of waves
- ξ_m Iribarren number = $\tan\alpha/s_m^{0.5}$
- s_m wave steepness = $2\pi H_s/gT_m^2$, and T_m is the mean period;

and the transition from plunging to surging is given by a critical value of ξ_m:

$$\xi_m = (6.2 \, P^{0.31} \, (\tan\alpha)^{0.5})^{1/(P+0.5)} \tag{2c}$$

Damage to armour on a range of core / underlayer configurations were analysed. Values of P given by Van de Meer vary from 0.1 for armour on underlayer over an impermeable slope, to 0.6 for a homogeneous mound of armour, with intermediate values of 0.4 and 0.5 also described. These formulae were derived for normal wave attack, and do not include corrections for roundheads or junctions.

3. DESIGN OF MODEL STUDIES

3.1 Test structures and facility

The objective of the physical model tests was to quantify the stability of rock armour on four typical rock armoured structures on a 1:7 beach slope:
- a) a breakwater or groyne roundhead, Type 2, (Fig 2);
- b) an L-shaped groyne, formed from a) above;
- c) an inclined groyne, Type 1, (Fig 3); and
- d) a simple 1:2 rubble sea wall slope.

The 1:2 sea wall section was tested primarily as a control structure, damage to which could be compared directly with predictions by the existing design formulae.

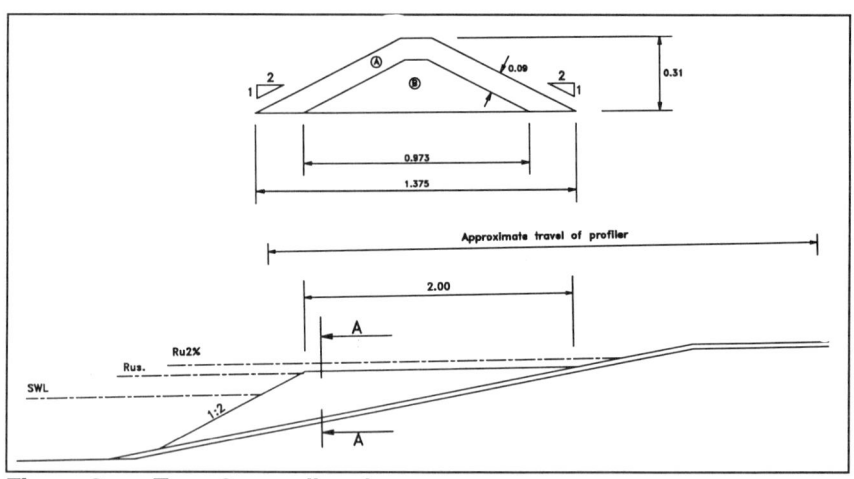

Figure 2 Type 2 roundhead groyne

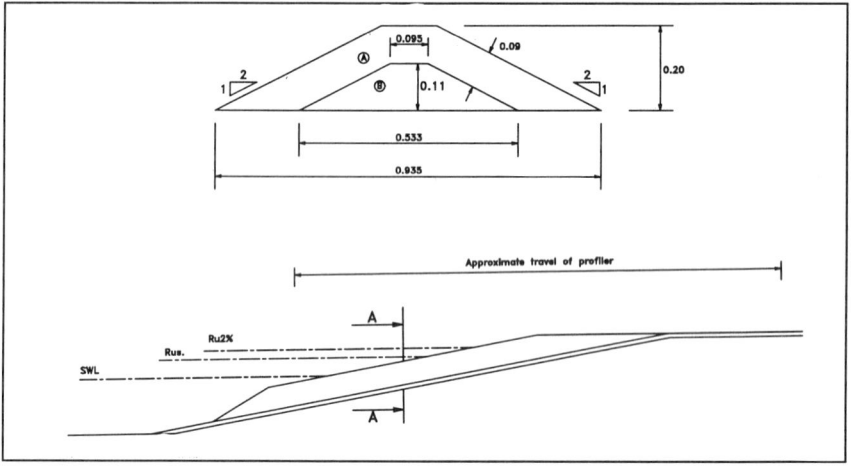

Figure 3 Type 1 groyne

Early in the research, a review of prototype structures suggested that the crest level should allow overtopping, and often fell between significant and 2% run up levels. The 2% run-up level was calculated using empirical prediction methods in CIRIA (1991). The principal test parameters may be summarised:

Water and bed levels	0.0m; -0.7m
Target wave height	$H_s = 0.13$m, ($H_s/\Delta D_{n50} = 1.7$)
Mean sea steepnesses	$s_m = 0.02$ and 0.04
Test duration	$N_z = 1000$ and 3000 waves
Armour and core size	$D_{n50} = 0.045$m; $D_{n50} = 0.024$m
Side slope angles	Cot $\alpha = 2.0$
Toe and crest levels	-0.36m, ($-8D_{n50}$); +0.22m, ($+5D_{n50}$)

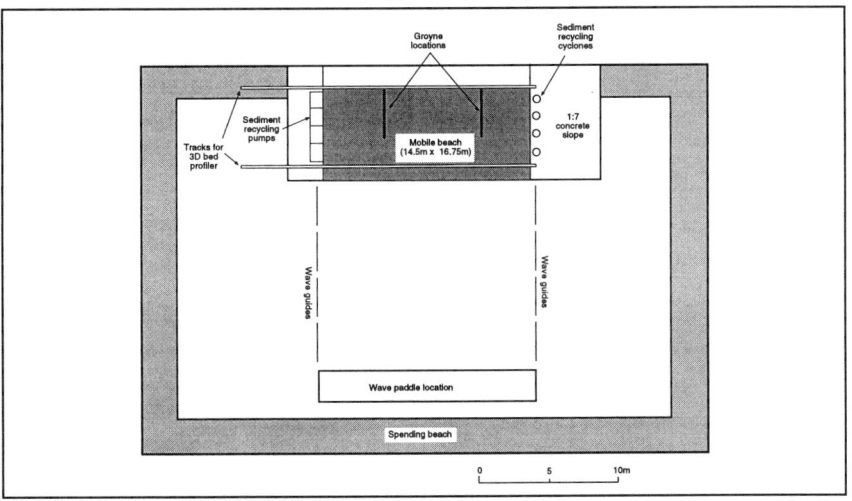

Figure 4   Roundhead Basin test facility

The structures were constructed and tested in a large wave basin which measured 40m by 27m (Fig 4). The models were not to a particular scale, so analysis of measurements used dimensionless terms, but the structures may be viewed as modelled at ratios of between 1:10 and 1:20. Two structures were tested at one time. Type 1 and 2 groynes were tested in the first series. The Type 1 groyne was then replaced by the sea wall, and the Type 2 groyne was modified to the "L" shape (Fig 5).

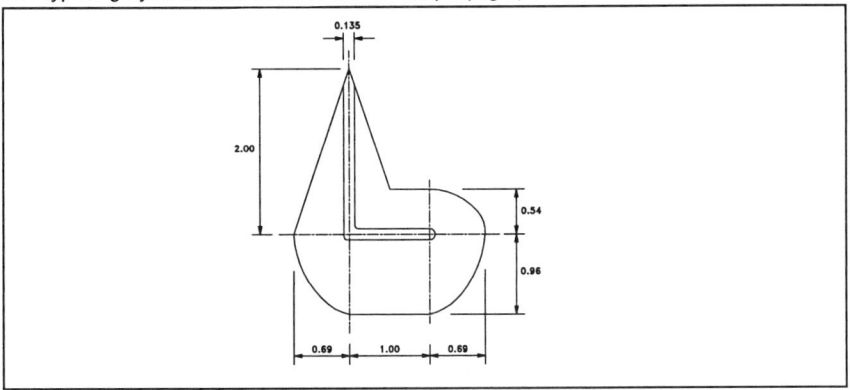

Figure 5   Plan of L-shaped groyne

The 1:2 slope sea wall section was 1.0m wide (equivalent to 22 D_{n50}). The toe was set at the same level as the toe of the "L" shaped and roundhead groynes.

The low level groyne, Type 1, was 4.7m long (105 D_{n50}), 0.94m (21 D_{n50}) wide at the base, and reached a height of 0.2m (4.4 D_{n50}) above the beach. The armour was laid to a thickness $t_a = 2D_{n50}$, slopes of 1:2, and crest width of $3D_{n50}$.

The high level roundhead groyne, Type 2, was shorter and wider than the Type 1 groyne at 2.95m (66 D_{n50}) long and 1.38m (31 D_{n50}) wide at the widest part of the roundhead. The crest was 0.31m (6.9 D_{n50}) above the beach at the seaward end.

The "L" shaped groyne was formed by extending from the roundhead of the Type 2 groyne. The groyne was 2.96m (66 D_{n50}) long, with a side limb of 2.38m (53 D_{n50}). A straight section of 1.0m (22 D_{n50}) formed the "L" shape.

Three 5m random wave paddles produced normal wave attack for these tests. The 1:7 slope beach used by Coates (1994) measured 12.5m by 8m, and this area was covered by an automatic three-axis bed profiler. Profile measurements of the structures were recorded on a personal computer, allowing armour displacements to be quantified by comparing profile lines over selected areas of each model. The measured damage was compared with damage predicted by the Van der Meer method for simple sections.

3.2 Test conditions

These tests were designed to give useful data over a wide range of structural and environmental variables. The experiments were intended to give intermediate armour damage at wave heights less than the maximum possible in the basin, equivalent to $H_s/\Delta D_{n50}=2.6$ for armour of $D_{n50}=0.045$m. The "target" wave height was equivalent to $H_s/\Delta D_{n50}=1.7$, and this wave height was used in the design of the model structures.

Around European coastlines, storm wave conditions are generally steep and narrow-banded, often described well by the JONSWAP spectra. Storm waves are relatively steep, but armour damage also depends upon sea steepness. These tests used two steepnesses, $s_m=0.02$ and 0.04 to explore the possible effects of longer waves. Each test was planned to be run for N=1000 and 3000 waves duration.

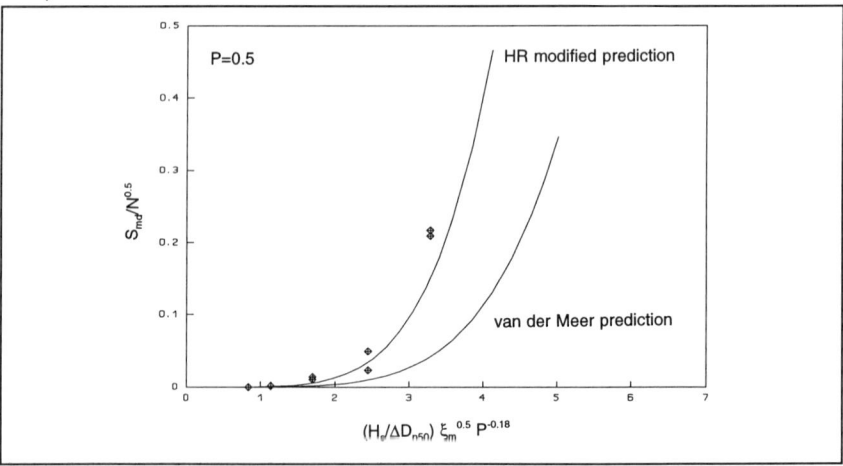

Figure 6 Sea wall damage, $s_m=0.04$, P=0.5

3.3 Test procedures

The first part of each test used 100 waves, during which close observations were made of the structures. If no movement was seen in these 100 waves, the test was stopped. If any damage (including rocking) was observed, then the test continued to 1000 waves when each profile line was re-surveyed. The test then continued for a further 2000 waves. After each test the structures were re-surveyed and photographed. If they had been damaged such that repair was necessary, the armour was re-built around the zones damaged. Testing was stopped when the structures needed re-building.

Profiling covered set lines over each of the groynes under test. The profiler incorporated a touch-sensitive foot of diameter equivalent to $0.8D_{n50}$, and took observations at $0.5D_{n50}$ intervals along each profile line. The profile results were used to calculate the area of erosion on each profile line, A_e, and hence the damage parameter S defined in section 2.

4. TEST RESULTS

4.1 Simple slopes

The test results were initially very surprising, as virtually all of the profiles showed significantly more damage than predicted by conventional formulae (Fig 6). Some local increases in damage had been expected, but not the consistently greater damage found here. Analysis attention was focused first on the 1:2 sea wall slope, and equivalent section on the L-shaped groyne, but even these simple cases showed significantly greater damage than predicted by the Van der Meer equations. During the analysis period, additional data on the performance of armoured groynes on two bed slopes became available from tests at CEPYC in Madrid. Initial analysis of these had been presented by Baonza & Berenguer (1992), and their test data were further analyzed by Allsop & Franco (1992) as part of the EC MAST project G6-S. These data also indicated that the sea bed slope might be significant in increasing armour damage.

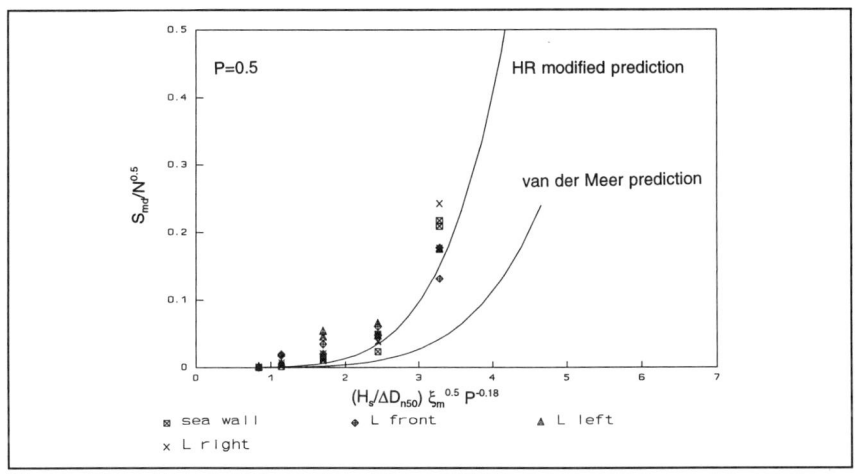

Figure 7   Sea wall and front face of "L" shaped groyne

Damage results for the sea wall section alone are summarised in Figure 6, using axes based on equation (2a) that allow results for each wave steepness and test duration to be presented together. The standard Van der Meer equation for plunging waves is shown, together with a version of the equation with a revised coefficient:

$$H_s/\Delta D_{n50} = 4.8\ P^{0.18}\ (S/\sqrt{N_z})^{0.2}\ \xi_m^{-0.5} \tag{3a}$$

The fit of the data to this modified equation is good over the area of main interest. Further support for the revised equation is given by the comparison of damage on the front face of the "L" shaped groyne with the sea wall, shown in Figure 7.

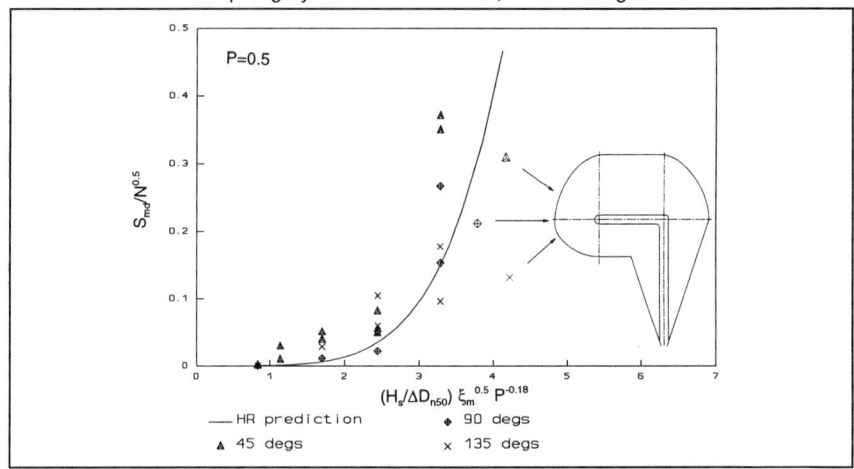

Figure 8   Damage to curved parts of "L" shaped groyne

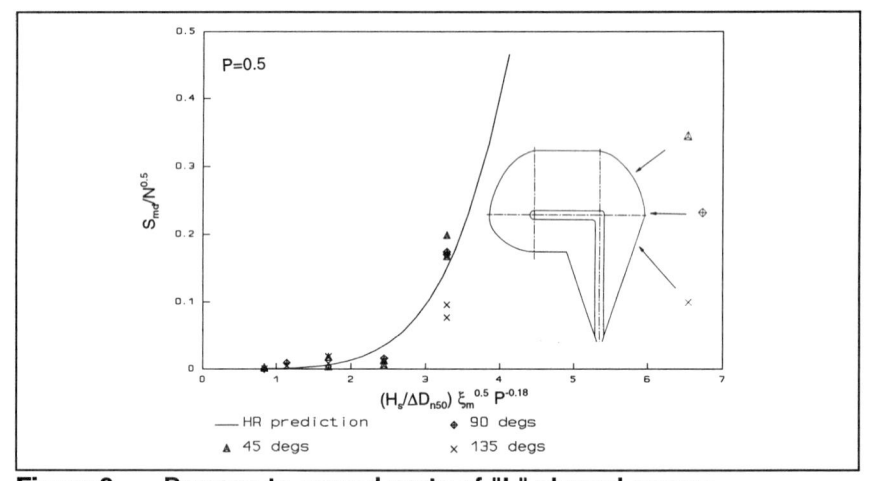

Figure 9   Damage to curved parts of "L" shaped groyne

The first result from this analysis was therefore the conclusion that damage is significantly increased by steep (local) beach slopes, even for simple slopes subject to normal wave attack. A modification to the Van der Meer equation for plunging waves is given in equation (3a), and a similar increase for surging waves would be given by:

$$H_s/\Delta D_{n50} = 0.77\ P^{-0.13}\ (S/\sqrt{N_z})^{0.2}\ \sqrt{\cot\alpha}\ \xi_m^P \tag{3b}$$

4.2 Roundhead and "L" shaped groynes

Once the effect of the steep beach slope had been accounted for, the test results suggest that some sections of the more complex 3-dimensional structures experience greater damage than the simple slopes under normal wave attack, but that the spread of damage spatially is somewhat variable. This is shown in Figures 8 and 9 where damage on the opposing sides of the "L" shaped groyne are contrasted with the new equation (3a). For most positions, the modified prediction method in eqn (3a) gives a reasonable estimate of the damage, but for the zones shown in Figure 8, damage at larger wave heights is still greater than would be predicted by the new method. In itself, this is not surprising, as it is well known that breakwater roundheads require larger armour units, and/or shallower slopes for the same stability as trunk sections.

4.3 Inclined, Type 1, groyne

Damage to the inclined groyne, Type 1, varied along its length, with the location of greatest damage depending on wave height and period. The mean level of damage taken over the active length of the groyne, and derived by averaging the erosion areas from each profile, fits the general prediction given by eqn (3a), and is summarised in Figure 10.

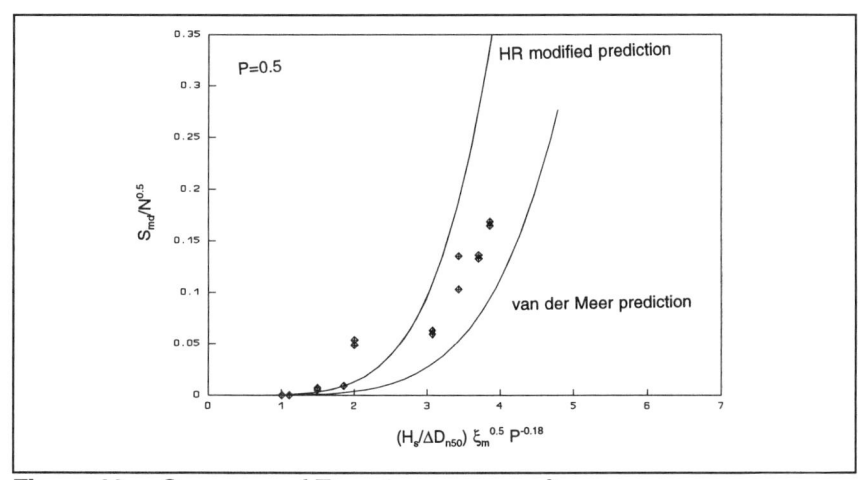

Figure 10 Summary of Type 1 groyne results

Peak values of (local) damage along the length of the groyne however often reached twice the mean, often at the wave run-up and run-down limits along the groyne, and this is illustrated in Figures 11-13 which show local levels of damage plotted against position along the groyne from the landward end for increasing relative wave heights. For $H_s/\Delta D_{n50}$ up to 1.72 (Figures 11-13), damage only exceeds S=5 at 1000 waves over small regions. For $H_s/\Delta D_{n50}$=2.16 (Fig 14) however, damage over most of the length of the structure has exceeded this criterion.

8 CONCLUSIONS

These tests have shown that the stability of armour on beach control structures depends critically on the local sea bed slope. Results from tests using a beach of 1:7 were used to develop a modified coefficient for use in the Van der Meer equation. The changes to the coefficients are equivalent to increasing the mean armour mass by a factor of 2.2 to maintain armour stability.

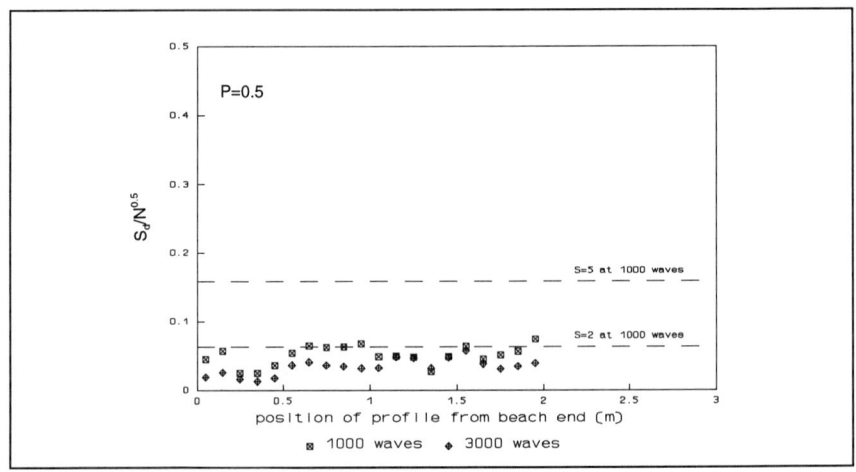

Figure 11   Type 1 groyne, $s_m=0.04$, $H_s/\Delta D_{n50}=0.83$

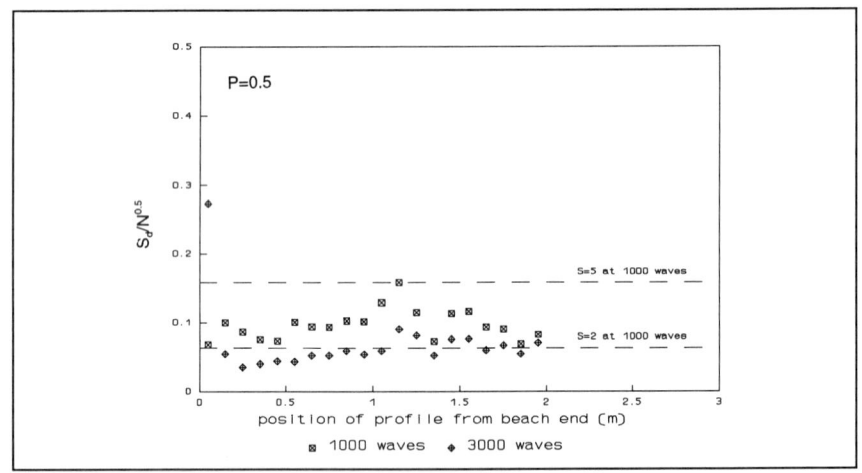

Figure 12   Type 1 groyne, $s_m=0.04$, $H_s/\Delta D_{n50}=1.12$

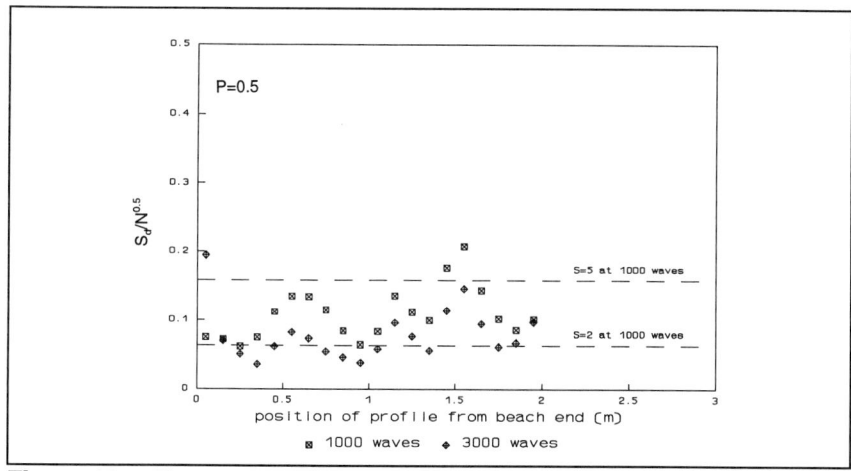

Figure 13   Type 1 groyne, $s_m=0.04$, $H_s/\Delta D_{n50}=1.72$

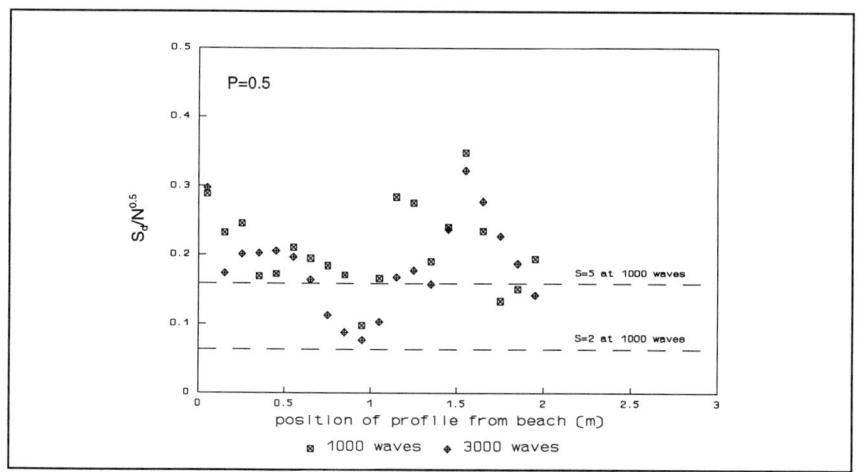

Figure 14   Type 1 groyne, $s_m=0.04$, $H_s/\Delta D_{n50}=2.16$

This study has identified some unusual effects, and some unexpected conclusions. The tests suggest that steep beach slopes may change the form of wave breaking such that the performance of structures in or behind steep (beach) slopes should be reviewed to identify whether a more general effect may lead to problems.

ACKNOWLEDGEMENTS

The laboratory studies covered in this paper were supported by the UK Ministry of Agriculture, Fisheries, and Food under Research Commissions on Flood Defence. Additional support was given by the European Union under MAST I project G6-S, and by HR Wallingford. The hydraulic model tests covered in section 5 were conducted by RJ

Jones assisted by MK Reeves and P Bona in the Coastal Group of HR Wallingford, and supervised by NWH Allsop. The authors are most grateful for the assistance of Andrew Bradbury of the Coast Protection Unit of New Forest District Council for valuable advice in the analysis of the laboratory tests, and for the additional data from Hurst Spit; to CEPYC and to C Franco of University of Rome for assistance in the MAST project G6-S topic 3R2.

Preparation of this paper was supported by the University of Sheffield and HR Wallingford.

REFERENCES

Allsop NWH (1994) "Design of rock armoured beach control structures" Paper 2.1 in Proceedings of 29th Conference of River and Coastal Engineers, MAFF, Loughborough, July 1994

Allsop NWH & Franco C (1992) "MAST G6-S Coastal Structures Topic 3R: Performance of rubble mound breakwaters singular points" Paper 3.12 to G6-S Final Overall Workshop, Lisbon, November 1992

Allsop NWH, Jones RJ & Bradbury AP (1995) "Design of beach control structures on shingle beaches" Paper to Conference on Coastal Structures and Breakwaters, Institution of Civil Engineers, London, April 1995

Baonza A & Berenguer JM (1992) "Experimental research on groyne stability under very oblique wave action" Proc. Conf. Civil Engineering in the Oceans V, ASCE, Texas, 1992

CIRIA (Simm JD, Ed.) (1991) "Manual on the use of rock in coastal and shoreline engineering" CIRIA Special Publication 83, London, November 1991

Coastal Engineering Research Centre, CERC (1984) "Shore Protection Manual", Vols I-II, US Gov Printing Off, Washington, 4th edition 1984

Coates TT (1994) "Physical modelling of the response of shingle beaches in the presence of control structures" Proc Conf Coastal Dynamics '94, Universtat Politecnica de Catalunya, Barcelona, February 1994

Jones RJ & Allsop NWH (1993) "Stability of rock armoured beach control structures" HR report SR 289, HR Wallingford, March 1993

Van der Meer JW. (1988) "Rock slopes and gravel beaches under wave attack", PhD thesis Delft University of Technology, April 1988. (available as Delft Hydraulics Communication 396)

CHAPTER 85

Effect on Roughness to Irregular Wave Run-up

Jea-Tzyy Juang*

ABSTRACT

A study of the irregular wave runup and rundown phenomenon on rough impermeable slope is conducted in order to investigate the effects of various surface roughness to the wave runup-rundown activities based on various kind of dyke and incident wave conditions. The roughness of dyke surface was defined as f_r which is functions of shape of the slender rectangular obstacles and its displacement distance. Results shows that the correlation between the relative significant runup height (R_{us}/H_s) and the incident wave steepness (H_s/L) have the same tendency either in regular or in irregular waves. The relative runup height has its maximum value when the dyke slope ($\cot\theta$) near 2. Next, the runup characteristics in regular wave was agree well with the Rayleigh distribution but it is a little bit smaller than that the Rayleigh's in irregular wave. Besides, the strong correlation between the wave runup and the surf similarity parameter ξ was found. The decrease of the relative runup height with the increase of the surface roughness was got also. It may be useful to applied into the design work of the seawall.

INTRODUCTION

Wave run-up on coastal structures such as seawalls, dykes and so forth, is an important factor in the design of the height of the structures; therefore, many studies on the wave runup phenomenon and its characteristics have been carried out. However, the studies of the effect of roughness on the dyke to the run-up height are still scarce to find. Therefore, a study of the irregular wave run-up and run-down phenomenon on rough impermeable slope was conducted here in order to investigate the effect of various surface roughness to the wave runup-rundown activities based on the different kind of the dyke and the incident wave conditions.

---

* Deputy Director, Institute of Harbour & Marine Technology Wu-Chi, Taichung Hsien, Taiwan, ROC.

ANALYTICAL CONSIDERATION

Based on the concept of control volume, the correlation among the incident wave energy, the reflected wave energy and the energy dissipation during the wave run-up and overtopping (as shown in Fig. 1) can be written as (Cross, 1972)

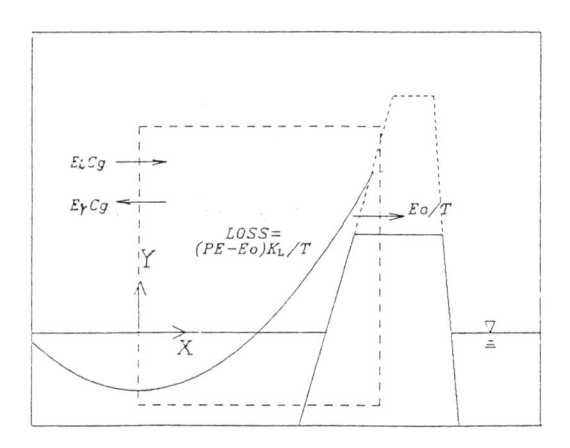

Fig.1 Definition sketch of runup and overtopping

$$E_i \cdot C_g - E_r \cdot C_g - \frac{E_o}{T} = \frac{PE - E_o}{T} \cdot K_1 \tag{1}$$

where E_i: Incident wave energy density
  E_r: Reflected wave energy density
  C_g: Group velocity
  E_o: Overtopping energy
  PE: Potential energy of the entire runup wedge above SWL
  K_1: Rundown energy loss coefficient
  T : Wave period

By condition of no overtopping, the above equation can be simplified as

$$E_i \cdot C_g - E_r \cdot C_g = \frac{PE}{T} \cdot K_1 \tag{2}$$

The potential energy PE can be computed (Juang, 1992) by the following equation with all notations was shown in Fig. 2.

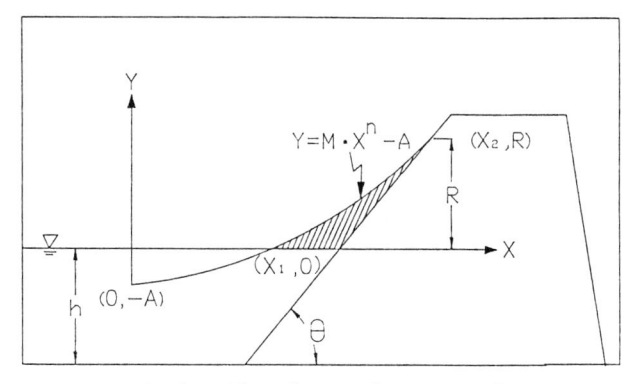

Fig.2 The shape of runup wedge

$$PE = \rho g \left\{ \int_{X_1}^{X_2} \frac{Y^2}{2} dx - \frac{R^3}{6\alpha} \right\}$$

$$= \frac{1}{2} \rho g \left\{ \frac{M^2}{2n+1}(X_2^{2n+1} - X_1^{2n+1}) - \frac{2 \cdot MA}{n+1}(X_2^{n+1} - X_1^{n+1}) \right.$$

$$\left. + A^2(X_2 - X_1) - \frac{R^3}{3\alpha} \right\} \qquad (3)$$

in which Y is the water surface elevation above the still water level at maximum runup. Its corresponding equation is

$$Y = M \cdot X^n - A \qquad (4)$$

where X is the distance from the trough shoreward; A is the amplitude at the trough and M, n the coefficients that was function of surf parameter ξ (=$\tan\theta /\sqrt{H/L}$) as shown in Table 1.

By using the linear wave theory, the wave height in front of the inclining dyke becomes

Table 1   Values of M and n

$\cot\theta$	0.5	1.0	2.0	3.0
M	$54.228\xi^{-3.5}$	$5.993\xi^{-3.68}$	$1.748\xi^{-4.67}$	$0.437\xi^{-3.19}$
n	$1.0075\xi^{0.19}$	$0.608\xi^{0.59}$	$0.858\xi^{0.64}$	$0.873\xi^{0.59}$

$$H = H_i + H_r = H_i(1 + K_r) \tag{5}$$

where H_i: incident wave height
 H_r: reflected wave height
 K_r: reflection coefficient

Due to the wave energy was proportional to the wave height, therefore we can have

$$E_r = K_r^2 \cdot E_i \tag{6}$$

Substitute eq.(6) into eq.(2), the equation becomes

$$E_i(1 - K_r^2) \cdot C_g \cdot T = PE \cdot K_1$$

$$\text{or} \quad PE = \frac{1 - K_r^2}{K_1} \cdot E_i \cdot C_g \cdot T$$

DEFINITION OF ROUGHNESS COEFFICIENT

The roughness of the dike surface in this experiment study was made of small rectangle slender obstacles on the dyke as shown in Fig.3. The idea of that frame was come from the stream flow. Because the roughness coefficient at the bottom of the river was proportional to the diameter of the bed particles. Therefore, the roughness coefficient f_r in this study was defined by the following step.

(1) $f_r \propto D$; the higher the height of the slender obstacle, the rougher the roughness coefficient.
(2) $f_r \propto 1/W$; the wider the width of the slender obstacle, the smoother the roughness coefficient.
(3) $f_r \propto A_r = BD/D_m(B+W_m)$; the bigger the ratio of the effective cross section A_r to the maximum cross section, the rougher the roughness coefficient.

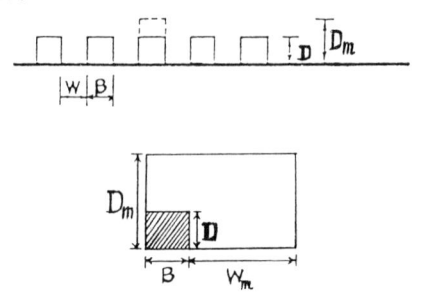

Fig.3 Definition diagram of the roughness coefficient

To view the whole situation that metioned above, we can defined the roughness coefficient as follow

$$f_r = \frac{BD^2}{WD_m(B + W_m)}$$

Various roughness coefficient (include $f_r=0$) with different ratio of D/B and W/B was shown in Table 2.

Table 2   Contrast table among D/B, W/B and f_r

D/B	0	0.25	0.25	0.25	0.5	1.0	1.0	1.0
W/B	0	4.5	3.0	1.5	1.5	4.5	3.0	1.5
f_r	0	0.0025	0.0038	0.0076	0.0303	0.0404	0.0606	0.1212

EXPERIMENT STUDY

A series of tests was carried out in a 100m long, 1.5m wide and 2m height wave flume. A random wave generator of servo-controlled electro-hydraulic system that was made by Danish Hydraulic Institute in Denmark is installed at one end. Artificial dyke model of various rough surface (include smooth) with four different kind of dyke slopes ($\cot\theta$ = 0.5, 1, 2, 3) were installed at the other end where was 45m from the wave generator. The flume was divided into two sections. One equal 100 cm and the other 40 cm. The later (smaller) channel was used for measure the incident wave which will not to interfere by the reflected wave. All the experimental apparatus is shown in Fig.4.

Several wave gauge (ch.1 to 5) of capacitance type was used for measured the incident waves (Ch.1 to 4) and the run-up (Ch.5). The experiment was complete in conditions on fixed water depth (40cm) and incident wave period (1.2sec). The incident significant wave height are approximate to 3.5, 5.8, 6.8, 7.5, 8.0 and 8.8cm separately. The random waves used for tests were simulated to have Pierson-Moskowitz type spectra. Both of water surface variations of incident wave and run-up height were recorded simultaneously by an analog data recorder. The recorders were digitized by an A-D converter at a sampling interval of 0.025 sec. The measuring duration is about 100 sec.

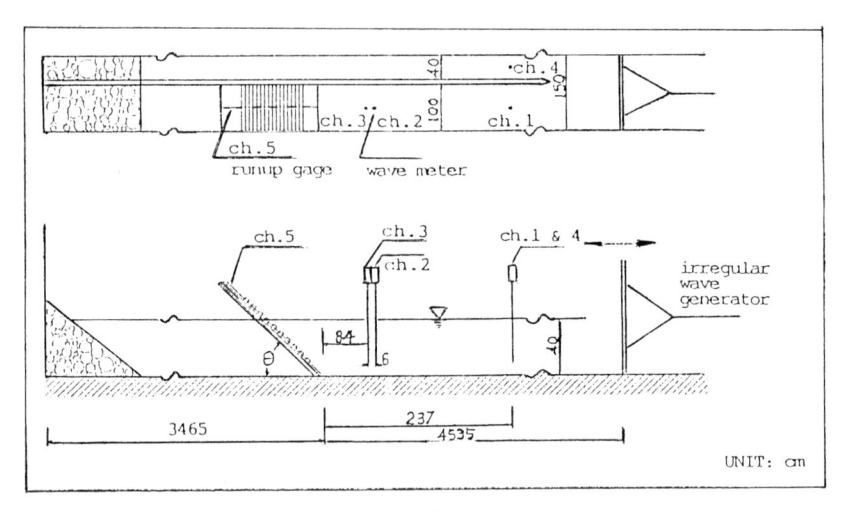

Fig.4 Layout of the test flume

For data analysis, at the begining, there have two ways to analyze the characteristics of incident and run-up waves. One is the statistics method to count out the significant wave. The other is the spectra analysis method to calculate the spectra energy then the significant wave height. Due to wave height and period are very important factors in wave structure interaction such as wave run-up. Therefore, the dimensionless distribution of relative wave height (H/H_m) to wave period (T/T_m) of the experimental waves was compared as shown in Fig.5. In the figure, it can be seen that the relative wave period (T/T_m) increases with the increase of wave height in the smaller waves ($H<1.3H_m$), but wave periods are distributed around $T/T_m=1.3$ in the wave field higher than arbitrary critical value ($H>1.3H_m$). Meanwhile, from the data shown in the figure, we have

$H_{1/10} = 1.97\ H_m$
$H_{1/3} = 1.56\ H_m$
$T_{1/3} = T_{1/10} = T_{max} = 1.3\ T_m$

It is identified from the above results that the joint distribution of wave height and period has a good agreement with the analyzed results of field data by Goda (1985). In other words, it can be stated that the simulated irregular waves used in model tests represent random ocean waves fairly well.

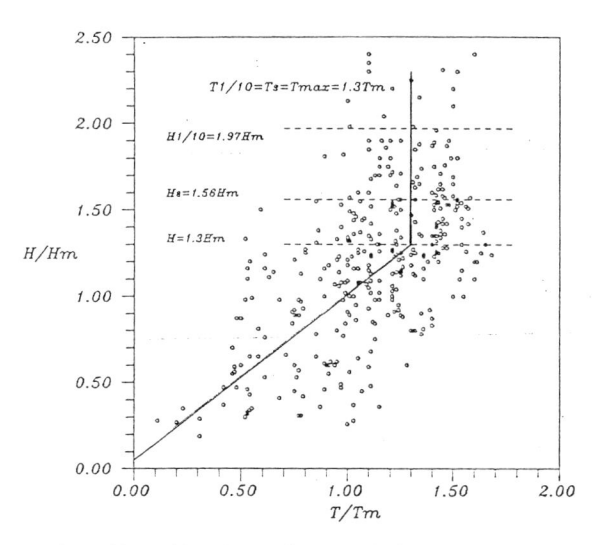

Fig.5 Joint distribution of wave height and wave period

In order to investigate the distribution status of the run-up waves of the irregular wave. The computational method of statistical significant wave height was used for counting the significant runup height (R_{us}) also. Part of the results ($f_r=0$ & $f_r=0.1212$) was shown in Fig.6 and Fig.7 respectively. From those figures, we can find the distribution of irregular wave runup heights (R_{us} to R_{um}) provides a good approximation to the Rayleigh distribution on the smooth surface dyke. But it will be overestimated when the dyke slope milder than 1 to 2 in rough surface ($f_r=0.1212$) test.

Secondly, all the experiment data which shows the relationship between the significant relative run-up and run-down height with the surf parameter was shown in Fig.8. In the figure, the envelope lines (1) and (2) indicate the extreme value of the relative run-up and run-down height respectively. Which was got from the experiment result by X_{ue} (1991) in condition of smooth dyke slope.

As to the effect of roughness to irregular wave runup, the relation diagram between the significant relative runup height and the surf similarity parameter in different kind of the surface roughness ($f_r= 0 \sim 0.1212$) was shown in Fig.9. From those figures, we can find that the effect of roughness to the relative runup height was certainly. If we sum up those correlation curves toge-

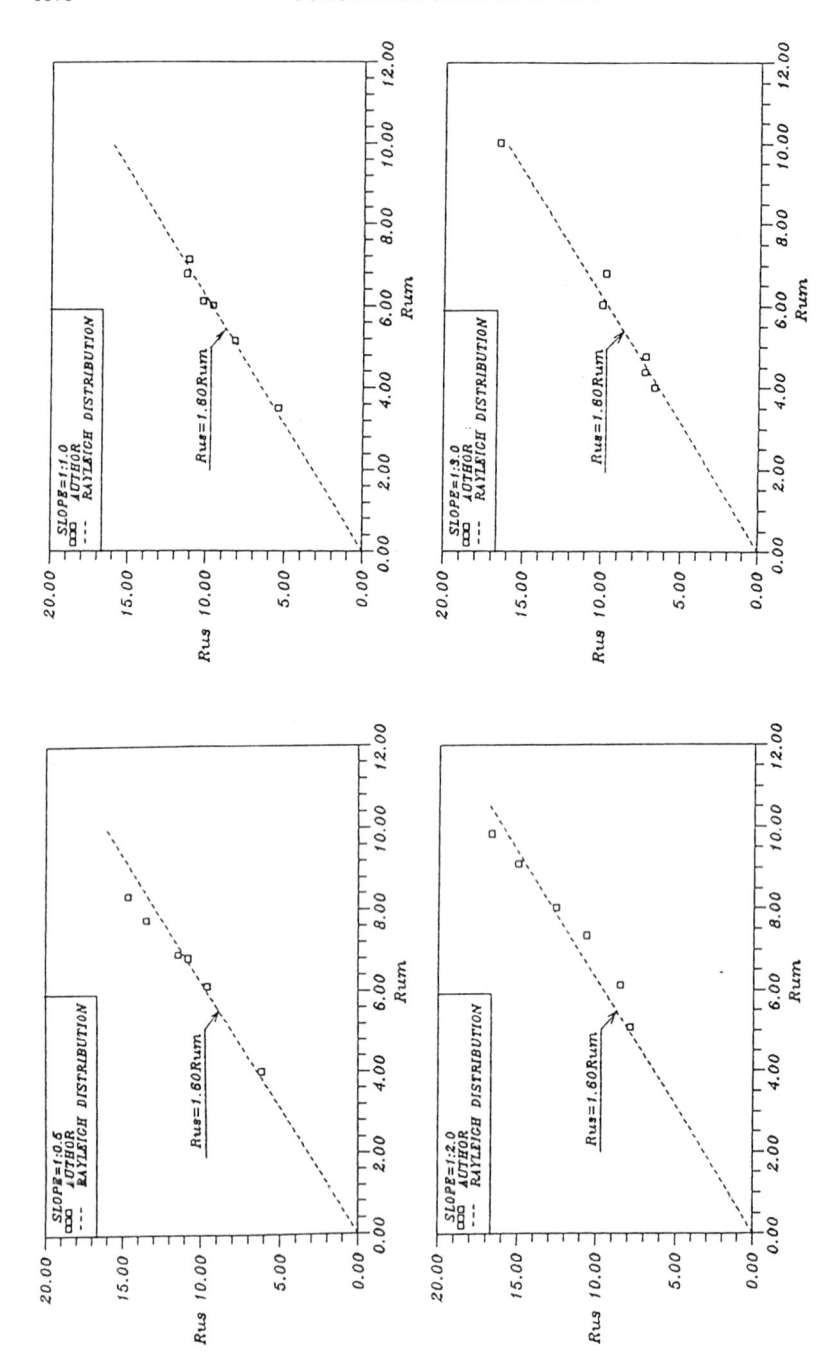

Fig.6 Relationship between significant runup height and mean runup height in smooth dike

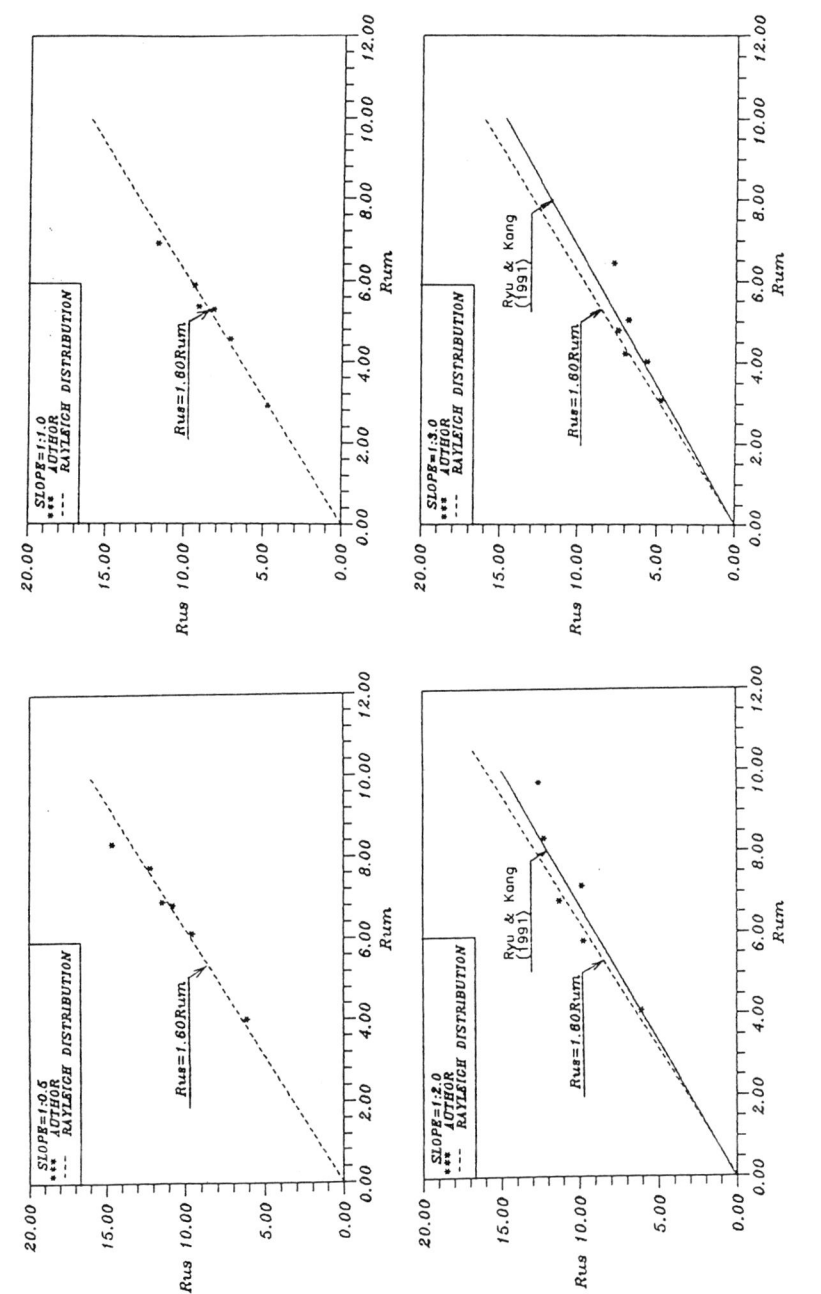

Fig.7 Relationship between significant runup height and mean runup height in rough dike

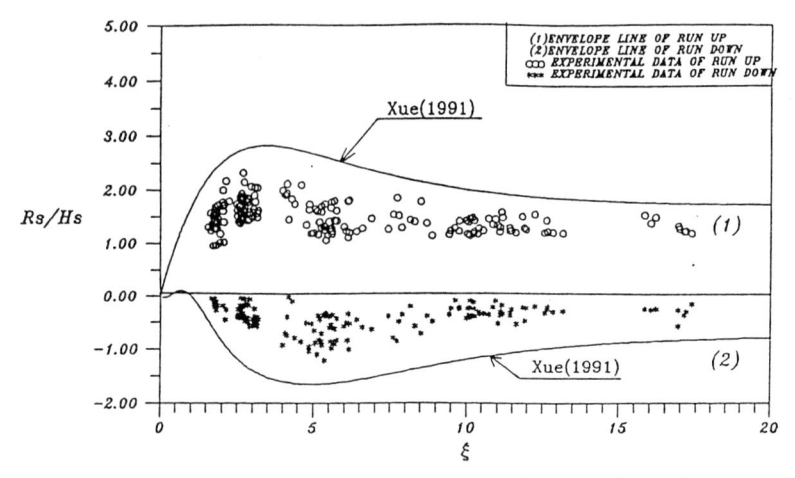

Fig.8 Relationship between run-up-down height and surf parameter

ther, we can find the rougher the surface of the dyke slope, the lower the relative runup height decreased as shown in Fig.10. When the roughness coefficient approach to 0.1212, the corretation curve between the relative runup and the surf parameter was very similar with those results which was presented by Ryu in 1990 in rubble mound experiment.

At last, the normalized relationship between the relative run-up and rundown height and the surf parameter was shown in Fig.11 and 12 respectively. The meaning of normalized stand for the ratio of relative run-up(down) height in rough dyke compared with those in smooth dyke. From the figures, we can see that when the surf parameter becomes bigger, that means the dyke slope approach to vertical, the influence of roughness to wave run-up-down height will vanished.

CONCLUSIONS

1. The distribution of the irregular wave runup heights provides a good approximation to the Rayleigh distribution on the smooth surface dyke. But it will be overestimated when the dyke slope milder than 1 to 2 in rough surface test.
2. The effect of roughness to the relative runup height was certainly and the rougher the surface of the dyke slope, the lower the relative runup height will decreased.
3. When the roughness coefficient approach to 0.1212, the relation

IRREGULAR WAVE RUN-UP

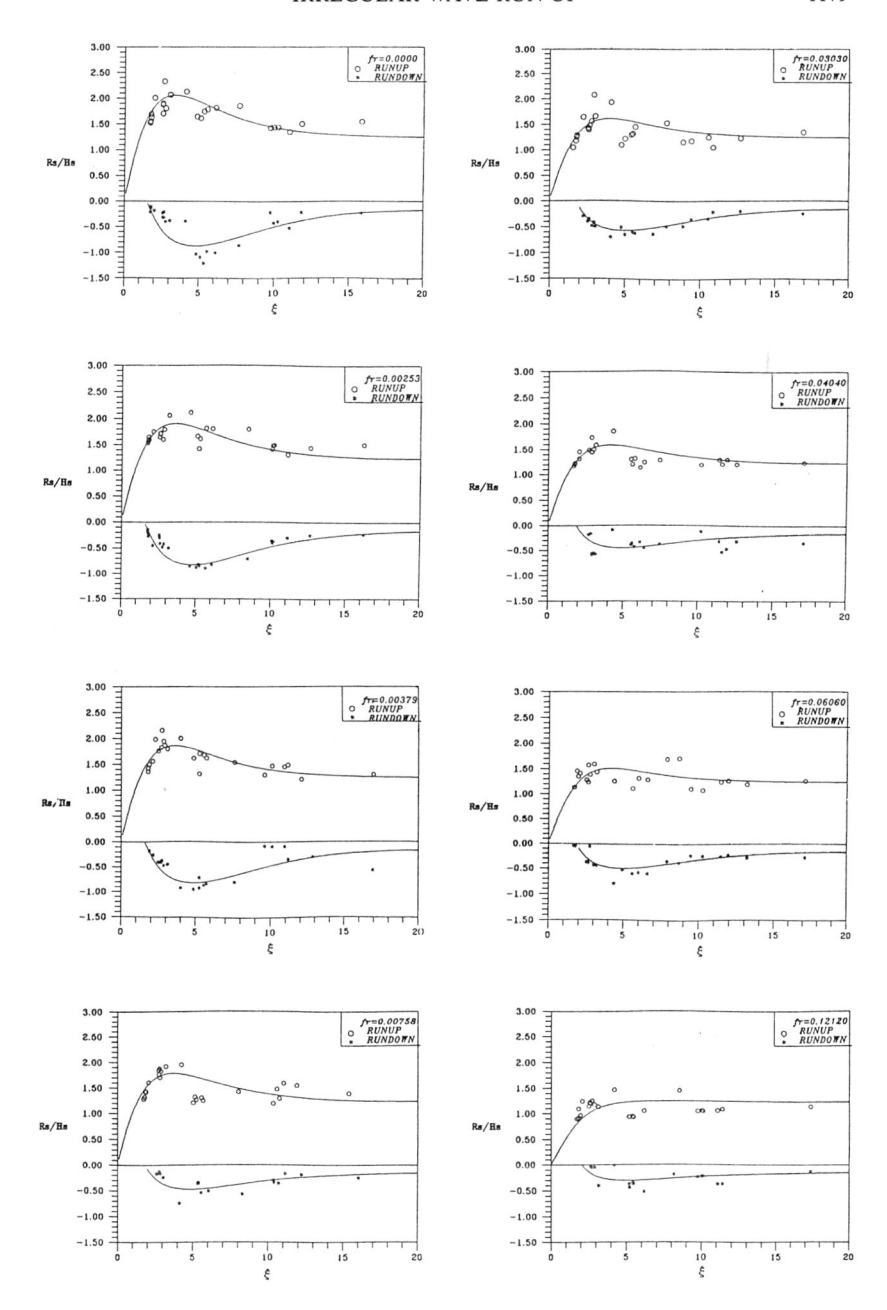

Fig.9 Relationship between significant relative runup height and the surf parameter

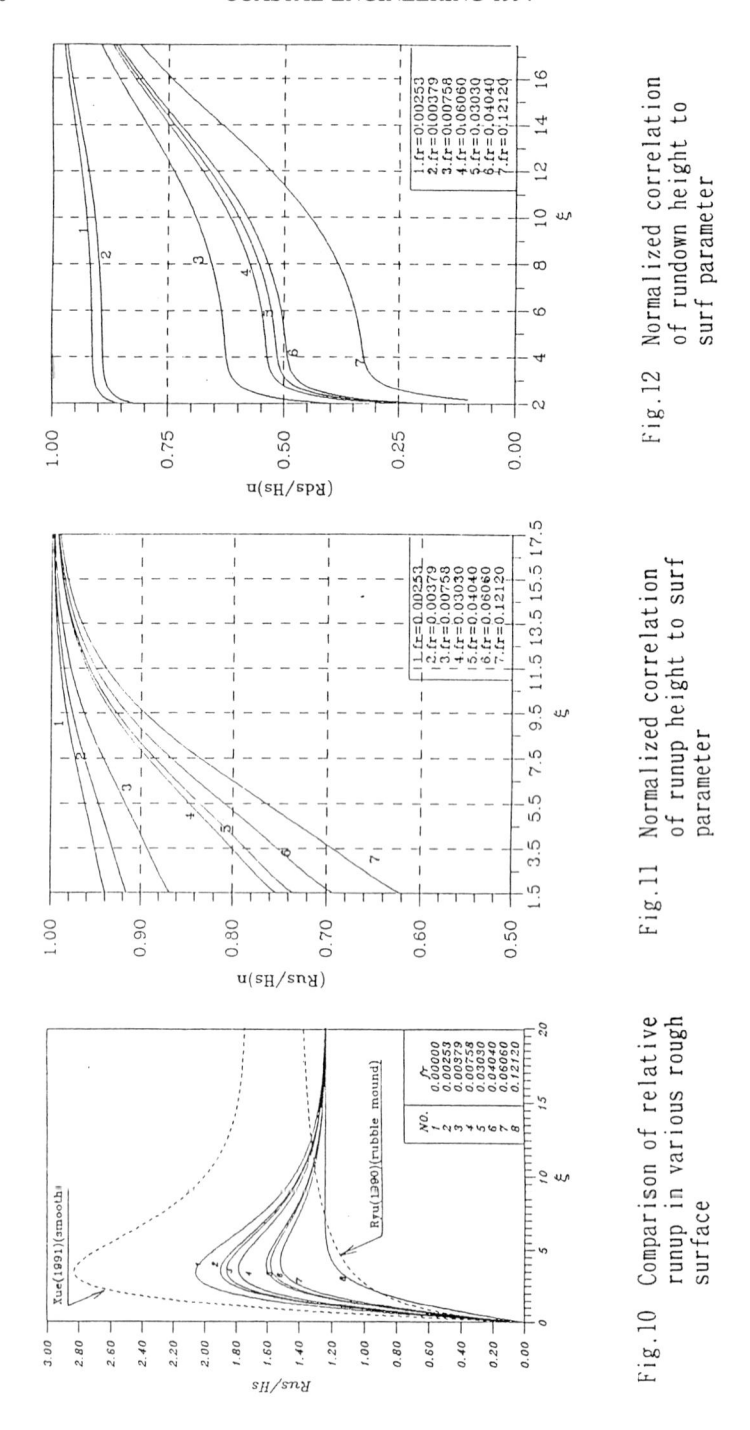

Fig.12 Normalized correlation of rundown height to surf parameter

Fig.11 Normalized correlation of runup height to surf parameter

Fig.10 Comparison of relative runup in various rough surface

curve between the relative runup height and the surf parameter was quite similar with those results in rubble mound experiment.
4. When the surf parameter becomes bigger, that's the dyke slope approach to vertical, the influence of roughness to wave run-up-down height will vanished.

ACKNOWLEDGEMENTS

This study was sponsored by the National Science Council of ROC under Grant NSC 81-0209-E124-01. The author would like to thank Mr. J. M. Chou, C. C. Chien and Miss E. J. Chien for their assistance in experiment and preparing the manuscript.

REFERENCES

Cross, R.H. and C.K.Sollitt(1972): Wave transmission by overtopping, Journal of Waterway, Port and Coastal Engineering, Vol.98, No. WW3, PP.295-309.

Goda, Y.(1985): Random seas and design of maritime structures, University of Tokyo Press, PP.1-323.

Juang, J.T.(1992): Effect on wind speed to wave runup, Coastal Engineering, ASCE, Vol.2, PP.1245-1257.

Kobayashi, N., D. T. Cox and A. Wurjanto (1990): Irregular wave reflection and runup on rough impermeable slopes, Journal of Waterway, Port, Coastal and Ocean Engineering, Vol.116, No.6, PP.708-726.

Mase, H.(1989):Random wave runup height on gentle slope, Journal of Waterway, Port, Coastal and Ocean Engineering, Vol.115, No.5, PP. 649-661.

Meer, J.W. and C.M.Stam(1992): Wave runup on smooth and rock slopes of coastal structures, Journal of Waterway, Port, Coastal and Ocean Engineering, Vol.118, No.5, PP.534-550.

Ryu, C.R. and H.Y. Kang(1990): A prediction model of irregular wave runup height on coastal structures, Proc. Coastal engineering, ASCE, Vol.1, PP.371-383.

Xue, H.C., D. Guo, S.H. Pan and H.S.Zhong(1991): Wave runup-rundown amplitude on slopes, China Ocean Engineering, Vol.5, No.1, PP.39 -49.

CHAPTER 86

Wave Overtopping of Breakwaters under Oblique Waves

Jørgen JUHL[1] and Peter SLOTH[1]

Abstract

A series of hydraulic model tests has been carried out in a wave basin with the aim of studying the effect of oblique waves on the wave overtopping of traditional rubble mound breakwaters without superstructure. The model tests concentrated on measuring the mean overtopping discharge for wave attacks varying from 0° (perpendicular to the structure) up to 50°. Analyses of the overtopping results were made with respect to the significant wave height, wave steepness, crest free board, crest width and angle of wave attack. The paper describes the influence of these parameters on the mean overtopping discharge for a traditional rubble mound breakwaters with an armour layer slope of 1:2.0.

Introduction

Wave overtopping of coastal structures is influenced by a large number of parameters related to breakwater geometry, construction materials, and hydrographic data. Some of the main parameters are listed below:

Geometrical parameters:
free board, crest configuration and width, slope of armour layer (irregular slope), and water depth

Construction material parameters:
porosity, stone shape and diameter (artificial blocks)

Hydrographic parameters:
wave height, wave period, angle of wave attack, wave steepness, spreading, wave sequences, wind conditions, and water level

---

[1] Danish Hydraulic Institute, Agern Allé 5, DK-2970 Hørsholm, Denmark

Wave overtopping is normally studied under perpendicular wave attack in wave flumes. Jensen and Juhl (1987) have shown overtopping data from model testing of rubble mound breakwaters and sea dikes. The paper mainly concentrates on mean overtopping discharges, but also includes a description of the horizontal distribution of the overtopping behind a breakwater, individual wave overtoppings and the influence of wind on wave overtopping, and a comparison to prototype measurements.

Franco et al (1994) have established a formula for the mean overtopping discharge for vertical breakwaters exposed to perpendicular wave attack. The influence of various geometrical types of breakwaters is taken into account using influence factors in connection with the general formula for a fully vertical breakwater. Further, a prediction formula for the probability distribution of individual overtopping volumes is presented. The effect of overtopping volumes on persons and cars behind a crown wall of a vertical breakwater was assessed by Franco (1993), and a set of critical overtopping discharges were proposed (safety criteria).

Only a little research has been made to study the influence of the angle of wave attack on the amount of overtopping water. De Wall and Van der Meer (1992) have carried out tests on the influence on wave run-up and overtopping on smooth slopes. The angle of wave attack, β, was varied from 0° up to 80°, and tests were performed with both long-crested and short-crested waves. For long-crested waves, a few tests showed larger run-up for angles between 10° and 30° than for perpendicular waves, but on average no increase was found. This also applies to the average measured overtopping discharges. For perpendicular wave attack, no difference in wave overtopping was measured between tests with long-crested and short-crested waves, whereas for oblique waves the influence of the angle of wave attack was less for short-crested waves. A reduction in the mean overtopping discharge of about 40 per cent was found for long-crested waves with an angle of 50° and of about 15 per cent for short-crested waves.

Galland (1994) has measured the number of waves overtopping rubble mound breakwaters exposed to oblique wave attack. The model tests were made with four different types of armour units, ie quarry stones, accropodes, antifer cubes and tetrapodes. In the case with quarry stones, the test results for long-crested waves showed a significant decrease in the percentage of waves overtopping the crest by increasing the angle of wave attack. For a dimensionless free board, R_c/H_s, higher than 2.0, no overtopping waves were measured, and for $R_c/H_s = 1.0$ the percentage of overtopping waves was about ten per cent for perpendicular waves which was reduced to no overtopping waves for an angle of 75°.

Model Set-up and Test Programme

Physical model tests have been carried out in a wave basin at the Danish Hydraulic Institute with the aim of measuring mean overtopping discharges defined as the volumes of wave overtopping per unit length of the breakwater per unit time. The basin was equipped with a movable wave generator in order to study the effect for different angles of wave attack, see **Fig 1**. The tests were carried out using long-crested irregular waves generated on basis of a Pierson-Moskowitz spectrum.

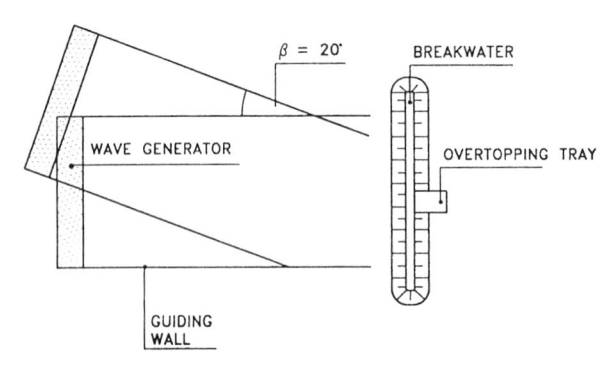

Fig. 1 Model plan for the basin tests showing the set-up for perpendicular and 20° wave attack.

The modelled structure had a total length of about seven metres and was constructed as a traditional rubble mound breakwater with a core, filter layer and armour layer of quarry stones, see **Fig. 2**. The size of the armour stones was selected not to allow for significant damage during testing, ie a nominal diameter, $D_{n,50}$, of about 0.04 m. The model consisted of a horizontal seabed which together with the use of a fixed breakwater height of 0.45 m means that variations in the crest free board were obtained by changes in the water level.

The wave conditions in the model were measured by seven resistance type wave gauges located in front of the breakwater. For perpendicular wave attack, the incident wave conditions and the reflection coefficients have been calculated using a multi-gauge technique. The significant wave height was calculated as $4 \times \sqrt{m_0}$, where m_0 is the zeroth moment of the spectral energy density function.

The overtopping water was collected in a 0.6 m wide tray located immediately behind the breakwater in a level corresponding to the crest elevation of the breakwater. This means that the recorded wave overtopping refer to water passing the rear edge of the breakwater crest. By measuring the total amount of overtopping water after each test with a duration of 600 to 1800 seconds, the mean overtopping discharge, q, was calculated.

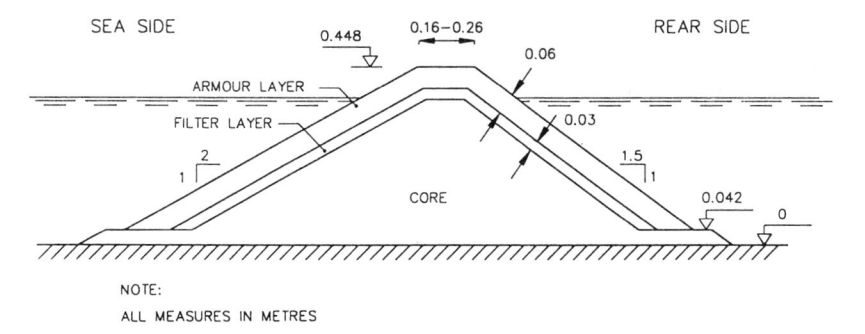

Fig. 2 Typical cross-section of breakwater used in the overtopping tests.

The following ranges of parameters were tested in the model study (all measures are in model measures):

- Significant wave height, H_s: 0.05 to 0.11 m
- Peak wave period, T_p: 0.8 to 2.0 s
- Wave steepness, s_{0p}: 0,018, 0,025, 0,030 and 0,045
- Crest free board, R_C: 0.050, 0.075 and 0.100 m
- Width of crest, B: 0.16 (4·D_{n50}), 0.21 and 0.26 m
- Slope angle, cotα: 2.0
- Angle of wave attack, β: 0° to 50° in steps of 10°

The wave steepness is given by the ratio between the significant wave height and the deep water wave length calculated on basis of the peak wave period:

$$s_{0p} = H_s/L_{0p} = 2\pi/g \cdot H_s/T_p^2$$

A parameter often used in the research on coastal structures is the surf similarity parameter given as:

$$\xi_{0p} = \tan\alpha/\sqrt{s_{0p}}$$

The model tests were run in test series with fixed wave steepness, ie a fixed ratio between the significant wave height and the deep water wave length. Thus all tests were made with a surf similarity parameter larger than 2 (two), which means that the wave conditions can be characterised as non-breaking waves.

The dimensionless free board, defined as R_c/H_s, varied between 0.5 and 2.0, which means that the tests covered both low and high crested breakwaters.

Test Results

This section includes analyses of the influence of the various tested parameters (wave height, wave steepness, crest free board, crest width and angle of wave attack) on the mean overtopping discharge. The influence of each of the parameters is described in the following.

Previous studies and the analysis of the test results showed that it is important to distinguish between situations with a large amount of water passing the breakwater crest, 'green water', and situations with a small amount of water passing the crest, 'spray'. Observations in the model showed that a rule of thumb to distinguish the two types of wave overtoppings is the dimensionless free board, R_c/H_s. It was found that for R_c/H_s larger than unity, the major part of the wave overtopping will occur as spray, whereas for R_c/H_s less than unity green water is dominant.

Influence of Wave Height

Previous research on wave overtopping has shown that the wave overtopping increases almost exponentially with the wave height, which is confirmed by the present model tests. An example of the influence of the significant wave height on the overtopping volume is shown in **Fig. 3** for an angle of wave attack of 40°. It should be noted that also the wave period changes due to the fixed wave steepnesses. It is observed that for perpendicular waves, the wave steepness has some influence on the overtopping. This influence became smaller for oblique waves, and for 40° no influence is found.

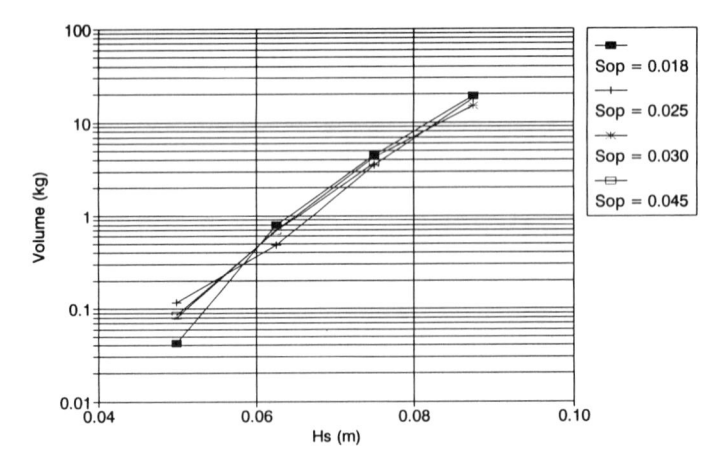

Fig. 3 Plot of the overtopping volume as function of the significant wave height, H_s. Crest free board, $R_c=0.05$ m. Crest width, B=0.16 m. Angle of wave attack, $\beta=40°$.

Influence of Wave Steepness

The influence of the wave steepness is shown in **Fig. 4** for two crest free boards, ie a low-crested and a high-crested breakwater. From the results, it is found necessary to distinguish between two different cases, ie $H_s > R_c$ and $H_s < R_c$. There is a tendency for decreasing overtopping volumes by an increase in the wave steepness for the case of $H_s > R_c$, whereas there is a tendency for increasing overtopping volumes for the case of $H_s < R_c$. The test results show that the influence of the wave steepness is decreasing for oblique waves, and generally no influence is found for an angle of wave attack of 30° as shown in **Fig. 7**.

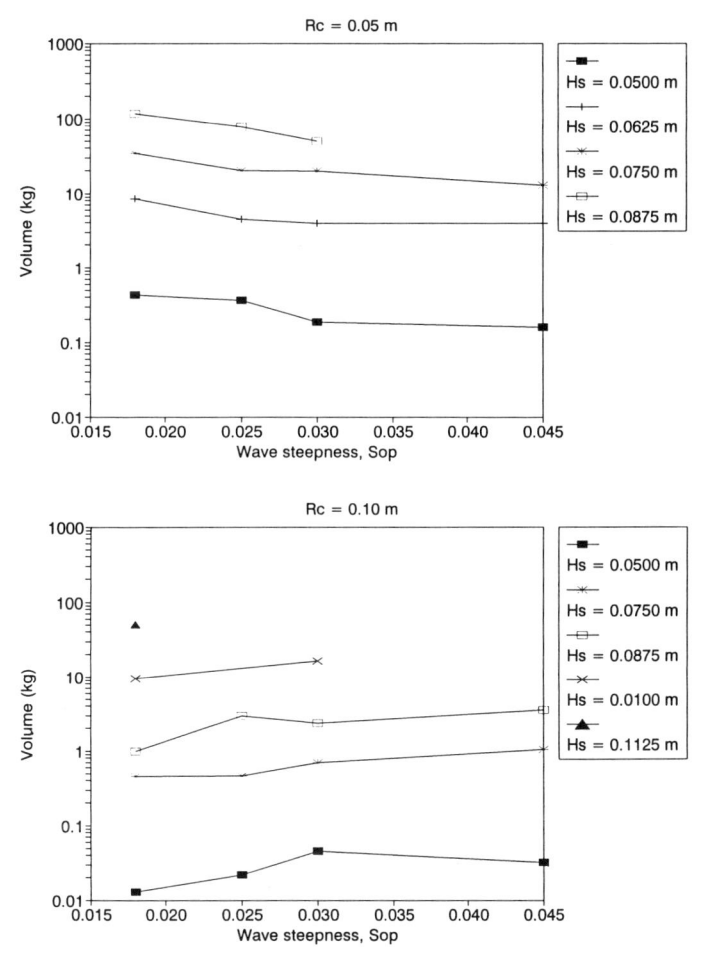

Fig. 4 Plots of the overtopping volume as function of the wave steepness, s_{op}. Crest width, B=0.16 m. Angle of wave attack, $\beta=0°$.

Influence of Crest Free Board

Tests were made with three different crest free boards, which resulted in a dimensionless freeboard, R_c/H_s, ranging from 0.5 up to 2.0. The crest freeboard together with the significant wave height are the major parameters governing the amount of water overtopping a breakwater. Examples of the influence of the crest free board are shown in **Fig. 5**. Comparison of the result for perpendicular wave attack and for an angle of 20° shows that the influence of an increase of the crest free board is larger for the latter case, which is due to an increased distance from the intersection of the still water level with the main armour layer to the rear side of the crest (the location of the measuring tray).

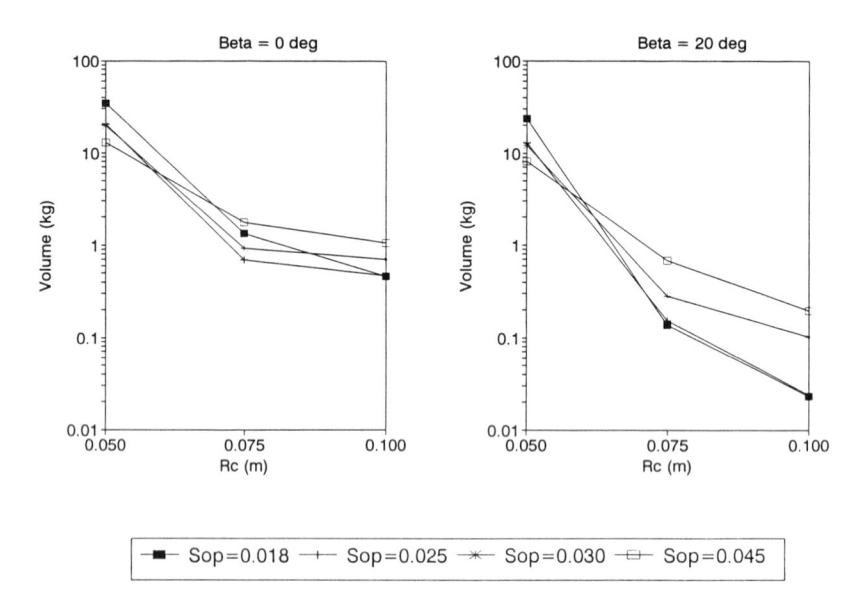

Fig. 5 Plots of the overtopping volume as function of the crest free board, R_c. $H_s = 0.075$ m. Crest width, $B = 0.16$ m.

Influence of Crest Width

Three crest widths were studied for perpendicular wave attack. Generally, it was found that the overtopping volume is decreasing with increasing crest width. However, the influence is smaller than the influence of the significant wave height and the crest free board. An example of the test results is presented in **Fig. 6**.

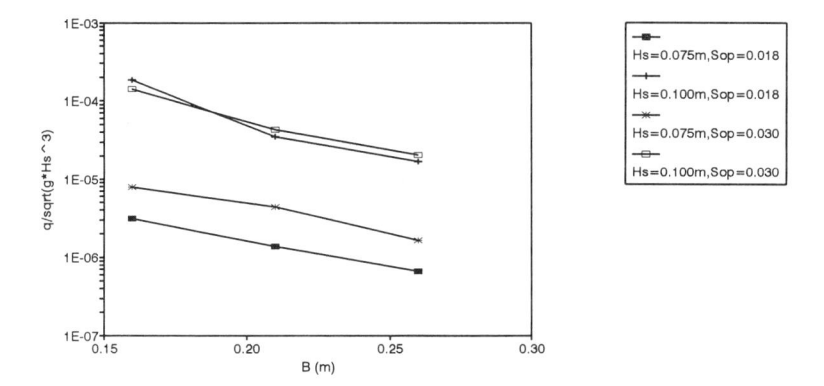

[Figure: chart with legend showing Hs=0.075m,Sop=0.018; Hs=0.100m,Sop=0.018; Hs=0.075m,Sop=0.030; Hs=0.100m,Sop=0.030. Y-axis: q/sqrt(g*Hs^3), X-axis: B (m)]

Fig. 6 Plots of the overtopping volume as function of the crest width, B. Crest free board, $R_c=0.10$ m. Angle of wave attack, $\beta=0°$.

Influence of Angle of Wave Attack

Test were carried out with angles of wave attack varying from 0° (perpendicular waves) up to 50° in steps of 10°. Typical examples of the influence of the angle are shown in **Fig. 7**, which includes the dependency of the wave steepness (it should be noted that the overtopping volumes are plotted in a linear scale). A pronounced characteristic in some of the cases is a maximum in the amount of wave overtopping for an angle of 10°.

In order to study in more details the influence of the angle of wave attack, dimensionless plots of the test data are shown in **Fig. 8**. These plots present the ratio between the overtopping volume for oblique waves and for perpendicular waves as function of the angle of wave attack for each of the three tested crest free boards. This ratio corresponds to a reduction factor taking into account wave obliquity. The plots show some scatter, but general trends can be recognised.

In the case with the smallest tested free board ($R_c=0.05$ m), the average of the tests shows a maximum in the mean overtopping discharge for an angle of 10°. A pronounced decrease in the overtopping is found increasing the angle of wave attack to 20° and 30°. For an angle of about 50°, the amount of overtopping water is on average reduced by 90 per cent compared to perpendicular wave attack, ie a reduction factor of 0.1. This reduction in the wave overtopping for an angle of wave attack of 50° is significantly higher than the reduction of about 40 per cent found for smooth slopes in De Wall and Van der Meer (1992).

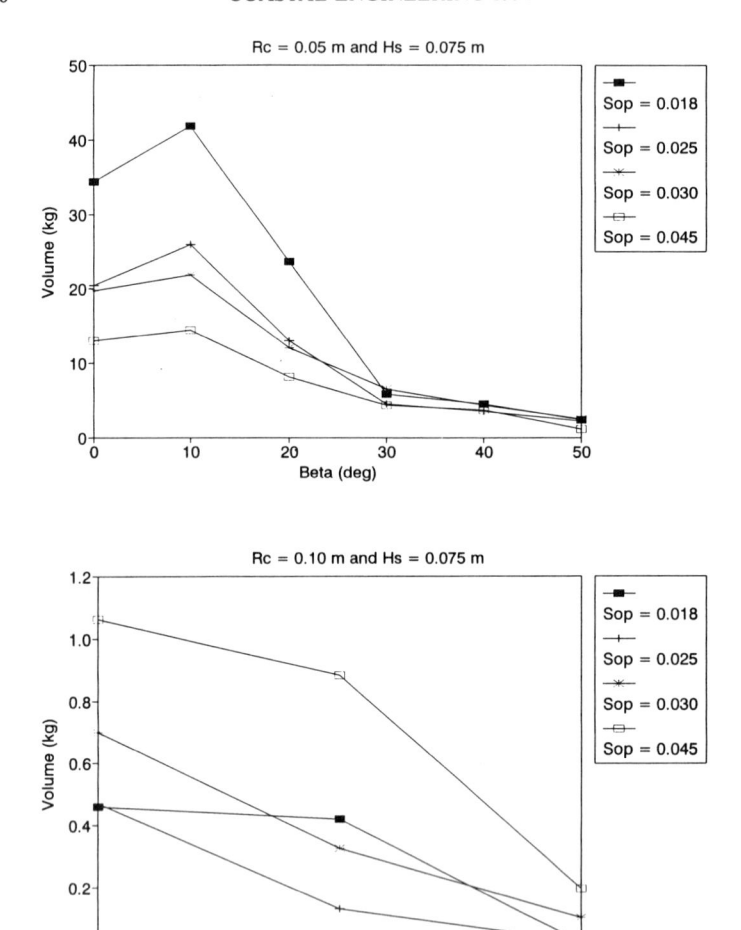

Fig. 7 Plots of the overtopping volume as function of the angle of wave attack, β. The plots include the influence of the wave steepness, s_{0p}. Crest width, B=0.16 m.

For the cases with higher free boards, only a few tests show a maximum in the overtopping for an angle of 10°, and on average a decrease is found. Comparing the three plots, it is found that the influence of the angle is getting more pronounced for the cases with the higher free boards. For the highest tested free board (R_c=0.10 m), a reduction factor of about 0.2 is found for an angle of 20°.

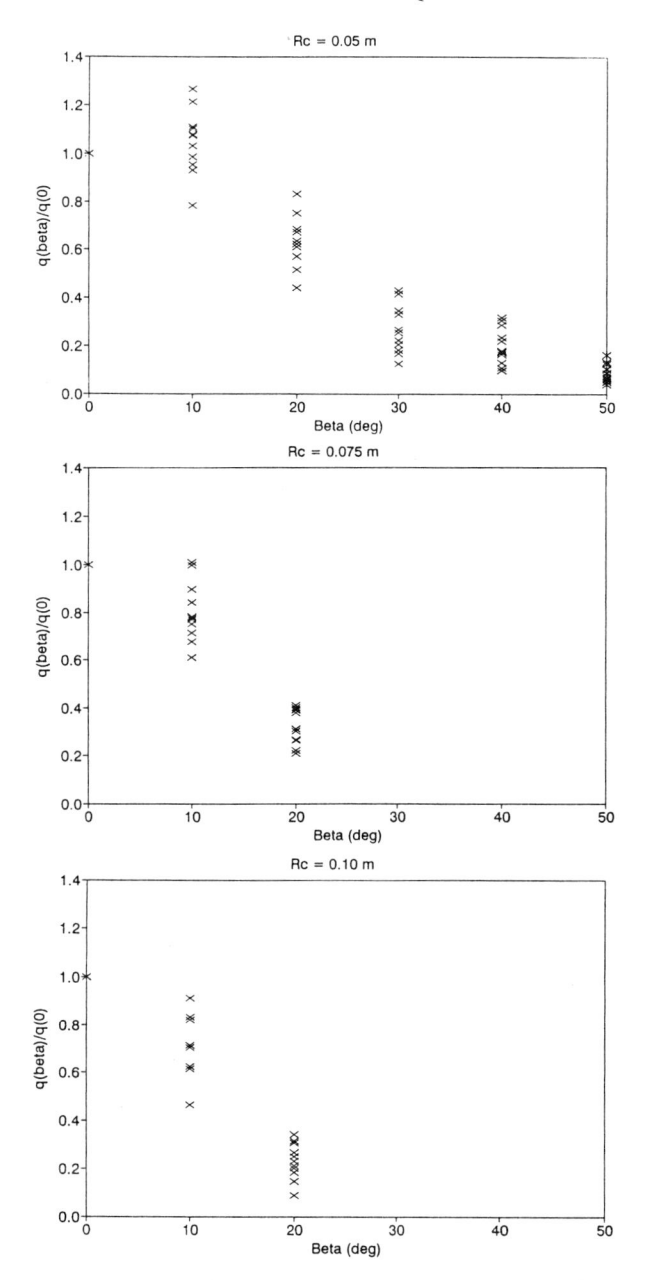

Fig. 8 Plots of the ratio between the overtopping discharge for oblique waves, q_β, and for perpendicular waves, q_0, as function of the angle of wave attack.

The overtopping volumes in De Wall and Van der Meer (1992) refer to the amount of water passing the crest of a slope, which is a relevant measure in the case of an impermeable slope. In the case of a permeable rubble mound breakwater, the amount of overtopping water will be dependant on the location of the measurement, ie the volume will be different when measuring at the front side edge or rear side edge of the crest due to the porous crest of the breakwater. The difference in the reference location of the overtopping measurements will have an effect on the influence of the angle of wave attack, as the distance from the front edge to the rear edge of the breakwater will increase with the angle of wave attack. Further, in the case of a rubble mound breakwater, the permeable layers will result in a faster decrease of the overtopping volume than for an impermeable slope.

Dimensionless Presentations

Through the years, overtopping results have been presented in numerous ways, including dimensionless plots. The most used dimensionless parameters are the dimensionless overtopping discharge, $Q = q/\sqrt{g \cdot H_s^3}$, and the dimensionless free board, $R = R_c/H_s$.

Examples of dimensionless plots of the test results obtained for perpendicular waves and for waves with an angle of 50° are shown in **Fig. 9**. The plots show that the test data for dimensionless free boards less than about 1.5 can be fitted to a straight line, ie Q can be described by an exponential function of R_c/H_s.

The model tests carried out with different crest widths showed that this parameter has an influence on the overtopping, and it is found that a combination of the crest free board and width can be used for describing the combined influence of these two parameters. For a fixed wave steepness, the dimensionless mean overtopping discharge can be fitted to an exponential function using $(2R_c + 0.35B)/H_s$ as parameter.

The test results show that the influence of the wave steepness is small compared to the influence of the other tested parameters, and that it is decreasing for oblique waves. A dimensionless plot excluding the influence of the wave steepness is presented in **Fig. 10** for all the tests made with perpendicular waves. The overtopping data are found to fit reasonable to the dimensionless parameter, $(2R_c + 0.35B)/H_s$, ranging between about 1.5 and 4.0.

Dividing the dimensionless overtopping discharge with average reduction factors for wave obliquity (see the plots presented in **Fig. 8**), all the overtopping data are presented in a dimensionless plot in **Fig. 11**. The figure presents the dimensionless overtopping discharge divided by the influence factor as function of the established dimensionless parameter, $(2R_c + 0.35B)/H_s$, taking into account both the crest free board and width. It is concluded that the influence of oblique waves on the wave overtopping discharge can be described by an influence factor for wave obliquity. A slight increase in the scatter is found including the data for oblique waves, see **Figs. 10 and 11**.

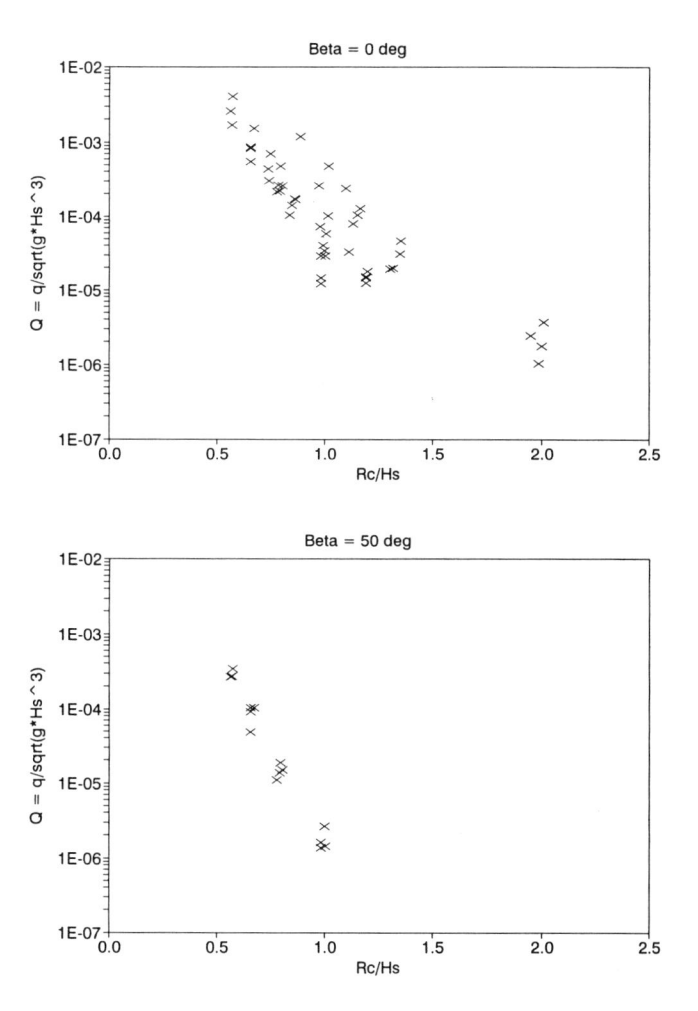

Fig. 9 Dimensionless overtopping discharge, Q, as function of dimensionless free board, R_c/H_s. Crest width, B=0.16 m.

1194　　　　　　　　　COASTAL ENGINEERING 1994

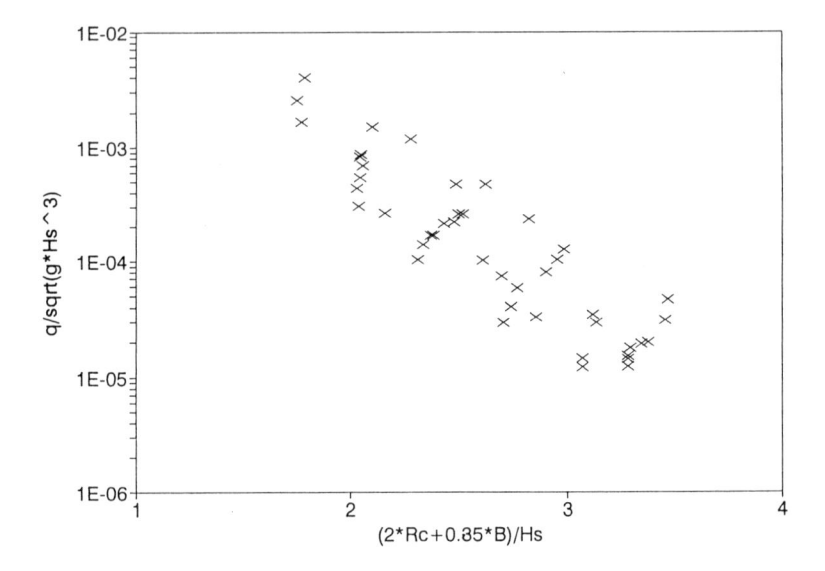

Fig. 10 Dimensionless overtopping discharge, Q, as function of the dimensionless parameter, $(2R_c+0.35B)/H_s$. Angle of wave attack, $\beta=0°$.

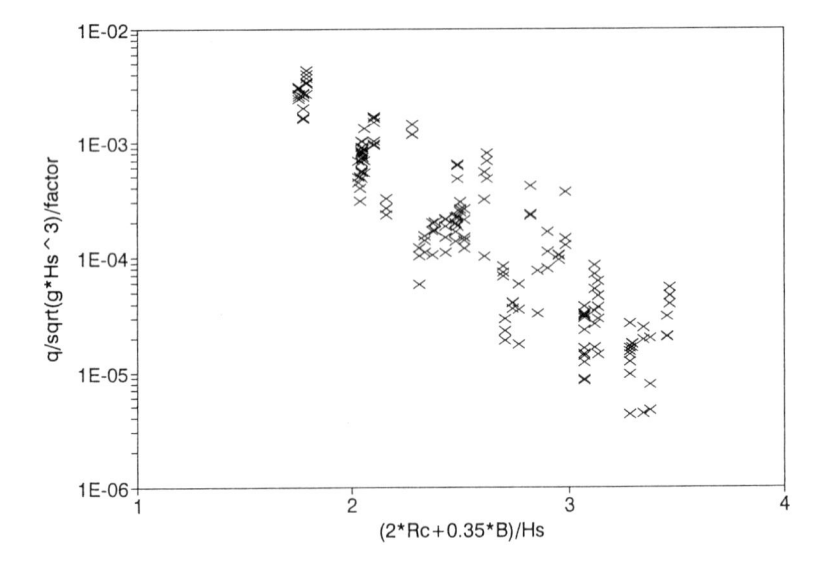

Fig. 11 Dimensionless overtopping discharge including an influence factor to take into account wave obliquity, Q/factor, as function of the dimensionless parameter, $(2R_c+0.35B)/H_s$.

Conclusions

A wide range of parameters influence wave overtopping of coastal structures. The present research study on overtopping of a traditional rubble mound breakwaters with an armour layer slope of 1:2.0 has concentrated on the angle of wave attack, the crest free board, crest width, and the general wave parameters (H_s, T_p, and s_{0p}).

Model tests were carried out in a wave basin for measuring the mean overtopping discharge, q (m³/m/s), and covered angles of wave attack ranging from 0° (perpendicular waves) up to 50° in step of 10°.

The significant wave height and the crest free board are the most important parameters with respect to wave overtopping. The crest width is found to have some smaller influence.

Testing with four wave steepnesses show that this parameter has an influence on the wave overtopping for perpendicular waves, whereas the influence is decreasing for oblique waves. However, the influence is small compared to the influence of the other tested parameters.

From analyses of the test results, it can be concluded that the reduction factor for wave obliquity (described by the ratio between the overtopping for oblique waves and perpendicular waves) is dependent on the crest free board. In case of a low-crested breakwater, the average of the tests show a maximum in the mean overtopping discharge for an angle of 10°. A pronounced decrease is found increasing the angle to 20° and 30°. For an angle of 50°, an average reduction factor of 0.1 is estimated. For breakwaters with a higher crest free board, the influence of the angle is larger, eg for the high-crested breakwater a reduction factor of about 0.2 is found for 20°.

The dimensionless overtopping discharge data for perpendicular waves (including all four wave steepnesses) show to fit well when plotted against a dimensionless parameter combining the influence of the crest free board and width $(2R_c+0.35B)/H_s$. Dividing the dimensionless overtopping discharges with average reduction factors for wave obliquity, it is found that the same dimensionless parameter gives a reasonable fit also for oblique waves. This means that for rubble mound breakwaters with a slope of 1:2.0, the dimensionless overtopping discharge including a reduction factor for wave obliquity can be described by an exponential function of $(2R_c+0.35B)/H_s$.

In order to establish a general overtopping formula for rubble mound breakwaters, it will be required to include the influence of, for instance, alternative breakwater geometries, water depth, artificial armour units, and superstructures.

Acknowledgement

The research has been funded by the Danish Research Council and the Danish Hydraulic Institute. The authors would like to thank Ms Kirsten Bossen, Mr Christian Paulsen, Mr Andreas Roulund and Mr Micael Aamann for carrying out the model tests and making the preliminary analysis, which were part of their B.Sc. dissertations at the Danish Academy of Engineering.

References

De Wall, J.P. and J.W. van der Meer. (1992). *Wave run-up and overtopping on coastal structures*. Proceedings of 23rd International Conference on Coastal Engineering, Venice, Italy.

Galland, J-C. (1994). *Rubble mound breakwaters stability under oblique waves: an experimental study*. Proceedings of 24th International Conference on Coastal Engineering, Kobe, Japan.

Jensen, O.J. and J. Juhl. (1987). *Wave overtopping on breakwaters and sea dikes*. Proceedings of Second International Conference on Coastal and Port Engineering in Developing Countries, Beijing, China.

Franco, L. (1993). *Overtopping of vertical face breakwaters: results of model tests and admissible overtopping rates*. Proceedings MAST2-MCS, 1st project workshop, Madrid, Spain.

Franco, L.; M. de Gerloni and J.W. van der Meer. (1994). *Wave overtopping at vertical and composite breakwaters*. Proceedings of 24th International Conference on Coastal Engineering, Kobe, Japan.

Van der Meer, J.W. and C-J.M. Stam. (1992). *Wave run-up on smooth and rock slopes of coastal structures*. Journal of Waterway, Port, Coastal, and Ocean Engineering, Vol 118, No 5, 1992.

SUBJECT INDEX
Page number refers to first page of paper

Aeration, 1895
Aerial photographs, 1998
Air entrainment, 1496
Airport runways, 2683
Aquatic plants, 142
Armor units, 918, 958, 986, 1001, 1061, 1090, 1143, 1157, 1227, 1388, 1412, 1426, 1439, 1511, 1541, 1568, 1625, 1641, 1713
Artificial islands, 1568

Barrier islands, 181, 2222, 2417, 2886, 3251, 3491
Bathymetry, 871
Beach erosion, 315, 609, 1197, 1880, 1934, 2070, 2252, 2340, 2380, 2434, 2449, 2571, 2683, 2726, 2943, 3208, 3393, 3478, 3491
Beach nourishment, 1359, 1797, 1934, 2100, 2222, 2395, 2668, 2886, 3016, 3208, 3491, 3507, 3548, 3564, 3579
Beaches, 540, 1157, 1782, 1812, 1880, 2115, 2325, 2395, 2449, 3167, 3378
Bed load, 3360
Bed load movement, 2513
Bed ripples, 635, 1975, 2013, 2043, 2070, 2140
Bed roughness, 300
Bedforms, 2185
Benefit cost analysis, 3237
Berms, 1075, 1343, 1625, 2252, 2712, 3594
Blocks, 932
Boundary conditions, 272, 442, 594, 886, 1770
Boundary element method, 871, 1241, 1700
Boundary layer, 594, 1827
Boundary layer flow, 384, 2527
Breaking waves, 219, 330, 399, 412, 511, 525, 594, 609, 1101, 1255, 1312, 1496, 1553, 1672, 1739, 1895, 1961, 2252, 2282, 2365, 2461, 2503, 2557, 2583, 2813, 2856, 3167
Breakwaters, 272, 525, 791, 1016, 1030, 1046, 1101, 1255, 1269, 1298, 1343, 1359, 1373, 1397, 1412, 1426, 1439, 1454, 1580, 1595, 1625, 1657, 1700, 1754, 1880, 2583, 2653, 2668, 2871, 3167, 3208, 3420, 3564, 3608
Buoyant jets, 3045
Buried pipes, 1553, 2571, 3099

Caissons, 1030, 1046, 1255, 1298, 1373, 1580
Calibration, 207
Case reports, 3491, 3507, 3564, 3579, 3594
Channel bends, 128
Channels, waterways, 128, 3002, 3060, 3139, 3533
Circular channels, 1373
Climatic changes, 3193, 3251, 3462
Coastal engineering, 1, 1837
Coastal management, 3579
Coastal morphology, 142, 1782, 1797, 1837, 1906, 2100, 2207, 2222, 2252, 2340, 2417, 2513, 2871, 3126, 3223, 3251
Coastal processes, 57, 2380, 2406, 3237, 3335
Coastal structures, 511, 2476, 3154, 3167, 3522, 3564
Cohesionless sediment, 2043, 2406
Cohesive sediment, 2004, 2058, 3060
Collisions, 3030
Comparative studies, 3208
Composite structures, 1030
Computer analysis, 1090
Computer models, 2311
Concrete, 1426, 1641
Concrete blocks, 1269
Conical bodies, 1595
Consolidation, soils, 2004, 2902
Construction, 3594
Coral reefs, 609
Cost effectiveness, 1412
Cross sections, 1130
Currents, 113, 384, 565, 624, 1484, 2476, 2503
Cylinders, 973, 1115, 1212

Volume 1 1-1196 **Volume 2** 1197-2416 **Volume 3** 2417-3614

Dam breaches, 2755
Damage assessment, 2197
Damage estimation, 1412
Damage patterns, 1001, 1090, 1157, 1657
Deep water, 12, 412, 455, 579, 973, 1343, 3608
Deltas, 2542
Design, 1327, 1359
Digital mapping, 1998
Dikes, 1075, 1169, 2197, 2639, 2755, 3350
Dilution, 3045
Discrete elements, 192
Dispersion, 3071
Displacements, 1255, 1625
Distribution, 3086
Distribution patterns, 192
Dolos, 958, 1388, 1426, 1511, 1641
Drainage systems, 1568, 2571
Dredge spoil, 2928, 3305, 3507, 3579
Dredging, 2972, 3002, 3016, 3060, 3139, 3393, 3533
Dunes, 1934, 2028, 2197, 2434, 2488, 2755, 2770

Earthquakes, 886
Eddy viscosity, 384
Embankments, 1130
Energy dissipation, 635, 1454, 2557
Energy losses, 761
Entrainment, 3071
Entropy, 232, 340
Environmental impacts, 3178
Erosion, 2058, 2170, 2488, 2542, 2639, 3360
Erosion control, 1880, 1934, 2070, 2155, 2683, 2712, 3193, 3208, 3378, 3491
Estuaries, 2004, 3060, 3178, 3281, 3408, 3533

Failure modes, 1388, 1713
Failures, 1526, 1657, 1754, 3350
Field investigations, 540, 719, 1946, 2028, 2513, 3522
Field tests, 665, 689, 945, 1269, 2115, 2610, 2799
Filters, digital, 168
Finite element method, 746, 871, 1388
Fish habitats, 3420
Fisheries, 1484, 3447, 3478

Flocculation, 3060
Flood control, 3126, 3193
Flooding, 2197, 2755, 3154
Flow characteristics, 27
Flow measurement, 1975
Flow patterns, 1770
Flow rates, 98, 1975, 2058, 2085
Flow simulation, 2140
Flow visualization, 861, 3045
Fluid flow, 98
Fluid-structure interaction, 1115
Flumes, 86, 1641, 1880, 2843, 2913
Free surfaces, 86
Frequency analysis, 12
Frequency response, 207
Friction, 1983
Friction factor, 565, 986

Geomorphology, 1837
Global warming, 3462
Grasses, 2639
Gravel, 1880
Gravity waves, 86, 579, 635, 1919, 3071
Greens function, 442
Groins, structures, 1157, 1327, 1782, 2668, 3564

Harbor engineering, 791, 806
Harbor structures, 3507
Harbors, 871, 2987
Hydraulic models, 1001, 1182, 1197, 1269, 2987, 3447
Hydraulic performance, 918, 1373
Hydraulic properties, 1454
Hydraulic roughness, 2871
Hydraulics, 2902
Hydrodynamic pressure, 973, 1484
Hydrodynamics, 861, 1849, 1919, 2282, 2340, 2741, 3099, 3432
Hydrologic aspects, 3251

Impact loads, 958, 3522
Impulsive loads, 1580
Indonesia, 609, 821
Inlets, waterways, 2943, 3432
Instrumentation, 42

Japan, 1, 247, 674, 886, 1016, 1046
Jetties, 2799

Kinematics, 540

Laboratory tests, 42, 98, 128, 168, 650, 689, 776, 1075, 1101, 1143, 1157, 1212, 1241, 1284, 1610, 1641, 1782, 1946, 2115, 2325, 2449, 2461, 2476, 2488, 2843, 2856, 3295
Lagoons, 2928, 3223
Lagrangian functions, 2828
Lakes, 2380, 3281
Land reclamation, 2972, 3178, 3305
Landslides, 821
Layered soils, 3360
Lift, 1115, 3099
Linear analysis, 285
Liquefaction, 2698, 3350
Littoral currents, 27, 1895, 1919, 1983, 2237, 2267, 2282, 2297, 2503, 2741
Littoral deposits, 1327
Littoral drift, 2513, 2595, 2625, 2726, 2799, 2943, 2972, 3478, 3507
Long waves, 455, 791, 847, 886, 1961

Marshes, 3548
Mass transport, 86, 2828, 2913
Mathematical models, 1241, 1727, 2125, 2197, 2350, 2755, 2928, 3223
Measuring instruments, 42, 207
Meteorological data, 79
Mining, 3335, 3393
Mixing, 3071, 3281
Model accuracy, 57, 2595
Model analysis, 57
Model studies, 1880, 2770
Model tests, 1030, 1061, 1182, 1197, 1298
Model verification, 901, 1961, 2595, 2610, 2785, 3223
Monitoring, 1511, 1812, 2100, 2222, 2799, 2886, 3522, 3579
Monsoons, 3378
Mooring, 791, 847
Movable bed models, 300, 1782, 1906, 2972
Mud, 2004, 2913, 3266, 3360, 3408

Navier-Stokes equations, 511, 1739, 2297
Navigation, 3139, 3533
Nearshore circulation, 68, 207, 272, 330, 1895, 2237, 2267, 2340, 2365, 2461, 2583, 2610, 2741, 3420

Netherlands, 2886, 3193, 3208, 3350
Nonlinear analysis, 157, 285, 427, 467, 1961
Normal stress, 986
North Sea, 261, 1359, 1797
Numerical calculations, 427
Numerical models, 27, 79, 113, 157, 192, 272, 442, 467, 511, 689, 746, 776, 806, 821, 871, 886, 1016, 1672, 1700, 1739, 1880, 1919, 1946, 1983, 2043, 2070, 2140, 2155, 2237, 2311, 2434, 2503, 2542, 2557, 2610, 2653, 2741, 2785, 2843, 2871, 3086, 3295, 3305, 3420

Ocean mining, 3393
Ocean waves, 232, 340, 806, 1526
Offshore engineering, 2928
Offshore structures, 1115
Oscillations, 847, 1255, 2987
Oscillatory flow, 1541, 1553, 1975, 1983, 3408
Outfall sewers, 1553, 3045, 3071, 3564
Overtopping, 511, 918, 1016, 1030, 1046, 1130, 1182, 1373, 1568, 1687, 1700, 2028, 2639, 3154
Oxygen content, 3447
Oxygenation, 3167

Particle motion, 2406
Permeability, 1101
Perturbation, 746, 973, 1469
Photogrammetry, 1998
Photographic analysis, 1511
Plates, 1312, 1454
Plunging flow, 1739
Polders, 2197
Pollution, 3071
Pore pressure, 932, 1727, 2698, 3111
Pore water pressure, 3030, 3369
Porous media, 635, 1770
Porous media flow, 1739
Ports, 1197, 3608
Portugal, 3608
Pressure distribution, 1770
Probabilistic methods, 1754, 3154
Probability density functions, 482
Probability distribution, 247, 356, 482, 497
Progressive waves, 86

Prototypes, 1016
Public safety, 1016, 1046

Random waves, 192, 272, 285, 300, 370, 412, 427, 482, 553, 565, 579, 704, 719, 1130, 1157, 1610, 2125, 2252, 2843
Recreation, 2395
Recreational facilities, 2668
Reefs, 1484
Regression models, 247
Rehabilitation, 3594, 3608
Research, 1
Residence time, 553
Residual strength, 2639
Return flow, 1101
Revetments, 932, 1001, 1269, 1687
Rip currents, 2583
Riprap, 3320
Risk analysis, 3154
Rivers, 3002, 3126, 3295
Rotational flow, 861
Rubble-mound breakwaters, 918, 958, 986, 1001, 1061, 1090, 1182, 1227, 1397, 1511, 1526, 1610, 1713, 1727, 3594

Salinity, 2928, 3533
Sand, 3360
Sand transport, 2557
Sandbars, 157, 1837, 1865, 2222, 2311, 2325, 2571, 2712, 2770, 2856, 3126, 3548
Scale effect, 1143, 2668, 2770
Scour, 1284, 1595, 3320
Scouring, 1212
Sea cliffs, 2170
Sea floor, 3111, 3139, 3320, 3369
Sea level, 2115, 2311, 2340, 3193, 3462
Sea state, 247
Sea walls, 1046, 1130, 1169, 1496, 1568, 1672, 1812, 3154, 3335
Sediment concentration, 1827, 2085, 2125, 2476, 2813
Sediment deposits, 2185, 3305
Sediment transport, 27, 98, 142, 300, 370, 525, 1212, 1327, 1770, 1827, 1837, 1849, 1865, 1906, 1919, 1946, 1975, 2028, 2043, 2070, 2125, 2140, 2185, 2282, 2406, 2503, 2513, 2527, 2595, 2625, 2770, 2785, 2799, 2843, 2871, 2902, 2928, 2928, 2972, 3295, 3360, 3478
Sedimentation, 2004, 3002, 3139
Sensors, 2185, 2799
Settling velocity, 3408
Shallow water, 261, 330, 370, 455, 467, 482, 665, 704, 731, 761, 1212, 1343, 1849, 2350, 2928, 3281
Shape, 1343, 1388, 1439, 1625
Shear stress, 98, 565, 594, 986, 1595, 2058
Shear waves, 1919
Shellfish, 3086, 3420
Ship motion, 791, 847, 871
Ships, 3030
Shoaling, 12, 467, 594, 2365
Shock, 1075
Shore protection, 142, 3237, 3335
Shoreline changes, 1327, 1782, 1812, 1837, 1906, 1946, 1998, 2380, 2610, 2625, 2653, 2683, 2726, 2755, 2785, 2813, 2943, 3178, 3295, 3335, 3378, 3393, 3462, 3478
Silts, 2902, 3060
Simulation models, 821, 3266
Sliding, 1580
Slope stability, 1553
Slopes, 315, 746, 1075, 1568, 1865, 2325, 2350, 3167
Soil mechanics, 2902, 3030
Soil permeability, 2902, 3111
South Africa, 1511
Spatial distribution, 579, 674
Spectral analysis, 12, 68, 384, 731
Spits, coastal, 2380, 2726, 3478
Stability, 918, 932, 1061, 1143, 1227, 1397, 1426, 1439
Stability analysis, 1865
Standing waves, 3369
State-of-the-art reviews, 1, 832
Statistical analysis, 674
Statistical data, 832
Statistics, 79
Stones, 1553, 1625
Storm surges, 901, 2170, 2434
Storms, 79, 261, 1526, 1849, 1934, 2843, 3491, 3608
Stratification, 3360
Stratified flow, 3447

Strength, 1388
Stress distribution, 1713
Structural response, 832, 1412, 1641
Structural strength, 958
Submerged discharge, 3320
Subsidence, 1541
Surf beat, 399, 689, 1961
Surf zone, 98, 315, 399, 553, 1895, 1934, 2085, 2155, 2237, 2267, 2282, 2350, 2503, 2513, 2542, 2712, 2813, 2856
Surface roughness, 1169
Surface waves, 467, 635, 3111
Suspended load, 3360
Suspended sediments, 1827, 1849, 1865, 2013, 2043, 2085, 2125, 2406, 2476, 2527, 2770, 2813, 3002, 3016, 3281
Suspended solids, 3408

Tensile stress, 1713
Three-dimensional flow, 370
Three-dimensional models, 871, 901, 1227, 2267, 2297, 2340, 2461, 2741, 3432
Tidal currents, 2282, 3281
Tidal hydraulics, 861
Tidal marshes, 3251
Tidal waters, 181, 2417, 3533
Tides, 3462
Time dependence, 272, 285, 553, 635
Tin, 3393
Tombolo, 1197, 2653
Topography, 272, 746, 1212, 1865, 1946, 2325, 2557, 2610, 3126
Transport rate, 2785, 2828, 2913, 2972
Tsunamis, 821, 886
Turbidity, 3016
Turbulence, 98, 861, 901, 2712, 3281
Turbulent boundary layers, 300, 384
Turtles, 3491
Two-dimensional analysis, 3447
Two-dimensional flow, 27
Two-dimensional models, 1030, 2155
Typhoons, 1, 219, 674, 2434

Ultrasonic testing, 624
Undertow, 330, 399, 2125, 2610, 2712, 2785
Uplift pressure, 932, 1298

Velocity profile, 86

Vertical cylinders, 1284, 1469, 1595
Vibration, 1115
Vibration analysis, 455
Viscoelasticity, 2913
Viscoplasticity, 2913
Viscosity, 3408
Vortices, 113, 525, 1115, 1975, 2140, 2856

Water content, 3266
Water depth, 261, 1343
Water flow, 1373
Water level fluctuations, 2170
Water levels, 2365, 3533
Water pollution, 3178
Water pressure, 1541
Water quality, 2395
Water surface profiles, 455, 624, 704, 1469
Water table, 2115, 2449, 2571
Water waves, 455
Waterfront facilities, 901
Wave action, 315, 918, 1001, 1061, 1212, 1227, 1269, 1739, 1934, 2013, 2058, 2639, 2698, 2712, 3016, 3045, 3099, 3266, 3522
Wave attenuation, 1312, 2913
Wave climatology, 181, 247, 1526, 2311
Wave crest, 356, 497, 540
Wave damping, 142, 609
Wave diffraction, 128, 285, 442, 973
Wave energy, 609, 761, 791, 806, 1454, 2698
Wave forces, 330, 1046, 1075, 1115, 1298, 1312, 1439, 1469, 1484, 1496, 1580, 1754
Wave generation, 68, 168, 650, 704, 776
Wave groups, 689, 776, 832, 847, 945, 1526, 1568, 1827, 1961, 2828
Wave height, 79, 181, 192, 247, 261, 356, 370, 399, 412, 497, 579, 609, 674, 746, 1016, 1101, 1182, 1241, 1397, 1657, 1754, 2155, 2207, 2311, 2350, 2365, 2488, 2698, 2828, 2987, 3266, 3608
Wave measurement, 42, 57, 207, 232, 624, 731, 776, 806, 832, 1610, 1906, 2207
Wave pressure, 1312, 1541
Wave propagation, 157, 168, 181, 232, 665, 1241

Wave reflection, 128, 168, 442, 719, 945, 1241, 1610, 2434
Wave refraction, 68, 285, 442
Wave runup, 27, 315, 399, 553, 689, 761, 776, 821, 886, 1169, 1687, 1727, 2115, 2155, 2170, 2207, 2325, 2488, 2557, 2571, 2583
Wave spectra, 12, 42, 68, 232, 340, 370, 384, 467, 482, 497, 624, 665, 719, 731, 806, 2843
Wave tanks, 128, 650, 1469, 1880, 1906, 2207, 2449
Wave velocity, 427, 525, 540, 565, 986, 1046, 1454, 2527
Waves, 113, 2476
Wind, 57
Wind direction, 79, 3447
Wind speed, 79, 261, 330, 1687
Wind waves, 12, 68, 181, 232, 330, 665, 731, 761, 901, 945, 2350

AUTHOR INDEX
Page number refers to first page of paper

Ahn, K., 482
Akeda, S., 3420
Alexis, A., 2902
Alexis, Alain, 2004
Allsop, N. W. H., 918, 1130, 1157, 3154
Alvarez, J., 3579
An, Nguyen Ngoc, 2913
Andrassy, Christopher J., 2100
Anglin, C. D., 2380
Aono, Toshio, 12
Asai, Tadashi, 232, 847
Asano, Toshiyuki, 27
Aydin, Ismail, 1770

Badiei, Peyman, 1782
Baird, W. F., 3608
Bakker, Willem T., 1797
Bakker, Wim T., 2197
Balzano, Andrea, 2928
Barnes, T. C. D., 776
Barthel, Volker, 791
Basco, David R., 1812
Battjes, J. A., 157
Bedford, K., 1827
Beji, Serdar, 427
Bellomo, Doug, 1812
Benoit, Michel, 42, 1610
Bergmann, Hendrik, 553, 1075
Bertotti, Luciana, 57, 79
Besley, P., 918
Bezuijen, Adam, 932
Blomgren, Sten H., 1327
Boccotti, Paolo, 945
Bodge, Kevin R., 2943
Booij, N., 68, 261
Bouws, E., 261
Brandt, Günther, 181
Briggs, Michael J., 806
Brøker, I., 2871
Brøker, Ida, 2958
Bruce, T., 1975
Burcharth, Hans F., 958

Caballería, Miquel, 1983
Carnero, Ovidio Varela, 1657
Cavaleri, Luigi, 57, 79
Chacaltana, J. T. Aquije, 973
Chang, C., 1700
Chang, Chen-Yue, 3045

Chang, Hsien-Kuo, 2972
Chaplin, John R., 1115
Chen, Xinjian, 3281
Chen, Yih-Far, 3045
Chen, Z., 3223
Chilo, Bruno, 1090
Chin, Ikuo, 3478
Chisholm, T. A., 1849
Chou, C. R., 2987
Christensen, E. Damgaard, 1865
Christensen, Erik Damgaard, 1919
Christensen, Morten, 168
Christiansen, N., 1595
Chyan, Jih-Ming, 3045
Cieślikiewicz, Witold, 86
Coates, T. T., 1880
Collado, F. R., 2542
Cornett, Andrew, 986
Cornett, Andrew M., 1001
Coussirat, M. G., 2542
Covarsi, Manuel F., 3564
Cox, Daniel T., 98
Cummins, I., 113

da Silva, A. F. Teles, 973
Daemrich, Karl-Friedrich, 2828
Dally, William R., 1895
Dalrymple, Robert A., 128
Damgaard, Jesper Svarrer, 1919
d'Angremond, K., 1713, 1754
Daniil, E. I., 3167
Davidson, D. D., 3608
Davies, Michael H., 1001
de Gerloni, M., 1030
De Groot, M. B., 1727
de Groot, Maarten B., 3350
de Quirós, Fernando Bernaldo, 3237
de Ronde, J. G., 261
de Ronde, John G., 761
de Ruig, J. H. W., 3208
de Vriend, Huib J., 594
Dean, Robert G., 1906, 2449, 3491
Deguchi, Ichiro, 370, 2476, 3002
Deigaard, R., 1865, 2043
Deigaard, Rolf, 1919, 2583
Dette, Hans H., 1934
Dette, Hans-H., 2843
Dette, Hans-Henning, 553
Dibajnia, Mohammad, 1946

Dijkman, Michiel, 3522
Dodd, N., 1880
Dodd, Nicholas, 1961
Dowd, Millard, 3491
Drago, Michele, 540
Dubi, Alfonse, 142

Earnshaw, H. C., 1975
Easson, W. J., 1975
Easson, William J., 525
Edge, B. L., 3608
Edge, Billy, 1687
Edge, Billy L., 3491
Eldeberky, Y., 157, 261
Endo, Taiji, 2395
Endoh, Kimihiko, 1016, 1046

Falqués, Albert, 1983
Ferier, P. G. P., 261
Fernandez, Joaquin, 3594
Fisher, John S., 1998, 2488
Foda, Mostafa A., 3099
Fontijn, Henri L., 1553
Franco, C., 918
Franco, L., 1030
Fredsøe, J., 1595, 1865, 2043
Fredsøe, Jørgen, 1919
Frigaard, Peter, 168
Fujima, Koji, 1115
Fukushima, Masahiro, 2571
Furuta, Goichi, 861

Gal, J. Andorka, 261
Galland, J.-C., 1061
Gallois, Stéphane, 2004
Gärtner, Joachim, 181
Giménez, Marcos H., 192
Glaser, Detlef, 181
Goda, Yoshimi, 1241
Goldenbogen, Roland, 3251
Gomez, Daniel, 3594
Gomez-Pina, G., 3579
Gómez-Pina, Gregorio, 3507
Goto, Chiaki, 12
Gotoh, H., 1541
Gotoh, Hitoshi, 2013
Gracia, V., 2542
Grass, Tony J., 565
Greated, C. A., 1975
Greated, Clive A., 540

Grote, Wout V., 1553
Grüne, Joachim, 181, 1075
Gudmestad, Ove T., 86
Guiducci, F., 1343, 1625
Guiducci, Franco, 1090

Hamilton, David G., 1782
Han, W. Y., 2987
Hancock, Mark W., 2028
Hanes, Daniel M., 3016
Hansen, E. A., 2043
Hansen, N.-E. Ottesen, 3030
Hansen, S. B., 1595
Hanslow, David J., 207, 2115
Hanson, Hans, 1327
Hanzawa, Minoru, 1143
Hara, Koji, 2461
Hardaway, C. Scott, 2653
Harkins, Gordon S., 806
Harris, J. M., 300
Harshinie, Karunarathna G., 689
Hashida, Misao, 219
Hashimoto, Noriaki, 232, 624, 847
Hashimoto, O., 1197
Hashimoto, Seiya, 2070
Hata, Sadakatsu, 2571
Hatada, Yoshio, 247, 674
Hattori, Masataro, 1101
Hattori, Takeshi, 704
Hayashi, Kenjirou, 1115
Hazelton, John M., 1812
Herbert, D. M., 1130
Hibbert, Kevin, 207
Higaki, Futoshi, 1115
Higano, Junya, 3086
Hoekstra, Piet, 2222
Holthuijsen, L. H., 68, 261
Holtzhausen, A., 1511
Holtzhausen, A. H., 1388
Hosoi, Yoshihiko, 1454
Houwing, Erik-Jan, 2058
Houwman, Klaas, 2222
Hsu, Tai-Wen, 2972
Hujii, Atsushi, 1046
Hwang, Kyu-Nam, 2488
Hwung-Hweng, Hwung, 3045

Ikeno, Masaaki, 3320
Imamura, Fumihiko, 821, 886
Iranzo, Vicente, 1983

AUTHOR INDEX

Irie, Isao, 2070
Ishii, Toshimasa, 272
Isobe, Masahiho, 719
Isobe, Masahiko, 272, 285, 2785, 3266
Ito, Kazunori, 847
Ito, Masahiro, 1143
Ito, Yoshiki, 624
Itoh, Sadahiko, 1454
Iwagaki, Yuichi, 1, 1143
Iwata, Koichiro, 1212, 1439
Izumi, Tatsuhisa, 3378

Jaffe, Bruce E., 2085
Janssen, Hans, 3522
Jensen, Anders, 2958
Jensen, Frerk, 181
Johnsen, John, 2958
Johnson, Hakeem, 2871
Johnson, Patrick, 3491
Jones, R. J., 918, 1157
Juang, Jea-Tzyy, 1169
Juhl, J., 1754
Juhl, Jørgen, 1182

Kabiling, Michael B., 2557
Kaczmarek, L. M., 300
Kaihatsu, Sumio, 1269
Kaiser, Ralf, 181
Kamata, A., 1197
Kamphuis, J. William, 1782
Kana, Timothy W., 2100
Kang, Hong-Yoon, 2115
Karjadi, Entin A., 2155
Katoh, Kazumasa, 315
Katopodi, Irene, 2125, 2527
Kawaguchi, T., 1197
Kawata, Yoshiaki, 330
Kersting, Nico F., 1797
Kheyruri, Z., 1727
Kim, Chang-Je, 1212
Kim, Hyeon-Ju, 1526
Kim, Hyoseob, 2140
Kim, Kyu Han, 2476
Kim, Taerim, 340
Kimura, A., 497
Kimura, Akira, 356
Kimura, Katsutoshi, 1227
Kirby, James T., 128
Kitou, Nikos, 2125
Kittitanasuan, Wudhipong, 1241

Kiyokawa, Tetsushi, 650
Klammer, P., 1255
Klammer, Peter, 1298
Klatter, Leo, 3522
Kobayashi, Akio, 847
Kobayashi, Masonori, 1269
Kobayashi, Nobuhisa, 98, 1373, 2028, 2155
Kobayashi, Tomonao, 861, 1284
Koelewijn, Ria, 2527
Kohlhase, Søren, 1298
Kojima, Haruyuki, 1312
Komar, P. D., 2170
Koole, R., 3071
Koontanakulvong, Sucharit, 3002
Kortenhaus, Andreas, 1298
Kos'Yan, R. D., 2185
Kraak, A. W., 3208
Kraak, Arie, W., 2197
Kraus, Nicholas C., 1327
Kriebel, D. L., 2325
Kriebel, David L., 2207
Kroon, Aart, 2222
Kung, Chen-Shan, 1837
Kunisu, Hiroshi, 1568
Kunz, H., 3533
Kunz, Hans, 3251
Kuriyama, Yoshiaki, 2237
Kusuda, Tetsuya, 3408
Kuwabara, S., 3420
Kuwahara, Hisami, 3086
Kwon, J. G., 370

Lamberti, A., 1343, 1625
Larson, Magnus, 2252
Larsonneur, Claude, 2282
Laustrup, Christian, 1359
Law, Adrian W. K., 3099
Lee, Dal Soo, 1373
Lee, J., 1827
Lee, J. J., 1700
Lee, Jung Lyul, 2267
Lee, Tsong-Lin, 3369
Levoy, Franck, 2282
Li, Bin, 2297, 3548
Liberatore, Gianfranco, 525
Lin, Li-Hwa, 340, 2770
Lindenberg, Jaap, 3350
Lindo, Mark H., 3594
Lintrup, Morten, 2928

Liu, Shuxue, 731
Liu, Zhou, 958
Longo, Sandro, 2527
Luger, S., 1511
Luger, S. A., 1388

MacDonald, N. J., 2311
Maddrell, Roger, 3548
Maddrell, Roger J., 2297
Madsen, Holger Toxvig, 1359
Madsen, O. S., 1849
Madsen, Ole Secher, 384
Madsen, P. A., 399
Madsen, Per A., 2583
Magda, Waldemar, 3111
Magoon, O. T., 3608
Mano, A., 3126
Mansard, E., 3608
Mansard, E. P. D., 832, 1397
Mansard, Etienne, 791, 986
Mansard, Etienne P. D., 1001
Manzenrieder, Helmut, 3139
Mase, Hajime, 635
Matsumi, Y., 1397
Matsumoto, Akira, 442
Matsunaga, Nobuhiro, 219
Matsuoka, Michio, 442
Matsutomi, Hideo, 886
Mayer, R. H., 2325
Mayerle, Roberto, 2340
McDougal, W. G., 2170
Meadowcroft, I. C., 3154
Medina, Josep R., 192, 1412
Melby, Jeffrey A., 1426, 1641
Memos, Constantine D., 2350
Miura, H., 2698
Miyaike, Yoshihito, 1212
Miyamoto, Y., 3420
Mizui, Hiroyuki, 219
Mizumoto, T., 1197
Mizutani, Norimi, 1439
Mocke, G. P., 2365, 3335
Møller, Jacob Steen, 2958
Monfort, Olivier, 2282
Mori, Nobuhito, 412
Moutzouris, C. I., 3167
Mulder, J. P. M., 2886
Murakami, Hiroshi, 1143
Murakami, Hitoshi, 1454
Murakami, Keisuke, 1469, 2070

Muraoka, Kohji, 3178
Muraoka, Kouji, 3447

Nadaoka, Kazuo, 427, 650
Nagai, Toshihiko, 232, 624, 847
Nagao, M., 3126
Nairn, R. B., 2380
Nakagawa, Hiroji, 2013
Nakagawa, Yasuyuki, 427
Nakamura, Satoshi, 2070
Nakamura, Tetuya, 1312
Nakatsuji, Keiji, 3178, 3447
Nemoto, Kenji, 1143
Nielsen, Peter, 207, 2115, 2406
Niemeyer, Hanz D., 181, 1797, 2417
Nishi, Ryuichiro, 2434
Nishida, H., 1484
Nishihira, F., 1484
Nishimura, Hitoshi, 442
Nishimura, Tsukasa, 861
Nochino, Masao, 455
Nwogu, Okey, 467

Ochi, M. K., 482
O'Connor, B. A., 300, 2311
O'Connor, Brian A., 2140
Oda, Kazuki, 3305
Oda, Kenji, 1284
Oh, Tae-Myoung, 1906, 2449
Ohhama, Hoiku, 2395
Ohta, T., 497
Ohta, Takao, 356
Ohyama, Takumi, 650, 871
Okayasu, Akio, 98, 2461
O'Neil, S., 1827
Ono, Masanobu, 2476, 3002
Ono, Nobuyuki, 2070
Oshiro, Shinichi, 609
Osiecki, Daniel A., 1895
Osmond, Bill, 3548
Oumeraci, H., 1255, 1672
Oumeraci, Hocine, 1298
Overton, Margery F., 1998, 2488
Owen, M. W., 1130

Park, Woo Sun, 1373
Partenscky, H.-W., 1255, 1672
Péchon, Philippe, 2503
Peerbolte, E. Bart, 3193
Peña, Carlos, 3564

Peña-Santana, Patricia G., 2395
Peregrine, D. H., 776, 1496
Perrier, G., 2043
Petit, H. A. H., 511, 1739
Petti, Marco, 525, 540
Phelp, D., 1511
Phelp, D. T., 1388
Pluijm, M., 3208
Podymov, I. S., 2185
Presti, A. Lo, 2625
Prieto, J., 2542
Pruszak, Zbigniew, 2513
Putrevu, U., 2741

Quinn, Paul A., 525, 540

Ramadan, Khaled A. H., 2527
Ramírez, Jose L., 3507
Raudkivi, Arved J., 1934
Reeve, D. E., 3154
Ribberink, Jan S., 2527
Ris, R. C., 68
Rodriguez, A., 2542
Roelvink, J. A., 3223
Rousset, Hélène, 2282
Rubin, David M., 2085
Ruessink, Gerben, 2222
Rufin, Teofilo Monge, Jr., 1439
Ruggiero, P., 2170
Ruiz, Luis Felipe Vila, 3237
Rutledge, J., 1397
Ryu, Cheong-Ro, 1526

Sakai, Hiroyuki, 1101
Sakai, T., 1541
Saleh, Wameidh M., 565
Sallenger, Asbury, Jr., 2085
Sánchez-Arcilla, A., 2542, 2625
Sánchez-Carratalá, Carlos R., 192
Sancho, F. E., 2741
Sand, S. E., 832
Sasaki, Hiroshi, 624
Sato, Michio, 2434, 2571
Sato, Shinji, 2557
Sawamoto, M., 3126
Sawamura, Yoshiyuki, 1454
Sawaragi, Toru, 2476, 3002
Schäffer, H. A., 399
Schäffer, Hemming A., 2583
Schiereck, Gerrit J., 1553

Schmeltz, Edward J., 3594
Schoonees, J. S., 2595, 2668
Schroeder, Ernst, 3251
Schröter, Andreas, 2340
Schüttrumpf, Holger, 553
Scott, R. D., 2380
Seijffert, J. W. W., 2639
Sekimoto, Tsunehiro, 1568
Shen, Daoxian, 3266
Sheng, Y. P., 3432
Sheng, Y. Peter, 3281
Shiba, Kazuhiko, 2785
Shibata, Takao, 847
Shibayama, Tomoya, 2461, 2813, 2913, 3295
Shigematsu, Takaaki, 3305
Shih, S.-M., 2170
Shim, Youngbo, 2140
Shimizu, Takao, 3320
Shimizu, Takuzo, 1946, 2610
Shimosako, Kenichiro, 1580
Shinoda, Seirou, 704
Shiraishi, Naofumi, 2395
Shuto, Nobuo, 821, 886
Sierra, J. P., 2625
Sigemura, Toshiyuki, 1115
Simons, Richard R., 565
Simonsen, B. C., 3030
Sistermans, Paul G. J., 1553
Sloth, Peter, 1182
Smit, F., 2365
Smith, G. G., 3335
Smith, G. M., 2639
Sørensen, O. R., 399
Sørensen, Ole R., 2583
Stansberg, Carl Trygve, 579
Steetzel, Henk J., 2197
Sterndorff, M. J., 3030
Stive, Marcel J. F., 594, 761
Stive, Marcer, 1837
Stoutjesdijk, Theo P., 3350
Sueyoshi, Toshiaki, 3178
Sugihara, Yuji, 219
Suh, Kyung Duck, 2653
Sulaiman, Dede M., 609
Sumer, B. M., 1595
Sunamura, Tsuguo, 2856
Svendsen, I. A., 2741
Swan, C., 113, 3071
Swart, D. H., 2668, 3335

Syamsudin, Abdul R., 2683

Tada, A., 1484
Takahashi, S., 2698
Takahashi, Shigeo, 1016, 1046, 1227, 1580
Takahashi, Tomoharu, 624
Takahashi, Tomoyuki, 821, 886
Takayama, Tomotsuka, 624
Takeba, Ken, 635
Tamashita, Takao, 901
Tanaka, Masahiro, 650
Tanimoto, Katsutoshi, 689, 1227, 1580
Tehrani, Mehrdad M., 565
Teisson, Charles, 42, 1610, 2503
Theron, A. K., 2595
Thomas, P., 2902
Thomas, Pierre, 2004
Tillotson, K. J., 2170
Ting, Francis C. K., 2712
Toba, E., 3579
Tobikik, Isao, 847
Togashi, Hiroyoshi, 746
Tolman, H. L., 261
Tomasicchio, G. R., 1343, 1625
Toms, Geffery, 1837
Tonder, A. van, 1388, 1511
Tönjes, P., 511, 1739
Tøorum, Alf, 142
Topliss, M. E., 1496
Torfs, Hilde, 3360
Toue, Takao, 847
Tsai, Ching-Piao, 3369
Tsuchida, Mitsuru, 871
Tsuchiya, Toshito, 609
Tsuchiya, Yoshito, 2683, 3378, 3478
Tsujimoto, Tetsuro, 2013
Tsuru, Masahito, 2610
Tsutsui, Shigeaki, 609
Turk, George F., 1426, 1641

Uda, Takaaki, 2726
Ujiie, Hisayoshi, 3320

Valdecantos, Vicente Negro, 1657
van de Graaff, Jan, 2197
van de Kreeke, J., 2886
van den Bosch, P., 511, 1739
van der Lem, J. C., 3208
van der Meer, J. W., 1030, 1713, 1754, 2639
Van Dongeren, A. R., 2741
Van Evra, R., 1827
van Gent, M. R. A., 511, 1727, 1739
van Nes, C. P., 1713
van Rijn, Leo C., 2058
van Vessem, P., 2886
van Vledder, Gerbrant Ph., 761
Verhagen, L. A., 665
Villaret, C., 2043
Visser, Paul J., 2197, 2755
von Lany, P. H., 3154
Vongvisessomjai, Suphat, 3393

Walstra, D. J. R., 3223
Wang, Hsiang, 340, 2267, 2434, 2770, 2843
Wang, Xu, 2770
Ward, Donald L., 1687
Watanabe, Akira, 272, 719, 1946, 2610, 2785, 3266
Watson, G., 776
Weggel, J. R., 3608
Whalin, R. W., 3608
White, Thomas E., 2799
Wibner, Christopher G., 1687
Williams, Greg, 1812
Wind, Herman G., 3193
Winyu, Rattanapitikon, 2813
Woltering, Stefan, 2828
Wright, L. D., 1849
Wu, N. T., 1672
Wu, Yongjun, 2843

Yamada, Akiko, 3295
Yamada, Minoru, 3478
Yamaguchi, Masataka, 247, 674
Yamamoto, Koji, 2726
Yamamoto, Masato, 1143
Yamamoto, S., 2698
Yamamoto, T., 1541
Yamamoto, Yoshimichi, 689
Yamanishi, Hiroyuki, 3408
Yamashita, Takao, 609, 2683, 3378
Yamauchi, Kazuaki, 3320
Yamazaki, H., 1727
Yamazaki, Tsuyoshi, 1568
Yan, Yixin, 3060
Yang, David W., 3594
Yano, K., 3420